Omics for Personalized Medicine

Debmalya Barh • Dipali Dhawan
Nirmal Kumar Ganguly
Editors

Omics for Personalized Medicine

Editors
Debmalya Barh
Centre for Genomics and Applied Gene Technology
Institute of Integrative Omics and Applied Biotechnology (IIOAB)
Nonakuri, Purba Medinipur,
West Bengal, India

Nirmal Kumar Ganguly
Policy Centre for Biomedical Research
Translational Health Science and Technology Institute
(Department of Biotechnology Institute, Government of India)
Office@National Institute of Immunology
New Delhi, India

Dipali Dhawan
B.V. Patel Pharmaceutical Education and Research Development Centre
Ahmedabad, Gujarat, India

Institute of Life Sciences
Ahmedabad University
Ahmedabad, Gujarat, India

ISBN 978-81-322-1183-9 ISBN 978-81-322-1184-6 (eBook)
DOI 10.1007/978-81-322-1184-6
Springer New Delhi Heidelberg New York Dordrecht London

Library of Congress Control Number: 2013949007

Printed on acid-free paper

Springer is part of Springer Science+Business Media (www.springer.com)

Purnendu Bhusan Barh
(22nd Feb 1940–28th Feb 2008)

This book is dedicated to ***Purnendu Bhusan Barh***
(S/O Ambika Charan Barh and F/O Debmalya Barh), an eminent academician, philosopher, career master, and transformator who is the soul and inspiration behind the Institute of Integrative Omics and Applied Biotechnology (IIOAB) and all its activities.

Foreword

This is the right time to clarify the issues that emerge from "omics sciences" (genomics, pharmacogenomics, transcriptomics, proteomics, metabolomics, connectomics, nutrigenomics, ecotoxicogenomics, environmental genomics, etc.) in terms of hopes and limitations. Of course, several hurdles are ahead. Given the ongoing additional developments in omics technologies assisted by powerful bioinformatic tools, optimism is in place to deliver the expectations raised by the genomic revolution. The ambition of "omics sciences" is to characterize a large (if not complete) set of molecules within a biological sample and their orchestral interactions in fine-tuning the phenotype.

The aim of these "pluriomics" is to get full insight into the complex biological systems often designated as "integrative systems biology." The semantic bioinformatic-based linking process of multitude omics sciences has been named "interactomics" which is not limited to protein-protein interactions but englobes all cellular processes. Presently, high-throughput technologies have generated more data than that we can handle. Harmonization and standardization of complex omics data itself is a great challenge to understand the dynamics of biological networks in space and time and also, through metadata analysis, to recover meaningful biological response to our questions in health and disease. In the omics ocean, it is quite difficult to navigate even for the experienced biologists and clinicians. To fill this gap and to soothen the annoyment of health-care scientists, "*Omics for Personalized Medicine*" provides a detailed description of concepts and technologies in the first section and the present "state-of-the-art" applications in the remaining six sections. The editors, led by the eminent physician *cum* biomedical scientist Prof. N.K. Ganguly and coordinated by the principal member of the editorial board, Debmalya Barh along with Dipali Dhawan, have brought a distinguished list of contributors from around the world to achieve the formidable task of explaining what is "omics" and how their concepts can be applied to personalized medicine. Since the fields of applications of omics are quite large, for reasons of clarity, six sections are introduced (cancer/autoimmune, neuropsychiatric, cardiovascular/pulmonary, metabolic, and infectious disorders without ignoring the implications in ortho, gyneco, and opthalmo medical specialities as well as in cell-based therapies). The concept of "personalized medicine" is not novel to physicians who have since long practiced it with some subjective appreciations and approximations, while the ambition of present concept of "personalized prescription" aims to achieve personalized "drug selection and drug dose" and intends to be evidence-based through omics knowledge. These approaches need not necessarily be based on patient's genetic makeup as we recognize more and more the importance of epigenetic modifications in drug tolerance, response, and toxicity which is also highlighted in the first section. After all, gene expression is influenced by xenobiotics including drugs, and the destiny of xenobiotics in our body is dictated by organism's gene expression network. We are in the process of searching the road map, and the "omics for personalized medicine" intends to be your GPS ("genomic positioning system").

Dr. R. Krishnamoorthy
Former Director of Research Pharmacogenomics Section
Institut National de la Santé et de la Recherche Médicale (INSERM)
Paris, France

R. Krishnamoorthy, Ph.D., D.Sc., is an eminent scientist and former director of research, INSERM, Paris, France. A specialist in the fields of pharmacogenetics, pharmacogenomics, hematological genetic disorders, and immunogenetics among others, Dr. Krishnamoorthy is dynamically engaged in several international committees and advisory boards and presently is a consultant for research in molecular hematology for a number of institutions in different countries. He works in collaboration with top research institutes from India, Brazil, the USA, the UK, Germany, Oman, Mauritius, Bahrain, Qatar, and Maldives and has more than 260 international publications. He had peer-reviewed for major journals like Am J Hum Gen, Blood, Nat Reviews, and PLoS One and other PLoS series.

Preface

The goal of humankind is to alleviate the human suffering through the gain of knowledge. In the last few decades, the rapid advancement of "omics" technologies has shown us a way to understand the genetic causes of diseases that has led the way for better treatment options and management for the diseases that afflict humankind. The sequencing of human genome was the first major step towards the realization and advancement of personalized medicine. Researchers all over the world are in unison in their opinion that this is just the beginning and as more and more innovations occur, omics technologies will gradually become the part of regular treatment and will be adopted by both physicians and patients alike.

The major need of the hour is the dissemination of knowledge of the latest technological development among students of science, medical professionals, policy makers, and educated public, in general, so that these revolutionary technologies are adopted more quickly and bring about positive changes in public health scenario. This book is a sincere effort by researchers all over the world to bring that knowledge and development trends in pharmacogenomics to the public.

The book "Omics for Personalized Medicine" as the title implies presents a comprehensive analysis of several aspects of pharmacogenomics and personalized medicine. The book has been divided into seven sections comprising 34 chapters. Section I contains the general aspects of the omics technologies that have been developed and also the future innovations that will transform public health. Sections II–VII comprise chapters on personalized medicine covering several diseases.

The book chapters have been written by scientists from all over the world and with a focus on present and future global trends in the field of pharmacogenomics and personalized medicine. Each chapter is comprehensively written and includes figures and illustrations for the benefit of the readers. The book has been written and complied for over a year ensuring that all the latest aspects of pharmacogenomic research trends have been incorporated in the book. The efforts of authors are being reflected in the considerable amount of information incorporated in each of the chapter of the book.

The book begins with a chapter entitled "Pharmaco-Geno-Proteo-Metabolomics and Personalized Medicine: An overview" by Drs. Raghavachari and Gucek, wherein they give an overview of the different omics technologies. The second chapter by Dr. Mitra and colleagues describes the current status and future of pharmacogenomics along with the issues that need to be resolved. The third chapter by Dr. Hong and colleagues is about next-generation sequencing including protocols, algorithms, and software packages for analyzing the data generated. In Chap. 4, Dr. Gunduz and colleagues elaborate on a novel discipline of pharmacoepigenomics and the epigenetic markers that can help in understanding interpersonal variations in response to therapeutics. In Chap. 5, Dr. Hassan explores the relationship between dietary factors and prevention of disorders like tumors, diabetes, and obesity. In Chap. 6, Son, Tuzmen, and Dr. Hizel talk about the designing and potential applications of studies related to pharmacogenomics. In Chap. 7, Drs. Sharma and Munshi describe the opportunities and applications of personalized medicine in clinical trials. In Chap. 8, Drs. Munshi and Ahuja talk about ethics, a very important concept in pharmacogenomics. In Chap. 9, Dr. Hernández-Lemus outlines some emerging ideas in the field of computational pharmacogenomics. In Chap. 10, Ms. Pasha and Dr. Scaria have highlighted the online resources available in personal genomics.

Sections II–VII of this book highlight the applications of personalized medicine in different fields. Drs. Dhawan and Padh provide an overview on the well-studied molecular markers for efficacy/toxicity and also discuss the evolution of companion diagnostics in cancer in Chap. 11. Dr. Gunduz and colleagues elaborate on the recent advances in the field of personalized medicine in acute myeloid leukemia in Chap. 12. Drs. George, Selvarajan, and Srinivasamurthy discuss the molecular markers in autoimmune disorders in Chap. 13. In Chap. 14, Drs. Archer and Fredriksson focus on studies related to personalized medicine in Parkinson's disease. In Chap. 15, Drs. Archer and Johansson highlight the progress achieved in the field of Alzheimer's disease in relation to personalized medicine. In Chap. 16, Drs. Archer, Ricci, and Ricciardi talk about the relevance of pharmacogenomics in mood disorders. Drs. Callejas and Garre discuss the association of gene polymorphisms with opioid and alcohol addiction in Chap. 17. Dr. Blum and colleagues highlight on an important link of neurogenetics and nutrigenomics with accurate genetic diagnosis and dopamine D2 agonist therapy to induce dopaminergic activation in Chap. 18. Dr. Yiannakopoulou focuses on variations that affect response on analgesic treatments in terms of efficacy and safety in Chap. 19. In Chap. 20, Dr. Chatterjee discusses personalized medicine in cardiovascular disease which is the leading cause of death in the world. In Chap. 21, Dr. Jiménez-Varo and colleagues have elaborated on the studies of oral anticoagulants especially coumarin derivatives in relation to pharmacogenetics. In Chap. 22, Drs. Xie and Zhang focus on efficacy of dual antiplatelet therapy in terms of tailored medicine. Dr. Crotti and colleagues highlight the importance of polymorphisms associated with therapy resistance in people with respiratory diseases in Chap. 23. Dr. Munshi and colleagues discuss the significant heterogeneity in the efficacy and adverse drug reactions of antiallergic and antiasthmatic drugs in Chap. 24. In Chap. 25, Dr. Selvarajan and colleagues extensively discuss the application of personalized medicine in drug response pertaining to type 2 diabetes mellitus and other metabolic disorders. In Chap. 26, Dr. Aguilera and colleagues have focused on clinical biomarkers which impact obese individuals mainly those that detect host biomarkers and also gut microbiota biomarkers. In Chap. 27, Drs. Ganguly and Saha elaborately discuss the biomarkers for infectious diseases in the view of efficacy and safety of tailored medicine. In Chap. 28, Dr. Dwivedi and colleagues focus on understanding the genetic makeup and progression of disorders and treatment in various viral diseases. Dr. Venegas and colleagues present concerns related to pathologies most prevalent in the elderly population like osteoporosis in connection to personalized medicine in Chap. 29. Dr. Kaur and colleagues focus on clinical relevance of genetic variations in terms of response to therapeutics in gynecological disorders in Chap. 30. Dr. Majumdar and colleagues have discussed the lesser studied aspect of pharmacogenomics in the field of infertility in Chap. 31. Dr. Schwartz and colleagues talk about personalized medicine in open-angle glaucoma and age-related macular degeneration, two important causes of visual loss and well studied in opthalmology in Chap. 32. In Chap. 33, Dr. Girija Sanal gives an extensive overview of the field of stem cell biology, genetic engineering, and regenerative medicine in terms of tailored therapy for future applications in the clinic. In the last chapter, Dr. Chemello and colleagues have discussed chronic kidney diseases and calcineurin inhibitors as immunosuppressive therapy.

The book will be a very useful and comprehensive information source as well as good scientific read for both the undergraduate and postgraduate students of medicine and researchers of the allied fields who have keen interest in the field of pharmacogenomics and personalized medicine.

We further welcome the suggestions and comments from the readers of the book for the future editions of the book.

(Editors)
Purba Medinipur, West Bengal, India — Debmalya Barh
Ahmedabad, Gujarat, India — Dipali Dhawan
New Delhi, India — Nirmal Kumar Ganguly

About the Book

"Omics for Personalized Medicine" will give to its prospective readers the insight of both the current developments and the future potential of personalized medicine. The book brings into light how the pharmacogenomics and *omics* technologies are bringing a revolution in transforming the medicine and the health care sector for the better. Students of biomedical research and medicine along with medical professionals will benefit tremendously from the book by gaining from the diverse fields of knowledge of new age personalized medicine presented in the highly detailed chapters of the book. The book chapters are divided into seven sections for convenient reading with the first section covering the general aspects of pharmaocogenomic technology that includes latest research and development in *omics* technologies. The first section also highlights the role of *omics* in modern clinical trials and even discusses the ethical consideration in pharmocogenomics. The remaining six sections focus on the development of personalized medicine in several areas of human health. The topics covered range from metabolic and neurological disorders to non-communicable as well as infectious diseases, and even explores the role of pharmacogenomics in cell therapy and transplantation technology. Thirty-four chapters of the book cover several aspects of pharmacogenomics and personalized medicine and have taken into consideration the varied interest of the readers from different fields of biomedical research and medicine. Advent of pharmacogenomics is the future of modern medicine, which has resulted from culmination of decades of research and now is showing the way forward. The book is an honest endeavour of researchers from all over the world to disseminate the latest knowledge and knowhow in personalized medicine to the community health researchers in particular and the educated public in general.

Contents

Section II Personalized Medicine: Cancer and Autoimmune Disorders

Section III Personalized Medicine: Neuro and Psychiatric Disorders

Section IV Personalized Medicine: Cardiovascular and Pulmonary Disorders

Section I

General Aspects of Pharmacogenomics

Pharmacogenomics, Pharmacoproteomics, and Pharmacometabolomics and Personalized Medicine: An Overview

1

Nalini Raghavachari and Marjan Gucek

Abstract

The mapping of the human genome has been an important milestone in understanding the interindividual differences in genetic predisposition to diseases and individuals' responsiveness to drugs. These factors have now begun to revolutionize the clinical landscape with new therapeutic strategies defined as "personalized medicine." Personalized medicine is believed to transform the traditional "one size fits all" model of medicine by applying individual gene-based information to better manage a patient's disease or predisposition toward a disease and to tailor strategies for the prevention and treatment of diseases. In this context, recent explosion in the omics tools and technologies is believed to generate valuable pharmacogenomic, proteomic, and metabolic information of patients which would serve as potential accelerating factors for the development of personalized medicine. Personalized medicine has the ability to change the overall landscape of medicine from diagnosis and treatment to prevention. The success of personalized therapy in the future will depend on scientific advances in pharmacogenomics, proteomics, and metabolics and on a systems biology approach in the diagnosis and treatment of complex diseases. This chapter discusses the basic concepts and advancements in the fields of pharmacogenomics, pharmacoproteomics, and pharmacometabolomics and their applications in personalized medicine.

N. Raghavachari (✉)
Division of Geriatrics & Clinical Gerontology
National Institute of Aging, NIH, Bethesda,
MD 20892, USA

DNA Sequencing and Genomics Core Facility – Genetics and Developmental Biology Center, National Heart Lung and Blood Institute, National Institutes of Health, Bethesda, MD 20892, USA
e-mail: nalini.raghavachari@nih.gov

M. Gucek
Proteomics Core Facility – Systems Biology Center, National Heart Lung and Blood Institute, National Institutes of Health, Bldg 10, Center Drive, R, 8C 103, Bethesda, MD 20892, USA
e-mail: Marjan.gucek@nih.gov

D. Barh et al. (eds.), *Omics for Personalized Medicine*,
DOI 10.1007/978-81-322-1184-6_1, © Springer India 2013

1 Personalized Medicine

1.1 Basic Concepts

Declare the past, diagnose the present, and foretell the future.

–Hippocrates

The art of medicine, so defined, is a matter of prudence, because it is the combination of careful history taking, diligent examination, and diagnosis followed by skillful treatment and prognosis (Hippocrates 1993). With the advancement of tools and technologies in the omics field, the practice of these acts in clinical medicine has now reached an exciting period wherein prediction, diagnosis, and treatment of diseases can be safely practiced with tailored individualized therapy.

In this context, the human genome project has made available the complete human genome sequence which is currently being analyzed at an accelerating rate to unravel the impact of variations of the human genome sequence on the pathogenesis of complex diseases such as cancer, heart diseases, and diabetes and the response to drug therapy in each of these conditions. The rapid accumulation of knowledge on genome–disease and genome–drug interactions from these analyses has led to the emergence of personalized medicine in the clinical landscape. This field though young is a rapidly advancing field of healthcare that is informed by each person's unique clinical, genetic, genomic, proteomic, and metabolomic information (Blum 2011; Farooqi et al. 2012). Because these factors are different for every person, the nature of diseases including their onset, their course, and how they might respond to drugs or other interventions is as individual as the people who have them. Thus, personalized medicine involves identifying precisely the molecular makeup of each individual information that allows accurate predictions to be made about a person's susceptibility of developing disease, the course of disease, and its response to treatment (Davis et al. 2009; Sadee 2011).

In current practice, personalized medicine consists of trying to integrate comprehensive knowledge on individual factors such as genetic makeup or environmental conditions in diagnostic decisions or therapeutic interventions. In addition, it contributes to discovery and clinical testing of new drugs, biomarkers, and other medical products. Though sometimes personalized medicine is still described as a phenomenon of the future, it is believed to have begun its impact on patient treatments.

Few applications where molecular testing is being used include identification of breast and colon cancer patients who are likely to benefit from new treatments and newly diagnosed patients with early-stage invasive breast cancer to determine the likelihood of recurrence (Longo et al. 2010; Chouchane et al. 2011).

In another example, a genetic test for patients with an inherited cardiac condition aims to help their physicians determine which course of hypertension treatment to prescribe in order to maximize benefit and minimize serious side effects (Bochud et al. 2011). In this respect, while genomic information has been well utilized, pharmacoproteomics and pharmacometabolomics are only starting to emerge in the medical and proteomics field with a goal to bring understanding to the role of proteins in drug development and disease. Thus personalized medicine promises many medical innovations and has the potential to change the way clinical medicine is being practiced now (Mini and Nobili 2009; Chouchane et al. 2011; Mehta et al. 2011).

1.2 Benefits and Challenges

Personalized medicine promises significant benefits to patients, physicians, and pharmaceuticals and also has few challenges. Major benefits include:

(a) *Early diagnoses and therapy interventions.* Molecular analysis could determine precisely which variant of a disease a person has, or whether an individual is susceptible to drug toxicities, to help guide treatment choices. In healthy individuals such analysis could improve the ability to identify which individuals are predisposed to develop a particular condition and guide decisions about interventions that might prevent it, delay its onset, or reduce its impact (Panareo et al. 2011; Sadee 2011; Sudhindra et al. 2011).

(b) *Efficient drug development.* A better understanding of genetic variations could help scientists identify new disease subgroups or their associated molecular pathways and design drugs that target them. Molecular analysis could also help select patients for inclusion in, or exclusion from, late-stage clinical trials – helping gain approval for drugs that might otherwise be abandoned because they appear to be ineffective in the larger cohort of patient population (Hong and Oh 2010; Mocellin et al. 2010).

(c) *Effective therapies.* Currently, physicians often have to use trial and error to find the most effective medication for each patient. With the knowledge obtained from GWAS on molecular variations, it will be easier to best predict how a patient will react to a treatment, develop accurate and cost-effective tests, and determine which medications are likely to work best (Vermeire et al. 2010; Awada et al. 2012) for each individual.

Although the benefits with the practice of personalized medicine are numerous, it also has inherent challenges especially in healthcare policy and ethics in clinical trials, intellectual property rights, reimbursement policies, and patient privacy and confidentiality. Given the array of issues, it is extremely important that a broad spectrum of life science companies, healthcare providers, and policymakers participate in shaping the successful application of this powerful new opportunity to prevent, diagnose, and manage diseases (Qattan et al. 2012; van Rooij et al. 2012). Nevertheless, it can be confidently said that with the adaptation of an integrated policy framework that balances new tools and technologies and patient, industry, and scientific interests, personalized medicine would be able to fulfill its promise as rapidly as possible and accomplish the major goal of the human genome project/genomic medicine. With these advancements in science, medical practice in the future will be customized to each individual as depicted in Fig. 1.1.

2 Pharmacogenomics and Personalized Medicine

2.1 Basic Concepts

Pharmacogenomics as a recently emerged discipline stems from the fusion of two scientific fields pharmacogenetics and genomics (Newman 2012). The term pharmacogenomics is also intertwined with the term personalized medicine and in general refers to the study of how genes affect the way a patient responds to medication. Genomic differences can influence the efficacy of medications, can be the source of serious drug side effects, and can increase the risk of drug-to-drug interactions (Whirl-Carrillo et al. 2012). By having an evidence-based report of a patient's genomic drug suitability profile, a clinician can better understand how their specific patients may react to a medication. The rapid accumulation of knowledge on genome–disease and genome–drug interactions has also impelled the transformation of pharmacogenetics into a new entity of human genetics, namely, pharmacogenomics. Enabled by high-throughput technologies in DNA analysis, genomics introduces a further dimension to individualized predictive medicine. Determining an individual's unique genetic profile in respect to disease risk and drug response will have a profound impact on understanding the pathogenesis of disease, and it may enable truly personalized therapy (Ma et al. 2012). This concept of therapy with the right drug at the right dose in the right patient has emerged as an urgent requirement in a recent survey of studies conducted in the USA on adverse drug effects in hospitalized patients (Nielsen and Borregaard 2009; Rio et al. 2009; Roederer and McLeod 2010; Madian et al. 2012). Thus, it is anticipated that pharmacogenomics will play an integral role in disease assessment, drug discovery and development, and selection of the type of drug. Moreover, it may provide information useful to the selection of dosage regimen for an individual patient. Pharmacogenomics holds great promise to shed scientific light on the costly process of drug

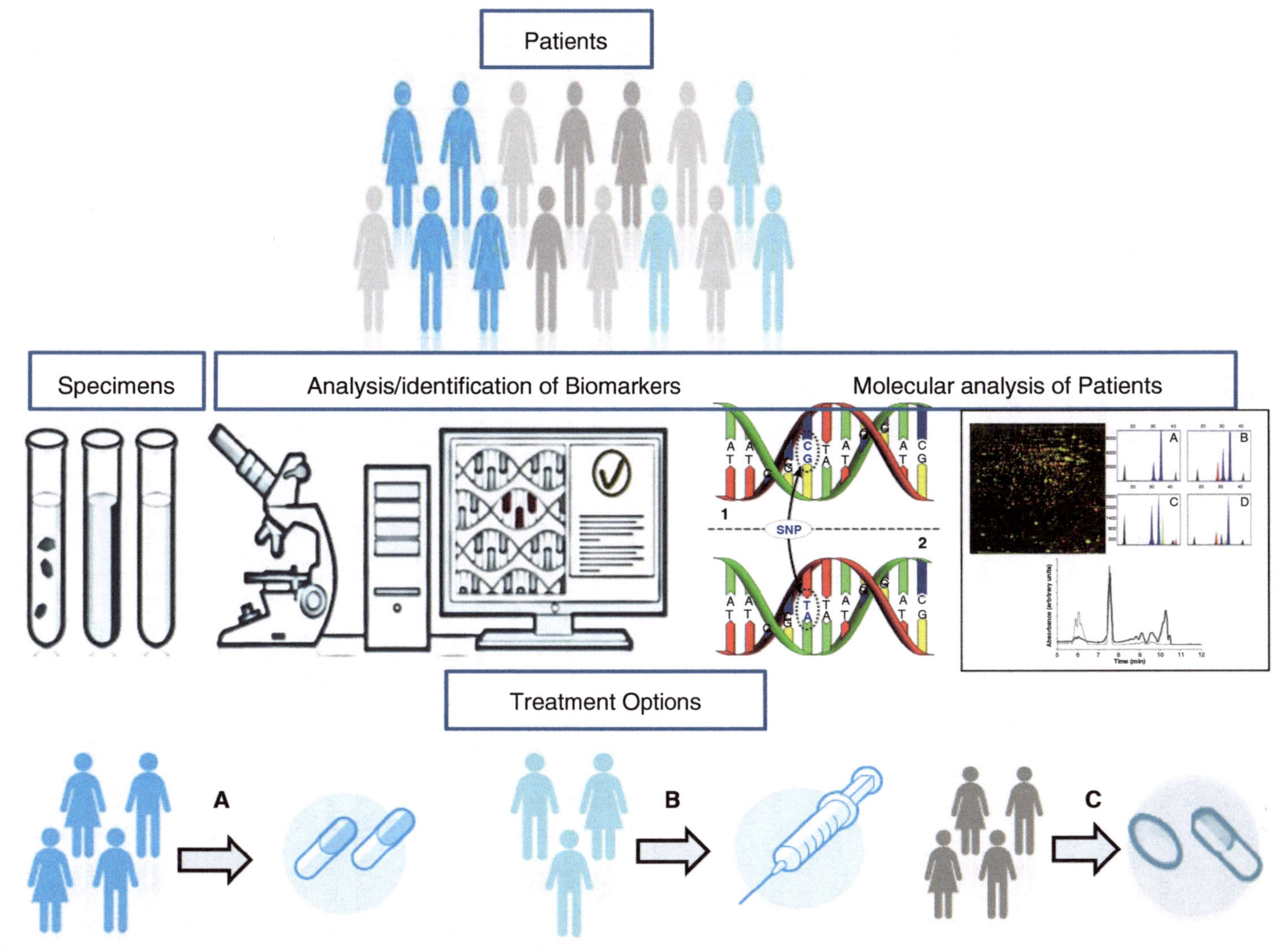

Fig. 1.1 The future of medicine

development and to provide greater confidence about the risks and benefits of drugs in specific populations.

2.2 Tools and Technologies

In the last decade, examination of human diseases for biomarker discovery and disease mechanism has advanced significantly with high-throughput technologies such as high-density microarrays manufactured in companies such as Affymetrix, Illumina, and Agilent (Hardiman 2008; Bhasker and Hardiman 2010) and deep sequencing of the genomes of humans and other species (Matkovich et al. 2010). Technological progress in analyzing millions of genes and intergenic variants in the form of single nucleotide polymorphisms (SNP) and copy number variants per individual has accelerated our comprehension of individual differences in genetic makeup. Genome-wide association studies (GWAS) have successfully identified common genetic variations associated with numerous complex diseases. However, most common genetic variants have been shown to confer only modest risk, and their association accounts for a small portion of the total heritability of disease variance. This has led to reexamination of the contribution of the environment, gene–environment interaction, epigenetics, and rare genetic variants in complex diseases. Next-generation (Illumina, SOLiD, 454, Ion torrent) or third-generation sequencing technologies (PacBio, Nanopore) have now emerged (Pareek et al. 2011) and new more efficient and cost-effective tools are evolving rapidly. With the significant drop in cost for sequencing, it is now possible to perform whole exome sequencing to identify rare variants and to use novel exome arrays to genotype these rare variants in the population at an acceptable cost and time. Recent advances in the analysis of regulatory elements such as miRNA and long noncoding RNA and epigenetic changes that regulate to a large extent modulations in expressed genes have resulted in an explosion of genomic data that can be well utilized for unraveling novel disease biomarkers and disease mechanisms which in turn could result in novel drug discovery (Costa 2009; Bhartiya et al. 2012). However, the analysis of rare variants is still challenging since the methods employed are underpowered and the amount of information generated is overwhelming, frequently exceeding available computational capacity. Rapid advances in genomic technologies are being quickly adopted by researchers and are expected to fuel great progress in this area of pharmacogenomics.

2.3 Applications in Medicine

This evolving field of pharmacogenomics is at an intersection of pharmaceuticals and genetics and specifically studies the variability in drug response due to heredity. Pharmacogenomics is gaining a lot of momentum with the advent of new methods and technologies for genome analysis and is widely believed to play a major role in predictive and personalized medicine (Roden et al. 2006). The way a person responds to a drug (in both positive and negative manner) is a complex trait that is influenced by many different genes. Without knowing all of the genes involved in drug response, scientists have found it difficult to develop genetic tests that could predict a person's response to a particular drug. The inherited variations in genes that dictate drug response and explore the ways of these variations can be used to predict whether a patient will have a good response or a bad response or no response at all to a particular drug. Most importantly, the knowledge on the influence of genetic variation on drug response in patients by correlating gene expression or single nucleotide polymorphisms with a drug's efficacy or toxicity is believed to be immensely helpful in reducing drug-caused morbidity and mortality. Pharmacogenomics will have high impact in areas such as oncology where many therapies are currently available, but each one works only for a small percentage of cancer patients. Genetic polymorphisms of proteins involved in drug targeting (i.e., pharmacodynamics) and drug metabolism and transport (i.e., pharmacokinetics) are the most important causes of individual variability in drug safety and efficacy. Some genetic variations can affect them by

Table 1.1 Few examples of individualized treatment of patients with some of the commonly used drugs

Biomarker	Disease	Drug	Application
Her-2/neu receptor	Breast cancer	Herceptin	Select Herceptin (trastuzumab) for breast cancer
BRCA1/2	Breast/ovarian	Tamoxifen	Breast and ovarian cancer inherited risk, prophylactic tamoxifen and surgery
CYP2D6/CYP2D19	Cancer	Cancer drugs	Guide prescribing/adjust dose of ~25 % of commonly used drugs
VKOR/CYP2C9	Blood disorders	Warfarin	Dosing of warfarin
CYP2C19	CVD	Clopidogrel	Dosing for variability in platelet aggregation
HLA-B	HIV	Abacavir	Genotyping to determine hypersensitivity
B lactam	Bacterial infection	Flucloxacillin	To determine sensitivity
CYP450	Hypercholesterolemia	Statins	Risk of myopathy

changing the biological context or environmental sensitivity of the drug response.

A classic example of the use of pharmacogenomics is in the use of the drug warfarin. With respect to the drug targets for the drug warfarin, the commonly used anticoagulant, polymorphisms located in the enzyme vitamin K epoxide reductase complex subunit 1 (VKORC1) result in differential response of patients to the drug (Hall and Wilkins 2005). Inhibition of VKORC1 by warfarin leads to vitamin K depletion and, consequently, the production of coagulation factors with reduced activities. Several mutations in the coding region of VKORC1 have been identified (Joffe et al. 2004) and have been found to be associated with resistance to warfarin, necessitating larger warfarin doses. Similarly, Cytochrome P450 family members are important enzymes that metabolize most, if not all, clinical drugs (McGraw and Waller 2012). With CYP450 polymorphisms, subjects may process the medication too rapidly (ultra-metabolizers), rendering it ineffective, or too slowly (poor metabolizers), so that drug concentrations will build up in blood, potentially causing adverse reactions and, in contrast, in case of prodrugs, ineffective activation (McGraw and Waller 2012). Another classic example of pharmacogenomics is the usage of tamoxifen, best-known cancer drug, and the level of CYP2D6, one of the best-studied drug-metabolizing enzymes in each individual cancer patient (Beverage et al. 2007). About 10 % of the general population has a slow-acting form of this enzyme, and 7 % have a superfast-acting form. In total, 35 % are carriers of an abnormally functional 2D6 allele. This thereby suggests that prior knowledge on the expression of CYP2D6 would be highly beneficial for tailored treatment of cancer patients (Beverage et al. 2007).

Similarly, statins that are widely used in the treatment of hypercholesterolemia provide a practical application for pharmacogenomics. The mechanism of statin-induced myopathy is still unclear and generally rare: 1/10,000 cases per year on standard therapy (Donnelly et al. 2011). However, the incidence increases with higher statin doses or concomitant use of other drugs, such as cyclosporine or amiodarone. A common SNP of SLCO1B1 an influx transporter on the basolateral membrane of hepatocytes that facilitates the drug uptake has been shown to markedly affect individual variation of statin pharmacokinetics in healthy volunteers (Mangravite et al. 2008; Superko et al. 2012). It was found that that genotyping SLCO1B1 could be useful to identify persons with abnormal SLCO1B1 and thereby increase the safety and efficacy of statin therapy (Donnelly et al. 2011). These applications are illustrated in Table 1.1.

The above mentioned examples of successful pharmacogenetic testing, unraveling the basis of some individual drug responses by single-gene polymorphisms, can guide future research in pharmacogenomics and its application. In addition, identification of gene pathways and networks; the integration of genetics and genomics, proteomics, metabolomics, and epigenetics; and noncoding RNA derived from hypothesis-free investigation in GWAS could immensely help practicing physicians in providing a tailored successful treatment regimen for patients in a timely and cost-effective manner. Pharmacogenomics is expected to help physicians and patients by enabling pharmaceutical companies to bring more drugs into market that are targeted at those patients who are more likely to benefit from the drugs.

2.4 Challenges

While technological advancements have accelerated the identification of biomarkers of disease susceptibility, treatment, and progression, the challenge that we now face is in the implication of these discoveries in clinical practice. This requires replication of the finding in an independent population followed by pathway elucidation and validation of efficacy and clinical utility in at-risk populations. Most importantly, it has to be borne in mind that establishing relationships between numerous genomic markers and clinical phenotypes is not an easy task. Large-scale, prospective clinical studies are difficult but much needed to demonstrate the utility of genotyping as a basis for individualized medicine.

As personalized medicine enters mainstream medical practice, physicians and other healthcare providers will have to administer the application of growing numbers of genomic tests, pharmacogenomically guided therapies, and treatment decisions based on predictive evidence and risk estimation. Future advances in human genomics and their application to pharmacogenomics are expected to bring continued advances to this important area of translational medicine and revolutionize the practice of medicine in future.

3 Pharmacoproteomics

3.1 Basic Concepts

Pharmacoproteomics is the application of proteomics technologies to drug discovery and development (Jain 2004). In contrast to pharmacogenomics, which is a well-established concept, pharmacoproteomics is only starting to emerge in the medical and proteomics field with a goal to bring understanding to the role of proteins in drug development and disease. The roles of proteins are very diverse, and mass spectrometry-based proteomics has established sophisticated tools and instruments that can identify proteins and measure the changes in protein levels, posttranslational modifications (e.g., kinase signaling), localization, and protein–protein or drug–protein interactions.

The key to a successful drug development is a complete understanding of the disease and its pathophysiology. Proteomics is only one of the pieces of the big puzzle, with other omics approaches bringing other pieces to form the full picture. Such a complex network of genes, proteins, and small molecules yields a functional phenotype, with perturbations to the network representing pathophenotype (Loscalzo 2011). Nevertheless, there is a lot of variability among individuals in their genome, proteome, or metabolome; that is why "personalized medicine" tries to measure and tailor the drugs specific to an individual (Bencharit 2012). The high-throughput omics technologies are on the verge of being able to deliver on the promise of personalized medicine with proteomics being almost there.

The term is seldom used in the scientific literature; most of the citations are coming in from the recent years: a PubMed search for pharmacoproteomics shows 63 papers as of September 2012 and only 18 for pharmacometabolomics, whereas generic "proteomics" search term yields 35,853 and the search term "pharmacogenomics" yields 12,866 results (PubMed). Given that the proteomics field has seen a dramatic growth only in the last 10 years, it is only now that sub-proteomics fields are emerging (such as organellar proteomics, serum proteomics, disease proteomics, pharmacoproteomics (Witzmann and Grant 2003). For the purpose of this chapter, we are going to focus on proteomics as a tool to understand disease and its markers and on proteomics as a platform for personalized medicine.

In the past decade, the new proteomics technological advances were driven by biomarker discovery, especially for early detection and treatment of diseases such as cancer and cardiovascular diseases. The discovery phase is usually less challenging as compared to bringing these biomarkers into clinical setting which is a more daunting task. For example, there are currently just over 200 protein targets in plasma proteome that are measured clinically, which only represents less than 1 % of the human proteome (Anderson 2010). Additionally, the rate of validation of biomarkers from discovery phase into clinical use has been extremely low. There is still a lot of work to be done to bring the findings from basic discovery studies into clinical setting.

3.2 Tools and Technologies

Proteomics studies proteins and their posttranslational modifications, protein interactions, and their localization. Mass spectrometry has been the method of choice for identification and quantification of proteins since the discovery of electrospray and MALDI ionizations which enabled to charge the biomolecules in vacuum (Fenn et al. 1989; Cho and Normile 2002). There has been a steady stream of improvements in sensitivity and speed of mass spectrometers enabling to detect lower amounts of peptides/proteins, with a greater reliability. The current state of the art is more than 10,000 proteins identified in a cell line (Nagaray et al. report identification of 10,255 proteins in HeLa cell line (Nagaraj et al. 2011), whereas Beck et al. report that U2OS cells express at least 10,000 proteins; for 7,300 proteins they also estimated their cellular concentrations (Beck et al. 2011)). Once we take into account the protein posttranslational modifications (PTMs), the numbers increase greatly, especially with the most widely studied phosphorylation. Huttlin et al. carried out phosphoproteomic characterizations of nine mouse tissues and identified nearly 36,000 phosphorylation sites in over 6,000 proteins (Huttlin et al. 2010). The identifications of novel sites will likely to continue since there are up to 500,000 possible phosphorylation sites in the human proteome (including all serines, threonines, and tyrosines).

All these numbers are impressive; however, they pose a significant bioinformatics challenge to extract some meaningful hypothesis from this vast information pool. Systems biology strives to organize these data in order to understand the internal workings in a cell, its signaling, and network interactions, either among proteins, or proteins–metabolites, or proteins–genes. As a consequence of the developments in systems biology approaches, there is an emerging paradigm to view human disease as a complex network of proteins, genes, and metabolites that all contribute to the disease phenotype (Loscalzo 2011; Chan et al. 2012). With the advent of all the high-throughput omics approaches, it is now possible to embark onto personalized screening of an individual's genome, proteome, and metabolome and compile and compare the data with systems biology approaches, all in order to understand the disease and/or potential risk for developing a disease. Chen et al. used such an approach in a groundbreaking paper which brings the personalized medicine into reality and proves that systems biology approach can work in medicine (Chen et al. 2012).

Mass spectrometry-based proteomics is used to identify and quantify proteins and their posttranslational modifications, it can be also utilized to study protein–protein or protein–drug interactions, and it can be used to follow temporal protein changes during progression or treatment of a disease. Although mass spectrometry was invented more than 100 years ago (Downard 2007), its application in the arena of large biomolecules is fairly recent. The novel instrument designs (dual-pressure ion trap) and mass analyzers (orbitrap) (Hardman and Makarov 2003) have revolutionized the way we identify and quantify proteins. These new instruments have sensitivities and speeds that were almost unimaginable a decade ago. Coupling these capable instruments with new approaches in sample preparation (e.g., enrichment of various posttranslational modifications, antibody-based protein/peptide enrichment) and fractionation (e.g., isoelectric focusing, strong cation exchange chromatography) led to a wealth of proteomics data that can be mined using systems biology approaches. Simultaneously, relative quantification of proteins has become easier using many new approaches and labeling

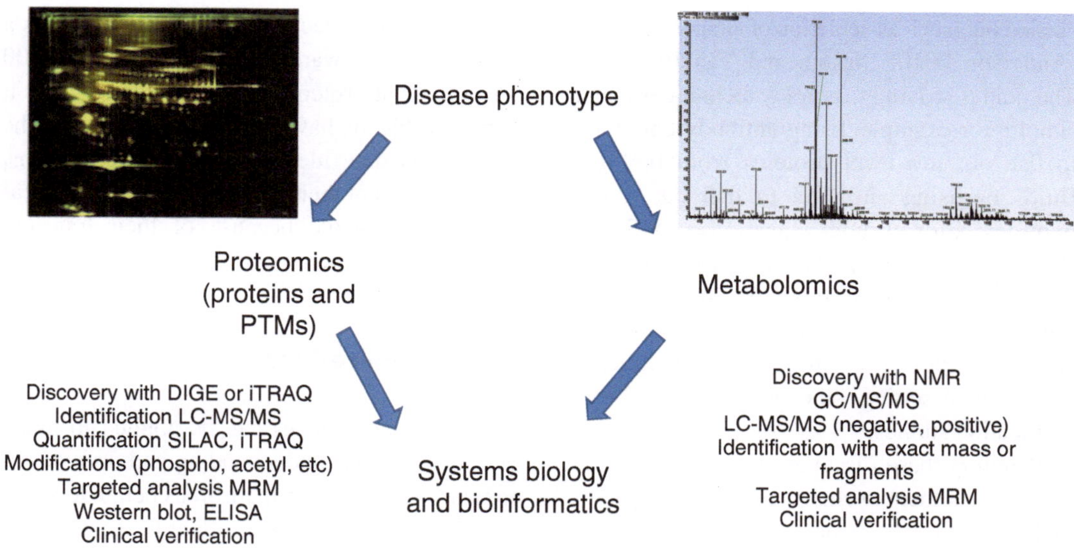

Fig. 1.2 Proteomics and metabolomics tools for disease biomarker discovery

chemistries (label-free, SILAC, iTRAQ, TMT) (Boja and Rodriguez 2012; Paulo et al. 2012). Nevertheless, traditional DIGE (2D fluorescence difference gel electrophoresis) approach is still well suited for clinical samples. In this type of experiment, the two protein samples (control, drug treated) are labeled with fluorescent dyes, mixed together, and run in a single gel to see differences in protein levels and/or posttranslational modifications (see Fig. 1.2 for overview of mass spectrometry tools and approaches).

Posttranslational modifications are attachments of small groups onto amino acids and can have a profound effect onto activity, localization, and folding of proteins (although they tend to be decorative sometimes). The most widely studied are the following (listed are also some recent reviews on respective PTMs): phosphorylation (Macek et al. 2009; Grimsrud et al. 2010; Nita-Lazar 2011), acetylation (Choudhary et al. 2009), S-nitrosylation (Gucek and Murphy 2010; Murphy et al. 2012), glycosylation (An et al. 2009), ubiquitination (Shi et al. 2011; Vertegaal 2011), etc. Because all of these moieties are usually substoichiometric (individual amino acids are only modified up to just a few percentages), there are enrichment strategies unique for each PTM that take advantage of physicochemical properties or are antibody based with the antibody targeting a specific modified amino acid. An example for the former is enrichment of phosphorylated proteins and peptides using immobilized metal affinity chromatography (IMAC) and an example for the latter is antibody-based enrichment of acetylated peptides using anti-acetyllysine antibody (Kim et al. 2006). Phosphoproteomics is able to identify thousands and thousands of modified sites by exploiting the physicochemical characteristics of the phosphate group to interact with positively charged metal ions: immobilized metal affinity chromatography (IMAC, gallium or iron based), metal oxide affinity chromatography (MOAC, titanium dioxide based), and strong cation exchange (SCX). (A comparison of phosphopeptide and/or phosphoprotein enrichment strategies is given by Grimsrud et al. (Grimsrud et al. 2010).)

Pharmacoproteomics faces some challenges that are connected with the sample types that are usually used when, e.g., studying protein–drug interactions in biological fluids (serum, saliva, tissue specimens). Such samples are difficult to solubilize, have a wide protein dynamic range, and contain abundant proteins that need to be depleted before proteomics analysis can be carried out (such as depletion of albumin in serum) (Anderson 2010; Stastna and Van Eyk 2012). The field is striving to employ techniques that are simple, for example, using antibody enrichment to fish out low-level proteins from biological fluids or using antibody to fish out specific peptides after tryptic digestion – SISCAPA (Kuhn et al. 2012). These approaches have proven to be viable in biomarker discovery with reproducibility of determination – which is a crucial component of any protein quantification in a clinical setting – becoming acceptable for verification studies.

Bringing these discovery tools into a clinical setting is another big challenge. Protein separation tools and mass spectrometric detection are well suited for the discovery process of biomarkers or measuring relative changes in protein levels and their PTM changes in disease or drug treatment. The next step is usually a verification step where several clinical samples (up to 100) are analyzed to verify that a given protein biomarker is actually uniformly changing across these specimen. The methods mostly used can be traditional (western blots and ELISA) or sophisticated mass spectrometric assays using triple quadrupole mass spectrometers (MRM – multiple reaction monitoring assays) (Picotti and Aebersold 2012). These types of instruments are capable of reliably quantifying hundreds of analytes in a single, fast separation run with minimal sample preparation. The idea is to bring these methods later into a routine use in a clinical setting. The technology for protein/PTM quantification has progressed enormously in the recent years: faster, high-pressure peptide/protein separation (UPLC), fast-scanning triple quadrupole instruments, and powerful data processing.

On the other hand, the use of proteomics in personalized medicine has to satisfy some criteria that are at this stage still difficult to meet – it has to be fast (proteomics usually employs a lot of laborious sample preparation), comprehensive (it is impossible to identify all proteins in a sample without extensive fractionation and prolonged mass spectrometry time), and cheap (the most powerful mass spectrometers cost well over $500 k, not including reagents and expertise). Especially, the aspect of comprehensiveness is a big bottleneck toward the ultimate goal "$1,000 per individual proteome." At the current stage it is impossible to have it all – even though the instruments may identify most of the proteins, identifying all of their posttranslational modifications is impossible because of their transient nature and substoichiometric ratios.

3.3 Applications

Proteomics approaches can be applied to many aspects of disease understanding and drug development, with the three main areas being development of protein biomarkers, identification of drug targets, and understanding of the molecular mechanism of drug action (Yu 2011).

A major drive for proteomics research in the last decade was biomarker discovery for early detection and diagnosis in cancer and cardiovascular diseases. For example, prostate-specific antigen (PSA) was introduced 25 years ago as a protein marker to track progression of prostate cancer and has been very successful to detect early stages of prostate cancer, but it has been also attributed to overdiagnosis in some cases (Prensner et al. 2012).

Proteomics technologies can be successfully applied in cardiovascular field, for example, Addona et al. developed and validated a pipeline to identify early biomarkers of cardiac injury from the blood of patients undergoing a therapeutic, planned myocardial infarction (PMI) for treatment of hypertrophic cardiomyopathy (Addona et al. 2011). Another very promising protein biomarker, especially for diagnosis of acute myocardial infarction, is troponin (Jaffe 2012). The cardiac troponin consists of three different proteins (encoded by three different genes) that as a complex play a regulating role in cardiac contraction. The sensitivity of troponin determination has increased significantly in the last years, enabling now to measure the levels of troponin in healthy individuals and thus set a benchmark values for healthy and acute states. There are other promising cardiovascular biomarkers, such as C-reactive protein (CRP) – a 25 kDa protein associated with increased cardiovascular risk in several studies (Hochholzer et al. 2010).

There have been many studies that utilized proteomics approaches to detect therapeutic responses and drug actions. Specifically, proteomics analyses of drug effects on, e.g., a cell line or an animal has benefitted a lot from recent developments in relative quantification techniques (such as SILAC or iTRAQ). In such studies, a drug is added to one sample for a period of time, and then the protein pool is compared to the control, to find the differences either in protein levels or in the levels of specific posttranslational modifications (such as phosphorylation, acetylation). Some techniques (iTRAQ and TMT) also enable measuring time-dependant protein profiles in up to eight samples at a time.

Malki et al. used a pharmacoproteomic approach to study the response to antidepressants for two inbred mouse strains with the goal to reveal molecular mechanisms underlying antidepressant action and identify new biomarkers for two antidepressants with proven efficacy in the treatment of depression but divergent mechanisms of action (Malki et al. 2012). The study was able to identify three proteins involved in serotonergic (PXBD5, YHWAB, SLC25A4) and one in noradrenergic antidepressant action (PXBD6).

Apart from studying proteomics differences in "western medicine," there is a vast interest in the usage of natural herbal products for therapeutic applications in treatment of diseases and disorders that are little responsive to "western drugs." For example, Manavalan et al. investigated the effects of tianma (a traditional Chinese herbal medicine often used for the treatment of cerebrovascular diseases) on the brain protein metabolism by quantitative proteomics to gain evidence for a direct relationship between tianma treatment and brain functions (Manavalan et al. 2012). They found out that long-term treatment with tianma effects the brain by downregulating the expressions of various proteins, such as Gnao1 and Dctn2, which are related to neuronal growth cone control and synaptic activities.

Another way that quantitative proteomics can be used is to study the drug action on thousands of peptides which are posttranslationally modified (e.g., phosphorylation sites). In a recent study, triple labeling by stable isotope labeling by amino acids in cell culture (SILAC) was used to quantify the effect of kinase inhibitors on phosphorylation sites in growth factor-stimulated cells (Pan et al. 2009). The authors found out that dasatinib, a clinical drug directed against BCR–ABL, which is the cause of chronic myelogenous leukemia, affected nearly 1,000 phosphopeptides. A similar study was conducted on lysine acetylation and the authors were able to identify 3,600 lysine acetylation sites on 1,750 proteins and quantified acetylation changes in response to the deacetylase inhibitors suberoylanilide hydroxamic acid and MS-275 (Choudhary et al. 2009). These examples of drug action on the protein levels or the levels of PTMs are a showcase that pharmacoproteomics is technologically mature to reveal the intricacies of drug actions in an organism or cell. Some selected applications of pharmacoproteomics are listed in Table 1.2.

3.4 Challenges

There was an unprecedented technological advancement in proteomics in the last decade, both on the instrumental side and the sample preparation side. Pharmacoproteomics is slowly emerging as an important tool in understanding the disease, one of the important omics approaches in the systems biology view of modern medicine. The main challenge is still the sample preparation (proteins in serum), comprehensiveness of the proteomics analysis, and data analysis; however, making sense of all the proteins and their modifications is probably the biggest challenge of them all.

4 Pharmacometabolomics

4.1 Basic Concepts

Metabolomics is an emerging omics tool with the goal to analyze (identify and quantify) endogenous small molecules in a biological system (Patti et al. 2012). It has the potential to

Table 1.2 Selected applications of pharmacometabolomics

Disease marker/drug studied	Sample source	Notes on dose and drug response	References
Protein changes with nortriptyline (antidepressant)	Mouse brain	Dose, 5 mg/kg Change in PRDX6 only	Malki et al. (2012)
Protein changes with escitalopram (antidepressant)	Mouse brain	Dose, 4 mg/kg Change in PRDX5, SLC25A4, and YWHAB	Malki et al. (2012)
Tianma effect on brain functions	Rat brain	Dose, 2.5 g/kg/day Downregulation in Gnao1 and Dctn2	Manavalan et al. (2012)
Dasatinib effect on protein phosphorylation	Growth factor-stimulated cells	Dose, 10 mM dasatinib Effect on downstream MAPK pathways	Pan et al. (2009)
MS-275 effect on protein acetylation	Whole-cell lysates of MV4-11 cells	Dose, 5 μM Regulation of cyclin-dependent kinase Cdc28	Choudhary et al. (2009)

describe any biological state with thousands of metabolites, their networks, and interactions. A metabolomics study often comprises of a relative quantitation on sample sets from a normal state and a treated state, where the treatment can be of any nature, such as genetic knockout, administration of a drug, or change in diet. Metabolomics is relevant clinically, especially with promising discoveries of metabolic biomarkers that are a direct signature of biological functions and are thus easier to track back to the phenotype (Kaddurah-Daouk et al. 2008). Genes and proteins, on the other hand, are subjected to epigenetic regulation for genes and posttranslational modification for proteins. Serum metabolomics is becoming a tool of choice in many clinical diagnostics applications where a metabolite can be measured for early detection of disease (Zhang et al. 2012b). Furthermore, metabolomics profiles after drug administration can be used as a valuable predictor of a drug's efficacy.

As it is the case in pharmacoproteomics, the term pharmacometabolomics is being gradually used more and more. It is especially useful with regard to evaluating the drug response phenotypes by measuring the metabolites before and after drug administration in order to highlight the biochemical pathways that may be targets for the drugs (Kaddurah-Daouk et al. 2008). With the impressive advancements in technology, these types of studies will become more common.

4.2 Tools and Technologies

Discovery metabolomics is as challenging as the proteomics counterpart. In fact, some of the instrumentation can be used for both, such as mass spectrometers can be utilized for metabolomics or proteomics work. Again, the bioinformatics challenge in metabolomics is even more profound – there are potentially tens of thousands small molecules in any biological system. Identifying those and assembling their interactions into a meaningful network is a huge task.

The technologies mostly used for metabolomics are NMR and mass spectrometry-based platforms. NMR is very comprehensive and can easily deduce structure of small molecules. Its disadvantages are relatively low sensitivity and a high cost associated with the purchase and maintaining of the instruments (Serkova and Brown 2012). Mass spectrometry-based platforms are used both in the discovery phase and in the routine applications. Metabolites are separated either by gas chromatography (GC-MS) or by liquid chromatography (LC-MS) (Zhang et al. 2012a). GC-MS has been used for analysis of small molecules for a long time, and it can be very sensitive and reproducible and uses less expensive instrumentation; however, it is only applicable to volatile compounds unless we are willing to perform derivatization which can increase the cost of analyses. LC-MS-based techniques are more suited for polar molecules which can be analyzed in positive or negative electrospray mode (modern instruments enable also rapid switching between the positive and negative mode, thus collecting valuable information on all the types of metabolites). In the discovery metabolomics, time-of-flight instruments are mostly employed, which can measure accurate metabolite mass (up to a few ppm) and provide structural information. On the other hand, in the clinical setting with targeted metabolomics, MRM (multiple reaction monitoring) is mostly used with triple quadrupole instruments.

Data analysis in the discovery metabolomics often helps in finding the difference in metabolite levels between two or more samples (e.g., healthy vs. control) and then identifies such a compound. Another feature of data analysis is network analysis and integrating the metabolic information with protein and gene information to get the networks of interconnected small molecules, proteins, and genes, thus the systems biology picture of a biological system. And this is where the biggest challenges are (Korman et al. 2012; Ram et al. 2012).

4.3 Applications

Advances in technology (both mass spectrometry and NMR based) enable metabolomics to have a great impact on pharmacology. It's becoming easier to answer questions about a drug's effect on metabolites, its toxicity, and efficacy. Furthermore, metabolomics can predict response of an individual to a drug before the drug is administered (Nicholson et al. 2011), leading to the ultimate goal of individualized therapy.

One of the fundamental examples of pharmacometabolomics in practice was work by Clayton et al. who used NMR to study metabolomics profile of urine before paracetamol (acetaminophen) was administered to rats (Clayton et al. 2006). They showed that they can predict the extent of liver damage sustained after paracetamol administration on the basis of predose urinary composition. The authors envisioned that this technology "might provide the basis of a future population-screening tool for selecting individuals according to their suitability for treatment with particular drugs, drug classes, or drug doses." Nevertheless, when a similar study was carried out in humans, the predose metabolomic urine profile was not able to predict which patients would develop mild liver injury (Winnike et al. 2010). However, the authors were able to draw such a prediction from comparing the urine metabolite profiles before and shortly after dosing.

Changes in urine metabolite patterns as a molecular marker have been also utilized by Klawitter et al. to assess the effect of cyclosporine on the kidney after a single dose in healthy humans (Klawitter et al. 2010). The study used both NMR and LC-MS to detect changes in urine metabolite profile as early as 4 h after taking an oral dose of cyclosporine. The major metabolite changes were in the levels of citrate, hippurate, lactate, TMAO, creatinine, and phenylalanine.

In the field of cancer, metabolomics profiling can be used to qualitatively measure cancer progression. For example, sarcosine (an N-methyl derivative of the amino acid glycine) is a metabolite and a potential biomarker for early detection of prostate cancer (Sreekumar et al. 2009). The authors used a combination of GC-MS and LC-MS to profile more than 1,100 metabolites across 262 clinical samples related to prostate cancer. They discovered that as prostate cancer progresses toward metastatic disease, amino acid metabolism along the nitrogen breakdown pathways increases. As a consequence, the authors believe that the levels of sarcosine increase with the severity of prostate cancer. Some selected applications of pharmacometabolomics are listed in Table 1.3.

4.4 Challenges

The increasing sensitivity in metabolite detection and quantification has helped in identifying more small molecules and their patterns that can be correlated with a disease or the drug's effect on the organism. Pharmacometabolomics has a potential to play a key role in personalized medicine, especially because of the fact that metabolites are downstream from genes and proteins, thus reflecting the changes in phenotype

Table 1.3 Selected applications of pharmacometabolomics

Disease marker/drug studied	Sample source	Notes on dose and drug response	References
Effect of acetaminophen on urine metabolites (rats)	Rat urine	Dose, 600 mg/kg The ratio of paracetamol glucuronide to paracetamol as predictive marker	Clayton et al. (2006)
Effect of galactosamine hydrochloride on urine metabolites (rats)	Rat urine	Dose, 800 mg/kg The metabolites classify animals to "responders" and "nonresponders"	Clayton et al. (2006)
Acetaminophen effect on urine metabolites (human)	Human urine	Dose, 4 g acetaminophen a day for 7 days Predictive metabolites are derived from NAPQI	Winnike et al. (2010)
Cyclosporine effect on urine metabolites	Human urine	Dose, 5 mg/kg single Predictive metabolites are citrate, hippurate, lactate, TMAO, creatinine, phenylalanine	Klawitter et al. (2010)
Sarcosine as a metabolite marker in prostate cancer	Tissue, urine, and plasma in prostate cancer patients	Dose, no treatment Predictive metabolite is sarcosine which increases with cancer progression	Sreekumar et al. (2009)

most rapidly. The ideal outcome of all the omics approaches would be a personal genome, proteome, and metabolome which can all be used by a physician to make a decision on treatment.

5 Future Perspectives

Systems biological approach by integrating genomic, proteomic, and metabolomic information from each patient is transforming the era of "one size fits all" therapy concepts into a new era of personalized medicine that already has emerged into the clinical landscape. The individualization of medicine and healthcare appears to be a promising trend, wherein right therapy for the patient at the right dose and time can be implemented successfully. The future of personalized medicine should therefore rely on the methods of systems biology to develop more efficient and effective diagnostic, prognostic, and predictive biomarkers.

Acknowledgements The authors gratefully acknowledge the support from the Intramural Research Program of NHLBI/NIH.

References

Addona TA, Shi X et al (2011) A pipeline that integrates the discovery and verification of plasma protein biomarkers reveals candidate markers for cardiovascular disease. Nat Biotechnol 29(7):635–643

An HJ, Froehlich JW et al (2009) Determination of glycosylation sites and site-specific heterogeneity in glycoproteins. Curr Opin Chem Biol 13(4): 421–426

Anderson NL (2010) The clinical plasma proteome: a survey of clinical assays for proteins in plasma and serum. Clin Chem 56(2):177–185

Awada A, Vandone AM et al (2012) Personalized management of patients with solid cancers: moving from patient characteristics to tumor biology. Curr Opin Oncol 24(3):297–304

Beck M, Schmidt A et al (2011) The quantitative proteome of a human cell line. Mol Syst Biol 7:549

Bencharit S (2012) Progresses and challenges of omics studies and their impacts in personalized medicine. J Pharmacogenomics Pharmacoproteomics 3(1):e105

Beverage JN, Sissung TM et al (2007) CYP2D6 polymorphisms and the impact on tamoxifen therapy. J Pharm Sci 96(9):2224–2231

Bhartiya D, Kapoor S et al (2012) Conceptual approaches for lncRNA drug discovery and future strategies. Expert Opin Drug Discov 7(6):503–513

Bhasker CR, Hardiman G (2010) Advances in pharmacogenomics technologies. Pharmacogenomics 11(4):481–485

Blum HE (2011) Personalized medicine. Praxis 100(3):159–166

Bochud M, Burnier M et al (2011) Top three pharmacogenomics and personalized medicine applications at the nexus of renal pathophysiology and cardiovascular medicine. Curr Pharmacogenomics Person Med 9(4):299–322

Boja ES, Rodriguez H (2012) Mass spectrometry-based targeted quantitative proteomics: achieving sensitive and reproducible detection of proteins. Proteomics 12(8):1093–1110

Chan SY, White K et al (2012) Deciphering the molecular basis of human cardiovascular disease through network biology. Curr Opin Cardiol 27(3):202–209

Chen R, Mias GI et al (2012) Personal omics profiling reveals dynamic molecular and medical phenotypes. Cell 148(6):1293–1307

Cho A, Normile D (2002) Nobel Prize in chemistry. Mastering macromolecules. Science 298(5593):527–528

Chouchane L, Mamtani R et al (2011) Personalized medicine: a patient-centered paradigm. J Transl Med 9:206

Choudhary C, Kumar C et al (2009) Lysine acetylation targets protein complexes and co-regulates major cellular functions. Science 325(5942):834–840

Clayton TA, Lindon JC et al (2006) Pharmaco-metabonomic phenotyping and personalized drug treatment. Nature 440(7087):1073–1077

Costa FF (2009) Non-coding RNAs and new opportunities for the private sector. Drug Discov Today 14(9–10):446–452

Davis JC, Furstenthal L et al (2009) The microeconomics of personalized medicine: today's challenge and tomorrow's promise. Nat Rev Drug Discov 8(4):279–286

Donnelly LA, Doney AS et al (2011) Common nonsynonymous substitutions in SLCO1B1 predispose to statin intolerance in routinely treated individuals with type 2 diabetes: a go-DARTS study. Clin Pharmacol Ther 89(2):210–216

Downard KM (2007) Historical account: Francis William Aston: the man behind the mass spectrograph. Eur J Mass Spectrom (Chichester, Eng) 13(3):177–190

Farooqi AA, Rana A et al (2012) NutriTRAILomics in prostate cancer: time to have two strings to one's bow. Mol Biol Rep 39(4):4909–4914

Fenn JB, Mann M et al (1989) Electrospray ionization for mass spectrometry of large biomolecules. Science 246(4926):64–71

Grimsrud PA, Swaney DL et al (2010) Phosphoproteomics for the masses. ACS Chem Biol 5(1):105–119

Gucek M, Murphy E (2010) What can we learn about cardioprotection from the cardiac mitochondrial proteome? Cardiovasc Res 88(2):211–218

Hall AM, Wilkins MR (2005) Warfarin: a case history in pharmacogenetics. Heart 91(5):563–564

Hardiman G (2008) Applications of microarrays and biochips in pharmacogenomics. Methods Mol Biol 448:21–30

Hardman M, Makarov AA (2003) Interfacing the orbitrap mass analyzer to an electrospray ion source. Anal Chem 75(7):1699–1705

Hippocrates C (1993) The sources of medical ethics: Hippocrates and his disciples. Rev Infirm 43(8):12–14

Hochholzer W, Morrow DA et al (2010) Novel biomarkers in cardiovascular disease: update 2010. Am Heart J 160(4):583–594

Hong KW, Oh B (2010) Overview of personalized medicine in the disease genomic era. BMB Rep 43(10):643–648

Huttlin EL, Jedrychowski MP et al (2010) A tissue-specific atlas of mouse protein phosphorylation and expression. Cell 143(7):1174–1189

Jaffe AS (2012) Troponin – past, present, and future. Curr Probl Cardiol 37(6):209–228

Jain KK (2004) Role of pharmacoproteomics in the development of personalized medicine. Pharmacogenomics 5(3):331–336

Joffe HV, Xu R et al (2004) Warfarin dosing and cytochrome P450 2C9 polymorphisms. Thromb Haemost 91(6):1123–1128

Kaddurah-Daouk R, Kristal BS et al (2008) Metabolomics: a global biochemical approach to drug response and disease. Annu Rev Pharmacol Toxicol 48:653–683

Kim SC, Sprung R et al (2006) Substrate and functional diversity of lysine acetylation revealed by a proteomics survey. Mol Cell 23(4):607–618

Klawitter J, Haschke M et al (2010) Toxicodynamic effects of ciclosporin are reflected by metabolite profiles in the urine of healthy individuals after a single dose. Br J Clin Pharmacol 70(2):241–251

Korman A, Oh A et al (2012) Statistical methods in metabolomics. Methods Mol Biol 856:381–413

Kuhn E, Whiteaker JR et al (2012) Interlaboratory evaluation of automated, multiplexed peptide immunoaffinity enrichment coupled to multiple reaction monitoring mass spectrometry for quantifying proteins in plasma. Mol Cell Proteomics 11(6):M111 013854

Longo R, D'Andrea M et al (2010) Pharmacogenetics in breast cancer: focus on hormone therapy, taxanes, trastuzumab and bevacizumab. Expert Opin Investig Drugs 19(Suppl 1):S41–S50

Loscalzo J (2011) Systems biology and personalized medicine: a network approach to human disease. Proc Am Thorac Soc 8(2):196–198

Ma JD, Lee KC et al (2012) Clinical application of pharmacogenomics. J Pharm Pract 25(4):417–427

Macek B, Mann M et al (2009) Global and site-specific quantitative phosphoproteomics: principles and applications. Annu Rev Pharmacol Toxicol 49:199–221

Madian AG, Wheeler HE et al (2012) Relating human genetic variation to variation in drug responses. Trends Genet 28(10):487–495

Malki K, Campbell J et al (2012) Pharmacoproteomic investigation into antidepressant response in two mouse inbred strains. Proteomics 12(14):2355–2365

Manavalan A, Feng L et al (2012) New insights into the brain protein metabolism of Gastrodia elata-treated rats by quantitative proteomics. J Proteomics 75(8): 2468–2479

Mangravite LM, Wilke RA et al (2008) Pharmacogenomics of statin response. Curr Opin Mol Ther 10(6):555–561

Matkovich SJ, Zhang Y et al (2010) Deep mRNA sequencing for in vivo functional analysis of cardiac transcriptional regulators: application to Galphaq. Circ Res 106(9):1459–1467

McGraw J, Waller D (2012) Cytochrome P450 variations in different ethnic populations. Expert Opin Drug Metab Toxicol 8(3):371–382

Mehta R, Jain RK et al (2011) Personalized medicine: the road ahead. Clin Breast Cancer 11(1):20–26

Mini E, Nobili S (2009) Pharmacogenetics: implementing personalized medicine. Clin Cases Miner Bone Metab 6(1):17–24

Mocellin S, Shrager J et al (2010) Targeted Therapy Database (TTD): a model to match patient's molecular profile with current knowledge on cancer biology. PLoS One 5(8):e11965

Murphy E, Kohr M et al (2012) S-nitrosylation: a radical way to protect the heart. J Mol Cell Cardiol 52(3):568–577

Nagaraj N, Wisniewski JR et al (2011) Deep proteome and transcriptome mapping of a human cancer cell line. Mol Syst Biol 7:548

Newman WG (2012) Pharmacogenetics: transforming clinical medicine. J R Coll Physicians Edinb 42(3): 244–247

Nicholson JK, Wilson ID et al (2011) Pharmacometabonomics as an effector for personalized medicine. Pharma-cogenomics 12(1):103–111

Nielsen FC, Borregaard N (2009) Pharmacogenetics and tailored drug therapy. Ugeskr Laeger 171(10):790–794

Nita-Lazar A (2011) Quantitative analysis of phosphorylation-based protein signaling networks in the immune system by mass spectrometry. Wiley Interdiscip Rev Syst Biol Med 3(3):368–376

Pan C, Olsen JV et al (2009) Global effects of kinase inhibitors on signaling networks revealed by quantitative phosphoproteomics. Mol Cell Proteomics 8(12):2796–2808

Panareo S, Rossi R et al (2011) A practical method for the estimation of therapeutic activity in the treatment of Graves' hyperthyroidism. Q J Nucl Med Mol Imaging 55(5):576–585

Pareek CS, Smoczynski R et al (2011) Sequencing technologies and genome sequencing. J Appl Genet 52(4):413–435

Patti GJ, Yanes O et al (2012) Innovation: metabolomics: the apogee of the omics trilogy. Nat Rev Mol Cell Biol 13(4):263–269

Paulo JA, Kadiyala V et al (2012) Mass spectrometry-based proteomics for translational research: a technical overview. Yale J Biol Med 85(1):59–73

Picotti P, Aebersold R (2012) Selected reaction monitoring-based proteomics: workflows, potential, pitfalls and future directions. Nat Methods 9(6):555–566

Prensner JR, Rubin MA et al (2012) Beyond PSA: the next generation of prostate cancer biomarkers. Sci Transl Med 4(127):127rv123

PubMed from http://www.ncbi.nlm.nih.gov/pubmed/

Qattan M, Demonacos C et al (2012) Roadmap to personalized medicine. Croat Med J 53(4):294–297

Ram PT, Mendelsohn J et al (2012) Bioinformatics and systems biology. Mol Oncol 6(2):147–154

Rio J, Comabella M et al (2009) Predicting responders to therapies for multiple sclerosis. Nat Rev Neurol 5(10):553–560

Roden DM, Altman RB et al (2006) Pharmacogenomics: challenges and opportunities. Ann Intern Med 145(10): 749–757

Roederer MW, McLeod HL (2010) Applying the genome to national drug formulary policy in the developing world. Pharmacogenomics 11(5):633–636

Sadee W (2011) Genomics and personalized medicine. Int J Pharm 415(1–2):2–4

Serkova NJ, Brown MS (2012) Quantitative analysis in magnetic resonance spectroscopy: from metabolic profiling to in vivo biomarkers. Bioanalysis 4(3):321–341

Shi Y, Xu P et al (2011) Ubiquitinated proteome: ready for global? Mol Cell Proteomics 10(5):R110 006882

Sreekumar A, Poisson LM et al (2009) Metabolomic profiles delineate potential role for sarcosine in prostate cancer progression. Nature 457(7231):910–914

Stastna M, Van Eyk JE (2012) Secreted proteins as a fundamental source for biomarker discovery. Proteomics 12(4–5):722–735

Sudhindra A, Ochoa R et al (2011) Biomarkers, prediction, and prognosis in non-small-cell lung cancer: a platform for personalized treatment. Clin Lung Cancer 12(6):360–368

Superko HR, Momary KM et al (2012) Statins personalized. Med Clin North Am 96(1):123–139

van Rooij T, Wilson DM et al (2012) Personalized medicine policy challenges: measuring clinical utility at point of care. Expert Rev Pharmacoecon Outcomes Res 12(3):289–295

Vermeire S, Van Assche G et al (2010) Role of genetics in prediction of disease course and response to therapy. World J Gastroenterol 16(21):2609–2615

Vertegaal AC (2011) Uncovering ubiquitin and ubiquitin-like signaling networks. Chem Rev 111(12):7923–7940

Whirl-Carrillo M, McDonagh EM et al (2012) Pharmacogenomics knowledge for personalized medicine. Clin Pharmacol Ther 92(4):414–417

Winnike JH, Li Z et al (2010) Use of pharmaco-metabonomics for early prediction of acetaminophen-induced hepatotoxicity in humans. Clin Pharmacol Ther 88(1):45–51

Witzmann FA, Grant RA (2003) Pharmacoproteomics in drug development. Pharmacogenomics J 3(2):69–76

Yu LR (2011) Pharmacoproteomics and toxicoproteomics: the field of dreams. J Proteomics 74(12): 2549–2553

Zhang A, Sun H et al (2012a) Modern analytical techniques in metabolomics analysis. Analyst 137(2): 293–300

Zhang A, Sun H et al (2012b) Serum metabolomics as a novel diagnostic approach for disease: a systematic review. Anal Bioanal Chem 404(4):1239–1245

2 Personalized Genome, Current Status, and the Future of Pharmacogenomics

Rohan Mitra, Mohan Lal Gope, and Rajalakshmi Gope

Abstract

Adverse drug reactions (ADRs) are one of the most dreadful medical conditions that affect a considerable number of individuals when they are taking single or multiple prescription drugs. Often these adverse reactions can occur with specialized drugs that are used to treat more serious disorders. Seldom ADRs can also occur due to intake of even simpler drugs such as penicillin and aspirin. In spite of volumes of data on ADRs, at present we still go through "one size fits all" model in dealing with prescription drugs. This scenario could change due to the emergence of new ways to overcome or minimize ADRs. Pharmacogenomics is one such ways to overcome many horrors of side effects caused by drugs, including ADRs. Pharmacogenomics is the combination of pharmacy and the patient's genetic composition which interact in an intricate manner to produce positive as well as negative drug reactions. When positive, it is for the betterment of patients, and when negative it leads to ADRs which oftentimes is fatal. Pharmacogenomics is an emerging field of science which is still in its infancy. Technologies that were developed along with the Human Genome Project (HGP), such as faster DNA sequencing protocols and efficient data handling softwares would help in the rapid advancement of pharmacogenomics in the near future. In addition, the reduced cost to obtain complete sequence of individual genome would provide data on single nucleotide polymorphisms (SNPs) and haplotype map (HapMap). These data would provide pattern of individual genetic variations which could be useful in managing diseases and treating patients effectively. In this chapter we will look at the current status and the future of pharmacogenomics which will aid in the development of personalised care. We will also discuss some of the obstacles that would have to be dealt with in achieving such target.

R. Mitra • R. Gope (✉)
Department of Human Genetics, NIMHANS,
Bangalore 560029, India
e-mail: rlgope@gmail.com

M.L. Gope
B-701 Ajmera Infinity, Electronic City Phase – 1,
Bangalore 560100, India

D. Barh et al. (eds.), *Omics for Personalized Medicine*,
DOI 10.1007/978-81-322-1184-6_2, © Springer India 2013

1 Introduction

The study of genetics refers to the genetic composition of any living cell or organism. Pharmacogenetics is a study of differential response of individuals to various pharmacological agents which could directly be correlated to the genes of an individual that are inherited primarily from the parents. Simply put, pharmacogenomics is the evaluation of how an individual's genetic composition affects the physiological response to drugs. In a broad sense, the term "pharmacogenomics" comes from the combination of "pharmacology" and "genomics" and is therefore considered as the combination of pharmaceuticals and genetics. Hence, pharmacogenomics refers to the study of genes, both inherited and acquired by an individual. These genes determine drug behavior such as dosage, metabolism, side effect, drug interactions, toxicity and half-life. It is generally believed that in future drugs could be customized and personalized to individual needs. It would be based on patients' genetic informations as well as other factors such as environment, occupation, diet, age, lifestyle and state of health, that could influence the outcome (Lagoumintzis et al. 2010; Ma and Lu 2011). In pharmacogenomics the overall genetic and genomic informations of the patients are collected and evaluated. Data on single nucleotide polymorphisms (SNPs), HapMap and genome-wide association studies (GWAS) and data on RNA and protein are used for this purpose. The results obtained from such evaluations are used to design custom-made drugs for individual patients. These results can also be used to evaluate the efficacy of novel drugs and patients' response to existing drugs (Laing et al. 2011). Overall the study of pharmacogenomics involves why or how the individual's genetic composition is responsible for the variation to drug response. It is a combination of many areas of science such as pharmacology, pharmaceutics, toxicology, population genetics, genomics, pharmacology, and biostatistics (Lagoumintzis et al. 2010; Sato et al. 2010; Ma and Lu 2011). Many scientists envisage that pharmacogenomics would eventually help pharmacists to determine how an individual would react to a particular medication. Such protocols will be based on genetic tests done at caregiver's premises in a cost-effective manner. They will be done by individuals with training in handling user-friendly instruments, computers, and analytical softwares. The final outcome would be to minimize and eventually eliminate ADRs.

2 Adverse Drug Reaction and the Need for Personalized Medicine

In general, the response to any drug by an individual is inherited from his/her parents. Variations in the DNA sequence and the genetic code determine the differential response by individuals for the same drug. Harmful side effects of certain drugs cause adverse drug reaction which is unintended. ADRs are one of the leading causes of fatalities in many countries. It is important to understand that more than 50 % ADRs is generally preventable. In certain instances the cost of managing ADRs exceeds the cost of medications. Therefore, it is important to develop medications that are safe to the individual patients with low or no ADRs. ADRs generally occur with the dose that is normally prescribed for all individuals with similar disease conditions and symptoms. The Journal of American Medical Association reported more than 2.2 million adverse drug reactions and more than 100,000 hospital deaths for the year 1994 in the United States alone, making it one of the serious issues to be dealt with in managing and treating patients (Willcox et al. 1994). ADRs are reported to be one of the leading causes of hospitalization and death in the USA, and it is also reported to be the 4th leading cause of death. An international system for monitoring ADR was established in 1971 by the World Health Organization (WHO). WHO oversees the policy matters, and the actual operation is done by the WHO Collaborating Centre for International Drug Monitoring, Uppsala Monitoring Centre (UMC), in Uppsala, Sweden. The WHO database currently consists of more than three million suspected ADR cases (Lindquist 2008; Clark 2010). A vigilant system is generally in place from the time the drug is formulated, manufactured and tested in a variety of experimental set-ups. Then it is tested

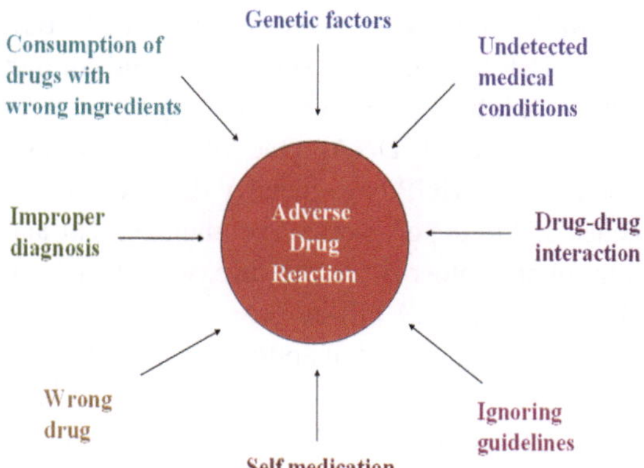

Fig. 2.1 Factors involved in adverse drug reactions (ADR)

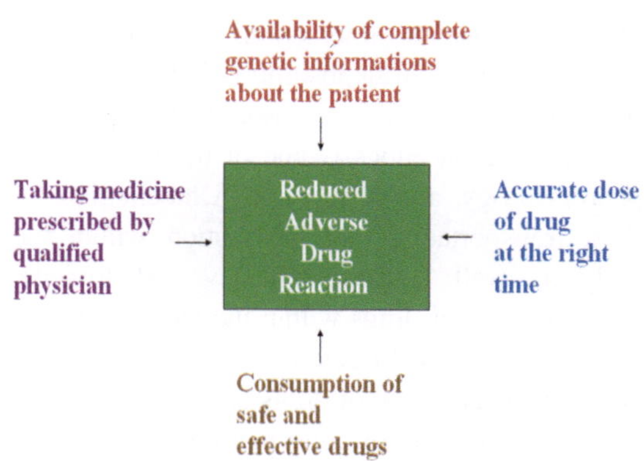

Fig. 2.2 Precautionary measures to avoid adverse drug reactions

in human beings in its pre-approval stages and is finally approved for human consumption. Some of the examples of ADRs include (a) consumption of thalidomide to control morning sickness in expectant mothers which leads to disfigured newborn, or (b) routinely used oral contraceptives which can cause thromboembolism or blood clot, or (c) statins which are used to regulate the level of cholesterol but can also cause degeneration of muscle cells. It is important to understand that no drug is safe and all drugs come with certain risk factors and some of them fatal. Some of the ADRs are due to a variety of reasons such as (a) improper diagnosis of the patient's disease condition, (b) prescription of a wrong dosage of the right medicine, (c) consumption of medication without a prescription, (d) not following the guidelines of the medications, (e) drug-drug interaction which could occur due to any of the following: (i) simultaneous consumption of multiple prescription drugs; (ii) consumption of traditional medicines along with prescription drugs; (iii) interaction of prescription drug with environment, (f) consumption of fake medication in the market which does not have any active ingredient or wrong ingredient that does not meet the scientific requirement, and (g) undetected medical condition, unidentified genetic factors, or allergic conditions of the patient (Fig. 2.1). Almost all medicines have both beneficial and harmful effects, and no medicine is totally safe. However, the side effects can be minimized by taking certain precautions such as (a) consuming only good-quality medicines that are prescribed by a qualified doctor, (b) taking drugs that are known to be safe and effective, and (c) specific drugs given to the right patient in the accurate dose at the appropriate time (Fig. 2.2). The knowledge of ADRs, availability of human genome sequence, and individual variations in SNPs and HapMaps point towards the fact that one drug does not fit all. It has also created the need for the development and use of safe drugs for individual patients.

3 Pharmacology

Drugs exert their functions by physically interacting with specific proteins. These proteins are encoded by specific genes. In general drug action involves either enhancement or suppression of certain protein functions. Pharmacology deals with medicine and biology in order to evaluate mechanism of drug actions with protein(s). The drugs can be any man-made or naturally occurring molecules. Once these molecules are in the intracellular environment, they can exert biochemical, physiological effect on the living cells, organs, and eventually the whole organism. Therefore, pharmacology is the study that deals with the communication between the chemicals and the living organism. It includes a variety of related subjects such as substances with medicinal values, drug composition, properties, interaction, therapeutic values, medical applications, anti-pathogenic actions, and toxicology. Pharmacology can be divided into two broad areas of biological chemistry, one dealing with the dynamics of and the other dealing with the kinetics of drug actions and interactions. It also involves studies related to drug absorption, effect, metabolism, interaction, and excretion from the living system. Pharmaceutics is one of the branches of pharmacology, and it deals with mechanism of drug transportation and distribution within the body as well as drug efficacy and toxicity. Concentration of drugs within the various organs, tissue, and cell types determines the site of drug action. Concentration of drugs at various sites can vary from the time of consumption till the metabolic products are flushed out of the system. The physicochemical properties of the drug determine its action which in turn depends on the genetics of the patients. In addition, the toxic nature of the drugs is also exerted by its interaction with specific proteins encoded by specific genes. Hence, the study of pharmacology involves direct patient care. However, the study of pharmacy includes drug discovery and testing, evaluation of dose, efficiency and toxicity and drug trials (Hollinger 2003; Nagle and Nagle 2005; Ma and Lu 2011).

4 Genetics

The study of genetics deals with the molecular structure and functions of genes within the cell and in an organism. The living things inherit genes from the parents which determine the health and disease process in individual's lifetime. The science of modern genetics began with the discovery of laws of heredity by Gregor Mendel in 1865. Later, DNA was identified as the genetic material and its structure and function were determined. Subsequently, a range of protocols were developed to clone and sequence genes and to identify their functions. These steps greatly advanced the field of molecular genetics (Weiling 1991).

5 Human Genome Project (HGP)

The advancement of human molecular genetics paved ways to the commencement of the Human Genome Project (HGP). The main task of HGP was to identify and map human genes and to obtain the entire sequence of human DNA. HGP identified 30,000 to 50,000 functional human genes each consisting of approximately 10,000 base of DNA. One of the important objectives of HGP is to identify disease-causing genes and to develop effective management and treatment protocols for the diseases. Data from HGP revealed 2.9 billion base pairs of human DNA which consisted of approximately 3 million variables. Among the three million variable base pairs, approximately 100,000 are confined to full human variation. It appears that approximately 10,000 base pairs could have relevance to pharmacology. Hence, among the 2.9 billion base pairs, only 0.1 % accounts for individual variations, and 3 % of which encodes for genes (Venter et al. 2001; Lander 2011). Technological advancement in human genome sequencing can make this process more affordable. Then the individual variations in DNA sequences could be identified quickly in a cost-effective manner at the point of caregiver. Robust computational and analytical softwares could compile and analyse large volumes of data with little assistance from experts. Thus, it would be easier for health care providers to formulate effective treatment protocols.

6 HapMap

Variations between individuals and between their drug responses correspond to and exist within the DNA sequence in specific locations, and they can be mapped accurately. The international HapMap project officially started in the year 2002. It was aimed at finding the haplotype map of human genome which determines genetic variations. A haplotype is defined a set of variations within the individual's DNA sequence or polymorphism that has the tendency to be inherited together. It can also be a set of combination of alleles or SNPs that are found on the same chromosome. The HapMap data provide the pattern of genetic variation between individuals and between groups of individuals. The genetic variations together with environmental conditions influence the state of health and disease conditions and ADRs. The

Phase I data from the HapMap project was released in 2005; the phase II and III datasets were published in 2007 and 2009 respectively. It is a collaborative effort of multiple scientific and academic research centers from many countries such as the USA, Canada, the United Kingdom, Nigeria, Japan, and China. All the data from the HapMap project such as SNP frequency, haplotype, and genotype are freely available for the public, and it can also be viewed through Haploview program (Barrett et al. 2005; International HapMap Consortium 2010).

Many human diseases have involvement of more than one gene and gene products. Some of them have added environmental influences as well. The genetic and environmental factors have significant role in the development and progression of disorders such as diabetes, cancer, heart disease, schizophrenia, hypertension, and arthritis. They also have influence on the individual response to the pharmacological agents used to manage or treat these disorders. The HapMap project identified approximately ten million common variants, mainly SNPs in limited sets of samples (Goldstein and Cavalleri 2005; Hinds et al. 2005). The integrated data sets of common and rare alleles are included in HapMap public release version 3 (HapMap3). Consequently, HapMap3 includes SNPs and single nucleotide polymorphism data base (dbSNP). The dbSNP contains short deletions and insertions, micro satellite markers, short tandem repeats (STRs), heterozygous sequences, copy number polymorphisms (CNP) and other minor genetic variations (The SNP Consortium LTD – SNP search). HapMap data is an important link between the human diseases and treatment or management protocols (Database of Genotype and Phenotype (dbGaP) located at http://www.ncbi.nim.nih.gov/entrez/query.fcgi?db=gap.

7 SNPs

The observed genetic variations among various groups of populations are not random. It is evident that certain combinations of genetic variations are more common than other.

Approximately 99.5 % of the DNA sequence is shared by any two unrelated human beings. The remaining 0.5 % may harbor changes at a single nucleotide at a particular region of the gene on a particular human chromosome which are known as single nucleotide polymorphisms (SNPs). SNPs are found in the coding and noncoding regions of the chromosomes. They are also found in the regulatory regions and intron regions of the genes. When they do not alter the amino acid sequence of the protein, SNPs generally have no influence on the functions of proteins coded by the genes in which they are found. SNPs are acquired during evolution and passed on to subsequent generations. Human beings have diploid number of chromosomes and therefore two sets of alleles for each gene, except the male Y chromosome. If the SNP in one allele is identified, the same in the other allele can be predicted (Fig. 2.3). The international HapMap project identified only the common SNPs. These common SNPs occur in each allele at least in 1% of the population. When they are located in any gene of importance or disease genes, the SNPs can be used as "tags". These SNP tags can be used to study that particular disease in terms of inheritance and progression of the disease. These are known as tagSNPs and they correlate well with neighboring SNPs. The alleles of tagSNPs will provide information of individual haplotype. By screening haplotype of individuals with and without disease, the disease-related tagSNPs and haplotype can be identified. Frequently the SNPs lead to change in the restriction enzyme recognition sites which can help identify loss of heterozygosity (LOH). LOH can be used to identify their specificity in the initiation and progression of diseases such as cancer (McCarthy and Hilfiker 2000; Laing et al. 2011; Varela and Amos 2010; American Association for Cancer Research Cancer Concepts Factsheet on SNPs; The SNP Consortium LTD).

SNPs can determine the initiation and progression of human diseases. They can also influence individuals' response to infectious diseases, environment, drugs, vaccines, and other agents. SNPs have a key role in the field of pharmacogenomics and in the development of personal-

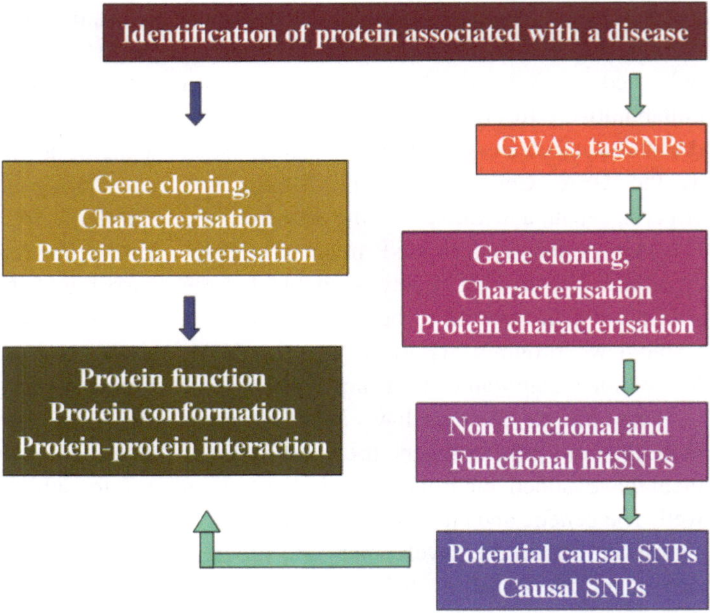

Fig. 2.3 Use of GWA data and SNPs in pharmacogenomics

ized medicine. One of the most important applications of SNPs is the comparison of genomic region harboring SNPs between individuals with and without any specific disease. It is also important within the group of individuals with the disease where many parameters such as the onset, progression, pathology, and response to drugs vary widely. This is particularly important in multiple gene disorders where the data from genome-wide association studies (GWAS) are important in understanding, treating, and managing these diseases (Laing et al. 2011; Varela and Amos 2010; The SNP Consortium LTD – SNP search).

For more information on SNPs, please refer to the websites given at the end of references.

8 Genome-Wide Association Studies (GWAS) Data in Pharmacogenomics

Data from the GWAS are important in assessing the specific drug performance, drug interaction, and efficacy of the existing management and treatment protocols. Association of phenotype of a disease to a single gene is found only for a very few human disease condition. Most human disorders do not follow Mendelian pattern of inheritance. Most human diseases, especially the psychological and psychiatric disorders, involve more than one gene and interplay between gene(s) or family of genes as well as environment. Data on GWAS play a central role in identifying the causative agent in the diagnosis and development of management/treatment protocols for some of these diseases even before the onset of the disease or manifestation of associated symptoms (Borobia et al. 2012). It appears that some of the SNPs identified in patients with psychological and psychiatric disorders are variants that are not necessarily associated with disease per se. In such cases functional analysis is required to identify the actual involvement of such variations, hitSNPs are the ones which are initially identified to be associated with a disease condition. If the hitSNPs are located within the protein coding regions or regulatory regions, it affects the gene expression. If the hitSNPs happen to be not related to the disease condition or if it is localized in the noncoding regions of the genes or intergenic, resequencing of those genes could be performed to rule out involvement of genes with the hitSNPs (Borobia et al. 2012). It is also important rule out noninvolvement of hitSNPs in the disease condition by analyzing their effect on posttranscriptional and posttranslational modifications of mRNA and protein, respectively. Posttranscriptional modifications could affect mRNA processing, stability and splicing. mRNA splice variants could produce protein isoforms with altered activity or function. Moreover, functional analysis of protein requires the informations on posttranslational modification (PTM) of proteins. Because PTMs determine location of proteins within the cell, protein-protein interaction, protein half-life, higher-order structure of proteins, ubiquitin system, etc. (Fig. 2.3).

9 Clinical Use of Genomes in Pharmacogenomics

Pharmacogenomics is aimed at targeted treatment for specific diseases which uses a variety of sets of database such as DNA sequence, SNPs, HapMap of patients, GWAS data for gene expressions, environment, and drug sensitivity (Fig. 2.4). Recently, "The Personal Genome Project" (PGP) permitted patients to give "open consent" for a pilot study and publish the data from ten pilot participant genomes called as "PGP-10." In this study the Genome-Environment-Trait Evidence (GET-Evidence) system was introduced. In GET-Evidence, the software will help identify previously unidentified sequence variants. The data from this study is automatically processed to reveal published and novel variants of healthy individuals for interpretation of disease process. This study would predict the onset of disease and disease progression. GET-Evidence project has so far identified 3.2 million variants per person. Now PGP is planning to analyze genome from as many as 10,000 who would waive the privacy rights. PGP may not be useful for late-onset diseases (Ball et al. 2012). Similar studies should contribute to better understand pharmacogenomics. It would also contribute towards better management of diseases, simultaneously reducing cost.

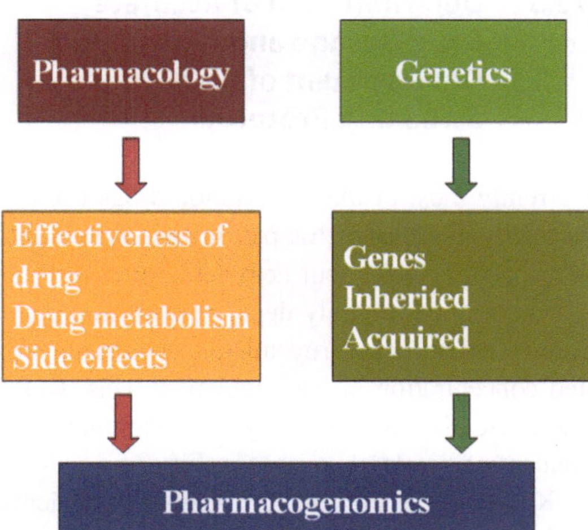

Fig. 2.4 Definition of pharmacogenomics which is the combination of pharmacology and genetic composition of an individual

10 Benefits of Pharmacogenomics

10.1 Development of More Safe and Potent Medicines

Innumerable benefits are attributed to pharmacogenomic studies, one of them being novel drug discovery (Fig. 2.5). The drug discovery procedure could draw attention to a particular disease and to individual patients. The drug companies will be able to create custom-made therapeutics based on the genetic informations of a patient. It would also be based on the gene expression profile and the levels of functional proteins, and their conformations. In addition, status of regulatory molecules such as micro RNA, siRNA and noncoding RNA would also be considered for this purpose. Drugs can also be developed based on age of onset of disease, severity of the disease, patients' hormonal status, and a variety of other biochemical

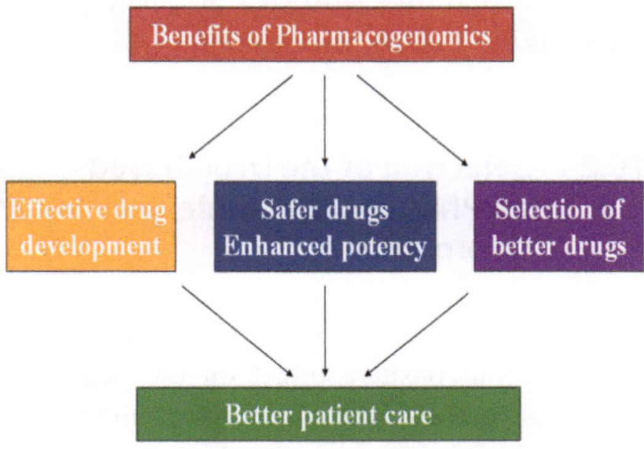

Fig. 2.5 Immediate advantages of pharmacogenomics

parameters. Drugs can be designed depending on the proteins, enzymes, and their conformations associated with the diseases. This procedure will assist pharmaceutical companies to design and produce targeted therapeutics which will have maximum efficacy. Such precision in the drug development and drug design process will greatly reduce the severity of side effects and ADRs. In addition, the precision in drug design and targeted therapeutics will help decrease damage to the nearby cells and other organs.

Based on the pharmacogenomic profile, doctors can formulate the best therapeutic protocols and prescribe the best available drugs right from the beginning. By doing so, guess work and trial-and error-based treatments can be avoided. Some of the treatments based on guesswork could be useless, wasteful, or at times could cause damage to the patients' overall health irreparably. Formulation of patient-specific therapeutic will also lead to a speedy recovery of the patients from the disease conditions with minimal side effects. The end result could be reduced mortality due to adverse drug response (Lazarou et al. 1998).

Therapeutic antibodies could be developed based on genetic material such as DNA or RNA, and they should benefit the patients in enhancing their immune system. This would eliminate risks such as infection and other complications associated with certain immunization procedures based on live viral vaccines. The nucleic acid-based antibodies will be relatively less expensive to produce, more stable than the conventional antibodies raised against protein antigens, and therefore will be easy to store. These antibodies can also be simultaneously raised against mixture of DNA/RNA of many pathogenic strains (Leitner et al. 1999).

10.2 Selection of the Drug Based on Pharmacogenomic Informations

In general, drug approval procedures are based on its overall positive effect on the health of human population and not based on individual benefits. Pharmacogenetics is the study of how genes influence an individual's response to drugs. A part of the pharmacogenomic study involves evaluation of pharmacogenetic data of the patients as well (Fig. 2.5). Use of pharmacogenetic data appears to be a recent trend; however, in reality it is about half a century old. In the 1950s, scientists first identified deficiencies in enzymes which could be responsible for adverse reactions to drugs, and it was also identified that this trait could be inherited. For example, an antimalarial drug could cause anemia in African American soldiers, and the same showed no side effect in soldiers of Caucasian origin. Later it was discovered that the adverse side effect was caused by a variation the glucose-6-phosphate dehydrogenase (G6PD) gene which was found only in individuals of African descent and not in those of Caucasian origin. It was further shown that the normal form of G6PD enzyme has protective effect on the red blood cells (RBCs). The variable form lacks the same leading to damage to RBCs in individuals carrying the variant G6PD gene (Beutler 1996; Weber 1997). Later, a variety of other enzyme variants that cause adverse side effect have been identified by trial-and-error method. To identify the enzyme variants, drugs were first administered, and the metabolism of the drug is tracked by monitoring the by-product in blood and urine.

10.3 Determination of Accurate Drug Dosage and Development of Modern Screening Protocols

Currently most of the prescription drugs use the standard parameters such as patients' age and weight. Pharmacogenomics will help determine the drug dosage more accurately depending on important parameters such as drug tolerance, metabolism, and concentration in the circulation. This would maximize the drug efficacy and minimize the chances of accidental overdose (Fig. 2.5).

Knowledge on the genetic makeup of the patients will give options to individuals to make enough changes in lifestyle and adjust to the environmental conditions at younger age in order to minimize severity of a genetic disease. Similarly, advanced knowledge of disease susceptibility will help patients and doctors to be more attentive so that the onset and the severity of disease can be monitored closely. These procedures will help choose the right stage and time for drug intervention which could maximize the positive outcome.

Currently the patients are given one of many suitable, available drugs by their doctors. An optimum dose is calculated based on the standard parameters such as age, sex, weight, hormonal status, other prescription medication that the patients are taking at any given time, stage of the disorder, and many other parameters. This dosage is adjusted later depending on the patients' response side effects. Oftentimes it also depends on the effectiveness of the drugs on other patients. The pharmacogenomic data of the patients will help doctors to avoid unnecessary trial-and-error method of drug prescription. Instead, the doctors will have the patients' genetic profile and based on that the best available drugs can be prescribed at the beginning stages of the treatment protocol. This also will help doctors to avoid any guesswork and provide the right pharmacological care to the patients. Thus, the disease can be managed safely, with little side-effects which perhaps can lead to speedy recovery. Pharmacogenomics in this regard will become a remarkable tool to minimize hospitalization and fatalities due to drug overdosage and unfavorable drug interactions (Grant 2001; Roberts et al. 2012).

10.4 Evolution of Drug Development and Approval and Reduced Cost of Health Care

It will be relatively easier for the pharmaceutical industries to develop effective, genome-specific drugs. The initial trials for these drugs can be carried out in selected population with certain trait instead of general population. Such procedures will be cost effective and it will also minimize the risk during drug trials (Fig. 2.5).

Speedy, genome-based drug development and population-based drug trials and approval would reduce the overall number of cases experiencing harmful drug interactions. Availability of more potent drugs together with novel genome-based early diagnostic procedures would reduce the dose and course of medication. This would reduce the cost of treatment which in turn could make the overall health care cost effective.

Pharmacogenomics would lead to emergence of new class of drugs such as antisense RNA, siRNA, oligonucleotide-based drugs, novel immunotherapeutics, and other drugs based on evaluation of individual genetic risk assessment. In such process it could channel the drug development process from curative to preventive direction. Development of preventive therapeutics could involve other businesses such as nutrition and fitness industries and formation of a new branch of preventive caregivers or group of consultants for healthy individuals. Thus, pharmacogenomics has the potential to force new business drive based on preventive therapy for patients with genetic predisposition to specific diseases. Combining genetics, disease diagnosis, and drug prescription will produce new industries (Grant 2001; Roberts et al. 2012; Rodon et al. 2012). Overall, emergence of such industries could promote development and availability of safe and efficient drugs with less toxicity for an affordable price.

11 Pharmacogenomics Today

To a limited extent pharmacogenomics is currently being used. Cytochrome P450 (CYP) is a group of enzymes which catalyze oxidation of organic substance (Guengerich 2006). The substrates for these enzymes are metabolic intermediates, xenobiotic substance, and toxic chemicals. The family of enzymes, CYP, breaks down more than 30 different groups of drugs. The efficiency of these enzymes and their ability to metabolize certain drugs depend on the DNA variations in genes that code for these enzymes (Ingelman-Sundberg 2004). Less effective or inactive forms of these enzymes would result in drug accumulation eventually drug overdose leading to adverse side effects. Currently, genetics screening is used in clinical trials to evaluate the cytochrome P450 variants in patients and their ability to metabolize a particular drug. DNA microarray tests for 29 different CYP2D6 genetic variants and tests to identify polymorphism for the CYP2C19 are currently available (Collins and McKusick 2001; Brockmoller et al. 2000, Cytochrome P450 dissertations, Human Cytochrome P450 (CYP) 2010, http://www.cypalleles.ki.se/). Results from these tests help predict poor, intermediate, good, or best form of CYPs, and this test is approved by Federal Drug Agency (USA).

The enzyme thiopurine methyltransferase (TPMT) breaks down the thiopurine drug which is a chemotherapeutic agent for childhood leukemia. A small percentage of patients of Caucasian origin have genetic variants that prevent them from producing an active form of TPMT enzyme, and these patients tend to accumulate toxic levels of the drug. Currently, doctors can use genetic screening to identify patients for the TPMT deficiency and adjust the drug dosage accordingly (Adams 2008).

The enzyme UDP-glucuronosyltransferase is involved in the metabolism of irinotecan, a drug used to treat metastatic colorectal cancer. The variations of the gene for this enzyme, UGT1A1 (*28 variants), are found among population. These individuals express the UGT1A1 enzyme in the liver. Individuals homozygous for UGT1A1 gene are recommended for reduced drug dose. A test which detects genetic marker that will identify patients who would have adverse reaction due to irinotecan-induced toxicity has been approved by FDA. Irinotecan is one of the widely used chemotherapeutic agents whose dose is determined according to the patients' genotype (Desai et al. 2003).

Warfarin is used to prevent blood clot, and variants of the genes CYP2C9 and VKORC1 (vitamin K epoxide reductase) are involved in its efficacy as an anticoagulant. Warfarin is metabolized in the liver by oxidation. The CYP2C9 protein is involved in the metabolism of warfarin and exerts its anticoagulant effect by inhibiting the activity of VKORC1. Three SNPs, two in the CYP2C9 gene and one in the VKORC1 gene, have been identified to play key roles in determining the effect of warfarin as an anticoagulant (Rieder et al. 2005; Hillman et al. 2004; Anderson et al. 2007; Limdi et al. 2008). A diagnostic test to identify patients with variants for these SNPs has been approved by FDA. Results from this test help in determining the warfarin dose so that the patients will not suffer from bleeding in the head or gastrointestinal tract. This test prevents warfarin-induced bleeding in patients and consequently decreases damage to neighboring healthy cells.

12 Future Pharmacogenomic Strategies in the Management of Brain Tumors

Human brain tumors account for less than 2 % of all human neoplasms, but they cause a disproportionately larger cancer-related morbidity and mortality. Brain tumors are the second leading cause of cancer deaths in men aged 20–39 years of age and fifth leading cause of death in women in the same age group. Glioblastoma multiforme is one of the most aggressive form of human brain tumors. The median survival period of patients diagnosed with these tumors is less than 1 year. This survival period has not changed over several decades in spite of intense research in this area. A few patients with anaplastic astrocytomas and glioblastomas survive for 2 years after diagnosis. Meningiomas are slow-growing tumors which occur in older patients (Kaye and Laws Jr 2001; Weinberg 2007).

The human brain tumors consist of a variety of neoplasms including glioma, meningioma, ependymoma, and vestibular schwannoma (VS). The VS tumors occur in both sporadic and familial forms (Kaye and Laws Jr 2001; Martuza and Eldridge 1988; Brackmann and Kwartler 1990; Weinberg 2007). The sporadic forms are generally unilateral, and familial forms are almost always bilateral. Sporadic VS are known to occur mostly in middle age in the fifth or sixth decade. The familial form generally occurs at a younger age with peak prevalence in third decade. The unilateral sporadic VS tumors are predominant in males, and the bilateral familial forms are

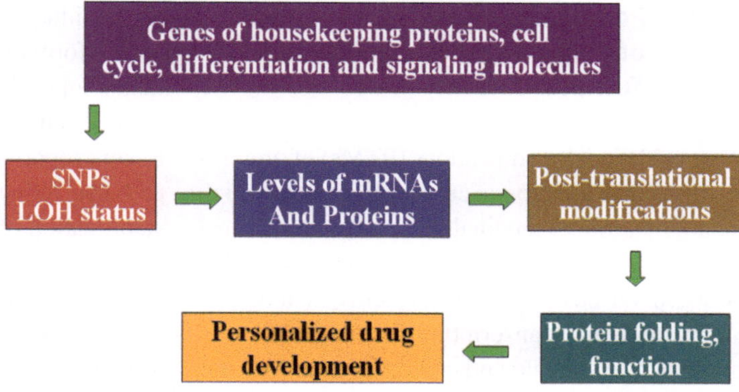

Fig. 2.6 Future of pharmacogenomics. Use of combined data on genes, proteins, protein modifications and function of proteins in brain tumor management

predominant in females (Martuza and Eldridge 1988; Brackmann and Kwartler 1990).

Alterations at the p53 gene locus have been reported in more than 50 % of gliomas, and they are not common in meningiomas. p53 regulates various genes involved in cell cycle, apoptosis, development, differentiation, DNA repair, aging and senescence (Hainaut and Weiman 2009; Hanahan and Weinberg 2011). Loss of heterozygosity at p53 gene locus is reported for human brain and nervous system tumors (Thomas et al. 2005; Dayalan et al. 2006a, b; Mathivanan et al. 2007a, b; Rohini et al. 2007, 2008). The retinoblastoma gene, RB1, is an important tumor suppressor gene with multiple functions such as cell cycle regulation, cell differentiation, and apoptosis (Chen et al. 1989; Weinberg 1991, 1995, 2007; Hanahan and Weinberg 2011).

12.1 SNPs and LOH Status in Human Brain Tumors

SNPs occur throughout the genome. If the SNPs create restriction enzyme recognition sites, it can be used for analysis of loss of heterozygosity (LOH). Analysis of structure of RB1 and p53 gene in human vestibular schwannomas (VS) and other intracranial tumors with polymorphic probes showed LOH at the intron 1 locus of RB1 and p53 gene loci. The human tumor suppressor genes, RB1 and p53, are not known to be mutated in human vestibular schwannomas (VS). However, there are reports showing loss of heterozygosity (LOH) in the 5′ intron, that is, intron 1 locus of RB1 and p53 gene in approximately 25 and 54 % of informative cases, respectively, in human VS tumors (Thomas et al. 2005; Dayalan et al. 2006a, b, c). The LOH was analyzed using the change in restriction enzyme BamH1 or Bgl II recognition site at the intron 1 region of RB1 and p53 gene, respectively. The 5′ introns are reported to be rich in regulatory elements which influence gene expression. Approximately 19 and 63 % of brain tumors showed LOH at the intron 1 locus of RB1 and p53 genes, respectively, in human brain tumors (Mathivanan et al. 2007a, b; Rohini et al. 2007, 2008). The presence of LOH is indicative of genetic instability in patients with VS tumors. LOH can cause haploinsufficiency for proteins involved in cell cycle and DNA repair pathway which could lead to genomic instability and carcinogenesis. Human intracranial tumors are heterogeneous disorder with variable age of onset, speed of disease progression, extent of metastasis, and survival period after the diagnosis of the disease. SNPs and LOH status in RB1 and p53 genes could influence these parameters. Therefore, LOH status of these two important tumor suppressor genes in human brain tumors could have an important role in the management of patients with brain tumors. Currently various drugs are available for a variety of tumors including brain tumors, and some of which are dependent on the mutational status of p53 gene. Therefore, it is important to evaluate all types of alterations at the p53 locus in brain tumors so that these drugs can be prescribed for patients depending on the mutational status of p53 (Fig. 2.6).

12.2 Posttranslational Modification of Proteins in Human Brain Tumors

Posttranslational modifications (PTMs) of proteins are important function in eukaryotic system which in turn determines the final shape of the functional protein molecule. The PTMs are interconnected between various regulatory proteins such as signaling proteins, transcription factors, regulatory proteins of cell cycle and cell differentiation, and proteins involved in cell death and apoptosis. In general these modifications include phosphorylation, acetylation, glycosylation, methylation, and carboxylation, just to mention a few. Among the PTMs, phosphorylation is the most predominant one, and it occurs at serine, threonine, and tyrosine residues of the protein molecule (Minguez et al. 2012).

Human p53 protein is known to undergo posttranslational modification in many serine and threonine sites. Some of the posttranslational modifications stabilize the p53 protein through disrupting the binding to MDM2. MDM2 binding is necessary for ubiquitin-mediated proteosomal degradation of p53 protein. Presence of a higher percentage of Ser 392 phosphorylated p53 protein is reported in the VS tumors of the young patients of 35 years or less as compared to the older patients who are above 35 years of age (Dayalan et al. 2006c). It appears that in the young patients the Ser 392 phosphorylation could alter the conformation of p53 protein which could hinder binding to MDM2 resulting in decreased degradation and accumulation of higher level of p53. In the older patients the p53 protein may form complex with MDM2 but could have defective intermediates that could lead to accumulation of p53 (Dayalan et al. 2006a). Increased percentage of Ser 392 phosphorylated form of p53 protein is found in brain tumors of higher histological grade, that is, grade IV of glioblastoma multiforme which is highly malignant, and it is biologically the most aggressive form (Rohini et al. 2007). Ser 392 phosphorylated p53 protein is suggested to have an important role in the aggressive behavior of these tumors perhaps due to reduced affinity to MDM2 binding. Phosphorylation at the Ser 392 residue of wild type p53 protein could alter its conformation thus altering its function as a tumor suppressor. Increased level of p53 isoforms is reported in the high-grade human gliomas. Recent report shows that Ser 392 phosphorylated p53 has reduced affinity to MDM2 leading to stabilization of p53 protein in sarcomas cell lines (Piccinin et al. 2012). It is important to evaluate the posttranslational modification of p53 protein in tumors before applying any therapeutic protocols that are developed based on only the point mutation status of p53 gene.

The RB1 gene product pRb undergoes phosphorylation at the early G1 phase of the cell cycle, and it is dephosphorylated once the cell cycle is completed. There are 16 serine/threonine sites which can be phosphorylated by the cyclin-CDK (cyclin-dependent kinases) complexes, cyclin D/CDK4, cyclin E/CDK2, and cyclin A/CDK2, when there is mitogenic signal available in extracellular milieu (Chen et al. 1989; Weinberg 2007; Hanahan and Weinberg 2011). Higher level of phosphorylated forms of pRb protein is present in human VS tumors and in other human brain tumors such as gliomas and meningiomas. The Ser 567 that is located at the A domain of the pocket region of pRb protein is inefficiently phosphorylated during normal cell cycle. Efficient phosphorylation of this site is reported in human VS tumors which could aid in tumor growth (Mitra et al. 2012). It is important to note that the dephosphorylation of Ser 567 of pRb is the initial event in sodium butyrate-induced cell death of VS tumors in vitro (Mitra et al. 2012). Therefore, it is essential to identify the phosphorylation status of both pRb and p53 tumor suppressor proteins in order to develop effective management protocol for patients with brain tumors (Fig. 2.6).

12.3 Tumor Heterogeneity and Pharmacogenomics

A variety of data on chromosome alterations, tumor ploidy, immunohistochemistry, mutation status and mRNA profile were analyzed from a primary renal cell carcinoma and its associated metastatic tumors. The exome sequencing data from these tumors were also used for this analysis. Results from this study showed heterogeneity within the tumor. Further, the data revealed branched evolution of tumor during growth and metastasis. The observed intra-tumor heterogeneity also included loss of function of multiple tumor suppressor genes. The gene expression profile at various regions within the tumor showed regions with good and bad prognosis. These variations can mislead investigators and clinicians who develop treatment protocols based on data from single biopsy sample from one or two regions within the tumor (Gerlinger et al. 2012). These data also points to the fact that the intra-tumor heterogeneity could aid and promote tumor evolution and variable adaptation of tumors via Darwinian selection. Existence of such intra-tumor heterogeneity could complicate development and testing of personalized medicine for treatment.

12.4 Duplex Sequencing

In order to develop personalized medicine for tumors, we need the patient's DNA sequence and the sequence of the tumor DNA. Due to the tumor heterogeneity, it is important to get DNA sequence of tumor from various regions to get an overall assessment of the molecular changes that have contributed in the tumor development. In general DNA sequencing has an error rate of approximately 1 % which would limit the accuracy of the sequences. Deep sequencing is generally applied to genetically heterogeneous samples, and this technique could produce errors which may not be suitable if the management and treatment protocols are based on DNA sequences, SNPs, HapMaps, GWAS, etc. Recently duplex sequencing is developed which is reported to greatly reduce the errors. In this method each DNA strand is independently tagged, and each of the two strands is sequenced. Because of the complementary nature of DNA, the true alterations are expected on both strands in the same locations. Theoretically, duplex sequencing is estimated to have low error rate of less than one artifactual mutation per billion nucleotide sequence. In addition, if the mutation is detected in only one of the strands, it can be used to identify site specific DNA damage in duplex. The duplex sequencing would have wide application in pharmacogenomics as the DNA sequence plays an important role in this area of medicine (Schmitta et al. 2012).

Currently a few management protocols are being used to care for patients with brain tumors (Weinberg 2007). However, targeted therapeutic for cancer patients is shifting towards the pharmacogenomic informations and patient-specific biomarkers. Any initial drug trial at phase 1 provides the basic informations such as toxicity, dosage, and efficacy for appropriate patient population. These basic informations together with pharmacogenomic data of patient and control groups from various geographic and ethnic background could help in the development of specific treatment protocols. The types of cancers, the demography of the patients, and eventually the individual pharmacogenomic profile of the patient could be helpful in developing patient-specific treatment protocols. In addition, SNP, LOH, and PTM data combined with analysis of tumor heterogeneity and duplex sequencing could provide a better platform to develop, test, and provide patient-specific care for brain tumor patients.

13 Ethical Issues

Opponents of human genome project believe that it is a waste of time and resources. Many also believe that pharmacogenomics, like other new areas of sciences which branched out of human genome project, represents misallocation of funds. These individuals believe that time and resources should be spent on solving other major problems facing the survival of humanity, such as famine, flood, epidemics due to infectious diseases, and providing clean water to everyone.

Others believe that pharmacogenomics has the potential to benefit millions of patients as well as economic benefits for the health care industry.

Fig. 2.7 Ethics in pharmacogenomics

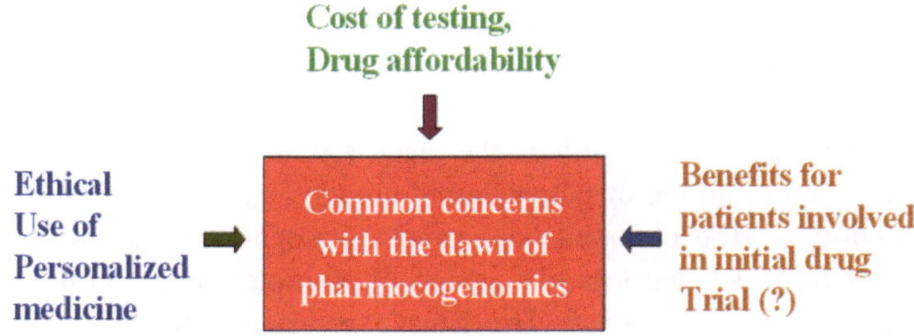

The arguments for the advancement of pharmacogenomics include the mortality due to adverse drug reaction which amounts to tens of thousands or even more. In USA alone, drug reaction is thought to kill and hospitalize approximately 100,000 patients. Many of these deaths are due to individual genetic variants and therefore can be prevented by obtaining pharmacogenomics data of these patients prior to drug administration. However, the field of pharmacogenomics is still in its infancy. The science and technology associated with pharmacogenomics has to evolve in order for it to be used effectively (Rothstein and Epps 2001).

In general, the drugs are prescribed based on certain fundamental data on the patients such as age, gender, symptoms, and other clinical criteria. These criteria may not be sufficient to ensure the safety of the patients. Each year more than two million people have fatal drug reaction, and another two million have serious, but nonfatal drug reaction. In the absence of pharmacogenomics data, it is impossible for the physicians to know in advance the side effects of drugs prescribed to their patients. Pharmacogenomics would help in the development of safe and effective drugs and determination of appropriate dosage. Thus, it could eventually eliminate the current practice of "one size fits all" method of drug prescription.

13.1 Individuals' Affordability and Patient Benefits of Pharmacogenomics

Currently the cost to obtain individual pharmacogenomic data appears to be enormous. However, new efficient procedures which would make it affordable to all are being developed. At the initial stages it may appear that only the rich will benefit from customized medicines, but it should become affordable to all once the procedures are developed, standardized, and put into practice. Some of the novel and affordable protocols may be patented by a few companies who could fix the rates keeping the profit in mind, and it would become unaffordable for the masses (Rothstein and Epps 2001). Researchers who have shares in these companies may also include many unnecessary procedures which would make the cost of pharmacogenomic testing higher than the actual cost. Under these circumstances individual health could be overlooked for the sake of monetary benefits (Fig. 2.7).

13.2 Benefits to the Patients Involved in Initial Drug Trials

Historically it has been observed that the population who initially participated in research and development and drug trials did not receive the anticipated health benefits. This is due to lack of health insurance to cover the high cost of treatments once they are commercialized. For example, the patients with Gaucher disease who participated in the development of safe and effective treatment were denied access to it by their insurance companies due to high cost of the new treatment, and patients could not afford to bear the cost on their own. Another example is the participation of Canavan's Disease Support Group who helped in raising funds for research and recruited willing patients to participate in research but was denied continuous participation in the further development of treatment procedures (Rothstein and Epps 2001; Neil and Carigie 2004;

Peterson-Iyer 2008). This group which originally supported the research lab in many ways is suing the lab for abrupt exclusion (Fig. 2.7).

13.3 Ethical Use of Individualized Medicine

Advanced knowledge of the individuals' response to a particular medicine should help doctors to avoid prescribing ineffective or dangerous medications. However, availability of such personal informations could also be misused deliberately in order to prevent patients from getting the right treatment. Or under extreme conditions patients may be deliberately given medications that are ineffective for them (Rothstein and Epps 2001; Neil and Carigie 2004; Peterson-Iyer 2008).

Pharmacogenomic profiling is not the only way to design patient-specific drugs. Environmental factors, lifestyle behaviors, diet, and many other factors affect the efficacy and safety of prescription drugs. Variations in patients' response to drugs also depend on many factors other than SNPs, such as drug-drug interaction, drug to food interaction, and sunlight exposure of the patients. Variations in the manner in which the drugs are metabolized by the individual patient depend mostly on the pharmacogenomic data. Therefore, the potency of drugs appears to be dependent on pharmacogenomics components (Fig. 2.7).

14 Pharmacogenomics and Protection of Individual Rights

A child enrolled in genetic research study giving consent. The father of this individual's medical records was obtained by a lab without his consent. The father contacted the Office of Human Research Protections (OHRP) of the US Department of Health and Human Services and indicated that his right to privacy is violated. OHRP sided with the father, prohibited the child from using father's medical informations or getting access to any further informations on his father citing the father's right to privacy in this regard. It is a little confusing situation regarding whose right is violated and whose right is more important. In this case does the child has the right to use fathers medical informations to further medical research or to get the right evaluation of his genetic condition which he could pass on to his offspring. What about the right of the child and his offspring's? Under HIPPA it is unclear if the fathers' right is protected or not.

15 What Is Health Insurance Portability and Accountability Act (HIPPA), USA

The major goal of this act is protect individual's right regarding their medical history and health information while allowing flow of personal medical conditions to advance medical research aimed towards better health care to the public. HIPPA protects (a) individuals present and past physical and mental health conditions; (b) provision of health care to individual; and (c) past, present, and future provisional health care payments to an individual. The individual information can be reidentified and the medical informations can be disclosed only if necessary.

16 Risk of Pharmacogenomics

16.1 Risk of Psychological Harm

One of the major ethical issues is regarding testing for disease predisposition status of an individual. Certain genetic disorders have late onset, and knowing these informations could psychologically harm the individual years ahead of the onset of actual disease. Diseases such as Huntington's disease and Alzheimer's disease have late onset, and effective treatments do not exist for these. It is difficult to establish ethical standards for low and high risk in certain diseases which is further complicated by competing interest (Williams-Jones and Corrigan 2003; Foster and Sharp 2005; Shastry 2006). In such circumstances an apparent low risk can be made to look like very high risk and vice versa (Fig. 2.8).

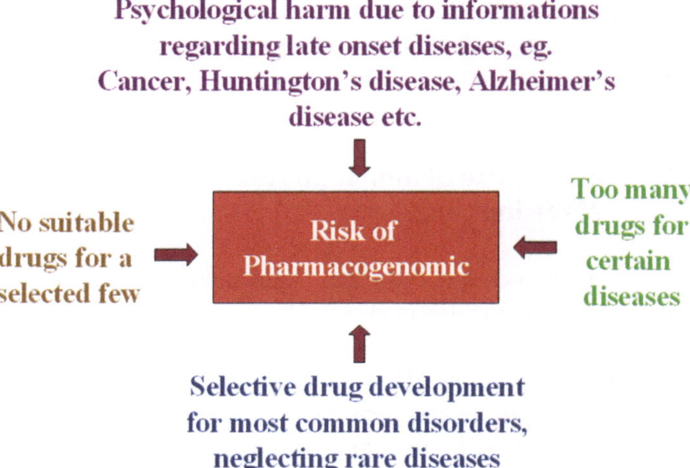

Fig. 2.8 Risks associated with pharmacogenomics

16.2 Insufficient Drug Alternatives for a Selected Group of Patients

Detailed pharmacogenomic data on an individual could sometimes be disadvantageous if these individuals already have genetic disorders or could develop one in the latter years. Some of these rare disorders may have only a few approved drugs available for treatments. If the genetic variants of the patients prevent the use of the existing drugs, they may be left with no treatment options available to them. Such scenario should be given serious consideration as it could cause severe psychological problems to the patients' years before the onset of the actual disease (Williams-Jones and Corrigan 2003; Foster and Sharp 2005; Shastry 2006) (Fig. 2.8).

16.3 Obstructions in the Development of Pharmacogenomics-Based Patient Management

The task involved in identifying the gene(s) and family of genes involved in multiple gene disorders is highly complex. There are three billion bases in human genome which are placed in a sequential manner, and the SNPs are reported to be present in approximately every 300–400 bases. Initially all the SNPs will have to be identified and analyzed, and then their involvement in a particular disorder will have to be evaluated in order to use this data for patient management. This step has to be repeated for each individual which could be time consuming and very expensive. This procedure also can have negative impact as some patients may be sensitive to all the available drugs and they may be left without any treatment options. The pharmaceutical companies may produce drugs which are effective in most patients. This could leave a small but a significant group of patients who suffer from drug sensitivity without any drugs. The other extreme scenario could be presence of multiple drugs for a single disorder in which case the health care provider may have to run extra diagnostic procedures to select the ones that is best suited for each patient (Williams-Jones and Corrigan 2003; Foster and Sharp 2005; Shastry 2006). Such extra steps could also be time consuming and expensive.

17 Conclusion

Pharmacogenomics is an emerging field and it is still in its infancy. It appears that the cost of whole genome sequencing could become affordable for each individual and that would help designer drugs to suit individual needs. Health care providers will have more understanding about their patients and will be able to offer better care. With the aid of pharmacogenomic informations, appropriate dose of drug can be provided to patients and that would reduce the cost of treatment considerably. The more the number of patients screened for pharmacogenomic information, the better the understanding of the possible role of SNPs and HapMaps. This could lead to better understanding of human beings globally, and it could also throw light into human evolution and adaptation to various environmental conditions. At this juncture we do not know if pharmacogenomics will be accessible to all individuals. It would serve its purpose only if and when it helps all human beings equal access to better health care.

Acknowledgement The authors thank National Institute of Mental Health and Neurosciences (NIMHANS), Bangalore 560029, India, for logistic support in the preparation of manuscript. Dr. Mohan L. Gope is a retired Professor.

References

Adams JU (2008) Pharmacogenomics and personalized medicine. Nat Educ 1:1

Anderson JL, Horne BD, Stevens SM et al (2007) Randomized trial of genotype-guided versus standard warfarin dosing in patients initiating oral anticoagulation. Circulation 116:2563–2570

Ball MP, Thakuria JV, Zaranek AW et al (2012) A public resource facilitating clinical use of genomes. Proc Natl Acad Sci U S A 109:11920–11927

Barrett JC, Fry B, Maller J, Daly MJ (2005) Haploview: analysis and visualization of LD and haplotype maps. Bioinformatics 21(2):263–265

Beutler E (1996) G6PD: population genetics and clinical manifestations. Blood Rev 10:45–52

Borobia AM, Lubomirov R, Ramırez E et al (2012) An acenocoumarol dosing algorithm using clinical and pharmacogenetic data in Spanish patients with thromboembolic disease. PLoS One 7:e41360. doi:10.1371/journal.pone.0041360

Brackmann DE, Kwartler JA (1990) A review of acoustic tumors: 1983–1988. Am J Otol 11:216–232

Brockmoller J, Kirchheiner J, Meisel C et al (2000) Pharmacogenetic diagnostics of cytochrome P450 polymorphisms in clinical drug development and in drug treatment. Pharmacogenomics 1:125–151

Chen PL, Scully P, Shew JY, Wang JYJ, Lee WH (1989) Phosphorylation of the retinoblastoma gene product is modulated during cell cycle and cellular differentiation. Cell 58:1193–1198

Clark D (2010) Expecting the worst – a publication from the Uppsala Monitoring Centre. Drug Saf 33:1135–1136

Collins FS, McKusick VA (2001) Implications of the human genome project for medical science. J Am Med Assoc 285:540–544

Dayalan AHPP, Keshava R, Mathivanan J et al (2006a) The p53 gene and human vestibular schwannomas. Ann Neurosci 13:77–91

Dayalan AHPPD, Thomas R, Mathivanan M et al (2006b) The tumor suppressor gene retinoblastoma (RB1) in human vestibular schwannomas. Ann Neurosci 13:113–124

Dayalan AHPP, Mathivanan J, Keshava R et al (2006c) Age dependent phosphorylation and deregulation of p53 in human vestibular schwannomas. Mol Carcinog 45:38–46

Desai AA, Innocenti F, Ratain MJ (2003) Pharmacogenomics: road to anticancer therapeutics nirvana? Oncogene 22:6621–6628

Foster MW, Sharp RR (2005) Will investments in biobanks, prospective cohorts, and markers of common patterns of variation benefit other populations of drug response and disease susceptibility gene discovery? Pharmacogenomics J 5:75–80

Gerlinger M, Rowan AJ, Horswell S et al (2012) Intratumor heterogeneity and branched evolution revealed by multiregion sequencing. N Engl J Med 366:883–892

Goldstein DB, Cavalleri GL (2005) Genomics: understanding human diversity. Nature 437:1241–1242

Grant SFA (2001) Pharmacogenetics and pharmacogenomics: tailored drug therapy for the 21st century. Trends Pharmacol Sci 22:3–4

Guengerich FP (2006) Cytochrome P450s and other enzymes in drug metabolism and toxicity. AAPS J 8:E101–E111

Hainaut P, Weiman KG (2009) 30 years and a long way into p53 research. Lancet Oncol 10:913–919

Hanahan D, Weinberg RA (2011) Hallmark of cancer: the next generation. Cell 144:646–674

Hillman MA, Wilke RA, Caldwell MD et al (2004) Relative impact of covariates in prescribing warfarin according to CYP2C9 genotype. Pharmacogenetics 14:539–547

Hinds DA, Stuve LL, Nilsen GB et al (2005) Whole genome patterns of common DNA variation in three human populations. Science 307:1072–1079

Hollinger MA (2003) Introduction to pharmacology. CRC Press, Boca Raton, Florida, USA, ISBN 0-415-28033-8

Human Cytochrome P450 (CYP) Allele Nomenclature Committee. Home Page of the Human Cytochrome

P450 (CYP) Allele Nomenclature Committee. http://www.cypalleles.ki.se/. Accessed 25 Mar 2010

Ingelman-Sundberg M (2004) Pharmacogenetics of cytochrome P450 and its applications in drug therapy: the past, present and future. Trends Pharmacol Sci 25:193–200

International HapMap Consortium (2010) Integrating common and rare genetic variation in diverse human populations. Nature 467:52–58

Kaye AH, Laws ER Jr (eds) (2001) Brain tumors: an encyclopedic approach, 2nd edn. © Harcourt Publishers Limited, London

Lagoumintzis G, Poulas K, Patrinos GP (2010) Genetic data bases and their potential in pharmacogenomics. Curr Pharm Des 16:2223–2231

Laing RE, Hess P, Shen Y et al (2011) The role and impact of SNPs in pharmacogenomics and personalized medicine. Curr Drug Metab 12:460–486

Lander ES (2011) Initial impact of the sequencing of the human genome. Nature 470:187–197

Lazarou J, Pomeranz BH, Corey PN (1998) Incidence of adverse drug reactions in hospitalized patients: a meta-analysis of prospective studies. JAMA 279: 1200–1205

Leitner WW, Ying H, Restifo NP (1999) DNA and RNA based vaccines: principles, progress and prospects. Vaccine 18:765–777

Limdi NA, McGwin G, Goldstein JA et al (2008) Influence of CYP2C9 and VKORC1 1173C/T genotype on the risk of hemorrhagic complications in African-American and European-American patients on warfarin. Clin Pharmacol Ther 8(83):212–221

Lindquist M (2008) Vigibase, the WHO global ICSR database system: basic facts. Drug Inf J 42:409–419

Ma Q, Lu AY (2011) Pharmacogenetics, pharmacogenomics, and individualized medicine. Pharmacol Rev 63:437–459

Martuza RL, Eldridge R (1988) Neurofibromatosis 2 (bilateral acoustic neurofibromatosis). N Eng J Med 318:684–688

Mathivanan J, Rohini K, Gope ML et al (2007a) Altered structure and deregulated expression of the tumor suppressor gene retinoblastoma (RB1) in human brain tumors. Mol Cell Biochem 302:67–77

Mathivanan J, Rohini K, Gope ML et al (2007b) Possible role of the tumor suppressor gene retinoblastoma (RB1) in human brain tumor development. Ann Neurosci 14:72–82

McCarthy JJ, Hilfiker R (2000) The use of single-nucleotide polymorphism maps in pharmacogenomics. Nat Biotechnol 18:505–508

Minguez P, Parea L, Diella F et al (2012) Deciphering a global network of functionally associated post-translational modifications. Mol Syst Biol 8:599

Mitra R, Indira Devi B, Gope ML et al (2012) Sodium Butyrate modulates pRb phosphorylation and induces cell death in human vestibular schwannomas *in vitro*. Indian J Exp Biol 50:19–27

Nagle H, Nagle N (2005) Pharmacology: an introduction. McGraw Hill, Boston. ISBN 0-07-312275-0

Neil D, Carigie J (2004) The ethics of pharmacogenomics. Monash Bioeth Rev 23:9–20

Peterson-Iyer K (2008) Pharmacogenomics, ethics and public policy. Kennedy Inst Ethics J 18:35–56

Piccinin S, Tonin E, Sessa S et al (2012) A "twist box" code of p53 inactivation: twist box:p53 interaction promotes p53 degradation. Cancer Cell 22:404–415

Read more: http://www.ukdissertations.com/dissertations/sciences/genetic-polymorphism-governing-the-cyp2d6-cytrochrome-p450-enzyme-subfamily-in-drug-metabolism.php#ixzz26i3mqjcg

Rieder MJ, Reiner AP, Gage BF et al (2005) Effect of VKORC1 haplotypes on transcriptional regulation and warfarin dose. N Engl J Med 352:2285–2293

Roberts NJ, Vogelstein JT, Parmigiani G et al (2012) The predictive capacity of personal genome sequencing. Sci Transl Med 4:135lr3

Rodon J, Saura C, Dienstmann R et al (2012) Molecular prescreening to select patient population in early clinical trials. Nat Rev Clin Oncol 9:359–366

Rohini K, Mathivanan J, Dayalan AHPP et al (2007) Loss of heterozygosity of the p53 gene and deregulated expression of its mRNA and protein in human brain tumors. Mol Cell Biochem 300:101–111

Rohini K, Mathivanan J, Gope ML et al (2008) Functional modulation of the p53 gene and its protein in human brain tumors. Ann Neurosci 15:75–86

Rothstein MA, Epps PG (2001) Ethical and legal implications of pharmacogenomics. Nat Rev Genet 2:228–231

Sato Y, Laird M, Yoshida T (2010) Biostatistic tools in pharmacogenomics – advances, challenges, potential. Curr Pharm Des 16:2232–2240

Schmitta MW, Kennedya SR, Salka JJ et al (2012) Detection of ultra-rare mutations by next-generation sequencing. Proc Natl Acad Sci U S A 109:14508–14513

Shastry BS (2006) Pharmacogenetics and the concept of individualized medicine. Pharmacogenomics J 6:16–21

Thomas R, Dayalan AHPP, Mathivanan J et al (2005) Altered structure and expression of RB1 gene and increased phosphorylation of pRb in human vestibular schwannomas. Mol Cell Biochem 271:113–121

Varela MA, Amos W (2010) Heterogeneous distribution of SNPs in the human genome: microsatellites as predictors of nucleotide diversity and divergence. Genomics 95:151–159

Venter JC et al (2001) The sequence of the human genome. Science 291:1304–1351

Weber WW (1997) Pharmacogenetics, Oxford University Press, New York

Weiling F (1991) Historical study: Johann Gregor Mendel 1822–1884. Am J Med Genet 40:1–25, discussion 26

Weinberg RA (1991) Tumor suppressor genes. Science 254:1138–1146

Weinberg RA (1995) The retinoblastoma protein and cell cycle control. Cell 81:323–330

Weinberg RA (2007) Biology of cancer. Garland Sciences, Taylor & Francis Group, LLC, New York, NY, USA

Williams-Jones B, Corrigan OP (2003) Rhetoric or hype. Where's the ethics in pharmacogenomics. Am J Pharmacogenomics 3:375–383

Willcox SM, Himmelstein DU, Woolhandler S (1994) Inappropriate drug prescribing for the community-dwelling elderly. J Am Med Assoc 272:292–296

Websites on SNPs, SNPs in Cancer, GWAS

American Association for Cancer Research Cancer Concepts Factsheet on SNPs

Database of Genotype and Phenotype (dbGaP) located at: http://www.ncbi.nlm.nih.gov/entrez/query.fcgi?db=gap

The SNP Consortium LTD – SNP search

3 Next-Generation Sequencing (NGS): A Revolutionary Technology in Pharmacogenomics and Personalized Medicine

Huixiao Hong, Wenqian Zhang, Zhenqiang Su, Jie Shen, Weigong Ge, Baitang Ning, Hong Fang, Roger Perkins, Leming Shi, and Weida Tong

Abstract

Personalized medicine can improve healthcare by selecting treatments that are more efficacious or induce less adverse responses in stratified cohorts sharing differentiating genetic traits. Personalized medicine has advanced quickly, providing both opportunities and challenges for the pharmaceutical industry and regulatory agencies in the twenty-first century. Pharmacogenomics is the key to the identification of personalized medicine biomarkers useful for efficacy and safety that can ultimately be clinically applied for diagnosis, prognosis, and treatment selection. The requisite technologies and approaches needed for pharmacogenomics have steadily advanced over more than a decade in terms of both capability and cost. In 2005, 454 Life Sciences announced their sequencing-by-synthesis technology, the first next-generation sequencing (NGS) platform, proclaiming the breakthrough in sequencing technology. NGS is revolutionizing pharmacogenomics and personalized medicine, with several NGS platforms commercially available. Illumina, Roche 454, and Applied Biosystems are the current major vendors. This chapter will characterize the technical assessments of NGS, including comparative analyses across platforms, experimental protocols, algorithms for mapping short reads to reference

The views presented in this article do not necessarily reflect those of the US Food and Drug Administration.

H. Hong (✉) • Z. Su • J. Shen • W. Ge • R. Perkins
L. Shi • W. Tong
Division of Bioinformatics and Biostatistics, National Center for Toxicological Research, US Food and Drug Administration, 3900 NCTR Road, Jefferson, AR 72079, USA
e-mail: huixiao.hong@fda.hhs.gov

W. Zhang
Beijing Genomic Institute, Beishan Industrial Zone, Yantian District, Shenzhen, Guangdong 518083, China

B. Ning
Division of Systems Biology, National Center for Toxicological Research, US Food and Drug Administration, 3900 NCTR Road, Jefferson, AR 72079, USA

H. Fang
Office of Scientific Coordination, National Center for Toxicological Research, US Food and Drug Administration, 3900 NCTR Road, Jefferson, AR 72079, USA

D. Barh et al. (eds.), *Omics for Personalized Medicine*,
DOI 10.1007/978-81-322-1184-6_3,

genomes, strategies for quantitatively measuring expression levels, and methods for detecting single nucleotide polymorphisms (SNPs). Different pipelines and software packages for analyzing NGS data will be reviewed. Examples and a prospective outlook on applications of NGS in pharmacogenomics and personalized medicine will be given.

1 Introduction

1.1 Personalized Medicine

The term "personalized medicine" has been used for some 13 years (Langreth and Waldholz 1999) in the context widely understood today of "the right drug for the right patient with the right dose at the right time through the right route." The President's Council of Advisors on Science Technology (PCAST) has a more comprehensive definition for personalized medicine: "'Personalized medicine' refers to the tailoring of medical treatment to the individual characteristics of each patient. It does not literally mean the creation of drugs or medical devices that are unique to a patient, but rather the ability to classify individuals into subpopulations that differ in their susceptibility to a particular disease or their response to a specific treatment. Preventive or therapeutic interventions can then be concentrated on those who will benefit, sparing expense and side effects for those who will not" (President's Council of Advisors on Science Technology 2008).

The rapid increase in the number of published articles per year referring to personalized medicine, shown in Fig. 3.1, is a testament to the substantial and accelerating scientific interest (Jorgensen 2009) that is driven by real medical needs and fostered by rapid advances in information-rich and high-throughput technologies. While technology trends and palpable scientific excitement has moved personalized medicine from futuristic to realistic, a lack of demonstrable progress in the clinical appearance of personalized medicine related biomarkers portends substantial challenges remain for both the pharmaceutical industry and regulatory agencies.

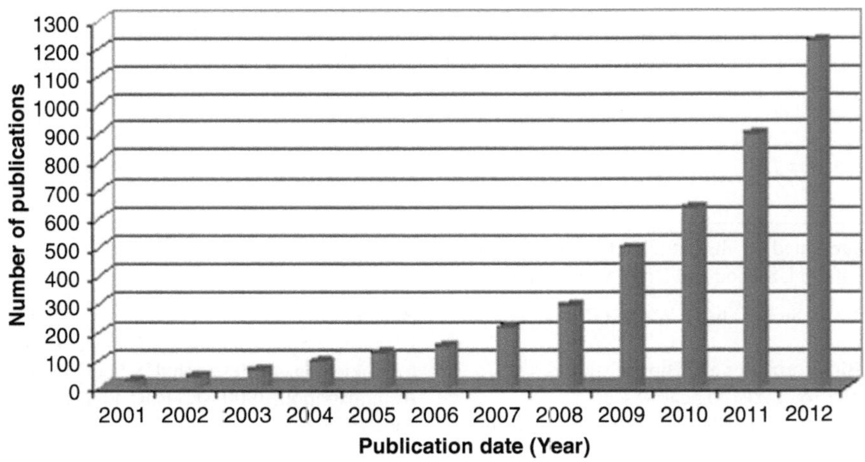

Fig. 3.1 Annual number of publications related to personalized medicine from 2000 to 2011 based on a keyword search in PubMed. Keyword used: personalized medicine. Fields searched: title and abstract (Search was conducted on August 31, 2012. The number of publications in 2012 was projected from 815 for the first 8 months to 1,219 for the whole year)

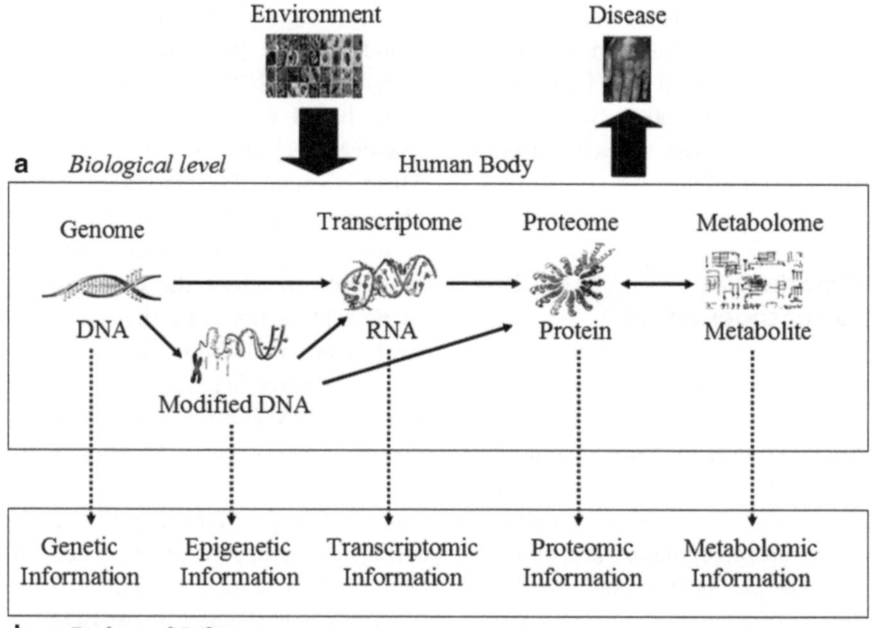

Fig. 3.2 Taxonomy of molecules used for biomarkers. *Box a* depicts the biological levels and associated molecules that can be identified as biomarkers with the appropriate technologies. *Box b* gives the corresponding categories of the molecular biomarkers based on the type of molecule class

1.2 Pharmacogenomics

Adequate individuals' biological and lifestyle (environment) information is an essential prerequisite for realizing personalized medicine applications. The biological information comprises the molecules and their measurements for individuals that genetically distinguish a subpopulation and can therefore be translated into clinical practices for personalized medicine. Figure 3.2 provides a taxonomy of molecular classes pertinent for personalized medicine. Most active pharmacogenomics research fully utilizes these molecular classes based on DNA and RNA measurements yielding genetic, epigenetic, and transcriptomic information.

Pharmacogenomics is a scientific research field that attempts to explain how genomic and genetic variations affect a patient's response to a drug. Motulsky first reported a pharmacogenomics study in 1957 (Motulsky 1957). That study found enzymes for the metabolism of a number of drugs exhibited variation in activity among patients that were correlated with the adverse reactions to those drugs. Since then, pharmacogenomics has enormously advanced in accordance with technologies, leading to the discovery of genetic variants that can be used as markers to monitor the efficacy and safety of drugs on the market (Hong et al. 2010).

Some 3.1 million common SNPs in human populations have been identified by the HapMap project (The International HapMap Consortium 2007). In concert, state-of-art high-throughput SNP genotyping technology enables simultaneously genotyping of hundreds of thousands of SNPs, making genome-wide association studies (GWAS) a promising pharmacogenomics research field for linking genetics with biological response. Some GWAS have found associations between genetic variations of patients and their therapeutic responses to drugs such as thiazide diuretic (Turner et al. 2008), warfarin (Takeuchi et al. 2009), and iloperidone (Lavedan et al. 2008). Other GWAS have found associations between genetic variants and the adverse events caused by drugs such as the elevation of serum alanine aminotransferase by ximelagatran (Kindmark et al. 2008) and drug-induced liver injury due to flucloxacillin (Daly et al. 2009). Both types of studies are important as risk assessment considers both efficacy and safety.

1.3 Biomarkers in Pharmacogenomics

In this book chapter, a biomarker in pharmacogenomics is a molecule or a set of molecules of DNA or RNA that can be measured and used for purposes such as characterizing financial risk in pharmaceutical drug development, quantifying risk or benefit to inform regulatory decisions, and ultimately predicting clinical response (including selecting clinical treatment or tailoring dose to patient) (Hong et al. 2010). Here we classify pharmacogenomics biomarkers into three types: genetic biomarkers, epigenetic biomarkers, and transcriptomic biomarkers, as depicted in Fig. 3.2.

1.3.1 Genetic Biomarkers

Genetic biomarkers are used in personalized medicine practices mainly for selecting those patients who benefit more or have less risk of adverse drug reactions (ADRs) from a particular drug. For example, human leukocyte antigen (HLA) allelic marker, HLA-B*5701, is a genetic biomarker that can be used to predict increased risk of developing hypersensitivity reactions to the antiretroviral drug, abacavir. Thus, it is recommended that the marker's presence is assessed in patients with human immunodeficiency virus (HIV) prior to treatment (Lucas et al. 2007; Mallal et al. 2008; US Food and Drug Administration 2012).

Investigation of biomarkers in drug-metabolizing genes may lead to a better understanding of drug efficacy. Drug-metabolizing genes encode proteins responsible for metabolizing exogenous toxicants including drugs; genetic polymorphisms are prevalent in these genes. Gene deletions, missense, nonsense, and splice site mutations can alter or abolish enzyme activity, whereas mutations causing amino acid substitutions can lead to markedly modified enzyme action. A good example is the anticoagulant warfarin (a commonly used blood thinner). Warfarin is metabolized by the enzyme CYP2C9 that exhibits wide genetically based activity variation among patients (Sanderson et al. 2005; Takahashi et al. 2006). There is a poor efficacy to safety margin with this drug and, since individual response to warfarin varies so greatly, that patients have to be closely monitored to adjust the dose during treatment to prevent hemorrhage that can result in fatal strokes, etc. In addition, the target of warfarin is vitamin K epoxide reductase and polymorphisms in its gene (VKORC1) account for some of the dosing variation among individuals, with certain haplotype groups requiring larger doses and other haplotype groups requiring lower doses, to reach safe therapeutic levels. Studies have identified variants associated with warfarin adverse reactions that resulted in the FDA updating the label for warfarin, stating that "the patient's *CYP2C9* and *VKORC1* genotyping information, when available, can assist in selection of starting dose" (Kim et al. 2009).

Another example of biomarkers enabling personalized medicine is the genetic testing of mutations in the KRAS gene (Kirsten *ras*) to avoid treatment of patients with metastatic colorectal cancer who cannot benefit from the use of panitumumab (Vectibix®) and cetuximab (Erbitux®). KRAS is a protein encoded by the KRAS gene (McGrath et al. 1983); mutations in the KRAS gene make potent oncogenes and the protein products of oncogenes play a role in many cancers (Kranenburg 2005). Pharmacogenetics studies revealed that the presence of mutations in the KRAS gene was associated with the poor response to panitumumab or cetuximab therapy in patients with colorectal cancer. Accordingly, drug labels were updated by the US Food and Drug Administration (FDA) and suggest that when these two anti-epidermal growth factor receptor (EGFR) antibody drugs are used for the treatment of patients with metastatic colorectal cancer, gene tests for *KRAS* mutations are recommended (Hong et al. 2010).

Table 3.1 lists the genetic biomarkers mentioned in some drug labels approved by the FDA. In addition, to promote drug efficacy and drug

Table 3.1 Genetic biomarkers mentioned in labels of drug products approved by FDA (Hong et al. 2010)

Biomarker	Drug	Section in label
CYP2C19	Plavix®	Clinical pharmacology, precautions, dosage and administration
	Vfend®	Clinical pharmacology
	Effient®	Use in specific populations, clinical pharmacology, Clinical Studies
	Celebrex®	Clinical pharmacology
CYP2C9	Celebrex®	Clinical pharmacology
	Effient®	Use in specific populations, clinical pharmacology, clinical studies
	Coumadin®	Clinical pharmacology, precautions
CYP3A4	Celebrex®	Clinical pharmacology
	Codeine sulfate	Drug interactions, clinical pharmacology
CYP3A5	Effient®	Use in specific populations, clinical pharmacology, clinical studies
CYP2B6	Effient®	Use in specific populations, clinical pharmacology, clinical studies
CYP2D6	Strattera®	Dosage and administration, warnings and precautions, drug interactions, clinical pharmacology
	Prozac®	Clinical pharmacology, precautions
	Codeine sulfate tablets	Warnings and precautions, drug interactions, use in specific populations, clinical pharmacology
VKORC1	Coumadin®	Clinical pharmacology, precautions
UGT1A1	Camptosar®	Clinical pharmacology, warnings, dosage and administration
	Tasigna®	Drug interactions, clinical pharmacology
HLA-B*1502	Tegretol®	Warnings, precautions
HLA-B*5701	Ziagen®	Warnings and precautions
Deletion 5q	Revlimid®	Hematologic toxicity, clinical studies, precautions, adverse reactions

safety, FDA maintains a database (http://www.fda.gov/drugs/scienceresearch/researchareas/pharmacogenetics/ucm083378.htm) of genetic variants that affect the treatment outcomes of some drugs. Included are genetic biomarkers and related advices/warnings that are listed on the drug labels that indicate efficacy differences and possible adverse reactions among patient with certain genotypes.

1.3.2 Epigenetic Biomarkers

While DNA sequence of an organism may not be changed, nongenetic factors, environmental, and lifestyle-related influences such as nutrition and exposure to stress can induce epigenetic alteration that causes the organism's genes to behave (or express themselves) differently (Bird 2007). The genetic background of a person can provide information at the individual level on the possibility of developing some diseases and responses (efficacy and safety) to some drugs, while epigenetic alterations might be more directly associated with the phenotype observed. Consequently, epigenetic biomarkers hold great promise for personalized medicine.

Numerous mechanisms are involved in gene regulation, among which, chromatin remodeling (such as DNA methylation and histone methylation), microRNAs (miRNAs, the term will be used hereafter), and other noncoding RNAs are the most studied (Bird 2007; Berger et al. 2009). Chromatin remodeling is an approach of epigenetic alteration that is achieved either by the post translational modification of the amino acids that make up histone proteins or by adding methyl groups to the DNA (most likely at CpG sites) to convert cytosine to 5-methylcytosine. Though epigenetic biomarkers of histone modification (Kondo et al. 2008) and noncoding RNAs (He et al. 2005; Lu et al. 2005) have been reported, the most common ones are the biomarkers identified based on DNA methylation. The epigenetic biomarkers being pursued so far are mainly utilized for the diagnosis and prognosis of cancers, some of which will be discussed in detail below.

Cancer epigenetic biomarkers can be simply categorized as diagnostic and prognostic biomarkers based on their clinical applications.

Table 3.2 Fluids and tissues that have been used in discovery of epigenetic biomarkers

Fluid/tissue	Cancer type	Epigenetic biomarkers (methylated genes)
Biopsy	Prostate	GSTP1, RARβ2, APC, TIG1
Ejaculate	Prostate	GSTP1
Nipple fluid	Breast	RASSF1
Peritoneal	Ovary	BRCA1, RASSF1
	Breast	RASSF1, APC, DAPK
Serum	Colorectum	SEPT9, TMEEF2, NGFR
	Ovary	BRCA1, RASSF1
Sputum	Lung	$P16^{INK4A}$, RASSF1, MGMT
Stool	Colorectum	SFRP2, CDKN2A, hMLH1
Urine	Bladder	RASSF1, APC, $p14^{ARF}$, DAPK, BCL2, TERT
	Prostate	GSTP1, RARβ2, APC, RASSF1

It has been observed that DNA methylation patterns are often altered in early-stage tumors (Weisenberger et al. 2006; Irizarry et al. 2009). For example, methylation detection in conjunction with a cervical Pap test is more sensitive to diagnose lethal cervical cancers than the Pap test alone (Kahn et al. 2008). For the most part, to be potentially useful in clinical practice, diagnostic epigenetic biomarkers identified from easily obtainable and readily available body fluids and tissues are the most valuable. Table 3.2 lists some fluids and tissues that have been used in epigenetic biomarkers development (Duffy et al. 2009). However, one of the challenges for developing diagnostic biomarkers is the requirement of low false positive rates when biomarkers are used for screening in an overall healthy population since a moderately low false positive rate can lead to a substantial number of unnecessary follow-up examinations in healthy individuals. In contrast, epigenetic biomarkers used as prognostic indicators do not present the same concern in terms of specificity. In addition, many epigenetic biomarkers were also developed for classifying disease subtypes or disease stages among already diagnosed individuals (Weisenberger et al. 2006).

1.3.3 Transcriptomic Biomarkers

DNA microarrays were introduced for analyzing gene expression profiles approximately 17 years ago (Schena et al. 1995; Lockhart et al. 1996). They enable measuring expression levels of thousands of genes in a single experiment. This technology has become ubiquitously applied across research and drug development. The basic principle behind DNA microarrays experiments is the base-pairing hybridization between two DNA strands as complementary nucleic acid sequences form hydrogen bonds between complementary nucleotide base pairs. A typical experiment for measuring gene expressions using DNA microarrays consists of the following steps for a two color array: mRNA molecules in the cells/tissues of interest are extracted and collected, the mRNA molecules are labeled by attaching fluorescent nucleotides to the cDNAs that are generated by using a reverse transcriptase enzyme (usually samples labeled with special fluorescent dyes), the labeled cDNAs are added onto a DNA microarray slide and hybridized to their complementary DNAs attached on the microarray slide, and finally, a special scanner is used to measure the fluorescent intensity for each spot/areas on the microarray slide. In this manner, DNA microarrays represent a mature technology for measuring transcriptome expression for specific cell/tissue and differential expressions between disease and healthy populations. Microarrays can thus be applied to discover transcriptomic biomarkers for personalized medicine.

Some transcriptomic biomarkers are validated for the application in clinical diagnosis and risk assessments. For example, in 2007, FDA approved a DNA microarray-based diagnostic kit (MammaPrint®) that measures the transcription level of 70 genes in breast cancer patients (Van't Veer et al. 2002; van de Vijver et al. 2002; Glas et al. 2006; Buyse et al. 2006). The profiles of the 70-gene signature are usable by physicians as a prognosis test but only for breast cancer patients who are less than 61 years old with Stage I or Stage II disease, tumor size ≤5.0 cm, and who are lymph node negative. Another example of transcriptomic biomarkers is Onco*type* DX®, a 21-gene assay (Paik et al. 2004; Goldstein et al. 2008) approved by the FDA in 2005 and currently widely used for lymph node negative breast cancer patients whose tumors tested positive for hormone receptors (Harris et al. 2007). Based on

Table 3.3 Representative technologies used in pharmacogenomics

Type	Technology	Detection	Principal	Application
LCA	PCR-RFLP	Gel-based	Restriction site	Genotyping
	AS-PCR	Gel-based and homogeneous fluorescence	AS amplification	Genotyping
	TaqMan-PCR	Homogeneous fluorescence	FRET quenched hydrolysis probes	Genotyping and gene expression
HCA	Microsphere array	Flow cytometry fluorescence	Bead-immobilized oligonucleotide	Genotyping and gene expression
	DNA microarray	Fluorescence	Oligonucleotide hybridization	Genotyping and gene expression
	NGS	Fluorescence	Pyrosequencing and sequencing-by-synthesis or ligation	DNA and RNA sequencing

Abbreviations: *LCA* low content analysis, *HCA* high content analysis, *PCR* polymerase chain reaction, *RFLP* restriction fragment length polymorphism, *AS* allele-specific, *NGS* next-generation sequencing

results from multi-institutional clinical trials, the 21 genes (*MK167*, *STK15 (AURKA)*, *BURC5*, *CCNB1*, *MYBL2*, *MMP11*, *CTSL2*, *GRB7*, *HER2 (ERBB2)*, *GSTM1*, *CD68*, *BAG1*, *ESR1*, *PGR*, *BCL2*, *SCUBE2*, *ACTB*, *GAPDH*, *RPLPO*, *GUS*, and *TFRC*) are validated to be able to predict recurrences, deaths, and responses to chemotherapy for patients with estrogen-positive breast cancer (Paik et al. 2004).

Given the amount of resources (including current RNA sequencing efforts) being applied to find transcriptomic biomarkers, it is expected that the number of qualified biomarkers will grow.

1.4 Technologies for Pharmacogenomics

Rapid advances in technologies for measuring genomic variations that affect patients' responses to drugs have been responsible for accelerating pharmacogenomics research that is potentially extensible to personalized medicine. The technologies that have been used in pharmacogenomics can be conveniently divided into two types based on the number of prospective genomic or genetic markers that can be simultaneously analyzed: low content analysis (LCA) and high content analysis (HCA) technologies. Table 3.3 summarizes a few of the representative technologies; there was no intent to provide a complete list.

LCA technologies are used when a few SNPs within a single gene (or a few genes) need to be genotyped or the gene expression levels for a few genes need to be measured. Typical LCA technologies include methods such as PCR-restriction fragment length polymorphism (PCR-RFLP), allele-specific PCR (AS-PCR), TaqMan-PCR, hybridization probe-melting analysis, and oligonucleotide ligation-PCR reactions (Newton et al. 1989; Syvanen et al. 1990; Holland et al. 1991; Barany 1991).

PCR is a biochemical technique that has been used for some 30 years to generate millions of copies from a single DNA molecule or from a few DNA molecules. A pair of primers, complementary to the 3′ ends on each of the two strands of the DNA molecule, is used in the amplification. A typical PCR experiment consists of 15–40 cycles, with each cycle having several temperature changes. The amplification in PCR is achieved in three major steps: denaturation, annealing, and elongation. First, denaturation is conducted by heating the reaction to a preset temperature in a short time to melt the DNA template by breaking the hydrogen bonds. In the annealing step, the temperature of the reaction is lowered to a preset temperature (depending the primers used) quickly (usually about 15–40 s) to anneal the primers. In the elongation (or extension) step, a new DNA molecule complementary to the DNA template sequence from the primers is synthesized using an enzyme (Tag DNA polymerase) by adding dNTPs (deoxyribonucleotide triphosphate) to the template in the 5′–3′ direction. PCR is a relatively mature and sensitive technology for measuring expression levels of genes. It has been broadly applied in biomedical research and diagnosis of diseases.

PCR-restriction fragment length polymorphism (PCR-RFLP) technology was an old fashion technology that was used for genotyping common SNPs within candidate genes, especially when a variant naturally occurs at a restriction enzyme cutting site (Goldstein and Blaisdell 1996). This technology is labor intensive and difficult to automate. More importantly, errors happen frequently due to incomplete restriction digestion. It has been replaced, generally, by other more efficient and accurate genotyping methods.

In allele-specific PCR (AS-PCR), two homologous primers that vary only in their 3′ ends (one is complementary to the normal allele and another is complementary to the variant allele) are differentially amplified (Newton et al. 1989). Gel or homogeneous fluorescent dyes are then used to detect the PCR reaction products (Higuchi et al. 1992). In the PCR amplification, undesired side reactions such as primer dimerization can generate noise that confounds the signal from the fluorescent dye bound to the expected reaction products. Thus, optimizing PCR amplification selectivity is vital to data quality. However, the homogeneous real-time fluorescent dye-based detection is easier to automate than the gel-based electrophoretic analysis.

Many genotyping methods use direct probe, as opposed to the complement, sequence that is homogeneously amplified for detecting variants. The most popular one is the TaqMan-PCR that uses 5′-nuclease hydrolysis probes (i.e., the TaqMan probes) (Holland et al. 1991). These are designed in a way that a dye in the native oligonucleotide probe quenches the reporter fluorophore. In amplification, a TaqMan probe hybridizes to its complementary oligonucleotide and is cleaved by the 5′-nuclease action of thermostable DNA polymerases in the primer extension and, thus, the quencher and fluorescent reporters are separated and released for the next cycle. The fluorescent signal of the reporter dye from the TaqMan probes can be quantified in real time. The measured fluorescent intensity steadily strengthens with each PCR cycle and reaches a threshold that is preset for determining whether homozygous or heterozygous alleles are contained in the genotyping sample. The number of SNPs that can be simultaneously genotyped in a single TaqMan-PCR is determined by fluorescent signals that can be measured within a single PCR reaction. The TaqMan-PCR instruments on the market are capable of simultaneously genotyping four to six SNPs.

LCA technologies have their limitations in time, cost, and material efficiency. HCA technologies are the solution when a large number of genomic or genetic markers need to be interrogated. HCA technologies are usually based on hybridization or single-base extension technologies, which makes high-throughput SNP genotyping and gene expression profiling possible. The early stage of HCA technologies were array-based assays, such as oligonucleotide (Cronin et al. 1996) and bead-based microarrays (Armstrong et al. 2000) for simultaneously genotyping a large number of genetic variants.

The microsphere array is a bead-based genotyping technology that became available early in the last decade. The technology is based on attaching oligonucleotide anti-tag sequences to polystyrene microspheres with different dyes that can be read using a laser instrument. Tag complementary oligonucleotides covalently combined with allele-specific primer extensions can be detected. This technology allows tens of SNPs to be simultaneously genotyped in a single tube.

GWAS became a promising research approach for pharmacogenomics 5 years ago when the genotypes of more than 3.1 million common SNPs in human populations were determined by the HapMap project (The International HapMap Consortium 2007), and high-throughput SNP genotyping technology was advanced enough to enable simultaneous genotyping of hundreds of thousands of SNPs using high-density oligonucleotide microarrays. However, the common genetic variants that were identified by GWAS typically contributed only to a small portion of the total variation in the phenotype (Frazer et al. 2009). Rare genetic variants having high penetrance were also found to contribute to the phenotypes of interest, e.g., blood pressure (Ji et al. 2008). Consequently, the best way for obtaining biological information on DNA is to have the single-base resolution for the DNA sequence of an individual, which can be used to interrogate all genetic variants, both common and rare. Next-generation sequencing is currently the best choice for most applications and will be discussed below.

2 Next-Generation Sequencing

2.1 Background

The Sanger method was invented in 1977 (Sanger et al. 1977). The method involves DNA synthesis in the presence of chain-terminating inhibitors followed by electrophoresis. It is noted for excellent accuracy and reasonable read length but very low throughput and high expense, rendering it unsuitable for deciphering the human genome, crucial information for realizing personalized medicine. In 2005, 454 Life Sciences announced their breakthrough sequencing-by-synthesis technology, the first next-generation sequencing (NGS) platform (Margulies et al. 2005). Since then there has been remarkable advances in DNA sequencing technologies known as NGS or as massively parallel sequencing (Shendure and Ji 2008; Reis-Filho 2009; Ansorge 2009; Voelkerding et al. 2009; Metzker 2010). Currently, the Illumina HiSeq-2000 and HiScan, the Roche 454 GS-FLX, and the Applied Biosystems SOLiD Analyzer 5500xl are the most used and commercially available platforms, albeit the Illumina platforms dominate the market. In addition to the platforms that already exist on the market, new NGS platforms are under development and mainly adopt single DNA molecule sequencing technology (e.g., nanotechnology and electron microscopy) which can read through DNA templates in real time without amplification, and thus could be more accurate with potentially longer reads (e.g., Pacific BioSciences RS system produces reads of >1,000 bp; nanoAnalyzer from BioNanomatrix, now BioNano Genomics, generates reads of around 400,000 bp (Das et al. 2010)).

2.2 NGS Workflow

To date, NGS has been successfully applied to different aspects of applications, such as RNA sequencing for profiling gene expression (also called RNA-seq or whole transcriptome sequencing) (Mortazavi et al. 2008; Wang et al. 2008a, Nagalakshmi et al. 2008; Cloonan et al. 2008), chromatin immunoprecipitation followed by sequencing (ChIP-seq) for identifying DNA-binding sites of proteins (Johnson et al. 2007; Park 2009; Schmidt et al. 2010), sequencing to identify methylated DNA (methyl-seq) (Brunner et al. 2009; Hormozdiari et al. 2009) for DNA methylation analysis, whole human genome sequencing (Wheeler et al. 2008; Wang et al. 2008b) for determination of genetic variants, and targeted sequencing of specific candidate genes or the entire human exome in large numbers of individuals (Hodges et al. 2007). The common workflow for studies using NGS technologies is depicted in Fig. 3.3, with main aspects of the process described below.

2.2.1 Library Generation

The first step in NGS is to make a library from a DNA or a cDNA sample. Different NGS platforms use divergent protocols and reagent kits for the library preparation. In practice, the instruction from a specific NGS vendor should be strictly followed to pursue the quality of the library. Here we briefly review the principle and procedure of the library preparation for the Illumina HiSeq-2000 platform, the most popular platform on the market, though readers are recommended to follow Illumina's instruction for detailed procedures for library preparation.

The Illumina HiSeq-2000, an upgrade of the Illumina GA system that was the first short-read sequencing platform, currently dominates the NGS platform market (Metzker 2010). Figure 3.4 depicts the principle and procedure of library

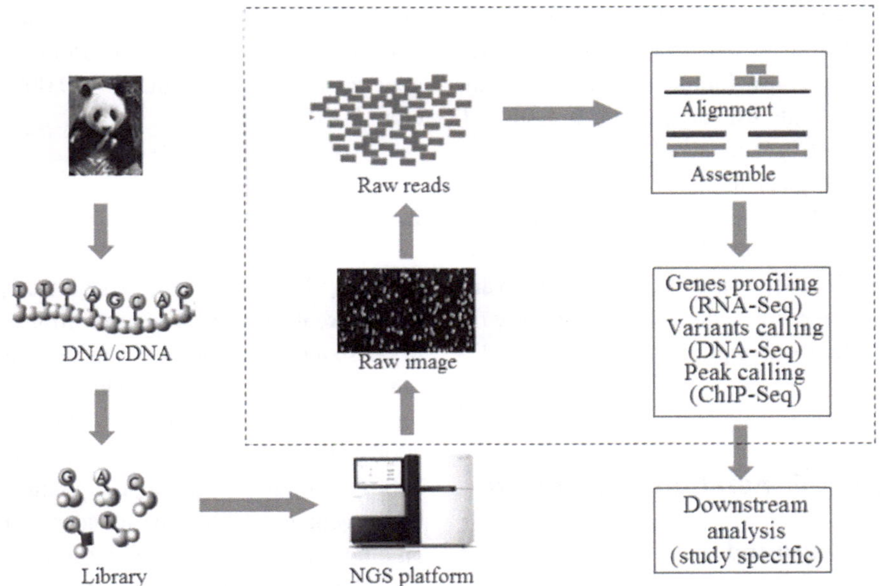

Fig. 3.3 Overview of a common NGS workflow

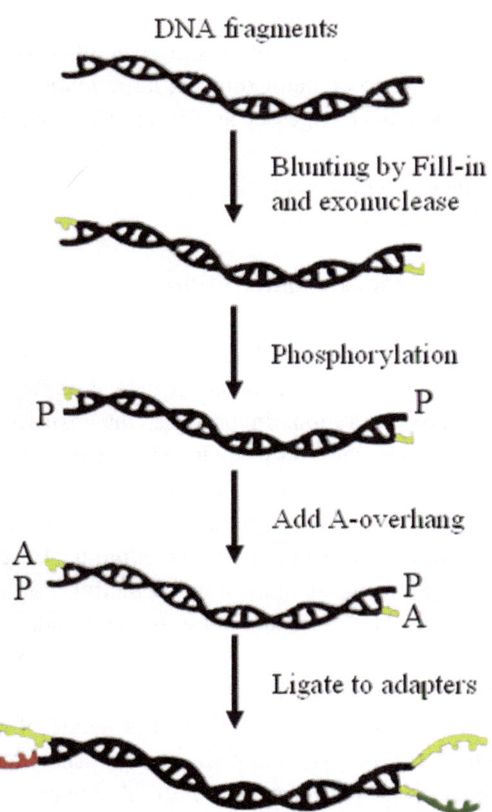

Fig. 3.4 Library preparation of Illumina NGS platform

preparation for the Illumina platform. DNA samples are first sheared into fragments that are then end-repaired to generate 5′-phosphorylated blunt ends. The Klenow fragment of DNA polymerase is then used to attach a single "A" base to the 3′ end of each DNA fragments, which enables the DNA fragments for ligation to oligonucleotide adapters. After ligation to adapters at both ends, the DNA fragments are denatured, and single-stranded DNA fragments are attached to reaction chambers that are optically transparent solid surfaces called a flow cell. The attached DNA fragments are extended and amplified by bridge PCR amplification.

2.2.2 Sequencing

Different chemical reactions for sequencing are used in current NGS platforms. The main features of three most popular NGS platforms are compared in Table 3.4. It is important to note that these metrics are constantly changing when newer models of the same platforms are released. Both the Illumina HiSeq-2000 system and the SOLid 5500xl system use the short-read sequencing technologies, while the Roche 454 FLX system provides relatively longer reads that could be advantageous for de novo sequencing of new genomes.

Table 3.4 Comparison of NGS platforms

	Platform		
	Illumina	Roche 454	SOLid
Sequencing reaction	Sequencing- by-synthesis	Pyrosequencing	Ligation-based sequencing
Amplification	Bridge PCR	Emulsion PCR	Emulsion PCR
Read length	~100 bp	~700 bp	~75 bp
Paired ends/separation	Yes/200 bp	Yes/3,000 bp	Yes/3,000 bp
Reads per run (in millions)	3,000	1	1,500
Run time	11 days	23 h	8 days
Comments	Most widely used	Longer reads, fast run, higher cost	Good data quality

The Illumina HiSeq-2000 system currently is the most widely used short-read sequencing platform. It uses the sequencing-by-synthesis (SBS) method. By using the SBS technology, the bases of a DNA fragment are determined using a proprietary reversible terminator-based method when they are synthesized to "grow" the DNA strands. Since many synthetic reactions are carried out simultaneously in a reaction chamber (flow cell), the sequences of millions of DNA fragments are determined in parallel. To be specific, when synthesizing a base, a fluorescence-labeled terminator-bound dNTP is added to the strand and imaged. Then, the terminator is cleaved for the synthesis of the next base. Since there are four reversible terminator-bound dNTPs (dATP, dCTP, dGTP, and dTTP) in each sequencing cycle, natural competition minimizes incorporation bias. During each sequencing cycle, bases are called directly from the measured fluorescence intensities of each base. The final data obtained are the base-by-base information of the fragments.

The Roche 454 Genome Sequencer FLX is based on pyrosequencing technology. DNA fragments are sheared into shorter segments that are then ligated to specific oligonucleotide adapters for amplification by emulsion PCR on the surfaces of agarose beads. The current read length of DNA sequencing produced by the Roche 454 platform is about 700 bp and is the longest read among the three short-read NGS platforms. Therefore, the Roche 454 could be more suitable for applications requiring longer reads, such as RNA isoform identification in RNA-seq and de novo assembly of microbes in metagenomics (Mocali and Benedetti 2010).

The Applied Biosystems SOLiD 5500xl sequencer technology is based on the principle of sequencing-by-ligation. In this method, sheared DNA fragments are amplified by an emulsion PCR approach with small magnetic beads. DNA fragments on the surface of each magnetic bead are then "sequenced" by detection of the oligonucleotide ligation. One notable drawback of this platform is its tedious and time-consuming procedures for DNA library preparation prior to sequencing, currently amounting to some 5 days.

2.2.3 Data Analysis

In the workflow of an NGS project (Fig. 3.3), once the samples are sequenced using a sequencing platform, the task becomes one for bioinformaticians. Therefore, NGS data analysis (the rectangle in dash line, Fig. 3.3) is a crucial step in an NGS project.

Mapping or Assembling Short Reads to a Reference Genome

The first step in NGS data analysis is to align or to assemble the huge amount of short reads to a reference genome. Although NGS is a powerful sequencing tool, the short length of the reads generated from NGS technology limits its biological applications (Li and Homer 2010). Therefore, accurate alignment or assembly of the millions of short reads is an important determinate of experiment success. A variety of algorithms and software packages have been specifically developed for the task (Chistoserdova 2010), some of the most popular of which are listed in Table 3.5.

Table 3.5 Short-read sequence alignment tools

Name	Description
Bowtie	Uses a Burrows–Wheeler transform to create a permanent, reusable index of the genome; faster run for short sequence alignment to reference genome
BWA	Slower than bowtie but allows indels in alignment
MAQ	Performs only un-gapped alignments and allows up to three mismatches
SeqMap	Up to five mixed substitutions and insertions/deletions. Various tuning options and input/output formats
SOAP	Allows up to three gaps and mismatches. SOAP2 uses bidirectional BWT to build the index of references and increases the running speed
TopHat	Splice junction mapper for RNA-Seq reads

Bowtie (http://bowtie.cbcb.umd.edu/) is a program for aligning short read sequences to a genome that is very fast without burdensome amounts of computer memory (Langmead et al. 2009). The reference genome is first indexed using a scheme based on the Burrows–Wheeler index; the short reads are then mapped to the indexed genome, making its memory footprint small. Backtracking and double indexing are the two major algorithmic strategies Bowtie uses to rapidly align short reads to a genome. Bowtie allows mismatches and favors high-quality alignments through backtracking, while the double indexing strategy makes Bowtie avoid excessive backtracking. One of the drawbacks of this program is that gaps are not allowed in alignment. However, its new version, Bowtie 2, overcomes this and can do gapped alignment.

BWA (http://maq.sourceforge.net/) is a fast "lightweight" software package for aligning short sequencing reads against a large reference sequence such as the human genome (Li and Durbin 2009). This alignment algorithm uses a backward search strategy with the Burrows–Wheeler transform (BWT) algorithm to align sequences to the reference genome. It allows mismatches and gaps in mapping. BWA can align both single-end and paired-end reads to a reference. It uses a mapping quality index for assessing the goodness of a read aligned with a specific region in the reference genome. Therefore, it provides choices for a read mapping to the best location in the reference or to multiple regions if requested.

MAQ (http://maq.sourceforge.net/) is a freely available aligner for rapidly mapping short reads from NGS to a reference genome. It generates genotype calls according to the consensus sequence of a diploid genome by using quality scores (Li et al. 2008b). MAQ searches for the un-gapped match with lowest mismatch score. This program calculates a quality score for each alignment to measure the probability that the true alignment differs from the one the program found. It then searches for the alignment with the lowest score. It uses un-gapped mapping. Users can use MAQ for aligning reads, calling consensus sequences such as SNPs and indel variants, simulating diploid genomes, and processing alignment results in various ways.

SeqMap (http://www.stanford.edu/group/wonglab/jiangh/seqmap/) is an efficient sequence mapping program for mapping millions of short reads to a reference genome (Jiang and Wong 2008). It finds all possible locations in a whole reference genome for each of the short read sequences from NGS. SeqMap allows up to five substitutions and insertions/deletions. The FASTA file format is used for inputting data and alignment results and output is in various formats such as the SAM. SeqMap supports parallel computing and, thus, can be run on a cluster of computers. This program is very efficient, usually taking just a few hours on a desktop PC for a typical alignment job of NGS data. One distinct feature of this aligner is that it indexes and hashes the input short reads before mapping them to the reference genome.

SOAP (http://soap.genomics.org.cn/) is efficient for aligning short read sequences from NGS to a reference genome (Li et al. 2008a). It allows both gapped and un-gapped alignments and can align both single-read and paired-end reads. The seed-and-hash lookup table algorithm is used to accelerate its alignment process. Briefly, a two-bits-per-base encoding strategy is used to transform the short read sequences and the reference sequence into a numeric format. The base difference between a short read sequence and the reference sequence is then calculated using the

Table 3.6 Programs for de novo assembly

Name	Developer	Website
ABySS	Simpson JT et al.	www.bcgsc.ca/platform/bioinfo/software/abyss
ALLPATHS	Butler J et al.	ftp.broadinstitute.org/pub/crd/ALLPATHS/
Cufflinks	Trapnell C et al.	cufflinks.cbcb.umd.edu/
Edena	Hernandez D et al.	www.genomic.ch/edena.php
MIRA	Chevreux B et al.	sourceforge.net/apps/mediawiki/mira-assembler
SOAPdenovo	Li R et al.	soap.genomics.org.cn/soapdenovo.html

lookup table. SOAP is a command-driven program and easily used for batch processing with a user-specific script. It also provides a multithread option and thus supports parallel computing. It accepts FASTA format for the reference genome and both FASTA and FASTQ formats for the input short reads.

TopHat (http://tophat.cbcb.umd.edu/) is designed to align short reads from RNA-Seq to a reference genome (Trapnell et al. 2009). TopHat is designed mainly for finding exon–exon splicing junctions, although it can be used for other tasks. In the TopHat pipeline, the input short read sequences are first aligned to the reference genome using Bowtie. A short read can contiguously align to more than one exon in the genome since many exons are shorter than the length of the reads obtained from current NGS. Therefore, TopHat splits an input short read into shorter segments, and then independently aligns these segments to different exons. The aligned exons are connected together to generate the complete alignment. While mainly developed for aligning short reads from an Illumina platform such as HiSeq-2000 (~100 bp in length), it can align reads up to 1,024 bp. It can handle both single-end and paired-end reads. Users need to be aware of the parameters used in the program. The default values are optimized for alignment of short reads to the mammalian genomes. When aligning to other genomes, the parameters should be reset to suitable values.

De Novo Assembly of Short Reads

De novo assembly refers to the process of assembling short reads to an unknown genome or to a much larger reference sequence by directly "sewing" the sequences of short reads that are produced using an NGS platform. De novo assembly of NGS data is different from reference-based assembly, and it has more challenges than mapping assemblies, in terms of complexity and time consuming. Current de novo assembly algorithms have difficulty assembling large genomes, and are generally more suitable for small genomes such as found in bacteria (Chaisson et al. 2009). Table 3.6 lists some widely used de novo assembly tools. In this section, we briefly describe them.

The ABySS program (Simpson et al. 2009) is written in C++ and performs de novo sequence assembly from short paired-end reads. It has two versions, the single-processor version is designed for assembly small genomes with a size up to 100 Mbp. The parallel version is designed for assembly of large genomes with a size greater than 100 Mbp, and it is implemented using MPI (message passing interface).

ALLPATHS (Butler et al. 2008) is a de novo assembler that assembles short sequences such as the reads generated from NGS platforms into high-quality genome assemblies. One distinct feature of ALLPATHS is that its assemblies can be nonlinear. Actually, the assembling results are presented in graphs that can retain ambiguities in the input reads arising from polymorphism, sequencing errors, unresolved repeats, etc. It works better on the reads produced by Illumina platform such as HiSeq-2000 than on other platforms. ALLPATHS is not the best choice for assembling long reads such as those from the Roche 454 FLX. The program requires high sequence coverage of the genome in order to compensate for the short length of the reads. The required depth of raw reads coverage, before any error correction or filtering, depends on both the length and quality of the paired-end reads. As a rule of thumb, the required coverage for a human genome is one flow cell of data produced by the Illumina HiSeq-2000.

Cufflinks (Trapnell et al. 2010) is a program designed for multiple functions: assembling transcripts, estimating abundances of assembled transcripts, and quantifying differential expression of the transcripts based on RNA-Seq data. It assembles the short reads into a parsimonious set of transcripts and then estimates the relative abundances of these transcripts, taking into account the biases in library preparation protocols. Cufflinks can ignore fragments that map to the genome more than a specified number of times, but by default it uses all fragments in the alignment file. This program fills gaps smaller than 50 bp and joins the "transfrags" in coverage when assembling transcripts. Before testing for differential expression or regulation of genes and transcripts, Cuffdiff, another program in the package, checks goodness of the variance model for the gene or transcript. Because a positional bias correction was reducing accuracy on certain data sets in some genes, it was modified with an option to model sequence-specific bias. Its new library size normalization model is based on the geometric mean. It uses Eigen package for matrix operations that makes good use of vector registers in modern processors, speeding up the numerical routines used during abundance estimation.

Edena (Exact DE Novo Assembler) (Hernandez et al. 2008) is mainly designed for assembling short reads generated from the Illumina NGS platforms, including HiSeq-2000. It is based on the traditional overlap-layout-consensus paradigm. The program is still in the development stage, thus some features are incomplete and new functions continue to be added. Currently, both Linux (64 and 32) and Windows versions are available.

MIRA (Mimicking Intelligent Read Assembly) (Chevreux et al. 2004) was developed for the Linux operating system. This program was designed for hybrid de novo assembling short reads generated from different NGS platforms such as Roche 454 and Illumina. Therefore, it assembles reads instead of a mix of (eventually shredded) consensus sequence and reads. One restriction is the length of reads that must be less than 15 kbp.

SOAPdenovo (Li et al. 2010a) performs de novo assembly of large genomes from short reads generated from NGS platforms. This program uses the De Bruijn graph in which each node is a k-mer (the suggested k-mer is 25-mer, but it supports up to 127-mer to utilize long reads). The longer the k-mer that is used, the lower the quantity of nodes is obtained, and the more the memory is consumed. Dijkstra's algorithm is used to detect bubbles that are then merged into a single path if the sequences of the parallel paths are very similar. It produces data by its Data Preparation Module that is necessary to run the "map" and "scaff" steps using the contigs generated by SOAPdenovo or other assemblers. At beginning, the program was specially designed to assemble short reads from Illumina NGS platform such as HiSeq-2000, but now it works for short reads from other NGS platforms.

Visualizing and Annotating the Mapped or Assembled Results

Once mapping or assembly is completed, the immediate and imperative task is visualizing mapping and assembly results that are in the form of huge text or binary files. The complex and rich information requires graphical display commensurate with the task of interrogating and interpreting the data, and much effort has been expended by the scientific community toward this end. Table 3.7 lists some popular visualization and annotation tools for NGS data analysis. Below we will briefly introduce EagleView (Huang and Marth 2008) and IGV (Robinson et al. 2011).

EagleView (http://bioinformatics.bc.edu/marthlab/EagleView) is a genome assembler viewer and can be used for data integration. It can display many types of information such as base qualities and genome feature annotations. Users can easily use it to visually examine the quality of a genome assembly and to validate polymorphism candidate sites (e.g., SNPs) identified by polymorphism discovery programs. EagleView can be used to interpret assembly results and to produce hypotheses. It is written in C++ and is available for Windows, Linux, and Mac operating systems. EagleView accepts different types of data files such as the standard ACE genome assembly file and optional READS, EGL, and MAP files. It uses the optional files to visualize a genome

Table 3.7 Programs for visualizing and annotating mapping and assuming results

Name	Description and reference
EagleView	EagleView is developed at Boston College (Huang and Marth 2008). It is an information-rich genome assembler viewer that can display a dozen different types of information including base quality
IGV	IGV (Integrative Genomics Viewer) developed at MIT (Robinson et al. 2011). It is a high-performance visualization tool for interactive exploration of large, integrated genomic data sets. It supports a wide variety of data types, including next-generation sequence data, and genomic annotations
SeqTools	SeqTools was developed at Wellcome Trust Sanger Institute and can be obtained from www.broadinstitute.org/software/igv/home (Sonnhammer and Hollich 2005). It contains three tools: Blixem, Dotter, and Belvu which can be used independently and called from other tools as part of a software pipeline. In a typical application, Blixem is called to analyze a set of alignments in more detail, and Dotter is called within Blixem to give a graphical representation of a particular alignment
MapView	MapView developed at Sichuan University in China (Bao et al. 2009). It is a short-reads alignment viewer with genetic detection capability for NGS data analysis. It supports a compact alignment view for both single-end and paired-end short reads, multiple navigation and zoom modes, and multithread processing. It can handle large-scale data with high computational efficiency. Moreover, it offers automated genetic variation detection
SAM	SAM (Sequence Assembly Manager) is a WGA (whole genome assembly) management and visualization tool developed at Canada's Michael Smith Genome Sciences Centre (Warren et al. 2005). It provides a generic platform for manipulating, analyzing, and viewing WGA data, regardless of input type (www.bcgsc.ca/platform/bioinfo/software/sam)

assembly using the additional information contained in the standard files. Users can define the features to be visualized using EagleView by adding the features in the data files easily as the input files are simple text format (EagleView documentation contains the detailed format for each type file). EagleView automatically detects and opens the associated optional files from the same directory when it opens an ACE file. The key features of EagleView include (1) fast and efficient memory usage; (2) platform-specific signals pinpoint views of base quality; (3) pinpoints views of read identifier and strand distinct marks for discrepancy sites; (4) provides genome annotation; (5) navigation by read ID, contig ID, genomic features, or user defined locations, and both unpadded and padded positions; (6) data utility tools; and (7) customizable font and color.

IGV (http://www.broadinstitute.org/software/igv/home) provides concurrent visualization of multiple data types of samples, as well as a correlation of integrated data sets with clinical and phenotypic data. Sample annotations and associations with data tracks can be defined using a tab-delimited text file format in which sample identifier (used to link different types of data for the same sample), phenotype, outcome, cluster membership, and other clinical data can be the content. IGV uses the annotations that are visualized as a heatmap to group, sort, filter, and overlay diverse data types to generate a comprehensive picture of the integrated data set. Diverse genomic data types such as aligned sequence reads, identified mutations and copy numbers, RNA interference screens, gene expression, methylation, and genomic annotations can be easily integrated in IGV. Users can navigate a data set in a way similar to Google Maps: the genome can be zoomed and paned at any level of detail from whole genome to a single-base pair. One useful feature of IGV is that users can visualize their own genomic data sets alongside publicly available data because data sets can be input from a user's computer or loaded remotely from online resources such as cloud-based repositories. This makes IGV particularly effective for collaborative projects as it allows collaborators to share data remotely over the Internet. The current IGV version utilizes Java 6 or later that has to be installed on a user's computer. Several genomes are hosted in IGV genome server, including human (UCSC hg16–19 from UCSC Genome Bioinformatics, genome.ucsc.edu; Assembly b37 from 1000 Genomes, www.1000genomes.org), mouse (UCSC mm7–9 from UCSC Genome Bioinformatics), and rat (UCSC Baylor 3.4/rn4 based on version 3.4 produced by the Atlas group at Baylor Human Genome). Users can load a hosted genome into IGV by selecting the genome from the drop-down list in the tool bar. IGV provides the igvtools utility that contains several tools for preprocessing data. (1) tool *tile* can be used to convert a sorted data input file to a binary tiled data (.tdf) file and to preprocess large data sets for improved IGV performance (supported input file formats are .wig, .cn, .snp, .igv, .res, and .gct). (2) Tool *count* can be used to calculate average alignment or feature density for over a specified window size across the genome and to create a track that can be displayed in IGV, for example, as a bar chart (supported input file formats are .sam, .bam, .aligned, .psl, .pslx, and bed). (3) Tool *index* can be used to convert an ASCII alignment file or feature file to an index file that is required for loading alignment files into IGV. Index files can significantly improve performance for large feature files (supported input file formats are .sam, .aligned, .vcf, .psl, and .bed). (4) Tool *sort* that can be used to sort the input file by start position and to prepare data files for tools that required sorted input files (supported input file formats are .cn, .igv, .sam, .aligned, .psl, .bed, and .vcf).

3 NGS Applications

NGS technology is used to determine sequences of short DNA fragments in a very high-throughput manner. Theoretically, there is no limitation for application of NGS to scientific research so long as short oligonucleotide fragments are of high quality. NGS has already been widely applied to many scientific research areas, such as determination of sequences of genomes of many species (Harris et al. 2008; Chaisson and Pevzner 2008; Li et al. 2010b); identification of genetic variants by re-sequencing whole genome or targeted genes (Yeager et al. 2008; Yi et al. 2010); characterization of transcriptomes for cells, tissues, and organisms by sequencing their mRNA content (RNA-Seq) (Denoeud et al. 2008; Pan et al. 2008; Hiller et al. 2009; Guttman et al. 2010; Trapnell et al. 2010); identification of DNA-binding proteins and epigenetic markers by ChIP-Seq (Robertson et al. 2007; Kharchenko et al. 2008; Kaufmann et al. 2010); and pinpointing of the genomic content in a complex sample by metagenomics (Turnbaugh et al. 2010; Fierer et al. 2010). Many comprehensive review articles on NGS technologies and their applications have been published (Wang et al. 2009; Kato 2009; Voelkerding et al. 2009). This section does not intend to cover a comprehensive of applications of NGS technologies, and rather focuses on NGS technologies with the most potential applications in pharmacogenomics.

3.1 RNA Sequencing

Gene expression profiles of patients can be used to interrogate their responses to drugs for both efficacy and safety considerations. Identification of a gene or a panel of genes that are associated with a patient's response to a specific drug is an important topic in pharmacogenomics. NGS is thought to be more accurate and precise in measuring expression levels of genes or transcripts, compared to other technologies such as PCR and DNA microarray. Therefore, it is expected that NGS will accelerate pharmacogenomics research and translation of the findings in pharmacogenomics into personalized medicine.

Directly sequencing RNAs/cDNAs offers an alternative approach for high-throughput transcriptome analysis (Marioni et al. 2008; Wilhelm et al. 2010). RNA sequencing (simplified as RNA-Seq) is revolutionary in its abilities to precisely measure expression of genes at the whole genome level (Li et al. 2010c). Its higher resolution output makes NGS a promising tool for discovery of novel transcripts, exploration of allele-specific expression, and identification of alternative splice variants and posttranscriptional mutations and isoforms compared to the conventional Sanger sequencing and microarray-based approaches (Sultan et al. 2008; Chepelev et al. 2009; Perkins et al. 2009; Tang et al. 2009; Hittinger et al. 2010). RNA-Seq has been used to characterize RNA populations and provided more complicated pictures of RNA regulation and expression through alternative splicing, alternative polyadenylation, and RNA editing (Nagalakshmi et al. 2008; Guttman et al. 2009; Li et al. 2009a). NGS has expanded our understanding of the extent and complexity of gene expression and the mechanisms of RNA expression regulation in both eukaryotic and prokaryotic genomes (Jacquier 2009; Sorek and Cossart 2010; Licatalosi and Darnell 2010).

The most active field of applying RNA-seq in pharmacogenomics is the discovery of cancer-related genomic biomarkers. Since the Human Genome Project finished sequencing the human genome and published the first human genome map draft (a landmark in genomics) (International Human Genome Sequencing Consortium 2001; Venter et al. 2001), the understanding and discovery of cancer genomic biomarkers became feasible. The recent advances in RNA-seq technology have made it possible to more precisely profile genes at the whole genome level and discover new RNA molecules such as new gene transcripts, small RNAs, alternative splicing products, and gene fusions products (Lipson et al. 2012). Sinicrop et al. developed and optimized the RNA-Seq library chemistry as well as bioinformatics methods for whole transcriptome profiling (Sinicropi et al. 2012). This work led to not only the re-discovery of RNA biomarkers for disease recurrence risk that were previously identified by RT-PCR analysis of a cohort of 136 patients but also the identification of a new group of recurrence risk biomarkers that were not previously discovered using DNA microarrays in a separate cohort of patients (Sinicropi et al. 2012).

3.2 DNA Sequencing

Identifying genetic variants (such as SNPs and mutations) is possible after successfully aligning (or assembling) NGS short reads to a genome. However, distinguishing the causal variants for a phenotype from the large number of apparently novel genetic variants present by chance in any human genome can be difficult. Identification of rare mutations introduced into the population that differ from low-frequency alleles descendent from ancient ancestors is still more difficult when analyzing DNA sequencing data. One of the key factors for the success of application of DNA sequencing in pharmacogenomics is the reliability of software packages for detecting genetic variants such as SNPs from the NGS data.

Many software packages for detecting SNPs from DNA sequencing data have been developed. For example, SOAPsnp (Li et al. 2009b) is a member of the Short Oligonucleotide Analysis Package (http://soap.genomics.org.cn/soapsnp.html). SOAPsnp is a re-sequencing utility that can be used to assemble a consensus sequence for a genome after the raw reads are aligned to the reference genome. It identifies SNPs and determines their genotypes by comparing the aligned reads with the reference genome using the consensus sequence. SOAPsnp is a command line driven program written in C/C++ that generally runs under the 64-bit Linux system. However, this program has been tested on other operation systems. It only needs ~0.5 GB or even less memory to run but its output might be very large because it consumes a lot of hard disk space when memory is small. In its text output mode, the output file may be 60 times the genome size (e.g., 180G free space is required to run a human genome). The program was evaluated using the Asian genome re-sequencing project data, based on the Illumina HapMap 1 M BeadChip Duo genotyping sites. Over 99 % of the known SNPs were identified and 99.9 % of the genotypes for the covered SNPs are consistent, indicating its reliability for SNP calling.

DNA sequencing has been actively applied in pharmacogenomics and many interesting findings have been obtained. Some representative studies and their major finding are revisited below. Pleasance et al. sequenced the genomes of a malignant melanoma and a lymphoblastoid cell line obtained from the same patient to provide somatic mutation information from cancer cells (Pleasance et al. 2010). Sequence data analysis detected 33,345 somatic base substitutions, 680 small deletions, 303 small insertions, 51 somatic rearrangements, and all well-validated somatic copy number alterations and regions of loss of heterozygosity. Lee et al. sequenced a genome from a primary lung tumor and a normal one from adjacent normal tissue of the same patient (Lee et al. 2010). Comparative analysis identified more than 50,000 high-confidence single nucleotide variants (SNV) as somatic variants. Shah et al. sequenced the genome and transcriptome of an estrogen-receptor-positive metastatic lobular breast cancer (Shah et al. 2009). DNA sequence data analysis led to the identification of 32 somatic non-synonymous coding mutations. Tran et al. sequenced the blood, tumor biopsy, and archived tumor samples of 50 patients collected from four different cancer centers by using targeted exon sequencing (Tran et al. 2013). Combining with multiplex somatic mutation genotyping using Sequenom MassARRAY and Sanger sequencing, they assessed the feasibility of incorporating real-time analysis of somatic mutations within exons of 19 genes into patient management. Moreover, the identified actionable mutations were similar between archival and biopsy samples, implying that the cancer mutations do not change much across clinical stages making them good predictors of drug response. Their results demonstrated that the use of next-generation sequencing for real-time genomic profiling in advanced cancer patients is feasible.

Some well-known mutations in BRCA1 and BRCA2 (BRCA1/2) genes can markedly elevate risk of developing breast and ovarian cancers. However, mutations in unscreened regions of BRCA1/2 and other predisposition genes need identification. Ozcelik et al. recently applied long-range PCR and NGS for BRCA1/2 mutation analysis (Ozcelik et al. 2012). By sequencing the genomic DNA from 12 cancer patients, all 19 distinct (51 total) BRCA1 and 35 distinct (63 total) BRCA2 sequence alterations detected by the Sanger sequencing were confirmed, with no false-negatives. Moreover, variants from introns and untranslated regions were identified, demonstrating that NGS can provide accurate and more comprehensive genetic information.

These demonstrable successes portend that advances in DNA sequencing will further enable molecular diagnostics, accelerate pharmacogenomics research, and expand translation of genetic biomarkers into personalized medicine.

3.3 miRNA Sequencing

miRNAs have been found to be involved in many diseases (Jiang et al. 2009), including diabetes (Hennessy and O'Driscoll 2008), cardiomyopathies (van Rooij et al. 2006), psychiatric disorders such as schizophrenia (Barbato et al. 2008; Beveridge et al. 2009), and cancer (Medina and Slack 2008). High-throughput sequencing has been applied for profiling known and novel small RNAs such as miRNAs that regulate gene expression and play a major role in most biological processes. They could be used as biomarkers and as therapeutic agent targets (Fasanaro et al. 2010; Dreyer 2010; Nana-Sinkam and Croce 2011) for drug development. Consequently, identifying miRNAs and measuring their expression using NGS has been a recent pharmacogenomics research field.

Theoretically, the expression of a therapeutic target protein involved in a disease can be blocked by inhibitors that repress the protein itself or by particular miRNAs that repress the expression of the protein. Thus, administration of miRNAs mimetics can simulate the endogenous miRNAs population repressing a detrimental gene and its protein products. Therefore, miRNA inhibitors and some miRNA mimics can be used as therapeutic candidates for various diseases, including cardiovascular disease, neurological disorders, and viral infection (Nana-Sinkam and Croce 2011). Drugs that target miRNAs are in the discovery and development pipelines of many pharmaceutical companies. Table 3.8 lists some miRNAs that are used as the targets of a number of new drug products in preclinical studies and in clinical trials.

Several miRNA biomarkers for diagnosis and prognosis of disease have recently been identified using NGS technologies. One example is the differential expression of the oncogenic miRNAs of the *miR17-92* cluster and the *miR-181* family among different types of neuroblastoma patients, which was measured using the SOLid NGS platform. The expression levels of those miRNAs were higher in the five unfavorable neuroblastoma patients. In contrast, the expression levels of the tumor suppressive miRNAs of *miR-542-5p* and

Table 3.8 Examples of miRNAs in drug development

Company	Generic name	Target	Indication	Status
Santaris Pharma	Miravirsen	miR-122	Hepatitis C virus	Phase IIA
Regulus Therapeutics	Unspecified	miR-21	Fibrosis	Clinical
Regulus Therapeutics	Unspecified	miR-21	Cancer	Clinical
Regulus Therapeutics	Unspecified	mi-R122	Hepatitis C virus	Clinical
Regulus Therapeutics	Unspecified	mi-155	Inflammation	Preclinical
Regulus Therapeutics	Unspecified	miR-33a	Metabolic diseases	Preclinical
miRagen Therapeutics	MGN-9103	miR-208/499	Chronic Heart Failure	Preclinical
miRagen Therapeutics	MGN-1374	miR-15/195	Post-MI Remodeling	Preclinical
miRagen Therapeutics	MGN-4893	miR-451	Polycythemia vera	Preclinical
miRagen Therapeutics	MGN-4420	miR-29	Cardiac fibrosis	Lead optimization
Mirna Therapeutics	Unspecified	Let-7	Lung cancer	Preclinical
Mirna Therapeutics	Unspecified	miR-34	Prostate cancer	Preclinical
Rosetta Genomics	TCDD	miR-191	Hepatocellular carcinoma	Preclinical
Rosetta Genomics	Unspecified	miR-34a	Liver cancer	Preclinical

miR-628 measured using NGS were much higher in the five favorable neuroblastoma patients compared to the five unfavorable neuroblastoma patients whose expressions of these two miRNAs were virtually absent (Schulte et al. 2010).

4 Future Perspectives

GWAS are based on the "common disease–common variants hypothesis" and had a great wave of research activities 6 years ago. The huge prospect for GWAS to advance pharmacogenomics and personalized medicine has so far largely proved disappointing. Fortunately, next-generation sequencing is fundamentally changing the way in which genomic information of individuals at the base level is being obtained for a holistic understanding of the human genome. Sequencing target genes of diseases or whole genomes of patients is expected to reveal additional instances of lower-frequency, higher-penetrance alleles and is renewing prospects for more effective personalized therapy. It is also expected that more and more genomic biomarkers will be identified and used in the development of drug products in the future. The challenges in the future are not only the discovery of genomic biomarkers, but also how to incorporate them into drug development, regulatory decision making, and personalized medicine. To enhance FDA's scientific base for regulatory decision making on pharmacogenomics findings from NGS for personalized medicine, FDA initiated SEQC (sequence quality control) project, also known as MAQC-III (http://www.fda.gov/ScienceResearch/BioinformaticsTools/MicroarrayQualityControlProject/default.htm#MAQC-IIIalsoknownasSEQC) to address various issues on the reliability of NGS technologies for pharmacogenomics.

The fields of pharmacogenomics and personalized medicine are undergoing a revolution today because of the advances in NGS technologies. Both common genetic variants that have been explored by GWAS and rare genetic variants that have not been explored can be identified using NGS. The authors believe this approach will drive the next wave of pharmacogenomics and personalized medicine advances. We expect NGS to make significant contributions to our understanding of pharmacogenomics and will redefine the field of personalized medicine. As the pharmacogenomics knowledge base expands, translation of pharmacogenomics findings into personalized medicine will be within reach.

References

Ansorge WJ (2009) Next-generation DNA sequencing techniques. Nat Biotechnol 25:195–203

Armstrong B, Stewart M, Mazumder A (2000) Suspension arrays for high throughput, multiplexed single

nucleotide polymorphism genotyping. Cytometry 40:102–108
Bao H, Guo H, Wang J, Zhou R, Lu X, Shi S (2009) MapView: visualization of short reads alignment on a desktop computer. Bioinformatics 25:1554–1555
Barany F (1991) Genetic disease detection and DNA amplification using cloned thermostable ligase. Proc Natl Acad Sci USA 88:189–193
Barbato C, Giorge C, Catalanotto C, Cogoni C (2008) Thinking about RNA? MicroRNAs in the brain. Mamm Genome 19:541–551
Berger SL, Kouzarides T, Shiekhattar R, Shilatifard A (2009) An operational definition of epigenetics. Genes Dev 23:781–783
Beveridge NJ, Gardiner E, Carroll AP et al (2009) Schizophrenia is associated with an increase in cortical microRNA biogenesis. Mol Psychiatry 15:1176–1189
Bird A (2007) Perceptions of epigenetics. Nature 447: 396–398
Brunner AL, Johnson DS, Kim SW et al (2009) Distinct DNA methylation patterns characterize differentiated human embryonic stem cells and developing human fetal liver. Genome Res 19:1044–1056
Butler J, MacCallum I, Kleber M, Shlyakhter IA, Belmonte MK, Lander ES, Nusbaum C, Jaffe DB (2008) ALLPATHS: de novo assembly of whole-genome shotgun microreads. Genome Res 18:810–820
Buyse M, Loi S, Van't Veer L et al (2006) Validation and clinical utility of a 70-gene prognostic signature for women with node-negative breast cancer. J Natl Cancer Inst 98:1183–1192
Chaisson MJ, Pevzner PA (2008) Short read fragment assembly of bacterial genomes. Genome Res 18:324–330
Chaisson MJ, Brinza D, Pevzner PA (2009) De novo fragment assembly with short mate-paired reads: does the read length matter? Genome Res 19:336–345
Chepelev I, Wei G, Tang Q, Zhao K (2009) Detection of single nucleotide variations in expressed exons of the human genome using RNA-Seq. Nucleic Acids Res 37:e106
Chevreux B, Pfisterer T, Drescher B, Driesel AJ, Müller WEG et al (2004) Using the miraEST assembler for reliable and automated mRNA transcript assembly and SNP detection in sequenced ESTs. Genome Res 14:1147–1159
Chistoserdova L (2010) Recent progress and new challenges in metagenomics for biotechnology. Biotechnol Lett 32:1351–1359
Cloonan N, Forrest AR, Kolle G et al (2008) Stem cell transcriptome profiling via massive-scale mRNA sequencing. Nat Methods 5:613–619
Cronin MT, Fucini RV, Kim SM, Masino RS, Wespi RM, Miyada CG (1996) Cystic fibrosis mutation detection by hybridization to light-generated DNA probe arrays. Hum Mutat 7:244–255
Daly AK, Donaldson PT, Bhatnagar P et al (2009) HLA-B*5701 genotype is a major determinant of drug induced liver injury due to flucloxacillin. Nat Genet 41:816–819
Das SK, Austin MD, Akana MC, Deshpande P, Cao H, Xiao M (2010) Single molecule linear analysis of DNA in nano-channel labeled with sequence specific fluorescent probes. Nucleic Acids Res 38:e177
Denoeud F, Aury JM, Da Silva C et al (2008) Annotating genomes with massive-scale RNA sequencing. Genome Biol 9:R175
Dreyer JL (2010) New insights into the roles of microRNAs in drug addiction and neuroplasticity. Genome Med 2:92
Duffy MJ, Napieralski R, Martens JW et al (2009) Methylated genes as new cancer biomarkers. Eur J Cancer 45:335–346
Fasanaro P, Greco S, Ivan M, Capogrossi MC, Martelli F (2010) MicroRNA: emerging therapeutic targets in acute ischemic diseases. Pharmacol Ther 125:92–104
Fierer N, Lauber C, Zhou N et al (2010) Forensic identification using skin bacterial communities. Proc Natl Acad Sci USA 107:6477–6481
Frazer KA, Murray SS, Schork NJ, Topol EJ (2009) Human genetic variation and its contribution to complex traits. Nat Rev Genet 10:241–251
Glas AM, Floore A, Delahaye LJ et al (2006) Converting a breast cancer microarray signature into a high-throughput diagnostic test. BMC Genomics 7:278
Goldstein JA, Blaisdell J (1996) Genetic tests which identify the principal defects in CYP2C19 responsible for the polymorphism in mephenytoin metabolism. Methods Enzymol 272:210–218
Goldstein LJ, Gray R, Badve S et al (2008) Prognostic utility of the 21-gene assay in hormone receptor-positive operable breast cancer compared with classical clinicopathologic features. J Clin Oncol 26:4063–4071
Guttman M, Amit I, Garber M et al (2009) Chromatin signature reveals over a thousand highly conserved large non-coding RNAs in mammals. Nature 458:223–227
Guttman M, Garber M, Levin JZ et al (2010) Ab initio reconstruction of cell type-specific transcriptomes in mouse reveals the conserved multi-exonic structure of lincRNAs. Nat Biotechnol 28:503–510
Harris L, Fritsche H, Mennel R et al (2007) American Society of Clinical Oncology 2007 update of recommendations for the use of tumour markers in breast cancer. J Clin Oncol 25:5287–5312
Harris TD, Buzby PR, Babcock H et al (2008) Single-molecule DNA sequencing of a viral genome. Science 320:106–109
He L, Thomson JM, Hemann MT et al (2005) A microRNA polycistron as a potential human oncogene. Nature 435:828–833
Hennessy E, O'Driscoll L (2008) Molecular medicine of microRNAs: structure, function, and implications for diabetes. Expert Rev Mol Med 10:e24
Hernandez D, François P, Farinelli L, Osterås M, Schrenzel J (2008) De novo bacterial genome sequencing: millions of very short reads assembled on a desktop. Genome Res 18:802–809
Higuchi R, Dollinger G, Walsh PS, Griffith R (1992) Simultaneous amplification and detection of specific DNA sequences. Biotechnology (N Y) 10:413–417

Hiller D, Jiang H, Xu W, Wong WH (2009) Identifiability of isoform deconvolution from junction arrays and RNA-Seq. Bioinformatics 25:3056–3059

Hittinger CT, Johnston M, Tossberg JT, Rokas A (2010) Leveraging skewed transcript abundance by RNA-Seq to increase the genomic depth of the tree of life. Proc Natl Acad Sci USA 107:1476–1481

Hodges E, Xuan Z, Balija V et al (2007) Genome-wide in situ exon capture for selective re-sequencing. Nat Genet 39:1522–1527

Holland PM, Abramson RD, Watson R, Gelfand DH (1991) Detection of specific polymerase chain reaction product by utilizing the 5′–3′ exonuclease activity of *Thermus aquaticus* DNA polymerase. Proc Natl Acad Sci USA 88:7276–7280

Hong H, Goodsaid F, Shi L, Tong W (2010) Molecular biomarkers: a US FDA effort. Biomarkers Med 4:215–225

Hormozdiari F, Alkan C, Eichler EE, Sahinalp SC (2009) Combinatorial algorithms for structural variation detection in high-throughput sequenced genomes. Genome Res 19:1270–1278

Huang W, Marth G (2008) EagleView: a genome assembly viewer for next-generation sequencing technologies. Genome Res 18:1538–1543

International Human Genome Sequencing Consortium (2001) Initial sequencing and analysis of the human genome. Nature 409:860–921

Irizarry RA, Ladd-Acosta C, Wen B et al (2009) The human colon cancer methylome shows similar hypo- and hypermethylation at conserved tissue-specific CpG island shores. Nat Genet 41:178–186

Jacquier A (2009) The complex eukaryotic transcriptome: unexpected pervasive transcription and novel small RNAs. Nat Rev Genet 10:833–844

Ji W, Foo JN, O'Roak BJ et al (2008) Rare independent mutations in renal salt handling genes contribute to blood pressure variation. Nat Genet 40:592–599

Jiang H, Wong WH (2008) SeqMap: mapping massive amount of oligonucleotides to the genome. Bioinformatics 24:2395–2396

Jiang Q, Wang Y, Hao Y et al (2009) miR2Disease: a manually curated database for microRNA deregulation in human disease. Nucleic Acids Res 37:D98–D104

Johnson DS, Mortazavi A, Myers RM, Wold B (2007) Genome-wide mapping of *in vivo* protein-DNA interactions. Science 316:1497–1502

Jorgensen JT (2009) New era of personalized medicine: a 10-year anniversary. Oncologist 14:557–558

Kahn SL, Ronnett BM, Gravitt PE, Gustafson KS (2008) Quantitative methylation-specific PCR for the detection of aberrant DNA methylation in liquid-based Pap tests. Cancer 114:57–64

Kato K (2009) Impact of the next generation DNA sequences. Int J Clin Exp Med 2:193–202

Kaufmann K, Muino JM, Osteras M et al (2010) Chromatin immunoprecipitation (ChIP) of plant transcription factors followed by sequencing (ChIP-SEQ) or hybridization to whole genome arrays (ChIP-CHIP). Nat Protoc 5:457–472

Kharchenko PV, Tolstorukov MY, Park PJ (2008) Design and analysis of ChIP-seq experiments for DNA-binding proteins. Nat Biotechnol 26:1351–1359

Kim MJ, Huang SM, Meyer UA, Rahman A, Lesko LJ (2009) A regulatory science perspective on warfarin therapy: a pharmacogenetic opportunity. J Clin Pharmacol 49:138–146

Kindmark A, Jawaid A, Harbron CG et al (2008) Genome-wide pharmacogenetic investigation of a hepatic adverse event without clinical signs of immunopathology suggests an underlying immune pathogenesis. Pharmacogenomics J 8:186–195

Kondo Y, Shen L, Cheng AS et al (2008) Gene silencing in cancer by histone H3 lysine 27 trimethylation independent of promoter DNA methylation. Nat Genet 40:741–750

Kranenburg O (2005) The KRAS oncogene: past, present, and future. Biochim Biophys Acta 1756:81–82

Langmead B, Trapnell C, Pop M, Salzberg SL (2009) Ultrafast and memory-efficient alignment of short DNA sequences to the human genome. Genome Biol 10:R25

Langreth R, Waldholz M (1999) New era of personalized medicine: targeting drugs for each unique genetic profile. Oncologist 4:426–427

Lavedan C, Licamele L, Volpi S et al (2008) Association of the *NPAS3* gene and five other loci with response to the antipsychotic iloperidone identified in a whole genome association study. Mol Psychiatry 14:804–819

Lee W, Jiang Z, Liu J et al (2010) The mutation spectrum revealed by paired genome sequences from a lung cancer patient. Nature 465:473–477

Li R, Li Y, Kristiansen K, Wang J (2008a) SOAP: short oligonucleotide alignment program. Bioinformatics 24:713–714

Li H, Ruan J, Durbin R (2008b) Mapping short DNA sequencing reads and calling variants using mapping quality scores. Genome Res 18:1851–1858

Li H, Durbin R (2009) Fast and accurate short read alignment with Burrows-Wheeler transform. Bioinformatics 25:1754–1760

Li JB, Levanon EY, Yoon JK et al (2009a) Genome-wide identification of human RNA editing sites by parallel DNA capturing and sequencing. Science 324:1210–1213

Li R, Li Y, Fang X, Yang H, Wang J, Kristiansen K, Wang J (2009b) SNP detection for massively parallel whole-genome re-sequencing. Genome Res 19:1124–1132

Li H, Homer N (2010) A survey of sequence alignment algorithms for next-generation sequencing. Brief Bioinform 11:473–483

Li R, Zhu H, Ruan J, Qian W, Fang X, Shi Z et al (2010a) De novo assembly of human genomes with massively parallel short read sequencing. Genome Res 20:265–272

Li R, Fan W, Tian G et al (2010b) The sequence and de novo assembly of the giant panda genome. Nature 463:311–317

Li B, Ruotti V, Stewart RM, Thomson JA, Dewey CN (2010c) RNA-Seq gene expression estimation with read mapping uncertainty. Bioinformatics 26:493–500

Licatalosi DD, Darnell RB (2010) RNA processing and its regulation: global insights into biological networks. Nat Rev Genet 11:75–87

Lipson D, Capelletti M, Yelensky R et al (2012) Identification of new ALK and RET gene fusions from colorectal and lung cancer biopsies. Nat Med 18:382–384

Lockhart DJ, Dong H, Byrne MC et al (1996) Expression monitoring by hybridization to high-density oligonucleotide arrays. Nat Biotechnol 14:1675–1680

Lu J, Getz G, Miska EA et al (2005) MicroRNA expression profiles classify human cancers. Nature 435:834–838

Lucas A, Nolan D, Mallal S (2007) HLA-B*5701 screening for susceptibility to abacavir hypersensitivity. J Antimicrob Chemother 59:591–593

Mallal S, Phillips E, Carosi G et al (2008) HLA-B*5701 screening for susceptibility to abacavir hypersensitivity. N Engl J Med 358:568–569

Margulies M, Egholm M, Altman WE et al (2005) Genome sequencing in microfabricated high-density picolitre reactors. Nature 437:376–380

Marioni JC, Mason CE, Mane SM, Stephens M, Gilad Y (2008) RNA-seq: an assessment of technical reproducibility and comparison with gene expression arrays. Genome Res 18:1509–1517

McGrath JP, Capon DJ, Smith DH et al (1983) Structure and organization of the human Ki-ras protooncogene and a related processed pseudogene. Nature 304:501–506

Medina PP, Slack FJ (2008) MicroRNAs and cancer: an overview. Cell Cycle 7:2485–2492

Metzker ML (2010) Sequencing technologies – the next generation. Nat Rev Genet 11:31–46

Mocali S, Benedetti A (2010) Exploring research frontiers in microbiology: the challenge of metagenomics in soil microbiology. Res Microbiol 161:497–505

Mortazavi A, Williams BA, McCue K, Schaeffer L, Wold B (2008) Mapping and quantifying mammalian transcriptomes by RNA-Seq. Nat Methods 5:621–628

Motulsky AG (1957) Drug reactions enzymes and biochemical genetics. J Am Med Assoc 165:835–837

Nagalakshmi U, Wang Z, Waern K, Shou C, Raha D, Gerstein M, Snyder M (2008) The transcriptional landscape of the yeast genome defined by RNA sequencing. Science 320:1344–1349

Nana-Sinkam SP, Croce CM (2011) MicroRNAs as therapeutic targets in cancer. Transl Res 157:216–225

Newton CR, Graham A, Heptinstall LE et al (1989) Analysis of any point mutation in DNA. The amplification refractory mutation system (ARMS). Nucleic Acids Res 17:2503–2516

Ozcelik H, Shi X, Mc C et al (2012) Long-range PCR and next-generation sequencing of Brca1 and Brca2 in breast cancer. J Mol Diagn 14:467–475

Paik S, Shak S, Tang G et al (2004) A multi-gene assay to predict recurrence of tamoxifen-treated, node-negative breast cancer. N Engl J Med 351:2817–2826

Pan Q, Shai O, Lee LJ, Frey BJ, Blencowe BJ (2008) Deep surveying of alternative splicing complexity in the human transcriptome by high-throughput sequencing. Nat Genet 40:1413–1415

Park PJ (2009) ChIP–seq: advantages and challenges of a maturing technology. Nat Rev Genet 10:669–680

Perkins TT, Kingsley RA, Fookes MC et al (2009) A strand-specific RNA-Seq analysis of the transcriptome of the typhoid bacillus Salmonella typhi. PLoS Genet 5:e1000569

Pleasance ED, Cheetham RK, Stephens PJ et al (2010) A comprehensive catalogue of somatic mutations from a human cancer genome. Nature 463:191–196

President's Council of Advisors on Science Technology (2008) Priorities for personalised medicine. http://www.whitehouse.gov/files/documents/ostp/PCAST/pcast_report_v2.pdf. Accessed 28 Aug 2012

Reis-Filho JS (2009) Next-generation sequencing. Breast Cancer Res 11(Suppl 3):S12

Robertson G, Hirst M, Bainbridge M et al (2007) Genome-wide profiles of STAT1 DNA association using chromatin immunoprecipitation and massively parallel sequencing. Nat Methods 4:651–657

Robinson JT, Thorvaldsdóttir H, Winckler W, Guttman M, Lander ES, Getz G, Mesirov JP (2011) Integrative genomics viewer. Nat Biotechnol 29:24–26

Sanderson S, Emery J, Higgins J (2005) CYP2C9 gene variants, drug dose, and bleeding risk in warfarin-treated patients: a HuGEnet™ systemic review and meta-analysis. Genet Med 7:97–104

Sanger F, Nicklen S, Coulson AR (1977) DNA sequencing with chain-terminating inhibitors. Proc Natl Acad Sci USA 74:5463–5467

Schena M, Shalon D, Davis RW, Brown PO (1995) Quantitative monitoring of gene expression patterns with a complementary DNA microarray. Science 270:467–470

Schmidt D, Wilson MD, Ballester B et al (2010) Five-vertebrate ChIP–seq reveals the evolutionary dynamics of transcription factor binding. Science 328:1036–1040

Schulte JH, Marschall T, Martin M et al (2010) Deep sequencing reveals differential expression of microRNAs in favorable versus unfavorable neuroblastoma. Nucleic Acids Res 38:5919–5928

Shah SP, Morin RD, Khattra J et al (2009) Mutational evolution in a lobular breast tumour profiled at single nucleotide resolution. Nature 461:809–813

Shendure J, Ji H (2008) Next-generation DNA sequencing. Nat Biotechnol 26:1135–1145

Simpson JT, Wong K, Jackman SD, Schein JE, Jones SJ, Birol I (2009) ABySS: a parallel assembler for short read sequence data. Genome Res 19:1117–1123

Sinicropi D, Qu K, Collin F et al (2012) Whole transcriptome RNA-Seq analysis of breast cancer recurrence risk using formalin-fixed paraffin- embedded tumor tissue. PLoS One 7:e40092

Sonnhammer EL, Hollich V (2005) Scoredist: a simple and robust protein sequence distance estimator. BMC Bioinformatics 6:108

Sorek R, Cossart P (2010) Prokaryotic transcriptomics: a new view on regulation, physiology and pathogenicity. Nat Rev Genet 11:9–16

Sultan M, Schulz MH, Richard H et al (2008) A global view of gene activity and alternative splicing by deep sequencing of the human transcriptome. Science 32:956–960

Syvanen AC, Aalto-Setala K, Harju L, Kontula K, Soderlund H (1990) A primer-guided nucleotide incorporation assay in the genotyping of apolipoprotein E. Genomics 8:684–692

Takahashi H, Wilkinson GR, Nutescu EA et al (2006) Different contributions of polymorphisms in VKORC1 and CYP2C9 to intra- and inter-population differences in maintenance doses of warfarin in Japanese, Caucasians and African Americans. Pharmacogenet Genomics 16:101–110

Takeuchi F, McGinnis R, Bourgeois S et al (2009) A genome-wide association study confirms *VKORC1*, *CYP2C9*, and *CYP4F2* as principal genetic determinants of warfarin dose. PLoS Genet 5, E1000433

Tang F, Barbacioru C, Wang Y et al (2009) MRNA-Seq whole-transcriptome analysis of a single cell. Nat Methods 6:377–382

The International HapMap Consortium (2007) A second generation human haplotype map of over 3.1 million SNPs. Nature 449:851–862

Tran B, Brown AM, Bedard PL et al (2013) Feasibility of real time next generation sequencing of cancer genes linked to drug response: results from a clinical trial. Int J Cancer 132:1547–1555. doi:10.1002/ ijc.27817

Trapnell C, Pachter L, Salzberg SL (2009) TopHat: discovering splice junctions with RNA-Seq. Bioinformatics 25:1105–1111

Trapnell C, Williams BA, Pertea G et al (2010) Transcript assembly and quantification by RNA-Seq reveals unannotated transcripts and isoform switching during cell differentiation. Nat Biotechnol 28:511–515

Turnbaugh PJ, Quince C, Faith JJ et al (2010) Organismal, genetic, and transcriptional variation in the deeply sequenced gut microbiomes of identical twins. Proc Natl Acad Sci USA 107:7503–7508

Turner ST, Bailey KR, Fridley BL et al (2008) Genomic association analysis suggests chromosome 12 locus influencing antihypertensive response to thiazide diuretic. Hypertension 52:359–365

US Food and Drug Administration (2012) Table of valid genomic biomarkers in the context of approved drug labels. http://www.fda.gov/Drugs/ScienceResearch/ResearchAreas/Pharmacogenetics/ucm083378.htm. Accessed 28 Aug 2012

Van't Veer LJ, Dai H, van de Vijver MJ et al (2002) Gene expression profiling predicts clinical outcome of breast cancer. Nature 415:530–536

van de Vijver MJ, He YD, Van't Veer LJ et al (2002) A gene-expression signature as a predictor of survival in breast cancer. N Engl J Med 347:1999–2009

van Rooij E, Sutherland LB, Liu N et al (2006) A signature pattern of stress-responsive microRNAs that can evoke cardiac hypertrophy and heart failure. Proc Natl Acad Sci USA 103:18255–18260

Venter JC, Adams MD, Myers EW et al (2001) The sequence of the human genome. Science 291: 1304–1351

Voelkerding KV, Dames SA, Durtschi JD (2009) Next-generation sequencing: from basic research to diagnostics. Clin Chem 55:641–658

Wang ET, Sandberg R, Luo S, Khrebtukova I, Zhang L, Mayr C, Kingsmore SF, Schroth GP, Burge CB (2008a) Alternative isoform regulation in human tissue transcriptomes. Nature 456:470–476

Wang J, Wang W, Li R et al (2008b) The diploid genome sequence of an Asian individual. Nature 456:60–65

Wang Z, Gerstein M, Snyder M (2009) RNA-Seq: a revolutionary tool for transcriptomics. Nat Rev Genet 10:57–63

Warren RL, Butterfield YS, Morin RD, Siddiqui AS, Marra MA, Jones SJM (2005) Management and visualization of whole genome shotgun assemblies using SAM. Biotechniques 38:715–720

Weisenberger DJ, Siegmund KD, Campan M et al (2006) CpG island methylator phenotype underlies sporadic microsatellite instability and is tightly associated with BRAF mutation in colorectal cancer. Nat Genet 38:787–793

Wheeler DA, Srinivasan M, Egholm M et al (2008) The complete genome of an individual by massively parallel DNA sequencing. Nature 452:872–876

Wilhelm BT, Marguerat S, Goodhead I, Bahler J (2010) Defining transcribed regions using RNA-seq. Nat Protoc 5:255–266

Yeager M, Xiao N, Hayes RB et al (2008) Comprehensive re-sequence analysis of a 136 kb region of human chromosome 8q24 associated with prostate and colon cancers. Hum Genet 124:161–170

Yi X, Liang Y, Huerta-Sanchez E et al (2010) Sequencing of 50 human exomes reveals adaptation to high altitude. Science 329:75–78

4 Pharmaco-epigenomics

Mehmet Gunduz, Muradiye Acar, Kubra Erdogan,
Elif Nihat Cetin, and Esra Gunduz

Abstract

Epigenetic modifications are defined as the study of heritable changes in phenotype that do not involve alterations in the DNA sequence. DNA methylation, posttranslational modifications of the histone proteins, and miRNAs are regulating the expression of genes as well as drug-metabolizing genes. Epigenetic regulation is essential for normal developmental and cellular processes. Conversely, abnormal epigenetic regulation is a character of complex diseases, including cancer, hematological malignancies, psychiatric disorders, and other diseases. Pharmaco-epigenomics is a novel discipline and involves the study of epigenetic factors in the interpersonal variation to drugs. Epigenetic biomarkers can be used to diagnose disease, estimate disease progression, or predict interpersonal variations in response to therapy. Unlike genetic alterations, changes in epigenetic machinery are reversible, and this reversible characteristic makes them an attractive therapeutic targets.

1 Epigenetics and Pharmaco-epigenomics

Epigenetics can be defined as the study of heritable changes in phenotype that do not involve alterations in the DNA sequence (Hamm and Costa 2011). Epigenetics contains three molecular mechanisms which interact with each other; these are DNA methylation, modification of histones, and RNA-mediated regulation of gene expression (Peedicayil 2006). Epigenetics offers to the researchers field of a new playground for their study and modify interindividual variation in clinical effects, variation in drug response and toxicity, and new targets for drug therapy (Mateo Leach et al. 2010). This is the branch of genetics that also offers a new viewpoint to understanding the role of the environment's interactions with the genome in causing disease and in modulating those interactions to improve human health (Feinberg 2007). Epigenomic profiles have shown varieties from cell type to cell type and may alter over time and in response to physiologic, pathologic, and pharmacologic triggers (Mateo Leach et al. 2010).

Pharmaco-epigenomics contains the study of the roles of epigenomic modifications in intrapersonal and interpersonal variations in response of individuals to drugs, in the effects of drugs on

M. Gunduz (✉) • M. Acar • K. Erdogan • E.N. Cetin
E. Gunduz
Department of Medical Genetics, Faculty of Medicine, Turgut Ozal University, Ankara, Turkey
e-mail: mehmet.gunduz@gmail.com

D. Barh et al. (eds.), *Omics for Personalized Medicine*,
DOI 10.1007/978-81-322-1184-6_4, © Springer India 2013

gene expression profiles, in the mechanism of action of drugs and adverse drug reactions, and in the discovery of new drug target therapy (Peedicayil 2008). While the genotype varies only between individuals, epigenotypes may vary in three distinct situations: between cell types; within a cell type across individuals; and as a function of a time, during development and aging (Bjornsson et al. 2008). Furthermore, the epigenotype is sensitive to modification by environmental factors, including pharmacologic agents (Stamatoyannopoulos and Dunham 2008).

Describing of epigenomic variation at the cellular and individual level will ensure a critical backdrop for understanding its potential variability under circumstances of pharmacologic and environmental exposure (Stamatoyannopoulos and Dunham 2008). Recent studies have shown us that not only the genetic changes are responsible from heritable phenotypic variation but that epigenetic changes also play an important role in the variation of tendency to disease and to drug response (Mateo Leach et al. 2010). In controlling gene expression, epigenetic variations can play a crucial role and therefore offer another mechanism clarifying to interindividual variation (Dennis 2003).

In drug response, the role of epigenetics has just perceived, but its impending influence on drug metabolism and disposition promises to be important (Mateo Leach et al. 2010). The important point in the area of pharmacogenomics from 2000 was to explain the most important polymorphic alleles and their functional conclusions, partly as conclusions of the completion of the human genome sequence. The effect of genetic polymorphisms on drug response and efficacy was not fully known; there was no major attempt to explain epigenetic factors for the interindividual differences in, for example, the activity and function of drug-metabolizing enzymes and transporters (Ingelman-Sundberg and Gomez 2010) (Fig. 4.1). Actually, pharmaco-epigenomics has an encouraging future. For example, an attractive phase of epigenetics that would really be of excellent benefit to further understanding interindividual variation in drug response is its plasticity (Mateo Leach et al. 2010).

2 Epigenetic Disease Mechanism

At least there are three different types of epigenetic modifiers. These are DNA methylation, histone modifications, and ncRNAs. The first type of modification called DNA methylation affects the genomic DNA directly. The second type of modification called posttranslational histone modification affects DNA indirectly (Hamm and Costa 2011). And the third modification includes different classes of ncRNAs, which can physically bind to the DNA, change its conformation, and, in the case of microRNAs, silence genes by posttranslational control (Costa 2008).

These epigenetic modifications may be different among tissues and individuals, but they also may alter in time during aging or as a result of environmental interactions or different disease. Because the epigenome plays a important role in programming the expression of the genome, differences in gene expression among individuals that affect the response to drugs may be modified by epigenomic variations on top of nucleotide mutationsorpolymorphisms(Stamatoyannopoulos and Dunham 2008).

In contrast to genetic changes, the epigenetic changes can be reversed by the inhibition of DNA methylation and histone deacetylation (Sebova and Fridrichova 2010). Despite their heritability, these changes could be reversible, increasing the possibility of a new therapeutic target (Hamm and Costa 2011). Also epigenetic modifications influence the expression of drug-metabolizing enzymes, drug transporters, and nuclear receptors that settle the expression of various genes and ultimately affect the response to drugs (Gomez and Ingelman-Sundberg 2009).

Proper epigenetic regulation is crucial for normal differentiation in embryogenesis and development. Conversely, abnormal epigenetic regulation is a character of complex diseases, involving cancer, diabetes, heart disease, and other diseases. Understanding the effects of epigenetic modifications in normal development and disease consequently leads to the future development of more effective epigenetic-based therapies (Hamm and Costa 2011). In brief, epigenetic

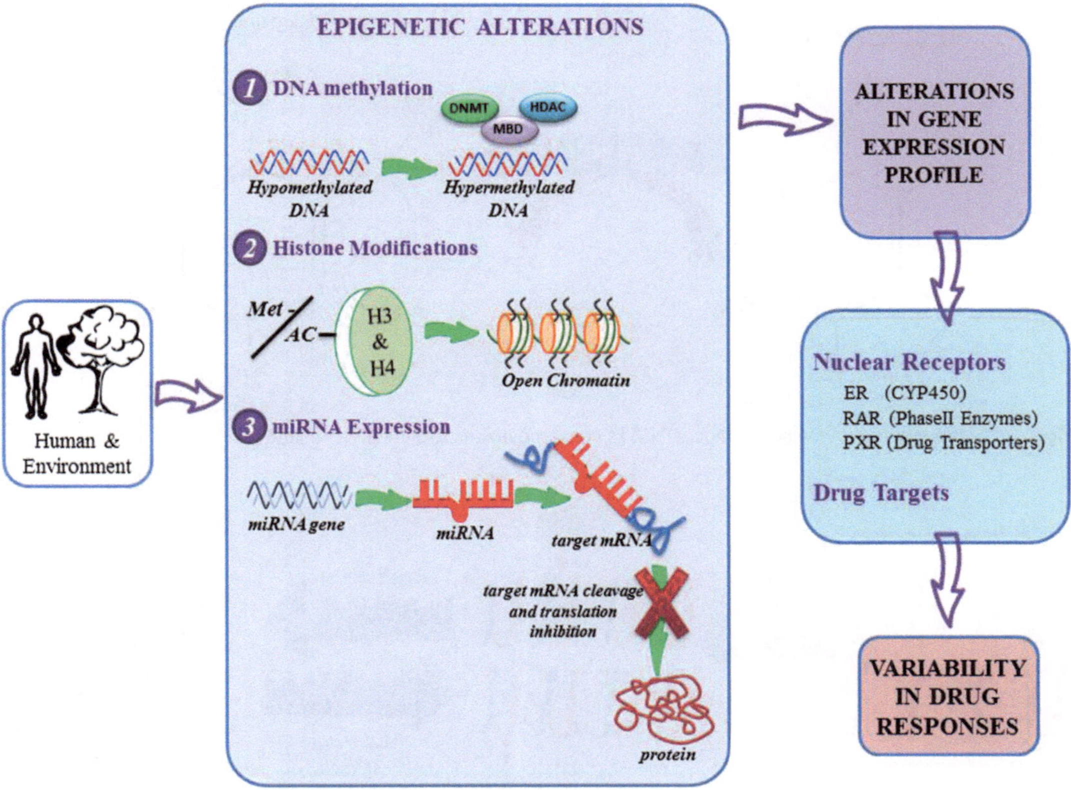

Fig. 4.1 Association between epigenetic modifications and drug response

processes are natural and crucial to the function of organisms, but if they occur inconveniently, there can be big adverse health and behavioral effects (Csoka and Szyf 2009).

2.1 DNA Methylation

DNA methylation is defined as the transfer of a methyl group (CH3) from the universal methyl donor S-adenosylmethionine to the carbon in the fifth position of cytosine in the CpG dinucleotide sequence (Fog et al. 2007) (Fig. 4.2). DNA methylation exist in mammals mainly as the sequence 5′-CG-3′, which also is referred to as a CpG dinucleotide; approximately 70 % of all CpGs in humans are methylated (Issa 2004). On the other hand, in the 5′ regulatory region of many genes there are clusters of unmethylated CpGs known as CpG islands. This frequency of CpG dinucleotides in CpG islands is higher than the other DNA regions. Particularly, differential methylation of CpG islands is part of the epigenetic variation found in humans (Peaston and Whitelaw 2006).

Three main DNA methyltransferases are mediated to transferring the methyl groups. These are DNMT1, DNMT3A, or DNMT3B (Hermann et al. 2004). All DNMTs include multiple domains and are able to form variable complexes with other corepressors. DNMT1 is known as a maintenance enzyme that preferably binds to hemimethylated DNA and copies methylation patterns from the parent strand to the newly synthesized strand during the cell cycle S-phase (Mattson 2004). DNMT3a and DNTM3b, named as de novo methyltransferases, can methylate unmethylated DNA sequences (Toyota and Issa 1999) (Fig. 4.3). DNMT2 and DNMT3L are the other two kinds of methyltransferases which manifest very little or no DNMT activities, respectively (Li et al. 1999).

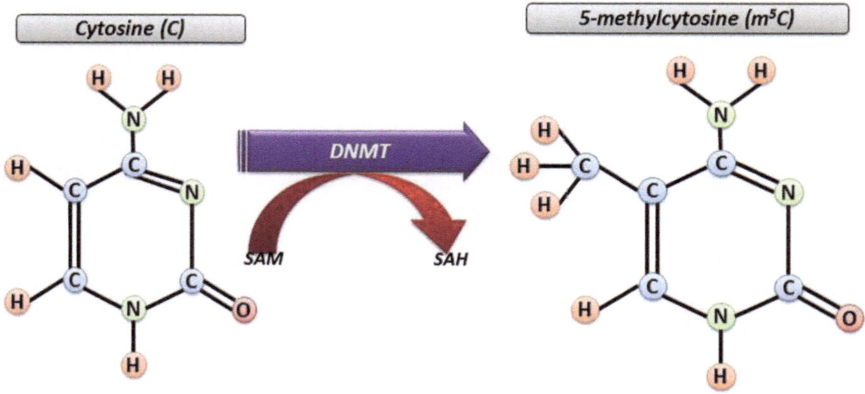

Fig. 4.2 Methylation of cytosine in DNA. DNMTs convert cytosine to 5-methylcytosine

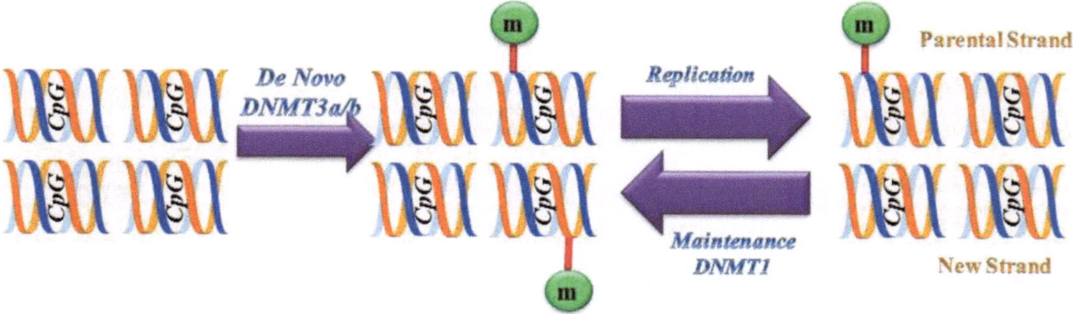

Fig. 4.3 De novo and maintenance methylation patterns

Researches on cancer developments and recent findings have indicated that development and progression of tumors are associated significantly with epigenome modifications. Epigenetic therapy can be possible with reversibility of epigenetic alterations. Tissues of cancers when compared with normal tissues demonstrate widespread epigenetic changes in DNA methylation such as global DNA hypomethylation and hypermethylation of promoters in many cancer-related genes (Esteller 2007).

CYP gene expression levels can be influenced by the epigenetic alterations and particularly DNA methylation (Baer-Dubowska et al. 2011). No critical functional polymorphisms have been described in CYP1A1 gene. Despite that, there are some epigenetic mechanisms that have been documented for the regulation of CYP1A1 expression (Rodriguez-Antona et al. 2010). In prostate cancer cells, DNA methylation contributes to the regulation of CYP1A1 gene (Okino et al. 2006). Smoking is induced to the CYP1A2, and CYP1A2 affected by diet is involved in the metabolism of several drugs and interferon. Methylation of a CCGG site is apparently related to CYP1A2 expression (Hammons et al. 2001). Believed to play an important role for CYP2A13 is initiation of carcinogenesis. 5-aza-2-deoxycytidine (a DNA-demethylating agent) and trichostatin A (an inhibitor of HDAC) are used for treating human lung cancer cells and showed a ~10-fold increase in the level of *CYP2A1* expression (Ling et al. 2007).

DNA methylation is affecting to the expression of drug transporter proteins (Baer-Dubowska et al. 2011). Two major superfamilies of membrane transporter proteins are the ATP-binding cassette (ABC) and soluble carrier (SLC) proteins. This ABC transporters are often related with decreasing cellular accumulation of anticancer drugs and multidrug resistance of tumors (Sugars and Rubinsztein 2003). Some SLC transporters such as folate, nucleoside, and amino acid transporters usually enlarge chemosensitivity by mediating the cellular uptake of hydrophilic drugs such as gemcitabine and other nucleoside analogues (Candelaria et al. 2010). Epigenetic status of the *MDR1* (human multidrug resistance gene 1) locus detects its expression following treatment with chemotherapeutic drugs like daunorubicin and etoposide, so that drugs activate *MDR1* transcription only when the promoter region is significantly hypomethylated. Histone modification can induce to upregulation of MDR1. MDR1 upregulation correlates with increasing histone 3 acetylation and histone 3 methylation at lysine 4 (K4) (Baker et al. 2005). The gene for solute carrier family 5 (iodide transporter) member 8 (*SPC5A5*) has been characterized as a tumor-suppressor gene and is downregulated by promoter methylation in pancreatic and prostatic cancers. This gene expression was avoided by treatment with DNA methylation inhibitors (Pandey and Picard 2009). Other transporters have also been proved to be epigenetically downregulated. Abnormal hypermethylation pattern of the reduced folate carrier (*RFC*) gene has been associated with resistance to methotrexate in cancer cell lines, primary osteosarcomas, lymphoproliferative disorders (Kastrup et al. 2008), and breast cancer (Worm et al. 2001). These examples pointed that circumvention in drug transporter expression at the epigenetic level may stand for one way to overcome drug resistance (Baer-Dubowska et al. 2011).

Nuclear receptors including the transcription control of genes encoding drug transporters and enzymes. DNA methylation promotes the expression of members of the retinoic acid receptor family. Loss of retinoic acid receptor α 2 expression in head and neck squamous cell carcinomas was associated with its hypermethylation, which occurs early in head and neck carcinogenesis (Youssef et al. 2004).

2.2 Histone Modification

The eukaryotic genome is compacted with histone proteins. In the nucleosomes, genomic DNA is folded and compacted around core histone proteins (two copies of each of the core histones H2A, H2B, H3, and H4), forming the basic repeat units of chromatin. Four different core histone proteins have an amino-terminal tail that sticks out from the chromatin fiber and is thought to interact with DNA or other histones or proteins (Mano 2008). Posttranslational modification of histone proteins can change the electrostatic interaction between the histones and adjacent DNA (Ray-Gallet and Almouzni 2010). The interaction of histone proteins with DNA has a major impact on the accessibility of transcriptional machinery to their target DNA sequences and therefore regulates transcriptional activity. Through this crucial mechanism, nucleosomes carry epigenetically inherited information in the form of covalent modifications of their core histone proteins (Mano 2008).

A myriad of histone-modifying enzymes has been identified in recent years. Histone acetyltransferases (HAT) and histone deacetylases (HDACs) are the most studied enzyme groups (Kouzarides 2007). Acetylation of histones by histone acetyltransferases promotes gene transcription by creating a more accessible chromatin structure, whereas HDAC-induced deacetylation diminishes histone-DNA and histone-nonhistone protein interactions, impairing transcription of genes (Sasaki and Matsui 2008). As such, transcriptionally silent genes are commonly associated with deacetylation of histone H3 and H4 (Ballestar et al. 2003; Jones and Baylin 2007). Histone methyltransferase and histone demethylase enzymes regulate methylation of specific amino acid residues resulting in activation of gene expression (histone H3 lysine 4 [H3K4]) or repression of gene expression (histone H3 lysine 9 [H3K9]) (Richards and Elgin 2002).

DNA methylation and histone modifications are frequently disrupted in cancer cells (Claes et al. 2010). Both histone methyltransferase and histone demethylase enzymes are misregulated in cancer. Also, expression level of HDACs is increased and there is a global reduction in histone acetylation levels (Shi 2007).

2.3 Noncoding RNA (ncRNA)

Human genome (approximately 98 %) is included of noncoding DNAs. These regions do not encode to proteins but produce functional RNAs (Mattick 2004). These noncoding RNAs (ncRNAs) comprised of transfer RNAs, ribosomal RNAs, miRNAs, small nuclear RNAs (snRNAs), and small nucleolar RNAs (snoRNAs), which have important functions in gene regulation (Ambros 2004; Hobert 2008). miRNAs (MicroRNAs or miR) are short (18–25 nt) double-stranded RNAs that originate from nuclear and cytoplasmic larger precursor molecules. miRNAs play a crucial role in the regulations of gene expression at a posttranscriptional level. miRNAs degraded or silenced their targeted mRNAs by binding to partly complementary base pairs in 3 untranslated regions of mRNAs. miRNA regulation of protein expression is very complicated because several genes may be targeted by the same miRNA and several miRNAs can target the same gene (Schroen and Heymans 2009).

The expression levels of miRNAs are tissue and disease specific (Mateo Leach et al. 2010). Abnormal miRNA expression is known to conduce to the pathogenesis of diseases, and hereby, miRNAs may serve as a novel target for therapy (Zhang and Farwell 2008).

Also miRNAs play a critical role in posttranscriptional regulation of ADME genes (Yu 2009; Nakajima and Yokoi 2011). Drugs administered orally undergo a series of processes, namely, absorption, distribution, metabolism, and excretion (ADME). Drug-metabolizing enzymes and transporters are involved in ADME of drugs. Among them especially, xenobiotic-metabolizing enzymes such as cytochrome P450 (CYP or P450) isoforms play a crucial role in metabolic elimination of drugs (Guengerich 2006), and ABC and SLC transporters have high effect on drug absorption, distribution, and excretion (Giacomini et al. 2010). Recent studies show that genetic polymorphism has an important function in inhibition and induction of ADME enzymes and transporters, while others such as posttranscriptional regulation of ADME genes are under active study. Posttranscriptional regulation of ADME genes are controlled by miRNAs. Change of ADME gene expression by miRNAs may alter ADME gene expression, change the capacity of drug metabolism and transport, and influence the sensitivity of cells to xenobiotics (Yu and Pan 2012).

miRNAs are direct impact on mRNA transcription, but some miRNAs, epi-miRNAs, have an indirect influence on gene transcription by affecting the epigenetic machinery, comprehending DNA methyltransferases, histone deacetylases, and polycomb repressive complex genes (Fabbri and Calin 2010).

3 Disease and Pharmaco-epigenomics

Many genes encoding transcription factors, cell cycle regulators, nuclear receptors, drug transporters, and enzymes are controlled by epigenetic mechanisms (Baer-Dubowska et al. 2011). Several modifications can influence the epigenetic condition of a particular locus in the genome. Epigenetic modifications are fundamental for normal development, and they continue to have a role in gene regulation and genome stability throughout the lifespan of an organism (Pogribny and Beland 2009). Genomic instability is a crucial phenotype that permits cancer cells to generate oncogenic translocations, inactivate tumor-suppressor genes, and amplify oncogenes and drug-resistance genes (Toyota et al. 2009).

Given the prevalence and importance of epigenetic changes throughout development, it is not surprising that epigenetic aberrations conduce to the pathogenesis of cancer. Abnormal epigenetic changes were first described in colon cancer (Feinberg and Vogelstein 1983), but subsequent research in the field of oncology demonstrated that epigenetic changes might be one of the most common abnormities in cancers (Pogribny and Beland 2009). In other words, epigenetic abnormalities have been determined in several other human diseases and disorders such as cancer, hematologic malignancies, psychiatric disorders and other diseases (Hamm and Costa 2011).

DNA methyltransferase inhibitors and histone deacetylase inhibitors are current epigenetic therapies. DNA methylation inhibitors and HDAC inhibitors promote gene expression by promoting a more open chromatin structure (Hamm and Costa 2011). These inhibitors have been FDA approved for several years and are clinically successful (Kelly et al. 2010).

The goal of epigenetic therapy is to reverse epigenetic silencing and reactivate various genes expectation for a therapeutic effect such as growth arrest, cell differentiation, or apoptosis. In vitro studies have shown that pharmacologic inhibition of DNMT or HDAC resulting in reactivation of silenced genes either physiologically or pathologically (Bender et al. 1998; Marks et al. 2001).

3.1 Inhibitors of DNMT

Many genes that are hypermethylated in cancer and other diseases can be reactivated with DNA methyltransferase (DNMT) inhibitors (Baer-Dubowska et al. 2011). Some drugs that inhibit DNA methyltransferases have been demonstrated to reactivate silenced genes and induce apoptosis or differentiation of malignant cells (Donkena et al. 2010).

DNMTs are overexpressed in almost all types of tumors. Consequently, the promoter regions of numerous silenced tumor-suppressor genes (TSGs) are hypermethylated. Some drugs include 5-azacytidine (5AC, Vidaza), 5-aza-2-deoxycytidine (DAC, Dacogen, 5-aza-CdR), and 1-(b-D-ribofuranosyl)-1,2-dihydropyrimidin-2-one (ZEB, Zebularine) have the ability to reverse aberrant DNA hypermethylation in silent TSGs. During the RNA synthesis or DNA replication, these compounds are able to incorporate into RNA (5AC), or DNA (5AC, DAC), sequences as cytidine analogues, respectively (Gowher and Jeltsch 2004). DNA methylation is prevented by covalent binding between the cytidine analogues incorporated in the DNA chain and DNMTs (Juttermann et al. 1994).

5-AC is a nucleoside inhibitor that is incorporated into DNA. DNA methyltransferase methylate both cytosine residues and 5-azacytosine residues in the DNA chain. However, 5-azacytosine inhibits the resolution of a covalent reactive intermediate. This event contributes to the DNA methyltransferase being trapped and inactivated in the form of a covalent protein-DNA adduct, which results in depletion of cellular DNA methyltransferase (Santi et al. 1984). In vitro studies have demonstrated that a lower dose of 5-AC induces apoptosis and cell differentiation by promoting the expression of crucial genes that are silenced by hypermethylation machinery (Jones and Taylor 1980). Two inhibitors of DNA methyltransferases, Vidaza and decitabine have already been approved by the FDA as effective drugs for treatment of myelodysplastic syndromes (MDS) (Mack 2006).

Decitabine, the deoxyribose analogue of 5-azacytidine, demonstrates more specificity with greater inhibition of DNA methylation status and less toxicity than 5-AC. However, it also has significant toxic effects (Billam et al. 2010). One potential adverse impact is activation of genes that are normally epigenetically silenced, so negatively affect disease outcome. DNA hypomethylation can promote tumorigenesis (Eden et al. 2003; Gaudet et al. 2003), and treatment of tumor cells with a global DNA hypomethylating agent, DAC, can promote tumor progression and cancer (Hamm et al. 2009).

Zebularine is more stable than 5-AC or decitabine cytidine analogue (Billam et al. 2010). Decitabine and zebularine can decrease vessel formation and circumvent proliferation of tumor-conditioned endothelial cells by reactivation of growth-inhibiting genes, such as JUNB, THBS-1, and IGFBP3, known to be silenced in tumor-conditioned endothelial cells (Hellebrekers et al. 2006).

Also DNA methylation is inhibited by some nonnucleoside compounds. EGCG, the main polyphenol compound in green tea, binds and blocks the active site of DNMT1 enzyme (Fang et al. 2003). RG108, a small-molecule inhibitor directly and specifically inhibits DNMT1 with low toxicity than other compounds (Brueckner et al. 2005). MG98, a specific antisense oligonucleotide, represents another class of DNA methyltransferase inhibitors. MG98 has displayed antitumor activity in preclinical trials and is recently being tested in a phase II clinical trials (Donkena et al. 2010).

3.2 Inhibitors of HDAC

HDAC inhibitors can be classified into four main families. These families are the hydroxamic acids (trichostatin A and suberoylanilide hydroxamic acid), the short-chain fatty acids (sodium butyrate, phenylbutyrate, and valproic acid), the epoxy ketones (trapoxin), and the benzamides. The most widely studied inhibitors are trichostatin A, suberoylanilide hydroxamic acid, sodium butyrate, and phenylbutyrate. HDAC inhibitors prevent growth, induce differentiation and apoptosis, and have efficient anticancer activities (Carey and La Thangue 2006; Minucci and Pelicci 2006).

3.3 Pharmaco-epigenomics in Cancer

During the carcinogenesis, epigenetic regulation is disrupted, leading to the abnormal expression pattern of many fundamental genes. TSGs are inhibited after the hypermethylation of their promoter sequences in tumors. On the other hand, oncogenes can be activated by the sequence-specific decrease of DNA methylation (Peedicayil 2006). Epigenetic changes can lead to carcinogenesis by silencing or activating essential genes (Jones and Baylin 2002).

Given the critical function of epigenetic changes in cancer, similar to genetic changes, they can be used for the diagnosis and molecular classification of cancer, and to predict cancer progression or response to therapy. Also epigenetic changes can be used for cancer detection (Claes et al. 2010). Glutathione S-transferase 1 gene is hypermethylated in 80–90 % of prostate cancer patients, whereas benign hyperplastic prostate tissue is not hypermethylated (Esteller et al. 1998; Jeronimo et al. 2001). Intriguing evidences indicate that epigenetic changes occur early in cancer development before other biomarkers are detectable (Claes et al. 2010). For example, hypermethylation of the tumor-suppressor gene CDKN2A can already be identified in bronchial preneoplastic epithelium of smokers (Belinsky et al. 1998).

Targeting of epigenetic alterations is a very promising strategy in cancer therapy or chemoprevention (Hauser and Jung 2008). In non-small-cell lung cancer (NSCLC), an unmethylated insulin-like growth factor-binding protein 3 (IGFBP3) promoter is indicative of responsiveness to cisplatin-based chemotherapy and predictor of the chemotherapy outcome (Ibanez de Caceres et al. 2010). Also epigenetic modifications can be used to evaluate treatment efficiency and disease progression. For example, methylation of paired-like homeodomain transcription factor 2 (PITX2) can be used to estimate outcome of early breast cancer patients following adjuvant tamoxifen therapy (Martens et al. 2009). Patients with *p16* hypermethylation had lower bladder cancer recurrence rates after IL-2 treatment compared to patients with no *p16* hypermethylation after IL-2 treatment (Jarmalaite et al. 2010).

Colorectal cancer patients with *K-ras* mutations reportedly do not respond to treatment with cetuximab or panitumumab (Freeman et al. 2008; Khambata-Ford et al. 2007). This mutation may permit cells to escape the anti-EGFR activity of cetuximab and panitumumab (Toyota et al. 2009). These patients show methylation of multiple CpG islands and DNA methylation of specific genes may contribute to resistance to these anti-EGFR antibodies (Toyota et al. 2000).

3.4 Pharmaco-epigenomics for Hematological Disorders

Alterations in epigenetic regulation can cause some cells to develop several abnormalities, while other cells of the same type remain normal (Mund et al. 2006). Epigenetic regulation is part of physiologic processes and becomes abnormal in neoplasia, where silencing of critical genes by DNA hypermethylation or histone deacetylation can contribute to leukemogenesis as an alternative to genetic changes (Oki and Issa 2010).

p15, p73, E-cadherin, ID4, and RARb2 genes demonstrate abnormal DNA methylation pattern in acute myelogenous leukemia (AML) (Oki and Issa 2010). In AML, promoter methylation of $p15^{INK4B}$ has been known for many years (Herman et al. 1994), and the methylation of this gene in MDS appears to be associated with poor prognosis and a higher chance of developing AML (Uchida et al. 1997). Frequently methylated genes in AML are HIC1 (Melki et al. 1999), ID4 (Yu et al. 2005), SHP1 (Chim et al. 2004), MYOD1, PITX2 (Toyota et al. 2001), and calcitonin (Melki et al. 1999). A histone methyltransferase, MLL gene is located on chromosome 11q23, is often fused with other genes leading to leukemogenesis through chromatin modulation (Cimino et al. 1991). The N-terminal domain of MLL contains a transcriptional repression domain (Zeleznik-Le et al. 1994). This domain recruits HDAC1 and HDAC2 (Xia et al. 2003). One of the crucial functions of MLL gene is the maintenance of expression of HoxA9 and HoxC8 genes by binding to their promoter DNA and keeping the chromatin open for transcription machinery (Milne et al. 2002; Nakamura et al. 2002). The C-terminus of MLL includes a transcriptional-activation domain that binds to a HAT, CBP (Ernst et al. 2001).

Widespread translocations in AML result in fusion proteins that affect gene transcription by HAT/HDAC recruitment (Oki and Issa 2010). AML1 normally interacts with a co-activator complex containing p300, which has HAT activity, and will result in acetylation of histone proteins and transcriptional activation. In translocation of t(8;21)(q22;q22), AML1/ETO translocation has HDAC activity instead of HAT, and results in transcriptional repression. Thus, this activity prevents myeloid differentiation (Durst and Hiebert 2004; Scandura et al. 2002).

The goal of epigenetic therapy is to reverse epigenetic silencing and reactivate various genes to induce a therapeutic effect such as differentiation, growth arrest, or apoptosis (Oki and Issa 2010). Meantime, it has been established that certain tumor types respond well to HDAC and DNMT inhibitor treatments, with the best clinical efficacy seen in hematologic disorders (Claes et al. 2010). DNMT inhibitor decitabine was initially evaluated twenty years ago in cancer treatment and has shown important antitumor activity in hematologic disorders (Aparicio and Weber 2002). Decitabine is approved for the treatment of patients with AML or MDS (Hackanson et al. 2005; Silverman and Mufti 2005). Vorinostat (suberoylanilide hydroxamic acid) is one of the HDAC inhibitors and used for treatment of cutaneous T-cell lymphoma in patients with progressive, persistent, or recurrent disease (Khan and La Thangue 2008). DNMT inhibitor azacitidine used at relatively high dose (600–1,500 mg/m^2/course) in refractory AML resulted in about 45 % overall responses (Glover et al. 1987).

3.5 Pharmaco-epigenomics in Psychiatry

Epigenetic modifications is critical to cellular differentiation, development, and behavior, including learning and memory (Wood et al. 2006; Levenson and Sweatt 2005). The epigenetic informations are responsive to synaptic activity and provide a link between experience, genetic predisposition, and changes in neural function. Valproic acid is the anticonvulsant and mood-stabilizing drug in the nervous system. Also it is an inhibitor of HDAC1 and its antiepileptic effects are connective to changes in histone acetylation. Studies in recent years have revealed that inhibitors of class 1 and 2 HDACs demonstrate novel therapeutic approaches to treat neurodegenerative disorders, depression and anxiety, and the cognitive deficits that accompany many neurodevelopmental disorders (Abel and Zukin 2008).

The environmental factors can effect on an individual's genes to cause depression and related disorders (Poulter et al. 2008). The antidepressant, imipramine, can induce the acetylation of histones, thereby reverse depression-induced repressive chromatin modifications (Tsankova et al. 2006).

Epigenetic abnormalities are a common topic in disorders of synaptic plasticity, and cognition comprises neurodevelopmental disorders (Rubinstein-Taybi syndrome, Rett syndrome, and fragile X syndrome), neurodegenerative disorders (Huntington's disease, Parkinson's disease and ischemia), and mood disorders (depression and anxiety) (Abel and Zukin 2008).

3.6 Pharmaco-epigenomics for Neurodegenerative Diseases

Epigenetic abnormalities play a major role in the pathogenesis of cancer and some multifactorial diseases such as schizophrenia and bipolar disorder, both of which are due to epigenetic defects rather than genetic effects (Peedicayil 2006).

Neurological disorders in which an epigenetic gene is mutated comprise α-thalassemia/mental retardation X-linked syndrome (ATRX), Rett syndrome, Rubinstein-Taybi, and Coffin-Lowry syndromes. ATRX and Rett syndromes are both X-linked disorders caused by mutations in a chromatin remodeling complex protein and in methyl-CpG binding domain protein 2 (MeCP2), respectively. Rubinstein-Taybi syndrome is associated with the dysfunction of a HAT, while Coffin-Lowry syndrome is caused by deficiencies in a histone phosphorylase (Urdinguio et al. 2009).

Alzheimer's disease (AD) is defined by aberrant CREB/CBP signaling. AD is a neurodegenerative disorder that currently affects nearly 2 % of the population in industrialized countries; the risk of AD dramatically increases in individuals beyond the age of 70. Neuropathological specialty of AD comprises extracellular plaques and intracellular tangles. Endoproteolytic cleavage of the transmembrane amyloid precursor protein (APP) composes β-amyloid peptides, which aggregate to form plaques. A common thema in neurodegenerative disorders (HD, PD, and AD) is the idea that intraneuronal aggregates such as plaques interfere with transcription machinery and cause deficiency in plasticity and cognition (Mattson 2004).

Hypermethylation in certain genes can be observed in normal tissues during the process of aging (Issa 1999). HDAC inhibitors used to ameliorate the cognitive and motor deficits characteristic of Huntington's disease (HD). HD is an inherited, late-onset autosomal-dominant neurodegenerative disorder defined by progressive motor, psychiatric, and cognitive decline (Bates 2001). HD is caused by a polyglutamine (CAG) expansion in the 5′coding region of the huntingtin (htt) gene. Transcriptional abnormalities play a major role in the pathogenesis of HD. Mutant htt localizes primarily to the nucleus where it forms aggregates of mutant polyQ protein, which bind and functionally circumvent transcription factors and co-activators such as CBP (Sugars and Rubinsztein 2003). Loss of accessible CBP leads to dysregulation of CBP-/CREB-mediated gene expression profile, histone deacetylation, and ultimately neuronal death (Abel and Zukin 2008).

4 Conclusion

During the development of a disease, epigenetic regulation is disrupted, leading to the abnormal expression of many crucial genes. To use epigenomic therapies efficiently, it will be necessary to understand the epigenetic modifications in both normal and disease states. Pharmacoepigenomics' aim is to provide a new individualized treatment and discover new drug targets. In the future, pharmacoepigenetics and pharmacoepigenomics will play an essential role in pharmacology and clinical medicine.

References

Abel T, Zukin RS (2008) Epigenetic targets of HDAC inhibition in neurodegenerative and psychiatric disorders. Curr Opin Pharmacol 8(1):57–64

Ambros V (2004) The functions of animal microRNAs. Nature 431(7006):350–355

Aparicio A, Weber JS (2002) Review of the clinical experience with 5-azacytidine and 5-aza-2′-deoxycytidine in solid tumors. Curr Opin Investig Drugs 3(4):627–633

Baer-Dubowska W, Majchrzak-Celinska A, Cichocki M (2011) Pharmocoepigenetics: a new approach to predicting individual drug responses and targeting new drugs. Pharmacol Rep 63(2):293–304

Baker EK, Johnstone RW, Zalcberg JR et al (2005) Epigenetic changes to the MDR1 locus in response to chemotherapeutic drugs. Oncogene 24(54):8061–8075

Ballestar E, Paz MF, Valle L et al (2003) Methyl-CpG binding proteins identify novel sites of epigenetic inactivation in human cancer. EMBO J 22(23):6335–6345

Bates GP (2001) Huntington's disease. Exploiting expression. Nature 413(6857):691, 693–694

Belinsky SA, Nikula KJ, Palmisano WA et al (1998) Aberrant methylation of p16(INK4a) is an early event in lung cancer and a potential biomarker for early diagnosis. Proc Natl Acad Sci USA 95(20):11891–11896

Bender CM, Zingg JM, Jones PA (1998) DNA methylation as a target for drug design. Pharm Res 15(2):175–187

Billam M, Sobolewski MD, Davidson NE (2010) Effects of a novel DNA methyltransferase inhibitor zebularine on human breast cancer cells. Breast Cancer Res Treat 120(3):581–592

Bjornsson HT, Sigurdsson MI, Fallin MD et al (2008) Intra-individual change over time in DNA methylation with familial clustering. JAMA 299(24):2877–2883

Brueckner B, Garcia Boy R, Siedlecki P et al (2005) Epigenetic reactivation of tumor suppressor genes by a novel small-molecule inhibitor of human DNA methyltransferases. Cancer Res 65(14):6305–6311

Candelaria M, de la Cruz-Hernandez E, Perez-Cardenas E et al (2010) Pharmacogenetics and pharmacoepigenetics of gemcitabine. Med Oncol 27(4):1133–1143

Carey N, La Thangue NB (2006) Histone deacetylase inhibitors: gathering pace. Curr Opin Pharmacol 6(4):369–375

Chim CS, Wong AS, Kwong YL (2004) Epigenetic dysregulation of the Jak/STAT pathway by frequent aberrant methylation of SHP1 but not SOCS1 in acute leukaemias. Ann Hematol 83(8):527–532

Cimino G, Moir DT, Canaani O et al (1991) Cloning of ALL-1, the locus involved in leukemias with the t(4;11)(q21;q23), t(9;11)(p22;q23), and t(11;19)(q23;p13) chromosome translocations. Cancer Res 51(24):6712–6714

Claes B, Buysschaert I, Lambrechts D (2010) Pharmaco-epigenomics: discovering therapeutic approaches and biomarkers for cancer therapy. Heredity 105(1):152–160

Costa FF (2008) Non-coding RNAs, epigenetics and complexity. Gene 410(1):9–17

Csoka AB, Szyf M (2009) Epigenetic side-effects of common pharmaceuticals: a potential new field in medicine and pharmacology. Med Hypotheses 73(5):770–780

Dennis C (2003) Epigenetics and disease: altered states. Nature 421(6924):686–688

Donkena KV, Young CY, Tindall DJ (2010) Oxidative stress and DNA methylation in prostate cancer. Obstet Gynecol Int 2010, Article ID 302051, doi:10.1155/2010/302051

Durst KL, Hiebert SW (2004) Role of RUNX family members in transcriptional repression and gene silencing. Oncogene 23(24):4220–4224

Eden A, Gaudet F, Waghmare A et al (2003) Chromosomal instability and tumors promoted by DNA hypomethylation. Science 300(5618):455

Ernst P, Wang J, Huang M et al (2001) MLL and CREB bind cooperatively to the nuclear coactivator CREB-binding protein. Mol Cell Biol 21(7):2249–2258

Esteller M (2007) Cancer epigenomics: DNA methylomes and histone-modification maps. Nat Rev Genet 8(4):286–298

Esteller M, Corn PG, Urena JM et al (1998) Inactivation of glutathione S-transferase P1 gene by promoter hypermethylation in human neoplasia. Cancer Res 58(20):4515–4518

Fabbri M, Calin GA (2010) Epigenetics and miRNAs in human cancer. Adv Genet 70:87–99

Fang MZ, Wang Y, Ai N et al (2003) Tea polyphenol (−)-epigallocatechin-3-gallate inhibits DNA methyltransferase and reactivates methylation-silenced genes in cancer cell lines. Cancer Res 63(22):7563–7570

Feinberg AP (2007) Phenotypic plasticity and the epigenetics of human disease. Nature 447(7143):433–440

Feinberg AP, Vogelstein B (1983) Hypomethylation distinguishes genes of some human cancers from their normal counterparts. Nature 301(5895):89–92

Fog CK, Jensen KT, Lund AH (2007) Chromatin-modifying proteins in cancer. APMIS 115(10):1060–1089

Freeman DJ, Juan T, Reiner M et al (2008) Association of K-ras mutational status and clinical outcomes in patients with metastatic colorectal cancer receiving panitumumab alone. Clin Colorectal Cancer 7(3):184–190

Gaudet F, Hodgson JG, Eden A et al (2003) Induction of tumors in mice by genomic hypomethylation. Science 300(5618):489–492

Giacomini KM, Huang SM, Tweedie DJ et al (2010) Membrane transporters in drug development. Nat Rev Drug Discov 9(3):215–236

Glover AB, Leyland-Jones BR, Chun HG et al (1987) Azacitidine: 10 years later. Cancer Treat Rep 71(7–8):737–746

Gomez A, Ingelman-Sundberg M (2009) Pharmacoepigenetics: its role in interindividual differences in drug response. Clin Pharmacol Ther 85(4):426–430

Gowher H, Jeltsch A (2004) Mechanism of inhibition of DNA methyltransferases by cytidine analogs in cancer therapy. Cancer Biol Ther 3(11):1062–1068

Guengerich FP (2006) Cytochrome P450s and other enzymes in drug metabolism and toxicity. AAPS J 8(1):E101–E111

Hackanson B, Robbel C, Wijermans P et al (2005) In vivo effects of decitabine in myelodysplasia and acute myeloid leukemia: review of cytogenetic and molecular studies. Ann Hematol 84(Suppl 1):32–38

Hamm CA, Costa FF (2011) The impact of epigenomics on future drug design and new therapies. Drug Discov Today 16(13–14):626–635

Hamm CA, Xie H, Costa FF et al (2009) Global demethylation of rat chondrosarcoma cells after treatment with 5-aza-2′-deoxycytidine results in increased tumorigenicity. PLoS One 4(12):e8340

Hammons GJ, Yan-Sanders Y, Jin B et al (2001) Specific site methylation in the 5′-flanking region of CYP1A2 interindividual differences in human livers. Life Sci 69(7):839–845

Hauser AT, Jung M (2008) Targeting epigenetic mechanisms: potential of natural products in cancer chemoprevention. Planta Med 74(13):1593–1601

Hellebrekers DM, Jair KW, Vire E et al (2006) Angiostatic activity of DNA methyltransferase inhibitors. Mol Cancer Ther 5(2):467–475

Herman JG, Latif F, Weng Y et al (1994) Silencing of the VHL tumor-suppressor gene by DNA methylation in renal carcinoma. Proc Natl Acad Sci USA 91(21): 9700–9704

Hermann A, Gowher H, Jeltsch A (2004) Biochemistry and biology of mammalian DNA methyltransferases. Cell Mol Life Sci 61(19–20):2571–2587

Hobert O (2008) Gene regulation by transcription factors and microRNAs. Science 319(5871):1785–1786

Ibanez de Caceres I, Cortes-Sempere M, Moratilla C et al (2010) IGFBP-3 hypermethylation-derived deficiency mediates cisplatin resistance in non-small-cell lung cancer. Oncogene 29(11):1681–1690

Ingelman-Sundberg M, Gomez A (2010) The past, present and future of pharmacoepigenomics. Pharmacogenomics 11(5):625–627

Issa JP (1999) Aging, DNA methylation and cancer. Crit Rev Oncol Hematol 32(1):31–43

Issa JP (2004) CpG island methylator phenotype in cancer. Nat Rev Cancer 4(12):988–993

Jarmalaite S, Andrekute R, Scesnaite A et al (2010) Promoter hypermethylation in tumour suppressor genes and response to interleukin-2 treatment in bladder cancer: a pilot study. J Cancer Res Clin Clin Oncol 136(6):847–854

Jeronimo C, Usadel H, Henrique R et al (2001) Quantitation of GSTP1 methylation in non-neoplastic prostatic tissue and organ-confined prostate adenocarcinoma. J Natl Cancer Inst 93(22):1747–1752

Jones PA, Baylin SB (2002) The fundamental role of epigenetic events in cancer. Nat Rev Genet 3(6):415–428

Jones PA, Baylin SB (2007) The epigenomics of cancer. Cell 128(4):683–692

Jones PA, Taylor SM (1980) Cellular differentiation, cytidine analogs and DNA methylation. Cell 20(1):85–93

Juttermann R, Li E, Jaenisch R (1994) Toxicity of 5-aza-2′-deoxycytidine to mammalian cells is mediated primarily by covalent trapping of DNA methyltransferase rather than DNA demethylation. Proc Natl Acad Sci USA 91(25):11797–11801

Kastrup IB, Worm J, Ralfkiaer E et al (2008) Genetic and epigenetic alterations of the reduced folate carrier in untreated diffuse large B-cell lymphoma. Eur J Haematol 80(1):61–66

Kelly TK, De Carvalho DD, Jones PA (2010) Epigenetic modifications as therapeutic targets. Nat Biotechnol 28(10):1069–1078

Khambata-Ford S, Garrett CR, Meropol NJ et al (2007) Expression of epiregulin and amphiregulin and K-ras mutation status predict disease control in metastatic colorectal cancer patients treated with cetuximab. J Clin Oncol 25(22):3230–3237

Khan O, La Thangue NB (2008) Drug Insight: histone deacetylase inhibitor-based therapies for cutaneous T-cell lymphomas. Nat Clin Pract Oncol 5(12):714–726

Kouzarides T (2007) Chromatin modifications and their function. Cell 128(4):693–705

Levenson JM, Sweatt JD (2005) Epigenetic mechanisms in memory formation. Nat Rev Neurosci 6(2):108–118

Li Q, Ahuja N, Burger PC et al (1999) Methylation and silencing of the Thrombospondin-1 promoter in human cancer. Oncogene 18(21):3284–3289

Ling G, Wei Y, Ding X (2007) Transcriptional regulation of human CYP2A13 expression in the respiratory tract by CCAAT/enhancer binding protein and epigenetic modulation. Mol Pharmacol 71(3):807–816

Mack GS (2006) Epigenetic cancer therapy makes headway. J Natl Cancer Inst 98(20):1443–1444

Mano H (2008) Epigenetic abnormalities in cardiac hypertrophy and heart failure. Environ Health Prev Med 13(1):25–29

Marks P, Rifkind RA, Richon VM et al (2001) Histone deacetylases and cancer: causes and therapies. Nat Rev Cancer 1(3):194–202

Martens JW, Margossian AL, Schmitt M et al (2009) DNA methylation as a biomarker in breast cancer. Future Oncol 5(8):1245–1256

Mateo Leach I, van der Harst P, de Boer RA (2010) Pharmacoepigenetics in heart failure. Curr Heart Fail Rep 7(2):83–90

Mattick JS (2004) RNA regulation: a new genetics? Nat Rev Genet 5(4):316–323

Mattson MP (2004) Pathways towards and away from Alzheimer's disease. Nature 430(7000):631–639

Melki JR, Vincent PC, Clark SJ (1999) Cancer-specific region of hypermethylation identified within the HIC1 putative tumour suppressor gene in acute myeloid leukaemia. Leukemia 13(6):877–883

Milne TA, Briggs SD, Brock HW et al (2002) MLL targets SET domain methyltransferase activity to Hox gene promoters. Mol Cell 10(5):1107–1117

Minucci S, Pelicci PG (2006) Histone deacetylase inhibitors and the promise of epigenetic (and more) treatments for cancer. Nat Rev Cancer 6(1):38–51

Mund C, Brueckner B, Lyko F (2006) Reactivation of epigenetically silenced genes by DNA methyltransferase inhibitors: basic concepts and clinical applications. Epigenetics 1(1):7–13

Nakajima M, Yokoi T (2011) MicroRNAs from biology to future pharmacotherapy: regulation of cytochrome P450s and nuclear receptors. Pharmacol Ther 131(3): 330–337

Nakamura T, Mori T, Tada S et al (2002) ALL-1 is a histone methyltransferase that assembles a supercomplex of proteins involved in transcriptional regulation. Mol Cell 10(5):1119–1128

Oki Y, Issa JP (2010) Epigenetic mechanisms in AML – a target for therapy. Cancer Treat Res 145:19–40

Okino ST, Pookot D, Li LC et al (2006) Epigenetic inactivation of the dioxin-responsive cytochrome P4501A1 gene in human prostate cancer. Cancer Res 66(15): 7420–7428

Pandey DP, Picard D (2009) miR-22 inhibits estrogen signaling by directly targeting the estrogen receptor alpha mRNA. Mol Cell Biol 29(13):3783–3790

Peaston AE, Whitelaw E (2006) Epigenetics and phenotypic variation in mammals. Mamm Genome 17(5):365–374

Peedicayil J (2006) Epigenetic therapy – a new development in pharmacology. Indian J Med Res 123(1):17–24

Peedicayil J (2008) Pharmacoepigenetics and pharmacoepigenomics. Pharmacogenomics 9(12):1785–1786

Pogribny IP, Beland FA (2009) DNA hypomethylation in the origin and pathogenesis of human diseases. Cell Mol Life Sci 66(14):2249–2261

Poulter MO, Du L, Weaver IC et al (2008) GABAA receptor promoter hypermethylation in suicide brain: implications for the involvent of epigenetic processes. Biol Psychiatry 64(8):645–652

Ray-Gallet D, Almouzni G (2010) Nucleosome dynamics and histone variants. Essays Biochem 48(1):75–87

Richards EJ, Elgin SC (2002) Epigenetic codes for heterochromatin formation and silencing: rounding up the usual suspects. Cell 108(4):489–500

Rodriguez-Antona C, Gomez A, Karlgren M et al (2010) Molecular genetics and epigenetics of the cytochrome P450 gene family and its relevance for cancer risk and treatment. Hum Genet 127(1):1–17

Santi DV, Norment A, Garrett CE (1984) Covalent bond formation between a DNA-cytosine methyltransferase and DNA containing 5-azacytosine. Proc Natl Acad Sci USA 81(22):6993–6997

Sasaki H, Matsui Y (2008) Epigenetic events in mammalian germ-cell development: reprogramming and beyond. Nat Rev Genet 9(2):129–140

Scandura JM, Boccuni P, Cammenga J et al (2002) Transcription factor fusions in acute leukemia: variations on a theme. Oncogene 21(21):3422–3444

Schroen B, Heymans S (2009) MicroRNAs and beyond: the heart reveals its treasures. Hypertension 54(6):1189–1194

Sebova K, Fridrichova I (2010) Epigenetic tools in potential anticancer therapy. Anticancer Drugs 21(6):565–577

Shi Y (2007) Histone lysine demethylases: emerging roles in development, physiology and disease. Nat Rev Genet 8(11):829–833

Silverman LR, Mufti GJ (2005) Methylation inhibitor therapy in the treatment of myelodysplastic syndrome. Nat Clin Pract Oncol 2(Suppl 1):S12–S23

Stamatoyannopoulos JA, Dunham I (2008) Epigenomics at the tipping point. Foreword. Pharmacogenomics 9(12):1781–1783

Sugars KL, Rubinsztein DC (2003) Transcriptional abnormalities in Huntington disease. Trends Genet 19(5): 233–238

Toyota M, Issa JP (1999) CpG island methylator phenotypes in aging and cancer. Semin Cancer Biol 9(5):349–357

Toyota M, Ohe-Toyota M, Ahuja N et al (2000) Distinct genetic profiles in colorectal tumors with or without the CpG island methylator phenotype. Proc Natl Acad Sci USA 97(2):710–715

Toyota M, Kopecky KJ, Toyota MO et al (2001) Methylation profiling in acute myeloid leukemia. Blood 97(9):2823–2829

Toyota M, Suzuki H, Yamashita T et al (2009) Cancer epigenomics: implications of DNA methylation in personalized cancer therapy. Cancer Sci 100(5):787–791

Tsankova NM, Berton O, Renthal W et al (2006) Sustained hippocampal chromatin regulation in a mouse model of depression and antidepressant action. Nat Neurosci 9(4):519–525

Uchida T, Kinoshita T, Nagai H et al (1997) Hypermethylation of the p15INK4B gene in myelodysplastic syndromes. Blood 90(4):1403–1409

Urdinguio RG, Sanchez-Mut JV, Esteller M (2009) Epigenetic mechanisms in neurological diseases: genes, syndromes, and therapies. Lancet Neurol 8(11): 1056–1072

Wood MA, Hawk JD, Abel T (2006) Combinatorial chromatin modifications and memory storage: a code for memory? Learn Mem 13(3):241–244

Worm J, Kirkin AF, Dzhandzhugazyan KN et al (2001) Methylation-dependent silencing of the reduced folate carrier gene in inherently methotrexate-resistant human breast cancer cells. J Biol Chem 276(43): 39990–40000

Xia ZB, Anderson M, Diaz MO et al (2003) MLL repression domain interacts with histone deacetylases, the polycomb group proteins HPC2 and BMI-1, and the corepressor C-terminal-binding protein. Proc Natl Acad Sci USA 100(14):8342–8347

Youssef EM, Lotan D, Issa JP et al (2004) Hypermethylation of the retinoic acid receptor-beta (2) gene in head and neck carcinogenesis. Clin Cancer Res 10(5):1733–1742

Yu AM (2009) Role of microRNAs in the regulation of drug metabolism and disposition. Expert Opin Drug Metab Toxicol 5(12):1513–1528

Yu A-M, Pan Y-Z (2012) Noncoding microRNAs: small RNAs play a big role in regulation of ADME? Acta Pharm Sin B 2(2):93–101

Yu L, Liu C, Vandeusen J et al (2005) Global assessment of promoter methylation in a mouse model of cancer identifies ID4 as a putative tumor-suppressor gene in human leukemia. Nat Genet 37(3):265–274

Zeleznik-Le NJ, Harden AM, Rowley JD (1994) 11q23 translocations split the “AT-hook” cruciform DNA-binding region and the transcriptional repression domain from the activation domain of the mixed-lineage leukemia (MLL) gene. Proc Natl Acad Sci USA 91(22):10610–10614

Zhang B, Farwell MA (2008) microRNAs: a new emerging class of players for disease diagnostics and gene therapy. J Cell Mol Med 12(1):3–21

Personalized Nutrition and Drug–Nutrient Interactions

5

Yousef I. Hassan

Abstract

Nutrition is the cornerstone of health. This chapter explores the relationship between dietary factors and the development of modern concepts in controlling and preventing chronic health issues such as tumors, diabetes, and obesity. While recent advances in molecular biology enhanced our understanding of the root causes of such diseases, future implementation of personalized-nutrition and medicine could generate dietary recommendations and interventions that will achieve optimum health standards and provide maximum disease prevention and control.

The conventional pharmacokinetics approach completely ignores environmental (such as intestinal microbial flora) or lifestyle factors (such as dietary choices) that influence many physiological parameters. These factors in fact are essential in dictating the bioavailability, metabolism, and degradation, hence the efficacy of most of the modern drugs. The alternative approach offered by modern nutrigenomics and pharmacogenomics, accounts for such variables in order to intervene in more predictable fashion with better prognosis outcomes.

1 Introduction

Early in the nineteenth and twentieth centuries, most of the scientific groundwork conducted in the field of nutrition was to elucidate nutritional factors, their chemical nature, dietary sources, upper and lower intake limits, and how to prevent deficiencies in both animals and human models (Zempleni et al. 2008). Tremendous advancements in the nutrition field were pioneered by early biochemists, biologists, and nutritionists leading to the transformation of this branch of knowledge into a solid evidence-based practice instead of the mysterious witchcraft-filled one that dominated once in the early history of mankind (Stover and Garza 2006). Most of the nutrition-related discoveries were later utilized by governmental agencies, regulatory bodies, and food processors and manufactures to enhance the quality of food products served to public or to eliminate actual or possible deficiencies of micronutrients such as

Y.I. Hassan (✉)
Division of Cancer Biology and Genetics,
Queen's University Cancer Research Institute,
Kingston, ON, Canada
e-mail: youhassan@yahoo.com

D. Barh et al. (eds.), *Omics for Personalized Medicine*,
DOI 10.1007/978-81-322-1184-6_5, © Springer India 2013

ascorbic acid, folic acid, and iron (Almeida and Cardoso 2010; Caudill et al. 1997). It can be said with confidence that most of the early nutritional recommendations were exclusively based on the upper and lower limits of intake and were established to match the appearance or disappearance of some diagnostic symptoms of nutrient deficiencies. More importantly, any recommended supplementation was designed to target the whole studied population without considering any differences in the nutritional needs among different subgroups within that population. While this approach proved to be successful in eradicating deficiencies in high-income countries and decreasing the reported numbers of such cases in low-income countries, it was proven to have some side effects related to the incorrectness of the imbedded assumption – which assumes that all individuals within a certain population are homogeneous in their need for nutrients – and that other possibly or even better scenarios should be explored (Vakili and Caudill 2007; Kaput 2008). The second stage of nutrition-field evolution came to light, after almost two decades of molecular biologists establishing the individualistic molecular differences among human beings. While we all share 99.99 % of our genome sequences with other human fellows and comrades around the globe, we still have our unique fingerprints that distinguish us from the rest. Signature differences such as single-nucleotide polymorphisms (SNPs) do exist between individuals even within the same family affecting how cells transcribe, translate, and use the genetic information in an amazing fashion (Cui and He 2005).

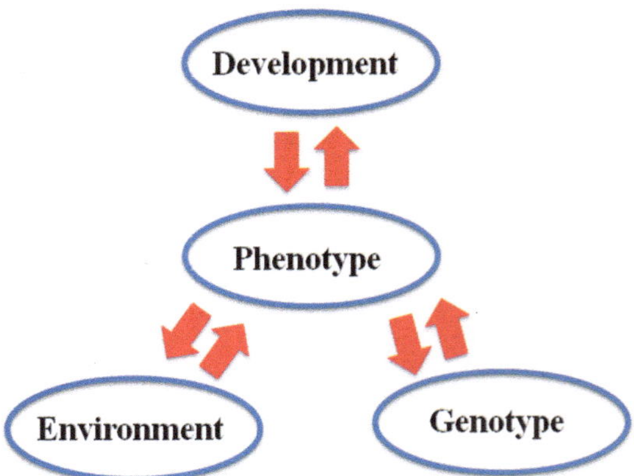

Fig. 5.1 Vibrant and dynamic relationships are found between genes and environment that influence organism's phenotype throughout early development and in later stages of life

Generally speaking, our phenotype (a collection of traits) is affected by our genes as well as by the surrounding environmental factors. Variations in any trait/phenotype are acceptably expressed as the result of the genotype and environment interactions ($P = G \times E$) (Hassan and Zempleni 2010). Historically, most of our understanding and research efforts went around the environmental factors. This was mainly due to the lack of proper research techniques and methods, which were needed to explore the genetic information variability. While these techniques were out of our grasps earlier, the picture is dramatically changing nowadays, and the latest developments of the twentieth and twenty-first centuries are enabling us to fully explore genetically and environmentally encoded factors. A better understanding of the relationship between environmental factors and genetic determinants is developing day by day with the help of such a large arsenal of molecular techniques available for researchers at very low costs. Scientist also started to recognize that phenotypes of living organisms reflect a dynamic and vibrant nature of interactions rather than a rigid and static one. Figure 5.1 depicts the interchangeability of the genetic and environmental factors influencing our phenotype and development. In essence, our genetic makeup defines our susceptibility to a disease or condition, but only the environmental factors (such as diet, sleep, and exercise) determine who among those with such genetic susceptibility will later develop the disease (Simopoulos 2010). In theory, by exploring and knowing our risk/susceptibility factors ahead of time and modifying and tailoring our lifestyles, it is possible to avoid or delay the onset of most of the chronic diseases that prevail in our modern societies.

This chapter will introduce the readers to the principles of diversity among humans and human populations from a nutritional prospective, the molecular and genetic bases of nutritional diversity, the principles of nutrigenetics and nutrigenomics, the importance of personalized-nutrition, and finally comment on the future practice of personalized nutrition.

2 The Diversity of Human Populations and Their Nutritional Condition

Deep understanding of how our bodies respond to nutrients intake has twofolds: first, those factors that are intrinsic to food itself. Example of the intrinsic factors include nutrient content of each serving/portion of the consumed food, nutrient bioavailability, presence of anti-nutritional factors, and how the food was stored, handled, and prepared. These intrinsic factors influence the final dose that we get from that specific nutrient.

A second set of factors that affect our exposure levels to nutrients are imbedded in our ability to digest, absorb, and utilize specific nutrients at the cellular level. The way our cells express different receptors, transports, and hormones shapes the fashion of how we respond to our diet. The nutritional needs of humans differ from one individual to another (based on their genomic makeup), growth stage (childhood, adulthood, etc.), and physiological state (pregnancy, lactation, etc.). In essence, all these biological factors together dictate the ultimate needs of different nutrients and the required balance among these nutrients during consumption.

In reality, not only biology influences our diet or how we respond to nutrients. Behavioral and socio-economical factors do in a similar way. Our lifestyle, food prices, seasonal availability, personal perception, taste, sedentary behavior, meal frequency, sleep cycle alteration, temporal variation, and religion all together influence our dietary choices and intakes adding what might be considered as a third level of variance and complexity to how individuals utilize and respond to food. Furthermore, some exogenous and random factors such as exposure to sunlight, toxins, allergens, or bacterial inoculum may exert effects on how individuals respond to dietary components (Fay and German 2008).

Diet does not only shape our short life spans contributing to our health and disease status but also forms a major determinant of how we – as humans – have evolved. During the evolution and emergence of complex life forms, food and nutrients stood on the top of environmental-pressures list (Hinde and German 2012). Recent advances in understanding the role of nutrition in human genome evolution suggest that part of the genetic variations among humans is caused by adaptation to the available diet (Ye and Gu 2011). In fact, modern findings are showing that even variables such as taste – which we did not have any idea about its biological base until the recent past – can now be tracked and attributed to specific epigenetic elements that are believed to form the foundation of how environmental factors impact our cells and their physiological status. Molecular and genetic bases of environmental and nutritional diversity are discussed later in this chapter.

From the evolutionary point of view, nutritional sciences, in general, study only short-term regulatory responses (spanning months to years) that modulate the state of organisms in relation to the environment and nutrient homeostasis. On the other hand, nutritional evolution focuses on the longer-term developmental responses trying to detect any accumulative genetic changes that take place in a Darwinian fashion. Searching and archiving nutritional factors that affect our evolution has an immense practical implication on our strive for better health. For example, and as life expectancies and aging populations are on the rise in our modern societies, tailored and age-specific nutritional interventions might be advantageous in our efforts to eliminate or reduce the numbers of age-related chronic diseases. Yet the data obtained by most of the current studies in this regard is lacking a mechanistic approach of how this could be possibly achieved. One recent analysis of such research shortcomings attributed our lack of success to the failure of understanding the evolutionary context of aging and longevity (McDonald and Ruhe 2011). Current protocols and research designs neglect to acknowledge that longevity is only a by-product for genes that were merely intended to make organisms survive until the age of reproduction and that they are not products of random evolution/selective pressure that is supposed to start in early years of active reproduction. Hence, comparing differences between young (control/normal) groups with old (abnormal/experimental) groups is faulty premise by itself. To ratify such faulty premises, more and more efforts are being invested in order to identify human-specific selection events that distinguished human from nonhuman primates in addition to local adaptation events that contributed to human diversity. Significant interdisciplinary efforts with the help of anthropologists, social, and environmental scientists are aiming nowadays at elucidating the underlying mechanisms behind such differences and divergences.

Finally, as humans adapt to their environment and evolve, it will be very interesting to see the impact of our current energetically dense and easily digestible diet on future generations (Hinde and German 2012). Only limited a number of generations have been subjected to such high caloric abundance leading to the excessive weight gain that we witness nowadays in our societies. Consequences of such nutrient overconsumption are subject of concern from the human evolutionary context. Unfortunately, it is predicted that such practices are more likely to have a severely adverse nature.

3 Molecular and Genetic Bases of Nutritional Diversity

Nutrition is an environmental factor of paramount importance. Despite the early documented facts of nutrients being major regulators of our physiological status (glucose is one of the best examples that jump into mind here), scientists did not have clear and well-established mechanisms of how nutrients (such as vitamin, minerals, fatty acids) influence such large array of biological processes. These processes span gene expression, protein translation and trafficking, and enzyme functions. Figure 5.2 sets examples of vitamins that are involved in some important biological processes. It was only lately when we started discovering the bases of how food constitutions have the capacity to activate/inactivate gene expression, influence mRNA translation, and impact protein folding. For instance, most of the vitamins were thought earlier to be merely as enzyme cofactors that aid to accomplish one or a limited number of functions usually taking place in the cytosol. Eventually, it was discovered that

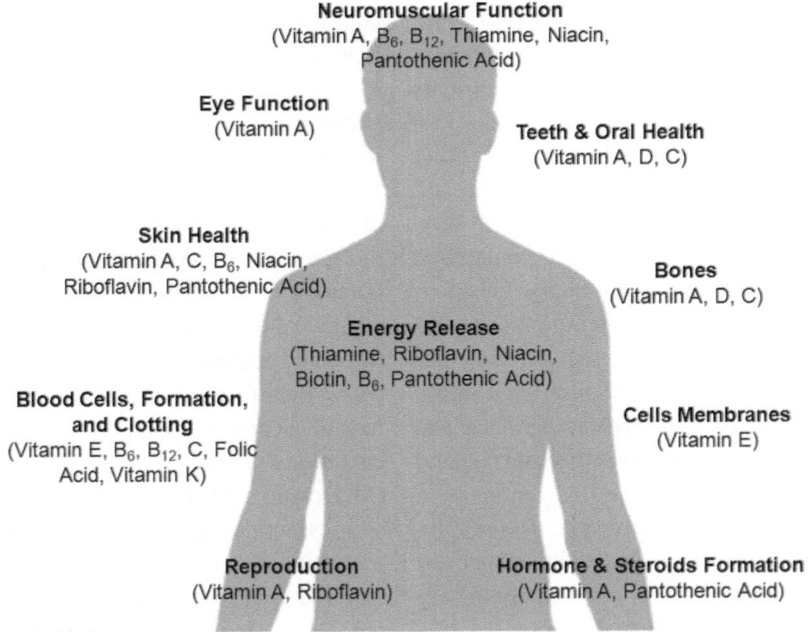

Fig. 5.2 Micronutrients and vitamins are involved in vital biological processes spanning functions in the neuromuscular system to reproduction

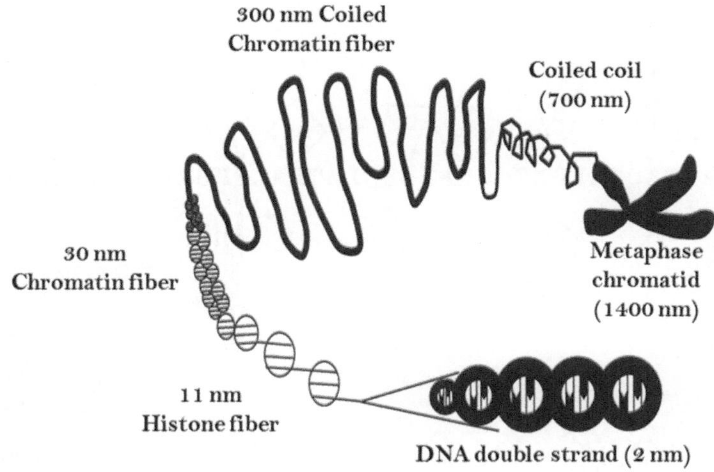

Fig. 5.3 Different classes of histones help DNA condensation and the assembly of the chromatin fiber inside cell's nucleus

some vitamins have a global influence on gene transcription and genome stability (such as vitamin A and folic acid). Most of the current reported mechanisms of nutrient-action on gene expression and protein translation imply the involvement of DNA, histones, or both. Histones are basic proteins that associate with DNA in the nucleus of the cell (Bustin et al. 1969). They are rich in basic amino acids (around 20 % of their amino acid compositions are arginines and lysines). Electrostatic interactions between the positive charges of these amino acids and the negative charges of phosphate groups of the DNA backbone mediate packaging of DNA into chromatin. Two copies of each of the four core histones (H2A, H2B, H3, and H4) form an octamer (Bottomley 2004). DNA (around 146 base pairs) is wrapped around this octamer to produce a nucleosomal core particle. The second level of organizing chromatin comes with what is known as a linker region, where 20–60 bp of DNA links one nucleosome to another. Each linker region is occupied by a single molecule of histone H1, giving a "beads on a string" appearance (Cheung et al. 2000). This 11-nm histone fiber is then further packed into an irregular 30-nm chromatin fiber structure that is coiled into even more complex structures, which eventually form the chromosome (Fig. 5.3). Binding of histones to DNA does not depend on a particular nucleotide sequence in the DNA, but it does critically depend on the amino acid sequence of histones.

Binding of transcription factors to gene promoters may be inhibited if the promoter is blocked by a nucleosome and transcription is usually associated with sliding nucleosomes along the DNA molecule, exposing the gene's promoter so that the transcription factors can access that region. Transcription of protein-coding genes is carried out by RNA polymerase II (RNAP II). In order for the polymerase to travel along the DNA, a complex of proteins removes the nucleosomes in front of RNAP II and then replaces them after RNAP II has transcribed the sequence. This removal of nucleosomes in front of RNA polymerases and putting them back after transcription is completed, is known as a "chromatin-remodeling" event.

Around 75–80 % of the histone's amino acids make up the core, and only the N-terminal tails of histones protrude from the nucleosomal surface. Through modifications of the side chains of different amino acids on the N-terminus (exposed side of the nucleosomes) of histones, the chromatin structure can be controlled and part of the DNA sequence can be actively expressed or silenced. Chemical modifications of amino acid residues in histones include the covalent attachment of acetyl groups (CH_3CO–) to lysines, phosphate groups to serines and threonines, methyl groups to lysines and arginines, biotin to lysine groups, ubiquitinylation and sumoylation of lysine residues, and poly-ADP-ribosylation of glutamic or aspartic acid residues. Most of these

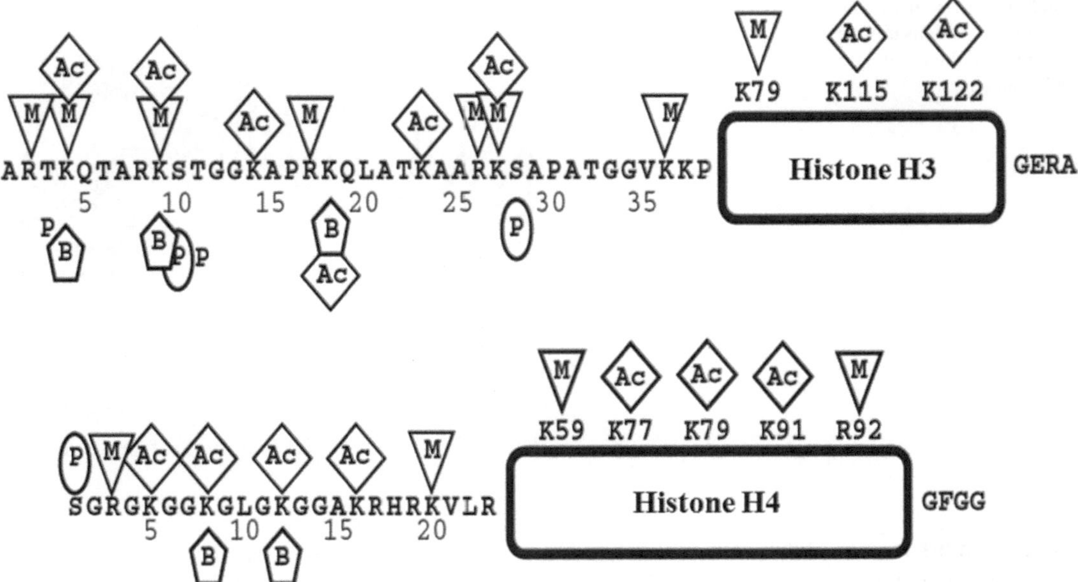

Fig. 5.4 Simplified map for some of the well-established modifications of N-termini of both histone H4 and H3 in mammalian cells (*P* phosphorylation, *Ac* acetylation, *M* methylation, and *B* biotinylation)

modifications occur on H3 and H4 histone tails (Fig. 5.4). The most important feature of these modifications is that they are reversible. For example, acetyl groups are added by enzymes called histone acetyltransferases (HATs) and removed by histone deacetylases (HDACs) (Allfrey et al. 1964). Acetylation of histone tails occurs in regions of chromatin that become active in gene transcription. Adding acetyl groups neutralizes the positive charges on lysines, thus reducing the strength of the association between the negatively charged DNA and the positively charged histones. It is now clear that histones are a dynamic component of chromatin and not simply inert DNA-packing material. All of these modifications are part of what is known as the "histone code."

Chromatin remodeling and epigenetic changes are implicated in gene expression regulation, mammalian cell development, embryogenesis, and environmental origins of chronic diseases such as diabetes mellitus and cancer (Chan et al. 2010; Oommen et al. 2005; Finnell et al. 2008; Ho and Zempleni 2009; Swanson et al. 2009; Wadhwa et al. 2009). It is now accepted that these genome and epigenome wide changes can alter gene expression levels, fix, or determine future set-points of chronic diseases with lifelong consequences (Arai and Kanai 2010; Bird 2007; Aagaard-Tillery et al. 2008).

The second level of epigenetic changes that connects nutrients to gene expression takes place at the DNA level through CpG islands. "CpG" simply stands for a cytosine "C" and guanine "G" connected by a phosphodiester bond "p." CpG islands are repetitive elements found in the promoter region of most genes (60 % of human genes have CpG islands) (Bird 1986; Antequera and Bird 1993). Cytosine in these regions can be modified by methyl group attachment ($-CH_3$) to both DNA strands rendering the genes inactive (known as hyper-methylated). CpG islands are one of the best examples where DNA methylation plays an important role in regulation of gene expression. For example, most of the changes that take place in folate deficiency point out to the involvement of this water-soluble vitamin in epigenetic changes taking place at DNA and histone levels through a methylation process. Folate is responsible for a sustained supply of methyl groups for histone and DNA methyltransferases. In other words, it is required for activating genes that need

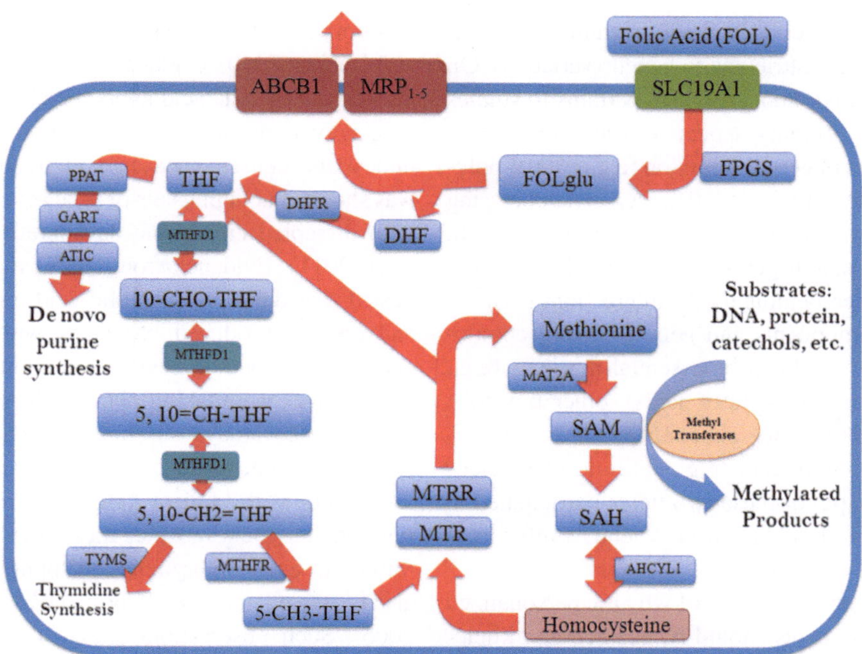

Fig. 5.5 Folate cellular cycle and folate-dependent modification of DNA and histones (*SLC19A1* solute carrier family 19 member 1, *FPGS* folylpolyglutamate synthase, *FOLglu* polyglutamylated folic acid, *DHF* dihydrofolate, *DHFR* dihydrofolate reductase, *THF* tetrahydrofolic acid, *MTHFD1* methylenetetrahydrofolate dehydrogenase, *MTHFR* 5,10-methylenetetrahydrofolate reductase, *MTR* methionine synthase, *MTRR* methionine synthase reductase, *MAT2A* methionine adenosyltransferase 2A, *SAM* S-adenosylmethionine, *SAH* S-adenosylhomocysteine, *AHCYL1* S-adenosylhomocysteine hydrolase-like 1, *TYMS* thymidylate synthase, *PPAT* phosphoribosylpyrophosphate amidotransferase, *GART* glycinamide ribonucleotide formyltransferase, *ATIC* aminoimidazole carboxamide ribonucleotide transformylase/inosine monophosphate cyclohydrolase, *MRP* multidrug resistance protein, *ABCB* ATP-binding cassette subfamily B)

to be active and for silencing genes that need to be silenced. S-adenosylmethionine (SAM) is the universal cellular source of methyl groups for DNA and histone methyltransferases, and it is provided through the folate and methionine pathways. These pathways are dependent on dietary sources of folic acid, choline, methionine, and vitamin B_{12} (Fig. 5.5). Low levels of DNA methylation are associated with active gene transcription, while hyper-methylation is associated with gene silencing. On the other hand, histone methylation, which neutralizes the charge on lysines (and arginines), can either stimulate or inhibit gene transcription in the same region. Methylation of lysine-4 in H3 is associated with active genes, while methylation of lysine-9 in H3 is associated with inactive genes (Bharathy and Taneja 2012; Ea et al. 2012).

Understanding how vitamins – and nutrients in general – regulate gene expression takes a pivotal importance in any recommended nutritional intervention that might be useful in preventing or fighting chronic diseases. In this section the reader was introduced to a mechanistic explanation of nutrient actions at the cellular levels and the diversity established by such mechanisms. The practical importance of such mechanistic understanding will be more clear at the end of this chapter.

4 Nutrigenetics, Nutrigenomics, and the Importance of Personalized Nutrition

Nutrigenetics correlates to the development of concepts, practices, and research on the genetic variation in response to dietary intake (e.g., individuals responding differently to the same diet such as attaining different levels of serum cholesterol/blood pressure attributed to genetic variation). On the other hand, nutrigenomics pertains to studies on the evolutionary aspects of diet and the role of nutrients in gene expression (e.g., role of polyunsaturated fatty acids (PUFA) in suppressing fatty acid synthase (mRNA) gene expression). In other words, nutrigenomics is defined as a mechanistic understanding for the interactions between dietary components and genes. Dietary components of interest can be essential nutrients (e.g., vitamins, minerals, fatty acids), other bioactive substances (e.g., phytochemicals), or metabolites of food components (e.g., retinoic acid, eicosanoids). Both of these two fields participated in creating and branding what scientists refer to as "personalized-nutrition."

A very good example of why the personalized nutrition concepts should be adapted and implemented in nutritional interventions as soon as feasible is set by folic acid (also known as B_9). This water-soluble vitamin is found in various food types including liver, orange juice, strawberries, dark green leafy vegetables, peanuts, and dried beans (especially black and kidney beans). Moreover, a large number of food products available in the market nowadays, such as infant formulas, meal replacement, nutritional bars, ready-to-eat breakfast cereals, and enriched cereal grains, are fortified to provide levels of 90–110 μg of folic acid per serving.

Folate mandatory fortification was established as early as 1998 in order to reduce incidences of neural tube defect (NTD), a preventable disease composed of a spectrum of congenital abnormalities arising from the failure of normal fusion of the fetal neural tube. NTD has detrimental effects on pregnancy outcomes leading to pregnancy termination, miscarriage, neonatal death, and lifelong disabilities (Almeida and Cardoso 2010; Lin and Birmingham 2010). Folate as a vitamin was given a lot of attention in the last two decades, and much of the data collected show its involvement as a biofactor in the epigenetic events taking place at both the DNA and histone levels (Lucock 2006; Stover and Garza 2006; Boyles et al. 2008; Finnell et al. 2008).

Even though folate mandatory supplementation proved to reduce incidences of *spina bifida* in modern populations (Alasfoor et al. 2010; Amarin and Obeidat 2010), evidence is accumulating that high intake of folic acid alters the phenotype and epigenotype of individuals which may lead to undesirable consequences later in life. Recently, it was shown that high levels of folic acid correlated with disrupted embryonic development (Pickell et al. 2011). Evidence from rodent models suggested that diets high in folic acid, methionine, and choline produced progenies with different coat colors or with kinked tails implying that certain changes were taking place at the epigenome level. A recent study even suggested that too much folic acid might result in nervous tissue damage associated with autism (Beard et al. 2011). Further, it is believed that these effects are not limited to early embryonic development but can also be passed onto the offspring (Waterland et al. 2006). Such observations raised a very critical question: Are we harming parts of our population by forcing them to consume too much folate?

Folate homeostasis and sensing is part of a complicated and tightly regulated process. Cells try to coordinate cytoplasmic folate concentration with chromatin-remodeling events to a very large degree. Major problems can appear when we flood our cellular systems with folate, choline, methionine, and vitamin B_{12}. With increased availability of S-adenosylmethionine (SAM) as a methyl donor for both DNA and histone methyltransferases, this could lead to silencing of genes that should always be active (such as proto-oncogene) or activating genes that always need to be inactive (oncogene). Luckily, mammalian cells possess mechanisms to reduce adverse effects of high intracellular concentrations of folate by effluxing it back to the extracellular space. The nature of the effluxing system is not clear yet, but it involves a family of ATP-dependent transporters. These transports are also known for their ability to transport different groups of drugs in a mechanism that principally evolved to protect cells from toxicity. At least 12 members of the multidrug resistance protein (MRP) family are reported to exist. These proteins are clinically relevant as they can efflux methotrexate, a folate analog that is commonly used as a genotoxic agent during cancer therapy. Generally speaking, the information available about how these effluxing proteins function in regard to folate homeostasis is very scarce. Further research is required to establish a correlation between this defense mechanism, folate dietary supplementation, and epigenetic markers of healthy and non-healthy cells.

Folic acid supplementation in order to prevent NTD is highly recommended, and it is obvious that the benefits of such practice outweigh the risks connected with it. At the time of conception and early pregnancy, there is no doubt that using and prescribing folate supplementation is acceptable. On the other hand, the questions that need to be answered sooner than later are as follows: What about the other subgroups of populations that are already fulfilling their daily folate requirements through dietary intake and then getting additional amounts through other enriched products? Does the hidden risk of folate fortification outweigh the foreseen benefits in these cases? Are we predisposing large numbers of people to factors that lead to chronic health issues later in life? Athletes, for example, tend to consume well above the Recommended Dietary Allowance (RDA) for most micronutrients either through diet or supplements. In some extreme cases, a simultaneous use of ten different supplements was reported, resulting in excess intake above tolerable upper limits of folate, vitamin E, and zinc (Carlsohn et al. 2011). What about the long-term effect of such practices combined with excess intake through dietary sources? What about combining this with faulty practices of labeling or manufacturing? One recent study examined folate content in 95 products that would be considered as sources of folate in Canadian markets and concluded that the analyzed folate values were higher than those reported on labels for "breads," "rolls and buns," and "ready-to-eat cereals" in 141, 118, and 237 %, respectively (Shakur et al. 2009).

It is extremely difficult to suggest a simple solution at least on the population level for such a complicated issue that fits all groups. More efforts and resources should be invested in order to scrutinize whether mandatory folate supplementation is a 100 % successful intervention at the population level or not. A much more realistic way to overcome the above disadvantages of folate supplementation would be focusing on adapting the personalized nutrition concept. For granted it would not be an easy journey, and most likely such adaptation will take time and efforts in order to shape this field and establish it in the clinical practice. New disciplines such as nutrigenomics, proteomics, and metabolomics will contribute to the rapid development of personalized nutrition practice. Proteomics tools will help us to study the full set of proteins encoded and expressed by a particular genome (belongs either to a model organism or a particular individual). Speaking at large, proteomics will identify a vast number of proteins in these organisms, map their interactions, and analyze their biological activities. On the other hand, metabolomics and metabolite-profiling will aid in measuring the real outcome of the potential changes suggested by genomics and proteomics. Metabolomics will investigate the physiological regulation and metabolic fluxes of individual cells or cell types. The full power and capacity in addition to the integration of all these fields will highlight the minor differences between individuals, between different physiological states, and more importantly between health and disease.

At a simplified level, nutrigenomics will be helpful in elucidating how certain dietary components affect protein profile of individuals; proteomics will give ideas about how altered protein profiles affect biological systems of such individuals; and finally metabolomics will show the cellular response to such changes. Overall, all these tools will pour in the direction of understanding the direct and indirect influence of nutrients on the physiological process (Fig. 5.6) and utilizing them in the personalized-nutrition approach.

5 Single-Nucleotide Polymorphisms (SNPs) in Nutrition and Drug Metabolism

The interaction between pharmacology and nutrition fields is in steady increase. Scientists on both sides are recognizing that simple solutions such as one disease–one target–one drug will not provide much success on the longer term in our battle against chronic diseases. But rather compound combinations

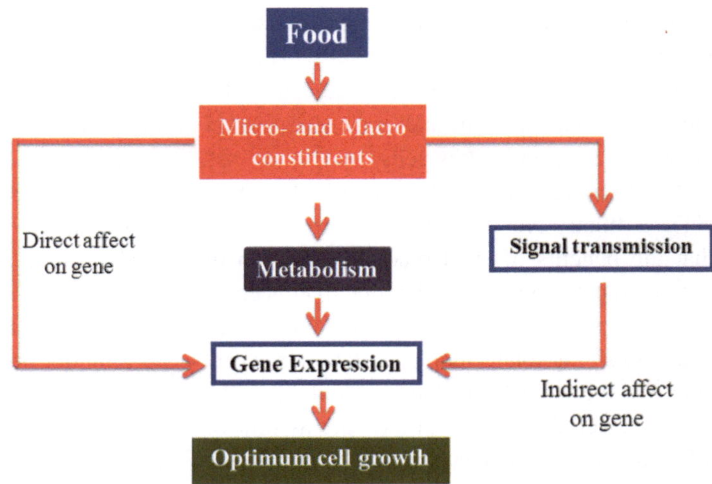

Fig. 5.6 Diet and food constituents can influence mammalian gene expression levels in more than one route (direct or indirect effects through cellular-signaling pathways)

of drugs/nutrients that involve multi-targets and affect multi-pathways are required in order to overcome such complicated conditions.

Generally speaking, single-nucleotide polymorphisms (SNPs) not only affect enzyme activities or protein functions leading to varied conversion rates of substrates (either nutrients or medication) and metabolites in the body but can also influence chemical and hormonal receptors leading to enhanced or attenuated signaling (Touhara 2002; Hasin-Brumshtein et al. 2009). This phenomenon is very important in understanding human behavior when it comes to the process of making dietary choices where factor such as taste and smell plays a pivotal role. McRae et al. (2012) reported a SNP that resides within the *OR2J3* odorant receptor which can influence individual's ability to detect the "freshly cut grass" smell. In this study an amino acid substitution, T113A, abolished the ability of OR2J3 receptor to respond to the chemical compound (namely, cis-3-hexen-1-ol) responsible for this odor. Later an extensive survey of genetic variations in human populations, found that the haplotypes and allele-frequencies of different variants of OR2J3 differ among populations around the globe. African populations (including African-Americans) for example have a relatively high proportion of the T113A nonresponsive haplotype. This study was among the few that explored the contributions of genotype variation to the human olfaction variation within different human societies.

Earlier studies tracked how individuals differ in their perception to the same chemical. For example, androstenone, a steroidal odorous material, is variously perceived by different individuals spanning from either being an offensive "sweaty and urinous" to pleasant "sweet and floral" to "odorless" (Wysocki and Beauchamp 1984; Bremner et al. 2003; Keller et al. 2007). In this case, the human odorant receptor OR7D4 was found to be responsible for this variation in odor perception. The receptor contains two non-synonymous SNPs resulting in two amino acid substitutions (R88W and T133M). Human subjects with RT/WM or WM/WM genotypes were found to be less sensitive to androstenone and rated the odor to be less unpleasant compared to subjects with the RT/RT genotype (Keller et al. 2007). This study was able to successfully link a genotypic variation to intensity variance in odor perception.

What is true for the smell is true also for the taste. SNPs affect most of our perception of taste (including but not limited to bitterness, sweetness, and umami). The variation in the perceived bitterness was linked to making food preferences over 50 years ago (Fischer et al. 1961, 1963). Evolutionary speaking, the variation in bitterness perception among populations reflects a local adaptation to the avoidance of plant toxins found in a particular geographic environment where that population inhibited (Kim et al. 2005). Among the 25 known human bitterness receptors and their respective genes (TAS2Rs), Hayes et al.

(2011) recently studied the allelic variation of some of these genes and showed their influence on responding to bitter compounds found in the food supply. A single-nucleotide polymorphism at the TAS2R19 receptor was connected with decreased liking of grapefruit juice and increased sensation of bitterness. An amino acid change from the less sensitive Arg299 to the more sensitive Cys299 explained the reported observation. Furthermore, variability in espresso-coffee bitterness was associated with a haplo-block containing TAS2R3, TAS2R4, and TAS2R5 genes.

Glutamate, mainly found in protein-rich foods (such as meat and cheese) and frequently used as a flavor enhancer directly added to selected cuisines, is considered as umami taste perception stimulus. Most recently, variations in human glutamate umami taste perception were correlated with variations in the human TAS1R3 gene (Chen et al. 2009). A subset of human subjects displayed extreme sensitivity (doubling of umami ratings) was mapped to multiple single-nucleotide polymorphisms – R757C, A5T, and R247H – in TAS1R3 receptor.

The above examples of single-nucleotide polymorphisms explain how we as humans perceive environmental factors differently and based on this perception we pick our dietary choices. The map of human SNPs is a good candidate for growing and exploration, and thousands of SNPs will be definitely added before we completely comprehend our dietary behavior under the lights of molecular biology concepts. But once such understanding crystallizes, it will revolutionize our approach for attaining better health standards with better dietary guidelines and better tools to tackle problems related to nutritional choices. In a simplified example, if a group of people have the natural tendency to avoid food “X”, just because they have a genetic disposition that intensifies the taste or odor of one constituent of that food, then this food could be replaced with another one that does not contain that particular constituent and does not produce the same adverse perception, yet still contains all the beneficiary nutrients.

Absorption and conversion rates of drugs are generally influenced by the chemistry of that particular drug, physiological status of the patient and his genetic map, and finally the environmental factors that influence the drug at the time of administration (Murray 2012). While pharmaceutical companies do their best in optimizing the chemical nature of most drugs by enhancing stability, solubility, and absorption; little can be done in regard to patient’s characteristics. What might be considered as an optimum dose for one patient with normal metabolic rates can be toxic for another patient with reduced or slow metabolic rates or can be ineffective in individuals privileged with high clearing and metabolizing rates. For that reason understanding how a drug will behave in all exposed subgroups is central. As it is the case of nutrients, drug metabolism is essentially influenced by single-nucleotide polymorphisms also. A new branch of knowledge known as pharmacogenomics is interested in answering some of these concerns related to drug efficacy in different populations. This field is involved in studying the genetic basis for the difference noticed between individuals in response to drug administration and in tailoring drug prescriptions to match individual genotypes. It uses genetic maps containing the most recent SNPs to distinguish between patients who might benefit from a particular drug and those who might not (Verschuren et al. 2012; Voora and Ginsburg 2012). One good example of such a case, is the recent study that was conducted in China correlating genetic polymorphisms of carboxylesterase 1 gene (CES1) to the susceptibility of antituberculosis drug-induced hepatotoxicity (Wu et al. 2012). Different genetic polymorphisms of CES1 were tracked in 473 tuberculosis patients with or without hepatotoxicity after antituberculosis chemotherapy. The results indicated significant associations between some single-nucleotide polymorphisms – such as rs8192950 AC genotype and rs1968753 GG genotype – with antituberculosis drug-induced hepatotoxicity. As tracking of these genotypes is as simple as running an array of polymerase chain reactions (PCRs), such drug–genotype correlations will prompt the future screening of any patient being treated from tuberculosis – for an extended time – to predict and prevent side effects.

Drug–nutrient interaction is another factor that is being scrutinized closely for the involvement of SNPs. Medications interact with foods and nutrients in several and diverse ways. Medications in some cases can decrease appetite or change the way a nutrient is being absorbed, metabolized, or excreted. On the other hand, a food–drug interaction is the effect of food (or a nutrient in that specific food) on a medication. Dietary nutrients can affect medications by altering their absorption or metabolism. The food could make the medications work faster or slower or even prevent it from being absorbed at all (Rodriguez-Fragoso et al. 2011; Boullata and Hudson 2012; Yasuji et al. 2012). Such interactions raise concerns that medications may lead to nutritional deficiencies or that diet may change how a medication works. The pronounced effect might range from a minor chance of adverse nutritional effects – especially if the medication is being used for a short time, such as ten days treatment – to a complicated and serious nutritional deficiency when some medications are used for months or years (Hines and Murphy 2011; Bartal et al. 2012). Children, older adults, pregnant women, people who are poorly nourished, and people with a chronic disease are at greater risk of medications affecting their nutritional health. In such cases, dietary changes to include more rich foods (in vitamins and minerals) are highly preferred (Ruggiero et al. 2012).

Last but not the least and since clinics around the world became aware of the overlapping issues between drugs and nutrients and in order to improve the therapeutic outcomes of their practices, a couple of comprehensive databases were assembled aiming at making the access of drug/nutrient interaction-related information easier and more reliable. Two databases to be noted over here are the PharmGKB (http://www.pharmgkb.org) and DrugBank (http://www.drugbank.ca). The PharmGKB database (Fig. 5.7) contains information related to more than 20,000 genes, 3,000 diseases (spanning cardiovascular, pulmonary, cancer, pathways, metabolic, and transporter domains), and 2,500 drugs in a searchable format. More than 460 genetic variants and SNPs that affect drug metabolism have been deposited in this database. Indeed, this database is foreseen as the foundation of a future central database which will encompass the genetic, genomic, molecular, and cellular phenotypes and any clinical information related to pharmacogenomics studies. On the other hand, the DrugBank database relates more to the bioinformatics and cheminformatics of drugs. It details drug's chemical, pharmacological, and pharmaceutical data with comprehensive drug-targets (such as protein sequence, protein structure, and metabolic pathways). To date, this database contains more than 4,800 drug entries. In addition, DrugBank contains extensive SNP-drug data that is useful for pharmacogenomics studies.

Over the past few years, an extensive amount of research has been conducted on the molecular characterization of target proteins in individuals and populations, leading to a clear understanding of the physiological significance of a number of these proteins and any single-nucleotide polymorphisms affecting them. Increasing evidence suggests that altered drug–nutrient interactions result from changes in activity of these proteins affected by those SNPs which differ from the interactions noticed among the general population. Accordingly, evaluation of the clinical relevance of such SNPs/interactions should be an integral part of risk assessment during drug development and commercialisation processes (Morlighem et al. 2011; Ovsyannikova and Poland 2011; Almal and Padh 2012).

6 Personalized Nutrition in Practice

The ultimate goal of personalized nutrition is to tweak the diet and adjust its components to match the personal needs and preferences of individuals based on their genetic makeups (Ganesh and Hettiarachchy 2012; Rubio-Aliaga et al. 2012). Some of the hurdles that we are facing nowadays in implementing such a concept are technical in nature. For example, the cost and time involved in sequencing whole genomes of individuals is still considered a daunting task (Almasy 2012; Peters et al. 2012). This is despite the technological advancements that we witnessed during the last three decades in this

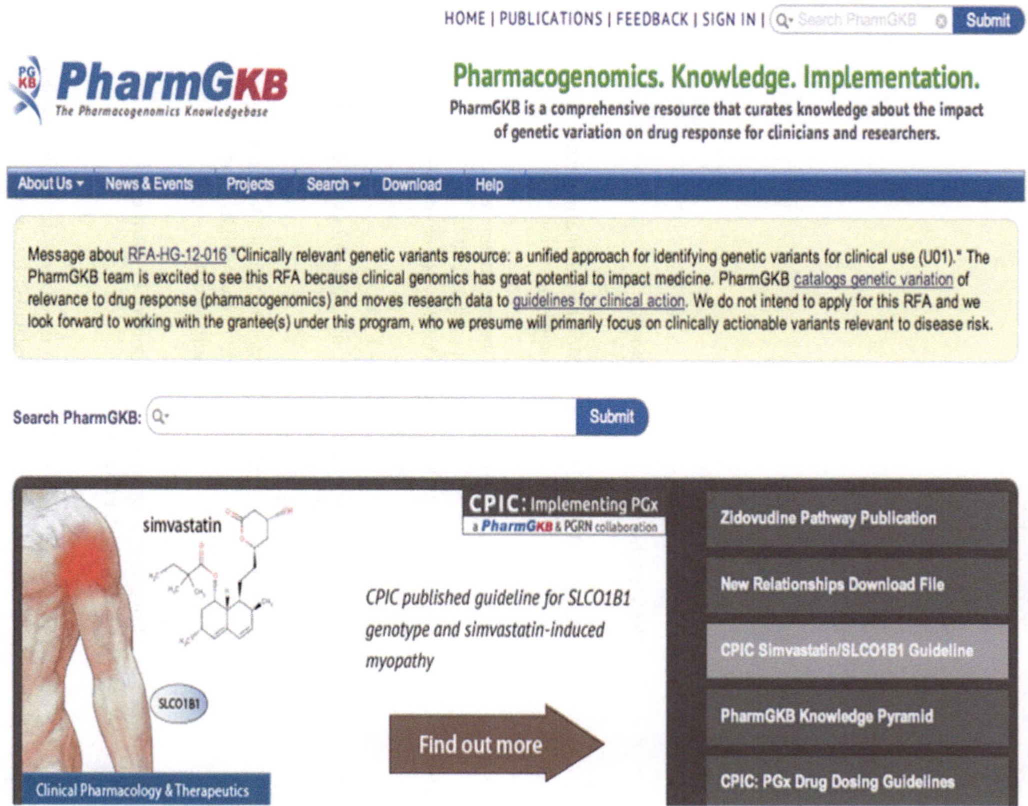

Fig. 5.7 The PharmGKB database contains information related to more than 20,000 genes, 3,000 diseases, and 2,500 drugs in a searchable format

regard, leading to the drop of sequencing-costs of model organisms from thousands of dollars per organism to less than hundreds of dollars range. For example, in a wide range of cancers it was reported that individualistic and genome-wide RNA expression profiling can be used as a powerful tool with a valuable prognostic and predictive value that could help in deciding the appropriate course of treatment. Both World Health Organization (WHO) and US Food and Drug Administration (FDA) adopted two RNA expression classifiers for breast cancer patients. Unfortunately and due to the costs and technical complexity of such tools, they did not reach the clinical implementation phase yet (Smeets et al. 2011). In reality, improving the existing sequencing methods to yield reliable reads in a feasible time frame might take some times (Donley et al. 2012; Roberts et al. 2012). This is true especially when we are talking about affordable costs that will cover the sequencing of entire genomes for individuals/patients.

Another catch for the personalized nutrition challenge comes from the goal that we are trying to achieve by itself. "health" and "healthy" are vague definitions, which indicate in most of the cases "the absence of a sickness/disease and its characteristic symptoms." When it comes to measure "individual's/patient's healthy state," we are still lacking meaningful markers that reflect how the optimum state should look like. Yes we do have so many biological parameters (such as cholesterol levels, HDL, LDL, glucose) with lower and higher numerals, but these usually tend to have wide range of values. In addition and due to the inconsistent genetic variation among individuals even within the same population and under the same environmental factors,

Table 5.1 Examples of monogenic and polygenic diseases that are affected by SNPs, nutrients, and gene–nutrient interaction

Single-nucleotide polymorphism (SNP)	Gene	Effect	References	Notes
(a) *Cardiovascular diseases* (*CVD*)				
rs964184, rs2108622, and rs7834588	ZNF259, CYP4F2, and NKAIN3	Influence serum α-tocopherol concentrations after vitamin E supplementation	Major et al. (2012)	Monitor long-term vitamin E supplementation
rs12676	Choline dehydrogenase	Increased susceptibility to dietary choline deficiency	Johnson et al. (2012)	Observe choline status and may be use supplementation
rs993609	FTO	Risk of being overweight/obese	Frayling et al. (2007)	Monitor carbohydrates intake
rs429358 and rs7412	ApoE	Higher plasma concentration of LDL-C	Minihane et al. (2007)	–
rs1801282	PPAR-γ	Increased adiposity and insulin resistance	Gonzalez Sanchez et al. (2002), Knouff and Auwerx (2004)	Limit high carbohydrates/fat intakes
rs2943634	Noncoding region of chromosome 2q36.3	Ischemic stroke	Arregui et al. (2012)	Monitor plasma levels of high-density lipoprotein (HDL), cholesterol, and adiponectin
rs28362491	NF-κB	Inflammatory bowel disease	Andersen et al. (2011)	–
rs533556	ApoAI	Higher cholesterol levels in response to PUFA intakes	Lovegrove and Gitau (2008), Engler (2009)	Control PUFA intake
(b) *Neurological disorders*				
rs111033744	Galactose-1-phosphate uridylyltransferase (GALT)	Galactosemia	Tyfield et al. (1999), Singh et al. (2011)	A monogenic disease treated with galactose-free diet

rs5030853	Phenylalanine hydroxylase (PAH)	Phenylketonuria	Mutch et al. (2005), Mutze et al. (2012)	A monogenic disease treated with phenylalanine-restricted tyrosine-supplemented diet
rs4895171	Insulin-degrading enzyme (IDE)	Late-onset sporadic Alzheimer disease	Bian et al. (2004)	–
(c) *Neoplastic diseases*				
rs249954 and rs120963	PALB2	An increase of breast cancer risk	Chen et al. (2008)	–
rs1801133 and rs1801131	Methylenetetrahydrofolate reductase (MTHFR)	Increased postmenopausal breast cancer risk	Ericson et al. (2009)	Pay attention to plasma folate concentrations
rs297299, rs3797310, rs12055266, rs7579, and rs3805435	SEPP1	Increased risk of colorectal cancer	Peters et al. (2008)	Monitor selenium status and increase consumption of foods rich in selenium
rs34713741	SELS	Elevated colorectal and gastric cancer risks	Sutherland et al. (2010)	Boost intake of foods rich in selenium
rs1230025	Alcohol dehydrogenase, ADH1	Increased risk of gastric cancer	Duell et al. (2012)	Reduce alcohol consumption
(d) *Metabolic disorders*				
rs363717	ABCA1	Hyper absorption of cholesterol and hypoalphalipoproteinemia	Bodzioch et al. (1999), Mutch et al. (2005)	–
rs11674694 and rs11678405	MAP4K4	Associated with type 2 diabetes	Sartorius et al. (2012)	–
rs4068	ADIPQO	Metabolic syndrome	Khabour et al. (2012)	–
rs3743462	NR2F2	Influence whole-body insulin sensitivity	Boutant et al. (2012)	–
rs1137100	Leptin receptor (LEPR)	Obesity and type 2 diabetes	Park et al. (2006)	–
rs2270188	CAV2	Type 2 diabetes	Fisher et al. (2011)	Decrease fat consumption

establishing a baseline for such "health" markers is still considered a very challenging mission.

The final hurdle about the personalized nutrition approach relates to the targeted population who will use these advances and recommendations in practice. In other words, scientist can make the genomic-related information available for public, but it's the public's responsibility to implement such recommendations and take them to the kitchen. Some of the critical questions over here are: How easy would it be for patients to understand the information that they are being offered? Would it be more confusing for them than the conventional concepts that we already have today? Will they understand that such information is specifically tailored for them and that even a patient with the same disease and the same symptoms might get a different set of recommendation? Indeed, we have a lot of questions that need to be answered.

Despite the aforementioned three major challenges, sincere efforts are being made to explore and implement the principle of personalized medicine and nutrition in today's current practices. Table 5.1 lists examples of SNPs, their association with monogenic and polygenic diseases, and some recommended steps of intervention (when attainable).

A recent pilot human study was also conducted in Canada with the help of healthy human subjects to explore their willingness and readiness to incorporate the genomic information and at the same time address some of the values and potential risks connected with the personalized nutrition approach (Nielsen and El-Sohemy 2012). In this well-designed study, a comparison was made between providing the subjects with a genotype-based dietary advice or providing them with a general dietary advice. The results of this study were very encouraging. Higher percentage of participants in the intervention group agreed that they understood the dietary advice that they were given. Less than 9 % of the participants in the intervention group felt that the genetic information that they were offered was ambiguous and uneasy to grasp. One issue against this study was that most of the incorporated subjects were among the young generations (20–35 years old) which is more likely to facilitate their adaptation to the genomic and personalize nutrition concepts compared to older ones.

7 Concluding Remarks

Preliminary data that involve gene–diet interactions of most chronic diseases (such as cardiovascular, diabetes, and cancer) are very promising, but mostly inconclusive. Challenges facing any advancement in the prevention and treatment of such diseases are similar to those encountered earlier in any drug discovery and development. The majority of these diseases are not caused by only one genetic variation within a single target gene but rather a result of complex interactions among different genes, with onset times influenced by environmental and lifestyle factors.

Nutrigenetics, nutrigenomics, proteomics, metabolomics, and personalized-nutrition can provide the needed bases for more specific dietary recommendations that match every individual's genetic map. Within this context, personalized-nutrition and diets tailored to meet the individualistic needs have the potential of revolutionizing the health sector in the same way as the industrialization of food processing did in the past. The improved microbiological and hygienic standards combined with nutrient retention through nutrient enrichment and fortification played a key role in the public health success with the noticed increased longevity rates in modern societies. The current difficulties that we face in measuring and evaluating how environmental and lifestyle factors contribute to disease risks will be soon addressed, opening the door for a more feasible and reliable prevention and treatment options.

References

Aagaard-Tillery KM, Grove K et al (2008) Developmental origins of disease and determinants of chromatin structure: maternal diet modifies the primate fetal epigenome. J Mol Endocrinol 41(2):91–102

Alasfoor D, Elsayed MK et al (2010) Spina bifida and birth outcome before and after fortification of flour with iron and folic acid in Oman. East Mediterr Health J 16(5):533–538

Allfrey V, Faulkner RM et al (1964) Acetylation and methylation of histones and their possible role in the regulation of RNA synthesis. Proc Natl Acad Sci USA 51:786–794

Almal SH, Padh H (2012) Implications of gene copy-number variation in health and diseases. J Hum Genet 57(1):6–13

Almasy L (2012) The role of phenotype in gene discovery in the whole genome sequencing era. Hum Genet 131(10):1533–1540

Almeida LC, Cardoso MA (2010) Recommendations for folate intake in women: implications for public health strategies. Cad Saude Publica 26(11):2011–2026

Amarin ZO, Obeidat AZ (2010) Effect of folic acid fortification on the incidence of neural tube defects. Paediatr Perinat Epidemiol 24(4):349–351

Andersen V, Christensen J et al (2011) Polymorphisms in NF-kappaB, PXR, LXR, PPARgamma and risk of inflammatory bowel disease. World J Gastroenterol 17(2):197–206

Antequera F, Bird A (1993) Number of CpG islands and genes in human and mouse. Proc Natl Acad Sci USA 90(24):11995–11999

Arai E, Kanai Y (2010) Genetic and epigenetic alterations during renal carcinogenesis. Int J Clin Exp Pathol 4(1):58–73

Arregui M, Fisher E et al (2012) Significant associations of the rs2943634 (2q36.3) genetic polymorphism with adiponectin, high density lipoprotein cholesterol and ischemic stroke. Gene 494(2):190–195

Bartal A, Matrai Z et al (2012) Novel oral anticancer drugs: a review of adverse drug reactions, interactions and patient adherence. Orv Hetil 153(2):66–78

Beard CM, Panser LA et al (2011) Is excess folic acid supplementation a risk factor for autism? Med Hypotheses 77(1):15–17

Bharathy N, Taneja R (2012) Methylation muscles into transcription factor silencing. Transcription 3(5):215–220

Bian L, Yang JD et al (2004) Insulin-degrading enzyme and Alzheimer disease: a genetic association study in the Han Chinese. Neurology 63(2):241–245

Bird AP (1986) CpG-rich islands and the function of DNA methylation. Nature 321(6067):209–213

Bird A (2007) Perceptions of epigenetics. Nature 447(7143):396–398

Bodzioch M, Orso E et al (1999) The gene encoding ATP-binding cassette transporter 1 is mutated in Tangier disease. Nat Genet 22(4):347–351

Bottomley MJ (2004) Structures of protein domains that create or recognize histone modifications. EMBO Rep 5(5):464–469

Boullata JI, Hudson LM (2012) Drug-nutrient interactions: a broad view with implications for practice. J Acad Nutr Diet 112(4):506–517

Boutant M, Ramos OH et al (2012) Glucose-dependent regulation of NR2F2 promoter and influence of SNP-rs3743462 on whole body insulin sensitivity. PLoS One 7(5):e35810

Boyles AL, Wilcox AJ et al (2008) Folate and one-carbon metabolism gene polymorphisms and their associations with oral facial clefts. Am J Med Genet A146A(4):440–449

Bremner EA, Mainland JD et al (2003) The prevalence of androstenone anosmia. Chem Senses 28(5):423–432

Bustin M, Rall SC et al (1969) Histone structure: asymmetric distribution of lysine residues in lysine-rich histone. Science 163:391–393

Carlsohn A, Cassel M et al (2011) How much is too much? A case report of nutritional supplement use of a high-performance athlete. Br J Nutr 105:1724–1728

Caudill MA, Cruz AC et al (1997) Folate status response to controlled folate intake in pregnant women. J Nutr 127(12):2363–2370

Chan D, Cushnie DW et al (2010) Strain-specific defects in testicular development and sperm epigenetic patterns in 5,10-methylenetetrahydrofolate reductase-deficient mice. Endocrinology 151(7):3363–3373

Chen P, Liang J et al (2008) Association of common PALB2 polymorphisms with breast cancer risk: a case-control study. Clin Cancer Res 14(18):5931–5937

Chen QY, Alarcon S et al (2009) Perceptual variation in umami taste and polymorphisms in TAS1R taste receptor genes. Am J Clin Nutr 90(3):770S–779S

Cheung P, Allis CD et al (2000) Signaling to chromatin through histone modifications. Cell 103:263–271

Cui XY, He GZ (2005) Single nucleotide polymorphisms, inflammation and nutrition of genes. Zhongguo Yi Xue Ke Xue Yuan Xue Bao 27(6):790–792

Donley G, Hull SC et al (2012) Prenatal whole genome sequencing: just because we can, should we? Hastings Cent Rep 42(4):28–40

Duell EJ, Sala N et al (2012) Genetic variation in alcohol dehydrogenase (ADH1A, ADH1B, ADH1C, ADH7) and aldehyde dehydrogenase (ALDH2), alcohol consumption and gastric cancer risk in the European Prospective Investigation into Cancer and Nutrition (EPIC) cohort. Carcinogenesis 33(2):361–367

Ea CK, Hao S et al (2012) EHMT1 binds to NF-κB p50 and represses gene expression. J Biol Chem 287(37): 31207–31217

Engler MB (2009) Nutrigenomics in cardiovascular disease: implications for the future. Prog Cardiovasc Nurs 24(4):190–195

Ericson UC, Ivarsson MI et al (2009) Increased breast cancer risk at high plasma folate concentrations among women with the MTHFR 677T allele. Am J Clin Nutr 90(5):1380–1389

Fay LB, German JB (2008) Personalizing foods: is genotype necessary? Curr Opin Biotechnol 19(2): 121–128

Finnell RH, Shaw GM et al (2008) Gene-nutrient interactions: importance of folic acid and vitamin B12 during early embryogenesis. Food Nutr Bull 29(2 Suppl): S86–S98; discussion S99–100

Fischer R, Griffin F et al (1961) Taste thresholds and food dislikes. Nature 191:1328

Fischer R, Griffin F et al (1963) Taste thresholds, cigarette smoking, and food dislikes. Med Exp Int J Exp Med 9:151–167

Fisher E, Schreiber S et al (2011) A two-step association study identifies CAV2 rs2270188 single nucleotide polymorphism interaction with fat intake in type 2 diabetes risk. J Nutr 141(2):177–181

Frayling TM, Timpson NJ et al (2007) A common variant in the FTO gene is associated with body mass index and predisposes to childhood and adult obesity. Science 316(5826):889–894

Ganesh V, Hettiarachchy NS (2012) Nutriproteomics: a promising tool to link diet and diseases in nutritional research. Biochim Biophys Acta 1824(10):1107–1117

Gonzalez Sanchez JL, Serrano Rios M et al (2002) Effect of the Pro12Ala polymorphism of the peroxisome proliferator-activated receptor gamma-2 gene on adiposity, insulin sensitivity and lipid profile in the Spanish population. Eur J Endocrinol 147(4):495–501

Hasin-Brumshtein Y, Lancet D et al (2009) Human olfaction: from genomic variation to phenotypic diversity. Trends Genet 25(4):178–184

Hassan YI, Zempleni J (2010) Molecular aspects of physical performance and nutritional assessment. In: Driskell JA, Wolinksy I (eds) Nutritional assessment of athletes, 2nd edn. Taylor and Francis, Inc, Boca Raton, pp 212–232

Hayes JE, Wallace MR et al (2011) Allelic variation in TAS2R bitter receptor genes associates with variation in sensations from and ingestive behaviors toward common bitter beverages in adults. Chem Senses 36(3):311–319

Hinde K, German JB (2012) Food in an evolutionary context: insights from mother's milk. J Sci Food Agric 92(11):2219–2223

Hines LE, Murphy JE (2011) Potentially harmful drug-drug interactions in the elderly: a review. Am J Geriatr Pharmacother 9(6):364–377

Ho E, Zempleni J (2009) Overview to symposium "Nutrients and epigenetic regulation of gene expression". J Nutr 139(12):2387–2388

Johnson AR, Lao S et al (2012) Choline dehydrogenase polymorphism rs12676 is a functional variation and is associated with changes in human sperm cell function. PLoS One 7(4):e36047

Kaput J (2008) Nutrigenomics research for personalized nutrition and medicine. Curr Opin Biotechnol 19(2): 110–120

Keller A, Zhuang H et al (2007) Genetic variation in a human odorant receptor alters odour perception. Nature 449(7161):468–472

Khabour OF, Wehaibi SH et al (2012) Association of adiponectin with hypertension in type 2 diabetic patients: the gender effect. Clin Exp Hypertens: 1–6

Kim U, Wooding S et al (2005) Worldwide haplotype diversity and coding sequence variation at human bitter taste receptor loci. Hum Mutat 26(3):199–204

Knouff C, Auwerx J (2004) Peroxisome proliferator-activated receptor-gamma calls for activation in moderation: lessons from genetics and pharmacology. Endocr Rev 25(6):899–918

Lin KW, Birmingham E (2010) Folic acid for the prevention of neural tube defects. Am Fam Physician 82(12):1533

Lovegrove JA, Gitau R (2008) Nutrigenetics and CVD: what does the future hold? Proc Nutr Soc 67(2):206–213

Lucock MD (2006) Synergy of genes and nutrients: the case of homocysteine. Curr Opin Clin Nutr Metab Care 9(6):748–756

Major JM, Yu K et al (2012) Genome-wide association study identifies three common variants associated with serologic response to vitamin E supplementation in men. J Nutr 142(5):866–871

McDonald RB, Ruhe RC (2011) Aging and longevity: why knowing the difference is important to nutrition research. Nutrients 3(3):274–282

McRae JF, Mainland JD et al (2012) Genetic variation in the odorant receptor OR2J3 is associated with the ability to detect the "grassy" smelling odor, cis-3-hexen-1-ol. Chem Senses 37(7):585–593

Minihane AM, Jofre-Monseny L et al (2007) ApoE genotype, cardiovascular risk and responsiveness to dietary fat manipulation. Proc Nutr Soc 66(2):183–197

Morlighem JE, Harbers M et al (2011) DNA amplification techniques in pharmacogenomics. Pharmacogenomics 12(6):845–860

Murray JF (2012) Personalized medicine: been there, done that, always needs work! Am J Respir Crit Care Med 185(12):1251–1252

Mutch DM, Wahli W et al (2005) Nutrigenomics and nutrigenetics: the emerging faces of nutrition. FASEB J J19(12):1602–1616

Mutze U, Beblo S et al (2012) Metabolomics of dietary fatty acid restriction in patients with phenylketonuria. PLoS One 7(8):e43021

Nielsen DE, El-Sohemy A (2012) A randomized trial of genetic information for personalized nutrition. Genes Nutr 7(4):559–566

Oommen AM, Griffin JB et al (2005) Roles for nutrients in epigenetic events. J Nutr Biochem 16:74–77

Ovsyannikova IG, Poland GA (2011) Vaccinomics: current findings, challenges and novel approaches for vaccine development. AAPS J J13(3):438–444

Park KS, Shin HD et al (2006) Polymorphisms in the leptin receptor (LEPR)–putative association with obesity and T2DM. J Hum Genet 51(2):85–91

Peters U, Chatterjee N et al (2008) Variation in the selenoenzyme genes and risk of advanced distal colorectal adenoma. Cancer Epidemiol Biomarkers Prev 17(5):1144–1154

Peters BA, Kermani BG et al (2012) Accurate whole-genome sequencing and haplotyping from 10 to 20 human cells. Nature 487(7406):190–195

Pickell L, Brown K et al (2011) High intake of folic acid disrupts embryonic development in mice. Birth Defects Res A Clin Mol Teratol 91(1):8–19

Roberts NJ, Vogelstein JT et al (2012) The predictive capacity of personal genome sequencing. Sci Transl Med 4(133):133ra158

Rodriguez-Fragoso L, Martinez-Arismendi JL et al (2011) Potential risks resulting from fruit/vegetable-drug interactions: effects on drug-metabolizing enzymes and drug transporters. J Food Sci 76(4):R112–R124

Rubio-Aliaga I, Kochhar S et al (2012) Biomarkers of nutrient bioactivity and efficacy: a route toward personalized nutrition. J Clin Gastroenterol 46(7):545–554

Ruggiero A, Cefalo MG et al (2012) The role of diet on the clinical pharmacology of oral antineoplastic agents. Eur J Clin Pharmacol 68(2):115–122

Sartorius T, Staiger H et al (2012) Association of common genetic variants in the MAP4K4 locus with prediabetic traits in humans. PLoS One 7(10):e47647

Shakur YA, Rogenstein C et al (2009) How much folate is in Canadian fortified products 10 years after mandated fortification? Can J Public Health 100(4):281–284

Simopoulos AP (2010) Nutrigenetics/nutrigenomics. Annu Rev Public Health 31:53–68

Singh R, Kaur G et al (2011) A case of classical galactosemia: identification and characterization of 3 distinct mutations in galactose-1-phosphate uridyl transferase (GALT) gene in a single family. Indian J Pediatr 78(7):874–876

Smeets SJ, Harjes U et al (2011) To DNA or not to DNA? That is the question, when it comes to molecular subtyping for the clinic! Clin Cancer Res 17(15):4959–4964

Stover PJ, Garza C (2006) Nutrition and developmental biology – implications for public health. Nutr Rev 64(5 Pt 2):S60–S71; discussion S72–S91

Sutherland A, Kim DH et al (2010) Polymorphisms in the selenoprotein S and 15-kDa selenoprotein genes are associated with altered susceptibility to colorectal cancer. Genes Nutr 5(3):215–223

Swanson JM, Entringer S et al (2009) Developmental origins of health and disease: environmental exposures. Semin Reprod Med 27(5):391–402

Touhara K (2002) Odor discrimination by G protein-coupled olfactory receptors. Microsc Res Tech 58(3):135–141

Tyfield L, Reichardt J et al (1999) Classical galactosemia and mutations at the galactose-1-phosphate uridyl transferase (GALT) gene. Hum Mutat 13(6): 417–430

Vakili S, Caudill MA (2007) Personalized nutrition: nutritional genomics as a potential tool for targeted medical nutrition therapy. Nutr Rev 65(7):301–315

Verschuren JJ, Trompet S et al (2012) A systematic review on pharmacogenetics in cardiovascular disease: is it ready for clinical application? Eur Heart J J33(2): 165–175

Voora D, Ginsburg GS (2012) Clinical application of cardiovascular pharmacogenetics. J Am Coll Cardiol 60(1):9–20

Wadhwa PD, Buss C et al (2009) Developmental origins of health and disease: brief history of the approach and current focus on epigenetic mechanisms. Semin Reprod Med 27(5):358–368

Waterland RA, Dolinoy DC et al (2006) Maternal methyl supplements increase offspring DNA methylation at axin fused. Genesis 44(9):401–406

Wu XQ, Zhu DL et al (2012) The relationship between carboxylesterase 1 gene polymorphisms and susceptibility to antituberculosis drug-induced hepatotoxicity. Zhonghua Nei Ke Za Zhi 51(7):524–530

Wysocki CJ, Beauchamp GK (1984) Ability to smell androstenone is genetically determined. Proc Natl Acad Sci USA A81(15):4899–4902

Yasuji T, Kondo H et al (2012) The effect of food on the oral bioavailability of drugs: a review of current developments and pharmaceutical technologies for pharmacokinetic control. Ther Deliv 3(1):81–90

Ye K, Gu Z (2011) Recent advances in understanding the role of nutrition in human genome evolution. Adv Nutr 2(6):486–496

Zempleni J, Hassan YI et al (2008) Biotin and biotinidase deficiency. Expert Rev Endocrinol Metab 3(6):715–724

Designing and Implementing Pharmacogenomics Study

6

Yeşim Aydın Son, Şükrü Tüzmen, and Candan Hızel

Abstract

Pharmacogenomics of today has its origins in the 1950s with pioneering studies of monogenic variations in drug metabolism and pharmacokinetics. With the completion of the Human Genome Project in 2003 and the advances in genomics such as the high-throughput genomics technologies, we are now in the postgenomics era. This transition is increasingly marked with study of polygenic and multifactorial traits such as common complex human diseases as well as pharmacodynamic differences among populations. Changes that emerge from postgenomics medicine are not, however, limited to seismic shifts in scale and scope of pharmacogenetics research. Importantly, many low- and middle-income countries (LMICs) of the South, Asia-Pacific, Eastern Mediterranean, and the Middle East are becoming notable contributors with rapid globalization of science and increasing access to genomics technologies. This brings about, in parallel, an acute demand for regional capacity building in LMICs so that the future evaluation and implementation of postgenomics technologies in personalized medicine take place in an integrated, sustainable, and equitable manner. This chapter aims to highlight the potential applications and opportunities as well as technical and strategic issues that this field offers to influence medical care.

Y. Aydın Son
Department of Health Informatics, Informatics Institute, Middle East Technical University (METU), Ankara 06800, Turkey

GENformatik, CSO, METU Technokent, Ankara 06800, Turkey
e-mail: yesim@metu.edu.tr; yesim.aydinson@gmail.com

Ş. Tüzmen
Department of Biological Sciences, Faculty of Arts and Sciences, Eastern Mediterranean University (EMU), Famagusta, North Cyprus

Pharmaceutical Genomics Division, Translational Genomics Research Institute (TGen), 13208 East Shea Blvd., Suite 110, Scottsdale, AZ 85259, USA
e-mail: sukru.tuzmen@emu.edu.tr

C. Hızel (✉)
Faculty of Pharmacy, Anadolu University, Eskişehir, Turkey

C2H-VichyGenomics, Vichy 03200, France
e-mail: hchizel@anadou.edu.tr; candan.hizel@gmail.com

D. Barh et al. (eds.), *Omics for Personalized Medicine*,
DOI 10.1007/978-81-322-1184-6_6, © Springer India 2013

1 Introduction

The era of postgenomics medicine arrived with the completion of the Human Genome Project (HGP) in 2003, exactly 50 years after the discovery of DNA by Watson and Crick. An important conceptual shift in this period is a greater emphasis on "prediction/prevention" of future health outcomes (e.g., disease susceptibility, response to health interventions) with use of individual genetic/genomics information. A corollary is that preventive and customized interventions and diagnostic tests may now be conceptualized (and in some cases implemented) during the presymptomatic phase of a disease or before pharmacotherapy is initiated. The concept of prevention represents the next step in the development of the "predictive medicine" as conceived by Jean Dausset, one of the three winners of the Nobel Prize in Physiology/Medicine in 1980. Dausset has suggested the term "predictive medicine" as a prerequisite step for preventive medicine (Hizel et al. 2009). Subsequently in 1993, Jacques Ruffie offered a more comprehensive definition for the term and laid the philosophical basis for this new field in a book entitled "Naissance de la Médecine Prédictive" (Birth of Predictive Medicine) (Ruffié 1993). Today, the term "predictive medicine" is replaced by one that is more precise: "personalized medicine", enabled by the introduction and availability of high-throughput genomics technologies. In the present postgenomics era, we are witnessing the increasing applications of genomics technologies in multiple sectors of health. The genomics science has now left the laboratory space and rapidly diffusing to many segments of the global society not only in North America and Western Europe but also in many other global regions and countries (Zimmern 2009; Pang 2009; Normile 2007; Sgaier et al. 2007; Suarez-Kurtz 2004). Importantly, a number of low- and middle-income countries (LMICs) are firmly engaged in large-scale genotyping projects to discern the genetic diversity among their populations and apply such knowledge to discovery of health interventions in the form of novel pharmaceuticals, vaccines, and companion diagnostics to customize their targeted optimal use (Zimmern 2009; Pang 2009; Normile 2007; Sgaier et al. 2007; Suarez-Kurtz 2004). Pharmacoge-nomics studies facilitate valuable understanding of the genomic characteristics that regulate an individual's response to particular drugs. Utilization of genomic information to personalize treatment is being examined in many clinical trials and is promising to become an integral factor in the application of individualized medicine (Tuzmen et al. 2011). It is anticipated that both discovery and development of medicines and their postmarketing pharmacovigilance can substantially benefit from the availability of mechanistic biomarkers of drug treatment outcomes (Frueh et al. 2008; Hizel et al. 2009).

2 Appropriateness and Validation of Pharmacogenomics

2.1 Genomics for Personalized Medicine/Healthcare: Pharmacogenomics vs. Pharmacogenetics

The real breakthrough made in genomics will have an important long-term impact on the development and utilization of drugs for treatment and prevention. Genomic medicine which is the use of genomic information to guide medical decision is an essential component of the broader personalized medicine/healthcare concept (Willard et al. 2005; Ginsburg and Willard 2009). Since the completion of Human Genome Project in 2003, symbolically announced postgenomic era, tremendous progress has been done during last several years in modern medicine due to rapid development of molecular medicine specifically genetics, informatics, and other high-throughput technologies (e.g., nanotechnology and proteomics) which has accelerated the use of predictive and personalized medicine including utilization of drugs for treatment in routine medical and pharmacy practice for more efficient treatment of individuals (Ozdemir et al. 2009; Sheffield and Phillimore 2009). Building on globalization of postgenomics R&D and rapid advances in genomics technologies, the predictive, preventive, personalized, and participatory (P4) medicine to which pharmacogenomics is a concept directly related to has come to represent the basic tenets of twenty-first-century healthcare in developed and developing countries alike (Gurwitz and Lunshof 2011). Importantly, there is growing anticipation among developing countries for P4 medicine and knowledge-based bioeconomy drawing from data-intensive OMICS technologies such as genomics and metabolomics (Hizel et al. 2009; Pang 2009; Ozdemir et al. 2012a).

In spite of tremendous progresses in large genome-scale sequencing and improvements in bioinformatics during the last several years, hospitalizing problem and many avoidable deaths due to adverse drug reactions (ADRs) are becoming real public health problem. In addition, inefficient treatment costs a lot of money. A 1998 meta-analysis of 39 prospective studies in US hospitals estimated that ADRs are the fourth to sixth leading cause of death in the USA with more than 100,000 deaths per year and 2.2 million serious adverse reactions per year (Lazarou et al. 1998). All antidepressants and antipsychotic medicines are processed by enzymes with a high incidence of poor metabolizers (PM), and hospitalized psychiatric patients who are poor metabolizers cost $4,000–6,000 more in medical care compared to patients with an average metabolizer genotype (Chou et al. 2000). ADRs are. The scientific evidence showed that adverse effect is because of 42 % dosing error and 50 % genetic factors (Phillips et al. 2001).

While the field of pharmacogenetics has a history dating back to the 1950s, the term pharmacogenomics first appeared in the literature in 1997 (Meyer 2004). Terms such as pharmacogenetics, pharmacogenomics, personalized medicine, and stratified medicine have all been used interchangeably over the past few years to describe a revolution that is occurring in medicine (Lindpaintner 2002). Even though pharmacogenetics and pharmacogenomics are two terms used interchangeably, pharmacogenetics is the study of how genetic differences influence the variability of individual patient responses to drugs, aims to distinguish responders from nonresponders, and predicts those in whom toxicity, whereas pharmacogenomics as the whole-genome application of pharmacogenetics takes a much more global approach to the impact of variation in genetic information on drug response. Moreover, pharmacogenomic analysis can identify disease susceptibility genes representing potential new molecular drug targets for therapeutic intervention by correlating gene expression or single-nucleotide polymorphisms (SNPs) with a drug's efficacy or toxicity (Sheffield and Phillimore 2009; Milos and Seymour 2004; Wang 2010) (Fig. 6.1). By doing so, pharmacogenomics aims to develop rational means to optimize drug therapy, with respect to the patients' genotype, to ensure maximum efficacy with minimal adverse effects (Becquemont 2009). Such approaches stand in contrast to the current "one size fits all" approach to drug prescription and dosing (Wolf et al. 2000; Swen et al. 2007; Squassina et al. 2010; Bartlett et al. 2012) and promise the advent of "personalized medicine" in which drugs and drug combinations are optimized for each individual's unique genetic makeup. To this end pharmacogenomics is also increasingly being explored in the primary care context (Grice et al. 2006; Bartlett et al. 2012).

Overall, the field of pharmacogenomics is promising a new era of personalized interventions based on the person's genotype, disease subtypes and characteristics, and other forms of personalization. Such a field promises the delivery of the right drug to the right person at the right dose and at the right time, in order to maximize effectiveness and minimize side effects (Khoury 2009).

2.2 Application and Implementation of Pharmacogenomics for Clinical Practice: Analytic Validity, Clinical Validity, and Clinical Utility

Even though pharmacogenomics is expected to determine the right drug at the right dose during the pharmacotherapy by identifying individuals at risk and decrease adverse events, its appropriate use in

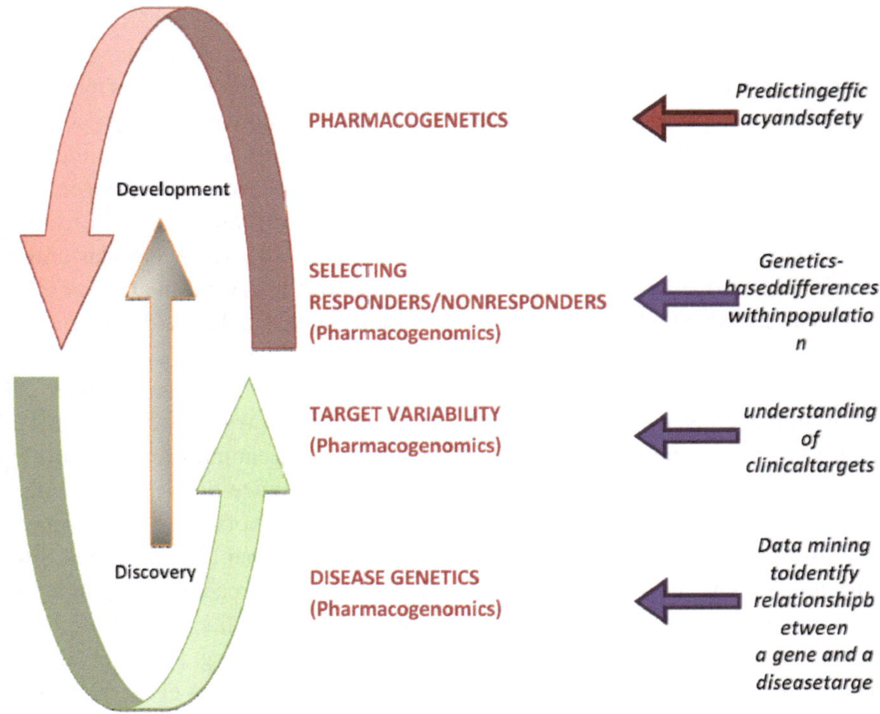

Fig. 6.1 Application of pharmacogenomics/genetics for clinical use (Adapted from Milos and Seymour 2004)

clinical uptake and its implementations in medical and pharmacy practice is an important issue (Reiss and American Pharmacists Association 2011; Mrazek and Lerman 2011). However, translation of pharmacogenomics studies into routine practice whether in medicine or pharmacy has confronted certain barriers such as identification of genetic markers associated with drug response which does not always equate to clinically useful predictors of adverse outcomes, limited availability of test, the high cost of genetic testing, disagreements about interpretation of results, and even lack of understanding about pharmacogenomics in general. To this end, pharmacogenomic discoveries need clinical/analytical validation and demonstration of clinical utility before acceptance into healthcare (Xie and Frueh 2005; Squassina et al. 2010; Scott 2011; O'Donnell et al. 2012) (Fig. 6.2).

Since the completion of Human Genome Project (HGP), progresses in genomics will have an important long-term impact on the development and utilization of drugs for treatment in the context of prediction and prevention of ADRs. In recent years, commercially available pharmacogenetic tests have been approved (recommended but not mandated) by the Food and Drug Administration (FDA) (http://www.fda.gov) and pharmacogenetic information was added to several drug labels, but their application in patient care remains very limited (Xie and Frueh 2005; Jennings et al. 2009). Even though many pharmacogenomic tests are available, these test have not yet been recommended for most drugs, and large-scale trials are needed to show that routine testing could improve patient outcomes in pharmacotherapy, such as CYPD6 polymorphism on tamoxifen treatment (Lee et al. 2010; Abraham et al. 2010; Kitzmiller et al. 2011). Probably the best examples of pharmacogenomic applications in routine practice are HER2 testing for breast cancer treatment (Phillips 2008), UGT1A1 variant associated with Gilbert's disease in patient receiving irinotecan (Palomaki et al. 2009; Marsh and Hoskins 2010; Evrard and Mbatchi 2012), HLA testing for Abacavir for HIV management (Lai-Goldman and Faruki 2008; Faruki et al. 2007; Khoury 2009), thiopurine methyltransferase (TPMT) polymorphism in patients with leukemia

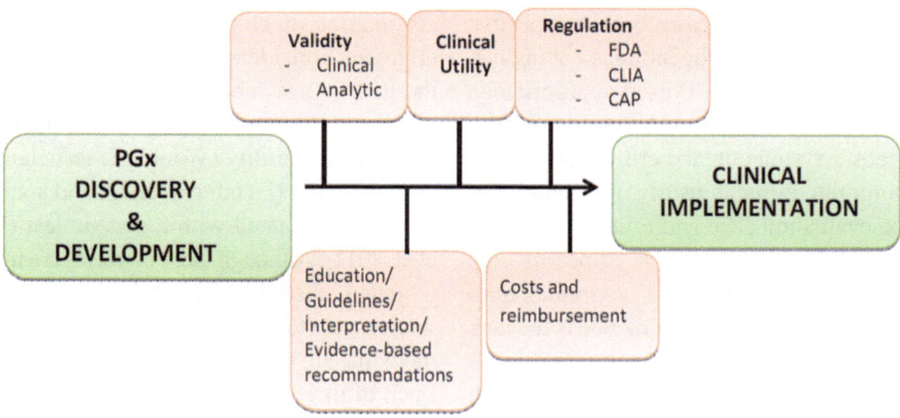

Fig. 6.2 The current challenges and barriers to clinical implementation of pharmacogenomics (Adapted from Scott 2011)

who are treated with thiopurines (Krynetski et al. 1996; McLeod et al. 2000; Evrard and Mbatchi 2012; Khoury 2009), and variations in two genes encoding CYP2C9, the enzyme primarily responsible for warfarin metabolism, and VKORC1 which is the site of action for warfarin (Seip et al. 2010; Donohue and Tirschwell 2011; Ozer et al. 2013). To be clinically useful, a pharmacogenetic test must predict the outcome of drug treatment or guide dose adjustment. The value of pharmacogenetics is heavily dependent on the identification of useful "biomarkers." Biomarkers are indicators that mark the presence of a potential gene-drug interaction or that measure response to therapeutic activity (Kitzmiller et al. 2011). However, not every drug needs a biomarker for predicting response. Biomarkers predicting drug response are especially useful when the drug's effect cannot be easily monitored or when treatment delay has important clinical consequences, or when it predicts the occurrence of severe side effects. Biomarkers can be used in clinical trials to stratify patients who respond to a new potential medicine appropriately, adversely, or not at all. In clinical practice, genetic biomarkers can be developed into diagnostic tests to measure the potential efficacy or toxicity of a particular therapy (Frueh et al. 2008). The FDA provides information on validated pharmacogenomic biomarker tests at its Web site, www.fda.gov. In pharmacotherapy for a genotype-specific population, these tests may be necessary to define which patients will benefit from the treatment and which will not (Frueh et al. 2008; Kitzmiller et al. 2011; La Thangue and Kerr 2011; Sadee 2011).

For personalized medicine/healthcare approach to select the right drug at the right dose for the right person and so to improve benefit/risk ratio for existing and common drug during pharmacotherapy, a prospective screening of a large stratified patient population with relevant genetic markers for a given disease is needed (Huang and Ratain 2009). Prospective pharmacogenomic analysis is a crucial step toward the validation of genetic markers as predictors of clinical management "in real-life" setting (Grossman 2007; Xie and Frueh 2005). Such prospective studies provide important evidence for regulatory authorities if it is necessary to modify the labelling of specific settled drugs in order to improve clinical practice and patient healthcare during pharmacotherapy. Due to the chances of high cost of such trials being very high, there is a lack of large-scale prospective studies to determine genetic variation in drug disposition and response (Xie and Frueh 2005).

While there are some examples in which prospective approach has been successfully applied, because of the high cost of such trials, most of them lack large-scale prospective studies to determine genetic variation in drug disposition and response, and so pharmacogenomics studies are currently performed retrospectively in an individual patient to identify reasons for treatment failure or unexpected ADRs (Huang and Ratain 2009; Hughes et al. 2008; Rauch et al. 2006; Xie and Frueh 2005). In any case, both prospective and retrospective pharmacogenomics testing can provide useful information if used appropriately. Hence, the decision to perform pharmacogenomics testing needs a systematic and critical evaluation of the potential clinical utility for genotype-phenotype evaluation (Xie and Frueh 2005).

The most important factor impeding the appropriate clinical uptake of genomics into routine medical and pharmacy practice is the lack of evidences for its clinical utility (i.e., evidence that use of a genomic technology improves health outcomes). Hence, pharmacogenomic applications have to be subjected to principles of evidence-based medicine and comparative effectiveness research to evaluate the benefit/risk ratio of drug use in practice (Khoury et al. 2009). Due to new pharmacogenomic testing becoming accessible, decision makers need reliable information on evidence of efficacy and cost-effectiveness of these tests in order to translate pharmacogenomic research into practice health (Khoury et al. 2009).

To develop and disseminate an integrated knowledge base on the clinical validity and utility of pharmacogenomic applications and to assure the implementation of evidence-based recommendations in practice by evaluating their impact on population health, public health is crucial. So, genomics can play a kind of honest broker role for this process (Khoury et al. 2009). Evaluation of Genomic Applications in Practice and Prevention (EGAPP) group launched in 2004 by the CDC National Office of Public Health Genomics is this kind of effort in the USA. This organization supports a well-coordinated, systematic process for evaluating genetic tests including pharmacogenomics applications that are in transition from research to clinical and public health practice in the USA (Teutsch et al. 2009; Khoury et al. 2009).

Since availability of pharmacogenomics data, alone, is not a sufficient reason for its use in inefficacy monitoring, pharmacogenomics requires an evidentiary framework for its use in decision making or genetic testing. In this sense, the ACCE framework is one of the highly prominent approaches for evaluation of genetic tests in a context of public health (Sanderson et al. 2005). As metrics of clinical biomarker value, ACCE approach considers the (AV) analytic validity (Is the genetic test robust in the lab?), (CV) clinical validity (Does the test predict a clinical event?), (CU) clinical utility (Would the test change what you do?), and (E) ethical, legal, and social issues (ELSIs) associated with a genetic test (Ozdemir et al. 2011). Although these concepts are imperative to genetic testing, some of their exact definitions with respect to pharmacogenetics and personalized medicine are not always clear (Lesko et al. 2010). Each of these components focuses on a different aspect of test performance and applicability in real-life settings, starting at the laboratory performing the test, through clinical studies proving its benefits, to more subjective evaluation of managed care burden associated with introduction of test to routine practice and reimbursement frameworks (CDC 2006).

Another important point is the necessity of the integration of pharmacogenomics building infrastructures for large-scale genome-wide association studies (GWAS) in order to identify and validate the biomarkers that reinforce the use of pharmacogenomics in clinical setting (see Sects. 4.3 and 4.4). Despite many successful and encouraging GWAS for pharmacogenomic traits have been performed to identify genes affecting drug response or susceptibility to ADRs used in the medical field (Crowley et al. 2009; Daly 2010; Stuart 2011), most drugs are not tested using GWAS (Stuart 2011). Extensively and publicly used personalized medicine/healthcare will lower the cost of the pharmacotherapy due to ADRs and prescription of drugs that have been proven inefficient in certain genotypes. It is very costly when a clinical trial is put to a stop by licensing authorities because of the small population who experiences ADRs. The new drives in pharmacogenetics make possible to develop and license a drug specifically designed for those who are not the small population genetically at risk for adverse side effects (Corrigan 2011).

The implementation of genetic data for a better prediction of response to medications and ADRs is becoming a reality in some clinical settings due to well-defined phenotypes being critical in the success of clinical validation affecting the interpretation of the genotype-phenotype association in pharmacogenomic studies. Hence, it has been submitted that the application of pharmacogenomic testing will be more appropriate and efficient if integrated with routine therapeutic drug monitoring (TDM) (Albers and Ozdemir 2004; Gervasini et al. 2010).

Finally, genomics and so pharmacogenomics on the methodology of drug development in preclinical and clinical trials should be viewed through a lens of population and public health to move beyond a simplistic focus on biotechnology. Hence, expertise in both genomics technologies as well as regional capacity to design rational pharmacogenomics studies and lack of qualification of information concerning interpretation of genomics tests result with precise personalized interpretation will be crucial (Hizel et al. 2009). Because both physicians and pharmacists might find it difficult to interpret the clinical value of pharmacogenetic test results, guidelines that link the result of a pharmacogenetic test to therapeutic recommendations might help to overcome these problems, but such guidelines are only sparsely available (Kirchheiner et al. 2005). In addition, as emphasized during the "1st International Certificate Program on Predictive & Personalized Medicine in Daily Modern Medicine and Pharmacy – P4 Medicine Meeting" (http://p4certificate.anadolu.edu.tr/p4/?q=en/home), education of scientists, healthcare professionals, and publics in genetics is crucial for appropriate application of pharmacogenetics to integrate into healthcare system.

3 Methodology, Population, and Methodological Biases

Until recently, genome-scale technologies were not widely available to exhibit connections between drug response and gene function phenotypes. Hence, functional translation of such information has been the rate-limiting step in obtaining insight into the underlying molecular mechanisms and identifying potential connections to drug response. This bottleneck was addressed by the introduction of the high-throughput RNAi-based applications (Tuzmen et al. 2011). Personalized medicine is facilitated and driven by the development and application of new genomic approaches to rapidly depict genetic sequence variations and gene expression profiles on a large scale. These types of experiments are typically accompanied by extensive genome-wide association studies aimed at identifying genes or gene sets that predict drug response. The associations are then extensively validated and evaluated for their efficacy to predict response. Often, the function of these associated genes and their potential role in drug response is unknown. Identification of the genes that are not only associated with drug response but also causally involved would add tremendous value to pharmacogenomics data, enabling a more intelligent utilization of such genes as diagnostic markers and as putative drug targets to ameliorate drug response (Tuzmen et al. 2011). In addition to accelerating the functional validation of gene lists from association-based pharmacogenomics data, high-throughput screening approaches for genome-scale functional analysis can potentially be a more relevant starting point for pharmacogenomics analysis. In such strategies, genes that are causally involved in modulating drug response are discovered, and their putative clinical association with drug response is later validated. With this principle, global phenotype analysis enables the examination of a wide range of genetic factors including associations with specific gene and transcript sequence variations, protein modifications, expression of mRNA or protein, and many other genetic and epigenetic variations that are often not used in primary pharmacogenomics analysis. Further, this approach can identify functionally associated genes, which after undergoing in-depth functional validation may also have utility as drug targets. Consequently, there is clearly an unmet need to develop and apply genome compatible strategies and methodologies to unveil functional modulators of drug response (Tuzmen et al. 2011).

The function of a gene could be determined by experimentally interfering with its expression. The resulting phenotype or effect on molecular endpoints and signaling pathways can then be examined.

3.1 The Muting Apparatus: RNA Interference (RNAi)

RNA interference (RNAi) is a naturally occurring phenomenon of gene expression control, triggered by the introduction of double-stranded RNA into a cell. This mechanism can be synthetically employed to downregulate expression of specific genes by transfecting mammalian cells with synthetic short interfering RNAs (siRNAs) (Fire et al. 1998; Tuschl 2001; Elbashir et al. 2002; Hannon 2002; Nakayashiki and Nguyen 2008; Travella and Keller 2009; Tuzmen et al. 2011).

When a long double-stranded RNA (~>200 bp) is introduced into a cell, it is recognized and cleaved by the enzyme Dicer resulting in 21–23 nucleotide long double-stranded RNAs referred to as siRNAs, which mediate gene silencing through an interaction with a protein complex called RNA-induced silencing complex (RISC) (Tuzmen et al. 2011). The RNAi technology involves the use of siRNAs as powerful mediators of a naturally occurring posttranscriptional gene-silencing mechanism. This mechanism is highly conserved among multicellular organisms as diverse as flies, yeast, plants, worms, and humans (Huppi et al. 2005; Tomari et al. 2007; Matranga and Zamore 2007; Zamore 2004; Roguev et al. 2008; Volpe et al. 2003; Hamilton et al. 2005; Caplen and Mousses 2003; Haley et al. 2003; Stroschein-Stevenson et al. 2009; Wendler et al. 2010; Saleem et al. 2012). In *Caenorhabditis elegans*, RNAi has been applied to generate "somatic knockdowns" for the functional analysis of thousands of genes (Sonnichsen et al. 2005; Kim et al. 2005; Grishok et al. 2008; Rocheleau 2012; Schwarz et al. 2012; Schøler et al. 2012). In mammalian cells, siRNAs can induce RNAi without activation of nonspecific dsRNA-dependent pathways (Caplen and Mousses 2003; Huppi et al. 2005; Sun, X et al. 2008; Sun, TM et al. 2008; Mariotti et al. 2009; Schnettler et al. 2012; Holland et al. 2012; Chalupnikova et al. 2013). Enzymatically or chemically synthesized siRNAs, as well as vector-based short hairpin RNAs (shRNAs), have been thoroughly applied to the study of function of individual mammalian genes (McManus and Sharp 2002; Huppi et al. 2005; Aleku et al. 2008; Rines et al. 2008; Sibley and Wood 2011; Zhou et al. 2012; Pieraets et al. 2012). SiRNAs act as a guide for the RNA-induced silencing complex (RISC) (Huppi et al. 2005; Filipowicz 2005; Ohrt et al. 2008; MacRae et al. 2008; Lima et al. 2012; Kubo et al. 2012; Derrien et al. 2012). These siRNAs can be designed to silence the expression of specific genes carrying a particular target sequence in high-throughput (HT) siRNA experimental systems. This may potentially be offered as a therapeutic strategy for inhibiting transcriptional regulation of genes, which in such cases can constitute a strategy that can inhibit targets that are not accommodating by small molecule compounds. Large-scale experiments using low-dose drug exposure combined with siRNA also represent a promising discovery strategy for the purpose of identifying synergistic targets that facilitate synthetic lethal combination phenotypes (Tuzmen et al. 2011). Considering such powerful applications, siRNA technology provides an exciting opportunity for relatively rapid and revolutionary approaches to cancer drug design. Small RNAs that harness the RNAi machinery may become the next new class of drugs for treating a variety of diseases (Tuzmen et al. 2011; Petrocca and Lieberman 2011).

In this section, we provide examples on the application of RNAi, with particular focus on HT siRNA phenotype profiling, technical requirements, methodological challenges, and population and methodological biases. Through specific examples, we also illustrate how such research can be used to advance and accelerate drug target discovery and clinical drug development to support cellular pharmacogenomics.

3.2 Resources for High-Throughput RNAi (HT-RNAi) Research

Thus far many attempts have driven the development of large siRNA and shRNA libraries targeting human genome utilizing RNAi (Aleku et al. 2008; Berns et al. 2004; Ganesan et al. 2008; Schlabach et al. 2008; Silva et al. 2008; Paddison and Hannon

2003). Large genome-wide siRNA libraries as well as focused libraries are commercially available from suppliers including Qiagen, Ambion, Life Technologies, Sigma, and Dharmacon. Similarly, shRNA libraries are commercially available from sources including Sigma Open Biosystems, System Biosciences, and GeneCopoeia.

3.3 Problems Faced by RNAi as Therapeutics: Could This Originate a Mainspring of Population and Methodological Biases?

Despite siRNAs proving to be very powerful inhibitors of gene regulation and enabling the elucidation and better comprehension of gene functions in variety of cell lines and organisms, several challenges of the siRNA technology remain to be resolved. *Sequence specificity*, *delivery*, *off-target effects*, and *alternative splice forms of genes* constitute some of the major problems associated with RNAi analysis. Hence, development of RNAi-based therapeutics can still be considered as being at its infancy.

3.3.1 siRNA Sequence Specificity

Amarzguioui (Amarzguioui et al. 2003; Amarzguioui and Prydz 2004) demonstrated that changes at the 3′ end of the guide siRNA strand are well tolerated. However, variations at the 5′ end and in the middle of the siRNAs have adverse effects, and these observations have also been validated by other scientists in the siRNA field (Boden et al. 2003; Das et al. 2004; Gitlin et al. 2002). Sano et al. (2008) recently showed that the character of the 2-nt 3′-overhang of siRNAs is the predominant determinant of which strand participates in the RNAi pathway. They illustrated that siRNAs with a unilateral 2-nt 3′-overhang on the antisense strand are more potent than siRNAs with 3′-overhangs at both ends, due to preferential loading of the antisense strand into the RISC (Sano et al. 2008). Several groups have proposed a set of guidelines that seek to have a focused choice of siRNAs that could potentially knockdown gene expression (Elbashir et al. 2002; Paddison and Hannon 2003). There are also web-based online software systems available for computing highly functional siRNA sequences with maximal target specificity in mammalian RNAi (Naito et al. 2004; Ui-Tei et al. 2007; Theis and Buchholz 2010; Kittler et al. 2007). It is recommended that systematic testing be performed to verify that the siRNA sequence under consideration targets a single gene as well as the correct splice variant of the gene of interest. A BLAST search of the selected sequence should be performed against sequence databases such as EST and/or Unigene libraries using the National Center for Biotechnology Information (NCBI) website (Dykxhoorn et al. 2003). These studies should include at least two carefully BLASTED siRNA sequences as well as scrambled and mismatched controls, and all the siRNAs should be used at the lowest active concentrations.

Furthermore, siRNAs yield mRNA knockdowns rather than knockouts, and care must be taken in the interpretation of negative outcomes given that even minute and undetectable changes in the level of protein expression might still be sufficient to stimulate a cellular functional response (Jones et al. 2004; Jones and Lindsay 2004). On the other hand, plasmid-based RNAi expression systems (shRNAs) can be considered as an alternative/addition to the synthetic siRNAs, but there are both advantages and disadvantages associated with this system (Rossi 2008). For example, chemically synthesized siRNA transfection is more efficacious than plasmid-based shRNA transfection as more cells will experience gene expression downregulation after siRNA transfection. The initiation of siRNA-transfected knockdown is immediate as opposed to shRNA-based strategies, which would require transcription and dicer processing. Plasmid-based RNAi expression systems permit plasmids to be readily regenerated and the duration of silencing can be extended using this system. However, the utility of plasmid-based strategies is limited in cell lines that are difficult to transfect and that cannot be grown for long periods of time in culture including primary cells (Dykxhoorn et al. 2003).

3.3.2 siRNA Delivery

Efficient, reproducible and rapid siRNA delivery is necessary for effective siRNA library screening. This creates a major obstacle especially for siRNA-directed silencing by lipid-based transfections. This is mainly because siRNAs cannot readily cross the mammalian cell membrane. Protocols often need to be optimized individually for the efficient delivery of siRNAs into different cell lines. There are a multitude of reagents available today for the transfection of siRNAs into mammalian cells. The number of commercial vendors has significantly increased over a short period of time and continues to grow today as more researchers are beginning to utilize a functional genomics approach (Tuzmen et al. 2011). The most common is lipid-based (cationic and polyamine) transfection. New cationic-based delivery methods have recently been tested (Han et al. 2008). Furthermore, newer technology for siRNA transfection has emerged, such as amphiphilic chemicals, nanoparticles, and microelectrode arrays (Sun, X et al. 2008; Sun, TM et al. 2008; Medarova et al. 2007, 2008; Patel and Muthuswamy 2012; Castillo et al. 2012; Toriyabe et al. 2013). Both lipid-based transfection and electroporation have been adapted for high-throughput siRNA delivery. Simultaneous transfection and plating of cells (reverse transfection) (Tuzmen et al. 2011) permit large number of sample plate handling in a short period of time (Kuuselo et al. 2007). Electroporation can deliver hundreds of siRNAs quickly when performed with multiwell plate electroporators (Ovcharenko et al. 2005; Arabsolghar and Rasti 2012). However, while electroporation can deliver siRNAs quicker in multiwell format, it has a major drawback of toxicity from the process itself. It may not be consistent from well to well, thus introducing a noise factor in the experiment that may make it hard to reproduce (personal communication). The necessity of initial optimization of transfection conditions for each cell line, and protocol optimization for individual targets, causes the high-throughput target identification and validation to be very challenging. Nevertheless, once the selection of the best delivery method (a transfection reagent and its particular conditions or electroporation conditions to increase the uptake of siRNAs while maintaining high cell viability) for the chosen cell type is identified, these challenges can be defeated (Tuzmen et al. 2011).

3.3.3 siRNA Off-Target Effects

Although RNAi offers specificity as one of its greatest advantages, off-target effects appear as a potential disadvantage. Consequently, a stringent screening strategy has to be accomplished for all candidate siRNAs in order to avoid or at least limit off-target effects (Anderson et al. 2008; Dua et al. 2011; Shum et al. 2013; Nolte et al. 2013). Several researchers have illustrated in their microarray studies that siRNAs can have off-target effects by silencing unintended genes (Jackson et al. 2003; Scacheri et al. 2004; Anderson et al. 2008). Additionally, other studies showed that some cationic liposomes induce or trigger interferon response (Heidel et al. 2004; Ma et al. 2005; Kim et al. 2007). Similarly, any viral vector harboring an shRNA expression cassette could stimulate an unintended immune response (Uprichard 2005). Also, the nature of the synthetic siRNAs may trigger the induction of the dsRNA cellular defense mechanism. It has been stated that interferon induction might be beneficial for therapy in some cases, but cytotoxicity may be an outcome of this uncontrolled induction of the innate defense mechanism (Uprichard 2005; Tuzmen et al. 2011). These studies made it possible and practical to chemically synthesize siRNAs, which could then be applied as a molecular biology tool for gene silencing in mammalian cells.

3.3.4 Alternative Splice Forms of Genes

Over 60 % of protein-coding genes in vertebrates express mRNAs that undergo alternative splicing. Alternative splicing of precursor messenger RNAs (pre-mRNAs) is one of the most important sources of protein diversity in vertebrates. The resulting collection of transcript isoforms presents significant challenges for contemporary biological assays. For example, real-time quantitative PCR (qPCR) validation of gene expression microarray results may be unsuccessful if the two technologies target different splice variants. Effective use of sequence-based technologies requires knowledge of the specific splice variant(s) that are targeted. Additionally, the critical roles of alternative splice forms in biological function and in disease suggest that assay results may be more informative if analyzed in the context of the targeted splice variant (Ryan et al. 2008). Defects in alternatively spliced process are linked to numerous human genetic diseases and various forms of cancer (Gaur 2006). Alternatively spliced forms of genes may reflect disease diversity and population specificity, which can deliver biased outcome of the results depending on what methodologies are used (Tian et al. 2011; Wegmann et al. 2008; Tucker et al. 2011).

3.4 RNAi and Microarray Technology

The emergence of novel technologies in molecular biology enabled the identification of new targets that have been correlated with human disease. Technical advances in RNAi field and the completion of Human Genome Project have also enabled the high-throughput, genome-wide RNAi analysis of various organisms. Several groups have utilized high-throughput RNAi (HT-RNAi) technology to systematically study hundreds or thousands of genes for knockdown phenotypes (Ganesan et al. 2008; Kimura et al. 2008). HT-RNAi is also routinely being used for rapid, genome-wide screening for genes involved in specific biological processes. The integration of this technology, with platforms such as microarray-based analysis, has enhanced the speed, accuracy, and throughput of such genome-wide screens (Mousses et al. 2003; Semizarov et al. 2004; Vanhecke and Janitz 2004; Wheeler et al. 2005; Ortega-Paino et al. 2008; Wilson and Plucinski 2011; Wu et al. 2011; McConnell et al. 2011; Tuzmen et al. 2011; Vainio et al. 2012).

3.5 High-Throughput RNAi Drug Sensitivity Screening

The value of high-throughput cell-based screening technology is that it not only enables a fast screening method for the phenotype associated with silencing of individual genes, but also it enables the researchers to study gene interactions in parallel. The establishment of an efficient HT-RNAi platform involves successful development of several stages including assay development, assay validation, high-throughput screening, and data analysis (Tuzmen et al. 2011). Once established, HT-RNAi can be used to identify and functionally validate target hits. One application of HT-RNAi is to conduct multidimensional analysis of drug modulator genes across multiple inputs (different chemical or drug exposures), in various systems (different cell lines or cell line variants), and across multiple outputs (cellular phenotypes or molecular endpoints). When each of these dimensions is multiplied, we achieve highly multidimensional RNAi data which can be extremely valuable in not only identifying chemo-modulators but also gaining some understanding about the chemo-selectivity of chemotype modulation and its relationship to various genomic parameters and contexts (Tuzmen et al. 2011).

High-throughput screening technologies are providing the means with which to perform large-scale and whole-genome RNAi phenotype studies with massive amounts of data (Rines et al. 2008; Ganesan et al. 2008). In addition to pushing the envelope of our current understanding of global gene function and regulation, it will also present substantial challenges in the areas of data management and analysis (Tuzmen et al. 2011). For this reason, it becomes important to focus on effective informatics methods in order to make novel biological conclusions (Zhang et al. 2006).

3.6 Drug Target Validation vs. Drug Discovery

Genomic technologies have massively accelerated the identification of genes that are linked with physiological processes and pathological states. The challenge has traditionally been to functionally evaluate these candidate genes to determine which ones would make appropriate and pharmacologically relevant drug targets. In this context, a gene is first associated with a disease process, or with response to therapy, and then the expression or function of a gene is perturbed experimentally to determine if a functional and causal link exists between the gene and signaling pathways, and the desired pharmacological outcome or phenotype (Thomas et al. 2006). Hence, RNAi technology is now being widely applied as a readily available research tool to accelerate the functional validation of the multitude of new candidate drug targets emerging from "omics" data (Ghosh and Poisson 2009; Navratil et al. 2011; Daka and Peer 2012). Our own laboratory at the Translational Genomics Research Institute (TGen) has used this approach to evaluate candidate drug targets which were found to be amplified and overexpressed genes in cancer genomes through comparative genomic hybridization (CGH) microarray and cDNA microarray analysis. In such experiments, small focused siRNA libraries (i.e., *Cancer Gene Library* of 278 siRNAs targeting 139 classic oncogenes and tumor suppressor genes) were created against prioritized candidates and used to rapidly determine the ones that were also necessary for cell growth and survival (Tuzmen et al. 2007). The functional data adds substantial value to the genomic information and permits the rapid prioritization of pharmacologically vulnerable targets to be distinguished from disease-associated genes that would not make appropriate drug targets. This approach however is limited in that it focuses on genes that are associated with the disease or with some related process. Although disease-related genes make sense as candidate drug targets, it is also possible that certain genes and pathways which are not involved in the etiology of the disease may also represent points of vulnerability that can be therapeutically exploited. The hypothesis behind this concept is that certain gene targets would become vulnerable under certain contexts, such as exposure to a drug or in the context of a cell in a pathological state, but would not be associated with the disease state (Tuzmen et al. 2011). Despite a very large body of knowledge about the genes that are in some way associated with cancer initiation, promotion, and progression, targeted cancer therapies have had limited success.

We propose an alternative approach that consists of focusing on vulnerabilities in cancer cells that arise from the cancer process. In this paradigm, the most important aspect is not the disease-causing genes, but the specific modifications to the normal genetic regulatory network that occur in the context of a cancer cell with a highly altered genome (Tuzmen et al. 2011). With HT-RNAi technology, it is possible to rapidly analyze genes in the genome to determine if they are needed for growth and survival in cancer cells but are not critical for survival of normal cells. Such genes may not be directly involved in the cancer process itself but nevertheless represent selectively vulnerable drug targets for therapeutic intervention. In a similar fashion, the cellular response to a specific drug may be controlled and modulated differently in various cell types, under different physiological and pathological conditions (Tuzmen et al. 2011). Systematic RNAi phenotype screening can reveal the genes that functionally regulate the response to specific drugs in cancer cells providing a means for development of new combination chemotherapies that exploit these chemo-selective vulnerabilities (Tuzmen et al. 2011).

3.7 Key Benefits of RNAi

- Relative to other functional methods, RNAi provides several advantages for studying the function of genes, gene-gene interactions, and interactions with gene-environment.
- Knocking down gene expression provides a more biologically relevant experimental outcome for analyzing gene function than overexpression of a gene in the context, where it is usually not functional.
- Utilization of siRNA to trigger RNAi has proven to be more potent, highly specific, and more reliable than other approaches such as antisense technology.
- The ease of use and high success rate in designing effective and specific siRNA reagents enabled this technology to be compatible with the creation of large, genome-wide libraries for systematic and global scale studies (Tuzmen et al. 2011).

3.8 Pharmacogenomics Applications of HT-RNAi

Genome-scale analysis of mRNA and protein expression has produced enticing clues about which genes may be associated with drug response and therapy failure in the clinic. Unfortunately, regulatory genetic networks that control drug response are highly complex, heterogeneous, and virtually impossible to model based on our current understanding. Therefore, most attempts to develop response biomarkers are currently based primarily on associative data from genomics analysis, with little regard to the functional role that genes have in modulating drug response. Consequently, these genes and gene signatures may initially show extremely strong associations with drug response, but when applied to large clinical trials, they often do not have sufficient predictive power to be useful (Tuzmen et al. 2011).

As an alternative, we have approached the problem from a systems analysis perspective, assuming a "black-box" model for the regulatory genetic network of a cell, and applied genome-wide RNAi for multidimensional analysis to reveal insightful functional relationships between synthetic genotypes and chemical compounds in cancer cells (Tuzmen et al. 2011). This approach is based on recent advances in RNAi technology and requires extensive expertise, infrastructure, and resources to perform HT-RNAi on a large scale (Tuzmen et al. 2011).

HT-RNAi screening has emerged to offer tremendous insights and knowledge behind the mechanisms and contexts of vulnerability of any particular drug therapeutic. Use of this information can then be applied to better position the therapeutic drug clinically by suggesting new drug combinations that may have enhance effectiveness or by identifying biomarkers that will indicate patient groups that will likely respond better to the therapeutic (Tuzmen et al. 2011). Consequently, it is very critical that the acquisition of the data from RNAi studies be as accurate and genuine. The use of appropriate siRNA controls will also provide the ability to reduce the selection of false-positive results from these screens. Hence, potential transient variables such as transfection reagent cytotoxicity that may contribute to false-positive selection can be identified and filtered out during the analysis (Echeverri et al. 2006; Zhang 2010; Das et al. 2012). These controls can then also be used to assess the quality of the HT-RNAi screen over a large genome-wide library. Plate-to-plate and run-to-run variations within the screen provide an overall assessment of the robustness of the screen.

Application of HT-RNAi to functional chemogenomics is a powerful strategy for "predictive pharmacogenomics" since it permits the discovery of candidate biomarkers that are causally related to drug response (Tuzmen et al. 2011).

4 Power Calculations and Statistical Analysis

The power of a statistical test can be described as the probability that it will give a statistically significant result when the null hypothesis is false. Fisher's and Neyman-Pearson's theories for hypothesis testing have become the two most popular of the statistical tests in biomedical sciences since they have been described (Fisher 1958; Neyman and Pearson 1933). In order to face the challenges of the biomedical data and clinical studies, strengths of both tests have been utilized by a combination approach where both a significance of the statistics can be provided with the power of the test (Lehmann 1993; Altman and Goodman 1994). Details of the theories and the applications of each test have been covered in many publications to date. In this chapter, our goal is neither a comprehensive review nor teaching basics of statistics but to address how power calculations can be effectively adapted to genome-wide studies and the statistical analysis approaches used for genome-wide association studies (GWAS) that are the backbone of pharmacogenomics.

4.1 Power Calculations

In statistical analysis the main parameters effecting the outcome can be listed as (1) the statistical model chosen for testing, (2) effect size and its variability, (3) sample size, (4) type I error (significance, α), and (5) type II error (power of the test, 1-β). Among all five parameters sample size is the one that researchers have the most control over and by determining the appropriate sample size for the study, significance of the results and the power of the test can be assured. Power calculations aim to determine the sample size for the desired statistical significance at the design stage of the studies. The smallest study size that can detect the appropriate effect size with desired power can be calculated through power calculations. Main goal of the power calculations is to design efficient studies where enough samples are analyzed without need for excessive data collection.

In general, biomedical study designs can vary from a basic candidate gene-finding study or association of genetic variations with a phenotype and drug response to genome-wide association studies to determine risk SNPs. In any of the study designs, case-control, cross-sectional, or prospective sampling strategies can be used for the selection of study population. The goal of the study and selection method of study population mostly determines the parameters that should be considered while calculating the sample size. Typically type I error of $\alpha=0.05$ (significance) and type II error of $\beta=0.20$ (power of 80 %) are accepted. In a classical genetic epidemiological study like candidate gene finding, the factors that have to be determined before any power calculation to find the appropriate sample size are the prevalence or the relative risk of the condition under study to determine effect size, and the ratio of cases vs controls as recruiting more controls can increase the power. In order to identify the association of genetic variations, additional parameters are necessary. Most important parameters are the mode of inheritance (multiplicative, additive, recessive, or dominant), linkage disequilibrium (LD), and marker allele frequency (maf). In either study the Pearson's chi-squared test is suggested for case-control design (Sasieni 1997). If a trio study is being conducted where three members of a family (child, mother, and father) are recruited, then effect size of the study cannot be represented with the prevalence or relative risk of the population. Advantages of a trio study are control for the population stratification and detection of parent of origin and/or any genotyping errors. Trio studies are preferred to study genetic variation of rare diseases. The Spielman's transmission disequilibrium test (TDT) is available for the analysis of trio studies where the number of trio families is used as the number of participants while calculating the expected power (Spielman et al. 1993).

When any of the studies described above is conducted at genomic level, like genotyping hundreds of thousands up to millions of single-nucleotide polymorphisms (SNPs) in a single study, to detect changes in the minor allele frequencies, its results represent genome-wide associations. Genome-wide studies are mainly designed for genomic marker discovery for chronic complex diseases, drug efficacy, and side-effect studies and require sample sizes of thousands for the accurate interpretation of the results. The details of the statistical design and analysis issues of genome-wide studies especially in pharmacogenomics studies will be covered in the following sections.

4.2 Tools for Calculating Sample Size

There is variety of software tools available to aid researchers with the power calculations at the design phase of their studies. The study designs that can be assessed with these tools range from candidate gene association studies to gene-gene interaction or gene-environment interactions for case-control or cross-sectional studies.

The most popular power calculation tools can be listed as:

(a) *CaTS*: Power **ca**lculator for **t**wo-**s**tage association studies is a genetic power calculator for case-control association studies including two-stage genome-wide association studies. It has been developed by Skol et al., and the software is available for download through http://www.sph.umich.edu/csg/abecasis/CaTS/index.html. Information on the prevalence, relative risk, allele frequency, mode of inheritance, case-control ratios, as well as the expected significance level can be determined using the sliders on the graphical interphase of the software. Tutorials provided with example data provides further training on the software. The details on the calculations implemented in the software are described in Skol et al. (2006).

(b) *Quanto* is a software that computes sample size or power for association studies of gene, gene-gene, or gene-environment interactions. Studies with matched case-control, case-sibling, case-parent, and case-only designs can be evaluated with Quanto through the graphical interface that allows users to select/change models to observe the results without restarting. Program runs on windows platforms and analysis results can be exported as a text file. The software can be downloaded through http://hydra.usc.edu/gxe/.

(c) *GPC*: Interactive **G**enetic **P**ower **C**alculator is a website that provides an interphase for various power calculations including discrete and quantitative trait TDT analysis (Purcell et al. 2003). The program can be reached from http://pngu.mgh.harvard.edu/~purcell/gpc/ where links to different analysis options and tutorials are available. After the statistical model of the study is selected, parameters such as prevalence, relative risk, and allele frequencies are asked. According to the expected type I and type II errors, the analysis returns appropriate sample sizes.

(d) *PAWE*: *P*ower for *a*ssociation *w*ith *e*rror computes power and sample size calculations for genetic case-control association studies in presence of genotyping errors and also offers the option to determine power for a fixed sample size (Gordon et al. 2002). In case of no genotyping errors, the calculation results will match other power calculators listed above. PAWE program also runs through its website http://linkage.rockefeller.edu/pawe. In the first screen analysis for either power calculations for a fixed sample size or sample size calculation for a fixed power can be selected. Then, the other parameters such as case-control ratios are entered, and the calculation results are returned on the screen.

4.3 Pharmacogenomics and Genome-Wide Association Study Design

Post-Human Genome technologies and knowledge shifted how we study genetics at molecular level from single gene approach to genomic level studies. Pharmacogenomics approach, where variation in drug effects are studied as a function of genetic variation, has been utilized to carry pharmacogenetic studies at genomic level. Genome-wide association studies (GWAS) are a fundamental approach in pharmacogenomics with their potential to reveal molecular basis of variations to drug efficacy and side effects within populations through analysis of polymorphisms (Gurwitz and McLeod 2009).

Genome-wide association studies are hypothesis-free, unbiased studies for discovery of novel variants with the capability to test few thousands to up to million markers at a time. In order to identify small changes with high confidence, a study design with large group of subjects is essential. Due to the challenges with analysis of appropriate case numbers, so far outcomes of completed GWAS for pharmacogenomics were only able to define major effects of genomic variations on drug response and susceptibility to side effects (Duncan 2009).

In GWAS design, large sample size is required to be able to detect rare variations and effects and also to be able to handle increasing number of false associations as hundreds of thousands of markers are studied simultaneously (Reich and Lander 2001; Wang et al. 2005). GWAS face a challenge on both side of the study design. Conducting studies at the required size increases the analysis cost, while recruiting enough participants might also raises its own problems. On the other hand, if these challenges are not met, overall study might fail to discover associations at a significant level due to the low power of the study.

A GWAS can be designed as a single stage or two stage depending on the goal of the analysis, which mainly affects the number of participants needed and the cost of the studies. Single-stage cohort studies aim to collect baseline information on the participants and monitor the changes in a prospective study. Even though single-stage cohort studies require the largest number of individual and have the highest cost, novel and predictive biomarkers and rare associations for complex diseases within populations can be discovered. Most of these studies are conducted by governmental support or funded by big pharmacological companies.

Single-stage case-control studies are preferred widely for academic and clinical research. Case-control studies have the strength of comparing allele frequencies in patients compared to the control group. In GWAS with case-control design, the sample size requirement is usually at a much lower scale than cohort studies. Investigations on the power of GWAS suggests around 1,000 participants each for both case and control groups for the single-stage or the initial stage of a two-stage study to be able to address the issue of increased number of false positives due to the genome-scale multiple marker analysis (Eberle 2007). Still many GWAS that have been completed are done with fewer participants than the statistically required number. These studies were able to discover the major associations between the variations, and the effects have been described even though there were limitations due to the sample size (Crowley et al. 2009).

In early 2000s when the idea of GWAS was new and studies were much costly, different approaches with two-stage design have been proposed. A usual two-stage design involves genome-wide analysis of the initial case-control group and then replicating the results for small number of associated markers on other case-control groups. Another example to a two-stage design would be using pooled sample of DNAs in the first stage to be able to survey higher number of individuals by following up with individual genotyping of few. Especially, the two-stage dependent design, in which the participants in the second stage are also sampled in the first stage, has been shown to be the more cost-effective even when compared to one-stage design (Wang et al. 2009). Availability of low-cost commercial genotyping platforms, along with the opportunity for researcher to test multiple hypotheses with multiple end points, is now the main driving cause behind selecting one-stage designs (Duncan 2009).

In general, sample size and power estimations options are compared to decide on the optimal study design which will provide the maximum power for the total cost of the study. In some cases, the maximum power can also be optimized against other parameter such as sample size, genotype cost, and type I error.

4.4 Analysis of Genome-Wide Association Studies

Genome-wide association studies (GWAS) aim to discover genomic level variance among individuals or different case-control groups within a population in a holistic and agnostic manner. The case-control studies mainly focus on identification of significant variations that are likely to be associated with phenotypes, susceptibilities, disease conditions, and drug efficacy or side effects. Following the advancements in the genomics-based technologies, the single-stage case-control design has become much cost-effective and overwhelmingly preferred as mentioned in the previous sections. Thus, this section will focus on the overall statistical analysis of GWAS with single-stage case-control design for discovering polymorphic associations. Also, further approaches that need to be employed before the validation studies will be discussed.

4.4.1 Association Analysis and Multiple Testing

In a typical GWAS after genotyping is completed, allelic frequencies of all control and case subjects are compared and the significant variations between cases and controls are reported. A widely used software PLINK is designed to perform basic, large-scale statistical analysis of GWAS. PLINK provides a comprehensive toolkit solely for the whole-genome association analysis step of genotype-phenotype data, so it can handle GWAS data only after the genotyping is completed. PLINK is developed by Shaun Purcell and his colleagues at the Center for Human Genetic Research Massachusetts General Hospital and the Broad Institute of Harvard and MIT (Purcell et al. 2007). It is free, open-source software basically operated through command line, available for download at http://pngu.mgh.harvard.edu/~purcell/plink. The graphical user interphase and visualization of analysis results and is supported by third-party tools like gPLINK and Haploview.

PLINK offers all the basic association testing models available for the analysis of GWAS analysis of case-control studies including standard allelic test, Fisher's exact test, Cochran-Armitage trend test, Mantel-Haenszel, and Breslow-Day tests for stratified samples, dominant, recessive, and general models and model comparison tests. Multi-marker predictors, haplotypic tests, copy number variant analysis, and TDT analysis for family-based studies are additional analysis options offered by PLINK.

GWAS analysis provides testing for multiple hypotheses as hundreds of thousands of SNP markers are tested simultaneous. That raises the issue of high number of false-positive results that would pass the thresholds of statistical significance purely by chance. The multiple-hypothesis correction tests just as Bonferroni and false discovery rate (FDR) have been suggested, but these generic multi-hypothesis correction tests do not account for the association of SNPs to each other on the same haplotype through LD and are found to be too conservative for GWAS analysis (Storey and Tibshirani 2003).

The resulting *p*-values of associations should be interpreted with utmost caution for GWAS analysis. The *p*-value indicates the "strength of the evidence found against the null hypothesis." It has been shown that the hypothesis testing results with widely excepted p-value 0.05 may not be the conventional threshold to support evidence against the null hypothesis in case of GWAS analysis (Azuaje 2010.) The genome-wide significance threshold can be estimated by accounting for (1) ratio of markers selected to the total number of markers investigated in a GWAS. As the total number of selected SNPs is small enough compared to up to millions of SNPs, it can be ignored in the calculations. So, the problems present itself as the ratio of genome-wide error to total number of markers in the study. Then, in a genome-wide study of millions of SNP markers with the estimated genome-wide type I error of 0.05, the *p*-value threshold of $0.5 \cdot 10^{-7}$ would be acceptable to determine statistical significance of the identified associations (Panagiotou and Ioannidis 2012).

PLINK also offers the meta-analysis option, through which the results of independent studies, with same research hypothesis, are combined. It does not require access to the original data, but the analysis results are collected and a consensus is presented. Meta-analysis has been accepted as an appropriate method to increase the statistical power in genetic analysis (Sagoo et al. 2009; Yesupriya et al. 2008).

4.4.2 SNP Subset Selection

Next stage in the GWAS analysis after determination of statistically associated SNPs is the replication and/or validation of the study for selected markers through independent studies. Focusing on a smaller subset of SNPs from the initial study is recommended in order to decrease the burden on the cost and statistical analysis. The subset of SNPs can be selected through various approaches depending on the choice of the researcher. The most direct forward approach is to use the calculated *p*-values by setting an appropriate threshold for the study including all SNPs with a *p*-value smaller than the cutoff in the validation phase. Many alternative strategies also have been suggested such as population attributable risk, the false-positive report probability, Bayes factors or *q*-values, empirical Bayes estimates of effect size, and multi-marker methods like a local scan statistic where a specific variable is used to compare the significance assigned to the SNPs in the study. The specific variable that carries prior information about the markers has been shown to improve the subset selection, allowing the researchers to focus on SNPs with higher rate of true associations (reviewed in Duncan et al. 2009).

This observation has paved the road to the development of further methodologies for the selection of SNPs, based on multiple parameter criteria rather than single prior information.

4.4.3 Prioritization of Associated SNPs

The linkage disequilibrium (LD) between SNPs is not considered during the statistical analysis, while the *p*-value associations are calculated for SNP biomarker discovery in a GWAS setting; therefore, concentrating solely on the statistical evidence alone is not an adequate approach. So, it is highly unlikely that the true causal polymorphisms would be identified by using a single threshold for the *p*-values of association as the genotypes will be very similar. Additionally, simulations showed that even when there is weak predictive prior knowledge and even the ones that are truly predictive are unclear, the prioritization through hierarchical Bayes ranking utilizing multiple features led to better power than just relying of *p*-value thresholds (Duncan et al. 2009).

The necessity of prioritizing SNPs for subsequent GWAS and/or functional experiments leads the search for optimum prioritization method for SNPs. Alongside with statistical evidence, use of other available information such as biological/functional effects of SNPs and other contributing factors to the disease susceptibility like conservation, natural selection, and microRNA binding has been suggested (reviewed in Üstünkar 2011). Today many software tools are available for the systematic prioritization of SNPs based on the various of biological information available in public databases such as SPOT (Saccone 2010), FunctSNP (Goodswen et al. 2010), FASTSNP (Yuan 2006), SNPLogic (Pico 2009), SNPinfo (Xu and Taylor 2009), and SNPit (Shen 2009).

Recently, using meta-analysis is also suggested for prioritizing genes and SNPs for subsequent studies after GWAS. Besides combining test statistics from comparable studies, other statistical approaches like Fisher's method of combining *p*-values are proposed (Fisher 1948). Another method favored by researchers is converting the test statistics into *z* scores and using odds ratios and regression coefficients. There are various number studies for SNP prioritization utilizing meta-analysis results (reviewed in Üstünkar 2011). Some of them are based mainly on meta-analysis of the *p*-values, whereas others like METU-SNP utilize predetermined decision rules that are based on available evidence in addition to meta-analysis results to guide the decision (Üstünkar and Aydın Son 2011).

For many of the GWAS results, however, the underlying mechanisms, such as the effect of the genetic variant on regulation, function and structure of the protein product, or transcription, remain unclear, and overall associations are able to explain only a limited amount of heritability. These issues may be clarified by more detailed investigations on the prioritized SNPs in combination with the analyses of rare variants at genomic and epigenomic levels. Further studies on wide range of populations would validate the findings of GWAS and ultimately will provide the evidence needed for developing predictive and personalized approaches for pharmacogenomics applications and maintaining public health.

5 Conclusion

At the twenty-first century, we are fortunate to witness the revolutionary changes in pharmaceutical and medical science due to finalization of Human Genome Project (HGP) in 2003. In the current postgenomic era, there is rapid increase in the amount of genomic knowledge, which could be utilized to improve pharmacotherapy via pharmacogenomics by identifying *patients for ADR risk*. Pharmacogenomics studies of variability in drug safety and efficacy using information from the entire genome of a patient without a priori hypotheses specific to a candidate gene is the cornerstone of P4 medicine (Gurwitz and Lunshof 2011). However, pharmacogenomics is still at its infancy. Even though clinically meaningful associations have been established for few drugs, and various technologies that are now being used such as gene expressions, genome-wide associations studies, epigenomics, proteomics, and metabolomics, it rarely provides simple, clear-cut result

(Kirchheiner et al. 2005; Swen et al. 2007; Crowley et al. 2009; Daly 2010; Stuart 2011; Clayton et al. 2006, 2009; Ingelman-Sundberg and Gomez 2010). Prior to pharmacogenetic testing being routinely available in clinical settings, there is an urgent need to have more prospective evidence to support its clinical utility.

Recently developed pharmacogenomics tests and emerging technologies should be carefully implemented in clinical practice as soon as they are available, for the public to appreciate the benefits of personalized medicine/healthcare. Teaching the current, continuously updated knowledge of pharmacogenomics should not be postponed until the new paradigm arrives. Education of healthcare providers, graduate medical and pharmacy students, policymakers, the public, and governments about genomics medicine and bioeconomy contributes to "collective learning" essential in the postgenomic era for integration of pharmacogenomics as a part of P4 medicine into mainstream healthcare (Hizel et al. 2009). Legal aspects should also be considered in terms of the problem of availability and the degree of qualification of commercial drug tests on the market (Williams-Jones and Corrigan 2003; Ozdemir 2010; Ozdemir et al. 2011, 2012b; Ozdemir and Knoppers 2010).

References

Abraham JE, Maranian MJ, Driver KE, Platte R, Kalmyrzaev B, Baynes C, Luccarini C, Shah M, Ingle S, Greenberg D, Earl HM, Dunning AM, Pharoah PD, Caldas C (2010) CYP2D6 gene variants: association with breast cancer specific survival in a cohort of breast cancer patients from the United Kingdom treated with adjuvant tamoxifen. Breast Cancer Res 12:R64

Albers LJ, Ozdemir V (2004) Pharmacogenomic-guided rational therapeutic drug monitoring: conceptual framework and application platforms for atypical antipsychotics. Curr Med Chem 11:297–312

Aleku M, Schulz P, Keil O, Santel A, Schaeper U, Dieckhoff B, Janke O, Endruschat J, Durieux B, Roder N, Loffler K, Lange C, Fechtner M, Mopert K, Fisch G, Dames S, Arnold W, Jochims K, Giese K, Wiedenmann B, Scholz A, Kaufmann J (2008) Atu027, a liposomal small interfering RNA formulation targeting protein kinase N3, inhibits cancer progression. Cancer Res 68:9788–9798

Altman DG, Goodman SN (1994) SN transfer of technology from statistical journals to the biomedical literature. Past Trends Future Predict 272:129–132

Amarzguioui M, Prydz H (2004) An algorithm for selection of functional siRNA sequences. Biochem Biophys Res Commun 316:1050–1058

Amarzguioui M, Holen T, Babaie E, Prydz H (2003) Tolerance for mutations and chemical modifications in a siRNA. Nucleic Acids Res 31:589–595

Anderson E, Boese Q, Khvorova A, Karpilow J (2008) Identifying siRNA-induced off-targets by microarray analysis. Methods Mol Biol 442:45–63

Arabsolghar R, Rasti M (2012) Optimal electroporation condition for small interfering RNA transfection into MDA-MB-468 cell line. Iran J Med Sci 37:187–193

Azuaje F (2010) Bioinformatics and biomarker discovery: omic data analysis for personalized medicine. Wiley, Chichester

Bartlett G, Zgheib N, Manamperi A, Wang W, Hizel C, Kahveci R, Yazan Y (2012) Pharmacogenomics in primary care: a crucial entry point for global personalized medicine? Curr Pharmacogenomics Person Med 10:101–105

Becquemont L (2009) Pharmacogenomics of adverse drug reactions: practical applications and perspectives. Pharmacogenomics 10:961–969

Berns K, Hijmans EM, Mullenders J, Brummelkamp TR, Velds A, Heimerikx M, Kerkhoven RM, Madiredjo M, Nijkamp W, Weigelt B, Agami R, Ge W, Cavet G, Linsley PS, Beijersbergen RL, Bernards R (2004) A large-scale RNAi screen in human cells identifies new components of the p53 pathway. Nature 428:431–437

Boden D, Pusch O, Lee F, Tucker L, Shank PR, Ramratnam B (2003) Promoter choice affects the potency of HIV-1 specific RNA interference. Nucleic Acids Res 31:5033–5038

Caplen NJ, Mousses S (2003) Short interfering RNA (siRNA)-mediated RNA interference (RNAi) in human cells. Ann N Y Acad Sci 1002:56–62

Castillo B, Bromberg L, López X, Badillo V, González Feliciano JA, González CI, Hatton TA, Barletta G (2012) Intracellular delivery of siRNA by polycationic superparamagnetic nanoparticles. J Drug Deliv 2012:218940

CDC (2006) Evaluation of genetic testing. ACCE: A CDC-sponsored project carried out by the Foundation of Blood Research. http://www.cdc.gov/genomics/gtesting/ACCEhtm

Chalupnikova K, Nejepinska J, Svoboda P (2013) Production and application of long dsRNA in mammalian cells. Methods Mol Biol 942:291–314. doi:10.1007/978-1-62703-119-6_16

Chou WH, Yan FX, de Leon J, Barnhill J, Rogers T, Cronin M, Pho M, Xiao V, Ryder TB, Liu WW, Teiling C, Wedlund PJ (2000) Extension of a pilot study: impact from the cytochrome P450 2D6 polymorphism on outcome and costs associated with severe mental illness. J Clin Psychopharmacol 20:246–251

Clayton TA, Lindon JC, Cloarec O, Antti H, Charuel C, Hanton G, Provost JP, Le Net JL, Baker D, Walley RJ, Everett JR, Nicholson JK (2006) Pharmacometabonomic phenotyping and personalized drug treatment. Nature 440:1073–1077

Clayton TA, Baker D, Lindon JC, Everett JR, Nicholson JK (2009) Pharmacometabonomic identification of a significant host-microbiome metabolic interaction affecting human drug metabolism. Proc Natl Acad Sci U S A 106:14728–14733 [Epub 2009 Aug 10]

Corrigan OP (2011) Personalized medicine in a consumer Age. Curr Pharmacogenomics Personal Med 9:168–176

Crowley JJ, Sullivan PF, McLeod HL (2009) Pharmacogenomic genome-wide association studies: lessons learned thus far. Pharmacogenomics 10:161–163

Daka A, Peer D (2012) RNAi-based nanomedicines for targeted personalized therapy. Adv Drug Deliv Rev 64:1508–1521

Daly AK (2010) Genome-wide association studies in pharmacogenomics. Nat Rev Genet 11:241–246

Das AT, Brummelkamp TR, Westerhout EM, Vink M, Madiredjo M, Bernards R, Berkhout B (2004) Human immunodeficiency virus type 1 escapes from RNA interference-mediated inhibition. J Virol 78:2601–2605

Das S, Ghosal S, Kozak K, Chakrabarti J (2012) An siRNA designing tool with a unique functional off-target filtering approach. J Biomol Struct Dyn [Epub ahead of print] PMID: 23140209

Derrien B, Baumberger N, Schepetilnikov M, Viotti C, De Cillia J, Ziegler-Graff V, Isono E, Schumacher K, Genschik P (2012) Degradation of the antiviral component ARGONAUTE1 by the autophagy pathway. Proc Natl Acad Sci U S A 109:15942–15946

Donohue MM, Tirschwell DL (2011) Implications of pharmacogenetic testing for patients taking warfarin or clopidogrel. Curr Neurol Neurosci Rep 11:52–60

Dua P, Yoo JW, Kim S, Lee DK (2011) Modified siRNA structure with a single nucleotide bulge overcomes conventional siRNA-mediated off-target silencing. Mol Ther 19:1676–1687

Duncan TC (2009) Genome-wide association studies for discrete traits. Genet Epidemiol 33(Suppl 1):1098–2272

Duncan TC et al (2009) Methodological issues in multistage genome-wide association studies. Stat Sci 24:414–429

Dykxhoorn DM, Novina CD, Sharp PA (2003) Killing the messenger: short RNAs that silence gene expression. Nat Rev Mol Cell Biol 4:457–467

Eberle MA (2007) Power to detect risk alleles using genome-wide tag SNP panels. PLoS Genet 3(10):e170

Echeverri CJ, Beachy PA, Baum B, Boutros M, Buchholz F, Chanda SK, Downward J, Ellenberg J, Fraser AG, Hacohen N, Hahn WC, Jackson AL, Kiger A, Linsley PS, Lum L, Ma Y, Mathey-Prevot B, Root DE, Sabatini DM, Taipale J, Perrimon N, Bernards R (2006) Minimizing the risk of reporting false positives in large-scale RNAi screens. Nat Methods 3:777–779

Elbashir SM, Harborth J, Weber K, Tuschl T (2002) Analysis of gene function in somatic mammalian cells using small interfering RNAs. Methods 26:199–213

Evrard A, Mbatchi L (2012) Genetic polymorphisms of drug metabolizing enzymes and transporters: the long way from bench to bedside. Curr Top Med Chem 12:1720–1729

Faruki H, Heine U, Brown T, Koester R, Lai-Goldman M (2007) HLA-B*5701 clinical testing: early experience in the United States. Pharmacogenet Genomics 17:857–860

Filipowicz W (2005) RNAi: the nuts and bolts of the RISC machine. Cell 122:17–20

Fire A, Xu S, Montgomery MK, Kostas SA, Driver SE, Mello CC (1998) Potent and specific genetic interference by double-stranded RNA in Caenorhabditis elegans. Nature 391:806–811

Fisher R (1948) Combining independent tests of significance. Am Stat 2(Suppl):30

Fisher R (1958) Statistical methods for research workers. Hafner, New York

Frueh FW, Amur S, Mummaneni P, Epstein RS, Aubert RE, DeLuca TM, Verbrugge RR, Burckart GJ, Lesko LJ (2008) Pharmacogenomic biomarker information in drug labels approved by the United States food and drug administration: prevalence of related drug use. Pharmacotherapy 28:992–998

Ganesan AK, Ho H, Bodemann B, Petersen S, Aruri J, Koshy S, Richardson Z, Le LQ, Krasieva T, Roth MG, Farmer P, White MA (2008) Genome-wide siRNA-based functional genomics of pigmentation identifies novel genes and pathways that impact melanogenesis in human cells. PLoS Genet 4:e1000298

Gaur RK (2006) RNA interference: a potential therapeutic tool for silencing splice isoforms linked to human diseases. Biotechniques 40(Suppl):15–22

Gervasini G, Benítez J, Carrillo JA (2010) Pharmacogenetic testing and therapeutic drug monitoring are complementary tools for optimal individualization of drug therapy. Eur J Clin Pharmacol 66:755–774

Ghosh D, Poisson L (2009) "Omics" data and levels of evidence for biomarker discovery. Genomics 93:13–16

Ginsburg GS, Willard HF (2009) Genomic and personalized medicine: foundations and applications. Transl Res 154:277–287

Gitlin L, Karelsky S, Andino R (2002) Short interfering RNA confers intracellular antiviral immunity in human cells. Nature 418:430–434

Goodswen S et al (2010) FunctSNP: an R package to link SNPs to functional knowledge and dbAutoMaker: a suite of Perl scripts to build SNP databases. BMC Bioinformatics 11(1):311

Gordon D et al (2002) Power and sample size calculations for case-control genetic association tests when errors present: application to single nucleotide polymorphisms. Hum Hered 54:22–33

Grice GR, Seaton TL, Woodland AM et al (2006) Defining the opportunity for pharmacogenetic intervention in primary care. Pharmacogenomics 7(1):61–65

Grishok A, Hoersch S, Sharp PA (2008) RNA interference and retinoblastoma-related genes are required for repression of endogenous siRNA targets in Caenorhabditis elegans. Proc Natl Acad Sci U S A 105:20386–20391

Grossman I (2007) Routine pharmacogenetic testing in clinical practice: dream or reality? Pharmacogenomics 8:1449–1459

Gurwitz D, Lunshof JE (2011) Personalized participatory medicine: sharing knowledge and uncertainty. Genome Med 3:69

Gurwitz D, McLeod HL (2009) Genome-wide association studies: powerful tools for improving drug safety and efficacy. Pharmacogenomics 10:157–159

Haley B, Tang G, Zamore PD (2003) In vitro analysis of RNA interference in Drosophila melanogaster. Methods 30:330–336

Hamilton B, Dong Y, Shindo M, Liu W, Odell I, Ruvkun G, Lee SS (2005) A systematic RNAi screen for longevity genes in C. elegans. Genes Dev 19:1544–1555

Han SE, Kang H, Shim GY, Suh MS, Kim SJ, Kim JS, Oh YK (2008) Novel cationic cholesterol derivative-based liposomes for serum-enhanced delivery of siRNA. Int J Pharm 353:260–269

Hannon GJ (2002) RNA interference. Nature 418:244–251

Heidel JD, Hu S, Liu XF, Triche TJ, Davis ME (2004) Lack of interferon response in animals to naked siRNAs. Nat Biotechnol 22:1579–1582

Hizel C, Gök S, Sardas S, Bernard-Gallon D, Maugard C, Genç E, on behalf of the Istanbul Working Group on Personalized Medicine (2009) Personalized and predictive medicine in Turkey: a symposium report of the Istanbul Working Group on Personalized Medicine, Istanbul, Turkey, September 10–12, 2009. Curr Pharmacogenomics Person Med 7:297–301

Holland AJ, Fachinetti D, Han JS, Cleveland DW (2012) Inducible, reversible system for the rapid and complete degradation of proteins in mammalian cells. Proc Natl Acad Sci U S A 109:E3350–E3357

Hong-Guang Xie, Frueh FW (2005) Pharmacogenomics steps toward personalized medicine. Personal Med 2:325–337 http://www.fda.gov/Drugs/ScienceResearch/ResearchAreas/Pharmacogenetics/ucm083378.htm

Huang RS, Ratain MJ (2009) Pharmacogenetics and pharmacogenomics of anticancer agents. CA Cancer J Clin 59:42–55

Hughes S, Hughes A, Brothers C, Spreen W, Thorborn D, CNA106030 Study Team (2008) PREDICT-1 (CNA106030): the first powered, prospective trial of pharmacogenetic screening to reduce drug adverse events. Pharm Stat 7:121–129

Huppi K, Martin SE, Caplen NJ (2005) Defining and assaying RNAi in mammalian cells. Mol Cell 17:1–10

Ingelman-Sundberg M, Gomez A (2010) The past, present and future of pharmacoepigenomics. Pharmacogenomics 11:625–627

Jackson AL, Bartz SR, Schelter J, Kobayashi SV, Burchard J, Mao M, Li B, Cavet G, Linsley PS (2003) Expression profiling reveals off-target gene regulation by RNAi. Nat Biotechnol 21:635–637

Jennings L, Van Deerlin VM, Gulley ML, College of American Pathologists Molecular Pathology Resource Committee (2009) Recommended principles and practices for validating clinical molecular pathology tests. Arch Pathol Lab Med 133:743–755

Jones SW, Lindsay MA (2004) Overview of target validation and the impact of oligonucleotides. Curr Opin Mol Ther 6:546–550, Review

Jones SW, Souza PM, Lindsay MA (2004) siRNA for gene silencing: a route to drug target discovery. Curr Opin Pharmacol 4:522–527

Khoury MJ (2009) Interview: Dr. Muin J. Khoury discusses the future of public health genomics and why it matters for personalized medicine and global health. Curr Pharmacogenomics Person Med 7:158–163

Khoury MJ, Rich EC, Randhawa G, Teutsch SM, Niederhuber J (2009) Comparative effectiveness research and genomic medicine: an evolving partnership for 21st century medicine. Genet Med 11:707–711

Kim JK, Gabel HW, Kamath RS, Tewari M, Pasquinelli A, Rual JF, Kennedy S, Dybbs M, Bertin N, Kaplan JM, Vidal M, Ruvkun G (2005) Functional genomic analysis of RNA interference in C. elegans. Science 308:1164–1167

Kim JY, Choung S, Lee EJ, Kim YJ, Choi YC (2007) Immune activation by siRNA/liposome complexes in mice is sequence- independent: lack of a role for Toll-like receptor 3 signaling. Mol Cells 24(2):247–254

Kimura J, Nguyen ST, Liu H, Taira N, Miki Y, Yoshida K (2008) A functional genome-wide RNAi screen identifies TAF1 as a regulator for apoptosis in response to genotoxic stress. Nucleic Acids Res 36:5250–5259

Kirchheiner J, Fuhr U, Brockmöller J (2005) Pharmacogenetics-based therapeutic recommendations – ready for clinical practice? Nat Rev Drug Discov 4:639–647

Kittler R, Surendranath V, Heninger AK, Slabicki M, Theis M, Putz G, Franke K, Caldarelli A, Grabner H, Kozak K, Wagner J, Rees E, Korn B, Frenzel C, Sachse C, Sönnichsen B, Guo J, Schelter J, Burchard J, Linsley PS, Jackson AL, Habermann B, Buchholz F (2007) Genome-wide resources of endoribonuclease-prepared short interfering RNAs for specific loss-of-function studies. Nat Methods 4(4):337–344

Kitzmiller JP, Groen DK, Phelps MA, Sadee W (2011) Pharmacogenomic testing: relevance in medical practice: why drugs work in some patients but not in others. Cleve Clin J Med 78:243–257

Krynetski EY, Tai HL, Yates CR, Fessing MY, Loennechen T, Schuetz JD, Relling MV, Evans WE (1996) Genetic polymorphism of thiopurine S-methyltransferase: clinical importance and molecular mechanisms. Pharmacogenetics 6:279–290

Kubo T, Yanagihara K, Takei Y, Mihara K, Sato Y, Seyama T (2012) SiRNAs conjugated with aromatic compounds induce RISC-mediated antisense strand selection and strong gene-silencing activity. Biochem Biophys Res Commun 426(4):571–577

Kuuselo R, Savinainen K, Azorsa DO, Basu GD, Karhu R, Tuzmen S, Mousses S, Kallioniemi A (2007) Intersex-like (IXL) is a cell survival regulator in pancreatic cancer with 19q13 amplification. Cancer Res 67:1943–1949

La Thangue NB, Kerr DJ (2011) Predictive biomarkers: a paradigm shift towards personalized cancer medicine. Nat Rev Clin Oncol 8:587–596. doi:10.1038/nrclinonc.2011.121

Lai-Goldman M, Faruki H (2008) Abacavir hypersensitivity: a model system for pharmacogenetic test adoption. Genet Med 10:874–878

Lazarou J, Pomeranz B, Corey P (1998) Incidence of adverse drug reactions in hospitalized patients. JAMA 279:1200–1205

Lee KC, Ma JD, Kuo GM (2010) Pharmacogenomics: bridging the gap between science and practice. J Am Pharm Assoc 50:e1–e14; quiz e15–17

Lehmann EL (1993) The Fisher, Neyman-Pearson theories of testing hypotheses: one theory or two? J Am Stat Assoc 88:1241–1249

Lesko LJ, Zineh I, Huang SM (2010) What is clinical utility and why should we care? Clin Pharmacol Ther 88:729–733

Lima WF, Prakash TP, Murray HM, Kinberger GA, Li W, Chappell AE, Li CS, Murray SF, Gaus H, Seth PP, Swayze EE, Crooke ST (2012) Single-stranded siRNAs activate RNAi in animals. Cell 150(5):883–894

Lindpaintner K (2002) The impact of pharmacogenetics and pharmacogenomics on drug discovery. Nat Rev Drug Discov 1:463–469

Ma Z, Li J, He F, Wilson A, Pitt B, Li S (2005) Cationic lipids enhance siRNA-mediated interferon response in mice. Biochem Biophys Res Commun 330:755–759

MacRae IJ, Ma E, Zhou M, Robinson CV, Doudna JA (2008) In vitro reconstitution of the human RISC-loading complex. Proc Natl Acad Sci U S A 105:512–517

Mariotti M, Castiglioni S, Maier JA (2009) Inhibition of T24 human bladder carcinoma cell migration by RNA interference suppressing the expression of HD-PTP. Cancer Lett 273:155–163

Marsh S, Hoskins JM (2010) Irinotecan pharmacogenomics. Pharmacogenomics 11:1003–1010

Matranga C, Zamore PD (2007) Small silencing RNAs. Curr Biol 17:R789–R793

McConnell KI, Schweller RM, Diehl MR, Suh J (2011) Live-cell microarray surface coatings supporting reverse transduction by adeno-associated viruses. BioTechniques 51(4):255–258

McLeod HL, Krynetski EY, Relling MV, Evans WE (2000) Genetic polymorphism of thiopurine methyltransferase and its clinical relevance for childhood acute lymphoblastic leukemia. Leukemia 14:567–572

McManus MT, Sharp PA (2002) Gene silencing in mammals by small interfering RNAs. Nat Rev Genet 3:737–747

Medarova Z, Pham W, Farrar C, Petkova V, Moore A (2007) In vivo imaging of siRNA delivery and silencing in tumors. Nat Med 13:372–377

Medarova Z, Kumar M, Ng SW, Yang J, Barteneva N, Evgenov NV, Petkova V, Moore A (2008) Multifunctional magnetic nanocarriers for image-tagged siRNA delivery to intact pancreatic islets. Transplantation 86:1170–1177

Meyer UA (2004) Pharmacogenetics – five decades of therapeutic lessons from genetic diversity. Nat Rev Genet 5:669–676

Milos PM, Seymour AB (2004) Emerging strategies and applications of pharmacogenomics. Hum Genomics 1:444–455

Mousses S, Caplen NJ, Cornelison R, Weaver D, Basik M, Hautaniemi S, Elkahloun AG, Lotufo RA, Choudary A, Dougherty ER, Suh E, Kallioniemi O (2003) RNAi microarray analysis in cultured mammalian cells. Genome Res 13:2341–2347

Mrazek DA, Lerman C (2011) Facilitating clinical implementation of pharmacogenomics. JAMA 306:304–305

Naito Y, Yamada T, Ui-Tei K, Morishita S, Saigo K (2004) siDirect: highly effective, target-specific siRNA design software for mammalian RNA interference. Nucleic Acids Res 32(Web Server issue):W124–W129

Nakayashiki H, Nguyen QB (2008) RNA interference: roles in fungal biology. Curr Opin Microbiol 11:494–502

Navratil V, de Chassey B, Combe CR, Lotteau V (2011) When the human viral infectome and diseasome networks collide: towards a systems biology platform for the aetiology of human diseases. BMC Syst Biol 5:13

Neyman J, Pearson E (1933) On the problem of the most efficient tests of statistical hypotheses. Philos Trans R Soc A 231:289–337

Nolte A, Ott K, Rohayem J, Walker T, Schlensak C, Wendel HP (2013) Modification of small interfering RNAs to prevent off-target effects by the sense strand. N Biotechnol 30:159–165

Normile D (2007) Consortium hopes to map human history in Asia. Science 306(5702):1667

O'Donnell PH, Bush A, Spitz J, Danahey K, Saner D, Das S, Cox NJ, Ratain MJ (2012) The 1200 patients project: creating a new medical model system for clinical implementation of pharmacogenomics. Clin Pharmacol Ther 92(4):446–449

Ohrt T, Mutze J, Staroske W, Weinmann L, Hock J, Crell K, Meister G, Schwille P (2008) Fluorescence correlation spectroscopy and fluorescence cross-correlation spectroscopy reveal the cytoplasmic origination of loaded nuclear RISC in vivo in human cells. Nucleic Acids Res 36:6439–6449

Ortega-Paino E, Fransson J, Ek S, Borrebaeck CA (2008) Functionally associated targets in mantle cell lymphoma as defined by DNA microarrays and RNA interference. Blood 111:1617–1624

Ovcharenko D, Jarvis R, Hunicke-Smith S, Kelnar K, Brown D (2005) High-throughput RNAi screening in vitro: from cell lines to primary cells. RNA 11:985–993

Ozdemir V (2010) Pharmacogenomics: reflecting on the old and new social, ethical and policy issues in post-genomics medicine. In: Schwab M, Kaschka WP, Spina E (eds) Pharmacogenomics in psychiatry, vol 25, Advances in biological psychiatry. Karger, Basel, pp 12–29

Ozdemir V, Knoppers BM (2010) One size does not fit all: toward "upstream ethics"? Am J Bioeth 10:42–44

Ozdemir V, Husereau D, Hyland S, Samper S, Salleh MZ (2009) Personalized medicine beyond genomics: new technologies, global health diplomacy and anticipatory governance. Curr Pharmacogenomics Person Med 7:225–230

Ozdemir V, Joly Y, Knoppers BM (2011) ACCE, pharmacogenomics, and stopping clinical trials: time to extend the CONSORT statement? Am J Bioeth 11:11–13

Ozdemir V, Fisher E, Dove ES, Burton H, Wright GEB, Masellis M, Warnich L (2012a) End of the beginning and public health pharmacogenomics: knowledge in 'mode 2' and P5 medicine. Curr Pharmacogenomics Person Med 10:1–6

Ozdemir V, Joly Y, Dove ES, Karalis A, Avard D, Knoppers BM (2012b) Are we asking the right ethics questions on drug shortages? Suggestions for a global and anticipatory ethics framework. Am J Bioeth 12:13–15

Özer M, Demirci Y, Hizel C, Sarikaya S, Karalti İ, Kaspar Ç, Alpan S, Genç E (2013) Impact of genetic factors (CYP2C9, VKORC1 and CYP4F2) on warfarin dose requirement in the Turkish population. Basic Clin Pharmacol Toxicol 112:209–214

Paddison PJ, Hannon GJ (2003) siRNAs and shRNAs: skeleton keys to the human genome. Curr Opin Mol Ther 5:217–224

Palomaki GE, Bradley LA, Douglas MP, Kolor K, Dotson WD (2009) Can UGT1A1 genotyping reduce morbidity and mortality in patients with metastatic colorectal cancer treated with irinotecan? An evidence-based review. Genet Med 11:21–34

Panagiotou AO, Ioannidis JPA (2012) What should the genome-wide significance threshold be? Empirical replication of borderline genetic associations. Int J Epidemiol 41(1):273–286

Pang T (2009) Pharmacogenomics and personalized medicine for the developing world – too soon or just-in-time? A personal view from the World Health Organization. Curr Pharmacogenomics Person Med 7:149–157

Patel C, Muthuswamy J (2012) High efficiency, site-specific transfection of adherent cells with siRNA using microelectrode arrays (MEA). J Vis Exp (67). PII:4415. doi:10.3791/4415

Petrocca F, Lieberman J (2011) Promise and challenge of RNA interference-based therapy for cancer. J Clin Oncol 29(6):747–754, Review

Phillips KA (2008) Closing the evidence gap in the use of emerging testing technologies in clinical practice. JAMA 300:2542–2544

Phillips KA, Veenstra DL, Oren E, Lee JK, Sadee W (2001) Potential role of pharmacogenomics in reducing adverse drug reactions: a systematic review. JAMA 286:2270–2279

Pico AR (2009) SNPLogic: an interactive single nucleotide polymorphism selection, annotation, and prioritization system. Nucleic Acids Res 37:D803–D809

Pieraets S, Cox L, Gielen O, Cools J (2012) Development of a siRNA and shRNA screening system based on a kinase fusion protein. RNA 18(6):1296–1306 [Epub 2012 Apr 26]

Purcell S, Cherny SS, Sham PC (2003) Genetic power calculator: design of linkage and association genetic mapping studies of complex traits. Bioinformatics 19(1):149–150

Purcell S et al (2007) PLINK: a toolset for whole-genome association and population-based linkage analyses. Am J Hum Genet 81:559–575

Rauch A, Nolan D, Martin A, McKinnon E, Almeida C, Mallal S (2006) Prospective genetic screening decreases the incidence of abacavir hypersensitivity reactions in the Western Australian HIV cohort study. Clin Infect Dis 43:99–102

Reich DE, Lander ES (2001) On the allelic spectrum of human disease. Trends Genet 17:502–510

Reiss SM, American Pharmacists Association (2011) Integrating pharmacogenomics into pharmacy practice via medication therapy management. J Am Pharm Assoc 51:e64–e74

Rines DR, Gomez-Ferreria MA, Zhou Y, DeJesus P, Grob S, Batalov S, Labow M, Huesken D, Mickanin C, Hall J, Reinhardt M, Natt F, Lange J, Sharp DJ, Chanda SK, Caldwell JS (2008) Whole genome functional analysis identifies novel components required for mitotic spindle integrity in human cells. Genome Biol 9:R44

Rocheleau CE (2012) RNA interference: systemic RNAi SIDes with endosomes. Curr Biol 22(20):R873–R875. doi:10.1016/j.cub.2012.08.039

Roguev A, Bandyopadhyay S, Zofall M, Zhang K, Fischer T, Collins SR, Qu H, Shales M, Park HO, Hayles J, Hoe KL, Kim DU, Ideker T, Grewal SI, Weissman JS, Krogan NJ (2008) Conservation and rewiring of functional modules revealed by an epistasis map in fission yeast. Science 322:405–410

Rossi JJ (2008) Expression strategies for short hairpin RNA interference triggers. Hum Gene Ther 19:313–317

Ruffié J (1993) Naissance de la Médecine Prédictive. Odile Jacob, Paris

Ryan MC, Zeeberg BR, Caplen NJ, Cleland JA, Kahn AB, Liu H, Weinstein JN (2008) SpliceCenter: a suite of web-based bioinformatic applications for evaluating the impact of alternative splicing on RT-PCR, RNAi, microarray, and peptide-based studies. BMC Bioinformatics 9:313

Saccone SF (2010) SPOT: a web-based tool for using biological databases to prioritize SNPs after a genome-wide association study. Nucleic Acids Res 38:W201–W209

Sadee W (2011) Genomics and personalized medicine. Int J Pharm 415:2–4

Sagoo GS, Little J, Higgins JPT (2009) Systematic reviews of genetic association studies. Human Genome Epidemiology Network [Makale]. PLoS Med 6(3):e28

Saleem S, Schwedes CC, Ellis LL, Grady ST, Adams RL, Johnson N, Whittington JR, Carney GE (2012) Drosophila melanogaster p24 trafficking proteins have vital roles in development and reproduction. Mech Dev 129(5–8):177–191

Sanderson S, Zimmern R, Kroese M, Higgins J, Patch C, Emery J (2005) How can the evaluation of genetic tests be enhanced? Lessons learned from the ACCE framework and evaluating genetic tests in the United Kingdom. Genet Med 7:495–500

Sano M, Sierant M, Miyagishi M, Nakanishi M, Takagi Y, Sutou S (2008) Effect of asymmetric terminal structures of short RNA duplexes on the RNA interference activity and strand selection. Nucleic Acids Res 36:5812–5821

Sasieni PD (1997) From genotypes to genes: doubling the sample size. Biometrics 53:1253–1261

Scacheri PC, Rozenblatt-Rosen O, Caplen NJ, Wolfsberg TG, Umayam L, Lee JC, Hughes CM, Shanmugam KS, Bhattacharjee A, Meyerson M, Collins FS (2004) Short interfering RNAs can induce unexpected and divergent changes in the levels of untargeted proteins in mammalian cells. Proc Natl Acad Sci U S A 101:1892–1897

Schlabach MR, Luo J, Solimini NL, Hu G, Xu Q, Li MZ, Zhao Z, Smogorzewska A, Sowa ME, Ang XL, Westbrook TF, Liang AC, Chang K, Hackett JA, Harper JW, Hannon GJ, Elledge SJ (2008) Cancer proliferation gene discovery through functional genomics. Science 319:620–624

Schnettler E, Sterken MG, Leung JY, Metz SW, Geertsema C, Goldbach RW, Vlak JM, Kohl A, Khromykh AA, Pijlman GP (2012) Noncoding flavivirus RNA displays RNA interference suppressor activity in insect and Mammalian cells. J Virol 86:13486–13500

Schøler LV, Møller TH, Nørgaard S, Vestergaard L, Olsen A (2012) Isolating genes involved with genotoxic drug response in the nematode Caenorhabditis elegans using genome-wide RNAi screening. Methods Mol Biol 920:27–38. doi:10.1007/978-1-61779-998-3_3

Schwarz EM, Kato M, Sternberg PW (2012) Functional transcriptomics of a migrating cell in Caenorhabditis elegans. Proc Natl Acad Sci U S A 109(40): 16246–16251. doi:10.1073/pnas.1203045109 [Epub 2012 Sep 18]

Scott SA (2011) Personalizing medicine with clinical pharmacogenetics. Genet Med 13:987–995. doi:10.1097/GIM.0b013e318238b38c

Seip RL, Duconge J, Ruaño G (2010) Implementing genotype-guided antithrombotic therapy. Future Cardiol 6:409–424

Semizarov D, Kroeger P, Fesik S (2004) siRNA-mediated gene silencing: a global genome view. Nucleic Acids Res 32:3836–3845

Sgaier SK, Jha P, Mony P, Kurpad A, Lakshmi V, Kumar R, Ganguly NK (2007) Biobanks in developing countries: needs and feasibility. Science 318: 1074–1075

Sheffield LJ, Phillimore HE (2009) Clinical use of pharmacogenomic tests in. Clin Biochem Rev 30:55–65

Shen TH (2009) Carlson CS and Tarczy-Hornoch P SNPit: a federated data integration system for the purpose of functional SNP annotation. Comput Methods Programs Biomed 95(2):181–189

Shum D, Bhinder B, Ramirez CN, Radu C, Calder PA, Beauchamp L, Farazi T, Landthaler M, Tuschi T, Magdaleno S, Djaballah H (2013) An arrayed RNA interference genome-wide screen identifies candidate genes involved in the MicroRNA 21 biogenesis pathway. Assay Drug Dev Technol 11:191–205

Sibley CR, Wood MJ (2011) Identification of allele-specific RNAi effectors targeting genetic forms of Parkinson's disease. PLoS One 6(10):e26194 [Epub 2011]

Silva JM, Marran K, Parker JS, Silva J, Golding M, Schlabach MR, Elledge SJ, Hannon GJ, Chang K (2008) Profiling essential genes in human mammary cells by multiplex RNAi screening. Science 319:617–620

Skol AD et al (2006) Joint analysis is more efficient than replication-based analysis for two-stage genome-wide association studies. Nat Genet 38:209–213

Sonnichsen B, Koski LB, Walsh A, Marschall P, Neumann B, Brehm M, Alleaume AM, Artelt J, Bettencourt P, Cassin E, Hewitson M, Holz C, Khan M, Lazik S, Martin C, Nitzsche B, Ruer M, Stamford J, Winzi M, Heinkel R, Roder M, Finell J, Hantsch H, Jones SJ, Jones M, Piano F, Gunsalus KC, Oegema K, Gonczy P, Coulson A, Hyman AA, Echeverri CJ (2005) Full-genome RNAi profiling of early embryogenesis in Caenorhabditis elegans. Nature 434:462–469

Spielman RS, McGinnis RE, Ewens WJ (1993) Transmission test for linkage disequilibrium: the insulin gene region and insulin-dependent diabetes mellitus (IDDM). Am J Hum Genet 52:506–516

Squassina A, Manchia M, Manolopoulos VG, Artac M, Lappa-Manakou C, Karkabouna S, Mitropoulos K, Del Zompo M, Patrinos GP (2010) Realities and expectations of pharmacogenomics and personalized medicine: impact of translating genetic knowledge into clinical practice. Pharmacogenomics 11:1149–1167

Storey JD, Tibshirani R (2003) Statistical significance for genomewide studies. Proc Natl Acad Sci U S A 100(16):9440–9445

Stroschein-Stevenson SL, Foley E, O'Farrell PH, Johnson AD (2009) Phagocytosis of Candida albicans by RNAi-treated Drosophila S2 cells. Methods Mol Biol 470:347–358

Stuart AS (2011) Personalizing medicine with clinical pharmacogenetics. Genet Med 13:987–995

Suarez-Kurtz G (2004) Pharmacogenomics in admixed populations: The Brazilian Pharmacogenetics/pharmacogenomics network–REFARGEN. Pharmacogenomics J 4:347–348

Sun TM, Du JZ, Yan LF, Mao HQ, Wang J (2008) Self-assembled biodegradable micellar nanoparticles of amphiphilic and cationic block copolymer for siRNA delivery. Biomaterials 29:4348–4355

Sun X, Rogoff HA, Li CJ (2008) Asymmetric RNA duplexes mediate RNA interference in mammalian cells. Nat Biotechnol 26:1379–1382

Swen JJ, Huizinga TW, Gelderblom H, de Vries EG, Assendelft WJ, Kirchheiner J, Guchelaar HJ (2007) Translating pharmacogenomics: challenges on the road to the clinic. PLoS Med 4:e209

Teutsch SM, Bradley LA, Palomaki GE, Haddow JE, Piper M, Calonge N, Dotson WD, Douglas MP, Berg AO, EGAPP Working Group (2009) The Evaluation of Genomic Applications in Practice and Prevention (EGAPP) initiative: methods of the EGAPP Working Group. Genet Med 11:3–14

Theis M, Buchholz F (2010) MISSION esiRNA for RNAi screening in mammalian cells. J Vis Exp (39). PII:2008. doi:10.3791/2008

Thomas RK, Weir B, Meyerson M (2006) Genomic approaches to lung cancer. Clin Cancer Res 12:4384s–4391s

Tian Y, Liao IH, Zhan X, Gunther JR, Ander BP, Liu D, Lit L, Jickling GC, Corbett BA, Bos-Veneman NG, Hoekstra PJ, Sharp FR (2011) Exon expression and alternatively spliced genes in Tourette Syndrome. Am J Med Genet B Neuropsychiatr Genet 156B(1):72–78

Tomari Y, Du T, Zamore PD (2007) Sorting of Drosophila small silencing RNAs. Cell 130:299–308

Toriyabe N, Hayashi Y, Harashima H (2013) The transfection action activity of R8-modified nanoparticles and siRNA condensation using pH sensitive stearylated-octahistidine. Biomaterials 34:1337–1343

Travella S, Keller B (2009) Down-regulation of gene expression by RNA-induced gene silencing. Methods Mol Biol 478:185–199

Tucker BA, Scheetz TE, Mullins RF, DeLuca AP, Hoffmann JM, Johnston RM, Jacobson SG, Sheffield VC, Stone EM (2011) Exome sequencing and analysis of induced pluripotent stem cells identify the cilia-related gene male germ cell-associated kinase (MAK) as a cause of retinitis pigmentosa. Proc Natl Acad Sci U S A 108(34):E569–E576

Tuschl T (2001) RNA interference and small interfering RNAs. Chembiochem 2:239–245

Tuzmen S, Kiefer J, Mousses S (2007) Validation of short interfering RNA knockdowns by quantitative real-time PCR. Methods Mol Biol 353:177–203

Tuzmen S, Tuzmen P, Arora S, Mousses S, Azorsa D (2011) RNAi-based functional pharmacogenomics. In: DiStefano JK (ed) Disease gene identification: methods and protocols, Part 4, vol 700, Methods in molecular biology. Springer, New York, pp 271–290

Ui-Tei K, Naito Y, Saigo K (2007) Guidelines for the selection of effective short-interfering RNA sequences for functional genomics. Methods Mol Biol 361:201–216

Uprichard SL (2005) The therapeutic potential of RNA interference. FEBS Lett 579:5996–6007

Üstünkar G (2011) An integrative approach to structured SNP prioritization and representative SNP selection for genome-wide association studies: algorithms and systems. PhD thesis, Middle East Technical University, Ankara

Üstünkar G, Aydın Son Y (2011) METU-SNP: an integrated software system for SNP-complex disease association analysis. J Integr Bioinform 8:187

Vainio P, Wolf M, Edgren H, He T, Kohonen P, Mpindi JP, Smit F, Verhaegh G, Schalken J, Perälä M, Iljin K, Kallioniemi O (2012) Integrative genomic, transcriptomic, and RNAi analysis indicates a potential oncogenic role for FAM110B in castration-resistant prostate cancer. Prostate 72(7):789–802

Vanhecke D, Janitz M (2004) High-throughput gene silencing using cell arrays. Oncogene 23:8353–8358

Volpe T, Schramke V, Hamilton GL, White SA, Teng G, Martienssen RA, Allshire RC (2003) RNA interference is required for normal centromere function in fission yeast. Chromosome Res 11:137–146

Wang L (2010) Pharmacogenomics: a systems approach. Wiley Interdiscip Rev Syst Biol Med 2:3–22

Wang W, Barratt B, Clayton D, Todd J (2005) Genome-wide association studies: theoretical and practical concerns. Nat Rev Genet 6:109–118

Wang J, Liang H, Zou G (2009) Optimal 2-stage design with given power in association studies. Biostat 10(2):324–326

Wegmann D, Dupanloup I, Excoffier L (2008) Width of gene expression profile drives alternative splicing. PLoS One 3(10):e3587

Wendler F, Gillingham AK, Sinka R, Rosa-Ferreira C, Gordon DE, Franch-Marro X, Peden AA, Vincent JP, Munro S (2010) A genome-wide RNA interference screen identifies two novel components of the metazoan secretory pathway. EMBO J 29(2): 304–314

Wheeler DB, Carpenter AE, Sabatini DM (2005) Cell microarrays and RNA interference chip away at gene function. Nat Genet 37(Suppl):S25–S30

Willard HF, Angrist M, Ginsburg GS (2005) Genomic medicine: genetic variation and its impact on the future of health care. Philos Trans R Soc Lond B Biol Sci 360:1543–1550

Williams-Jones B, Corrigan OP (2003) Rhetoric and hype: where's the 'ethics' in pharmacogenomics? Am J Pharmacogenomics 3:375–383

Wilson PA, Plucinski M (2011) A simple Bayesian estimate of direct RNAi gene regulation events from differential gene expression profiles. BMC Genomics 12:250

Wolf CR, Smith G, Smith RL (2000) Science, medicine, and the future. Pharmacogenetics 320:987–990

Wu N, Castel D, Debily MA, Vigano MA, Alibert O, Mantovani R, Iljin K, Romeo PH, Gidrol X (2011) Large scale RNAi screen reveals that the inhibitor of DNA binding 2 (ID2) protein is repressed by p53 family member p63 and functions in human keratinocyte differentiation. J Biol Chem 286(23): 20870–20879

Xu Z, Taylor JA (2009) SNPinfo: integrating GWAS and candidate gene information into functional SNP selection for genetic association studies. Nucleic Acids Res 37:W600–W605

Yesupriya A et al (2008) The continued need to synthesize the results of genetic associations across multiple studies. Genet Med 10(8):633–635

Yuan H (2006) FASTSNP: an always up-to-date and extendable service for SNP function analysis and prioritization. Nucleic Acids Res 34:W635–W641

Zamore PD (2004) Plant RNAi: How a viral silencing suppressor inactivates siRNA. Curr Biol 14:R198–R200

Zhang XD (2010) An effective method for controlling false discovery and false nondiscovery rates in genome-scale RNAi screens. J Biomol Screen 15(9): 1116–1122

Zhang X, Yang X, Chung N, Gates A, Stec E, Kunapuli P, Holder D, Ferrer M, Espeseth A (2006) Robust statistical methods for hit selection in RNA interference high-throughput screening experiments. Pharmacogenomics 7:299–309

Zhou H, DeLoid G, Browning E, Gregory DJ, Tan F, Bedugnis AS, Imrich A, Koziel H, Kramnik I, Lu Q, Kobzik L (2012) Genome-wide RNAi screen in IFN-γ-treated human macrophages identifies genes mediating resistance to the intracellular pathogen Francisella tularensis. PLoS One 7(2):e31752 [Epub 2012 Feb 16]

Zimmern RL (2009) Genomic medicine: the future is now. Lessons for Hong Kong from the House of Lords Report on Genomic Medicine. Hong Kong Med J 15:324–325

Omics Approaches and Applications in Clinical Trials

7

Sanjeev Sharma and Anjana Munshi

Abstract

Clinical trials are aimed at compiling evidence of efficacy and safety through the use of drugs in a large number of patients. Large numbers are required to overcome the issues such as disease heterogeneity, partial understanding of the underlying disease mechanisms, and variability in drug response adverse drug reactions. Due to poor efficacy and suboptimal safety, the failure rate of potential products in development is more than 90 %. This pipeline attrition has an enormous cost – both financially and time-wise.

Novel approaches and tools of omics, especially "pharmacogenomics," have enabled exploration of the molecular mechanisms underlying differences in drug response and thereby reduce attrition. After the sequencing of human genome, genomic information has benefitted the mankind in a number of ways. The stratification of patients based on their genotype, thus prevents the occurrence of severe drug reactions and better clinical outcome in susceptible patients, is one of the few examples of use of genomics in the area of health care. Pharmacogenomics identifies safety biomarkers and provides an opportunity to rescue a compound that would otherwise fail to be commercialized due to safety concerns. Further the knowledge of pharmacogenomics along with bioinformatics during different phases of clinical trials will formulate the drugs with more therapeutic effectiveness and with minimum adverse effects. This will also reduce the cost of adverse drug reactions being experienced by the patients and chances of withdrawal of the drug molecule from the market.

S. Sharma (✉)
Department of Clinical Pharmacology,
Apollo Health City, Jubilee Hills, Hyderabad,
Andhra Pradesh, India
e-mail: sanjeevsham@yahoo.co.uk

A. Munshi
Department of Molecular Biology, Institute of
Genetics and Hospital for Genetic Diseases,
Osmania University, Begumpet, Hyderabad 500016,
Andhra Pradesh, India

D. Barh et al. (eds.), *Omics for Personalized Medicine*,
DOI 10.1007/978-81-322-1184-6_7, © Springer India 2013

1 Introduction

A clinical trial is defined as any investigation in human subjects intended to discover or verify the clinical, pharmacological, and/or other pharmacodynamic effects of an investigational product(s) and/or to identify any adverse reactions to an investigational product(s) and/or to study absorption, distribution, metabolism, and excretion of an investigational product(s) with the object of ascertaining its safety and/or efficacy (Aronson 2004).

The evolution of clinical research has been a long and fascinating journey. Evidences from history suggest that drug discoveries are the result of serendipity and a product of trial and error method "guided as much by the intuition and serendipity of chemists, biologists and physicians as by any rational linear process" (Papanikolaw 1999). Prior to the existence of pharmaceutical industry, drugs were discovered by accident and their uses passed down by verbal and written records (Ratti and Trist 2001). People have found berries, roots, and barks that could be used to treat or to alleviate symptoms of illness (Boa 2003). One of the best examples is of willow bark containing salicin, which was used to reduce fever, and of cinchona bark from which quinine was discovered (Boa 2003).

In the twentieth century, when the pharmacological basis of disease and drugs were defined, the drug discovery process was based on the conventional methods of hit and trial (Papanikolaw 1999). The pharmaceutical industry was initiating screening of known compounds or randomly testing any available molecules. A few examples of successful drugs, which were discovered in this way, are chlorpromazine and meprobamate benzodiazepines (Ratti and Trist 2001). In this process expected potential lead molecules were obtained from serendipity or from screening of the chemical diversity (Papanikolaw 1999). These molecules were further optimized by medicinal chemist and were passed to the process of development and eventually into the market. However, this approach suffered from a drawback of lack of sufficient molecules with high enough structural diversity and efficacy.

In the 1980s, more rational approach was developed using in vitro assays (rather than previously used in vivo assays) on animal tissues which became a central part in the process for giving valuable information on structure–activity relationships. This also helped in pharmacophore construction. Therefore, if the lead molecule fails, then there is sufficient information on structure and activity in order to allow the cause for failure to be extricated from the molecule (Papanikolaw 1999).

From the very first recorded trial of legumes in biblical times to the first randomized controlled trial of streptomycin in 1946, the history of clinical trial is full of examples (Bhatt 2010). In 1747, scurvy trial was conducted by James Lind which contained most of the elements of a controlled trial. The UK Medical Research Council's (MRC) trial of patulin for the treatment of common cold in year 1943 was the first double-blind controlled trial. These studies have paved the way for conducting the first randomized control trial of streptomycin in pulmonary tuberculosis which was carried out in 1946 by MRC of the UK.

Over the years, as the discipline of clinical trials grew in sophistication and influence, a number of clinical trials have come into existence with groundbreaking results. The clinical trials consist of four phases: phase I is conducted on healthy volunteers and mainly include pharmacokinetic studies, phase II is designed to explore the therapeutic efficacy and tolerability in diseased individuals, phase III trials are conducted with large sample sizes and produce confirmatory results, whereas phase IV trials assess the long-term drug efficacy and also include post market surveillance studies.

The most common cause of termination of a clinical trial or failure of a drug molecule is lack of desired therapeutic efficacy followed by safety concerns (McCarthy et al. 2005). This occurs in phase II and phase III trials. Phase III trials utilize a very large share of the resources and are conducted on a large population group. Therefore, the unsuccessful ending of a trial is unacceptable to a pharmaceutical firm in terms of time loss and financial loss incurred. The complete process of drug development is extremely expensive. It has been found that industry-funded sources put the cost at approximately €500–800 million per marketed drug (Anon 2001; DiMasi 2002; DiMasi et al. 2003). Few suggest that the figure is considerably lower (Henry and Lexchin 2002; Public Citizen 2001). After discovery of the compound, it takes approximately 10–15 years for a drug to reach to the market (Anon 2001). In addition, there is a high attrition rate with only one out of every 5,000 chemical compounds considered to have a therapeutic potential of being successfully developed for clinical use.

Before the advent of pharmacogenomics and proteomics, efficacy and safety concerns were poorly determined in a clinical trial leading to higher attrition rate. Nowadays, the predictability of safety and efficacy of a drug has increased to a significant level as these are influenced by genomic profile of an individual which can be assessed by pharmacogenomic studies (Surendiran et al. 2008).

The last decade has seen remarkable technological developments in many sectors of biological and medical research. The sequencing of Human Genome and International HapMap projects has increased the expectations in health-care field by generating the tools which are aimed at identifying and characterizing human diversity (Shyh-Yoh et al. 2012). These projects have provided a large resource of single nucleotide polymorphisms (SNPs), explaining the variations between different individuals and different ethnic groups. Now there is lot of literature available explaining contribution of genetic variability in drug-metabolizing enzymes, transporters, and drug targets toward interindividual differences in efficacy and toxicity of many medicines. Pharmacogenomics broadly explains the interindividual variability related to drugs prescribed on the basis of individual's genomic profile. Till now focus of pharmacogenomics was on studying the relationship between DNA variants – mainly in the genes encoding for drug absorption, drug metabolism, and excretion system. Pharmacokinetics refers to the analysis of how drug molecules available in the blood stream, transported to the target site with the help of transporter proteins, binds to the target and subsequently metabolized and excreted. Pharmacodynamics refers to the target receptor binding and subsequent signaling or metabolic events that determine the therapeutic efficacy and is the area of focus of pharmacogenomics recently (Shyh-Yoh et al. 2012). The addition of pharmacogenomic information has improved the understanding of interindividual variability in pharmacokinetic and pharmacodynamic profile of a drug molecule administered in a subgroup of population.

Despite the increase in investment in research almost twofold, the industry has submitted 50 % fewer new drug applications to the US Food and Drug Administration (FDA) in comparison with applications filed in the years of 1997–1998 (Lesko and Woodcock 2004). Moreover, number of new applications filed to regulatory agencies for drug approval has shown a reduction almost every year over the last decade.

2 History of Pharmacogenomics

Numerous efforts have focused on identifying the mechanisms underlying the variability in drug response from individual to individual. Initially, disease heterogeneity, age, gender, diet, coadministration of drugs, renal function, and hepatic function are various causes identified to modify drug response. A large body of literature governs the role of genetic factors in the pathogenesis of various diseases as well as in interindividual variability toward drug response. Genetic factors can account for 20–95 % of the variability in drug therapies (Evans and McLeod 2003). Throughout the history, the art of prescribing medicines has evolved from advances in various fields of science and technology. One of the major areas which has changed the view of researchers and clinicians is pharmacogenomics/pharmacogenetics. The history of pharmacogenetics dates back to 510 BC, when Pythagoras observed that the ingestion of fava beans results in potentially fatal condition in some individuals but not in others (Bhatt 2010). Since then there have been a number of landmarks in the field of drug discovery and development, for example, a link between the inability to taste phenylthiocarbamide (PTC) and an autosomal recessive trait was the first report of pharmacogenomics study in 1932. This report demonstrated that certain chemicals react differently based

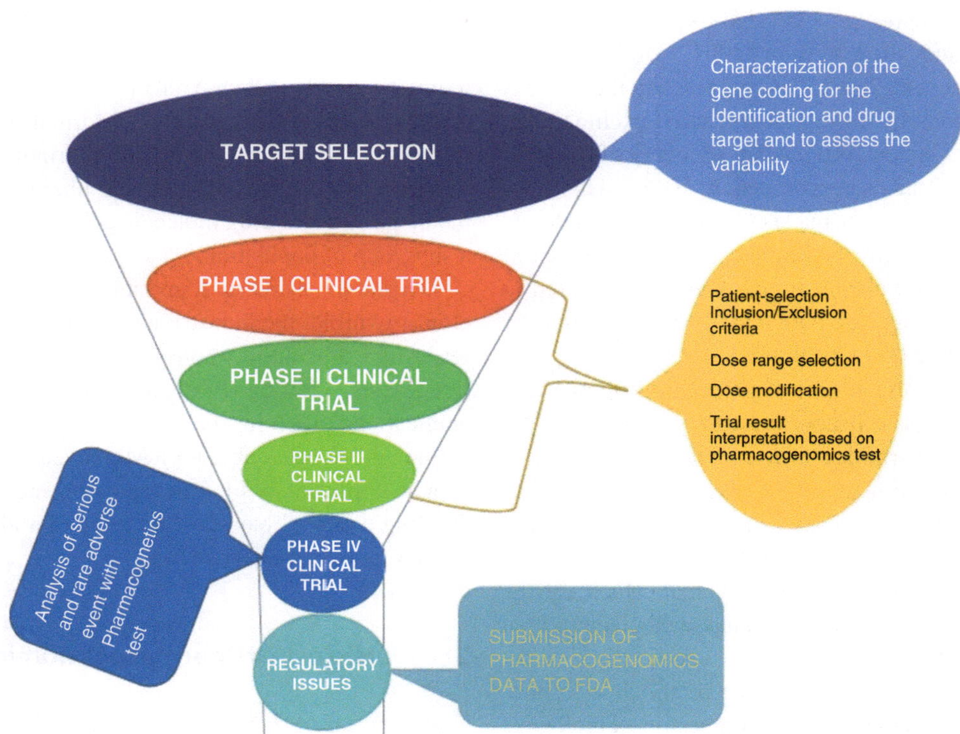

Fig. 7.1 Figure depicting role of pharmacogenomics in various stages of drug development

on genetic differences (Mancinelli et al. 2000). Another example of genetic differences in drug metabolism was the varied response of patients to succinylcholine, described by Kalow in the 1950s (Sweeney 2005). A deficiency in plasma cholinesterase activity (an enzyme) affecting approximately 0.03 % of the population, which was due to an inherited abnormality of succinylcholine metabolism resulted in prolonged paralysis. The term pharmacogenetics was coined in 1959 by Friedrich Vogel to describe the examining of inherited differences in response to drugs (Vogel 1959). Basically the variation in genes in the form of single nucleotide polymorphisms, copy number variation, and others has been found to be implicated in interindividual variability toward drug response and drug-associated adverse effects leading to poor clinical outcome. The rapidly emerging fields of pharmacogenomics and proteomics have become a key phase of new drug development, examples of which can be observed in medications used in pain management, chemotherapeutics, epilepsy, hypertension, cardiovascular disorders, cancer, metabolic disorders, and various other diseases. Therefore, the incorporation of pharmacogenomics in clinical practice has the potential to improve efficacy and reduce toxicity by providing the right drug to the right person in right dose (Pirmohmad and Lewis 2004). The process of drug development includes target identification, preclinical drug development including screening studies using animal models, clinical drug development, and marketing (phase I, II, III, IV). Figure 7.1 outlines the use of pharmacogenomics in drug development process.

3 Target Selection

The process of drug development starts with the identification of a potential target at which a drug can initiate expected effect in a particular disease. Researchers try to find/select a biological target such as a particular metabolizing enzyme, a transporter, a receptor or ion channel, and a protein in signal transduction, to develop the expected drug molecule. The selection of drug or lead molecule involves finding a group of molecule/single molecule which might have the potential to target the protein or enzyme and can modulate its activity. Lead candidates thus selected are chemically modified and pharmacologically characterized to obtain a candidate molecule with desirable pharmacokinetic and pharmacodynamic properties and subjected to preclinical development. It helps to find the right drugs for the right patients so as to avoid undesirable adverse drug effects. A number of research tools are available for exploring new methods for preventing, diagnosing, and treating disease which have replaced the traditional methods of target selection and drug development process. At present the target selection is carried out by the medical genomics, transcriptomics, and proteomics. DNA microarrays and microfluidic devices, high-throughput screening, are the few examples of advanced tools of genetics which are used in pharmacogenomic profiling. The massive amounts of sequence data is handled by using improved genome-based research strategy, informatics and analytical methods, and automated instruments. Genomic techniques have helped us in identifying new gene targets and their association between established biomarkers and drug responses. The Pharmacogenetics Research Network (PGRN), The SNP Consortium (TSC), and the International HapMap Project are large-scale collaborations in pharmacogenomics area. Efforts from the PGRN have already contributed to the development of Pharmacogenomics Knowledge Base program (PharmGKB), by Stanford University. This database evaluates the functions of proteins, identifies polymorphisms, and assesses the relationship of genetic variants to clinical drug responses. The database is made up of patient information such as medical history, drug responses, side effects, and DNA sequences. Currently it contains information on 608 genes, 540 drugs, 523 diseases, and 27 annotated pharmacogenomics genes (Klein et al. 2001; Stanford University 2007). TSC was a nonprofit partnership between major pharmaceutical, technological companies, and academic research centers, formed in year 1999. The aim of TSC was to identify 300,000 single nucleotide polymorphisms (SNPs) of biomedical interest.

The HapMap Project, aimed to identify common genetic variations in humans, was started in 2002 (International HapMap Project 2007). HapMap is a useful tool not only for conducting association studies but also allowing researchers to link variations to predict the risk for specific diseases (Ahn 2007). HapMap will enable the rapid dissection of genetic contributions to drug responses used in the treatment of various diseases. The identification of the haplotype of vitamin K reductase 1 gene (VKORC1) in correlation with warfarin sensitivity was based data generated by HapMap (Geisen et al. 2005). Similarly the role of human leukocyte antigen (HLA) haplotypes in determining the drug-associated adverse drug reactions was another contribution of HapMap (Hung et al. 2005).

4 Preclinical Drug Development

Preclinical phase of drug development process involves the laboratory-based analysis of a potential drug molecule. This phase consists of a wide array of chemical and biochemical assays, animal models, and cell culture methods. Mainly pharmacological studies (pharmacokinetic and pharmacodynamic) and toxicological studies (acute, subacute, chronic, reproductive, and mutagenic) are performed to check the lead molecule. Pharmacogenetics has great impact on preclinical phase of drug development. Interindividual variability in metabolism of drugs is known for years now. The identification of the molecular abnormalities underlying phenotypic variability has led to the development of in vitro screens. For instance, a major advance has been the development of cell lines expressing drug-metabolizing enzymes, e.g., the cytochrome P450 enzymes. The CYP450 group of enzymes is the most versatile group of biological catalysts known to exist in nature and is involved in the metabolism of majority of the available drugs. This allows the consideration of interaction of a drug with a particular enzyme (CYP450 enzyme) at an early stage of development and the subsequent prediction of polymorphic metabolism in human with the possibility of drug–drug interactions (Park and Pirmohamed 2001). Further development of a molecule is abandoned, once it is clear from preclinical studies that a drug is a substrate for a polymorphic metabolizing enzyme, in case, if the molecule is developed, an appropriate warning for prescriber is placed on drug label. This type of screening may be extended to the protein targets on which drugs act, e.g., ion channels and receptors. Further development in pharmacogenomics has been concentrated on the use of gene expression profiling in order to predict toxicity and drug-related adverse effects. A large amount of money is being spent for developing databases of gene expression profiles in relation with known toxicants with the hope that databases would allow future candidate gene selection and reduce attrition rates later in the drug development processes. Although this may help in certain situations, where the adverse effect depends on an idiosyncratic feature found only in a small proportion of patients, it is also important to note that such screens will not replace animal experimentation as they are not absolute predictive (Lindpaintner 2002). However, it is possible that these screens will allow more focused animal experimentation, leading to a reduction in the total number of animals tested, and thereby saving time and cost because of their high-throughput nature (Pirmohmad and Lewis 2004).

5 Phase I–IV Studies

These clinical studies, ranging from "first in man" kinetic and tolerability studies (phase I) in small numbers of healthy volunteers to the large randomized clinical trials designed to assess the efficacy of a compound (phase II and III), provide the basis for regulatory approval. These trials are based on a statistical model of frequency, aimed at compiling evidence of efficacy and safety through the use of the drug in large numbers of subjects (McCarthy et al. 2005). These large number of study subjects are needed to overcome many issues such as disease heterogeneity, partial understanding of underlying disease mechanisms, variability in drug responses, and placebo effects. Currently, the failure rate of potential products in development has been found to be more than 90 %. The main causes for this have been supposed to be poor efficacy and suboptimal safety. This "pipeline attrition" has been estimated to bear enormous cost – both financially and time-wise. The average cost of developing a market product is approximately more than $800 million (DiMasi et al. 2003). The estimated capitalized phase cost for a new chemical entity is about US$1.6 million for animal model testing, 15.2 million for phase I trial, 16.7 million for phase II, and US$27.1 million for phase III trials. The average time from first screening to marketing varies between 8 and 15 years (DiMasi et al. 2003). The large-scale phase III studies consume the majority of these resources and also represent up to 50 % of the overall attrition (Gilbert 2003). Novel approaches and tools such as pharmacogenomics may enable exploration of the pathophysiological mechanisms underlying differences in drug response and reduce attrition (Lesko et al. 2003). Pharmacogenomics can be applied either retrospectively or prospectively in the trials. When we look back over the results of clinical trials, using genomic data to create insights into issues such as the kinetic and dynamic properties of drugs, efficacy, and adverse events, it is termed as retrospective pharmacogenomics. Prospective pharmacogenomics allows proactive identification of patient subgroups (poor metabolizers or rapid metabolizers) that would be predictive of either positive or negative responses to a drug. If this data would be available before or between phase IIa and IIb trials, this would significantly shorten and simplify phase III and increase the probability of success (Lesko and Woodcock 2004). In addition, the ability to prospectively identify subgroups of patients by therapeutic response during early phase II would permit the progression of multiple compounds that can treat overlapping groups of patients with the same disease subcomponent (Roses 2004).

Phase IV refers to the period after the licensing and marketing of a drug. Phase IV studies include hypothesis generating, spontaneous reporting, hypothesis testing, and pharmacoepidemiological analysis. Since phase IV involves exposure of large number of patients to the drug, detection of rare adverse events usually occurs in this phase. Storage of DNA samples from patients treated with the drug in this phase may allow pharmacogenomic testing and identification of genetic predisposing factors. The best example is of abacavir hypersensitivity, where post-marketing studies have found a major genetic predisposing factor in the major histocompatibility complex (MHC) locus (Mallal et al. 2002; Hetherington et al. 2002). For the identification of rare and long-term drug toxicities, reduction in total number of patients may require more structured phase IV trial (Pirmohmad and Lewis 2004).

Further, prospective collection of DNA samples in phase IV might be expensive as the cost of this may have to be borne by the pharmaceutical industry (Roses 2000). This may result in the development of more expensive product, and therefore, this type of a shift in cost, to the field of health care, is unclear at present.

6 Role of Pharmacogenomics in Pharmacokinetics

Pharmacokinetics deals with absorption, distribution, metabolism, and excretion of drugs. Interindividual variation in metabolism appears to limit the rate at which medicines are cleared from the body thus producing toxic effect. From the year 1950 onward, pharmacogenomics is used to explore genetic variation affecting metabolism (Daly 2003). The principle underlying the design of such pharmacogenomic studies has remained essentially the same for many years (McCarthy et al. 2005).

Recently with the availability of more sophisticated tools, many enzymes and transporters involved in the drug disposition and clearance have come under focus. Absorption and distribution are also better understood and the relevant alleles are well characterized, along with their frequency distribution in different groups across the world. Studies on drug clearance on different individuals have revealed that at least 40 % of drug metabolism is via polymorphic CYP450 enzymes (Ingleman-Sundberg 2004). While these data may appear straightforward, we need to be aware of complexities (confounding factors) that may impact the interpretation or analysis in a study. For example, environmental confounders (such as smoking), grapefruit juice, and drug–drug interaction can affect CYP expression and change the metabolic route of a number of drugs which are the substrates for CYP450 enzymes. It has been estimated that 6 % of patients on two medications, 50 % of those on five medications, and nearly 100 % of those on ten medications experience adverse drug reactions (ADRs) because of drug–drug interaction (McCarthy et al. 2005). Interference in the metabolism of drugs through CYPP450 pathways is the major cause of these ADRs. For example, a person who is having CYP2D6 genotype responsible for extensive metabolizer was treated with the potent CYP2D6 inhibitor drug, bupropion, became a phenotypically poor metabolizer. The extensive metabolizers in this enzyme are the result of some of the SNPs in the gene encoding for the enzyme with altered metabolic activity. This example clearly demonstrates the complexity of assessing the exact causes of ADRs based on CYP2D6 genotype (Guzey et al. 2002).

It is well established that CYP2D6 poor metabolizers have increased biological exposure and increased risk of adverse events to a variety of commonly prescribed medicines. Despite the extensive research of literature, there is a shortage of clinical outcome studies demonstrating the clinical benefit of such interventions. This lack of clinical utility data makes it difficult for physicians to apply this information in their prescribing decisions on routine bases.

Although prescribing practices have not yet been significantly altered from the traditional approach of "one size fits all," pharmacogenomic information has already been appeared in the label information provided when the drug is dispensed (McCarthy et al. 2005). In the USA, labels for atomoxetine and 6-mercaptopurine provide information to the physician regarding the metabolism of these medicines by polymorphic enzymes

Table 7.1 DNA-based biomarkers of enzyme activities considered as valid biomarkers

Enzyme	Drugs	Outcome measures	Study results
TPMT	6-MP	Dose-limiting hematopoietic toxicity	More in TPMT deficiency or heterozygosity
UGT1A1	Irinotecan	Grade 3/4 neutropenia, pharmacokinetic parameters (AUC ratio of SN38G/SN38)	UGT1A1 7/7 and 6/7 more frequent than 6/6, UGT1A1*28 and *6 with reduce ratios
CYP2C9	Warfarin	Maintenance dose time to reach stable dosing	Patients with *2 and *3 maintained with lower doses and took longer time to reach stable dosing
CYP2D6	Codeine	Morphine formation, analgesic effects	Higher in extensive metabolizer

Hung et al. (2005)

(CYP2D6 and thiopurine methyltransferase, respectively). Physicians have utilized this information to identify poor metabolizers by doing genotypic tests. But the prescreening of patients by the physicians is not required prior to prescribing these medicines (Lesko et al. 2003).

An updated list of CYP450 enzymes and literature references for in vitro or in vivo activities for various alleles is available online (http://www.imm.ki.se/cypalleles/). In addition to polymorphisms in metabolizing enzymes, there are polymorphisms in transporters, receptors, and other therapeutic target receptor genes. Various reviews have focused on extent to which the metabolizing enzyme genotypes have significantly influenced the drug response in people belonging to different ethnic population (Xie et al. 2001; Evans and McLeod 2003; Weinshilboum 2003; Pauli-Magnus and Kroetz 2004). There are several enzymes that are considered "established" biomarkers based on the criteria described in a recently released guidance on pharmacogenomic data submission (FDA 2005, http://www.fda.gov/cder/guidance/6400fnl.pdf). These valid biomarkers are defined as being measured in an analytical test system with well-established performance characteristics and for which there is evidence about the physiological, toxicological, pharmacological, or clinical significance of the results (FDA 2005). Table 7.1 includes several drug-metabolizing enzymes considered valid biomarkers and summarizes the published correlation data between the genotypes and outcome measures (e.g., clinical efficacy, ADR, doses, PK, and PD) for some drugs. Table 7.2 includes pharmacogenomic biomarkers in drug labels approved by the FDA.

7 Pharmacogenomics of Drug Efficacy

Variable efficacy is an important issue to be considered for drug development, which has direct clinical impact of a drug on patients and health-care systems. The most common reason for terminating the development of medicines is failure to show efficacy in phase II studies. The large-scale phase III studies required to establish safety and efficacy are very expensive; therefore, if more information could be extracted from smaller phase II studies to define the efficacy of a candidate drug molecule more clearly, then valuable time and resources would be saved during phase III. In fact, the variable efficacy of medicines, even in apparently homogeneous patient groups recruited in phase II studies, can obscure true and significant efficacy in a subset of patients, thus leading to inappropriate termination of the drug (McCarthy et al. 2005). Efficacy prediction is a very promising area for pharmacogenomics. Based on genetic and other validated biomarkers, we can identify subgroup of patients in phase II studies, in which new drugs are effective. Compounds or the drug molecule that are effective in patient subgroups may be subjected to further stages of drug development, and this will significantly increase the delivery of new medicines to meet the unmet patient needs and thus increasing the productivity of pharmaceutical research and development. The critical issue is whether the phase II studies are appropriate to generate robust pharmacogenomics data that can be reliably used to support the further development of compounds. Over half a dozen products

Table 7.2 Pharmacogenomic biomarkers in drug labels approved by the FDA

Drug	Therapeutic area	Biomarker	Label sections
Celecoxib	Analgesics	CYP2C9	Dosage and administration, drug interactions, use in specific populations, clinical pharmacology
Codeine	Analgesics	CYP2D6	Warnings and precautions, use in specific populations, clinical pharmacology
Tramadol and acetaminophen	Analgesics	CYP2D6	Clinical pharmacology
Quinidine	Antiarrhythmics	CYP2D6	Precautions
Terbinafine	Antifungals	CYP2D6	Drug interactions
Voriconazole	Antifungals	CYP2C19	Clinical pharmacology, drug interactions
Chloroquine	Anti-infectives	G6PD	Precautions
Rifampin, isoniazid, and pyrazinamide	Anti-infectives	NAT1; NAT2	Adverse reactions, clinical pharmacology
Abacavir	Antivirals	HLA-B*5701	Boxed warning, contraindications, warnings and precautions, patient counseling information
Boceprevir	Antivirals	IL28B	Clinical pharmacology
Maraviroc	Antivirals	CCR5	Indications and usage, warnings and precautions, clinical pharmacology, clinical studies, patient counseling information
Peginterferon alfa-2b	Antivirals	IL28B	Clinical pharmacology
Telaprevir	Antivirals	IL28B	Clinical pharmacology
Carvedilol	Cardiovascular	CYP2D6	Drug interactions, clinical pharmacology
Clopidogrel	Cardiovascular	CYP2C19	Boxed warning, dosage and administration, warnings and precautions, drug interactions, clinical pharmacology
Isosorbide and hydralazine	Cardiovascular	NAT1; NAT2	Clinical pharmacology
Metoprolol	Cardiovascular	CYP2D6	Precautions, clinical pharmacology
Prasugrel	Cardiovascular	CYP2C19	Use in specific populations, clinical pharmacology, clinical studies
Propafenone	Cardiovascular	CYP2D6	Clinical pharmacology
Propranolol	Cardiovascular	CYP2D6	Precautions, drug interactions, clinical pharmacology
Ticagrelor	Cardiovascular	CYP2C19	Clinical studies
Cevimeline	Dermatology and dental	CYP2D6	Drug interactions
Dapsone	Dermatology and dental	G6PD	Indications and usage, precautions, adverse reactions, patient counseling information
Fluorouracil	Dermatology and dental	DPD	Contraindications, warnings
Tretinoin	Dermatology and dental	PML/RARα	Boxed warning, dosage and administration, precautions
Dexlansoprazole (1)	Gastroenterology	CYP2C19	Clinical pharmacology, drug interactions
Dexlansoprazole (2)	Gastroenterology	CYP1A2	Clinical pharmacology
Esomeprazole	Gastroenterology	CYP2C19	Drug interactions, clinical pharmacology
Omeprazole	Gastroenterology	CYP2C19	Dosage and administration, warnings and precautions, drug interactions

(continued)

Table 7.2 (continued)

Drug	Therapeutic area	Biomarker	Label sections
Pantoprazole	Gastroenterology	CYP2C19	Clinical pharmacology, drug interactions, special populations
Rabeprazole	Gastroenterology	CYP2C19	Drug Interactions, clinical pharmacology
Sodium phenylacetate and sodium benzoate	Gastroenterology	UCD (NAGS; CPS; ASS; OTC; ASL; ARG)	Indications and usage, description, clinical pharmacology
Sodium phenylbutyrate	Gastroenterology	UCD (NAGS; CPS; ASS; OTC; ASL; ARG)	Indications and usage, dosage and administration, nutritional management
Eltrombopag (1)	Hematology	Factor V Leiden (FV)	Warnings and precautions
Eltrombopag (2)	Hematology	Antithrombin III deficiency (SERPINC1)	Warnings and precautions
Lenalidomide	Hematology	Chromosome 5q	Boxed warning, indications and usage, clinical studies, patient counseling
Warfarin (1)	Hematology	CYP2C9	Dosage and administration, precautions, clinical pharmacology
Warfarin (2)	Hematology	VKORC1	Dosage and administration, precautions, clinical pharmacology
Atorvastatin	Metabolic and endocrinology	LDL receptor	Indications and usage, dosage and administration, warnings and precautions, clinical pharmacology, clinical studies
Pravastatin	Metabolic and endocrinology	APOE2	Clinical studies, use in specific populations
Carisoprodol	Musculoskeletal	CYP2C19	Clinical pharmacology, special populations
Carbamazepine	Neurology	HLA-B*1502	Boxed warning, warnings and precautions
Clobazam	Neurology	CYP2C19	Clinical pharmacology, dosage and administration, use in specific populations
Dextromethorphan and quinidine	Neurology	CYP2D6	Clinical pharmacology, warnings and precautions
Galantamine	Neurology	CYP2D6	Special populations
Phenytoin	Neurology	HLA-B*1502	Warnings
Tetrabenazine	Neurology	CYP2D6	Dosage and administration, warnings, clinical pharmacology
Arsenic trioxide	Oncology	PML/RARα	Boxed warning, clinical pharmacology, indications and usage, warnings
Brentuximab vedotin	Oncology	CD30	Indications and usage, description, clinical pharmacology
Busulfan	Oncology	Ph chromosome	Clinical studies
Capecitabine	Oncology	DPD	Contraindications, precautions, patient information
Cetuximab (1)	Oncology	EGFR	Indications and usage, warnings and precautions, description, clinical pharmacology, clinical studies
Cetuximab (2)	Oncology	KRAS	Indications and usage, dosage and administration, warnings and precautions, adverse reactions, clinical pharmacology, clinical studies
Cisplatin	Oncology	TPMT	Clinical pharmacology, warnings, precautions

(continued)

Table 7.2 (continued)

Drug	Therapeutic area	Biomarker	Label sections
Crizotinib	Oncology	ALK	Indications and usage, warnings and precautions, adverse reactions, clinical pharmacology, clinical studies
Dasatinib	Oncology	Ph chromosome	Indications and usage, clinical studies, patient counseling information
Denileukin diftitox	Oncology	CD25	Indications and usage, warnings and precautions, clinical studies
Erlotinib	Oncology	EGFR	Clinical pharmacology
Everolimus	Oncology	HER2/NEU	Indications and usage, boxed warning, adverse reactions, use in specific populations, clinical pharmacology, clinical studies
Exemestane	Oncology	ER and/or PGR	Indications and usage, dosage and administration, clinical studies, clinical pharmacology
Fulvestrant	Oncology	ER	Indications and usage, patient counseling information
Gefitinib	Oncology	EGFR	Clinical pharmacology
Imatinib (1)	Oncology	C-Kit	Indications and usage, dosage and administration clinical pharmacology, clinical studies
Imatinib (2)	Oncology	Ph chromosome	Indications and usage, dosage and administration, clinical pharmacology, clinical studies
Imatinib (3)	Oncology	PDGFR	Indications and usage, dosage and administration, clinical studies
Imatinib (4)	Oncology	FIP1L1-PDGFRα	Indications and usage, dosage and administration, clinical studies
Irinotecan	Oncology	UGT1A1	Dosage and administration, warnings, clinical pharmacology
Lapatinib	Oncology	HER2/NEU	Indications and usage, clinical pharmacology, patient counseling information
Letrozole	Oncology	ER and/or PGR	Indications and usage, adverse reactions, clinical studies, clinical pharmacology
Mercaptopurine	Oncology	TPMT	Dosage and administration, contraindications, precautions, adverse reactions, clinical pharmacology
Nilotinib (1)	Oncology	Ph chromosome	Indications and usage, patient counseling information
Nilotinib (2)	Oncology	UGT1A1	Warnings and precautions, clinical pharmacology
Panitumumab (1)	Oncology	EGFR	Indications and usage, warnings and precautions, clinical pharmacology, clinical studies
Panitumumab (2)	Oncology	KRAS	Indications and usage, clinical pharmacology, clinical studies
Pertuzumab	Oncology	HER2/NEU	Indications and usage, warnings and precautions, adverse reactions, clinical studies, clinical pharmacology
Rasburicase	Oncology	G6PD	Boxed warning, contraindications

(continued)

Table 7.2 (continued)

Drug	Therapeutic area	Biomarker	Label sections
Tamoxifen (1)	Oncology	ER	Indications and usage, precautions, medication guide
Tamoxifen (2)	Oncology	Factor V Leiden (FV)	Warnings
Tamoxifen (3)	Oncology	Prothrombin mutations (F2)	Warnings
Thioguanine	Oncology	TPMT	Dosage and administration, precautions, warnings
Tositumomab	Oncology	CD20 antigen	Indications and usage, clinical pharmacology
Trastuzumab	Oncology	HER2/NEU	Indications and usage, precautions, clinical pharmacology
Vemurafenib	Oncology	BRAF	Indications and usage, warning and precautions, clinical pharmacology, clinical studies, patient counseling information
Aripiprazole	Psychiatry	CYP2D6	Clinical pharmacology, dosage and administration
Atomoxetine	Psychiatry	CYP2D6	Dosage and administration, warnings and precautions, drug interactions, clinical pharmacology
Chlordiazepoxide and amitriptyline	Psychiatry	CYP2D6	Precautions
Citalopram (1)	Psychiatry	CYP2C19	Drug interactions, warnings
Citalopram (2)	Psychiatry	CYP2D6	Drug interactions
Clomipramine	Psychiatry	CYP2D6	Drug interactions
Clozapine	Psychiatry	CYP2D6	Drug interactions, clinical pharmacology
Desipramine	Psychiatry	CYP2D6	Drug interactions
Diazepam	Psychiatry	CYP2C19	Drug interactions, clinical pharmacology
Doxepin	Psychiatry	CYP2D6	Precautions
Fluoxetine	Psychiatry	CYP2D6	Warnings, precautions, clinical pharmacology
Fluoxetine and olanzapine	Psychiatry	CYP2D6	Drug interactions, clinical pharmacology
Fluvoxamine	Psychiatry	CYP2D6	Drug interactions
Iloperidone	Psychiatry	CYP2D6	Clinical pharmacology, dosage and administration, drug interactions, specific populations, warnings and precautions
Imipramine	Psychiatry	CYP2D6	Drug interactions
Modafinil	Psychiatry	CYP2D6	Drug interactions
Nefazodone	Psychiatry	CYP2D6	Drug interactions
Nortriptyline	Psychiatry	CYP2D6	Drug interactions
Paroxetine	Psychiatry	CYP2D6	Clinical pharmacology, drug interactions
Perphenazine	Psychiatry	CYP2D6	Clinical pharmacology, drug interactions
Pimozide	Psychiatry	CYP2D6	Warnings, precautions, contraindications, dosage and administration
Protriptyline	Psychiatry	CYP2D6	Precautions
Risperidone	Psychiatry	CYP2D6	Drug interactions, clinical pharmacology
Thioridazine	Psychiatry	CYP2D6	Precautions, warnings, contraindications
Trimipramine	Psychiatry	CYP2D6	Drug interactions

(continued)

Table 7.2 (continued)

Drug	Therapeutic area	Biomarker	Label sections
Valproic acid	Psychiatry	UCD (NAGS; CPS; ASS; OTC; ASL; ARG)	Contraindications, precautions, adverse reactions
Venlafaxine	Psychiatry	CYP2D6	Drug interactions
Indacaterol	Pulmonary	UGT1A1	Clinical pharmacology
Ivacaftor	Pulmonary	CFTR (G551D)	Indications and usage, adverse reactions, use in specific populations, clinical pharmacology, clinical studies
Drospirenone and ethinyl estradiol	Reproductive	CYP2C19	Precautions, drug interactions
Clomiphene	Reproductive and urologic	Rh genotype	Precautions
Tolterodine	Reproductive and urologic	CYP2D6	Clinical pharmacology, drug interactions, warnings and precautions
Azathioprine	Rheumatology	TPMT	Dosage and administration, warnings and precautions, drug interactions, adverse reactions, clinical pharmacology
Flurbiprofen	Rheumatology	CYP2C9	Clinical pharmacology, special populations
Mycophenolic acid	Transplantation	HGPRT	Precautions

www.fda.gov/drugs/scienceresearch/researchareas/pharmacogenetics/. Accessed on 22 Oct 2012

have been approved in the USA with either mandatory or recommended genetic testing associated with use of the drugs.

The example of Herceptin (trastuzumab) highlights how specifically a drug can be prescribed using pharmacogenomic information and can be used to progress the medicines through the research and development pipeline. Overexpression of the ErbB2 gene is associated with increased tumor aggressiveness and poorer prognosis. Herceptin – a humanized monoclonal antibody against the ErbB2 receptor – has been approved for the treatment of breast cancer (Vogel and Franco 2003; Noble et al. 2004). Retrospective examination of the clinical trials of Herceptin showed in patient with tumor overexpressing ErbB2 protein, a positive response was more likely. Therefore, the measurement of ErbB2 overexpression conducted by methods using immunohistochemistry or fluorescent in situ hybridization (FISH) has been used to assess the appropriateness of treatment with Herceptin in patients with breast cancer. The availability of a test for a subgroup of patients with higher probability of responding to treatment with Herceptin has allowed this drug to progress through further studies of approval and reach to the clinic from the manufacturing pharmaceutical firm.

The same paradigm has been applied to understand the response of lung cancer patients to Iressa or gefitinib, where a positive response has been found to be closely associated with the presence of active mutations in epidermal growth factor receptor (EGFR) gene, the main target of the drug used in the lung tumors. Tests for EGFR expression, such as Genzyme's EGFR Mutation Assay, have helped to predict the patients which may respond better to gefitinib (Lynch et al. 2004). This drug treatment was approved by the FDA in 2003 and was developed by AstraZeneca (Kirk et al. 2008).

A phase II study of a GlaxoSmithKline anti-obesity compound which included analysis of all of the patients treated has showed less efficacy in contrast to the reports available in the literature for the current "gold standard" therapies. However, pharmacogenomics analysis using candidate genes based on the compound's target and putative mechanism of action showed association between three pharmacogenomic markers and weight loss during the study. Thirty-six percent of the patients could be clustered to show significantly greater weight loss when subgroup was identified by using one of these three alleles (McCarthy et al. 2005).

Gleevec (imatinib), a targeted cancer drug, has got FDA approval in 2001 and was developed by Novartis (Kirk et al. 2008). The drug acts by inhibiting the specific tyrosine kinase enzyme (BCR-ABL) and is used to treat multiple cancers including chronic myelogenous leukemia and gastrointestinal stromal tumors. This abnormal tyrosine kinase is created by the Philadelphia chromosome, a translocation between chromosome 9 and 22. Quantification of p210 expression levels is a method to monitor response to the drug (Kirk et al. 2008).

Another example of FDA-approved drug is Tarceva (erlotinib), a drug developed in 2005 by Genentech, which targets the EGFR tyrosine kinase domain. It is used to treat non-small cell lung cancer and pancreatic cancer. Genzyme Mutation Assay can be used to predict patient's response to erlotinib (Kirk et al. 2008).

Erbitux (cetuximab) is an EGFR inhibitor used to treat metastatic colorectal cancer and head and neck cancer. EGFR immunostain performed on tumor tissue is used to predict patient's response to cetuximab. The drug label requires the EGFR test. Vectibix (panitumumab) is designed to treat EGFR expressing metastatic colorectal cancer in patients who have received prior treatment. Vectibix differs from Erbitux in its isotype (IgG2 for Vectibix and IgG1 for Erbitux) and potentially in its mechanism of action.

Irinotecan is a semisynthetic analogue of camptothecin and requires metabolic activation by carboxyl esterase to form the active metabolite 7-ethyl-10-hydroxycamptothecin (SN-38), which in turn inhibits topoisomerase-I (Rothenberg et al. 1993). SN-38 is further detoxified via formation of SN-38 glucuronide (SN38G). Irinotecan has potent antitumor activity against a wide range of tumors, and it is one of the most commonly prescribed chemotherapy agents. Diarrhea and myelosuppression are the dose-limiting toxicities of irinotecan and these toxicities limit optimal utilization of irinotecan (Lee et al. 2005). These toxicities have been found to be associated with increased levels of SN-38 (due to accumulation). Polymorphisms in UDP-glucuronosyltransferase 1A1 (UGT1A1), the enzyme responsible for glucuronidation of SN-38 is the focus of pharmacogenomic studies of irinotecan (Gupta et al. 1994). Variations in UGT1A1 activity most commonly arise from polymorphisms in the UGT1A1 promoter region that contains several repeating TA elements. The presence of seven tendom (TA) repeats (referred to as UGT1A1*28), instead of the wild-type number of six, results in reduced UGT1A1 expression and activity (Bosma et al. 1995). Accordingly, UGT1A1*28 has been found to be associated with reduced glucuronidation of SN-38 leading to increased exposure to SN-38 and increased clinical toxicity for patients (Bosma et al. 1995; Iyer et al. 1998, 1999, 2002; Innocenti et al. 2004). UGT1A1*28 alleles are present in approximately 35 % of Caucasians and African Americans, while their frequency has been found to be much lower in Asians (Beutler et al. 1998; Innocenti et al. 2002). In addition, common UGT1A1 promoter variants are in linkage disequilibrium, and the haplotype structure of the promoter appears to differ among different ethnic groups (Innocenti et al. 2002).

In 2005, the FDA has recommended the labeling of drugs and to add the information that patients with polymorphisms in UGT1A1 gene, particularly the TA7/*28 variant, should receive lower doses of the drug to avoid drug toxicity (Yates et al. 1997). Irinotecan is the first chemotherapeutic agent to be dosed according to a patient's genotype. Therefore, we are able to achieve optimal therapeutic response of many drugs by predicting an individual's response to a drug based on the genotypic information.

The crizotinib trial in the area of oncology was unusual because it did not contain a control arm for placebo treatment. The anticancer drug crizotinib is a potent inhibitor of a fusion protein ALK-EML4 which results from chromosomal rearrangement observed in about 4 % lung cancer patients. This drug acts by shrinking of tumors visibly and immediately, in some cases within 48 h. In this trial, clinicians enrolled patients which were positive for ALK biomarker. Even though only 82 patients were enrolled, the results were found to be statistically significant (Vaidyanathan 2012).

8 Pharmacogenomics in Drug Safety

A key objective of development of clinical trials producing new drug molecule/molecules is not only establishing the drug efficacy but also to define potential safety issues associated with new drugs. ADRs cause approximately 106,000 deaths each year in the USA (Lazarou et al. 1998; Rau et al. 2004). Increasing evidences have shown that genetic variations can predispose individual risk to ADRs. For example, nearly 30 % of patients who had ADRs due to antidepressants had CYP2D6 genotype encoding for enzyme which makes them poor metabolizers. This genotype has been found to be present in only 7 % of Caucasian population (Rau et al. 2004).

Pharmacogenomics-based clinical trials significantly contribute in understanding of rare adverse events. Sufficient cases of ADRs may only be reported once a medicine is registered and more widely available. Therefore, retrospective studies including may be required along with various approaches to accelerate the efficient collection of such cases and controls in order to evaluate the pharmacogenomic analysis of ADRs.

Serious and fatal hypersensitivity reactions have been associated with abacavir (antiretroviral drug). Patients bearing the HLA-B*5701 allele are at high risk for experiencing hypersensitivity reaction to abacavir. Therefore, prior to initiating treatment with abacavir, screening for the HLA-B*5701 allele is recommended; this approach has been found to decrease the risk of a hypersensitivity reaction (http://dailymed.nlm.nih.gov/dailymed). Screening is also recommended prior to re-initiation of abacavir in patients of unknown HLA-B*5701 status who have previously tolerated abacavir. For HLA-B*5701-positive patients, treatment with an abacavir-containing regimen is not recommended and should be considered only with close medical supervision and under exceptional circumstances when the potential benefit outweighs the risk.

Candidate gene analysis identified polymorphisms in the human leukocyte antigen (HLA) region on chromosome 6 with strong association of drug-induced hypersensitivity reactions (Hetherington et al. 2002). The association between certain markers in the HLA region was reported to be strongly present among Caucasians, weakly in Hispanics, and was not detected in Blacks (Hetherington et al. 2002). Stevens–Johnson Syndrome (SJS) is a rare, serious cutaneous reaction that is occasionally seen in response to a range of medicines such as antiepileptic drug carbamazepine, which was reported to have some genetic association with HLA markers (Chung et al. 2004). It has been found that genetic variant HLA-B*1502 was found to be strongly associated with carbamazepine-induced SJS in the Taiwanese population. Carbamazepine-induced SJS produces another incidence explaining the rare adverse event where pharmacogenomics technology has provided significant new insights. The above mentioned examples of adverse events clearly indicate that association of a strong genetic component associated with the risk of developing these drug-associated adverse events of some people which was impossible to predict before the genomic era came into existence. Moreover, this type of clinical trials have led to remarkable findings in the field of health care.

9 Mitigating Risk in Drug Development

Potential safety signals can be apparent in early phase II studies and can have a significant (negative) impact on the risk for further development of a drug molecule. For example, if reversible changes in liver function tests are seen in a small subset of patients in a phase II study, it would be difficult to assess the importance of this in relation to drug metabolism alone. A number of medicines have failed either in late development or after launch due to a subset of patients who exhibited liver function changes and subsequently developed severe liver failure (McCarthy et al. 2005).

If high-risk patients could be identified before initiating the drug, the overall safety of a medicine in use would increase considerably and this could avoid failure of drug to reach the market.

Pharmacogenomics is expected to demonstrate the basis of safety concerns such as liver function changes seen during the process of drug development. For example, in recent clinical studies on tranilast (a product under development for reducing restenosis after coronary angioplasty), approximately 8 % of individuals have shown elevated levels of unconjugated serum bilirubin, which dissipated on termination of drug treatment. This phenotype showed some similarity to Gilbert's syndrome. Gilbert's syndrome is a condition which is characterized by episodic increase in unconjugated bilirubin levels but not associated with long-term impact on liver functions (McCarthy et al. 2005).

Genetic analysis of Gilbert's syndrome patients has revealed a strong genetic susceptibility marker in the promoter region of the gene uranyl glucosyltransferase (UGT-1A), where a variable tandem repeat (TA) repeat is located. "Wild-type" or normal allele activity is associated with six copies of the TA repeat, whereas seven copies of the TA repeat have been found to be associated with reduced expression of UGT-1A and increased propensity to Gilbert's syndrome. This example strongly supports the hypothesis that the observed hyperbilirubinemia can be due to tranilast-induced Gilbert's syndrome (McCarthy et al. 2005).

Another example of genomic information predicting the drug response and toxicity can be explained with the help of cancer chemotherapy. Pharmacogenomics approaches have been applied to many existing chemotherapeutic agents in an effort to identify relevant genetic variations which may better predict the response of a patient toward prescribed chemotherapeutic agent (Lee et al. 2005). 6-Mercaptopurine (6-MP) is a purine antimetabolite used in the treatment of leukemia. The antitumor activity of 6-MP is exerted via the inhibition of the formation of nucleotides necessary for DNA and RNA synthesis. Thiopurinemethyltransferase (TPMT) catalyzes the S-methylation of 6-MP to form inactive metabolites. Genetic variations in the TPMT gene have been found to have profound effects on the bioavailability and toxicity of 6-MP. It has been demonstrated that about 1 in 300 individuals inherit TPMT deficiency as an autosomal recessive trait (Lee et al. 2005). Patients who carry TPMT polymorphisms are at higher risk for severe hematologic toxicities when treated with 6-MP because these polymorphisms lead to a decrease in the rate of 6-MP metabolism (Evans et al. 1991; Lennard et al. 1993). The molecular basis for polymorphic TPMT activity has been well defined. In around 95 % of observed cases, TPMT deficiency is due to three particular TPMT alleles, designated as TPMT*2, TPMT*3A, and TPMT*3C (Innocenti et al. 2000). Appropriate 6-MP dose reductions for TPMT-deficient patients have allowed for similar toxicity and survival outcomes as patients with normal TPMT levels (Relling et al. 1999). Genotyping methods have been established for the molecular diagnosis of TPMT deficiency and can assist determining a safe starting dose for 6-MP therapy (Yates et al. 1997). TPMT testing is now being used for dose optimization in children before 6-MP therapy is initiated.

10 Rare Adverse Events and Collaborative Models

The feasibility of clinically validating the predictive power of pharmacogenomics test to predict rare safety event is questionable. The number of rare ADEs in a typical clinical development program of 5,000–10,000 patients will be approximately 5–10 cases (Frank 2008). For rare adverse events, different post-marketing surveillance scenarios have been described with regard to validating test findings. Pharmacogenomic test will be required for assessing the seriousness and frequency of the ADEs. Pharmacogenomics will help in explaining the adverse drug reactions. The industry is now realizing to join efforts in a collaborative mode to pool resources to make significant headway in the area of rare but serious adverse events. One example is the recent creation of the Serious Adverse Events Consortium (SAEC), which is a nonprofit partnership of industry, academia, and government. It is an effort to identify the genetic markers related to harmful drug reactions (Frank 2008).

The short-term priorities of the SAEC are drug-induced liver injury and serious skin rash and QT prolongation. This is a major step forward in understanding the basic science underlying drug-induced toxicity and, if successful, will benefit the pharmaceutical industry overall but more importantly benefit patients and society at large.

11 Pharmacogenomics in Pediatrics

According to National Health Institute, approximately 70 % of the medicines given to children have only been tested in adults. But there are a number of anatomical, physiological, and genetic differences among children and adults. The gene products and their expression in particular phenotype–genotype correlations observed in neonates may be markedly different from the one observed in trials on adults (Leeder et al. 2010). Further, evidences have suggested that the response to some medications may differ in children relative to adults. The extent of pharmacokinetic variability for a drug which is metabolized by a polymorphic enzyme may be same, increased, or decreased in children based on relative development of other enzymes or transporters involved in absorption, distribution, metabolism, and excretion of the drug (Leeder et al. 2010). Thus, genotype-based pharmacokinetic studies involving poor metabolizers and extensive metabolizers/ultrarapid metabolizers at different ages or developmental stages will help in revealing any age-related differences in drug disposition and the dose–exposure relationship that can then be incorporated in the efficacy studies. The role of pharmacogenomic-based pharmacokinetic studies is to establish the range of drug exposures that might be present in a pediatric population at a given age or developmental stage. The range of doses required to achieve the most appropriate drug exposure with which to conduct efficacy studies can be established. Similarly, there may be differences in receptor (and pathway involved) response and hence differences in exposure–response relationships in children, compared with adults, that can be identified if dosing strategies in pediatric studies are designed to achieve exposures known to be associated with the desired therapeutic response in adults (Leeder et al. 2010). The pediatric drug development may be influenced by adult experience of clinical trials but depends on increased knowledge of both pharmacogenomics and ontogeny in children. The determinants of drug disposition and action are multifaceted and are polygenic in their origin. Therefore, while designing a clinical trial with children as participants, genes expressed or not expressed according to their age and developmental stages influencing the drug kinetics and dynamics should be considered and taken care. The quantity and type of procedures used in a pediatric trial should be related to minimize discomfort, pain, and fright, and invasive procedures should be used only when clinically necessary. The increased use of established biomarkers is also an appropriate way to minimize invasive procedures in such trials in order to prove the efficacy of drug treatment. Another sensitive issue is the frequency of withdrawal of blood sample to isolate deoxyribonucleic acid (DNA) and to analyze the serum samples for biochemical assays, which should be minimized during study period. A clinical trial's safety and efficacy endpoints might need to be adjusted for pediatric clinical trials before implementation.

Efficacy endpoints commonly used in adult studies should not be translated directly to the study recruiting children. Apart from this, safety follow-up must always be larger and longer in order to detect any adverse effects on development as the children grow in later stages of study period. The route of drug administration and type of drug formulation may vary from adults to children because adults can be given tablets or capsules which is not appropriate in case of small children. Given that many factors complicate the pharmacogenomics in pediatric clinical trial design, utmost care should be taken by sponsors for protocol design, drug formulation, consent, enrollment, ethnicity, age group, and many other factors to enhance the efficacy and benefits of therapeutic molecule.

12 Pharmacogenomics and Regulatory Issues

The function of regulatory agencies such as the European Medicines Agency (EMEA) and Food and Drug Administration dealing with the whole area of pharmacogenomics in terms of clinical trials, licensing, and labeling is not clear at present. As the predictive power of pharmacogenetic testing increases, labeling is likely to become more prescriptive (Robertson et al 2002). An important issue in drug development is new technology not encompassed in many guidelines. Thus, new guidelines will have to be developed. The FDA and EMEA are actively supporting the introduction of pharmacogenetics-based therapy (Lesko and Woodcock 2004), with regular meetings with industry and discussion of "safe harbor"-type arrangements to encourage joint discussion of the "meaning" and interpretation of pharmacogenetics data. According to the FDA, it is necessary to submit pharmacogenomic data by pharmaceutical firms which are involved in these type of trials (Surendiran et al. 2008).

There are many drugs currently in the market that have information displayed on their labels indicating the possible adverse drug effects and mandatory genetic testing for their prescribing. When a drug has been shown to be efficacious in patients with a certain genotype, the indication in the label should reflect this. Therefore, the drug will be licensed not only for a particular condition but will also be recommended for use in patients with certain genotypes. This has been already exemplified by the label for trastuzumab (Herceptin) used in the treatment of breast cancer. Any prescription for this drug without the particular genotype test of patients will therefore have to be considered to be outside the licensed indication. This may lead to issues how instructions for use will be encouraged or enforced. Doctors currently have the right to prescribe "off-label," and it is difficult to see how this would change (Pirmohmad and Lewis 2004). How regulatory agencies will ensure appropriate prescribing of medicines based on pharmacogenetic principles remains an outstanding question to be considered. It is unlikely that pharmacogenetics testing will become part of regulatory requirements for all drugs. The need for pharmacogenomic testing for a particular drug will depend upon many factors, in particular on the genetic factors determining its disposition, pharmacodynamic characteristics, and its therapeutic index. Therefore, a drug that has a high degree of efficacy in a large percent of the population shows little interindividual variability in kinetics and dynamics and has a wide therapeutic index should not necessarily require pharmacogenetic testing.

In contrast, a drug that has narrow therapeutic index and is efficacious in 30 % of the population, for example, some antipsychotics, should be subject to pharmacogenetic testing prior to prescription. Preclinical studies should be planned in order to identify the routes of metabolism and disposition of a particular drug and its mechanism of action. If any of these parameters are subject to genetic polymorphism that could theoretically or in practice effect the response to the drug, then pharmacogenetic/genomic testing should be encouraged.

Another major issue is some patients having an "orphan genotype," i.e., a genotype that cannot be treated with currently available drugs because those patients have been classed as non-responders or as susceptible to particular adverse reactions. This will represent a small proportion of the population. Pharmaceutical companies may be reluctant to develop new medicines for this group because of their potential unprofitability. In such cases, further regulatory measures are needed so that these orphan genotypes are treated in the same way as orphan diseases are treated at present (Motl et al 2003). This may provide financial incentives for the pharmaceutical industry to develop medicines for small patient populations with these orphan genotypes. Alternatively, as already mentioned, smaller genomics/drug development companies may enter such markets, in a similar way to the orphan drug market.

Since many more medicines could fall into the category of "orphan" products because of a reduced market, pharmaceutical companies might seek extension of orphan drug legislation in order to obtain development subsidies.

13 Statistical Methods

Pharmacogenomics is a rapidly developing field of research with highest potential for immediate benefit of patient and society at large. To reap the full benefits of the data generated from pharmacogenomics trials and studies, a careful consideration should be given to the study population under investigation, the phenotype being examined and the statistical methodology used in data analysis (Turner et al. 2010). The US FDA has been a strong proponent of the use of pharmacogenomics in clinical trials, but still there is a scope to implement its mission in a better way and to stimulate the use of pharmacogenomics information (Kirk et al. 2008; Kalow et al. 1998). The type of statistical analysis required to recognize retrospective analysis of pharmacogenomic data may replace or supplement prospective pharmacogenomic studies and is worthy of consideration. The classic frequentist statistical approaches have long been favored in clinical trials. A frequentist statistical approach uses a sequential monitoring of data and provides the probability that a drug is effective (Kirk et al. 2008).

The US FDA is not willing to allow drug sponsors the opportunity of "cherry picking" clinical endpoints after the data is known (Kirk et al. 2008). With very high drug candidate failure rate (95 %) and lot of biomarker data of significant power, in phase III studies, any method that may improve the insight into valid biomarker relationship related to efficacy or safety should be utilized. Recently instead of classical frequentist statistical approach Bayesian statistics have gained more attention in clinical trials. Bayesian trials allow for continuous monitoring of data and provide the probability of effectiveness of drug. Bayesian methods compare the probability to prior probabilities and updated posterior probabilities. The drawback in implementing Bayesian statistics is logistical and computational complexity. However, logistical capability and the availability of computing power have advanced quite considerably in recent years.

Pfizer has published results of their Acute Stroke Therapy by Inhibition of Neutrophils (ASTIN) trial in 2003. This Bayesian trial offered adaptive allocation of patients to different dose groups. In case of strong efficacy, a seamless transition to phase III was possible (Lowe 2006). In addition to these logistical challenges and conducting the study in 100 centers worldwide and other problems of more study subjects recruitment, the trial accomplished its goal and the drug was promptly terminated by Pfizer. Although modern computing has made calculations more feasible, a significant amount of work remains to be done for implementing Bayesian statistics in clinical trials (Kirk et al. 2008).

Strict adherence to prospective study designs will also hinder widespread use of pharmacogenomics, e.g., the approval process for Amgen's Vectibix has clearly illustrated this hurdle. Amgen's retrospective study examination showed that in phase III study patients with wild-type KRAS genotype were more likely to respond to Vectibix in comparison with mutated KRAS genotype. The metastatic colorectal cancer drug was approved by the EMEA on retrospective data. Later on the FDA determined that a prospectively designed study must be submitted before the KRAS biomarker inhibitor (Amgen's Vectibix) could be added to the USA.

In March 2008, the FDA's Pharmaceutical Science and Clinical Pharmacology Advisory Committee debated a preference for collecting DNA samples from all patients in clinical trials (Usdin 2008). Such steps may ultimately lead to improving conditions for pharmacogenomics discovery that may otherwise be too difficult to implement in earlier phase trials.

14 Patent Prosecution in Pharmacogenomics

Patent law represents a delicate threshold for promoting innovative and lifesaving discoveries (Kirk et al. 2008). Patent system is getting weakened due to hurdles imposed by courts, Congress, and the US Patent and trademark Office (USPTO). This weakened patent system has major negative effect on pharmaceutical and genomic innovator companies (Mancinelli et al. 2000). Changes in subject matter patentability analysis have already increased the considerable burden for innovations directed to human genetic material and biotechnology. Once an invention crosses the initial threshold of utility, it is subject to the rigors of novelty, non-obviousness, written description, and enablement (Kirk et al. 2008). Any further degradation of patent rights available to innovator companies in the area of gene therapy and molecular diagnostics could prove fatal to groundbreaking research in diseases such as cancer, diabetes, depression, and schizophrenia. Research, development, and funding of new genetic tests based on isolated genomic information may move therapeutics toward "personalized medicine" from "one size fits all" approach. This is possible only if these genetic tests are eligible for the necessary patent protection to promote investment (Council for International Organizations of Medical Sciences 2005). Trade secret protection is largely impractical for biotechnology and genetic material due to the stringent regulatory environment demanding transparency, and these products can be easily reverse engineered. Patent reforms are need of hour for promoting pharmacogenomics. The challenges to pharmacogenomics patents are still evolving. Because of their direct application to biological life on earth, pharmacogenomics and genomics patents are subject to intense scrutiny by the various patent offices. Knowledge of the challenges in the field pharmacogenomics patent process would lead to the development of more skillful prosecution system and more rapid innovation overall (www.iploft.com/NU-Pharmaco).

15 Conclusion

Pharmacogenomic is being used globally for assessing the safety profile of drugs. The pharmacogenomics in pharmaceutical industry is a powerful tool, but its maximum benefit is yet to be achieved. Analysis of multiple genes, haplotype analysis, genome wide scans, and drug designing can be utilized by pharmaceutical industry for drug development process in various phases of clinical trials. As a consequence an individual's genomic profile can be used to predict the safety and efficacy of a pharmacological agent. The present "one size fits all" paradigm in clinical trials will shift toward evidence-based efficacy models with improved efficacy and reduced economic burden to the ultimate customer or the patient.

The integration of other omics-based technologies such as proteomics, transcriptomics, metabolomics, and epigenomics in different phases of clinical trials would improve the drug discovery and development. The omics approaches have strong potential to screen the drug molecule at different levels and to develop novel drug targets, exploring the mechanism of action of drugs, identifying ADRs and enabling its translation from benchside to bedside.

References

Ahn C (2007) Pharmacogenomics in drug discovery and development. Genomics Inform 5(2):41–45

Anon (2001) Tufts Centre for the Study of Drug Development pegs cost of a new prescription medicine at $802 million. Press Release, Tufts Centre for the Study of Drug Development, Tufts University. 30 November. http://www.tufts.edu/med/csdd/

Aronson JK (2004) What is a clinical trial? Br J Clin Pharmacol 58(1):1–3

Beutler E, Gelbart T, Demina A (1998) Racial variability in the UDP-glucuronosyltransferase 1 (UGT1A1) promoter: a balanced polymorphism for regulation of bilirubin metabolism? Proc Natl Acad Sci U S A 95:8170–8174

Bhatt A (2010) Evolution of clinical research: a history before and beyond James Lind. Perspect Clin Res 1:6–10

Boa AN (2003) Introduction to drug discovery. Department of Chemistry, University of Hull, Hull

Bosma PJ, Chowdhury JR, Bakker C, Gantla S, de Boer A, Oostra BA, Lindhout D, Tytgat GN, Jansen PL, Oude Elferink RP et al (1995) The genetic basis of the reduced expression of bilirubin UDP-glucuronosyltransferase 1 in Gilbert's syndrome. N Engl J Med 333:1171–1175

Chung WH, Hung SI, Hong HS, Hsih MS, Yang LC, Ho HC, Wu JY, Chen YT (2004) Medical genetics: a marker for Stevens–Johnson syndrome. Nature 428:486

Council for International Organizations of Medical Sciences (CIOMS) (2005) Pharmacogenetics: towards improving treatment with medicines. CIOMS, Geneva

Daly AK (2003) Pharmacogenetics of the major polymorphic metabolizing enzymes. Fundam Clin Pharmacol 17:27–41

DiMasi JA (2002) The value of improving the productivity of the drug development process: faster times and better decisions. Pharmacoeconomics 3:1–10

DiMasi JA, Hansen RW, Grabowski HG (2003) The price of innovation: new estimates of drug development costs. J Health Econ 22(2):151–185

Evans WE, McLeod HL (2003) Pharmacogenomics – drug disposition, drug targets, and side effects. N Engl J Med 348(6):538–549

Evans WE, Horner M, Chu YQ, Kalwinsky D, Roberts WM (1991) Altered mercaptopurine metabolism, toxic effects, and dosage requirement in a thiopurine methyltransferase-deficient child with acute lymphocytic leukemia. J Pediatr 119:985–989

FDA CDER/CBER/CDRH (2005) Guidance for industry: pharmacogenomic data submission. Posted on March 22, 2005: http://www.fda.gov/cder/guidance/6400fnl.pdf

FDA. Table of pharmacogenomic biomarkers in drug labels. http://www.fda.gov/drugs/scienceresearch/researchareas/pharmacogenetics

Frank M (2008) Application of pharmacogenomics in clinical trials. In: Cohen N (ed) Pharmacogenomics and personalized medicine, Ith edn. Humana Press, Totowa

Geisen C, Watzka M, Sittinger K, Steffens M, Daugela L, Seifried E, Muller CR, Wienker TF, Oldenburg J (2005) VKORC1haplotypes and their impact on the inter-individual and interethnical variability of oral anticoagulation. Thromb Haemost 94:773–779

Gilbert J (2003) Rebuilding big PhRMA's business model. In Vivo 21:73–80

Gupta E, Lestingi TM, Mick R, Ramirez J, Vokes EE, Ratain MJ (1994) Metabolic fate of ironectin in humans correlation of glucuronidation with diarrhea. Cancer Res 54:3723–3725

Guzey C, Norstrom A, Spigset O (2002) Change from the CYP2D6 extensive metabolizer to the poor metabolizer phenotype during treatment with bupropion. Ther Drug Monit 24:436–437

Henry D, Lexchin J (2002) The pharmaceutical industry as a medicines provider. Lancet 360(9345):1590–1595

Hetherington S, Hughes AR, Mosteller M, Shortino D, Baker KL, Spreen W, Lai E, Davies K, Handley A, Dow DJ, Fling ME, Stocum M, Bowman C, Thurmond LM, Roses AD (2002) Genetic variations in HLA-B region and hypersensitivity reactions to abacavir. Lancet 359:1121–1122

Hung SI, Chung WH, Liou LB, Chu CC, Lin M, Huang HP, Lin YL, Lan JL, Yang LC, Hong HS, Chen MJ, Lai PC, Wu MS, Chu CY, Wang KH, Chen CH, Fann CS, Wu JY, Chen YT (2005) HLA-B*5801 allele as a genetic marker for severe cutaneous adverse reactions caused by allopurinol. Proc Natl Acad Sci U S A 102:4134–4139

Ingleman-Sundberg M (2004) Human drug metabolizing cytochrome P450 enzymes: properties and polymorphisms. Naunyn Schmiedebergs Arch Pharmacol 369:89–104

Innocenti F, Iyer L, Ratain MJ (2000) Pharmacogenetics: a tool for individualizing antineoplastic therapy. Clin Pharmacokinet 39:315–325

Innocenti F, Grimsley C, Das S, Ramírez J, Cheng C, Kuttab-Boulos H, Ratain MJ, DiRienzo A (2002) Haplotype structure of the UDP-glucuronosyltransferase 1A1 promoter in different ethnic groups. Pharmacogenetics 12:725–733

Innocenti F, Undevia SD, Iyer L, Chen PX, Das S, Kocherginsky M, Karrison T, Janisch L, Ramírez J, Rudin CM, Vokes EE, Ratain MJ (2004) Genetic variants in the UDP-glucuronosyltransferase 1A1 gene predict the risk of severe neutropenia of irinotecan. J Clin Oncol 22:1382–1388

International HapMap Project (2007) HapMap homepage. Retrieved February 1, 2008 from http://www.hapmap.org

Iyer L, King CD, Whitington PF, Green MD, Roy SK, Tephly TR, Coffman BL, Ratain MJ (1998) Genetic predisposition to the metabolism of irinotecan (CPT-11). Role of uridine diphosphate glucuronosyltransferase isoform 1A1 in the glucuronidation of its active metabolite (SN-38) in human liver microsomes. J Clin Invest 101:847–854

Iyer L, Hall D, Das S, Mortell MA, Ramírez J, Kim S, Di Rienzo A, Ratain MJ (1999) Phenotype-genotype correlation of in vitro SN-38 (active metabolite of irinotecan) and bilirubin glucuronidation in human liver tissue with UGT1A1 promoter polymorphism. Clin Pharmacol Ther 65:576–582

Iyer L, Das S, Janisch L, Wen M, Ramírez J, Karrison T, Fleming GF, Vokes EE, Schilsky RL, Ratain MJ (2002) UGT1A1*28 polymorphism as a determinant of irinotecan disposition and toxicity. Pharmacogenomics J 2:43–47

Kalow W, Tang BK, Endrenyi I (1998) Hypothesis: comparisons of inter- and intra-individual variations can substitute for twin studies in drug research. Pharmacogenetics 8:283–289

Kirk RJ, Hung JL, Horner SR, Perez JT (2008) Implications of pharmacogenomics for drug development. Exp Biol Med (Maywood) 233(12):1484–1497

Klein TE, Chang JT, Cho MK, Easton KL, Fergerson R, Hewett M, Lin Z, Liu Y, Liu S, Oliver DE, Rubin DL, Shafa F, Stuart JM, Altman RB (2001) Integrating genotype and phenotype information: an overview of the PharmGKB Project. Pharmacogenomics J 1:167–170

Lazarou J, Pomeranz BH, Corey PN (1998) Incidence of adverse drug reactions in hospitalized patients: a meta-analysis of prospective studies. J Am Med Assoc 279:1200–1205

Lee W, Lockhart AC, Kim RB, Rothenberg ML (2005) Cancer pharmacogenomics: powerful tools in cancer chemotherapy and drug development. Oncologist 10(2):104–11

Leeder JS, Kearns GL, Spielberg SP, van den Anker J (2010) Understanding the relative roles of pharmacogenetics and ontogeny in pediatric drug development and regulatory science. J Clin Pharmacol 50(12):1377–1387

Lennard L, Gibson BE, Nicole T, Lilleyman JS (1993) Congenital thiopurinemethyltransferase deficiency and 6-mercaptopurine toxicity during treatment for acute lymphoblastic leukaemia. Arch Dis Child 69:577–579

Lesko LJ, Woodcock J (2004) Translation of pharmacogenomics and pharmacogenetics: a regulatory perspective. Nat Rev Drug Discov 3:763–769

Lesko LJ, Salerno RA, Spear BB, Anderson DC, Anderson T, Brazell C, Collins J, Dorner A, Essayan D, Gomez-Mancilla B, Hackett J, Huang SM, Ide S, Killinger J, Leighton J, Mansfield E, Meyer R, Ryan SG, Schmith V, Shaw P, Sistare F, Watson M, Worobec A (2003) Pharmacogenetics and pharmacogenomics in drug development and regulatory decision making: report of the first FDA-PWG-PhRMA-DruSafe Workshop. J Clin Pharmacol 43:342–358

Lindpaintner K (2002) Pharmacogenetics and the future of medical practice. Br J Clin Pharmacol 54:221–230

Lowe D (2006) What you need to know about adaptive trials. Pharm Exec Jul 1

Lynch TJ, Bell DW, Sordella R, Gurubhagavatula S, Okimoto RA, Brannigan BW, Harris PL, Haserlat SM, Supko JG, Haluska FG, Louis DN, Christiani DC, Settleman J, Haber DA (2004) Activating mutations in the epidermal growth factor receptor underlying responsiveness of non-small-cell lung cancer to gefitinib. N Engl J Med 350:2129–2139

Mallal S, Nolan D, Witt C, Masel G, Martin AM, Moore C, Sayer D, Castley A, Mamotte C, Maxwell D, James I, Christiansen FT (2002) Association between presence of HLA-B*5701, HLA-DR7, and HLA-DQ3 and hypersensitivity to HIV-1 reverse transcriptase inhibitor abacavir. Lancet 359:727–732

Mancinelli L, Cronin M, Sade'e W (2000) Pharmacogenomics: the promise of personalized medicine. AAPS PharmSci 2(1):1–13. http://aapspharmsci.org/articles/ps0201/ps020104/ps020104.pdf

McCarthy AD, Kennedy JL, Middleton LT (2005) Pharmacogenetics in drug development. Philos Trans R Soc B 360:1579–1588

Motl S, Miller SJ, Burns P (2003) Programs established by FDA to expedite patient access. Am J Health Syst Pharm 60(4):339–345

Noble ME, Endicott JA, Johnson LN (2004) Protein kinase inhibitors: insights into drug design from structure. Science 303:1800–1805

Papanikolaw J (1999) Bioinformatics emerges as key technology for developing new drugs. Chem Mark Report 255(21):ABI/INFORM 22

Park BK, Pirmohamed M (2001) Toxicogenetics in drug development. Toxicol Lett 120(1–3):281–291

Patent Prosecution in Pharmacogenomics. More information is available on http://www.iploft.com/

Pauli-Magnus C, Kroetz DL (2004) Functional implications of genetic polymorphisms in the multidrug resistance gene MDR1 (ABCB1). Pharm Res 21:904–913

Pirmohmad M, Lewis G (2004) The implications of pharmacogenetics and pharmacogenomics for drug development and health care. In: Mossialos E, Mrazek M, Walley T (eds) Regulating pharmaceuticals in Europe: striving for efficiency, equity and quality. Open University Press, Maidenhead

Public Citizen (2001) Rx R&D myths: the case against the drugs industry "scare card". http://www.citizen.org/publications/release.cfm?ID=7065

Ratti E, Trist D (2001) Continuing evolution of the drug discovery process in the pharmaceutical industry. Pure Appl Chem 73(1):67–75

Rau T, Wohlleben G, Wuttke H, Thuerauf N, Lunkenheimer J, Lanczik M, Eschenhagen T (2004) CYP2D6 genotype: impact on adverse effects and nonresponse during treatment with antidepressants – a pilot study. Clin Pharmacol Ther 75:386–393

Relling MV, Hancock ML, Rivera GK, Sandlund JT, Ribeiro RC, Krynetski EY, Pui CH, Evans WE (1999) Mercaptopurine therapy intolerance and heterozygosity at the thiopurine S-methyltransferase gene locus. J Natl Cancer Inst 91:2001–2008

Robertson JA, Brody B, Buchanan A, Kahn J, McPherson E (2002) Pharmacogenetic challenges for the health care system. Health Aff 21(4):155–167

Roses AD (2000) Pharmacogenetics and the practice of medicine. Nature 405:857–865

Roses AD (2004) Pharmacogenetics and drug development: the path to safer and more effective drugs. Nat Rev Drug Discov 3:645–656

Rothenberg ML, Kuhn JG, Burris HA 3rd, Nelson J, Eckardt JR, Tristan-Morales M, Hilsenbeck SG, Weiss GR, Smith LS, Rodriguez GI et al (1993) Phase I and Pharmacokinetic trial of weekly CPT-11. J Clin Oncol 11:2194–2204

Shyh-Yoh L, Stringer F, Hirayama M (2012) The impact of pharmacogenomics research on drug development. Drug Metab Pharmacokinet 27(1):2–8

Stanford University (2007) PharmGKB: the pharmacogenetics and pharmacogenomics knowledge base. Retrieved January 31, 2008 from http://www.pharmgkb.org

Surendiran A, Pradhan SC, Adithan C (2008) Role of pharmacogenomics in drug discovery and development. Indian J Pharmacol 40(4):137–143

Sweeney BP (2005) Pharmacogenomics: the genetic basis for variability in drug response. In: Cashman JN, Grounds RM (eds) Recent advances in anaesthesia and intensive care, vol 23. Cambridge University Press, Cambridge, pp 1–34

Turner SD, Crawford DC, Ritchie MD (2010) Methods for optimizing statistical analyses in pharmacogenomics research. Expert Rev Clin Pharmacol 2(5):559–570, Author manuscript; available in PMC

Usdin S (2008) Proof is in the pudding. Biocentury 16(14):A1–A6

Vaidyanathan G (2012) Redefining clinical trials: the age of personalized medicine. Cell 148:1079–1080

Vogel F (1959) Moderneprobleme der humangenetik. Ergeb Inn Med Kinderheilkd 12:52–125

Vogel CL, Franco SX (2003) Clinical experience with trastuzumab (herceptin). Breast J 9(6):452–462

Weinshilboum R (2003) Inheritance and drug response. N Engl J Med 348:529–537

Xie HG, Kim RB, Wood AJ, Stein CM (2001) Molecular basis of ethnic differences in drug disposition and response. Annu Rev Pharmacol Toxicol 41:815–850

Yates CR, Krynetski EY, Loennechen T, Fessing MY, Tai HL, Pui CH, Relling MV, Evans WE (1997) Molecular diagnosis of thiopurine S-methyltransferase deficiency: genetic basis for azathioprine and mercaptopurine intolerance. Ann Intern Med 126(8):608–614

Ethical Considerations in Pharmacogenomics

8

Anjana Munshi and Yog Raj Ahuja

Abstract

The science of pharmacogenomics holds the promise to change the way in which clinical drug trials are conducted as well as the prescription of drugs on a routine basis. However, there are a number of ethical and regulatory challenges such as subject recruitment, privacy, sample collection and storage, and confidentiality, which are of major concern for conducting genomics-based clinical studies. Stratification of clinical study subjects into subgroups on the basis of genotype is another problem in designing clinical studies which might lead to the subject selection biases. This could also lead to spurious interpretations of statistical analysis. Subtype stratification can also result in scientific challenges and data analysis because of penetrance. The variable degree of clinically relevant phenotypic expression of genetic variation can lead to false positives. Moreover, the discrimination by job providers and insurance firms are other issues of legal considerations and need to be answered. The pharmacogenomics can also lead to the development of orphan phenotype. If a new drug molecule is being developed only for a subpopulation of patients, then issues such as distributive justice and fairness to accept the new drug also need to be considered. No doubt pharmacogenomics is emerging as a boon for medical fraternity; however, the translation of pharmacogenomics into clinical practice requires to frame and address the legal and ethical issues along with incentives to overcome these roadblocks.

A. Munshi (✉)
Department of Molecular Biology, Institute of Genetics and Hospital for Genetic Diseases, Osmania University, Begumpet, Hyderabad 500016, Andhra Pradesh, India
e-mail: anjanadurani@yahoo.co.in

Y.R. Ahuja
Department of Genetics and Molecular Medicine, Vasavi Medical and Research Centre, Khairatabad, Hyderabad 500004, Andhra Pradesh, India

D. Barh et al. (eds.), *Omics for Personalized Medicine*,
DOI 10.1007/978-81-322-1184-6_8, © Springer India 2013

Abbreviations

ADRs	Adverse Drug Reactions
DNA	Deoxyribonucleic acid
FDA	Food and Drug Administration
A-HeFT	African-American Heart Failure Trial
BiDil	Isosorbide dinitrate and hydralazine
CYP2D6	Cytochrome P450, family 2, subfamily D, polypeptide 6
G6PD	Glucose 6-phosphate dehydroge-nase
NADPH	Nicotine-adenine dinucleotide phosphate
NAT-2	*N*-acetyl transferase-2
HLA	Human leukocyte antigen
ALOX-5	Arachidonate-5 lipoxygenase enzyme
UNESCO	United Nations Educational Scientific and Cultural Organization
HUGO	Human Genome Organization
PharmaGKB	Pharmacogenomics knowledge base
UK	United Kingdom
US	United States
VGDS	Voluntary Genomic Data Submission
EMEA	European Medicines Agency
COMT	Catechol-*O* methyltransferase
CYP2C9	Cytochrome P450, family 2, subfamily C, polypeptide 9
HER-2	Human Epidermal Growth Factor Receptor 2, also known as Neu
TPMT	Thiopurine methyltransferase
PON-1	Paraoxonase gene-1

1 Introduction

Pharmacogenomics is a nascent field that encompasses the study of genetic variation undergoing individual differences in drug responses. The existing system of health care involves "trial and error" method for prescription of drugs, which contributes approximately three million ineffective prescriptions per year (personalized medicine: the emerging pharmacogenomics revolution 2005). The drug-associated adverse effects (ADRs) have been reported to affect around 2.2 million people and 106,000 deaths in the USA. Moreover, the economic burden and drug withdrawal cases are increasing due to ADRs (Ernst and Grizzle 2001). Pharmacogenomics may reduce the health-care costs by précising and optimizing the drug treatment. It paves the way to develop the novel and better targeted therapeutic agents from the very first stage of designing a clinical trial to its induction into the market.

The pharmacogenomics also holds the promise of streamlining the clinical trials with the use of genotyping so as to enable researchers to "rescue" drugs that could not be approved under conventional models of clinical trials (Rothstein and Epps 2001). In other words, drugs which were rejected earlier on the basis of conventional trials with higher adverse effects will become more acceptable because of few or no adverse effects when tested in a subgroup of the population with a particular genotype. It will also reduce the time and money spend on the development of a drug molecule. Pharmacogenomics is at the stage of infancy with the availability of very few examples, which are established in the clinical practice. The pharmacogenomics will also provide greater efficiency in the allocation of resources for the development of new drug molecules.

The translation of pharmacogenomics into routine clinical practice although bears a scientific promise but also raises some ethical concerns for the conduct of human research. Researchers have raised a number of general, ethical, social, and legal problems for translation of pharmacogenomics into clinical practice. A few among those are informed consent form for the use of deoxyribonucleic acid (DNA), issues of confidentiality and privacy, storage and use of genetic information, and potential for discrimination and stigmatization (Rothstein and Epps 2001; Clarke et al. 2001; Robertson 2002; Holtzman and Watson 1997; Human Genetic Commission 2002).

In this chapter, we have tried to highlight the key concerns associated with the practice of pharmacogenomics and emphasized that consideration of ethical and regulatory issues will provide an effective framework for the implementation of pharmacogenomics into clinical practice under the following subheadings:

2 Clinical Study Design

The practice of evidence-based clinical medicine is purely dependent on the careful designing and conduct of the clinical trial. A clinical trial is a controlled prospective study involving human subjects or participants which is designed to determine the efficacy of a therapeutic intervention, preventive measures, diagnostic procedure, and use of the medical or surgical device (Friedman et al. 1996; Issa 2002). The clinical trials are divided into four phases: phase I, phase II, phase III, and phase IV. In phase I, a new drug molecule is tested for the first time in a small group of people (20–80) to evaluate its efficacy, safety, side effects, and dosage range. In phase II, the drug to be studied is given to a large group of people (100–300) to evaluate its efficacy. This phase determines the broader range of toxicities. Phase III confirms the effectiveness and side effects of the drug in 1,000–3,000 study subjects and compares it to commonly used treatments. After the completion of phase III, new drug application is filed with the drug regulatory authority, which will consider all the drug-related data from preclinical and clinical studies to approve or reject the new drug for marketing. Phase IV consists of post marketing surveillance studies. The fundamental characteristics of a trial design are sufficient sample size, statistical power, and control of sampling bias (Issa 2002). The randomized clinical trial and double-blind randomized clinical trial (in which both investigator and subject do not know about the trial) are considered a gold standard for determining treatment efficacies (Friedman et al. 1996). The objective of randomization is to reduce the bias and variability as much as possible (Issa 2002).

2.1 Incorporation of Genomic Information into Clinical Trial Design

In the twenty-first century, the clinical trials are designed incorporating pharmacogenomic profiling into trials either prospectively or retrospectively, in clinical trials (Issa 2002). Retrospective trials use the conventional methods and high-throughput sequencing techniques to identify genotypes so as to correlate them with disease susceptibility and drug response (Issa 2002). In case of prospective studies, the genomic profiling is used to reduce the incidences of ADRs due to variability in the pharmacokinetic profile of a drug (Issa 2002). At present, the pharmacogenomic profiling is mainly used in designing of phase I trials. The individuals are recruited on the basis of their genotypic profile either to predict metabolism (slow or rapid) and to prevent ADRs (Issa 2002).

In phase II trials, candidate gene approach is used to identify the efficacy of a drug according to genotypes of the subjects. Phase III trials, involve pharmacogenomic profiling for identifying the responders and nonresponders. Sample size, allele frequency, and gene-effect size are some of the parameters which should be taken into consideration in designing and conducting of pharmacogenomic study (Issa 2002).

In the USA, Federal regulations that govern human subjects research include "Risks to subjects should be minimized by using procedures with sound research design…" (Levine 1993; Tri Council Policy Statement: Ethical Council for Research involving Humans 1998; World Medical Association Declaration of Helsinki Ethical Principles for Medical Research involving Human Subjects 2000). Since the goal of pharmacogenomic research is to focus on interindividual drug variability, therefore the pharmacogenomic trials start with a different assumption about the participating subjects – that interindividual heterogeneity is inherent (Issa 2002).

The candidate gene studies and genome-wide association studies are mainly used to design a pharmacogenomic trial. According to an estimate, genome-wide association trials require at least a threefold increase in sample size as compared to candidate gene approach (Cardon et al. 2000). Major technical issues have been found to surround the reliability and validity of clinical trials, designed and streamlined according to the genomic criteria (Martin and Morrison 2006). These issues are statistical problems, e.g., study with a small subgroup of patients, decreased chances of detecting rare ADRs (ADRs), sample bias, lack of consistency in controlling pharmacogenomics variables, and the difficulty in replication of genetic association studies (Martin and Morrison 2006). Therefore, the regulatory authorities are taking steps to redesign the trials, because this type of practice can increase the possibility of unsafe medicines reaching the market (Martin and Morrison 2006) The Food and Drug Administration (FDA) has already indicated that if a trial contains any element of selection which is on the basis of genotypic information, there is greater need for tight clinical governance and improved post marketing surveillance (Martin and Morrison 2006).

3 Privacy and Confidentiality of Recruited Subjects

The genomic information is inherently personal; at the same time it is familial and (Issa 2000; Nuffield Council of Bioethics 2003; Buchanan et al. 2002). There are serious risks of discrimination and loss of privacy which are needed to be addressed in a pharmacogenomic study. Genomic information can stigmatize a person as it predicts the person's health (Patowary 2005). The pharmacogenomic information will reveal the status of individuals as responders and nonresponders. Therefore, the patient's willingness to undergo genetic testing is an important issue to be considered during the design of a clinical study. A number of studies have documented that general lack of knowledge and awareness creates problems in understanding the benefits of genetic testing especially in people with lower socioeconomic status (Hughes et al. 1997; Lipkus et al. 1999; Kinney et al. 2001; Shields 2001; Peters et al. 2004; Singer et al. 2004; Bates et al. 2005; Murphy et al. 2009; Suther and Kiros 2009) However, it was found that even the subjects with higher educational background express fewer concerns about possible misuse of genomic information (Suther and Kiros 2009).

There could be psychological implications for individuals in case the genomic information, labels them as "difficult to treat" (Issa 2000). No one will like that this information should be provided to others especially in the context of marriage and childbearing issues.

The personal information of the recruited study subjects may also serve as a ground for job discrimination by employers and insurance companies (Patowary 2005). Individuals will find it more difficult to find affordable health insurance based on their genomic information. Insurance companies will ask to pay higher premiums because of non-responsiveness to therapeutic outcome regardless of whether or not they develop the disease for which the treatment should be used (Issa 2000). At present, moratorium exists for the use of genomic information in setting insurance premiums only in a few countries (Ho et al. 2010).

The genomic information can be misused where there is strong social and cultural tradition of individual's reproductive freedom. Women are more vulnerable to suffer from this information especially in developing countries, where there is bias and discrimination against women (Patowary 2005; Mastroianni et al. 1994).

Pharmaceutical companies or researchers often collect and store samples from study subjects. The risk of storage of these samples is either to replicate or dispute findings that other researchers might present at a later date (Issa 2000). These can also be used for further studies. Therefore, taking informed consent from the recruited subjects is very important prior to their recruitment for sample collection. The informed consent form should be designed in a way so as to inform all the risks (including ADRs) and benefits of the study. The informed consent is the main vehicle by which the patient is empowered in the system (Verma et al. 2011). Obtaining an informed consent from the patients is a challenging issue as it can limit the pharmacogenomic trials in many ways (Anderson et al. 2003).

The storage of DNA samples in a DNA bank is another important issue in a pharmacogenomic study protocol. It can ensue a number of ethical issues, e.g., who will own the DNA samples for the future analysis. Usually, this information is entered and stored in computerized databases. But, this can impose serious issues related to autonomy, privacy of recruited subjects that warrant clarification, discussion, and immediate action (Issa 2002). Genomic information is more vulnerable to violations of privacy as it consists of an individual's "diary" of predicting the future (Annas 1998).

Another question is of feedback given to the study subjects. Careful consideration is therefore required to be given to the level of anonymity that should be applied to genomic information which has been stored as a part of research (Issa 2000). Each country has its own regulations for the protection of personal data. In some countries, rules allow the study subjects/patients to access all the medical and genetic information or individual feedback. In that case, samples will not be anonymized for research purpose. Therefore, the research study aimed at identifying genomic profiles in context to disease susceptibility and drug response should only be conducted on subjects who are aware of all the risks and benefits of research and voluntarily accept it.

4 Subject Stratification

Drug development strategies which are based on pharmacogenomic profiling assume that certain genetic variations will be identified affecting the response of the drug. This leads to the stratification of clinical trial subjects into subgroups based on their genotype (Issa 2002). Stratification leads to a number of ethical challenges. Firstly, genotyping as criteria of inclusion and exclusion of research participants might lead to the loss of benefits of research participation or to unfair representation in the trials (Issa 2002). This genotype-based stratification could lead to development of the subject selection biases. Excluding individuals from the trial on the basis of genotypic information could lead to the loss of benefits that they might achieve by participating in the trial (Issa 2002).

Stratification of study subjects into smaller subgroups can confound the statistical analysis and interpretation of the results (Becker 2001). For example, if 100 study subjects are stratified into groups of 10 each, it is evident that statistically significant differences might be perceived. It is important to question whether these statistical differences are clinically relevant because studies can be statistically significant but clinically insignificant (Sterne and Davey 2001; Becker 2001).

Pharmacogenomic data which is obtained as a result of stratification of research subjects into smaller groups will probably increase the spurious interpretation of statistical analysis (Sterne and Davey 2001; Issa 2002). Subject stratification can also lead to scientific and ethical challenges in data analysis due to the concept of penetrance. The variability in the degree of clinically relevant phenotypic expression because of genetic variation can lead to false-positive results (Zhao 2000; Issa 2002). As a consequence of this some of the participants might not be offered the trial medication in case the given polymorphism is highly penetrant. The clinical consequences of penetrance also pose a challenge for regulatory authorities, e.g., USFDA in labeling of drugs (Issa 2002). More research work needs to be carried out to predict the effects of penetrance on drug response and adverse effects.

5 Race and Ethnicity

Pharmacogenomic information varies according to race and ethnicity. Significant disparities exist in health status, health-care utilization, and clinical outcomes among different racial and ethnic groups (Harty et al. 2006). The recent approval by the FDA, of a combination of drugs, isosorbide dinitrate and hydralazine (BiDil) in heart failure, has created much discussion among researchers (Branca 2005; FDA News 2005). The combination of two generic drugs BiDil has got the approval for heart failure in self-identified Blacks, a step toward personalized medicine (FDA News 2005). The approval in part was on the basis of African-American Heart failure trial (A-HeFT), which showed approximately 43 % reduction in death as compared to placebo, a significant improvement over previous trials among unselected groups of patients (Taylor et al. 2004). The efficacy of BiDil, slowing down the progression of heart disease, provides a strong evidence that a particular genotype prevailing in a racial and ethnic subgroup of the population would be helpful in guiding the prescription of drugs.

A considerable debate is going on the issue of race and ethnicity in pharmacogenomics trials. The argument is based on either to continue the use of race and ethnicity or to abandon it altogether (Fullilove 1998; Rivara and Finberg 2001; Schwartz 2001; Burchard et al. 2003; Risch et al. 2002). This can be better explained with the help of an example, enzyme cytochrome P450 subfamily 2D6 (CYP2D6) is one of the oxidizing enzymes responsible for the metabolism of a number of drugs. Single nucleotide polymorphisms in the gene encoding for this enzyme can vary between different racial groups (Lipton 2003). The impaired activity of this enzyme is responsible for the altered pharmacokinetic profile of a number of drugs metabolized by the enzyme. It has been reported that among Caucasian, about 7 % population has a genetic variant which contributes to the reduced activity of CYP2D6 enzyme (Lipton 2003). As a result, this group of population is poor metabolizer for many of the medicines which are the substrates of CYP2D6 enzyme. Ethiopians and Saudi Arabians have high frequency of another genetic variant of CYP2D6 which makes them good responders because of increased activity of encoded enzyme (Weber 1997). Different variants of glucose 6-phosphate dehydrogenase (G6PD), an enzyme responsible for generation of nicotine-adenine dinucleotide phosphate (NADPH) in mature red blood cells, are found in very high frequency in African, Mediterranean, and Asiatic populations. Some of the genetic variants of G6PD have been found to disrupt the normal functions of the enzyme (loss of function). The pharmacological agents which are the substrates of G6PD, e.g., antimalarial drug, primaquine, have been found to induce hemolytic anemia in patients with impaired function of G6PD enzyme (Beutler 2007).

Another example is of antituberculosis drug, isoniazid, which is inactivated by acetylation. The impaired metabolism of the drug via slow acetylation due to genetic variation in one of key enzyme *N*-acetyl transferase-2 (NAT-2) results in accumulation of toxic levels of isoniazid (Weber 1997). Therefore, the variation in NAT-2 gene directly influences the metabolism (either rapid or slow) of isoniazid as well as other drugs and carcinogenic compounds (Grant et al. 1997). Studies have identified that approximately 50 % of Caucasian populations are genotypically slow metabolizers (acetylators) of isoniazid. More than 80 % of individuals in certain Middle Eastern populations and about 20 % Japanese are also slow metabolizers (Weber 1997).

Some of the genetic markers predicting the ADRs of certain drugs, e.g., human leukocyte antigen (HLA), has been found to be present in high frequency among Han Chinese population (8 %) as compared to rest of the world (Geer et al. 1998). In India, a 0–6 % prevalence has been reported (Shankarkumar 2004; Rajalingam et al. 2004; Rani et al. 2007). Studies have documented a significant high association of HLA*B 1502 variant with the development of serious adverse effects such as Stevenson-Johnson syndrome and toxic epidermal necrolysis in epileptic patients on carbamazepine (an antiepileptic drug) (Mehta et al. 2009; Chung et al. 2004; Chang et al. 2011).

These examples identify the significant role of genomic variation and ethnicity in pharmacogenomics research. However, population descriptors, i.e., race and ethnicity, are crude measures having useful predictive values at the level of individual patients. The use of both the terms race and ethnicity without accurate evaluation of other contributing factors has been observed to raise the risk of reinforcing and creating inequities in the field of health care (Harty et al. 2006; Jones 2001; Kahn 2005). However, this cannot be overlooked because each ethnic group might have specific genotypes responsible for ADRs and interindividual variation in a high frequency, and, therefore, this information needs to be taken into consideration for prescribing a particular drug in a particular ethnic group.

6 Orphan Drug Development

In certain cases, drug-associated severe adverse effects are the major problems encountered by physicians and patients. Therefore, the major ethical issue is that whether the new drug developed is meant only for the most common genotype or for the group identified as good responders. And this new drug will not be useful for nonresponders because this group is too small to attract investment from the pharmaceutical companies. According to Orphan Drug Act, 1983 (passed in the USA), drugs are considered orphan drugs under two circumstances: first, if the condition treated by drug affects fewer than 20,000 patients and, second, if there is no reasonable expectation that the cost of developing the drug will be recovered from its sales (FDA 2001; Rai 2002). According to this law, an "orphan disease" is a condition that affects fewer than 200,000 people. This act provides financial incentives which include tax credits, for pharmaceutical firms to direct their efforts for the developing products or drugs to treat orphan conditions/diseases. Companies could generate a demand for a prescribed drug by offering tests to identify responders and nonresponders. As a result of this, the entire population might be given minimum attention in this type of market driven drug development process resulting in "Orphan Population" (Issa 2002).

The problem has a high magnitude in case of developing countries with poor population. To overcome this issue, serious planning and significant public sector investment in research, increased pricing, new policies similar to orphan drug legislation in the USA and Europe might be required to provide the resources and financial incentives to encourage the private sectors in the pharmacogenomic studies (Robertson 2002; Danzon and Towse 2002). Researchers have already identified the situation where genetic variations do appear to predict the drug response in a manner that can create orphan drug population. For example, in patients with asthma, the presence of a particular type of single-nucleotide polymorphism in the promoter region of arachidonic acid converting enzyme, i.e., ALOX5, is responsible for lack of response of anti-asthmatic drug treatment (Rai 2002).

7 Unequal Distribution of Benefits

Provision of resource allocation is a big issue when the money is spent for research and design. Problems regarding equity and distributive justice are a central part in pharmacogenomic research (Beauchamp and Childress 2001). It is believed by some that research involving tools of genomics, e.g., Human Genomic Project, is the poor allocation of resources when there are majority of people suffering from lack of food, safe drinking, water, housing, and other basic amenities of life in developing countries (Patowary 2005). In that case it seems to be a luxury to invest on pharmacogenomics which is going to benefit only the rich and urbanized people. Contrary to this it has been observed that every year hundreds of people die due to hypersensitivity reactions or ADRs induced by drugs which can be avoided by testing the genotype before prescribing the available drugs. Further this will also reduce the frequency of visits to clinicians and recovery time, thereby decreasing the cost of hospitalization (Patowary 2005). The pharmacogenomics may also be costly and will be accessible to only rich. The excellent example in this concern is of patients of Gaucher's disease. The sufferers of this disease helped the pharmaceutical companies to develop effective treatment and unfortunately were denied access to the treatment by insurance companies due to higher cost.

An economic analysis of industry pharmacogenomics indicates that whether this new technology of personalized medicine is useful or good investment for capitalists (Todd 1999). Market pressures and excessive enthusiasm for conducting trials including pharmacogenomic protocols can result into study subjects being exposed to risks in trials which might be abandoned due to volatile market trends and uncertain longevity of specific patent holders (Issa 2002).

8 Conflict of Interest in Business and Research

Conflict of interest is developed among researchers who have invested their time and knowledge in conducting pharmacogenomic research for the drug companies interested in commercialization of their product (Agnew 2000). The chances of ignorance of their research interests are more after the companies have achieved their goal (Patowary 2005). Because the companies have to invest huge amount of money in developing a product and bringing it to the market, the question is whether these companies will be ready to develop a drug for a small subset of population with rare genotype.

Clinical research involving pharmacogenomic studies creates a close relationship between industry and academics with large amount of money involved in it (Patowary 2005; Corrigan and Williams-Jones 2006). These conflicts can be controlled by a process of systemic disclosure to both the ethics committees and research subjects involved in the trials/studies. In some cases it may also include the modifications in the protocol or to the research team (Howard et al. 2011). The concept of benefit sharing introduced by the United Nations Educational Scientific and Cultural Organization (UNESCO) and HUGO provides avenues for exploring to establish a more equitable relation between study participants, researchers, and sponsors in the field of pharmacogenomic studies (UNESCO 2003; HUGO 2000).

The recommendations in this concern include that there should be a declaration of conflict of interest regularly updated and reported by the principal investigators (PIs) and co-PIs of all the pharmacogenomic trials (Howard et al. 2011). The recruited study subjects should be informed that they will not be entitled to receive any commercial or financial benefits or future intellectual property rights. The researchers involved in pharmacogenomic studies expecting to commercialize their findings should consider a benefit sharing plan within the research protocols (Howard et al. 2011).

9 Data Sensitivity

Pharmacogenomic studies require a large-scale use of genetic information and individual's drug response profile to compare and evaluate the efficacy and drug-associated adverse effects, thus raising the issues about storage, security, and access to genetic information (Lipton 2003). Therefore, there should be an efficient database storage system. The development of such a database should be sensitive enough to overcome the issues of the significant issues of privacy and confidentiality (Robertson 2002). One such example of a database is Pharmacogenomics Knowledge Base (PharmaGKB) which is based on a project entitled the "Pharmacogenomics Research Network." This data can be shared between different people or researchers. Data sharing can be done without obtaining the informed consent from the research participants (Howard et al. 2011). Disregard of the consent form of research in data sharing is unethical because it violates the rights of confidentiality and integrity of the research participants (Joly et al. 2005).

Registration of clinical trials is a must for the researchers and companies running clinical trials in every nation. The USFDA has adopted some of the measures in this concern so as to regulate and secure the pharmacogenomics data of human subjects involved in studies (FDA News 2005). These include guidance for publication which is a crucial first step to overcome the problem of uncertainty and other concerns including confidentiality and intellectual property (UK Pharmacogenetics Study Group 2006). Secondly, there should be a framework for co-development of drug-diagnostic protocols/techniques (www.fda.gov/coder/genomics/pharmacoconceptfn.pdf). Thirdly, the Voluntary Genomic Data Submission (VGDS) initiative has been introduced to encourage pharmaceutical companies involved in pharmacogenomic trials and to submit exploratory pharmacogenomic information (www.fda.gov/coder/genomics/VGDS.htm). In Europe, the pharmacogenetic terminology has been defined properly by European Medicines Agency (EMEA). EMEA has also introduced a system of brief meetings in order to allow sponsors and regulators to discuss pharmacogenetic data informally at an early stage of study (CHMP 2005, Draft Guideline on Pharmacogenetics Briefing Meetings, EMEA/CHMP/20227/2004).

10 Social and Economic Concerns

Social and economic considerations have a greater impact in directing the pharmacogenomic research and its translation into the clinical field. A number of issues such as access to insurance, employment, and health-care resources can also lead to discrimination among individuals. In case the disease susceptibility is known for an individual, other family members related to the affected individual may feel at an increased risk of becoming ill and having fewer therapeutic alternatives, even when the risk has not been scientifically quantified (Martin et al. 2000). A social issue that can arise from pharmacogenomic-based therapeutics is a paradigm shift in therapeutic choices, caused by the perception of "uniqueness" based on the genomic profile of an individual (Issa 2002). For example, genetic variants affecting responsiveness to a particular asthma medication or antihypertensive can result in subtle subconscious "shifting of blame" by the principal investigator or the concerned physician involved in a clinical setting (Issa 2002).

Focusing on another issue of race and ethnicity, the pharmacogenomics information can also lead to a neglect of socioenvironmental factors which might lead to population-based differences in health status (Lee 2005). For example, many drug-metabolizing enzymes have qualitative and quantitative differences between different racial groups. Catechol-*O* methyltransferase (COMT) is an enzyme involved in the metabolism of drugs levodopa and methyldopa used in the treatment of Parkinson's diseases and hypertension, respectively. This enzyme also influences the production of estrogen metabolites associated with breast cancer disease. It has been suggested that the low-activity allele for enzyme COMT is less frequent among African and East Asian populations showing a clear evidence of risk in therapeutic benefit among these racial groups (McLeod et al. 1998; Syvanen et al. 1998).

Overlapping as well as distinguished concerns might arise in front of health-care providers, patients, and pharmaceutical companies (Goldstein et al. 2003). The genomic information could offer significant diagnostic value about the patient's response toward a pharmacological agent to the health-care providers. Therefore, economically both the health-care providers as well as the health-care receivers or patients will be benefited in terms of developing an exclusive system of giving the precise medication to the patients and avoiding the unnecessary wastage of money due to the bypass of drug-associated adverse effects (Goldstein et al. 2003). According to an estimate by You et al. (2004), the cost of major adverse effects (bleeding due to warfarin sensitivity) of the anticoagulant drug warfarin prescribing was found to be approximately US$6,000, which could be averted by prospective CYP2C9 genotyping (You et al. 2004; Caraco et al. 2008).

Studies have found that the economic calculations for the pharmaceutical industry are more complicated because of market segmentation (Shah 2003). Few drugs which work well are rejected or withdrawn from the market because of rare but ADRs in a subgroup of the population. This results in a huge loss of money from the viewpoint of a pharmaceutical firm, because the development of a drug molecule crosses a number of phases before coming into the market involving lots of expenditure and time. The antiepileptic drug felbamate, antipsychotic clozapine, and most of drugs have been withdrawn owing to QT-interval-associated arrhythmias (e.g., Terfenadine). If pharmacogenomic or pharmacogenetic predictors of adverse drug events could prevent the exposure of susceptible patients on the basis of genetic information, the costs of developing a single drug molecule on a large-scale effort could be fully recovered (Goldstein et al. 2003).

The development of genotype-based treatment will potentially result in a shift of costs of drug development from the pharmaceutical sector to the market or clinical sector (Pirmohamed and Lewis 2004). For example, the cost of the development of a drug from the process of manufacturing in a company and reaching up to licensing authority will be reduced via a more efficient streamlined process. But the use of drug after getting licensed will incur the combined cost of drug and pharmacogenetic test. As a consequence of this, there is a risk of new drugs developed with the aim of being prescribed to certain or only to rare genotype. Further, this will reduce the market size for that drug and will be too expensive for the patients (Pirmohamed and Lewis 2004).

There is no doubt that pharmacogenomic information has created a revolution in the field of clinical medicine, with few examples such as Herceptin for the patients with breast cancer who overexpress HER2 and Gleevec for patients with chronic myeloid leukemia that results from Philadelphia chromosome (Roses 2000; Goldstein et al. 2003). However, the tailor-made medicines or the personalized medicine will be created only for a few people or a subgroup of people therefore will be more expensive as compared to "bulk drugs" (Bolt et al. 2008). The main concern is that should these pharmacogenomic applications be reimbursed by health insurance companies or the cost will be paid by the patient itself (Bolt et al. 2008). This situation can lead to inequity among people those who can pay for these and the one who cannot.

The implications for providers of health care arise particularly if a patient would like to go for a genomic test (Lipton 2003). The concerned clinician might or might not be willing to prescribe a medicine on the basis of genomic information. Moreover, what will happen in case of nonresponders with no other option of treatment available. Ultimately, who will be deciding that whether or not a patient should take a tailored medicine against the doctor's advice (Lipton 2003). These are a few concerns which need to be resolved by formulating drug formularies with particular genomic information and generating the awareness among patients and doctors about genetic knowledge. The reason is that our academic profile of medicines does not provide sufficient knowledge and training except few introductory chapters in the curriculum. There is a need to develop the policies and to create funding incentives in order to promote equitable access to the benefits of pharmacogenomics and to protect the patients, their families, and disadvantaged social groups (Pirmohamed and Lewis 2004).

11 Regulatory Issues

The enormous potential of pharmacogenomics providing therapeutic benefits in terms of both patient and health provider heavily outweighs the risks. Providing such a powerful information in the hands of health-care providers and the ones who are involved in the discovery of novel approaches to disease treatment or prevention offers promise to the betterment of the society (McLeod and Evans 2001). At the same time, a number of other issues can be regulated by generating a better framework under the rules and regulations so as to minimize the chances of abuse or misuse of genomic information. According to "Title VII of the Civil Rights Act of 1964," discrimination on the basis of genetic information is the clear violation of the law because this will fall along social, racial, and ethnic lines (Nebert and Bingham 2001). A classic example of this type of discrimination appeared in the 1970s, when the sickle cell anemia was used as a criteria for denying from employment in a number of jobs (Nebert and Bingham 2001).

Antidiscrimination legislations are required so as to prevent the misuse of personal genomic data and to avoid stigmatization of a particular subgroup (Pirmohamed and Lewis 2004). To prevent the development of orphan drug and orphan genotype, public sector investment is required in the private sector. Cost-effective and cost-benefit analyses are required to be carried out so as to ensure fair pricing of pharmacogenomic products and diagnostics developed from pharmacogenomics trials.

The European Agency for the Evaluation of Medicinal Products, UK's Medicines and Healthcare Products Regulatory Agency, Japan's Pharmaceutical Affairs Bureau, and the FDA are the examples of regulatory agencies controlling the research studies and clinical trials. An evidence of the regulatory agency USFDA has been observed during the translation of genomic medicine into medical care in August 2007, when USFDA announced that it was bringing to doctors' attention, the potential usefulness of getting a patient's genetic profile before prescribing the drug "warfarin." Warfarin (used as anticoagulant agent) is a drug of choice in various cardiovascular treatments (Pirmohamed and Lewis 2004). The drug-metabolizing enzyme cytochrome P450 2C9 (CYP2C9) has been found to have a number of genetic variants affecting the pharmacokinetic of warfarin to a marked level. Therefore, the persons with a particular genetic variant affecting drug response should be screened prior to prescribing, as indicated on the label of the drug. The FDA has the responsibility of regulating genetic tests to ensure scientific validity and utility. In addition to the genetic variants, the interaction of these variants with the environment or food also needs to be considered.

There are some countries in which legal systems are based on common law (court decisions by judges); physicians and pharmacists are having liability under the theories of negligence which involves the violation of a duty based on a wrong interpretation of genetic tests.

It is the duty of physicians to prescribe the right drug in the right dose, to the right patient at the right time. Pharmacists are also liable because of dispensing the drugs. The responsibility of the pharmacist is to warn the patient about the ADRs as well as to provide other necessary information such as dosage of drug, drug and food interaction, and drug-drug interaction. The pharmacist dispenses the drugs according to the instructions of doctor, but there are some jurisdictions which have imposed liability on pharmacists for the harm that result from a prescribed drug that was properly not dispensed in accordance to the prescription (Rothstein and Epps 2001). PharmGened is a program (2008–2011) designed to provide physicians, pharmacists, students, and other health-care professionals to access evidence-based pharmacogenomic information and to improve their knowledge (Lee 2005).

The generics of drugs already exist in the market; prescribing drugs on the basis of genomic information could impact the sale of that drug and can also put a question mark for it to be tested again or reevaluation for its efficacy and side effects. Pharmacogenomics can also impose liability on pharmaceutical firms for not conveying proper information of risks and associated adverse effects on the label. It is not possible in a clinical trial to predict all the adverse effects owing to the number of genes involved in affecting the response of a drug along with consideration of environmental factors (Wood and Woosley 1998). The example of SmithKline Beecham will give a meaning to the above statement. In year 2000, four persons had filed a class action lawsuit against SmithKline Beecham, alleging that the manufacturer of a vaccine for Lyme disease knew that some individuals would be susceptible to arthritis on exposure to the vaccine because of their genotype, but did not provide this important information on the label of the drug (Cassidy vs. SmithKline Beecham 1999). The pharmaceutical company has stopped selling the vaccine in February 2002 and agreed in 2003 to an out of court settlement of more than one million dollars to the patients (Vaccination News 2003).

Regulatory policies which can successfully achieve the dual objectives of ensuring public safety and promoting innovation in health technology should be imperative in order to enjoy the fruits of this emerging new field (Issa 2002). In year 2003, the FDA had issued a draft guidance regarding the submission of pharmacogenomic data, in order to encourage drug companies to conduct pharmacogenomic tests during drug development and to maintain an appropriate level of safety and accuracy for consumers (FDA drug safety newsletter).

12 Pharmacogenomics into Clinic

The sequencing of the human genome in 2003 has raised the expectations of researchers in the use of genomic information in response to individual variability toward prescribed medicines. The era of personalized medicine is

Table 8.1 FDA-approved few drug labels including pharmacogenomic data

Name of the drugs	Disease in which the drug is prescribed	Genotypic information/tests available
6-mercaptopurine	Leukemia	Patients with inherited little or no thiopurine-*S*-methyltransferase (TPMT) activity due to polymorphism are at increased risk for severe mercaptopurine toxicity from conventional doses of mercaptopurine and generally require substantial dose reduction
		Laboratory tests are available to determine the TPMT status
Azathioprine	Organ transplant, immunosuppressive, prodrug	Laboratory tests are available to determine the TPMT status so as to avoid adverse drug reactions
Atomoxetine	ADHD	Laboratory tests are available to identify
		CYP2D6 poor metabolizers
Irinotecan	Colon cancer	Heterozygous patients (carriers of one variant allele and one wild-type/normal allele which results in intermediate UGT1A1 activity) may be at an increased risk for neutropenia
Carbamazepine	Epilepsy	Patients from at risk populations should be screened for the presence of HLA*B1502 prior to initiating treatment with carbamazepine. Patients testing positive for the allele should not be treated with the drug
Warfarin	Cardiovascular diseases	Genotyping for VKORC1 and CYP2C9 genes is advised to avoid adverse drug effects
Herceptin	Breast cancer	Herceptin treatment can lead to ventricular dysfunction and congestive cardiac failure among 25–30 % patients. Therefore, patients should be tested for over-expression of Her-2/neu2protein before prescribing

going to provide the best and precise therapeutic interventions. Current methods of prescribing medicines based on weight, age, conditions of liver and kidney, presence of any other disease, and concomitant medicines are no more successful because of drug-associated adverse effects in some of the patients. Therefore, the tailored medicines based on the individual's genomic profile will guide the concerned clinician to prescribe the right medicine in the right dose for the right person. There are a few examples of translation of pharmacogenomic information into medical practice. The common gene variant (c 94C>A) of thiopurine methyltransferase (TPMT) – an enzyme affecting the metabolism of cancer drugs, azathiopurine and mercaptoprine (which are used in acute lymphoblastic childhood leukemia), is being used for the prediction of right dose of these two medicines (Meyer 2000; McCarthy and Hilfiker 2000). Herceptin has created another breakthrough in the treatment of breast cancer, and patients are getting benefits from it, e.g., Table 8.1.

Different variants of cytochrome 2D6 (CYP2D6) gene have been found to be associated with debrisoquine-sparteine drug oxidation and are correlated with gastrointestinal, liver, and lung cancer (Clapper 2000). The genetic variants of paraoxonase gene-1 (PON-1) gene have been found to be correlated with the risk of developing not only cardiovascular disease but also predicts the toxicity of organophosphates such as pesticides, parathion chlorpyrifos and diazinon, and nerve agents sarin and soman (Nebert and Bingham 2001; Brophy et al. 2001).

13 Conclusions

The pharmacogenomics is in its infancy and what is already known about pharmacogenomics is just the tip of the iceberg, a greater understanding of the way in which patients with a particular genotypic profile response to a drug allows manufacturer to identify subtypes of a population that will benefit most from a particular drug. The pharmacogenomics is in its inception; therefore, concerned ethical, social, and legal issues need to addressed and formulated carefully. To obtain maximum benefit from pharmacogenomics, we need to achieve legitimate concerns and safeguard against malpractices in clinical medicine. There should be right formulations of policy issues along with the combination of constraints and incentives so as to protect and promote the patient and society so that the pharmacogenomics would be widely introduced and accepted into the routine clinical practice. The goal of pharmacogenomics is to speed up the process of drug development with reduced failure, more safety, and therapeutic values. The inclusion of pharmacogenomics into clinical practice will be determined by their scientific values and validity with respect to human values; there is an urgent need for researchers, health-care providers, government, and pharmaceutical organization to work hard in hand in order to formulate an international guideline/regulation for the use of pharmacogenomics produce safer and more effective therapeutic strategies.

References

Agnew B (2000) Studies trace patchwork of conflict policies. Science 290:1873

Anderson JL, Carlquist JF, Horne BD, Muhlestein JB (2003) Cardiovascular pharmacogenomics: current status, future prospects. J Cardiovasc Pharmacol Therapeut 8(1):71–83

Annas GJ (1998) Some choice: law, medicine and the market. Oxford University Press, New York

Bates BR, Lynch JA, Bevan JL, Condit CM (2005) Warranted concerns, warranted outlooks: a focus group study of public understandings of genetic research. Soc Sci Med 60(2):331–344

Beauchamp TL, Childress JF (2001) Principles of biomedical ethics, 5th edn. Oxford University Press, New York

Becker RE (2001) Modifying clinical trial designs to test treatments for clinical significance in individual patients. Clin Drug Investig 21:727–733

Beutler E, Duparc S, G6PD Deficiency Working Group (2007) Glucose-6-phosphate dehydrogenase deficiency and antimalarial drug development. Am J Trop Med Hyg 77(4):779–789

Bolt ILLE, Kalis A, Derijks J, van Delden JJM (2008) Ethical questions in the field of pharmacogenetics. EJHP Pract 14:30–34

Branca MA (2005) BiDil raises questions about race as a marker. Nat Rev Drug Discov 4:615–616

Brophy VH, Hastings MD, Clendenning JB, Richter RJ, Jarvik GP, Furlong CE (2001) Polymorphisms in the human paraoxonase promoter. Pharmacogenetics 11:77–84

Buchanan A, Califano A, Kahn J, McPherson E, Robertson J, Brody B (2002) Pharmacogenetics: ethical issues and policy options. Kennedy Inst Ethics J 12(1):1–15

Burchard EG, Ziv E, Coyle N, Gomez SL, Tang H, Karter AJ, Mountain JL, Pérez-Stable EJ, Sheppard D, Risch N (2003) The importance of race and ethnic background in biomedical research and clinical practice. N Eng J Med 348:1170–1175

Caraco Y, Blotnick S, Muszkat M (2008) CYP2C9 genotype-guided warfarin prescribing enhances the efficacy and safety of anticoagulation: a prospective randomized controlled study. Clin Pharmacol Ther 83(3):460–470

Cardon LR, Idury RM, Harris TJR (2000) Testing drug response in the presence of genetic information: sampling issues for clinical trials. Pharmacogenetics 10:503–510

Cassidy v. Smithkline Beecham (Pa.Chester County Dec. 14, 1999) 99-10423

Chang CC, Too CL, Murad S, Hussein S (2011) Association of HLAB*1502 with carbamazepine-induced toxic epidermal necrolysis and Stevens-Johnson syndrome in Malaysian population. Int J Dermatol 50(2):221–224. doi:10.1111/j.1365-4632.2010.04745.x

CHMP (2005) Draft guideline on pharmacogenetics briefing meetings, EMEA/CHMP/20227/2004 (17 March). Guideline on pharmacogenetics briefing meetings, EMEA/CHMP/PGxWP/20227/2004 (27 April 2006). www.emea.eu.int/pdfs/human/pharmacogenetics/2022704en.pdf

Chung WH, Hung SI, Hong HS, Hsih MS, Yang LC, Ho HC, Wu JY, Chen YT (2004) Medical genetics: a marker for Stevens-Johnson syndrome. Nature 428(6982):486

Clapper ML (2000) Genetic polymorphism and cancer risk. Curr Oncol Rep 2:251–256

Clarke A, English V, Harris H, Wells F (2001) Report on ethical considerations. Int J Pharmceut Med 15:89–94

Corrigan OP, Williams-Jones B (2006) Pharmacogenetics: the bioethical problem of DNA investment banking. Stud Hist Philos Biol Biomed Sci 37:550–565

Danzon P, Towse A (2002) The economics of gene therapy and of pharmacogenetics. Value Health 5(1):5–13

Food and Drug Administration (2001) Miscellaneous provisions relating to the Orphan Drug Act. Food and Drug Administration htpp://www.fda.gov/opacom/laws/orphandg.htm

Ernst FR, Grizzle AJ (2001) Drug-related morbidity and mortality: updating the cost-of-illness model. J Am Pharm Assoc (Wash) 41:192–199

European Agency for the Evaluation of Medicinal Products, Position paper on terminology in pharmacogenetics (2002) EMEA/CPMP/3070/01, London 1–6

FDA News (2005) FDA approves BiDil heart failure drug for Black patients. June 23, 2005. Food and Drug Administration. http://www.fda.gov/bbs/topics/NEWS/2005/NEW1190html

Friedman LM, Furberg CD, DeMets DL (1996) Fundamentals of clinical trials. Wright, Boston

Fullilove MT (1998) Comment: abandoning "race" as a variable in public health research – an idea whose time has come. Am J Public Health 88(9):1297–1298

Geer L, Terasaki PI, Gjertson DW (1998) HLA frequency. In: Gjertson DW, Terasaki PI (eds) HLA. American Society for Histocompatibility and Immunogenetics, Lenexa, pp 327–363

Goldstein DB, Tate SK, Sisodiya SM (2003) Pharmacogenetics goes genomic. Nat Rev 4:937–947

Grant DM, Hughes NC, Janezic SA, Goodfellow GH, Chen HJ, Gaedigk A, Yu VL, Grewal R (1997) Human acetyltransferase polymorphisms. Mutat Res 376:61–70

Harty L, Johnson K, Power A (2006) Race and ethnicity in the era of pharmacogenomics. J Clin Pharmacol 46:405–407

Ho MK, Goldman D, Heinz A, Kaprio J, Kreek MJ, Li MD, Munafò MR, Tyndale RF (2010) Breaking barriers in the genomics and pharmacogenetics of drug addiction. Clin Pharmacol Ther 88(6):779–791

Holtzman NA, Watson MS (1997) Promoting safe and effective genetic testing in the United States. The National Human Genome Research Institute, Bethesda

Howard HC, Joly Y, Avard D, Laplante N, Philips M, Tardif JC (2011) Informed consent in the text of pharmacogenomic research: ethical considerations. Pharmacogenomics J 11:155–161

Hughes C, Gomez-Caminero A, Benkendorf J, Kerner J, Isaacs C, Barter J, Lerman C (1997) Ethnic differences in knowledge and attitudes about BRCA1 testing in women at increased risk. Patient Educ Couns 32(1–2):51–62

Human Genetic Commission (2002) http://genome.wellcome.ac.uk/doc_WTD021009.html

Human Genome Organization (2000) Statement on benefit sharing. Human Genome Organization, London

Issa AM (2000) Ethical considerations in clinical pharmacogenomics research. Trends Pharmacol Sci 21:247–249

Issa AM (2002) Ethical perspectives on pharmacogenomic profiling in the drug development process. Nat Rev Drug Discov 4(1):300–308

Joly Y, Knoppers BM, Nguyen MT (2005) Stored tissue samples: through the confidentiality maze. Pharmacogenomics J 5(1):2–5

Jones CP (2001) Invited commentary: "race," racism, and the practice of epidemiology. Am J Epidemiol 154(4):299–304; 305–306

Kahn J (2005) Misreading race and genomics after BiDil. Nat Genet 37:655–656

Kinney AY, Croyle RT, Dudley WN, Bailey CA, Pelias MK, Neuhausen SL (2001) Knowledge, attitudes, and interest in breast-ovarian cancer gene testing: a survey of a large African-American kindred with a BRCA1 mutation. Prev Med 33(6):543–551

Lee SS (2005) Racializing drug design: implications of pharmacogenomics for health disparities. Am J Public Health 95(12):2133–2138

Levine RJ (1993) New international ethical guidelines for research involving human subjects. Ann Intern Med 119(4):339–341

Lipkus IM, Iden D, Terrenoire J, Feaganes JR (1999) Relationships among breast cancer concern, risk perceptions, and interest in genetic testing for breast cancer susceptibility among African-American women with and without a family history of breast cancer. Cancer Epidemiol Biomark Prev 8(6):533–539

Lipton P (2003) Pharmacogenetics: the ethical issues. Pharmacogenomics J 3:14–16

Martin LF, Robinson A, Moore BJ (2000) Socioeconomic issues affecting the treatment of obesity in the new millennium. Pharmacoeconomics 18(4):335–353

Martin P, Morrison M (2006) Realizing the potential of genomic medicine. A research report published by Royal Pharmaceutical Society of Great Britain

Mastroianni AC, Faden R, Federman D (1994) Women and health research: ethical and legal issues of including women in clinical studies. Institute of Medicine. National Academy Press, Washington, DC

McCarthy JJ, Hilfiker R (2000) The use of single-nucleotide polymorphism maps in pharmacogenomics. Nat Biotechnol 18:505–508

McLeod HL, Syvanen AC, Githang'a J, Indalo A, Ismail D et al (1998) Ethnic differences in catechol O-methyltransferase pharmacogenetics: frequency of the codon 108/158 low activity allele is lower in Kenyan than Caucasian or SouthWest Asian individuals. Pharmacogenetics 8:195–199

McLeod HL, Evans WE (2001) Pharmacogenomics: unlocking the human genome for better drug therapy. Annu Rev Pharmacol Toxicol 41:101–121

Mehta TY, Prajapati LM, Mittal B, Joshi CG, Sheth JJ, Patel DB, Goyal RK, Dave DM (2009) Association of HLAB*1502 allele and carbamazepine induced Stevens-Johnson syndrome among Indians. Indian J Dermatol Venereol Leprol 75:579–582

Meyer U (2000) Pharmacogenetics and adverse drug effects. Lancet 356(9242):1667–1671

Murphy EJ, Wickramaratne P, Weissman MM (2009) Racial and ethnic differences in willingness to participate in psychiatric genetic research. Psychiatr Genet 19(4):186–194

Nebert DW, Bingham E (2001) Pharmacogenomics: out of the lab and into the community. Trends Biotechnol 19(12):519–523

Nuffield Council of Bioethics (2003) Pharmacogenetics: ethical issues. NCB: London. bioethics@nuffieldfoundation.org

Patowary S (2005) Pharmacogenomics – therapeutic and ethical issues. Kathmandu Univ Med J 12(4): 428–430

(2005) Personalized medicine: the emerging pharmacogenomics revolution. PricewaterhouseCoopers, New York. http://www.pwc.com/techforecast/pdfs/pharmaco-wb-x.pdf

Peters N, Rose A, Armstrong K (2004) The association between race and attitudes about predictive genetic testing. Cancer Epidemiol Biomark Prev 13(3):361–365

Pirmohamed M, Lewis G (2004) In: Mossialos E, Mrazek M, Walley T (eds) Regulating pharmaceuticals in Europe: striving for efficiency, equity and quality. Open University Press, Maidenhead, pp 279–296

Rai AK (2002) Pharmacogenetic interventions, orphan drugs, and distributive justice: the role of cost-benefit analysis. Soc Philos Policy 19(2):246–270

Rajalingam R, Parham P, Mehra NK (2004) HLA-A, -B, -Cw, -DQA1, -DQB1 and -DRB1 alleles and KIR alleles in a Hindu population from Punjab, India. Hum Immunol 65:958–960

Rani R, Marcos C, Lazaro AM, Zhang Y, Stastny P (2007) Molecular diversity of HLA-A, -B and C alleles in a North Indian population as determined by PCR-SSOP. Int J Immunogenet 34:201–208

Risch N, Burchard E, Ziv E, Tang H (2002) Categorization of humans in biomedical research: genes, race and disease. Genome Biol 3(7):comment 2007.1–2007.12

Rivara F, Finberg L (2001) Use of the terms race and ethnicity. Arch Pediatr Adolesc Med 155(2):119

Robertson JA (2002) Consent and privacy in pharmacogenetic testing. Nat Genet 28(3):207–209

Roses AD (2000) Pharmacogenetics and practice of medicine. Nature 405:857–865

Rothstein MA, Epps PG (2001) Ethical and legal implications of pharmacogenomics. Nat Rev Genet 2:228–231

Schwartz RS (2001) Racial profiling in medical research. N Engl J Med 344(18):1392–1393

Shah J (2003) Economic and regulatory considerations in pharmacogenomics for licensing and healthcare. Nat Biotechnol 21(7):747–753

Shankarkumar U (2004) HLA-A, -B and -Cw allele frequencies in a Bhil population from Western India. Hum Immunol 65:960–961

Shields AE (2001) Ethical concerns related to developing pharmacogenomic treatment strategies for addiction. Addict Sci Clin Pract 6:32–43

Singer E, Antonucci T, Van Hoewyk J (2004) Racial and ethnic variations in knowledge and attitudes about genetic testing. Genet Test 8(1):31–43

Sterne JA, Davey SG (2001) Sifting the evidence – what's wrong with significance tests? Br Med J 322(7280): 226–231

Suther S, Kiros GE (2009) Barriers to the use of genetic testing: a study of racial and ethnic disparities. Genet Med 11(9):655–662

Syvanen AC, Githangga J, Indalo A, Ismail D, Dewar K, Ulmanen I, Sludden J (1998) Ethnic differences in catechol O-methyltransferase pharmacogenetics: frequency of the codon 108/158 low activity allele is lower in Kenyan than Caucasian or South-west Asian individuals. Pharmacogenetics 8(3):195–199

Taylor AL, Ziesche S, Yancy C, Carson P, D'Agostino R Jr, Ferdinand K, Taylor M, Adams K, Sabolinski M, Worcel M, Cohn JN, African-American Heart Failure Trial Investigators (2004) Combination of isosorbide dinitrate and hydralazine in blacks with heart failure. N Eng J Med 351(20):2049–2057

Todd JA (1999) Interpretation of results from genetic studies of multifactorial diseases. Lancet 354(1):15–16

Tri-Council Policy Statement: Ethical Conduct for Research Involving Humans (1998) A report by a working group of the three federal funding councils: Medical Research Council of Canada, Natural Sciences and Engineering Research Council of Canada, and Social Sciences and Humanities Research Council of Canada

UK Pharmacogenetics Study Group (2006) Policy issues in pharmacogenetics: a policy briefing from the UK Pharmacogenetics Study Group. http://www.york.ac.uk/res/pgx/publications/PGxpolicyissues2006

UNESCO (2003) International Bioethics Committee. Declaration on Human Genetic Data, Paris

Vaccination News (2003) Glaxo settles Lyme disease vaccine suit 9 July 2003. http//vaccinationnews.com/dailynews/2003/july09/GlaxoSettles9.htm

Verma A, Khurana T, Bharti SK (2011) Pharmacogenomics in clinical research and practice: an ethical consideration. Pharmacology (Online) 1:453–461

Weber W (1997) Pharmacogenetics. Oxford University Press, Oxford

Wood AJJ, Woosley R (1998) Making medicines safer – the need for an independent drug safety board. N Engl J Med 339:1851–1853

World Medical Association Declaration of Helsinki Ethical Principles for Medical Research Involving Human Subjects, Adopted by the 18th WMA General Assembly Helsinki, Finland, 1964, and amended by the 52nd WMA General Assembly, Edinburgh (2000)

You JH, Chan FW, Wong RS, Cheng G (2004) The potential clinical and economic outcomes of pharmacogenetics-oriented management of warfarin therapy – a decision analysis. Thromb Hemost 92:590–597

Zhao HY (2000) Family-based association studies. Stat Methods Med Res 9:563–587

Computational Pharmacogenomics

9

Enrique Hernández-Lemus

Abstract

After the advent of high-throughput research in the biomedical sciences, greatly enhanced with the rise of genomics, proteomics, and metabolomics, it is a recognized fact that the science of pharmacology will also experiment dramatic changes. The search for therapeutic targets has become far too complex to be guided, either by an elevated form of *educated guesses* or by the expert opinion of principal investigators, no matter how rich a wealth of experience and how high the level of knowledge they may possess. Organismal response to drugs depends ultimately on a series of extremely entrenched molecular *pulls and triggers* encoded on a complex web behind the regulatory mechanisms in gene expression, cell signaling, and metabolic control. Pharmaceutical genomics or pharmacogenomics is thus related with the discovery and ultimately the clinical application of such enormous bodies of information. Massive data analysis, classification techniques, network reconstruction, and dynamical modeling are thus at the core of pharmacogenomics. In this chapter we will outline some of the main ideas and techniques in the emerging field of computational pharmacogenomics, about their use in research, development, and clinical application and also as diagnostic/prognostic tools, and in the field of targeted therapeutics and personalized medicine.

1 Introduction

Contemporary drug discovery and design is no longer possible without resorting to sophisticated computational mathematical modeling techniques. In this chapter we will review some new tenets and opportunities in the computational modeling of the complex biological process behind disease characterization at the genomic level and its relation with modern techniques of drug discovery. We will be discussing how these computational methods may impact in such disparate areas of pharmacology such as prediction of efficacy of drugs (and also synergistic effects in *drug cocktails*) by means of computational modeling of molecular pathways (Adams et al. 2009;

E. Hernández-Lemus (✉)
Computational Genomics Department, National Institute of Genomic Medicine, México City, Mexico
e-mail: ehernandez@inmegen.gob.mx

D. Barh et al. (eds.), *Omics for Personalized Medicine*,
DOI 10.1007/978-81-322-1184-6_9, © Springer India 2013

Arikuma et al. 2008; Azuaje et al. 2011), using computer-aided text mining (Almeida 2010; Chiang and Butte 2009) to search for drug leads in the ever-growing online biomedical literature (Garten and Altman 2009; Harpaz et al. 2012), and applying genetic overlap between various disease-associated genes to predict alternative drug use (Overby et al. 2010).

By combining computational genomics calculations in data from genotyping (Danecek et al. 2011; Guo et al. 2006; Johnson 2007), DNA and RNA sequencing (Kahvejian et al. 2008; Li et al. 2010; Nekrutenko and Taylor 2012), MS proteomics (Deutsch et al. 2008), gene expression microarrays (Chengalvala et al. 2007; Baca-López et al. 2009), and small molecule analysis (Adams et al. 2009) with integrative modeling and bioinformatics approaches (Eissing et al. 2011), it may be possible to overcome the bottleneck of nearly all pharmacological studies (the one that makes FDA approval a long and hard race) which is the scarce knowledge that we posses of the extremely complex molecular mechanisms underlying human phenotypes, in particular those for common diseases such as cancer, diabetes, as well as metabolic, autoimmune, and infectious diseases. The use of high-performance computing methods (Hernández-Lemus 2011; Gentleman et al. 2005; Okimoto et al. 2009) holds the potential for achieving a substantial reduction in the costs of drug development by reducing the length and scope of empirical tests but also by broadening the set of viable pharmacological targets and by making possible to identify novel therapeutic strategies in silico.

Computational pharmacogenomics is an upcoming science that is nourished in many well-established disciplines:

Molecular modeling or structural biology (Krissinel 2007, 2010) is perhaps one of the most mature areas in computational biology. It is aimed at predicting the main structural properties of biological targets (Krissinel and Henrick 2004, 2007) (in some cases even modeling them in a block-by-block *Lego-brick fashion*). Molecular models base their discoveries in powerful techniques blending computational quantum chemistry with biophysics (Krissinel 2010), along with experimental crystallography and informatics. It is precisely with the aid of chemo-informatics that it is possible to model characteristic features of promissory drug leads and inferring their physical interaction with targets.

Sequence analysis includes a number of mathematical an computational methods (Lipman and Pearson 1985; Pearson and Lipman 1988; Smith and Waterman 1981) used to compare nucleotide and amino acid sequences for a variety of phenotypes (Krissinel 2007), thus allowing the imputation of functional roles to gene variants by considering evidence from homologous genes, in similar phenotypes, cases/controls, intraspecies, and even in different biological systems. Sequence analysis has a foundational role in target identification, in particular in the earlier stages of drug discovery.

Computational biology Former applications of computational sciences to biology dealt with data storage, retrieval, and analysis, quite often without any form of data management or modification. This discipline is what we currently name bioinformatics. Computational biology, on the other hand, deals with the quantitative analysis, probabilistic modeling, statistical characterization and even computational process of biological (mostly molecular) data (Gentleman et al. 2005). These methods arise since raw information although permits some degree of understanding is not enough to realize the enormous complexity linked with biological processes.

Data integration is a series of computer-based methods whose goal is the effective analysis of datasets coming from fusing multiple data types (Eissing et al. 2011; Huson et al. 2011). These may include hypothesis established from literature mining, gene expression data from microarrays and RNA-seq experiments, protein-protein interactions from yeast two-hybrid experiments and immunoassays, and even clinical patient records. Data integration involves the estimation of statistical measures of bias, uncertainty, and cross correlation in the data between and within sources (Szklarczyk et al. 2011). Under proper methodological designs, data integration may increasingly turn important for drug discovery, even more so since availability of high-throughput data for a variety of experimental platforms in diverse biosystems that are relevant to human disease grows at an unprecedented fast pace.

Pattern recognition and classification methods are intended to imitate the way that we perceived things happening in the real world. Related perception tasks such as face recognition, graphology, shape detection algorithms, and discrimination between levels of relevance of text in search queries and others are extremely difficult to implement by computers even when they are considered almost trivial human activities. However, a series of algorithms, most of them based in the tenets of statistical- and machine-learning theories, artificial intelligence and neural networks (Hernández-Lemus and Rangel-Escareño 2011) have been developed. In biomedicine and drug design, pattern recognition methods have been used to discover multimolecular processes and build models based on complex databases often containing molecular, environmental, and clinical information. Information theory and signal processing are other computer science approaches to pattern identification (Hernández-Lemus et al. 2009; Hernández-Lemus 2011).

Network biology started as a visualization tool but has turned into an extremely powerful analysis framework for complex bio-processes (Arrell and Terzic 2010; Faustino and Terzic 2008, Hernández-Lemus and Rangel-Escareño 2011). If we think about molecules inside the cell – say genes, proteins, RNAs, or even exogenous small molecules – as the vertices on a graph, the physical or functional interactions between such molecules as directed links, we can view cells as molecular networks. Complex networks are dynamic objects whose functions (local and global) may be shaped and reshaped by inner and outer processes. In this sense, drugs may *interact* with the networks representing molecular processes driving the cell's functions (Schlitt and Brazma 2007). Thus, by studying the nature of biological network, one may have a clearer picture of synergies, side effects, and feedback loops that may determine whether a drug is effective to reach its target, but even more, whether its use may produce secondary effects by hitting other targets or by affecting other pathways in otherwise unpredictable ways.

In brief, computational approaches for pharmacogenomics may include not only database design and implementation, data sharing strategies, statistical analysis, and computational methods development but also methodological approaches akin to systems biology, in particular, mathematical/computational models of molecular interaction networks for genotype to phenotype, gene regulatory interactions from whole-genome expression data, population genomics approaches, and multiscale drug characterization. All of these studies rely on heavy computational implementations that may be fully integrated with traditional pharmacology and genetics, in order to ease the transition of designed pharmaceuticals from bench to bed.

2 Bioinformatics

2.1 Bioinformatics Applications in Pharmacogenomics

The role of bioinformatics applications in modern biomedicine has evolved from being a mere companion to a quite central role. If we want to study the ways by which normal cell functions differ from those of diseased conditions, biological datasets may be handled in order to form a comprehensive picture of the underlying phenomena. Bioinformatics involves the analysis and interpretation of various types of data, such as nucleotide and amino acid sequences, protein domains, and protein structures (Altschul et al. 1997). In order to do this, bioinformatics is related with the development and implementation of tools that make possible a fast, efficient access to biological information. It also encompasses the use and management of these information sources (Warren et al. 2005). Bioinformatics is also in charge of the development of new algorithms and statistical models used to assess (and sometimes even infer) relationships among members of large datasets by focusing on the development and application of computationally intensive techniques (Gentleman et al. 2005). As stated these methods deal with pattern recognition, data mining, machine-learning algorithms, and visualization with particular application to alignment of both nucleic acid and amino acid sequences, gene discovery, and whole-genome assembly but also in drug design and discovery, protein structure alignment, protein structure prediction, as well as gene expression and protein-protein interaction profiling, genome-wide association studies for genome variants, etc (Ouzounis 2012).

Bioinformatics techniques may be applied both in basic research settings and within the clinical/pharmaceutical R & D efforts. In the last case, the discipline termed *translational bioinformatics* (TB) (Altman 2012; Butte and Ito 2012) aims to narrow the gap between identifying pathogenic pathways and diseased phenotypes on the one hand and guiding molecular experiments to improve target identification, drug selection, and clinical trial design (Eng et al. 2011). Computational approaches of this kind may turn into an effective way to solve these problems, without *drowning in data* (Hoppe et al. 2011). TB should allow decision makers to take a stand in relevant manners by means of complementary information sources, way beyond than it could be achieved by resorting to the capabilities human memory, intuition, and pattern thinking alone (Fernald et al. 2011; Harpaz et al. 2012). Bioinformatics analyses have been implemented, for instance, to predict drug interactions and look up for adverse effects and synergy in drug cocktails (Arikuma et al. 2008; Sarkar et al. 2011).

With the same mind-frame as TB (i.e., clinical and pharmacological application of basic biomedical research), a field is now emerging under the name computational pharmacogenomics (CPG). CPG is an innovative and evolving science that assembles and mines disparate datasets curated (and contextualized in terms of extensive databases and hypothesis) from the domains of human biology, biochemical pathways, and drug-targeting mechanisms of action (Thorn et al. 2010). It borrows principles from computational biology, informatics, and information technology as well as from genomic platforms and classic pharmacology. In silico software platforms consist of multiple linked databases that can include structure-activity relationships, animal disease models, drug chemistry, toxicology, gene expression and genomic microarrays, sequence and functional annotations, and phenotypes of disease and drug response. CPG implementation and use in drug development has been limited in most disease areas. Oncology is the exception where the combination of modern tumor phenotyping and bioinformatics has successfully delineated mechanistic details in tumors such as gene signatures and protein networks. In brief, CPG involves the development of algorithms to analyze basic molecular and cellular data with the goal of optimizing drug design and clinical care (Holford and Karlsson 2007).

Technologies to measure DNA sequences are under constant development and evolution. Now, we are able to measure not only the genetic variations with extreme accuracy at lower costs but the sequencing of whole genomes and exomes, which has become quite cheap (about $4,000 USD or less per sample) and with unprecedented quality. The pace of increase in our ability to sequence DNA has in fact surpassed the growth in computer power for the past several years (the so-called Moore's law), and therefore, DNA sequencing has begun to further great demands on the computer data infrastructure, in terms of both storage and analysis. Hence, the challenges in CPG are moving fast toward gargantuan computation facilities, powerful algorithms, and intelligent experimental designs and strategies that will no doubt pave the way to completely new forms of drug discovery.

2.2 Database Archiving

Electronic repositories for biological data and information (in particular for molecular biology and "omics") are becoming an important tool to help us analyze an enormous number of biological phenomena, in such disparate fields as structural biology, protein science, the architecture of metabolic pathways, gene regulation, epigenomics, and pharmacogenomics (Adams et al. 2009; Arikuma et al. 2008; Flicek et al. 2011; Fujita et al. 2011; He et al. 2010; Kent et al. 2002; Overby et al. 2010). This immense wealth of knowledge may be applied to improve our ability to treat diseases and design better medications. The tasks of designing, developing, and managing biological databases are central to bioinformatics and computational biology. Data contents are quite diverse: gene sequences, textual descriptions, categorized attributes of molecules, pathways and processes – also called ontology classifications – literature citations, and quantitative data (such as the results of full genome sequencing, RNA expression, and protein contents for many phenotypic conditions). Biological knowledge is so diverse that it may be archived in both general (e.g., NCBI, EMBL) and specialized databases (UniProt, GeneCards, COSMIC, etc.). By this very reason, there is a need for means to manage and integrate the data of several sources in a consistent manner.

Biological databases are large, organized bodies of data, managed by interfaces designed to update, query, and retrieve components of the data stored. Common standards for managing information are still been developed. Hence, each database often works under their proprietary standard of data formatting, storage, and management. Since often large amounts of data are not easily managed in standard personal computers, databases with thousands (or millions) of entries may become too large to handle for non-trained users (Hernández-Lemus 2011).

Another approach to deal with large databases is to store all the information in a *relational database*. Relational databases have connections or pointers to additional data in other databases or tables (think of these as hyperlinks in webpages). This is the approach followed by major generalistic database engines such as NCBI, EMBL, and UCSC Genome Browser. Most people's first encounters with bioinformatics datasets are with these extremely useful resources that, however, by their generalistic web-based nature are limited when treating with large amounts of quite specific data.

In addition to storing structural data such as sequences, protein structures, and so on, biological databases are also useful to retrieve the information about functional features of condition-specific experiments. The explosion of molecular measurements generated in experiments using high-throughput technologies has also brought a change in the philosophy for sharing the associated data, a change also forwarded by economic reasons associated with today's size of experimental projects. For this reason, now it is common that many journals, especially those with *high impact*, require these data to be archived in public-access databases before a manuscript is even reviewed. Several international public data repositories, such as NCBI's *Gene Expression Omnibus* (GEO) and the European Bioinformatics Institute's *ArrayExpress*, have been created as means to store and distribute these data.

Gene Expression Omnibus contains millions samples coming from thousands of experiments. In order to ease interpretation and analysis, such enormous databases are complemented by so-called knowledge bases of functional annotation. This *annotation files* describe the biological processes (sometimes even the specific signaling pathways) in which these molecules are known to be involved. Integration of this knowledge with the data presents unprecedented opportunities to accelerate and improve our understanding of biology as well as a potential opportunity for data-driven drug design and development.

2.3 Data-Mining Techniques

Data mining (sometimes called text mining) is the generic name of a series of methods and algorithms aimed to extract phrases and statements from texts and databases by analyzing electronically stored versions of them (Almeida 2010). The goal is that by having this information assembled and analyzed, it is possible to generate new knowledge. Strange as this may sound: this is exactly what scholars do. When we (humans) read and analyze series of data sources (say books), we are able to find patterns that are not necessarily contained in either of these sources, at least not in an evident way. Data-mining techniques use methods derived from computational linguistics, natural language processing, data mining, and artificial intelligence (Chiang and Butte 2009). In the context of biomedical research, *literature mining* (i.e., data mining in bibliographic collections) has been used to look up in thousands of articles by searching biological terms such as molecule names and molecular interactions to further improve hypothesis generation (Garten and Altman 2009).

Recent progress in computational analysis of texts has produced algorithms that can recognize even complex biological entities. Gene names are notoriously difficult to identify because they often involve several otherwise common words strung together (e.g., *sonic hedgehog* is a gene involved in cell signaling) (Garten et al. 2010). Literature-mining tools may be used also in the primary research behind drug design: they may help us to infer links between molecular entities and clinical entities, thus enabling to highlight unnoticed connections across the information in diverse subdisciplines. We can envision the day when connecting phenotypical descriptions – even at the individual patient level – with controlled measurements of lab values, but also with imaging studies, and coded diagnoses may guide decisions about clinical responses to diseases and the dream of *individualized therapies* (Harpaz et al. 2012; Rinaldi et al. 2012).

The way in which data mining works is as follows: by analyzing large and complex arrays of data, statistical and mathematical techniques are used to discover *hidden patterns* which we may loosely unanticipated regularities. For instance, the fact that published genetic studies in both malaria and sickle cell anemia (two quite different diseases in origin, prevalence, and phenomenology) point out to similar genome-associated regions or even to the same set of molecules. This may lead to a scientific discovery (Yao et al. 2009), even before actually performing any experiment. Strange as this may seem, this is a widely used strategy to *score* for new drugs in combinatorial medicinal chemistry. Some drug cocktails were even designed under these tenets. Data-mining techniques have also enabled the identification of patterns in clinical patient records that have pointed out to new therapeutics. The combination of literature and database mining with individualized genomic and clinical data should further our understanding of the molecular mechanisms behind drug success and failure. Table 9.1 displays a variety of bioinformatics, database, and data-mining tools.

3 Computational Sequence Analysis

3.1 Sequencing Applications in Pharmacogenomics

Applications of nucleic acid and amino acid sequencing have been thoroughly used to unveil and understand genotype/phenotype relationships in relation with drug responses (Lahti et al. 2012; Guo et al. 2006). Drug delivery and metabolism designs have particularly improved with such computational interventions (Adams et al. 2009; Boran and Iyengar 2010a, b), but also adverse effects (Chiang and Butte 2009; Hammann et al. 2010) and drug sensitivity (Cohen et al. 2011) methodologies now rely on knowledge of the sequence/function computational relationship discovery. The next subsection will discuss further details on these methods.

3.2 Sequence Alignment

The human genome has been often referred to as the *Book of Life*. Such description goes well beyond the obvious parallels. Nucleic acid and protein sequences are well suited to be analyzed as texts, even by resorting to the tools of analytical and computational linguistics (Altschul et al. 1997). Here we will be talking about algorithms for the alignment of sequences of symbols, as well as quantitative measures (scores) used to determine the best alignment. We will discuss common measures of sequence similarity (that we may then assimilate with biological closeness). By referring to these so-called sequence alignment algorithms that are able to include calculation methods to determine the statistical significance of such sequence similarity scores resulting from a database search (Lipman and Pearson 1985; Pearson and Lipman 1988; Smith and Waterman 1981), molecular biologists usually think of biological similarity in terms of chemical similarity. Thus, one may expect mutations that cause large changes in chemical properties to be uncommon since they may end up destroying

Table 9.1 Bioinformatics, database, and data-mining tools

Tool	Main use	URL	References
BLAST[a,b]	Sequence alignment and similarity search	http://blast.ncbi.nlm.nih.gov/	Smith and Waterman (1981) and Altschul et al. (1997)
FASTA[a,b]	Sequence alignment and similarity search (includes protein/DNA search)	http://www.ebi.ac.uk/Tools/sss/fasta/	Lipman and Pearson (1985) and Pearson and Lipman (1988)
CLUSTAL[a,b]	Multiple sequence alignment	http://www.ebi.ac.uk/Tools/msa/clustalw2/	Thompson et al. (1997), Chenna et al. (2003), and Larkin et al. (2007)
Genome browser[b]	Comprehensive genome annotation database	http://genome.ucsc.edu/	Kent et al. (2002) and Fujita et al. (2011)
Ensembl[b]	Genome annotation database including comparative genomics tools	http://www.ensembl.org/index.html	Flicek et al. (2011) and Stabenau et al. (2004)
PDBeFold[a,b]	Three-dimensional protein structures comparisons	http://www.ebi.ac.uk/msd-srv/ssm/ssmstart.html	Krissinel and Henrick (2004) and Krissinel (2007)
PDBePISA[a,b]	Quaternary protein structure prediction, macromolecular (protein, DNA/RNA, and ligand) interfaces	http://www.ebi.ac.uk/msd-srv/prot_int/pistart.html	Krissinel and Henrick (2007) and Krissinel (2010)
BioPerl[a,c]	Collection of computational biology analysis suite in Perl with data-mining-enhanced capabilities	http://www.bioperl.org/wiki/Main_Page	Stajich et al. (2002)
EMBOSS[a,c]	Collection of computational biology analysis suite	http://emboss.sourceforge.net/	Rice et al. (2000)
Bioconductor[a]	[R]-programming language-based tools for the analysis and comprehension of high-throughput genomic data: sequencing, microarrays, annotation, high-throughput assays	http://www.bioconductor.org/	Gentleman et al. (2005)
EBIMed[c]	Information retrieval and extraction from biomedical databases	http://www.ebi.ac.uk/Rebholz-srv/ebimed/index.jsp	Rebholz-Schuhmann et al. (2007)

[a]Computational tool
[b]Database
[c]Data-mining tool

the protein's three-dimensional structure, hence its biological function. In the other hand, changes between *chemically close* amino acids should happen more often. These comparison strategies may lead us to the multiple sequence analysis scenario.

3.2.1 Multiple Sequence Alignment

Performing multiple (possibly related) sequence alignments is a step that we may think as the starting point of many bioinformatics analyses. Constructing a multiple alignment of a number of homologous nucleotide or protein sequences, despite their frequent use, remains one of the algorithmically most challenging areas in bioinformatics and computational biology research (Chenna et al. 2003).

The problem of multiple sequence alignment (MSA) consists in finding out hypothesis about the dynamics of substitutions, insertions, and deletions and how they shape the actual (mostly primary) structure in genes and protein sequences (Huson et al. 2011). The usual input in MSA algorithms is a set of related (homologous) sequences that may be phylogenetically related and are suspected to have biomolecular functions in common. These methods produce a matrix or table where each row corresponds to an input sequence and each column corresponds to a position in the alignment. Columns represent sequence regions that have all diverged due to genome dynamics. Gaps correspond to insertions/deletions and thus do not have common counterparts in all sequences.

A MSA serves as a starting point in the search to find out phylogenetic distances within sequences that may be investigated by tree-building methods based on the alignment. This way one may be aware of sequence similarities, useful, for instance, to build animal models for pharmaceutical research and preclinical trials. Maybe the best possible example in pharmacogenomics research is in the study of the family of *CYP2D6*-related molecules that are known to play an important role in drug metabolism. Conserved domains, which may indicate functionally important sites, such as binding sites, active sites, or sites related to other key functions, can be identified by homology relations in an MSA. The optimal solution to a pairwise alignment problem often can be computationally found within reasonable time. However, in practice MSA is by large more complex. In order to deal with this enormous computational burden in practical times, heuristic algorithms are used for construction and analysis of MSA. The pharmaceutical R & D community is especially enthusiastic in the development of such heuristics due to the intrinsic complexity associated with drug-target finding within large sequences.

3.3 Next-Generation Sequencing

Biological studies at the genomic level can now be analyzed at unprecedented levels of accuracy. No doubt, new sequencing techniques have thus not only reshaped the scope of genome sequencing but also helped to impact a whole generation of basic, clinical, and translational research scientists moving into a new (and largely unfamiliar to them) world (Kahvejian et al. 2008). New kinds of questions relating structure-function of genes may be answered with precision and a breadth of scope that seemed unattainable just a few years ago. One single run on ABI's next-generation SOLiD platform, for example, is able to yield about 6 gigabases of sequence. The competition is hard with the 10 terabytes for a single Genome Analyzer run making the "mere" 12 gigabytes that Roche 454 sequencer may produce seem like *peanuts* (Li et al. 2008, 2010). Data management of such amounts of data is just the beginning. These requirements pale since significant computational infrastructure is also required to reassemble the short reads from many of these instruments into genomic scaffolds or exons. Instrument manufacturers are only providing the software to analyze what comes off the sequencer under a restricted set of applications: say de novo genome sequencing, gene expression analysis, or perhaps targeted resequencing. But further computational biology and bioinformatics analysis are out of the scene (Nekrutenko and Taylor 2012).

In the immediate future, next-generation sequencing platforms may remain out of the sight of researchers and companies without a strong bioinformatics platform. Computational methods for the analysis of NGS data have been developed (Shendure and Ji 2008). These algorithms are aimed at alignment of sequence reads to a reference; base-calling and/or polymorphism detection; *de novo assembly*, from paired or unpaired reads; and genome browsing and annotation. NGS alignment and assembly represent particularly conflicting problems due to the small size of the reads. This is so because alignment algorithms (e.g., BLAST) are largely adequate for long reads – generated by *past-generation* sequencing – but are unlikely to be the best algorithms for handling short-read sequence data. New ad hoc alignment tools have been developed with view to a fast alignment of large sets of short reads of NGS. These methods are also able to check for mismatches and/or gaps. Some of these tools are based in classic alignment algorithms, but new algorithms specifically designed to deal with large amounts of short reads are under development (Simpson et al. 2009). In Table 9.2 we can see a review of tools for sequence alignment and NGS analysis.

4 Genotyping and Genome-Wide Association Studies

Genome-wide association studies (also termed GWAS) are the common denomination of a series of statistical genetic procedures aimed to link (associate) common genetic variation (often in the form of single-nucleotide polymorphisms or SNPs as measured by genotyping platforms) in the human genome with differences in the allelic frequency between cases and controls (Li et al. 2011). The main assumption made in GWAS is that large frequency differences point out statistical association of given genomic regions influencing (in an undisclosed manner) the healthy or diseased phenotype. High-depth GWAS are often based in the implementation of linkage disequilibrium (LD) mapping, since it is likely that mutations shared by large proportion of affected (or unaffected) individuals will be located in the vicinity of other shared alleles (due to specific patterns of genomic recombination), thus forming a haplotype in the region in which the mutation was originated. In other words, since point mutations do not occur really at random, their probability distribution will be differentially affected by functional features, possibly related with the affected phenotype or disease (Marchini et al. 2007).

GWAS emerge as a data-driven alternative to the traditional approach of *candidate gene* search to look up for functional regions in the genome associated with complex phenotypes (in particular disease phenotypes) suited to perform better in replication and clinical trial studies. For instance, before the advent of GWAS, less than 1 % of candidate gene studies were replicated. Even if GWAS have showed a number of limitations, they also have advantages, such as being hypothesis-free, and presenting a higher statistical power and best cost/benefit than linkage studies. However, as opposed to candidate gene and linkage studies, GWAS involve a big deal of computational and statistical analysis (Patterson et al. 2006). These analyses include genotyping data quality controls, Hardy-Weinberg equilibrium assessment for each marker (recall that today's standards range in few millions of markers along the genome), missing genotype calculations, identity by descend and identity by state analysis of the individuals, linkage and population stratification corrections, multidimensional scaling and principal component visualization to detect substructure, significance tests, and case/control statistical association calculations. Additional calculations may be customized to include copy number variant assessment, SNP imputation, and multi-allelic marker associations (Marchini et al. 2007; Purcell et al. 2007).

4.1 Genotyping Applications in Pharmacogenomics

One important aspect that has been studied deeply within the pharmacogenomics arena is that of individual response to drugs, since this is the basis of personalized medicine (Fernald et al. 2011). Genomic variation at the individual level has been analyzed under the GWAS paradigm (Li et al. 2011) in recent years (Guo et al. 2006).

Table 9.2 High-throughput and NGS sequence alignment tools

Tool	Main use	URL	References
ABySS	Parallel, paired-end sequence assembly of short-read, de Novo assembly	http://www.bcgsc.ca/platform/bioinfo/software/abyss	Simpson et al. (2009)
Velvet	De novo assembly tool using graphs		Zerbino and Birney (2008)
SOAP	Gapped and ungapped alignment of short oligonucleotides onto reference sequences	http://soap.genomics.org.cn/	Li et al. (2008)
SOAPdenovo	Short-read de novo assembly tool for assembling short oligonucleotide into contigs and scaffolds	http://soap.genomics.org.cn/soapdenovo.html	Li et al. (2010)
Galaxy	Genomic analysis suite for the analysis of multiple alignments, comparison of genomic annotations, profiling of metagenomic samples, etc	http://wiki.g2.bx.psu.edu/	Nekrutenko and Taylor (2012)
SAM	Sequence Assembly Manager platform for manipulating, analyzing, and viewing whole genome assembly data	http://www.bcgsc.ca/platform/bioinfo/software/sam	Warren et al. (2005)
AMOS	Modular open-source whole-genome assembly software	http://sourceforge.net/projects/amos	–
MEGAN	Metagenomics analysis tools and whole-genome taxonomy	http://ab.inf.uni-tuebingen.de/software/megan/welcome.html	Huson et al. (2011)

Particular attention has been given to drug metabolism and how this may affect the transition of chemical therapeutics from the lab to the patient (Lesko 2012). Dose adjustment and computational prescription have been improved by considering individual genomic variations (Linares and Linares 2011; Guo et al. 2006). Some tools for the analysis of high-throughput genotyping data and genome-wide association studies are presented in Table 9.3.

5 Whole-Genome Gene Expression Analysis

5.1 Gene Expression Analysis Applications in Pharmacogenomics

Genome-wide transcriptional profiling (or gene expression analysis (GEA)) is a technology that has provided biomedical researchers with global information about functional genomics at the mRNA expression level and its regulation (Chengalvala et al. 2007). GEA has deepened our understanding about the dynamics of the different processes involved in the extremely complex phenomena behind gene regulation unveiling previously unappreciated biomolecular relations, for instance, the coupling between nuclear and cytoplasmic transcription and metabolic processes. Gene expression analysis made evident the intricate communication existent within regulatory units, for example, in the organization of transcription factors into regulatory motifs. It is these kinds of regulatory interactions what one is looking for in studies of molecular specificity in drug targets (Herwig and Lehrach 2006; Linares and Linares 2011). Since complex biochemical mechanisms of therapeutic intervention may be better dissected by considering a comprehensive genomic scenario, transcription factor patterns must be accounted for (Wei et al. 2005).

5.2 Statistical Data Preprocessing

A strong challenge in GEA is related to how to effectively dissociate actual gene expression values from experimental noise. The hybridization noise has been found to have very strong dependence on the expression level, with different characteristics for the low and high expression values. It is essential to develop statistical methods that can discern fluctuations caused solely by the measurement processes. A related source of undesired variation is that, on increasing mRNA concentration levels, the distance of the average gene intensity to the background noise increases. The levels of background intensity could thus mask the effects of some mean-valued expression levels, i.e., the average shining effect could hide a relatively important signal (Baca-López et al. 2009).

Also, in the vast majority of the applications of GeneChips, one wishes to learn how mRNA concentration profiles differ in response to genetic, cellular, and environmental differences. One important instance is when large (or small) expression of a given gene or set of genes may cause an illness (such as cancer), thus resulting in variation between diseased and normal tissues (a so-called case–control comparison). However, observed intensity levels also depend on sample preparation, manufacture of the arrays, and lab processing of such arrays (dye labeling, hybridization, and scanning). These are called sources of obscuring variation. Due to these facts, unless arrays are correctly normalized, comparing data from different arrays can lead to misleading results.

Several methods have been proposed to normalize the arrays, and it has been found that quantile normalization (QN) performs best. The goal of QN is to make the distribution of probe intensities the same for arrays within a given category (quantile); in such case a quantile-quantile plot (QQ plot) will be given as an identity line. A generalization for the usual QQ plot for n data vectors could be given in terms of a projection operator formalism. In brief, we quantile-normalize the arrays to the same distribution by taking the mean quantile and substituting it instead of the data item in the original dataset.

A general scheme has been proposed, called the Robust Multi-array Average (RMA) algorithm which (1) background-corrects the arrays using an expectation-of-the-signal transformation (a so-called B-transform), (2) normalizes the arrays by a QN, and (3) fits a linear model to

Table 9.3 Genotyping and genome-wide association studies tools

Tool	Main use	URL	References
PLINK	Whole-genome association analysis toolset: data management, quality control, population stratification, and case/control association studies from genotyping data	http://pngu.mgh.harvard.edu/~purcell/plink/	Purcell et al. (2007)
EIGENSOFT	Population structure analysis and correction in whole-genome genotyping data	http://genepath.med.harvard.edu/~reich/Software.htm	Patterson et al. (2006)
IMPUTE	Inference of unobserved genotypes in SNP association studies.	https://mathgen.stats.ox.ac.uk/impute/impute.html	Marchini et al. (2007)
Chiamante	Joint genotype-calling algorithm for array and sequence data	http://www.well.ox.ac.uk/~jared/chiamante/	O'Connell and Marchini (2012)
MACH	Estimates long haplotypes or infers missing genotypes in samples of unrelated individuals	http://www.sph.umich.edu/csg/abecasis/mach/	–
BIMBAM	Bayesian imputation association mapping for complex diseases and traits	http://stephenslab.uchicago.edu/software.html#bimbam	Servin and Stephens (2007)
QUICKTEST	Statistical methods for association testing using uncertain (imputed) genotypes	http://toby.freeshell.org/software/quicktest.shtml	Johnson (2007)

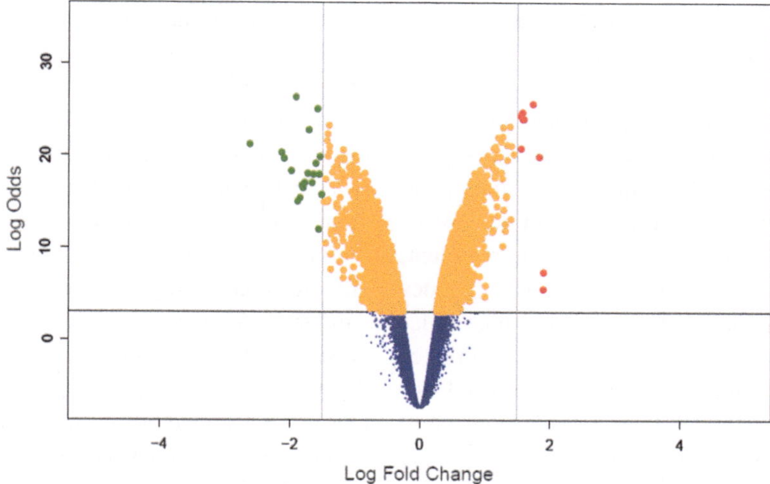

Fig. 9.1 A volcano plot displays information about changes in gene expression for whole genome experiments based in calculations for the size of differential expression (the log-fold change) and its statistical significance (the log-odds). *Red dots* are statistically significant over-expressed genes while *green dots* are statistically significant under-regulated genes

summarize the probe intensities for each probe set (Gentleman et al. 2005).

RMA is less noisy than all other measures at lower mRNA concentrations, has a smaller spread (and thus is better tailored to detect differentially expressed probe sets), and has greater sensitivity. For all the reasons above, we decided to perform RMA correction and preprocessing of the raw gene expression data before performing any additional analyses (Baca-López et al. 2009).

5.3 Differential Expression and Statistical Significance

GEA when applied in a whole-genome basis can provide us with genome-wide patterns of gene expression across many phenotypes. In principle, detailed analyses may reveal links between patterns of mRNA expression and their corresponding phenotypes in what has been called a *genomic signature*. The basic level of analysis of these patterns consists in categorizing whether observed differences in expression possess statistical significance, that is, if we are really able to discern between different patterns with a degree of certainty. However, under common experimental designs in contemporary genomics, traditional statistical methods fail to overcome the challenges imposed by features such as low signal-to-noise ratios in the experimental measurements (due to technological reasons), high biological variability of the samples, and an extremely small number of these as compared with the number of variables (i.e., probes in a gene expression microarray).

Due to this fact, Baldi and Long, for instance, developed a Bayesian probabilistic framework for MA data analysis. Baldi-Long analysis (also called a CyberT-test) consists in modeling log-expression values by independent normal distributions, parameterized by corresponding means and variances with hierarchic prior distributions. These point estimates, combined with a statistical *t*-test (a regularized *t*-test), provide a systematic inference algorithm that compares favorably with the more widely used simple *t*-test or fold methods and partly compensate for the lack of replication within the actual GEA framework (Baca-López et al. 2009). Statistical results of log-fold change and statistical significance in differential gene expression are often displayed in the form of a so-called volcano plot. An example of a volcano plot can be seen in Fig. 9.1.

5.4 Clustering and Classification of Expression Profiles

Clustering methods are computational procedures aimed to identify genes that behave in a similar way over a range of circumstances, biological conditions, or samples (e.g., a pathological or diseased condition, pharmacological treatment, and so on). The goal is to find out groups of genes based in the assumption that genes showing similar expression patterns share other characteristics such as common regulatory elements, related biological functions, or cell lineage. This (apparently naïve) assumption has been verified with the successful application of clustering algorithms for the analysis of cellular functions, both in *minor* species (such as yeast) and in mammals and humans. A variety of data-clustering methods have been used to analyze microarray data, included in this list are hierarchic clustering, k-means clustering, deterministic annealing, self-organizing maps, combinatorial and graph-theoretical approaches, etc. A common good practice is to apply more than one clustering algorithm to a given dataset and then analyze reproducibility in group assignations or clustering between different methods.

5.4.1 Neighborhood Analysis

These methods are based in the implementation of iterative procedures looking for clusters in terms of their *central points* or mean values. Once we have defined the center of a set of clusters, each gene is then assigned to the cluster with the closest center to this gene. The clustering algorithm then adjusts the new centers of the clusters in order to minimize the sum of the distances of every gene in a cluster to its center. With this, a new set of group centers is obtained, genes can then be reassigned, and the process is repeated iteratively until convergence criteria are satisfied.

5.4.2 Hierarchic Clustering

Hierarchic clustering is an agglomerative procedure in which one-member *groups* of data points (say for gene expression profiles) are fused together to form bigger-and-bigger groups. There are many different techniques; however, most of the times the process starts by calculating a *distance matrix* among the pairs of data points for every gene. Genes with the smallest distance of separation are clustered together. Once a new group has been formed by fusing two groups, the distance matrix is updated to calculate the distance between this and other clusters (Hammann et al. 2010). This procedure leads to a hierarchic graph or *dendrogram* in which multiple groups are clustered – together according with their similarity, giving rise to a hierarchic tree in which close branches represent sets of genes with coordinated patterns of gene expression. Hierarchic clustering is aimed at forming sets of data points (or classes) that possess common features in such a way that the information obtained by analyzing the classes may reveal internal pattern structures useful for a clear-cut biological interpretation that may lead to clinical and pharmacological applications.

Most clustering techniques can be expressed in terms of the so-called Lance-Williams algorithm as follows:

1. Calculate the proximity graph.
2. Combine the closest (more similar groups).
3. Update the proximity (distance) matrix to reflect the new distance between groups.
4. Repeat steps 2 and 3 until just one group remains.

One visual tool to analyze the results of hierarchic clustering is the *heatmap* plot which displays genes in samples as clustered by similarity in their expression patterns; an example of a heatmap plot showing changes in global gene expression due to the action of a chemotherapeutic is shown in Fig. 9.2.

5.4.3 Neural Networks

In these methods, an artificial neural network (ANN) is used to model gene expression. ANNs are computational algorithms based in a simplified model of human brain neurons. ANNs can be *trained* to recognize and categorize complex patterns. Pattern recognition is attained by means of setting ANN parameters by means of an error minimization procedure that is tuned by means of *learning by experience* (just like the human brain). The correlation between the expression patterns for different genes is used to define the *weight matrix* in the propagation function (akin to synaptic propagation in biological neurons) of

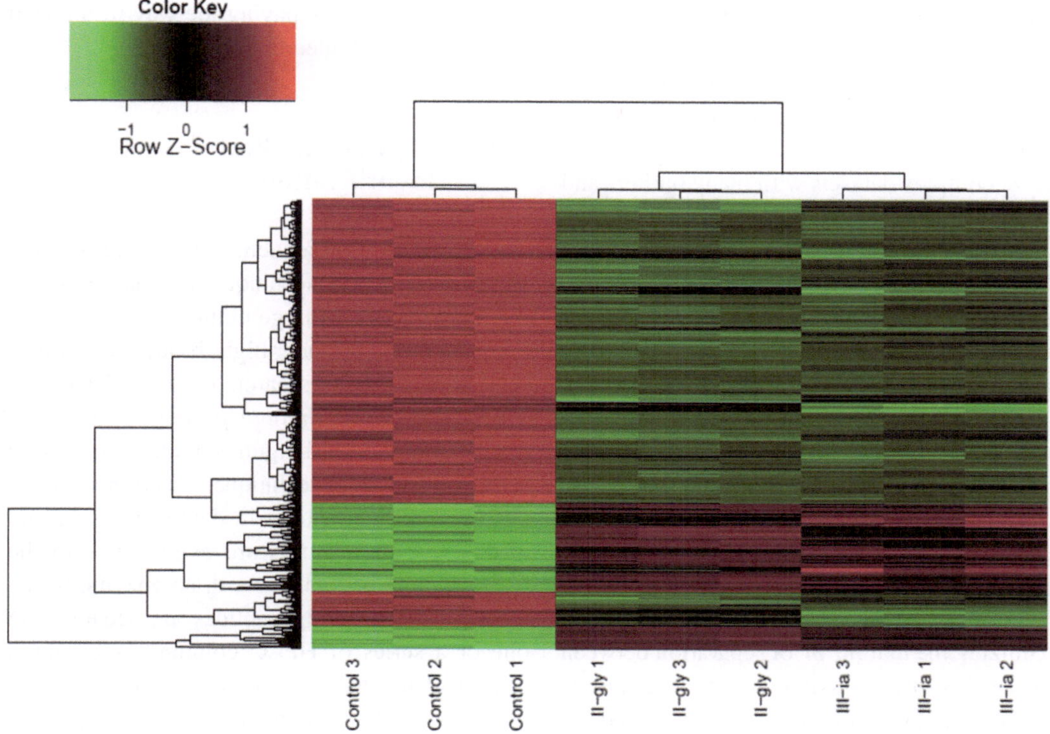

Fig. 9.2 A heatmap plot shows a hierachic clustering in the gene expression patterns of three distinctive groups (each one having three samples): a control group and treatment with two different drugs. Genes/samples are clustered together according with their expression profiles –here color-coded: from *red* (overexpression) to *green* (subexpression)

an ANN, whereas the activation module (playing the role of dendrite activation) updates the *membership* of genes to their respective clusters. Common ANN methods used in gene expression analysis are the *perceptron*, *multilayered perceptron*, *Cauchy machines*, and *Kohonen networks* (also related with self-organizing maps).

Statistical Learning

In the theory of statistical learning, there are two basic types of problems:

- *Supervised learning:* in which there is a result or outcome that we want to predict in terms of characteristic features or quantities
- *Unsupervised learning:* in which we want to find patterns in the data, as to how are these organized, classified, clustered in groups, and so on, relying on little or no a priori information

In order to do this, we create statistical models that are then fed with training datasets to obtain response variables.

ANNs may give us two types of response variables:

- *Quantitative response:* In this we obtain a numerical value of the response.
- *Qualitative response:* In this we obtain a set of tags or classes.

The learning problem is called a *regression model* when the response variables are quantitative and a *classification* when the response variables are qualitative.

Learning task: Given an entrance value X, a learning task produces a prediction $\hat{Y}$ by using a prediction rule on a set of observed data L called a *training dataset.*

5.4.4 Self-Organizing Maps

Self-organizing maps (SOMs) are the number given to the family of classification methods based on iterative processes in which the datasets (say gene expression vectors) in each group are trained to find the greatest difference in the given clusters, i.e., the system is given an imposed partial structure in the start-up stage that is later modified as *driven by the data*. SOMs perform better than other classification methods to analyze complex datasets with many outliers and dummy parameters.

5.4.5 Support Vector Machines

The computational theory of support vector machines (SVMs) is a relatively new technique based on the idea of structural flow minimization. In many applications, SVMs have shown a greater performance than other learning machines (such as ANNs) and have been introduced as powerful tools to solve classification problems [?]. An SVM maps the entry points to a higher-dimensional space (e.g., $\Re^2 \rightarrow \Re^3$ and so on) and finds a hyperplane that divides the data while maximizing the margin m of separation between classes in this space. Maximizing m is a quadratic programming problem that may be solved by Lagrange multipliers. SVMs then find the optimal hyperplane by using metric functions called the *kernel* of the SVM. Such hyperplane may be represented by the combination of a few data entry points called the *support vectors*.

6 Proteomics

6.1 Proteomics Applications in Pharmacogenomics

Ultimately, the cell's biological function depends on the action of proteins and multiprotein assemblies. Hence, pharmacological interventions often rely on knowledge of protein abundance, composition of the proteome, and protein interactions. Due to their chemical nature (less stable than nucleic acids), however, the large-scale systematic studies of the protein composition and conformation in diverse phenotypes have been somewhat delayed in comparison with their nucleic acid counterpart. Nevertheless due to advance in biotechnological and analytical tools, now it is possible (and will be more so in the near future) to take a look at the full arsenal of proteins in an organism, thus paving the way to directed pharmacological interventions.

6.2 Peptide and Protein Identification

The application of sequential (or tandem) mass spectrometry (TMS) techniques has made possible the identification of large numbers of proteins in complex mixtures under high-throughput settings typical of today's research in "omic" biology. TMS (also called *shotgun proteomics* for their similar approach to shotgun genomic sequencing) as previous "omic" techniques mentioned in this chapter depends heavily in data processing and analysis. A protein mixture extracted from the tissue sample is digested with an enzyme (often trypsin); the resulting peptides are separated in one or a series of HPLC columns. A common analytical technique is protein fractionation by means ion exchange to reduce sample complexity. The isolated peptides are then injected into the MS device, either in a fluid phase by means of *electro-spray* ionization from a reverse-phase HPLC column or as particles taken by laser pulses from a solid matrix – the so-called MALDI plate – onto which the molecules have been spotted. Then the mass spectrometer selects significant peaks to be isolated and subject to further collision-induced fragmentation, hence the name TMS or MS/MS. A mass spectrum of the fragment ions (more properly a mass-charge spectrum), known as a *tandem mass spectrum*, is obtained for each selected precursor peptide (Deutsch et al. 2008).

The ultimate analysis for these mass spectra uses a clever combination of computational biology sorting algorithms (for distributions of m/z time-of-flight measurements) and informatics techniques such as data mining (to look up for the most likely peptides compliant with the m/z spectrum just calculated). The output of such analysis is the desired list of significant proteins in the sample. This list is inferred from a combinatorial search and reconstruction of proteins in the MS spectra; starting from the more likely peptide list – which is also reconstructed but this one from the MS/MS spectra. This list of proteins may also present either quantitative or semiquantitative measures of abundance as well as indicators of statistical significance for the data-mining inference, peptide reconstruction, and protein reconstruction (Eng et al. 2011; Frickenschmidt et al. 2008; Günther et al. 2006).

6.3 Protein-Protein Interactions

In order to reach a systems-level understanding of cellular functions, it is mandatory to decipher all functional interactions among proteins in the cell. A remarkable progress along these lines has been made recently in terms of experimental measurements but quite especially with regard to computational prediction and assessment methods. Protein interaction information databases have still not reach the level of comprehensiveness of their genomic counterparts, since protein-protein interaction information is more prone to errors, thus requiring a considerably bigger effort to be effectively annotated (Krissinel and Henrick 2007).

Due to their chemical properties, proteins may play many functional roles when interacting with each other: they may form stable complexes, partake in biochemical reactions to create metabolic pathways, and arrange themselves by myriads of direct and indirect regulatory interactions. These connections can be also be properly conceptualized as networks. Considering the size and extremely complex organization of these, we can view the genome (and its associated proteome) as something more rich than just a catalogue of genetic functions: a complex dynamic entity that we may call the *interactome* (He et al. 2010; Sun et al. 2012). Protein network calculations help to increase the statistical power in genetic studies, particularly fitted for drug discovery (Vicini 2010). For instance, the knowledge of the interactome gives us a better, more comprehensive understanding of enzyme metabolism, hence enabling us to predict subtle relations between phenotypes and gene functions.

The functional annotation and database curation of protein interactions (either protein-protein or protein-nucleic acid) is far more challenging than other types of "omic" data due to the fact that functional interactions between proteins span through a large spectrum of mechanisms, sometimes quite subtle and thus often presenting large experimental error rates. Another issue is that often these interactions are context specific, depending in the cellular environment, tissue type, and other dynamically changing phenotypical conditions. For these reasons, a great deal of information is needed to properly describe the various aspects of a given protein interaction. However, the role that interactomics will play – in combination with other sources of "omic" data – in the development of modern pharmacology is, no doubt, quite outstanding. This is so due to the contextuality of protein interactions, since is in this specificity that phenotypic individuality (e.g., in patient-specific drug metabolism, pharmacokinetics, and pharmacodynamics, as well as in possible drug-drug interactions) that lies the power of pharmacogenomic approaches. Table 9.4 displays a collection of tools related to proteomics, protein-protein interaction databases, and computational annotation tools.

7 Gene Regulatory Networks

7.1 Biological Networks Applications in Pharmacogenomics

Once a genetic regulatory network (GRN) is known for a certain phenotype, too much in silico pharmacological analysis can be done (Arrell and Terzic 2010). For instance, if we know that a certain protein, say a transcription factor, is a drug target, knowledge of all their target genes may help us to prevent undesired secondary effects even before preclinical studies, thus helping us to save too much money and time invested in the research, design, and testing of drugs that may not even reach the clinic – or even worst that may enter the market only to be discontinued or discredited, even harming patients (Faustino and Terzic 2008).

Table 9.4 Proteomics, Protein-protein interaction databases, and computational annotation tools

Tool	Main use	URL	References
PeptideProphet	Validation of peptide assignments to MS/MS spectra made by database search programs	http://peptideprophet.sourceforge.net/	Deutsch et al. (2008)
ProteinProphet	Computational protein assignment	http://proteinprophet.sourceforge.net/	Deutsch et al. (2008)
EMBOSS	Integrative omics analysis suite	http://emboss.sourceforge.net/	Rice et al. (2000)
GenMAPP 2	Pathway analysis and interaction networks	http://www.genmapp.org/	Salomonis et al. (2007)
STRING	Database of known and predicted protein interactions	http://string-db.org/	Szklarczyk et al. (2011)
DTome	Computational framework for construction of drug-target interactome	http://bioinfo.mc.vanderbilt.edu/DTome/	Sun et al. (2012)

7.2 Inference Methods of Gene Regulatory Networks

The probabilistic inference of GRNs based in "omic" data has become a fundamental problem in contemporary computational biology. A comprehensive GRN leads us to understand the interplay between thousands of genes and other molecules such as transcription factors and promoters. The associated challenges are threefold: technological, biological, and computational: measurement processes (GeneChips, RNA-sequencing techniques, shotgun proteomics, etc.) often generate results with a low signal-to-noise ratio; also there are too many variables involved (number of genes and interactions among them) commonly thousand – or at least hundreds – of times as many as experimental samples. In addition, the highly nonlinear biochemical dynamics underlying genomic control make most statistical techniques useless (Hernández-Lemus and Rangel-Escareño 2011).

However, since GRNs are extremely well-suited theoretical models to describe the *complete* state of a cell under certain biological conditions at a given time, much effort has been put into treating the problem of their inference with new, more powerful computational techniques developed ad hoc. The inference of these interaction connectivity networks is mathematically related with the solution of a so-called inverse problem, i.e., you know the effect – a molecular phenotype – but you don't know the causes, the set of molecular interactions in the network. A large number of computational algorithms to solve this inference problem, commonly termed *reverse engineering* methods, exist (Hernández-Lemus et al. 2009). As we have seen, there are two major shortcomings related to the computational inference of GRNs: (1) nonlinearity and (2) large number of variables. Computational methods have been developed to deal with issues (1) and (2). It can be seen that most of these methods rely on some information-theoretical measure of dependency. Such measures have been used to define (and quantify) relevance, redundancy, and interaction in such large noisy datasets by capturing nonlinear dependencies. These methods have been designed to provide a the cellular network topology from the transcriptional interactions as revealed by, say, gene expression measurements that are then treated as samples from a joint probability distribution. Validation of network against available real-life data is also an important stage in the discovery of reliable GRNs (Schlitt and Brazma 2007). An example of a gene regulatory network is shown in Fig. 9.3.

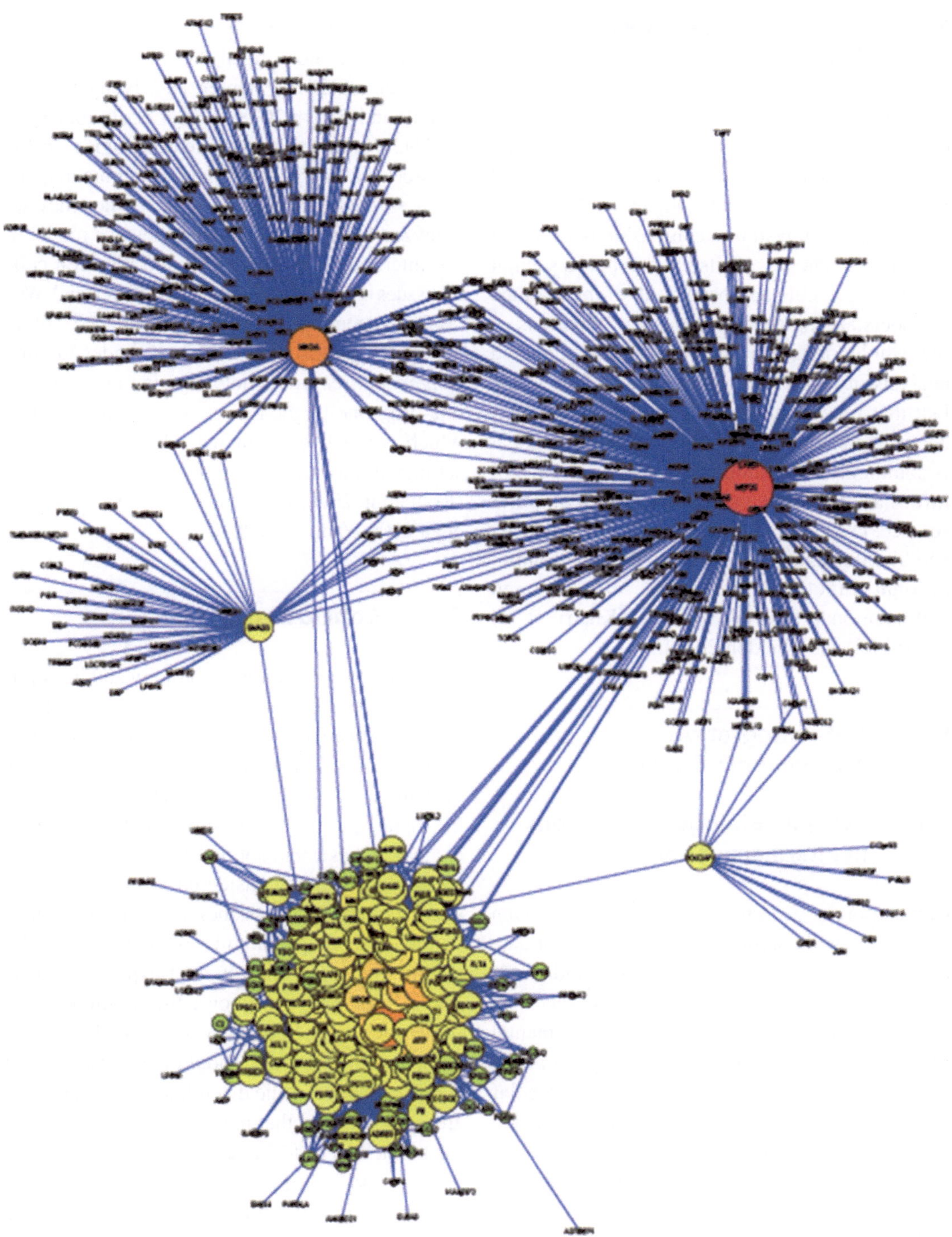

Fig. 9.3 Gene regulatory network displaying direct transcriptional interactions of genes acting as master regulators that may be potential drug-targets. Genes are color and size- coded according with their connectivity (*bigger red circles* correspond with highly connected genes, whereas *smaller green* ones are lowly connected genes)

7.3 Methods for Networks Analysis

Functional genomics deals with the large wealth of information/data produced by high-throughput "omic" research to describe gene functions and interactions. A principal subject of functional genomics is to unveil the relationship between an organism's genome and its phenotype; as such, it has many possible approaches, all of them aimed at understanding the properties and functions of the whole organism in terms of the detailed analysis of genes and gene products. Due to the fact that functional genomics involves the management and careful dissection of vast datasets, it is commonly based in the application of tools of information theory, statistics, and computer science. Data mining, computational intelligence, and bioinformatics are the usual tools of computational functional genomics, also deemed computational systems biology (Khatri et al. 2011; Pleil and Sheldon 2011).

8 Data Integration and System's Biology

8.1 System's Biology Applications in Pharmacogenomics

Contemporary genomic research is now driven toward a new theoretical framework, that of data integration. The goal of such new paradigm is being able to make sense of huge datasets with different (often disparate) kinds of experimental and analytical data produced by the upcoming of new techniques in high-throughput molecular biology (Atkinson and Lyster 2010). Biological data integration when taken into the extreme gives rise to the discipline called systems biology that focuses on understanding the behavior of entire biological systems, not just individual components (Navlakha and Bar-Joseph 2011; Mitra et al. 2009). Methods in systems biology broadly cover: qualitative network diagrams linking molecular and phenotypic entities, probabilistic graphical models such as Bayesian networks, prediction methods based on differential equations, and state space models based in optimization and control theory, among others (Berg et al. 2010; Berger and Iyengar 2011; Harrill and Rusyn 2008; Tegnér et al. 2008).

These complementary methodologies enable the study of emergent biological phenomena, i.e., features that become apparent only when they are analyzed by means of a comprehensive model of the interactions within a cell and between cells in tissues (Rodriguez et al. 2010; Silver and Way 2007). Systems pharmacology is founded within the same philosophy aimed at understanding drug responses as a whole, moving away from a one-drug-one-target model to a systemic view which is much more fit with the actual clinical settings in modern multidimensional therapeutics (Kohl et al. 2010; Mac Gabhann et al. 2010; Wist et al. 2009; Yang et al. 2010).

9 Conclusions

The study of computational biology and its applications to pharmacology is a wide emerging field. It is also a deep, highly multidisciplinary, and fast evolving one. Due to these facts, it is extremely difficult to present a comprehensive review on the matters, useful for different kind of researchers in both basic and applied sciences. This chapter may be better considered as a glimpse on some interesting approaches and tools of computational pharmacogenomics quite biased to the author's personal interests and experience. It is like a brief tour that shows the highlights, but sometimes, out of necessity, neglects the details. After reading this, one should have just a broad and fuzzy scope of the discipline. However, there are some sections that may result even more useful to the interested reader: mainly Tables 9.1, 9.2, 9.3, and 9.4, describing some useful tools for the practicing researcher but quite especially the references. It is here that one may find further inside in more specialized instances.

The cooperation between quantitative sciences like mathematics, physics, and computer science, the biochemical sciences, medicine, and pharmacology has resulted in a complex but quite intriguing avenue of research, one that has generated new avenues to transit both in basic investigations and especially in translational efforts to take the extremely large wealth of biological knowledge and make it useful for the design and implementation of therapeutic interventions and drug design.

References

Adams JC, Keiser MJ, Basuino L, Chambers HF, Lee D-S et al (2009) A mapping of drug space from the viewpoint of small molecule metabolism. PLoS Comput Biol 5(8):e1000474

Almeida JS (2010) Computational ecosystems for data-driven medical genomics. Genome Med 2(9):67

Altman RB (2012) Translational bioinformatics: linking the molecular world to the clinical world. Clin Pharmacol Ther 91(6):994–1000

Altschul SF, Madden TL, Schäffer AA, Zhang J, Zhang Z, Miller W, Lipman DJ (1997) Gapped BLAST and PSI-BLAST: a new generation of protein database search programs. Nucleic Acids Res 25:3389–3402

Arikuma T, Yoshikawa S, Azuma R, Watanabe K, Matsumura K, Konagaya A (2008) Drug interaction prediction using ontology-driven hypothetical assertion framework for pathway generation followed by numerical simulation. BMC Bioinformatics 9(Suppl 6):S11

Arrell DK, Terzic A (2010) Network systems biology for drug discovery. Clin Pharmacol Ther 88(1):120–125

Atkinson AJ Jr, Lyster PM (2010) Systems clinical pharmacology. Clin Pharmacol Ther 88(1):3–6

Azuaje FJ, Zhang L, Devaux Y, Wagner DR (2011) Drug-target network in myocardial infarction reveals multiple side effects of unrelated drugs. Sci Rep 1:52

Baca-López K, Hernández-Lemus E, Mayorga M (2009) Information-theoretical analysis of gene expression data to infer transcriptional interactions. Rev Mex Fís 55(6):456–466

Berg JM, Rogers ME, Lyster PM (2010) Systems biology and pharmacology. Clin Pharmacol Ther 88:1

Berger SI, Iyengar R (2011) Role of systems pharmacology in understanding drug adverse events. Wiley Interdiscip Rev Syst Biol Med 3(2):129–135

Boran AD, Iyengar R (2010a) Systems pharmacology. Mt Sinai J Med 77(4):333–344

Boran AD, Iyengar R (2010b) Systems approaches to polypharmacology and drug discovery. Curr Opin Drug Discov Devel 13(3):297–309

Butte AJ, Ito S (2012) Translational bioinformatics: data-driven drug discovery and development. Clin Pharmacol Ther 91(6):949–952

Chengalvala MV, Chennathukuzhi VM, Johnston DS, Stevis PE, Kopf GS (2007) Gene expression profiling and its practice in drug development. Curr Genomics 8(4):262–270

Chenna R, Sugawara H, Koike T, Lopez R, Gibson TJ, Higgins DG, Thompson JD (2003) Multiple sequence alignment with the clustal series of programs. Nucleic Acids Res 31:3497–3500

Chiang AP, Butte AJ (2009) Data-driven methods to discover molecular determinants of serious adverse drug events. Clin Pharmacol Ther 85(3):259–268

Cohen AL, Soldi R, Zhang H, Gustafson AM, Wilcox R, Welm BE, Chang JT, Johnson E, Spira A, Jeffrey SS, Bild AH (2011) A pharmacogenomic method for individualized prediction of drug sensitivity. Mol Syst Biol 7:513

Danecek P, Auton A, Abecasis G, Albers CA, Banks E, DePristo MA, Handsaker RE, Lunter G, Marth GT, Sherry ST, McVean G, Durbin R (2011) The variant call format and VCFtools. Bioinformatics 27:2156–2158

Deutsch EW, Lam H, Aebersold R (2008) Data analysis and bioinformatics tools for tandem mass spectrometry in proteomics. Physiol Genomics 33:18

Eissing T, Kuepfer L, Becker C, Block M, Coboeken K, Gaub T, Goerlitz L, Jaeger J, Loosen R, Ludewig B, Meyer M, Niederalt C, Sevestre M, Siegmund HU, Solodenko J, Thelen K, Telle U, Weiss W, Wendl T, Willmann S, Lippert J (2011) A computational systems biology software platform for multiscale modeling and simulation: integrating whole-body physiology, disease biology, and molecular reaction networks. Front Physiol 2:4

Eng L, Ibrahim-zada I, Jarjanazi H, Savas S, Meschian M, Pritchard KI, Ozcelik H (2011) Bioinformatic analyses identifies novel protein-coding pharmacogenomic markers associated with paclitaxel sensitivity in NCI60 cancer cell lines. BMC Med Genomics 4:18

Faustino RS, Terzic A (2008) Bioinformatic networks: molecular reticles for pinpointing pharmacological target selection. Clin Pharmacol Ther 84(5):543–545

Fernald GH, Capriotti E, Daneshjou R, Karczewski KJ, Altman RB (2011) Bioinformatics challenges for personalized medicine. Bioinformatics 27(13):1741–1748

Flicek P, Amode MR, Barrell D, Beal K, Brent S, Chen Y, Clapham P, Coates G, Fairley S, Fitzgerald S, Gordon L, Hendrix M, Hourlier T, Johnson N, Kähäri A, Keefe D, Keenan S, Kinsella R, Kokocinski F, Kulesha E, Larsson P, Longden I, McLaren W, Overduin B, Pritchard B, Riat HS, Rios D, Ritchie GR, Ruffier M, Schuster M, Sobral D, Spudich G, Tang YA, Trevanion S, Vandrovcova J, Vilella AJ, White S, Wilder SP, Zadissa A, Zamora J, Aken BL, Birney E, Cunningham F, Dunham I, Durbin R, Fernández-Suarez XM, Herrero J, Hubbard TJ, Parker A, Proctor G, Vogel J, Searle SM (2011) Ensembl 2011. Nucleic Acids Res 39(Database issue):D800–D806, 1111111

Frickenschmidt A, Frohlich H, Bullinger D, Zell A, Laufer S, Gleiter CH, Liebich H, Kammerer B (2008) Metabonomics in cancer diagnosis: mass spectrometry-based profiling of urinary nucleosides from breast cancer patients. Biomarkers 13(4):435–449

Fujita PA, Rhead B, Zweig AS, Hinrichs AS, Karolchik D, Cline MS, Goldman M, Barber GP, Clawson H, Coelho A, Diekhans M, Dreszer TR, Giardine BM, Harte RA, Hillman-Jackson J, Hsu F, Kirkup V, Kuhn RM, Learned K, Li CH, Meyer LR, Pohl A, Raney BJ, Rosenbloom KR, Smith KE, Haussler D, Kent WJ (2011) The UCSC Genome Browser database: update 2011. Nucleic Acids Res 39(Database issue): D876–D882

Garten Y, Altman RB (2009) Pharmspresso: a text mining tool for extraction of pharmacogenomic concepts and relationships from full text. BMC Bioinformatics 10(Suppl 2):S6

Garten Y, Coulet A, Altman RB (2010) Recent progress in automatically extracting information from the pharmacogenomic literature. Pharmacogenomics 11(10):1467–1489

Gentleman R, Carey V, Huber W, Irizarry R, Dudoit S (2005) Bioinformatics and computational biology solutions using R and bioconductor. Springer, New York

Günther S, Senger C, Michalsky E, Goede A, Preissner R (2006) Representation of target-bound drugs by computed conformers: implications for conformational libraries. BMC Bioinformatics 9(7):293

Guo Y, Weller P, Farrell E, Cheung P, Fitch B, Clark D, Wu SY, Wang J, Liao G, Zhang Z, Allard J, Cheng J, Nguyen A, Jiang S, Shafer S, Usuka J, Masjedizadeh M, Peltz G (2006) In silico pharmacogenetics: warfarin metabolism. Nat Biotechnol 24(5):531–536

Hammann F, Gutmann H, Vogt N, Helma C, Drewe J (2010) Prediction of adverse drug reactions using decision tree modeling. Clin Pharmacol Ther 88(1):52–59

Harpaz R, DuMouchel W, Shah NH, Madigan D, Ryan P, Friedman C (2012) Novel data-mining methodologies for adverse drug event discovery and analysis. Clin Pharmacol Ther 91(6):1010–1021

Harrill AH, Rusyn I (2008) Systems biology and functional genomics approaches for the identification of cellular responses to drug toxicity. Expert Opin Drug Metab Toxicol 4(11):1379–1389

He Z, Zhang J, Shi X-H, Hu L-L, Kong X et al (2010) Predicting drug-target interaction networks based on functional groups and biological features. PLoS One 5(3):e9603

Hernández-Lemus E (2011) On the application of compressive sampling techniques to high throughput data in computational genomics. Theor Appl Inform 23(3–4):177–192

Hernández-Lemus E, Rangel-Escareño C (2011) The role of information theory in gene regulatory network inference. In: Deloumeaux P, Gorzalka JD (eds) Information theory: new research, Mathematics research developments series. Nova Publishing, New York

Hernández-Lemus E, Velázquez-Fernández D, Estrada-Gil JK, Silva-Zolezzi I, Herrera-Hernández MF, Jiménez-Sánchez G (2009) Information theoretical methods to deconvolute genetic regulatory networks applied to thyroid neoplasms. Physica A 388: 5057–5069

Herwig R, Lehrach H (2006) Expression profiling of drug response–from genes to pathways. Dialogues Clin Neurosci 8(3):283–293

Holford N, Karlsson MO (2007) Time for quantitative clinical pharmacology: a proposal for a pharmacometrics curriculum. Clin Pharmacol Ther 82(1):103–105

Hoppe R, Brauch H, Kroetz DL, Esteller M (2011) Exploiting the complexity of the genome and transcriptome using pharmacogenomics towards personalized medicine. Genome Biol 12(1):301

Huson DH, Mitra S, Weber N, Ruscheweyh H, Schuster SC (2011) Integrative analysis of environmental sequences using MEGAN4. Genome Res 21:1552–1560

Johnson T (2007) Bayesian method for disease gene detection and mapping, using a case and control design and DNA pooling. Biostatistics 8:546–565

Kahvejian A, Quackenbush J, Thompson JF (2008) What would you do if you could sequence everything? Nat Biotechnol 26(10):1125–1133

Kent WJ, Sugnet CW, Furey TS, Roskin KM, Pringle TH, Zahler AM, Haussler D (2002) The human genome browser at UCSC. Genome Res 12(6):996–1006

Khatri P, Sarwal MM, Butte AJ (2011) Applications of translational bioinformatics in transplantation. Clin Pharmacol Ther 90(2):323–327

Kohl P, Crampin EJ, Quinn TA, Noble D (2010) Systems biology: an approach. Clin Pharmacol Ther 88(1):25–33

Krissinel E (2007) On the relationship between sequence and structure similarities in proteomics. Bioinformatics 23:717–723

Krissinel E (2010) Crystal contacts as nature's docking solutions. J Comput Chem 31(1):133–143

Krissinel E, Henrick K (2004) Secondary-structure matching (PDBeFold), a new tool for fast protein structure alignment in three dimensions. Acta Cryst D60:2256–2268

Krissinel E, Henrick K (2007) Inference of macromolecular assemblies from crystalline state. J Mol Biol 372:774–797

Lahti JL, Tang GW, Capriotti E, Liu T, Altman RB (2012) Bioinformatics and variability in drug response: a protein structural perspective. J R Soc Interface 9(72):1409–1437

Larkin MA, Blackshields G, Brown NP, Chenna R, McGettigan PA, McWilliam H, Valentin F, Wallace IM, Wilm A, Lopez R, Thompson JD, Gibson TJ, Higgins DG (2007) Clustal W and Clustal X version 2.0. Bioinformatics 23:2947–2948

Lesko LJ (2012) Drug research and translational bioinformatics. Clin Pharmacol Ther 91(6):960–962

Li R, Li Y, Kristiansen K, Wang J (2008) SOAP: short oligonucleotide alignment program. Bioinformatics 24(5):713–714

Li R, Zhu H, Ruan J, Qian W, Fang X, Shi Z, Li Y, Li S, Shan G, Kristiansen K, Li S, Yang H, Wang J, Wang J (2010) De novo assembly of human genomes with massively parallel short read sequencing. Genome Res 20(2):265–272

Li Y, Sidore C, Kang HM, Boehnke M, Abecasis G (2011) Low coverage sequencing: implications for the design of complex trait association studies. Genome Res 21:940–951

Linares OA, Linares AL (2011) Computational opioid prescribing: a novel application of clinical pharmacokinetics. J Pain Palliat Care Pharmacother 25(2):125–135

Lipman DJ, Pearson WR (1985) Rapid and sensitive protein similarity searches. Science 227(4693): 1435–1441

Mac Gabhann F, Annex BH, Popel AS (2010) Gene therapy from the perspective of systems biology. Curr Opin Mol Ther 12(5):570–577

Marchini J, Howie B, Myers S, McVean G, Donnelly P (2007) A new multipoint method for genome-wide association studies via imputation of genotypes. Nat Genet 39:906–913

Mitra S, Klar B, Huson DH (2009) Visual and statistical comparison of metagenomes. Bioinformatics 25(15):1849–1855

Navlakha S, Bar-Joseph Z (2011) Algorithms in nature: the convergence of systems biology and computational thinking. Mol Syst Biol 7:546

Nekrutenko A, Taylor J (2012) Next-generation sequencing data interpretation: enhancing reproducibility and accessibility. Nat Rev Genet 13(9):667–672

O'Connell J, Marchini J (2012) Joint genotype calling with array and sequence data. Genet Epidemiol 36(6):527–537. doi:10.1002/gepi.21657

Okimoto N, Futatsugi N, Fuji H, Suenaga A, Morimoto G et al (2009) High-performance drug discovery: computational screening by combining docking and molecular dynamics simulations. PLoS Comput Biol 5(10):e1000528

Ouzounis CA (2012) Rise and demise of bioinformatics? Promise and progress. PLoS Comput Biol 8(4):e1002487

Overby CL, Tarczy-Hornoch P, Hoath JI, Kalet IJ, Veenstra D (2010) Feasibility of incorporating genomic knowledge into electronic medical records for pharmacogenomic clinical decision support. BMC Bioinformatics 11(Suppl 9):S10

Patterson N, Price AL, Reich D (2006) Population structure and eigenanalysis. PLoS Genet 2(12):e190

Pearson WR, Lipman DJ (1988) Improved tools for biological sequence comparison. Proc Natl Acad Sci U S A 85(8):2444–2448

Pleil JD, Sheldon LS (2011) Adapting concepts from systems biology to develop systems exposure event networks for exposure science research. Biomarkers 16(2):99–105

Purcell S, Neale B, Todd-Brown K, Thomas L, Ferreira MAR, Bender D, Maller J, Sklar P, de Bakker PIW, Daly MJ, Sham PC (2007) PLINK: a toolset for whole-genome association and population-based linkage analysis. Am J Hum Genet 81(3):559–575

Rebholz-Schuhmann D, Kirsch H, Arregui M, Gaudan S, Riethoven M, Stoehr P (2007) EBIMed–text crunching to gather facts for proteins from Medline. Bioinformatics 23(2):e237–e244

Rice P, Longden I, Bleasby A (2000) EMBOSS: the European molecular biology open software suite. Trends Genet 16(6):276–277

Rinaldi F, Clematide S, Garten Y, Whirl-Carrillo M, Gong L, Hebert JM, Sangkuhl K, Thorn CF, Klein TE, Altman RB (2012) Using ODIN for a PharmGKB revalidation experiment. Database (Oxf) 2012:bas021

Rodriguez B, Burrage K, Gavaghan D, Grau V, Kohl P, Noble D (2010) The systems biology approach to drug development: application to toxicity assessment of cardiac drugs. Clin Pharmacol Ther 88(1):130–134

Salomonis N, Hanspers K, Zambon AC, Vranizan K, Lawlor SC, Dahlquist KD, Doniger SW, Stuart J, Conklin BR, Pico AR (2007) GenMAPP 2: new features and resources for pathway analysis. BMC Bioinformatics 8:217

Sarkar IN, Butte AJ, Lussier YA, Tarczy-Hornoch P, Ohno-Machado L (2011) Translational bioinformatics: linking knowledge across biological and clinical realms. J Am Med Inform Assoc 18:e354–e357

Schlitt T, Brazma A (2007) Current approaches to gene regulatory network modeling. BMC Bioinformatics 8(Suppl 6):S9

Servin B, Stephens M (2007) Imputation-based analysis of association studies: candidate regions and quantitative traits. PLoS Genet 3(7):e114

Shendure J, Ji H (2008) Next-generation DNA sequencing. Nat Biotechnol 26(10):1135–1145

Silver PA, Way JC (2007) Molecular systems biology in drug development. Clin Pharmacol Ther 82(5): 586–590

Simpson JT, Wong K, Jackman SD, Schein JE, Jones SJ, Birol I (2009) ABySS: a parallel assembler for short read sequence data. Genome Res 19(6):1117–1123

Smith TF, Waterman MS (1981) Identification of common molecular subsequences. J Mol Biol 147(1):195–197

Stabenau A, McVicker G, Melsopp C, Proctor G, Clamp M, Birney E (2004) The Ensembl core software libraries. Genome Res 4(5):929–933

Stajich J, Block D, Boulez K, Brenner S, Chervitz S, Dagdigian C, Fuellen G, Gilbert J, Korf I, Lapp H, Lehväslaiho H, Matsalla C, Mungall C, Osborne B, Pocock M, Schattner P, Senger M, Stein L, Stupka E, Wilkinson M, Birney E (2002) The Bioperl toolkit: Perl modules for the life sciences. Genome Res 12(10):1611–1618

Sun J, Wu Y, Xu H, Zhao Z (2012) DTome: a web-based tool for drug-target interactome construction. BMC Bioinformatics 13(Suppl 9):S7

Szklarczyk D, Franceschini A, Kuhn M, Simonovic M, Roth A, Minguez P, Doerks T, Stark M, Muller J, Bork P, Jensen LJ, von Mering C (2011) The STRING database in 2011: functional interaction networks of proteins, globally integrated and scored. Nucleic Acids Res 39(Database issue):D561–D568

Tegnér JN, Compte A, Auffray C, An G, Cedersund G, Clermont G, Gutkin B, Oltvai ZN, Stephan KE, Thomas R, Villoslada P (2008) Computational disease modeling – fact or fiction? BMC Syst Biol 3:56

Thompson JD, Gibson TJ, Plewniak F, Jeanmougin F, Higgins DG (1997) The ClustalX windows interface: flexible strategies for multiple sequence alignment aided by quality analysis tools. Nucleic Acids Res 25:4876–4882

Thorn CF, Klein TE, Altman RB (2010) Pharmacogenomics and bioinformatics: PharmGKB. Pharmacogenomics 11(4):501–505

Vicini P (2010) Multiscale modeling in drug discovery and development: future opportunities and present challenges. Clin Pharmacol Ther 88(1):126–129

Warren RL, Butterfield YS, Morin RD, Siddiqui AS, Marra MA, Jones SJM (2005) Management and visualization of whole genome shotgun assemblies using SAM. Biotechniques 38:715–720

Wei T, Liao B, Ackermann BL, Jolly RA, Eckstein JA, Kulkarni NH, Helvering LM, Goldstein KM, Shou J, Estrem ST, Ryan TP, Colet JM, Thomas CE, Stevens JL, Onyia JE (2005) Data-driven analysis approach for biomarker discovery using molecular-profiling technologies. Biomarkers 10(2–3):153–172

Wist AD, Berger SI, Iyengar R (2009) Systems pharmacology and genome medicine: a future perspective. Genome Med 1(1):11

Yang R, Niepel M, Mitchison TK, Sorger PK (2010) Dissecting variability in responses to cancer chemotherapy through systems pharmacology. Clin Pharmacol Ther 88(1):34–38

Yao L, Evans JA, Rzhetsky A (2009) Novel opportunities for computational biology and sociology in drug discovery. Trends Biotechnol 27(9):531–540

Zerbino DR, Birney E (2008) Velvet: algorithms for de novo short read assembly using de Bruijn graphs. Genome Res 18(5):821–829

Pharmacogenomics in the Era of Personal Genomics: A Quick Guide to Online Resources and Tools

10

Ayesha Pasha and Vinod Scaria

Abstract

The post-HapMap era has seen a spurt in the resources available for genome analysis and their interpretation along with databases and consortiums working on personal genomes. The advent of next-generation sequencing has provided an impetus to personal genomics while at the same time giving an indirect boost to concepts such as pharmacogenomics. Modern-day pharmacogenomics (or PGx) has evolved from the mundane – highlighting variations in common genes for drugs used the most, to the correlative study of the influence of a single gene on multiple drugs or combinatorial drugs on other drugs used concomitantly and so on. The vast outburst in data has led to a simultaneous growth in the databases and consortiums working on PGx as well as curating the data generated. This review covers those online resources and tools which are instrumental to the interpretation of personal genomes and contain pharmacogenetically relevant data. Further, it has been divided into five main categories depending upon the content and utility of the resource into (1) pharmacogenomics databases, (2) variation databases, (3) tools/resources for analysing PGx data, (4) community efforts and consortia and (5) standards for data representation.

1 Introduction

Pharmacogenomics (PGx) is the study of the genetic basis of interindividual variability in drug response (Lee et al. 2009; Kitzmiller et al. 2011). Pharmacogenomics has recently emerged into one major area of human genetics, thanks to the availability of technology which enables genome-scale analysis of variations for association testing and better molecular understanding of disease processes. The recent years have seen a spurt of

A. Pasha
Open Source Drug Discovery Unit, Council of Scientific and Industrial Research, Anusandhan Bhavan, New Delhi 110001, India

V. Scaria (✉)
GN Ramachandran Knowledge Center for Genome Informatics, CSIR Institute of Genomics and Integrative Biology, Mall Road, Delhi, India
e-mail: vinods@igib.res.in

D. Barh et al. (eds.), *Omics for Personalized Medicine*,
DOI 10.1007/978-81-322-1184-6_10, © Springer India 2013

interest in understanding genetic variations and their effect on drug action and/or adverse events. At present, there are over thousands of reported variants in the human genome with known pharmacogenetic effects. Of the different types of genetic variations that could potentially contribute to variability in drug response, single-nucleotide variations are by large, the type which has been extensively explored. Single-nucleotide variations in genes coding for enzymes, carriers, transporters and targets, which are inherently linked with the pharmacodynamics and pharmacokinetics of a drug, may result in altered responses to drugs and in some cases severe adverse effects (Marsh and McLeod 2006; Pirmohamed 2011). Apart from genetic variations involved in drug transport, metabolism and action, the genetic variations in immune genes have also been explored with respect to their relation with a subtype of adverse drug reactions which are idiosyncratic in nature.

One of the major reasons for the increasing relevance of pharmacogenetics is the incidence of adverse drug reactions or ADRs. Several studies have indicated that the incidences of ADRs are about 1.6–41.4 % in patients undergoing therapy with approximately $17–29 billion spent annually on avoidable adverse events. In the USA, ADRs are responsible for ~100,000 deaths annually, while in the United Kingdom alone, they have been associated with a 25 % increase in the length of hospitalisation (Cano and Rozenfeld 2009). A separate study also reported that 56.3 % of ADRs cause no or minor disability, 7.0 % ADRs may be associated with permanent disability and 7.4 % ADRs may result in death (de Vries et al. 2008). Pharmacogenomics offers immense promise in this area, as a number of adverse events could be potentially prevented by adequate and appropriate genetic testing and prescriptions based on the genetic testing. The relevance of pharmacogenetic testing may further be emphasised by the increasing number of drugs for which genetic testing has been recommended by the FDA. At present, the FDA recommends genetic tests in the drug labels of 103 drugs for 113 biomarkers (US-FDA). Hopefully, the widespread adoption of pharmacogenomics testing in clinical practice would significantly improve patient health-care options to ensure clinical efficacy and cost-effectiveness (Yang et al. 2011; McLeod and Isaacs 2011; Deverka et al. 2010).

Despite the rapid advancements in methodology for clinical testing and a number of markers with pharmacogenetic associations made available, only a very small number of pharmacogenetic tests have made it into actual clinical practice (Flynn 2011). This limitation is primarily due to the lack of replication of genetic markers which show moderate effects and lack of biological understanding of the pathways and mechanisms of many drugs and their action. The rise of pharmacogenomics has been partially fuelled by the Human Genome Project and other associated projects including the HapMap project and now the 1000 Genomes Project which aims at deciphering the landscape of genetic variations in world populations.

The Human Genome Project and tremendous advancements in technology and throughput of genome sequencing have enabled the sequencing of a number of personal genomes, including genomes from different populations of the world. Further improvements in the technology and throughput promise personal genomics to be commonplace in clinical settings. Apart from the technological advancements in the throughput and scale of genomic sequencing, the realisation of personal genomics in clinical settings would only be possible with the availability of better, faster and efficient bio-computational pipelines and workflows to effectively mine relevant information from genomes. In addition, efficient high-quality curated datasets of genetic variants also need to be available to make mining meaningful and clinically relevant. In addition, these resources should be user-friendly to offer the most appropriate information to the clinician in the time of need. The current review focuses on the major online resources created over the years (Fig. 10.1) for pharmacogenomics along with their distinct features and applications. While an attempt has been made to cover as many resources as possible, a review on all of them is beyond the scope of this article. Presently, the datasets and resources available on pharmacogenomics are scattered in a variety of databases, tools and

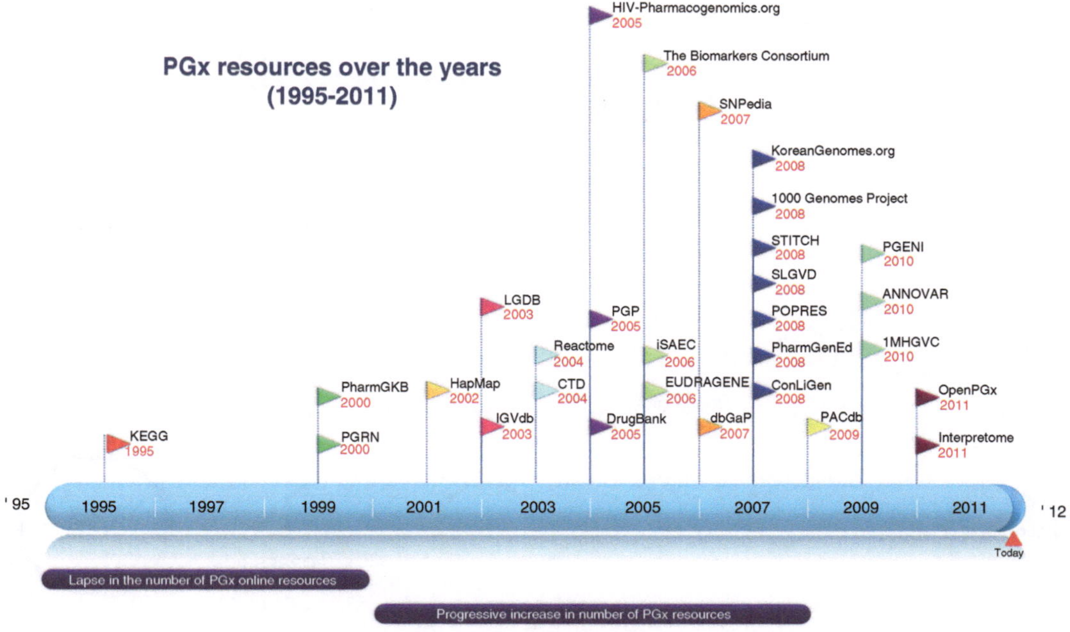

Fig. 10.1 Growth of pharmacogenomics resource over the years (1995–2011)

resources and lack a common or unified protocol. In this review, we discuss a non-exhaustive list of databases, resources and computational tools which would enable the easy access of relevant pharmacogenomics information. Figure 10.2 covers the interlink of these resources to each other which highlights the process of data exchange and generation as well as their interdependent relationships. The online resources listed within this article have been divided into four main classes depending upon their broad role.

2 Pharmacogenomics Databases

2.1 General Databases

2.1.1 Pharmacogenomics Knowledge Base (PharmGKB)

PharmGKB (Hernandez-Boussard et al. 2008; Giacomini et al. 2007; Kawamoto et al. 2009; Klein and Altman 2004; Owen et al. 2008; Thorn et al. 2005) is a web-based, public portal of genotype-phenotype information relevant to pharmacogenetics. PharmGKB editors collect, curate and disseminate data on human genetic variations which impact drug response. The resource presents information relevant to pharmacogenomics in the form of Variant Annotations, Clinical Annotations and Very Important Pharmacogene (VIP) Summaries. It also makes available clinically relevant information in the form of Pharmacogenomics-Based Drug-Dosing Guidelines, Drug Labels with Pharmacogenomic Information. The resource also hosts information on drug-centred pathways.

PharmGKB has four levels of data collection, management and implementation which are collectively known as the PharmGKB Knowledge Pyramid:

1. Knowledge extraction, which involves the manual curation of PGx data from publications in PubMed along with the development of Natural Language Processing techniques to efficiently mine this data.
2. Knowledge annotation, aggregation and integration that establishes associations between genetic variants and drugs reported in literature (Variant Annotations) creates pathways for a

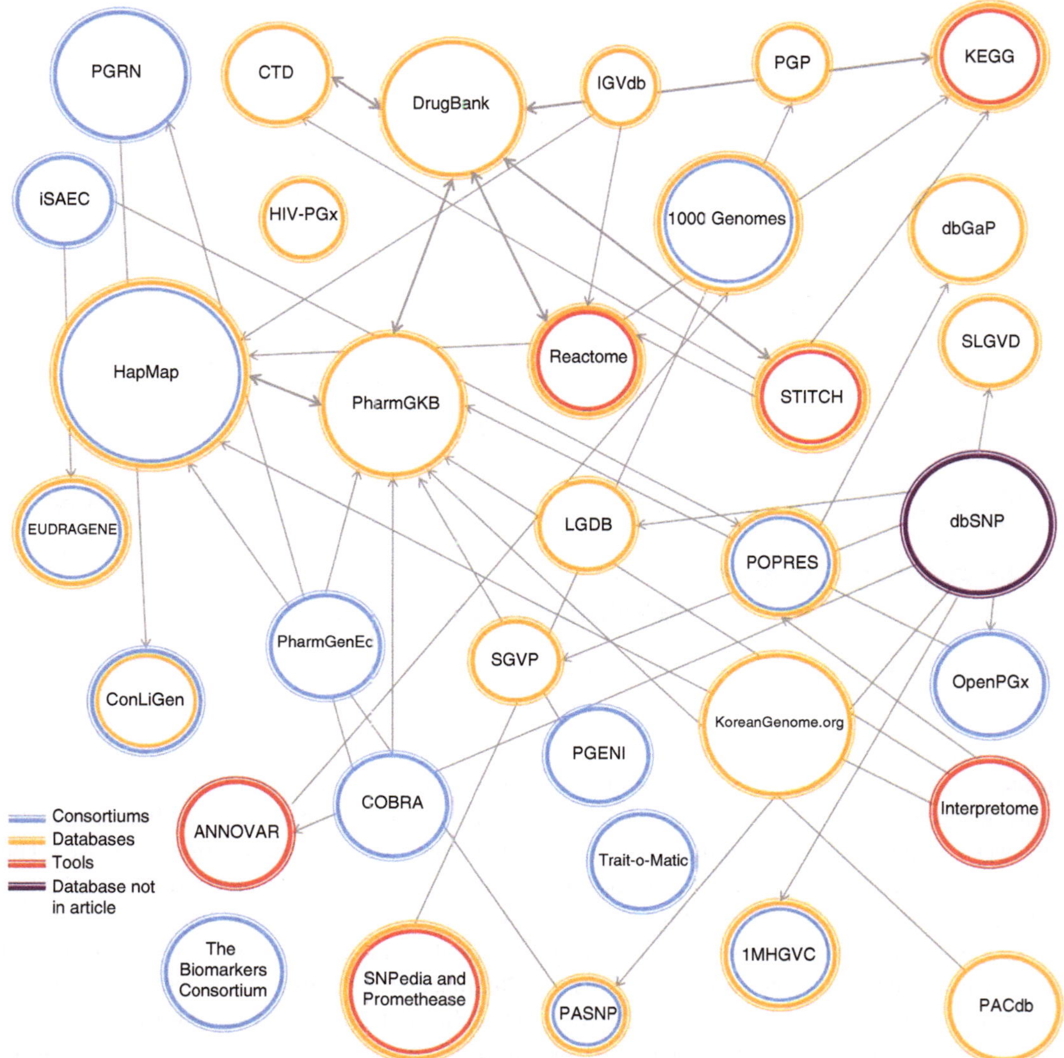

Fig. 10.2 Interlinks between different resources. *One-sided arrows* signify outward links from one resource to another, and *two-sided arrows* signify resources that are linked to each other

drug which has pharmacogenetic association (both pharmacodynamic and pharmacokinetic pathways) and provides summaries on genes which have important roles in drug response (Very Important Pharmacogene (VIP)).

3. Clinical interpretation, which includes summaries on the association between a drug and each genotype of genetic variants in the form of Clinical Annotations, created by aggregating the Variant Annotations for a particular genetic variant – drug association pair. Each Clinical Annotation is supplemented by literature evidence and may move up or down the scale depending on further publications supporting or contradicting the associations.
4. Clinical implementation is the last level within the PharmGKB Knowledge Pyramid and includes clinically relevant data such as FDA Drug Label with PGx information, genetic tests for specific variants which influence treatment response and Drug-Dosing Guidelines.

Based on these levels of data organisation, PharmGKB collects and organises data into five pertinent categories: (1) clinical outcomes (CO), (2) pharmacodynamics and drug response (PD), (3) pharmacokinetics (PK), (4) molecular and cellular functional assays (FA) and (5) genotype (GN). Four out of the five categories analyse the influence of genetic variations on clinical/medical outcomes (CO), biological and physiological responses to drugs which are not strong enough to alter treatment regimes (PD), levels or concentration of drugs or their metabolites at their site of action (PK) and results of molecular and cellular functional assays (FA). The fifth data category, i.e. GN, deals with the genetic variations independent of individual drugs that are measured as sequence variations in individual genes and may form the basis for variations in response to drugs. Data on PharmGKB may also be searched according to genes, drugs, diseases and pathways which include entire datasets in an A–Z index that may be searched alphabetically. PharmaGKB also closely works with several consortiums, projects and implementation partnerships such as the Clinical Pharmacogenetics Implementation Consortium (CPIC), International Clopidogrel Pharmacogenomics Consortium (ICPC), Inter national SSRI Pharmacogenomics Consortium (ISPC), International Tamoxifen Phar macogenomics Consortium (ITPC), International Warfarin Pharmacogenetics Consortium (IWPC), International Warfarin Pharmacogenetics Con sortium-Genome Wide Association Studies (IWPC-GWAS), Translational Pharmacogenetics Project (TPP), University of Florida & Shands Personalized Medicine Program and 1200 Patients Project to facilitate the implementation and development of the concepts of PGx.

2.1.2 DrugBank

DrugBank (Wishart et al. 2006, 2008; Knox et al. 2011) is an online resource which has been developed to link detailed drug data, i.e. chemical, pharmaceutical and pharmacological data, with comprehensive drug target information including the sequence, pathway and structure of these targets. It was designed as a comprehensive in silico resource which could be used for a wide range of applications in drug discovery including cheminformatics. DrugBank contains 6,711 drug entries comprising of 1,447 FDA-approved small molecule drugs, 131 FDA-approved biotech (protein/ peptide) drugs, 85 nutraceuticals and 5,080 experimental drugs which are linked to 4,227 nonredundant proteins, i.e. drug targets, enzymes, carriers and transporters. The drugs and proteins listed within these entries are obtained from several databases and include the unique IDs assigned to them by other databases (such as the UniProt ID and KEGG ID) providing users access/links to expanded information contained within other databases. The information on each drug in DrugBank is organised as a DrugCard. Each DrugCard entry consists of more than 150 data fields with almost half the components dedicated to drug/chemical data and the other half to drug protein data. Users may visualise data by clicking on the "Browse" button which would present a tabulated format of the DrugBank information or they may browse by drug name (DrugBrowse), its pharmaceutical application (PharmaBrowse), genes/SNPs associated with drugs (GenoBrowse), drug pathways (Pathway Browse), drug class depending upon whether they are organic or inorganic (ClassBrowse) and the associated targets/carriers/enzymes/transporters (AssociationBrowse). DrugBank can also be mined for data depending upon the "Search" query used. There are five search queries users can employ: (1) ChemQuery through which users can draw or write a chemical structure and search DrugBank for drugs/compounds which are similar to or identical to the structure being studied; (2) Text Query which allows users to search through the database using a sophisticated text search option based on the Lucene query language that supports Boolean Search (AND, OR, NOT) and can search terms that would match to specific terms within the database; (3) Interaction Search, which provides users three search options depending upon the type of interaction, i.e. drug interaction lookup (used to search drug-drug interactions for one or more drugs at one time), Multi Search (can be used to determine drug-drug interactions between two lists of drugs entered) and Food Interaction Lookup (that lists the drug-food interactions for the list of drugs submitted); (4) SequenceSearch that can be used for BLASTP (protein) sequence search from amongst the 18,000 sequences listed in DrugBank and presents sequences in the FASTA format; and (5) Data Extractor, which allows users to extract data from DrugBank by selecting DrugCard fields listed within the database and providing complex or constrained queries, the results of which are presented in HTML, CSV or printable list format. DrugBank also includes external links to almost all major bioinformatics and biomedical databases, such as GenBank, SwissProt/UniProt, PDB, ChEBI, KEGG, Pub Chem and PubMed; numerous drug and pharmaceutical databases such as RxList, PharmGKB and FDA labels; and reciprocal links to SwissProt/UniProt, Wikipedia, BioMOBY and PubChem. DrugBank is a comprehensive database which includes drugs, targets, enzymes, transporters and carriers and can be of consequential relevance in determining PGx associations.

2.1.3 Database of Genotypes and Phenotypes (dbGaP)

dbGaP (Mailman et al. 2007) is a publicly available database maintained by NCBI that archives the results of genotype-phenotype association studies and at present archives data on 285 studies, 131,482 variables, 3,052 documents, 3,519 analysis studies and 2,267 datasets. Users can access data through two options: (1) open, which includes the nonsensitive data that can be viewed by anyone, and (2) controlled, which is the individual-level data or in other words information on the personal, phenotypic and genotypic features of the individuals participating in the study and requires authorisation for access. Open access data can further be divided into four categories: (1) study data which includes all relevant information such as the study description, protocol documents, data collection instruments and other experimental facets; (2) phenotypic data for the variables studied in the form of summaries and individual-level data; (3) genotypic data consisting of the individual genotypes, pedigree information, fine-mapping results, re-sequencing traces, etc.; and (4) statistical data which includes data and results from association studies and linkage analysis if available. Users may also browse through the top studies within the database by searching through five data categories, i.e. Studies (which includes all aspects of a study along with the different studies included within the database), Variables (which consists of a list of variables studied that can be individually searched through), Analyses (includes all analyses archived within the database and their results), Documents and Datasets associated with studies. dbGaP also includes the Phenotype-Genotype Integrator (PhenGenI) tool which integrates the NHGRI genome-wide association study (GWAS) catalogue data with other databases within NCBI such as Gene, dbGaP, OMIM, GTEx and dbSNP. PhenGenI tool allows users to select and search either a phenotype or a genotype and present the results on the corresponding genotypic and phenotypic associations, respectively. dbGaP serves as an important database in determining PGx associations through the use of tools or studying the data included within the resource.

2.1.4 Comparative Toxicogenomics Database (CTD)

CTD (Davis et al. 2009, 2011a, b; Mattingly et al. 2004) is an online resource that curates and extracts data from several sources to aid in the discovery of novel relationships between toxicants and in human health, i.e. in other words to elucidate the influence of environmental agents on human health. The data within the database is manually curated into three types of interactions, i.e. chemical-gene, chemical-disease and gene-disease relationships, which together form the core data and are integrated to generate the novel chemical-gene-disease networks which illuminate the underlying mechanisms for variable susceptibility and environmentally influenced diseases. CTD has 11 data categories: chemicals, genes, chemical-gene/protein interactions, diseases, gene-disease associations, chemical-disease associations, references, organisms, gene ontology, pathways and exposures. Out of all these datasets, the chemicals, genes, chemical-gene/protein interactions, pathways, gene-disease associations and chemical-disease association datasets include information relevant to PGx, while the remaining datasets contain information substantiating the data within the database and may be pertinent to pharmacogenomics. CTD also includes analysis tools which may be used to analyse and interpret data including the (1) Batch Query which allows users to download entire datasets associated with a list of diseases, genes, chemicals, gene ontology terms and pathways; (2) Gene Set Enricher that can be used to find GO functional annotations or pathways for a set of genes; (3) MyGeneVenn which can be used to compare the user's gene list with those genes that have up to two associated chemicals or diseases; (4) MyVenn, used for viewing relationships between lists of CTD chemicals, diseases, genes, GO terms or pathways or any other data; and (5) VennViewer that can be used to compare datasets for up to three chemicals, diseases and genes. Thus, CTD is a vast database for toxicogenomic associations which has a consequential role in its applicability to PGx research.

2.2 Specific Databases

These databases include data that is specific either to a particular disease (e.g. CancerResource) or to a drug (e.g. ConLiGen). While the entire data or focus of these databases may not lie completely in pharmacogenomics, they do contain information relevant to PGx associated with that particular disease or drug extracted from literature, other PGx databases and in some cases experimental analysis. The following section provides a preview on these databases.

2.2.1 HIV-Pharmacogenomics.org

HIV-Pharmacogenomics.org (http://www.hiv-pharmacogenomics.org/) is a comprehensive resource which has been created to assist researchers and health-care professionals studying the genetic factors influencing the clinical outcomes following drug therapy in HIV patients. Antiviral drugs used in the treatment of HIV not only cause severe adverse reactions and long-term drug toxicity, but also given the duration and combinatorial nature of therapeutics in HIV, a carefully selected and planned therapeutic regime is required. The database is a highly relevant resource for pharmacogenetics of HIV therapeutics. Users may search the database by drug name, gene or toxicity induced. Further, the genes may also be searched according to their pharmacogenetic associations, i.e. genes that alter drug metabolism, genes altering drug transport, genes inducing drug-associated toxicity, genes altering drug therapy and those that are natural history modifiers. Natural history modifiers include those genes and their variants, the presence or absence of which alter HIV progression, susceptibility and therapeutics. Pharmacogenetic associations may be easily visualised by following a step by step approach at the end of which the association if present will be available to the user.

2.2.2 ConLiGen

The International Consortium on Lithium Genetics (ConLiGen) (Schulze et al. 2010, http://www.conligen.org/imprint.html) was established to study the pharmacogenetics of lithium. Lithium though toxic has a central role in the psychotherapeutics of bipolar disorder. The pharmacogenomic data on lithium toxicity is sparse. This has been primarily because of the lack of large sample numbers and high level of interindividual variability which precludes genome-wide association studies on smaller number of individuals. ConLiGen consortium was formed with the aim of performing a genome-wide association study (GWAS) of stringently defined response to lithium and identifying the phenotype of lithium response with the eventual aim of determining all aspects of lithium therapy including the genetic factors predisposing to adverse effects of lithium. The consortium assembled a cohort of over 1,200 patients with bipolar disorder, who were further categorised into responders and nonresponders and studied to determine the pharmacogenetic associations to lithium therapy in bipolar disorder.

2.3 Others

2.3.1 PACdb

PACdb (Gamazon et al. 2010, http://www.pacdb.org/index.html) or the Pharmacogenomics and

Cell Database is an online PGx resource which curates and provides information on the relationships between SNPs, gene expression and cellular sensitivity to drugs analysed in cell-based models. The results of these cell model experiments are used to identify genetic variants which affect drug responses. PACdb utilises the lymphoblastoid cell line derived from different target populations as the experimental cell model due to the availability of extensive genotypic data and the potential of these cell lines in deducing haematological toxicities and other types of toxicities. An additional area of focus for PACdb is the development of computational workflows and pipelines which would help in the management of the vast amounts of data generated and its interpretation. The database has three distinct logical layers, i.e. the front end, middle tier and data storage, which enables the development of the three layers independently. The resource also allows users to input their own data as well as post their own queries. The middle tier of the database comprises of data relevant to solving the queries posted by the user such as the drug, SNP, gene, expression, phenotype and population. The last layer, i.e. the data storage component, simply stores all data in a relational database based on MySQL. PACdb provides whole genome analysis of the relationship between genotype and drug-induced toxicity, transcript clustering, the relationship between the gene expression and drug phenotype and other functional and physical annotations. The webpage of the database provides an external link to a SNP database SCAN which has two categories of SNP annotations: (1) physical annotations which classify SNPs according to their position relative to the gene and its Linkage Disequilibrium (LD) and (2) functional annotations wherein SNPs are categorised on the basis of their influence on gene expression. All the SNP genotype information within the database is extracted from HapMap for two population types, i.e. CEU and YRI. At present, the PACdb contains information only on the following drugs: carboplatin, cisplatin, etoposide, daunorubicin, busulfan, ara-C, doxyfluridine (capecitabine), hydroxyurea and pemetrexed, but foresees further incorporation of pharmacogenetic data, particularly that which has been established in the LCL models and a linkup with PharmGKB to facilitate the utilisation of the data in clinical and experimental settings.

3 Variation Databases

3.1 World Population Databases

3.1.1 HapMap

HapMap (Deloukas and Bentley 2004; O'Shaughnessy 2006; Zhang et al. 2008; http://hapmap.ncbi.nlm.nih.gov/index.html.en) or the International HapMap Project is a multinational endeavour established with the aim of identifying and cataloguing the genetic variations between individuals, whether they be similarities or dissimilarities, in order to evaluate the effects of genes and genetic variations on response to drugs, diseases and human health. While it does not establish definite links between genetic variants and their effects, it does provide information which may be used for this purpose by other researchers. The HapMap collaboration includes scientists and funding agencies from Japan, the United Kingdom, Canada, China, Nigeria and the United States. DNA samples from populations of European, African and Asian descent were analysed to identify haplotypes within them and determine the links for these to diseases found within the population or in general. A total of 270 individuals were genotyped, and these individuals were either completely unrelated or consisted of "trios" of both parents and one adult child. The database is built on samples collected from the Yoruba people from Nigeria (30 sets of trios); individuals from Tokyo, Japan (45, unrelated); Beijing, China (45, unrelated); and the USA (30 sets of trios from individuals with Northern or Western European ancestry). HapMap data can be searched graphically by feeding the chromosome number directly to the search engine or by using the Genome Browser. Users may also browse using the different datasets available, i.e. Genotype (individual genotypes submitted to HapMap), Frequencies (allele and genotype frequencies obtained from genotype data), LD Data

(Linkage Disequilibrium (LD) properties), Phasing data (data obtained by utilising the PHASE software), Allocated SNPs (dbSNP reference SNP clusters), CNV Genotypes (CNV data), Recombination rates and Hotspots, SNP Assays (compiled data on SNP assays), Perlegen amplicons (mapping of Perlegen amplicons to HapMap assayLSID), Raw data on signal intensities obtained from genotypes, Inferred genotypes and Mitochondrial and chrY Haplogroups. Using TagSNPs, the haplotypes are identified along with their frequencies, and at present, more than 3.5 million SNPs have been identified by HapMap with the eventual aim of identifying the 10 million or more SNPs in the human genome. The project has been divided into three phases at present: Phase 1 aimed at the genotyping of one common SNP per 5,000 bases and identification of more than 1 million SNPs, phase 2 targeted the identification of 2 million additional SNPs, and phase 3 has increased the number of DNA samples genotyped from 270 in the initial two phases to 1,301 samples in HapMap3. The data obtained may be used to analyse population, gender and other individual responses to drug therapy, propensity for diseases, etc., highlighting the importance of HapMap in realising PGx studies and targets.

3.1.2 1000 Genomes Project

The 1000 Genomes Project (Via et al. 2010; Wise 2008; 1000 Genomes Consortium 2010) is an international consortium including research foundations and institutes from the United Kingdom, the USA, China and Germany, created with the target of identifying genetic variations having population frequencies of at least 1 %. The main aim of the project is to collect 2,500 DNA samples from ethnically diverse populations and sequence their genomes at a 4X coverage. The first portion of the project included genotypic data from 1,167 samples across 13 populations (2010–2011), and the second portion aimed at the genotyping of 633 samples from 7 populations (early 2011) and the third set (late 2011) at the genome sequencing of 700 samples. To aid the overall sequencing process and to ensure the sufficiency of the protocol to be followed, several pilot studies were initially conducted including the sequencing of the genome of 180 individuals at 2–4X coverage, 2 trios at 20–60X coverage and 1,000 gene regions in 900 samples at 50X coverage. The data available at 1000 genomes may be directly downloaded from the mirror "ftp" sites of the database. The genome information so amassed foresees wide applications in patient health care and therapeutics as it may provide details on the genes/genetic variants associated with diseases, drug response phenotypes and other traits as well as improve the applicability of genetic tests.

3.1.3 POPulation REference Sample (POPRES)

POPRES (Nelson et al. 2008) is a DNA resource created from a large number of individuals (~6,000) participating in multiple studies from around the world. PCA analysis was used to confirm the overall integrity and elucidate the important features in the genetic structure of these diverse populations. POPRES was established to enable the extensive genotyping of DNA samples in order to facilitate PGx research. DNA samples were collected from ten ethnically diverse populations of African-American, East Asian, South Asian, Mexican and European origin, and genotype data obtained from a genome-wide panel of SNPs (500,000) attempted on all participants was carefully evaluated to identify potential markers and subjects which might have utility in a wide range of applications. The data so obtained is freely available through dbGaP for application in research and is useful in terms of generating contrast groups for PGx studies such as ADR studies where the treatment regimes remain the same, but patient variability in drug sensitivity owing to causative genetic variants may be identified.

3.2 Population-Specific Datasets

3.2.1 Pan-Asian SNP Database/ Pan-Asian SNP Consortium (PASNP)

The Pan-Asian SNP database (Ngamphiw et al. 2011; http://www4a.biotec.or.th/PASNP) is a comprehensive dataset of more than 50,000 SNPs identified in 1,808 DNA samples covering 73 Asian populations. Users may visualise the data either in the "Summary" format which provides an overview of the entire SNP and CNV data or through the "CNV" option which allows users to search CNVs on entire chromosomes, using sampleIDs or through CNV type. The data obtained has also been analysed to elucidate the migratory patterns of the East and Southeast Asian populations with the database having external links to other SNP databases for the comparison of SNPs and CNVs in these repositories to those identified in the 73 populations.

3.2.2 Indian Genome Variation Database (IGVdb)

The Indian Genome Variation Database (Indian Genome Variation Consortium 2005; http://www.igvdb.res.in/index.php) has been developed with the aim of studying the underlying genetics in four major linguistic groups of India, i.e. the Dravidian, Indo-European, Austro-Asiatic and Tibeto-Burman populations. The database serves as resource of SNPs and CNVs and repeats within the Indian subpopulation genomes in more than 1,000 genes at present which could have consequential PGx impact. It also provides a visual representation of the genic location of the SNPs and allows users to upload their own data on novel SNP's. IGVdb aims at the utilisation of this data for the construction of haplotype maps, their application in disease association studies and PGx and analysing the functional relevance of biomarkers identified in the Indian health-care system.

3.2.3 EUDRAGENE

EUDRAGENE (Molokhia and McKeigue 2006; https://www.eudragene.org/index.php) is a European Union (EU) initiative to collect DNA samples to study and identify genes which may be associated with major/severe drug side effects. During the first phase, researchers at various centres associated with the consortium aim to study an initial set of important ADRs that cause severe effects in patients on use of the associated drugs (which normally would be the best treatment alternative) or present symptoms that are usually not consistent with the disease diagnosed. The second phase aims at expanding the number of ADRs studied. The data generated through these studies along with the samples analysed (EUDRAGENE would collect 100 samples of each ADR type along with an equal number of healthy controls) would be freely available to researchers through a database maintained by the coordinating members of the consortium. Apart from this, physicians may upload their case studies on patients who have shown ADRs following drug therapy. Similarly, data may also be entered by the patient, with the information divided into sections outlining drug reaction, medical history, alcohol consumption, family background, ADR drug reaction data, additional questionnaire and eligibility criteria. The database can only be accessed by registering with the consortium and downloading completely as a CSV file. Alternatively, data may be filtered according to drug reaction, sex and age group before downloading. EUDRAGENE has already established collaborations with several research groups belonging to Austria, Denmark, Canada, Belgium, France, Italy, the Netherlands, Spain, Sweden and the UK which would work together to facilitate pharmacogenomic studies.

3.2.4 Other Population-Specific Genomes

There are several databases which include the whole genome and genome-wide association studies based on specific populations. While the world population databases include genotypic information for multiple populations, the population-specific databases focus entirely on elucidating the genotype-phenotype associations. Other than IGVdb and EUDRAGENE, other population-specific databases include:

A. *Koreangenome.org or the Korean Genome Project* (http://koreangenome.org/index.php/Main_Page) which includes the personal genomes of individuals of Korean nationality. It is an open access endeavour to collect, analyse and distribute Korean genomes. The second phase of this project known as the Korean Personal Genome Project will provide an open, interactive interface for common people and researchers interested in the field of personal genomics. Data may be obtained by downloading from the mirror "ftp" site of the Koreangenome.org.

B. *Genome Database of Latvian Population (LGDB)* (http://www.p3gobservatory.org) is a repository of all genetic research done in Latvia. The main aim of this project is to facilitate the application of genetic data in pharmacogenetics and genetic epidemiology. At present, it is generating GWA data from 95 individuals of Latvian origin.

C. *Sri Lankan Genome Variation Database* (Samarakoon et al. 2011; http://hgucolombo.org/SLGVD.aspx) is a repository of the SNPs identified in three major ethnic communities of Sri Lanka, i.e. the Sinhalese, Sri Lankan Tamils and Moors. The database may be searched by gene name, gene or SNP ID and allows users to submit data as well.

D. *1Malaysia Human Genome Variation Consortium (1Mhgvc)* (http://1mhgvc.kk.usm.my/) is a project which has been established to map the genome variations in the ethnically diverse populations of Malaysia. It aims to determine the migratory history of the country's populations, the genetic similarities between them and the implications of these variations on the various facets of research including pharmacogenomics.

E. *Singapore Genome Variation Project* (http://www.statgen.nus.edu.sg/~SGVP/; Teo et al. 2009) aims at the identification of the common variants across 1 million SNPs in the three ethnic populations of Singapore, i.e. Chinese (99 samples), Indians (95 samples) and Malays (98 samples). The data may be downloaded as bulk data under eight categories or browsed through using the "GBrowser" option. At present, the database includes ~1.6 million SNPs assayed across 268 individuals from these three populations.

3.3 Personal Genome Datasets

3.3.1 Personal Genome Project

The Personal Genome Project (Church 2005; http://www.personalgenomes.org/) is a Harvard Medical School initiative which has been established with the aim of implementing the data obtained from personal genome sequencing. PGP is a follow-up on the Human Genome Project focused at improving the understanding of genetic and environmental factors contributing to human traits by collecting biological samples for DNA/genome sequencing, cell line creation, etc. from volunteers and making the data obtained freely accessible to researchers worldwide. PGP targets at the dissipation of information on the utility of personal genomes in health care and disease, along with the development of tools to facilitate the analysis and interpretation of genotypic information as well as the ethical, social and legal issues (ELSI) associated with personal genome sequencing. The project has been developed targeting its application in science and technology, ELSI, health care, personal knowledge and development of products based on genetic information. Apart from the genetic data of an individual, PGP also collects personal and medical data so as to provide a comprehensive overview which may be used in PGx studies.

4 Tools/Resources for Analysing PGx Data

Several tools and resources have been developed which may be used to curate, analyse and interpret the data obtained from pharmacogenomic studies. These tools may be divided into two categories: one which may be used for visualisation of the drug pathways/interactions and those which are used for the interpretation of the data.

4.1 Resources Which Aid in the Visualisation of Drug Pathways and Interactions

4.1.1 STITCH

STITCH or Search Tool for Interacting Chemicals (Kuhn et al. 2010, 2012; http://stitch.embl.de/) is an online resource which has been developed to identify known and predicted associations between two chemicals and chemical-proteins based on data obtained from experiments, databases and literature. The current available version of STITCH is

3.1 and includes data on interactions between 300,000 small molecules and 2.6 million proteins from 1,133 organisms. Compared to STITCH 2, the latest version of STITCH has an almost four-fold increase in the number of chemicals with interactions (74,000 in STITCH 2 to 312,000 in STITCH 3.1). Four data sources have been used to construct these interactions: (1) databases containing experimental information, i.e. ChEMBL, PDSP Ki Database, BindingDB and PDB; (2) sources for drug targets, i.e. DrugBank, GLIDA, Matador, TTD and CTD; (3) repositories containing pathways, i.e. KEGG, NCI/Nature Pathway Interaction Database (http://pid.nci.nih.gov), Reactome and BioCyc; and (4) by mining literature using both manual curation and Natural Language Processing (NPL). Users can browse through the interaction networks using any one of the five search modes, i.e. by name, chemical structure, protein sequence, multiple names or multiple sequences.

STITCH provides four types of views by which users can visualise the interaction networks. These include the confidence view (represents the degree of confidence between interacting moieties), evidence view (represents the types of evidence whether from databases, experiments, etc. which support the associations within the network), actions view (that provides a view of the nature of interaction between moieties, i.e. inhibition, activation, expression and posttranscriptional modification) and the interactive view (which represents all types of interactions between the moieties). Apart from these, stereoisomers of a drug may be viewed or combined and chemical links to other chemical structures may also be visualised. The interface of the database also includes search options that may be used to view the experiments, databases and literature reviewed to construct the network. The database allows users to view the associations of a particular drug not only to other proteins but also with other chemical structures, presenting a complex interaction network of all associations possible for a chemical structure.

4.1.2 Reactome

Reactome (Vastrik et al. 2007; Matthews et al. 2009; http://www.reactome.org/ReactomeGWT/entrypoint.html) is an open access pathway database which has been manually curated and annotated from literature by several experts as well as cross-referenced to several other databases such as HapMap, Ensembl and UniProt. The database includes visual representations of the data curated as simple as classic metabolic pathways and comprehensive as the signalling pathways. The core of the pathway model is the "reaction" along with the "entities" such as drug molecules and nucleic acids, which participate in them. Users may use the database to (1) browse pathways directly, (2) map IDs of compounds or entities existing within Reactome by integrating data or by inputting data and identifying their function in existing or novel pathways, (3) compare entities present in humans to those in other organisms through orthologies and (4) analyse gene expression data to study their effect on pathways and reactions in organisms. The data within the database may be downloaded in BioPax, SBML, MySQL and PSI-MITAB formats as well as through web-based APIs. Reactome also includes tools for data analysis and interpretation such as BioMart, PathFinder and Cytoscape Plugins. The pathways within the database especially those pertaining to those in humans may be used for research and analysis, pathways modelling, systems biology as well as PGx applications to analyse effects of drug pathway alterations on drug response and phenotypes.

4.1.3 Kyoto Encyclopedia of Genes and Genomes (KEGG)

KEGG (Ogata et al. 1999; Kanehisa and Goto 2000; Kanehisa et al. 2012; http://www.kegg.jp/kegg/) is a knowledge base which has been developed to help facilitate the systemic analysis of gene function in terms of the networks of genes and molecules with the aim of understanding the utilities and functions of biological systems. It is a collection of graphical diagrams representing the biochemical pathways created by integrating data on genes, proteins and chemicals with molecular wiring diagrams of interaction, reaction and relational networks. KEGG is essentially an integrated resource of 17 databases which are broadly categorised into three major classes, that is, systems information, genomic information and chemical information. The systems information databases contain data on the mechanisms by which molecules or genes are networked, complementary to the information for individual molecules and genes in existing molecular biology databases. The systems information databases are KEGG Pathways, KEGG Brite, KEGG Module, KEGG Disease, KEGG Drug and KEGG Environ. The genomic information databases include gene catalogues for organisms whose entire or partial genomes have been sequenced as well as links for individual genes to components within the KEGG biochemical pathways. These include KEGG Orthology, KEGG Genes, KEGG Genome, KEGG DGenes and KEGG SSDB. The chemical information databases contain entire lists of chemical elements, compounds as well as other substances in living organisms under the broad category of KEGG LIGAND and links individuals chemical entities to components within the KEGG biochemical pathways. KEGG LIGAND includes KEGG Compounds, KEGG Glycans, KEGG Reaction, KEGG RPair, KEGG RClass and KEGG Enzyme. The database uses a unique colour coding to distinguish between the different datasets and individual components (known as KEGG objects) along with a KEGG object identifier consisting of a database-dependent prefix and a five-digit number. Considering the milieu of genetic, molecular and chemical information present within the database, it could serve as an important PGx resource.

4.2 Tools for the Interpretation of Personal Genome Data

4.2.1 Interpretome

Interpretome (Karczewski et al. 2012; www.interpretome.com) is an interpretation engine which has been developed to help users interpret their own personal genomes. The web-based resource allows users to upload their personal genomes, which are provided by the direct-to-consumer (DTC) companies in a tab-limited file, and interprets their genome in a comprehensive and client-friendly manner. Interpretome guarantees client confidentiality and does not allow individuals other than the customer to access their genomes. The interface of the engine has four utilities: Lookup, Explore, Clinical and Ancestry. The Lookup utility looks up the SNPs within the user's personal genome, and if it does not find any SNPs that match, it imputes data from HapMap. In case of imputing SNPs from HapMap, the utility relies on the "phases" provided by HapMap for a SNP to identify the presence of a particular variation. The Explore utility allows users to explore their genomes to determine similarities between them and existing exercises present on the engine. For example, users can explore their genomes to determine the presence of certain SNPs which are indicative of disease probability, exceptional longevity, height, etc. The Ancestry utility can be used to determine the user's ancestry on the basis of their ancestral and individual similarity and chromosome painting. The Clinical utility is the most relevant with respect to the field of PGx. This utility analyses the user's genomes to find SNPs which have been determined as being associated with diseases such as diabetes, to calculate the appropriate warfarin dose as well as find biomarkers within the personal genome which have been found to have a role in PGx determinations. The engine links with PharmGKB to analyse a user's genome for these biomarkers.

4.2.2 SNPedia and Promethease

SNPedia (Cariaso and Lennon 2012; http://www.snpedia.com) is a freely accessible, Wiki resource which lists all the medical, genealogical, phenotypic and forensic associations of genetic variations, i.e. SNPs. The inclusion of variants is dependent upon the level of scientific significance associated with it in terms of the number of individuals used in the study, medical history, method of analysis, etc. The SNPs are denoted by their respective "rsIDs," and users may search through by genes (by clicking on a specific gene name within the list which produces all the SNPs reportedly associated with that gene and a brief description), genosets (phenotypes caused by a set of variants listed together), genomes (whole genomes with their SNP listings), medicines (list of drugs with the SNPs responsible for altered phenotypes/response) and medical condition (listing the SNPs that may be linked to a disorder). Clicking on the rsID opens a "variant page" which includes the entire information available on that particular SNP along with external links to other databases such as dbSNP, HapMap, Ensembl, PharmGKB, NextBio and 23andMe. SNPedia also includes a tool "Promethease" which builds a personal genome report using data from SNPedia and a file of genotypes and may be employed by users to analyse their genomes. Users may also submit data to the database including novel SNPs and whole genomes sequenced by external sources.

4.2.3 Trait-O-Matic

Trait-O-Matic (http://snp.med.harvard.edu/) is an online resource for finding and classifying genetic variants in whole genomes. While it does not provide any whole genome sequencing facilities, users either may submit queries under confidentiality to determine genetic variants and their associated roles or may download the Trait-O-Matic tool to analyse "at home" the genetic variants in their genomes. It also includes the whole genomes of individuals with the entire genotypic data such as variants and their phenotype (diseases, adverse events, etc.), chromosome number, position and frequencies. Cell lines have also been developed from the samples provided by certain individuals. It is a useful resource for PGx analysis of personal genomes.

4.2.4 ANNOVAR

ANNOVAR (Wang et al. 2010; http://www.openbioinformatics.org/annovar/) is a functional annotation, command line-driven, open source software tool which may be used for annotating genetic variants detected from diverse organisms such as mouse, worm, fly, yeast and human genomes hg18 and hg19 using certain PERL scripts. It has been developed to identify novel SNPs in the vast amounts of data generated by next-generation sequencing technologies, shifting focus from few common SNPs to previously unknown SNPs, overcoming the limitations associated with the use of SIFT and PolyPhen2 in SNP identification and extension of annotation to organisms other than humans. In order to annotate variants, ANNOVAR needs to be provided with the chromosome number, start position, end position, reference nucleotide and observed nucleotides of the variants and based on these performs gene-based, region-based, filter-based and other types of annotations. Gene-based annotations identify those variations which may result in an altered state of protein coding as well as the amino acid affected; region-based annotations may be used to identify those SNPs which are present in specific regions of the genome and the filter-based annotations which identify variants by filtering them on the basis of their listing in dbSNP, subsets of SNPs in the 1000 Genomes Project (MAF > 1 %), non-synonymous SNPs with a SIFT score of >0.05 and other annotations on mutations. Apart from this ANNOVAR may also be used for other utilities such as the identification of SNPs within 1000 genomes that are in strong LD with a GWAS hit and other datasets. The computational pipeline of ANNOVAR has the capability to perform gene-based functional annotations in ~4 min and "variants reduction" procedure linking SNPs to diseases and traits in ~15 min, making it easier and practical to handle vast amount of data per day.

5 Community Efforts and Consortia

Consortiums are an essential part of all fields of scientific research as they not only dissipate information to increase awareness but also form collaborations between researchers from different parts of the world, aiding research and its advancement. Most consortiums also include databases on relevant results obtained by researchers associated with them or a comprehensive database on all data mined from literature. Access to these consortiums or their databases usually requires registration but no monetary remuneration; however, memberships to them do require certain fees to be paid. The following section is a review on some of the major PGx consortiums and initiatives.

5.1 Pharmacogenomics Research Network (PGRN)

PGRN (Giacomini et al. 2007; www.pgrn.org) is a NIH initiative which comprises of a consortium of researchers working on diverse fields but having the ultimate goal of establishing correlations between drug response and genetic variations. The scientific data obtained by these research groups is stored and annotated within PharmGKB. PGRN has been established with the objective of implementing PGx studies in clinical practice, research, etc.; developing new tools and experimental methodologies for implementation in PGx research; enabling sharing of data as well as biological samples; creating cross-disciplinary groups to facilitate the discovery and dissemination of new data; building alliances and partnerships; and collaborating with related disciplines to advance PGx knowledge and its applications. The research groups within the network may focus either on drugs which can be used to treat disorders or on groups of proteins which interact with drugs.

5.2 PGENI

PGENI or the Pharmacogenetics for Every Nation Initiative (http://www.pgeni.org/) works to facilitate the implementation of pharmacogenomics in the health-care systems of both developed and developing nations. The consortium aims at the collection of blood samples from major ethnic groups of every country and genotyping to identify genetic variants and their patterns in the population. Apart from this, it has compiled a list of drugs that are used the most in developing nations from the "14th Essential Medicines" list published by WHO and would incorporate the entire list of drugs in the future with the aim of identifying PGx associations. PGENI would also identify those populations that are at a high risk for toxicity or poor drug response and suggest alternative therapies for these populations. Both PubMed and PharmGKB were used for determining genes/SNPs influencing drug response and currently include 154 markers for 206 drugs. At present, PGENI is a consortium of 104 countries with five major coordinating centres, Brazil, China, Greece, Mexico and South Africa, and aims to bring about the integration of PGx in public health-care decisions to improve the overall quality of health-care services provided as well increasing the understanding and outreach of PGx in developing nations.

5.3 PharmGenEd

PharmGenEd (http://pharmacogenomics.ucsd.edu/about-us/pharmgened-objectives.aspx) is a pharmacogenomics education programme developed by the University of San Diego, California, to facilitate the implementation of pharmacogenomics in clinical settings and to provide elementary evidence-based knowledge on PGx to pharmacists and physicians, pharmacy and medical students as well as other health-care professionals. The programme collaborates with several research experts in the field of PGx, clinicians, health professionals and professional representatives to enable the dissemination of pharmacogenetic information and design educational materials. PharmGenEd states four major objectives or programme components: (1) to continue education programmes for health-care practitioners, specifically pharmacists and clinicians using live presentations, online modules and written articles; (2) to provide resources and training for faculty members in the form of modules on PGx for implementation in course curriculum, which may be disseminated to students and other faculty trainers alike; (3) to provide resources and training for health-care practitioners on PGx through online modules and live presentations by other health-care professionals well versed with the concepts and field of PGx; and (4) to establish online PGx resources in collaboration with other individuals in the field in the form of virtual communities and pubcasts. PharmGenEd has also collaborated with several other organisations and PGx communities such as PharmGKB to enable a process of continued implementation and development of the concepts of pharmacogenomics.

5.4 OpenPGx

The Open Personal Genomics Consortium (OpenPGx) (http://www.openpgx.org/) is a global, open access collaborative initiative created to facilitate the development and application of PGx in patient health care. The main targets of OpenPGx are (1) to create an open access database of all biomarkers which may be used in clinical setting; (2) to develop novel algorithms and tools to facilitate data mining and interpretation of personal genomes; (3) to provide an interactive platform for students, researchers, health-care professionals and experts from the field of PGx; and (4) to establish evidence-based guidelines for the application of pharmacogenomic data to patient health care. OpenPGx database will include data manually curated from literature on pharmacogenomic associations between drugs and genetic polymorphs. It would present a comprehensive overview of all information pertaining to the drug-gene interaction which would include the drug indication, gene name, rsID of the risk allele, statistical variables used to validate the associations (hazards ratio, odds ratio, *P* value, etc.), study details such as the population studied, allele frequencies and the PubMed IDs of the studies supporting an interaction. The database would also include a map representing all reported associations/interactions between drugs, targets, carriers, enzymes and transporters obtained from several major databases such as DrugBank, PharmGKB and UniProt and from literature. The pathway map has been developed using CellDesigner and is available in the SBML format to enable export to other visualisation softwares. OpenPgx has also established collaborations with several laboratories (at present within South Asia) to facilitate data exchange and exhaustive study on genetic variants and their influence on drug response.

5.5 The Biomarkers Consortium

The Biomarkers Consortium (http://www.biomarkersconsortium.org/) is an open collaboration between several researchers, created with the aim of identifying and qualifying novel or promising biomarkers to facilitate their application in drug therapeutics and hence improving the overall quality of human therapeutics. The main research focus of the consortium lies in four fields, cancer, immunity and inflammation, metabolic disorders and neuroscience, with several pioneering projects active in each of these. Apart from the above mentioned, the Biomarker Consortium also develops and works on projects based on other diseases, therapeutic areas and overlapping fields. Researchers can join the consortium either as scientific members (those who are associated with scientific organisations and whose area of research is related to the use of biomarkers) or supporting members (those who support the consortiums' activities but do not wish to provide any scientific inputs) on payment of annual or triennial dues. Researchers can also submit projects to the consortium which may be in any one of the main research areas of the consortium or any other associated field, and the acceptability of these depends on the fulfilment of the target requirements of the consortium. Thus, the broad goals of the consortium are (1) to develop and qualify biomarkers through the use of existing and newly developed techniques; (2) to enunciate the role of biomarkers in diagnosing diseases, improving clinical practice and therapeutic responses; (3) to make available the results obtained from all projects within the consortium to the entire scientific community; and (4) to assist in the regulatory decision making pertaining to these biomarkers.

5.6 International Serious Adverse Events Consortium (iSAEC)

The International Serious Adverse Events Consortium (http://www.saeconsortium.org/) is a collaboration between the pharmaceutical industry and the FDA, established to identify and qualify genetic variants which may be useful in predicting drug-induced severe adverse reactions. Apart from this, researchers from both academics and industries involved in the development of novel genetic and computational techniques for application in whole genome sequencing/mapping to identify and qualify the genetic variants associated with adverse events may also join the consortium. The data collected from the pharmaceutical industry and academic institutions/networks is made available through the data portal which may be accessed only after agreeing to the restrictions posed by the iSAEC and signing the data use agreement. The research initiatives within the consortium have been divided into two phases:

1. Phase 1 Research Plan: During the first phase, the consortium's main research focus lay in two fields, both involving the identification of common DNA variants associated with drug-induced liver diseases and serious skin rashes, i.e. Stevens-Johnson syndrome ("SJS") and toxic epidermal necrolysis ("TEN"). This phase essentially involved the collection of genotypic and clinical data relevant to these areas. Apart from this, the build-up of the core operational processes and other computational techniques was also completed within this phase.
2. Phase 2 Research Plan: While phase 1 focused on the identification of the common variants which may have a role in drug-induced liver diseases and serious skin rashes, phase 2 aims at identifying the rare variants which may be involved with these conditions with the help of next-generation sequencing technology.

Keeping the present and future research focuses of the consortium in mind, the main aims established by iSAEC are (1) to identify the DNA variants which may be associated with drug-induced severe adverse reactions and other similar research, (2) to make available publicly the results of the research conducted and (3) to educate both the scientific and non-scientific communities about affairs related to severe ADRs.

5.7 Consortium on Breast Cancer Pharmacogenomics (COBRA)

COBRA (http://medicine.iupui.edu/clinpharm/cobra/) has been established to study the influence of multiple genetic factors on the clinical pharmacology of tamoxifen and aromatase inhibitors, with the eventual extension of the results obtained for tamoxifen to other drugs within the same class. The aims of the consortium are (1) to identify the common genetic variants of the human oestrogen receptor, its nuclear coactivators and repressors through the use of bio-computational and sequencing techniques; (2) to analyse through in vitro assays whether the genetic variants of the two ER receptor subtypes (ERα and ERβ) alter gene expression and function; (3) to determine if the genetic variants of ERα and ERβ identified through the first two objectives along with the genotype/haplotype information alter the response to tamoxifen; (4) to ascertain the role of human, genetically polymorphic, drug-metabolising enzymes in the metabolism of aromatase inhibitors; and (5) to analyse if the variants in candidate genes identified through all the studies conducted are associated with curated phenotypic outcomes such as the oestrogen metabolite concentrations, pharmacokinetics, bone metabolism and serum lipid subfractions in breast cancer patients receiving exemestane and letrozole. The consortium webpage also provides external links to PGRN, PharmGKB and a table listing all the drug interactions with cytochrome P450 isoforms. The complete table may be visualised directly on the COBRA webpage or extracted in the PDF format. Furthermore, an additional table on the clinically relevant cytochrome P450s with their interacting drugs is also available.

6 Standards for Data Representation

These are languages which have been developed to aid the exchange of data between visualisation softwares and researchers without changing the file format for each specific software.

6.1 Genomic Sequence Variation Markup Language (GSVML)

GSVML (Nakaya et al. 2010) has been developed to enable the exchange of genomic information especially that pertaining to SNPs in an internationally standardised and defined format. It was created using eight steps following case analysis and domain evaluations and allows users to interact with each other irrespective of the Markup Language or data format used. GSVML is human health oriented with its entire focus on the exchange of clinical and omics data. It consists of three data categories, variation data, direct annotations and indirect annotations, out of which the variation data is a compulsory and the remaining two are optional fields. The variation data describes the allele, its type, position, region and length; the direct annotation data deals with the associated gene, experimental assay and other genetic information, and the indirect annotation provides pertinent omics, clinical and environmental data. While there are several Markup Languages available which can be used for data exchange, GSVML is unique as it has been created specifically for the exchange of data on human genome sequence variations. Considering the monumental increase in genomic data, especially the ever-expanding interest and data generation on the human genetic variations, GSVML aims at providing a communicative platform for this data to be suitably mined, analysed and interpreted. GSVML has been developed in collaboration with Health Level Seven Clinical Genomics Special Interest Group (HL7 CG SIG) and in accordance with the requirements of the International Organisation of Standardisation (ISO), which is essential for its clinical and omic application. Further, all three data categories are essential and need to be used in tandem for its apt use in clinical and omic settings.

6.2 SBML

The Systems Biology Markup Language (SBML) (Hucka et al. 2003; http://www.sbml.org) is an XML-based representation format used for the export and storage of computational models of biological processes. It has a standard structure which includes specific components such as compartment, species, reaction, parameters, unit definitions, rules, initial assignment and event, which enable the transport of models from one visualisation software to the other. The main purpose of SBML is to enable the use of several softwares without the need for writing and creating separate models for every tools file format, to enable export and utilisation of files by other researchers irrespective of the tool used and to extend the lifetime of models beyond that of the tool used in its creation/writing. Due to the restrictions posed by certain tools in the characters allowed in symbolic names, SBML restricts the characters' types such that names may only start with an underscore ("_") followed by letters, digits, etc. or with a letter, should be at least two characters long, may be upper case or lower case and must only be limited to the plain ASCII characters in order to ensure compatibility with existing softwares. Apart from this, each entity has annotations attached with it that link entities within a model or to those in other databases as well. The current available versions of SBML are Level 3 Version 1 Core, Level 2 Version 4 Release 1 and Level 1 Version 2.

7 Conclusion

Pharmacogenomics holds the keys to the next generation in therapeutics and patient health care. The prospect of individualising therapy attracts not only researchers but clinicians and patients as well. This is evident in the increasing number of individuals who have made available their whole genome through open sources and consortiums. The fast growth has further been abetted by the outcrop of databases and online resources created to facilitate pharmacogenomics research (Synopsis of online resources in Table 10.1). Some of these work as consortiums which annotate and curate PGx data and make them available through open databases as in the case of PharmGKB, OpenPGx, dbGaP, etc. Others such as PGENI, PGRN and iSEAC bring together researchers in the field to facilitate PGx research and its implementation, creating awareness of its implications in health care. These consortiums/projects have become quintessential in the exploration and implementation of PGx. Collaborations established through them help in the build-up of

Table 10.1 Online resources for pharmacogenomics

Resource	Type	Salient features	URL
Pharmacogenomics Knowledge Base (PharmGKB)	Database	It is an open access knowledge base containing genotype-phenotype information relevant to pharmacogenetics. The database distinguishes itself as one of the few dedicated resources carrying specifically pharmacogenetic information	http://www.pharmgkb.org/
DrugBank	Database	This online resource is a detailed repository of drug data pertaining to their chemical, pharmacological and pharmaceutical properties linked to comprehensive drug target information and other proteins involved in their ADME	http://www.drugbank.ca/
Database of Genotypes and Phenotypes (dbGaP)	Database	It is an NCBI initiative which archives the results on genotype-phenotype association studies published to date and contains a database listing this data	http://www.ncbi.nlm.nih.gov/gap
Comparative Toxicogenomics Database (CTD)	Database	CTD is a repository containing information on the influence of environmental agents on human health not restricted to drugs only along with datasets of genes, diseases and chemicals	http://ctdbase.org/
HIV-Pharmacogenomics.org	Database	A database containing pharmacogenetic information on genes influencing the response of drugs used in the therapy of HIV	http://www.hiv-pharmacogenomics.org/
International Consortium on Lithium Genetics (ConLiGen)	Database and consortium	ConLiGen has been established with the aim of performing GWAS in order to elucidate the pharmacogenomic associations of lithium used in the therapy of bipolar disorder	http://www.conligen.org/
Pharmacogenomics and Cell Database (PACdb)	Database	A database containing results of the association studies between SNPs, gene expression and cellular sensitivity to drugs evaluated on the lymphoblastoid cell line derived from different target populations	http://www.pacdb.org/
International HapMap Project	Database and consortium	An international collaboration to identify and catalogue genetic variants in individuals which may influence drug response, disease propensity and human health care	http://hapmap.ncbi.nlm.nih.gov/
1000 Genomes Project	Database and consortium	1000 Genomes is an multinational endeavour to sequence the whole genomes of DNA samples from 2,500 individuals with the aim of identifying genetic variants having population frequencies of at least 1 %	http://www.1000genomes.org/
Population Reference Sample (POPRES)	Database and consortium	It is a DNA resource created by analysing samples collected from ~6,000 individuals to identify potential genetic markers that may be used in a variety of applications including PGx	http://www.ncbi.nlm.nih.gov/projects/gap/cgi-bin/study.cgi?study_id=phs000145.v1.p1
Pan Asian SNP Database (PASNP)	Database and consortium	It is a repository of more than 50,000 SNPs and CNVs identified from 73 populations of Asian descent in an attempt to elucidate pharmacogenetically relevant biomarkers for the Asian population	http://www4a.biotec.or.th/PASNP

(continued)

Table 10.1 (continued)

Resource	Type	Salient features	URL
Indian Genome Variation Database (IGVdb)	Database	It is a database of SNPs, CNVs and repeats in more than 1,000 genes identified in the genomes of individuals from four major linguistic groups of India with potential application in haplotype analysis, PGx, etc.	http://www.igvdb.res.in/
EUDRAGENE	Database and consortium	It is an European Union (EU) initiative to identify genes associated with ADRs using DNA samples collected from individuals from several European nations to elucidate PGx associations	https://www.eudragene.org/
Koreangenome.org/the Korean Genome Project (KPGP)	Database	KPGP is a collection of personal genomes from individuals of Korean descent	http://kpgp.kr/index.php/Main_Page
Genome Database of Latvian Population (LGDB)	Database	It is a repository of the genetic research done in Latvia with the aim of its eventual application in PGx and will also include GWAS data of 95 individuals of Latvian descent in the future	http://www.p3gobservatory.org/catalogue.htm;jsessionid=ACE6E593F10B80573D64E965FA2DB3D8?measureId=18
Sri Lankan Genome Variation Database	Database	SLGVD is a database of the SNPs identified in the genomes of three major ethnic communities of Sri Lanka	http://www.hgucolombo.net/slgv/rsOrder.php
1Malaysia Human Genome Variation Consortium (1MHGVC)	Database and consortium	It is a resource containing the genetic variants identified in the ethnically diverse populations of Malaysia established to study the migratory patterns, PGx applications, etc. of these variants with respect to the Korean population	http://1mhgvc.kk.usm.my/
Singapore Genome Variation Project (SGVP)	Database	SGVP is an initiative aimed at identifying the common variants across 1 million SNPs in the genomes of three major ethnically diverse populations of Singapore and evaluating their PGx roles	http://www.statgen.nus.edu.sg/~SGVP/
Personal Genome Project (PGP)	Database	A follow-up to the Human Genome Project, it has been established to collect DNA samples in order to facilitate the understanding of the role of genetic and environmental factors in human phenotypes as well as their utility in health care and disease	http://www.personalgenomes.org/
Search Tool for Interacting Chemicals (STITCH)	Tool and database	It is an online resource developed to identify interactions between two chemicals or proteins and chemicals using published data and other databases. At present, it includes interactions between 300,000 chemicals and 2.6 million proteins in 1,133 organisms that may be analysed to decipher novel PGx associations	http://stitch.embl.de/
Reactome	Tool and database	Reactome is a pathway database containing visual representations of the metabolic, signalling and other reactions associated with chemicals, curated from literature	http://www.reactome.org/ReactomeGWT/entrypoint.html
Kyoto Encyclopedia of Genes and Genomes (KEGG)	Tool and database	It is a graphical database which has been created to enable the systematic analysis of gene function, integrating data on genes, proteins and chemicals to produce molecular diagrams of interaction	http://www.genome.jp/kegg/

Interpretome	Tool	It is a free, online, interpretation tool which may be used by individuals to interpret their own personal genomes including identifying SNPs which may be associated with diabetes, warfarin dosing and other biomarkers	http://esquilax.stanford.edu/
SNPedia and Promethease	Tool and database	SNPedia is a online, Wiki resource containing genes and their variants along with their medical, genealogical, phenotypic and forensic associations. It also includes the tool Promethease which may be used for personal genome interpretation based on the data within SNPedia	http://www.snpedia.com/index.php/SNPedia
Trait-O-Matic	Tool	It is an online tool which may be used for identifying genetic variants in personal genomes which may be done by submitting whole genomes to the web-tool or by downloading the tool for self-use	http://snp.med.harvard.edu/
ANNOVAR	Tool	It is a functional annotation, command line-driven, open source software tool that may be used to annotate genetic variants in genomes of diverse organisms	http://www.openbioinformatics.org/annovar/
Pharmacogenomics Research Network (PGRN)	Consortium	PGRN is a consortium of several research groups working on different aspects of pharmacogenomics, with the study results published through PharmGKB	http://www.pgrn.org/display/pgrnwebsite/PGRN+Home
Pharmacogenetics for Every Nation Initiative (PGENI)	Consortium	PGENI is a multinational initiative to facilitate the implementation of pharmacogenomics in the health-care systems of developed and developing nations. It also carries out research by collecting blood samples from different ethnic communities to identify genetic biomarkers	http://www.pgeni.org/
PharmGenEd	Consortium	It is an education programme developed for health-care practitioners and patients to make them aware of the concepts and impact of PGx. It has also established collaborations with other datasets and resources to facilitate PGx research	http://pharmacogenomics.ucsd.edu/about-us/pharmgened-objectives.aspx
Open Personal Genomics Consortium (OpenPGx)	Consortium	OpenPGx is an open access, collaborative initiative to facilitate PGx research and includes researchers from most South Asian countries. It also includes a database of genetic variants that have been found to be associated with drug phenotypes manually curated from literature	http://www.openpgx.org/
The Biomarkers Consortium	Consortium	It is an open collaboration between several researchers to identify novel biomarkers influencing drug response in four major disease categories: cancer, immunity and inflammation, metabolic disorders and neuroscience	http://www.biomarkersconsortium.org/
International Serious Adverse Events Consortium (iSAEC)	Consortium	iSAEC is a collaborative initiative between the FDA and pharmaceutical industries to identify novel genetic variants which may influence drug response. The data generated may be accessed via the data portal of the iSAEC	http://www.saeconsortium.org/

(continued)

Table 10.1 (continued)

Resource	Type	Salient features	URL
Consortium on Breast Cancer Pharmacogenomics (COBRA)	Consortium	It is a consortium cum database of the genetic variants influencing tamoxifen and aromatase inhibitors drug responses. COBRA is also involved in research to identify novel variants which may influence response to drugs used in the therapy of breast cancer	http://medicine.iupui.edu/clinpharm/cobra/
Genomic Sequence Variation Markup Language (GSVML)	Language software	GSVML is a representation format that has allows for exchange of genetic information especially that pertaining to SNPs	
Systems Biology Markup Language	Language software	SBML is the standard representation format that has been created to facilitate the exchange of computational models of biological processes between different visualisation softwares without the need for writing or creating separate models for each tool	http://sbml.org/Main_Page

research, overcoming the limitations associated with small sample sizes, analysing allele frequencies in different populations and identifying genetic markers relevant for one population or in common. Further, databases such as DrugBank, ConLiGen and CTD act as ancillary resources and help disseminate data that aids in PGx research as well as serve as a source for future, novel studies. At present, there are more than 30 databases and consortiums working on and containing pharmacogenomics data, and many more will continue to come up in the future. It is essential to mine these resources to the full extent and make available freely all data for scientific analysis and implementation in patient health care so as to make real the concept of "personalised medicine."

Apart from databases and consortiums, a large number of tools and languages have been generated which help facilitate the analysis, interpretation and visualisation of data generated. Some such as KEGG, Reactome and STITCH mainly deal with the visual representation of PGx data as pathways, while others such as Interpretome, Trait-O-Matic and ANNOVAR may be used for personal genome analysis to identify novel variations (SNPs or CNVs) or common, known variants which may be pharmacogenetically relevant. Finally, languages such as SBML and GSVML were also developed to allow for the interchange and exchange of data between researchers.

The upsurge of databases, consortiums and tools has immensely expanded the horizons of PGx research and paved the way for its widespread clinical application. As more and more health-care personnel and patients become acquainted with the application and prospects of PGx, the focus has shifted from single genes to whole genomes. While, in some cases, the whole genome sequencing has been undertaken by large-scale endeavours such as the 1000 Genomes Project, in other situations, consensual genome sequencing of volunteering individuals has been done and applied in PGx evaluations. Another crucial development has been the passing of the Genetic Information Nondiscrimination Act (GINA) by the US Congress in 2008, which mandates that there should be no discrimination between individuals on the basis of their genetic make-up and propensity for diseases. Pharmacogenomics has taken vast leaps from the researchers table to the clinical environ, but still leaves a lot to be accomplished. Research in pharmacogenomics may be considered unidirectional with the minutiae of most common variants studied but leaving the larger part of the genome unexplored. An additional glaring fact at present is that of all the publications made in the field, approximately half are reviews (Pirmohamed 2011), making it evident that there is a swift and essential need to shift focus from reviews to pioneering and consequential research.

With several new drugs being approved by the FDA, especially those which have been approved in view of their beneficial role in spite of critical adverse effects even in clinical trials, PGx needs to evolve at a much faster rate so as to ensure that patients reap the best benefits of drug therapeutics and eliminate the chances and severity of ADRs. While consortiums are working to facilitate PGx research, there is the need to increase the role of health-care practitioners in these as well, so as to ensure that the benefits of research conducted actually reach the patient at the grass-root level.

References

1000 Genomes Project Consortium (2010) A map of human genome variation from population-scale sequencing. Nature 467(7319):1061–1073

1Malaysia Human Genome Variation Consortium. http://1mhgvc.kk.usm.my/

ANNOVAR. http://www.openbioinformatics.org/annovar/.

Cano FG, Rozenfeld S (2009) Adverse drug events in hospitals: a systematic review. Cad Saude Publica 25:S360–S372

Cariaso M, Lennon G (2012) SNPedia: a wiki supporting personal genome annotation, interpretation and analysis. Nucleic Acids Res 40(Database issue):D1308–D1312

Church GM (2005) The personal genome project. Mol Syst Biol 1:2005.0030

Consortium on Breast Cancer Pharmacogenomics (COBRA). http://medicine.iupui.edu/clinpharm/cobra/

Davis AP, Murphy CG, Saraceni-Richards CA, Rosenstein MC, Wiegers TC, Mattingly CJ (2009) Comparative toxicogenomics database: a knowledgebase and discovery tool for chemical-gene-disease networks. Nucleic Acids Res 37(Database issue):D786–D792

Davis AP, King BL, Mockus S, Murphy CG, Saraceni-Richards C, Rosenstein M, Wiegers T, Mattingly CJ

(2011a) The comparative toxicogenomics database: update 2011. Nucleic Acids Res 39(Database issue):D1067–D1072

Davis AP, Wiegers TC, Murphy CG, Mattingly CJ (2011b) The curation paradigm and application tool used for manual curation of the scientific literature at the Comparative Toxicogenomics Database. Database (Oxford). 2011:bar034.

de Vries EN, Ramrattan MA, Smorenburg SM, Gouma DJ, Boermeester MA (2008) The incidence and nature of in-hospital adverse events: a systematic review. Qual Saf Health Care 17(3):216–223

Deloukas P, Bentley D (2004) The HapMap project and its application to genetic studies of drug response. Pharmacogenomics J 4(2):88–90

Deverka PA, Vernon J, McLeod HL (2010) Economic opportunities and challenges for pharmacogenomics. Annu Rev Pharmacol Toxicol 50(1):423–437

EUDRAGENE. https://www.eudragene.org/index.php.

Flynn AA (2011) Pharmacogenetics: practices and opportunities for study design and data analysis. Drug Discov Today 16(19–20):862–866

Gamazon ER, Duan S, Zhang W, Huang RS, Kistner EO, Dolan ME, Cox NJ (2010) PACdb: a database for cell-based pharmacogenomics. Pharmacogenet Genomics 20(4):269–273

Genome Database of Latvian Population [LGDB]. http://www.p3gobservatory.org

Giacomini KM, Brett CM, Altman RB, Benowitz NL, Dolan ME, Flockhart DA, Johnson JA et al (2007) The pharmacogenetics research network: from SNP discovery to clinical drug response. Clin Pharmacol Ther 81(3):328–345

Hernandez-Boussard T, Whirl-Carrillo M, Hebert JM, Gong L, Owen R, Gong M, Gor W, Liu F, Truong C, Whaley R, Woon M, Zhou T, Altman RB, Klein TE (2008) The pharmacogenetics and pharmacogenomics knowledge base: accentuating the knowledge. Nucleic Acids Res 36(Database issue):D913–D918

HIV-Pharmacogenomics.org. http://www.hiv-pharmacogenomics.org/

Hucka M, Finney A, Sauro HM, Bolouri H, Doyle JC, Kitano H, Arkin AP et al (2003) The systems biology markup language (SBML): a medium for representation and exchange of biochemical network models. Bioinformatics 19(4):524–531

Indian Genome Variation Consortium (2005) The Indian Genome Variation database (IGVdb): a project overview. Hum Genet 118(1):1–11

Indian Genome Variation Database (IGVdb). http://www.igvdb.res.in/index.php

International Consortium on Lithium Genetics (ConLiGen). http://www.conligen.org/imprint.html

International HapMap Project. http://hapmap.ncbi.nlm.nih.gov/index.html.en

International Serious Adverse Effects Consortium (iSAEC). http://www.saeconsortium.org/

Interpretome. www.interpretome.com

Kanehisa M, Goto S (2000) KEGG: Kyoto encyclopedia of genes and genomes. Nucleic Acids Res 28(1):27–30

Kanehisa M, Goto S, Sato Y, Furumichi M, Tanabe M (2012) KEGG for integration and interpretation of large-scale molecular data sets. Nucleic Acids Res 40(Database issue):D109–D114

Karczewski KJ, Tirrell RP, Cordero P, Tatonetti NP, Dudley JT, Salari K, Snyder M, Altman RB, Kim SK (2012) Interpretome: a freely available, modular, and secure personal genome interpretation engine. Pac Symp Biocomput 2012:339–350

Kawamoto K, Orlando LA, Voora D, Lobach DF, Joy S, Cho A, Ginsburg GS (2009) Evaluation of the PharmGKB knowledge base as a resource for efficiently assessing the clinical validity and utility of pharmacogenetic assays. AMIA Annu Symp Proc 2009:307–311

Kitzmiller JP, Groen DK, Phelps MA, Sadee W (2011) Pharmacogenomic testing: relevance in medical practice: why drugs work in some patients but not in others. Cleve Clin J Med 78(4):243–257

Klein TE, Altman RB (2004) PharmGKB: the pharmacogenetics and pharmacogenomics knowledge base. Pharmacogenomics J 4(1):1

Knox C, Law V, Jewison T, Liu P, Ly S, Frolkis A, Pon A, Banco K, Mak C, Neveu V, Djoumbou Y, Eisner R, Guo AC, Wishart DS (2011) DrugBank 3.0: a comprehensive resource for 'omics' research on drugs. Nucleic Acids Res 39(Database issue):D1035–D1041

Korean Genome Project (KPGP). http://koreangenome.org/index.php/Main_Page

Kuhn M, Szklarczyk D, Franceschini A, Campillos M, von Mering C, Jensen LJ, Beyer A, Bork P (2010) STITCH 2: an interaction network database for small molecules and proteins. Nucleic Acids Res 38(Database issue):D552–D556

Kuhn M, Szklarczyk D, Franceschini A, von Mering C, Jensen LJ, Bork P (2012) STITCH 3: zooming in on protein-chemical interactions. Nucleic Acids Res 40(Database issue):D876–D880

Kyoto Encyclopedia of Genes and Genomes (KEGG). http://www.kegg.jp/kegg/

Lee KC, Ma JD, Kuo GM (2009) Pharmacogenomics: bridging the gap between science and practice. J Am Pharm Assoc (2003). 50(1):e1–e14; quiz e15-7

Mailman MD, Feolo M, Jin Y, Kimura M, Tryka K, Bagoutdinov R, Hao L, Kiang A, Paschall J, Phan L, Popova N, Pretel S, Ziyabari L, Lee M, Shao Y, Wang ZY, Sirotkin K, Ward M, Kholodov M, Zbicz K, Beck J, Kimelman M, Shevelev S, Preuss D, Yaschenko E, Graeff A, Ostell J, Sherry ST (2007) The NCBI dbGaP database of genotypes and phenotypes. Nat Genet 39(10):1181–1186

Marsh S, McLeod HL (2006) Pharmacogenomics: from bedside to clinical practice. Hum Mol Genet. 15 Spec No. 1:R89–R93

Matthews L, Gopinath G, Gillespie M, Caudy M, Croft D, de Bono B, Garapati P, Hemish J, Hermjakob H, Jassal B, Kanapin A, Lewis S, Mahajan S, May B, Schmidt E, Vastrik I, Wu G, Birney E, Stein L, D'Eustachio P (2009) Reactome knowledgebase of human biological pathways and processes. Nucleic Acids Res D619(Database issue)

Mattingly CJ, Colby GT, Rosenstein MC, Forrest JN, Boyer JL (2004) Promoting comparative molecular studies in environmental health research: an overview of the comparative toxicogenomics database (CTD). Pharmacogenomics J 4(1):5–8

McLeod HL, Isaacs KL (2011) Preemptive pharmacogenetic testing: insufficient data equal unsatisfactory guidance. Ann Intern Med 154(12):842–844

Molokhia M, McKeigue P (2006) EUDRAGENE: European collaboration to establish a case-control DNA collection for studying the genetic basis of adverse drug reactions. Pharmacogenomics 7(4):633–638

Nakaya J, Kimura M, Hiroi K, Ido K, Yang W, Tanaka H (2010) Genomic sequence variation markup language (GSVML). Int J Med Inform 79(2):130–142

Nelson MR, Bryc K, King KS, Indap A, Boyko AR, Novembre J, Briley LP et al (2008) The population reference sample, POPRES: a resource for population, disease, and pharmacological genetics research. Am J Hum Genet 83(3):347–358

Ngamphiw C, Assawamakin A, Xu S, Shaw PJ, Yang JO, Ghang H, Bhak J, Liu E, Tongsima S, Pan-Asian HUGO, Consortium SNP (2011) PanSNPdb: the Pan-Asian SNP genotyping database. PLoS One 6(6):e21451

O'Shaughnessy KM (2006) HapMap, pharmacogenomics, and the goal of personalized prescribing. Br J Clin Pharmacol 61(6):783–786

Ogata H, Goto S, Sato K, Fujibuchi W, Bono H, Kanehisa M (1999) KEGG: Kyoto encyclopedia of genes and genomes. Nucleic Acids Res 27(1):29–34

Open Personal Genomics Consortium (OpenPGx). http://www.openpgx.org/

Owen RP, Altman RB, Klein TE (2008) PharmGKB and the International Warfarin Pharmacogenetics Consortium: the changing role for pharmacogenomic databases and single-drug pharmacogenetics. Hum Mutat 29(4):456–460

Pan-Asian SNP Database. http://www4a.biotec.or.th/PASNP

Personal Genome Project. http://www.personalgenomes.org/

Pharmacogenomics and Cell Database. http://www.pacdb.org/index.html

Pharmacogenomics Education Program (PharmGenEd). http://pharmacogenomics.ucsd.edu/about-us/pharmgened-objectives.aspx

Pharmacogenomics for Every Nation Initiative (PGENi). http://www.pgeni.org/

Pharmacogenomics Research Network (PGRN). www.pgrn.org

Pirmohamed M (2011) Pharmacogenetics: past, present and future. Drug Discov Today 16(19–20):852–861

Reactome. http://www.reactome.org/ReactomeGWT/entrypoint.html

Samarakoon PS, Jayasekara RW, Dissanayake VHW (2011) The Sri Lankan genome variation database. Sri Lanka J Biomed Inform 2(1):9–20

Schulze TG, Alda M, Adli M, Akula N, Ardau R, Bui E, Chillotti C et al (2010) The International Consortium on Lithium Genetics (ConLiGen): an initiative by the NIMH and IGSLI to study the genetic basis of response to lithium treatment. Neuropsychobiology 62(1):72–78

Search Tool for Interacting Chemicals (STITCH). http://stitch.embl.de/

SNPedia and Promethease. http://www.snpedia.com

Sri Lankan Genome Variation Database (SLGVD). http://hgucolombo.org/SLGVD.aspx

Systems Biology Markup Language (SBML). http://www.sbml.org

Teo YY, Sim X, Ong RT, Tan AK, Chen J, Tantoso E, Small KS, Ku CS, Lee EJ, Seielstad M, Chia KS (2009) Singapore genome variation project: a haplotype map of three Southeast Asian populations. Genome Res 19(11):2154–2162

The Biomarkers Consortium. http://www.biomarkersconsortium.org/

Thorn CF, Klein TE, Altman RB (2005) PharmGKB: the pharmacogenetics and pharmacogenomics knowledge base. Methods Mol Biol 311:179–191

Trait-O-Matic. http://snp.med.harvard.edu/

Vastrik I, D'Eustachio P, Schmidt E, Joshi-Tope G, Gopinath G, Croft D, de Bono B, Gillespie M, Jassal B, Lewis S, Matthews L, Wu G, Birney E, Stein L (2007) Reactome: a knowledge base of biologic pathways and processes. Genome Biol 8(3):R39

Via M, Gignoux C, Burchard EG (2010) The 1000 Genomes Project: new opportunities for research and social challenges. Genome Med 2(1):3

Wang K, Li M, Hakonarson H (2010) ANNOVAR: functional annotation of genetic variants from high-throughput sequencing data. Nucleic Acids Res 38(16):e164

Wise J (2008) Consortium hopes to sequence genome of 1000 volunteers. BMJ 336(7638):237

Wishart DS, Knox C, Guo AC, Shrivastava S, Hassanali M, Stothard P, Chang Z, Woolsey J (2006) DrugBank: a comprehensive resource for in silico drug discovery and exploration. Nucleic Acids Res 34(Database issue):D668–D672

Wishart DS, Knox C, Guo AC, Cheng D, Shrivastava S, Tzur D, Gautam B, Hassanali M (2008) DrugBank: a knowledgebase for drugs, drug actions and drug targets. Nucleic Acids Res 36(Database issue):D901–D906

Yang YT, Wiley E, Leppard J (2011) Individualized medicine and pharmacogenomics: ethical, legal and policy challenges. J Med Pers 9:48–57

Zhang W, Ratain MJ, Dolan ME (2008) The HapMap resource is providing new insights into ourselves and its application to pharmacogenomics. Bioinform Biol Insight 2:15–23

Section II

Personalized Medicine: Cancer and Autoimmune Disorders

Pharmacogenomics and Personalized Medicine for Cancer

11

Dipali Dhawan and Harish Padh

Abstract

Cancer pharmacogenomics is a growing field with new avenues being explored and applied to the routine practice when validated. There is an enormous amount of data present in literature regarding association studies that have been undertaken around the world with different cancer patient populations. However, some association studies have contradictory results or no association with the studied molecular markers. Many of the genetic tests developed have been translated to the clinic and are being recommended by the US Food and Drug Administration (FDA) as well for a better outcome of the patients when prescribed these medicines. In this chapter we provide an overview of the well-studied molecular markers for efficacy and toxicity of cancer therapeutics and also discuss the evolution of the field of companion diagnostics in cancer.

1 Introduction

The therapeutic window for anticancer agents is often narrow, necessitating that the drug concentration required to produce a therapeutic effect be near the concentration that leads to significant toxicities. Given the toxic effects of these drugs, the potential therapeutic regimen is hence limited to a certain cumulative dose, and the patients are treated at maximum dose levels that can be tolerated by them. Inherited genetic variations in drug-metabolism genes lead to differential treatment outcomes and severe toxicities in different patient subtypes; hence, pharmacogenomics enables in selecting the right dose for the right patient and also facilitates in titrating the dose for each patient. Pharmacogenetics studies of anticancer agents are potentially difficult by somatic polymorphisms in the cancer mass, even though this is not likely to impact toxicity.

Researchers around the world have studied the effect of several genetic variations including single nucleotide polymorphisms (SNPs), copy number variations (CNVs), deletions and insertions on

D. Dhawan (✉)
Department of Cellular and Molecular Biology, B.V. Patel Pharmaceutical Education and Research Development Centre, Ahmedabad 380054, Gujarat, India

Institute of Life Sciences, Ahmedabad University, Ahmedabad 380054, Gujarat, India
e-mail: dipali.dhawan@gmail.com

H. Padh
Department of Cellular and Molecular Biology, B.V. Patel Pharmaceutical Education and Research Development (PERD) Centre, Thaltej-Gandhinagar Highway, Thaltej, Ahmedabad 380054, Gujarat, India

D. Barh et al. (eds.), *Omics for Personalized Medicine*,
DOI 10.1007/978-81-322-1184-6_11, © Springer India 2013

the metabolism of drugs prescribed to cancer patients of different ethnicity. It has been observed that frequency and effect of polymorphisms vary with the group of population studied. Hence it is beneficial to study each population and understand the effect of the polymorphisms in that particular population before concluding any association study. Here we review some of the majorly studied polymorphic drug-metabolizing enzymes, drug targets, and drug transporters that affect the metabolism of cancer therapeutics widely prescribed to different cancer patients. The manuscript also gives an overview of companion diagnostics approved by the Food and Drug Administration (FDA), USA.

2 Genetic Variations Affecting Drug Response and Toxicity with Cancer Chemotherapy

2.1 Polymorphisms in Drug-Metabolizing Enzymes

There are a number of Phase I and Phase II drug-metabolizing enzymes that play an important role in detoxification of the cancer therapeutics (Dhawan and Padh 2013). This section gives an overview of the current status of the extensively studied polymorphic drug-metabolizing enzymes like thiopurine S-methyltransferase, UDP-glucuronosyltransferase 1A1, dihydropyrimidine dehydrogenase, and cytochrome P450 2D6.

2.1.1 Thiopurine S-Methyltransferase (TPMT) and 6-Mercaptopurine

Thiopurine S-methyltransferase (TPMT) catalyzes the S-methylation of 6-mercaptopurine (6-MP), its oral prodrug azathioprine, and 6-thioguanine (6-TG) into inactive metabolites (Lennard 1992). These drugs are prescribed for chronic inflammatory disorders, transplantation, and hematological diseases like childhood and adult ALL, childhood acute myeloid leukemia, and childhood non-Hodgkin lymphoma (Sahasranaman et al. 2008). Thiopurine being an inactive drug gets converted to its active metabolite 6-thioguanine nucleotide (6-TGN). This activated metabolite gets incorporated into DNA and causes cell-cycle arrest and apoptosis (Lennard 1992; Swann et al. 1996). Myelosuppression is one of the major serious clinical repercussions of thiopurines. It has been reported that in general, about 90 % of the population are extensive TPMT metabolizers, 10 % are intermediate metabolizers, and 0.3 % are poor metabolizers (Krynetski et al. 1996; McLeod et al. 1994, 2000). Patients with a low TPMT activity are prone to cellular accumulation of 6-TGN and develop severe toxicity even at standard doses of thiopurine drugs (Krynetski et al. 1996; McLeod et al. 1994, 2000; Krynetski and Evans 1999).

The lower TPMT activity can be majorly attributed to the polymorphisms reported in *TPMT* gene. According to literature, *TPMT*2* (238G>C; Ala80Pro), *TPMT*3A* (460G>A; Ala154Thr and 719G>A; Tyr240Cys), and *TPMT*3C* (719G>A; Tyr240Cys) account for about 80–95 % of lower TPMT activity (Otterness et al. 1997; Yates et al. 1997; Spire-Vayron de la Moureyre et al. 1998; Tai et al. 1996, 1997; Gardiner et al. 2000). It has been observed that patients heterozygous or homozygous for *TPMT*2*, *TPMT*3A*, or *TPMT*3C* result in intermediate or poor TPMT activity. There is convincing evidence for severe toxicity from thiopurine drugs observed in heterozygous and homozygous patients polymorphic for *TPMT*2*, *TPMT*3A*, or *TPMT*3C* as compared to wild-type patients. Based on evidence from literature, doses need to be titrated for intermediate and poor metabolizers where doses might be reduced to 60 and 10 % of the conventional dose, respectively, to avoid severe toxicity in these patients (Kirchheiner et al. 2005). *TPMT* genotyping assays have been designed to be used on a routine basis in the clinic to avoid toxicity, save time, and be cost-effective for the patient (Nasedkina et al. 2006; Schaeffeler et al. 2008; Litos et al. 2007; Davison et al. 2006). *TPMT* is the most well-studied example of the effect of genotype on drug metabolism in patients and has helped in dose reduction in patients polymorphic for this gene.

2.1.2 UDP-Glucuronosyltransferase 1A1 (UGT1A1) and Irinotecan

More than 60 different polymorphisms in *UGT1A1* have been identified, *UGT1A1*28* being the most clinically relevant polymorphism according to different studies (Nagar and Remmel 2006). This variant allele is a microsatellite polymorphism which consists of an additional TA (thymine-adenine) tandem repeat in the regulatory TATA box located in the promoter region of *UGT1A1* (Bosma et al. 1995). The addition makes it seven TA repeats ($A(TA)_7TAA$), instead of six ($A(TA)_6TAA$), in the wild-type sequence (Bosma et al. 1995; Monaghan et al. 1996). Some less frequent variants also occur with five and eight TA repeats. It has been reported that the glucuronidation activity of UGT1A1 is associated inversely with the number of TA repeats (Bosma et al. 1995; Beutler et al. 1998; Iyer et al. 1999; Raijmakers et al. 2000).

UGT1A1 is mainly involved in the metabolism of irinotecan which is a derivative of camptothecin. Irinotecan is activated to SN-38 through hydrolysis mediated by carboxyesterases. SN-38 is the activated form which is converted by UGT1A1 to SN-38-glucuronide (SN-38G), which is the inactivated form (Ciotti et al. 1999; Garcia-Carbonero and Supko 2002). In patients prescribed with irinotecan with *UGT1A1*28* polymorphism, the glucuronidation activity is reduced; hence, the ratio of SN-38/SN-38G is greater, leading to severe toxicity in these patients. Many studies around the world have confirmed this effect of *UGT1A1*28* causing higher toxicity in patients prescribed with irinotecan (Marcuello et al. 2004a, b; Toffoli et al. 2006; Innocenti et al. 2004; Kweekel et al. 2008; Côté et al. 2007; Pillot et al. 2006; Hoskins et al. 2007).

It has been observed that patients homozygous for *UGT1A1*28* show severe hematological toxicities like neutropenia and nonhematological toxicities like diarrhea after irinotecan therapy (McLeod et al. 2010; Iyer et al. 2002; Marcuello et al. 2004a, b; Toffoli et al. 2006; Ando et al. 2000; de Jong et al. 2006; Innocenti et al. 2004; Kweekel et al. 2008; Ferraldeschi et al. 2009; Rouits et al. 2004; Glimelius et al. 2011). Some studies however show contradictory results with no association of *UGT1A1*28* with toxicity observed (Stewart et al. 2007; Côté et al. 2007; Han et al. 2006; Pillot et al. 2006) which might be due to the small sample size or low frequency of the polymorphism. Hence it would be imperative to study this polymorphism in each population with a good sample size to elucidate the actual scenario.

The dose administered to the patients is also an important factor in determining the toxicity as observed in a meta-analysis including 821 patients on irinotecan treatment regime. It was concluded in this study that homozygosity for *UGT1A1*28* was a risk factor for hematologic grade 3–4 toxicity, with a high dose (>300 mg/m^2) as well as a medium dose (~180 mg/m^2); however, with a low dose (~100 mg/m^2), the risk for severe toxicity was similar to that of wild-type patients (Hoskins et al. 2007). Nevertheless, more studies of this kind are warranted for a better understanding of the effect of genotype and dose on toxicity in each population. Hence, if the genotype of the patient is known in advance, it would be of assistance in titrating the dose for that patient, to avoid toxicity and have a beneficial effect of the therapy.

FDA has approved the identification of the *UGT1A1*28* genotype by a rapid testing kit (Invader® UGT1A1). It is well reported that the other *UGT1A1* alleles can also result in a reduced enzyme activity, however; these variants have not been included on the irinotecan package insert or the FDA-approved assay.

Some other less studied polymorphisms in *UGT1A1* include -3156G>A (*UGT1A1*93*) which is also present in the promoter region (Innocenti et al. 2004) and 211G>A [Gly71Arg] (*UGT1A1*6*) which is located in exon 1 and leads to lower enzymatic activity (Sai et al. 2004).

2.1.3 Dihydropyrimidine Dehydrogenase (DPD) and 5-FU

DPD is the main enzyme in 5-FU metabolism and is responsible for inactivation and elimination of 5-FU in the liver (>80 %). Orotate phosphoribosyltransferase (OPRT) catalyzes the conversion of 5-FU to fluorouridine monophosphate (FUMP), which is subsequently phosphorylated to activate fluorouridine triphosphate (FUTP). FUTP incorporates into RNA and thereby compromises RNA processing and function. The *DPYD* gene is located in human chromosome 1p22, comprising of 23 exons and is approximately 950 kilobase pairs (Wei et al. 1998). More than 30 SNPs and deletion mutations have been identified within *DPYD*; however, only few of the variants have functional effect on enzymatic activity. The most important variant IVS14+1G>A (*DPYD*2A*, rs3918290) is a splice site variant, and has been observed in about 40–50 % of patients who developed grade 4 neutropenia, and shows association with DPD deficiency (Van Kuilenburg et al. 2002). Different ethnic groups have shown variation in DPD enzymatic activity, with higher mean DPD activity observed in Koreans while lower activity reported in African Americans compared to Whites (Yen and McLeod 2007). Morel and colleagues have reported that the sensitivity, specificity, and positive and negative predictive values of the detection of the three major SNPs (IVS14+1G>A, 2846A>T and 1679T>G) in *DPYD* as factors predicting 5-FU toxicity were 0.31, 0.98, and 0.62 and 0.94, respectively (Morel et al. 2006). They also observed that only 50–60 % of patients with genetic variations in *DPYD* develop severe 5-FU toxicity (Morel et al. 2006). Many researchers around the globe have developed several clinical assays for assessing DPD activity, mRNA expression, and metabolite formation, as well as SNPs within *DPYD* (Yen and McLeod 2007).

2.1.4 Cytochrome P450 2D6 (CYP2D6) and Tamoxifen

Several *CYP2D6* allelic variants have been reported and are classified by enzymatic activity into poor, intermediate, extensive, and ultrarapid metabolizers. There is a lot of ethnic variation in the frequency of the variants for *CYP2D6* (Ingelman-Sundberg et al. 2007). The most important *CYP2D6* alleles are wild-type *CYP2D6*1* (fully functional enzyme); *CYP2D6*3–*8*, **11–*16*, **18–*20*, **38*, **40*, **42*, and **44* (null alleles with essentially no enzymatic activity); *CYP2D6*9*, **10*, **17*, **29*, **36*, **37*, and **41* (reduced-activity alleles); and **1XN*, **2XN*, **35XN*, and **41XN* (amplified alleles with numerous copies of the gene).

Numerous studies have shown the effect of *CYP2D6* genotype in tamoxifen metabolism. Women on tamoxifen therapy, with two or additional completely functional copies of *CYP2D6*, have elevated plasma endoxifen levels as compared to patients with one or additional nonfunctional alleles (*CYP2D6*3* to **6*) or those taking CYP2D6 inhibitors (Jin et al. 2005), which means that lowered CYP2D6 activity results in decreased metabolism of tamoxifen. The most common null allele *CYP2D6*4* was associated with poor metabolism in a retrospective study of 223 postmenopausal women (Goetz et al. 2005). Patients homozygous for *CYP2D6*4* had poorer outcomes in terms of lesser time to recurrence and poor disease-free survival than women with *CYP2D6*4/*1* or *CYP2D6*1/*1* genotypes. It was observed that patients with reduced-activity *CYP2D6* alleles (**4*, **5*, **10*, **41*) had significantly poorer outcome in terms of higher relapse rates, lesser times to recurrence of disease, and worst survival than carriers of functional alleles.

One of the most frequent adverse effects of tamoxifen treatment is hot flashes which are treated by antidepressants like selective serotonin reuptake inhibitors (Fisher et al. 1998). It is well known that some antidepressants are inhibitors of CYP2D6 activity; for example, coadministration of paroxetine or fluoxetine during tamoxifen therapy causes a significant decrease in the levels of endoxifen in extensive metabolizers as compared to patients not taking CYP2D6 inhibitors (Jin et al. 2005; Borges et al. 2006). FDA has recommended that the tamoxifen package should have enclosures stating that patients who are CYP2D6 poor metabolizers are at high risk for relapse of the breast cancers if treated with tamoxifen and further that coadministration of certain serotonin reuptake inhibitors which can inhibit CYP2D6 can influence the metabolism of tamoxifen (Young 2006). It has been reported that based on the patient's genotype, an alternative therapy like aromatase inhibitors can be prescribed. However, more studies are needed for premenopausal women as most of the data available is on postmenopausal women.

2.2 Polymorphisms Affecting Targeted Therapies

There are a number of targeted therapies that are being used in the clinic nowadays, which are directed towards some specific molecule to be inhibited in a pathway. Polymorphisms in the target or its downstream effector pathway can lead to a change in the outcome in patients; hence the targeted therapies can also be titrated according to the effect of polymorphisms in these patients. Here we discuss an example of tyrosine kinase inhibitors which are routinely used in the clinic.

2.2.1 Epidermal Growth Factor Receptor (EGFR) and Tyrosine Kinase Inhibitors

Gefitinib and erlotinib are classified as small molecular tyrosine kinase inhibitors (TKIs) of EGFR. Mutations in the *EGFR* gene have been identified and have been reported to be associated with response (Lynch et al. 2004; Paez et al. 2005). In-frame deletions and missense mutations in the *EGFR* tyrosine kinase domain were observed in eight out of nine patients responsive to gefitinib. In order to confirm if these mutations were present only in the tumor tissue of responding patients, matched normal tissue of the same patients and tumor tissue of nonresponding patients were also checked for the presence of these mutations. As expected these mutations were not observed in the later samples which confirmed that these mutations are responsible for a favorable response to gefitinib (Lynch et al. 2004). These results were also confirmed in another study involving nine tumor patients of whom five were responding and had mutations and four were nonresponding and had no mutations (Paez et al. 2005). The most well-studied mutations are in-frame deletion in exon 19 and a missense mutation at codon 858 which leads to an arginine to leucine substitution (L858R) (Sequist et al. 2007). In the presence of epidermal growth factor (EGF), the mutated *EGFR* had significantly increased and prolonged activation compared to wild type as observed in the in vitro studies (Lynch et al. 2004).

A number of studies have attempted to define the response rates and clinical outcomes of EGFR TKIs in patients with and without *EGFR* mutations. Response rates of 75–90 % with a median progression-free survival (PFS) of 7.7–11.5 months were observed in Japanese patients on gefitinib treatment (Asahina et al. 2006; Yoshida et al. 2007; Inoue et al. 2006; Tamura et al. 2008), whereas North American patients showed response rates of 55 % with median PFS of 9.2 months (Sequist et al. 2008). Although mutations in *EGFR* lead to a higher response in patients, there are also issues with acquired resistance. There are a number of probable mechanisms leading to resistance including a secondary *EGFR* mutation T790M which causes a threonine to methionine change (Kobayashi et al. 2005) and another being MET amplification (Engelman et al. 2007; Bean et al. 2007).

Other functional polymorphisms in *EGF* and *EGFR* have been identified that affect gene expression, promoter activity, and protein production (Araujo et al. 2007; Gebhart et al. 1999; Buerger et al. 2000; Liu et al. 2005; Moriai et al. 1994; Bhowmick et al. 2004; Zhang et al. 2007). The first one includes CA simple sequence repeat 1 (*cA-SSR1*) which is highly polymorphic and is located in intron 1 of *EGFR*. It contains 14–21 *CA* dinucleotide repeats (Araujo et al. 2007; Desai et al. 2003). It has been reported that the length of the CA repeat is inversely associated with *EGFR* gene transcription (Gebhart et al. 1999). The second one is a *-16G/T* polymorphism (rs712829) which is located in the promoter region of *EGFR* gene (Liu et al. 2005). The third one is *-91C/A* polymorphism (rs712830) and is also located in the *EGFR* promoter region, and the variant may lead to a differential protein production (Nomura et al. 2007; Liu et al. 2003). The last one is *R497K* polymorphism (rs1154348) which causes an A>G alteration, and the Lys allele leads to decreased activity of EGFR (Moriai et al. 1994).

There have been a number of studies to evaluate the clinical outcomes in lung cancer patients and the effect of polymorphisms affecting the activity of EGFR. One of the studies involves the retrospective analysis of 137 Japanese patients on gefitinib (Ichihara et al. 2007). It was observed that the common mutations in *EGFR* caused sensitivity to gefitinib; however, none of the studied polymorphisms were associated with response or survival (Ichihara et al. 2007). Another study with 86 Korean patients treated with gefitinib studied *EGFR* mutation and CA repeat polymorphism and observed that the common mutations were significantly correlated with response and survival (Han et al. 2007a, b). This study showed that the shorter the CA repeats, the better the response which was confirmed by another study by Nie and colleagues where an analysis on 70 Chinese patients treated with gefitinib showed similar results (Nie et al. 2007). However, more studies need to be undertaken with a larger patient population for a better understanding of the effect of these polymorphisms on treatment outcome in different cancers.

2.3 Polymorphisms in Drug Transporters

The ATP-binding cassette (ABC) family of transmembrane proteins is one of the largest protein families that actively transport a variety of substrates across cell membranes and are commonly called ABC transporters (Dassa and Bouige 2001). Numerous ABC proteins are multidrug efflux pumps and play an important role in the uptake and distribution of drugs as well as protect the body from exogenous toxins (Schinkel and Jonker 2003). Varying levels of ABC transporters have been responsible for different responses to drug therapy in different populations, making it inevitable to study polymorphisms in these populations (Choudhuri and Klaassen 2006). Many single nucleotide polymorphisms (SNPs) result in varying protein expression level and transport function, thereby affecting drug absorption, plasma concentration, distribution, and finally elimination from the body.

2.3.1 P-Glycoprotein

P-glycoprotein (P-gp) has a substrate affinity towards many anticancer drugs like doxorubicin, vinblastine, paclitaxel, docetaxel, etoposide, and irinotecan (Schwab et al. 2003). *ABCB1*, the gene-encoding P-gp, is reported to have numerous functional polymorphisms that vary in allele frequencies among different ethnic groups (Schwab et al. 2003; Hoffmeyer et al. 2000; Cascorbi et al. 2001; Kim et al. 2001; Sai et al. 2003). In the case of *ABCB1*, 3435C>T (Ile1145Ile; *ABCB1*6*) is the most well studied and is observed to be in strong linkage disequilibrium with a further SNP 1236C>T (Gly412Gly; *ABCB1*8*) and the triallelic variant 2677G>T/A (Ala893Ser/Thr) (Kim et al. 2001; Kroetz et al. 2003). The above mentioned three SNPs are together designated as *P-gp*2* (Kim et al. 2001). There are contradictory reports regarding the functional effect of 3435C>T on mRNA stability or lower mRNA expression (Hoffmeyer et al. 2000; Siegsmund et al. 2002; Hitzl et al. 2001; Wang et al. 2005) or increased expression levels (Kim et al. 2001; Nakamura et al. 2002; Illmer et al. 2002).

The homozygous *P-gp*2* variant has been reported to be correlated with poorer renal clearance of irinotecan and its active metabolite SN-38 (Sai et al. 2003). Also, 3435TT showed a significant association with grade 3 diarrhea in 107 patients on irinotecan and cisplatin therapy, suffering with non-small cell lung cancer (NSCLC) (Han et al. 2007a, b). In the case of NSCLC patients on docetaxel and cisplatin, 33 % of 3435TT allele carriers experienced grade ≥ 2 diarrhea as compared to 4 % heterozygous and 11 % wild-type patients (Isla et al. 2004). The polymorphisms in *ABCB1* have shown contradictory results in response to treatment outcome in different populations. This might be due to ethnic differences and type of treatment prescribed in the studied population. Additional studies for the polymorphisms in *ABCB1* are required for the better understanding of the role of these polymorphisms in treatment outcome and further in facilitating personalized anticancer therapy.

2.3.2 BCRP (ABCG2)

Anticancer therapeutics like methotrexate, topotecan, imatinib, and gefitinib are substrates for ABCG2 (Cusatis and Sparreboom 2008). Many studies have reported the effect of polymorphisms in *ABCG2* on its activity (Bäckström et al. 2003; Bosch et al. 2005; Zamber et al. 2003). 421C>A (Gln141Lys) and 376C>T (Gln126stop) are some of the well-studied SNPs. The 421C>A SNP, which is higher in Japanese (30 %) than in Whites (10 %), affects the efficiency of translation of *ABCG2* and results in lower expression in the placenta (Imai et al. 2002; Kobayashi et al. 2005). Other additional studies with White (de Jong et al. 2004) and Asian (Jada et al. 2007) patients on irinotecan therapy, 421C>A did not significantly influence irinotecan pharmacokinetics (de Jong et al. 2004). There are contradictory reports of the effect of 421C>A on imatinib and gefitinib. Eighty-two patients on imatinib, heterozygous for 421C>A, showed no significant pharmacokinetic outcome in one of the studies (Gardner et al. 2006), while in the other study 67 heterozygous patients showed a 22 % poorer clearance of imatinib (Petain et al. 2008). Further, in the case of patients treated with gefitinib, the same SNP was associated with an increased accumulation of gefitinib (Li et al. 2007) and diarrhea (grade 1 or 2) (Cusatis et al. 2006). This data suggests that the expression of ABCG2 might vary among different populations; hence, further studies are warranted to get a better idea about the effect of SNPs and also the effect of haplotypes in these populations.

2.4 Polymorphisms in Drug Targets

There are a large number of studies focusing on polymorphic drug-metabolizing enzymes and their association with drug response; however, the number of studies on the effect of polymorphic drug targets on drug response in the clinic is very few. The most well-studied example reviewed here is about thymidylate synthase which is a drug target for 5-fluorouracil.

2.4.1 Thymidylate Synthase (TYMS) and 5-FU

Thymidylate synthase is important in DNA synthesis and repair and cell proliferation. It is the main enzyme in the de novo synthesis of thymidylate (dTMP) which is needed for synthesis of DNA. Anticancer drugs targeting TYMS include fluoropyrimidines, methotrexate, and many others. 5-Fluorodeoxyuridine monophosphate, which is the key active metabolite of fluoropyrimidines, forms a ternary complex with TYMS and methylenetetrahydrofolate (5,10-CH_2FH_4). This leads to inhibition of TYMS and ultimately arrest of DNA synthesis followed by cell death (Grem 2000; Rustum et al. 1997; Welsh et al. 2000). The antitumor activity of 5-fluorouracil (5-FU) is inversely proportional to the expression level of TYMS; cells that have low expression level are more sensitive to 5-FU and vice versa (Leichman et al. 1997; Nishimura et al. 1999). Preclinical studies in cancer cell lines have revealed that overexpression of *TYMS* is linked with resistance to 5-FU (Copur et al. 1995). The most well-studied polymorphism in *TYMS* is a tandem repeat in the 5′ promoter enhancer region of the *TYMS* (*TSER*) which affects the expression of the gene. The tandem repeat is of a 28-bp sequence repeated different number of times: 2 (*TSER**2), 3 (*TSER*3*), 4 (*TSER*4*), 5 (*TSER*5*), and 9 (*TSER*9*), although *TSER*2* and *TSER*3* are the most frequent (Horie et al. 1995; Kaneda et al. 1987; Luo et al. 2002; Marsh et al. 2000). As the number of repeats increases, the expression level of TYMS increases (Kaneda et al. 1987). A G>C SNP occurs within the second 28-bp repeat of TSER*3 which is denoted as *TSER*3G/TSER*3C* wherein *TSER*3C* has a lower expression than *TSER*3G* and is comparable to that of *TSER*2* (Kawakami and Watanabe 2003; Mandola et al. 2003; Marcuello et al. 2004a, b; Kawakami et al. 1999; Morganti et al. 2005). The activity of TYMS can be divided into three based on these alleles: low (**2/*2* or **2/*3C* or **3C/*3C*), intermediate (**2/*3G* or **3C/*3G*), and high (**3G/*3G*). Numerous studies have analyzed the effect of genetic polymorphism in *TYMS* on clinical outcome with fluoropyrimidines wherein gastrointestinal or breast cancer patients on fluoropyrimidines with high TYMS expression showed lower clinical activity than in patients with low TYMS expression (Marcuello et al. 2004a, b; Pullarkat et al. 2001; Lecomte et al. 2004; Park et al. 2002; Villafranca et al. 2001; Ruzzo et al. 2006; Goekkurt et al. 2006; Largillier et al. 2006). It is however thought that a meta-analysis would give a better insight into the predictive value and clinical significance of this polymorphism. A study analyzing tumor tissue DNA of colorectal cancer patients on 5-FU showed that **3/*3* patients had no long-term benefit of survival with chemotherapy, while those with **2/*2* or **2/*3* genotype had significantly longer survival with chemotherapy (Iacopetta et al. 2001). Another interesting observation in one of the studies was that a loss of heterozygosity at the *TYMS* locus in **2/*3* patients may either become **2/loss* or **3/loss*. The patients with tumor **2/loss* had a superior response to S-1 (an oral fluoropyrimidine) of 80 % versus 14 % for the ones with **3/loss* (Uchida et al. 2004).

Another common polymorphism in *TYMS* is the deletion of six base pairs in its 3′ UTR region (Ulrich et al. 2000), which leads to lower mRNA stability and protein expression of *TYMS* (Mandola et al. 2004). Clinically there is less association of this polymorphism with toxicity than the 28-bp tandem repeat in the case of colorectal (Lecomte et al. 2004; Stoehlmacher et al. 2004; Sharma et al. 2008), gastric (Goekkurt et al. 2009), or breast (Largillier et al. 2006) cancer treated with 5-FU-based chemotherapeutics. However, one study reported a longer progression-free survival observed only in patients on 5-FU therapy and not on capecitabine therapy (Martinez-Balibrea et al. 2008).

3 Companion Diagnostics in Cancer Therapy

One of the recently developed areas is of companion diagnostics where a diagnostic is developed along with a drug to screen for patients for predictable clinical outcome. The combination of diagnostic test and the drug can be codeveloped and co-marketed. A number of pharmaceutical companies like Abbott, Pfizer, Bristol-Myers Squibb (BMS), Boehringer Ingelheim, GlaxoSmithKline, and AstraZeneca have developed companion diagnostics (Dhawan and Padh 2011). The concept of companion diagnostics holds immense potential to assist in making correct treatment decisions. Codevelopment of a companion diagnostic can reduce the time taken for a novel therapeutic to be approved as well as avoid adverse drug reactions for patients with dreadful diseases like cancer where time is very critical for the patients. The US FDA has realized the urge for companion diagnostics in a recent draft guidance which summarizes their inclination for concurrently approving a drug along with its companion diagnostic test (FDA 2011). Another approach being taken up is to codevelop companion diagnostics within the company, together with the development of the corresponding therapeutic. This is being followed by big pharma companies, like Abbott, Johnson & Johnson, and Roche. They have all developed in-house capabilities to develop companion diagnostics (Naylor and Cole 2010). The advantage of this kind of a strategy is to streamline the development of both pharmaceutical products, as the knowledge gained during the drug discovery process can be applied to the development of the companion diagnostics. A timeline of development of biomarkers in oncology is depicted in Figure 11.1 which illustrates the progress in the field of pharmacogenomics in oncology.

3.1 FDA-Approved Companion Diagnostics on the Market

There is a lot of activity in companion diagnostics in the area of oncology. A number of companion diagnostics markers have been developed for targeted therapeutics like Herceptin (Genentech), Tarceva (OSI Pharmaceuticals/Genentech), Iressa (AstraZeneca), Erbitux (ImClone/Bristol-Myers Squibb), and Vectibix (Amgen). There is a considerable progress in the use of companion diagnostic markers to predict toxicity, efficacy and drug dosage to meet the critical endpoints. The list of FDA-approved companion diagnostics is summarized in Table 11.1.

The TheraGuide® 5-FU test has been marketed by Myriad Genetics Inc. to predict toxicity to 5-FU (and capecitabine). This test comprises of sequencing the *DPYD* gene; the identification of known mutations including IVS14+1G>A, D949V, and I560S; and PCR amplification of *TYMS* 5′-UTR tandem repeats. The test is based on reports from several clinical studies (Morel et al. 2006; Lecomte et al. 2004; Ichikawa et al. 2006; Pullarkat et al. 2001). The test has reported a high technical specificity and sensitivity, available at http://www.myriadtests.com/provider/doc/TheraGuide-5-FU-Technical-Specifications.pdf (accessed on September 3, 2012). The commonly

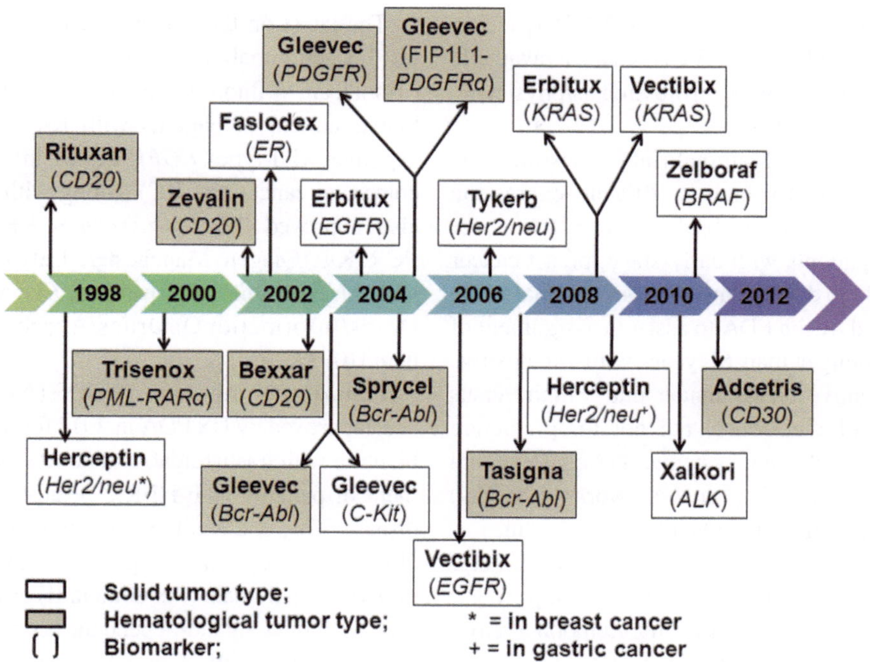

Fig. 11.1 Timeline representative of biomarker development in cancer

Table 11.1 FDA-approved companion diagnostic tests for cancer

Cancer type	Diagnostic manufacturer	Target	Method	Pharmaceutical company	Therapeutic
Breast cancer	Dako	HER2/neu	IHC	Genentech	Herceptin (trastuzumab)
Colorectal cancer	Qiagen	EGFR	RT-PCR	Bristol-Myers Squibb	Erbitux (cetuximab)
Leukemia	Quest Diagnostics	ABL kinase domain mutation in CML	Nested PCR, RT-PCR, sequencing	Novartis	Tasigna (nilotinib)
Leukemia	Genzyme	ABL kinase domain mutation in CML	PCR, sequencing	Bristol-Myers Squibb	Sprycel (dasatinib)
Lung cancer	Abbott Molecular	ALK	FISH	Pfizer	Xalkori (crizotinib)
Melanoma	Cobas	BRAF V600E	RT-PCR	Genentech	Zelboraf (vemurafenib)

Adapted from Naylor and Cole (2010), Hoggatt (2011)

IHC immunohistochemistry, *RT-PCR* real time-polymerase chain reaction, *FISH* fluorescence in situ hybridization

used predictors of breast cancer prognosis include progesterone receptor, estrogen receptor, and HER-2/neu. Oncotype DX®, a diagnostic assay, quantifies the likelihood of breast cancer relapse and assesses the benefit from chemotherapy in women with newly diagnosed, early-stage breast cancer (for both tamoxifen and CMF (cyclophosphamide, methotrexate, 5-FU) or methotrexate/5-FU/leucovorin chemotherapy). This test has been recommended in the National Comprehensive

Cancer Network (NCCN) Clinical Practice in Oncology Guidelines – Breast Cancer (available at http://www.nccn.org/professionals/physician_gls/PDF/breast.pdf).

Another assay, MammaPrint®, uses microarray technology to analyze 70-gene expression profiles from frozen breast tumor tissue for selecting patients with early-stage, breast cancer more likely to develop metastasis. MammaPrint® is approved by the FDA to assist in assignment of women younger than 61 years with ER-positive or ER-negative breast cancer into a high versus low risk for recurrence, but not for predicting benefit from adjuvant systemic therapy (National Comprehensive Cancer Network (NCCN) Clinical Practice in Oncology Guidelines – Breast Cancer, available at http://www.nccn.org/professionals/physician_gls/PDF/breast.pdf).

A companion diagnostic for Zelboraf (vemurafenib) for treating late-stage melanoma and a further for Xalkori (crizotinib) for treating late-stage lung cancer join the list of approvals for companion diagnostics by US FDA in 2011. Zelboraf, the first important "personalized" medicine approval of FY 2011, is specially indicated for the treatment of patients with melanoma whose tumors express a gene mutation called BRAF V600E. Zelboraf has been approved with a unique genetic test called the "cobas 4800 BRAF V600 Mutation Test", a companion diagnostic that will enable in determining whether a patient's melanoma cells have the BRAF V600E mutation (http://www.fda.gov/AboutFDA/CentersOffices/OfficeofMedicalProductsandTobacco/CDER/ucm268301.htm). Xalkori, the second important "personalized" medicine approval of FY 2011, was approved with a companion diagnostic test that will allow the drug to be beneficial to a particular group of patients. The genetic test, a first-of-a-kind genetic test called Vysis ALK Break Apart FISH Probe Kit, helps determine if a patient has the abnormal ALK gene. The approval of Xalkori with the diagnostic test enables the selection of patients who are more likely to respond to this drug. This may help in targeted therapy reducing adverse drug reactions in these patients (http://www.fda.gov/AboutFDA/CentersOffices/OfficeofMedicalProductsandTobacco/CDER/ucm270058.htm).

The latest on the list of approvals by FDA in 2012 is cetuximab in combination with FOLFIRI (irinotecan, 5-fluorouracil, leucovorin) for first-line treatment of patients with K-ras mutation-negative (wild-type), *EGFR*-expressing metastatic colorectal cancer (mCRC). Along with this FDA also approved the *Therascreen*® KRAS RGQ PCR Kit (Qiagen Manchester, Ltd) concurrent with this cetuximab approval (http://www.fda.gov/Drugs/InformationOnDrugs/ApprovedDrugs/ucm310933.htm).

Pertuzumab injection (PERJETA) has also been approved by US FDA in 2012 for use in combination with trastuzumab and docetaxel for treatment of patients with HER-2 positive metastatic disease. This is a recombinant humanized monoclonal antibody that targets the extracellular dimerization domain (Subdomain II) of HER2 and thereby blocks ligand-dependent heterodimerization of HER2 with other HER family members (http://www.fda.gov/Drugs/InformationOnDrugs/ApprovedDrugs/ucm307592.htm). Although the list of companion diagnostics is growing by the day, there are some advantages and certain limitations of these which are summarized in Table 11.2.

4 Pharmacogenomic Biomarkers in Drug Labels

It is very evident from the examples mentioned above that pharmacogenomics can enable in stratifying the population of patients into responders and nonresponders, in turn avoiding adverse drug response and titrating drug dose. Drug labels can impart information on the effect of biomarkers to adverse reactions, genotype-specific dose titration, mechanisms of action, as well as the polymorphic drug target and polymorphic metabolizing enzymes.

The US Food and Drug Administration (FDA) has recommended a label change for tamoxifen in 2006 including a mention of *CYP2D6* genotype testing as an option for women before they are prescribed tamoxifen. Also, the AmpliChip technology has enabled the identification of the CYP2D6 genotype and reports the individuals as poor, intermediate, extensive, or ultrarapid metabolizers. This test was approved by US FDA

Table 11.2 Advantages and disadvantages of pairing therapeutics with diagnostics

Advantages	Disadvantages
Application: development	
Increased chances of meeting efficacy endpoints and lower adverse drug reactions with a group of patients	Difficulty in codevelopment due to regulatory processes and timing
Potential for reduced numbers of failures at a later stage	Requirement for strong early data of biomarker association to support patient stratification during late-stage clinical trials
Possibility of reduction in length of clinical trials	Association of biomarker with response is not complete
Minimized risk of subsequent discovery of biomarker which allows more accurate assessment of potential market size	
Application: commercialization	
Provides increased accuracy in assessment of target patient population	Diagnostic inconsistencies may impact the brand
Possibility of shorter clinical trials, hence earlier launch	Stratification of potential market
Higher efficacy in a particular group of population, hence increase in market value and early reimbursement of investment	
Marketing message of improvements over standards of care in a stratified population can result in pulling market from patients and physicians	

Adapted from Khosrow-Shahi et al. (2009)

on December 24, 2004, and was adapted by many commercial companies involved with genotyping (e.g., DNAdirect) (DNAdirect). In 2003, the addition of pharmacogenetic information concerning *TPMT* polymorphisms and treatment toxicity to the drug label for 6-mercaptopurine was recommended by the US FDA. Hence in 2004, the label for 6-mercaptopurine was changed with inclusion of *TPMT* testing and dosage recommendations provided for TPMT-deficient patients. Further the US FDA has also approved the addition of a warning to the irinotecan label as well as marketing of the Invader® *UGT1A1* Molecular Assay (Third Wave Technologies) for the detection of *UGT1A1*28* (Ratain 2006; Maitland et al. 2006; Hasegawa et al. 2004). Over 30 oncology drugs provide information on pharmacogenomic biomarkers on their labels and have been summarized in Table 11.3.

5 Conclusion

With the advancement in identification of new molecular markers that can help in predicting the response to drugs, there is a hope that the patients will benefit from the tailored medicine which is prescribed to them. Further, the US FDA recommends that the approval of new drugs becomes faster if accompanied by genetic tests that can predict the drug response. It would also save the drug from being taken off the market if adverse drug reactions are observed after it has been marketed. Hence, these molecular marker-based tests not only are beneficial to the patients but also save the pharmaceutical company's time and money that it has invested in developing a new drug. In the case of cancer, where the therapeutic index of the drugs is narrow and the time of treatment is critical, chances of drug toxicity are high; the area of cancer pharmacogenomics will enable the patients to be benefited with the available therapeutics by titrating the medication to the individual patients. The most common pharmacogenomic biomarkers for prevention of toxicities in cancer patients are summarized in Table 11.4.

6 Future Prospects

The field of cancer pharmacogenomics is evolving every day with newer insights about association of genetic variations with response to therapeutics. Numerous studies have been carried out to establish an association of the commonly observed genetic variations with efficacy and/or toxicity to the prescribed medicines in cancer

Table 11.3 Pharmacogenomic biomarkers in oncology drug labels

Drug	Cancer	Biomarker	Label sections
Arsenic trioxide	Some types of leukemia	PML/RARα	Boxed warning, clinical pharmacology, indications and usage, warnings
Brentuximab vedotin	Hodgkin lymphoma, anaplastic large cell lymphoma (ALCL)	CD30	Indications and usage, description, clinical pharmacology
Busulfan	Chronic myelogenous leukemia, some lymphomas	Ph chromosome	Clinical studies
Capecitabine	Breast, colorectal	DPD	Contraindications, precautions, patient information
Cetuximab (1)	Head and neck, colorectal	EGFR	Indications and usage, warnings and precautions, description, clinical pharmacology, clinical studies
Cetuximab (2)	Head and neck, colorectal	KRAS	Indications and usage, dosage and administration, warnings and precautions, adverse reactions, clinical pharmacology, clinical studies
Cisplatin	Testicular, bladder, ovarian, lung	TPMT	Clinical pharmacology, warnings, precautions
Crizotinib	Advanced non-small cell lung cancer (having mutation in ALK gene)	ALK	Indications and usage, warnings and precautions, adverse reactions, clinical pharmacology, clinical studies
Erlotinib	Pancreatic cancer, non-small cell lung cancer	EGFR	Clinical pharmacology
Everolimus	Kidney cancer, advanced breast cancer	Her2/neu	Indications and usage, boxed warnings, adverse reactions, use in specific populations, clinical pharmacology, clinical studies
Exemestane	Breast cancer (in postmenopausal women)	ER and PgR receptor	Indications and usage, dosage and administration, clinical studies, clinical pharmacology
Fulvestrant	Advanced breast cancer (in postmenopausal women)	ER receptor	Indications and usage, patient counseling information
Gefitinib	Some types of non-small cell lung cancer	EGFR	Clinical pharmacology
Imatinib (1)	Chronic myelogenous leukemia (CML) and acute lymphocytic leukemia (ALL)	C-Kit	Indications and usage, dosage and administration, clinical pharmacology, clinical studies
Imatinib (2)	Chronic myelogenous leukemia (CML) and acute lymphocytic leukemia (ALL)	Ph chromosome	Indications and usage, dosage and administration, clinical pharmacology, clinical studies
Imatinib (3)	Chronic myelogenous leukemia (CML) and acute lymphocytic leukemia (ALL)	PDGFR	Indications and usage, dosage administration, clinical studies

(continued)

Table 11.3 (continued)

Drug	Cancer	Biomarker	Label sections
Imatinib (4)	Chronic myelogenous leukemia (CML) and acute lymphocytic leukemia (ALL)	FIP1L1-PDGFRα	Indications and usage, dosage administration, clinical studies
Irinotecan	Colorectal	UGT1A1	Dosage and administration, warnings, clinical pharmacology
Lapatinib	Some types of breast cancer	Her2/neu	Indications and usage, clinical pharmacology, patient counseling information
Letrozole	Breast cancer (in postmenopausal women)	ER and PgR receptor	Indications and usage, adverse reactions, clinical studies, clinical pharmacology
Mercaptopurine	Leukemia	TPMT	Dosage and administration, contraindications, precautions, adverse reactions, clinical pharmacology
Nilotinib (1)	Chronic myelogenous leukemia (CML)	Ph chromosome	Indications and usage, patient counseling information
Nilotinib (2)	Chronic myelogenous leukemia (CML)	UGT1A1	Warnings and precautions, clinical pharmacology
Panitumumab (1)	Colorectal	EGFR	Indications and usage, warnings and precautions, clinical pharmacology, clinical studies
Panitumumab (2)	Colorectal	KRAS	Indications and usage, clinical pharmacology, clinical studies
Pertuzumab	Certain types of breast cancer	Her2/neu	Indications and usage, warnings and precautions, adverse reactions, clinical studies, clinical pharmacology
Rasburicase	A number of cancers	G6PD	Boxed warning, contraindications
Tamoxifen	Breast	ER receptor	Indications and usage, precautions, medication guide
Thioguanine	Some kinds of acute leukemia	TPMT	Dosage and administration, precautions, warnings
Tositumomab	Certain types of non-Hodgkin lymphoma	CD20 antigen	Indications and usage, clinical pharmacology
Trastuzumab	Certain types of breast cancer and stomach cancer	Her2/neu	Indications and usage, precautions, clinical pharmacology
Vemurafenib	Advanced melanomas (with certain mutation in BRAF gene)	BRAF	Indications and usage, warning and precautions, clinical pharmacology, clinical studies, patient counseling information

Adapted from http://www.fda.gov/drugs/scienceresearch/researchareas/pharmacogenetics/ucm083378.htm

Table 11.4 Common pharmacogenomic biomarkers and their impact on toxicities in cancer patients

Biomarkers	Polymorphisms	Therapeutic agents (brand)	Cancer	Pharmacokinetic/ pharmacodynamic impact	Frequency of variant poor metabolizer phenotype	Clinical impact
Uridine diphosphate-glucuronyltransferase (UGT1A1)	$A(TA)_7TAA$, 3156G>A, 211G>A	Irinotecan (Camptosar), nilotinib (Tasigna)	Breast, leukemia, and lymphoma	Increased systemic exposure to SN-38 with UGT1A1*28	Deficiency of the enzyme may occur in 35 % of Caucasians and African Americans	Homozygous UGT1A1*28 genotype leads to risk of severe diarrhea, neutropenia at doses >200 mg/m^2
Dihydropyrimidine dehydrogenase (DPD)	IVS14+1G>A, 2846A>T, 1679T>G	Fluorouracil (5-FU), capecitabine (Xeloda)	Breast, head and neck, colorectal	Mutation causes systemic increase in 5-FU	3 % of the Caucasian population may have deficiency of this enzyme	Deficiency can lead to fatal neurologic and hematologic toxicities. Grade 3 diarrhea and hand-foot syndrome linked with 5-FU plasma levels >3 mg/L in males
Thiopurine methyltransferase (TPMT)	238G>C, 460G>A, 719G>A	6-Mercaptopurine (6-MP, Purinethol)	Leukemia and lymphoma	Inactivates 6-MP; low or absent TPMT activity increases systemic drug exposure	~10 % of Caucasians are poor metabolizers while 0.3 % of patients have complete deficiency of the enzyme	Patients with low or absent TPMT activity have an increased risk of developing severe, life-threatening myelotoxicity
CYP3A4/3A5	>30 polymorphisms	Cyclophosphamide (Cytoxan)	Breast, neuroblastoma, retinoblastoma, leukemia, and lymphoma	Activates the prodrug to its active form	45 % of African Americans and 9 % of Caucasians have a variant allele	Patients with deficiency of this enzyme have variability in efficacy and increased toxicity
CYP2B6	>30 polymorphisms	Cyclophosphamide, ifosfamide	Breast, neuroblastoma, retinoblastoma, leukemia, and lymphoma	Major metabolizing enzyme for cyclophosphamide and ifosfamide	>45 % of individuals are polymorphic for this enzyme	Patients with dysfunctional CYP2B6 may have an increased risk of nephrotoxicity

CYP2D6	>50 polymorphisms	Tamoxifen (Nolvadex)	Breast	Converts tamoxifen to endoxifen which is the potent active metabolite	Nonfunctional variants are observed in 7 % of Caucasians, <3 % of African Americans, and 1 % of Asians	Deficiency of CYP2D6 causes reduction of endoxifen and may change toxicity
Glutathione S-transferase alpha 1 (GST1)	Null mutations	Cyclophosphamide, busulfan (Myleran)	Chronic myelogenous leukemia, some lymphomas	A family of enzyme detoxifying electrophilic groups of some chemotherapeutics	Deficiency of this enzyme is observed in ~57 % of Caucasians and 13 % of African Americans	Deficiency of this enzyme may result in increased toxicity
Excision repair cross-complementing rodent repair deficiency group 1 and 2 (ERCC1, ERCC2)	ERCC1-118C>T ERCC2-8092C>A	Cisplatin, carboplatin, oxaliplatin	Bladder, ovarian, lung, testicular, head and neck	An endonuclease repairing damaged DNA segments	No data	Not conclusive; high gene expression associated with inferior outcomes for bladder cancer
Multidrug resistance protein [ABCB1 (MDR1), ABCB2 (MRP2)]	ABCB1-3435C>T, 1236C>T, 2677G>T/A ABCB2-421C>A, 376C>T	Paclitaxel (Taxol, Onxol)	Lung, ovarian, breast, head and neck	Encodes for P-gp responsible for efflux of drugs from the cell	More than 50 SNPs and 3 insertion/deletion polymorphisms have been reported	Overexpression of ABCB1 leads to drug resistance; wild-type patients demonstrated reduced neuropathy

Feng et al. (2012)

patients. However, some of the association studies need to be validated on a large number of patient population before the results can be translated to bedside. There is enormous amount of data on ethnic differences observed in terms of frequency of the genetic variations present in the studied population, which might affect the metabolism of drugs to an extent and in turn affect response in the clinic. This would help in tailoring drugs for each study population and as an individual.

The field of companion diagnostics has already grown tremendously and will continue to grow with an increase in partnerships between pharmaceutical companies and companies developing genetic tests to evaluate drug response. This will enable a better future for the cancer patients as it will save time that is spent in empirically deciding the drugs currently prescribed to them; it will also save them from adverse drug reactions and also help in tailoring the dose for each patient. A lot of research needs to be done to enable the complete technology transfer of the field of pharmacogenomics from bench side to the clinic for routine use. So far SNP is the widely studied variant. We have recently learned that other types of variants like CNV, INDELs, etc. could also affect physiology and pathophysiology (Dhawan and Padh 2009; Almal and Padh 2012). We need to wait until we learn about their frequency and implication of new variants in health and disease. It may turn out that various combinations of all variants may be the strongest biomarker for disease processes, and then we shall realize the true potential of pharmacogenetics.

References

Almal SH, Padh H (2012) Implications of gene copy-number variation in health and diseases. J Hum Genet 57(1):6–13

Ando Y, Saka H, Ando M et al (2000) Polymorphisms of UDP-glucuronosyltransferase gene and irinotecan toxicity: a pharmacogenetic analysis. Cancer Res 60:6921–6926

Araujo A, Ribeiro R, Azevedo I et al (2007) Genetic polymorphisms of the epidermal growth factor and related receptor in non-small cell lung cancer – a review of the literature. Oncologist 12:201–210

Asahina H, Yamazaki K, Kinoshita I et al (2006) A phase II trial of gefitinib as first-line therapy for advanced non-small cell lung cancer with epidermal growth factor receptor mutations. Br J Cancer 95:998–1004

Bäckström G, Taipalensuu J, Melhus H et al (2003) Genetic variation in the ATP-binding cassette transporter gene ABCG2 (BCRP) in a Swedish population. Eur J Pharm Sci 18:359–364

Bean J, Brennan C, Shih JY et al (2007) MET amplification occurs with or without T790M mutations in EGFR mutant lung tumors with acquired resistance to gefitinib or erlotinib. Proc Natl Acad Sci USA 104(52):20932–20937

Beutler E, Gelbart T, Demina A (1998) Racial variability in the UDP-glucuronosyltransferase 1 (UGT1A1) promoter: a balanced polymorphism for regulation of bilirubin metabolism? Proc Natl Acad Sci USA 95:8170–8174

Bhowmick DA, Zhuang Z, Wait SD, Weil RJ (2004) A functional polymorphism in the *EGF* gene is found with increased frequency in glioblastoma multiforme patients and is associated with more aggressive disease. Cancer Res 64:1220–1223

Borges S, Desta Z, Li L, Skaar TC, Ward BA, Nguyen A, Jin Y, Storniolo AM, Nikoloff DM, Wu L, Hillman G, Hayes DF, Stearns V, Flockhart DA (2006) Quantitative effect of CYP2D6 genotype and inhibitors on tamoxifen metabolism: implication for optimization of breast cancer treatment. Clin Pharmacol Ther 80:61–74

Bosch TM, Kjellberg LM, Bouwers A et al (2005) Detection of single nucleotide polymorphisms in the ABCG2 gene in a Dutch population. Am J Pharmacogenomics 5:123–131

Bosma PJ, Chowdhury JR, Bakker C et al (1995) The genetic basis of the reduced expression of bilirubin UDP-glucuronosyltransferase 1 in Gilbert's syndrome. N Engl J Med 333:1171–1175

Buerger H, Gebhardt F, Schmidt H et al (2000) Length and loss of heterozygosity of an intron 1 polymorphic sequence of EGFR is related to cytogenetic alterations and epithelial growth factor receptor expression. Cancer Res 60:854–857

Cascorbi I, Gerloff T, Johne A et al (2001) Frequency of single nucleotide polymorphisms in the P-glycoprotein drug transporter MDR1 gene in white subjects. Clin Pharmacol Ther 69:169–174

Choudhuri S, Klaassen CD (2006) Structure, function, expression, genomic organization, and single nucleotide polymorphisms of human ABCB1 (*MDR1*), ABCC (*MRP*), and ABCG2 (*BCRP*) efflux transporters. Int J Toxicol 25:231–259

Ciotti M, Basu N, Brangi M et al (1999) Glucuronidation of 7-ethyl-10-hydroxycamptothecin(SN-38) by the human UDP-glucuronosyltransferases encoded at the UGT1 locus. Biochem Biophys Res Commun 260:199–202

Copur S, Aiba K, Drake JC, Allegra CJ, Chu E (1995) Thymidylate synthase gene amplification in human colon cancer cell lines resistant to 5-fluorouracil. Biochem Pharmacol 49:1419–1426

Côté JF, Kirzin S, Kramar A et al (2007) UGT1A1 polymorphism can predict hematologic toxicity in patients treated with irinotecan. Clin Cancer Res 13:3269–3275

Cusatis G, Sparreboom A (2008) Pharmacogenomic importance of ABCG2. Pharmacogenomics 9:1005–1009

Cusatis G, Gregorc V, Li J et al (2006) Pharmacogenetics of ABCG2 and adverse reactions to gefitinib. J Natl Cancer Inst 98:1739–1742

Dassa E, Bouige P (2001) The ABC of ABCs: a phylogenetic and functional classification of ABC systems in living organisms. Res Microbiol 152:211–229

Davison JE, McMullin MF, Catherwood MA (2006) Genotyping of thiopurine methyltransferase in patients with acute leukemia using LightCycler PCR. Leuk Lymphoma 47:1624–1628

de Jong FA, Marsh S, Mathijssen RH et al (2004) ABCG2 pharmacogenetics: ethnic differences in allele frequency and assessment of influence on irinotecan disposition. Clin Cancer Res 10:5889–5894

de Jong FA, Kehrer DF, Mathijssen RH et al (2006) Prophylaxis of irinotecan-induced diarrhea with neomycin and potential role for UGT1A1*28 genotype screening: a double-blind, randomized, placebo-controlled study. Oncologist 11:944–954

Desai AA, Innocenti F, Ratain MJ (2003) Pharmacogenomics: road to anticancer therapeutics nirvana? Oncogene 22:6621–6628

Dhawan D, Padh H (2009) Pharmacogenetics: technologies to detect copy number variations. Curr Opin Mol Ther 11(6):670–680

Dhawan D, Padh H (2011) Omics approaches in cancer biomarker and targeted anticancer drug discovery. In: Barh D, Blum K, Madigan MA (eds) OMICS: biomedical perspectives and applications. CRC Press, Taylor & Francis LLC, Boca Raton

Dhawan D, Padh H (2013) Genetic variability and chemotoxicity of 5-Fluorouracil and cisplatin in head and neck cancer patients: a preliminary study. Indian J Med Res 137:125–129

DNAdirect. Your genes. Your health. Your choices. http://www.dnadirect.com/. Accessed 3 Sept 2012

Engelman JA, Zejnullahu K, Mitsudomi T et al (2007) MET amplification leads to gefitinib resistance in lung cancer by activating ERBB3 signaling. Science 316:1039–1043

FDA-Medical devices (Guidance documents) (2011) Draft guidance for industry and food and drug administration staff: in vitro companion diagnostic devices 14 July 2011 (Online)

Feng X, Pearson D, Listiawan M et al (2012) Pharmacogenomic biomarkers for toxicities associated with chemotherapy. US Pharm 37(Oncology suppl):2–7

Ferraldeschi R, Minchell LJ, Roberts SA et al (2009) UGT1A1*28 genotype predicts gastrointestinal toxicity in patients treated with intermediate-dose irinotecan. Pharmacogenomics 10:733–739

Fisher B, Costantino JP, Wickerham DL, Redmond CK, Kavanah M, Cronin WM, Vogel V, Robidoux A, Dimitrov N, Atkins J, Daly M, Wieand S, Tan-Chiu E, Ford L, Wolmark N (1998) Tamoxifen for prevention of breast cancer: report of the National Surgical Adjuvant Breast and Bowel Project P-1 Study. J Natl Cancer Inst 90:1371–1388

Garcia-Carbonero R, Supko JG (2002) Current perspectives on the clinical experience, pharmacology, and continued development of the camptothecins. Clin Cancer Res 8:641–661

Gardiner SJ, Begg EJ, Barclay ML et al (2000) Genetic polymorphism and outcomes with azathioprine and 6-mercaptopurine. Adverse Drug React Toxicol Rev 19:293–312

Gardner ER, Burger H, van Schaik RH et al (2006) Association of enzyme and transporter genotypes with the pharmacokinetics of imatinib. Clin Pharmacol Ther 80:192–201

Gebhart F, Zanker KS, Brandt B et al (1999) Modulation of epidermal growth factor receptor gene transcription by a polymorphic dinucleotide repeat in intron 1. J Biol Chem 274:13176–13180

Glimelius B, Garmo H, Berglund A et al (2011) Prediction of irinotecan and 5-fluorouracil toxicity and response in patients with advanced colorectal cancer. Pharmacogenomics J 11:61–71

Goekkurt E, Hoehn S, Wolschke C et al (2006) Polymorphisms of glutathione S-transferases (GST) and thymidylate synthase (TS) – novel predictors for response and survival in gastric cancer patients. Br J Cancer 94:281–286

Goekkurt E, Al-Batran SE, Mogck U et al (2009) Pharmacogenetic analyses of hematotoxicity in advanced gastric cancer patients receiving biweekly fluorouracil, leucovorin, oxaliplatin and docetaxel (FLOT): a translational study of the Arbeitsgemeinschaft Internistische Onkologie (AIO). Ann Oncol 20(3):481–485

Goetz MP, Rae JM, Suman VJ, Safgren SL, Ames MM, Visscher DW, Reynolds C, Couch FJ, Lingle WL, Flockhart DA, Desta Z, Perez EA, Ingle JN (2005) Pharmacogenetics of tamoxifen biotransformation is associated with clinical outcomes of efficacy and hot flashes. J Clin Oncol 23:9312–9318

Grem JL (2000) 5-Fluorouracil: forty-plus and still ticking. A review of its preclinical and clinical development. Invest New Drugs 18:299–313

Han JY, Lim HS, Shin ES et al (2006) Comprehensive analysis of UGT1A polymorphisms predictive for pharmacokinetics and treatment outcome in patients with non-small-cell lung cancer treated with irinotecan and cisplatin. J Clin Oncol 24:2237–2244

Han JY, Lim HS, Yoo YK et al (2007a) Associations of ABCB1, ABCB2, and ABCG2 polymorphisms with irinotecan-pharmacokinetics and clinical outcome in patients with advanced non-small cell lung cancer. Cancer 110:138–147

Han SW, Jeon YK, Lee KH et al (2007b) Intron 1 *CA* dinucleotide repeat polymorphism and mutations of epidermal growth factor receptor and gefitinib responsiveness in non-small cell lung cancer. Pharmacogenet Genomics 17:313–319

Hasegawa Y, Sarashina T, Ando M et al (2004) Rapid detection of UGT1A1 gene polymorphisms by newly developed invader assay. Clin Chem 50:1479–1480

Hitzl M, Drescher S, van der Kuip H et al (2001) The C3435T mutation in the human MDR1 gene is associated with altered efflux of the P-glycoprotein substrate rhodamine 123 from CD56+ natural killer cells. Pharmacogenetics 11:293–298

Hoffmeyer S, Burk O, von Richter O et al (2000) Functional polymorphisms of the human multidrug-resistance gene: multiple sequence variations and correlation of one allele with P-glycoprotein expression and activity in vivo. Proc Natl Acad Sci USA 97:3473–3478

Hoggatt J (2011) Personalized medicine-trends in molecular diagnostics. Mol Diagn Ther 15:53–55

Horie N, Aiba H, Oguro K et al (1995) Functional analysis and DNA polymorphism of the tandemly repeated sequences in the 5′-terminal regulatory region of the human gene for thymidylate synthase. Cell Struct Funct 20:191–197

Hoskins JM, Goldberg RM, Qu P et al (2007) UGT1A1*28 genotype and irinotecan-induced neutropenia: dose matters. J Natl Cancer Inst 99:1290–1295

Iacopetta B, Grieu F, Joseph D et al (2001) A polymorphism in the enhancer region of the thymidylate synthase promoter influences the survival of colorectal cancer patients treated with 5-fluorouracil. Br J Cancer 85:827–830

Ichihara S, Toyooka S, Fujiwara Y et al (2007) The impact of epidermal growth factor receptor gene status on gefitinib-treated Japanese patients with non-small cell lung cancer. Int J Cancer 120:1239–1247

Ichikawa W, Takahashi T, Suto K et al (2006) Orotate phosphoribosyltransferase gene polymorphism predicts toxicity in patients treated with bolus 5-fluorouracil regimen. Clin Cancer Res 12(13):3928–3934

Illmer T, Schuler US, Thiede C et al (2002) MDR1 gene polymorphisms affect therapy outcome in acute myeloid leukemia patients. Cancer Res 62:4955–4962

Imai Y, Nakane M, Kage K et al (2002) C421A polymorphism in the human breast cancer resistance protein gene is associated with low expression of Q141K protein and low-level drug resistance. Mol Cancer Ther 1:611–616

Ingelman-Sundberg M, Sim SC, Gomez A, Rodriguez-Antona C (2007) Influence of cytochrome P450 polymorphisms on drug therapies: pharmacogenetic, pharmacoepigenetic, and clinical aspects. Pharmacol Ther 116:496–526

Innocenti F, Undevia SD, Iyer L et al (2004) Genetic variants in the UDP-glucuronosyltransferase 1A1 gene predict the risk of severe neutropenia of irinotecan. J Clin Oncol 22:1382–1388

Inoue A, Suzuki T, Fukuhara T et al (2006) Prospective phase II study for chemotherapy naive patients with advanced non-small cell lung cancer with epidermal growth factor receptor gene mutations. J Clin Oncol 24(21):3340–3346

Isla D, Sarries C, Rosell R et al (2004) Single nucleotide polymorphisms and outcome in docetaxel-cisplatin-treated advanced non-small-cell lung cancer. Ann Oncol 15:1194–1203

Iyer L, Hall D, Das S et al (1999) Phenotype-genotype correlation of in vitro SN-38 (active metabolite of irinotecan) and bilirubin glucuronidation in human liver tissue with UGT1A1 promoter polymorphism. Clin Pharmacol Ther 65:576–582

Iyer L, Das S, Janisch L et al (2002) UGT1A1*28 polymorphism as a determinant of irinotecan disposition and toxicity. Pharmacogenomics J 2:43–47

Jada SR, Lim R, Wong CI et al (2007) Role of UGT1A1*6, UGT1A1*28 and ABCG2 c.421C_A polymorphisms in irinotecan-induced neutropenia in Asian cancer patients. Cancer Sci 98:1461–1467

Jin Y, Desta Z, Stearns V, Ward B, Ho H, Lee KH, Skaar T, Storniolo AM, Li L, Araba A, Blanchard R, Nguyen A, Ullmer L, Hayden J, Lemler S, Weinshilboum RM, Rae JM, Hayes DF, Flockhart DA (2005) CYP2D6 genotype. Antidepressant use, and tamoxifen metabolism during adjuvant breast cancer treatment. J Natl Cancer Inst 97:30–39

Kaneda S, Takeishi K, Ayusawa D et al (1987) Role in translation of a triple tandemly repeated sequence in the 5′-untranslated region of human thymidylate synthase mRNA. Nucleic Acids Res 15:1259–1270

Kawakami K, Watanabe G (2003) Identification and functional analysis of single nucleotide polymorphism in the tandem repeat sequence of thymidylate synthase gene. Cancer Res 63:6004–6007

Kawakami K, Omura K, Kanehira E et al (1999) Polymorphic tandem repeats in the thymidylate synthase gene is associated with its protein expression in human gastrointestinal cancers. Anticancer Res 19:3249–3252

Khosrow-Shahi F, Doyle K, Naigamwalla D (2009) Detecting the future: a perspective on the rising value of diagnostics in oncology therapeutics. Oncol Bus Rev March:8–17

Kim RB, Leake BF, Choo EF et al (2001) Identification of functionally variant MDR1 alleles among European Americans and African Americans. Clin Pharmacol Ther 70:189–199

Kirchheiner J, Fuhr U, Brockmoller J (2005) Pharmacogenetics-based therapeutic recommendations – ready for clinical practice? Nat Rev Drug Discov 4:639–647

Kobayashi D, Ieiri I, Hirota T et al (2005) Functional assessment of ABCG2 (BCRP) gene polymorphisms to protein expression in human placenta. Drug Metab Dispos 33:94–101

Kroetz DL, Pauli-Magnus C, Hodges LM et al (2003) Sequence diversity and haplotype structure in the human ABCB1 (MDR1, multidrug resistance transporter) gene. Pharmacogenetics 13:481–494

Krynetski EY, Evans WE (1999) Pharmacogenetics as a molecular basis for individualized drug therapy: the thiopurine S-methyltransferase paradigm. Pharm Res 16:342–349

Krynetski EY, Tai HL, Yates CR et al (1996) Genetic polymorphism of thiopurine S-methyltransferase: clinical importance and molecular mechanisms. Pharmacogenetics 6:279–290

Kweekel DM, Gelderblom H, Van der Straaten T et al (2008) UGT1A1*28 genotype and irinotecan dosage in patients with metastatic colorectal cancer: a Dutch Colorectal Cancer Group study. Br J Cancer 99:275–282

Largillier R, Etienne-Grimaldi MC, Formento JL et al (2006) Pharmacogenetics of capecitabine in advanced breast cancer patients. Clin Cancer Res 12:5496–5502

Lecomte T, Ferraz J, Zinzindohoué F et al (2004) Thymidylate synthase gene polymorphism predicts toxicity in colorectal cancer patients receiving 5-fluorouracil-based chemotherapy. Clin Cancer Res 10:5880–5888

Leichman CG, Lenz HJ, Leichman L et al (1997) Quantitation of intratumoral thymidylate synthase expression predicts for disseminated colorectal cancer response and resistance to protracted-infusion fluorouracil and weekly leucovorin. J Clin Oncol 15:3223–3229

Lennard L (1992) The clinical pharmacology of 6-mercaptopurine. Eur J Clin Pharmacol 43:329–339

Li J, Cusatis G, Brahmer J et al (2007) Association of variant ABCG2 and the pharmacokinetics of epidermal growth factor receptor tyrosine kinase inhibitors in cancer patients. Cancer Biol Ther 6:432–438

Litos IK, Emmanouilidou E, Glynou KM et al (2007) Rapid genotyping of CYP2D6, CYP2C19 and TPMT polymorphisms by primer extension reaction in a dipstick format. Anal Bioanal Chem 389:1849–1857

Liu W, Innocenti F, Chen P et al (2003) Interethnic difference in the allelic distribution of human epidermal growth factor receptor intron 1 polymorphism. Clin Cancer Res 9:1009–1012

Liu W, Innocenti F, Wu MH et al (2005) A functional common polymorphism in a Sp1 recognition site of the epidermal growth factor receptor. Cancer Res 65:46–53

Luo HR, Lü XM, Yao YG et al (2002) Length polymorphism of thymidylate synthase regulatory region in Chinese populations and evolution of the novel alleles. Biochem Genet 40:41–51

Lynch TJ, Bell DW, Sordella R et al (2004) Activating mutations in the epidermal growth factor receptor underlying responsiveness of non-small cell lung cancer to gefitinib. N Engl J Med 350:2129–2139

Maitland M, Vasisht K, Ratain M (2006) TPMT, UGT1A1 and DPYD: genotyping to ensure safer cancer therapy? Trends Pharmacol Sci 27(8):432–437

Mandola MV, Stoehlmacher J, Muller-Weeks S et al (2003) A novel single nucleotide polymorphism within the 5′ tandem repeat polymorphism of the thymidylate synthase gene abolishes USF-1 binding and alters transcriptional activity. Cancer Res 63:2898–2904

Mandola MV, Stoehlmacher J, Zhang W et al (2004) A 6 bp polymorphism in the thymidylate synthase gene causes message instability and is associated with decreased intratumoral TS mRNA levels. Pharmacogenetics 14:319–327

Marcuello E, Altés A, del Rio E et al (2004a) Single nucleotide polymorphism in the 5′ tandem repeat sequences of thymidylate synthase gene predicts for response to fluorouracil-based chemotherapy in advanced colorectal cancer patients. Int J Cancer 112:733–737

Marcuello E, Altés A, Menoyo A et al (2004b) UGT1A1 gene variations and irinotecan treatment in patients with metastatic colorectal cancer. Br J Cancer 91:678–682

Marsh S, Ameyaw MM, Githang'a J et al (2000) Novel thymidylate synthase enhancer region alleles in African populations. Hum Mutat 16:528

Martinez-Balibrea E, Abad A, Aranda E et al (2008) Pharmacogenetic approach for capecitabine or 5-fluorouracil selection to be combined with oxaliplatin as first-line chemotherapy in advanced colorectal cancer. Eur J Cancer 44:1229–1237

McLeod HL, Lin JS, Scott EP et al (1994) Thiopurine methyltransferase activity in American white subjects and black subjects. Clin Pharmacol Ther 55:15–20

McLeod HL, Krynetski EY, Relling MV et al (2000) Genetic polymorphism of thiopurine methyltransferase and its clinical relevance for childhood acute lymphoblastic leukemia. Leukemia 14:567–572

McLeod HL, Sargent DJ, Marsh S et al (2010) Pharmacogenetic predictors of adverse events and response to chemotherapy in metastatic colorectal cancer: results from North American Gastrointestinal Intergroup Trial N9741. J Clin Oncol 28:3227–3233

Monaghan G, Ryan M, Seddon R et al (1996) Genetic variation in bilirubin UPD-glucuronosyltransferase gene promoter and Gilbert's syndrome. Lancet 347:578–581

Morel A, Boisdron-Celle M, Fey L et al (2006) Clinical relevance of different dihydropyrimidine dehydrogenase gene single nucleotide polymorphisms on 5-fluorouracil tolerance. Mol Cancer Ther 5:2895–2904

Morganti M, Ciantelli M, Giglioni B et al (2005) Relationships between promoter polymorphisms in the thymidylate synthase gene and mRNA levels in colorectal cancers. Eur J Cancer 41:2176–2183

Moriai T, Kobrin MS, Hope C, Speck L, Korc M (1994) A variant epidermal growth factor receptor exhibits altered type-α transforming growth factor binding and transmembrane signaling. Proc Natl Acad Sci USA 91:10217–10221

Nagar S, Remmel RP (2006) Uridine diphosphoglucuronosyltransferase pharmacogenetics and cancer. Oncogene 25:1659–1672

Nakamura T, Sakaeda T, Horinouchi M et al (2002) Effect of the mutation (C3435T) at exon 26 of the MDR1 gene on expression level of MDR1 messenger ribonucleic acid in duodenal enterocytes of healthy Japanese subjects. Clin Pharmacol Ther 71:297–303

Nasedkina TV, Fedorova OE, Glotov AS et al (2006) Rapid genotyping of common deficient thiopurine S-methyltransferase alleles using the DNA microchip technique. Eur J Hum Genet 14:991–998

Naylor S, Cole T (2010) Companion diagnostics in the pharmaceutical industry part II: business models. Drug Discov World 11(3):61–68

Nie Q, Wang Z, Zhang GC et al (2007) The epidermal growth factor receptor intron 1 (*CA*)n microsatellite polymorphism is a potential predictor of treatment outcome in patients with advanced lung cancer treated with gefitinib. Eur J Pharmacol 570:175–181

Nishimura R, Nagao K, Miyayama H et al (1999) Thymidylate synthase levels as a therapeutic and prognostic predictor in breast cancer. Anticancer Res 19:5621–5626

Nomura M, Shigematsu H, Lin L et al (2007) Polymorphisms, mutations, and amplification of the *EGFR* gene in non-small cell lung cancers. PLoS Med 4(4):E125

Otterness D, Szumlanski C, Lennard L et al (1997) Human thiopurine methyltransferase pharmacogenetics: gene sequence polymorphisms. Clin Pharmacol Ther 62:60–73

Paez JG, Janne PA, Lee JC et al (2005) *EGFR* mutations in lung cancer: correlation with clinical response to gefitinib therapy. Science 304:1497–1500

Park DJ, Stoehlmacher J, Zhang W et al (2002) Thymidylate synthase gene polymorphism predicts response to capecitabine in advanced colorectal cancer. Int J Colorectal Dis 17:46–49

Petain A, Kattygnarath D, Azard J et al (2008) Population pharmacokinetics and pharmacogenetics of imatinib in children and adults. Clin Cancer Res 14:7102–7109

Pillot GA, Read WL, Hennenfent KL et al (2006) A phase II study of irinotecan and carboplatin in advanced non-small cell lung cancer with pharmacogenomic analysis: final report. J Thorac Oncol 1:972–978

Pullarkat S, Stoehlmacher J, Ghaderi V et al (2001) Thymidylate synthase gene polymorphism determines response and toxicity of 5-FU chemotherapy. Pharmacogenomics J 1:65–70

Raijmakers MT, Jansen PL, Steegers EA et al (2000) Association of human liver bilirubin UDP-glucuronyltransferase activity with a polymorphism in the promoter region of the UGT1A1 gene. J Hepatol 33:348–351

Ratain MJ (2006) From bedside to bench to bedside to clinical practice: an odyssey with irinotecan. Clin Cancer Res 12:1658–1660

Rouits E, Boisdron-Celle M, Dumont A et al (2004) Relevance of different UGT1A1 polymorphisms in irinotecan-induced toxicity: a molecular and clinical study of 75 patients. Clin Cancer Res 10:5151–5159

Rustum YM, Harstrick A, Cao S et al (1997) Thymidylate synthase inhibitors in cancer therapy: direct and indirect inhibitors. J Clin Oncol 15:389–400

Ruzzo A, Graziano F, Kawakami K et al (2006) Pharmacogenetic profiling and clinical outcome of patients with advanced gastric cancer treated with palliative chemotherapy. J Clin Oncol 24:1883–1891

Sahasranaman S, Howard D, Roy S (2008) Clinical pharmacology and pharmacogenetics of thiopurines. Eur J Clin Pharmacol 64:753–767

Sai K, Kaniwa N, Itoda M et al (2003) Haplotype analysis of ABCB1/MDR1 blocks in a Japanese population reveals genotype-dependent renal clearance of irinotecan. Pharmacogenetics 13:741–757

Sai K, Saeki M, Saito Y et al (2004) UGT1A1 haplotypes associated with reduced glucuronidation and increased serum bilirubin in irinotecan-administered Japanese patients with cancer. Clin Pharmacol Ther 75:501–515

Schaeffeler E, Zanger UM, Eichelbaum M et al (2008) Highly multiplexed genotyping of thiopurine S-methyltransferase variants using MALD-TOF mass spectrometry: reliable genotyping in different ethnic groups. Clin Chem 54:1637–1647

Schinkel AH, Jonker JW (2003) Mammalian drug efflux transporters of the ATP binding cassette (ABC) family: an overview. Adv Drug Deliv Rev 55:3–29

Schwab M, Eichelbaum M, Fromm MF (2003) Genetic polymorphisms of the human MDR1 drug transporter. Annu Rev Pharmacol Toxicol 43:285–307

Sequist LV, Bell DW, Lynch TJ, Haber DA (2007) Molecular predictors of response to epidermal growth factor receptor antagonists in non-small cell lung cancer. J Clin Oncol 25(5):587–595

Sequist LV, Martins RG, Spigel D et al (2008) First-line gefitinib in patients with advanced non-small cell lung cancer harboring somatic *EGFR* mutations. J Clin Oncol 26(15):2442–2449

Sharma R, Hoskins JM, Rivory LP, Zucknick M, London R, Liddle C, Clarke SJ (2008) Thymidylate synthase and methyltetrahydrofolate reductase gene polymorphisms and toxicity to capecitabine in advanced colorectal cancer patients. Clin Cancer Res 14:817–825

Siegsmund M, Brinkmann U, Schaffeler E et al (2002) Association of the P-glycoprotein transporter MDR1 (C3435T) polymorphism with the susceptibility to renal epithelial tumors. J Am Soc Nephrol 13:1847–1854

Spire-Vayron de la Moureyre C, Debuysère H, Sabbagh N et al (1998) Detection of known and new mutations in the thiopurine S-methyltransferase gene by single-strand conformation polymorphism analysis. Hum Mutat 12:177–185

Stewart CF, Panetta JC, O'Shaughnessy MA et al (2007) UGT1A1 promoter genotype correlates with SN-38 pharmacokinetics, but not severe toxicity in patients receiving low-dose irinotecan. J Clin Oncol 25:2594–2600

Stoehlmacher J, Park DJ, Zhang W et al (2004) A multivariate analysis of genomic polymorphisms: prediction of clinical outcome to 5-FU/oxaliplatin combination chemotherapy in refractory colorectal cancer. Br J Cancer 91:344–354

Swann PF, Waters TR, Moulton DC et al (1996) Role of postreplicative DNA mismatch repair in the cytotoxic action of thioguanine. Science 273:1109–1111

Tai HL, Krynetski EY, Yates CR et al (1996) Thiopurine S-methyltransferase deficiency: two nucleotide transitions define the most prevalent mutant allele associated with loss of catalytic activity in Caucasians. Am J Hum Genet 58:694–702

Tai HL, Krynetski EY, Schuetz EG et al (1997) Enhanced proteolysis of thiopurine S-methyltransferase (TPMT) encoded by mutant alleles in humans (TPMT*3A, TPMT*2): mechanisms for the genetic polymorphism of TPMT activity. Proc Natl Acad Sci USA 94:6444–6449

Tamura K, Okamoto I, Negoro S et al (2008) Multicentre prospective phase II trial of gefitinib for advanced non-small cell lung cancer with epidermal growth factor receptor mutations: results of the West Japan Thoracic Oncology Group trial (WJTOG0403). Br J Cancer 98:907–914

Toffoli G, Cecchin E, Corona G et al (2006) The role of UGT1A1*28 polymorphism in the pharmacodynamics and pharmacokinetics of irinotecan in patients with metastatic colorectal cancer. J Clin Oncol 24:3061–3068

Uchida K, Hayashi K, Kawakami K et al (2004) Loss of heterozygosity at the thymidylate synthase (TS) locus on chromosome 18 affects tumor response and survival in individuals heterozygous for a 28-bp polymorphism in the TS gene. Clin Cancer Res 10:433–439

Ulrich CM, Bigler J, Velicer CM et al (2000) Searching expressed sequence tag databases: discovery and confirmation of a common polymorphism in the thymidylate synthase gene. Cancer Epidemiol Biomarkers Prev 9:1381–1385

Van Kuilenburg AB, Meinsma R, Zoetekouw L, Van Gennip AH (2002) Increased risk of grade IV neutropenia after administration of 5-fluorouracil due to a dihydropyrimidine dehydrogenase deficiency: high prevalence of the IVS14+1g>a mutation. Int J Cancer 101:253–258

Villafranca E, Okruzhnov Y, Dominguez MA et al (2001) Polymorphisms of the repeated sequences in the enhancer region of the thymidylate synthase gene promoter may predict downstaging after preoperative chemoradiation in rectal cancer. J Clin Oncol 19:1779–1786

Wang D, Johnson AD, Papp AC et al (2005) Multidrug resistance polypeptide 1 (MDR1, ABCB1) variant 3435C>t affects mRNA stability. Pharmacogenet Genomics 15:693–704

Wei X, Elizondo G, Sapone A et al (1998) Characterization of the human dihydropyrimidine dehydrogenase gene. Genomics 51:391–400

Welsh SJ, Titley J, Brunton L et al (2000) Comparison of thymidylate synthase (TS) protein up-regulation after exposure to TS inhibitors in normal and tumor cell lines and tissues. Clin Cancer Res 6:2538–2546

Yates CR, Krynetski EY, Loennechen T et al (1997) Molecular diagnosis of thiopurine S-methyltransferase deficiency: genetic basis for azathioprine and mercaptopurine intolerance. Ann Intern Med 126:608–614

Yen J, McLeod H (2007) Should DPD analysis be required prior to prescribing fluoropyrimidines? Eur J Cancer 43:1011–1016

Yoshida K, Yatabe Y, Park JY et al (2007) Prospective validation for prediction of gefitinib sensitivity by epidermal growth factor receptor mutation in patients with non-small cell lung cancer. J Thorac Oncol 2(1):22–28

Young D (2006) Genetics examined in tamoxifen's effectiveness: recurrence warning urged for labeling. Am J Health Syst Pharm 63:2286–2296

Zamber CP, Lamba JK, Yasuda K et al (2003) Natural allelic variants of breast cancer resistance protein (BCRP) and their relationship to BCRP expression in human intestine. Pharmacogenetics 13:19–28

Zhang W, Weissfeld JL, Romkes M, Land SR, Grandis JR, Siegfried JM (2007) Association of the *EGFR* intron 1 *CA* repeat length with lung cancer risk. Mol Carcinog 46(5):372–380

Pharmacogenomics in Acute Myeloid Leukemia

12

Omer Faruk Hatipoglu, Onur Bender, Esra Gunduz, and Mehmet Gunduz

Abstract

Acute myeloid leukemia (AML) is a cancer of the myeloid line of blood cells. In AML, the bone marrow makes many immature cells called blasts, which do not mature and cannot fight infections. AML is a heterogeneous neoplasm with several pathological, genetic, and molecular subtypes. Combinations of various doses and schedules of drugs have been the majority of treatment for all types of AMLs in adult patients. However, not all patients have the same response to these treatments, some of which are adverse responses that are potentially life threatening. Because interindividual responses to AML medications can vary considerably, the potential for genetic contributions to variable drug responses is significant. The pharmacogenomics approach tries to find prognostic and predictive biomarkers permitting to identify patients who could benefit from a particular treatment or those exhibiting higher risks of toxicity. Pharmacogenomics is a rapidly improving science with the potential to revolutionize drug discovery/development and offers one possibility for rationalizing therapy/dose selection. It combines many different fields such as genetics, genomics, molecular biology, pharmacology, pharmaceutics, and population biology. This chapter focuses on treatment of AML, genetic and clinical prognostic markers, and recent advances in the field of pharmacogenomics in AML.

1 Introduction

Acute myeloid leukemia (AML), also known as acute myelogenous leukemia or acute nonlymphocytic leukemia, is a clonal hematopoietic disorder that may be derived from either a hematopoietic stem cell or a lineage-specific progenitor cell (Lane et al. 2009; Jordan et al. 2006). AML is one of the four types of leukemia (Table 12.1). According to the American Cancer Society in 2010, approximately 13,000 new cases

O.F. Hatipoglu • O. Bender • E. Gunduz
M. Gunduz (✉)
Department of Medical Genetics,
Faculty of Medicine, Turgut Ozal University,
Ankara, Turkey
e-mail: mehmet.gunduz@gmail.com

D. Barh et al. (eds.), *Omics for Personalized Medicine*,
DOI 10.1007/978-81-322-1184-6_12, © Springer India 2013

Table 12.1 Classification of leukemias

	Acute	Chronic
Lymphoid origin	Acute lymphoblastic leukemia (ALL) – mostly in young children	Chronic lymphocytic leukemia (CLL) – mostly over the age of 55
Myeloid origin	Acute myeloid leukemia (AML) – occurs in both adults and children	Chronic myeloid leukemia (CML) – mainly in adults

of AML are diagnosed in the United States, and it is commonly a disease of elderly people with a median age of diagnosis of approximately 72 years old (Juliusson et al. 2009). In most cases the etiology for AML is unknown. However, several known risk factors have been identified. Environmental risk factors, including exposure to ionizing radiation, petroleum, benzene, and benzene-containing compounds, can result in bone marrow damage leading to AML (Levine and Bloomfield 1992; Austin et al. 1988). Patients who develop AML may have an antecedent predisposing disease, such as aplastic anemia and myeloma. A number of inherited conditions, for example, Down syndrome, Bloom syndrome, and others, carry an increased risk of AML (Wiernik 1997; Miller 1971; Zipursky et al. 1992; Bartsocas and Loukopoulos 1992; Horwitz 1997). AML is classified morphologically according to the French-American-British (FAB) classification subtypes of M0 to M7 (Bennett et al. 1976). Recently there is a massive progression in new prognostic factors in AML, which is driving our understanding of disease biology and also the development of new therapeutic targets. Many new important prognostic factors have been identified including gene mutations in nucleophosmin1 (NPM1), internal tandem duplications in the fms-related tyrosine kinase-3 (FLT3-ITD), and CCAAT/enhancer-binding protein-α (CEBPA). Such gene mutations have attracted considerable attention in AML with normal karyotype.

2 Molecular Alterations in Acute Myeloid Leukemia

Studies showed that approximately 45 % of adult AML patients had no detectable chromosomal aberrations until recently but the availability of new genetic and molecular prognostic markers in AML has grown considerably. Patients with AML are generally categorized into three risk groups: favorable, intermediate, and adverse (Byrd et al. 2002; Estey and Dohner 2006) (Table 12.2). In recent years, a number of prognostically important gene mutations or deregulated gene expressions have been identified in all subtypes of AML. For example, internal tandem duplication (ITD) in the fms-like tyrosine kinase-3 gene (FLT3), a hematopoietic growth factor receptor, has been associated with poor treatment outcome. Another molecular abnormality that is frequently detected in AML is the presence of a partial tandem duplication of the mixed lineage leukemia gene (MLL-PTD); patients with an MLL-PTD mutation have in general a relative worse prognosis (Zorko et al. 2012; Kindler et al. 2010). Most frequent prognostically important molecular abnormalities and their biological/clinical features in AML are depicted in Table 12.3.

Multistep pathogenesis of AML model hypothesizes that AML is the consequence of a collaboration between two broadly defined complementation groups (Kelly and Gilliland 2002). Group one (class I) mutations consist of activated signal transduction pathways, such as BCR/ABL, TEL/PDGFßR, N-RAS, K-RAS, or activated FLT3, resulting in enhanced proliferation and/or survival of leukemic progenitor cells. These mutant genes affect a CML-like disease when expressed alone. Another group (class II) comprises mutations, which result in loss of function of transcription factors or components of the transcriptional co-activation complex. t(8;21), inv(16)/t(16;16), and t(15;17) or mutations in CEBPA, MLL, and NPM1 would be considered class II mutations. These mutations have both emerged as important prognostic and predictive markers and targets for novel therapies.

Table 12.2 Prognostic subgroups of AML based on presenting cytogenetics and genetic lesions

Cytogenetic subgroup	% of patient	10 year survival (%)	Cytogenetic findings	Molecular abnormalities
Favorable	23	69	t(15;17) PML	NPM1, CEBPA
			t(8; 21) AML1	Non FLT3-ITD (normal cytogenetics)
			inv (16) CBFβ	
Intermediate	17	33	Normal cytogenetics	c-KIT mutation with:
			+8, t(3;5)4, t(9;11)(p22q23)	t(8;21)(q22;q22), or
				inv(16)(p13.q22), t(16;16) (p13.q22)
Unfavorable	17	12	Complex (≥3 abnormalities)	FLT3-ITD
			inv(3)(q21q26)/t(3;3)(q21;q26)	Non NPM1 mutation
			t(11q23) [non t(9;11) (p21 ~ 22;q23), t(11;19)(q23;p13)]	(Normal cytogenetics)
			(−5, 5q-), (MK+), (−7, 7q-)	High EVI1 expression
			abn(3q) [non t(3;5) (q21 ~ 25;q31 ~ 35)], −7, add(7q)/del(7q), t(10;11) (p11 ~ 13;q23), t(9;22)(q34;q11), add(5q), del(5q), −5, t(6;11) (q27;q23), −17/abn(17p)	
(Normal karyotype)	43	38		

Adapted from Byrd et al. (2002), Estey and Dohner (2006), Foran (2010), Grimwade et al. (1998, 2010), Vardiman et al. (2009), Dohner et al. (2010)

Table 12.3 Mutated genes in AML

Gene	Biological/clinical features	References
NPM1	Encodes a phosphoprotein with pleiotropic functions	Schnittger et al. (2009)
	NPM1 mutations found in 25–35 % of adult AML, in 45–64 % of NC-AML	Mrozek et al. (2007), Dohner et al. (2005)
FLT3	Mutations are found in 30 % of NC-AML	Chang et al. (2010)
	Member of the class III receptor tyrosine kinase family	Stirewalt and Radich (2003)
CEBPA	Encodes a master regulatory transcription factor in hematopoiesis	Preudhomme et al. (2002)
	Mutations are found in 15–20 % of NC-AML	

2.1 Mutations in NPM1

Mutations in nucleophosmin (NPM1), also known as nucleolar phosphoprotein B23 or numatrin, are the most frequently acquired molecular abnormalities in AML (Thiede et al. 2006). Falini and colleagues described a set of common mutations within the final exon of the NPM1 gene in primary NK-AML patients, which alter the N-terminal domain nuclear localization signal leading to abnormal cytoplasmic accumulation of the NPM1 phosphoprotein (Falini et al. 2005). NPM1 encodes a phosphoprotein and monitoring can be performed by quantitative PCR (Schnittger et al. 2009; Grisendi et al. 2006). NPM1 mutations were found in 25–35 % of adult AML and in 45–64 % of NC-AML (Mrozek et al. 2007; Dohner et al. 2005). Nucleophosmin has multiple functions, some of which are as follows: (a) encodes a phosphoprotein with pleiotropic functions, (b) genomic stability and DNA repair, (c) endoribonuclease activity, (d) centrosome duplication during cell cycle, (e) regulation of ARF-p53 tumor suppressor pathway, (f) histone chaperones, and (g) prevents apoptosis (Falini et al. 2005; Lindstrom 2011).

2.2 Mutations in FLT3

FLT3 (fms-like tyrosine kinase-3) is a receptor tyrosine kinase, located on chromosome 13q12, with important roles in hematopoietic stem/progenitor cell survival and proliferation. It regulates cell proliferation and differentiation, and FLT3 mutation results in constitutively active FLT3 protein that promotes Stat5 phosphorylation, leading to uncontrolled hematopoietic cell proliferation. Same large-scale studies have shown that FLT3/ITD mutations, found in about 30 % of patients with AML, are strongly associated with a poor prognosis and a high leukemia cell count in patients with AML, suggesting that FLT3 mutations are involved in disease progression (Chang et al. 2010; Stirewalt and Radich 2003).

2.3 Mutations in CEBPA

The CEBPA (CCAAT/enhancer-binding protein-alpha) is an essential transcription factor for granulocytic differentiation, and mutation leads to failure of granulocytic differentiation and its null mice lack neutrophils and eosinophils but retain monocytes (Leroy et al. 2005; Zhang et al. 1997). Mutations of CEBPA are found in 15–20 % of NC-AML and it encodes a master regulatory transcription factor in hematopoiesis (Preudhomme et al. 2002). Two major types of heterozygous CEBPA mutations have been identified in AML. The most frequent mutation is combination of N-terminal mutations and bZIP mutations. When both mutations exist, they are usually located in different mutated alleles, and CEBPA wild-type protein is not expressed (Wouters et al. 2009).

3 Gene Expression Profiling and SNPs for Diagnosis and Prognosis

In AML, gene expression profiling has been used to provide prognostic information and to discriminate between different histological subtypes. Recently, large clinical trials have demonstrated that distinct molecular aberrations in AML such as mutations of the CEBPA and NPM1 genes can be reliably diagnosed based on resulting characteristic gene expression signatures. Such gene expression can also provide additional molecular insights as exemplified by the identification of the role of deregulated FLT3 expression in MLL-rearranged leukemia (Haferlach et al. 2005; Armstrong et al. 2002). Many studies have reported the association of gene expression profiles with prognosis in cancer patients. Haferlach et al. have performed the clinical utility of gene expression profiling in AML in 11 laboratories across three continents and included more than 3,000 patients. This study has shown the technology robust for the diagnosis of hematologic malignancies with high accuracy. Briefly, gene expression profiling can be used as a powerful tool in investigating a variety of transcriptional phenomena in AML (Van't Veer et al. 2002; Schramm et al. 2005; Haferlach et al. 2010).

3.1 Array Technology

DNA microarray technology has made it possible to simultaneously analyze the expression of thousands of genes in a small sample of tumor tissue, and this technology might permit the identification of novel markers for early detection of disease and provide insights into the mechanisms of cancer growth and chemotherapy resistance (Liotta and Petricoin 2000).

miRNAs are small, noncoding RNAs that downregulate gene expression by binding to the 3′UTR of the target gene with partial complementarity, which leads to an inhibition of translation and facilitated degradation of the target mRNA, at the posttranscriptional level (Filipowicz et al. 2005; Bartel 2004). Recent studies have shown that many miRNAs are implicated in common human cancers through a variety of mechanisms and miRNAs have the potential to be used in disease prognosis and diagnosis. Polymorphisms in the miRNA pathway (miR-polymorphisms) are also emerging as powerful tools to understand the biology of a disease (Mishra and Bertino 2009; Rukov et al. 2011). The first report linking miRNA and lymphocytic leukemia involved the deletion or the downregulation of miR-15 and miR-16 (Calin et al. 2002; Cimmino et al. 2005). Cytogenetically, more than 200 chromosomal abnormalities identified with AML. Some recent studies have shown using microRNA expression profiling that several of cytogenetic subtypes of AML are associated with microRNA signatures (Mrozek et al. 2004).

3.2 Single-Nucleotide Polymorphism

Single-nucleotide polymorphism (SNP) is the most common polymorphism, which is due to single base pair difference, and SNP arrays allow in determining disease susceptibility and for measuring the efficacy of drug therapies designed specifically for individuals (Sachidanandam et al. 2001). Two studies of AML indicate the value of paired samples in SNP array analysis. The first study analyzed 86 adult AML cases using ultra-high resolution Affymetrix 6.0 SNP arrays with paired normal DNA in all cases, and the second one studied 111 pediatric AML cases with high-resolution SNP arrays using paired samples for 65 patients (Walter et al. 2009; Radtke et al. 2009). SNP array to identify somatic genetic changes linked to transformation deletion, and amplification, will prove superior to standard cytogenetic analysis for classification of AML cases.

4 Pharmacogenomics in AML

4.1 Gene Expression Related to Drug Resistance in AML

Drug resistance constitutes one of the problems in therapy of AML. In the AML patients, the occurrence of resistance to structurally and functionally unrelated chemotherapeutic agents is called multidrug resistance (MDR). MDR is caused by multidrug resistance protein1 (MDR1) gene product, a 170–180 kDa glycoprotein known as P-glycoprotein (Pgp) (Hirose et al. 2003). Pgp has been shown to be associated with poor treatment outcome in AML patients and overexpression of Pgp in cell lines and offers cellular resistance to a wide variety of anticancer drugs, including many agents used in the treatment of AML (van der Kolk et al. 2002). MRP1 homologues MRP2, MRP3, MRP5, and MRP6 have been shown to be expressed at variable levels in AML patient cells. In addition Badrul et al. have found different polymorphisms of the MDR1 gene in many populations, and it was significantly influenced by ethnicities (Badrul Hisam et al. 2006). The list of some polymorphisms in drug transporter genes in AML is outlined in Table 12.4 (van den Heuvel-Eibrink et al. 2001; Kim et al. 2006a, b; Illmer et al. 2002; Gradhand and Kim 2008; Schuetz et al. 1999; Abla et al. 2008; Huang and Sadee 2006).

Eisele et al. performed gene expression patterns in 14 patients with de novo AML, for the purpose of identification of the genes, which might be related to drug resistance. The result has shown that between the complete remission and blast persistence groups with regard to nucleotide metabolism, apoptosis, and reactive oxygen species metabolism, the drug resistance in AML is a heterogenous phenomenon and that might be better focusing on the alteration of the expression of different genes (Eisele et al. 2007).

Novel approaches for drug resistance are the combined use of FLT3 inhibitors with inhibitors of viability signaling or components of PI3K/Akt, MAPK, and JAK/STAT pathway signaling. However, several FLT3 inhibitors are presently being used in clinical trials with sufficient efficacy. In addition signaling pathways in AML may in the future help rationally selected targeted therapies in individual patients.

4.2 Drug Transporters and Drug Delivery in AML

Strategies of drug delivery have largely depended on therapeutic efficacy of drugs in AML. Molecular targeted therapy, with the potential for increased selectivity and fewer adverse effects, holds promise in the treatment of AML. One of the candidate targets is the FLT3 and has led to

Table 12.4 Polymorphisms in drug transporter genes in AML

Gene name	Transporter name	SNP	Treatment	Reference
MDR1	MDR1	G2677T	Etoposide, mitoxantrone, idarubicin	van den Heuvel-Eibrink et al. (2001)
MDR1	MDR1	G2677A	Cyclosporine	Kim et al. (2006a)
MDR1	MDR1	G2677T/A + C3435T	Idarubicin, cytarabine	Kim et al. (2006b)
MDR1	MDR1	C3435T	Etoposide, mitoxantrone, daunorubicin	Illmer et al. (2002)
ABCC4, ABCC5	MRP4, MRP5	Gly187Trp, Gly487Glu	Azidothymidine, mercaptopurine, thioguanine, abacavir	Gradhand and Kim (2008), Schuetz et al. (1999), Abla et al. (2008)
SLC28A1, SLC28A2, SLC28A3	CNT1, CNT2, CNT3	Unclear relevance	Didanosine, idoxuridine, zidovudine	Huang and Sadee (2006)

Table 12.5 Drug transporters and inhibitor relevant to AML

Name	Inhibitors	Reference
ABCC1(MRP-1)	Daunorubicin, mitoxantrone, etoposide	Abla et al. (2008)
ABCC2(MRP-2)	Vincristine, mitoxantrone, etoposide	Huang and Sadee (2006)
ABCC3(MRP-3)	Etoposide, teniposide, methotrexate	Eisele et al. (2007)
ABCB1	Etoposide, daunorubicin	Smith et al. (2004)
ABCG2	Mitoxantrone, methotrexate, cladribine, topotecan, imatinib	Rodriguez-Ariza et al. (2011)

the approval of a molecularly based therapy in AML. Several drugs such as MLN 518, PKC 412, and CEP-701 have been developed to target FLT3 activity in AML. Phase 1 and 2 studies in patients with poor-risk AML have been completed; PKC 412 and CEP 701 can show biological activity and inhibit autophosphorylation in the target (Smith et al. 2004; Stone et al. 2004). The other target is vascular endothelial growth factor (VEGF) pathway, which is a key regulator of angiogenesis. It has led to the development of several VEGF-targeted approaches (Rodriguez-Ariza et al. 2011).

Expression of the ATP-binding cassette (ABC) transporter P-glycoprotein (Pgp) and expression of the ABC transporter multidrug resistance protein (MRP) and the vault-transporter lung resistance protein (LRP) have generally been reported to correlate with prognosis in AML (List et al. 1996; Leith et al. 1999). Transport proteins play an important role in targeted therapy of AML, such as ABCC1 (Burcu et al. 2008), ABCC2 (van der Kolk et al. 1998), ABCC3 (De Jonge et al. 2010), and ABCB1 (Green et al. 2012) (Table 12.5). ABCG2 encodes a transporter protein that is associated with multidrug-resistant phenotypes in AML. High levels of expression of ABCG2 are generally associated with a poor prognosis. Clinical interest has focused on its role in multidrug resistance in human cancers (Robey et al. 2007, 2009).

4.3 Genetic Polymorphisms of Drug-Metabolizing Enzymes and Transporters in AML

There are over 30 families of drug-metabolizing enzymes in humans. Most of them have genetic variants which cause functional changes in the proteins encoded and thereby change the metabolism of drugs (Evans and Relling 1999).

The superfamily of cytochrome P450 enzymes has a crucial role in the metabolism of drugs; so far 17 families of CYPs with about 50 isoforms have been characterized in the human genomes, and these enzymes show extensive structural polymorphism. The most abundant cytochrome

P450 family is CYP3A, and the two key members of this family CYP3A4 and CYP3A5 catalyze the metabolism of a substantial percentage of medications (Relling and Dervieux 2001). Molecular epidemiological studies have recently shown associations between these enzyme variants and altered risk of variety of cancers such as ovarian, bladder, and breast (Spurdle et al. 2001; Brockmoller et al. 1996; Loktionov et al. 2001). CYP2E1, CYP2D6, and GSTM1 genes have shown significant correlation in the etiology of acute leukemias. Rollinson et al. reported an association between GSTT1 and GSTM1 null genotypes and the risk of AML (Rollinson et al. 2000), Davies et al. reported GSTM1 null genotype has an increased risk for AML in children (Davies et al. 2000), and Sayitoglu et al. showed statistically significant association of adult AML but failed to link CYP2D6*4 with AML incidence (Aydin-Sayitoglu et al. 2006).

The genetic polymorphism of thiopurine methyltransferase (TPMT) is one of the most well-known examples of clinical pharmacogenomics. TPMT catalyzes the S-methylation of the thiopurine drugs such as 6-mercaptopurine, azathioprine, and thioguanine (Peng et al. 2008). Genetic polymorphism of TPMT can negatively influence tolerance to thiopurine-based drugs, causing dosage-related hematopoietic toxicity (Dubinsky et al. 2002). Niedzielska et al. have shown that SNPs of TPMT are associated with lower risk of recurrence of acute leukemia in the pediatric population (Niedzielska and Chybicka 2011).

4.4 Genetic Alterations Influencing Anti-AML Drug Response and Adverse Effects

4.4.1 Cytarabine (Ara-C)

Cytarabine, ara-C, is a chemotherapy agent used mainly in the treatment of AML and non-Hodgkin lymphoma. Ara-C kills cancer cells by interfering with DNA synthesis (Wang et al. 1997). Key enzymes within the ara-C metabolic pathway are the deoxycytidine kinase (dCK) which plays an important role in ara-C resistance in vitro. Inactivated dCK transcripts due to alternative splicing were indicated in AML patients resistant to chemotherapy(Veuger et al. 2000). Jing et al. identified two single-nucleotide polymorphisms in the regulatory region at dCK (−201C>T and −360C>G) (Shi et al. 2004).

Galmarini et al. reported that expression of 5′-nucleotidase (5NT) or deficiency of human equilibrative nucleoside transporter 1 (hENT1) involved in resistance to ara-C in patients with AML (Galmarini et al. 2002). It is reported that the AA and GG genotypes of Janus kinase-2 (JAK2), member of a family of tyrosine kinases, A830G are important markers for therapy outcomes in AML patients in a Chinese population (Zhong et al. 2010). Yue et al. collected cDNAs from 52 leukemia/lymphoma samples and analyzed by direct sequencer. They reported that a population characterized with 208A genotype for human cytidine deaminase (HDCA), which leads one more sensitive to ara-C treatment than prototype (Yue et al. 2003). Ara-C metabolism and key enzymes are outlined in Fig. 12.1.

4.4.2 Doxorubicin and Daunorubicin

Daunorubicin and doxorubicin, anti-AML drugs, belonging to the anthracycline group, are widely used in human cancer chemotherapy (Weiss 1992). The anthracyclines are substrates for GSTs, and those enzymes play an important role in the detoxification of chemotherapeutic agents (Michael and Doherty 2005). Studies reported that deletions in GSTM1 and GSTT1, GST subfamily genes, influence the outcome of treatment with induction therapy in AML patients (Voso et al. 2002). Lack of enzymatic activity of GSTs by polymorphisms will lead to reduced detoxification. Studies reported that a remarkable relation exists between risk for death due to toxicity and GSTT1 null genotype in AML patients. Davies et al. showed that GSTT1 null genotype was associated with a worse prognosis in children with AML and reduced survival after standard induction therapy compared to children with at least one GSTT1 allele (Davies et al. 2001).

Doxorubicin is a known MDR1 substrate and inducer, and its cytostatic efficacy is thus limited by MDR1 overexpression. Daunorubicin is used in the treatment of hematopoietic malignancies.

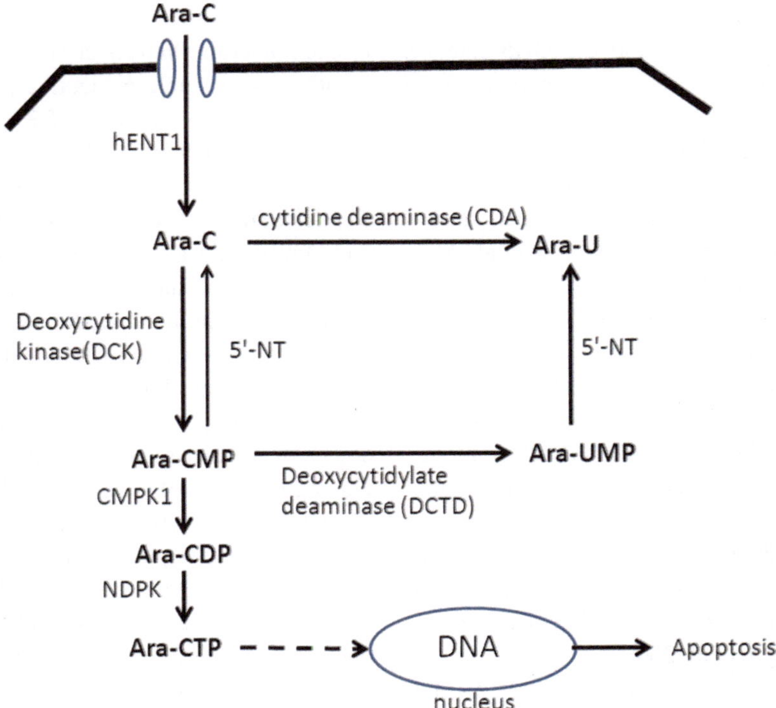

Fig. 12.1 Overview of ara-C metabolism. *CMPK1* cytidine monophosphate (UMP-CMP) kinase 1, *hENT1* equilibrative nucleoside transporter 1, *CDA* cytidine deaminase, *DCK* deoxycytidine kinase, *DCTD* deoxycytidylate deaminase, *5′-NT* 5′-nucleotidase, *NDPK* nucleoside diphosphate kinase

Carbonyl reductase1 (CBR1) is the important enzyme involved in metabolism of daunorubicin in AML (Bogason et al. 2010). Varatharajan et al. reported the wide interindividual variation in the RNA expression of CBR1 and CBR3. They have found the polymorphisms of CBR1 SNP rs25678 in CBR1 and CBR3 on plasma pharmacokinetics of daunorubicin in primary cells from AML patients (Varatharajan et al. 2012).

5 Conclusion and Future Aspects

The field of pharmacogenetics has promised a new way to approach the treatment of groups of patients who share important genetic elements. Through the discovery of new genetic targets, pharmacogenomics has the potential to improve quality of life and reduce health-care costs by decreasing the number of treatment failures and adverse drug reactions. In common with many new technologies, the generalizability and clinical application of pharmacogenomics have proved more challenging than expected. Pharmacogenomics-based personalized medicine holds the promise of reducing the number of adverse drug events and drug failures, thereby vastly improving quality of life while reducing health-care costs.

Combined with proteomic profiling, these technologies have helped identify new targets and signaling pathways and may soon help to identify individual patients likely to benefit from specific therapies. Genetic predictive testing for AML dose adjustment of avoidance of adverse events requires large-scale studies taking into account the complexities of the AML phenotype.

NPM1, FLT3, and CEBPA mutations in AML are attractive targets for molecularly targeted therapy. Those mutational screening should become part of the initial work-up of a newly diagnosed AML. In addition better understanding of mechanisms of those mutations will aid in the discovery and design of more effective treatment strategies. Narrowing the phenotype by focusing on pathways will make this genetically tractable and lead to successful clinical translation. The discovery of new drug targets is no less important, and the large-scale efforts in whole genome mapping of AML traits will be key in this aspect. It is essential that prospective studies are undertaken, prior to any treatment modification, to assess the true effects of those polymorphisms and mutations and determine whether the effect is drug specific or disease related. In summary, the discovery of new prognostic and predictive markers has increased significantly, improving our understanding of AML biology.

References

Abla N, Chinn LW, Nakamura T, Liu L, Huang CC, Johns SJ, Kawamoto M, Stryke D, Taylor TR, Ferrin TE, Giacomini KM, Kroetz DL (2008) The human multidrug resistance protein 4 (MRP4, ABCC4): functional analysis of a highly polymorphic gene. J Pharmacol Exp Ther 325(3):859–868

Armstrong SA, Staunton JE, Silverman LB, Pieters R, den Boer ML, Minden MD, Sallan SE, Lander ES, Golub TR, Korsmeyer SJ (2002) MLL translocations specify a distinct gene expression profile that distinguishes a unique leukemia. Nat Genet 30(1):41–47

Austin H, Delzell E, Cole P (1988) Benzene and leukemia. A review of the literature and a risk assessment. Am J Epidemiol 127(3):419–439

Aydin-Sayitoglu M, Hatirnaz O, Erensoy N, Ozbek U (2006) Role of CYP2D6, CYP1A1, CYP2E1, GSTT1, and GSTM1 genes in the susceptibility to acute leukemias. Am J Hematol 81(3):162–170

Badrul Hisam HR, Ros M, Aziz A, Narazah M (2006) Screening for 3435C>T and 2677G>T/A polymorphisms of multi-drug resistance (MDR1) gene in Malay patients with leukemia. Malays J Biochem Mol Biol 14(1)

Bartel DP (2004) MicroRNAs: genomics, biogenesis, mechanism, and function. Cell 116(2):281–297

Bartsocas CS, Loukopoulos D (1992) Bloom's syndrome: incidence, age of onset, and types of leukemia in the Bloom's syndrome registry. In: Bartsocas CS, Loukopoulos D (eds) Genetics in hematologic disorders. Hemisphere, Washington, DC, pp 241–258

Bennett JM, Catovsky D, Daniel MT, Flandrin G, Galton DA, Gralnick HR, Sultan C (1976) Proposals for the classification of the acute leukaemias. French-American-British (FAB) co-operative group. Br J Haematol 33(4):451–458

Bogason A, Masquelier M, Lafolie P, Skogastierna C, Paul C, Gruber A, Vitols S (2010) Daunorubicin metabolism in leukemic cells isolated from patients with acute myeloid leukemia. Drug Metab Lett 4(4):228–232

Brockmoller J, Cascorbi I, Kerb R, Roots I (1996) Combined analysis of inherited polymorphisms in arylamine N-acetyltransferase 2, glutathione S-transferases M1 and T1, microsomal epoxide hydrolase, and cytochrome P450 enzymes as modulators of bladder cancer risk. Cancer Res 56(17):3915–3925

Burcu M, O'Loughlin KL, Ford LA, Baer MR (2008) Amonafide L-malate is not a substrate for multidrug resistance proteins in secondary acute myeloid leukemia. Leukemia 22(11):2110–2115

Byrd JC, Mrozek K, Dodge RK, Carroll AJ, Edwards CG, Arthur DC, Pettenati MJ, Patil SR, Rao KW, Watson MS, Koduru PR, Moore JO, Stone RM, Mayer RJ, Feldman EJ, Davey FR, Schiffer CA, Larson RA, Bloomfield CD (2002) Pretreatment cytogenetic abnormalities are predictive of induction success, cumulative incidence of relapse, and overall survival in adult patients with de novo acute myeloid leukemia: results from Cancer and Leukemia Group B (CALGB 8461). Blood 100(13):4325–4336

Calin GA, Dumitru CD, Shimizu M, Bichi R, Zupo S, Noch E, Aldler H, Rattan S, Keating M, Rai K, Rassenti L, Kipps T, Negrini M, Bullrich F, Croce CM (2002) Frequent deletions and down-regulation of micro- RNA genes miR15 and miR16 at 13q14 in chronic lymphocytic leukemia. Proc Natl Acad Sci U S A 99(24):15524–15529

Chang P, Kang M, Xiao A, Chang J, Feusner J, Buffler P, Wiemels J (2010) FLT3 mutation incidence and timing of origin in a population case series of pediatric leukemia. BMC Cancer 10:513

Cimmino A, Calin GA, Fabbri M, Iorio MV, Ferracin M, Shimizu M, Wojcik SE, Aqeilan RI, Zupo S, Dono M, Rassenti L, Alder H, Volinia S, Liu CG, Kipps TJ, Negrini M, Croce CM (2005) miR-15 and miR-16 induce apoptosis by targeting BCL2. Proc Natl Acad Sci U S A 102(39):13944–13949

Davies SM, Robison LL, Buckley JD, Radloff GA, Ross JA, Perentesis JP (2000) Glutathione S-transferase polymorphisms in children with myeloid leukemia: a Children's Cancer Group study. Cancer Epidemiol Biomarkers Prev 9(6):563–566

Davies SM, Robison LL, Buckley JD, Tjoa T, Woods WG, Radloff GA, Ross JA, Perentesis JP (2001) Glutathione S-transferase polymorphisms and outcome of chemotherapy in childhood acute myeloid leukemia. J Clin Oncol 19(5):1279–1287

De Jonge HJ, Valk PJ, Veeger NJ, ter Elst A, den Boer ML, Cloos J, de Haas V, van den Heuvel-Eibrink MM, Kaspers GJ, Zwaan CM, Kamps WA, Lowenberg B, de Bont ES (2010) High VEGFC expression is associated with unique gene expression profiles and predicts adverse prognosis in pediatric and adult acute myeloid leukemia. Blood 116(10):1747–1754

Dohner K, Schlenk RF, Habdank M, Scholl C, Rucker FG, Corbacioglu A, Bullinger L, Frohling S, Dohner H (2005) Mutant nucleophosmin (NPM1) predicts favorable prognosis in younger adults with acute myeloid leukemia and normal cytogenetics: interaction with other gene mutations. Blood 106(12):3740–3746

Dohner H, Estey EH, Amadori S, Appelbaum FR, Buchner T, Burnett AK, Dombret H, Fenaux P, Grimwade D, Larson RA, Lo-Coco F, Naoe T, Niederwieser D, Ossenkoppele GJ, Sanz MA, Sierra J, Tallman MS, Lowenberg B, Bloomfield CD (2010) Diagnosis and management of acute myeloid leukemia in adults: recommendations from an international expert panel, on behalf of the European LeukemiaNet. Blood 115(3):453–474

Dubinsky MC, Yang H, Hassard PV, Seidman EG, Kam LY, Abreu MT, Targan SR, Vasiliauskas EA (2002) 6-MP metabolite profiles provide a biochemical explanation for 6-MP resistance in patients with inflammatory bowel disease. Gastroenterology 122(4):904–915

Eisele L, Klein-Hitpass L, Chatzimanolis N, Opalka B, Boes T, Seeber S, Moritz T, Flasshove M (2007) Differential expression of drug-resistance-related genes between sensitive and resistant blasts in acute myeloid leukemia. Acta Haematol 117(1):8–15

Estey E, Dohner H (2006) Acute myeloid leukaemia. Lancet 368(9550):1894–1907

Evans WE, Relling MV (1999) Pharmacogenomics: translating functional genomics into rational therapeutics. Science 286(5439):487–491

Falini B, Mecucci C, Tiacci E, Alcalay M, Rosati R, Pasqualucci L, La Starza R, Diverio D, Colombo E, Santucci A, Bigerna B, Pacini R, Pucciarini A, Liso A, Vignetti M, Fazi P, Meani N, Pettirossi V, Saglio G, Mandelli F, Lo-Coco F, Pelicci PG, Martelli MF (2005) Cytoplasmic nucleophosmin in acute myelogenous leukemia with a normal karyotype. N Engl J Med 352(3):254–266

Filipowicz W, Jaskiewicz L, Kolb FA, Pillai RS (2005) Post-transcriptional gene silencing by siRNAs and miRNAs. Curr Opin Struct Biol 15(3):331–341

Foran JM (2010) New prognostic markers in acute myeloid leukemia: perspective from the clinic. Hematology Am Soc Hematol Educ Program 2010:47–55

Galmarini CM, Thomas X, Calvo F, Rousselot P, El Jafaari A, Cros E, Dumontet C (2002) Potential mechanisms of resistance to cytarabine in AML patients. Leuk Res 26(7):621–629

Gradhand U, Kim RB (2008) Pharmacogenomics of MRP transporters (ABCC1-5) and BCRP (ABCG2). Drug Metab Rev 40(2):317–354

Green H, Falk IJ, Lotfi K, Paul E, Hermansson M, Rosenquist R, Paul C, Nahi H (2012) Association of ABCB1 polymorphisms with survival and in vitro cytotoxicity in de novo acute myeloid leukemia with normal karyotype. Pharmacogenomics J 12(2):111–118

Grimwade D, Walker H, Oliver F, Wheatley K, Harrison C, Harrison G, Rees J, Hann I, Stevens R, Burnett A, Goldstone A (1998) The importance of diagnostic cytogenetics on outcome in AML: analysis of 1,612 patients entered into the MRC AML 10 trial. The Medical Research Council Adult and Children's Leukaemia Working Parties. Blood 92(7):2322–2333

Grimwade D, Hills RK, Moorman AV, Walker H, Chatters S, Goldstone AH, Wheatley K, Harrison CJ, Burnett AK (2010) Refinement of cytogenetic classification in acute myeloid leukemia: determination of prognostic significance of rare recurring chromosomal abnormalities among 5876 younger adult patients treated in the United Kingdom Medical Research Council trials. Blood 116(3):354–365

Grisendi S, Mecucci C, Falini B, Pandolfi PP (2006) Nucleophosmin and cancer. Nat Rev Cancer 6(7):493–505

Haferlach T, Kohlmann A, Schnittger S, Dugas M, Hiddemann W, Kern W, Schoch C (2005) Global approach to the diagnosis of leukemia using gene expression profiling. Blood 106(4):1189–1198

Haferlach T, Kohlmann A, Wieczorek L, Basso G, Kronnie GT, Bene MC, De Vos J, Hernandez JM, Hofmann WK, Mills KI, Gilkes A, Chiaretti S, Shurtleff SA, Kipps TJ, Rassenti LZ, Yeoh AE, Papenhausen PR, Liu WM, Williams PM, Foa R (2010) Clinical utility of microarray-based gene expression profiling in the diagnosis and subclassification of leukemia: report from the International Microarray Innovations in Leukemia Study Group. J Clin Oncol 28(15):2529–2537

Hirose M, Hosoi E, Hamano S, Jalili A (2003) Multidrug resistance in hematological malignancy. J Med Invest 50(3–4):126–135

Horwitz M (1997) The genetics of familial leukemia. Leukemia 11(8):1347–1359

Huang Y, Sadee W (2006) Membrane transporters and channels in chemoresistance and -sensitivity of tumor cells. Cancer Lett 239(2):168–182

Illmer T, Schuler US, Thiede C, Schwarz UI, Kim RB, Gotthard S, Freund D, Schakel U, Ehninger G, Schaich M (2002) MDR1 gene polymorphisms affect therapy outcome in acute myeloid leukemia patients. Cancer Res 62(17):4955–4962

Jordan CT, Guzman ML, Noble M (2006) Cancer stem cells. N Engl J Med 355(12):1253–1261

Juliusson G, Antunovic P, Derolf A, Lehmann S, Mollgard L, Stockelberg D, Tidefelt U, Wahlin A, Hoglund M (2009) Age and acute myeloid leukemia: real world data on decision to treat and outcomes from the Swedish Acute Leukemia Registry. Blood 113(18):4179–4187

Kelly LM, Gilliland DG (2002) Genetics of myeloid leukemias. Annu Rev Genomics Hum Genet 3:179–198

Kim DH, Park JY, Sohn SK, Lee NY, Suh JS, Lee KB (2006a) The association between multidrug resistance-1 gene polymorphisms and outcomes of allogeneic HLA-identical stem cell transplantation. Haematologica 91(6):848–851

Kim DH, Park JY, Sohn SK, Lee NY, Baek JH, Jeon SB, Kim JG, Suh JS, Do YR, Lee KB (2006b) Multidrug resistance-1 gene polymorphisms associated with treatment outcomes in de novo acute myeloid leukemia. Int J Cancer 118(9):2195–2201

Kindler T, Lipka DB, Fischer T (2010) FLT3 as a therapeutic target in AML: still challenging after all these years. Blood 116(24):5089–5102

Lane SW, Scadden DT, Gilliland DG (2009) The leukemic stem cell niche: current concepts and therapeutic opportunities. Blood 114(6):1150–1157

Leith CP, Kopecky KJ, Chen IM, Eijdems L, Slovak ML, McConnell TS, Head DR, Weick J, Grever MR, Appelbaum FR, Willman CL (1999) Frequency and clinical significance of the expression of the multidrug resistance proteins MDR1/P-glycoprotein, MRP1, and LRP in acute myeloid leukemia: a Southwest Oncology Group Study. Blood 94(3):1086–1099

Leroy H, Roumier C, Huyghe P, Biggio V, Fenaux P, Preudhomme C (2005) CEBPA point mutations in hematological malignancies. Leukemia 19(3):329–334

Levine EG, Bloomfield CD (1992) Leukemias and myelodysplastic syndromes secondary to drug, radiation, and environmental exposure. Semin Oncol 19(1):47–84

Lindstrom MS (2011) NPM1/B23: a multifunctional chaperone in ribosome biogenesis and chromatin remodeling. Biochem Res Int 2011:195209

Liotta L, Petricoin E (2000) Molecular profiling of human cancer. Nat Rev Genet 1(1):48–56

List AF, Spier CS, Grogan TM, Johnson C, Roe DJ, Greer JP, Wolff SN, Broxterman HJ, Scheffer GL, Scheper RJ, Dalton WS (1996) Overexpression of the major vault transporter protein lung-resistance protein predicts treatment outcome in acute myeloid leukemia. Blood 87(6):2464–2469

Loktionov A, Watson MA, Gunter M, Stebbings WS, Speakman CT, Bingham SA (2001) Glutathione-S-transferase gene polymorphisms in colorectal cancer patients: interaction between GSTM1 and GSTM3 allele variants as a risk-modulating factor. Carcinogenesis 22(7):1053–1060

Michael M, Doherty MM (2005) Tumoral drug metabolism: overview and its implications for cancer therapy. J Clin Oncol 23(1):205–229

Miller RW (1971) Deaths from childhood leukemia and solid tumors among twins and other sibs in the United States, 1960–67. J Natl Cancer Inst 46(1):203–209

Mishra PJ, Bertino JR (2009) MicroRNA polymorphisms: the future of pharmacogenomics, molecular epidemiology and individualized medicine. Pharmacogenomics 10(3):399–416

Mrozek K, Heerema NA, Bloomfield CD (2004) Cytogenetics in acute leukemia. Blood Rev 18(2):115–136

Mrozek K, Dohner H, Bloomfield CD (2007) Influence of new molecular prognostic markers in patients with karyotypically normal acute myeloid leukemia: recent advances. Curr Opin Hematol 14(2):106–114

Niedzielska E, Chybicka A (2011) Risk of recurrence of acute leukemia in children depending on the polymorphism for TPMT, CYP2C9 and CYP2C19. Onkol Pol 14(2):68–70

Peng Y, Feng Q, Wilk D, Adjei AA, Salavaggione OE, Weinshilboum RM, Yee VC (2008) Structural basis of substrate recognition in thiopurine s-methyltransferase. Biochemistry 47(23):6216–6225

Preudhomme C, Sagot C, Boissel N, Cayuela JM, Tigaud I, de Botton S, Thomas X, Raffoux E, Lamandin C, Castaigne S, Fenaux P, Dombret H (2002) Favorable prognostic significance of CEBPA mutations in patients with de novo acute myeloid leukemia: a study from the Acute Leukemia French Association (ALFA). Blood 100(8):2717–2723

Radtke I, Mullighan CG, Ishii M, Su X, Cheng J, Ma J, Ganti R, Cai Z, Goorha S, Pounds SB, Cao X, Obert C, Armstrong J, Zhang J, Song G, Ribeiro RC, Rubnitz JE, Raimondi SC, Shurtleff SA, Downing JR (2009) Genomic analysis reveals few genetic alterations in pediatric acute myeloid leukemia. Proc Natl Acad Sci U S A 106(31):12944–12949

Relling MV, Dervieux T (2001) Pharmacogenetics and cancer therapy. Nat Rev Cancer 1(2):99–108

Robey RW, Polgar O, Deeken J, To KW, Bates SE (2007) ABCG2: determining its relevance in clinical drug resistance. Cancer Metastasis Rev 26(1):39–57

Robey RW, To KK, Polgar O, Dohse M, Fetsch P, Dean M, Bates SE (2009) ABCG2: a perspective. Adv Drug Deliv Rev 61(1):3–13

Rodriguez-Ariza A, Lopez-Pedrera C, Aranda E, Barbarroja N (2011) VEGF targeted therapy in acute myeloid leukemia. Crit Rev Oncol Hematol 80(2): 241–256

Rollinson S, Roddam P, Kane E, Roman E, Cartwright R, Jack A, Morgan GJ (2000) Polymorphic variation within the glutathione S-transferase genes and risk of adult acute leukaemia. Carcinogenesis 21(1):43–47

Rukov JL, Vinther J, Shomron N (2011) Pharmacogenomics genes show varying perceptibility to microRNA regulation. Pharmacogenet Genomics 21(5):251–262

Sachidanandam R, Weissman D, Schmidt SC, Kakol JM, Stein LD, Marth G, Sherry S, Mullikin JC, Mortimore BJ, Willey DL, Hunt SE, Cole CG, Coggill PC, Rice CM, Ning Z, Rogers J, Bentley DR, Kwok PY, Mardis ER, Yeh RT, Schultz B, Cook L, Davenport R, Dante M, Fulton L, Hillier L, Waterston RH, McPherson JD, Gilman B, Schaffner S, Van Etten WJ, Reich D, Higgins J, Daly MJ, Blumenstiel B, Baldwin J, Stange-Thomann N, Zody MC, Linton L, Lander ES, Altshuler D (2001) A map of human genome sequence variation containing 1.42 million single nucleotide polymorphisms. Nature 409(6822):928–933

Schnittger S, Kern W, Tschulik C, Weiss T, Dicker F, Falini B, Haferlach C, Haferlach T (2009) Minimal residual disease levels assessed by NPM1 mutation-specific RQ-PCR provide important prognostic information in AML. Blood 114(11):2220–2231

Schramm A, Schulte JH, Klein-Hitpass L, Havers W, Sieverts H, Berwanger B, Christiansen H, Warnat P, Brors B, Eils J, Eils R, Eggert A (2005) Prediction of clinical outcome and biological characterization of neuroblastoma by expression profiling. Oncogene 24(53):7902–7912

Schuetz JD, Connelly MC, Sun D, Paibir SG, Flynn PM, Srinivas RV, Kumar A, Fridland A (1999) MRP4: a previously unidentified factor in resistance to nucleoside-based antiviral drugs. Nat Med 5(9):1048–1051

Shi JY, Shi ZZ, Zhang SJ, Zhu YM, Gu BW, Li G, Bai XT, Gao XD, Hu J, Jin W, Huang W, Chen Z, Chen SJ (2004) Association between single nucleotide polymorphisms in deoxycytidine kinase and treatment response among acute myeloid leukaemia patients. Pharmacogenetics 14(11):759–768

Smith BD, Levis M, Beran M, Giles F, Kantarjian H, Berg K, Murphy KM, Dauses T, Allebach J, Small D (2004)

Single-agent CEP-701, a novel FLT3 inhibitor, shows biologic and clinical activity in patients with relapsed or refractory acute myeloid leukemia. Blood 103(10):3669–3676

Spurdle AB, Webb PM, Purdie DM, Chen X, Green A, Chenevix-Trench G (2001) Polymorphisms at the glutathione S-transferase GSTM1, GSTT1 and GSTP1 loci: risk of ovarian cancer by histological subtype. Carcinogenesis 22(1):67–72

Stirewalt DL, Radich JP (2003) The role of FLT3 in haematopoietic malignancies. Nat Rev Cancer 3(9):650–665

Stone RM, De Angelo J, Galinsky I, Estey E, Klimek V, Grandin W, Lebwohl D, Yap A, Cohen P, Fox E, Neuberg D, Clark J, Gilliland DG, Griffin JD (2004) PKC 412 FLT3 inhibitor therapy in AML: results of a phase II trial. Ann Hematol 83(Suppl 1):S89–S90

Thiede C, Koch S, Creutzig E, Steudel C, Illmer T, Schaich M, Ehninger G (2006) Prevalence and prognostic impact of NPM1 mutations in 1485 adult patients with acute myeloid leukemia (AML). Blood 107(10):4011–4020

van den Heuvel-Eibrink MM, Wiemer EA, de Boevere MJ, van der Holt B, Vossebeld PJ, Pieters R, Sonneveld P (2001) MDR1 gene-related clonal selection and P-glycoprotein function and expression in relapsed or refractory acute myeloid leukemia. Blood 97(11): 3605–3611

van der Kolk DM, de Vries EG, Koning JA, van den Berg E, Muller M, Vellenga E (1998) Activity and expression of the multidrug resistance proteins MRP1 and MRP2 in acute myeloid leukemia cells, tumor cell lines, and normal hematopoietic CD34+ peripheral blood cells. Clin Cancer Res 4(7):1727–1736

van der Kolk DM, de Vries EG, Muller M, Vellenga E (2002) The role of drug efflux pumps in acute myeloid leukemia. Leuk Lymphoma 43(4):685–701

Van't Veer LJ, Dai H, van de Vijver MJ, He YD, Hart AA, Mao M, Peterse HL, van der Kooy K, Marton MJ, Witteveen AT, Schreiber GJ, Kerkhoven RM, Roberts C, Linsley PS, Bernards R, Friend SH (2002) Gene expression profiling predicts clinical outcome of breast cancer. Nature 415(6871):530–536

Varatharajan S, Abraham A, Zhang W, Shaji RV, Ahmed R, Abraham A, George B, Srivastava A, Chandy M, Mathews V, Balasubramanian P (2012) Carbonyl reductase 1 expression influences daunorubicin metabolism in acute myeloid leukemia. Eur J Clin Pharmacol 68(12):1577–1586

Vardiman JW, Thiele J, Arber DA, Brunning RD, Borowitz MJ, Porwit A, Harris NL, Le Beau MM, Hellstrom-Lindberg E, Tefferi A, Bloomfield CD (2009) The 2008 revision of the World Health Organization (WHO) classification of myeloid neoplasms and acute leukemia: rationale and important changes. Blood 114(5):937–951

Veuger MJ, Honders MW, Landegent JE, Willemze R, Barge RM (2000) High incidence of alternatively spliced forms of deoxycytidine kinase in patients with resistant acute myeloid leukemia. Blood 96(4):1517–1524

Voso MT, D'Alo F, Putzulu R, Mele L, Scardocci A, Chiusolo P, Latagliata R, Lo-Coco F, Rutella S, Pagano L, Hohaus S, Leone G (2002) Negative prognostic value of glutathione S-transferase (GSTM1 and GSTT1) deletions in adult acute myeloid leukemia. Blood 100(8):2703–2707

Walter MJ, Payton JE, Ries RE, Shannon WD, Deshmukh H, Zhao Y, Baty J, Heath S, Westervelt P, Watson MA, Tomasson MH, Nagarajan R, O'Gara BP, Bloomfield CD, Mrozek K, Selzer RR, Richmond TA, Kitzman J, Geoghegan J, Eis PS, Maupin R, Fulton RS, McLellan M, Wilson RK, Mardis ER, Link DC, Graubert TA, DiPersio JF, Ley TJ (2009) Acquired copy number alterations in adult acute myeloid leukemia genomes. Proc Natl Acad Sci U S A 106(31):12950–12955

Wang WS, Tzeng CH, Chiou TJ, Liu JH, Hsieh RK, Yen CC, Chen PM (1997) High-dose cytarabine and mitoxantrone as salvage therapy for refractory non-Hodgkin's lymphoma. Jpn J Clin Oncol 27(3):154–157

Weiss RB (1992) The anthracyclines: will we ever find a better doxorubicin? Semin Oncol 19(6):670–686

Wiernik PH (1997) Leukemias and plasma cell myeloma. Cancer Chemother Biol Response Modif 17:390–407

Wouters BJ, Lowenberg B, Erpelinck-Verschueren CA, van Putten WL, Valk PJ, Delwel R (2009) Double CEBPA mutations, but not single CEBPA mutations, define a subgroup of acute myeloid leukemia with a distinctive gene expression profile that is uniquely associated with a favorable outcome. Blood 113(13):3088–3091

Yue L, Saikawa Y, Ota K, Tanaka M, Nishimura R, Uehara T, Maeba H, Ito T, Sasaki T, Koizumi S (2003) A functional single-nucleotide polymorphism in the human cytidine deaminase gene contributing to ara-C sensitivity. Pharmacogenetics 13(1):29–38

Zhang DE, Zhang P, Wang ND, Hetherington CJ, Darlington GJ, Tenen DG (1997) Absence of granulocyte colony-stimulating factor signaling and neutrophil development in CCAAT enhancer binding protein alpha-deficient mice. Proc Natl Acad Sci U S A 94(2):569–574

Zhong Y, Chen B, Feng J, Cheng L, Li Y, Qian J, Ding J, Gao F, Xia G, Chen N, Lu Z (2010) The associations of Janus kinase-2 (JAK2) A830G polymorphism and the treatment outcomes in patients with acute myeloid leukemia. Leuk Lymphoma 51(6):1115–1120

Zipursky A, Poon A, Doyle J (1992) Leukemia in Down syndrome: a review. Pediatr Hematol Oncol 9(2):139–149

Zorko NA, Bernot KM, Whitman SP, Siebenaler RF, Ahmed EH, Marcucci GG, Yanes DA, McConnell KK, Mao C, Kalu C, Zhang X, Jarjoura D, Dorrance AM, Heerema NA, Lee BH, Huang G, Marcucci G, Caligiuri MA (2012) Mll partial tandem duplication and Flt3 internal tandem duplication in a double knock-in mouse recapitulates features of counterpart human acute myeloid leukemias. Blood 120(5):1130–1136

Pharmacogenomics of Autoimmune Diseases

13

Melvin George, Sandhiya Selvarajan, and Suresh Kumar Srinivasamurthy

Abstract

Autoimmune diseases such as rheumatoid arthritis (RA), systemic lupus erythematosus (SLE), and psoriasis cause a considerable degree of morbidity worldwide. Although the treatment of these conditions has shown progress over the last decade with the steady trickle of new drug molecules, drug therapy is far from satisfactory due to the reduced efficacy and maximal toxicity in certain patients. Several factors are known to influence the efficacy and toxicity of these drugs such as age, gender, liver and kidney function, and concomitant drug therapy. Another crucial factor influencing drug response of the patient is the genetic constitution of the patient. For example, polymorphisms in the gene MTHFR such as *677T>C* can increase methotrexate serum levels and lead to toxicity in a patient with RA. Similarly polymorphisms in the drug transporter ABCB1 are associated with decreased efficacy to methotrexate. Polymorphisms within the *TNF* promoter region have been shown to modify the clinical efficacy and toxicity of anti-TNF therapy in RA patients. SLE patients with polymorphic *TPMT* gene may require a reduced dose of azathioprine to circumvent the catastrophe of fatal bone marrow suppression. Polymorphisms in the *TYMS* gene could lead to reduced efficacy with methotrexate in psoriatic arthritis patients. Although our understanding of autoimmune diseases has improved considerably over the last decade and several studies in pharmacogenomics of autoimmune diseases have been carried out, the only clinical application is *TPMT* testing for azathioprine. Yet with improved methodology adopted in pharmacogenomics studies coupled with novel technologies, the field of pharmacogenomics does appear to offer significant promise in the coming years towards the dream of personalized medicine in autoimmune diseases.

M. George, MD., DM. (✉)
Department of Cardiology, SRM Medical College and Hospital Research Centre, Kattankulathur, Kancheepuram District, Tamil Nadu, India
e-mail: melvingeorge2003@gmail.com

S. Selvarajan • S.K. Srinivasamurthy
Division of Clinical Pharmacology, JIPMER, Pondicherry, India
e-mail: sandhiyaselvarajan@yahoo.com; sandhiyaselvarajan@gmail.com

D. Barh et al. (eds.), *Omics for Personalized Medicine*,
DOI 10.1007/978-81-322-1184-6_13, © Springer India 2013

1 Introduction

Autoimmune diseases are one of the major causes of morbidity in developed and developing countries among young and middle-aged women (Cooper and Stroehla 2003). In women, autoimmune diseases constitute one of the top ten causes of death. Most autoimmune diseases have an unequal preponderance in females, being almost 65 %. Rheumatoid arthritis, psoriasis, systemic lupus erythematosus, Hashimoto's thyroiditis, Sjogren's syndrome, scleroderma, Wegener's granulomatosis, and systemic vasculitis are some of the major autoimmune diseases. The annual direct healthcare costs related to these diseases are more than hundred billion US dollars. As our understanding of these diseases has increased over the last few decades, it has also led to a number of new molecules entering the market. Although several new drugs have entered the market which target specific pathways of these diseases, the treatment of these disease conditions is far from satisfactory owing to the frequency of adverse drug reactions and variable and limited efficacy in different population. Several factors contribute to the variation of the drug response seen among these patients such as age, gender, presence of earlier drug therapy or concomitant drug therapy, stage of disease, and the genetic status of the individual. Pharmacogenetics is the branch of medicine which deals with the study of how genetic factors contribute to the variation in drug response of the individual. Apart from improving our ability to predict the variability of drug response and toxicity, pharmacogenomics also helps us to develop new biomarkers and new targets for drug development (Fig. 13.1). This chapter attempts to highlight the recent advances in the pharmacogenetics of some of the common autoimmune diseases.

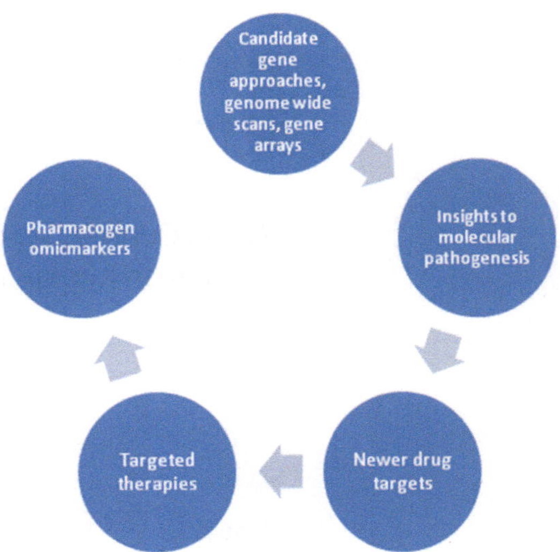

Fig. 13.1 Differing facet of pharmacogenomics in autoimmune diseases which includes development of newer targeted therapies with pharmacogenomic markers

2 Pharmacogenomics of Rheumatoid Arthritis

Rheumatoid arthritis is one of the most common autoimmune diseases worldwide with its prevalence hovering between 0 and 1 % in different regions. The disease causes untold suffering to millions of people every year. The symptoms of the disease, such as joint swelling and tenderness involving more than five joints, which are accentuated in the early hours of the day cause tremendous morbidity in these patients. Rheumatoid arthritis is also known to cause increased mortality due to the increased association with cardiovascular disease, malignancy, and infections in these patients (Owlia et al. 2012). The diminished quality of life and the added economic burden in terms of hospitalizations make this an important health problem to contend with. Some of the common drugs for rheumatoid arthritis include corticosteroids, methotrexate, sulfasalazine, and TNF-alpha inhibitors such as infliximab, etanercept, and adalimumab. Some of the new additions to this array of molecules which have been recently approved by US FDA include golimumab, certolizumab pegol, and tocilizumab (Reichert 2012). Although these new molecules appear promising in phase 2 and 3 clinical trials, as demonstrated by favorable safety and efficacy data, they are yet to become common place in drug therapy of rheumatoid arthritis owing to their prohibitive cost and limited availability in most nations (Hoebert et al. 2012). Hence, it is not surprising that the older drugs

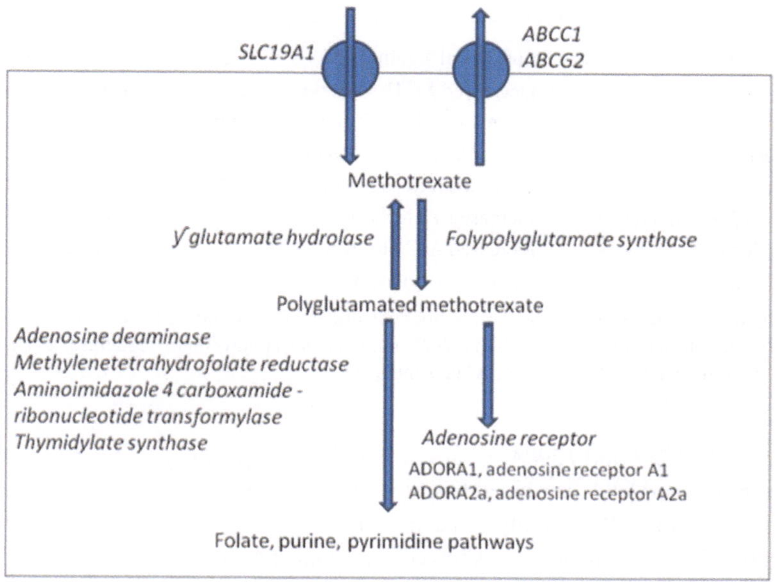

Fig. 13.2 Metabolism of methotrexate within the cell

such as methotrexate still occupy a pivotal role in the management of rheumatoid arthritis. The major adverse effects of methotrexate include neurologic toxicity, gastrointestinal complications including nausea, vomiting and diarrhea, liver dysfunction, hematologic abnormalities, rash, stomatitis, and alopecia (Mittal et al. 2012).

2.1 Effect of Methotrexate on Folate and Other Pathways

The entry of methotrexate into the cell is directed by a solute carrier family protein such as SLC19A1, while the efflux of the drug is mediated by ATP-binding cassette family transporters, also termed as MDR transporters. The enzyme folylpolyglutamate synthase helps in converting the methotrexate into its different polyglutamate forms (MTXPG), which helps in the retention of intracellular methotrexate (Fig. 13.2). This reaction can also be reversed by the enzyme gamma-glutamyl hydrolase. The MTXPG that are formed will in turn inhibit dihydrofolate reductase that converts dihydrofolate to tetrahydrofolate. Tetrahydrofolate further undergoes methylation to result in the formation of 5-methyltetrahydrofolate which acts as a carbon donor for several reactions such as conversion of homocysteine into methionine (Davila and Ranganathan 2011).

Methotrexate also has additional effects by acting on the pyrimidine and purine synthesis. It inhibits the thymidylate synthetase thereby preventing conversion of deoxyuridylate to deoxythymidylate. It also inhibits the enzyme AICAR transformylase that leads to intracellular accumulation of aminoimidazole carboxamide adenosine ribonucleotide (AICAR). AICAR and its metabolites inhibit the enzymes involved in the metabolism of adenosine such as adenosine deaminase and AMP deaminase, resulting in a rise in adenosine that has a powerful anti-inflammatory effect (Davila and Ranganathan 2011).

2.2 Genetic Polymorphisms and Methotrexate in RA

Methotrexate is well known to exhibit variable efficacy and toxicity in the treatment of rheumatoid arthritis. Several studies have been done to illustrate the role of genetic polymorphisms in influencing the response of methotrexate in terms of both efficacy and toxicity (Table 13.1). For example, polymorphisms of the gene

Table 13.1 Effect of genetic polymorphisms on methotrexate efficacy and toxicity in rheumatoid arthritis

Gene	Polymorphism	Functional significance	Clinical outcome
MTHFR	*677C>T*	Decreased MTHFR enzyme levels	Increased toxicity
	1298A>C	Decreased MTHFR enzyme activity	Increased efficacy and toxicity
SLC19A1	*80G>A*	Higher levels of MTXPG	Increased efficacy
ABCB1	*3425C>T*	Increased intracellular uptake of MTX	Decreased efficacy
TYMS	*5-UTR repeat element*	Increased TYMS activity	Decreased efficacy
ATIC	*347C>G*	Increased AICAR levels	Increased toxicity
IL-1R	*IL-1RN*3*	Alters IL-1 synthesis	Decreased efficacy

Abbreviations: *AICAR* aminoimidazole carboxamide ribonucleotide, *ATIC* aminoimidazole carboxamide ribonucleotide transformylase, *IL-1R* interleukin 1 receptor, *MTHFR* methylenetetrahydrofolate reductase, *MTXPG* methotrexate polyglutamate, *SLC* solute carrier, *TYMS* thymidylate synthase

encoding *MTHFR*, *677C>T* and *1298A>C*, are in linkage disequilibrium and lead to reduced levels, and thus activity, of MTHFR. A study done in 125 RA patients of European descent showed that RA patients with *DHFR-317AA* genotype had less favorable response to MTX as measured by DAS (Disease Activity Scoring) (Milic et al. 2012). A study in US population revealed that an SNP in the *ATIC* gene, rs4673993 was associated with low disease activity in patients on MTX (Lee et al. 2009). Methotrexate-induced liver dysfunction was found to be significantly correlated with non-*TT* genotype polymorphism at *GGH T16C* gene in a study done in Japanese children with juvenile idiopathic arthritis (Yanagimachi et al. 2011). A meta-analysis was done to determine the influence of pharmacogenetics on the efficacy and toxicity of methotrexate. Although as many as 12 genetic polymorphisms in the methotrexate pathway were studied, sufficient data was available only with respect to two polymorphisms, namely, *C677T* and *A1298C* in *MTHFR* gene. The *C677T* polymorphism was found to be significantly associated with methotrexate toxicity, while *A1298C* was not associated with toxicity. No polymorphisms could be positively correlated with efficacy in this meta-analysis (Fisher and Cronstein 2009). Besides these polymorphisms in recent years, there have also been studies which have attempted to build a model that combines clinical parameters with genetic polymorphisms. For example, clinical parameters such as rheumatoid factor status, smoking status, gender, and disease activity were compared with polymorphisms in adenosine pathway genes and folate pathway genes to predict efficacy of methotrexate. The *MTHFR 677C>T* variant has been found to be the most widely reported polymorphisms that is positively correlated with methotrexate toxicity (Schmeling et al. 2005; Aggarwal et al. 2006; Wessels et al. 2006; Ranganathan et al. 2008; Hughes et al. 2006). Yet quite number of studies have not shown an association between these polymorphisms and methotrexate efficacy and toxicity (Taraborelli et al. 2009a, b). A meta-analysis failed to prove a positive association between *C677T* and *A1298C* polymorphisms of MTHFR and the toxicity and efficacy of methotrexate in RA (Lee and Song 2010).

2.3 Genetic Polymorphisms and Sulfasalazine in RA

This drug has been used in RA treatment for more than 30 years. The drug is activated into 5-aminosalicylic acid and sulfapyridine after ingestion. Sulfapyridine is metabolized by N-acetyltransferase 2. Based on polymorphisms in *NAT2* gene, individuals can be classified as rapid and slow acetylators. Studies have demonstrated that toxicity with sulfasalazine such as headache, nausea, abdominal discomfort, and rash is more frequent in patients who are slow acetylators (Davila and Ranganathan 2011). Those patients without the wild-type haplotype at *NAT2* were more likely to experience adverse events by sulfasalazine (Taniguchi et al. 2007).

2.4 Genetic Polymorphisms and Leflunomide in RA

Leflunomide is metabolized into an active metabolite A771726 that causes reversible inhibition of the rate-limiting enzyme in pyrimidine synthesis, namely, dihydroorotate dehydrogenase (DHODH). Thus, a missense polymorphism in the human *DHODH* gene will reduce the DHODH enzyme activity (Davis et al. 1996). In a study carried out in 147 RA patients, remission was found to be more common in patients with *C* allele rather than *A* allele. However, this polymorphism has also found to be linked with a sevenfold increased risk of adverse events from leflunomide such as hepatotoxicity and gastrointestinal and mucosal toxicity (Pawlik et al. 2009). Based on in vitro studies, estrogen was found to interfere in the suppression of cytokine production by leflunomide. This led to exploration of the association between estrogen receptor polymorphisms and leflunomide response. Polymorphisms in the estrogen receptor could potentially alter the estrogen receptor expression. In a prospective study done in 115 patients with RA, the *ESR1* rs9340799 *AA* and rs2234693 *TT* genotype were associated with better response to leflunomide therapy. Leflunomide is converted into its active metabolite by CYP1A2. *CYP1A2*1F* polymorphism was found to be correlated with leflunomide toxicity as it accentuates the conversion of leflunomide to its active metabolite (Grabar et al. 2009; Dziedziejko et al. 2011).

2.5 Genetic Polymorphisms and Tumor Necrosis Factor Antagonists in RA

Polymorphisms in the *TNF* gene locus such as *–308G>A* and *–238A>G* influence TNF production. A study reported increased efficacy of TNF-alpha antagonists in patients having *–308G>A* polymorphism (Cuchacovich et al. 2004). A meta-analysis of 13 studies revealed an association between treatment response to infliximab and the TNF-alpha *–238 A/G* polymorphism, but no associations between treatment response and the TNF-alpha *–308 A/G* polymorphism (Lee et al. 2010). Polymorphism in *TNFRSF1B 196T>G* is associated with decreased efficacy as it influences receptor shedding and ligand binding. DNA microsatellites which are nothing but repeat sequences of A and T in the intronic portions of DNA can sufficiently influence gene transcription. There are five microsatellites in the TNF locus, namely, TNFa to TNFe, and they influence the production of TNF production. There have been studies that have shown TNFd and TNFa2 increase TNF production while low level of TNF is associated with TNFa6. However, a study done in 457 patients with RA showed that response to etanercept was not dependent on any TNF microsatellite markers (Criswell et al. 2004).

TNF antagonists are antibodies, and the Fcδ portion binds to the Fcδ receptors. Polymorphisms in the FcδR mediate antibody-dependent cellular cytotoxicity (ADCC). For instance, an SNP in FCGR3A which encodes a Val158Phe variant will influence the binding affinity of IgG1 and promote increased ADCC and apoptosis resulting in drug toxicity. The TNF promoter *–308* allele was found to be a selective marker which influences the response to TNF antagonists such as etanercept, infliximab, and adalimumab in RA patients. In a meta-analysis of TNF-alpha promoter *–308 A/G* polymorphism and responsiveness to anti-TNF therapy, individuals with RA who carry the *A* allele did not respond as well to TNF antagonists as those with the *G* allele (Lee et al. 2006). A meta-analysis of polymorphisms in *TNFAIP3* revealed that these patients were at a greater predilection to develop rheumatoid arthritis (Lee et al. 2012).

Besides TNF, there are other genes that have also been explored to look for the influence of their polymorphic variants that affect response to biologic agents. The PTPRC is a CD45 tyrosine protein phosphatase C that regulates BCR and TCR signaling which influence the secretion of several cytokines. In a prospective study of 1,283 patients with rheumatoid arthritis, the *rs10919563* variant was found to be associated with favorable response to biologic therapy (Cui et al. 2010). The MAPK14 is an important signaling molecule involved in the production of several proinflammatory cytokines and matrix metalloproteinases. Patients with RA having SNPs in *MAPK14* gene were found to have improved efficacy with infliximab and adalimumab. However, there was no improved response with etanercept in these patients with variant polymorphisms in the MAPK14 pathway (Coulthard et al. 2011).

An important confounding factor in the interpretation and application of pharmacogenetic studies is that the same genetic variants which are markers of severe disease are also found to be associated with better response to anti-TNF therapy. This can complicate the predictive value of pharmacogenetics markers in autoimmune diseases. An ideal tissue to use in pharmacogenetics of rheumatic diseases is to obtain the synovial tissue. However, this is seldom accessible and so one has to contend with the less reliable DNA from peripheral blood. Another limiting factor in most pharmacogenetic studies is the lack of power due to smaller sample size and racially homogenous nature of the population making it difficult to generalize the results in the other populations. So there is a definite onus on pharmacogenetic researchers to carry out larger prospective, multicentric, and multiethnic studies to overcome these problems. Although there has been an increasing trend in the rise of genome-wide association studies, this approach is equally daunting with its own potential limitations.

One of the reasons that has been argued for the variable results in correlation between efficacy and different polymorphisms is that rheumatoid arthritis is a waxing and waning disease, and since most of the studies that are carried out are snapshot studies, they may not reflect the true picture as to whether a patient is a responder or a nonresponder to drug therapy. A better way of investigating this correlation would be to perform long-term longitudinal studies to correlate efficacy variables with polymorphisms (de Rotte et al. 2010).

3 Pharmacogenomics of SLE

SLE is an autoimmune, multisystemic disease which is defined by the presence of autoantibodies to self-antigens and to antigens present inside the nucleus which leads to progressive destruction of several organs. The disease is more commonly seen in Asians, African Americans, and Hispanics. The various features of the disease include skin rashes, arthralgia, and hematologic, renal, and immunologic abnormalities. In SLE both the innate and acquired immune responses are deranged resulting in impaired T cell response, production of autoantibodies, and apoptosis. The disease is characteristically associated with exacerbations and remissions. The disease is conventionally treated with NSAIDs, antimalarials, corticosteroids, high-dose immunoglobulins, and immunosuppressants such as azathioprine, cyclophosphamide, methotrexate, and mycophenolic acid.

3.1 Genetic Polymorphisms and Hydroxychloroquine in SLE

Polymorphisms in TNF and IL-10 can influence the response to hydroxychloroquine in SLE patients. The basal production of IL-10 can be influenced by three polymorphisms (*IL-10-1082A>G*, *-IL-10-819C>T*, *−592C>A*). Those patients having the genotype which causes high TNF production and low IL-10 production were found to have better response to hydroxychloroquine (Davila and Ranganathan 2011). A study done in patients with discoid lupus erythematosus showed that the response of patients to hydroxychloroquine was not influenced by *CYP2C8* or *CYP2D6* polymorphisms (Wahie et al. 2011). Yet the data for the application of pharmacogenetics in guiding the use of hydroxychloroquine for SLE remain sparse (Table 13.2).

3.2 Genetic Polymorphisms and Cyclophosphamide in SLE

Cyclophosphamide is a DNA alkylating agent used in the treatment of SLE and glomerulonephritis secondary to SLE. It is broken down by the enzyme CYP2C19 in the liver. One of the adverse effects known to occur in patients with cyclophosphamide is ovarian failure. In a study done in Thai patients with SLE, polymorphism in *CYP2C19*

Table 13.2 Genetic polymorphisms affecting drug therapy in SLE

Gene	Polymorphism	Clinical outcome
CYP2C19	*CYP2C19*2*	Reduced toxicity with cyclophosphamide
CYP2B6	*CYP2B6*5*	Reduced toxicity with cyclophosphamide
GSTP	*Ile 105Val*	Increased toxicity with cyclophosphamide
IL-10	*1082A>G*	Increased efficacy with hydroxychloroquine
	819C>T	
	592C>A	
NAT2	*NAT2*4*	Increased toxicity in slow acetylators to sulfasalazine
TNF	*308A>G*	Increased efficacy with hydroxychloroquine

Abbreviations: *CYP* cytochrome P450 enzyme system, *GSTP* glutathione *S*-transferase, *Il-10* interleukin 10, *NAT N*-acetyl transferase, *TNF* tumor necrosis factor

such as *CYP2C19**2 reduced the likelihood of ovarian toxicity as compared to those with wild-type *CYP2C19*1*1* (Ngamjanyaporn et al. 2011). Patients with SLE who were heterozygous for this polymorphism also were observed to have greater risk of developing cyclophosphamide-induced end-stage renal disease complicating the lupus nephritis. Polymorphisms of *GSTP* gene were also said to play a major role in the metabolism of cyclophosphamide in the liver. In a study performed in 102 patients, individuals with GSTP polymorphisms were noticed to have a greater risk of developing adverse effects to cyclophosphamide. The *GSTP1* variant gene polymorphisms in *105* codon due to a substitution of isoleucine to valine results in a reduction in the catalytic activity of GST protein as well as its inherent stability. Myelotoxicity and gastrointestinal toxicity were more common in patients with *GSTP1*- *105 I/V or *GSTP1*-105*V*/*V* genotype than in patients with the *GSTP1* wild-type genotype (Zhong et al. 2006).

3.3 Genetic Polymorphisms and Rituximab in SLE

Rituximab is a monoclonal antibody that targets CD20 B cells and mediates antibody-dependent cellular cytotoxicity. The *FcγR IIIA Val158Phe* polymorphism was helpful in predicting the response to drug therapy in one study. But the sample size for this study was small, and these findings were not replicated by any other investigators. Although the number of studies which dealing with SLE pharmacogenomics are scant, the data on the role of genetic factors in determining the prevalence of the disease is abundant (Davila and Ranganathan 2011).

4 Pharmacogenomics of Psoriasis

Psoriasis is a complex genetic disorder which has diverse manifestations from skin to joints that affects 2–3 % of Caucasian population. The skin lesions of psoriasis are distinct and very well circumscribed, circular, red papules or plaques with a grey or silvery-white, dry scale, which are typically distributed symmetrically on the scalp, elbows, knees, lumbosacral area, and in the body folds (Langley et al. 2005). Patients with psoriatic arthritis start with oligoarticular disease and progress on to severe polyarticular disease. Distal joints are frequently affected in psoriatic arthritis. A ray pattern of distribution is observed in which all the joints of a single digit are more likely to get affected than the same joints on both sides, which is typical of RA. Other classical features of psoriatic arthritis include erythema over affected joints, the presence of spinal involvement, the presence of enthesitis (inflammation at the point where tendon gets attached to bone), and lower level of tenderness (Gladman et al. 2005). The nails are additionally affected in several patients. The quality of life in patients with psoriasis is diminished when compared to the general population (Jankovic et al. 2011).

Genome-wide association studies have identified the loci involved in disease susceptibility. At present, 16 loci have been identified, as being associated with susceptibility to psoriasis (Hebert et al. 2012). A study showed that the combined genetic risk involving 10 loci could account for 11.6 % of the genetic variance in psoriasis (Chen et al. 2011). The *HLA-C*, *IL-12B*, *TRAF3IP2*, and *FBXL19* genes have been associated with psoriatic arthritis susceptibility in genome-wide association studies (Bluett and Barton 2012). Apart from genetic susceptibility, not much is explained about genetic heritability, which remains to be explored. There is emerging evidence that the different psoriasis subtypes have different genetic makeup such as increased association of *PSORS1* in psoriasis vulgaris and guttate psoriasis, which may be of pharmacogenetic importance (Vasku et al. 2007). Genome-wide approach may also be promising in identifying the genetic predictors of treatment response. Treatments available for psoriasis act mainly as immunosuppressants and inhibitors of inflammation. As pathogenesis and genetic predisposition varies from person to person, concept of personalized medicine is more relevant in management of psoriasis. Further currently available drugs mainly control the disease activity, and there is no evidence of change in long-term course of the disease with these drugs.

Variation in drug response is another prime concern in the management of psoriasis. Both lack of response and adverse drug reactions occur in significant proportion of patients. Therapies for psoriasis are broadly divided as topical, phototherapy, systemic agents, and biologics. The choice of drugs depends on severity, site, adverse drug reactions, and drug response in the patient. Thus, if genetic makeup of individual would help to tailor these drugs to the person to whom it is appropriately matched, then the overall cost and even adverse drug reactions could be prevented. It is reported that 15–30 % of interindividual variation in drug response can be attributed to the genes coding the proteins involved in pharmacokinetics and pharmacodynamics (Table 13.3) (Hebert et al. 2012).

Table 13.3 Genetic polymorphisms affecting drug therapy in psoriasis

Gene	Polymorphism	Clinical outcome
ABCC1	*1219–176T>C*	Improved methotrexate efficacy
	rs35592	
	3391–1960G>A	
	rs2238476	
	4009A>G	
	rs28364006	
ABCG2	*204–592C>T*	Improved methotrexate efficacy
	rs17731538	
	1194+928A>T	
	rs13120400	
ADORA2A	*4205975G>A*	Increased methotrexate toxicity
IL-6	*−174G/C*	Reduced efficacy to TNF-α blockers
MTHFR	*1289A>C*	Reduced toxicity to methotrexate
SLC19A1	*80G>A*	Increased methotrexate toxicity
TPMT	*TPMT *2, TPMT*3B, TPMT*3C*	Increased azathioprine toxicity
TYMS	*28bp repeat rs34743033*	Reduced efficacy to methotrexate
VDR	*Fok1 F, Taq1 T, A-1012G A*	Enhanced response to calcipotriol
VEGF	*+405, −460*	Reduced efficacy to TNF-α blockers

Abbreviations: *ABC* ATP-binding cassette protein, *ADOR* adenosine receptor, *IL-6* interleukin 6, *MTHFR* methylenetetrahydrofolate reductase, *SLC* solute carrier, *TPMT* thiopurine methyl transferase, *TYMS* thymidylate synthase, *VDR* vitamin D receptor, *VEGF* vascular endothelial growth factor

4.1 Genetic Polymorphisms and Vitamin D Analogues in Psoriasis

Topical vitamin D is well used in mild psoriasis. They exert antiproliferative, prodifferentiation of keratinocytes with localized immunosuppressive effects in dermis. This action is mainly by binding to vitamin D receptors. For unknown reasons, certain patients do not respond well to vitamin D. Since vitamin D acts through nuclear receptors, studies have explored the relationship between vitamin D receptor polymorphism and response to vitamin D. There was a significant difference in the genotypes *Ff*, *ff*, and *TT* between vitamin D3 therapy responders and nonresponders. Haplotype analysis of the results showed that certain haplotypes were less responsive to therapy (Acikbas et al. 2012). Three polymorphism of VDR gene *Fok1* F allele, *Taq1* T allele, and *A-1012G* A allele are identified to be associated with an enhanced response to calcipotriol (Halsall et al. 2005).

4.2 Genetic Polymorphisms and Methotrexate in Psoriasis

Methotrexate is an effective drug for moderate to severe psoriasis. Some patients are refractory to treatment and other sustains adverse drug reactions. Several studies have addressed the pharmacogenomics involved in methotrexate therapy of psoriasis. Common genetic polymorphisms of genes coding adenosine deaminase (ADA), 5-aminoimidazole-4-carboxamide ribonucleotide transformylase (ATIC), 5,10-methylenetetrahydrofolate reductase (MTHFR), solute carrier family 19, member 1 (SCL19A1), and thymidylate synthase (TS) were studied in retrospective cohort of 203 patients with psoriasis on methotrexate therapy. A specific polymorphism in *SLC19A1(80A allele)* which is involved in influx of methotrexate into cells and TS (3′ untranslated region six base pair deletion) involved in pyrimidine synthesis were associated with toxicity. *Methylenetetrahydrofolate reductase* polymorphism has been shown to predict methotrexate toxicity for psoriasis patients treated with methotrexate (Hebert et al. 2012). However, these findings were not replicated in another study that looked at single-nucleotide polymorphisms (SNPs) across four genes that are relevant to methotrexate metabolism [*folylpolyglutamate synthase* (*FPGS*), *gamma-glutamyl hydrolase* (*GGH*), *methylenetetrahydrofolate reductase* (*MTHFR*), and *5-aminoimidazole-4-carboxamide ribonucleotide transformylase* (*ATIC*)]. No significant association was found between clinical outcomes with methotrexate and the SNPs studied (Warren et al. 2009).

4.3 Genetic Polymorphisms and Cyclosporine in Psoriasis

Cyclosporine is metabolized by CYP3A4 and CYP3A5, and efflux transporter protein ABCB1 is involved in transporting out of cell. Pharmacogenetics of cyclosporine has been studied in renal transplant patients, which has shown no effect on drug response (Press et al. 2010). Pharmacogenetics of cyclosporine in psoriasis is yet to be studied. Recently, the study has identified 220 early response genes (day 14 post treatment) that were downregulated by cyclosporine in psoriasis. This also demonstrated modulation of genes from activated T cells and the "type 1" pathway, Th17 pathway. Cyclosporine also reduced TNF and inducible NO synthase in dendritic cells. It is also hypothesized that myeloid-derived genes contribute to pathogenic inflammation in psoriasis as cyclosporine modulated more myeloid-derived genes than activated T cell genes in responders (Haider et al. 2008).

4.4 Genetic Polymorphisms and Azathioprine in Psoriasis

The use of azathioprine in psoriasis has been superseded by other drugs. The active metabolite of azathioprine is 6 mercaptopurine which is deactivated by the intracellular enzyme thiopurine methyltransferase. Polymorphisms in *TPMT* such as *TPMT *2*, *TPMT*3B, and TPMT*3C* significantly vary the enzyme activity, such that it correlates inversely with myelotoxic adverse effects. Azathioprine in individuals with reduced/absent TPMT is associated with early myelosuppression. Thus, pretreatment assessment of *TPMT* in peripheral blood specimen guides the titration of azathioprine, such that azathioprine can be totally avoided in absent/very low TPMT activity.

4.5 Genetic Polymorphisms and Synthetic Retinoids in Psoriasis

Acitretin is a synthetic retinoid that is used in moderate to severe psoriasis. It is hypothesized that acitretin interacts with cytosolic proteins and nuclear receptors such as retinoic acid receptors and retinoid X receptors to alter gene transcription. Thus, it results in normalization of epidermal cell proliferation, differentiation, and cornification (Montrone et al. 2009; Young et al. 2006). The pharmacological relevance of the polymorphisms of these receptors with treatment response has not been investigated.

Psoriatic plaques have shown to have elevated levels of vascular endothelial growth factor (VEGF) which is involved in promoting angiogenesis. Retinoids inhibit VEGF production. Two *VEGF* gene polymorphisms, namely, *+405* and *–460* have been implicated in retinoid blockade of VEGF production. A study demonstrated increased prevalence of the *–460 TT* genotype among nonresponders (Young et al. 2006).

4.6 Genetic Polymorphisms and TNF Antagonists in Psoriasis

TNF antagonists are effective treatment for severe forms of psoriasis causing disease remission in 80 % of cases. Nearly 44 polymorphisms in the TNF gene have been identified. Two SNPs in the promoter region *G to A* transitions at the *–238* and *–308* sites appear to be functionally related to TNF expression in rheumatoid arthritis. Studies have shown many genetic loci as markers of TNF antagonist response, such as genes encoding transcription factors (AFF3), cell surface membrane proteins (CD226), and components of the Toll-like receptor and NF-kB pathways (Tan et al. 2010; Potter et al. 2010). Further, SNPs at the *–238* and *–308* TNF gene loci have been associated with disease susceptibility; however, there is need to explore the pharmacogenetic/genomic significance of these polymorphisms (Louis et al. 1998). In 15 psoriasis patients on etanercept therapy, patients were categorized as "responders" or "nonresponders" (Zaba et al. 2009).

Gene clusters expressed in the lesional biopsy were identified by gene arrays. The downregulation of genes involved in Th17 lymphocyte activity and IL-17 signaling was associated with treatment response. Thus, it is inferred that for etanercept response, downstream Th17 suppression is necessary, also increasing the understanding of the pathogenesis of psoriasis. Allelic variance at the genes *IL-12B* (encoding the p40 subunit common to IL-12 and IL-23), *IL-23A* (encoding IL-23 p19 subunit), and *IL-23R* (encoding IL-23 receptor subunit) is shown to be associated with susceptibility to psoriasis by genome-wide association scans (Cargill et al. 2007; Liu et al. 2008; Nair et al. 2009).

Increasing molecular and cellular evidence suggests that the IL-23/Th17 axis is key to the pathogenesis of psoriasis. This leads to the hypothesis of targeting common p40 subunit of the cytokines IL-12 and IL-23 to block their actions by monoclonal antibodies (Mak et al. 2009). Ustekinumab is the IL-12/IL-23 antagonist developed, and it has been shown to have very good anti-psoriatic efficacy (Leonardi et al. 2008). However, the interaction between polymorphisms identified at the *IL-12B*, *IL-23A*, and *IL-23R* genes and the efficacy/toxicity of IL-12/IL-23 antagonists remains to be explored. Pharmacogenomics in psoriasis is presently devoid of clinical application in spite of significant research in this area over the last decade. However, it is hoped that with improved understanding of the disease, larger prospective studies, and improved cost-effective technology, pharmacogenomics would find definite clinical application in psoriasis as in other diseases.

5 Pharmacogenomics of Inflammatory Bowel Disease

Inflammatory bowel disease with its two subtypes, namely, Crohn's disease and ulcerative colitis, has been observed to have greater prevalence in Europe and North America. But recent reports indicate that the disease is also increasing in prevalence in the developing regions such as Asia and the Middle East (Molodecky et al. 2012; Loftus 2004; Niriella et al. 2010). The pathogenesis and exact etiology of IBD is yet to be elucidated, but an interaction between genetic susceptibility, environmental factors, and the host immune response is said to play a major role in the pathophysiology of the disease (Niriella et al. 2010). Abdominal pain and diarrhea are the most common symptoms of IBD. Systemic complaints that may be associated with IBD include fever, weight loss, malaise, and arthralgia. The symptomatology of the disease being nonspecific may mimic irritable bowel syndrome and other intestinal disorders. Since the disease is associated with exacerbations and remissions, the two goals of therapy include achievement of remissions and prevention of disease flares. The common drugs used in the treatment of inflammatory bowel disease include corticosteroids, aminosalicylates, and biologic agents such as infliximab, adalimumab, and certolizumab pegol.

More than 30 susceptible novel IBD susceptible loci have been found in the last decade across several centers. Some of the genes include *NOD2, MUC3A, MST1, OCTN1, OCTN2, ABCB1, IRF5, IL-23R, and NKX2-3.* The first susceptibility gene described was the nucleotide-binding oligomerization domain 2 (*NOD2*) on chromosome 16 which encodes a protein involved in recognizing the muramyl dipeptide component of the peptidoglycan cell wall of bacteria. This leads to subsequent activation of NF-kB, an important mediator of the immune response. The discovery of NOD2 has led to the realization that the innate immunity is specifically disturbed in Crohn's disease. Polymorphisms in NOD2 such as *Arg702Trp*, *Gly908Arg*, and *Leu1007fsinsCNOD2* were found to impair NF-kB activation confer susceptibility to CD (Ishihara et al. 2009; Weizman and Silverberg 2012). In spite of this, only 22 % of patients with Crohn's disease carry a NOD polymorphism. Polymorphisms in IL-23R are associated with Crohn's disease and ulcerative colitis (Duerr et al. 2006). Although there is sufficient evidence for NOD2, it does not have sufficient sensitivity and specificity to enter routine clinical practice (Weizman and Silverberg 2012).

The most widely used application of pharmacogenomics in IBD is azathioprine use and *TPMT* polymorphisms in IBD similar to those described in earlier disorders (Table 13.4) (Chouchana et al. 2012). IBD patients with polymorphisms such as *Leu155His* in the NALP1 complex were found to develop steroid resistance more commonly than those without these polymorphisms (De et al. 2011). Individuals with variants in *IL-23R* were also associated with improved response to infliximab therapy (Jurgens et al. 2010). In a retrospective cohort of 700 patients, a genetic risk marker score based on 46 SNPs

Table 13.4 Genetic polymorphisms affecting drug therapy in inflammatory bowel disease

Gene	Polymorphism	Clinical outcome
IL-23R	*rs1004819*	Increased/decreased efficacy with infliximab
NALP1	*Leu155His*	Reduced efficacy to steroids
TPMT	*TPMT*2*, *TPMT*3B*, *TPMT*3C*	Increased azathioprine toxicity

Abbreviations: *Il-23R* interleukin 23 receptor, *NALP1* NACHT leucine-rich-repeat protein 1, *TPMT* thiopurine methyl transferase

was developed that predicted patients' need for surgery due to refractoriness to medical therapy. On the basis of this score, four groups were identified, and the risk of colectomy in these patient groups was 0, 17, 74, and 100 % (Haritunians et al. 2010).

6 Limitations of Pharmacogenetic Studies

The major problems associated with pharmacogenomics of autoimmune diseases are the limited sample size, high degree of discordance between studies, and the lack of application in clinical practice settings of most the variant polymorphisms that have been discovered in relation to rheumatic diseases pharmacotherapy. Poor methodological rigor followed in designing pharmacogenetic studies was cited as one of the major reasons for failure of replication of positive findings of initial investigators (Jorgensen and Williamson 2008). For example, use of relative risk may be an inferior measure of outcome differences than taking the absolute risk difference to determine if genotype does make a difference in drug response. Most pharmacogenetic studies are observational in nature, but in order to differentiate if the genotype is a prognostic factor or a true effect modifier of the outcome, one should perform randomized controlled trials and collect the DNA samples from patients in a prospective fashion. If they are collected later, it could lead to selection bias and selective loss of genotype if they are associated with a poor outcome. The presence of drug-gene interaction could complicate the analysis of the data gathered. Ideally drug-gene interactions are appreciated best when outcomes in the treatment and nontreatment group are compared within one genotype (Smits et al. 2005; Kelly et al. 2005; Cobos et al. 2011).

7 Conclusion

Rheumatic diseases such as rheumatoid arthritis, SLE, psoriasis, and other less common autoimmune diseases contribute to the increased morbidity and mortality worldwide. Although the pharmacological armamentarium to tackle these illnesses has steadily increased in the last two decades, some of these drugs are fraught with issues such as narrow therapeutic index, severe adverse reactions, and variable therapeutic response. Pharmacogenomic studies have appeared to elucidate some of the mechanisms that better predict an individual's response to these drugs. However, till date the only successful application for pharmacogenetic testing in clinical practice for autoimmune diseases is the *TPMT* testing for azathioprine in rheumatic diseases.

References

Acikbas I, Sanli B, Tepeli E, Ergin S, Aktan S, Bagci H (2012) Vitamin D receptor gene polymorphisms and haplotypes (Apa I, Bsm I, Fok I, Taq I) in Turkish psoriasis patients. Med Sci Monit 18:CR661–CR666

Aggarwal P, Naik S, Mishra KP, Aggarwal A, Misra R (2006) Correlation between methotrexate efficacy & toxicity with C677T polymorphism of the methylenetetrahydrofolate gene in rheumatoid arthritis patients on folate supplementation. Indian J Med Res 124:521–526

Bluett J, Barton A (2012) What have genome-wide studies told us about psoriatic arthritis? Curr Rheumatol Rep 14:364–368

Cargill M, Schrodi SJ, Chang M et al (2007) A large-scale genetic association study confirms IL12B and leads to the identification of IL23R as psoriasis-risk genes. Am J Hum Genet 80:273–290

Chen H, Poon A, Yeung C et al (2011) A genetic risk score combining ten psoriasis risk loci improves disease prediction. PLoS One 6:e19454

Chouchana L, Narjoz C, Beaune P, Loriot MA, Roblin X (2012) Review article: the benefits of pharmacogenetics for improving thiopurine therapy in inflammatory bowel disease. Aliment Pharmacol Ther 35:15–36

Cobos A, Sanchez P, Aguado J, Carrasco JL (2011) Methodological quality in pharmacogenetic studies with binary assessment of treatment response: a review. Pharmacogenet Genomics 21:243–250

Cooper GS, Stroehla BC (2003) The epidemiology of autoimmune diseases. Autoimmun Rev 2:119–125

Coulthard LR, Taylor JC, Eyre S et al (2011) Genetic variants within the MAP kinase signalling network and anti-TNF treatment response in rheumatoid arthritis patients. Ann Rheum Dis 70:98–103

Criswell LA, Lum RF, Turner KN et al (2004) The influence of genetic variation in the HLA-DRB1 and LTA-TNF regions on the response to treatment of early rheumatoid arthritis with methotrexate or etanercept. Arthritis Rheum 50:2750–2756

Cuchacovich M, Ferreira L, Aliste M et al (2004) Tumour necrosis factor-alpha (TNF-alpha) levels and influence of −308 TNF-alpha promoter polymorphism on the responsiveness to infliximab in patients with rheumatoid arthritis. Scand J Rheumatol 33:228–232

Cui J, Saevarsdottir S, Thomson B et al (2010) Rheumatoid arthritis risk allele PTPRC is also associated with response to anti-tumor necrosis factor alpha therapy. Arthritis Rheum 62:1849–1861

Davila L, Ranganathan P (2011) Pharmacogenetics: implications for therapy in rheumatic diseases. Nat Rev Rheumatol 7:537–550

Davis JP, Cain GA, Pitts WJ, Magolda RL, Copeland RA (1996) The immunosuppressive metabolite of leflunomide is a potent inhibitor of human dihydroorotate dehydrogenase. Biochemistry 35:1270–1273

de Rotte MC, Luime JJ, Bulatovic M, Hazes JM, Wulffraat NM, de Jonge R (2010) Do snapshot statistics fool us in MTX pharmacogenetic studies in arthritis research? Rheumatology (Oxford) 49:1200–1201

De IS, Stocco G, Martelossi S et al (2011) Genetic predictors of glucocorticoid response in pediatric patients with inflammatory bowel diseases. J Clin Gastroenterol 45:e1–e7

Duerr RH, Taylor KD, Brant SR et al (2006) A genome-wide association study identifies IL23R as an inflammatory bowel disease gene. Science 314:1461–1463

Dziedziejko V, Kurzawski M, Safranow K, Chlubek D, Pawlik A (2011) The effect of ESR1 and ESR2 gene polymorphisms on the outcome of rheumatoid arthritis treatment with leflunomide. Pharmacogenomics 12:41–47

Fisher MC, Cronstein BN (2009) Metaanalysis of methylenetetrahydrofolate reductase (MTHFR) polymorphisms affecting methotrexate toxicity. J Rheumatol 36:539–545

Gladman DD, Antoni C, Mease P, Clegg DO, Nash P (2005) Psoriatic arthritis: epidemiology, clinical features, course, and outcome. Ann Rheum Dis 64(Suppl 2): ii14–ii17

Grabar PB, Rozman B, Logar D, Praprotnik S, Dolzan V (2009) Dihydroorotate dehydrogenase polymorphism influences the toxicity of leflunomide treatment in patients with rheumatoid arthritis. Ann Rheum Dis 68:1367–1368

Haider AS, Lowes MA, Suarez-Farinas M et al (2008) Identification of cellular pathways of "type 1," Th17 T cells, and TNF- and inducible nitric oxide synthase-producing dendritic cells in autoimmune inflammation through pharmacogenomic study of cyclosporine A in psoriasis. J Immunol 180:1913–1920

Halsall JA, Osborne JE, Pringle JH, Hutchinson PE (2005) Vitamin D receptor gene polymorphisms, particularly the novel A-1012G promoter polymorphism, are associated with vitamin D3 responsiveness and non-familial susceptibility in psoriasis. Pharmacogenet Genomics 15:349–355

Haritunians T, Taylor KD, Targan SR et al (2010) Genetic predictors of medically refractory ulcerative colitis. Inflamm Bowel Dis 16:1830–1840

Hebert HL, Ali FR, Bowes J, Griffiths CE, Barton A, Warren RB (2012) Genetic susceptibility to psoriasis and psoriatic arthritis: implications for therapy. Br J Dermatol 166:474–482

Hoebert JM, Mantel-Teeuwisse AK, van Dijk L, Bijlsma JW, Leufkens HG (2012) Do rheumatoid arthritis patients have equal access to treatment with new medicines?: tumour necrosis factor-alpha inhibitors use in four European countries. Health Policy 104:76–83

Hughes LB, Beasley TM, Patel H et al (2006) Racial or ethnic differences in allele frequencies of single-nucleotide polymorphisms in the methylenetetrahydrofolate reductase gene and their influence on response to methotrexate in rheumatoid arthritis. Ann Rheum Dis 65:1213–1218

Ishihara S, Aziz MM, Yuki T, Kazumori H, Kinoshita Y (2009) Inflammatory bowel disease: review from the aspect of genetics. J Gastroenterol 44:1097–1108

Jankovic S, Raznatovic M, Marinkovic J et al (2011) Health-related quality of life in patients with psoriasis. J Cutan Med Surg 15:29–36

Jorgensen AL, Williamson PR (2008) Methodological quality of pharmacogenetic studies: issues of concern. Stat Med 27:6547–6569

Jurgens M, Laubender RP, Hartl F et al (2010) Disease activity, ANCA, and IL23R genotype status determine early response to infliximab in patients with ulcerative colitis. Am J Gastroenterol 105:1811–1819

Kelly PJ, Stallard N, Whittaker JC (2005) Statistical design and analysis of pharmacogenetic trials. Stat Med 24:1495–1508

Langley RG, Krueger GG, Griffiths CE (2005) Psoriasis: epidemiology, clinical features, and quality of life. Ann Rheum Dis 64(Suppl 2):ii18–ii23; discussion ii24–ii5

Lee YH, Song GG (2010) Associations between the C677T and A1298C polymorphisms of MTHFR and the efficacy and toxicity of methotrexate in rheumatoid arthritis: a meta-analysis. Clin Drug Investig 30:101–108

Lee YH, Rho YH, Choi SJ, Ji JD, Song GG (2006) Association of TNF-alpha −308 G/A polymorphism with responsiveness to TNF-alpha-blockers in rheumatoid arthritis: a meta-analysis. Rheumatol Int 27:157–161

Lee YC, Cui J, Costenbader KH, Shadick NA, Weinblatt ME, Karlson EW (2009) Investigation of candidate polymorphisms and disease activity in rheumatoid arthritis patients on methotrexate. Rheumatology (Oxford) 48:613–617

Lee YH, Ji JD, Bae SC, Song GG (2010) Associations between tumor necrosis factor-alpha (TNF-alpha) −308 and −238 G/A polymorphisms and shared epitope status and responsiveness to TNF-alpha blockers in rheumatoid arthritis: a metaanalysis update. J Rheumatol 37:740–746

Lee YH, Bae SC, Choi SJ, Ji JD, Song GG (2012) Associations between TNFAIP3 gene polymorphisms and rheumatoid arthritis: a meta-analysis. Inflamm Res 61:635–641

Leonardi CL, Kimball AB, Papp KA et al (2008) Efficacy and safety of ustekinumab, a human interleukin-12/23 monoclonal antibody, in patients with psoriasis: 76-week results from a randomised, double-blind, placebo-controlled trial (PHOENIX 1). Lancet 371:1665–1674

Liu Y, Helms C, Liao W et al (2008) A genome-wide association study of psoriasis and psoriatic arthritis identifies new disease loci. PLoS Genet 4:e1000041

Loftus EV Jr (2004) Clinical epidemiology of inflammatory bowel disease: incidence, prevalence, and environmental influences. Gastroenterology 126:1504–1517

Louis E, Franchimont D, Piron A et al (1998) Tumour necrosis factor (TNF) gene polymorphism influences TNF-alpha production in lipopolysaccharide (LPS)-stimulated whole blood cell culture in healthy humans. Clin Exp Immunol 113:401–406

Mak RK, Hundhausen C, Nestle FO (2009) Progress in understanding the immunopathogenesis of psoriasis. Actas Dermosifiliogr 100(Suppl 2):2–13

Milic V, Jekic B, Lukovic L et al (2012) Association of dihydrofolate reductase (DHFR) -317AA genotype with poor response to methotrexate in patients with rheumatoid arthritis. Clin Exp Rheumatol 30:178–183

Mittal N, Mittal R, Sharma A, Jose V, Wanchu A, Singh S (2012) Treatment failure with disease-modifying antirheumatic drugs in rheumatoid arthritis patients. Singap Med J 53:532–536

Molodecky NA, Soon IS, Rabi DM et al (2012) Increasing incidence and prevalence of the inflammatory bowel diseases with time, based on systematic review. Gastroenterology 142:46–54

Montrone M, Martorelli D, Rosato A, Dolcetti R (2009) Retinoids as critical modulators of immune functions: new therapeutic perspectives for old compounds. Endocr Metab Immune Disord Drug Targets 9:113–131

Nair RP, Duffin KC, Helms C et al (2009) Genome-wide scan reveals association of psoriasis with IL-23 and NF-kappaB pathways. Nat Genet 41:199–204

Ngamjanyaporn P, Thakkinstian A, Verasertniyom O et al (2011) Pharmacogenetics of cyclophosphamide and CYP2C19 polymorphism in Thai systemic lupus erythematosus. Rheumatol Int 31:1215–1218

Niriella MA, De Silva AP, Dayaratne AH et al (2010) Prevalence of inflammatory bowel disease in two districts of Sri Lanka: a hospital based survey. BMC Gastroenterol 10:32

Owlia MB, Mostafavi Pour Manshadi SM, Naderi N (2012) Cardiac manifestations of rheumatological conditions: a narrative review. ISRN Rheumatol 2012:463620. doi:10.5402/2012/463620, Epub 2012 Oct 17

Pawlik A, Herczynska M, Kurzawski M, Safranow K, Dziedziejko V, Drozdzik M (2009) The effect of exon (19C > A) dihydroorotate dehydrogenase gene polymorphism on rheumatoid arthritis treatment with leflunomide. Pharmacogenomics 10:303–309

Potter C, Cordell HJ, Barton A et al (2010) Association between anti-tumour necrosis factor treatment response and genetic variants within the TLR and NF{kappa}B signalling pathways. Ann Rheum Dis 69:1315–1320

Press RR, de Fijter JW, Guchelaar HJ (2010) Individualizing calcineurin inhibitor therapy in renal transplantation–current limitations and perspectives. Curr Pharm Des 16:176–186

Ranganathan P, Culverhouse R, Marsh S et al (2008) Methotrexate (MTX) pathway gene polymorphisms and their effects on MTX toxicity in Caucasian and African American patients with rheumatoid arthritis. J Rheumatol 35:572–579

Reichert JM (2012) Marketed therapeutic antibodies compendium. MAbs 4:413–415

Schmeling H, Biber D, Heins S, Horneff G (2005) Influence of methylenetetrahydrofolate reductase polymorphisms on efficacy and toxicity of methotrexate in patients with juvenile idiopathic arthritis. J Rheumatol 32:1832–1836

Smits KM, Schouten JS, Smits LJ, Stelma FF, Nelemans P, Prins MH (2005) A review on the design and reporting of studies on drug-gene interaction. J Clin Epidemiol 58:651–654

Tan RJ, Gibbons LJ, Potter C et al (2010) Investigation of rheumatoid arthritis susceptibility genes identifies association of AFF3 and CD226 variants with response to anti-tumour necrosis factor treatment. Ann Rheum Dis 69:1029–1035

Taniguchi A, Urano W, Tanaka E et al (2007) Validation of the associations between single nucleotide polymorphisms or haplotypes and responses to disease-modifying antirheumatic drugs in patients with rheumatoid arthritis: a proposal for prospective pharmacogenomic study in clinical practice. Pharmacogenet Genomics 17:383–390

Taraborelli M, Andreoli L, Archetti S, Ferrari M, Cattaneo R, Tincani A (2009a) Methylenetetrahydrofolate reductase polymorphisms in methotrexate treatment of rheumatoid arthritis patients. Review of the literature and personal experience. Reumatismo 61:98–106

Taraborelli M, Andreoli L, Archetti S, Ferrari M, Cattaneo R, Tincani A (2009b) Methylenetetrahydrofolate reductase polymorphisms and methotrexate: no association with response to therapy nor with drug-related adverse events in an Italian population of rheumatic patients. Clin Exp Rheumatol 27:499–502

Vasku V, Bienertova VJ, Pavkova GM, Vasku A (2007) Three retinoid X receptor gene polymorphisms in plaque psoriasis and psoriasis guttata. Dermatology 214:118–124

Wahie S, Daly AK, Cordell HJ et al (2011) Clinical and pharmacogenetic influences on response to hydroxychloroquine in discoid lupus erythematosus: a retrospective cohort study. J Investig Dermatol 131:1981–1986

Warren RB, Smith RL, Campalani E et al (2009) Outcomes of methotrexate therapy for psoriasis and relationship to genetic polymorphisms. Br J Dermatol 160:438–441

Weizman AV, Silverberg MS (2012) Have genomic discoveries in inflammatory bowel disease translated into clinical progress? Curr Gastroenterol Rep 14:139–145

Wessels JA, de Vries-Bouwstra JK, Heijmans BT et al (2006) Efficacy and toxicity of methotrexate in early rheumatoid arthritis are associated with single-nucleotide polymorphisms in genes coding for folate pathway enzymes. Arthritis Rheum 54:1087–1095

Yanagimachi M, Naruto T, Hara T et al (2011) Influence of polymorphisms within the methotrexate pathway genes on the toxicity and efficacy of methotrexate in patients with juvenile idiopathic arthritis. Br J Clin Pharmacol 71:237–243

Young HS, Summers AM, Read IR et al (2006) Interaction between genetic control of vascular endothelial growth factor production and retinoid responsiveness in psoriasis. J Investig Dermatol 126:453–459

Zaba LC, Suarez-Farinas M, Fuentes-Duculan J et al (2009) Effective treatment of psoriasis with etanercept is linked to suppression of IL-17 signaling, not immediate response TNF genes. J Allergy Clin Immunol 124:1022–1030

Zhong S, Huang M, Yang X et al (2006) Relationship of glutathione S-transferase genotypes with side-effects of pulsed cyclophosphamide therapy in patients with systemic lupus erythematosus. Br J Clin Pharmacol 62:457–472

Section III

Personalized Medicine: Neuro and Psychiatric Disorders

Pharmacogenomics and Personalized Medicine in Parkinsonism

14

Trevor Archer and Anders Fredriksson

Abstract

Pharmacogenetic-pharmacogenomic development from a single gene approach to incorporate pathway-based and genome-wide approaches has been benefitted from the emergence of several parallel technologies, such as genomics, transcriptomics, metabolomics, and proteomics which have contributed to and enhanced significantly propensities for generation and testing of pharmacogenomic hypotheses both paralleled and followed by associated developments in the clinical practice for treating Parkinson's disease (PD). The notion of "personalized medicine," incorporating the customization of healthcare, with decisions and practices that suited to each individual patient through application of genetic, biomarker, gene-environment interactive, or other information, involves principles through which drugs, drug combinations, and drug administration properties are optimized for each individual's unique genetic makeup. The personalized medication of antiparkinsonian drug therapy; the symptomatic and regional disruptions; genetic, epigenetic, and biomarkers of the disorder; and the pharmacogenomics of neuroleptic drug-induced parkinsonism provide outlets for eventual understanding and management. As a case study in personalized medicine in the laboratory, physical exercise combined with the electromagnetic wavelength treated *Saccharomyces cerevisiae* yeast, Milmed, was demonstrated to abolish the marked hypokinesia induced by the dopamine (DA) neurotoxin, MPTP, as well as the severe loss of DA in the striatal region of the C57/BL6 mice studied. The Exercise-Milmed coadministration induced also a profound increase in brain-derived neurotrophin levels (BDNF) in the mouse parietal cortex region that included the motor cortex.

T. Archer (✉)
Department of Psychology, University of Gothenburg, Box 500, Gothenburg 40530, Sweden
e-mail: Trevor.archer@psy.gu.se

A. Fredriksson
Department of Neuroscience Psychiatry, Uppsala University, Uppsala 751 85, Sweden

D. Barh et al. (eds.), *Omics for Personalized Medicine*,
DOI 10.1007/978-81-322-1184-6_14, © Springer India 2013

1 Introduction

Pharmacogenomics is defined by the branch of pharmacology which deals with the influence of genetic variation on drug response in patients, in the present case parkinsonian, by correlating gene expression or single-nucleotide polymorphisms with a drug's efficacy or toxicity (Wang 2010). Weinshilboum and Wang (2006) have defined pharmacogenomics as the study of the role of inherited and acquired genetic variation in response to drug interventions. A major focus of pharmacogenomics is the development of suitable methods that facilitate the efficacy of drug therapy, in association with individuals' genotype, so that maximum efficacy is balanced against minimal adverse effects thereby the attainment of viable therapeutic windows (Becquemont 2009). Current attempts at attainment of bioinformatics, including genomics, proteomics, and metabolomics, should expedite identification of proteins/enzymes, activated proteins, genes, and gene variations facilitating drug therapies (Becquemont et al. 2011), e.g., those that contribute to the etiopathogenesis of Parkinson's disease (PD) and related therapeutic measures (Clayton 2012; Kaiser et al. 2003; Wang et al. 2011). Pharmacogenomics facilitates the identification of biomarkers, not only to facilitate derivation of "disorder staging" but also to allow optimal possibilities for therapeutic drug selection and targets, dose windows, drug dispositions, treatment duration, and projections of adverse reactions that may be avoidable (Evans and McLeod 2003; Weinshilboum 2003; Woodcock 2010). The technologies involved have led greater effectiveness in drug selection and administration accompanied by reductions in adverse/side effect profiles through investigations of neuroscientific, regulatory, and neuropsychological agents (Wang et al. 2000).

PD is a relatively common, idiopathic neurodegenerative movement disorder characterized by impaired motor function, including resting tremors, rigidity, akinesia/bradykinesia, and postural instability as the cardinal symptoms (Gaggelli et al. 2006; Jankovic 2008; Lees et al. 2009). It is a progressive neurodegenerative disorder and, compared with familial forms, is associated most often with advanced age (>55 years of age). The pathophysiology of PD involves dopaminergic neuron death and accumulation of Lewy bodies associated with mutations in α-synuclein, a 14-kDa protein predominantly expressed in the brain and CNS (Rasia et al. 2005). PD patients show decreased levels of presynaptic dopamine (DA) neuron terminal markers in the basal ganglia (Felicio et al. 2009), consistent with loss of dopaminergic terminals due to degeneration of neuronal cell bodies in the substantia nigra pars compacta (Hattori et al. 2006). PD patients exhibit decreased levels of DA transporters (DATs) and vesicular monoamine transporter type 2 (VMAT2), as well as reduced activity of dopa decarboxylase, assessed by striatal conversion of L-dopa to DA, according to PET and SPECT analyses (Al Hadithy et al. 2008; Lewitt et al. 2012). Wu et al. (2012) using MRI showed that the substantia nigra pars compacta expressed a decreased connectivity with several regions, including the striatum, globus pallidus, subthalamic nucleus, thalamus, supplementary motor area, dorsolateral prefrontal cortex, insula, default mode network, temporal lobe, cerebellum, and pons in patients compared to controls. They found that L-dopa administration partially normalized the pattern of connectivity to a similarity such as that expressed by the healthy volunteers involving causal connectivity of basal ganglia networks from the substantia nigra pars compacta. Postsynaptic D_2 DA receptors (D_2Rs) are either unaffected or increased in the striatum of untreated PD patients (Antonini et al. 1994). Oxidative injury appears to be one effect of α-synuclein (α-Syn) aggregates and could ultimately produce neuronal cell death. α-Syn, a 140 residue, intrinsically disordered protein is localized in presynaptic terminals of DA neurons (Yang et al. 2010). Autonomic nervous system involvement occurs at early stages in both PD and incidental Lewy body disease and affects the sympathetic, parasympathetic, and enteric nervous systems. It has been proposed that α-Syn pathology in PD has a distal to proximal progression along autonomic pathways.

According to Braakian notions, the enteric nervous systems are affected before the dorsal motor nucleus of the vagus, and distal axons of cardiac sympathetic nerves degenerate before there is loss of paravertebral sympathetic ganglion neurons. Cersosimo and Benarroch (2012a) have shown that consistent with neuropathological findings, some autonomic manifestations, such as constipation or impaired cardiac uptake of norepinephrine precursors, occur at early stages of the disease even before the onset of motor symptoms (cf. Braak et al. 2007; Cersosimo and Benarroch 2012b; Hawkes et al. 2007).

The evolution of pharmacogenetics-pharmacogenomics from a single gene approach to incorporate pathway-based and genome-wide approaches has been described comprehensively (Wang 2009; Wang and Weinshilboum 2008). Prolifically, several parallel technologies, such as genomics, transcriptomics, metabolomics, and proteomics, have contributed to and enhanced significantly propensities for generation and testing of pharmacogenomic hypotheses followed by associated developments in clinical practice (Lesko and Woodcock 2004; Wang and Weinshilboum 2006). The incorporation of transcriptomic and metabolomic findings has offered an important tactic for assessing and predicting variation in drug-response phenotypes and "translational" variants (Dettmer and Hammock 2004; Hughes et al. 2009; Lindon et al. 2004; Mendes 2006). Studies that focus upon pharmacogenomics involve the rapid scanning of markers across the genome of individuals affected by a certain disorder, e.g., PD, or drug-response phenotype, in comparison with unaffected individuals, with tests for association that compare genetic variations in case–control settings (Manolio 2010). Several oxidative phosphorylation (OXPHOS) system complex activities and quantities are reduced in PD. Toxicogenomics, combining toxicology with genomics, describes the collection, interpretation, and storage of information about gene and protein activity within particular cell or tissue of an organism in response to toxic substances in order to elucidate molecular mechanisms evolved in the expression of toxicity and to derive molecular expression patterns (i.e., molecular biomarkers) that predict toxicity or the genetic susceptibility to PD. OXPHOS functioning is affected by the mutations of PD-linked nuclear genes (Bar-Yaacov et al. 2012), and inactivation of other nuclear genes related to mitochondrial DNA replication and expression leads to PD (Pennington et al. 2010; Orth and Schapira 2001). Lopez-Gallardo et al. (2011) have described the extent to which nuclear and mitochondrial genetic and environmental factors, primary through gene-environment interplay, induce additive/synergistic effects thereby elevating the risk for PD. Population polymorphisms pertaining to mitochondrial DNA replication and expression that influence interactions with different xenobiotics, substances in an individual but which are not normally produced or expected to be there, may present susceptibility factors that contribute to the etiopathogenesis of the disorder.

The notion of "personalized medicine," a model that outlines the customization of healthcare, with decisions and practices that suited to each individual patient through application of genetic, biomarker, gene-environment interactive, or other information (Shastry 2005, 2006), refers in which drugs and drug combinations are optimized for each individual's unique genetic makeup (Squassina et al. 2010). A multitude of factors influence the emergence and progression of parkinsonism, motor symptoms, disability, outcome prognosis, and drug fluctuations, in addition to the clinically significant non-motor features such as depression, anxiety, sleep disturbances, smell and taste loss, compulsive behaviors, and dementia. Disorder staging diagnostic intervention optimizes early identification and accurate diagnosis and management of patients. Therapy of PD is an achievable goal considering the recent advances in neuroimaging, genetic testing, and other evolving diagnostic measures. Successful intervention is facilitated by proteomics, the comprehensive analysis and characterization of proteins and protein isoforms encoded by the human genome. Important biological functions, such as growth and development of the brain and CNS involving migration, differentiation and synaptogenesis, neuronal death, cellular movement, localization and integrity, and stem cell differentiation, are controlled by signal transduction, an epigenetic process modulated by protein enzyme activity (Clayton et al. 2006). Seventeen regions of the genome are present with common variations that affect the risk of developing Parkinson's disease. Nine genes have been identified that, when mutated, may cause the disorder (Annesi et al. 2011; Cooper-Knock et al. 2012; Dumitriu et al. 2012; Kumar et al. 2012; Lachenmayer and Yue 2012; Maruyama and Naoi 2012). In PD, the range of personalized medicine, from physical exercise schedules (Archer et al. 2011a, b; Schenkman et al. 2012) to personalized deep brain stimulation (Wagle Shukla and Okun 2012) to identification of premotor populations (Streffer et al. 2012), continues to flourish.

2 Personalized Medication and Anti-PD Treatment

Anti-PD compounds, such as L-dopa and direct-acting DA agonists, show an efficacy dependent upon patient characteristics in reducing symptoms of movement disorder (Devos et al. 2009; van Hilten et al. 2000). The pharmacotherapy of PD has focussed upon dopaminergic compounds, largely the DA precursor, L-dopa (L-3,4-dihydroxyphenylalanine), which is very much the treatment of choice for the disorder (Fahn 1999). Orally administered L-dopa is absorbed by the intestine, enters the bloodstream, crosses the blood–brain barrier, and enters dopaminergic neurons in the brain where it is converted to DA through the action of the aromatic amino acid decarboxylase (AADC) enzyme. In general, it is administered in combination with a decarboxylase inhibitor to prevent conversion to DA peripherally. Large to very large fluctuations in the responses of individual patients to anti-PD drug medication have been observed (Fabbrini et al. 1988), not least due to the correlation between motor performance and plasma concentration of L-dopa (Jankovic and Stacy 2007; Pahwa and Lyons 2009). Similar extent of fluctuation has been described with regard to the development of motor complications, e.g., L-dopa-induced dyskinesia and side effects such as hallucinations and sleepiness, up to 45 % of users within 5 years (Graham et al. 1997). Additionally, the likelihood of ischemic complications presents another symptom for consideration (Arbouw et al. 2012). The notion of personalization medication emerges in the context of compounds designed to reduce fluctuations in PD: Lewitt et al. (2000) assessed the pharmacokinetic profile, efficacy, and safety of XP21279 administered with carbidopa (CD) in subjects with Parkinson's disease (PD) experiencing motor fluctuations and explore dose correspondence between CD-levodopa and XP21279 administered with carbidopa. They observed that XP21279 provided significantly less variability in LD concentration compared with carbidopa-levodopa in 10 PD patients presenting motor fluctuations, consistent with a lower peak-to-trough fluctuation for XP21279. The expressed patterns of percentage of patients' "OFF-periods" were consistent with the levodopa concentration-time profiles for each respective treatment. Compared with carbidopa-levodopa treatment, 6 of 10 study completers experienced reduction of 30 % or greater in average daily OFF time during the last 4 days in the XP21279 treatment period. XP21279 resulted in an increase in the time spent during "ON-periods" without functionally blocking dyskinesias, and the mean time to ON after the first morning XP21279 dose was not delayed, as compared with carbidopa-levodopa.

It is likely that genetic variations in gene coding for drug metabolism and drug availability contribute to a large extent towards the interindividual variability experienced in response to drugs of therapy (Swen et al. 2007), not least regarding the pharmacogenetics of anti-PD compounds (Arbouw et al. 2007). The genetic variability, together with gender and reproductive factors (Nicoletti et al. 2011), of each individual determines largely the interindividual variability in the responses to anti-PD drug therapy (Zappia et al. 2005). Although the findings emerging from several genetic association studies implying links between anti-PD drug-induced dyskinesias and polymorphisms are conflicting, several genes appear to be involved, including the *DRD2* gene, the *DRD4* gene, the DA transporter (*DAT*) gene, the μ1-opioid receptor (*OPRM1*) gene, the cholecystokinin (*CCK*) gene, the apolipoprotein E (*APOE*) gene, the preprohypocretin (*HCRT*) gene, and the catechol-o-transferase (*COMT*) gene (Arbouw et al. 2009; Liu et al. 2009a, b; Paus et al. 2008; Williams-Gray et al. 2008). Lin et al. (2007) in a study of 251 PD patients observed that the frequency of the angiotensin I-converting enzyme gene homozygote ACE-II genotype with L-dopa-induced psychosis was significantly higher than that in PD patients without the adverse effect. Additionally, the possible role of brain-derived neurotrophic factor (BDNF) in L-dopa-induced dyskinesias has been considered since the factor is involved in synaptogenesis, synaptic plasticity and efficacy (Chase 2004; Woo et al. 2005), modulation of receptor systems underlying L-dopa-induced dyskinesias (Guillin et al. 2003), and the pathogenesis of dopaminergic neurotransmission in PD (Fumagalli et al. 2006; Momose et al. 2002). Foltynie et al. (2009) studied the influence of a common functional polymorphism of the BDNF gene on the risk for development of L-dopa-induced dyskinesias in a cohort of 315 PD patients, independently and variably treated with L-dopa and/or other DA interventions. PD patients with the met allele of BDNF, linked with lower activity-dependent secretion of BDNF, presented significantly higher risk of developing dyskinesia earlier in the course of dopaminergic agent therapy. Pharmacogenetic-pharmacogenomic studies facilitate the description and definition of genetic variations in gene coding and the regulation of the proteins involved in the pathways underlying dyskinesic expressions evolving from the pathophysiology of PD (Arbouw et al. 2010).

Genetic predispositions modulating factors causing anti-PD drug therapy fluctuations bedevil both treatment prognosis and outcome appraisals. In order to investigate more closely whether or not genetic predispositions may contribute to the pathophysiological development of medication-related complications in PD, Paus et al. (2009) reassessed the impact of the *DRD3* Ser9Gly polymorphism on development of motor complications in a large-scale association study based on the gene bank of the German Competence Network on Parkinson's disease, using stepwise regression analysis. Despite incorporating established clinical risk factors to avoid overlooking an effect of genotype, no effect of DRD3 Ser9Gly on chorea, dystonia, or motor fluctuation expressions in PD was observed. They confirmed that duration of PD was confirmed as the most important clinical risk factor, followed by age of disease onset and female gender. Furthermore, it was not possible to identify any effect of *DRD3* Ser9Gly on tremor in PD, even when regarding various symptom combinations to avoid missing a weak effect on the phenotype (Paus et al. 2010). AADC provides the major pathway for decarboxylation of L-dopa to DA. Thus, the relationship between subregional AADC activity in the striatum and the PD symptoms, using high-resolution PET with an AADC tracer, 6-[^{18}F]fluoro-L-m-tyrosine (FMT), has been assessed (Asari et al. 2011). They found that FMT uptake was decreased in the posterior putamen regardless of predominant motor symptoms and disease duration in all 101 patients and that severity of bradykinesia, rigidity, and axial symptoms was correlated with the decrease of FMT uptake in the putamen, especially in the anterior part. L-Dopa is metabolized to 3-O-methyldopa by catechol-O-methyltransferase (COMT) under conditions of AADC inhibition that involve L-dopa complications (Alachkar et al. 2010). COMT inhibitors, e.g., entacapone, increase the duration of motor responding by PD patients 30–40 %, decreasing the motor fluctuations, and effectively elevating the ON period (Kaakkola 2010; Rinne et al. 1998; Ruottinen and Rinne 1996a, b). The extent and definition of COMT polymorphisms, particularly the *COMT* gene (rs4608) resulting in Val157Met, and related DA and L-dopa metabolism and symptom profiles have been described (e.g., Kiyohara et al. 2011; Vallelunga et al. 2012; Wu et al. 2012). Corvol et al. (2011) determined the consequences of COMT polymorphisms upon 58 PD patients' responses to entacapone (200 mg) coadministered with L-dopa (50 mg), with regard to high (Val/Val, *COMT*^{HH}), intermediate (Val/Met, *COMT*^{HL}), and low (Met/Met, *COMT*^{LL}) COMT activity (Hernán et al. 2002). They observed that the gain in the best ON-period time was higher in *COMT*^{HH} patients than in *COMT*^{LL} patients. Area under the concentration over time curve of L-dopa increased more after entacapone in *COMT*^{HH} patients than in *COMT*^{LL} patients, and COMT inhibition by entacapone was higher in *COMT*^{HH} patients than in *COMT*^{LL} patients. They concluded that the *COMT*^{HH} genotype enhanced the effect of entacapone on the pharmacodynamics and pharmacokinetics of L-dopa in PD patients.

It has been observed that the benefits of levodopa therapy become less marked over time, possibly due to the degeneration of nigrostriatal dopaminergic neurons inducing a progressive loss of AADC, the enzyme that converts levodopa into dopamine (Contin et al. 1994). Gene transfer of dopamine-synthesizing enzymes into the striatal neurons and/or neuroprotective interventions has led to behavioral recovery in animal models (Harms et al. 2011; Huo et al. 2012; Laganiere et al. 2010; Zhou et al. 2011). Muramatsu et al. (2010) have provided evidence for the safety and efficacy of AADC gene therapy in Phase I study of PD treatment. Using PET imaging with [(18)F]fluoro-L-m-tyrosine tracer was used for evaluation of AADC expression and the UPDRS; Mittermeyer et al. (2012) observed elevated PET signal in the first 12 months that persisted over 4 years in both dose (high and low) groups. The elevated PET value, compared with the presurgery baseline, was maintained over the 4-year monitoring period. The UPDRS off medication for 12 h improved in the first 12 months for all the patients, but deteriorated slowly in subsequent years. These studies implied that a therapy strategy involving manipulation of the AADC gene may prove a viable alternative. In a primate model of PD, intrastriatal infusion of an adeno-associated viral type 2 vector containing the human AADC gene (AAV-hAADC) results in robust response to low-dose levodopa without the side effects associated with higher doses. In a clinical trial, patients with moderately advanced PD received bilateral intraputaminal infusion of AAV-hAADC vector (Christine et al. 2009). Although gene therapy was well tolerated, 1 symptomatic and 2 asymptomatic intracranial hemorrhages followed the operative procedure. Total and motor rating scales improved in both cohorts. Motor diaries also showed increased on-time and reduced off-time without increased ON-period time dyskinesia. At 6 months, FMT PET showed a 30 % increase of putaminal uptake in the low-dose cohort and a 75 % increase in the high-dose cohort. It appears that bilateral intrastriatal infusion of adeno-associated viral type 2 vector containing the human AADC gene improves mean scores on the Unified Parkinson's Disease Rating Scale by approximately 30 % in the on and off states, but the surgical procedure may be associated with an increased risk of intracranial hemorrhage and self-limited headache.

3 Pharmacogenomics of Neuroleptic-Induced Parkinsonism

Neuroleptic-induced parkinsonism (NIP), secondary parkinsonism, presents a movement disorder occurring in 15–40 % of patients treated with antipsychotic medication, with accompanying adverse effects on drug compliance, self-esteem, and quality of life (Gerlach 1999; Hirose 2006). It tends to develop slowly, over days to weeks, expressing high levels of variability in individual sensitivity/susceptibility to all the extrapyramidal side effects as a function of the pharmacological and pharmacodynamic profiles of the compounds applied (Friedman 2006, 2010; Thomas and Friedman 2010). Several risk factors affect the predisposition to development of NIP, including advanced age, gender (female), type of neuroleptic drug, age at diagnosis, and dose levels of drugs applied (Caligiuri et al. 1999, 2000; Jabs et al. 2003). Nevertheless, the high levels of variation in incidence have prompted attempts to identify genetic predisposition (e.g., Lencer et al. 2004). Using logistic regression and controlling for population stratification, age, gender, Simpson-Angus scale score at baseline, and concomitant use of anticholinergic drugs, Alkelai et al. (2009) identified several single-nucleotide polymorphisms associated with NIP severity.

They identified a number of candidate genes that were likely to contribute to the pathophysiology of the syndrome.

Pharmacogenetic studies have generalized the several expressions of NIP, i.e., tardive dyskinesias, akathisia, and dystonia, into a single clinical syndrome (Gunes et al. 2007; Guzey et al. 2007; Nakazono et al. 2005). The etiopathogenesis could involve the antagonistic actions of neuroleptic compound upon dopamine DA D2 receptor gene, the DRD2 gene (Lidow 2000; Reynolds 2004; Mihara et al. 2000). However, Chong et al. (2003) failed to obtain evidence that the D2 genotype was involved in the pathophysiology of tardive dyskinesias in Chinese patients with schizophrenia. Instead, they pointed out that the association of tardive dyskinesias with the serine/serine genotype of the DRD3 may be an epiphenomenon of patients with a subtype of schizophrenic patients with greater exposure to neuroleptic drugs (see also Lee et al. 2008, 2010; Tan et al. 2003). Additionally, the modulating effects of the serotonin 2A and 2C receptor genes, HTR2A and HTR2C, respectively, have been considered (Hamdani et al. 2005; Lerer et al. 2005). Grønbaek et al. (2008) studied the association between polymorphisms for DRD3, HTR2A, and HTR2C and NIP, rigidity, bradykinesia, and rest tremor in 117 African-Caribbean inpatients at the D.R. Capriles clinic (Curacao, Netherlands Antilles). Inclusion criteria were (1) absence of organic and neurological disorders that could cause movement disorders, (2) a history of neuroleptic use over at least 3 months, and (3) informed consent. Determination of polymorphisms was performed according to standard protocols, the Unified Parkinson's Disease Rating Scale (UPDRS) for assessment of NIP, rigidity, bradykinesia, and rest tremor. In the male patients, significant associations between DRD2 (the *-141CDel*-allele carriership) and rigidity, and HTR2C (*23Ser*-allele carriership) and bradykinesia, were obtained. Their overall conclusions pertained to symptom-specific pharmacogenomic, personalized medicine analyses (see also Bakker et al. 2006).

4 Milmed-Exercise Combination as Personalized Intervention in PD

Physical exercise has been described as any and all activity that generates force through muscular activity that disrupts a homeostatic state (McArdle et al. 1974; Scheuer and Tipton 1977). Although daily physical activity holds benefits for general measures of function, quality of life, and physical strength, as well as increasing endurance (Dechamps et al. 2010; Marks et al. 2009, 2010), much evidence presents the manifest advantages for cerebral integrity during aging (Kramer et al. 1999; Lustig et al. 2009; Marks et al. 2011). Any bodily activity that enhances or maintains physical fitness implies the involvement of regular and frequent exercise. Morris and Schoo (2004) have defined exercise as a planned, structured physical activity with the purpose of improving one or more aspects of physical fitness and functional capacity. Physical exercise offers a nonpharmacologic, noninvasive intervention that enhances brain health and plasticity (Cotman and Berchtold 2002). It has been characterized on the basis of type, intensity, frequency, and duration, with either endurance or resistance as the training end point (Mougios 2010). Long-term exercise benefits brain functioning by increasing cerebral blood flow and oxygenation (Linkis et al. 1995), mobilizing growth factors and synaptic plasticity (Hunsberger et al. 2007), and facilitating performance through neurotransmitter release (Morishima et al. 2006; Waters et al. 2008). Regular physical exercise holds particular benefits for older individuals, whether under conditions of normal aging or affected by neurodegenerative disorders (Archer 2011; Archer et al. 2011a, b).

Repeated administration of 1-methyl-4-phenyl-1,2,3,6-tetrahydropyridine (MPTP) to C57/BL6 mice induces selective and long-lasting lesions of dopamine (DA) in nigrostriatal regions of the brain (Jackson-Lewis et al. 1995; Jones-Humble et al. 1994). The susceptibility of mice to the neurotoxic actions of MPTP can be quite variable, depending on gender and strain differences, expressed in functional, neurochemical, and histochemical analyses (Schwarting et al. 1999; Sedelis et al. 2000a, b, 2001, 2003). C57/BL6 and Swiss Webster strains were shown to differ in c-Jun N-terminal kinases (JNKs) and c-JUN activation in response to MPTP. JNKs, of the mitogen-activated protein kinase family, are responsive to stress stimuli, such as cytokines, ultraviolet irradiation, heat shock, and osmotic shock; c-Jun is the name of a gene and protein that, in combination with c-Fos, forms the AP-1 early response transcription factor. MPTP induced COX-2, an enzyme responsible for inflammation and pain, responding exclusively in C57/BL6 mice (Boyd et al. 2007). MPTP, administered systemically, induces parkinsonism in human and nonhuman primates (Langston 1985) that results in the loss of *substantia nigra* cells in the *pars compacta* of adult animals (Chiueh et al. 1985). It destroys selectively nigrostriatal neurons thereby inducing acute, subacute, long-lasting, and even permanent effects that resemble certain features of PD, particularly the hypokinesic effects (Schultz et al. 1989). Systemic administration of MPTP (2×40 mg/kg, s.c.) caused L-dopa reversible hypoactivity (Fredriksson et al. 1990; Sundström et al. 1990). A less rigorous dose regime, e.g., 2×20 or 25 or 30 mg/kg, of MPTP has been found not to reduce motility in the C57 black mice, although DA brain concentrations may indicate up to 50–80 % reductions (Heikkila et al. 1989; Sonsalla and Heikkila 1986), unless given much more repeatedly (cf. Kurz et al. 2007). The parameters of MPTP treatment neurotoxicity in mice are extremely long-lasting (up to and beyond 52 weeks after treatment) with strong correlations between the functional deficits, particularly hypokinesia, the main biomarker, severe DA depletions, and a dose- and time-dependent recovery of several parameters of motor behavior following treatment with the DA precursor, L-dopa (Archer and Fredriksson 2003; Fredriksson and Archer 1994; Fredriksson et al. 1999).

In the unilateral 6-hydroxydopamine rat model of Parkinson's disease (PD), Tillerson et al. (2001) abolished the lesion-induced motor asymmetry by forcing the rats to use affected (contralateral) limb, whereas forced nonuse exacerbated the injury (Tillerson et al. 2002). Both dopamine (DA) and 3,4-dihydroxyphenylacetic acid (Dopac) were elevated markedly in "casted" 6-OHDA-treated rats (forced to use the contralateral limb) compared with "non-casted" rats (Cohen et al. 2003). Archer and Fredriksson (2010) found that daily running-wheel activity attenuated the hypokinesic effects of MPTP in both a concentrated (2×40 mg/kg, 24-h interval) and progressive (1×40 mg/kg, weekly doses over 4 weeks) schedule with regard to spontaneous motor behavior and activity following a subthreshold dose of L-dopa. The loss of DA in each case was attenuated by exercise also (Experiment I, 61 % of control rather than 17 %; Experiment II, 24 % rather than 11 %). Using the progressive schedule of MPTP treatment and extending the exercise intervention from 7 to 14 weeks, it was shown that spontaneous motor activity after MPTP was close to restoration, whereas activity after subthreshold L-dopa was completely recovered (Fredriksson et al. 2011); DA levels were restored from 17 % (non-exercised) to 64 % in the 14-week exercise intervention, and levels of brain-derived neurotrophic factor were increased significantly. It was shown also that both the functional and DA deficits by MPTP were attenuated even by delayed introduction of exercise (Archer and Fredriksson 2012).

Brain-derived neurotrophic factor (BDNF) is a neurotrophin with widespread expression in the brain and is connected intimately with brain metabolism and homeostasis (Chaldakov 2011). It is associated with neurogenesis, neuronal survival, and neuroreparation in the brain and CNS (Cui 2006; Numakawa et al. 2010). Treatment interventions that enhance BDNF-related signaling have the potential to restore neural connectivity (Kaplan et al. 2010). Physical exercise induces improvements in motor ability and enhances BDNF expression (Macias et al. 2009). It is linked to elevated BDNF levels in the hippocampus (Neeper et al. 1996; Oliff et al. 1998). Voluntary running, as physical activity, amplifies the BDNF signal that augments neurogenesis through diverse molecular pathways (Stranahan et al. 2009). BDNF mediates several essential morphological changes at neuronal levels that include dendritic arborization (Imamura and Greer 2009; Zhou et al. 2008), axonal and dendritic remodeling (Jeanneteau et al. 2010; Menna et al. 2003), synaptogenesis (Liu et al. 2009b; Tchantchou et al. 2009), and synaptic efficacy (Boulanger and Poo 1999; Sallert et al. 2009). Faherty et al. (2005) have shown that a combination of exercise, social interactions and learning, or exercise alone during adulthood gave total protection against MPTP-induced parkinsonism. They found also that changes in mRNA expression suggested that increases in glial-derived neurotrophic factors, coupled with a decrease of dopamine-related transporters (e.g., dopamine transporter, DAT; vesicular monoamine transporter, VMAT2), contributed to the observed neuroprotection of dopamine neurons in the nigrostriatal system following MPTP exposure. Tajiri et al. (2010) observed that exercise induced behavioral recovery in an animal model of PD and caused increased BDNF and glial-derived neurotrophic factor (GDNF) in the striatum of 6-OHDA-treated rats.

The production of Milmed is a patent-protected treatment of yeast cultures, but for any synergistic antiparkinson effect, a regime of physical exercise must be incorporated: the basis of "personalized medicine" builds upon this particular combination, whereas the pharmacogenetic aspect involves the selective susceptibility of the C57/Bl6 mouse strain for the DA neurotoxin, MPTP. The yeast cultures, *Saccharomyces cerevisiae* or *Saccharomyces carlsbergensis*, have been utilized as industrially important cell factories (Nielsen and Jewett 2008), with many regulatory pathways conserved between these yeasts and humans (Zhang et al. 2010). Cell death studies using yeast apoptosis increasingly provide a model for analyzing the cascade of molecular events that contribute to neurodegenerative disorders (Carmona-Gutierrez et al. 2010; Petranovic et al. 2010). Several features of PD have been reproduced in yeast with cell death promotion in a concentration-dependent manner (Outeiro and Lindquist 2003) with possibilities for facilitating the development of both therapeutic targets and compounds (Braun et al. 2009; Teneiro and Outeiro 2010). This experiment illustrates, for the purposes of this review, that the treatment of yeast cell cultures themselves provides an agent that provides an antiparkinson effect. The treatment and preparation of *Saccharomyces cerevisiae* or *Saccharomyces carlsbergensis* with electromagnetic waves in the extreme high frequency (EHF) range of 30–300 GHz produces a treated yeast extract, given the name Milmed (i.e., Milmed®). This treatment was developed through the pioneering work of MB Golant (Golant 1994; Golant et al. 1994; Ragimov et al. 1991) upon the genesis and reparation of cells. The coadministration of Milmed with daily physical exercise has been reported to induce plasticity in attenuating MPTP-induced motor deficits (Oscarson et al. 2009).

This following experiment illustrates that, as an example of "personalized medicine," physical exercise in a running wheel combined with administration of the treated Milmed (yeast extract) under conditions where the extract was charged or uncharged would ameliorate the functional and DA deficits induced by MPTP under Milmed regimes. Additionally, BDNF levels in the parietal cortex (including the motor cortex) were assayed in order to assess the effects of MPTP, MPTP+Exercise, and MPTP+Exercise combined with charged or uncharged Milmed ("yeast extract"). Here, the notion of "personalized medicine" derives from the particular combination of physical exercise regime with twice weekly doses of Milmed. In the case of the mice studied here, daily 30-min bouts of running-wheel activity constituted the exercise regime.

4.1 Description of Exercise-Milmed Intervention

4.1.1 Animals

Male C57 Bl/6 mice were purchased from B&K, Sollentuna, Sweden, and were maintained, five to a cage, in plastic cages in a room at temperature of 22 ± 1 °C and a 12/12 h constant light/dark cycle (lights on between 06.00 and 18.00 h). They were placed and maintained in groups of four to six animals in a room maintained for male mice only following arrival at the laboratory for about 2 weeks in order to acclimatize. Free access to food and water was maintained throughout, except for the day previous to the initiation to wheel-running exercise which occurred at the end of the second week following arrival. They were housed in groups of six animals, wheel-running exercised and activity chamber tested only during the hours of light (08.00–15.00 h). All exercising and testing was performed in a normally lighted room. Half of the mice in each treatment condition (MPTP-Exer, MPTP-Exer-Milmed, and Vehicle) were given wheel-running exercise, whereas the other half were placed in a clean laboratory cage for the same period in a room in which the running wheels were placed. Motor activity was tested in a specially arranged test room. This test room, in which all 12 ADEA activity test chambers, each identical to the home cage, were placed, was well secluded and used only for this purpose. Each test chamber (i.e., motor activity test cage) was placed in a soundproofed wooden box with 12-cm-thick walls and front panels and a small double-glass window to allow observation; each box had a dimmed lighting.

Three weeks following arrival, four groups ($n=10$) of DSP4-treated and two groups of vehicle-treated mice were administered with either MPTP (2×40 mg/kg, s.c., 24 h between injections) or vehicle (0.9 % physiological saline injected s.c. in a volume of 2 ml/kg body weight). Milmed (see below for details of preparation) or vehicle was administered twice weekly.

Experiments were carried out in accordance with the European Communities Council Directive of 24 November 1986 (86/609/EEC) after approval from the local ethical committee (Uppsala University and Agricultural Research Council) and by the Swedish Committee for Ethical Experiments on Laboratory Animals (license S93/92 and S77/94, Stockholm, Sweden).

4.1.2 Drugs

MPTP (Research Biomedical Inc., MA, USA, 2×20 mg/kg or 2×40 mg/kg, s.c., with a 24-h interval between injections in each case) was dissolved in saline and administered s.c. in a volume of 2 ml/kg body weight. Milmed was obtained through treatment and preparation of Saccharomyces cerevisiae with electromagnetic waves in the extreme high frequency (EHF) range of 30–300 GHz to produce the treated yeast extract (cf. Golant 1994). Saline was used as vehicle in each case.

4.1.3 Behavioral Measurements and Apparatus

Activity test chambers. An automated device, consisting of macrolon rodent test cages (40×25×15 cm) each placed within two series of infrared beams (at two different heights, one low and one high, 2 and 8 cm, respectively, above the surface of the sawdust, 1 cm deep), was used to measure spontaneous motor activity (RAT-O-MATIC, ADEA Elektronic AB, Uppsala, Sweden). The distances between the infrared beams were as follows: the low-level beams were 73 mm apart lengthwise and 58 mm apart breadthwise in relation to the test chamber; the high-level beams, placed only along each long side of the test chamber, were 28 mm apart. According to the procedures described previously (Archer et al. 1986), the following parameters were measured: *locomotion* was measured by the low grid of infrared beams. Counts were registered only when the mouse in the horizontal plane is ambulating around the test cage. *Rearing* was registered throughout the time when at least one high-level beam was interrupted, i.e., the number of counts registered was proportional to the amount of time spent rearing. *Total activity* was measured by a sensor (a pickup similar to a gramophone needle, mounted on a lever with a counterweight) with which the test cage was constantly in contact. The sensor registered all types of vibration received from the test cage, such as those produced both by locomotion and rearing as well as shaking, tremors, scratching, and grooming. All three behavioral parameters were measured over three consecutive 20-min periods. The motor activity test room, in which all 12 ADEA activity test chambers, each identical to the home cage, were placed, was well secluded and used only for

Table 14.1 Chronological and experimental design for MPTP treatment, exercise schedule, and Milmed (charged/uncharged) administration over the 14-week experiment

Time and test	Day	Vehicle	MPTP	MPTP+Exer	MPTP+Exer+ Milm(1)	MPTP+Exer+ Milm(0)
	Monday	Cage	Cage	Exer	Exer	Exer
	Tuesday	Cage	Cage	Exer	Exer	Exer
Weeks 1–4	Wednesday	Cage	Cage	Exer	Exer	Exer
	Thursday	Cage	Cage	Exer	Exer	Exer
Tests 1–4[a]	Friday	Test+sal	Test+MPTP[b]	Test+MPTP[b]	Test+MPTP[b]	Test+MPTP[b]
	Monday	Cage	Cage	Exer	Exer	Exer
	Tuesday	Cage	Cage	Exer	Exer	Exer
Weeks 5–8	Wednesday	Cage	Cage	Exer	Exer	Exer
	Thursday	Cage	Cage	Exer	Exer	Exer
Tests 5–8[a]	Friday	Test+sal	Test	Test	Test	Test
	Monday	Cage	Cage	Exer	Exer	Exer
	Tuesday	Cage	Cage	Exer	Exer	Exer
Weeks 9–14	Wednesday	Cage	Cage	Exer	Exer	Exer
	Thursday	Cage	Cage	Exer	Exer	Exer
Tests 9–14[a]	Friday[c]	Test+sal	Test	Test	Test	Test

[a]Spontaneous activity over 60 min
[b]MPTP (40 mg/kg) injected during the first 4 weeks
[c]L-dopa (5 mg/kg, s.c.) tests after 60-min habituation to test cages 6, 8, 10, 12, and 14 weeks

this purpose. Each test chamber (i.e., activity cage) was placed in a soundproofed wooden box with 12-cm-thick walls and front panels and day lighting. Motor activity parameters were tested on one occasion only, over three consecutive 20-min periods, at the age of 3–4 months. Groups of mice were treated with MPTP and then given access to running wheels (30 min/day, 4 times/week), with or without concomitant treatment with Milmed ([Milm(1)-charged] or [Milm(0)]-uncharged = yeast itself), as displayed in Table 14.1.

4.1.4 Neurochemical Analysis

Mice were killed by cervical dislocation within 2 weeks of completion of behavioral testing. Determination of DA was performed using a high-performance liquid chromatograph with electrochemical detection (HPLC-EC), according to Björk et al. (1991), as modified by Ye Liu et al. (1995). Striatal regions were rapidly dissected out and stored at −80(C until neurochemical analysis. DA concentration was measured as follows: the frozen tissue samples were weighed and homogenized in 1 ml of 0.1 M perchloric acid, and alpha-methyl-5-hydroxytryptophan was added as an internal standard. After centrifugation (12,000 rpm, i.e., 18,600 g, 4 °C, 10 min) and filtration, 20 µl of the supernatant was injected into the HPLC-EC to assay DA. The HPLC system consisted of a PM-48 pump (Bioanalytical Systems, BAS) with a CMA/240 autoinjector (injection volume, 20 µl), a precolumn (15 × 3.2 mm, RP-18 Newguard, 7 µm), a column (100 × 4.6 mm, SPHERI-5, RP-18, 5 µm), and an amperometric detector (LC-4B, BAS, equipped with a Ag/AgCl reference electrode and a MF-2000 cell) operating at a potential of +0.85 V. The mobile phase, pH 2.69, consisted of K_2HPO_4 and citric acid buffer (pH 2.5), 10 % methanol, sodium octyl sulfate, 40 mg/l, and EDTA. The flow rate was 1 ml/min, and the temperature of the mobile phase was 35 °C.

4.1.5 BDNF Analysis

The methods and procedures described by Viberg et al. (2008) were maintained. Frontal cortex, parietal cortex, and hippocampus tissues from the mice in each group were sonicated in 20 volumes (w/v) of ice-cold lysis buffer (137 mM

NaCl; 20 mM Tris–HCl, pH 8.0; 1 mM phenylmethyl-sulfonyl fluoride; 10 lg/ml aprotinin; 1 lg/ml leupeptin). The homogenate was centrifuged for 20 min at 200,009 g at 4 °C, and the supernatant was acidified (pH 3) with HCl and neutralized back to pH 7.6 with NaOH. The Promega Emax TM ImmunoAssay System was used to determine the amount of BDNF in the samples according to the technical bulletin supplied by the distributor. Briefly, BDNF from each sample was captured with a monoclonal antibody (mAb) against BDNF; captured BDNF was then bound to a second specific polyclonal antibody (pAb) against BDNF. After washing, the amounts of specifically bound pAb were detected by using a specific anti-IgY antibody conjugated to horseradish peroxidase (HRP) as a tertiary reactant. Unbound conjugate was removed through washing, and after an incubation period with a chromogenic substrate, the color change was measured in a microplate reader at 450 nm. The amount of BDNF was proportional to the color change generated and compared with a standard curve. The cross-reactivity to other neurotrophic factors was less than 3 %, and the purity of the anti-BDNF antibodies was greater than 95 %.

4.2 Effects of Exercise-Milmed on MPTP-Induced Deficits

4.2.1 Spontaneous Motor Activity

Mice treated with MPTP showed a marked hypokinesic effect over all 14 Test days from Test day 2 onwards. Access to physical exercise (running wheel) retarded the onset of MPTP-induced hypokinesia until Test day 5 and then attenuated the hypokinesia throughout. Uncharged Milmed (Milm(0)) in combination with physical exercise also retarded and attenuated the hypokinesia induced by MPTP. Charged Milmed (Milm(1)) in combination with physical exercise abolished any sign of hypokinesia throughout. Split-plot ANOVA indicated a Treatment × Days interaction: ($F(52, 629) = 16.08$, $p < 0.0001$). Pairwise testing with Tukey's HSD indicated the following differences over all three motor activity parameters, locomotion, rearing, and total activity:

Vehicle, MPTP + Exer + Milm(1) > MPTP + Exer + Milm(0), MPTP + Exer > MPTP during Test days 2–14 (Fig. 14.1).

4.2.2 L-Dopa-Induced Activity

Mice treated with MPTP showed a marked hypokinesic effect over all five L-dopa-induced tests (Test days 6, 8, 10, 12, and 14). Both physical exercise, by itself, or combined with uncharged Milmed attenuated the loss of L-dopa-induced activity over Test days 6, 8, 10, and 12, but abolished this loss on Test day 14. Physical exercise combined with charged Milmed abolished MPTP-induced L-dopa activity deficits throughout. Split-plot ANOVA indicated a Treatment × Days interaction: $F(16, 224) = 9.65$, $p < 0.0001$. Pairwise testing with Tukey's HSD indicated the following differences over all three motor activity parameters, locomotion, rearing, and total activity:

Vehicle, MPTP + Exer + Milm(1) > MPTP + Exer + Milm(0), MPTP + Exer > MPTP during Test days 2–14 (Fig. 14.2).

4.2.3 Neurochemical Analysis

Mice treated with MPTP showed a marked loss of DA in the striatum (17 % of control values). This effect was attenuated strongly by physical exercise by itself (MPTP + Exercise = 64 % of control values) or combined with uncharged Milmed (MPTP + Exercise + Milmed[yeast](0) = 65 % of control values). The combination of physical exercise with charged Milmed (MPTP + Exercise + Milmed[yeast](1)) abolished completely any loss of DA (101 % of control values). One-way ANOVA indicated a significant group effect: $F(4, 30) = 47.27$, $p < 0.0001$. Tukey testing indicated the following differences:

MPTP + Exercise + Milmed[yeast](1), Vehicle > MPTP + Exercise + Milmed[yeast](0), MPTP + Exercise > MPTP (Fig. 14.3).

4.2.4 BDNF Analysis

Mice treated with MPTP showed a marked elevation of BDNF in the parietal cortex (including motor cortex) of the MPTP group (431 % compared with control values) compared with the

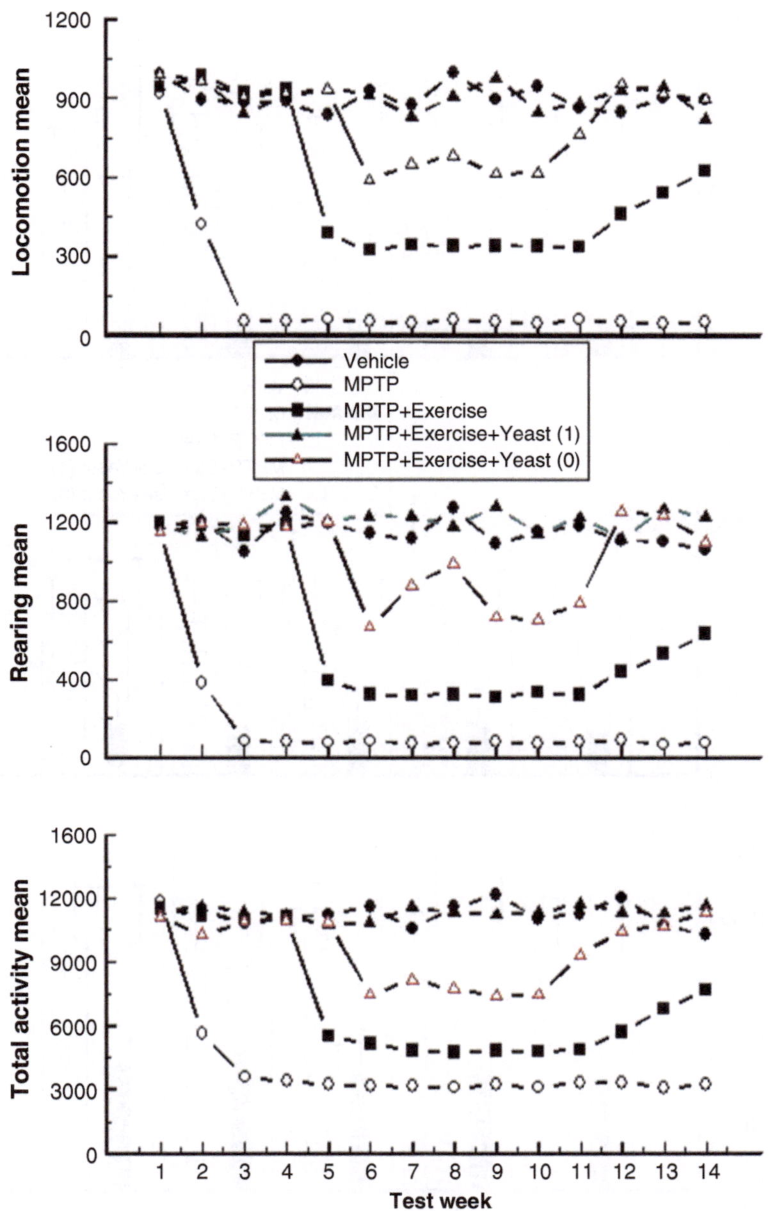

Fig. 14.1 Mean locomotion, rearing, and total activity during the spontaneous motor activity by Vehicle, MPTP, MPTP+Exercise, MPTP+Exercise+Milmed [yeast](1), and MPTP+Exercise+Milmed [yeast](0) over Test days 1–14

vehicle group. This elevation of BDNF was increased by both physical activity itself (528 % of control values) and in combination with uncharged Milmed (MPTP+Exercise+Milmed [yeast](0)=534 % of control values). BDNF elevation was greatest in the case of the charged Milmed (MPTP+Exercise+Milmed[yeast](1)) group (853 % of controls). One-way ANOVA indicated a significant group effect: $F(4, 30)=18.27$, $p<0.0001$. Tukey testing indicated the following differences:

MPTP+Exercise+Milmed[yeast](1)>MPTP +Exercise+Milmed[yeast](0), MPTP+Exercise >MPTP>Vehicle (Fig. 14.4).

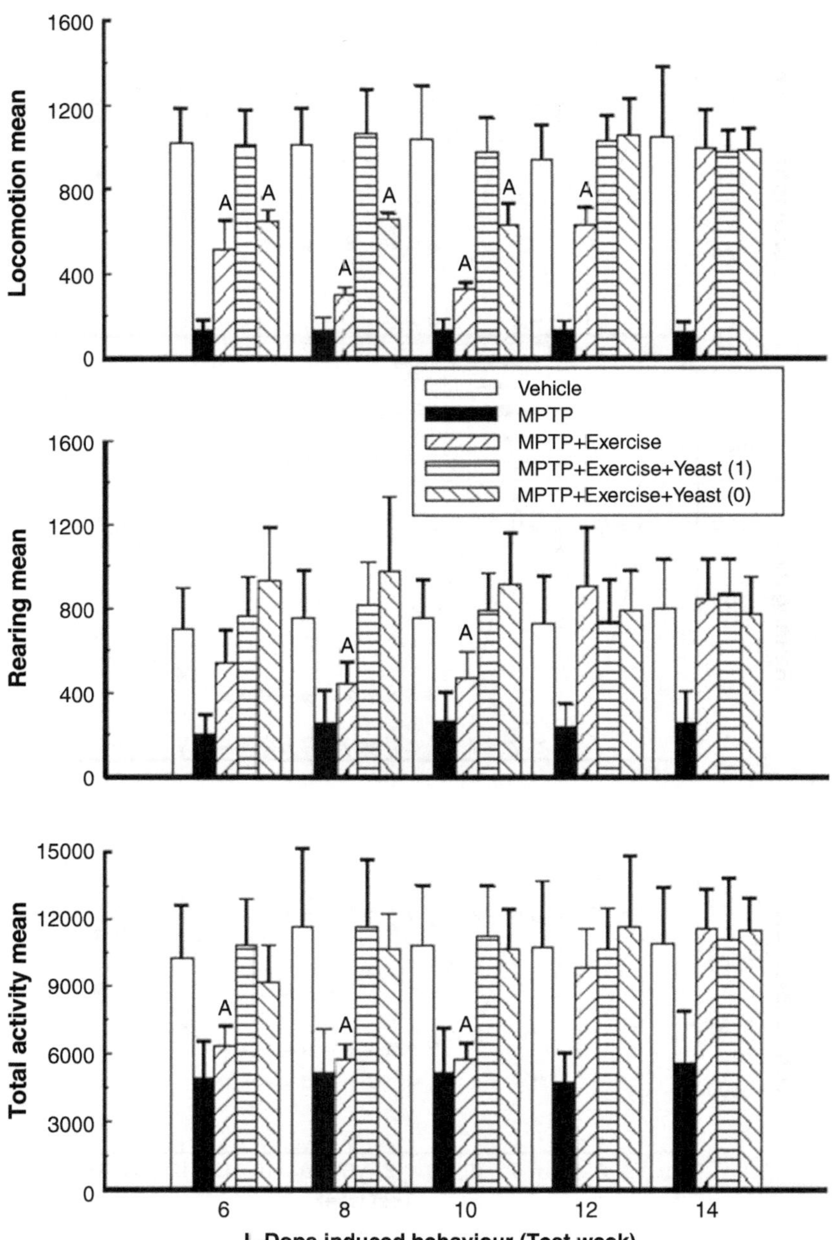

Fig. 14.2 Mean locomotion, rearing, and total activity during the L-dopa-induced motor activity by Vehicle, MPTP,MPTP+Exercise,MPTP+Exercise+Milmed[yeast](1), and MPTP+Exercise+Milmed[yeast](0) groups on Tests 1–5 over Test days 6, 8, 10, 12–14

4.3 Milmed-Exercise Synergism Abolishes MPTP-Induced Deficits

Physical exercise alleviates both the symptoms and the biomarkers (e.g., DA loss) of PD in the laboratory and in the clinical setting (Archer and Fredriksson 2010, 2012; Archer et al. 2011a, b; Fredriksson et al. 2011). The combination physical exercise and Milmed abolished MPTP-induced parkinsonism in the laboratory both functionally and neurochemically such that the antiparkinsonian effects were in excess of the summation of exercise and Milmed effects by

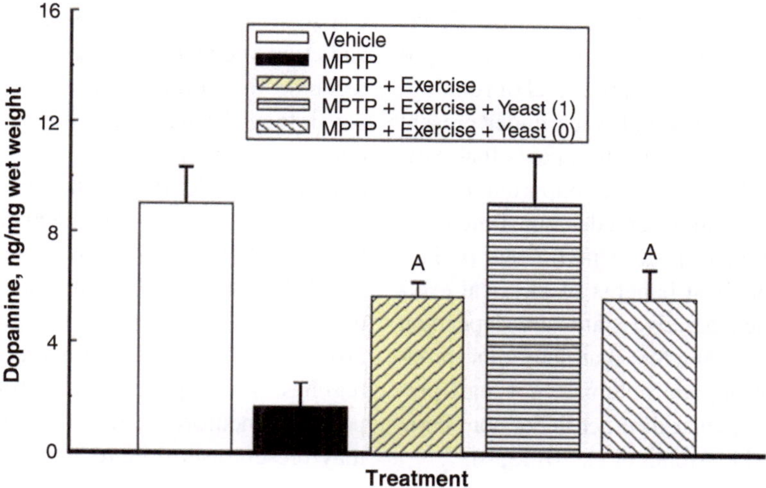

Fig. 14.3 Mean striatal dopamine concentrations in the Vehicle, MPTP, MPTP+Exercise, MPTP+Exercise+Milmed[yeast](1), and MPTP+Exercise+Milmed[yeast](0) groups

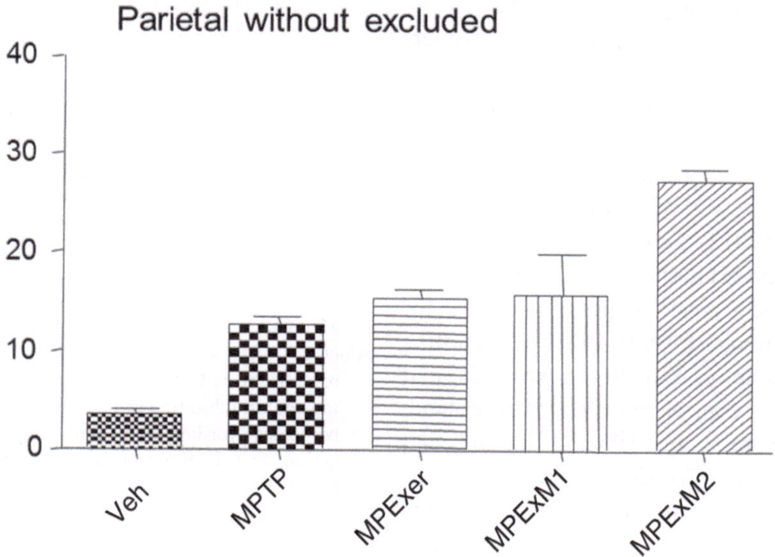

Fig. 14.4 Mean parietal cortex (including motor cortex) BDNF concentrations in the Vehicle, MPTP, MPTP+Exercise, MPTP+Exercise+Milmed[yeast](1), and MPTP+Exercise+Milmed[yeast](0) groups

themselves; in the absence of exercise, Milmed does not affect MPTP-induced hypokinesia or DA loss at all (Archer and Fredriksson, unpublished results). The notion that exercise-induced elevations in BDNF may be of significance for the treatment of aging disorders is not novel, since memantine, a medium-affinity uncompetitive *N*-methyl-D-aspartate receptor antagonist applied clinically as a neuroprotective agent to treat AD and PDs, increased BDNF mRNA levels markedly in the limbic cortex at clinically relevant doses (Marvanová et al. 2001). The present findings that (a) MPTP treatment induced a marked increase in parietal BDNF and (b) exercise over 14 weeks further increased levels of BDNF in the parietal cortex appear to lend credence for the involvement of BDNF in the exercise-induced recovery of function and DA innervation following repeated doses of MPTP. The lack of exercise-induced changes in hippocampal BDNF suggests that the level of running-wheel exercise per day and week under present conditions was insufficient. It appears that parietal cortex BDNF may exert an important mediatory role, hitherto unobserved, upon functional and biomarker recovery in experimental parkinsonism. The manifest benefits of physical exercise on neurodegenerative states are dependent on a variety of parameters that determine prognosis, intervention, and outcome, not least pertaining to the particular disorder under consideration (Archer 2011; Archer et al. 2011a, b).

5 Conclusions

Clinicogenetic trials have demonstrated that therapeutic drug efficacy or toxicity or susceptibility for adverse effects presents disorder-intervention features increasingly found to be governed by genetic and epigenetic principles (Kalinderi et al. 2011). In a large multicentered study to ascertain the frequency and pathogenicity of reported VPS35 variants worldwide, Sharma et al. (2012) sought to identify a mutation (p.Asp620Asn) in the vacuolar protein sorting 35 (VPS35) gene as possible cause for autosomal dominant form of PD. The identified pathogenic variant p. Asp620Asn was identified in 8 cases and 1 control from Italy, US, Poland, and Australia; three sites (Poland, Ireland, and US) detected p.Leu774Met variant in 6 cases and 4 controls; and one site (Norway) detects p.Gly1Ser variant in 3 cases and 1 control with two reported variants (p.Arg524Trp and p.Ile241Met) monomorphic. The overall analysis described an increased risk for PD for p.Asp620Asn and p.Leu774Met variants, respectively, in their cohort, thereby highlighting the role of rare variants in the complex PD condition. Current levels of information pertaining to notions of pharmacogenomics, epigenetics, and biomarkers that are modulated by interindividual variability affect the diagnosis, intervention, and prognosis of both PD disorder expression and therapeutic strategies. Symptom profiles and course of disease, etiopathological heterogeneity, and etiopathogenesis may be elucidated through recourse to a dimensional approach to pathophysiology through the distinguished endophenotypes and biomarkers of disorder progression (Archer et al. 2010). Much increasing evidence suggests that epigenetic mechanisms, such as DNA methylation, histone modifications, and small RNA-mediated mechanisms, may regulate the expression of PD-related genes (Coppedè 2012). Finally, the coadministration of exercise regimes with agents (like Milmed) offering potential neuroreparative/neurogenesis agency may present useful ingredients for personalized medicine.

References

Antonini A, Schwarz J, Oertel WH, Beer HF, Madeja UD, Leenders KL (1994) [11C]raclopride and positron emission tomography in previously untreated patients with Parkinson's disease: Influence of L-dopa and lisuride therapy on striatal dopamine D2-receptors. Neurology 44(7):1325–1329

Al Hadithy AF, Wilffert B, Stewart RE, Looman NM, Bruggeman R, Brouwers JR, Matroos GE, van Os J, van Harten PN (2008) Pharmacogenetics of parkinsonism, rigidity, rest tremor, and bradykinesia in African-Caribbean inpatients: differences in association with dopamine and serotonin receptors. Am J Med Genet B 147B:890–897

Alachkar A, Brotchie JM, Jones OT (2010) Locomotor response to L-DOPA in reserpine-treated rats following central inhibition of aromatic L-amino acid decarboxylase: further evidence for non-dopaminergic actions of L-DOPA and its metabolites. Neurosci Res 68:44–50

Alkelai A, Greenbaum L, Rigbi A, Kanyas K, Lerer B (2009) Genome-wide association study of antipsychotic-induced parkinsonism severity among schizophrenia patients. Psychopharmacology 206: 491–499

Annesi F, De Marco EV, Rocca FE, Nicoletti A, Pugliese P, Nicoletti G, Arabia G, Tarantino P, De Mari M, Lamberti P, Gallerini S, Marconi R, Epifanio A, Morgante L, Cozzolino A, Barone P, Torchia G, Zappia M, Annesi G, Quattrone A (2011) Association study between the LINGO1 gene and Parkinson's disease in the Italian population. Parkinsonism Relat Disord 17:638–641

Arbouw ME, van Vugt JP, Egberts TC, Guchelaar HJ (2007) Pharmacogenetics of antiparkinsonian drug treatment: a systematic review. Pharmacogenomics 8:159–176

Arbouw ME, Movig KL, Egberts TC, Poels PJ, van Vugt JP, Wessels JA, van der Straaten RJ, Neef C, Guchelaar HJ (2009) Clinical and pharmacogenetic determinants for the discontinuation of non-ergoline dopamine agonists in Parkinson's disease. Eur J Clin Pharmacol 65:1245–1251

Arbouw ME, Guchelaar HJ, Egberts TCG (2010) Novel insights in pharmacogenetics of drug response in Parkinson's disease. Pharmacogenomics 11:127–129

Arbouw ME, Movig KL, Guchelaar HJ, Neef C, Egberts TC (2012) Dopamine agonists and ischemic complications in Parkinson's disease: a nested case–control study. Eur J Clin Pharmacol 68:83–88

Archer T (2011) Physical exercise alleviates debilities of normal aging and Alzheimer's disease. Acta Neurol Scand 123:221–238

Archer T, Fredriksson A (2003) An antihypokinesic action of α2-adrenoceptors upon MPTP-induced behavior deficits in mice. J Neural Transm 110:183–200

Archer T, Fredriksson A (2010) Physical exercise attenuates MPTP-induced deficits in mice. Neurotox Res 18:313–327

Archer T, Fredriksson A (2012) Delayed exercise-induced functional and neurochemical partial restoration following MPTP. Neurotox Res 21:210–221

Archer T, Fredriksson A, Jonsson G, Lewander T, Mohammed AK, Söderberg U (1986) Central noradrenaline depletion antagonizes aspects of d-amphetamine-induced hyperactivity in the rat. Psychopharmacology 88:141–146

Archer T, Beninger RJ, Palomo T, Kostrzewa RM (2010) Epigenetics and biomarkers in the staging of neuropsychiatric disorders. Neurotox Res 18:347–366

Archer T, Johansson B, Fredriksson A (2011a) Exercise alleviates Parkinsonism: clinical and laboratory evidence. Acta Neurolog Scand. doi:10.1111/j.1600-0404.2010.01360.x

Archer T, Fredriksson A, Schütz E, Kostrzewa R (2011b) Influence of physical exercise on neuroimmunological functioning and health: aging and stress. Neurotox Res 20:69–83

Asari S, Fujimoto K, Miyauchi A, Sato T, Nakano I, Muramatsu S (2011) Subregional 6-[18F]fluoro-L-m-tyrosine uptake in the striatum in Parkinson's disease. BMC Neurol 11:35

Bakker PR, van Harten PN, van Os J (2006) Antipsychotic-induced tardive dyskinesia and the Ser9Gly polymorphism in the DRD3 gene: a meta analysis. Schizophr Res 83:185–192

Bar-Yaacov D, Blumberg A, Mishmar D (2012) Mitochondrial-nuclear co-evolution and its effects on OXPHOS activity and regulation. Biochim Biophys Acta 1819:1107–1111

Becquemont L (2009) Pharmacogenomics of adverse drug reactions: practical applications and perspectives. Pharmacogenomics 10:961–969

Becquemont L, Alfirevic A, Amstutz U, Brauch H, Jacqz-Aigrain E, Laurent-Puig P, Molina MA, Niemi M, Schwab M, Somogyi AA, Thervet E, Maitland-van der Zee AH, van Kuilenburg AB, van Schaik RH, Verstuyft C, Wadelius M, Daly AK (2011) Practical recommendations for pharmacogenomics-based prescription: 2010 ESF-UB conference on pharmacogenetics and pharmacogenomics. Pharmacogenomics 12:113–124

Björk L, Lindgren S, Hacksell U, Lewander T (1991) (S)-UH-301 antagonizes ®-8-OH-DPAT-induced cardiovascular effects in the rat. Eur J Pharmacol 199:367–370

Boulanger L, Poo MM (1999) Gating of BDNF-induced synaptic potentiation by cAMP. Science 284:1982–1984

Boyd JD, Jang H, Shepherd KR, Faherty C, Slack S, Jiao Y, Smeyne RJ (2007) Response to 1-methyl-4-phenyl-1,2,3,6-tetrahydropyridine (MPTP) differs in mouse strains and reveals a divergence in JNK signaling and COX-2 induction prior to loss of neurons in the substantia nigra pars compacta. Brain Res 1175:107–116

Braak H, Sastre M, Bohl JR, de Vos RA, Del Tredici K (2007) Parkinson's disease: lesions in dorsal horn layer I, involvement of parasympathetic and sympathetic pre- and postganglionic neurons. Acta Neuropathol 113(4):421–429

Braun RJ, Büttner S, Ring J, Kroemer G, Madeo F (2009) Nervous yeast: modeling neurotoxic cell death. Trends Biochem Sci 35:135–144

Caligiuri MP, Lacro JP, Jeste DV (1999) Incidence and predictors of drug-induced parkinsonism in older psychiatric patients treated with very low doses of neuroleptics. J Clin Psychopharmacol 19:322–328

Caligiuri MP, Jeste DV, Lacro JP (2000) Antipsychotic-induced movement disorders in the elderly: epidemiology and treatment recommendations. Drugs Aging 17:363–384

Carmona-Gutierrez D, Jungwirth H, Eisenberg T, Madeo F (2010) Cell cycle control of cell death in yeast. Cell Cycle 9:4046

Cersosimo MG, Benarroch EE (2012a) Autonomic involvement in Parkinson's disease: pathology, pathophysiology, clinical features and possible peripheral biomarkers. J Neurol Sci 313(1–2):57–63. doi:10.1016/j.jns.2011.09.030

Cersosimo MG, Benarroch EE (2012b) Pathological correlates of gastrointestinal dysfunction in Parkinson's disease. Neurobiol Dis 46(3):559–564. doi:10.1016/j.nbd.2011.10.014

Chaldakov G (2011) The metabotrophic NGF and BDNF: an emerging concept. Arch Ital Biol 149:257–263

Chase TN (2004) Striatal plasticity and extrapyramidal motor dysfunction. Parkinsonism Relat Disord 10:305–313

Chiueh CC, Burns RS, Markey SP, Jacobowitz DM, Kopin IJ (1985) Primate model of parkinsonism: selective lesion of nigrostriatal neurons by 1-methyl-4-phenyl-1,2,3,6-tetrahydropyridine produces an extrapyramidal syndrome in rhesus monkeys. Life Sci 36:213–218

Chong SA, Tan EC, Tan CH, Mythily, Chan YH (2003) Polymorphisms of dopamine receptors and tardive dyskinesia among Chinese patients with schizophrenia. Am J Med Genet B Neuropsychiatr Genet 116B: 51–54

Christine CW, Starr PA, Larson PS, Eberling JL, Jagust WJ, Hawkins RA, VanBrocklin HF, Wright JF, Bankiewicz KS, Aminoff MJ (2009) Safety and tolerability of putaminal AADC gene therapy for Parkinson disease. Neurology 73:1662–1669

Clayton TA (2012) Metabolic differences underlying two distinct rat urinary phenotypes, a suggested role for gut microbial metabolism of phenylalanine and a possible connection to autism. FEBS Lett 586:956–961

Clayton TA, Lindon JC, Cloarec O, Antti H, Charuel C, Hanton G, Provost JP, Le Net JL, Baker D, Walley RJ, Everett JR, Nicholson JK (2006) Pharmaco-metabonomic phenotyping and personalized drug treatment. Nature 440:1073–1077

Cohen AD, Tillerson JL, Smith AD, Schallert T, Zigmond MJ (2003) Neuroprotective effects of prior limb use in 6-hydroxydopamine-treated rats: possible role of GDNF. J Neurochem 85:299–305

Contin M, Riva R, Martinelli P, Cortelli P, Albani F, Baruzzi A (1994) Longitudinal monitoring of the levodopa concentration-effect relationship in Parkinson's disease. Neurology 44:1287–1292

Cooper-Knock J, Kirby J, Ferraiuolo L, Heath PR, Rattray M, Shaw PJ (2012) Gene expression profiling in human neurodegenerative disease. Nat Rev Neurol 8:518–530. doi:10.1038/nrneurol.2012.156

Coppedè F (2012) Genetics and epigenetics of Parkinson's disease. Sci World J 2012:489830

Corvol JC, Bonnet C, Charbonnier-Beaupel F, Bonnet AM, Fiévet MH, Bellanger A, Roze E, Meliksetyan G, Ben Djebara M, Hartmann A, Lacomblez L, Vrignaud C, Zahr N, Agid Y, Costentin J, Hulot JS, Vidailhet M (2011) The COMT Val158Met polymorphism affects the response to entacapone in Parkinson's disease: a randomized crossover clinical trial. Ann Neurol 69:111–118. doi:10.1002/ana.22155

Cotman CW, Berchtold NC (2002) Exercise: a behavioral intervention to enhance brain health and plasticity. Trends Neurosci 25:295–301

Cui Q (2006) Actions of neurotrophic factors and their signaling pathways in neuronal survival and axonal regeneration. Mol Neurobiol 33:155–179

Dechamps A, Diolez P, Thiaudière E, Tulon A, Onifade C, Vuong T, Helmer C, Bourdel-Marchasson I (2010) Effects of exercise programs to prevent decline in health related quality of life in highly deconditioned institutionalized elderly persons: a randomized controlled trial. Arch Intern Med 170:162–169

Dettmer K, Hammock BD (2004) Metabolomics – a new exciting field within the "omics" sciences. Environ Health Perspect 112:A396–A397

Devos D, French DUODOPA Study Group (2009) Patient profile, indications, efficacy and safety of duodenal levodopa infusion in advanced Parkinson's disease. Mov Disord 24:993–1000

Dumitriu A, Latourelle JC, Hadzi TC, Pankratz N, Garza D, Miller JP, Vance JM, Foroud T, Beach TG, Myers RH (2012) Gene expression profiles in Parkinson disease prefrontal cortex implicate FOXO1 and genes under its transcriptional regulation. PLoS Genet 8(6):e1002794

Evans WE, McLeod HL (2003) Pharmacogenomics – drug disposition, drug targets, and side effects. N Engl J Med 348:538–549

Fabbrini G, Mouradian MM, Juncos JL, Schlegel J, Mohr E, Chase TN (1988) Motor fluctuations in Parkinson's disease: central pathophysiological mechanisms, part I. Ann Neurol 24:366–371

Faherty CJ, Raviie Shepherd K, Herasimtschuk A, Smeyne RJ (2005) Environmental enrichment in adulthood eliminates neuronal death in experimental Parkinsonism. Brain Res Mol Brain Res 134:170–179

Fahn S (1999) Parkinson disease, the effect of levodopa, and the ELL-dopa trial. Earlier vs later L-DOPA. Arch Neurol 56:529–535

Felicio AC, Shih MC, Godeiro-Junior C, Andrade LA, Bressan RA, Ferraz HB (2009) Molecular imaging studies in Parkinson disease: reducing diagnostic uncertainty. Neurologist 15(l):6–16. doi:10.1097/NRL.0b013e318183fdd8

Foltynie T, Cheeran B, Williams-Gray CH, Edwards MJ, Schneider SA, Weinberger D, Rothwell JC, Barker RA, Bhatia KP (2009) BDNF val66met influences time to onset of levodopa induced dyskinesia in Parkinson's disease. J Neurol Neurosurg Psychiatry 80:141–144

Fredriksson A, Archer T (1994) MPTP-induced behavioral and biochemical deficits: a parametric analysis. J Neural Transm Park Dis Dement Sect 7:123–132

Fredriksson A, Plaznik A, Sundström E, Jonsson G, Archer T (1990) MPTP-induced hypoactivity in mice: reversal by L-dopa. Pharmacol Toxicol 67:295–301

Fredriksson A, Palomo T, Chase TN, Archer T (1999) Tolerance to a suprathreshold dose of L-dopa in MPTP mice: effects of glutamate antagonists. J Neural Transm 106:283–300

Fredriksson A, Stigsdotter IM, Hurtig A, Evalds-Kvist B, Archer T (2011) Running wheel activity restores MPTP-induced functional deficits. J Neural Transm 118:407–420

Friedman JH (2006) Atypical antipsychotics in the elderly with Parkinson disease and the "black box" warning. Neurology 67:564–566

Friedman JH (2010) Recommendations on the drug treatment of psychosis in Parkinson's disease. Am J Med 123:e19

Fumagalli F, Racagni G, Riva MA (2006) Shedding light into the role of BDNF in the pharmacotherapy of Parkinson's disease. Pharmacogenomics 6:95–104

Gaggelli E, Kozlowski H, Valensin D, Valensin G (2006) Copper homeostasis and neurodegenerative disorders (Alzheimer's, prion, and Parkinson's diseases and amyotrophic lateral sclerosis). Chem Rev 106:1995–2044

Gerlach J (1999) The continuing problem of extrapyramidal symptoms: strategies for avoidance and effective treatment. J Clin Psychiatry 60(Suppl 23):20–24

Golant MB (1994) Physical laws of medicine and their use in the realization of living organisms with EHF radiation. Radiophys Quantum Electron 37:45–47

Golant MB, Mudrik DG, Kruglyakova OP, Izvol'skaya VE (1994) Effect of EHF-radiation polarization on yeast cells. Radiophys Quantum Electron 37:82–84. doi:10.1007/BF01039307

Graham JM, Grunewald RA, Sagar HJ (1997) Hallucinosis in idiopathic Parkinson's disease. J Neurol Neurosurg Psychiatry 63:434–440

Grønbaek K, Treppendahl M, Asmar F, Guldberg P (2008) Epigenetic changes in cancer as potential targets for prophylaxis and maintenance therapy. Basic Clin Pharmacol Toxicol 0020103(5):389–396. doi:10.1111/j.1742-7843.2008.00325.x

Guillin O, Griffon N, Bezard E, Leriche L, Diaz J, Gross C, Sokoloff P (2003) Brain-derived neurotrophic factor controls dopamine D3 receptor expression: therapeutic implications in Parkinson's disease. Eur J Pharmacol 480:89–95

Gunes A, Scordo MG, Jaanson P, Dahl ML (2007) Serotonin and dopamine receptor gene polymorphisms and the risk of extrapyramidal side effects in perphenazine-treated schizophrenic patients. Psychopharmacology 190:479–484

Guzey C, Scordo MG, Spina E, Landsem VM, Spigset O (2007) Antipsychotic-induced extrapyramidal symptoms in patients with schizophrenia: associations with dopamine and serotonin receptor and transporter polymorphisms. Eur J Clin Pharmacol 63:233–241

Hamdani M, Bonniere M, Ades J, Hamon M, Boni C, Gorwood P (2005) Negative symptoms of schizophrenia could explain discrepant data on the association between the 5-HT2A receptor gene and response to antipsychotics. Neurosci Lett 377:69–74

Harms AS, Barnum CJ, Ruhn KA, Varghese S, Treviño I, Blesch A, Tansey MG (2011) Delayed dominant-negative TNF gene therapy halts progressive loss of nigral dopaminergic neurons in a rat model of Parkinson's disease. Mol Ther 19:46–52

Hattori N, Machida Y, Sato S, Noda K, lijima-Kitami M, Kubo S, Mizuno Y (2006) Molecular mechanisms of nigral neurodegeneration in Park2 and regulation of parkin protein by other proteins. J Neural Transm Suppl (70):205–208

Hawkes CH, Del Tredici K, Braak H (2007) Parkinson's disease: a dual-hit hypothesis. Neuropathol Appl Neurobiol 33(6):599–614

Heikkila RE, Sieber B-A, Manzino L, Sonsalla PK (1989) Some features of the nigrostriatal dopaminergic neurotoxin 1-methyl-4-phenyl-1,2,3,6-tetrahydropyridine (MPTP) in the mouse. Mol Chem Neuropathol 10:171–183

Hernán MA, Checkoway H, O'Brien R, Costa-Mallen P, De Vivo I, Colditz GA, Hunter DJ, Kelsey KT, Ascherio A (2002) MAOB intron 13 and COMT codon 158 polymorphisms, cigarette smoking, and the risk of PD. Neurology 58:1381–1387

Hirose G (2006) Drug-induced parkinsonism: a review. J Neurol 253(Suppl 3):22–24

Hughes AR, Brothers CH, Mosteller M, Spreen WR, Burns DK (2009) Genetic association studies to detect adverse drug reactions: abacavir hypersensitivity as an example. Pharmacogenomics 10:225–233

Hunsberger JG, Newton SS, Bennett AH, Duman CH, Russell DS, Salton SR, Duman RS (2007) Antidepressant actions of the exercise-regulated gene VGF. Nat Med 13:1476–1482

Huo LR, Liang XB, Li B, Liang JT, He Y, Jia YJ, Jia J, Gong XL, Yu F, Wang XM (2012) The cortical and striatal gene expression profile of 100 Hz electroacupuncture treatment in 6-hydroxydopamine-induced Parkinson's disease model. Evid Based Complement Altern Med 2012:908439

Imamura F, Greer CA (2009) Dendritic branching of olfactory bulb mitral and tufted cells: regulation by TrkB. PLoS One 4:e6729

Jabs BE, Bartsch AJ, Pfuhlmann B (2003) Susceptibility to neuroleptic-induced parkinsonism – age and increased substantia nigra echogenicity as putative risk factors. Eur Psychiatry 18:177–181

Jackson-Lewis V, Jakowec M, Burke RE, Przedborski S (1995) Time course and morphology of dopaminergic neuronal death caused by the neurotoxin 1-methyl-4-phenyl-1,2,3,6-tetrahydropyridine. Neurodegeneration 4:257–269

Jankovic J (2008) Parkinson's disease: clinical features and diagnosis. J Neurol Neurosurg Psychiatry 79(4):368–376. doi:10.1136/jnnp.2007.131045

Jankovic J, Stacy M (2007) Medical management of levodopa-associated complications in patients with Parkinson's disease. CNS Drugs 21:677–692

Jeanneteau F, Deinhardt K, Miyoshi G, Bennett AM, Chao MV (2010) The MAP kinase phosphatase MKP-1 regulates BDNF-induced axon branching. Nat Neurosci 13:1373–1379

Jones-Humble SA, Morgan PF, Cooper BR (1994) The novel anticonvulsant lamotrigine prevents dopamine depletion in C57 black mice in the MPTP animal model of Parkinson's disease. Life Sci 54:245–252

Kaakkola S (2010) Problems with the present inhibitors and a relevance of new and improved COMT inhibitors in Parkinson's disease. Int Rev Neurobiol 95:207–225

Kaiser R, Hofer A, Grapengiesser A, Gasser T, Kupsch A, Roots I, Brockmöller J (2003) L-Dopa-induced adverse effects in PD and dopamine transporter gene polymorphism. Neurology 60:1750–1755

Kalinderi K, Fidani L, Katsarou Z, Bostantjopoulou S (2011) Pharmacological treatment and the prospect of pharmacogenetics in Parkinson's disease. Int J Clin Pract 65:1289–1294

Kaplan GB, Vasterling JJ, Vedak PC (2010) Brain-derived neurotrophic factor in traumatic brain injury, post-traumatic stress disorder, and their comorbid conditions: role in pathogenesis and treatment. Behav Pharmacol 21:427–437

Kiyohara C, Miyake Y, Koyanagi M, Fujimoto T, Shirasawa S, Tanaka K, Fukushima W, Sasaki S, Tsuboi Y, Yamada T, Oeda T, Shimada H, Kawamura N, Sakae N, Fukuyama H, Hirota Y, Nagai M, Fukuoka Kinki Parkinson's Disease Study Group (2011)

Genetic polymorphisms involved in dopaminergic neurotransmission and risk for Parkinson's disease in a Japanese population. BMC Neurol 11:89

Kramer AF, Hahn S, Cohen NJ, Banich MT, McAuley E, Harrison CR, Chason J, Vakil E, Bardell L, Boileau RA, Colcombe A (1999) Ageing, fitness and neurocognitive function. Nature 400:418–419

Kumar KR, Lohmann K, Klein C (2012) Genetics of Parkinson disease and other movement disorders. Curr Opin Neurol 25:466–474

Kurz MJ, Pothakos K, Jamaluddon S, Scott-Pandorf M, Arellano C, Lau Y-S (2007) A chronic mouse model of Parkinson's disease has a reduced gait pattern certainty. Neurosci Lett 429:39–42

Lachenmayer ML, Yue Z (2012) Genetic animal models for evaluating the role of autophagy in eticpathogenesis of Parkinson disease. Autophagy 8(12), PMID: 22931754

Laganiere J, Kells AP, Lai JT, Guschin D, Paschon DE, Meng X, Fong LK, Yu Q, Rebar EJ, Gregory PD, Bankiewicz KS, Forsayeth J, Zhang HS (2010) An engineered zinc finger protein activator of the endogenous glial cell line-derived neurotrophic factor gene provides functional neuroprotection in a rat model of Parkinson's disease. J Neurosci 30:16469–16474

Langston JW (1985) MPTP neurotoxicity: an overview and characterization of phases of toxicity. Life Sci 36:201–206

Lee J, Poon LY, Chong SA (2008) Spontaneous dyskinesia in first-episode psychosis in a Southeast Asian population. J Clin Psychopharmacol 28:536–539

Lee J, Jiang J, Sim K, Chong SA (2010) The prevalence of tardive dyskinesia in Chinese Singaporean patients with schizophrenia: revisited. J Clin Psychopharmacol 30:333–335

Lees AJ, Hardy J, Revesz T (2009) Parkinson's disease. Lancet 373(9680):2055–2066. doi:10.1016/S0140-6736(09)60492-X

Lencer E, Eismann G, Kasten M, Kabakei K, Geithe V, Grimm J, Klein C (2004) Family history of primary movement disorders as a predictor for neuroleptic-induced extrapyramidal symptoms. Br J Psychiatry 185:465–471

Lerer B, Segman RH, Tan EC, Basile VS, Cavallaro R, Aschauer HN, Strous R, Chong SA, Heresco-Levy U, Verga M, Scharfetter J, Meltzer HY, Kennedy JL, Macciardi F (2005) Combined analysis of 635 patients confirms an age-related association of the serotonin 2A receptor gene with tardive dyskinesia and specificity for the non-orofacial subtype. Int J Neuropsychopharmacol 8:411–425

Lesko LJ, Woodcock J (2004) Translation of pharmacogenomics and pharmacogenetics: a regulatory perspective. Nat Rev Drug Discov 3:763–769

Lewitt PA, Ellenbogen A, Chen D, Lal R, McGuire K, Zomorodi K, Luo W, Huff FJ (2012) Actively transported levodopa prodrug XP21279: a study in patients with Parkinson disease who experience motor fluctuations. Clin Neuropharmacol 35(3):103–110. doi:10.1097/WNF.0b013e31824e4d7d

Lidow MS (2000) Neurotransmitter receptors in actions of antipsychotic medications. CRC Press, Boca Raton, p 256

Lin JJ, Yueh KC, Lin SZ, Harn HJ, Liu JT (2007) Genetic polymorphism of the angiotensin converting enzyme and L-dopa-induced adverse effects in Parkinson's disease. J Neural Sci 252:130–134

Lindon JC, Holmes E, Nicholson JK (2004) Metabonomics and its role in drug development and disease diagnosis. Expert Rev Mol Diagn 4:189–199

Linkis P, Jörgenson LG, Olesen HL, Madsen PL, Lassen NA, Secher NH (1995) Dynamic exercise enhances regional cerebral artery mean flow velocity. J Appl Physiol 78:12–16

Liu Y, Yu H, Mohell N, Nordvall G, Lewander T, Hacksell U (1995) Derivatives of cis-2-amino-8-hydroxy-1-methyltetralin: mixed 5-HT1A-receptor agonists and dopamine D2-receptor antagonists. J Med Chem 38:150–160

Liu X, Robinson ML, Schreiber AM, Wu V, Lavail MM, Cang J, Copenhagen DR (2009a) Regulation of neonatal development of retinal ganglion cell dendrites by neurotrophin-3 overexpression. J Comp Neurol 514:449–458

Liu YZ, Tang BS, Yan XX, Liu J, Ouyang DS, Nie LN, Fan L, Li Z, Ji W, Hu DL, Wang D, Zhou HH (2009b) Association of the *DRD2* and *DRD3* polymorphisms with response to pramipexole in Parkinson's disease patients. Eur J Clin Pharmacol 65:679–683

Lopez-Gallardo E, Iceta R, Iglesias E, Montoya J, Ruiz-Pesini E (2011) OXPHOS toxigenomics and Parkinson's disease. Mutat Res 728:98–106

Lustig C, Shah P, Seidler R, Reuter-Lorenz PA (2009) Aging, training, and the brain: a review and future directions. Neuropsychol Rev 19:504–522

Macias M, Nowicka D, Czupryn Sulejczak D, Skup M, Skangiel-Kramska J, Czarkowska-Bauch J (2009) Exercise-induced motor improvement after complete spinal cord transection and its relation to expression of brain-derived neurotrophic factor and presynaptic markers. BMC Neurosci 10:144

Manolio TA (2010) Genomewide association studies and assessment of the risk of disease. N Engl J Med 363:166–176

Marks BL, Katz L, Smith JK (2009) Exercise and the aging mind: buffing the baby boomer's body and brain. Phys Sports Med 37:119–125

Marks BL, Katz L, Styner M, Smith JK (2010) Aerobic fitness and obesity: relationship to cerebral white matter integrity in the brain of active and sedentary older adults. Br J Sports Med 45(15):1208–1215, PMID: 20558529

Marks BL, Katz LM, Styner M, Smith JK (2011) Aerobic fitness and obesity: relationship to cerebral white matter integrity in the brain of active and sedentary older adults. Br J Sports Med 45(15):1208–1215. doi:10.1136/bjsm.2009.068114

Maruyama W, Naoi M (2012) "70th Birthday Professor Riederer" induction of glial cell line-derived and brain-derived neurotrophic factors by rasagiline and

(−)deprenyl: a way to a disease-modifying therapy? J Neural Transm 120(1):83–89, PMID: 22892822

Marvanová M, Lakso M, Pirhonen J, Nawa H, Wong G, Castrén E (2001) The neuroprotective agent memantine induces brain-derived neurotrophic factor and trkB receptor expression in rat brain. Mol Cell Neurosci 18:247–258

McArdle WD, Katch FI, Katch VI (1974) Appendix C. In: Lupash E (ed) Essentials of exercise physiology. Lippincott Williams and Wilkens, Baltimore, p 723

Mendes P (2006) Metabolomics and the challenges ahead. Brief Bioinform 7:127

Menna E, Cenni MC, Naska S, Maffei L (2003) The anterogradely transported BDNF promotes retinal axon remodeling during eye specific segregation within the LGN. Mol Cell Neurosci 24:972–983

Mihara K, Suzuki A, Kondo T, Nagashima U, Ono S, Otani K, Kaneko S (2000) No relationship between Taq1 a polymorphism of dopamine D(2) receptor gene and extrapyramidal adverse effects of selective dopamine D(2) antagonists, bromperidol and nemonapride in schizophrenia: a preliminary study. Am J Med Genet 96:422–424

Mittermeyer G, Christine CW, Rosenbluth KH, Baker SL, Starr P, Larson P, Kaplan PL, Forsayeth J, Aminoff MJ, Bankiewicz KS (2012) Long-term evaluation of a phase 1 study of AADC gene therapy for Parkinson's disease. Hum Gene Ther 23:377–381

Momose Y, Murata M, Kobayashi K, Tachikawa M, Nakabayashi Y, Kanazawa I, Toda T (2002) Association studies of multiple candidate genes for Parkinson's disease using single nucleotide polymorphisms. Ann Neurol 51:133–136

Morishima M, Harada N, Hara S, Sano A, Takahashi A, Marita Y, Nakaya Y (2006) Monoamine oxidase A activity and norepinephrine level in hippocampus determine hyperwheel running in SPORTS rats. Neuropsychopharmacology 31:2627–2638

Morris M, Schoo A (2004) Optimizing exercise and physical activity in older adults. Butterworth Heinemann, Edinburgh

Mougios V (2010) Exercise metabolism. In: Bahrke MS (ed) Exercise biochemistry. Human Kinetics, Champaign, p 122

Muramatsu S, Fujimoto K, Kato S, Mizukami H, Asari S, Ikeguchi K, Kawakami T, Urabe M, Kume A, Sato T, Watanabe E, Ozawa K, Nakano I (2010) A phase I study of aromatic L-amino acid decarboxylase gene therapy for Parkinson's disease. Mol Ther 18: 1731–1735

Nakazono Y, Abe H, Murakami H, Koyabu N, Isaka Y, Nemoto Y, Murata S, Tsutsumi Y, Ohtani H, Sawada Y (2005) Association between neuroleptic drug-induced extrapyramidal symptoms and dopamine D2-receptor polymorphisms in Japanese schizophrenic patients. Int J Clin Pharmacol Ther 43:163–171

Neeper SA, Gomez-Pinilla F, Choi J, Cotman CW (1996) Exercise and brain neurotrophins. Nature 373:109

Nicoletti A, Nicoletti G, Arabia G, Annesi G, De Mari M, Lamberti P, Grasso L, Marconi R, Epifanio A, Morgante L, Cozzolino A, Barone P, Quattrone A, Zappia M (2011) Reproductive factors and Parkinson's disease: a multicenter case–control study. Mov Disord 26:2563–2566. doi:10.1002/mds.23951

Nielsen J, Jewett MC (2008) Impact systems biology on metabolic engineering of *Saccharomyces cerevisiae.* FEMS Yeast Res 8:122–131

Numakawa T, Suzuki S, Kumamaru E, Adachi N, Richards M, Kunagi H (2010) BDNF function and intracellular signaling in neurons. Histol Histopathol 25:237–258

Oliff HS, Berchtold NC, Isackson P, Cotman CW (1998) Exercise-induced regulation of brain-derived neurotrophic factor (BDNF) transcripts in the rat hippocampus. Brain Res Mol Brain Res 61:147–153

Orth M, Schapira AH (2001) Mitochondria and degenerative disorders. Am J Med Genet 106:27–36

Oscarson S, Archer T, Fredriksson A (2009) Physical exercise and Milmed: synergism and neuroreparation in antiparkinsonian effects. In: Abstracts of staging neuropsychiatric disorders: implications for etiopathogenesis and treatment, Fundation Cerebro y Mente, Mojacar, p 88

Outeiro TF, Lindquist S (2003) Yeast cells provide insight into alpha-synuclein biology and pathobiology. Science 302:1772–1775

Pahwa R, Lyons KE (2009) Levo-dopa-related wearing-off in Parkinson's disease: identification and management. Curr Med Res Opin 25:841–849

Paus S, Grünewald A, Klein C, Knapp M, Zimprich A, Janetzky B, Möller JC, Klockgether T, Wüllner U (2008) The *DRD2* TaqIA polymorphism and demand of dopaminergic medication in Parkinson's disease. Mov Disord 23:599–602

Paus S, Gadow F, Knapp M, Klein C, Klockgether T, Wüllner U (2009) Motor complications in patients form the German competence network on Parkinson's disease and the *DRD3* Ser9Gly polymorphism. Mov Disord 24:1080–1084

Paus S, Gadow F, Kaut O, Knapp M, Klein C, Klockgether T, Wüllner U (2010) Tremor in Parkinson's disease is not associated with the *DRD3* Ser9Gly polymorphism. Parkinsonism Relat Disord 16:381–383

Pennington K, Peng J, Hung CC, Banks RE, Robinson PA (2010) Differential effects of wild-type and A53T mutant isoform of alpha-synuclein on the mitochondrial proteome of differentiated SH-SY5Y cells. J Proteome Res 9:2390–2401

Petranovic D, Tyo K, Vemuri GN, Nielsen J (2010) Prospects of yeast systems biology for human health: integrating lipid, protein and energy metabolism. FEMS Yeast Res 10:1046–1059

Ragimov CR, Ter-Asaturov GP, Golant MB, Rogov KA, Balakireva LZ (1991) Stimulation of reparative osteogenesis by millimeter band electromagnetic radiation in experimental traumatic defects of the mandible. Bull Exp Biol Med 111:562–565

Rasia RM, Bertoncini CW, Marsh D, Hoyer W, Cherny D, Zwcckstetter M, Griesinger C, Jovin TM, Fernández CO (2005) Structural characterization of copper(II)

binding to alpha-synuclein: Insights into the bioinorganic chemistry of Parkinson's disease. Proc Natl Acad Sci USA 102(12):4294–4299

Reynolds GP (2004) Receptor mechanisms in the treatment of schizophrenia. J Psychopharmacol 18:340–345

Rinne UK, Larsen JP, Siden A, Worm-Petersen J (1998) Entacapone enhances the response to levodopa in parkinsonian patients with motor fluctuations. Nomecomt Study Group. Neurology 51:1309–1314

Ruottinen HM, Rinne UK (1996a) A double-blind pharmacokinetic and clinical dose–response study of entacapone as an adjuvant to levodopa therapy in advanced Parkinson's disease. Clin Neuropharmacol 19:283–296

Ruottinen HM, Rinne UK (1996b) Effect of one month's treatment with peripherally acting catechol-O-methyl transferase inhibitor, entacapone, on pharmacokinetics and motor response to levodopa in advanced parkinsonian patients. Clin Neuropharmacol 19:222–233

Sallert M, Rantamäki T, Vesikansa A, Anthoni H, Harju K, Ylikauhaluoma J, Taira T, Castren E, Lauri SE (2009) Brain-derived neurotrophic factor controls activity-dependent maturation of CA1 synapses by downregulating tonic activation of presynaptic kainate receptors. J Neurosci 29:11294–11303

Schenkman M, Hall DA, Barón AE, Schwartz RS, Mettler P, Kohrt WM (2012) Exercise for people in early- or mid-stage Parkinson disease: a 16-month randomized controlled trial. Phys Ther 92(11):1395–1410, PMID: 22822237

Scheuer J, Tipton CM (1977) Cardiovascular adaptations to physical training. Annu Rev Physiol 39:221–251

Schultz W, Scarnati E, Sundström E, Romo R (1989) Protection against 1-methyl-4-phenyl-1,2,3,6-tetrahydropyridine-induced parkinsonism by the catecholamine uptake inhibitor nomifensine: behavioral analysis in monkeys with partial striatal dopamine depletions. Neuroscience 31:219–230

Schwarting RK, Sedelis M et al (1999) Strain-dependent recovery of open-field behavior and striatal dopamine deficiency in the mouse MPTP model of Parkinson's disease. Neurotox Res 1:41–56

Sedelis M, Hofele K, Auburger GW, Morgan S, Huston JP, Schwarting RK (2000a) Evidence for resistance to MPTP in C57BL/6 mice × BALB/c F1 hybrids as compared to their progenitor strains. Neuroreport 11:1093–1096

Sedelis M, Hofele K, Auburger GW, Morgan S, Huston JP, Schwarting RK (2000b) MPTP susceptibility in the mouse: behavioural, neurochemical and histological analysis of gender and strain differences. Behav Genet 30:171–182

Sedelis M, Schwarting RK, Huston JP (2001) Behavioral phenotyping of the MPTP mouse model of Parkinson's disease. Behav Brain Res 125:109–125

Sedelis M, Hofele K, Schwarting RK, Huston JP, Belknap JK (2003) Chromosomal loci influencing the susceptibility to the parkinsonian neurotoxin 1-methyl-4-phenyl-1,2,3,6-tetrahydropyridine. J Neurosci 23:8247–8253

Sharma M, Ioannidis JP, Aasly JO, Annesi G, Brice A, Van Broeckhoven C, Bertram L, Bozi M, Crosiers D, Clarke C, Facheris M, Farrer M, Garraux G, Gispert S, Auburger G, Vilariño-Güell C, Hadjigeorgiou GM, Hicks AA, Hattori N, Jeon B, Lesage S, Lill CM, Lin JJ, Lynch T, Lichtner P, Lang AE, Mok V, Jasinska-Myga B, Mellick GD, Morrison KE, Opala G, Pramstaller PP, Pichler I, Park SS, Quattrone A, Rogaeva E, Ross OA, Stefanis L, Stockton JD, Satake W, Silburn PA, Theuns J, Tan EK, Toda T, Tomiyama H, Uitti RJ, Wirdefeldt K, Wszolek Z, Xiromerisiou G, Yueh KC, Zhao Y, Gasser T, Maraganore D, Krüger R, On behalf of the GEO-PD Consortium (2012) Large-scale replication and heterogeneity in Parkinson disease genetic loci. Neurology 79:659–667

Shastry BS (2005) Genetic diversity and new therapeutic concepts. J Hum Genet 50:321–328

Shastry BS (2006) Pharmacogenetics and the concept of individualized medicine. Pharmacogenomics J 6:16–21

Sonsalla PK, Heikkila RE (1986) The influence of dose and dosing interval on MPTP-induced dopaminergic neurotoxicity in mice. Eur J Pharmacol 129:339–345

Squassina A, Manchia M, Manolopoulos VG, Artac M, Lappa-Manakou C, Karkabouna S, Mitropoulos K, Del Zompo M, Patrinos GP (2010) Realities and expectations of pharmacogenomics and personalized medicine: impact of translating genetic knowledge into clinical practice. Pharmacogenomics 11:1149–1167

Stranahan AM, Zhou Y, Martin B, Maudsley S (2009) Pharmacomimetics of exercise: novel approaches for hippocampally-targeted neuroprotective agents. Curr Med Chem 16:4668–4678

Streffer JR, Grachev ID, Fitzer-Attas C, Gomez-Mancilla B, Boroojerdi B, Bronzova J, Ostrowitzki S, Victor SJ, Fontoura P, Alexander R (2012) Prerequisites to launch neuroprotective trials in Parkinson's disease: an industry perspective. Mov Disord 27:651–655. doi:10.1002/mds.25017

Sundström E, Fredriksson A, Archer T (1990) Chronic neurochemical and behavioral changes in MPTP-lesioned C57BL/6 mice: a model for Parkinson's disease. Brain Res 528:181–188

Swen JJ, Huizinga TW, Gelderblom H, de Vries EG, Assendelft WJ, Kirchheiner J, Guchelaar HJ (2007) Translating pharmacogenomics: challenges on the road to the clinic. PLoS Med 4:E209

Tajiri N, Yasuhara T, Shingo T, Kondo A, Yuan W, Kadota T, Wang F, Baba T, Tayra JT, Morimoto T, Jing M, Kikuchi Y, Kuramoto S, Agari T, Miyoshi Y, Fujino H, Obata F, Takeda I, Furuta T, Date I (2010) Exercise exerts neuroprotective effects on Parkinson's disease model of rats. Brain Res 1310:200–207

Tan EC, Chong SA, Mahendran R, Tan CH, Teo YY (2003) Mu opioid receptor gene polymorphism and neuroleptic-induced tardive dyskinesia in patients with schizophrenia. Schizophr Res 65:61–63

Tchantchou F, Lacor PN, Cao Z, Lao L, Hou Y, Cui C, Klein WL, Luo Y (2009) Stimulation of neurogenesis and synaptogenesis by bilobalide and quercetin via common final pathway in hippocampal neurons. J Alzheimers Dis 18:787–798

Teneiro S, Outeiro TF (2010) Simple is good: yeast models of neurodegeneration. FEMS Yeast Res 10:970–979

Thomas AA, Friedman JH (2010) Current use of clozapine in Parkinson disease and related disorders. Clin Neuropharmacol 33:14–16

Tillerson JL, Cohen AD, Philpower J, Miller GW, Zigmond MJ (2001) Forced limb-use effects on the behavioral and neurochemical effects of 6-hydroxydopamine. J Neurosci 21:4427–4435

Tillerson JL, Cohen AD, Philpower J, Miller GW, Zigmond MJ (2002) Forced nonuse in unilateral Parkinsonian rats exacerbates injury. J Neurosci 22:6790–6799

Vallelunga A, Flaibani R, Formento-Dojot P, Biundo R, Facchini S, Antonini A (2012) Role of genetic polymorphisms of the dopaminergic system in Parkinson's disease patients with impulse control disorders. Parkinsonism Relat Disord 18:397–399

van Hilten JJ, Ramaker C, Van de Beek WJ, Finken MJ (2000) Bromocriptine for levodopa-induced motor complications in Parkinson's disease. Cochrane Database Syst Rev 2000(2), CD001203

Viberg H, Mundy W, Eriksson P (2008) Neonatal exposure to decabrominated diphenyl ether (PBDE 209) results in changes inBDNF, CaMKII and GAP-43, biochemical substrates of neuronal survival, growth, and synaptogenesis. Neurotoxicology 29:152–159

Wagle Shukla A, Okun MS (2012) Personalized medicine in deep brain stimulation through utilization of neural oscillations. Neurology 78:1900–1901

Wang L (2009) Pharmacogenomics: a systems approach. Wiley Interdiscip Rev Syst Biol Med 2(1):3–22. doi:10.1002/wsbm.42

Wang L (2010) Pharmacogenomics: a systems approach. Wiley Interdiscip Rev Syst Biol Med 2:3–22

Wang GJ, Volkow ND, Fowler JS, Franceschi D, Logan J, Pappas NR, Wong CT, Netusil N (2000) PET studies of the effects of aerobic exercise on brain striatal dopamine release. J Nucl Med 41:1352–1356

Wang L, Weinshilboum RM (2006) Thiopurine S-methyltransferase (TPMT) pharmacogenetics: insights, challenges and future directions. Oncogene Res 25:1629–1638

Wang L, Weinshilboum RM (2008) Pharmacogenomics: candidate gene identification, functional validation and mechanisms. Hum Mol Genet 17:R174–R179

Wang L, McLeod HL, Weinshilboum RM, Feero WG, Guttmacher AE (2011) Genomics and drug response. N Engl J Med 364:1144–1153

Waters RP, Renner KJ, Pringle RB, Summers CH, Britton SL, Koch LG, Swallow JG (2008) Selection for aerobic capacity affects corticosterone, monoamines and wheel-running activity. Physiol Behav 18:1044–1054

Weinshilboum RM (2003) Inheritance and drug response. N Engl J Med 348:529–537

Weinshilboum RM, Wang L (2006) Pharmacogenetics and pharmacogenomics: development, science and translation. Ann Rev Genomics Hum Genet 7:223–245

Williams-Gray CH, Hampshire A, Barker RA, Owen AM (2008) Attentional control in Parkinson's disease is dependent on *COMT* VAL158Met genotype. Brain 131:397–408

Woo NH, Teng HK, Siao CJ, Chiaruttini C, Pang PT, Milner TA, Hempstead BL, Lu B (2005) Activation of p75NTR by proBDNF facilitates hippocampal long-term depression. Nat Neurosci 8:1069–1077

Woodcock J (2010) Assessing the clinical utility of diagnostics used in drug therapy. Clin Pharmacol Ther 88:765–773

Wu K, O'Keeffe D, Politis M, O'Keeffe GC, Robbins TW, Bose SK, Brooks DJ, Piccini P, Barker RA (2012) The catechol-O-methyltransferase Val158Met polymorphism modulates fronto-cortical dopamine turnover in early Parkinson's disease: a PET study. Brain 135(Pt 8):2449–2457

Yang ML, Hasadsri L, Woods WS, George JM (2010) Dynamic transport and localization of alpha-synuclein in primary hippocampal neurons. Mol Neurodegener 5(1):9. doi:10.1186/1750-1326-5-9

Zappia M, Annesi G, Nicoletti G, Arabia G, Annesi F, Messina D, Pugliese P, Spadafora P, Tarantino P, Carrideo S, Civitelli D, De Marco EV, Cirò-Candiano IC, Gambardella A, Quattrone A (2005) Sex differences in clinical and genetic determinants of L-dopa peak-dose dyskinesias in Parkinson's disease: an exploratory study. Arch Neurol 62:601–605

Zhang J, Vemuri G, Nielsen J (2010) Systems biology of energy homeostasis in yeast. Curr Opin Microbiol 13:382–388

Zhou XP, Wu KY, Liang B, Fu XQ, Luo ZG (2008) TrkB-mediated activation of geranylgeranyltransferase I promotes dendritic morphogenesis. Proc Natl Acad Sci USA 105:17181–17186

Zhou W, Bercury K, Cummiskey J, Luong N, Lebin J, Freed CR (2011) Phenylbutyrate up-regulates the DJ-1 protein and protects neurons in cell culture and in animal models of Parkinson disease. J Biol Chem 286:14941–14951

Pharmacogenomics and Personalized Medicine in Alzheimer's Disease

15

Trevor Archer and Boo Johansson

Abstract

Alzheimer's disease (AD) and related dementias, neurodegenerative disorders accompanied by progressive deterioration of cognitive capacity, every-day behavior abilities, and integrity of brain tissue, present an ever-growing, worldwide dilemma due to aging populations confronted by a related neuropathology. The "amyloid cascade hypothesis," that pathophysiology is driven by the ever-increasing burden of β-amyloid in the brains of afflicted patients, involves a poorly understood orchestration encompassing multitudes of enzymes and signaling pathways arranged in vast and diverse arrays of cellular processes, and vascular considerations all of which are under the control of predictive genes and susceptibility genes that describe genetic and genes x environmental epigenetic interactions. Genetic aspects of these disorders and the intricacies of pharmacogenomics implicated several neurotransmitter pathways, circuits and regional brain developments, and metabolism that reinforce the growing requirements for personalized medicine. The search for individual-based medication, in addition to genomic assay and biomarker identity, seeks to establish a "reregulation" of destructive β-amyloid pathways, an understanding and application of Aβ-linked immunotherapy, the initiation and formulation of pharmacogenetic/pharmacogenomics principles and methodologies, the emergence of the role of apolipoprotein (APOE) in therapeutic endeavor, the assessment and treatment of behavioral and psychological symptoms, the therapies focused upon frontotemporal dementia, and the interventions centered around instrumental activities of daily activities.

1 Introduction

Alzheimer's disease (AD) and related dementias present major health problems throughout the world, against a background of ever-older populations, with disability concerns in older adults. Prevalence increases exponentially from about 2 %

T. Archer (✉) • B. Johansson
Department of Psychology,
University of Gothenburg, Box 500,
SE 450 30 Gothenburg, Sweden
e-mail: trevor.archer@psy.gu.se

D. Barh et al. (eds.), *Omics for Personalized Medicine*,
DOI 10.1007/978-81-322-1184-6_15, © Springer India 2013

at 60–65 years of age to greater than 30–35 % in individuals aged more than 80 years, after 65 years of age doubling every 5-year age increase (Lobo et al. 2000). Over 35 million individuals were diagnosed with AD with an expectation of at least a doubling by 2030 (Alzheimer's disease International Consortium 1996, assessed 2012). Nevertheless, it is important to bear in mind that the onset and development of and associated disorders differs from individual to individual but that there exist a number of common symptoms between individuals. Early symptoms are often mistakenly thought to be "age-related" concerns or manifestations of stress (Waldemar 2007). AD (50–70 %), vascular dementia (30–40 %), and mixed dementia (15–20 %) account for more than 90 % of the most prevalent forms of dementia. AD presents the common type of dementia observed in 50–70 % of all individuals presenting the disorder in Western industrialized countries (Cacabelos 2008), with the neurodegenerative effects of the disease state associated with phenotypic expression of more than 200 genes (Cacabelos et al. 2012). In addition to the huge burden of responsibility placed upon caregivers and society, current estimates of treatment and mortality costs as well as the cost-effectiveness of disease-modifying drugs attest to the ever-increasing economic burden (Sköldunger et al. 2013). Early symptoms may interfere with the most complex daily living activities. The notion of staging is central to any discourse on the disorder. In the early stages, the most common symptom is difficulty in remembering recent events. Subtle problems with the executive functions of attentiveness, planning, flexibility, and abstract thinking, or impairments in semantic memory (memory of meanings and concept relationships) may be associated with the early stages of the disorder (Nygård 2003), as well as apathy which remains as the most characteristic symptom that persists throughout (Landes et al. 2001).

AD is a neurodegenerative disease accompanied by the progressive deterioration of cognitive capability and associated with the deposition of extracellular plaques, aggregated Aβ peptides, by intraneuronal neurofibrillary tangles and synapse and focal neuronal loss (Duykaerts et al. 2009). AD pathology is characterized by the accumulation of misfolded protein (beta-amyloid, a peptide of 36–43 amino acids processed from the amyloid precursor protein), oxidative damage, and inflammatory changes resulting in region-specific loss of synaptic connections and neuronal cell death (Querfurth and LaFerld 2010). When AD is suspected, the diagnosis is usually confirmed with tests that evaluate behavior and thinking abilities, often followed by brain imaging analyses if these techniques are available for application. With disease progression, several other symptoms may emerge, including states of confusion, irritability and aggression, fluctuations in mood states, problems in expressing language, and long-term memory loss. Withdrawal from family members, friends, and society should be expected as the patient declines and loses capacities. Eventually, bodily functions deteriorate, the patient requiring more and more care and supervision, and the continued loss ultimately leads to death. Due to the uniqueness of the disease state for each individual, any predictions as to how particular individuals may be affected are not only difficult but speculative, thereby emphasizing the extreme necessity for "personalized medication." The availability of current treatments for AD is severely limited with those drug interventions that are applicable only providing modest efficacy and symptom alleviation. AD symptoms and biomarkers progress for an unknown and variable amount of time before becoming fully apparent to family and friends; the individual's predicament may remain undiagnosed for years. The life expectancy following diagnosis has been estimated to be approximately 7 years, with less than 3 % of individuals surviving more than 14 years after diagnosis (Mölsä et al. 1986).

The "amyloid cascade hypothesis" of AD posits that the pathophysiology is driven by a progressively increasing burden of β-amyloid in the brains of afflicted patients, via the aberrant proteolysis of β-amyloid precursor protein (βAPP) into neurotoxic amyloid beta (Aβ) peptides that is central to the etiopathogenesis of AD. Nevertheless, the relative failure of the hypothesis, based upon the lack of sufficient therapeutic efficacy of β-amyloid-lowering agents, points towards a level of disorder complexity combined with a multiplicity of gene-environment interactions that have yet to be understood, although the failure of these agents clinically does necessarily discount the hypothesis. The evidence that β-amyloid processing is distributed over a multitude of enzymes and signaling pathways, all of which involve a vast and diverse array of cellular processes, suggests that the orchestration of AD pathophysiology is poorly understood. Alternatively, there is the possibility that neurotoxic β-amyloid levels had already wreaked sufficient destruction that causes irreversible damage to downstream cellular pathways that expressions of dementia had already been established. Teich and Arancio (2012) have argued that AD drug therapy would benefit from direct approaches through elucidation of mechanisms through which elevated β-amyloid disrupts synaptic and electrochemical physiology. Herrmann et al. (2011) indicate that not only amyloid-β aggregation but other possible neuronal mechanisms – such as hyperphosphorylated tau, neuroinflammation, and other processes – play important roles in the pathophysiology of this multifactorial disorder. Development of better disease models and biomarkers is essential for the advancement of knowledge of the disease mechanisms (see also Levy et al. 2011). Recent advances in cerebrospinal fluid (CSF) biomarkers and neuroimaging techniques, e.g., PET for Aβ imaging in brain, have facilitated the identification of "high-risk" patients for AD prior to onset of clinical signs and/or prodromal stages (Isaac et al. 2011). This category of patients may benefit from preventive therapies that retard progression to prodromal AD (cf. Archer et al. 2011c) or to AD dementia (Dubois et al. 2012). Vascular alterations predate AD neuropathology both in animal models of AD and in the development of cognitive deficits by "high-risk" AD patients.

Patients with a clinical history of stroke show increased risk for AD development (Honig et al. 2003), with "silent" brain infarcts (SBIs) more than doubling the risk for dementia, including AD (Vermeer et al. 2003, 2007); these AD with SBI patients present lower global function cognitive scores compared with "pure" AD patients (Song et al. 2007). A Japanese MRI study indicated that one third of AD-diagnosed patients were SBIs (Matsui et al. 2005). Dementia occurs in the presence of a lower number of AD lesions in patients presenting vascular lesions than those without (Zekry et al. 2002), and high vascular scores were linked to dementia and AD irrespective of type of vascular lesion (Strozyk et al. 2010), with strategic localization of lacunar infarcts and white matter lesions related to deficit profiles (Duering et al. 2011). Pimentel-Coelho and Rivest (2012) have reviewed the interactions between cerebrovascular diseases and AD brain pathophysiology, chronic cerebral hypoperfusion in AD and the mechanisms involved, arteriolosclerosis and cerebral amyloid angiopathy which constitute the most common forms of cerebral small-vessel disease of early origin (Deramecourt et al. 2012), impairments in cerebral autoregulation, vascular reactivity and cerebral blood flow, impaired angiogenesis in AD, and the possible links between vascular dysfunction and neurogenesis impairments in AD. They present a unifying vascular model hypothesis of interactions between cerebrovascular disease and pathogenic mechanisms in AD. The model outlines the notion that vascular risk factors and other AD-related genetic and environmental risk factors trigger a variety of events converging on neurovascular unit dysfunction and BBB damage (Zipser et al. 2007) that cause cerebral blood flow reductions in multiple brain regions prodromally and even prior to that. Disease progression is governed by synergistic interaction of these destructive forces resulting in synaptic impairments, neuronal dysfunction and neurodegeneration (Zlokovic 2011), cognitive deficits, and dementia. The progressive "derailing" of regenerative mechanisms, angiogenesis and neurogenesis, perpetuates the cycle of vascular and neuronal damage.

2 Genetic Aspects of AD

AD presents a genetically dichotomous disorder defined by (1) early-onset familial cases characterized by Mendelian inheritance (EO-FAD) and (2) late onset (>60 years) with no consistent mode of transmission (LOAD, see below; Bertram and Tanzi 2005, 2008, 2009, 2012; Bertram et al. 2007, 2008, 2010; Tanzi and Bertram 2005). EOAD (see below) occurs in (>60 years) AD cases of familial clustering with 80 % of cases involving inheritance of genetic factors (Gatz et al. 2006). Progressive cognitive deterioration, behavioral disturbances, and general functional decline are expressions of a polygenic condition in which hundreds of genes across the human genome may be associated with environmental inducers, cerebrovascular dysfunction, and epigenetic influences. Whereas "sporadic" AD (see below, linked with LOAD) is susceptible to genetic variants modulated through life environmental situation, EO-FAD is associated with rare, fully penetrant mutations in three different genes: (1) *APP* gene-amyloid-β protein precursor-chromosome 21q21-molecular phenotype increased $A\beta_{42}/A\beta_{40}$ ratio, (2) *PSEN1* gene-presenilin 1-chromosome14q24-molecular phenotype increased $A\beta_{42}/A\beta_{40}$ ratio, and (3) *PSEN2* chromosome 1q31-molecular phenotype increased $A\beta_{42}/A\beta_{40}$ ratio (Kounnas et al. 2010; Tanzi 2012). It seems that the only gene variant regarded as an established LOAD risk factor is the ε4 allele of *APOE* gene (see below, Strittmatter et al. 1993) on chromosome 19q13. The *APOE* ε4 allele appears neither necessary nor sufficient to cause AD, whereas the EO-FAD/EOAD mutations are sufficient but not necessary to cause AD, but poses a genetic risk factor reducing age of onset in a dose-dependent fashion with its role in lipid metabolism and transport secondary to Aβ clearance from brain (Roses et al. 2010). Although genome-wide association studies have identified 11 additional candidate genes (*CD33*, *CLU*, *CR1*, *PICALM*, *BIN1*, *ABCA7*, *CD2AP*, *EPHA1*, *MS4A6A*/*MS4A4E*, and *ATXN1*), while common gene polymorphisms, e.g., ε2 and ε4 variants of *APOE*, account for around 50 % of LOAD susceptibility, the four genes, *APP*, *PSEN1*, *PSEN2*, and *APOE*, cover about 30–50 % of AD heritability (Tanzi 2012; see below also). Elias-Sonnenschein et al. (2012) carried out a systematic, qualitative review covering 17 studies (that listed 47 polymorphisms in 26 genes), thereby investigating the potential relationship between the most compelling AD risk genes and markers for Aβ and tau in cerebrospinal fluid, PET imaging, and clinical neuropathological examination. Of all genes covered, only APOE and PICALM showed consistent effects on Aβ but not on tau, while no obvious effects were observed for CLU, CR1, ACE, SORL, and MAPT. Functional variants within genes at these loci need to be observed in order to confirm their role: Allen et al. (2012) have shown that CLU and MS4A4A eSNPs may to extent explain the LOAD risk association at these loci while CLU and ABCA7 may harbor additional strong eSNPs.

Five categories of genes are involved in pharmacogenomics for the most part: (1) genes involved in etiopathogenesis, (2) genes associated with the mechanism of action of a certain compound, (3) genes linked to phase I and phase II metabolic reactions, (4) genes associated with transporters, and (5) pleiotropic genes and/or genes associated with concomitant pathologies (Cacabelos et al. 2012), with APOE and CYP2D6 genes most extensively investigated (Albani et al. 2012; Kotze and van Rensburg 2012; Meda et al. 2012; Troyer et al. 2012; Wang et al. 2012; Zhong et al. 2013). It has been suggested that the relationship between the cognitive symptoms of AD and the classical biomarkers of the disorders, amyloid plaques and neurofibrillary tangles, may be tenuous (Cummings 2004), not least due to the presence of β-amyloid in normal older adults and AD patients (Alzenstein et al. 2008). Additionally, the lack of efficacy of therapies based on the amyloid hypothesis has initiated a reassessment of AD pathophysiology (Mangialasche et al. 2010). Despite these indications, much evidence implies that changes in the generation or the degradation of the amyloid-β peptide (Aβ) lead to the formation of aggregated structures that are the triggering molecular events in the pathogenic cascade of AD (Di Carlo et al. 2012). AD is associated with multiple genetic defects that are transmitted via either predictive (mutational) or susceptibility (risk) genes. Predictive (presymptomatic) genes allow the detection of gene mutations associated with disorders that appear after birth, often later in life, whereas different alleles (one or more forms of genetic sequence at a certain genome location) of susceptibility genes alter the risk of developing AD and/or AD severity. Personalized medicine is a medical model that proposes the customization of healthcare, with decisions and practices being tailored to the individual patient by use of genetic or other information. Mutations in genes predictive for AD increase likelihood for AD; on the other hand, genes endowing susceptibility appear not to be implicated in disorder causality but rather elevate individuals' predisposition/susceptibility for disorder.

Taken together with "disorder staging" (cf. Archer et al. 2011c), AD genetics seem to follow a dichotomy of early onset and late onset, with predictive genes (limited to about 5 %) accounting for the early-onset AD (EOAD) variant (Mastroeni et al. 2009, 2010), whereas late-onset AD (LOAD) with 95 % of cases "sporadic" is dependent on gene x environment interactions, with a variety of susceptibility genes, mainly apolipoprotein E (APOE), presents a substantial susceptibility risk (Muresan and Muresan 2012; Pedersen 2010). There exists a potentially protective effects of the apolipoprotein E (APOE) ε2 allele on cognitive functioning in individuals at risk for developing AD; Bonner-Jackson et al. (2012) have observed that the APOE ε2 allele provides a buffer against significant changes in daily functioning over time and is associated with better neuropsychological performance across a number of measures of cognitive performance. An outline of predictive and susceptibility genes has been accumulated by Alagiakrishnan et al. (2012) describing common mechanisms: for EOAD and LOAD, immune and inflammatory system-linked genes, cell membrane and synapse-linked genes, and cholesterol- and vascular-linked genes, for the pathophysiology of AD (see Table 15.1, Alagiakrishnan et al. 2012). Whereas EOAD is linked to predictive genes (APP, amyloid-β precursor protein; PSEN1 and 2, presenilin 1 and 2), LOAD is linked to both replicated (e.g., APOE; CLU, clusterin [apolipoprotein J]; B1N1, bridging integrator 1; PICALM, phosphatidylinositol-binding clathrin assembly protein; GAB2, glycine-rich protein-associated binding protein 2) and non-replicated susceptibility genes (e.g., A2M, alpha-2-macroglobulin; ACE, angiotensin-converting enzyme; ubiquitin 1; SORL1, sortilin-related receptor 1). The susceptibility genes are involved in several aspects of the etiopathogenesis of AD: clearance of amyloid, inflammatory proteins associated with the disorder, membrane synaptic transmission and removal of apoptotic cells, amyloid protein synthesis, neuroimmune functioning, lipid metabolism, and receptor-mediated endocytosis (Carrasquillo et al. 2010; DeMattos et al. 2004; Jun et al. 2010; Reiman et al. 2007; Rogers et al. 2010; Tanzi 2012) underlying the clinical characteristics, i.e., age of onset, accelerated cognitive decline, aggressive behavior, apathy, and poor response and adverse reactions to cholinesterase inhibitors (Barabash et al. 2009; Belbin et al. 2011; Hatanaka et al. 2000; Kolsch et al. 2009). Finally, several epigenetic mechanisms, including infections, stress, and nutritional and emotional factors (Chouliaras et al. 2010; Graff and Mansuy 2008; Nee and Lippa 1999; Urdinquio et al. 2009), whereas multiple vascular risk factors, mid-life hypertension, hypercholesterolemia, diabetes, and smoking, exert epigenetic influences elevating disorder risk (Chandra and Pandav 1998; Migliore and Coppede 2009).

AD patients present myelin loss in the cortical gray matter (GM) and white matter (WM), myelin breakdown (Bartzokis et al. 2003), and neuronal phenotypes (Roher et al. 2002; Vlkolinsky et al. 2001), together with reduction in size of corpus callosum (Hampel et al. 1998) and oligodendrocyte glial cell death (Lassmann et al. 1995). Taken together, these progressive damage areas reduce signal-to-noise ratios in the electrochemical signaling processes involved. On the other hand, oligodendrocyte progenitor/precursor cells (OPCs) which can be identified by their expression of a number of antigens, including the ganglioside GD3 (Curtis et al. 1988), the NG2 chondroitin sulfate proteoglycan (Tan et al. 2005), and the platelet-derived growth factor-alpha receptor subunit PDGF-alphaR (Pringle and Richardson 1993), that precede oligodendrocytes, are capable even of generating neurons and glial cells. OPCs expressing proteoglycan, a major component of the extracellular matrix, NG2, and transcription factor Olig2 are present in healthy human brains (Geha et al. 2010), representing the only proliferating

Table 15.1 Certain examples of underlying pathophysiology and expressions of AD in individuals with possible application of "personalized medicine"

Pathophysiology	Expression	Personalized medicine
Aβ deposition	↑Aβ peptide[a]	Modulation of βAPP processing
Loss of cognition	↓Memory performance	Anti-AchE[b], IADLs[c], exercise
Neuroinflammation	↓↑Cytokines	Aβ immunization (EB101)[d]
Insulin resistance	↑Blood sugar	*APOE* and *CYP2D6* manipulation
Gliosis	↓Myelination	BCHE-K[e] manipulation
Vascular burden	↑Hypertension	Exercise-induced angiogenesis
Free radical generation	↑Apoptosis	Chelation[f]

[a]Neurofibrillary tangles, neuron-synapsis loss, ↓brain volume
[b]Anti-acetylcholinesterase
[c]Instrumental activities of daily living
[d]EB101 vaccine
[e]Butyrylcholinesterase K
[f]Free radical scavengers

cell type in the adult parenchyma in both humans and mice (Dimou et al. 2008; Dimou and Götz 2012; Simon et al. 2011). In the healthy rodent brain, OPCs can generate myelinating oligodendrocytes (Kang et al. 2010; Zhu et al. 2011), neurorepair by differentiating into mature cells that remyelinate in mouse models of demyelination (Fancy et al. 2011; Islam et al. 2009; Zawadzka et al. 2010).

In mouse models, myelin integrity is associated with AD-linked mutations in neuron and oligodendrocyte lineage (3xTg) with presenilin mutations exerting effects upon OPCs/oligodendrocytes survival, differentiation, and myelination (Desai et al. 2009, 2010, 2011; Horiuchi et al. 2012). Behrendt et al. (2012) studied cells of the oligodendrocyte lineage in a mouse model of AD with chronic plaque deposition (APPPS1) and samples from human patients. In APPPS1 mice defects in myelin integrity, prevalent at 6 months of age, had normalized by 9 months of age concurrent with increases in proliferation and differentiation of OPCs; this observation implicated OPCs in the neuroreparation of myelin aberrations, but in postmortem human tissue, OLIGO2+ were decreased in the cortex. They concluded that the oligodendrocyte progenitors reacted specifically to amyloid plaque deposition in an AD-related mouse model and in human AD pathology but with distinct outcomes, with neuroreparation in the former that was limited severely in the latter. Microglia, principle immune effector cells in the brain, are involved also through interaction with fibrillar forms of Aβ via Toll-like receptors 2/4/6 and their co-receptors that are dependent upon interleukin receptor-associated kinases for signaling. Cameron et al. (2012) reported that microglial interleukin receptor-associated kinase4 was necessary for cascading activation of proinflammatory signaling pathways, p38, JNK (c-jun N-terminal kinase) and ERK (extracellular signal-regulated kinase), MAP (mitogen-activated protein) kinases (MAPKs), and reactive oxygen species (ROS); this loss of interleukin receptor-associated kinases was linked to diminished microgliosis and astrogliosis in aged mice. The MAPKs are implicated in directing cellular responses to a diverse array of stimuli, such as mitogens, osmotic stress, heat shock, and proinflammatory cytokines, as well as regulating cell proliferation, gene expression, differentiation, mitosis, cell survival, and apoptosis. Microglia analysis revealed altered patterns of gene expression associated with changes in microglial phenotypes in turn linked to IRF transcription factor expression govern those phenotypes. The loss of interleukin receptor-associated kinases promoted also amyloid clearance mechanisms that included elevated expression of insulin-degrading enzyme. It appears that interleukin receptor-associated kinase activation regulates microglial activation status and amyloid homeostasis under normal conditions but is dysregulated under pathophysiological conditions.

3 Pharmacogenomics of AD

Individual-specific therapeutic compounds are predicted to elevate treatment efficacy on the one hand and reduce adverse drug reaction on the other; pharmacogenomic determinations of symptom response to drug effects and potential toxic reactions are expected to influence the utility of both conventional and prospective pharmacotherapeutic interventions. It is likely that pharmacogenomic factors may account for 60–90 % of drug variability in drug disposition, metabolism, and pharmacodynamics. For example, 15 % of Caucasians are carriers of defective CYP2D6 polymorphisms linked to cholinesterase inhibitor lack of efficacy (Cacabelos 2005), whereas in APOE4 patients cholinesterase inhibitors induce high efficacy (Crentsil 2004). It is possible that around 50 % of adverse drug events in CNS disorders might be attributed to pharmacogenomic factors. Cacabelos et al. (2012) have produced a rationale that presents a practical approach to notions of pharmacogenomics and personalized therapeutics based on individual genomic profiles implies the management of different types of genes and their products including those (1) genes associated with the mechanism of action of psychotropic drugs (neurotransmitters, receptors, transporters), (2) genes encoding enzymes responsible for drug metabolism (phase I, phase II reactions), (3) disease-specific genes associated with a particular pathogenic cascade, and (4) pleiotropic genes (genes influencing multiple phenotypic traits) with multi-locational effects within metabolomic networks. The incorporation of genomic medicine procedures and pharmacogenomics into clinical practice, together with educational programs for the correct use of medication, will eventually revitalize current therapeutic notions applied to treat CNS disorders and in particular AD. It has been shown that about 57.76 % of patients with Alzheimer's disease are extensive metabolizers (EMs) for CYP2D6 enzymes, 31.06 % are intermediate metabolizers (IMs), 5.28 % are poor metabolizers (PMs), and 5.90 % were ultrarapid metabolizers: 73.71 % are CYP2C19-EMs, 25.12 % IMs, and 1.16 % PMs; 60.87 % are CYP2C9-EMs, 34.16 % IMs, and 4.97 % PMs; and 82.75 % presented CYP3A4/5-EMs, 15.88 % IMs, and 1.37 % UMs. It was observed that a trigenic cluster that integrated CYP2D6 + CYP2C19 + CYP2C9 polymorphic variants yielded 82 different haplotype-like profiles that represented 36 different pharmacogenetic phenotypes of which only 26.51 % of patients showed a pure 3EM phenotype. These data indicated directly that the incorporation of pharmacogenomic protocols to dementia research and clinical trials can foster therapeutics optimization by helping to develop cost-effective pharmaceuticals and improve drug efficacy and safety (Cacabelos 2009, 2010).

Paraoxonases are enzymes involved in the hydrolysis of organophosphates; paraoxonase 1 is an arylesterase with several biological activities. The three known genotypic forms are coded for by the PON set of genes: PON1 (an oxidant, prevents oxidation of LDL), PON2 (a ubiquitously expressed intracellular protein that can protect cells against oxidative damage), and PON3 (similar to PON1 in activity, but differing from it in substrate specificity) are all three located on the long arm of chromosome 7. Paraoxonase 1 (PON1) functions to protect the cholinergic system against nerve gases and the organophosphate family of pesticides. Leduc and Poirier (2008) observed that the frequency of the M55M genotype at the PON1 L55M locus was found to be significantly increased in AD patients relative to age-matched controls, and significant associations were observed between the PON1 L55M and Q192R polymorphisms and frontal cortex Abeta (amyloid-β) levels as well as ChAT (choline acetyltransferase) activity and nicotinic receptor density in the temporal cortex, thereby implying a major role in AD pathophysiology. Genetic variants in the paraoxonase (PON) gene cluster, particularly a single C/T promoter polymorphism (rs 705381) in the PON1 gene, with the T allele presenting increased risk of AD development, although Cellini et al. (2006) found no evidence of an interaction between the T risk allele and the ApoE epsilon4 allele status and no effect of the PON1 polymorphism on age at onset was detected. However, it was shown that subjects carrying the R allele were more likely to respond to cholinesterase inhibitors (Pola et al. 2005). Additionally, further implications of the involvement of the PON1 gene in AD etiopathology and responses to treatment have been forthcoming (Leduc et al. 2009, 2011). Borowczyk et al. (2012) have shown that PON1 protects mice against Hcy-thiolactone neurotoxicity by hydrolyzing it in the brain and suggest a mechanism by which PON1 can protect against neurodegeneration associated with hyperhomocysteinemia and Alzheimer's disease.

Cholinergic system dysfunctions due the neurodegenerative loss of cholinergic neurons are implicated in several conditions, such as chronic ethanol abuse (Arendt 1994), Creutzfeldt-Jakob disease (Arendt et al. 1984), Down's syndrome (Fodale et al. 2006), temporal lobe epilepsy (Friedman et al. 2007), as well as AD (Nardone et al. 2011; Rangani et al. 2012; Schliebs and Arendt 2006). For example, Marcone et al. (2012) assessed acetylcholinesterase activity with [^{11}C]-MP4A and PET using a maximum a posteriori Bayesian method (MAPB) based on a 2-tissue compartment-3-rate-constant reference region model. They obtained significant and widespread acetylcholinesterase activity reductions in several cortical regions and in the hippocampus in all probable AD subjects and amnesic mild cognitive impairment (MCI) subjects who progressed to AD (converters). Hippocampal acetylcholinesterase activity correlated significantly with long-term verbal and nonverbal memory in both amnestic MCI converters and probable AD. The comparable acetylcholinesterase activity reductions in probable AD and amnesic MCI converters indicate the presence of a widespread impairment of the cholinergic system already in the MCI phase. Neither the genetic, epigenetic, nor underlying molecular mechanisms in AD are understood despite some evidence that cholinergic pathways affect disorder onset and progression (Ellis 2005; Santibez et al. 2007). The cognitive deficits in MCI and moderate and severe AD are manifested concurrently with deficits in cholinergic markers, e.g., acetylcholinesterase activity, acetylcholine, muscarine, nicotine, acetylcholine receptors, nerve growth factors, and acetylcholine transferase activity (Ariga et al. 2010; Bierer et al. 1995; Gsell et al. 2004). Acetylcholinesterase may form a complex with amyloid precursor protein and presenilin 1 genes associated with AD pathology (Silveyra et al. 2008; Xia et al. 1997, 2000). Kim et al. (2011) studied whether or not the pharmacological inhibition of acetylcholinesterase activity in Drosophila larvae may induce synaptic defects. In ACE (angiotensin-converting enzyme gene) hypomorphic mutant larvae, the level of ACE mRNA and acetylcholinesterase activity was reduced compared to controls, thereby causing downregulation of branch length and number of boutons in Type 1 glutamatergic neuromuscular junctions. Controls that were given a sublethal dose of DDVP (2,2-dichlorovinyl dimethyl phosphate, organophosphorus insecticide) phenocopied the synaptic structural defects of the ACE hypomorphic mutant. It appears that the downregulation of acetylcholinesterase activity, whether genetic or pharmacological, alters synaptic architecture. Thus, the authors suggest the possibility that exposure to acetylcholinesterase inhibitors may alter synaptic architecture in AD patient brains with resultant loss of efficacy. Thus, it appears that several "prodromal" conditions, including MCI and/or conditions accelerating immunosenescence, contribute to the gradual deteriorations that may be diagnosed eventually as AD (Candore et al. 2010; Ciaramella et al. 2010).

4 Personalized Medicine in AD

The major neuropathological characteristics of AD include accumulation of Aβ peptide, neurofibrillary tangles composed of paired helical filaments with phosphorylated tau proteins, neuronal and synaptic loss (Selkoe 2000), reduced overall brain volume with specific damage to the entorhinal cortex and hippocampus (Blennow et al. 2006), neuroinflammation, gliosis and free radical generation (Nunomura et al. 2009), neurotransmitter and neurotrophic deficit, metabolic impairment and insulin resistance (Gandy 2012), and vascular burden (Lo et al. 2012). Aβ deposition antecedes cognitive impairment onset and in brain regions with greatest Aβ deposition, synaptic dysfunction may be imaged beginning at preclinical stages. In regions that are not identical with the ones with most Aβ deposition but heavily connected with them, regional atrophy and loss of white matter anisotropy can be detected later in the course of the disease, near the time when mild cognitive impairment is expressed. Brain pathology is marked also by other destructive forces: proteasome, for degradation of damaged/unneeded proteins, and chaperone, proteins assisting in the non-covalent folding or unfolding and the assembly or disassembly of other macromolecular structures but not occurring in these structures during the course of normal biological functions, dysregulations (Ambegaokar and Jackson 2012; Salminen et al. 2011). Lukiw (2013) has reviewed a compilation and update of current pharmacological strategies designed to downregulate Aβ42 peptide generation in an effort to ameliorate the tragedy of AD. Anti-acetylcholinesterase-, chelation-, *N*-methyl-D-ASPARTATE (NMDA) receptor antagonist-, statin-, Aβ immunization-, β-secretase-, γ-secretase-based, and other strategies to modulate βAPP processing have dominated pharmacological approaches directed against AD-type neurodegenerative pathology.

Aβ-linked immunotherapy offers a viable intervention (Lemere and Masliah 2010; Morgan 2006; Schenk 2002), and current immunotherapies are associated with autoimmunity-related adverse effects and mobilization of neurotoxic insoluble Aβ oligomers. Neuroimaging techniques have facilitated Aβ targeting and removal even during preclinical stages (Masdeu et al. 2012). Aβ-linked immunotherapy has provided a potentially promise option for treatment of AD deficits (Fu et al. 2010; Wilcock et al. 2009, 2011), including cognition (Dodart et al. 2002; Li et al. 2012; Maier et al. 2006). Liu et al. (2012) have described available evidence regarding the mechanisms of both endogenous and exogenous Aβ-specific antibodies, with a view to developing optimal immunotherapy based on peripheral Aβ clearance, targeting of the toxic domain of Aβ, and improvement of antibody specificity. As observed in AD patients (Mustafiz et al. 2011), transgenic mice expressing mutated forms of the human amyloid precursor protein gene, hAPP, show prematurely elevated levels of Aβ protein and deposition in the cerebral cortex and hippocampus (Sturchler-Pierrat et al. 1997). Presenilin 1 mutant transgenic mice indicate also elevated levels of Aβ protein (McGowan et al. 1999). William et al. (2012) observed that following monocular visual deprivation during the critical period, mice that express mutant alleles of amyloid precursor protein (APPswe) and presenilin 1 (PS1dE9), as well as mice that express APPswe alone, lacked ocular dominance plasticity in visual cortex. APPswe/PS1dE9 double transgenic mice have a prematurely accelerated accumulation of Aβ deposits (Jankowsky et al. 2004; Trinchese et al. 2004), thereby constituting a realistic model for testing potential therapeutic avenues (Malm et al. 2011). Carrera et al. (2012) targeted reduction of Aβ burden and deceleration of AD-like pathology, including inflammation, using the EB101 vaccine through induction of an anti-inflammatory T-helper-2 immune response. The EB101 vaccine consists of Aβ1-40 delivered in a novel immunogen-adjuvant composed of liposomes-containing sphingosine-1-phosphate. Treatment with EB101 induced a marked reduction of Aβ burden decreases in neurofibrillary tangle-like structure density and attenuation of astrocytosis. Thus, the immunization with EB101 prevented and reversed AD neuropathology by disrupting progression of the disorder. Targeting tau pathology with immunotherapy seems to arrest cognitive decline; Boutajangout et al. (2010) developed a new mouse model with accelerated tangle development, generated by crossing available strains that express all six human tau isoforms and the M146L presenilin mutation. They found that this strategy prevented completely severe cognitive impairment in three different tests.

Genomic medicine and pharmacogenetics introduce principles necessary for the treatment of neurodegenerative disorders (Vance and Tekin 2011). For instance, the therapeutic response to conventional compounds in AD patients is genotype specific with the CYP2D6-PMs, CYP2D6-Ums, and APOE-4/4 carriers presenting the worst responders (Chianella et al. 2011; Dergunov 2011; Klimkowicz-Mrowiec et al. 2011; Savino et al. 2011). *APOE* and *CYP2D6* may cooperate as pleiotropic genes in the metabolism of drugs and hepatic functions (Jann et al. 2002; Mahley and Huang 1999; Mahley and Ji 1999; Miksys et al. 2000). Cacabelos et al. (2012) have proposed several main conclusions to be drawn from the various notions and observations derived from the AD genomics and related pharmacogenomics: (1) the complexity of AD as a neurodegenerative disorder involves the variety of different gene clusters that are implicated and the expectation of further clusters; (2) that genes that have been screeded currently derive from different proteomic and metabolic pathway with the potential for affecting AD pathogenesis; (3) that the APOE gene appears to constitute a major risk factor for both degenerative and vascular dementia; (4) that the therapeutic response to conventional drugs is genotype specific, with CYP2D6-PMs, CYP2D6-Ums, and APOE-4/4 carriers predicting the worst levels of drug response; (5) that APOE and CYP2D6 may cooperate as pleiotropic genes in the metabolism of drugs and hepatic functions (see above); and (6) that the initiation and formulation of pharmacogenetic/pharmacogenomics methodologies and procedures into the personalized medication approach in AD holds promise for optimal interventional advantages.

The role of APOE in therapeutic endeavor is far-reaching: butyrylcholinesterase K (BCHE-K) is associated with increased risk for AD development in APOE ε4 carriers, whereas among APOE ε4 non-carriers, BCHE-K appears protective (Ferris et al. 2009; Patterson et al. 2011). The BCHE-K implication in AD varies with gender: in early AD, male BCHE-K-variant carriers with one or two APOE 4 alleles show prominent medial temporal atrophy, synaptic failure, cognitive decline, and accumulation of aggregated β-amyloid peptide. Elevations of synaptic acetylcholine in damaged but still functional cholinergic synapses improved cognitive symptoms, whereas increasing the ability of glia to support synapses and to clear β-amyloid peptide might be disease modifying (Lane and He 2013). On the other hand, chronic glial overactivation may drive degenerative processes, and in BCHE-K-variant-negative female individuals, generalized glial overactivation may constitute the main driver force graduating from mild cognitive impairment to AD. Females were more likely than males to have accelerated age-related myelin breakdown, more widespread white matter loss, loss of neural network connectivity, whole brain atrophy, and functional decline. Increasing extracellular acetylcholine levels blocks glial activation, reduces myelin loss and damage to neural network connectivity, and is disease modifying (ibid.). Darreh-Shori et al. (2012) coadministered BCHE-K pharmacogenetic observations in AD patients ($n = 179$) with proteomic and enzymatic analysis of plasma, cerebrospinal fluid, or both samples. They observed that the BCHE-K genotype was overrepresented among the patients, and plasma BuChE (pseudocholinesterase/butyrylcholinesterase) activity was gene dose dependently 20–50 % less among K carriers. Cerebrospinal fluid BuChE activity failed to show this robust K gene dose dependency as K homozygotes expressed 30–40 % less activity compared to both non-carriers and heterozygotes. CSF APOE protein expression too was altered by presence of the K allele. In APOE4 absence, cerebrospinal fluid BuChE activity was indistinguishable among the K carriers and non-carriers, while contrastingly in APOE4 presence, K carriers expressed K allele dose dependently a BuChE phenotype with reduced activity compared to non-carriers. Interestingly, patients' cognitive performance patterns were very similar to the BuChE variants. They concluded that the APOE4-dependent outcome of BCHE-K genotype as an AD risk factor arises through a differential phenotypic modulation of BuChE.

Throughout AD the progressive decline in cognitive ability and functional capacity combined with the incremental emergence of behavioral and psychological symptoms, referred to as behavioral and psychological symptoms of dementia (BPSD), afflicts family members and caregivers with a heavy burden, immense distress, and enormous social impact. During the course of disorder, >90 % of those with dementia develop at least one BPSD (Steinberg et al. 2008; Tschanz et al. 2011). A high proportion of AD patients exhibit aggression, about 20–25 % of those in contact with clinical services (Burns et al. 1990), or community dwelling (Lyketsos et al. 2000, 2005), and/or 40–60 % in care facilities (Ballard et al. 2009; Margallo-Lana et al. 2001). There is a small, yet significant, association between a dopamine D3 receptor (DRD3) polymorphism and psychosis-agitation in AD (Holmes et al. 2001). Sato et al. (2009) have observed that a significant association was observed between the presence of the DRD3 glycine allele and paranoid and delusional ideation, regardless of ApoE epsilon4. Nevertheless, it appears that serotonin (5-HT) 5-HT2A, 5-HT2C, and 5-HT transporter polymorphisms are implicated in AD neuropsychiatric symptoms (Craig et al. 2007; Pritchard et al. 2007, 2008, 2009). Angelucci et al. (2009) reported that the T allele of the 5-HT2A T102C polymorphism was associated with poorer responding to the antipsychotic compound, risperidone, treatment than the C allele. Ballard et al. (2011) have summarized results involving the efficacy and safety from short-term randomized controlled trials, case-registered studies, and long-term outcome studies. They concluded that atypical antipsychotic medications provide modest benefits that ought to be balanced against risk for serious adverse effects thereby underlining the requirement for further pharmacogenetic drug-oriented studies of BPSD in AD. In this regard, examining patients with AD who presented psychosis or agitation that had responded to risperidone therapy for 4–8 months, Devanand et al. (2012) found that discontinuation of risperidone was associated with an increased risk of relapse. Regarding cognitive deterioration, the applications of "cognition enhancers in both AD and among healthy individuals open an avenue that is not without complications" (Mohamed and Sahakian 2012).

Frontotemporal dementia (FTD), although distinction from AD, is not a sole entity encompassing a group of disorders unified by the proclivity for focal, asymmetrical cortical atrophy in anterior structures of the frontal and temporal and sparing parietal lobe structures that degenerate in AD (Mioshi et al. 2007; Rabinovici and Miller 2010; Rascovsky et al. 2007). It is a progressive behavioral condition characterized by apathy, disinhibition and loss of insight with loss of empathy, failure to perceive others' needs, and expression of misappropriate, maladaptive, even bizarre, behaviors, compounded by diminished capacity to think abstractly. In comparison with AD, FTD is associated with greater reductions of fractional anisotropy in frontal brain regions, whereas no region AD evidences greater reductions of fractional anisotropy when compared to FTD. The regional patterns of anisotropy reduction in FTD and AD compared to controls suggest a characteristic distribution of white matter degradation, more prominent in FTD (Zhang et al. 2009). Using a novel insight questionnaire, Hornberger et al. (2012) voxel-based morphometry analysis showed that overall loss of insight correlated with ventromedial-frontopolar prefrontal atrophy, with exception of social interaction and emotion insight loss, which additionally correlated with lateral temporal and amygdala atrophy, respectively. Their results imply that patients suffering under neurodegenerative conditions express variable loss of insight, with ventromedial and frontopolar cortex regions appearing to be particularly important for insight expression. Three major types of frontotemporal lobar degeneration (FTLD) have been categorized according to the main protein aggregate: (1) FTLD-tau, with aggregates of microtubule-associated protein tau predominating, (2) FTLD-TDP, with intraneuronal inclusion bodies containing ubiquitinated deposits of TAR DNA-binding protein, and (3) FTLD-fused in sarcoma protein (FUS) where similar ubiquitinated inclusion bodies instead contain the FUS. In a review of studies performed on applications of cholinesterase inhibitors and memantine in frontotemporal dementia, Kerchner et al. (2011) have ascertained that memantine, the glutamate blocker, may hold some promise but that the cholinesterase inhibitors, donepezil, galantamine, and rivastigmine, were ineffective, thereby indicating that rational, molecularly based pharmacotherapies should offer greater potential.

Instrumental activities of daily living (IADLs) involve a large variety of behaviors that determine the essential functionality of elder individuals and remain a central aspect of personalized medicine. An immense proportion of the elderly/elder aged dwell under different circumstances of community and monitored habitat: the instrumental activities of daily living (IADLs) present tasks that are necessary for independent community living (see Voigt-Radloff et al. 2012). Monaci and Morris (2012) showed that correlations between the measures of daily functioning and cognitive abilities and neuropsychiatric symptoms showed that initially neuropsychological test results tended to correlate with IADL rather than ADL measures. Neuropsychiatric symptoms were not correlated whether IADL or ADL. At follow-up, none of the neuropsychological function measures correlated with IADL or ADL, but neuropsychiatric symptoms were correlated with IADL. These tasks often require intact physical and cognitive function, the impairment of which may adversely affect health in older adults. In the current study, Bowling et al. (2012) have examined the associations between IADL impairment and incident heart failure (HF) in community-dwelling older adults. IADL impairment was also associated with all-cause mortality implying that among these community-dwelling older adults free of baseline heart failure, the IADL impairment is a strong and independent predictor of incident HF and mortality. Luttenberger et al. (2012) employed a 6-month intervention comprised of three components: motor stimulation, activities of daily living, and cognitive stimulation (MAKS), with groups of ten patients led by two therapists participating for 2 h, 6 days a week and controls received treatment as usual. At 6 months, their results showed improvement in overall dementia symptoms in the MAKS group and no change in the control group, with the largest effect on the social behavior and instrumental activity of daily living subscales. Saragat et al. (2012) observed that psycho-functional decline of patients with AD is related to body composition variations, with a relative increase of fat mass with respect to the muscle component. IADL interventions hold promise for individuals in community dwelling: tailored and activity-based leisure interventions exert positive impacts on caregiver satisfaction, and some interventions had positive results for clients' well-being and quality of life. Social participation interventions focused on people with dementia still able to engage in verbal social interactions (Letts et al. 2012).

5 Conclusion

Personalized medicine is advanced through optimal diagnosis; florbetaben ([(18)F]BAY 94–9172), a novel β-amyloid PET tracer current in global clinical development, provides diagnostic efficacy, pharmacokinetics, safety, and tolerability of florbetaben. Barthel et al. (2012) have shown that florbetaben to be a safe and efficacious β-amyloid-targeted tracer with favorable brain kinetics. Subjects presenting AD were found to be differentiated easily from healthy controls through both visual and quantitative assessment of the positron emission tomography data. This operator-independent, voxel-based analysis was shown to yield a whole brain β-amyloid load which appeared valuable as a surrogate marker of disorder severity, offering a predictive viability. Vanitallie (2012) has reviewed the evidence demonstrating that a proactive, neuroimaging and pharmacogenetic, personalized medicine approach, to which may be added neuroimmune functioning integrity, allows neurologic physician to reduce risk factors in AD-prone patients both by attending to such modifiable AD risk factors as hypertension, obesity, type 2 diabetes, insulin resistance, hypercholesterolemia, sedentary lifestyle, and current cigarette smoking and facilitating disorder staging and prognosis for therapy. Among younger patients, and other at-risk individuals, shown to express ε4 positivity, advice and coaching ought to cover not only avoidance in the participation in contact sports or other activities that expose them to risk of traumatic brain injury, but also recommended physical exercise health-promoting paradigms. Other aspects of personalized medicine in AD pertain to the manifest benefits of physical exercise for angiogenesis, mobilization of the efficacy of neuroimmune functioning, and quality of life

(Archer 2011; Archer et al. 2011a, b). Exercise intervention presents a unique, noninvasive treatment that incorporates different regimes, whether dynamic or static, endurance or resistance. Physical exercise intervention protects against vascular risk factors including hypertension, diabetes, cellular inflammation, aortic rigidity, and buttresses blood–brain barrier intactness (Archer 2012).

References

Alagiakrishnan K, Gill SS, Fagarasanu A (2012) Genetics and epigenetics of Alzheimer's disease. Postgrad Med J 88:522–529. doi:10.1136/postgradmedj-2011-130363

Albani D, Martinelli Boneschi F, Biella G, Giacalone G, Lupoli S, Clerici F, Benussi L, Ghidoni R, Galimberti D, Squitti R, Mariani S, Confaloni A, Bruno G, Mariani C, Scarpini E, Binetti G, Magnani G, Franceschi M, Forloni G (2012) Replication study to confirm the role of CYP2D6 polymorphism rs1080985 on donepezil efficacy in Alzheimer's disease patients. J Alzheimers Dis 30:745–749

Allen M, Zou F, Chai HS, Younkin CS, Crook J, Pankratz VS, Carrasquillo MM, Rowley CN, Nair AA, Middha S, Maharjan S, Nguyen T, Ma L, Malphrus KG, Palusak R, Lincoln S, Bisceglio G, Georgescu C, Schultz D, Malphrus KG, Palusak R, Lincoln S, Bisceglio G, Georgescu C, Schultz D, Malphrus KG, Palusak R, Lincoln S, Bisceglio G, Georgescu C, Schultz D, Rakhshan F, Kolbert CP, Jen J, Haines JL, Mayeux R, Pericak-Vance MA, Farrer LA, Schellenberg GD, Alzheimer's Disease Genetics Consortium, Petersen RC, Graff-Radford NR, Dickson DW, Younkin SG, Ertekin-Taner N (2012) Novel late-onset Alzheimer disease loci variants associate with brain gene expression. Neurology 79:221–228

Alzheimer's disease International Consortium (1996) Consensus guidelines for the clinical and pathologic diagnosis of dementia with Lewy bodies (DLB): report of the consortium on DLB international workshop. Neurology 47:1113–1124

Alzenstein HJ, Nebes RD, Saxton JA et al (2008) Frequent amyloid disposition without significant cognitive impairment among the elderly. Arch Neurol 65:1509–1517

Ambegaokar SS, Jackson GR (2012) The downward spiral of tau and autolysosomes: a new hypothesis in neurodegeneration. Autophagy 8:7, PMID: 22635052

Angelucci F, Bernardini S, Gravina P, Bellincampi L, Trequattrini A, Di Iulio F, Vanni D, Federici G, Caltagirone C, Bossù P, Spalletta G (2009) Delusion symptoms and response to antipsychotic treatment are associates with the 5-HT2A receptor polymorphism (102T/C) in Alzheimer's disease: a 3-year follow-up longitudinal study. J Alzheimer Dis 17:203–211

Archer T (2011) Physical exercise alleviates debilities of normal aging and Alzheimer's disease. Acta Neurol Scand 123:221–238

Archer T (2012) Influence of physical exercise on traumatic brain injury deficits: scaffolding effects. Neurotox Res. doi:10-1007/s12640-011-9297-0

Archer T, Fredriksson A, Johansson B (2011a) Exercise alleviates Parkinsonism: clinical and laboratory evidence. Acta Neurol Scand 123:73–84

Archer T, Fredriksson A, Schütz E, Kostrzewa RM (2011b) Influence of exercise on neuroimmunological functioning and health: aging and stress. Neurotox Res 20:69–83

Archer T, Kostrzewa RM, Beninger RJ, Palomo T (2011c) Staging neurodegenerative disorders: structural, regional, biomarker, and functional progressions. Neurotox Res 19:211–234

Arendt T (1994) Impairment in memory function and neurodegenerative changes in the cholinergic basal forebrain system induced by chronic intake of ethanol. J Neural Transm Suppl 44:173–187

Arendt T, Bibl V, Arendt A (1984) Neurone loss in the nucleus basalis of Meynert in Creutzfeldt-Jakob disease. Acta Neuropathol 65:85–88

Ariga T, Yanagisawa M, Wakade C, Ando S, Buccafusco JJ, McDonald MP, Yu RK (2010) Ganglioside metabolism in a transgenic mouse model of Alzheimer's disease: expression of Chol-1α antigens in the brain. ASN Neuro 2:e00044

Ballard C, Margallo-Lana M, O'Brien JT, James I, Howard R, Fossey J (2009) Top cited papers in International Psychogeriatrics: 6a. Quality of life for people with dementia living in residential and nursing home care: the impact of performance on activities of daily living, behavioral and psychological symptoms, language skills, and psychotropic drugs. Int Psychogeriatr 21:1026–1030

Ballard C, Creese B, Corbett A, Aarsland D (2011) Atypical antipsychotics for the treatment of behavioral and psychological symptoms in dementia, with a particular focus on longer term outcomes and mortality. Expert Opin Drug Saf 10:35–43. doi:10.1517/147403 38.2010.506711

Barabash A, Marcos A, Ancin I et al (2009) APOE, ACT and CHRNA 7 genes in the conversion from amnestic mild cognitive impairment to Alzheimer's disease. Neurobiol Aging 30:1254–1264

Barthel H, Luthardt J, Becker G, Patt M, Hammerstein E, Hartwig K, Eggers B, Sattler B, Schildan A, Hesse S, Meyer PM, Wolf H, Zimmermann T, Reischl J, Rohde B, Gertz HJ, Reininger C, Sabri O (2012) Individualized quantification of brain β-amyloid burden: results of a proof of mechanism phase 0 florbetaben PET trial in patients with Alzheimer's disease and healthy controls. Eur J Nucl Med Mol Imaging 38:1702–1714

Bartzokis G, Cummings JL, Sultzer D, Henderson VW, Neuchterlein KH, Mintz J (2003) White matter structural integrity in healthy aging adults and patients with Alzheimer's disease: a magnetic resonance imaging study. Arch Neurol 60:393–398

Behrendt G, Baer K, Buffo A, Curtis MA, Faull RL, Rees MI, Götz M, Dimou L (2012) Dynamic changes in myelin aberrations and oligodendrocyte generation in chronic amyloidosis in mice and men. Glia. doi:10.1002/glia.22432

Belbin O, Carrasquillo MM, Crump M et al (2011) Investigation of 15 of the top candidate genes for late-onset Alzheimer's disease. Hum Genet 129:273–282

Bertram L, Tanzi RE (2005) The genetic epidemiology of neurodegenerative disease. J Clin Invest 115: 1449–1457

Bertram L, Tanzi RE (2008) Thirty years of Alzheimer's disease genetics: systematic meta-analysis herald a new era. Nat Rev Neurosci 9:768–778

Bertram L, Tanzi RE (2009) Genome-wide association studies in Alzheimer's disease. Hum Mol Genet 18:R137–R145

Bertram L, Tanzi RE (2012) The genetics of Alzheimer's disease. Prog Mol Biol Transl Sci 107:79–100

Bertram L, McQueen MB, Mullin K, Blacker D, Tanzi RE (2007) Systematic meta-analyses of Alzheimer disease genetic association studies: the AlzGene database. Nat Genet 39:17–23

Bertram L, Lange C, Mullin K, Parkinson M, Hsiao M, Hogan MF, Schjeide BM, Hooli B, Divito J, Ionita I, Jiang H, Laird N, Moscarillo T, Ohlsen KL, Elliott K, Wang X, Hu-Lince D, Ryder M, Murphy A, Wagner SL, Blacker D, Becker KD, Tanzi RE (2008) Genome-wide association analysis reveals putative Alzheimer's disease susceptibility loci in addition to APOE. Am J Hum Genet 83:623–632

Bertram L, Lill CM, Tanzi RE (2010) The genetics of Alzheimer disease: back to the future. Neuron 68:270–281

Bierer LM, Haroutunian V, Gabriel S, Knott PJ, Carlin LS, Purohit DP, Perl DP, Schmeidler J, Kanof P, Davis KL (2005) Neurochemical correlates of dementia severity in Alzheimer's disease: relative importance of the cholinergic deficits. J Neurochem 64(2):749–760

Blennow K, de Leon MJ, Zetterberg H (2006) Alzheimer's disease. Lancet 368:387–403

Bonner-Jackson A, Okonkwo O, Tremont G, Alzheimer's Disease Neuroimaging Initiative (2012) Apolipoprotein E ε2 and functional decline in amnestic mild cognitive impairment and Alzheimer disease. Am J Geriatr Psychiatry 20:584–593

Borowczyk K, Shih DM, Jakubowski H (2012) Metabolism and neurotoxicity of homocysteine thiolactone in mice: evidence for a protective role of paraoxonase 1. J Alzheimers Dis 30:225–231

Boutajangout A, Quartermain D, Sigurdsson EM (2010) Immunotherapy targeting pathological tau prevents cognitive decline in a new tangle mouse model. J Neurosci 30:16559–16566

Bowling CB, Fonarow GC, Patel K, Zhang Y, Feller MA, Sui X, Blair SN, Alagiakrishnan K, Aban IB, Love TE, Allman RM, Ahmed A (2012) Impairment of activities of daily living and incident heart failure in community-dwelling older adults. Eur J Heart Fail 14:581–587

Burns A, Jacoby R, Levy R (1990) Psychiatric phenomena in Alzheimer's disease. IV. Disorders of behavior. Br J Psychiatry 157:86–94

Cacabelos R (2005) Pharmacogenomics and therapeutic prospects in Alzheimer's disease. Expert Opin Pharmacother 6:1967–1987

Cacabelos R (2008) Pharmacogenomics and therapeutic prospects in dementia. Eur Arch Psychiatry Clin Neurosci 258(Suppl 1):28–47

Cacabelos R (2009) Pharmacogenomics and therapeutic strategies for dementia. Expert Rev Mol Diagn 9:567–611

Cacabelos R (2010) Pharmacogenomic protocols in CNS disorders and dementia. Neurodegener Dis 7:167–169

Cacabelos R, Martinez-Bouza R, Carril JC, Fernandez-Novoa L, Lombardi V, Carrera I, Corzo L, McKay A (2012) Genomics and pharmacogenomics of brain disorders. Curr Pharm Biotechnol 13:674–725

Cameron B, Tse W, Lamb BT, Li X, Lamb BT, Landreth GE (2012) Loss of interleukin receptor-associated kinase 4 signaling suppresses amyloid pathology and alters microglial phenotype in a mouse model of Alzheimer's disease. J Neurosci 32:15112–15123. doi:10.1523/JNEUROSCI.1729-12.2012

Candore G, Bulati M, Caruso C, Castiglia L, Colonna-Romano G, Di Bona D, Duro G, Lio D, Matranga D, Pellicano M, Rizzo C, Scapagnini G, Vasto S (2010) Inflammation, cytokines, immune response, apolipoptotein E, cholesterol and, oxidative stress in Alzheimer disease: therapeutic implications. Rejuvenation Res 13:301–313

Carrasquillo MM, Belbin O, Hunter TA, Ma L, Bisceglio GD, Zou F, Crook JE, Pankratz VS, Dickson DW, Graff-Radford NR, Petersen RC, Morgan K, Younkin SG (2010) Replication of CLU, CR1 and PICALM associations with Alzheimer's disease. Arch Neurol 67:961–964

Carrera I, Etcheverria I, Fernandez-Novoa L, Lombardi V, Cacabelos R, Vigo C (2012) Vaccine development to treat Alzheimer's disease neuropathology in APP/PS1 transgenic mice. Int J Alz Dis Article ID 376138, 17. doi: 10.1155/2012/376138

Cellini E, Tedde A, Bagnoli S, Nacmias B, Piacentini S, Bessi V, Bracco L, Sorbi S (2006) Association analysis of the paraoxonase-1 gene with Alzheimer's disease. Neurosci Lett 408:199–202

Chandra V, Pandav R (1998) Gene-environment interaction in Alzheimer's disease: a potential role for cholesterol. Neuroepidemiology 17:225–232

Chianella C, Gragnaniello D, Maisano Delser P, Visentini MF, Sette E, Tola MR, Barbujani G, Fuselli S (2011) BCHE and CYP2D6 genetic variation in Alzheimer's disease patients treated with cholinesterase inhibitors. Eur J Clin Pharmacol 67:1147–1157

Chouliaras L, Rutten BP, Kenis G et al (2010) Epigenetic regulation in the pathophysiology of Alzheimer's disease. Prog Neurobiol 90:498–510

Ciaramella A, Bizzoni F, Salani F, Vanni D, Spaletta G, Sanarico N, Vendetti S, Caltagirone C, Bossu P (2010) Increased pro-inflammatory response by dendritic cells from patients with Alzheimer's disease. J Alzheimers Dis 19:559–572

Craig D, Donnelly C, Hart D, Carson R, Passmore P (2007) Analysis of the 5HT-2A T102C receptor polymorphism and psychotic symptoms in Alzheimer's disease. Am J Med Genet B Neuropsychiatr Genet 144B:126–128

Crentsil V (2004) The pharmacogenomics of Alzheimer's disease. Ageing Res Rev 3:153–169

Cummings JL (2004) Alzheimer's disease. N Engl J Med 351:56–67

Curtis R, Cohen J, Fok-Seang J, Hanley MR, Gregson NA, Reynolds R, Wilkin GP (1988) Development of macroglial cells in rat cerebellum. I. Use of antibodies to follow early in vivo development and migration of oligodendrocytes. J Neurocytol 17:43–54

Darreh-Shori T, Siawesh M, Mousavi M, Andreassen N, Nordberg A (2012)

Darreh-Shori T, Vijayaraghavan S, Aeinehband S, Piehl F, Lindblom RP, Nilsson B, Ekdahl KN, Långström B, Almkvist O, Nordberg A (2013) Functional variability in butyrylcholinesterase activity regulates intrathecal cytokine and astroglial biomarker profiles in patients with Alzheimer's disease. Neurobiol Aging. doi:10.1016/j.neurobiolaging.2013.04.027. pii: S0197-4580(13)00195-4

DeMattos RB, Cirrito JR, Parsadanian M et al (2004) ApoE and clusterin cooperatively suppress A beta levels and deposition: evidence that ApoE regulates extracellular A beta metabolism in vivo. Neuron 41:193–2002

Deramecourt V, Slade JY, Oakley AE, Perry RH, Ince PG, Maurage CA, Kalaria RN (2012) Staging and natural history of cerebrovascular pathology in dementia. Neurology 78:1043–1050

Dergunov AD (2011) Apolipoprotein E genotype as a most significant predictor of lipid response at lipid-lowering therapy: mechanistic and clinical studies. Biomed Pharmacother 65:597–603

Desai MK, Sudol KL, Janelsins MC, Mastrangelo MA, Frazer ME, Bowers WJ (2009) Triple-transgenic Alzheimer's disease mice exhibit region-specific abnormalities in brain myelination patterns prior to appearance of amyloid and tau pathology. Glia 57:54–65

Desai MK, Mastrangelo MA, Ryan DA, Sudol KL, Narrow WC, Bowers WJ (2010) Early oligodendrocyte/myelin pathology in Alzheimer's disease mice constitutes a novel therapeutic target. Am J Pathol 177:1422–1435

Desai MK, Guercio BJ, Narrow WC, Bowers WJ (2011) An Alzheimer's disease-relevant presenilin-1 mutation augments amyloid-beta-induced oligodendrocyte dysfunction. Glia 59:627–640

Devanand DP, Mintzer J, Schultz SK, Andrews HF, Sultzer DL, de la Pena D, Gupta S, Colon S, Schimming C, Pelton GH, Levin B (2012) Relapse risk after discontinuation of risperidone in Alzheimer's disease. N Engl J Med 367:1497–1507

Di Carlo M, Giacomazza D, San Biagio PL (2012) Alzheimer's disease: biological aspects, therapeutic perspectives and diagnostic tools. J Phys Condens Matter 24:244102

Dimou L, Götz M (2012) Shaping barrels: activity moves NG2+ glia. Nat Neurosci 15:1176–1178

Dimou L, Simon C, Kirchhoff F, Takebayashi H, Götz M (2008) Progeny of Olig2-expressing progenitors in the grey and white matter of the adult mouse cerebral cortex. J Neurosci 28:10434–10442

Dodart JC, Bales KR, Gannon KS, Greene SJ, DeMattos RB, Mathis C, DeLong CA, Wu S, Wu X, Holtzman DM, Paul SM (2002) Immunization reverses memory deficits without reducing brain Aβ burden in Alzheimer's disease model. Nat Neurosci 5:452–457

Dubois B, Feldman HH, Jacova C, Cummings JL, Dekosky ST, Barberger-Gateau P, Delacourte A, Frisoni G, Fox NC, Galasko D, Gauthier S, Hampel H, Jicha GA, Meguro K, O'Brien J, Pasquier F, Robert P, Rossor M, Salloway S, Sarazin M, de Souza LC, Stern Y, Visser PJ, Scheltens P (2012) Revising the definition of Alzheimer's disease: a new lexicon. Lancet Neurol 9:1118–1127

Duering M, Zieren N, Herve D, Jouvent E, Reyes S, Peters N, Pachai C, Opherk C, Chabriat H, Dichgans M (2011) Strategic role of frontal white matter tracts in vascular cognitive impairment: a voxel-based lesion-symptom mapping study in CADASIL. Brain 134:2366–2375

Duykaerts C, Delatour B, Potier MC (2009) Classification and basic pathology of Alzheimer's disease. Acta Neuropathol 118:5–36

Elias-Sonnenschein LS, Bertram L, Visser PJ (2012) Relationship between genetic risk factors and markers for Alzheimer's disease pathology. Biomark Med 6:477–495. doi:10.2217/bmm.12.56

Ellis JM (2005) Cholinesterase inhibitors in the treatment of dementia. J Am Osteopath Assoc 105:145–158

Fancy SP, Chan JR, Baranzini SE, Franklin RJ, Rowitch DH (2011) Myelin regeneration: a recapitulation of development? Ann Rev Neurosci 34:21–43

Ferris S, Nordberg A, Soininen H, Darreh-Shori T, Lane R (2009) Progression from mild cognitive impairment to Alzheimer's disease: effects of sex, butyrylcholinesterase genotype, and rivastigmine treatment. Pharmacogenet Genomics 19:635–646

Fodale V, Mafrica F, Caminiti V, Grasso G (2006) The cholinergic system in Down's Syndrome. J Intellect Disabil 10(3):261–274

Friedman A, Behrens CJ, Heineman U (2007) Cholinergic dysfunction in temporal lobe epilepsy. Epilepsia 48(s5):126–130

Fu HJ, Liu B, Frost JL, Lemere CA (2010) Amyloid-β immunotherapy for Alzheimer's disease. CNS Neurol Drug Targ 9:197–206

Gandy S (2012) Vascular burden and Alzheimer disease pathologic progression. Neurology 79:1349–1355

Gatz M, Reynolds CA, Fratiglioni L, Johansson B, Mortimer JA, Berg S, Fiske A, Pedersen NL (2006) Role of genes and environments for explaining Alzheimer disease. Arch Gen Psychiatry 63:168–174

Geha S, Pallud J, Junier MP, Devaux B, Leonard N, Chassoux F, Chneiweiss H, Daumas-Duport C, Varlet P (2010) NG2+/Olig2+ cells are the major cycle-related cell population of the adult human normal brain. Brain Pathol 20:399–411

Graff J, Mansuy IM (2008) Epigenetic codes in cognition and behaviour. Behav Brain Res 192:70–87

Gsell W, Jungkunz G, Riederer P (2004) Functional neurochemistry of Alzheimer's disease. Curr Pharm Des 10(3):265–293

Hampel H, Teipel SJ, Alexander GE, Horwitz B, Teichberg D, Schapiro MB, Rapoport SI (1998) Corpus callosum atrophy is a possible indicator of region- and cell type-specific neuronal degeneration in Alzheimer disease: a magnetic resonance imaging analysis. Arch Neurol 55:193–198

Hatanaka Y, Kamino K, Fukuo K et al (2000) Low density lipoprotein receptor-related protein gene polymorphisms and risk for late-onset Alzheimer's disease in a Japanese population. Clin Genet 58:319–323

Herrmann N, Chau SA, Kircanski I, Lanctôt KL (2011) Current and emerging drug treatment options for Alzheimer's disease: a systematic review. Drugs 71:2031–2065. doi:10.2165/11595870-000000000-00000

Holmes C, Smith H, Ganderton R, Arranz M, Collier D, Powell J, Lovestone S (2001) Psychosis and aggression in Alzheimer's disease: the effect of dopamine receptor gene variation. J Neurol Neurosurg Psychiatry 71:777–779

Honig LS, Tang MX, Albert S, Costa R, Luchsinger J, Manly J, Stern Y, Mayeux R (2003) Stroke and the risk of Alzheimer disease. Arch Neurol 60:1707–1712

Horiuchi M, Maezawa I, Itoh A, Wakayama K, Jin LW, Itoh T, Decarli C (2012) Amyloid beta1-42 oligomer inhibits myelin sheet formation in vitro. Neurobiol Aging 33:499–509

Hornberger M, Yew B, Gilardoni S, Mioshi E, Gleichgerrcht E, Manes F, Hodges JR (2012) Ventromedial-frontopolar prefrontal cortex atrophy correlates with insight loss in frontotemporal dementia and Alzheimer's disease. Hum Brain Mapp. doi:10.1002/hbm.22200

Isaac M, Vamvakas S, Abadie E, Jonsson B, Gispen C, Pani L (2011) Qualification opinion of novel methodologies in the predementia stage of Alzheimer's disease: cerebro-spinal-fluid related biomarkers for drugs affecting amyloid burden–regulatory considerations by European Medicines Agency focusing in improving benefit/risk in regulatory trials. Eur Neuropsychopharmacol 21:781–788

Islam MS, Tatsumi K, Okuda H, Shiosaka S, Wanaka A (2009) Olig2-expressing progenitor cells preferentially differentiate into oligodendrocytes in cuprizone-induced demyelinated lesions. Neurochem Int 54:192–198

Jankowsky JL, Fadale DJ, Anderson J, Xu GM, Gonzales V, Jenkins NA, Copeland NG, Lee MK, Younkin LH, Wagner SL, Younkin SG, Borchelt DR (2004) Mutant presenilins specifically elevate the levels of the 42 residue beta-amyloid peptide in vivo: evidence for augmentation of a 42-specific gamma secretase. Hum Mol Genet 13:159–170

Jann MW, Shirley KL, Small GW (2002) Clinical pharmacokinetics and pharmacodynamics of cholinesterase inhibitors. Clin Pharmacokinet 41:719–739

Jun G, Naj AC, Beecham GW et al (2010) Meta-analysis confirms CR1, CLU and PICALM as Alzheimer disease risk loci and reveals interaction with APOE genotypes. Arch Neurol 67:1473–1484

Kang SH, Fukaya M, Yang JK, Rothstein JD, Bergles DE (2010) NG2+ CNS progenitors remain committed to the oligodendrocyte lineage in postnatal life and following neurodegeneration. Neuron 68:668–681

Kerchner GA, Tartaglia MC, Boxer AL (2011) Abhorring the vacuum: use of Alzheimer's disease medications in frontotemporal dementia. Expert Rev Neurother 11:709–717

Kim W, Lee D, Choi J, Kim A, Han S, Park K, Choi J, Kim J, Choi Y, Lee SH, Koh YH (2011) Pharmacogenetic regulation of acetylcholinesterase activity in Drosophila reveals the regulatory mechanisms of AChE inhibitors in synaptic plasticity. Neurochem Res 36:879–893

Klimkowicz-Mrowiec A, Marona M, Spisak K, Jagiella J, Wolkow P, Szczudlik A, Slowik A (2011) Paraoxonase 1 gene polymorphisms do not influence the response to treatment in Alzheimer's disease. Dement Geriatr Cogn Disord 32:26–31

Kolsch H, Jessen F, Wiltfang J et al (2009) Association of SORL 1 gene variants with Alzheimer's disease. Brain Res 1264:1–6

Kotze MJ, van Rensburg SJ (2012) Pathology supported genetic testing and treatment of cardiovascular disease in middle age for prevention of Alzheimer's disease. Metab Brain Dis 27:255–266

Kounnas MZ, Danks AM, Cheng S, Tyree C, Ackerman E, Zhang X, Ahn K, Nguyen P, Comer P, Mao L, Yu C, Pleynet D, Digregorio PJ, Velicelebi G, Stauderman KA, Comer WT, Mobley WC, Li YM, Sisodia SS, Tanzi RE, Wagner SL (2010) Modulation of γ-secretase reduces β-amyloid deposition in a transgenic mouse model of Alzheimer's disease. Neuron 67:769–780

Landes AM, Sperry SD, Strauss ME, Geldmacher DS (2001) Apathy in Alzheimer's disease. J Am Geriatr Soc 49:1700–1707. doi:10.1046/j.1532-5415.2001.49282.x

Lane RM, He Y (2013) Butyrylcholinesterasegenotype and gender influence Alzheimer's disease phenotype. Alzheimers Dement. PMID: 22402324. 9(2):e1–73. doi: 10.1016/j.jalz.2010.12.005

Lassmann H, Bancher C, Breitschopf H, Wegiel J, Bobinski M, Jellinger K, Wisniewski HM (1995) Cell death in Alzheimer's disease evaluated by DNA fragmentation in situ. Acta Neuropathol 89:35–41

Leduc V, Poirier J (2008) Polymorphisms at the paraoxonase 1 L55M and Q192R loci affect the pathophysiology of Alzheimer's disease: emphasis on the cholinergic system and beta-amyloid levels. Neurodegener Dis 5:225–227

Leduc V, Théroux L, Dea D, Robitaille Y, Poirier J (2009) Involvement of paraoxonase 1 genetic variants in Alzheimer's disease neuropathology. Eur J Neurosci 30:1823–1830

Leduc V, Legault V, Dea D, Poirier J (2011) Normalization of gene expression using SYBR green qPCR: a case for paraoxonase 1 and 2 in Alzheimer's disease brains. J Neurosci Methods 200:14–19

Lemere CA, Masliah E (2010) Can Alzheimer disease be prevented by amyloid-beta immunotherapy? Nat Rev Neurol 6:108–119

Letts L, Edwards M, Berenyi J, Moros K, O'Neill C, O'Toole C, McGrath C (2012) Using occupations to improve quality of life, health and wellness, and client and caregiver satisfaction for people with Alzheimer's disease and related dementias. Am J Occup Ther 65:497–504

Levy K, Lanctôt KL, Farber SB, Li A, Herrmann N (2011) Does pharmacological treatment of neuropsychiatric symptoms in Alzheimer's disease relieve caregiver burden? Drugs Aging 29:167–179. doi:10.2165/11599140-000000000-00000

Li Y, Ma Y, Zong LX, Xing XN, Guo R, Jiang TZ, Sha S, Liu L, Cao YP (2012) Intranasal inoculation with an adenovirus vaccine encoding ten repeats of Aβ3-10 reduces AD-like pathology and cognitive impairment in Tg-APPswe/PSEN1dE9 mice. J Neuroimmunol 249:16–26

Liu YH, Giunta B, Zhou HD, Tan J, Wang YJ (2012) Immunotherapy for Alzheimer disease: the challenge of adverse effects. Nat Rev Neurol 8:465–469

Lo RY, Jagust WJ, For the Alzheimer's Disease Neuroimaging Initiative (2012) Vascular burden and Alzheimer disease pathologic progression. Neurology 79:1349–1355

Lobo A, Launer LJ, Fratiglioni L, Andersen K, Di Carlo A, Breteler MM, Copeland JR, Dartigues JF, Jagger C, Martinez-Lage J, Soininen H, Hofman A (2000) Prevalence of dementia and major subtypes in Europe: a collaborative study of population-based cohorts. Neurological Diseases in the Elderly Research Group. Neurology 54(11 Suppl 5):S4–S9

Lukiw WJ (2013) Amyloid beta (Aβ) peptide modulators and other current treatment strategies for Alzheimer's disease (AD). Expert Opin Emerg Drugs PMID: 22439907. 4:77. doi: 10.3389/fgene.2013.00077

Luttenberger K, Donath C, Uter W, Graessel E (2012) Effects of multimodal nondrug therapy on dementia symptoms and need for care in nursing home residents with degenerative dementia: a randomized-controlled study with 6-month follow-up. J Am Geriatr Soc 60(5):830–840

Lyketsos CG, Steinberg M, Tschanz JT, Norton MC, Steffens DC, Breitner JC (2000) Mental and behavioral disturbances in dementia: findings from the Cache County Study on Memory in Aging. Am J Psychiatry 157:708–714

Lyketsos CG, Toone L, Tschanz J, Rabins PV, Steinberg M, Onyike CU, Corcoran C, Norton M, Zandi P, Breitner JC, Welsh-Bohmer K, Anthony J, Østbye T, Bigler E, Pieper C, Burke J, Plassman B, Green RC, Steffens DC, Klein L, Leslie C, Townsend JJ, Wyse BW, Munger R, Williams M, Cache County Study Group (2005) Population-based study of medical comorbidity in early dementia and "cognitive impairment, no dementia (CIND)": association with functional and cognitive impairment: The Cache County Study. Am J Geriatr Psychiatry 13:656–664

Mahley RW, Huang Y (1999) Apolipoprotein E: from atherosclerosis to Alzheimer's disease and beyond. Curr Opin Lipidol 10:207–217

Mahley RW, Ji ZS (1999) Remnant lipoprotein metabolism: key pathways involving cell-surface heparan sulfate proteoglycans and apolipoprotein E. J Lipid Res 40:1–16

Maier M, Seabrook TJ, Lazo ND, Jiang L, Das P, Janus C, Lemere CA (2006) Short amyloid-beta (Abeta) immunogens reduce cerebral Abeta load and learning deficits in an Alzheimer's disease mouse model in the absence of an Abeta-specific cellular immune response. J Neurosci 26:4717–4728

Malm T, Koistinaho J, Kanninen K (2011) Utilization of APPswe/PS1dE9 transgenic mice in research of Alzheimer's disease: focus on gene therapy and cell-based therapy applications. Int J Alz Dis Article ID 517160, 8

Mangialasche F, Solomon A, Winblad B et al (2010) Alzheimer's disease: clinical trials and drug development. Lancet Neurol 9:702–716

Marcone A, Garibotto V, Moresco RM, Florea I, Panzacchi A, Carpinelli A, Virta JR, Tettamanti M, Borroni B, Padovani A, Bertoldo A, Herholz K, Rinne JO, Cappa SF, Perani D (2012) [(11)C]-MP4A PET cholinergic measurements in amnestic mild cognitive impairment, probable Alzheimer's disease, and dementia with Lewy bodies: a Bayesian method and voxel-based analysis. J Alzheimers Dis 31:387–399

Margallo-Lana M, Swann A, O'Brien J, Fairbairn A, Reichelt K, Potkins D, Mynt P, Ballard C (2001) Prevalence and pharmacological management of behavioral and psychological symptoms amongst dementia sufferers living in care environments. Int J Geriatr Psychiatry 16:39–44

Masdeu JC, Kreisl WC, Berman KF (2012) The neurobiology of Alzheimer disease defined by neuroimaging. Curr Opin Neurol 25(4):410–420. doi:10.1097/WCO.0b013e3283557b36

Mastroeni D, McKee A, Grover J et al (2009) Epigenetic differences in cortical neurons from a pair of monozygotic twins discordant for Alzheimer's disease. PLoS One 4:e6617

Mastroeni D, Grover J, Delvaux E et al (2010) Epigenetic changes in Alzheimer's disease: decrements in DNA methylation. Neurobiol Aging 31:2025–2037

Matsui T, Nemoto M, Maruyama M, Yuzuriha T, Yao H, Tanji H, Ootsuki M, Tomita N, Matsushita S, Higuchi S, Yoshida Y, Seki T, Iwasaki K, Furukawa K, Arai H (2005) Plasma homocysteine and risk of coexisting silent brain infarction in Alzheimer's disease. Neurodegener Dis 2:299–304

McGowan E, Sanders S, Iwatsubo T, Takeuchi A, Saido T, Zehr C, Yu X, Uljon S, Wang R, Mann D (1999) Amyloid phenotype characterization of transgenic mice overexpressing both mutant amyloid precursor protein and mutant presenilin 1 transgenes. Neurobiol Dis 6:231–244

Meda SA, Koran ME, Pryweller JR, Vega JN, Thornton-Wells TA, Alzheimer's Disease Neuroimaging Initiative (2012) Genetic interactions associated with 12-month atrophy in hippocampus and entorhinal cortex in Alzheimer's Disease Neuroimaging Initiative. Neurobiol Aging. pii: S0197-4580(12)00490-3. doi: 10.1016/j.neurobiolaging.2012.09.020

Migliore L, Coppede F (2009) Genetics, environmental factors and the emerging role of epigenetics in neurodegenerative diseases. Mutat Res 667:82–97

Miksys S, Rao Y, Sellers EM, Kwan M, Mendis D, Tyndale RF (2000) Regional and cellular distribution of CYP2D subfamily members in rat brain. Xenobiotica 30:547–564

Mioshi E, Kipps CM, Dawson K, Mitchell J, Graham A, Hodges JR (2007) Activities of daily living in frontotemporal dementia and Alzheimer disease. Neurology 68:2077–2084

Mohamed AD, Sahakian BJ (2012) The ethics of elective psychopharmacology. Int J Neuropsychopharmacol 15:559–571. doi:10.1017/S146114571100037

Mölsä PK, Marttila RJ, Rinne UK (1986) Survival and cause of death in Alzheimer's disease and multi-infarct dementia. Acta Neurol Scand 74:103–107. doi:10.1111/j.1600-0404.1986.tb04634.x

Monaci L, Morris RG (2012) Neuropsychological screening performance and the association with activities of daily living and instrumentalactivities of daily living in dementia: baseline and 18- to 24-month follow-up. Int J Geriatr Psychiatry 27:197–204

Morgan D (2006) Immunotherapy for Alzheimer's disease. J Alzheimer Dis 9(Suppl 3):425–432

Muresan V, Muresan Z (2012) A persistent stress response to impeded axonal transport leads to accumulation of amyloid-β in the endoplasmic reticulum, and is a probable cause of sporadic Alzheimer's disease. Neurodegener Dis 10:60–63

Mustafiz T, Portelius E, Gustavsson MK, Hölttä M, Zetterberg H, Blennow K, Nordberg A, Lithner CU (2011) Characterization of the brain β-amyloid isoform pattern at different ages of Tg2576 mice. Neurodegen Dis 8:352–363

Nardone R, Golaszewski S, Ladurner G, Tezzon F, Trinka E (2011) A review of transcranial magnetic stimulation in the in vivo functional evaluation of central cholinergic circuits in dementia. Dement Geriatr Cogn Disord 32:18–25

Nee LE, Lippa CF (1999) Alzheimer's disease in 22 twin pairs – 13-years follow-up: hormonal, infectious and traumatic factors. Dement Geriatr Cogn Disord 10:148–151

Nunomura A, Hofer T, Moreira PI, Castellani RJ, Smith MA, Perry G (2009) RNA oxidation in Alzheimer disease and related neurodegenerative disorders. Acta Neuropathol 118:151–166

Nygård L (2003) Instrumental activities of daily living: a stepping-stone towards Alzheimer's disease diagnosis in subjects with mild cognitive impairment? Acta Neurol Scand Suppl 107(179):42–46. doi:10.1034/j.1600-0404.107.s179.8.x

Patterson CE, Todd SA, Passmore AP (2011) Effect of apolipoprotein E and butyrylcholinesterase genotypes on cognitive response to cholinesterase inhibitor treatment at different stages of Alzheimer's disease. Pharmacogenomics J 11:444–450

Pedersen NL (2010) Reaching the limits of genome-wide significance in Alzheimer's disease: back to the environment. JAMA 303:1864–1865

Pimentel-Coelho PM, Rivest S (2012) The early contribution of cerebrovascular factors to the pathogenesis of Alzheimer's disease. Euro J Neurosci 35:1917–1937

Pola R, Flex A, Ciaburri M et al (2005) Responsiveness to cholinesterase inhibitors in Alzheimer's disease: a possible role for the 192 Q/R polymorphism of the PON-1 gene. Neurosci Lett 382:338–341

Pringle NP, Richardson WD (1993) A singularity of PDGF alpha-receptor expression in the dorsoventral axis of the neural tube may define the origin of the oligodendrocyte lineage. Development 117:525–533

Pritchard AL, Pritchard CW, Bentham P, Lendon CL (2007) Role of serotonin transporter mechanisms in the behavioral and psychological symptoms in probable Alzheimer disease patients. Dement Geriatr Cogn Disord 24:201–206

Pritchard AL, Harris J, Pritchard CW, Coates J, Haque S, Holder R, Bentham P, Lendon CL (2008) Role of 5HT 2A and 5HT 2C polymorphisms in behavioural and psychological symptoms of Alzheimer's disease. Neurobiol Aging 29:341–347

Pritchard AL, Ratcliffe L, Sorour E, Haque S, Holder R, Bentham P, Lendon CL (2009) Investigation of dopamine receptors in susceptibility to behavioural and psychological symptoms in Alzheimer's disease. Int J Geriatr Psychiatry 24:1020–1025

Querfurth HW, LaFerld FM (2010) Alzheimer's disease. N Engl J Med 362:329–344

Rabinovici GD, Miller BL (2010) Frontotemporal lobar degeneration: epidemiology, pathophysiology, diagnosis and management. CNS Drugs 24:375–398

Rangani RJ, Upadhya MA, Nakhate KT, Kokare DM, Subhedar NK (2012) Nicotine evoked improvement in learning and memory is mediated through NPY Y1 receptors in rat model of Alzheimer's disease. Peptides 33:317–328

Rascovsky KP, Hodges JR, Kipps CM et al (2007) Diagnostic criteria for the behavioral variant of frontotemporal dementia (bvFTD): current limitations and future directions. Alzheimer Dis Assoc Disorders 21:S14–S18

Reiman EM, Webster JA, Myers AJ et al (2007) GAB2 alleles modify Alzheimer's risk in APOE epsilon4 carriers. Neuron 54:713–720

Rogers J, Li R, Mastroenui D et al (2010) Peripheral clearance of amyloid beta peptide by complement

C3-dependent adherence to erythrocytes. Neurobiol Aging 27:1733–1739

Roher AE, Weiss N, Kokjohn TA, Kuo YM, Kalback W, Anthony J, Watson D, Luehrs DC, Sue L, Walker D, Emmerling M, Goux W, Beach T (2002) Increased A beta peptides and reduced cholesterol and myelin proteins characterize white matter degeneration in Alzheimer's disease. Biochemistry 41:11080–11090

Roses AD, Lutz MW, Amrine-Madsen H, Saunders AM, Crenshaw DG, Sundseth SS, Huentelman MJ, Welsh-Bohmer KA, Reiman EM (2010) A TOMM40 variable-length polymorphism predicts the age of late-onset Alzheimer's disease. Pharmacogenom J 10:375–384

Salminen A, Ojala J, Kaarniranta K, Hiltunen M, Soininen H (2011) Hsp90 regulates tau pathology through co-chaperone complexes in Alzheimer's disease. Prog Neurobiol 93:99–110

Santibez M, Bolumar F, Garca AM (2007) Occupational risk factors in Alzheimer's disease: a review assessing the quality of published epidemiological studies. Occup Environ Med 64:723–732

Saragat B, Buffa R, Mereu E, Succa V, Cabras S, Mereu RM, Viale D, Putzu PF, Marini E (2012) Nutritional and psycho-functional status in elderly patients with Alzheimer's disease. J Nutr Health Aging 16:231–236

Sato N, Ueki A, Ueno H, Shinjo H, Morita Y (2009) Dopamine D3 receptor gene polymorphism influences on behavioral and psychological symptoms of dementia (BPSD) in mild dementia of Alzheimer's type. J Alzheimers Dis 17:441–448

Savino M, Seripa D, Gallo AP, Garrubba M, D'Onofrio G, Bizzarro A, Paroni G, Paris F, Mecocci P, Masullo C, Pilotto A, Santini SA (2011) Effectiveness of a high-throughput genetic analysis in the identification of responders/non-responders to CYP2D6-metabolized drugs. Clin Lab 57:887–893

Schenk D (2002) Amyloid-β immunotherapy for Alzheimer's disease: the end of the beginning. Nat Rev Neurosci 3:824–828

Schliebs R, Arendt T (2006) The significance of the cholinergic system in the brain during aging and in Alzheimer's disease. J Neural Transm Suppl 113:1625–1644

Selkoe DJ (2000) Toward a comprehensive theory for Alzheimer's disease. Hypothesis: Alzheimer's disease is caused by cerebral accumulation and cytotoxicity of amyloid β-protein. Ann NJ Acad Sci 924:17–25

Silveyra M-X, Evin G, Montenegro M-F, Vidal CJ, Martinez S, Culvenor JG, Saez-Valero J (2008) Presenilin 1 interacts with acetylcholinesterase and alters its enzymatic activity and glycosylation. Mol Cell Biol 28:2908–2919

Simon C, Götz M, Dimou L (2011) Progenitors in the adult cerebral cortex: cell cycle properties and regulation by physiological stimuli and injury. Glia 59:869–881

Sköldunger A, Johnell K, Winblad B, Wimo A (2013) Mortality and treatment costs have a great impact on the cost-effectiveness of disease modifying drugs in Alzheimer's disease. Curr Alzheimer Res PMID: 23036018. 10(2): 207–216

Song IU, Kim JS, Kim YI, Eah KY, Lee KS (2007) Clinical significance of silent cerebral infarctions in patients with Alzheimer disease. Cogn Behav Neurol 20:93–98

Steinberg M, Shao H, Zandi P, Lyketsos CG, Welsh-Bohmer KA, Norton MC, Breitner JC, Steffens DC, Tschanz JT, Cache County Investigators (2008) Point and 5-year period prevalence of neuropsychiatric symptoms in dementia: the Cache County Study. Int J Geriatr Psychiatry 23(2):170–177

Strittmatter WJ, Saunders AM, Schmechel D, Pericak-Vance M, Enghild J, Salvesen GS, Roses AD (1993) Apolipoprotein E: high-avidity binding to β-amyloid and increased frequency of type 4 allele in late-onset familial Alzheimer disease. Proc Natl Acad Sci USA 90:1977–1981

Strozyk D, Dickson DW, Lipton RB, Katz M, Derby CA, Lee S, Wang C, Verghese J (2010) Contribution of vascular pathology to the clinical expression of dementia. Neurobiol Aging 31:1710–1720

Sturchler-Pierrat C, Abramowski D, Duke M, Wiederhold KH, Mistl C, Rothacher S, Ledermann B, Bürki K, Frey P, Paganetti PA, Waridel C, Calhoun ME, Jucker M, Probst A, Staufenbiel M, Sommer B (1997) Two amyloid precursor protein transgenic mouse models with Alzheimer disease-like pathology. Proc Natl Acad Sci USA 94:13287–13292

Tan AM, Zhang W, Levine JM (2005) NG2: a component of the glial scar that inhibits axon growth. J Anat 207:717–725

Tanzi RE, Bertram L (2005) Twenty years of the Alzheimer's disease amyloid hypothesis: a genetic perspective. Cell 120(4):545–555

Tanzi RE (2012) The genetics of Alzheimer's disease. Cold Spring Harb Perspect Med 2:a006296. doi:10.1101/cshperspect.a006296

Teich AF, Arancio O (2012) Is the amyloid hypothesis of Alzheimer's disease therapeutically relevant? Biochem J 446:165–177

Trinchese F, Liu S, Battaglia F, Walter S, Mathews PM, Arancio O (2004) Progressive age-related development of Alzheimer-like pathology in APP/PS1 mice. Ann Neurol 55:801–814

Troyer AK, Murphy KJ, Anderson ND, Craik FI, Moscovitch M, Maione A, Gao F (2012) Associative recognition in mild cognitive impairment: relationship to hippocampal volume and apolipoprotein E. Neuropsychologia. pii: S0028-3932(12)00456-3. doi: 10.1016/j.neuropsychologia.2012.10.018

Tschanz JT, Corcoran CD, Schwartz S, Treiber K, Green RC, Norton MC, Mielke MM, Piercy K, Steinberg M, Rabins PV, Leoutsakos JM, Welsh-Bohmer KA, Breitner JC, Lyketsos CG (2011) Progression of cognitive, functional, and neuropsychiatric symptom domains in a population cohort with Alzheimerdementia: the Cache County Dementia Progression Study. Am J Geriatr Psychiatry 19:532–542

Urdinquio RG, Sanchez-Mut JV, Esteller M (2009) Epigenetic mechanisms in neurological diseases: genes, syndromes and therapies. Lancet Neurol 8:1056–1072

Vance JM, Tekin D (2011) Genomic medicine and neurology. Continuum (Minneap Minn) 17 (2 Neurogenetics): 249–267

Vanitallie TB (2012) Preclinical sporadic Alzheimer's disease: target for personalized diagnosis and preventive intervention. Metabolism. pii: S0026-0495(12) 00328-9. doi: 10.1016/j.metabol.2012.08.024

Vermeer SE, Prins ND, den Heijer T, Hofman A, Koudstaal PJ, Breteler MM (2003) Silent brain infarcts and the risk of dementia and cognitive decline. N Engl J Med 348:1215–1222

Vermeer SE, Longstreth WT, Koudstaal PJ (2007) Silent brain infarcts: a systematic review. Lancet Neurol 6:611–619

Vlkolinsky R, Cairns N, Fountoulakis M, Lubec G (2001) Decreased brain levels of 2',3'-cyclic nucleotide-3'-phosphodiesterase in Down syndrome and Alzheimer's disease. Neurobiol Aging 22:547–553

Voigt-Radloff S, Leonhart R, Schützwohl M, Jurjanz L, Reuster T, Gerner A, Marschner K, van Nes F, Graff M, Vernooij-Dassen M, Rikkert MO, Holthoff V, Hüll M (2012) Interview for Deterioration in Daily LivingActivities in Dementia: construct and concurrent validity in patients with mild to moderate dementia. Int Psychogeriatr 24:382–390

Waldemar G (2007) Recommendations for the diagnosis and management of Alzheimer's disease and other disorders associated with dementia: EFNS guideline. Eur J Neurol 14:e1–e26. doi:10.1111/j.1468-1331.2006.01605.x

Wang L, Roe CM, Snyder AZ, Brier MR, Thomas JB, Xiong C, Benzinger TL, Morris JC, Ances BM (2012) Alzheimer disease family history impacts resting state functional connectivity. Ann Neurol 72:571–577

Wilcock DM, Gharkholonarehe N, Van Nostrand WE, Davis J, Vitek MP, Colton CA (2009) Amyloid reduction by amyloid-β vaccination also reduces mouse tau pathology and protects from neuron loss in two mouse models of Alzheimer's disease. J Neurosci 29:7957–7965

Wilcock DM, Zhao Q, Morgan D, Gordon MN, Everhart A, Wilson JG, Lee JE, Colton CA (2011) Diverse inflammatory responses in transgenic mouse models of Alzheimer's disease and the effect of immunotherapy on these responses. ASN Neuro 3:249–258. doi:10.1042/AN20110018

Williams AB, Li L, Nguyen B, Brown P, Levis M, Small D (2012) Fluvastatin inhibits FLT3 glycosylation in human and murine cells and prolongs survival of mice with FLT3/ITD leukemia. Blood 120(15):3069–3079. doi:10.1182/blood-2012-01-403493

Xia W, Zhang J, Perez R, Koo EH, Selkoe DJ (1997) Interaction between amyloid precursor protein and presenilins in mammalian cells: implications for the pathogenesis of Alzheimer's disease. Proc Natl Acad Sci USA 94:8208–8213

Xia W, Ray WJ, Ostaszewski BL, Rahmati T, Kimberly WT, Wolfe MS, Zhand J, Goate AM, Selkoe DJ (2000) Presenilin complexes with the C-terminal fragments of amyloid precursor protein at sites of amyloid β-protein generation. Proc Natl Acad Sci USA 97:9299–9304

Zawadzka M, Rivers LE, Fancy SP, Zhao C, Tripathi R, Jamen F, Young K, Goncharevich A, Pohl H, Rizzi M, Rowitch DH, Kessaris N, Suter U, Richardson WD, Franklin RJ (2010) CNS-resident glial progenitor/stem cells produce Schwann cells as well as oligodendrocytes during repair of CNS demyelination. Cell Stem Cell 6:578–590

Zekry D, Duychaerts C, Moulias R, Belmin J, Geoffre C, Herrmann F, Hauw JJ (2002) Degenerative and vascular lesions of the brain have synergistic effects in dementia of the elderly. Acta Neuropathol 103: 481–487

Zhang Y, Schuff N, Du AT, Rosen HJ, Kramer JH, Gorno-Tempini ML, Miller BL, Weiner MW (2009) White matter damage in frontotemporal dementia and Alzheimer's disease measured by diffusion MRI. Brain 132:2579–2592

Zhong Y, Zheng X, Miao Y, Wan L, Yan H, Wang B (2013) Effect of CYP2D6*10 and APOE polymorphisms on the efficacy of donepezil in patients with Alzheime's disease. Am J Med Sci PMID: 22986607. 345(3): 222–226. doi: 10.1097/MAJ.0b013e318255a8f9

Zhu X, Hill RA, Dietrich D, Komitova M, Suzuki R, Nishiyama A (2011) Age-dependent fate and lineage restriction of single NG2 cells. Development 138:745–753

Zipser BD, Johanson CE, Gonzalez L, Berzin TM, Tavares R, Hulette CM, Vitek MP, Hovanesian V, Stopa EG (2007) Microvascular injury and blood–brain barrier leakage in Alzheimer's disease. Neurobiol Aging 28:977–986

Zlokovic BV (2011) Neurovascular pathways to neurodegeneration in Alzheimer's disease and other disorders. Nat Rev Neurosci 12:723–738

Pharmacogenomics and Personalized Medicine in Mood Disorders

16

Trevor Archer, Serafino Ricci, and Max Rapp Ricciardi

Abstract

Pharmacogenomics and the search for personalized medicine focus on the attainment of individualized pharmacotherapies that cover genetic variation and target groups of patients that present neurodevelopmental aspects of symptom profiles and biomarkers underlying the pathophysiology of mood disorders. The identification of genetic biomarkers facilitates choice of treatment, prediction of response, and prognosis of outcome over a wide spectrum of symptoms associated with affective states thereby optimizing clinical practice procedures. Several strategies, under development and refinement, show the propensity for derivation of essential elements in the etiopathogenesis of disorder affecting drug efficacy, drug metabolism, and drug adverse effects, e.g., with regard to SSRIs; these include the following: transporter gene expression and genes encoding receptor systems, hypothalamic-pituitary-adrenal axis factors, neurotrophic factors, and inflammatory factors affecting neuroimmune function. Nevertheless, procedural considerations of pharmacogenetics presume the parallel investment of policies and regulations to withstand eventual attempts at misuse thereby ensuring patient integrity.

1 Introduction

Mood disorders are associated a group of diagnoses in the Diagnostic and Statistical Manual of Mental Disorders (DSM-IV-TR) classification system wherein a disturbance in the person's mood, or emotional or affective status, is considered to present the main underlying feature (Nettle and Bateson 2012). Both unipolar depression and bipolar disorder present clinically severe conditions characterized by recurring episodes of depressive symptom categories, and in the latter periods of mania, with a lifelong lasting prevalence (Ferrari et al. 2012; Merikangas et al. 2007; Pini et al. 2005). It has been suggested that whereas mood refers to the underlying or longitudinal emotional state, affect pertains to the external/visible expression of the individual observed

T. Archer (✉) • M.R. Ricciardi
Department of Psychology, University of Gothenburg, Box 500, SE-40530 Gothenburg, Sweden
e-mail: trevor.archer@psy.gu.se

S. Ricci
Department of Anatomy, Histology, Forensic Medicine and Orthopaedics, Sapienza University, Rome, Italy

D. Barh et al. (eds.), *Omics for Personalized Medicine*,
DOI 10.1007/978-81-322-1184-6_16, © Springer India 2013

by others (Sadock 2012). Unipolar depression and bipolar disorder, of the depressive disorder spectrum of mood disorders, present severe illnesses and are leading causes of disability and suffering among a large population of afflicted individuals. Mood disorders describe less severe forms of depressive disorders, yet although less extreme, dysthymic disorder induces long-lasting moodiness expressed through low, dark moods. Dysthymic disorder may occur by itself or in comorbid relation to other psychiatric, e.g., drug abuse, or mood disorders (Palomo et al. 2007a, b, c; Kaess et al. 2012). Both anxiety and depression are markedly comorbid and present strong relationships in continuous scale formats (Pollack 2005). These disorders are associated with marked negative effects upon work relationships and performance, attendance, daily functioning, and caregivers' situations, with overall increases in costs accumulating from loss of productivity, etc. (Aarø et al. 2011; Laxman et al. 2008; Veronese et al. 2012).

Several major aspects influence the eventual of individual developmental trajectories that possess an essential determinant modulating effect upon outcome of future intervention: (1) the type of agent that interferes with brain development, whether chemical, immune system activating, or conspicuous through absence; (2) the phase of brain development at which the agent exerts disruption (prenatal to gestational, postnatal to infancy, adolescent, or adult lifespan); (3) the age of expression of structural-functional abnormalities with emotional, cognitive, and everyday behavior domains; and (4) the particular pharmacogenomic-pharmacogenetic profiles mediating responses to drug therapies (Archer et al. 2010). Among the mood disorders, adolescent depression is considered relatively common with a prevalence ranges from 5 % (Bhatia and Bhatia 2007) to about 14–15 % in the United States of America (Merikangas et al. 2010; Office of Applied Studies 2005) and may predict adult depression (Aalto-Setala et al. 2002). Female sufferers from the disorder remain almost twice as many as male sufferers with the relative gender proportions evident already during adolescence (Cyranowski et al. 2000). Several factors including physiological changes experienced during puberty, experienced shift in social roles, affiliations and expectations regarding peers and adults, and transient affective status that may provide negative/stressful experiences (Angold and Costello 2006) are more or less environmental factors that may contribute to the prevalence of gender differences (Ge et al. 1994; Natsuaki et al. 2009; Silberg et al. 1999). Edwards et al. (2011a) have shown that the magnitude of environmental influences upon depressive symptoms during adolescence changes as a function of pubertal development, the timing of which differs across gender. Age may contribute a factor modulated extent of mood disorder: Among older females, Gillespie et al. (2012) obtained evidence that both anxiety and depression interact reciprocally with disrupted sleep, whereas among younger women, both anxiety and depression appear to have a causal impact on sleep. Finally, it seems the case that mood disorders genetically and environmentally correlated across adolescence (Edwards et al. 2011b, c).

"Anxiety sensitivity," a lowered threshold for expression of physical and emotional anxiety symptoms, is a risk factor for mood disorders in children and adults (Cox et al. 1999; Taylor et al. 1996), with multiple dimensions (Muris et al. 2001; Silverman et al. 2003). Factor analysis has implicated a hierarchical structure for anxiety sensitivity from which all its dimensions are derived from a higher-order, general anxiety-sensitivity factor. A hierarchical model consists of three dimensions: Physical, social, and mental anxiety-related incapacitation concerns, from a large factor analysis study of adolescents (Walsh et al. 2004), together with other observations have confirmed the anxiety-sensitivity model (Wright et al. 2010). Zinbarg et al. (2001) have provided results demonstrating that anxiety-sensitivity-physical concerns is the only one of the three anxiety-sensitivity group factors that contributes to relations with fear responses to these two challenges, whereas anxiety-sensitivity-mental incapacitation concerns produced a stronger positive linear association with depressed mood than did anxiety-sensitivity-physical concerns. In a self-report study of three test time points from adolescence to young adulthood with 2,651 participants from the

G1219 twin study, Brown et al. (2012) obtained a three-factor model that depicted the physical, social, and mental anxiety-related incapacitation concerns with higher levels of interpretability and parsimony. They found that multivariate genetic analyses supported a hierarchical structure with general genetic and non-shared environmental influences.

1.1 Symptom-Continuity Profile

The symptoms/expressions of anxiety and depression by individuals appear to be relatively stable over time (Foley et al. 2001; Lopez-Castroman et al. 2012; Lovibond 1998). The notion of genetic set point postulates that genetic factors determine the stable set points to which individuals will return eventually following the subsidence of an episode of anxiety and depression (mood disorder) that was induced by environmental factors (Kandler et al. 2010). Twin studies have shown that genetic factors influence markedly mood disorder levels and their temporal stability (Boomsma et al. 2005; Kendler et al. 2008; Gillespie et al. 2004). Using hierarchical linear modeling, Kendler et al. (2011) investigated individual differences and individual changes in symptom levels over time thereafter modeling the manner in which these components were altered over time. Within pairs, the predicted symptom levels for the twins diverged increasingly from childhood until late adulthood wherein divergence ceased. They observed that by middle adulthood, environmental influences contributed substantially to stable and predictable interindividual differences in mood disorder, anxiety, and depression levels (see also Tully et al. 2010). Zavos et al. (2012) studied both the direction of influences on longitudinal associations between anxiety and depression and anxiety sensitivity and the continuity of genetic and environmental influence on mood disorder for adolescent populations consisting of twin and sibling pairs. They found that all disorder traits were stable over time accounting for the largest proportion of variance at 2nd test time point but with evidence of reciprocal associations over time. Genetic effects were fairly stable across time with new genetic influences occurring at 2nd test time point; environmental effects appeared more time dependent.

1.2 Intervention Requirement

Untreated mood disorders in children with implications for later adolescence to young adult and adult development, career and employment, financial status, and health service requirements may exert unpleasant consequences (Knapp 2008). In all mammalian species, the peripartum period is characterized by multiple adaptations at neuroendocrine, molecular, brain circuitry, and behavioral levels that prepare the female for the challenges of motherhood. These changes have been well characterized ensuring the survival and nurturance of the offspring (). Hillerer et al. (2012) have described common peripartum adaptations and the postpartum impact of mood disorders, as well as the consequences of peripartum stress exposure on these maternal adaptations. Maternal psychopathology expressed by mood disorders, anxiety, and depression with impact on both mother and child presents a major public health concern (Goodman and Gotlib 1999; Talge et al. 2007). Prenatal depression was associated with adverse perinatal outcomes, including premature delivery and slower fetal growth rates (Diego et al. 2010). Children of depressed mothers not only have higher risk of depression but also may experience both elevated and continuing exposure to stressful experiences (Dougherty et al. 2011; Tarantino et al. 2011). Infants born to mothers with high versus low prenatal dopamine levels also had higher dopamine and serotonin levels as well as lower cortisol levels (Field et al. 2008). Thus, prenatal maternal dopamine levels were shown to be negatively related to prenatal depression scores and positively related to neonatal dopamine and behavioral regulation, although these effects are confounded by elevated serotonin levels. Hammen et al. (2012) have observed that youth of depressed mothers remain at risk for not only depression but also for continuing experiences of acute and chronic stress from childhood to age 20. The associations among depression and stress are bidirectional and portend continuing experiences of depression and further stress. Prenatally and postnatally, these mood disorders are linked with poor neurodevelopmental functions across emotional, daily behavior, and cognitive domains (Alaräisänen et al. 2012; Barker et al. 2012; Coiro et al. 2012). For example, prepartum depression is highly prevalent and is associated with negative outcomes in offspring (Davalos et al. 2012).

Prenatal depression presents a significant predictor of postpartum depression with detrimental consequences for fetal development. Much evidence links stress during pregnancy with offspring predisposition of ill-health, mood disorders, and maladjustment, following birth, during infancy, childhood, adolescence, and adulthood (Buss et al. 2010; Glover et al. 2010; Gluckman et al. 2005; Knorr et al. 2010; O'Connor et al. 2003; Pariante and Miller 2001). Barker et al. (2011) showed that maternal depression during both prenatal and postnatal periods exerted a wider impact on different types of child maladjustment than maternal anxiety which was associated with internalizing problems in the child. Prenatal risks were associated prospectively with child externalizing difficulties and verbal IQ, over and above prenatal-postnatal maternal anxiety, and depression influences. It appears that prevention of adverse outcomes in offspring may be efficaciously addressed through intervention treatments centered upon maternal anxiety-depression during the prenatal and postnatal phases. Parents presenting mood disorders and/or antisocial behaviors are impaired significantly on various components of parenting (Hoeve et al. 2009), displaying inadequate supervision and monitoring (Loeber and Stouthamer-Loeber 1986), parental negativity (Fienberg et al. 2007), and abuse and neglect (Eaves et al. 2010). These dysfunctional ingredients of parenting presage a wide range of disorders, which are more or less linked to mood disturbances, in the offspring (Cox et al. 2012a, b; Diler et al. 2011; Lewis et al. 2012; Nauta et al. 2012; Vandeleur et al. 2012). Duffy et al. (2012) have described the association of bipolar disorder with a number of genetic and epigenetic abnormalities linked to neurotransmitter, hormonal and immunologically mediated neurobiological pathways that implicate HPA axis, and immune system abnormalities in offspring. Long-standing maladaptive alterations may be mediated through changes in intracellular signaling pathways, oxidative stress, cellular energy metabolism, and apoptosis suggesting a substantial burden of illness. The "children-of-twins" method has been used to disaggregate family environment effects from genetic parent-offspring liabilities (Keller et al. 2010), and its application has demonstrated offspring psychopathology following parental anxiety (Narusyte et al. 2008) and depression (Silberg et al. 2010; Singh et al. 2011). Silberg et al. (2012) obtained distinctive patterns of transmission between parental antisocial behavior and offspring conduct, depression, and hyperactivity as juveniles. Genetic and family environmental factors accounted for parental antisocial behavior and offspring conduct disturbance. Family environmental factors explained the parental antisocial behavior – offspring depression association and genetic factors alone explained the impact of parental antisocial behavior on hyperactivity thereby underscoring genetic and epigenetic contributions. Although Lahey et al. (2011) found some common environmental influences, consistent with a "generalist genes and specialist environments" model, there existed limited sharing of environmental influences.

1.3 Gender Consideration

Women presenting mood disorders had higher cortisol levels, and their newborns had lower gestational age and birth weight (Field et al. 2010a). The cortisol effects in these studies were unfortunately confounded by low serotonin and low dopamine levels which in themselves could contribute to nonoptimal pregnancy outcomes. The negative effects of depression and cortisol were also potentially confounded by comorbid anxiety; by demographic factors including younger age, less education, and lower SES of the mothers; and by the absence of a partner or a partner who was unhappy about the pregnancy or a partner who was depressed. Substance use (especially caffeine use) was still another risk factor. All of these problems including prenatal depression, elevated cortisol, prematurity and low birth weight, and even postpartum depression have been reduced by prenatal massage therapy provided by the women's partners. Field et al. (2010b) compared the effects of comorbid depression and anxiety to the effects of depression alone and anxiety alone on pregnancy mood states and biochemistry and on neonatal outcomes in a large multiethnic sample. At the prenatal period, the comorbid and depressed groups had higher scores than the other groups on the depression measure. Nevertheless, the comorbid group had higher anxiety, anger, and daily hassle scores than the other groups, and they had lower dopamine levels. As compared to the nondepressed group, they also reported more sleep disturbances and relationship problems. The comorbid group also experienced a greater incidence of prematurity than the depressed, the high-anxiety, and the nondepressed groups. Although the comorbid and anxiety groups had lower birth weight than the nondepressed and depressed groups, the comorbid group did not differ from the depressed and anxiety groups on birth length. The neonates of the comorbid and depressed groups had higher cortisol and norepinephrine and lower dopamine and serotonin levels than the neonates of the anxiety and nondepressed groups as well as greater relative right frontal EEG. These data suggest that for some measures, comorbidity of depression and anxiety is the worst condition (e.g., incidence of prematurity), while for others, comorbidity is no more impactful than depression alone. Gender considerations affecting disorder and eventual intervention ought to be "captured" already during infancy: Luby et al. (2012) assessed (1) preschoolers (mean age = 4.6 years) with a current depressive syndrome and two nondepressed comparison groups and (2) longitudinal data pertaining to emotional reactivity at preschool age (mean age = 4.5 years) and later mood disturbance symptoms during the transition to primary school that allowed a retrospective determination of risk. They observed that depressed and at-risk boys displayed more anger than sadness in contrast to girls in the same groups and in contrast to no-disorder/low-risk controls.

As with depressive illness, mood/dysthymic disorders are considered to be more common among female individuals than among male individuals (Dell'osso et al. 2012; Vishnuvardhan and Saddichha 2012). Moderate to high levels of heritability in mood and affective disorders have been obtained from family, twin, and adoption studies (Boardman et al. 2011; Lubke et al. 2012; Smoller and Finn 2003; Williams et al. 2011), while Hasler and Northoff (2011) have described endophenotypes that modulate specificity, mood state independence, heritability, familiarity, clinical relevance, and possible associations with candidate genes in adults and children (Franić et al. 2010; Lazary 2010). Age at onset may be a considerable source of heterogeneity in mood disorders: Early onset of major depressive disorder (MDD), despite negative findings (Sullivan et al. 2000), is linked to increased disorder risk in first-degree relatives (Kupfer et al. 1989; Nierenberg et al. 2007; Tozzi et al. 2008). Age at onset may distinguish subtypes of depression with specific heritable components (Bergemann and Boles 2010; Hays 1976). Power et al. (2012) showed that analysis of age at onset, as a quantitative trait, indicated that, across all SNPs, common genetic variants explained a large proportion of the variance observed. The mood stabilizers, designed to treat the intense and sustained mood shifts of these disorders, offer effective management of bipolar disorder, mania, and depression, as well as maintenance (de Bartolomeis and Perugi 2012; Walpoth-Niederwanger et al. 2012). Squassina et al. (2010) have outlined the current state of pharmacogenomic research on long-term treatment with mood stabilizers indicating the necessity of pharmacogenetic implementations in clinical settings. In order to study molecular mechanisms underlying antidepressant action and identify new biomarkers to determine therapeutic response to these compounds, Malki et al. (2012) examined two efficacious ADs with divergent pharmacological profiles, noradrenergic and serotonergic, respectively. They were treated with either nortriptyline, pro-noradrenergic action, or citalopram, pro-serotonergic SSRI, or saline. Using quantitative proteomic analyses that identify specific drug response markers was undertaken on hippocampal tissue from a study design that used two inbred mouse strains, two depressogenic protocols and a control condition (maternal separation, chronic mild stress, control), two antidepressant drugs, and two dosing protocols which were assayed with 2DE and isobaric tandem mass tagging applied to the selected experimental groups. They found that significant strain- and stress-related differences occurred across both two-dimensional gel electrophoresis and tandem mass tagging data sets whereby they identified three gene products linked to serotonergic (PXBD5, YHWAB, SLC25A4) AD action and one linked to noradrenergic (PXBD6) AD action (see also Uher et al. 2009).

2 Pharmacogenetics of Treatment

Genetically mediated variability mapped onto trajectories of risk for psychopathology, especially that precipitated by environmental adversity, contributes to the understanding of biological pathways that mediate individual differences in behavior and in risk for psychopathology (Hariri 2010). Pharmacogenetics pursues the description and prediction of the extent to which the gene profiles of individuals affect their responses to therapeutic interventions, particularly drugs, designed to alleviate mood disorders (Mutsatsa and Currid 2012; Perlis 2007; Serretti et al. 2008), through applied notions of genes, proteins, and single-nucleotide polymorphisms (SNPs). Scharinger et al. (2011) have described comprehensive evidence on the influence of serotonergic genes (*SLC6A4*, *HTR1A*, *MAOA*, *TPH2*), and BDNF on the following neural intermediate phenotypes is displayed: amygdala reactivity, coupling of amygdala-anterior cingulate cortex (ACC) activity, ACC volume, hippocampal volume, and serotonin receptor 1A (5-HT1A) binding potential. Several factors contribute to the difficulties involving drug treatment efficacy, e.g., delay of onset of therapeutic effect and tolerance and compliance issues (Masand et al. 2002; Mitchell 2006). Pharmacogenetic studies of psychometric outcome measures of drug response are hampered by small effect sizes; these may be handled through intermediate endophenotypes of drug response, as by imaging studies suggest (Rabl et al. 2010), thereby strengthening the relationship between genes and drug response as well as providing new insights into the neurobiology of depression and individual drug responses. The pharmacogenetics of treatments for mood disorders may focus upon several aspects of drug action, including pharmacokinetics, neurotransmitter metabolism and metabolic enzymes, and transporter mechanisms. For example, Porcelli et al. (2011) have focused upon genes linked to pharmacodynamics and in the stratification of these identifications have indicated several inconsistencies across observations. Scharinger et al. (2010) have reviewed imaging genetic studies in mood disorders that apply more complex genetic disease models, such as epistasis and gene-environment interactions, and their impact on brain systems regulating emotion processing and interventional outcomes.

Therapygenetics occupies certain aspects of psychiatric genetic study that observes the relationship between specific genetic variants and differences in the level of success of biopsychological therapy (Eley et al. 2012) and links to the "differential susceptibility hypothesis," through which individuals vary in the extent that environmental experiences affect them (Belsky et al. 2009; Belsky and Pluess 2009; Ellis et al. 2011; van Ijzendoorn et al. 2012). Eley et al. (2012) collected DNA from 584 presenting anxiety disorder and undergoing manual-based cognitive-behavior therapy, all with four white European grandchildren. They tested whether or not treatment response was associated with the 5HTTLPR that was shown previously to moderate environmental influences upon depression (Ng et al. 2006). They observed that children with the short-short allele, genotype, were significantly more likely to respond to cognitive-behavior therapy than those children carrying a long allele. On the other hand, in a population of outpatients ($n = 330$) with PTSD who underwent 5HTTLPR genotyping, the long-long genotype was associated with greater responsiveness of PTSD to sertraline (100 mg/day) and with lower dropout due to adverse events, compared with the short-short and short-long 5HTTLPR genotypes (Mushtaq et al. 2012). Also, in another study with adult bulimia-mood disorder comorbidity patients (Richardson et al. 2008; Steiger et al. 2008), it was shown that the 5HTTPLR short allele predicted a poorer treatment response whether or not cognitive-behavior therapy or medication or a combined therapy was administered. Factors affecting therapeutic drug response rates include clinical heterogeneity, diagnostic uncertainty and variable symptom expression, environmental and social factors, comorbidity, gene-environment interactions, and genetic factors (Lohoff 2011).

Anxiety mood disorders, highly prevalent and persisting into adulthood (Merikangas et al. 2010; Weems 2008), often have childhood onset (Kessler et al. 2005), accompanied by several deficits/problems (Asendorpf et al. 2008; Erath et al. 2007; Owens et al. 2008) with risk for various states of future ill-health (Ehrensaft and Cohen 2012; Kim-Cohen et al. 2003). High rates of remission and treatment response are predicted by symptom severity (Hudson 2005), parental psychopathology (Rapee et al. 2009), and comorbid mood disorder (Ollendick et al. 2008). Meta-analyses from association data of mood disorders have indicated the role of particular genes in genetic risk, and the integration of association data from meta-analyses with differential expression data in brains of mood disorder patients could heighten the level of support for specific genes (Detera-Wadleigh and Akula 2011). Several lines of evidence imply mechanisms underlying the reported increase in anxiety-like behavior elicited by perturbation in brain-derived neurotrophic factor (BDNF) signaling (Daftary et al. 2012). The secretion of BDNF is activity dependent with reduced secretion linked to the effects of stress and mood disorders (cf. Kapczinski et al. 2008; Post 2010); AD treatments generally elevate BDNF secretion (Chen et al. 2001; Martinowich and Lu 2008). In the functional rs6265 (Val66Met) polymorphism, the Met allele is associated with decreased activity-dependent BDNF secretion (Egan et al. 2003), structural brain abnormalities in limbic regions (Gallinat et al. 2010), impaired hippocampal activity (Hariri et al. 2003), impaired associative fear learning (Hajcak et al. 2009), and defective BDNF secretion and increased anxiety-related behavior in "knock-in" mice (Chen et al. 2006). The Met allele decreases BDNF transport, contrary to the superior functioning of the BDNF polymorphism (Val(66)Met) Val allele, and has been associated with worsened performance on several cognitive domains in euthymic BD subjects and controls. Manic patients with the Val allele (Met-) had higher Barrow-Welsh Art Scale for creativity and neuropsychological test scores than Met+ carriers (Soeiro-de-Souza et al. 2012).

An association between Val66 allele and higher neuroticism has been found (Lang et al. 2005), whereas the Met allele was either linked to lower neuroticism (Frustaci et al. 2008) or had no association (Terracciano et al. 2010a, 2011). Nevertheless, significant associations have been reported between Met allele carriers and increased introversion (Terracciano et al. 2010b), increased harm avoidance (Montag et al. 2010), and significant gene-gene and gene-environment interactions pertaining to anxiety- and depression-linked endophenotypes (Gatt et al. 2009, 2010; Pezawas et al. 2008). Lester et al. (2012) report findings from a sample of 374 children with anxiety disorder of European ancestry (from clinics in Reading, UK, and Sydney, Australia), undergoing cognitive-behavior therapy, from whom DNA was collected from buccal cells with cheek swabs. Their treatment response was assessed at posttreatment and follow-up time points. No significant associations were observed between BDNF rs6265 and the response to psychotherapy; however, children with one or two copies of the T allele of NGF rs6330 showed a greater likelihood of relinquishing their primary anxiety diagnosis at follow-up. The recently discovered human BDNF Val66Met

(BDNF(Met)) polymorphism may play a role in stress vulnerability through pharmacogenetic influences affecting molecular and structural mechanisms underlying the interaction. Yu et al. (2012) observed that heterozygous BDNF(+/Met) mice displayed hypothalamic-pituitary-adrenal axis hyperreactivity, increased depressive-like and anxiety-like behaviors, and impaired working memory compared with WT mice after 7 days restraint stress. Also, BDNF(+/Met) mice exhibited more prominent changes in BDNF levels and apical dendritic spine density in the prefrontal cortex and amygdala after stress related to impaired working memory and elevated anxiety-like behaviors. These depressive-like behaviors in BDNF(+/Met) mice were reversed selectively with acute administration of desipramine, but not fluoxetine. These selective behavioral, molecular, and structural deficits appear to be part of the stress and human genetic BDNF(Met) polymorphism interaction. From an aspect of "personalized medicine" (see below) that desipramine but not fluoxetine exerted AD effects on BDNF(+/Met) mice implies that specific classes of ADs may be a more effective treatment option for depressive symptoms in humans with this genetic variant BDNF.

Pharmacometabolomics deals with the direct measurement of metabolites in an individual's bodily fluids, in the present case biomarkers within nervous tissues involving structure and function, in order to predict or assay the metabolism of pharmaceutical compounds by deriving information in combination with the other omics, genomics, transcriptomics, and proteomics (Kaddurah-Daouk et al. 2008). Using targeted metabolomics, Kaddurah-Daouk et al. (2012) compared biochemical profiles in the cerebrospinal fluid of unmedicated depressed or remitted MDD patients with those in healthy controls. They obtained differences in the rMDD group tryptophan and tyrosine metabolism compared to the other groups. The rMDD group presented higher methionine levels and larger methionine-to-glutathione ratios than the other groups, implicating methylation and oxidative stress pathways. These rMDD group reductions of metabolites of the tryptophan and tyrosine pathways may be associated with these individuals' vulnerability for developing mood disorder symptoms under tryptophan/catecholamine depletion. Abo et al. (2012) tested the hypothesis that pharmacometabolomic data could be efficiently merged with pharmacogenomic data by single-nucleotide polymorphism (SNP) imputation of metabolomic-derived pathway data on a "scaffolding" of genome-wide association SNP data to broaden and accelerate "pharmacometabolomic-informed pharmacogenomic" studies by eliminating the need for initial genotyping and by making broader SNP association testing possible. They had genotyped previously 131 tag SNPs for six genes encoding enzymes in the glycine synthesis and degradation pathway using DNA from 529 depressed patients treated with citalopram to pursue a glycine metabolomic "signal" associated with selective serotonin reuptake inhibitor (SSRI) responses. The authors identified a significant SNP in the glycine dehydrogenase gene. Subsequently, GWAS SNP data were generated for the same patients. In this study, we compared SNP imputation within 200 kb of these same six genes with the results of the previous tag SNP strategy as a rapid strategy for merging pharmacometabolomic and pharmacogenomic data. Imputed genotype data provided greater coverage and higher resolution than did tag SNP genotyping, with a higher average genotype concordance between genotyped and imputed SNP data for "1,000 genomes" (96.4 %) than HapMap 2 (93.2 %) imputation. Many low P-value SNPs with novel locations within genes were observed for imputed compared with tag SNPs, thus altering the focus for subsequent functional genomic studies. Their results implied that the administration of genome-wide association SNP data to impute SNPs for genes in pathways identified by other "omics" approaches makes it possible to rapidly and cost-efficiently identify SNP markers to "broaden" and accelerate pharmacogenomic studies. Table 16.1 provides a list of genes that are associated with pathophysiology or efficacy, metabolism, or availability of pharmacotherapeutic agents in mood disorders.

Table 16.1 The pharmacogenetics of certain genes associated with the pathophysiology or efficacy, metabolism, or availability of pharmacotherapeutic agents in mood disorders

Site of action	Gene	Promoter region	Clinical indication
Serotonin transporter	*SLC6A4*	5-HTTPLR	SSRI efficacy
Glycoprotein	*ABCB1*	Upstream/downstream promoters	ABC transporter[a]
CRH receptor of HPA axis[b]	*CRHR1*	Luciferase reporter plasmid	Suicidality
5-HT_{2A} receptor	*HTR2A*	−1438G/A (rs6311)	Overdensity
Glucocorticoid receptor	*NR3C1*	Luciferase reporter plasmid	Stress adaptation
BDNF neurotrophin	*BDNF*	Multiple promoters	AD-enhanced plasticity
AD drug action	*MAGI2*, *DTWD1*, *WDFY4*, and *CHL1*	Multiple promoters	Symptom exacerbation

[a]Transportation of a wide variety of substrates across extra- and intracellular membranes
[b]Corticotrophin-releasing hormone of the hypothalamic-pituitary-adrenal axis

3 ADs and Personalized Medicine

A multitude of factors influence the emergence and progression of the mood disorders (Zahn-Waxler et al. 2000). "Personalized medicine" involves the adaptation of therapies based on an individual's genetic and molecular profile. It may be achieved through the adaptation of therapy to each individual patient through the application of molecular tools based upon gene technologies (Maier and Zobel 2008). The notion of personalized medicine, a model that outlines the customization of healthcare, with decisions and practices that suited to each individual patient through application of genetic, biomarker, gene-environment interactive, or other information (Shastry 2005, 2006), refers in which drugs and drug combinations are optimized for each individual's unique genetic makeup (Archer 2012). Pharmacogenomic testing of patients differs from that of healthy controls with regard to clinical severity and treatment resistance, intolerance, presentation of higher self-reported anxiety, depression and mood fluctuations, greater likelihood of familial background of mood/anxiety disorder, and greater indications of prior ADs, mood stabilizers, and antipsychotic medication (Rundell et al. 2011). The identification of the particular relationships between genotype and drug response, including both the therapeutic effect and side-effect profile, will influence the medical practice of disorder intervention to a degree as yet impossible to assess. Despite the huge application of antidepressant (AD) compounds to afflicted individuals, only 60 % of those treated with these drugs show sufficient response to medication, and adverse effects are common while numerous pharmacogenetic studies point to the involvement of genetic factors. Gvozdic et al. (2012) have reviewed the available literature on pharmacogenetics of antidepressant response and side effects until summer of 2011, using the PubMed database. They observed that several variants in candidate genes involved in the pharmacokinetics or pharmacodynamics of antidepressants, including association findings in the serotonin transporter gene (*5-HTT*), serotonin receptor genes, a gene coding an efflux pump in the blood–brain barrier (*ABCB1*), and genes involved in the HPA axis. They concluded that future studies ought to investigate comprehensively the functional biomarker analyses and underlying pathophysiology in considerations of gene-gene and gene-environment interactions.

The identification of genetic biomarkers that predict antidepressant treatment response is likely to improve current clinical practice particularly since the therapeutic response to ADs is marked by interindividual variability, according to which a large proportion of MDD patients do not response adequately the 1st AD drug prescribed (Trivedi et al. 2006). Studies on antidepressant treatment response have focused on both aspects of pharmacogenetic research, identifying new candidate genes that may predict better treatment response for patients (Steimer et al. 2001) and taking into account the situation that AD drug response aggregates in families (Franchini et al. 1998). Narasimhan and Lohoff (2012) have reviewed recent findings on the pharmacogenetics of antidepressant drugs and future clinical applications. The individualization and optimization of treatment decisions for unipolar depression couched in terms of "the right drug/treatment for the right patient" remains restricted since current circumstances sufficiently powerful clinical or biological predictors which could help to achieve this goal are missing (Möller et al. 2012). The notion of personalized medicine has illustrated graphically by evidence emerging from studies of the fate of serotonin released into the synaptic cleft: Dysfunctions of serotonergic neurotransmission are involved in the physiopathogenesis of mood disorders. Serotonin concentration in the synaptic cleft is essentially regulated by the serotonin transporter (5-HTT), and in this regard, a length polymorphism repeat in the 5-HTT promoter region, termed 5-HTTLPR, has been linked to the disorder. Haenisch et al. (2012) detected a significant association between the TA haplotype (tagging the S allele of the 5-HTTLPR) and mood disorder, and consistent with previous findings of an association between the 5-HTTLPR S allele and mood disorder (see also Olgiati et al. 2012).

The burgeoning encroachments of pharmacogenomics-pharmacogenetics complemented with plasma concentration, therapeutic drug monitoring, and diagnostic staging analyses have contributed markedly to the search for personalized treatment of mood disorders. For example, it has been reported that the low-expressing short allele at the locus 5-HTTLPR (serotonin transporter gene promoter polymorphism) of the gene encoding the serotonin transporter is associated with a poor response to selective serotonin reuptake inhibitors (SSRIs), whereas the high-expressing long allele predicted an effective response to the ADs (Keers and Aitchison 2011), although replication has been questioned (Taylor et al. 2010; Zou et al. 2011). Porcelli et al. (2012) systematically reviewed 33 studies for a separate analysis of separately 19 studies performed on Caucasians and 11 on Asians. They tested two phenotypes – remission and response rates – and three genotype comparisons, ll versus ls/ss, ss versus ll/ls, and ll versus ss, using the Cochrane review manager and separate evaluations for SSRIs and mixed/other drugs. Divided by ethnic group, among Caucasians they found an association between l allele and both response and remission in the SSRI group. Only a marginal association between l allele and remission survived pooling together mixed antidepressant treatments. Among the Asians, a small effect of 5-HTTLPR on remission for mixed antidepressants was obtained. It was observed that gender, age, and age at onset modulated the association in Caucasians, whereas gender, age, and depression severity at baseline modulated the association in Asians. They concluded that whereas in Caucasians, 5-HTTLPR may be a predictor of antidepressant response and remission, among Asians it did not exert a major role. Dreimüller et al. (2012) have shown that among 5HTTLPR L (A) allele carriers, MDD patients presenting high AD concentrations gave a better response to treatment than patients with low serum concentrations.

Differences in depressive symptoms and mood disorders, pertaining to prevalence among adolescents, with higher prevalence among girls, between the genders emerge typically during adolescence, age at onset, and onset of puberty providing contributory factors (Kolltveit et al. 2012; Schuster et al. 2012; Zullig and Divin 2012). Edwards et al. (2011a) have showed that that pubertal development moderates environmental influences on depressive symptoms. Factors contributing to depressive symptoms during 14 years of age among more developed girls, relative to their less developed peers, were more critical but decreased in influence during age 17 with regard to depressive symptoms. The effects observed in boys were similar, but are delayed, paralleling the delay in pubertal development in boys compared to girls thereby supporting the premise that environmental influences on depressive symptoms during adolescence change as a function of pubertal development. Joinson et al. (2012) found that depressive symptoms among girls during mid-adolescence were more strongly influenced by breast stage than timing of menarche implies that the female rise in depression during adolescence may be due to increasing levels of estrogen and may account for the gender difference in rates of depression emerges at this stage. Nilsen et al. (2012) performed a systematic review of 32 anxiety studies and 13 depression studies that met predefined methodological criteria comprising client demographic factors such as age, gender, ethnicity, and IQ and clinical factors such as duration, type of diagnosis, pretreatment severity, and comorbidity. Most of the studies showed nonsignificant associations between demographic factors (gender and age) with treatment outcome for both the anxiety and the depression treatment trials. The anxiety studies showed mainly the lack of demographic or clinical factors predicting or moderating treatment outcome. In the case of depression studies, the findings implied that baseline symptom severity and comorbid anxiety may impact on treatment response. Gender differences in response to intervention other than medication may be revealing: Gender and crime victimization significantly modified treatment effects on distress and Behavioral Problems Index (Osypuk et al. 2012). Female adolescents in families without crime victimization benefited from moving-to-opportunity intervention, for all outcomes, distress, and major lifetime depressive disorder, whereas male adolescents in intervention families experiencing crime victimization expressed worse distress, more behavior problems, and higher, though nonsignificantly, major lifetime depressive disorder versus controls. Finally, in a community-based longitudinal sample of 309 adolescents reported depressive symptoms and negative life events at ages 11, 13, and 15. 5-HTTLPR and MAOA-uVNTR genotypes were ascertained via buccal swabs (Priess-Groben and Hyde 2012). They obtained significant four-way interaction between 5-HTTLPR, MAOA-uVNTR, NLE at age 13, and gender that predicted depressive symptoms at 15 years of age whereby girls were most likely to exhibit elevated depressive symptoms when experiencing negative life events if associated with low-expression MAOA-uVNTR alleles and short 5-HTTLPR alleles. For boys, low-expression MAOA-uVNTR alleles but long 5-HTTLPR alleles were implicated. Taken together, the existing observations of pretreatment patient variables as predictors and moderators of anxiety and depression treatment outcome provided little consistent knowledge concerning for what type of patients and under what conditions treatments work.

Keers (2012) has suggested that gene-environment interaction studies may provide an explanation for the above discrepancies regarding the 5-HTTLPR locus and the actions of SSRIs, particularly involving the interaction between stressors and 5-HTTLPR. Gene × environment interaction effects were observed for genes encoding components of the hypothalamic-pituitary-adrenal axis. The T allele of rs1360780 in FKBP5 increased the risk of posttraumatic stress disorder (PTSD) following childhood maltreatment and rs10402 (a single-nucleotide polymorphism in the gene encoding CRHR1) and moderated the effects of this maltreatment on several behavioral phenotypes, such as alcoholism, neuroticism, and depression, underlining the possibility that several polymorphisms have been shown to moderate the effects of environmental adversity on the development of depression and treatment response (Keers and Uher 2012). Additionally, it has been found that individuals possessing the S allele experienced more depressive symptoms, clinical depression, and suicide attempts following recent stressful events or childhood maltreatment/adversity than those individuals carrying the L allele (Uher and McGuffin 2010). Nevertheless, in a sample of 290 systematically recruited patients diagnosed with a single depressive episode, the outcome of antidepressant treatment and the presence of stressful life events during a 6-month period preceding onset of depression using structured interviews were assessed (Bukh et al. 2010). Nine polymorphisms in the genes encoding the serotonin transporter, brain-derived neurotrophic factor, catechol-*O*-methyltransferase, angiotensin converting enzyme, tryptophan hydroxylase, and the serotonin receptors 1A, 2A, and 2C were genotyped, but no evidence that the effects of the genetic polymorphisms on treatment outcome were dependent on stressful life events experienced by the individual prior to onset of depression was forthcoming (but see Nanni et al. 2011). Keers et al. (2011) observed that stressful/adverse life events predicted a marked more effective response to citalopram, but showed no effect in response to nortriptyline, variation in the 5-HTTLPR promoter region polymorphism, and another polymorphism in the gene, STin4, significantly modified these treatment effects.

Mania, a state of abnormally elevated or irritable mood, arousal, and/or energy levels, presents a volatile mood state that is characterized by faulty judgements with marked impulsiveness often causing patients to engage in dangerous and self-damaging behaviors (Quanbeck et al. 2003) with genetic and clinical connection (Mellerup et al. 2012). Together with early onset and delay of treatment (Post et al. 2010), risk of switches in mood polarity into hypomania or mania during acute and continuation trials of adjunctive AD treatment of bipolar depression (BP) poses a necessary consideration with an incidence of 17–31 % (Leverich et al. 2006). In selection of AD treatment for BP, there is risk for treatment-emergent/AD-induced manic/hypomanic switch in up to 40 % of patients (Truman et al. 2007), whereas placebo/course-of-illness switch rate was 4–5 % (Calabrese et al. 2005). The serotonin transporter gene, *SLC6A4*, encodes the protein responsible for serotonin reuptake from the synaptic cleft following release from serotonergic neurons. The association between AD-induced mania and candidate genetic variants, focusing upon the promoter polymorphism of *SLC6A4*, has been examined (Daray et al. 2010; de Aguilar Ferreira et al. 2010). Nevertheless, on the basis of a meta-analysis, Biernacka et al. (2012) in attempting to confirm an association between the serotonin transporter gene polymorphism, 5HTTLPR (see above), and AD-induced mania concluded that there was insufficient evidence.

Both the release of corticotrophin-releasing hormone (CRH) and arginine vasopressin (AVP) in the parvocellular neurons of the paraventricular nuclei (PVN) of the hypothalamus mediate parallel activation of the sympathetic nervous system and the hypothalamic-pituitary-adrenal (HPA) axis, in turn activating pro-opiomelanocortin (POMC) synthesis, processed to adrenocorticotrophin hormone (ACTH), which induces secretion of glucocorticoids from the adrenal cortex (de Kloet 2004; Herman and Cullinan 1997). Glucocorticoids act through mineralocorticoid (MR) and glucocorticoid (GR) receptors: the former, high-affinity receptors, implicated in the appraisal process and acute stress response onset and the latter, low-affinity receptors, promoting adaptation and recovery from stress (De Kloet et al. 1998). GR signaling of the negative feedback process involves a complex arrangement of agents involving the transcriptional regulation of target genes (Echeverria and Picard 2010). Preclinical and clinical studies point to impaired MR and CR signaling capacity coupled to CRH and AVP system overactivity (McEwen 2003). The overactivity of the HPA axis expressed by hypercortisolism, adrenal hyperplasia, and abnormalities in negative feedback characterizes the biological abnormality in melancholic depression. In depressive states, anterior pituitary CRH1 receptors are downregulated and response to CRH infusion is blunted, while, on the other hand, vasopressin V3 receptors in the anterior pituitary express enhanced responding to AVP stimulation which influences in maintaining HPA overactivity (Dinan and Scott 2005). Depressed patients showed elevated numbers of ACTH (Mortola et al. 1987) and cortisol (Rubin et al. 1987) secretory pulses as expressed through increased plasma and urinary free cortisol (Green and Kane 1983), changes accompanied by increased size of pituitary and adrenal glands (Axelson et al. 1992). During pregnancy, maternal cortisol promotes secretion of placental CRH (O'Keane et al. 2011). In a group of medication-free pregnant women presenting major depression ($n=27$) or not ($n=38$), O'Keane et al. (2011) found that maternal cortisol concentrations correlated highly with CRH secretion for all participants. Second trimester CRH concentrations and mean evening salivary concentrations were significantly higher in the depressed women.

Mood disorders are associated early adversity, often prenatal traumatic stress (Abe et al. 2007; Darnaudery and Maccari 2008; Nugent et al. 2011), and accompanied frequently by relative elevations of glucocorticoid stress hormones.

The "deregulation" and the irregularity of the HPA axis present a major aspect of symptom and biomarker profiles in depressive disorders (Abreu Feijo de Mello et al. 2003; Boyle et al. 2005; Holsboer and Ising 2008), focusing on the role of elevated cortisol (Schüle 2006) and the putative AD-induced normalization of HPA function (Burke et al. 2005; Pariante et al. 2004). The biological stress response exerts essential functions in coping with life events, differing widely between individuals with genetically and epigenetically determined set points during infancy and adolescence. It is possible the depressive spectrum disorders constitute an adaptive defense mechanism to excessive stress/distress with the HPA axis expressing a hub in brain stress circuits implicated in depressive subtypes (Bonfiglio et al. 2011). Nevertheless, both the hippocampus-amygdala-prefrontal cortex and the reciprocal monosynaptic cerebello-hypothalamic connections together with dense glucocorticoid binding sites play an important role in stress regulation and depressive disorder (Schutter 2012). Piwowarska et al. (2012) undertook to determine whether or not increased plasma concentrations occurred in MDD patients, using the Hamilton Depression Rating Scale (HDRS) and whether or not SSRI treatment with fluoxetine may "reregulate" cortisol levels in a study of 21 patients (14 female, 7 male, aged 29–75 years), with a mean score on HDRS, and 24 healthy controls. They observed that among patients responding to fluoxetine therapy (reduction of HDRS scores by at least 50 %), levels of cortisol were decreased. In mood disorders, higher mean cortisol levels and higher cortisol awakening rise indicate hyperactivity of the HPA axis and dysregulated GC sensitivity determined in part by polymorphisms in genes encoding receptors and proteins involved in HPA axis regulation (Bet et al. 2009; Russcher et al. 2005; Spijker and van Rossum 2009). Spijker and van Rossum (2012) have outlined both genetic and epigenetic changes influencing the set point and regulation of the HPA axis, with major effects upon mood states that could originate from traumatic experiences in utero and during infancy (McGowan et al. 2009; Navailles et al. 2010).

Generalized anxiety disorder (GAD), a highly prevalent chronic neuropsychiatric disorder with marked morbidity and mortality, is characterized by excessive, uncontrollable, and often irrational worry about everyday things that is disproportionate to the actual source of worry, with symptoms, that interfere with everyday behaviors, retained at least 6 months (Torpy et al. 2011). For both acute and chronic treatment, AD compounds with 40–70 % treatment response are prescribed (Baldwin and Nair 2005; Bandelow et al. 2012; García-Campayo et al. 2012). The 5-HT_{2A} receptor is expressed widely throughout the brain and CNS, particularly near most of the serotonergic terminal-rich areas, including neocortex (mainly prefrontal, parietal, and somatosensory cortex) and the olfactory tubercle, and is coded by the *HTR2A* gene. Links between the A-1438G (rs6311) polymorphism and mood disorders have been obtained (Chee et al. 2001), and several studies have found associations between the rs7997012 and rs17288723 SNPs and AD treatment outcome in patients presenting depression spectrum disorders (Horstmann et al. 2010; Lucae et al. 2010; McMahon et al. 2006). Venlafaxine is an SNRI (serotonin-norepinephrine reuptake inhibitor) for treatment of MDD, GAD, and comorbid indications. Lohoff et al. (2011) tested whether or not rs7997012 polymorphism predicted treatment outcome in 156 GAD patients in a 6-month open-label clinical trial administering venlafaxine XR with the Hamilton Anxiety Scale (HAM-A) and Clinical Global Expression of Improvement (CGI-I) score. The frequency of the G allele differed between responders (70 %) and nonresponders (56 %) at 6 months on HAM-A, and the G allele was associated with improvement. Similarly, Lohoff et al. (2012) studied the interaction between SLC6A4 5HTTLPR/rs25531 haplotype and rs7997012 polymorphism for venlafaxine XR in an 18-month relapse prevention trial comprising 112 patients. Patients with genotypes La/La+G/G or La/La+G/A (n=28) showed lower HAM-A scores than those with genotypes La/S+A/A or S/S+A/A at 6 months thereby concluding a gene-gene interaction between these markers.

The potential role of polymorphisms in genes regulating the HPA axis thereby affecting putatively AD drug efficacy has been addressed (Tyrka et al. 2009). GR is encoded by the NR3C1 gene, on chromosome 5, which has three protein domains: immunogenic, DNA, and ligand binding and several functional genetic polymorphisms (DeRijk et al. 2002). Relevant to mood disorders, SNPs in the region encoding the immunogenic domain involving changes in GR function, linked to "glucocorticoid resistance syndromes," have been identified, e.g., ER22/23EK (DeRijk et al. 2008), which induces loss of GR sensitivity (van Rossum et al. 2002). An overrepresentation of the ER22/23EK allele conferring GR resistance has been reported (van Rossum et al. 2006; van West et al. 2006). N363S and BclI polymorphisms are associated with hypersensitivity to glucocorticoids, whereas the ER22/23EK polymorphism is related to glucocorticoid resistance; both BclI and ER22/23EK polymorphisms were associated with susceptibility to develop major depression (van Rossum et al. 2006), while the ER22/23EK polymorphism was associated with a faster clinical response to antidepressant treatment. Horstmann and Binder (2011) have argued that despite the glucocorticoid measures and presence of polymorphisms, involving the stress hormone system, showing associations with response to ADs, necessary concurrent assessment of several clinical, biomarker, and pharmacokinetic variables is required, before a suitable level of predictability is achieved. Nevertheless, the structure-functional relationships of the HPA axis with regions involved in stress coping or non-coping and the dynamics of the glucocorticoid system are critical to notions concerning epigenetic influence on the etiopathogenesis of mood disorders (Schutter 2012) and predicting AD treatment response (Massart et al. 2012; Menke et al. 2012).

Meta-analysis of association data of mood disorders suggests the role of particular genes posing genetic risk with differential expression evidence in brains of mood disorder patients supporting the contributions of specific genes. The "neurotrophin hypothesis" of depression posits a role of brain-derived neurotrophic factor (BDNF) in depression, although it is unknown whether BDNF is more involved in the etiology of depression or in the mechanism of action of antidepressants. Accordingly, deficiency in neurotrophic support levels may underlie mood disorders such that elevation of neurotrophic status to normal levels engenders mood recovery. Castrén and Rantamäki (2010) have provided an account on the role of BDNF and its receptors in depression and the AD response presenting a model whereby the effects of AD treatments may occur via a reactivation of activity-dependent and BDNF-mediated cortical plasticity. Wolkowitz et al. (2011) observed that pretreatment, with SSRIs, BDNF levels were lower in the depressed subjects than the controls, but these levels did not correlate significantly with the pretreatment assessment of depression severity. Depression ratings improved with SSRI treatment, and BDNF levels increased with treatment. Changes in BDNF levels were not significantly correlated with changes in depression ratings. However, pretreatment BDNF levels were directly correlated with antidepressant responses, and "responders" to treatment (≥50 % improvement in depression ratings) had higher pretreatment BDNF levels than did "nonresponders." These results confirm low serum BDNF levels in unmedicated depressed subjects and confirm AD-induced elevations in BDNF levels but imply that ADs, in conjunction with correcting BDNF insufficiency, function through a permissive or facilitatory role of BDNF in the mechanism of action of ADs. In this context, network analysis of meta-analysis-generated candidate genes expressing differential expression in patient brains identified signaling pathways and functional clusters implicated in genetic risk for mood disorders (Detera-Wadleigh and Akula 2011).

In order to extract a suitable level of efficacy in the treatment of mood disorders, sufficient comprehension of individual differences in susceptibility to the adverse effects of AD compounds is required. Both evidences from clinical trials (Papakostas and Fava 2009) and general clinical experience (Souery et al. 2007) indicate that 50 % and less of patients presenting uncomplicated MDD will respond to any single AD. These realities combine with problems arising through adverse drug reactions, e.g., cardiac toxicity in certain individual (Cooke and Waring 2012), that lead to inability to tolerate side effects and termination of treatment (Bull et al. 2002; Mitchell 2006). Prediction of individual efficacy and side-effect profiles remains in the domain of "trial and error" thereby rendering both psychological and somatic distress. Several avenues imply that the markedly heritable quality of the AD response (Malhotra et al. 2004; McGuffin et al. 2010) should be particularly suitable for examination through applications of pharmacogenomic techniques (Tsai et al. 2011). The role of serotonergic function-related genes in AD-induced adverse effects has been the focus of pharmacogenetic investigation (Kato and Serretti 2010; Zobel and Maier 2010), with more or less serious disturbance of somatic functions that effectively exacerbated symptom profiles (Ayala 2009; Serretti and Chiesa 2009). Several aspects of AD mechanism of action have been observed through genome-wide associations with applications for drug response profiling in mood disorder (Adkins et al. 2010; Morag et al. 2011). Clark et al. (2012) performed genome-wide studies to identify genetic variation affecting susceptibility to adverse effect in mood disorder drugs, concluding that 10 out of the 34 SNPs satisfied the criterion for genome-wide significance: 10 SNPs mediated the effects of bupropion upon adverse effects for sexual behavior, particularly with regard to *MAGI2*, *DTWD1*, *WDFY4*, and *CHL1*.

4 Conclusions

The notion of personalized medicine for the development of efficacious and individualized pharmacotherapies taking into account genetic variation and target groups of patients who share biology, not just the symptoms, of mood disorders progresses steadily (McMahon and Insel 2012). One strategy in this enterprise has been to address gene polymorphisms in humans that affect SSRI efficacy; another focuses on the molecular targets affected by SSRIs in animal models of mood disorders. It is concluded that serotonin transporter gene variation affects the efficacy and adverse effects of SSRIs in humans, whereas SSRIs do not affect generally serotonin transport gene expression in animal studies, but rather alter mRNA levels of genes encoding serotonin receptors, components of other neurotransmitter systems, HPA axis-affecting factors, neurotrophic factors, and inflammatory factors (Kroeze et al. 2012). The role of physical exercise as personalized medication intervention for the alleviation of mood disorder symptoms remains only sparsely exploited (Archer et al. submitted). Exercise induces anti-inflammatory effects to both nondepressed subjects and those presenting negative affective and/or infected status (Archer et al. 2011). Following a 12-week exercise intervention program in patients presenting MDD, Rethorst et al. (2012) reported higher baseline levels of TNF-α that were associated with a greater decrease in depressive symptoms over the intervention period as well as a marked positive correlation between change in Il-1β and depressive symptom scores. Their results imply that high TNF-α was linked with prediction of differentially superior outcome with exercise treatment as opposed to AD medication wherein high TNF-α was linked to poor response. Finally, adverse, unfavorable early life conditions, particularly during perinatal stages and infancy, may lead to epigenetic regulation of genes involved in stress response, behavioral inhibition, and cognitive-emotional systems (Archer et al. 2012). The ultimate final outcome may be expressed through behaviors bedeviled by problems with impulse control, wrong decisions, eating disorders, alcoholism, indiscriminate social behavior, as well as mood disorders.

References

Aalto-Setala T, Marttunen M, Tuulio-Henriksson A, Poikolainen K, Lonnqvist J (2002) Depressive symptoms in adolescence as predictors of early adulthood depressive disorders and maladjustment. Am J Psychiatry 159:1235–1237

Aarø LE, Herbeć A, Bjørngaard JH, Mańczuk M, Zatoński WA (2011) Depressive episodes and depressive tendencies among a sample of adults in Kielce, south-eastern Poland. Ann Agric Environ Med 18:273–278

Abe H, Hidaka N, Kawagoe C, Odagiri K, Watanabe Y, Ikeda T, Nishimori T, Ishida Y (2007) Prenatal psychological stress causes higher emotionality, depression-like behavior, and elevated activity in the hypothalamic-pituitary-adrenal axis. Neurosci Res 59:145–151

Abo R, Hebbring S, Ji Y, Zhu H, Zeng ZB, Batzler A, Jenkins GD, Biernacka J, Snyder K, Drews M, Fiehn O, Fridley B, Schaid D, Kamatani N, Nakamura Y, Kubo M, Mushiroda T, Kaddurah-Daouk R, Mrazek DA, Weinshilboum RM (2012) Merging pharmacometabolomics with pharmacogenomics using '1000 Genomes' single-nucleotide polymorphism imputation: selective serotonin reuptake inhibitor response pharmacogenomics. Pharmacogenet Genomics 22: 247–253

Abreu Feijo de Mello A, Feijo de Mello M, Carpenter LL, Price LH (2003) Update on stress and depression: the role of the hypothalamic-pituitary-adrenal (HPA) axis. Revista Brasileria de Psiquiatry 25:231–238

Adkins DE, Aberg K, McClay JL, Hettema JM, Kornstein SG, Bukszar J, van den Oord EJ (2010) A genome-wide association study of citalopram response in major depressive disorder – a psychometric approach. Biol Psychiatry 68:e25–e27

Alaräisänen A, Miettunen J, Pouta A, Isohanni M, Räsänen P, Mäki P (2012) Ante- and perinatal circumstances and risk of attempted suicides and suicides in offspring: the Northern Finland birth cohort 1966 study. Soc Psychiatry Psychiatr Epidemiol 47(11):1783–1794, PMID: 22327374

Angold A, Costello EJ (2006) Puberty and depression. Child Adolesc Psychiatr Clin N Am 15(4):919–937, ix

Archer T (2012) Pharmacogenomics and personalized medicine in Parkinson's disease. In: Barh D (ed) Pharmacogenomics. Omics Publishing Group

Archer T, Josefsson T, Lindvall M (submitted) The effect of exercise on depressive symptoms and biomarkers: a review and integrated model. Submitted manuscript

Archer T, Kostrzewa RM, Beninger RJ, Palomo T (2010) Staging perspectives in neurodevelopment aspects of neuropsychiatry: agents, phases and ages at expression. Neurotox Res 18:287–305

Archer T, Fredriksson A, Schülz E, Kostrzewa RM (2011) Influence of physical exercise on neuroimmunological functioning and health: aging and stress. Neurotox Res 20:69–83

Archer T, Oscar-Berman M, Blum K, Gold M (2012) Neurogenetics and epigenetics in impulsive behavior: impact of reward circuitry. Genet Syndr Gene Ther 3:3. doi:10.4172/2157-7412.1000115

Asendorpf JB, Denissen JJA, van Aken MAG (2008) Inhibited and aggressive preschool children at 23 years of age: personality and social transitions into adulthood. Dev Psychol 44:997–1011

Axelson DA, Doraiswamy PM, Boyko OB, Rodrigo Escalona P, McDonald WM, Ritchie JC, Patterson LJ, Ellinwood EH Jr, Nemeroff CB, Krishnan KR (1992) In vivo assessment of pituitary volume with magnetic resonance imaging and systematic stereology: relationship to dexamethasone suppression test results in patients. Psychiatry Res 44:63–70

Ayala ME (2009) Brain serotonin, psychoactive drugs, and effects on reproduction. Cent Nerv Syst Agents Med Chem 9:258–276

Baldwin DS, Nair RV (2005) Escitalopram in the treatment of generalized anxiety disorder. Expert Rev Neurotherapeutics 5:443–449

Bandelow B, Sher L, Bunevicius R, Hollander E, Kasper S, Zohar J, Möller HJ, WFSBP Task Force on Mental Disorders in Primary Care, WFSBP Task Force on Anxiety Disorders, OCD and PTSD (2012) Guidelines for the pharmacological treatment of anxiety disorders, obsessive-compulsive disorder and posttraumatic stress disorder in primary care. Int J Psychiatry Clin Pract 16:77–84

Barker ED, Jaffee SR, Uher R, Maughan B (2011) The contribution of prenatal and postnatal maternal anxiety and depression to child maladjustment. Depress Anxiety 28:696–702

Barker ED, Copeland W, Maughan B, Jaffee SR, Uher R (2012) Relative impact of maternal depression and associated risk factors on offspring psychopathology. Br J Psychiatry 200:124–129

Belsky J, Pluess M (2009) Beyond diathesis stress: differential susceptibility to environmental influences. Psychol Bull 135:885–908

Belsky J, Jonassaint C, Pluess M, Stanton M, Brummett B, Williams R (2009) Vulnerability genes or plasticity genes? Mol Psychiatry 14:746–754

Bergemann ER, Boles RG (2010) Maternal inheritance in recurrent early-onset depression. Psychiatr Genet 20:31–34

Bet PM, Penninx BW, Bochdanovits Z, Uitterlinden AG, Beekman AT, van Schoor NM, Deeg DJ, Hoogendijk WJ (2009) Glucocorticoid receptor gene polymorphisms and childhood adversity are associated with depression: new evidence for a gene-environment interaction. Am J Med Genet B Neuropsychiatr Genet 150B:660–669

Bhatia SK, Bhatia SC (2007) Childhood and adolescent depression. Am Fam Physician 75:73–80

Biernacka JM, McElroy SL, Crow S, Sharp A, Benitez J, Veldic M, Kung S, Cunningham JM, Post RM, Mrazek D, Frye MA (2012) Pharmacogenomics of antidepressant induced mania: a review and meta-analysis of the serotonin transport gene (5HTTLPR) association. J Affect Disord 136:e21–e29

Boardman JD, Alexander KB, Stallings MC (2011) Stressful life events and depression among adolescent twin pairs. Biodemography Soc Biol 57:53–66

Bonfiglio JJ, Inda C, Refojo D, Holsboer F, Arzt E, Silberstein S (2011) The corticotropin-releasing hormone network and the hypothalamic-pituitary-adrenal axis: molecular and cellular mechanisms involved. Neuroendocrinology 94:12–20

Boomsma DI, van Beijsterveldt CE, Hudziak JJ (2005) Genetic and environmental influences on Anxious/Depression during childhood: a study from the Netherlands Twin Register. Genes Brain Behav 4:466–481

Boyle MP, Brewer JA, Funatsu M, Wozniak DF, Tsien JZ, Yukitoshi I, Muglia LJ (2005) Acquired deficit of forebrain glucocorticoid receptor produces depression-like changes in adrenal axis regulation and behavior. Proc Natl Acad Sci USA 102:473–478

Brown HM, Trzaskowski M, Zavos HMS, Rijsdijk FV, Gregory AM, Eley TC (2012) Phenotypic and genetic structure of anxiety sensitivity in adolescence and early adulthood. J Anxiety Disord 26:680–688

Bukh JD, Bock C, Vinberg M, Werge T, Gether U, Kessing LV (2010) No interactions between genetic polymorphisms and stressful life events on outcome of antidepressant treatment. Eur Neuropsychopharmacol 20:327–335

Bull SA, Hu XH, Hunkeler EM, Lee JY, Ming EE, Markson LE, Fireman B (2002) Discontinuation of use and switching of antidepressants: influence of patient-physician communication. JAMA 288:1403–1409

Burke HM, Davis MC, Otte C (2005) Depression and cortisol responses to psychological stress: a meta-analysis. Psychoneuroendocrinology 30:846–856

Buss C, Davis EP, Muftuler LT, Head K, Sandman CA (2010) High pregnancy anxiety during mid-gestation is associated with decreased gray matter density in 6–9-year-old children. Psychoneuroendocrinology 35(1):141–153. doi:10.1016/j.psyneuen.2009.07.010

Calabrese JR, Keck PEJ, MacFadden W, Minkwitz M, Ketter TA, Weisler RH, Cutler AJ, McCoy R, Wilson E, Mullen J (2005) A randomized, double-blind, placebo-controlled trial of quetiapine in the treatment of bipolar I or II depression. Am J Psychiatry 162:1351–1360

Castrén E, Rantamäki T (2010) The role of BDNF and its receptors in depression and antidepressant drug action: reactivation of developmental plasticity. Dev Neurobiol 70:289–297

Chee IS, Lee SW, Kim JL, Wang SK, Shin YO, Shin SC, Lee YH, Hwang HM, Lim MR (2001) 5-HT2A receptor gene promoter polymorphism -1438A/G and bipolar disorder. Psychiatr Genet 11:111–114

Chen B, Dowlatshahi D, MacQueen GM, Wang JF, Young LT (2001) Increased hippocampal bdnf immunoreactivity in subjects treated with antidepressant medication. Biol Psychiatry 50:260–265

Chen ZY, Jing D, Bath KG, Ieraci A, Khan T, Siao CJ, Herrera DG, Toth M, Yang C, McEwen BS, Hempstead BL, Lee FS (2006) Genetic variant BDNF (Val66Met) polymorphism alters anxiety-related behavior. Science 314:140–143

Clark SL, Adkins DE, Aberg K, Hettema JM, McClay JL, Souza RP, van den Oord EJCG (2012) Pharmacogenomic study of side-effects for antidepressant treatment options in STAR*D. Psychol Med 42:1151–1162

Coiro MJ, Riley A, Broitman M, Miranda J (2012) Effects on children of treating their mothers' depression: results of a 12-month follow-up. Psychiatr Serv 63:357–363

Cooke MJ, Waring WS (2012) Citalopram and cardiac toxicity. Eur J Clin Pharmacol 69(4):755–760, PMID: 22996077

Cox B, Borger S, Enns M (1999) Anxiety sensitivity and emotional disorders: psychometric studies and their theoretical implications. In: Anxiety sensitivity: theory, research, and treatment of the fear of anxiety. LEA Assoc, Mahwah, pp 115–148

Cox LJ, Stanley BH, Melhem NM, Oquendo MA, Birmaher B, Burke A, Kolko DJ, Zelazny JM, Mann JJ, Porta G, Brent DA (2012a) Familial and individual correlates of nonsuicidal self-injury in the offspring of mood-disordered parents. J Clin Psychiatry 73: 813–820

Cox LJ, Stanley BH, Melhem NM, Oquendo MA, Birmaher B, Burke A, Kolko DJ, Zelazny JM, Mann JJ, Porta G, Brent DA (2012b) A longitudinal study of nonsuicidal self-injury in offspring at high risk for mood disorder. J Clin Psychiatry 73:821–828

Cyranowski JM, Frank E, Young E, Shear MK (2000) Adolescent onset of the gender difference in lifetime rates of major depression: a theoretical model. Arch Gen Psychiatry 57:21–27

Daftary SS, Calderon G, Rios M (2012) Essential role of brain-derived neurotrophic factor in the regulation of serotonin transmission in the basolateral amygdala. Neuroscience 224:125–134, PMID: 22917617

Daray FM, Thommi SB, Ghaemi SN (2010) The pharmacogenetics of antidepressant-induced mania: a systematic review and meta-analysis. Bipolar Disord 12:702–706

Darnaudery M, Maccari S (2008) Epigenetic programming of the stress response in male and female rats by prenatal restraint stress. Brain Res Rev 57:571–585

Davalos DB, Yadon CA, Tregellas HC (2012) Untreated prenatal maternal depression and the potential risks to offspring: a review. Arch Womens Ment Health 15:1–14. doi:10.1007/s00737-011-0251-1

de Aguilar Ferreira A, Neves FS, Pimenta GJ, Mello MP, Miranda DM, Romano-Silva MA, De Marco LA, Corrêa H (2010) The role of genetic variation of BDNF gene in antidepressant-induced mania in bipolar disorder. Psychiatry Res 180(1):54–56. doi:10.1016/j.psychres.2010.04.036

de Bartolomeis A, Perugi G (2012) Combination of aripiprazole with mood stabilizers for the treatment of bipolar disorder: from acute mania to long-term maintenance. Expert Opin Pharmacother 13:2027–2036

De Kloet ER (2004) Hormones and the stressed brain. Ann NY Acad Sci 1018:1–15

De Kloet ER, Vreugbenhil E, Oitzl MS, Joels M (1998) Brain corticosteroid receptor balance in health and disease. Endocr Rev 19:269–301

Dell'osso L, Carmassi C, Musetti L, Socci C, Shear MK, Conversano C, Maremmani I, Perugi G (2012) Lifetime mood symptoms and adult separation anxiety in patients with complicated grief and/or post-traumatic stress disorder: a preliminary report. Psychiatry Res 198(3):436–440, PMID: 22436452

DeRijk RH, Schaaf M, de Kloet ER (2002) Glucocorticoid receptor variants: clinical implications. J Steroid Biochem Mol Biol 81:103–122

DeRijk RH, Van Leeuwen N, Klok MD, Zitman FG (2008) Corticosteroid receptor-gene variants: modulators of the stress-response and implications for mental health. Eur J Pharmacol 585:492–501

Detera-Wadleigh SD, Akula N (2011) A systems approach to the biology of mood disorders through network analysis of candidate genes. Pharmacopsychiatry 44(Suppl 1):S35–S42

Diego MA, Field T, Hernandez-Reif M, Schanberg S, Kuhn C, Gonzalez-Quintero VH (2010) Prenatal depression restricts fetal growth. Early Hum Dev 85:65–70

Diler RS, Birmaher B, Axelson D, Obreja M, Monk K, Hickey MB, Goldstein B, Goldstein T, Sakolsky D, Iyengar S, Brent D, Kupfer D (2011) Dimensional psychopathology in offspring of parents with bipolar disorder. Bipolar Disord 13:670–678

Dinan TG, Scott LV (2005) Anatomy of melancholia: focus on hypothalamic-pituitary-adrenal axis overactivity and the role of vasopressin. J Anat 207:259–264

Dougherty LR, Klein DN, Rose S, Laptook RS (2011) Hypothalamic-pituitary-adrenal axis reactivity in the preschool-age offspring of depressed parents: moderation by early parenting. Psychol Sci 22:650–658

Dreimüller N, Tadic A, Dragicevic A, Boland K, Bondy B, Lieb K, Laux G, Maier W, Müller MJ, Rao ML, Rietschel M, Röschke J, Zill P, Hiemke C (2012) The serotonin transporter promoter polymorphism (5-HTTLPR) affects the relation between antidepressant serum concentrations and effectiveness in major depression. Pharmacopsychiatry 45:108–113

Duffy A, Lewitzka U, Doucette S, Andreazza A, Grof P (2012) Biological indicators of illness risk in offspring of bipolar parents: targeting the hypothalamic-pituitary-adrenal axis and immune system. Early Interv Psychiatry 6:128–137

Eaves L, Prom E, Silberg J (2010) The mediating effect of parental neglect on adolescent and young adult antisociality: a longitudinal study of twins and their parents. Behav Genet 40:425–437

Echeverria PC, Picard D (2010) Molecular chaperones, essential partners of steroid hormone receptors for activity and mobility. Biochim Biophys Acta 1803:641–649

Edwards AC, Rose RJ, Kaprio J, Dick DM (2011a) Pubertal development moderates the importance of environmental influences on depressive symptoms in adolescent girls and boys. J Youth Adolesc 40(10): 1383–1393. doi:10.1007/s10964-010-9617-3

Edwards AC, Larsson H, Lichtenstein P, Kendler KS (2011b) Early environmental influences contribute to covariation between internalizing symptoms and alcohol intoxication frequency across adolescence. Addict Behav 36:175–182

Edwards AC, Sihvola E, Korhonen T, Pulkkinen L, Moilanen I, Kaprio J, Rose RJ, Dick DM (2011c) Depressive symptoms and alcohol use are genetically and environmentally correlated across adolescence. Behav Genet 41:476–487

Egan MF, Kojima M, Callicott JH, Goldberg TE, Kolachana BS, Bertolino A, Zaitsev E, Gold B, Goldman D, Dean M, Lu B, Weinberger DR (2003) The BDNF Val66Met polymorphism affects activity-dependent secretion of BDNF and human memory and hippocampal function. Cell 112:257–269

Ehrensaft MK, Cohen P (2012) Contribution of family violence to the intergenerational transmission of externalizing behavior. Prev Sci 13:370–383

Eley TC, Hudson JL, Creswell C, Tropeano M, Lester KJ, Cooper P, Farmer A, Lewis CM, Lyneham HJ, Rapee RM, Uher R, Zavos HM, Collier DA (2012) Therapygenetics: the 5HTTLPR and response to psychological therapy. Mol Psychiatry 17:236–241

Ellis BJ, Boyce WT, Belsky J, Bakermans-Kranenburg MJ, van Ijzendoorn MH (2011) Differential susceptibility to the environment: an evolutionary–neurodevelopmental theory. Dev Psychopathol 23:7–28

Erath S, Flenagan K, Bierman K (2007) Social anxiety and peer relations in early adolescence: behavioural and cognitive factors. J Abnorm Child Psychol 35:405–416

Ferrari AJ, Somerville AJ, Baxter AJ, Norman R, Patten SB, Vos T, Whiteford HA (2012) Global variation in the prevalence and incidence of major depressive disorder: a systematic review of the epidemiological literature. Psychol Med 25:1–11, PMID: 22831756

Field T, Diego M, Hernandez-Reif M, Figueiredo B, Deeds O, Ascencio A, Schanberg S, Kuhn C (2008) Prenatal dopamine and neonatal behavior and biochemistry. Infant Behav Dev 31:590–593

Field T, Diego M, Hernandez-Reif M (2010a) Prenatal depression effects and interventions: a review. Infant Behav Dev 33:409–418

Field T, Diego M, Hernandez-Reif M, Figueiredo B, Deeds O, Ascencio A, Schanberg S, Kuhn C (2010b) Comorbid depression and anxiety effects on pregnancy and neonatal outcome. Infant Behav Dev 33:23–29

Fienberg M, Button T, Neiderhiser JM, Reiss D, Hetheringtion EM (2007) Parenting and adolescent antisocial behavior and depression: evidence of genotype x parenting interaction. Arch Gen Psychiatry 64:457–465

Foley DL, Neale MC, Kendler KS (2001) Genetic and environmental risk factors for depression assessed by subject-rated symptom check list versus structured clinical interview. Psychol Med 31:1413–1423

Franchini L, Serretti A, Gasperini M, Smeraldi E (1998) Familial concordance of fluvoxamine response as a tool for differentiating mood disorder pedigrees. J Psychiatr Res 32:255–259

Franić S, Middeldorp CM, Dolan CV, Ligthart L, Boomsma DI (2010) Childhood and adolescent anxiety and depression: beyond heritability. J Am Acad Child Adolesc Psychiatry 49:820–829

Frustaci A, Pozzi G, Gianfranca F, Manzoli L, Boccia S (2008) Meta-analysis of the brain-derived neurotrophic factor gene (BDNS) Val66Met polymorphism in anxiety disorders and anxiety-related personality traits. Neuropsychobiology 58:163–170

Gallinat J, Schubert F, Brühl R, Hellweg R, Klär AA, Kehrer C, Wirth C, Sander T, Lang UE (2010) Met carriers of BDNF Val66Met genotype show increased N-acetylaspartate concentration in the anterior cingulate cortex. NeuroImage 49:767–771

García-Campayo J, Caballero F, Perez M, López V (2012) Prevalence and clinical features of newly diagnosed generalized anxiety disorder patients in Spanish primary care settings: the GADAP study. Actas Esp Psiquiatr 40:105–113

Gatt JM, Nemeroff CB, Dobson-Stone C, Paul RH, Bryant RA, Schofield PR, Gordon E, Kemp AH, Williams LM (2009) Interactions between BDNF Val66Met polymorphism and early life stress predict brain and arousal pathways to syndromal depression and anxiety. Mol Psychiatry 14:681–695

Gatt JM, Nemeroff CB, Schofield PR, Paul RH, Clark CR, Gordon E, Williams LM (2010) Early life stress combined with serotonin 3A receptor and brain-derived neurotrophic factor valine 66 to methionine genotypes impacts emotional brain and arousal correlates of risk for depression. Biol Psychiatry 68: 818–824

Ge X, Lorenz FO, Conger RD, Elder GH, Simons RL (1994) Trajectories of stressful life events and depressive symptoms during adolescence. Devel Psychol 30:467–483

Gillespie NA, Kirk KM, Evans DM, Heath AC, Hickie IB, Martin NG (2004) Do the genetic or environmental determinants of anxiety and depression change with age? A longitudinal study of Australian twins. Twin Res 7:39–53

Gillespie NA, Gehrman P, Byrne EM, Kendler KS, Heath AC, Martin NG (2012) Modeling the direction of causation between cross-sectional measures of disrupted sleep, anxiety and depression in a sample of male and female Australian twins. J Sleep Res 21(6):675–683. doi:10.1111/j.1365-2869.2012.01026.x

Glover V, O'Connor TG, O'Donnell K (2010) Prenatal stress and the programming of the HPA axis. Neurosci Biobehav Rev 35:17–22

Gluckman PD, Cutfield W, Hofman P, Hanson MA (2005) The fetal, neonatal, and infant environments – the long-term consequences for disease risk. Early Hum Dev 81:51–59

Goodman SH, Gotlib IH (1999) Risk for psychopathology in the children of depressed parents: a developmental approach to the understanding of mechanisms. Psychol Rev 106:458–490

Green HS, Kane JM (1983) The dexamethasone suppression test in depression. Clin Neuropharmacol 6:7–24

Gvozdic K, Brandl EJ, Taylor DL, Müller DJ (2012) Genetics and personalized medicine in antidepressant treatment. Curr Pharm Des 18(36):5853–5878, PMID: 22681166

Haenisch B, Herms S, Mattheisen M, Steffens M, Breuer R, Strohmaier J, Degenhardt F, Schmäl C, Lucae S, Maier W, Rietschel M, Nöthen MM, Cichon S (2012) Genome-wide association data provide further support for an association between 5-HTTLPR and major depressive disorder. J Affect Disord 146(3):438–440, PMID: 22921522

Hajcak G, Castille C, Olvet DM, Dunning JP, Roohi J, Hatchwell E (2009) Genetic variation in brain-derived neurotrophic factor and human fear conditioning. Genes Brain Behav 8:80–85

Hammen C, Hazel NA, Brennan PA, Najman J (2012) Intergenerational transmission and continuity of stress and depression: depressed women and their offspring in 20 years of follow-up. Psychol Med 42:931–942

Hariri AR (2010) Genetic polymorphisms: a cornerstone of translational biobehavioral research. Sci Transl Med 2(18):18ps6

Hariri AR, Goldberg TE, Mattay VS, Kolachana BS, Callicott JH, Egan MF, Weinberger DR (2003) Brain-derived neurotrophic factor val66met polymorphism affects human memory-related hippocampal activity and predicts memory performance. J Neurosci 23:6690–6694

Hasler G, Northoff G (2011) Discovering imaging endophenotypes for major depression. Mol Psychiatry 16:604–619

Hays P (1976) Etiological factors in manic-depressive psychoses. Arch Gen Psychiatry 33:1187–1188

Herman JP, Cullinan WE (1997) Neurocircuitry of stress: central control of the hypothalamic-pituitary-adrenocortical axis. Trends Neurosci 20:78–84

Hillerer KM, Neumann ID, Slattery DA (2012) From stress to postpartum mood and anxiety disorders: how chronic peripartum stress can impair maternal adaptations. Neuroendocrinology 95(1):22–38, PMID: 22042058

Hoeve M, Dubsd J, Eichelsheim V, van der Laan P, Smeenk W, Gerris J (2009) The relationship between parenting and delinquency: a meta-analysis. J Abnorm Child Psychol 37:749–775

Holsboer F, Ising M (2008) Central CRH system in depression and anxiety – evidence from clinical studies with CRH_1 receptor antagonists. Eur J Pharmacol 583:350–357

Horstmann S, Binder EB (2011) Glucocorticoids as predictors of treatment response in depression. Harv Rev Psychiatry 19:125–143

Horstmann S, Lucae S, Menke A, Hennings JM, Ising M, Roeske D, Müller-Myhsok B, Holsboer F, Binder EB (2010) Polymorphisms in GRIK4, HTR2A, and FKBP5 show interactive effects in predicting remission to antidepressant treatment. Neuropsychopharmacology 35:727–740

Hudson JL (2005) Efficacy of cognitive-behavioural therapy for children and adolescents with anxiety disorders. Behav Change 22:55–70

Joinson C, Heron J, Araya R, Paus T, Croudace T, Rubin C, Marcus M, Lewis G (2012) Association between pubertal development and depressive symptoms in girls from a UK cohort. Psychol Med 12:1–11, PMID: 22717026

Kaddurah-Daouk R, Kristal BS, Weinshilboum RM (2008) Metabolomics: a global biochemical approach to drug response and disease. Annu Rev Pharmacol Toxicol 48:653–683

Kaddurah-Daouk R, Yuan P, Boyle SH, Matson W, Wang Z, Zeng ZB, Zhu H, Dougherty GG, Yao JK, Chen G, Guitart X, Carlson PJ, Neumeister A, Zarate C, Krishnan RR, Manji HK, Drevets W (2012) Cerebrospinal fluid metabolome in mood disorders-remission state has a unique metabolic profile. Sci Rep 2:667, PMID: 22993692

Kaess M, Parzer P, Mattern M, Plener PL, Bifulco A, Resch F, Brunner R (2012) Adverse childhood experiences and their impact on frequency, severity, and the individual function of nonsuicidai self-injury in youth. Psychiatry Res 206(2–3):265–272. doi:10.1016/j.psychres.2012.10.012

Kandler C, Bleidorn W, Riemann R, Sinath FM, Thiel W, Angleitner A (2010) Sources of cumulative continuity in personality: a longitudinal multiple-rater twin study. J Pers Soc Psychol 98:995–1008

Kapczinski F, Frey BN, Andreazza AC, Kauer-Sant'Anna M, Cunha AB, Post RM (2008) Increased oxidative stress as a mechanism for decreased BDNF levels in acute manic episodes. Rev Bras Psiquiatr 30:243–245

Kato M, Serretti A (2010) Review and meta-analysis of antidepressant pharmacogenetic findings in major depressive disorder. Mol Psychiatry 15:473–500

Keers R (2012) Will gene-environment interactions explain differential antidepressant response? Pers Med 9:319–322

Keers R, Aitchison KJ (2011) Pharmacogenetics of antidepressant response. Exp Rev Neurother 11:101–125

Keers R, Uher R (2012) Gene-environment interaction in major depression and antidepressant treatment response. Curr Psychiatry Rep 14:129–137

Keers R, Uher R, Huezo-Diaz P, Smith R, Jaffee S, Rietschel M, Henigsberg N, Kozel D, Mors O, Maier W, Zobel A, Hauser J, Souery D, Placentino A, Larsen ER, Dmitrzak-Weglarz M, Gupta B, Hoda F, Craig I, McGuffin P, Farmer AE, Aitchison KJ (2011) Interaction between serotonin transporter gene variants and life events predicts response to antidepressants in the GENDEP project. Pharmacogenomics J 11:138–145

Keller M, Medland S, Duncan L (2010) Are extended twin family designs worth the trouble? A comparison of the bias, precision, and accuracy of parameters estimated in four twin family models. Behav Genet 40:377–393

Kendler KS, Gardner CO, Lichtenstein P (2008) A developmental twin study of symptoms of anxiety and depression: evidence for genetic innovation and attenuation. Psychol Med 38:1567–1575

Kendler KS, Eaves LJ, Loken EK, Pedersen NL, Middeldorp CM, Reynolds C, Boomsma D, Lichtenstein P, Silberg J, Gardner CO (2011) The impact of environmental experiences on symptoms of anxiety and depression across the life span. Psychol Sci 22:1343–1352

Kessler RC, Berglund P, Demler O, Jin R, Merikangas KR, Walters EE (2005) Lifetime prevalence and age-of-onset distributions of DSM-IV disorders in the national comorbidity survey replication. Arch Gen Psychiatry 62:593–602

Kim-Cohen J, Caspi A, Moffitt TE, Harrington H, Milne BJ, Poulton R (2003) Prior juvenile diagnoses in adults with mental disorder: developmental follow-back of a prospective-longitudinal cohort. Arch Gen Pstchiatry 60:709–717

Knapp M (2008) Health economics. In: Rutter M, Bishop D, Pine D, Scott S, Stevenson J, Taylor E, Thapar A (eds) Rutter's child and adolescent paychiatry. Wiley-Blackwell, Oxford, pp 123–133

Knorr U, Vinberg M, Kessing LV, Wetterslev J (2010) Salivary cortisol in depressed patients versus control persons: a systematic review and meta-analysis. Psychoneuroendocrinology 35:1275–1286

Kolltveit S, Lange-Nielsen II, Thabet AA, Dyregrov A, Pallesen S, Johnsen TB, Laberg JC (2012) Risk factors for PTSD, anxiety, and depression among adolescents in Gaza. J Trauma Stress 25:164–170

Kroeze Y, Zhou H, Homberg JR (2012) The genetics of selective serotonin reuptake inhibitors. Pharmacol Ther. doi:10.1016/j.pharmthera.2012.08.015

Kupfer DJ, Frank E, Carpenter LL, Neiswanger K (1989) Family history in recurrent depression. J Affect Disord 17:113–119

Lang UE, Hellweg R, Kalus P, Bajbouj M, Lenzen KP, Sander T, Kunz D, Gallinat J (2005) Association of a functional BDNF polymorphism and anxiety-related personality traits. Psychopharmacology (Berl) 180(1):95–99

Lahey BB, Van Hulle CA, Singh AL, Waldman ID, Rathouz PJ (2011) Higher-order genetic and environmental structure of prevalent forms of child and adolescent psychopathology. Arch Gen Psychiatry 68: 181–189

Laxman KE, Lovibond KS, Hassan MK (2008) Impact of bipolar disorder in employed populations. Am J Manag Care 14:757–764

Lazary J (2010) Serotonin transporter gene and negative life change events are associated with depressive phenotype. Neuropsychopharmacol Hung 12:347–354

Lester KJ, Hudson JL, Tropeano M, Creswell C, Collier DA, Farmer A, Lyneham HJ, Rapee RM, Eley TC (2012) Neurotrophic gene polymorphisms and response to psychological therapy. Transl Psychiatry 2:e108. doi:10.1038/tp.2012.33

Leverich GS, Altshuler LL, Frye MA, Suppes T, McElroy SL, Keck PE Jr, Kupka RW, Denicoff KD, Nolen WA, Grunze H, Martinez MI, Post RM (2006) Risk of switch in mood polarity to hypomania or mania in patients with bipolar depression during acute and continuation trials of venlafaxine, sertraline, and bupropion as adjuncts to mood stabilizers. Am J Psychiatry 163:232–239

Lewis KJ, Mars B, Lewis G, Rice F, Sellers R, Thapar AK, Craddock N, Collishaw S, Thapar A (2012) Do parents know best? Parent-reported vs. child-reported depression symptoms as predictors of future child mood disorder in a high-risk sample. J Affect Disord 141(2–3):233–236, PMID: 22609197

Loeber R, Stouthamer-Loeber M (1986) Family factors as correlates and predictors of juvenile conduct problems and delinquency. In: Tonry M, Norris N (eds) Crime and justice. University of Chicago Press, Chicago, pp 29–149

Lohoff FW (2011) The future of psychiatric pharmacogenomics. Pharmacogenomics 12:927–929

Lohoff FW, Aquino TD, Narasimhan S, Multani PK, Etemad B, Rickels K (2011) Serotonin receptor 2A (HTR2A) gene polymorphism predicts treatment response to venlafaxine XR in generalized anxiety disorder. Pharmacogenetics J_{19} 13(1):21–26. doi:1470-269X/11

Lohoff FW, Narasimhan S, Rickels K (2012) Interaction between polymorphisms in serotonin transport (SLC6A4) and serotonin receptor 2A (HTR2A) genes predict treatment response to venlafaxine XR in generalized anxiety disorder. Pharmacogenetics J 19: 1–6. doi: 1470-269X/12

Lopez-Castroman J, Galfalvy H, Currier D, Stanley B, Blasco-Fontecilla H, Baca-Garcia E, Harkavy-Friedman JM, Mann JJ, Oquendo MA (2012) Personality disorder assessments in acute depressive episodes: stability at follow-up. J Nerv Ment Dis 200:526–530

Lovibond PF (1998) Long-term stability of depression, anxiety, and stress syndromes. J Abnorm Psychol 107:520–526

Lubke GH, Hottenga JJ, Walters R, Laurin C, de Geus EJ, Willemsen G, Smit JH, Middeldorp CM, Penninx BW, Vink JM, Boomsma DI (2012) Estimating the genetic variance of major depressive disorder due to all single nucleotide polymorphisms. Biol Psychiatry 72(8):707–709, PMID: 22520966

Luby JL, Essex MJ, Armstrong JM, Klein MH, Zahn-Waxler C, Sullivan JP, Goldsmith HH (2012) Gender differences in emotional reactivity of depressed and at-risk preschoolers: implications for gender specific manifestations of preschool depression. J Clin Child Adolesc Psychol 38:525–537

Lucae S, Ising M, Horstmann S, Baune BT, Arolt V, Müller-Myhsok B, Holsboer F, Domschke K (2010) HTR2A gene variation is involved in antidepressant treatment response. Eur Neuropsychopharmacol 20:65–68

Maier W (2008 Mar) Zobel A (2008) Contribution of allelic variations to the phenotype of response to antidepressants and antipsychotics. Eur Arch Psychiatry Clin Nurosci 258(Suppl 1):12–20. doi:10.1007/s00406-007-1004-z

Malhotra AK, Murphy GM Jr, Kennedy JL (2004) Pharmacogenetics of psychotropic drug response. Am J Psychiatry 161:780–796

Malki K, Campbell J, Davies M, Keers R, Uher R, Ward M, Paya-Cano J, Aitchinson KJ, Binder E, Sluyter F, Kuhn K, Selzer S, Craig I, McGuffin P, Schalkwyk LC (2012) Pharmacoproteomic investigation into antidepressant response in two mouse inbred strains. Proteomics 12:2355–2365

Martinowich K, Lu B (2008) Interaction between BDNF and serotonin: role in mood disorders. Neuropsychopharmacology 33:73–83

Masand PS, Wang X, Gupta S, Schwartz TL, Virk S, Hameed A (2002) Comparison of risperidone and olanzapine in bipolar and schizoaffective disorders. Prim Care Companion J Clin Psychiatry 4(2):70–73

Massart R, Mongeau R, Lanfumey L (2012) Beyond the monoaminergic hypothesis: neuroplasticity and epigenetic changes in a transgenic mouse model of depression. Philos Trans R Soc Lond B Biol Sci 367:2485–2494

McEwen BS (2003) Early life influences on life-long patterns of behavior and health. Ment Retard Dev Disabil Res Rev 9(3):149–154

McGowan PO, Sasaki A, D'Alessio AC, Dymov S, Labonte B, Szyf M, Turecki G, Meaney MJ (2009) Epigenetic regulation of the glucocorticoid receptor in human brain associates with childhood abuse. Nat Neurosci 12:342–348

McGuffin P, Perroud N, Uher R, Butler A, Aitchison KJ, Craig I, Lewis C, Farmer A (2010) The genetics of affective disorder and suicide. Eur Psychiatry 25: 275–277

McMahon FJ, Insel TR (2012) Pharmacogenomics and personalized medicine in neuropsychiatry. Neuron 74:773–776

McMahon FJ, Buervenich S, Charney D, Lipsky R, Rush AJ, Wilson AF, Sorant AJ, Papanicolaou GJ, Laje G, Fava M, Trivedi MH, Wisniewski SR, Manji H (2006) Variation in the gene encoding the serotonin 2A receptor is associated with outcome of antidepressant treatment. Am J Hum Genet 78:804–814

Mellerup E, Andreassen O, Bennike B, Dam H, Durovic S, Hansen T, Melle I, Møller GL, Mors O, Koefoed P (2012) Connection between genetic and clinical data in bipolar disorder. PLoS One 7(9):e44623

Menke A, Klengel T, Binder EB (2012) Epigenetics, depression and antidepressant treatment. Curr Pharm Des 18(36):5879–5889, PMID: 22686347

Merikangas KR, Akiskal HS, Angst J, Greenberg PE, Hirschfeld RM, Petukhova M, Kessler RC (2007) Lifetime and 12-month prevalence of bipolar spectrum disorder in the National Comorbidity Survey replication. Arch Gen Psychiatry 64:543–552

Merikangas KR, He JP, Burstein M, Swanson SA, Avenevoli S, Cui L, Benjet C, Georgiades K, Swendsen J (2010) Lifetime prevalence of mental disorders in U.S. adolescents: results from the National Comorbidity Survey Replication–Adolescent Supplement (NCS-A). J Am Acad Child Adolesc Psychiatry 49:980–989

Mitchell AJ (2006) High medication discontinuation rates in psychiatry: how often is it understandable? J Clin Psychopharmacol 26:109–112

Möller HJ, Bitter I, Bobes J, Fountoulakis K, Höschl C, Kasper S, European Psychiatric Association (2012) Position statement of the European Psychiatric

Association (EPA) on the value of antidepressants in the treatment of unipolar depression. Eur Psychiatry 27:114–128

Montag C, Basten U, Stelzel C, Fiebach CJ, Reuter M (2010) The BDNF Val66Met polymorphism and anxiety: support for animal knock-in studies from a genetic association study in humans. Psychiatry Res 179:86–90

Morag A, Pasmanik-Chor M, Oron-Karni V, Rehavi M, Stingl JC, Gurwitz D (2011) Genome-wide expression profiling of human lymphoblastoid cells identifies CHL1 as a putative SSRI antidepressant response biomarker. Pharmacogenomics 12:171–184

Mortola JF, Liu JH, Gillin JC, Rasmussen DD, Yen SS (1987) Pulsatile rhythms of adrenocorticotrophin (ACTH) and cortisol in women with endogenous depression: evidence for increased ACTH pulse frequency. J Clin Endocrinol Metab 65:962–968

Muris P, Schmidt H, Merckelbach H, Schouten E (2001) Anxiety sensitivity in adolescents: factor structure and relationships to trait anxiety and symptoms of anxiety disorders and depression. Behav Res Ther 39:89–100

Mushtaq D, Ali A, Margoob MA, Murtaza I, Andrade C (2012) Association between serotonin transporter gene promoter-region polymorphism and 4- and 12-week treatment response to sertraline in posttraumatic stress disorder. J Affect Disord 136:955–962

Mutsatsa S, Currid TJ (2012) Pharmacogenetics: a reality or misplaced optimism? J Psychiatr Ment Health Nurs 20(4):314–320. doi:10.1111/j.1365-2850.2012.01910.x

Nanni V, Uher R, Danese A (2011) Childhood maltreatment predicts unfavorable course of illness and treatment outcome in depression. Am J Psychiatry 169:141–151

Narasimhan S, Lohoff FW (2012) Pharmacogenetics of antidepressant drugs: current clinical practice and future directions. Pharmacogenomics 13:441–464

Narusyte J, Neiderhiser J, D'Onofrio B (2008) Testing different types of genotype-environment correlation: an extended Children of Twins models. Devel Psychol 44:1591–1603

Natsuaki MN, Klimes-Dougan B, Ge X, Shirtcliff EA, Hastings PD, Zahn-Waxler C (2009) Early pubertal maturation and internalizing problems in adolescence: sex differences in the role of cortisol reactivity to interpersonal stress. J Clin Child Adolesc Psychol 38:513–524

Nauta MH, Festen H, Reichart CG, Nolen WA, Stant AD, Bockting CL, van der Wee NJ, Beekman A, Doreleijers TA, Hartman CA, de Jong PJ, de Vries SO (2012) Preventing mood and anxiety disorders in youth: a multi-centre RCT in the high risk offspring of depressed and anxious patients. BMC Psychiatry 12:31, PMID: 22510426

Navailles S, Zimnisky R, Schmauss C (2010) Expression of glucocorticoid receptor and early growth response gene 1 during postnatal development of two inbred strains of mice exposed to early life stress. Dev Neurosci 32:139–148

Nettle D, Bateson M (2012) The evolutionary origins of mood and its disorders. Curr Biol 22:R712–R721

Ng CH, Easteal S, Tan S, Schweitzer I, Ho BK, Aziz S (2006) Serotonin transporter polymorphisms and clinical response to sertraline across ethnicities. Prog Neuropsychopharmacol Biol Psychiatry 30:953–957

Nierenberg AA, Trivedi MH, Fava M, Biggs MM, Shores-Wilson K, Wisniewski SR, Balasubramani GK, Rush AJ (2007) Family history of mood disorder and characteristics of major depressive disorder: a STAR*D (sequenced treatment alternatives to relieve depression) study. J Psychiatr Res 41:214–221

Nilsen TS, Eisemann M, Kvernmo S (2012) Predictors and moderators of outcome in child and adolescent anxiety and depression: a systematic review of psychological treatment studies. Eur Child Adolesc Psychiatry 22(2):69–87, PMID: 22923065

Nugent NR, Tyrka AR, Carpenter LL, Price LH (2011) Gene-environment interactions: early life stress and risk for depressive and anxiety disorders. Psychopharmacology 214:175–196

O'Connor TG, Heron J, Golding J, Glover V (2003) Maternal antenatal anxiety and behavioural/emotional problems in children: a test of a programming hypothesis. J Child Psychol Psychiatry 44:1025–1036

Office of Applied Studies (2005) Institute of Mental Disorders and Drug Abuse, Bethesda, MD., USA

O'Keane V, Lightman S, Marsh M, Pawlby S, Papadopoulos AS, Moore R, Patrck K (2011) Increased pituitary-adrenal activation and shortened gestation in a sample of depressed pregnant women: a pilot study. J Affect Disord 130:300–3006

Olgiati P, Bajo E, Bigelli M, De Ronchi D, Serretti A (2012) Should pharmacogenetics be incorporated in major depression treatment? Economic evaluation in high- and middle-income European countries. Prog Neuro-Psychopharmacol Biol Psychiatry 36: 147–154

Ollendick TH, Jarrett MA, Grills-Taquecchel AE, Hovey JD, Wolff JC (2008) Comorbidity as a predictor and moderator of treatment outcome in youth with anxiety, affective, attention deficit/hyperactivity disorder, and oppositional/conduct disorders. Clin Psychol Rev 28:1447–1471

Osypuk TL, Schmidt NM, Bates LM, Tchetgen-Tchetgen EJ, Earls FJ, Glymour MM (2012) Gender and crime victimization modify neighborhood effects on adolescent mental health. Pediatrics 130:472–481

Owens M, Stevenson J, Norgate R, Hadwin J (2008) Processing efficiency theory in children: working memory as a mediator between trait anxiety and academic performance. Anxiety Stress Coping 21:417–430

Palomo T, Kostrzewa RM, Beninger RJ, Archer T (2007a) Treatment consideration and manifest complexity in comorbid neuropsychiatric disorders. Neurotox Res 12:43–60

Palomo T, Kostrzewa RM, Beninger RJ, Archer T (2007b) Genetic variation and shared biological susceptibility underlying comorbidity in neuropsychiatry. Neurotox Res 12:29–42

Palomo T, Archer T, Kostrzewa RM, Beninger RJ (2007c) Comorbidity of substance abuse with other psychiatric disorders. Neurotox Res 12:17–27

Papakostas GI, Fava M (2009) Does the probability of receiving placebo influence clinical trial outcome? A meta-regression of double-blind, randomized clinical trials in MDD. Eur Neuropsychopharmacol 19:34–40

Pariante CM, Miller AH (2001) Glucocorticoid receptors in major depression: relevance to pathophysiology and treatment. Biol Psychiatry 49:391–404

Pariante CM, Thomas SA, Lovestone S (2004) Do antidepressants regulate how cortisol affects the brain? Psychoneuroendocrinology 29:423–447

Perlis RH (2007) Pharmacogenetic studies of antidepressant response: how far from the clinic? Psychiatr Clin North Am 30(1):125–138

Pezawas L, Meyer-Lindenberg A, Goldman AL, Verchinski BA, Chen G, Kolachana BS, Egan MF, Mattay VS, Hariri AR, Weinberger DR (2008) Evidence of biologic epistasis between BDNF and SLC6A4 and implications for depression. Mol Psychiatry 13:709–716

Pini S, de Queiroz V, Pagnin D, Pezawas L, Angst J, Cassano GB, Wittchen HU (2005) Prevalence and burden of bipolar disorders in European countries. Eur Neuropsychopharmacol 15:425–434

Piwowarska J, Chimiak A, Matsumoto H, Dziklinska A, Radziwon-Zaleska M, Szelenberger W, Pachecka J (2012) Serum cortisol concentration in patients with major depression after treatment with fluoxetine. Psychiatry Res. doi:10.1016/j.psychres.2012.01.029

Pollack M (2005) Comorbid anxiety and depression. J Clin Psychiatry 66:22–29

Porcelli S, Drago A, Fabbri C, Gibiino S, Calati R, Serretti A (2011) Pharmacogenetics of antidepressant response. J Psychiatry Neurosci 36:87–113

Porcelli S, Fabbri C, Serretti A (2012) Meta-analysis of serotonin transporter gene promoter polymorphism (5-HTTLPR) association with antidepressant efficacy. Eur Neuropsychopharmacol 22:239–258

Post RM (2010) Mechanisms of illness progression in the recurrent affective disorders. Neurotox Res 18:256–271

Post RM, Leverich GS, Kupka RW, Keck PE Jr, McElroy SL, Altshuler LL, Frye MA, Luckenbaugh DA, Rowe M, Grunze H, Suppes T, Nolen WA (2010) Early-onset bipolar disorder and treatment delay are risk factors for poor outcome in adulthood. J Clin Psychiatry 71:864–872

Power RA, Keers R, Ng MY, Butler AW, Uher R, Cohen-Woods S, Ising M, Craddock N, Owen MJ, Korszun A, Jones L, Jones I, Gill M, Rice JP, Hauser J, Henigsberg N, Maier W, Zobel A, Mors O, Placentino AS, Rietschel M, Souery D, Kozel D, Preisig M, Lucae S, Binder EB, Aitchison KJ, Tozzi F, Muglia P, Breen G, Craig IW, Farmer AE, Müller-Myhsok B, McGuffin P, Lewis C (2012) Dissecting the genetic heterogeneity of depression through age at onset. Am J Med Genet B Neuropsychiatr Genet 159B:859–868. doi:10.1002/ajmg.b.32093

Priess-Groben HA, Hyde JS (2012) 5-HTTLPR X stress in adolescent depression: moderation by MAOA and gender. J Abnorm Child Psychol 41(2):281–294, PMID: 22836288

Quanbeck C, Frye M, Altshuler L (2003) Mania and the law in California: understanding the criminalization of the mentally ill. Am J Psychiatry 160:1245–1250

Rabl U, Scharinger C, Müller M, Pezawas L (2010) Imaging genetics: implications for research on variable antidepressant drug response. Expert Rev Clin Pharmacol 3:471–489

Rapee RM, Schniering CA, Hudson JL (2009) Anxiety disorders during childhood and adolescence: origins and treatment. Ann Rev Clin Psychol 5:311–341

Rethorst CD, Toups MS, Greer TL, Nakonezny PA, Carmody TJ, Grannemann BD, Huebinger RM, Barber RC, Trivedi MH (2012) Pro-inflammatory cytokines as predictors of antidepressant effects of exercise in major depressive disorder. Mol Psychiatry 45:1–6. doi: 10.1038/mp.2012.125

Richardson J, Steiger H, Schmitz N, Joober R, Bruce KR, Israel M, Gauvin L, Anestin AS, Dandurand C, Howard H, de Guzman R (2008) Relevance of the 5-HTTLPR polymorphism and childhood abuse to increased psychiatric comorbidity in women with bulimia-spectrum disorders. J Clin Psychiatry 69:981–990

Rubin RT, Poland RE, Lesser IM, Winston RA, Blodgett AL (1987) Neuroendocrine aspects of primary endogenous depression. I. Cortisol secretory dynamics in patients and matched controls. Arch Gen Psychiatry 44:328–336

Rundell JR, Staab JP, Shinozaki G, Saad-Pendergrass D, Moore K, McAlpine D, Mrazek D (2011) Pharmacogenomic testing in a tertiary care outpatient psychosomatic medicine practice. Psychosomatics 52:141–146

Russcher H, Smit P, van der Akker EL, van Rossum EF, Brinkmann AO, de Jong FH, Lamberts SW, Koper JW (2005) Two polymorphisms in the glucocorticoid receptor gene directly affect glucocorticoid-regulated gene expression. J Clin Endocrinol Metab 90: 5804–5810

Sadock BJ (2012) Inevitable suicide: a new paradigm in psychiatry. J Psychiatr Pract 18:221–224

Scharinger C, Rabl U, Sitte HH, Pezawas L (2010) Imaging genetics of mood disorders. NeuroImage 53:810–821

Scharinger C, Rabl U, Pezawas L, Kasper S (2011) The genetic blueprint of major depressive disorder: contributions of imaging genetics studies. World J Biol Psychiatry 12:474–488

Schüle C (2006) Neuroendocrinal mechanism of action of antidepressant drugs. J Neuroendocrinol 19:213–226

Schuster RM, Mermelstein R, Wakschlag L (2012) Gender-specific relationships between depressive symptoms, marijuana use, parental communication and risky sexual behavior in adolescence. J Youth Adolesc. PMID: 22927009

Schutter DJ (2012) The cerebello-hypothalamic-pituitary-adrenal axis dysregulation hypothesis in depressive disorder. Med Hypotheses. pii: S0306-9877(12)00404-5. doi: 10.1016/j.mehy.2012.08.027

Serretti A, Benedetti F, Mandelli L, Calati R, Caneva B, Lorenzi C, Fontana V, Colombo C, Smeraldi E (2008) Association between GSK-3beta-50T/C polymorphism and personality and psychotic symptoms in mood disorders. Psychiatry Res 158(2):132–140

Serretti A, Chiesa A (2009) Treatment-emergent sexual dysfunction related to antidepressants: a meta-analysis. J Clin Psychopharmacol 29:259–266

Shastry BS (2005) Genetic diversity and new therapeutic concepts. J Hum Genet 50:321–328

Shastry BS (2006) Pharmacogenetics and the concept of individualized medicine. Pharmacogenomics J 6:16–21

Silberg J, Pickles A, Rutter M, Hewitt J, Simonoff E, Maes H, Carbonneau R, Murrelle L, Foley D (1999) The influence of genetic factors and life stress on depression among adolescent girls. Arch Gen Psychiatry 56:225–232

Silberg J, Maes HH, Eaves L (2010) Genetic and environmental influences on the transmission of parental depression to children's depression and conduct disturbance. J Child Psychol Psychiatry 51:734–744

Silberg J, Maes H, Eaves LJ (2012) Unraveling the effect of genes and environment in the transmission of parental antisocial behavior to conduct disturbance, depression and hyperactivity. J Child Psychol Psychiatry 53:668–677. doi:10.1111/j.1469-7610.2011.02494.x

Silverman W, Goedhart A, Barrett P, Turner C (2003) The facets of anxiety sensitivity represented in the childhood anxiety sensitivity index: confirmatory analyses of factor models from past studies. J Abnorm Psychol 112:364–374

Singh AL, D'Onofrio BM, Slutske WS, Turkheimer E, Emery RE, Harden KP, Heath AC, Madden PA, Statham DJ, Martin NG (2011) Parental depression and offspring psychopathology: a children of twins study. Psychol Med 41:1385–1395

Smoller JW, Finn CT (2003) Family, twin, and adoption studies of bipolar disorders. Am J Med Genet 123:48–58

Soeiro-de-Souza MG, Post RM, de Sousa ML, Missio G, do Prado CM, Gattaz WF, Moreno RA, Machado-Vieira R (2012) Does BDNF genotype influence creative output in bipolar I manic patients? J Affect Disord 139:181–186

Souery D, Oswald P, Massat I, Bailer U, Bollen J, Demyttenaere K, Kasper S, Lecrubier Y, Montgomery S, Serretti A, Zohar J, Mendlewicz J (2007) Clinical factors associated with treatment resistance in major depressive disorder: results from a European multicentre study. J Clin Psychiatry 68:1062–1070

Spijker AT, van Rossum EFC (2009) Glucocorticoid receptor polymorphisms in major depression: focus on glucocorticoid sensitivity and neurocognitive functioning. Ann NY Acad Sci 117:9199–9215

Spijker AT, van Rossum EFC (2012) Glucocorticoid sensitivity in mood disorders. Neuroendocrinology 95: 179–186

Squassina A, Manchia M, Del Zompo M (2010) Pharmacogenomics of mood stabilizers in the treatment of bipolar disorder. Hum Genom Proteom 2010:159761. doi:10.4061/2010/159761

Steiger H, Joober R, Gauvin L, Bruce KR, Richardson J, Israel M, Anestin AS, Groleau P (2008) Serotonin-system polymorphisms (5-HTTLPR and -1438G/A) and responses of patients with bulimic syndromes to multimodal treatments. J Clin Psychiatry 69:1565–1571

Steimer W, Muller B, Leucht S, Kissling W (2001) Pharmacogenetics: a new diagnostic tool in the management of antidepressant drug therapy. Clin Chim Acta 308:33–41

Sullivan PF, Neale MC, Kendler KS (2000) Genetic epidemiology of major depression: review and meta-analysis. Am J Psychiatry 157:1552–1562

Talge NM, Neal C, Glover V (2007) Antenatal maternal stress and long-term effects on child neurodevelopment: how and why? J Child Psychol Psychiatry 48:245–261

Tarantino LM, Sullivan PF, Meltzer-Brody S (2011) Using animal models to disentangle the role of genetic, epigenetic, and environmental influences on behavioral outcomes associated with maternal anxiety and depression. Front Psychiatry 2:44

Taylor S, Koch WJ, Woody S, McLean P (1996) Anxiety sensitivity and depression: how are they related? J Abnorm Psychol 105:474–479

Taylor MJ, Sen S, Bhagwagar Z (2010) Antidepressant response and the serotonin transporter gene-linked polymorphic region. Biol Psychiatry 68:536–543

Terracciano A, Sanna S, Uda M, Deiana B, Usala G, Busonero F, Maschio A, Scally M, Patriciu N, Chen WM, Distel MA, Slagboom EP, Boomsma DI, Villafuerte S, Sliwerska E, Burmeister M, Amin N, Janssens AC, van Duijn CM, Schlessinger D, Abecasis GR, Costa PT (2010a) Genome-wide association scan for five major dimensions of personality. Mol Psychiatry 15:647–656

Terracciano A, Tanaka T, Sutin AR, Deiana B, Balaci L, Sanna S, Olla N, Maschio A, Uda M, Ferrucci L, Schlessinger D, Costa PT (2010b) BDNF Val66Met is associated with introversion and interacts with 5-HTTLPR to influence neuroticism. Neuropsychopharmacology 35:1083–1089

Terracciano A, Lobina M, Piras MG, Mulas A, Cannas A, Meirelles O, Sutin AR, Zonderman AB, Uda M, Crisponi L, Schlessinger D (2011) Neuroticism, depressive symptoms, and serum BDNF. Psychosom Med 73:638–642

Torpy JM, Burke AE, Golub RM (2011) JAMA patient page. Generalized anxiety disorder. JAMA 305:522

Tozzi F, Prokopenko I, Perry JD, Kennedy JL, McCarthy AD, Holsboer F, Berrettini W, Middleton LT, Chilcoat HD, Muglia P (2008) Family history of depression is associated with younger age of onset in patients with recurrent depression. Psychol Med 38:641–649

Trivedi MH, Rush AJ, Wisnieski SR, Nierenberg AA, Warden D, Ritz L, Norquist G, Howland RH, Lebowitz B, McGrath PJ, Shores-Wilson K, Biggs MM, Balasubramani GK, Fava M, STAR*D Study Team (2006) Evaluation of outcomes with citalopram for depression using measurement-based care in STAR*D:

implications for clinical practice. Am J Psychiatry 163:28–40

Truman CJ, Goldberg JF, Ghaemi SN, Baldassano CF, Wisniewski SR, Dennehy EB, Thase ME, Sachs GS (2007) Self-reported history of manic/hypomanic switch associated with antidepressant use: data from the Systematic Treatment Enhancement Program for Bipolar Disorder (STEP-BD). J Clin Psychiatry 68:1472–1479

Tsai SJ, Hong CJ, Liou YJ (2011) Recent molecular genetic studies and methodological issues in suicide research. Prog Neuropsychopharmacol Biol Psychiatry 35:809–817

Tully EC, Iacono WG, McGue M (2010) Changes in genetic and environmental influences on the development of nicotine dependence and major depressive disorder from middle adolescence to early adulthood. Dev Psychopathol 22:831–848

Tyrka AR, Price LH, Gelernter J, Schepker C, Anderson GM, Carpenter LL (2009) Interaction of childhood maltreatment with the corticotropin-releasing hormone receptor gene: effects on hypothalamic-pituitary-adrenal axis reactivity. Biol Psychiatry 66:681–685

Uher R, McGuffin P (2010) The moderation of the serotonin transporter gene of environmental adversity in the etiology of depression: 2009 update. Mol Psychiatry 15:18–22

Uher R, Maier W, Hauser J, Marusic A, Schmael C, Mors O, Henigsberg N, Souery D, Placentino A, Rietschel M, Zobel A, Dmitrzak-Weglarz M, Petrovic A, Jorgensen L, Kalember P, Giovannini C, Barreto M, Elkin A, Landau S, Farmer A, Aitchison KJ, McGuffin P (2009) Differential efficacy of escitalopram and nortriptyline on dimensional measures of depression. Br J Psychiatry 194:252–259

van Ijzendoorn MH, Belsky J, Bakermans-Kranenburg MJ (2012) Serotonin transporter genotype 5HTTLPR as a marker of differential susceptibility? A meta-analysis of child and adolescent gene-by-environment studies. Transl Psychiatry 2:e147

van Rossum EF, Koper JW, Huizenga NA, Uitterlinden AG, Janssen JA, Brinkmann AO, Grobbee DE, de Jong FH, van Duyn CM, Pols HA, Lamberts SW (2002) A polymorphism in the glucocorticoid receptor gene, which decreases sensitivity to glucocorticoids in vivo, is associated with low insulin and cholesterol levels. Diabetes 51:3128–3134

van Rossum EF, Binder EB, Majer M, Koper JW, Ising M, Modell S, Salyakina D, Lamberts SW, Holsboer F (2006) Polymorphisms of the glucocorticoid receptor gene and major depression. Biol Psychiatry 59:681–688

van West D, Van Den Eede F, Del-Favero J, Souery D, Norrback KF, Van Duijn C, Sluijs S, Adolfsson R, Mendlewicz J, Deboutte D, Van Broeckhoven C, Claes S (2006) Glucocorticoid receptor gene-based SNP analysis in patients with recurrent major depression. Neuropsychopharmacology 31:620–627

Vandeleur C, Rothen S, Gholam-Rezaee M, Castelao E, Vidal S, Favre S, Ferrero F, Halfon O, Fumeaux P, Merikangas KR, Aubry JM, Burstein M, Preisig M (2012) Mental disorders in offspring of parents with bipolar and major depressive disorders. Bipolar Disord 14:641–653

Veronese A, Ayuso-Mateos JL, Cabello M, Chatterji S, Nuevo R (2012) Work disability and depressive disorders: impact on the European population. Am J Phys Med Rehabil 91(13 Suppl 1):S62–S68

Vishnuvardhan G, Saddichha S (2012) Psychiatric comorbidity and gender differences among suicide attempters in Bangalore, India. Gen Hosp Psychiatry 34:410–414

Walpoth-Niederwanger M, Kemmler G, Grunze H, Wei U, Hörtnagl C, Strau R, Blasko I, Hausmann A (2012) Treatment patterns in inpatients with bipolar disorder at a psychiatric university hospital over a 9-year period: focus on mood stabilizers. Int Clin Psychopharmacol 27:256–266

Walsh T, Stewart S, McLaughlin E, Comeau N (2004) Gender differences in Childhood Anxiety Sensitivity Index (CASI) dimensions. J Anxiety Disord 18: 695–706

Weems CF (2008) Developmental trajectories of childhood anxiety: identifying continuity and change in anxious emotion. Dev Rev 28:488–502

Williams KE, Stemmle PG, Westphal LM, Rasgon NL (2011) Mood disorders in oocyte donor candidates: brief report and implications for future research. Hum Reprod 26:847–852

Wolkowitz OM, Wolf J, Shelly W, Rosser R, Burke HM, Lerner GK, Reus VI, Nelson JC, Epel ES, Mellon SH (2011) Serum BDNF levels before treatment predict SSRI response in depression. Prog Neuropsychopharmacol Biol Psychiatry 35:1623–1630

Wright K, Asmundson G, McCreary D, Stewart S, McLaughlin E, Comeau M (2010) Confirmatory factor analysis of the childhood anxiety sensitivity index: a gender comparison. Cogn Behav Ther 39:225–235

Yu H, Wang DD, Wang Y, Liu T, Lee FS, Chen ZY (2012) Variant brain-derived neurotrophic factor Val66Met polymorphism alters vulnerability to stress and response to antidepressants. J Neurosci 32:4092–4101

Zahn-Waxler C, Klimes-Dougan B, Slattery MJ (2000) Internalizing problems of childhood and adolescence: prospects, pitfalls, and progress in understanding the development of anxiety and depression. Dev Psychopathol 12:443–466

Zavos HMS, Frühling VR, Eley TC (2012) A longitudinal, genetically informative, study of associations between anxiety sensitivity, anxiety and depression. Behav Genet 42:592–602

Zinbarg RE, Brown TA, Barlow DH, Rapee RM (2001) Anxiety sensitivity, panic, and depressed mood: a reanalysis teasing apart the contributions of the two levels in the hierarchial structure of the Anxiety Sensitivity Index. J Abnorm Psychol 110:372–377

Zobel A, Maier W (2010) Pharmacogenetics of antidepressant treatment. Eur Arch Psychiatr Clin Neurosci 260:407–417

Zou YF, Wang F, Feng XL, Li WF, Tao JH, Pan FM, Huang F, Su H (2011) Meta-analysis of FKBP5 gene polymorphisms association with treatment response in patients with mood disorders. Neurosci Lett 484: 56–61

Zullig KJ, Divin AL (2012) The association between non-medical prescription drug use, depressive symptoms, and suicidality among college students. Addict Behav 37:890–899

Pharmacogenetics of Opioid and Alcohol Addiction

17

Desirée González Callejas, Marisa Cañadas Garre, Margarita Aguilera, Enrique Jiménez Varo, and Miguel A. Calleja Hernández

Abstract

Opiate and alcohol addiction are serious health and social problems, associated with high morbidity and mortality. The manifestation and development of these complex diseases involves genetic factors, environmental factors, and the pharmacokinetic and pharmacodynamic properties of the psychoactive substance.

Among the main therapeutic indications of opioids are substance abuse programs and treatment of pain. Classic welfare programs are aimed at drug abuse patient detoxification programs or opiate agonist maintenance, being the methadone maintenance program (MMP) the best known. However, for some of the most seriously affected drug users, these programs are ineffective, hence the existence of alternative therapies such as Heroin Prescription Program (HPP) in Switzerland and Holland. These programs are located within a therapeutic approach and integrated harm reduction in the public health network for drug abuse. In other countries, like Germany and Spain, several clinical trials with PPH have been successfully conducted.

The establishment of an individualized prescription of opioids safely and effectively is a difficult task for the physician, due to interindividual variability in the individual's response to the substance and the individual risk of the administered dose which is within therapeutic range.

In recent years, numerous studies have described several genes involved in opioid and alcohol dependence. Among genetic variations that affect opioid receptors, the SNP that affects the mu receptor in position 118

D.G. Callejas (✉) • M.C. Garre
E.J. Varo • M.A.C. Hernández
Pharmacogenetics Unit, Pharmacy Service, Hospital Universitario Virgen de las Nieves, Granada, Spain
e-mail: desiree.gonzalez.callejas@hotmail.com; marisacgarre@gmail.com

M. Aguilera
Pharmacogenetics Unit, Pharmacy Service, Hospital Universitario Virgen de las Nieves, Granada, Spain

Instituto de Nutrición y Tecnología de los Alimentos "José Mataix Verdú" de la Universidad de Granada (INYTA); Centro de Investigación Biomédica (CIBM), Granada, Spain
e-mail: maguiler@ugr.es

D. Barh et al. (eds.), *Omics for Personalized Medicine*,
DOI 10.1007/978-81-322-1184-6_17, © Springer India 2013

seems to be one of the most clinically relevant in alcohol and opioid addiction. Current pharmacotherapy targeted to alcohol addiction also has been associated to gene polymorphisms in opioid receptors.

Polymorphisms in the different isoenzymes of CYP450 (CYP3A4, CYP2D6, CYP2B6, CYP1A2, CYP2C9, and CYP2C19) also significantly influence in the pharmacokinetics and the effects of opioids and concomitant treatment of patients.

Pharmacogenetic analysis, focused on the opioid system gene polymorphisms and interactions related to CYP450, together with the study of clinical parameters may be a useful tool for the physician to adjust the appropriate pharmacotherapeutic dose in each particular case.

This chapter will review the most relevant gene polymorphisms related to alcohol and opioid addiction in the opioid system and the CYP450 complex.

1 Introduction

Abuse of central nervous system (CNS) depressant psychoactive substances, such as opiates and alcohol, causes serious health problems among consumers. The uncontrolled use of these substances originates manifestations such as habitual intoxication, drug tolerance, physical dependence, sensitization, and compulsive search or "craving" (Alvarez and Farre 2005).

1.1 Opioid Dependence

Opioid dependence is a complex disease which involves environmental factors, pharmacokinetic and pharmacodynamic properties of the psychoactive substances, and genetic factors (Alvarez and Farre 2005).

The pharmacological properties of opioids are of great therapeutic interest due to their analgesic effects; however, abuse of these substances causes serious organic, psychiatric, and social complications, making the approach for the treatment of these subjects very complex.

Currently, health services offer a wide range of alternative therapies to treat this disease. Methadone maintenance programs with opiate agonists are the most commonly used as they have shown their high efficiency and effectiveness (Marset 2005).

There is, however, a population of severely affected users, who escapes classic healthcare programs and for whom the heroin prescription programs (HPP) are destined. These programs are set within a therapeutic approach of harm reduction and integrated into the public healthcare network for drug addicts in countries like Switzerland and the Netherlands. In others, such as Spain, the clinical trial "PEPSA" took place in 2003, based on individualized protocolized prescription of opioid agonists, specifically diacetylmorphine (DAM) administered intravenously and methadone (MTD) orally. Establishing individualized doses of opiates, safely and effectively, is a difficult task due to interindividual differences in dose requirements of equivalent opioids among patients and the individual narrow therapeutical range.

1.2 Alcohol Dependence

Alcoholism etiology involves genetic, psychosocial, and environmental factors. The use and abuse of alcohol is one of the main factors related to the health of individuals, and their consequences have a major impact in terms of health and in social terms.

Alcohol dependence triggers craving caused by decreased dopamine levels, increased calcium channel, increased NMDA receptors, and decreased GABA.

There are several drug treatments available to approach this disease, based on the use of anti-craving drugs (acamprosate, naltrexone, or serotonergic drugs) or interdictors drugs (disulfiram or calcium cyanamide), along with psychotherapy and support groups.

In this chapter, we will focus on human studies showing the influence of certain gene polymorphisms in different molecular structures, mainly the opioid system and phase I metabolism (cytochrome P450) genes, on opioids and/or alcohol dependence.

2 Pharmacology of Psychoactive Substances

2.1 Pharmacotherapy in Opioid Dependence

Opiates that will be studied in this section are diacetylmorphine (DAM), morphine (MOR), and methadone (MET).

2.1.1 Diacetylmorphine

Indications

DAM is not available for therapeutic use in the USA, but is used in other countries, particularly the UK, as an alternative to morphine and other opioids for the treatment of severe pain (Micromedex® 2.0). In Spain, it is also approved for opioid abuse in the context of the PEPSA Program (Experimental Program for the Prescription of Stupefacients in Andalusia), an ongoing clinical trial (March et al. 2006).

Pharmacokinetics

Heroin (diacetylmorphine, DAM) reaches its maximum concentration 1–2 min after intravenous or smoked administration and after 5 min of intranasal or intramuscular administration.

The oral bioavailability of heroin and morphine is low, due to a first-pass hepatic metabolism.

Distribution of opioid is very good. It crosses the blood–brain barrier and the placenta. The volume of distribution for heroin is 2–5 L/kg and 3.3 L/kg for morphine. Heroin is more soluble in lipids than morphine and more quickly crosses the blood–brain barrier. Heroin hydrolyzes to monoacetylmorphine in 5–10 min, which will then be metabolized to morphine. Circulating morphine becomes morphine-3-glucuronide (inactive) and morphine-6-glucuronide (pharmacologically active). Codeine is transformed in part into morphine. Morphine is excreted in the bile, and a small proportion is eliminated in the feces. The fraction of free morphine in urine only reaches 10 % of the dose. Although heroin is not usually found in urine, the presence of monoacetylmorphine confirms heroin administration (Martindale: The Complete Drug Reference).

When diacetylmorphine reaches the brain, it interacts with opioid receptors, mainly mu receptors operating in the area of analgesia, and depresses respiration and delta receptors, which according to recent theories may be more associated with mood with analgesia. It reaches high concentrations in a short period of time, which provokes great pleasure in patients with opioid, the so-called flash effect, not experienced with other opioids.

DAM presents renal excretion. Between 42 and 70 % of a parenteral dose of DAM can be recovered in urine as morphine (unchanged and conjugated). After oral doses of diacetylmorphine, up to 77 % has been recovered in the urine as morphine in some studies. Elimination half-life is 3 min.

Consumption patterns are conditioned not only by kinetics and dynamics of the drug but also by cultural influences (fashion), social, availability or offering, and individual variables (motivations, diseases, etc.). Tolerance is high and develops relatively quickly. The same applies to psychological and physiological dependence.

Mechanism of Action

DAM is an opioid analgesic with pharmacological properties similar to morphine. DAM is essentially a prodrug, being converted to morphine and 6-acetylmorphine (active metabolite) after parenteral administration. After parenteral doses in patients with acute pain, the potency of diacetylmorphine on a weight basis is at least twice that of morphine (Wallenstein et al. 1990). DAM is completely converted to morphine due to first-pass metabolism after oral administration; it is less efficient than an equal dose of morphine in providing delivery of systemic morphine.

Similarly to other opioids, DAM exerts pharmacologic effects by acting as an agonist in the

CNS, primarily at mu receptors; however, there is also significant affinity for kappa and delta receptors. 6-Acetylmorphine and morphine have been demonstrated in vitro to bind to opiate receptors, with morphine being slightly more potent; in contrast, diacetylmorphine itself binds only weakly or not at all to opiate receptors (Micromedex® 2.0).

Interactions

Two minor interactions have been described for DAM, DAM-alprazolam, and DAM-nabilone. Other interactions that should be considered are those produced with morphine, the diamorphine metabolite (Micromedex® 2.0).

2.1.2 Morphine

MOR is normally administered intravenously, with onset of effect of less than 1 min and a peak showing analgesic effect at 20 min after injection. The duration of action is about 4 h. Morphine plasma levels do not correlate with pharmacological activity, reflecting a delay in the penetration of the morphine through the blood–brain barrier. The poor penetration of morphine in the CNS is due to its relatively poor solubility in lipids, 90 % of ionization at physiological pH, protein binding, and conjugation with glucuronic acid.

Indications

FDA-labeled indications for MOR include acute myocardial infarction (pain), anesthesia (adjunct), obstetric pain, chronic pain (moderate to severe, in patients requiring a continuous around-the-clock opioid analgesic for an extended period of time), patient-controlled analgesia, postoperative pain, premedication for anesthetic procedure, and acute pulmonary edema (adjunct) (Micromedex® 2.0).

Pharmacokinetics

Morphine is metabolized primarily by conjugation with glucuronic acid in the liver and other organs, especially in the kidneys. Approximately 5–10 % of morphine appears as morphine-6-glucuronide, an active metabolite that produces analgesia and depression of ventilation, accumulating in patients with renal or kidney failure.

The elimination half-life is 114 min for morphine and 173 min for morphine-3-glucuronide, an inactive and predominant metabolite.

Mechanism of Action

Morphine sulfate is a pure opioid agonist that binds selectively to the mu receptor. Primary actions are developed in the brain through transitory stimulation prior to depression. In the CNS, it promotes analgesia and respiratory depression by decreasing brain stem respiratory centers in response to carbon dioxide tension and electrical stimulation. It also decreases gastric, biliary, and pancreatic secretion, induces peripheral vasodilation, and promotes opioid-induced hypotension due to histamine release (Micromedex® 2.0).

Interactions

The concomitant use of MOR is contraindicated with naltrexone and monoamine oxidase inhibitors. Other serious interactions are produced with barbiturates, centrally acting muscle relaxants, opioid analgesics, benzodiazepines, opioid analgesics and opioid agonists/antagonists, cimetidine, chloral hydrate, ethchlorvynol, sodium oxybate, phenothiazines, and tapentadol (Micromedex® 2.0).

2.1.3 Methadone

Indications

FDA-labeled indications of MET are opioid abuse, maintenance therapy, and pain (moderate to severe) (Micromedex® 2.0).

Pharmacokinetics

MET is administered orally and has a prolonged half-life (18–24 h). It is subject to accumulation in repeated daily administrations. MET is rapidly absorbed after oral doses and has high oral bioavailability and acts 30–35 min after ingestion. Peak plasma concentrations have been reported 1–5 h after a single tablet. Its main effects can last 18–24 h, although some may last up to 36 h (Martindale: The Complete Drug Reference).

Methadone hydrochloride is readily absorbed from the gastrointestinal tract and after subcutaneous or intramuscular injections. It is widely distributed in the tissues, diffuses across the placenta, and is distributed into breast milk. MET is metabolized in the liver by hepatic microsomal enzymes, mainly by N-demethylation and cyclization, and the metabolites are excreted in the bile and urine. Metabolism is mainly catalyzed by CYP3A4, although other cytochrome P450 isoenzymes also contribute. MET is also able to accelerate its own metabolism by inducing the 3A4 isoenzyme of the cytochrome P450 system, responsible for demethylation, and inhibits 2D6 isoenzyme, responsible for the biotransformation system used by many drugs (Martindale: The Complete Drug Reference).

It undergoes considerable tissue distribution, and protein binding is reported to be 60–90 % with α1-acid glycoprotein being the main binding protein in plasma. Metabolism to the major metabolite *N*-demethylmethadone (2-ethylidine-1, 5-dimethyl-3,3-diphenylpyrrolidine) and the minor metabolite 2-ethyl-3,3-diphenyl-5-methylpyrrolidine, both of them inactive, occurs in the liver. These metabolites are excreted in the feces and urine with unmodified methadone. Other metabolites, including methadol and normethadol, have also been described. The liver may also serve as a major storage site of unchanged methadone which is taken up, bound nonspecifically by the liver, and released again mainly unchanged. Urinary excretion of methadone is pH-dependent, the lower the pH the greater the clearance (Martindale: The Complete Drug Reference).

In addition to marked interindividual variations, there are differences in the pharmacokinetics of methadone after single or multiple doses. Elimination half-lives vary considerably (a range of 15–60 h has been quoted) and may be much longer than the 18 h reported following a single dose. Careful adjustment of dosage is necessary with repeated doses.

Most studies have been in addicts. Plasma concentrations have been found to vary widely during methadone maintenance therapy with large differences between patients and wide fluctuations in individual patients. These variations in kinetics have also been seen in cancer patients. Furthermore, many factors can modify the pharmacological effectiveness of methadone, especially hepatopathies, very frequent in these patients, and chronic kidney disease, but also pharmacokinetic or pharmacodynamic interactions due to concomitant use of other drugs or medications for treating other diseases.

The tolerance induced by methadone is high and is developed rapidly inhibiting their subjective effects. It causes physical dependence and an abstinence syndrome that takes longer to appear than other opiates but is longer (about 2 weeks) and is more severe than withdrawal from other opiates, as methadone plasma concentrations remain more time in the body.

Mechanism of Action

Methadone hydrochloride, a synthetic opioid analgesic, is an agonist of the mu receptor. It has several actions qualitatively similar to those of morphine, mainly involving the CNS and organs composed of smooth muscles. Analgesia, sedation, and detoxification or maintenance in opioid addiction can be achieved with therapeutic use of methadone hydrochloride.

Interactions

There are numerous interactions between methadone and other drugs. The most severe and totally contraindicated are naltrexone, itraconazole, ziprasidone, dronedarone, thioridazine, sparfloxacin, mesoridazine, cisapride, rasagiline, pimozide, and posaconazole (Martindale: The Complete Drug Reference).

Most serious interactions may cause an increased risk of prolongation of QT interval. Opioid analgesics and opioid agonists/antagonists may lead to precipitation of withdrawal symptoms (abdominal cramps, nausea, vomiting, lacrimation, rhinorrhea, anxiety, restlessness, increased temperature, or piloerection). Some antiretrovirals, like lopinavir/ritonavir, didanosine, ritonavir, and zidovudine, cause moderate interactions, increasing or decreasing plasma levels of MET. Another moderate interaction occurs with the combined use of alcohol and methadone, increasing sedation (Martindale: The Complete Drug Reference).

2.2 Pharmacotherapy in Alcohol Addiction

Alcohol is absorbed orally, in the proximal portion of the small intestine (over 80 %) and stomach (up to 20 %). The rate of absorption of alcohol determines the magnitude of its concentrations and the intensity and duration of its pharmacological effects. This speed depends on many factors. So, it is faster if taken fasting or on an empty stomach (maximum concentration within 30–60 min) and slower in the presence of food.

The concentration of alcohol in the drink also influences, being faster absorption when your alcohol content is 20–30 % compared to 3–10 % drinks. If drinks administered are 40 % or more, it decreases gastric emptying. Alcoholic beverages containing carbonic gas (e.g., cava) or mixed with carbonated beverages (soda) have faster absorption.

Food delays absorption, resulting in much lower concentrations of ethanol in blood when taken on an empty stomach.

Alcohol is a highly soluble molecule and therefore is distributed throughout the body water, reaching similar concentrations in most tissues and organs well irrigated to those of the blood. It crosses the blood–brain and placental barrier and is excreted in breast milk. Because of its poor lipid solubility, it does not diffuse well into the fat. After administration of a single weight-adjusted dose, blood alcohol concentrations are higher in women than in men. This appears to be due to several factors. On one hand, women have less alcohol dehydrogenase in the stomach (see metabolism), and on the other they have a higher proportion of subcutaneous fat and a lower volume of blood.

Metabolic degradation of ethanol is essentially by oxidation in liver (90–98 %), and 2–10 % can be removed by secondary pathways, such as the kidney and the lung. Most alcohol is converted to acetaldehyde by the action of three enzymes:

- Alcohol dehydrogenase catalyzes the conversion of alcohol (oxidized form) into acetaldehyde (reduced form) with involvement of nicotinamide adenine dinucleotide (NAD) cofactor. This generates an excess of reducing equivalents in the cytosol, favoring hyperlactacidemia and the consequent alteration in the Krebs cycle with subsequent hypoglycemia.
- Cytochrome P450-dependent microsomal ethanol-oxidizing system (MEOS): This is a microsomal oxidase system present in the smooth endoplasmic reticulum of hepatocytes. It has greater activity in patients with chronic alcoholism.
- Catalases: These enzymes utilize hydrogen peroxide to convert ethanol into acetaldehyde. They are found in blood, bone marrow, liver, and kidney.

The rate of elimination of alcohol is 100 mg/kg/h. The elimination pathways are lung (50–60 %), enterohepatic (25–30 %), urine (5–7 %), sweat, tears, gastric juice, saliva, and breast milk.

The joint consumption of alcohol and opiate substances causes an increase of the sedative effects of both substances and affectation on psychomotor performance. Concomitant ingestion of alcohol can also increase opioid respiratory depression.

There are several drug treatments available to approach this disease, based on the use of anti-craving medications (acamprosate, naltrexone, or serotonergic drugs) or drugs interdictors (disulfiram or calcium carbimide), along with psychotherapy and support groups.

Currently, there are three drugs approved by the US Food and Drug Administration (FDA) indicated for the treatment of AD: naltrexone (oral and injectable extended-release formulations), acamprosate, and disulfiram (Table 17.1).

Other non-approved drugs include fluoxetine, nalmefene, ondansetron, and topiramate.

2.2.1 Naltrexone

Indications

FDA-labeled indications of naltrexone are alcohol dependence (maintenance of abstinence) and opioid dependence (relapse after detoxification and prophylaxis) (Micromedex® 2.0).

Pharmacokinetics

Naltrexone is well absorbed from the gastrointestinal tract but is subject to considerable first-pass metabolism and may undergo enterohepatic

Table 17.1 FDA-approved medications for the treatment of alcohol dependence

Drug	Nature	Dosage	Side effects	Contraindications (other than hypersensitivity to the drug)
Disulfiram	Aldehyde dehydrogenase inhibitor	Begin with 250 mg once per day; increase to 500 mg once per day	Disulfiram–alcohol interaction: palpitations, flushing, nausea, vomiting, headache	Alcohol, metronidazole, or Paraldehyde use; psychosis; cardiovascular disease
Naltrexone	μ-Opioid receptor antagonist	50 mg once per day	Nausea, headache, anxiety, sedation	Narcotic use, acute opioid withdrawal, acute hepatitis, liver failure
Acamprosate	*N*-Methyl-D-aspartate (NMDA) receptor inhibitor	333-mg enteric-coated tablets Adults >132 lb (60 kg): two tablets three times per day Adults <132 lb: two tablets with the morning meal, one with the midday meal, and one with the evening meal	Diarrhea, headache, flatulence, nausea, vomiting, dyspepsia	Severe renal impairment (creatinine clearance < 30 mL per minute [0.5 mL per second])

Modified from Williams (2005)

recycling. It is extensively metabolized in the liver, and the major metabolite, 6-β-naltrexol, may also possess weak opioid antagonist activity. Peak plasma concentrations of naltrexone and 6-β-naltrexol occur about 1 h after oral dosing. The elimination half-life of oral naltrexone is around 4 h and that of 6-β-naltrexol about 13 h. Plasma concentrations of a modified-release formulation of naltrexone peak about 2 h after intramuscular injection. This is followed by another peak around 2–3 days later, and then a slow decline in concentrations from about 14 days post-dose to at least 1 month. The elimination half-life of both naltrexone and 6-β-naltrexol from this formulation is about 5–10 days. Naltrexone is about 20 % bound to plasma proteins at therapeutic doses. Naltrexone and its metabolites are excreted mainly in the urine, with a small fraction in the feces. Less than 2 % of an oral dose of naltrexone is excreted unchanged (Martindale: The Complete Drug Reference).

Mechanism of Action

Naltrexone hydrochloride is a pure opioid antagonist. It blocks opioid effects through competitive binding at opioid receptors (Micromedex® 2.0).

Interactions

Concurrent use of naltrexone and opioids may result in precipitation of opioid withdrawal symptoms and decreased opioid effectiveness (Micromedex® 2.0). It has also been described as moderate interaction with yohimbine (Micromedex® 2.0).

2.2.2 Acamprosate

Indications

The indication approved by FDA for acamprosate is maintenance of abstinence in alcoholism (Micromedex® 2.0).

Pharmacokinetics

Absorption of acamprosate from the gastrointestinal tract is slow but sustained and is subject to considerable interindividual variation. Steady-state concentrations occur after dosage for 7 days. Bioavailability is reduced if given with food. Acamprosate is not protein bound, and although it is hydrophilic, it is reported to cross the blood–brain barrier. Acamprosate does not appear to be metabolized and is excreted unchanged in the urine. The elimination half-life after oral doses has been reported to be about 33 h (Martindale: The Complete Drug Reference).

Mechanism of Action

Acamprosate calcium helps maintain abstinence to alcohol through a mechanism that may involve an interaction with glutamate and GABA neurotransmitter systems centrally. It possesses dose-dependent reduction of alcohol intake specific for the type of alcohol and the mechanisms of dependence, without exhibiting anticonvulsant, antidepressant, or anxiolytic properties (Micromedex® 2.0).

Interactions

No interactions have been described for acamprosate (Micromedex® 2.0).

2.2.3 Disulfiram

Indications

The indication approved by FDA for disulfiram is alcoholism (Micromedex® 2.0).

Pharmacokinetics

Disulfiram is absorbed variably from the gastrointestinal tract and is rapidly reduced to diethyldithiocarbamate mainly by the glutathione reductase system in the erythrocytes; reduction may also occur in the liver. Time to peak concentration by oral administration is 3–8 h. It does not undergo protein binding. Diethyldithiocarbamate is metabolized in the liver to its glucuronide and methyl ester and to diethylamine, carbon disulfide, and sulfate ions. Metabolites are excreted mainly in the urine; carbon disulfide is exhaled in the breath. Elimination half-life is 20–33h (Martindale: The Complete Drug Reference).

Mechanism of Action

Disulfiram inhibits the oxidation of acetaldehyde, the main metabolite of alcohol (Micromedex® 2.0).

Interactions

Disulfiram is contraindicated in the concomitant treatment with amprenavir, metronidazole, and paraldehyde. Interactions occur with tranylcypromine, tinidazole, phenprocoumon, dicumarol, theophylline, fosphenytoin, phenytoin, diazepam, omeprazole, isoniazid, anisindione, warfarin, cannabis, amitriptyline, desipramine, imipramine, and chlordiazepoxide (Micromedex® 2.0).

2.2.4 Calcium Carbimide

Indications

The indication for disulfiram is alcoholism, although it is not yet approved by FDA (Micromedex® 2.0).

Pharmacokinetics

The oral bioavailability of carbimide is dose related; respective extents of systemic absorption after single oral doses of carbimide 0.3, 1, and 1.5 mg/kg (solution in water) are reportedly 50, 70, and 81 %. Available pharmacokinetic data (carbimide oral solution) suggest a first-order absorption process, with a simultaneous Michaelis–Menten first-pass effect that is saturable even at low doses (Martindale: The Complete Drug Reference).

Mechanism of Action

Carbimide is a potent inhibitor of aldehyde dehydrogenase (Micromedex® 2.0).

Interactions

Carbimide exacerbates alcohol reaction with metronidazole or other aldehyde dehydrogenase inhibitors. Carbimide may cause reaction with the alcohol contained in certain drugs or foods. Disulfiram treatment should be discontinued 10 days before starting treatment with carbimide. It is incompatible with aldehyde drugs such as paraldehyde and chloral derivatives (Micromedex® 2.0).

3 Pharmacogenetics of Drug Addiction Therapies

Advances in the field of genetics can drill down individual characteristics and vulnerability to develop opioid dependence and/or alcohol. Studies in families (twins and adopted children) also show that genetics play a major role in the development of these addictions.

3.1 The Opioid System

The endogenous opioid system is one of the many systems that constitute the central nervous system and primarily exerts a neuromodulation

Table 17.2 Main gene polymorphisms in genes of the opioid system

GENE	Gene polymorphism DNA sequence	Protein	Reference SNP
OPRM1	A118G	Asn40Asp	rs1799971
	C17T	Ala6Val	rs1799972
	C1031G	–	rs2075572
OPRD1	T921C	Gly307Gly	rs2234918
	G80T	Cys27Phe	rs1042114
	GCAACT haplotype		
OPRK1	G36T	Pro12Pro	rs1051660
	C846T	Val282Val	rs16918875
	A843G	Ala281Ala	rs702764
	GGCTTCT haplotype		

function. This system consists of three types of receptors: μ (mu), δ (delta), and κ (kappa). Among the variety of functions performed, probably the best known is restraining painful stimuli (in fact, the most potent analgesics that are currently available are opioid compounds), but is also involved in other important neurophysiological mechanisms as drug addiction, as it is related to the "enhancers" or pleasurable effects they provide. The opioid system is also involved in some responses of the organism to a stressful situation.

Natural and synthetic opiates and endogenous opioid peptides bind specifically and with high affinity to opioid receptors, which means that these substances are perfectly matched to opioid receptors. These receptors are often located in the end portion of the presynaptic axon of the nerve cell and modulate the release of neurotransmitters by inhibiting the entry into operation of the action potential, thereby decreasing the amount of released transmitter substance. The effect of the opioid receptor is very strong in the nerve cells that transmit pain, where the release of the transmitter substance or P substance pain is inhibited, which explains the analgesic effect of opioid receptors transmitters. Different opioids bind more or less strongly to the different types of opioid receptors: mu (μ), delta (δ), and kappa (κ). Patients with opiate addiction consume preferably opioids acting preferentially on μ opioid receptors which present greater analgesic effect.

The relative intrinsic activity (RIA) is defined as the ability of the ligands to produce a given effect on a receptor. MOR and MET have the same specificity for opioid receptor pure agonists and show the maximum AIR value, so they can produce the same effect.

3.2 Pharmacogenetics of Opioid Addiction

3.2.1 Mu-Opioid Receptor Gene (OPRM1)

This receptor is the main mediator of the reinforcing effects of opioids, making it a good candidate for a key role in opioid addiction. There are numerous studies based on the identification of single nucleotide polymorphisms (SNPs) on the mu-opioid gene (OPRM1) trying to prove its relation to opioid dependence.

The most studied opioid system polymorphisms are:

- OPRM1: A118G, C17T, and C1031G
- OPRD1: T921C, G80T, and GCAACT haplotype
- OPRK1: G36T, rs16918875, rs702764, and GGCTTCT haplotype

Table 17.2 shows the most important polymorphisms described in genes of the opioid system, with their correspondences in DNA sequence and protein. Table 17.3 shows the distribution of these gene polymorphisms in different populations.

A118G Polymorphism

Numerous polymorphisms have been identified on the mu-opioid receptor (OPRM1). The most interesting SNP related to opioid dependence is A118G, located at position 118 in exon

Table 17.3 Distribution of the genotype frequencies (%) of OPRM1, OPRD1, and OPRK1 gene polymorphisms in different ethnic populations

Population	Origin	Gene polymorphism		
		OPRM1		
		A118G		
		AA	AG	GG
HapMap CEU	European	70.8	27.4	1.8
HapMap HCB	Chinese	41.9	44.2	14.0
HapMap JPT	Japanese	33.7	39.5	26.7
HapMap YRI	Nigerian	100	0	0
		C17T		
CEU GENO PANEL	European	98.3	1.7	0
AAM GENO PANEL	African–American	73.8	23.0	3.3
CHB GENO PANEL	Asian	100	0	0
YRI GENO PANEL	Sub-Saharan African	50.0	43.3	6.7
JPT GENO PANEL	Asian	100	0	0
		C1031G		
		CC	CG	GG
HapMap CEU	European	26.7	61.7	11.7
HapMap HCB	Chinese	64.4	31.1	4.4
HapMap JPT	Japanese	64.4	28.9	6.7
HapMap YRI	Nigerian	16.7	43.3	40.0
		OPRD1		
		T921C		
		CC	CT	TT
NHLBI Exome Sequencing Project		31.0	48.9	20.1
CLINSEQ_SNP	European	26.9	44.6	28.4
CORNELL	Multiple	75.0	0	25.0
		G80T		
		GG	GT	TT
HapMap CEU	European	0.9	22.1	77.0
HapMap HCB	Chinese	0	0	100
HapMap JPT	Japanese	0	0	100
HapMap YRI	Nigerian	0	4.4	95.6
		OPRK1		
		G36T		
		GG	GT	TT
CAUC1	Caucasian	83.3	16.7	0
AFR1	African/African–American	87.0	13.0	0
HISP1	Hispanic	78.3	21.7	0
PAC1	Pacific RIM	66.7	33.3	0
			C846T	
		CC	CT	TT
HapMap CEU	European	91.2	8.8	0
HapMap HCB	Chinese	100	0	0
HapMap JPT	Japanese	100	0	0
HapMap YRI	Nigerian	73.5	23.9	2.7
			A843G	
		AA	AG	GG
HapMap CEU	European	73.5	24.8	1.8
HapMap HCB	Chinese	86.0	14.0	0
HapMap JPT	Japanese	87.2	11.6	1.2
HapMap YRI	Nigerian	15.0	42.5	42.5

Source: dbSNP

Table 17.4 Association studies of OPRM1 A118G polymorphism with drug dependence

Study	Year	N	Effect	Population origin
Bond	1998	113/39	G > no opioid dependence	European/American
				African/American/ Hispanic
Gelernter	1999	891	No differences in opioid and alcohol dependence	Caucasian
				African/American
				Hispanic
Szeto	2001	200/97	G > opioid dependence	Chinese
Franke	2001	287/221/365 Opioid/alcohol/controls	No differences in opioid and alcohol dependence	German
Shi	2002	145/383	No differences in opioid dependence	European/American
Schinka	2002	179/297	A > opioid and alcohol dependence	European/American
Tan	2003	87/404	A > opioid dependence	Indian
				Malay
				Chinese
Bart	2004	139/170	G > opioid dependence	Swedish
Bart	2005	389/170	G > alcohol dependence	Swedish
Drakenberg	2006	64/38	G > opioid dependence	European
Glatt	2007	126/156	No differences in opioid dependence	Chinese
Kapur	2007	126/156	G > opioid dependence	Indian
Kim	2009		G > alcohol dependence (only for women)	Korean
Nikolov	2011	1842/1451	No differences in opioid dependence	Bulgarian
				Italian
Nagaya	2012	55/51	G > opioid dependence	Malaysia
Koller	2012	1845/1863	G < alcohol dependence	German
Chen	2012	1900/2382	G > alcohol dependence in Caucasian	Asian
				Caucasian
Kumar	2012	130 opioid addicts	G > opioid and alcohol dependence	Indian
		110 alcohol addicts		
		200 controls addicts		

1 (118A/G) of OPRM1, involving the substitution of adenine for guanine in position 118 of the DNA sequence and resulting in the presence of aspartic acid instead of an asparagine at position 40 of the amino terminal of the receptor (Asp40). The OPRM1 encoded variant has been reported to bind the endogenous opioid peptide β-endorphin with threefold higher affinity receptor variants encoded by Asn40 (Bond et al. 1998; Zhang et al. 2006).

This change in the amino acid sequence of the protein suppresses a potential glycosylation site at the N-terminal region of the receptor. The presence of heterozygosity at this level (A/G) varies the affinity for β-endorphin in the receptor, significantly increasing its power and the action of this substance, which is tripled. Subjects expressing the Asp40 protein variant show a higher response of cortisol to receptor blockade by naloxone and lower agonist effect of the metabolite morphine-6-glucuronate (Zhang et al. 2006; Thorn et al. 2009). The allelic frequency varies depending on the population analyzed (Table 17.3).

There have been numerous studies focused on SNP A118G and opioid addiction, obtaining controversial results, as it will be described below. Table 17.4 shows the most relevant studies on association of OPRM1 A118G polymorphism with drug dependence.

An important factor to consider is the diversity of the population on which every study has been conducted. Bond et al. (1998) studied the association of A118G polymorphism in 113 opioid-dependent patients and 39 healthy individuals from different ethnical origins. Findings supported the hypothesis that heroin addiction could be due to a combination of genetic, environmental (including conduct) factors that contribute to the acquisition, persistence, relapse, and drug use. No significant differences in allele frequency for the SNP A118G were found; however, this SNP demonstrated a significant association to the population studied, showing higher frequency of the G allele among the Hispanic population controls. This variant acted as a protective factor against opiate consume in those patients (Bond et al. 1998).

In accordance with these results, the AA genotype has been found significantly associated to consume of substances of abuse, including alcohol and nicotine, in 179 dependent patients and 297 controls from Caucasian population (Schinka et al. 2002), and to heroin dependence in Indian population, as showed in a study on population from India, Malaya, and China in 87 subjects with heroin dependence and 404 healthy subjects (Tan et al 2003).

On the contrary, other study, conducted by Szeto et al. (2001) in Chinese population, compared 200 subjects with opioid dependence to 97 healthy subjects. They showed a higher frequency of the GG genotype and G allele among subjects with opioid dependence (39.5 vs. 29.4 %). The results of the study by Bart et al. (2004), conducted in 139 opioid-dependent patients and 170 nonaddicted subjects from Sweden, confirmed this association of the 118G allele in exon 1 of the gene OPRM1 with heroin addiction. These results also indicated that up to 21.0 % of the attributable risk of heroin addiction is mediated by the 118 allele G. A significant association between the 118G allele and opioid dependence was also found in Indian population (Kapur et al. 2007). The study of Drakenberg et al. (2006) on European population (64 opioid addicts and 38 controls) demonstrated the importance of mu-opioid receptors in heroin dependence. The results showed that about 90 % of the population with the 118G allele were heroin users, demonstrating that the A118G SNP of the mu-opioid receptor gene (OPRM1) was associated to abuse of this substance. On the contrary, the study by Glatt et al. (2007) on the prevalence of the Asn40Asp polymorphism did not show relevant differences among individuals with/without opioid dependence, despite 1,208 individuals from 473 Han Chinese families were analyzed.

On the other hand, case/control studies by Gelernter et al. (1999) in Caucasians, African–Americans, and Hispanics, Franke et al. (2001) in German–Caucasian population (287 heroin dependence, 221 alcohol addiction, and 365 controls), and Shi et al. (2002) in Chinese population (145 heroin-dependent subjects and 383 controls) showed no significant differences in the frequency of the different genotypes for OPRM1 A118G. The study by Gelernter et al. (1999) failed to find an association between allelic frequencies of A118G (Asp40Asn) y C17T (Ala6Val) variants despite the sample population was especially high (891 opioid/alcohol dependants) from Caucasian, Afro-American, and Hispanic origin.

More recently, Nikolov et al. (2011) conducted a study with a cohort of 1,842 (1,516 Bulgarian and 326 Italian) opiate-addicted subjects and 1,451 (1,178 Bulgarian and 273 Italian) controls, and the results suggested that there is no direct effect of 118G allele on the risk of opioid dependence among heroin-dependent subjects. These results are opposite to another case/control study in population from Malaysia, where a strong association between the G allele and opioid addiction was found (Nagaya et al. 2012). The frequencies found for alleles A and G in opiate dependents were 51 and 49 %, respectively, versus 73 and 27 % in healthy volunteers. Recent studies suggest the association of polymorphisms may be important in determining the risk profile for complex diseases such as addiction (Kumar et al. 2012). This study on Indian population recruited 130 heroin-dependent subjects, 110 alcohol-addicted subjects, and 100 control

Table 17.5 Association studies of OPRM1 C17T polymorphism with drug dependence

Study	Year	*N*	Effect	Population origin
Berretini	1997		T > opioid dependence	Caucasian
				African/American
Comptom	2003	50/59	No differences in opioid dependence	African–American
Rommelspacher	2001	327/340	T > alcohol dependence	German
Tan	2003	84/404	No differences in opioid dependence	Indian
				Malay
				Chinese
Kapur	2007	126/156	No differences	Indian

subjects, all males. A118G SNP on OPRM1 and three OPRK1 SNPs (rs16918875, rs702764, and rs963549) were identified. The results showed a significant association for A118G with heroin addiction and with alcohol addiction (Kumar et al. 2012).

C17T Polymorphism

C17T polymorphism is the second most common alteration mu-opioid receptor gene (OPRM1). Substitution of cytosine by thymine at position 17 causes a change in the primary structure of the resulting protein, with an alanine substituted in position 6 by a valine.

The frequency of this polymorphism varies depending on the population, from less than 2 % in the Caucasian population to over 20 % in African–Americans (Berretini et al. 1997). Table 17.3 shows the distribution of allelic frequencies in different populations.

Table 17.5 shows association studies of OPRM1 C17T polymorphism with drug dependence. The study conducted by Berretini et al. (1997) on African–American and Caucasian population found a higher frequency of the allelic variant C17T in the opioid dependence group.

On the other hand, in a study conducted in heroin-dependent subjects from Indian, Malay, and Chinese population (87 cases and 404 controls) to investigate the distribution of the three main polymorphisms of the mu-opioid receptor gene, only three subjects in the entire sample showed this polymorphism (Tan et al. 2003).

In other SNP association studies, C17T polymorphism showed no evidence in the populations studied (Bergen et al. 1997; Gelernter et al. 1999; Compton et al. 2003; Crowley et al. 2003; Luo et al. 2003; Tan et al. 2003; Zhang et al. 2006; Kapur et al. 2007). Compton et al (2003) showed no significant results in the association of A118G and C17T polymorphisms in opioid dependence, although in this case, the absence of association could be due to the small sample size and the heterogeneous ethnic backgrounds of subjects (50 opioid dependents in treatment with MET and 59 healthy controls).

C1031G Polymorphism

This SNP is located in intron 2 of the mu-opioid receptor gene (OPRM1).

Table 17.6 shows association studies carried to evaluate the influence of OPRM1 C1031G polymorphism on opioid and alcohol addiction. An increased frequency of the G allele in opioid-dependent patients compared to controls (15.5 vs. 8.3 %) was found by Szeto et al (2001) in a case/control study carried out on Chinese population. These results also showed a significant association for both A118G and C1031G polymorphisms and opioid dependence. The G allele was more frequent in the dependent heroin group (39.5 and 30.8 % for C1031G SNP and A118G, respectively) compared to controls (29.4 and 21.1 % for SNPs A118G and C1031G, respectively). This study suggested that the G allele variant of both A118G and C1031G polymorphisms can contribute to the vulnerability to heroin dependence.

The study conducted by Szeto et al. (2001) showed a significant association for both A118G and C1031G and opioid dependence. The G allele was more frequent in the heroin-dependent

Table 17.6 Association studies of OPRM1 C1031G polymorphism with drug dependence

Study	Year	*N*	Effect	Population origin
Szeto	2001	200/97	G>opioid dependence	Chinese
Tan	2003	97/404	No differences in opioid dependence	Indian 20/117 Malaya 25/131 Chinese 52/156
Zhang	2006	382/338	No differences in opioid dependence	European/American
Kumar	2012	130 opioid addicts 110 alcohol addicts 200 controls addicts	No differences in opioid or alcohol dependence	Indian

Table 17.7 Association of OPRD1 C921T polymorphism with drug dependence

Study	Year	*N*	Effect	Population origin
Mayer	1997	103/115	C>opioid dependence	German
Franke	1999	233 opioid addicts 262 alcohol addicts 173 healthy controls	No differences in opioid or alcohol dependence	German
Franke	1999	233 opioid+323 parents 262 alcohol+334 parents	No differences in opioid or alcohol dependence	German
Xu	2002	450/304	No differences in opioid dependence	Chinese
Zhang	2008	111 opioid addicts 225 cocaine addicts 557 alcohol addicts 443 healthy controls	No differences in opioid and alcohol dependence GCAACT Haplotype: opioid and alcohol dependence	European/American

group (39.5 and 30.8 % for C1031G polymorphism and A118G, respectively) compared with controls (29.4 and 21.1 % for A118G and C1031G polymorphisms, respectively). This study suggested that the G allele variant A118G and C1031G polymorphisms both can contribute to the vulnerability to heroin dependence (Szeto et al. 2001).

Other studies have failed to find the C1031G SNP of the mu-opioid receptor gene (OPRM1) associated to drug or alcohol dependence, as described by Zhang et al (2006) in European–American population and Tan et al (2003) for Asian population.

3.2.2 Delta-Opioid Receptor Gene (OPRD1)

This receptor is located on chromosome 1p34. It consists of three exons. Two polymorphisms in the coding sequence have been identified: C921T and G80T (Table 17.2).

C921T Polymorphism

This polymorphism involves a silent substitution of cytosine by thymine at codon 307 in position 921 in the DNA sequence. Numerous studies have been conducted, but yielding contradictory results (Table 17.7).

Mayer et al. (1997) conducted a study on 103 opioid-dependent patients and 115 controls from German population. Significant differences were found between the two groups, with the C allele and the CC genotype appeared more frequently in patients with opioid dependence. However, these findings could not be replicated later in two studies conducted by Franke et al. (1999) on German population. One of the studies was a case/control approach, with 233 opioid-dependent patients, 262 alcoholic dependents, and 173 healthy subjects; the other one was carried out in a family-based design which studied the 233 heroin addicts and 262 alcohol addicts and their parents (323 and 334, respectively).

Table 17.8 Association of OPRD1 G80T polymorphism with drug dependence

Study	Year	*N*	Effect	Population origin
Xu	2002	450/304	Absent in the population analyzed	Chinese
Zhang	2008	111 opioid addicts	G > opioid dependence	European/American
		225 cocaine addicts		
		557alcohol addicts		
		443 healthy controls		

In those studies, no evidence of any association between T921C delta-opioid receptor (OPRD1) gene polymorphism and opioid dependence or alcohol dependence was found (Franke et al. 1999).

Xu et al. (2002) attempted to demonstrate the association of this SNP and opioid dependence in Chinese population by conducting a case/control study with 450 opioid-dependent patients and 304 healthy subjects. The results obtained for genotype and allele frequencies of T921C SNP in the delta-opioid receptor gene (OPRD1) were not significantly different. Same result was reached in a larger study, which analyzed 11 OPRD1 SNPs in 1,063 European–American subjects, 620 cases (557 alcohol dependence, 225 cocaine dependence, and 111 opioid dependence), and 443 controls (Zhang et al. 2008). The results showed no association between the C921T variant and opioid dependence; however, it did show an association of a specific haplotype (GCAACT) in OPRD1 containing C921 and G80 alleles with dependence on alcohol, on cocaine, and especially on opiates.

G80T Polymorphism

G80T polymorphism on OPRD1 gene originates a nonsynonymous transversion in exon 1 causing the change of phenylalanine by cysteine in the amino acid sequence. This SNP is distributed very differently among different populations, and as in the previous polymorphisms, different studies have yielded contradictory results (Tables 17.2 and 17.7, respectively).

Xu et al. (2002) investigated the OPRD1 G80T SNP in a cases/control study in China (450 opioid-dependent patients and 304 healthy subjects), but they could not find any subject presenting this polymorphism (Table 17.8).

In the study by Zhang et al (2008), it was found that the G80T SNP may be associated with opioid dependence in European–American population. In the group of patients with opioid dependence, the G allele frequency was higher than in the control group (21 vs. 13.2 %); therefore, it could be considered a genetic risk factor for this disease (Zhang et al 2008).

3.2.3 Kappa-Opioid Receptor Gene (OPRK1)

The kappa-opioid receptor gene (OPRK1) appears to play a key role in the ability for stress response, in opiate withdrawal in response to psychostimulants, and in inhibition of the mesolimbic system (Rattan et al. 1992; Kreek 1997; Herz 1998). Studies on its association with opioid addiction are shown in Table 17.9.

rs16918875 and rs702764

In the study by Kumar et al. (2012), described above, three SNPs on OPRK1 were studied (rs16918875, rs702764, and rs963549). No association was found between these SNPs and opioid addiction, but two significant interactions between A118G in OPRM1 and two SNPS in OPRK1 (rs16918875 and rs702764) were identified. Subjects with heroin addiction who had the rs16918875 and A118G risk alleles presented increased odds in 2.51 times. There was also a significant interaction between GG/AG of OPRM1 A118G and GG of rs702764 in patients with opioid dependence (Kumar et al. 2012). However, when Zhang et al. (2008) analyzed seven SNPs on OPRK1, rs702764 SNP showed no association to opioid addiction (rs16918875 was not included in the panel assessed).

Table 17.9 Association of OPRK1 polymorphisms with drug dependence

Study	Year	N	Effect	Population origin
Gerra	2007	106/70	G36T: opioid dependence	European Caucasian
Yuferov	2004	145/146	G36T: association with opioid dependence	Hispanic
Zhang	2008	111 opioid addicts 225 cocaine addicts 557alcohol addicts 443 healthy controls	GGCTTCT Haplotype: associated with alcohol addiction rs702764: no association with opioid addiction	European/American
Kumar	2012	130 opioid addicts 110 alcohol addicts 200 controls addicts	rs16918875 and rs702764	Indian

G36T Polymorphism

In humans, the G36T gene polymorphism in OPRK1 gene has been associated with substance dependence; however, the results are not conclusive (Table 17.9). In the study of Gerra et al. (2007) conducted on 106 heroin dependents (Western Europe, Caucasian) and 70 healthy control subjects matched for race and sex with no history of substance use disorder were genotyped. The frequency of OPRK1 G36T gene polymorphism was significantly higher among heroin-dependent individuals compared with control subjects (Gerra et al. 2007).

In the study of Yuferov et al. (2004), the possible associations of variants in OPRK1 and individual vulnerability to develop opiate addiction were analyzed. They evaluated OPRK1 gene polymorphisms in 291 subjects from several ethnic groups (145 former heroin addicts and 146 controls). Twelve SNPs were identified in these patients, with nine novel variants and three previously reported SNPs (G36T, C459T, and A843G). Genotype and allele distribution of OPRK1 SNPs stratified by ethnicity and drug status were also described (Yuferov et al. 2004). Nine haplotypes were observed in African–Americans, six in Caucasians and five for Hispanics. The results in SNP frequency of G36T showed a significant association with opioid dependence (Yuferov et al. 2004).

GGCTTCT Haplotype

As described previously in the study of Zhang et al. (2008), seven SNPs of OPRK1 were analyzed in 390 cases (97 opioid, 327 alcohol, and 177 cocaine-dependent subjects) and 358 controls, but none showed association with substance dependence. However, the GGCTTCT haplotype (G90 and C921 alleles) showed to be associated to alcohol dependence (Zhang et al. 2008).

3.3 Importance of Drug Interactions in Treatments for Opioid Dependence Gene Polymorphisms of Drug Metabolism and Transport

Drug interactions can result in loss of efficiency or treatment toxicity. Pharmacological interactions are classified based on the pharmacokinetics or pharmacodynamics of the drug.

Pharmacokinetic interactions occur when a drug causes a change in absorption, distribution, metabolism, and/or elimination ("ADME") of another drug.

Pharmacodynamic interactions occur when two drugs are coadministered and the concentration-response curve of one or both drugs are altered without a change in pharmacokinetics.

Opioids are widely used drugs in the treatment of chronic pain and in the treatment of opioid dependence. These substances have a narrow therapeutic range and a life-threatening toxicity. There is a wide variability in the response to treatment, so that a usual dose in a patient tolerant to opioids may cause severe adverse reactions in a patient which does not have the same tolerance.

Therefore, in the therapeutic approach of opioid and alcohol dependence, it is necessary to know the concomitant pharmacotherapy for analyzing potential drug interactions with genes involved in phase I metabolism, especially CYP450.

3.3.1 Genes Involved in Phase I Metabolism: Cytochrome P450 Gene

CYP are proteins associated to the cytoplasmic membranes, mitochondria, and endoplasmic reticulum, where they act metabolizing hundreds of endogenous and exogenous substances.

Most CYPs act on several substrates, some of which can catalyze various types of reactions. In vivo, these substrates include numerous drugs or toxic components derived from metabolism, such as bilirubin. Cytochrome P450 enzymes are present in most tissues of the body and constitute the major enzyme complex involved in drug metabolism in the body and play a fundamental role in the oxidative metabolism phase (phase I metabolism).

Some of these drugs have the ability to increase or decrease the activity of enzymes (phenomena known as enzyme inhibition and enzyme induction, respectively). This is of crucial importance in the evaluation of drug interactions.

For example, when a drug which inhibits the enzyme which degrades a second drug in the presence of both, the second drug blood levels will increase, and subsequently the potential to originate an overdose. Conversely, if a drug induces metabolism of the second drug, concentrations decrease, still below therapeutic levels, a vital factor. This leads to a need for a complete knowledge of the enzymes involved in the metabolism of drugs used in humans to prevent errors in the therapeutic window or side effects.

Some opioids are metabolized by CYP450, like codeine, hydrocodone, oxycodone, methadone, tramadol, and fentanyl.

The CYP450 system comprises different isozymes that are responsible for drug metabolism. The main families related to drug metabolism are CYP1, CYP2, and CYP3. Genes encoding CYP450 isozymes are polymorphic. This is of key importance because certain allelic variants have an altered activity or completely lacking enzyme activity. Of these isoenzymes, CYP3A and CYP2D6 are primarily responsible for the metabolism of opioids through the CYP450 system. CYP2B6 also contributes to the metabolism of methadone.

CYP3A4 Gene

CYP3A metabolism is responsible for approximately 50 % of all currently available drugs. CYP3A4 mediates the metabolism of opioids. The activity of the enzymes of this subfamily has a high variability in the population, as they are very modular enzymes, either by other drugs (inducers or inhibitors), diseases, diet, or environmental factors, not forgetting the genetic factor. Numerous variants have been identified for CYP3A4, having different frequencies, especially between different ethnic groups.

The impact of certain genetic polymorphisms on variable responses to analgesics was reviewed by Samer et al. (2005), showing how pharmacogenetics can help optimizing analgesic therapy. According to this review, the administration of the standard dose may cause either toxicity or therapeutic ineffectiveness, depending on the substance and gene polymorphisms associated to molecular structures involved in its pharmacokinetics/pharmacodynamics. Drug interactions due to genetic variants mimicking the presence of inhibitors and inducers of CYP also contribute to the variable response to analgesics (Samer et al. 2005). Among the molecular structures to be considered in analgesia, CYP isoenzymes are the most important. CYP3A4 metabolizes strong opioids (buprenorphine, fentanyl, methadone), while weak opioids (codeine, tramadol), along with antidepressants and dextromethorphan, are metabolized by CYP2D6, and CYP2C9 is involved in NSAIDs metabolism (Samer et al. 2005).

CYP3A4 and CYP2B6 are the major isoforms involved in MET metabolism, and CYP2D6 contributes to a lesser extent, according to a study of 245 patients in methadone treatment enrolled to investigate the influence of CYP2B6, CYP2C9, and CYP2C19 gene polymorphisms on the pharmacokinetics of MET and the response to treatment (Crettol et al. 2006). In this study,

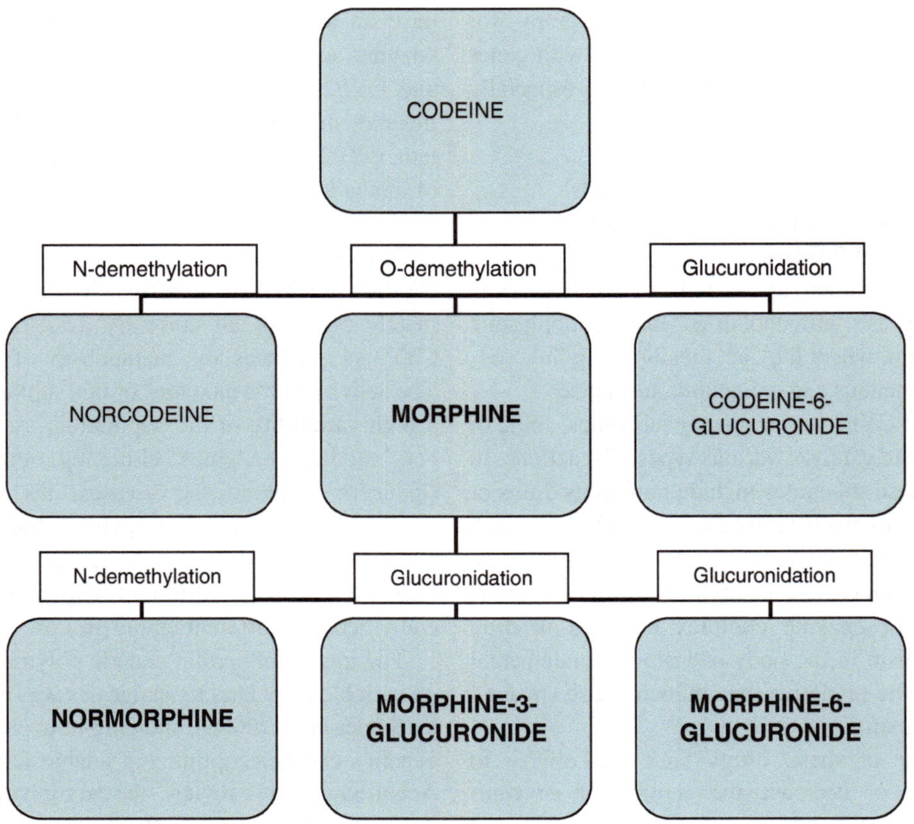

Fig. 17.1 Codeine metabolism

CYP2B6*6/*6 carriers and extensive or intermediate CYP2D6 metabolizers showed higher MET plasma levels, while CYP2C9 and CYP2C19 genotypes showed no influence on MET plasma levels (Crettol et al. 2006).

More recently, in a study conducted by Shiran et al. (2009) to analyze the influence of cytochrome P450 on the disposition of MET in 88 patients in MET maintenance therapy (MMT), CYP3A4, CYP2D6, and CYP1A2 activities were determined. Results demonstrated CYP3A activity had a moderate influence on methadone disposition, and therefore inhibitors and inducers in patients treated with methadone should be considered for possible interaction effects. No influence on MET disposition was found for CYP2D6 and CYP1A2 activities in these patients (Shiran et al. 2009).

Genetic variants in the CYP3A4 gene have been proposed as indicators to assess side effects and withdrawal symptoms of patients in methadone treatment (Chen et al. 2011). Particularly, genotypes and allele types of rs4646440 and rs2242480 in CYP3A4 were found significantly associated to the severity of withdrawal symptoms and side effects in 366 Han Chinese MMT patients (Chen et al. 2011).

CYP2D6 Gene

CYP2D6 is an enzyme involved in the metabolism of many important drugs, representing approximately 25 % of drugs metabolized by CYP450. It represents the best characterized cytochrome.

This enzyme catabolizes biotransformation through *O*-demethylation of opiates such as codeine (Fig. 17.1) and MET (Fig. 17.2). CYP2D6 has high phenotypic variability due to genetic polymorphisms. The metabolism rate is determined by the number of functional genes, as

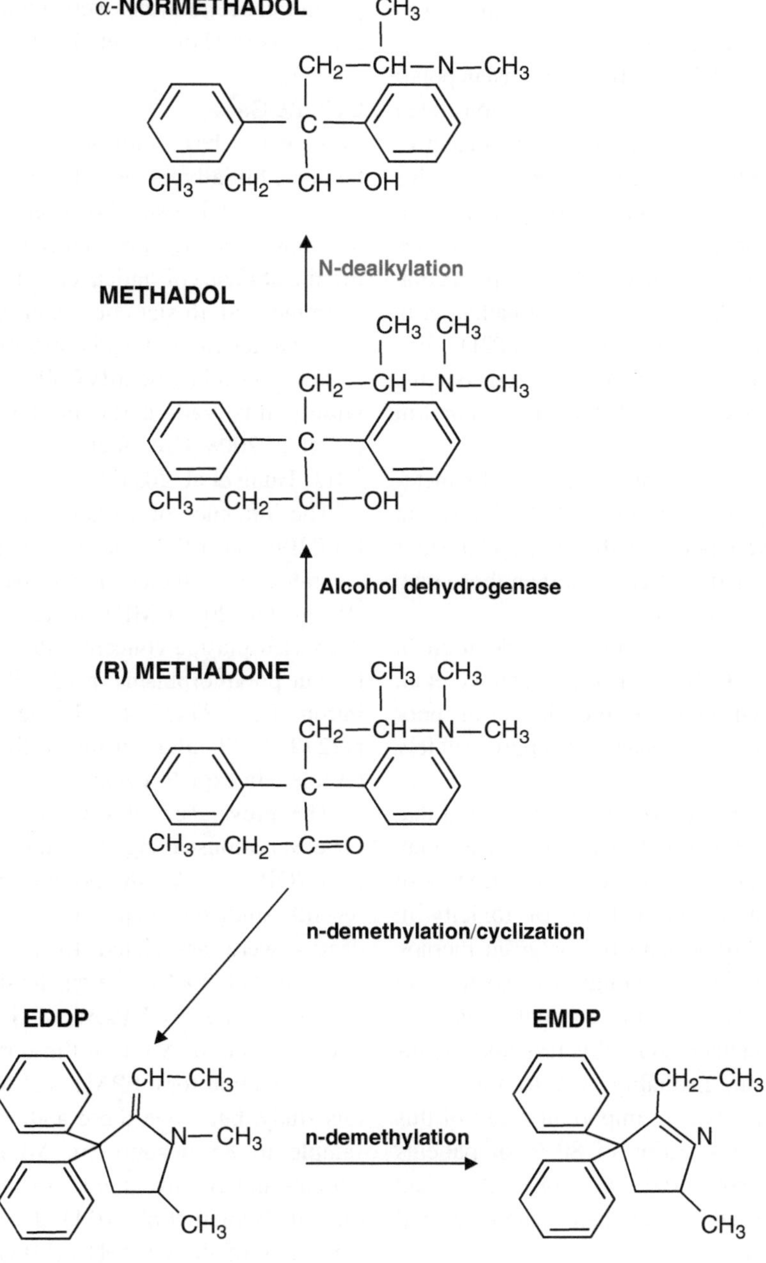

Fig. 17.2 Metabolism of methadone

well. CYP2D6 activity can be described under one of the following patterns:

- *Normal or extensive metabolizer*: Enzyme activity is normal. Patients present one or two functional genes.
- *Intermediate metabolizer*: Enzyme activity is slightly decreased.
- *Poor metabolizer*: Enzyme activity is greatly reduced or nonexistent. There is no specific functional gene. It produces a lower response to the pharmacological effects of the drugs.
- *Intensive or ultrarapid metabolizer*: These patients express three or more copies of the gene, so the activity is much higher

than normal. There is a greater sensitivity to the effects of the drugs.

Tyndale et al. (1997) study on Caucasian population (83 cases and 276 controls) demonstrated that subjects presenting the CYP2D6 defective genotype (CYP2D6*3 and CYP2D6*4 defective mutant alleles) showed protection against oral opiate dependence, mainly codeine. Other studies showed conflicting results regarding the protection of codeine dependence; Kathiramalainathan et al (2000) suggested that inhibition of CYP2D6 may decrease the likelihood of dependence on codeine; however, Fernandes et al (2002) could not confirm those results.

CYP2D6 has a fundamental role in the metabolism of methadone. Eap et al. (2001) found that patients who were poor metabolizers had a better response to methadone treatment than those who were ultrarapid metabolizers.

Another factor to take into consideration in the variability of CYP2D6 is race; Menoyo et al. (2006) observed that the prevalence of poor metabolizers in Caucasians is approximately 6–10 %.

Jannetto and Bratanow (2009) suggested that genotyping CYP2D6 and drug monitoring may improve efficiency in proper drug selection, dose optimization, and minimization of toxicity in patients with chronic pain or analgesic therapy treated with opioids. Although a trend to have the highest steady-state drug concentrations was found in poor metabolizers, this was not statistically significant, probably due to the small sample (61 patients). An important result of this study was the observation of 80 % of patients reporting adverse drug reactions also had impaired CYP2D6 metabolism (Jannetto and Bratanow 2009).

More recently, Fonseca et al. (2011) studied the allelic variants of genes encoding CYP3A5, CYP2D6, CYP2B6, CYP2C9, CYP2C19, and *P*-glycoprotein to investigate pharmacogenetic and pharmacokinetic factors involved in dose adjustment. A sample of 105 opioid-dependent patients of Caucasian origin, classified in responders and nonresponders, was analyzed. There were no differences in genotype frequencies between groups. Higher MET dose requirements and plasma concentrations were found in ultrarapid metabolizers (Fonseca et al. 2011).

CYP2B6 Gene

This gene has multiple polymorphisms in humans. Initially it was not considered important in drugs metabolism; however, recent research has shown the high prevalence of this enzyme in the metabolism of anticancer drugs (cyclophosphamide and ifosfamide), antiretroviral (efavirenz and nevirapine), and opiates (methadone).

Several studies identify CYP3A4 and CYP2B6 widely involved in the metabolism of MET (Li et al. 2008; Crettol et al. 2010; Bunten et al. 2011; Hung et al. 2011).

The influence of 10 gene polymorphisms on CYP2B6 on MET plasma concentration and clearance was studied in 366 patients in MMT (Wang et al. 2011). MET increased clearance and decreased plasma concentrations was associated to four polymorphisms: rs10403955 (T allele in intron 1), rs3745274 (G allele in exon 4), rs2279345 (T allele in intron 5), and rs707265 (A allele in exon 9) (Wang et al. 2011).

The prevalence of CYP2B6*4, *9, and *6 gene variations along with mu-opioid receptor gene (OPRM1) A118G polymorphism has been recently studied in 40 postmortem patients whose deaths were associated to MET dependence (Bunten et al. 2010). The results showed an association of these CYP2B6 alleles with higher plasma levels of MET in the population tested. No association for OPRM1 A118G was found in this study, but when these authors extended the sample to 84 postmortem MET-related death patients and 100 live nondrug-using control population (Bunten et al. 2011), they found that the frequency of the OPRM1 A118G variation was significantly higher in the control population. Apparently, the OPRM1 118G gene polymorphism conferred a protective mechanism against opioid toxicity. Therefore, these authors concluded that it would be useful to perform clinical screening to identify specific subjects carrying the CYP2B6*6 and OPRM1 118G variants before starting an MMT (Bunten et al. 2010, 2011).

Hung et al. (2011) assessed the impact of genetic polymorphisms on ABCB1, CYP2B6,

OPRM1, DRD2, and ANKK1 in MMT on Chinese population, showing that carriers of the ABCB1 3435C>T and CYP2B6 516G>T variants were more likely to require higher methadone doses than noncarriers.

CYP2C9 and CYP2C19 Genes

CYP2C9 participates in the metabolism of drugs like sulfonylurea and nonsteroidal anti-inflammatory drugs (NSAIDs). CYP2C19 is involved in drug metabolism of such as proton-pump inhibitors and antiepileptics. Although neither of these isoenzymes is involved in diacetylmorphine, morphine, MET, or alcohol metabolism, many studies related to opioid and alcohol dependence have considered them for the analysis of response to treatment. However, the influence of CYP2C9 and CYP2C19 genotype on MET plasma levels of patients treated with MET has not been elucidated yet (Crettol et al. 2005, 2010).

3.4 Pharmacogenetics of Alcohol Dependence

3.4.1 OPRM1

OPRM1 A118G

A selection of the most relevant studies conducted to establish an association between OPRM1 A118G polymorphism and the risk of alcohol dependence (AD) are described in Table 17.4.

A significant association between the OPRM1 118G allele and alcohol dependence was described in a case/control study of 559 patients (389 alcohol-dependent individuals and 170 population-based controls without drug or alcohol abuse or dependence), with an attributable risk for alcohol dependence in subjects with a 118G allele of 11.1 % (Bart et al. 2005). Other study performed in Korean men and women with AD compared to normal healthy individuals at a low risk for alcoholism reported a significantly AD higher risk for the 118G allele only for women (Kim et al. 2009).

A study conducted in 1,845 alcohol-dependent subjects and 1,863 controls from German population showed the association of A118G genotype to alcohol dependence (Koller et al 2012). The G allele was less frequent among individuals with alcohol dependence, in recessive (AA vs. GA/GG) and codominant (AA vs. GA) versus the control group (Koller et al. 2012).

The results of a meta-analysis conducted by Chen et al. (2012) which investigated the role of OPRM1 A118G polymorphism in alcohol dependence among Asian and Caucasian population, with a total of 1,900 cases and 2,382 controls studied, showed the 118G allele may be associated with alcohol dependence in Asians but not in Caucasians.

OPRM1 C17T

The study carried out by Rommelspacher et al. (2001) in 327 alcoholic dependents and 340 healthy control subjects showed an increase in the frequency of Val6 allele in subjects with alcohol dependence (Table 17.5).

3.4.2 OPRD1

OPRD1 C921T

As described previously, Franke et al. (1999) failed to find an association between T921C delta-opioid receptor (OPRD1) gene polymorphism and alcohol dependence in a sample of 262 alcoholic dependents and 173 healthy subjects (Table 17.7). A large study in 1,063 European–American subjects (557 alcohol dependence and 443 controls) did not find association between the C921T variant and alcohol dependence, but demonstrated an association of haplotype (GCAACT) in OPRD1 harboring C921 and G80 alleles to alcohol dependence (Zhang et al. 2008).

3.4.3 OPRK1

GGCTTCT Haplotype

The GGCTTCT haplotype on OPRK1 gene was found to be associated to alcohol dependence despite the seven SNPs of OPRK1 analyzed in 390 cases (97 opioid, 327 alcohol, 177 cocaine-dependent subjects) and 358 controls (Table 17.9) did not showed association with substance dependence (Zhang et al. 2008).

rs16918875 and rs702764

In the study by Kumar et al. (2012), described above, no association was observed between alcohol addiction and rs16918875, rs702764, and rs963549, but a significant interaction between A118G in OPRM1 and rs16918875 SNP in OPRK1 was found, being the odds 2.31 times higher in alcoholics (Table 17.9).

3.5 Pharmacogenetics of Alcohol Dependence Drugs

3.5.1 Naltrexone

Naltrexone is a mu-opioid receptor antagonist, although it can also bind, to a lesser degree, to kappa- and delta-opioid receptors in the brain (Shader 2003). The mechanism of action of naltrexone implicates the blockade of μ-opioid receptor, thereby attenuating the reinforcing effects of alcohol (Ray and Hutchison 2007). Naltrexone has demonstrated moderate efficacy in clinical trials.

The relation between the OPRM1 A118G polymorphism and pharmacological response to naltrexone has been widely investigated in the past few years (Oslin et al. 2003; McGeary et al. 2006; Gelernter et al. 2007; Ray and Hutchison 2007; Anton et al. 2008, 2012; Arias et al. 2008; Haile et al. 2008; O'Malley et al. 2008; Tidey et al. 2008; Kim 2009; Kim et al. 2009; Oroszi et al. 2009; Kranzler and Edenberg 2010; Ray et al. 2010a, b, 2012a, b; Wang et al. 2010; Kimura and Higuchi 2011; Arias and Sewell 2012; Ashenhurst et al. 2012; Chamorro et al. 2012). Evidence from these clinical trials and pharmacogenetic studies suggests that the presence of the variant 118G allele of OPRM1 may predict better treatment response to opioid receptor antagonists such as naltrexone (Oslin et al. 2003; McGeary et al. 2006; Ray and Hutchison 2007; Anton et al. 2008; Haile et al. 2008; O'Malley et al. 2008; Kim 2009; Oroszi et al. 2009; Ray et al. 2010a, b, 2012a, b; Arias and Sewell 2012; Ashenhurst et al. 2012; Chamorro et al. 2012). Particularly, the 118G allele significantly showed lower rates of relapse in alcohol-dependent individuals treated with naltrexone (Oslin et al. 2003; Haile et al. 2008; Kim 2009). Carriers of the 118G allele also showed increased urge for alcohol when treated with naltrexone (McGeary et al. 2006). Individuals with at least 1 copy of the G allele reported lower alcohol craving and higher efficacy of naltrexone on blunting alcohol-induced (Ray and Hutchison 2007; Ray et al. 2012a, b). In 2008, the results of the COMBINE study, from a clinical trial of 604 recently abstinent volunteers, confirmed and extended the observation that the functionally significant OPRM1 118G allele is able to predict naltrexone treatment response in alcoholic individuals (Anton et al. 2008). Results of this clinical trial showed that alcoholic subjects carrying the 118G allele had an increased percentage of days abstinent and a decreased percentage of heavy drinking if treated with naltrexone. The clinical outcome was good in 87.1 % of 118G carriers when treated with naltrexone, compared with only 54.8 % of individuals with the AA genotype (odds ratio, 5.75; confidence interval, 1.88–17.54) (Anton et al. 2008). Results of the COMBINE study also confirmed that the highest percent of good clinical outcome occurred in naltrexone-treated alcoholics carrying the 118G allele (Oroszi et al. 2009). Higher overall levels of vigor and lower levels negative mood after drinking have also been reported in carriers of the 118G allele in nontreatment-seeking heavy drinkers (Ray et al. 2010b). Naltrexone treatment has also shown to raise GABAergic neurosteroid levels only among carriers of the 118G allele, demonstrating a role of this polymorphism in the pharmacological effects of naltrexone (Ray et al. 2010a). Greater alcohol-induced sedation and subjective intoxication are features also associated to 118G allele in heavy drinkers treated with naltrexone (Ray et al. 2012a). Altogether, these results suggest that OPRM1 genotyping prior to treatment in alcoholic individuals might be useful to help in selecting pharmacotherapeutical alternatives.

Despite of all this increasing evidence, aiming to OPRM1 A118G polymorphism as a consistent predictor of naltrexone response in AD patients, some studies have failed to find an association of the OPRM1 118G polymorphism with naltrexone treatment response for AD (Gelernter et al. 2007; Tidey et al. 2008; Arias et al. 2008; Coller et al. 2011; Anton et al. 2012). However, the relevance of this polymorphism in the response to naltrexone becomes evident in a recent meta-analysis

Table 17.10 Non-FDA-approved medications for the treatment of alcohol dependence and associated gene polymorphisms

Drug	Nature	Gene	Polymorphism	References
Sertraline	Serotonin-specific reuptake inhibitor	SLC6A4	5-HTTPR	Kranzler et al. (2011)
Olanzapine	Antipsychotic	DRD4	DRD4 Exon 3 48-bp VNTR	Hutchison et al. (2003, 2006), Kranzler and Edenberg (2010), Arias and Sewell (2012)
Topiramate	Anticonvulsant	GRIK1	rs2832407	Kranzler et al. (2009), Ray et al. (2009)
Ondansetron	Serotonin 5-HT3 receptor antagonist	SLC6A4	5-HTTPR 3-HTT (rs1042173)	Johnson et al. (2000, 2002, 2003, 2011), Arias and Sewell (2012)

VNTR variable number of tandem repeats

that confirms that naltrexone-treated patients carrying the G allele present lower relapse rates than AA individuals (OR: 2.02, 95 % CI 1.26–3.22; $P=0.003$) (Chamorro et al. 2012).

The possible contribution to naltrexone pharmacogenetics of additional polymorphisms in the opioid receptor system, particularly in the kappa and delta (OPRK1 and OPRD1) genes, has been recently analyzed by Ashenhurst et al. (2012) in a double-blind placebo-controlled laboratory trial of naltrexone in 40 heavy drinkers. Multilevel models revealed significantly two associations, one between OPRK1 (rs997917) polymorphism and alcohol-induced sedation and other between OPRD1 (rs4654327) and alcohol-induced stimulation and craving. TT homozygotes for OPRK1 (rs997917) showed lower naltrexone-induced alcohol sedation as compared to carriers of the C allele, and carriers of the A allele at OPRK1 (rs997917) reported greater naltrexone-induced blunting of alcohol stimulation and alcohol craving compared to GG homozygotes. These findings support the involvement of the opioid receptor system in the response to naltrexone and suggest a potential accumulation of individual pharmacogenetic effects from different gene polymorphisms (Ashenhurst et al. 2012).

3.5.2 Acamprosate

Acamprosate is an inhibitor of the *N*-methyl-D-aspartate (NMDA) receptor that works through the glutamatergic/GABAergic negative reinforcement system.

Several gene polymorphisms have been described to be related to acamprosate pharmacological response in AD individuals (Ooteman et al. 2009; Kiefer et al. 2011).

Recently, the top 15 single nucleotide polymorphisms of a genome-wide association (GWA) by Bierut et al. (2010) were used to investigate relapse to alcohol drinking and treatment response in 374 alcohol-dependent subjects who underwent a randomized, double-blind, placebo-controlled trial with acamprosate, naltrexone, or placebo (Kiefer et al. 2011). The single nucleotide polymorphism, rs13273672, located in an intronic region of the GATA-binding protein 4 (GATA4) gene resulted to be associated with relapse in patients treated with acamprosate. The different GATA4 genotypes also showed a significant gene dose effect on the variance of atrial natriuretic peptide (ANP) plasma concentration (Kiefer et al. 2011).

3.5.3 Other Treatments for AD

Several non-FDA-approved drugs such as sertraline, olanzapine, topiramate, baclofen, and ondansetron have shown promising results in the investigational (and off-label) treatment of AD (Table 17.10) (Edwards et al. 2011; Arias and Sewell 2012). Heterogeneity in response, side effects, and dosage of these drugs has been tried to be explained by different gene polymorphisms.

Sertraline

Sertraline is a serotonin-specific reuptake inhibitor (SSRI) whose response has showed to be conditioned by a 44-bp repeat insertion polymorphism in the 5-HTT-linked promoter region (5-HTTLPR) of SLC6A4 gene that codes for the serotonin transporter (Lesch et al. 1996), resulting in long (L) and short (S) alleles (Kranzler et al. 2011). In AD, results from 134 patients from a prospective randomized placebo-controlled trial of the efficacy of sertraline in AD have showed that LL homozygotes are the most benefited by sertraline treatment (Kranzler et al. 2011).

Olanzapine

Olanzapine is an antipsychotic whose effects on craving for alcohol and drinking behavior have been associated to a variable number of tandem repeats (VNTR) gene polymorphism in the exon 3 of DRD4 gene, which encodes the dopamine D(4) receptor (Hutchison et al. 2003, 2006; Kranzler and Edenberg 2010). This variation in the DRD4 gene has also been proposed as a predictor of a better response to naltrexone and olanzapine (Arias and Sewell 2012).

Topiramate

Topiramate is an anticonvulsant with a promising future in the treatment of AD, although the optimal dose and treatment length have not been well determined yet (Edwards et al. 2011). Topiramate acts as a glutamate modulator that binds to the GluR5 of the kainic acid glutamate receptor subunit. Three SNPs in the GRIK1 gene (rs2832407 in intron 9, rs2186305 in intron 17, and rs2832387 in the 3'UTR), which encodes the GluR5 receptor subunit, have been associated to higher risk of AD in a study conducted in 1564 individuals (507 control subjects and 1,057 AD subjects), although only rs2832407 was significantly associated to phenotype (Kranzler et al. 2009). These three SNPs in the GRIK1 gene were also investigated as predictors of topiramate-induced side effects in 51 heavy drinkers, finding an association of one of the SNPs studied (rs2832407) with the severity of side effects and serum levels of topiramate (Ray et al. 2009).

Ondansetron

Ondansetron is a serotonin 5-HT3 receptor antagonist that has shown promising results in the treatment of AD (Johnson et al. 2011; Arias and Sewell 2012). Ondansetron was reported to reduce heavy drinking in individuals with early-onset alcoholism, mainly via reductions in craving and improvement of mood disturbances (Johnson et al. 2000, 2002, 2003). Response to ondansetron in AD treatment could be predicted by SLC6A4 (Johnson et al. 2011; Arias and Sewell 2012). 5-HTTLPR degenerate repeat polymorphic region in SLC6A4 gene has been proposed along with rs1042173 gene polymorphism in the UTR-3 region of this gene (3-HTT) as indicators of better response to ondansetron in alcoholics, especially for abstinence (Johnson et al. 2011). Particularly, individuals with the LL genotype who received ondansetron presented lower mean number of drinks per drinking day and a higher percentage of days abstinent in a randomized double-blind controlled trial conducted on 283 alcoholics (Johnson et al. 2011). In this study it also found a significant interaction between the 5-HTTLPR and rs1042173 variants, showing a better improvement for LL/TT individuals.

3.6 Importance of Drug Interactions in the Treatment of Alcohol Dependence

CYP2E1, as well as alcohol dehydrogenase and aldehyde dehydrogenase, is an important enzyme for the conversion of ethanol to acetaldehyde and acetate in humans.

The study of García-Bañuelos et al. (2012) carried out on 90 controls and 41 patients with alcohol addiction from Mexico showed that CYP2E1*c2 allele was associated with susceptibility to alcohol dependence.

Susceptibility to breast cancer mediated by alcohol consumption and genetic polymorphisms in the enzymes that metabolize alcohol, CYP2E1, and ALDH2 was examined in a Korean population (346 patients with histologically confirmed breast cancer and 377 controls without current or previous history of cancer) (Choi et al. 2003). The risk of developing breast cancer was 1.9 times higher in alcohol-addicted women carrying the CYP2E1 c2 allele than nonconsumers with the CYP2E1 c1/c1 genotype. This study therefore suggests that the CYP2E1*c2 allele may influence individual susceptibility to breast cancer in women who consume alcohol (Choi et al. 2003).

3.7 Final Considerations

After a thorough review of numerous studies concerning the involvement of the opioid system (OPRM1, OPRK1, and OPRD1) and cytochrome P450 gene (CYP3A4, CYP2B6, CYP2D6, CYP2C9, and CYP2C19) in the dependence on psychoactive substances, such as opiates and alcohol, a great variability in the results for both gene families is evident.

A118G polymorphism on OPRM1 gene is the most studied gene polymorphism in alcohol and opioid addiction and is probably the most promising biomarker of better response in these patients, despite discrepancies that have been manifested even in studies conducted on the same ethnicity. Many studies have failed to reproduce the positive results obtained by other authors, probably due to the need of a better definition of the phenotypes to study, considering the heterogeneity of these pathologies. Other factors, like the complexity of genetics in addiction, not only for the pathophysiology originated in different stages of addiction but also by the involvement of different genetic backgrounds, may influence in the heterogeneity of the population, along with the interaction between environmental factors and genes involved.

The influence of interactions between gene polymorphisms in different molecular structures is also an important question. Opioids are metabolized by phase I metabolism CYP450 isoenzymes, which show a high variability in the population. Therefore, modulation of these enzymes, through drug interactions simulating genetic defects (inhibitors/inducers), also contributes to the variable response to opioids hence the importance of knowing the concomitant drug treatment and use of other substances of abuse, the presence of pathologies, diet, environmental, and genetic factors.

Involvement in opiate dependence of other gene polymorphisms on the opioid receptor system, such as OPRM1 C17T and C1031G, OPRD1 T921C and G80T, and OPRK1 G36T, has also yielded inconclusive results. Individual and combined analyses have shown a great variability of results, therefore not demonstrating a clear involvement in opioid dependence.

In conclusion, A118G genotype is currently the most promising polymorphism that may help the clinician to adjust drug treatment to each individual. The existence of interactions between gene polymorphisms could also be important in determining the risk profile for complex diseases such as addiction.

Future prospects are based on the study of these gene polymorphisms in association with other molecular structures involved that allow replicating the results among individuals in a population, given that opioid/alcohol dependence is a disease very complex to approach due to the multiple factors involved individually, such as genetic factors common to the use of other substances (abuse drugs and/or concomitant drug therapy), specific genetic factors, environmental factors, and the intrinsic effects of the drug on the CNS.

References

Alvarez Y, Farre M (2005) Pharmacology of opioid drugs. In: Opiates monograph. Addictions Supplement 2(17):21–40. Socidrogalcohol Eds, Valencia

Anton RF, Oroszi G, O'Malley S, Couper D, Swift R, Pettinati H, Goldman D (2008) An evaluation of mu-opioid receptor (OPRM1) as a predictor of naltrexone response in the treatment of alcohol dependence: results from the Combined Pharmacotherapies and Behavioral Interventions for Alcohol Dependence (COMBINE) study. Arch Gen Psychiatry 65:135–144

Anton RF, Voronin KK, Randall PK, Myrick H, Tiffany A (2012) Naltrexone modification of drinking effects in a Subacute Treatment and Bar-Lab Paradigm: influence of OPRM1 and Dopamine Transporter (SLC6A3) Genes. Alcohol Clin Exp Res 36(11):2000–2007

Arias AJ, Sewell RA (2012) Pharmacogenetically driven treatments for alcoholism: are we there yet? CNS Drugs 26:461–476

Arias AJ, Armeli S, Gelernter J, Covault J, Kallio A, Karhuvaara S, Koivisto T, Mäkelä R, Kranzler HR (2008) Effects of opioid receptor gene variation on targeted nalmefene treatment in heavy drinkers. Alcohol Clin Exp Res 32:1159–1166

Ashenhurst JR, Bujarski S, Ray LA (2012) Delta and kappa opioid receptor polymorphisms influence the effects of naltrexone on subjective responses to alcohol. Pharmacol Biochem Behav 103(2):253–259

Bart G, Heilig M, LaForge KS, Pollak L, Leal SM, Ott J, Kreek MJ (2004) Substantial attributable risk related to a functional mu-opioid receptor gene polymorphism in association with heroin addiction in central Sweden. Mol Psychiatry 9:547–549

Bart G, Kreek MJ, Ott J, LaForge KS, Proudnikov D, Pollak L, Heilig M (2005) Increased attributable risk related to a functional mu-opioid receptor gene polymorphism in association with alcohol dependence in central Sweden. Neuropsychopharmacology 30:417–422

Bergen AW, Kokoszka J, Peterson R, Long JC, Virkkunen M, Linnoila M, Goldman D (1997) Mu opioid receptor gene variants: lack of association with alcohol dependence. Mol Psychiatry 2:490–494

Berretini WH, Hoehe MR, Ferrato TN, DeMaría PA, Gottheil E (1997) Human mu opioid receptor gene polymorphism and vulnerability to substance abuse. Addict Biol 2:303–308

Bierut LJ, Agrawal A, Bucholz KK, Doheny KF, Laurie C, Pugh E, Fisher S, Fox L, Howells W, Bertelsen S, Hinrichs AL, Almasy L, Breslau N, Culverhouse RC, Dick DM, Edenberg HJ, Foroud T, Grucza RA, Hatsukami D, Hesselbrock V, Johnson EO, Kramer J, Krueger RF, Kuperman S, Lynskey M, Mann K, Neuman RJ, Nöthen MM, Nurnberger JI Jr, Porjesz B, Ridinger M, Saccone NL, Saccone SF, Schuckit MA, Tischfield JA, Wang JC, Rietschel M, Goate AM, Rice JP, Gene, Environment Association Studies Consortium (2010) A genome-wide association study of alcohol dependence. Proc Natl Acad Sci USA 107:5082–5087

Bond C, LaForge KS, Tian M, Melia D, Zhang S, Borg L, Gong J, Schluger J, Strong JA, Leal SM, Tischfield JA, Kreek MJ, Yu L (1998) Single-nucleotide polymorphism in the human mu opioid receptor gene alters β-endorphin binding activity: possible implications for opiate addiction. Proc Natl Acad Sci USA 95:9608–9613

Bunten H, Liang WJ, Pounder DJ, Seneviratne C, Osselton D (2010) OPRM1 and CYP2B6 gene variants as risk factors in methadone-related deaths. Clin Pharmacol Ther 88:383–389

Bunten H, Liang WJ, Pounder DJ, Seneviratne C, Osselton D (2011) Interindividual variability in the prevalence of OPRM1 and CYP2B6 gene variations may identify drug-susceptible populations. J Anal Toxicol 35:431–437

Chamorro AJ, Marcos M, Mirón-Canelo JA, Pastor I, González-Sarmiento R, Laso FJ (2012) Association of μ-opioid receptor (OPRM1) gene polymorphism with response to naltrexone in alcohol dependence: a systematic review and meta-analysis. Addict Biol 17:505–512

Chen CH, Wang SC, Tsou HH, Ho IK, Tian JN, Yu CJ, Hsiao CF, Chou SY, Lin YF, Fang KC, Huang CL, Su LW, Fang YC, Liu ML, Lin KM, Hsu YT, Liu SC, Chen AC, Liu YL (2011) Genetic polymorphisms in CYP3A4 are associated with withdrawal symptoms and adverse reactions in methadone maintenance patients. Pharmacogenomics 12:1397–1406

Chen D, Liu L, Xiao Y, Peng Y, Yang C, Wang Z (2012) Ethnic-specific meta-analyses of association between the OPRM1 A118G polymorphism and alcohol dependence among Asians and Caucasians. Drug Alcohol Depend 123:1–6

Choi JY, Abel J, Neuhaus T, Ko Y, Harth V, Hamajima N, Tajima K, Yoo KY, Park SK, Noh DY, Han W, Choe KJ, Ahn SH, Kim SU, Hirvonen A, Kang D (2003) Role of alcohol and genetic polymorphisms of CYP2E1 and ALDH2 in breast cancer development. Pharmacogenetics 13:67–72

Coller JK, Cahill S, Edmonds C, Farquharson AL, Longo M, Minniti R, Sullivan T, Somogyi AA, White JM (2011) OPRM1 A118G genotype fails to predict the effectiveness of naltrexone treatment for alcohol dependence. Pharmacogenet Genomics 21:902–905

Compton P, Geschwind DH, Alarcón M (2003) Association between human mu-opioid receptor gene polymorphism, pain tolerance, and opioid addiction. Am J Med Genet B Neuropsychiatr Genet 121B:76–82

Crettol S, Déglon JJ, Besson J, Croquette-Krokkar M, Gothuey I, Hämmig R, Monnat M, Hüttemann H, Baumann P, Eap CB (2005) Methadone enantiomer plasma levels, CYP2B6, CYP2C19, and CYP2C9 genotypes, and response to treatment. Clin Pharmacol Ther 78:593–604

Crettol S, Déglon JJ, Besson J, Croquette-Krokar M, Hämmig R, Gothuey I, Monnat M, Eap CB (2006) ABCB1 and cytochrome P450 genotypes and phenotypes: influence on methadone plasma levels and response to treatment. Clin Pharmacol Ther 80:668–681

Crettol S, Petrovic N, Murray M (2010) Pharmacogenetics of phase I and phase II drug metabolism. Curr Pharm Des 16:204–219

Crowley JJ, Oslin DW, Patkar AA, Gottheil E, Demaria PA Jr, O'brien CP, Berrettini WH, Grice DE (2003) A genetic association study of the mu opioid receptor and severe opioid dependence. Psychiatr Genet 13:169–173

dbSNP database. National Center for Biotechnology Information (NCBI). Available at http://www.ncbi.nlm.nih.gov/SNP. Accessed Jun 2012

Drakenberg K, Nikoshkov A, Horváth MC, Fagergren P, Gharibyan A, Saarelainen K, Rahman S, Nylander I, Bakalkin G, Rajs J, Keller E, Hurd YL (2006) Mu opioid receptor A118G polymorphism in association with striatal opioid neuropeptide gene expression in heroin abusers. Proc Natl Acad Sci U S A 103:7883–7888

Eap CB, Broly F, Mino A, Hämmig R, Déglon JJ, Uehlinger C, Meili D, Chevalley AF, Bertschy G, Zullino D, Kosel M, Preisig M, Baumann P (2001) Cytochrome P450 2D6 genotype and methadone steady-state concentrations. J Clin Psychopharmacol 21:229–234

Edwards S, Kenna GA, Swift RM, Leggio L (2011) Current and promising pharmacotherapies, and novel research target areas in the treatment of alcohol dependence: a review. Curr Pharm 17:1323–1332

Fernandes LC, Kilicarslan T, Kaplan HL, Tyndale RF, Sellers EM, Romach MK (2002) Treatment of codeine dependence with inhibitors of cytochrome P450 2D6. J Clin Psychopharmacol 22:326–329

Fonseca F, de la Torre R, Díaz L, Pastor A, Cuyàs E, Pizarro N, Khymenets O, Farré M, Torrens M (2011) Contribution of cytochrome P450 and ABCB1 genetic variability on methadone pharmacokinetics, dose requirements, and response. PLoS One 6:e19527

Franke P, Nothen MM, Wang T, Neidt H, Knapp M, Lichtermann D, Weiffenbach O, Mayer P, Hollt V, Propping P, Marier W (1999) Human d-opioid receptor gene and susceptibility to heroin and alcohol dependence. Am J Med Genet 88:462–464

Franke P, Wang T, Nöthen MM, Knapp M, Neidt H, Albrecht S, Jahnes E, Propping P, Maier W (2001) Nonreplication of association between mu opioid receptor gene (OPRM1) A118G polymorphism and substance dependence. Am J Med Genet 105:114–119

García-Bañuelos J, Panduro A, Gordillo-Bastidas D, Gordillo-Bastidas E, Muñoz-Valle JF, Gurrola-Díaz CM, Sánchez-Enríquez S, Ruiz-Madrigal B, Bastidas-Ramírez BE (2012) Genetic polymorphisms of genes coding to alcohol-metabolizing enzymes in western Mexicans: association of CYP2E1*c2/CYP2E1*5B allele with cirrhosis and liver function. Alcohol Clin Exp Res 36:425–431

Gelernter J, Kranzler H, Cubells J (1999) Genetics of two mu opioid receptor gene (OPRM1) exon I polymorphisms: population studies, and allele frequencies in alcohol- and drug-dependent subjects. Mol Psychiatry 4:476–483

Gelernter J, Gueorguieva R, Kranzler HR, Zhang H, Cramer J, Rosenheck R, Krystal JH, VA Cooperative Study #425 Study Group (2007) Opioid receptor gene (OPRM1, OPRK1, and OPRD1) variants and response to naltrexone treatment for alcohol dependence: results from the VA Cooperative Study. Alcohol Clin Exp Res 31:555–563

Gerra G, Leonardi C, Cortese E, D'Amore A, Lucchini A, Strepparola G, Serio G, Farina G, Magnelli F, Zaimovic A, Mancini A, Turci M, Manfredini M, Donnini C (2007) Human kappa opioid receptor gene (OPRK1) polymorphism is associated with opiate addiction. Am J Med Genet B Neuropsychiatr Genet 144B:771–775

Glatt SJ, Bousman C, Wang RS, Murthy KK, Rana BK, Lasky-Su JA, Zhu SC, Zhang R, Li J, Zhang B, Li J, Lyons MJ, Faraone SV, Tsuang MT (2007) Evaluation of OPRM1 variants in heroin dependence by family-based association testing and meta-analysis. Drug Alcohol Depend 90:159–165

Haile CN, Kosten TA, Kosten TR (2008) Pharmacogenetic treatments for drug addiction: alcohol and opiates. Am J Drug Alcohol Abuse 34:355–381

Herz A (1998) Opioid reward mechanisms: a key role in drug abuse? Can J Physiol Pharmacol 76:252–258

Hung CC, Chiou MH, Huang BH, Hsieh YW, Hsieh TJ, Huang CL, Lane HY (2011) Impact of genetic polymorphisms in ABCB1, CYP2B6, OPRM1, ANKK1 and DRD2 genes on methadone therapy in Han Chinese patients. Pharmacogenomics 12:1525–1533

Hutchison KE, Wooden A, Swift RM, Smolen A, McGeary J, Adler L, Paris L (2003) Olanzapine reduces craving for alcohol: a DRD4 VNTR polymorphism by pharmacotherapy interaction. Neuropsychopharmacology 28:1882–1888

Hutchison KE, Ray L, Sandman E, Rutter MC, Peters A, Davidson D, Swift R (2006) The effect of olanzapine on craving and alcohol consumption. Neuropsychopharmacology 31:1310–1317

Jannetto PJ, Bratanow NC (2009) Utilization of pharmacogenomics and therapeutic drug monitoring for opioid pain management. Pharmacogenomics 10:1157–1167

Johnson BA, Roache JD, Javors MA, DiClemente CC, Cloninger CR, Prihoda TJ, Bordnick PS, Ait-Daoud N, Hensler J (2000) Ondansetron for reduction of drinking among biologically predisposed alcoholic patients: a randomized controlled trial. JAMA 284: 963–971

Johnson BA, Roache JD, Ait-Daoud N, Zanca NA, Velazquez M (2002) Ondansetron reduces the craving of biologically predisposed alcoholics. Psychopharmacology (Berl) 160:408–413

Johnson BA, Ait-Daoud N, Ma JZ, Wang Y (2003) Ondansetron reduces mood disturbance among biologically predisposed, alcohol-dependent individuals. Alcohol Clin Exp Res 27:1773–1779

Johnson BA, Ait-Daoud N, Seneviratne C, Roache JD, Javors MA, Wang XQ, Liu L, Penberthy JK, DiClemente CC, Li MD (2011) Pharmacogenetic approach at the serotonin transporter gene as a method of reducing the severity of alcohol drinking. Am J Psychiatry 168:265–275

Kapur S, Sharad S, Singh RA, Gupta AK (2007) A118g polymorphism in mu opioid receptor gene (oprm1): association with opiate addiction in subjects of Indian origin. J Integr Neurosci 6:511–522

Kathiramalainathan K, Kaplan HL, Romach MK, Busto UE, Li NY, Säwe J, Tyndale RF, Sellers EM (2000) Inhibition of cytochrome P450 2D6 modifies codeine abuse liability. J Clin Psychopharmacol 20:435–444

Kiefer F, Witt SH, Frank J, Richter A, Treutlein J, Lemenager T, Nöthen MM, Cichon S, Batra A, Berner M, Wodarz N, Zimmermann US, Spanagel R, Wiedemann K, Smolka MN, Heinz A, Rietschel M, Mann K (2011) Involvement of the atrial natriuretic peptide transcription factor GATA4 in alcohol dependence, relapse risk and treatment response to acamprosate. Pharmacogenomics J 11:368–374

Kim SG (2009) Gender differences in the genetic risk for alcohol dependence – the results of a pharmacogenetic study in Korean alcoholics. Nihon Arukoru Yakubutsu Igakkai Zasshi 44:680–685

Kim SG, Kim CM, Choi SW, Jae YM, Lee HG, Son BK, Kim JG, Choi YS, Kim HO, Kim SY, Oslin DW (2009) A micro opioid receptor gene polymorphism (A118G) and naltrexone treatment response in adherent Korean alcohol-dependent patients. Psychopharmacology (Berl) 201:611–618

Kimura M, Higuchi S (2011) Genetics of alcohol dependence. Psychiatry Clin Neurosci 65:213–225

Koller G, Zill P, Rujescu D, Ridinger M, Pogarell O, Fehr C, Wodarz N, Bondy B, Soyka M, Preuss UW (2012) Possible association between OPRM1 genetic variance at the 118 locus and alcohol dependence in a large treatment sample: relationship to alcohol dependence symptoms. Alcohol Clin Exp Res 36:1230–1236

Kranzler HR, Edenberg HJ (2010) Pharmacogenetics of alcohol and alcohol dependence treatment. Curr Pharm Des 16:2141–2148

Kranzler HR, Gelernter J, Anton RF, Arias AJ, Herman A, Zhao H, Burian L, Covault J (2009) Association of markers in the 3′ region of the GluR5 kainate receptor subunit gene to alcohol dependence. Alcohol Clin Exp Res 33:925–930

Kranzler HR, Armeli S, Tennen H, Covault J, Feinn R, Arias AJ, Pettinati H, Oncken C (2011) A double-blind, randomized trial of sertraline for alcohol dependence: moderation by age of onset [corrected] and 5-hydroxytryptamine transporter-linked promoter region genotype. J Clin Psychopharmacol 31:22–30

Kreek MJ (1997) Opiate and cocaine addictions: challenge for pharmacotherapies. Pharmacol Biochem Behav 57:551–569

Kumar D, Chakraborty J, Das S (2012) Epistatic effects between variants of kappa-opioid receptor gene and A118G of mu-opioid receptor gene increase susceptibility to addiction in Indian population. Prog Neuropsychopharmacol Biol Psychiatry 36:225–230

Lesch KP, Bengel D, Heils A, Sabol SZ, Greenberg BD, Petri S, Benjamin J, Müller CR, Hamer DH, Murphy DL (1996) Association of anxiety-related traits with a polymorphism in the serotonin transporter gene regulatory region. Science 274:1527–1531

Li Y, Kantelip JP, Gerritsen-van Schieveen P, Davani S (2008) Interindividual variability of methadone response: impact of genetic polymorphism. Mol Diagn Ther 12:109–124

Luo X, Kranzler HR, Zhao H, Gelernter J (2003) Haplotypes at the OPRM1 locus are associated with susceptibility to substance dependence in European-Americans. Am J Med Genet B Neuropsychiatr Genet 120:97–108

March JC, Oviedo-Joekes E, Perea-Milla E, Carrasco F, PEPSA Team (2006) Controlled trial of prescribed heroin in the treatment of opioid addiction. J Subst Abuse Treat 31:203–211

Marset M (2005) Heroin prescription programs. In: Opiates monograph. Addictions. Supplement 2(17): 21–40. Socidrogalcohol Eds, Valencia

Martindale: the complete drug reference. Accessed 15 June 2012. Available from URL: https://www medicinescomplete.com/mc/martindale/current/login.htm?uri=http%3A%2F%2Fwww.medicinescomplete.com%2Fmc%2Fmartindale%2Fcurrent%2F. Accessed Sep 2012

Mayer P, Rochlitz H, Rauch E, Rommelspacher H, Hasse HE, Schmidt S (1997) Association between a delta opioid receptor gene polymorphism and heroin dependence in man. Neuroreport 8:2547–2550

McGeary JE, Monti PM, Rohsenow DJ, Tidey J, Swift R, Miranda R Jr (2006) Genetic moderators of naltrexone's effects on alcohol cue reactivity. Alcohol Clin Exp Res 30:1288–1296

Menoyo A, del Río E, Baiget M (2006) Characterization of variant alleles of cytochrome CYP2D6 in Spanish populations. Cell Biochem Funct 24:381–385

MICROMEDEX® 2.0. Accessed 15 June 2012. Available from URL: http://www.thomsonhc.com/micromedex2/2.15.12/WebHelp/whnjs.htm. Accessed Sep 2012

Nagaya D, Ramanathan S, Ravichandran M, Navaratnam V (2012) A118G mu opioid receptor polymorphism among drug addicts in Malaysia. J Integr Neurosci 11:117–122

Nikolov MA, Beltcheva O, Galabova A, Ljubenova A, Jankova E, Gergov G, Russev AA, Lynskey MT, Nelson EC, Nesheva E, Krasteva D, Lazarov P, Mitev VI, Kremensky IM, Kaneva RP, Todorov AA (2011) No evidence of association between 118A>G OPRM1 polymorphism and heroin dependence in a large Bulgarian case-control sample. Drug Alcohol Depend 117(1):62–65

O'Malley SS, Robin RW, Levenson AL, GreyWolf I, Chance LE, Hodgkinson CA, Romano D, Robinson J, Meandzija B, Stillner V, Wu R, Goldman D (2008) Naltrexone alone and with sertraline for the treatment of alcohol dependence in Alaska natives and non-natives residing in rural settings: a randomized controlled trial. Alcohol Clin Exp Res 32:1271–1283

Ooteman W, Naassila M, Koeter MW, Verheul R, Schippers GM, Houchi H, Daoust M, van den Brink W (2009) Predicting the effect of naltrexone and acamprosate in alcohol-dependent patients using genetic indicators. Addict Biol 14:328–337

Oroszi G, Anton RF, O'Malley S, Swift R, Pettinati H, Couper D, Yuan Q, Goldman D (2009) OPRM1 Asn40Asp predicts response to naltrexone treatment: a haplotype-based approach. Alcohol Clin Exp Res 33:383–393

Oslin DW, Berrettini W, Kranzler HR, Pettinati H, Gelernter J, Volpicelli JR, O'Brien CP (2003) A functional polymorphism of the mu-opioid receptor gene is associated with naltrexone response in alcohol-dependent patients. Neuropsychopharmacology 28:1546–1552

Rattan AK, Koo KL, Tejwani GA, Bhargava HN (1992) The effect of morphine tolerance dependence and abstinence on immunoreactive dynorphin (1–13) levels in discrete brain regions, spinal cord, pituitary gland and peripheral tissues of the rat. Brain Res 584:207–212

Ray LA, Hutchison KE (2007) Effects of naltrexone on alcohol sensitivity and genetic moderators of medication response: a double-blind placebo-controlled study. Arch Gen Psychiatry 64:1069–1077

Ray LA, Miranda R Jr, MacKillop J, McGeary J, Tidey JW, Rohsenow DJ, Gwaltney C, Swift RW, Monti PM (2009) A preliminary pharmacogenetic investigation of adverse events from topiramate in heavy drinkers. Exp Clin Psychopharmacol 17:122–129

Ray LA, Hutchison KE, Ashenhurst JR, Morrow AL (2010a) Naltrexone selectively elevates GABAergic neuroactive steroid levels in heavy drinkers with the Asp40 allele of the OPRM1 gene: a pilot investigation. Alcohol Clin Exp Res 34:1479–1487

Ray LA, Miranda R Jr, Tidey JW, McGeary JE, MacKillop J, Gwaltney CJ, Rohsenow DJ, Swift RM, Monti PM (2010b) Polymorphisms of the mu-opioid receptor and dopamine D4 receptor genes and subjective responses to alcohol in the natural environment. J Abnorm Psychol 119:115–125

Ray LA, Bujarski S, Chin PF, Miotto K (2012a) Pharmacogenetics of naltrexone in Asian Americans: a randomized placebo-controlled laboratory study. Neuropsychopharmacology 37:445–455

Ray LA, Courtney KE, Bujarski S, Squeglia LM (2012b) Pharmacogenetics of alcoholism: a clinical neuroscience perspective. Pharmacogenomics 13:129–132

Rommelspacher H, Smolka M, Schmidt LG, Samochowiec J, Hoehe MR (2001) Genetic analysis of the mu-opioid receptor in alcohol-dependent individuals. Alcohol 24:129–135

Samer CF, Piguet V, Dayer P, Desmeules JA (2005) Genetic polymorphism and drug interactions: their importance in the treatment of pain. Can J Anaesth 52:806–821

Schinka JA, Town T, Abdullah L, Crawford FC, Ordorica PI, Francis E, Hughes P, Graves AB, Mortimer JA, Mullan M (2002) A functional polymorphism within the mu-opioid receptor gene and risk for abuse of alcohol and other substances. Mol Psychiatry 7:224–228

Shader RI (2003) Antagonists, inverse agonists, and protagonists. J Clin Psychopharmacol 23:321–322

Shi J, Hui L, Xu Y, Wang F, Huang W, Hu G (2002) Sequence variation in the mu opioid receptor gene (OPRM1) associated with human addiction to heroin. Hum Mutat 19:459–460

Shiran MR, Lennard MS, Iqbal MZ, Lagundoye O, Seivewright N, Tucker GT, Rostami-Hodjegan A (2009) Contribution of the activities of CYP3A, CYP2D6, CYP1A2 and other potential covariates to the disposition of methadone in patients undergoing methadone maintenance treatment. Br J Clin Pharmacol 67:29–37

Szeto CY, Tang NL, Lee DT, Stadlin A (2001) Association between mu opioid receptor gene polymorphisms and Chinese heroin addicts. Neuroreport 12:1103–1106

Tan EC, Tan CH, Karupathivan U, Yap EP (2003) Mu opioid receptor gene polymorphisms and heroin dependence in Asian populations. Neuroreport 14:569–572

Thorn CF, Klein TE, Altman RB (2009) Codeine and morphine pathway. Pharmacogenet Genomics 19: 556–558

Tidey JW, Monti PM, Rohsenow DJ, Gwaltney CJ, Miranda R Jr, McGeary JE, MacKillop J, Swift RM, Abrams DB, Shiffman S, Paty JA (2008) Moderators of naltrexone's effects on drinking, urge, and alcohol effects in non-treatment-seeking heavy drinkers in the natural environment. Alcohol Clin Exp Res 32:58–66

Tyndale RF, Droll KP, Sellers EM (1997) Genetically deficient CYP2D6 metabolism provides protection against oral opiate dependence. Pharmacogenetics 7:375–379

Wallenstein SL, Houde RW, Portenoy R, Lapin J, Rogers A, Foley KM (1990) Clinical analgesic assay of repeated and single doses of heroin and hydromorphone. Pain 41:5–13

Wang LL, Yang AK, He SM, Liang J, Zhou ZW, Li Y, Zhou SF (2010) Identification of molecular targets associated with ethanol toxicity and implications in drug development. Curr Pharm Des 16:1313–1355

Wang SC, Ho IK, Tsou HH, Tian JN, Hsiao CF, Chen CH, Tan HK, Lin L, Wu CS, Su LW, Huang CL, Yang YH, Liu ML, Lin KM, Chen CY, Liu SC, Wu HY, Chan HW, Tsai MH, Lin PS, Liu YL (2011) CYP2B6 polymorphisms influence the plasma concentration and clearance of the methadone S-enantiomer. J Clin Psychopharmacol 31:463–469

Williams SH (2005) Medications for treating alcohol dependence. Am Fam Physician 72:1775–1780

Xu K, Liu XH, Nagarajan S, Gu XY, Goldman D (2002) Relationship of the delta-opioid receptor gene to heroin abuse in a large Chinese case/control simple. Am J Med Genet 110:45–50

Yuferov V, Fussell D, LaForge KS, Nielsen DA, Gordon D, Ho A, Leal SM, Ott J, Kreek MJ (2004) Redefinition of the human kappa opioid receptor gene (OPRK1) structure and association of haplotypes with opiate addiction. Pharmacogenetics 14:793–804

Zhang H, Luo X, Kranzler HR, Lappalainen J, Yang BZ, Krupitsky E, Zvartau E, Gelernter J (2006) Association between two μ-opioid receptor gene (OPRM1) haplotype blocks and drug or alcohol dependence. Hum Mol Genet 15:807–819

Zhang H, Kranzler HR, Yang BZ, Luo X, Gelernter J (2008) The *OPRD1* and *OPRK1* loci in alcohol or drug dependence: *OPRD1* variation modulates substance dependence risk. Mol Psychiatry 13:531–543

Neurogenetics and Nutrigenomics of Reward Deficiency Syndrome (RDS): Stratification of Addiction Risk and Mesolimbic Nutrigenomic Manipulation of Hypodopaminergic Function

18

Kenneth Blum, David Han, John Giordano, Raquel Lohmann, Eric R. Braverman, Margaret A. Madigan, Debmalya Barh, John Femino, Mary Hauser, B.W. Downs, and Thomas Simpatico

Abstract

We have entered the genomics era with hope for the future of medicine including psychiatry. Understanding the role of DNA and polymorphic associations with brain reward circuitry has led to a new understanding of all addictive behaviors. We present here a brief review of the role of both neurogenetics and nutrigenomics as cornerstones that link more accurate genetic diagnosis and dopamine D2 agonist therapy to induce dopaminergic activation. Based on numerous experiments we are indeed proposing a novel approach. We challenge the entire recovery field to use these tools,

K. Blum (✉)
Department of Psychiatry, McKnight Brain Institute, College of Medicine, University of Florida, Gainesville, FL, USA

G and G Holistic Addiction Treatment Center, North Miami Beach, FL, USA

Department of Clinical Neurology, Path Foundation, New York, NY, USA

Institute of Integrative Omics and Applied Biotechnology, Nonakuri, Purba Medinipur, West Bengal, India

Dominion Diagnostics, LLC, North Kingstown, RI, USA

Global Integrated Services Unit, Center for Clinical and Translational Science, College of Medicine, University of Vermont, Burlington, VT, USA

Department of Addiction Research and Therapy, Malibu Beach Recovery Center, Malibu Beach, CA, USA

Department of Nutrigenomics, LifeGen, Inc., Austin, TX, USA
e-mail: drd2gene@gmail.com

D. Han
Department of Management Science and Statistics, University of Texas at San Antonio, San Antonio, TX, USA

J. Giordano
Department of Holistic Medicine, G and G Holistic Addiction Treatment Center, North Miami Beach, FL, USA

R. Lohmann
Department of Clinical Neurology, Path Foundation, New York, NY, USA

D. Barh et al. (eds.), *Omics for Personalized Medicine*,
DOI 10.1007/978-81-322-1184-6_18, © Springer India 2013

the result of years of scientific research into the nature of addiction, and to incorporate them into treatment programs for patients attending inpatient/outpatient addiction clinics, such as the Genetic Addiction Risk Score (GARS)™ for appropriate RDS diagnosis, Comprehensive Analysis of Reported Drugs (CARD)™ to determine both compliance and abstinence during treatment, natural D2 agonistic therapy (NAAT-KB220™), and, eventually, mRNA (patent pending) to determine pre- and post-candidate gene expressions in reward deficiency syndrome (RDS). We are, therefore, proposing a paradigm shift we have called "Reward Deficiency Solutions System (RDSS)™."

1 Introduction

Blum et al. (1996a) coined the term reward deficiency syndrome (RDS) to describe a dysfunction in the brain reward cascade (Blum and Kozlowski 1990) which directly links abnormal craving behavior with a deficit in a number of reward genes including dopaminergic, serotonergic, endorphinergic, GABAergic, adrenergic, opioidergic, cholinergic, as well as many second messengers. Dopamine controls feelings of well-being and is one of the most powerful neurotransmitters (Blum and Gold 2011; Tomasi and Volkow 2012). The interaction of other brain chemicals and neurotransmitters such as serotonin, the opioids (neuropeptides), and dopamine reduces stress and produces a sense of well-being. Good examples are the association of low levels of serotonin with depression and of high levels of opioids (the brains opium) with a sense of well-being (Pierce and Kumaresan 2006).

In order for one to feel good, these powerful neurotransmitters produced complex interactions in the brain and through a transcription process involving mRNA that regulate neurotransmitter receptor production. These processes ultimately regulate dopaminergic activity of the brain in the reward center, the mesolimbic system, and particularly in the nucleus accumbens (NAc). We provide a brief explanation of not only RDS but potential genetic predisposition diagnosis and the possible beneficial effects of natural D2 agonist therapy in substance use disorder (SUD) patients – see review Chen et al. (2011).

As stated earlier, RDS, which is emerging as an acceptable explanation of the interrelatedness of the impulsive, compulsive, and addictive behaviors including "process addictions," has

E.R. Braverman
Department of Psychiatry, McKnight Brain Institute, College of Medicine, University of Florida, Gainesville, FL, USA

Department of Clinical Neurology, Path Foundation, New York, NY, USA

M.A. Madigan • B.W. Downs
Department of Nutrigenomics, LifeGen, Inc., Austin, TX, USA

D. Barh
Centre for Genomics and Applied Gene Technology, Institute of Integrative Omics and Applied Biotechnology (IIOAB), Nonakuri, Purba Medinipur, WB, India

J. Femino
Meadows Edge Addiction Recovery Center, North Kingstown, RI, USA

Dominion Diagnostics, LLC, North Kingstown, RI, USA

M. Hauser
Dominion Diagnostics, LLC, North Kingstown, RI, USA

T. Simpatico
Dominion Diagnostics, LLC, North Kingstown, RI, USA

Global Integrated Services Unit, Center for Clinical and Translational Science, College of Medicine, University of Vermont, Burlington, VT, USA been the subject of concern by many notable investigators. The dopamine D2 receptor, part of the dopaminergic system, is involved in mesolimbic circuitry and brain reward mechanisms. Any dysfunction of the D2 dopamine receptors can result in aberrant seeking of substances such as drugs, alcohol, tobacco, and including food. Decades of research since the initial findings of Blum's group (Blum et al. 1990) indicate that genetics play a pivotal role in vulnerability to severe substance-seeking behavior. Blum et al. (1995a) utilized Bayes' theorem to predict future substance and process addictions. We proposed that DRD2 A1 allele variant of the D2 dopamine receptor gene is an important, common genetic determinant in predicting addictive disorders (Blum et al. 1995a, b, 1996a, b). In those studies, future RDS behaviors were predicted in subjects carrying the DRD2 Taq A1 allele at 74 %. Many neuroimaging studies have recently supported the RDS concept that links food craving and drug craving behavior (Burger and Stice 2012; Stice et al. 2010). While many genes are involved in RDS behaviors, it is apparent that a major role is played by the dopamine D2 receptor (Dunn et al. 2012). Johnson and Kenny (2012) found compulsive-like feeding behavior, measured as palatable food consumption that, like all addictions, resisted disruption by an aversive conditioned stimulus, in obese, but not lean, rats. Downregulation of striatal dopamine D2 receptors has been reported in both obese rats and drug-addicted humans. Moreover, the knockdown of striatal D2 receptors mediated by lentivirus, the development of reward deficits that were addiction-like, and the onset of compulsive-like food seeking were rapidly accelerated in rats when given extended access to palatable high-fat food. This data demonstrates that neuroadaptive responses in brain reward circuits that drive the development of compulsive eating are indeed triggered by overconsumption of palatable food. The authors suggest that obesity and drug addiction have common underlying hedonic mechanisms (Johnson and Kenny 2012).

It is noteworthy that others found that, in the ventromedial hypothalamus (VMH) of mice, selective BDNF depletion resulted in obesity due to hyperphagic behavior. Specifically, Cordeira and Rios (2011) found that in the ventral tegmental area of wild-type mice, BDNF and TrkB mRNA expression was influenced by consumption of palatable, high-fat food. It is particularly interesting that deficits have been noted in evoked release of dopamine in the dorsal striatum and NAc shell, but normal secretion in the NAc core was recorded in amperometric brain slices of mice depleted of central BDNF (Cordeira and Rios 2011). In addition, Lobo et al. (2012) recently showed that activation of D2+ neurons mimics the loss of TrkB and suppresses cocaine reward, while the activation of D1+ neurons induces the opposite effect. These insights into the molecular regulation of D1+ and D2+ neuronal activity at the circuit level identify the contribution these cell types make to cocaine reward Nestler (2010).

The reward gene DRD2 has been associated with pleasure (Blum et al. 1990), and the Taq1 A1 allele with reduced receptor density in the NAc (Noble et al. 1991). The evidence suggests that there is a three-part interaction that begins with dopamine receptor deficiency and results in a tendency to abuse alcohol and a reduced sensitivity to rewards. Based on these genetic characteristics, individuals from certain ethnic groups have a greater tendency toward alcoholism than others (Barr and Kidd 1993). The DRD2 has been one of the most widely studied genes in neuropsychiatric disorders in general and in alcoholism and other addictions in particular (PUBMED 3391 articles obtained 1/11/13). The Taq1 A1 allele of the dopamine D2 receptor gene has been associated with comorbid psychiatric conditions including antisocial personality disorder symptoms, high novelty seeking, obesity, spectrum disorders, gambling, and other process addictions (Palomo et al. 2007). The mesocorticolimbic dopaminergic pathway may be a common denominator for addictions and plays an especially prominent role in mediating reinforcement by abused drugs (Blum et al. 2012a, b, c).

When the mesocorticolimbic-dopamine reward system breaks down, perhaps caused by certain genetic variants, the result is RDS and subsequent drug-seeking behaviors. To reiterate, RDS refers to reward cascade dysfunction, and resultant aberrant behaviors, due to genetic and environmental influences (Blum et al. 1995b). Accordingly, neuronal release of brain dopamine that can satisfy unhealthy cravings and reduce negative feelings is activated by most positive reinforcers including alcohol and other drugs of abuse. Individuals are predisposed to a high risk for multiple addictive, impulsive, and compulsive behaviors if they have a deficiency of D2 receptors. Although other neurotransmitters, like glutamate, gamma-aminobutyric acid (GABA), and serotonin, may be necessary in stimulating the rewarding effects of ethanol and other drugs, initiating and reinstating drug use during protracted abstinence may be the providence of dopamine (Bowirrat and Oscar-Berman 2005; Wang et al. 2011). All roads do lead to dopamine.

Exploration of various treatment approaches utilizing Dominion's Comprehensive Analysis of Reported Drugs (CARD)™ found significant noncompliance to prescribed treatment medications and non-abstinence during outpatient/inpatient opioid treatment programs using Suboxone® and methadone (Blum et al. 2012d). Moreover, continued drug hunger and poor outcomes in terms of relapse prevention continue to plague the "recovery" field in general. Rather than correcting or compensating for pre-morbid dopamine system deficits, pharmacological therapies that focused on either interference with drug euphoria or harm reduction with opioid maintenance have had limited success (Blum and Gold 2011). In 2010 Blum and Gold proposed the incorporation of genetic testing and neuroadaptogen amino acid precursor enkephalinase–catecholamine–methyltransferase (COMT) inhibition therapy into residential, nonresidential, and aftercare addiction services. In order to treat the genetic basis of addiction, they proposed the identification of risk alleles for hypodopaminergic brain reward function (RDS) coupled with a natural but therapeutic nutraceutical formulation that could potentially produce dopamine release and cause the induction of D2-directed mRNA proliferation of D2 receptors in the human. They hypothesized that this proliferation of D2 receptors will induce the attenuation of drug-like craving behavior. These concepts await further confirmation via required neuroimaging studies. Very recent studies may shine some new light and result in potential therapeutic approaches that include, for example, targeting the heteromeric A(2A)–D(2) receptor complex (Filip et al. 2012).

2 Dopaminergic Activation Through Nutrigenomics

Although the use of nutritional approaches to treat the health-related consequences of addiction are well researched, there are not enough studies that involve treatment of addiction by the manipulation of neurotransmission in the brain reward system based on identified polymorphic genetics. In 2008 using single-gene candidates, microarray, and proteomics, Li et al. (2008) integrated 2,343 items of evidence that linked genes and chromosome regions to addiction. The evidence was collected from peer-reviewed articles published between 1976 and 2006. In these studies, they identified 1,500 human addiction-related genes and developed the first molecular database for addiction-related genes, KARG (http://karg.cbi.pku.edu.cn). They performed a meta-analysis of 396 genes. Of these genes, each supported by two or more items of evidence in the literature, they were able to identify 18 molecular pathways that were statistically significant and covered both upstream signaling events and downstream effects. They identified five molecular pathways as common pathways, found to have associations with all types of addictive drugs, which may be involved with addictive and rewarding actions. They connected the common pathways into a hypothetical common molecular network for addiction. Two new pathways, GnRH signaling pathway and gap junction, were identified. Clues to explain some of the irreversible features of addiction may be provided by their observation that fast and slow positive feedback loops were interlinked through CAMKII. The common thread involves dopaminergic and glutamatergic genes.

Interestingly the dopamine molecule promotes both "pleasure" and "stress-coping abilities." Wellness constitutes an enhanced state of pleasure with tranquility, while a major link to overeating is uncontrollable stress and carbohy-

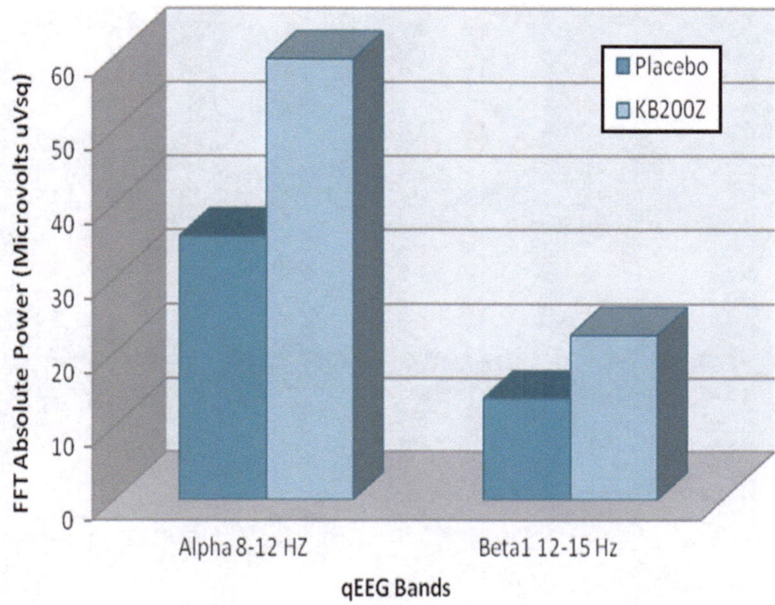

Fig. 18.1 KB220Z compared to placebo in psychostimulant abusers. KB220Z compared to placebo in psychostimulant abusers. Illustrates positive response of KB220Z compared to placebo in triple-blind, randomized, placebo-controlled study in psychostimulant abusers undergoing protracted abstinence (*Source*: Modified from Blum et al. (2010), with permission)

drate craving. We designed a study to evaluate the process of DNA customization of a nutritional solution for both wellness and weight management. For this study, we genotyped 1,058 subjects; these subjects were administered a patented nutraceutical based on polymorphic outcomes. In a subset, simple *t*-tests comparing a number of parameters, before treatment and after 80 days on the nutraceutical, were performed. Significant results were observed for appetite suppression, sugar craving reduction, weight loss, snack reduction, and reduction of late night eating ($P<0.01$); for enhanced quality of sleep, increased perception of overeating, and increased happiness ($P<0.05$); and for increased energy ($P<0.001$). Positive clinical parameters were tested by Blum et al. in 2008a against a number of genes (LEP, PPAR-gamma2, MTHFR, 5-HT2A, and DRD2 genes), and polymorphic correlates were obtained. In their study of all the polymorphisms and outcomes, only the DRD2 gene polymorphism (A1 allele) had the only significant Pearson correlation with days on treatment ($r=0.42$, $P=0.045$). This two-fold increase indicates that the A1 allele polymorphism is a very important genotype for compliance in treatment (Blum et al. 2008a).

3 Neuroimaging Evidence of NAAT in RDS

In a crossover, triple-blind, randomized, placebo-controlled qEEG study involving oral NAAT-KB220Z, increased low-beta activity and increased alpha wave activity were demonstrated, using quantitative electroencephalographic (qEEG) imaging, in the parietal brain loci. Using *t* statistics, significant differences between placebo and KB220Z were consistently observed during the first and then the subsequent, second week of analysis (Blum et al. 2010; Fig. 18.1).

Treatment of addictive behaviors with D2 agonist therapy is in agreement with NIDA scientists as well as other authors who concluded that dopamine/5-HT releasers and D2 agonists could be useful as therapeutic adjuncts in the treatment of psychiatric disorders: attention deficit disorder and depression as well as cocaine and alcohol addiction and obesity (Rothman 1994). In terms of relapse,

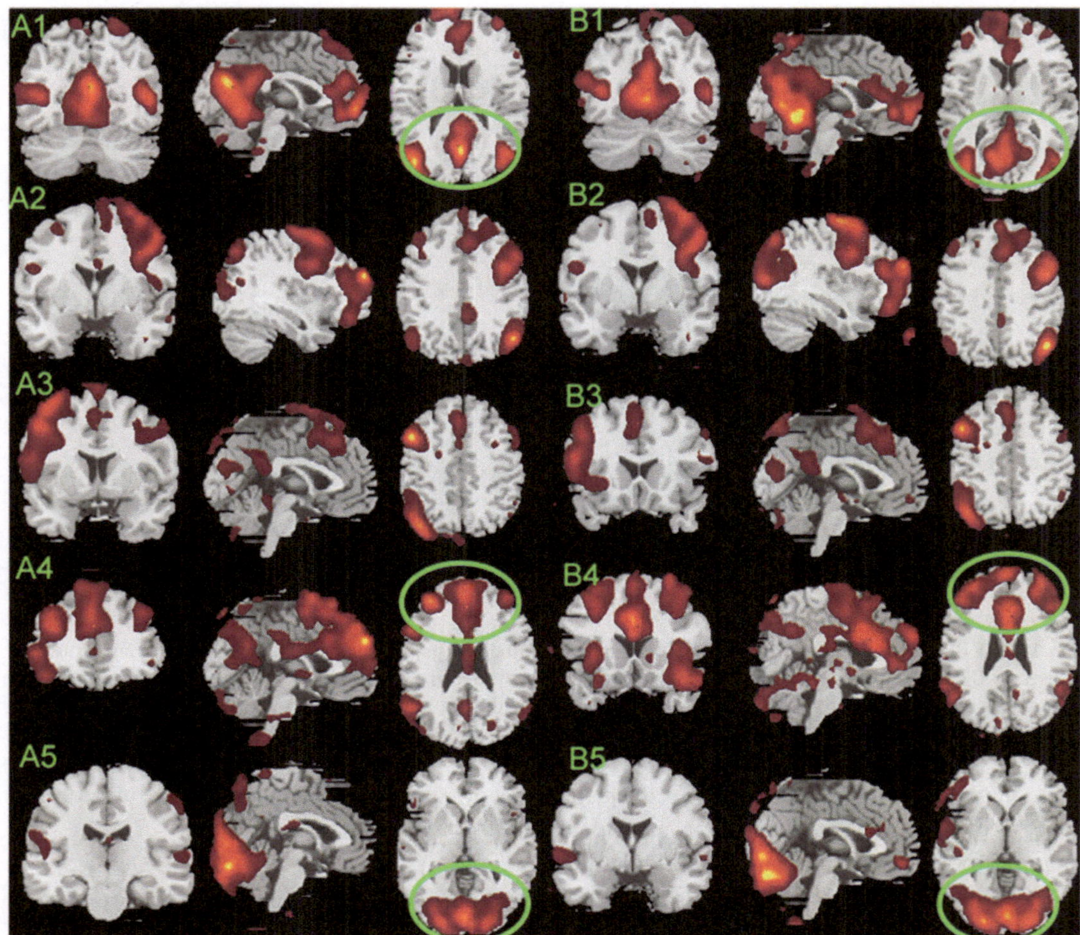

Fig. 18.2 One hour after KB220Z there is activation of dopaminergic pathways. Significant differences in BOLD activation of a number of prefrontal, cingulate gyrus, caudate–accumbens, and putamen brain regions. Brain regions A1–5 are pre-KB220Z and regions B1–5 are post-KB220Z (*Source*: Unpublished data Liu et al., with permission)

the above qEEG experiment showed the involvement of the prefrontal cortex, which was especially evident in dopamine D2 A1 allele subjects. This was the first time that any laboratory reported qEEG responses to a natural putative D2 agonist (Blum et al. 2010).

Similar success has been also obtained by showing that KB220Z also activated dopaminergic pathways using fMRI in Chinese heroin addicts undergoing protracted abstinence. Preliminary, unpublished results show significant differences in BOLD activation of a number of prefrontal, cingulate gyrus, and caudate–accumbens–putamen brain regions. This experiment suggests that 1 h after KB220Z there is activation of dopaminergic pathways (Fig. 18.2).

4 Reducing "Dopamine Resistance" Leads to Reduced Drug Seeking

Acute utilization of most psychoactive substances induces a feeling of well-being. Unfortunately if abuse is sustained and prolonged, a toxic feeling of pseudo well-being results in tolerance, discomfort, and dependence. The consequence of low dopamine receptor densities due to carrying the DRD2 A1 allelic genotype (30–40 % lower numbers) is excessive craving behavior, while normal or high dopamine receptor densities result in low craving-induced behavior (Noble et al. 1991). Prevention of excessive cravings might then be attained by the induction of dopamine D2 receptor proliferation in genetically prone individuals. Indeed experiments in vitro have shown that constant stimulation of the dopamine receptor system via a known D2 agonist in low doses results in significant proliferation of D2 receptors, in spite of genetic antecedents (Boundy et al. 1995; Bradaia et al. 2009). Negative feedback mechanisms are signaled by D2 receptor stimulation that induces mRNA expression and causes proliferation of D2 receptors in the mesolimbic system (Bradaia et al. 2009). Based on this molecular finding, constant natural induction of dopamine release may also support the production of D2-directed mRNA and result in the proliferation of D2 receptors in humans. The proliferation of D2 receptors will, in turn, attenuate human craving behavior. This has been proven using a form of gene therapy in work showing that DNA-directed overexpression of the DRD2 receptors results in a significant reduction in cocaine and alcohol craving-induced behavior in animals (Myers and Robinson 1999; Thanos et al. 2001, 2004, 2005, 2008).

5 Understanding the Role of mRNA Expression in Response to Drugs of Abuse and Reward Pathways

It is known that certain drugs of abuse and neuropathways interact in the genome to influence the biological function of mRNA as it relates to neurotransmission, enzymes involved in neurotransmitter metabolism, as well as specific neuronal receptors that are common in the production of feeling of well-being in the animal or human (Piechota et al. 2010). Hikida et al. found that the convergent input of dopaminergic modulation in the indirect striatopallidal pathways, and the direct striatonigral pathways, of the basal ganglia, are crucial in the rewarding and aversive learning involved in drug addiction. Hikida et al. (2010) explored basal ganglia information processing and integration through these two pathways. Using a reversible blocking technique, they blocked neurotransmission of each pathway selectively, using a doxycycline-dependent, specific expression of transmission-blocking tetanus toxin (Hikida et al. 2010). Two pathways with distinct but coordinated roles were found to be necessary for the modulation of dopamine-mediated acute psychostimulant actions. The direct pathway was prominent in reward learning and cocaine sensitization which is critical for making distinctions between associative and non-associative rewarding stimuli, while the indirect pathway was predominant in aversive behavior, used to avoid aversive stimuli for rapid memory formation (Hikida et al. 2010). While this is true, we ask: What is the effect of drugs of abuse on mRNA in these pathways? This concept is being explored by our laboratory and will be the subject of future papers.

6 Painting a Portrait of the Rational to Incorporate a "Neuroscience Module" in Treatment

6.1 The New Definition of Addiction

The American Society of Addiction Medicine (ASAM 2011), the US addiction specialty society of physicians, has recently redefined "addiction" as "addiction is a primary, chronic disease involving brain reward, motivation, memory and related circuitry. Dysfunction in these circuits leads to characteristic biological, psychological, social and spiritual manifestations. This is reflected in persons compulsively pursuing reward and/or relief by substance use and other behaviors. Addiction cannot be cured but can be brought into remission through a program of treatment, abstinence from all psychoactive substances, and supported recovery" (Smith 2012).

This new definition of addiction as a disease that involves brain reward, motivation, and memory establishes the work of our laboratory as a pioneer in the field. Dedicated scientific rigor during our 40-year sojourn into the mechanisms of the mesolimbic system has provided insight into the addictive brain and the neurogenetics involved in man's quest for happiness. To reiterate, feelings of well-being are experienced in the mesolimbic system of the brain (Ikemoto 2010). In the mesolimbic system, chemical messengers including serotonin, enkephalins, GABA, and dopamine work together for the release of dopamine in NAc where one

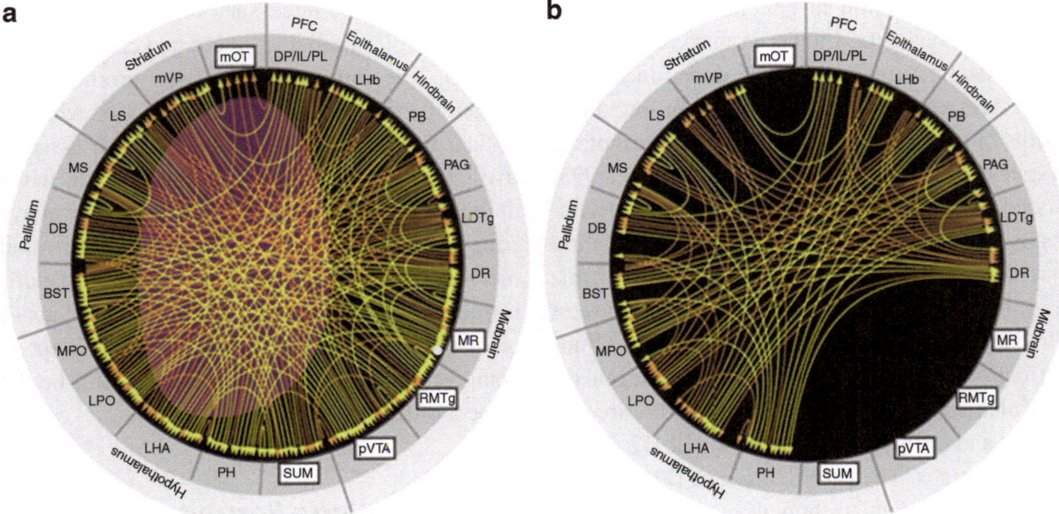

Fig. 18.3 Brain reward circuitry beyond the mesolimbic dopamine system. Hypothesized key components of brain reward circuitry and its organization. Afferents and efferents of key components of the circuitry are shown with *orange lines* for unidirectional connection and *yellow lines* for reciprocal connections. (**a**) The *purple* area corresponds to the medial forebrain bundle, at which electrical stimulation elicits vigorous self-stimulation. This depicts a tentative organization of brain reward circuitry at a macroscale level. (**b**) The connectivity of the circuitry after removing the trigger zones (Taken from: Ikemoto 2010, with permission)

experiences feelings of well-being. A series of neuronal events termed "brain reward cascade" was coined to describe the synthesis, vesicular storage, metabolism, and catabolism of neurotransmitters (Blum et al. 2008b, c; Figs. 18.3 and 18.4). These events are controlled by genes and polymorphic versions of these genes have certain variations which could lead to a breakdown of this cascade and ultimately dysregulation and dysfunction of the dopaminergic pathway. Although there is some debate (Salomone and Correa 2012), dopamine is generally regarded as the pleasure and antistress molecule. Any reduction in dopamine function could lead to unhappiness and reward deficiency and result in aberrant substance- and pleasure-seeking behavior (Blum et al. 2012a, b, c).

Homo sapiens are programmed to be motivated to drink, eat, have sex, and desire pleasurable experiences. Impairments in these natural reward mechanisms are governed by genetic polymorphic antecedents and can promote multiple impulsive, compulsive, and addictive behaviors in individuals. While many genetic variations are involved within the activity of the mesolimbic system, polymorphisms of genes that predispose individuals to excessive cravings and result in aberrant reward-seeking behaviors (Niehaus et al. 2009) include, among others, the dopamine D2 receptor (DRD2), dopamine D4 receptor (DRD4), dopamine transporter (DAT1), serotonergic-2A receptor (5-HTT2a), serotonergic transporter (5-HTTLPR), catechol-*O*-methyltransferase (COMT), monoamine oxidase (MOA), and GABA receptors (GABAB).

As stated, RDS is the umbrella term used to describe multiple impulsive, compulsive, and addictive behaviors with common genetic antecedents (Blum et al. 2011a). Individuals, especially in adolescence who have reduced numbers of dopaminergic and/or serotonergic receptors and an increased rate of synaptic dopamine catabolism, will, due to high MOA activity or high catabolic activity due to the COMT gene, have a predisposition to use any means, substance or behavior, to activate dopamine release. Dopamine release is activated by the use of glucose, alcohol, and drugs, like opiates, psychostimulants, and nicotine, and by process addictions, like gambling, sex, and even excessive internet gaming (Blum et al. 2011a; Wahlstrom et al. 2010).

Use of most drugs of abuse, including alcohol, is associated with dopamine release in the "reward pathway of the brain," the mesocorticolimbic

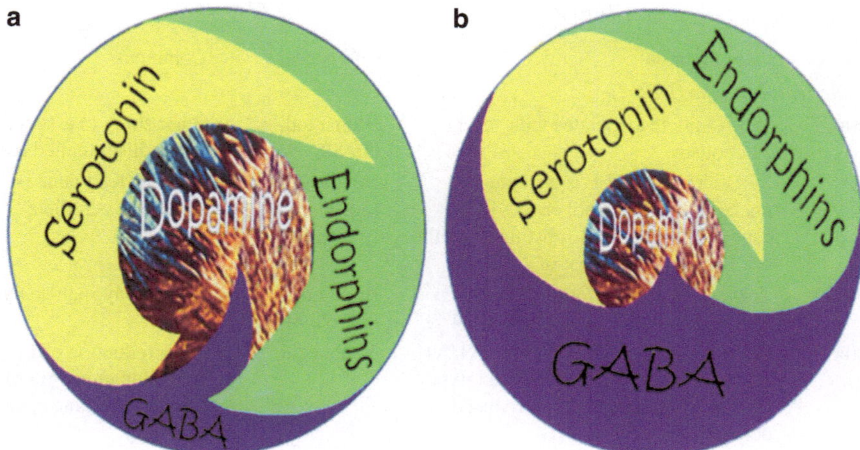

Fig. 18.4 Happy compared to unhappy brain. (**a**) Schematic represents the normal physiologic state of the neurotransmitter interaction at the mesolimbic region of the brain. Briefly in terms of the "brain reward cascade" first coined by Blum and Kozlowski (1990): serotonin in the hypothalamus stimulates neuronal projections of methionine enkephalin in the hypothalamus which in turn inhibits the release of GABA in the substania nigra, thereby allowing for the normal amount of dopamine to be released at the nucleus accumbens (reward site of brain). (**b**) Represents hypodopaminergic function of the mesolimbic region of the brain. It is possible that the hypodopaminergic state is due to gene polymorphisms as well as environmental elements including both stress and neurotoxicity from aberrant abuse of psychoactive drugs (i.e., alcohol, heroin, cocaine). Genetic variables could include serotonergic genes (serotonergic receptors [5HT2a], serotonin transporter 5HTlPR), endorphinergic genes (mu OPRM1 gene, proenkephalin (PENK), PENK polymorphic 3′ UTR dinucleotide (CA) repeats), GABAergic gene (GABRB3), and dopaminergic genes (ANKKI Taq A, DRD2 C957T, DRD4 7R, COMT Val/met substation, MAO-A uVNTR, and SLC6A3 9 or 10R). Any of these genetic and/or environmental impairments could result in reduced release of dopamine and/or reduced number of dopaminergic receptors (*Sources*: Blum et al. 2011b, with permission)

system, and a new mechanism involving local electrophysiology for dopamine release has been proposed (Kita et al. 2009). Activation of this dopaminergic system induces feelings of reward and pleasure (see Fig. 18.4, happy compared to unhappy brain); however, dopamine release in the ventral tegmental area (VTA) is also caused by aversive or painful stimuli like finding pleasure from bondage (Budygin et al. 2012).

6.2 Genetic Addiction Risk

The hypothesis is that regardless of the source, the presence of a hypodopaminergic state/trait is a primary cause of drug-seeking behavior. Certainly, polymorphisms that induce hypodopaminergic functioning are the mechanism of a genetic predisposition to chronic drug use and relapse (Blum and Gold 2011).

Numerous important studies that have explored the genetic risk for drug-seeking behaviors have reinforced the claim that individuals possessing dopaminergic gene variants are impulsive. These are association and linkage studies that implicate these alleles as risk antecedents that impact the mesocorticolimbic system. Dopaminergic gene variants include DRD2, DRD3, DRD4, DAT1, COMT, MOA-A, SLC6A4, Mu, and GABAB (see Table 18.1). Our laboratory in conjunction with Impact Genomics, Inc. and Dominion Diagnostics, Inc. is involved in development and research in 12 select centers across the United States to validate the first ever patented genetic test to determine a patient's genetic risk for RDS called Genetic Addiction Risk Score™ (GARS).

7 Relapse

Drug-seeking behavior can be triggered by reduced activity of the dopamine system (hypodopaminergic functioning) (Martinez et al. 2012). Hypodopaminergic functioning can be induced by variant alleles through reduced dopamine receptor density, blunted response to dopamine,

Table 18.1 Candidate reward genes and RDS (a sampling)

Polymorphism (s)	Study findings	References	Comments
(a) *D2 dopamine receptor gene* (*DRD2*)			
SNP rs: 1800497	Taq A1 allele associates with sever alcoholism	Blum et al. (1990)	First study to associate with alcoholism (called reward gene)
ANKKI-p. Glu713Lys	DRD2 Taq1A RFLP is a single-nucleotide polymorphism (SNP) that causes an amino acid substitution within the 11th ankyrin repeat of ANKK1	Neville et al. (2004)	The ANKKI gene is a reflection of DRD2 A1 allele
SNP rs: 1800497	This SNP has been found to predict future RDS behaviors as high as 74 %	Blum et al. (1996a, b)	Using Bayesian analysis
SNP rs: 1800497	Presence of the A1+ genotype (A1/A1, A1/A2) compared to the A− genotype (A2/A2) is associated with reduced density	Noble et al. (1991)	This reduction causes hypodopaminergic functioning in the dopamine reward pathway
SNP rs: 6277 at exon 7	T+ allele associates with alcohol dependence	Hill et al. (2008)	Associates with drug-seeking behavior and other RDS behaviors
SNP rs: 1800497	10-year follow-up that carriers of the DRD2 A1 allele have a higher rate of mortality compared to carriers of the A2 allele in alcohol-dependent individuals	Dahlgren et al. (2011)	*Taq1* A1 allele and a substantially increased relapse rate
DRD2-haplotypes I-C-G-A2 and I-C-A-A1	Confirmed the hypothesis that haplotypes, which are supposed to induce a low DRD2 expression, are associated with alcohol dependence	Kraschewski et al. (2009)	High frequency of haplotype was associated with Cloninger type 2 and family history of alcoholism
SNP rs: 1800497	Genotype analysis showed a significantly higher frequency for the Taq1A polymorphism among the addicts (69.9 %) compared to control subjects (42.6 %; Fisher's exact $\chi(2)$, $P<0.05$)	Teh et al. (2012)	The addicts had higher scores for novelty seeking (NS) and harm avoidance (HA) personality traits
(b) *D4 dopamine receptor gene* (*DRD4*)			
DRD4 – the 7-repeat (7R) VNTR	The length of the D4 dopamine receptor (DRD4) exon 3 variable number of tandem repeats (VNTR) affects DRD4 functioning by modulating the expression and efficiency of maturation of the receptor	Van Tol (1998)	The 7-repeat (7R) VNTR requires significantly higher amounts of dopamine to produce a response of the same magnitude as other size VNTRs
120 bp duplication, −616C/G, and −521C/T	Strong finding of −120 bp duplication allele frequencies with schizophrenia ($P=0.008$), −521 C/T polymorphism is associated with heroin addiction	Lai et al. (2010)	This reduced sensitivity or "dopamine resistance" leads to hypodopaminergic functioning. Thus, 7R VNTR has been associated with substance-seeking behavior
DRD4 7-repeat allele	A number of putative risk alleles using survival analysis revealed that by 25 years of age, 76 % of subjects with a DRD4 7-repeat allele were estimated to have significantly more persistent ADHD compared with 66 % of subjects without the risk allele	Biederman et al. (2009)	Findings suggest that the DRD4 7-repeat allele is associated with a more persistent course of ADHD

(continued)

Table 18.1 (continued)

Polymorphism (s)	Study findings	References	Comments
7-repeat allele of the dopamine D(4) receptor gene (DRD4)	Although the association between ADHD and DRD4 is small, these results suggest that it is real	Faraone et al. (2001)	For both the case-control and family-based studies, the authors found (1) support for the association between ADHD and DRD4, (2) no evidence that this association was accounted for by any one study, and (3) no evidence for publication bias
Dopamine D4 receptor (DRD4) exon 3 polymorphisms (48 bp VNTR)	Found significant differences in the short alleles (2–5 VNTR) frequencies between controls and patients with a history of delirium tremens and/or alcohol seizures ($P=0.043$)	Grzywacz et al. (2008)	A trend was also observed in the higher frequency of short alleles among individuals with an early age of onset of alcoholism ($P=0.063$)
Dopamine D4 receptor (DRD4) 7-repeat allele	Show that the 7-repeat allele is significantly overrepresented in the opioid-dependent cohort and confers a relative risk of 2.46	Kotler et al. (1997)	This is the first report of an association between a specific genetic polymorphism and opioid addiction
(c) *Dopamine transporter gene (DAT1)*			
Localized to chromosome 5p15.3. Moreover, within 3 noncoding region of DAT1 lies a VNTR polymorphism – 9-repeat (9R) VNTR	The 9-repeat (9R) VNTR has been shown to influence gene expression and to augment transcription of the dopamine transporter protein	Byerley et al. (1993)	Having this variant results in an enhanced clearance of synaptic dopamine, yielding reduced levels of dopamine to activate postsynaptic neurons
9-repeat (9R) VNTR	DAT1, genotype 9/9 was associated with early opiate addiction	Galeeva et al. (2002)	The combination of SERT genotype 10/10 with DAT1 genotype 10/10 was shown to be a risk factor of opiate abuse less than 16 years of age
Exon 15 rs27072 and VNTR (DAT), promoter VNTR and rs25531	The haplogenotypes 6-A-10/6-G-10 and 5-G-9/5-G-9 were more often present in type 2 alcoholics as compared with type 1 alcoholics [odds ratio (OR): 2.8] and controls (OR: 5.8), respectively	Reese et al. (2010)	In a typology proposed by Cloninger on the basis of adoption studies, a subgroup has been classified as type 2 with patients having high genetic loading for alcoholism, an early onset of alcoholism, a severe course, and coexisting psychiatric problems consisting of aggressive tendencies or criminality
VNTR polymorphism at the dopamine transporter locus (DAT1) 480 bp DAT1 allele	Using the haplotype-based haplotype relative risk (HHRR) method revealed significant association between ADHD/UADD and the 480 bp DAT1 allele (chi 2 7.51, 1 df, $P=0.006$)	Cook et al. (1995)	While there have been some inconsistencies associated with the earlier results, the evidence is mounting in favor of the view that the 10R allele of DAT is associated with high risk for ADHD in children and in adults alike

(continued)

Table 18.1 (continued)

Polymorphism (s)	Study findings	References	Comments
Dopamine transporter (DAT1) variable number tandem repeats (VNTR) genotypes – both 9- and 10-repeat alleles	The nonadditive association for the 10-repeat allele was significant for hyperactivity–impulsivity (HI) symptoms. However, consistent with other studies, exploratory analyses of the nonadditive association of the 9-repeat allele of DAT1 with HI and oppositional defiant disorder (ODD) symptoms also were significant	Lee et al.(2007)	The inconsistent association between DAT1 and child behavior problems in this and other samples may reflect joint influence of the 10-repeat and 9-repeat alleles
(d) *Catechol-O-methyltransferase* (*COMT*)			
COMT Val158Met and DRD2 Taq1A genotypes	COMT Val158Met and DRD2 Taq1A may affect the intermediate phenotype of central dopamine receptor sensitivity	Schellekens et al. (2012)	COMT Val158Met and DRD2 Taq1A may confer their risk of alcohol dependence through reduced dopamine receptor sensitivity in the prefrontal cortex and hindbrain, respectively
The functional polymorphism (COMT Val108/158Met) affects COMT activity with the valine (Val) variant associated with higher and the methionine (Met) variant with lower COMT activity	Male alcoholic suicide attempters, compared to male non-attempters, had the higher frequency of Met/Met genotype or Met allele and significantly (Kruskal–Wallis ANOVA on ranks and Mann–Whitney test) higher aggression and depression scores	Nedic et al. (2011)	These results confirmed the associations between Met allele and aggressive behavior or violent suicide attempts in various psychiatric diagnoses and suggested that Met allele of the COMT Val108/158 Met might be used as an independent biomarker of suicidal behavior across different psychopathologies
COMT Val(15) Met variation	Both controls and opiate users with Met/Met genotypes showed higher NS scores compared to those with the Val allele	Demetrovics et al. (2010)	Association of the COMT polymorphism and NS temperament scale has been shown for heroin-dependent patients and controls regardless of group status
A functional polymorphism (COMT Val158Met) resulting in increased enzyme activity has been associated with polysubstance abuse and addiction to heroin and methamphetamine	These results suggest a significant association between COMT Val158Met polymorphism and susceptibility to cannabis dependence	Baransel Isir et al. (2008)	Cannabis stimulates dopamine release and activates dopaminergic reward neurons in central pathways that lead to enhanced dependence. Catechol-*O*-methyltransferase (COMT) inactivates amplified extraneuronally released dopamine
(e) *Serotonin transporter gene*			
Serotonin transporter promoter polymorphism [5-HT transporter gene-linked polymorphic region (5-HTTLPR)]	5-HTTLPR had age-dependent effects on alcohol, tobacco, and drug use: substance use did not differ by genotype at age 9, but at age 15, the participants with the short (s)/s genotype had higher tobacco use, and at age 18, they were more active alcohol, drug, and tobacco users	Merenäkk et al. (2011)	Results reveal that expression of genetic vulnerability for substance use in children and adolescents may depend on age, gender, interaction of genes, and type of substance

(continued)

Table 18.1 (continued)

Polymorphism (s)	Study findings	References	Comments
The short (s), low-activity allele of a polymorphism (5-HTTLPR) in the serotonin transporter gene (SLC6A4) has been related to alcohol dependence	The 5-HTTLPR short allele predicted adolescent's growth (slope) in alcohol use over time. Adolescents with the 5-HTTLPR short allele showed larger increase in alcohol consumption than those without the 5-HTTLPR short allele	van der Zwaluw et al. (2010)	5-HTTLPR genotype was not related to the initial level (intercept) of alcohol consumption
Triallelic 5-HTTLPR genotype: SA/SA and SA/LG compared to LA/LA	Triallelic 5-HTTLPR genotype: SA/SA and SA/LG compared to LA/LA	Kosek et al. (2009)	Previously the 5-HTTLPR s allele has been associated with higher risk of developing chronic pain conditions, but in this study we show that the genotype coding for low 5-HTT expression is associated with a better analgesic effect of an opioid. The s allele has been associated with downregulation of 5-HT1 receptors, and we suggest that individuals with a desensitization of 5-HT1 receptors have an increased analgesic response to opioids during acute pain stimuli but may still be at increased risk of developing chronic pain conditions
(f) *Mu-opiate receptor* (*MOR*)			
A single-nucleotide polymorphism (SNP) in the human MOR gene (OPRM1 A118G) has been shown to alter receptor protein level in preclinical models and smoking behavior in humans	Independent of session, smokers homozygous for the wild-type OPRM1 A allele exhibited significantly higher levels of MOR BP (ND) than smokers carrying the G allele in bilateral amygdala, left thalamus, and left anterior cingulate cortex	Ray et al. (2011)	Among G allele carriers, the extent of subjective reward difference (denicotinized vs. nicotine cigarette) was associated significantly with MOR BP(ND) difference in right amygdala, caudate, anterior cingulate cortex, and thalamus
Polymorphism in A118G in exon 1 and C1031G in intron 2 of the MOR gene	Results showed a significant association for both A118G and C1031G polymorphisms and opioid dependence. The G allele is more common in the heroin-dependent group (39.5 and 30.8 % for A118G and C1031G polymorphisms, respectively) when compared to the controls (29.4 and 21.1 % for A118G and C1031G polymorphisms, respectively)	Szeto et al. (2001)	This study suggests that the variant G allele of both A118G and C1031G polymorphisms may contribute to the vulnerability to heroin dependence

(continued)

Table 18.1 (continued)

Polymorphism (s)	Study findings	References	Comments
A118G single-nucleotide polymorphism (SNP) in exon 1 of the MOR gene (OPRM1), which encodes an amino acid substitution, is functional, and receptors encoded by the variant 118G allele bind the endogenous opioid peptide beta-endorphin with threefold greater affinity than prototype receptors. Other groups subsequently reported that this variant alters stress responsivity in normal volunteers and also increases the therapeutic response to naltrexone (a mu-preferring opioid antagonist) in the treatment of alcohol dependence	There was a significant overall association between genotypes with a 118G allele and alcohol dependence ($P=0.0074$). The attributable risk for alcohol dependence in subjects with a 118G allele was 11.1 %	Bart et al. (2005)	There was no difference in A118G genotype between type 1 and type 2 alcoholics. In central Sweden, the functional variant 118G allele in exon 1 of OPRM1 is associated with an increased attributable risk for alcohol dependence
MOR gene knockout (KO) was examined in wild-type (+/+), heterozygote MOR-KO (+/–), and homozygote MOR-KO (–/–) mice on voluntary ethanol consumption	Heterozygous and homozygous MOR-KO mice consumed less ethanol than wild-type mice. These effects appeared to be greater in female KO mice than in male KO mice. MOR-KO mice, especially females, exhibited less ethanol reward in a conditioned place preference paradigm	Hall et al. (2001)	These data fit with the reported therapeutic efficacy of MOR antagonists in the treatment of human alcoholism. Allelic variants that confer differing levels of MOR expression could provide different degrees of risk for alcoholism
(g) *GABA beta subunit 3*			
GABA A receptor beta 3 subunit gene (GABRB3)	The G1– alleles of the GABRB3 in COAs were significantly higher than non-COAs	Namkoong et al. (2008)	In the same study the frequency of the A1+ allele at DRD2 in the COAs was significantly higher than non-COAs

(continued)

Table 18.1 (continued)

Polymorphism (s)	Study findings	References	Comments
Beta 3 subunit mRNAs	The levels of the beta 2 and beta 3 subunit mRNAs remain elevated at 24-h withdrawal from chronic ethanol. Chronic ethanol treatment increased the levels of both of these polypeptides in cerebral cortex	Mhatre and Ticku (1994)	Chronic ethanol administration produced an upregulation of the beta subunit mRNA and the polypeptide expression of these subunits in rat cerebral cortex
A1+ (A1A1 and A1A2 genotypes) and A1− (A2A2 genotype) alleles of the DRD2 and G1+ (G1G1 and G1 non-G1 genotypes) and G1− (non-G1 non-G1 genotype) alleles of the GABRB3 gene, study involved mood-related alcohol expectancy (AE) and drinking refusal self-efficacy (DRSE), were assessed using the Drinking Expectancy Profile	Patients with the DRD2 A1+ allele, compared with those with the DRD2 A1− allele, reported significantly lower DRSE in situations of social pressure. Similarly, lower DRSE was reported under social pressure by patients with the GABRB3 G1+ allele when compared to those with the GABRB3 G1− alleles. Patients with the GABRB3 G1+ allele also revealed reduced DRSE in situations characterized by negative affect than those with the GABRB3 G1− alleles. Patients carrying the GABRB3 G1+ allele showed stronger AE relating to negative affective change (e.g., increased depression) than their GABRB3 G1− counterparts	Young et al. (2004)	Molecular genetic research has identified promising markers of alcohol dependence, including alleles of the D2 dopamine receptor (DRD2) and the GABAA receptor beta 3 subunit (GABRB3) genes
A1+ (A1A1 and A1A2 genotypes) and A1− (A2A2 genotype) alleles of the DRD2 and G1+ (G1G1 and G1 non-G1 genotypes) and G1− (non-G1 non-G1 genotype) alleles of the GABRB3 gene, study involved mood-related alcohol expectancy (AE) and drinking refusal self-efficacy (DRSE), were assessed using the Drinking Expectancy Profile	Patients with the DRD2 A1+ allele, compared with those with the DRD2 A1− allele, reported significantly lower DRSE in situations of social pressure. Similarly, lower DRSE was reported under social pressure by patients with the GABRB3 G1+ allele when compared to those with the GABRB3 G1− alleles. Patients with the GABRB3 G1+ allele also revealed reduced DRSE in situations characterized by negative affect than those with the GABRB3 G1− alleles. Patients carrying the GABRB3 G1+ allele showed stronger AE relating to negative affective change (e.g., increased depression) than their GABRB3 G1− counterparts	Young et al. (2004)	Molecular genetic research has identified promising markers of alcohol dependence, including alleles of the D2 dopamine receptor (DRD2) and the GABAA receptor beta 3 subunit (GABRB3) genes

(continued)

Table 18.1 (continued)

Polymorphism (s)	Study findings	References	Comments
A1+ (A1A1 and A1A2 genotypes) and A1− (A2A2 genotype) alleles of the DRD2 and G1+ (G1G1 and G1 non-G1 genotypes) and G1− (non-G1 non-G1 genotype) alleles of the GABRB3 gene, study involved mood-related alcohol expectancy (AE) and drinking refusal self-efficacy (DRSE), were assessed using the Drinking Expectancy Profile	Patients with the DRD2 A1+ allele, compared with those with the DRD2 A1− allele, reported significantly lower DRSE in situations of social pressure. Similarly, lower DRSE was reported under social pressure by patients with the GABRB3 G1+ allele when compared to those with the GABRB3 G1− alleles. Patients with the GABRB3 G1+ allele also revealed reduced DRSE in situations characterized by negative affect than those with the GABRB3 G1− alleles. Patients carrying the GABRB3 G1+ allele showed stronger AE relating to negative affective change (e.g., increased depression) than their GABRB3 G1− counterparts	Young et al. (2004)	Molecular genetic research has identified promising markers of alcohol dependence, including alleles of the D2 dopamine receptor (DRD2) and the GABAA receptor beta 3 subunit (GABRB3) genes
Dinucleotide repeat polymorphisms of the GABA(A) receptor beta 3 subunit gene were compared to scores on the General Health Questionnaire-28 (GHQ)	Analysis of GHQ subscale scores showed that heterozygotes compared to the combined homozygotes had higher scores on the somatic symptoms ($P=0.006$), anxiety/insomnia ($P=0.003$), social dysfunction ($P=0.054$), and depression ($P=0.004$) subscales	Feusner et al. (2001)	The present study indicates that in a population of PTSD patients, heterozygosity of the GABRB3 major (G1) allele confers higher levels of somatic symptoms, anxiety/insomnia, social dysfunction, and depression than those found in homozygosity
GABRB3 major (G1) allele & DRD2 A1 allele	A significant progressive increase was observed in DRD2 A1 allelic prevalence ($P=3.1\times10(-6)$) and frequency ($P=2.7\times10(-6)$) in the order of nonalcoholics, less severe, and severe alcoholics. In severe alcoholics, compared to nonalcoholics, a significant decrease was found in the prevalence ($P=4.5\times10(-3)$) and frequency ($P=2.7\times10(-2)$) of the GABRB3 major (G1) allele. Furthermore, a significant progressive decrease was noted in G1 allelic prevalence ($P=2.4\times10(-3)$) and frequency ($P=1.9\times10(-2)$) in nonalcoholics, less severe, and severe alcoholics, respectively	Noble et al. (1998)	In sum, in the same population of nonalcoholics and alcoholics studied, variants of both the DRD2 and GABRB3 genes independently contribute to the risk for alcoholism, with the DRD2 variants revealing a stronger effect than the GABRB3 variants. However, when the DRD2 and the GABRB3 variants are combined, the risk for alcoholism is more robust than when these variants are considered separately

(continued)

Table 18.1 (continued)

Polymorphism (s)	Study findings	References	Comments
(h) *MOA-A*			
MAOA genotype	Significant three-way interactions, MAOA genotype by abuse and by sex, predicted dysthymic symptoms. Low-activity MAOA genotype buffered against symptoms of dysthymia in physically abused and multiply maltreated women. Significant three-way interactions, MAOA genotype by sexual abuse and by race, predicted all outcomes. Low-activity MAOA genotype buffered against symptoms of dysthymia, major depressive disorder, and alcohol abuse for sexually abused white participants. The high-activity genotype was protective in the nonwhite sexually abused group	Nikulina et al. (2012)	This prospective study provides evidence that MAOA interacts with child maltreatment to predict mental health outcomes
Low-repeat MAOA allele	Individuals with CUD had reductions in GMV in the orbitofrontal, dorsolateral, prefrontal, and temporal cortex and the hippocampus compared with controls. The orbitofrontal cortex reductions were uniquely driven by CUD with low-MAOA genotype and by lifetime cocaine use	Alia-Klein et al. (2011)	Long-term cocaine users with the low-repeat MAOA allele have enhanced sensitivity to gray matter loss, specifically in the orbitofrontal cortex, indicating that this genotype may exacerbate the deleterious effects of cocaine in the brain
MAOA u-VNTR	Girls, carrying the long MAOA u-VNTR variant showed a higher risk of being high alcohol consumers, whereas among boys, the short allele was related to higher alcohol consumption	Nilsson et al. (2011)	The present study supports the hypothesis that there is a relation between MAOA u-VNTR and alcohol consumption and that this relation is modulated by environmental factors
30 bp repeat in the promoter region of the monoamine oxidase-A gene (MAO-A)	Significant associations between cold pain tolerance and DAT1 ($P=0.008$) and MAO-A ($P=0.024$) polymorphisms were found. Specifically, tolerance was shorter for carriers of allele 10 and the rarer allele 11, as compared to homozygous for allele 9, and for carriers of allele 4 (MOA) as compared to homozygous for allele 3, respectively	Treister et al. (2009)	These results, together with the known function of the investigated candidate gene polymorphisms, suggest that low dopaminergic activity can be associated with high pain sensitivity, and vice versa
The Psychopathy Checklist-Revised (PCL-R) has shown a moderate association with violence and as such studied with MAOA-genotyped alcoholic offenders	The PCL-R total score predicts impulsive reconvictions among high-activity MAOA offenders (6.8 % risk increase for every one-point increase in PCL-R total score, $P=0.015$), but not among low-activity MAOA offenders, whereas antisocial behavior and attitudes predicted reconvictions in both genotypes (17 % risk increase among high-activity MAOA offenders and 12.8 % increase among low-activity MAOA offenders for every one-point increase in factor 2 score)	Tikkanen et al. (2011)	Results suggest that the efficacy of PCL-R is altered by MAOA genotype, alcohol exposure, and age, which seems important to note when PCL-R is used for risk assessments that will have legal or costly preventive work consequences

(continued)

Table 18.1 (continued)

Polymorphism (s)	Study findings	References	Comments
Genotyping of two functional polymorphisms in the promoter region of the serotonin transporter and monoamine oxidase-A, respectively (5-HTTLPR and MAOA-VNTR), was performed in a group of women with severe alcohol addiction	Within the group of alcoholics, when the patients with known comorbid psychiatric disorders were excluded, aggressive antisocial behavior was significantly linked to the presence of the high- activity MAOA allele	Gokturk et al. (2008)	The pattern of associations between genotypes of 5-HTTLPR and MAOA-VNTR in women with severe alcoholism differs from most corresponding studies on males
The MAOA gene presents several polymorphisms, including a 30 bp VNTR in the promoter region (MAOA-uVNTR). Alleles with 3.5 and 4 repeats are 2–10 times more efficient than the 3-repeat allele	The results suggest that the 3-repeat allele is associated to (1) alcohol dependence ($P<0.05$), (2) an earlier onset of alcoholism ($P<0.01$), (3) comorbid drug abuse among alcoholics ($P<0.05$), and (4) a higher number of antisocial symptoms ($P<0.02$)	Contini et al. (2006)	Results confirmed previous reports showing an association of the low-activity 3-repeat allele of MAOA-uVNTR polymorphism with substance dependence and impulsive/antisocial behaviors. These findings in a different culture further support the influence of the MAOA-uVNTR in psychiatric disorders
(i) *Dopamine D3*			
The genotypes of the BDNF Val66Met and DRD3 Ser9Gly polymorphisms. BDNF regulate expression of D3	Logistic regression analysis showed a significant main effect for the Val/Val genotype of the BDNF Val66Met polymorphism ($P=0.020$), which predicted bipolar II patients. Significant interaction effects for the BDNF Val66Met Val/Val genotype and both DRD3 Ser9Gly Ser/Ser and Ser/Gly genotypes were found only in bipolar II patients ($P=0.027$ and 0.006, respectively)	Lee et al. (2012)	Evidence that the BDNF Val66Met and DRD3 Ser9Gly genotypes interact only in bipolar II disorder (hypomania) and that bipolar I (mania) and bipolar II may be genetically distinct
D3R KO mice	The possible interaction between morphine-induced tolerance and D3 receptors has not been investigated. Compared with wild-type (WT) mice, the dopamine D3 receptor knockout (D3R KO) mice showed pronounced hypoalgesia. The D3R KO mice clearly developed lower morphine-induced tolerance and showed attenuated withdrawal signs compared with the WT mice	Li et al. (2012)	These results suggest that D3 receptors regulate basal nociception and are involved in the development of morphine-induced tolerance and withdrawal
DNA microarrays of two different alcohol-preferring rat lines (HAD and P) and D3 receptors	Data revealed an upregulation of the dopamine D3 receptor (D3R) after 1 year of voluntary alcohol consumption in the striatum of alcohol-preferring rats that was confirmed by qRT-polymerase chain reaction	Vengeliene et al. (2006)	Long-term alcohol consumption leads to an upregulation of the dopamine D3R that may contribute to alcohol seeking and relapse. We therefore suggest that selective antagonists of this pharmacological target provide a specific treatment approach to reduce alcohol craving and relapse behavior

(continued)

Table 18.1 (continued)

Polymorphism (s)	Study findings	References	Comments
Gly9 homozygotes in comparison to Ser9 carriers of D3 receptor gene	German descent and have found Diminished parietal and increased frontal P300 amplitudes in Gly9 homozygotes in comparison to Ser9 carriers were found in 124 unrelated healthy subjects of German descent. Further studies should address the direct role of the DRD3 Ser9Gly polymorphism in attenuated P300 amplitudes in psychiatric disorders like schizophrenia or alcoholism	Mulert et al. (2006)	An important reason for the interest in P300 event-related potentials are findings in patients with psychiatric disorders like schizophrenia or alcoholism in which attenuations of the P300 amplitude are common findings
Dopamine receptor D3 gene Bal I polymorphism	Patients above the median value for cognitive impulsiveness (one of the three dimensions of the Barratt scale) were more frequently heterozygous than both alcohol-dependent patients with lower impulsiveness (OR = 2.51, $P = 0.019$) and than 71 healthy controls (OR = 2.32, $P = 0.025$)	Limosin et al. (2005)	The D3 receptor gene has been associated with addictive behaviors especially impulsiveness
Bal I polymorphism at the DRD3 gene	Patients with a sensation-seeking score above 24 were more frequently homozygotes for both alleles than patients with a sensation-seeking score under 24 ($P = 0.038$) or controls ($P = 0.034$)	Duaux et al. (1998)	These results suggest that the DRD3 gene may have a role in drug-dependence susceptibility in individuals with high sensation-seeking scores
mRNA of both DRD2 and DRD3 gene expression	After a chronic schedule of intermittent bingeing on a sucrose solution, mRNA levels for the D2 dopamine receptor and the preproenkephalin and preprotachykinin genes were decreased in dopamine-receptive regions of the forebrain, while D3 dopamine receptor mRNA was increased. The effects of sugar on mRNA levels were of greater magnitude in the nucleus accumbens than in the caudate–putamen	Spangler et al. (2004)	Striatal regions of sugar-dependent rats show alterations in dopamine and opioid mRNA levels similar to morphine-dependent rats

and/or enhanced dopamine catabolism in the reward pathway (Yacubian et al. 2007). Possibly, cessation of chronic drug use also can induce a hypodopaminergic state that reinstates drug seeking in an attempt to address the withdrawal state (Zhang et al. 2012).

Paloyelis et al. (2010) lent support to the impulsive nature of individuals possessing RDS genetics in a recent article that suggests that variants in the COMT gene can predict impulsive choice behavior and could be an indicator for treatment targets. Those people with addictive disorders are at risk of relapse; returning to drug use even after prolonged periods of abstinence is well known and the subject of considerable research. In animal models of relapse, drug-seeking behavior can be triggered by priming with injections of the drugs, by drug-associated environmental stimuli, and by stress (foot-shock stress). According to Self and Nestler (1998), the neural mechanisms that are the basis of relapse can be viewed in general terms as drug-like, mini-withdrawal processes (Blum et al. 2008b). Bossert et al. (2006) demonstrated that we cannot ignore the importance of neurochemical mechanisms involved in drug-induced relapse behavior. These investigators used a drug relapse model, previously shown to induce relapse by reexposing rats to heroin-associated contexts. After the extinction of drug-reinforced responding, they found that exposure to different heroin-associated contexts reinstated heroin seeking. Reinstatement was reduced when glutamate transmission was inhibited in components of the mesolimbic dopa-

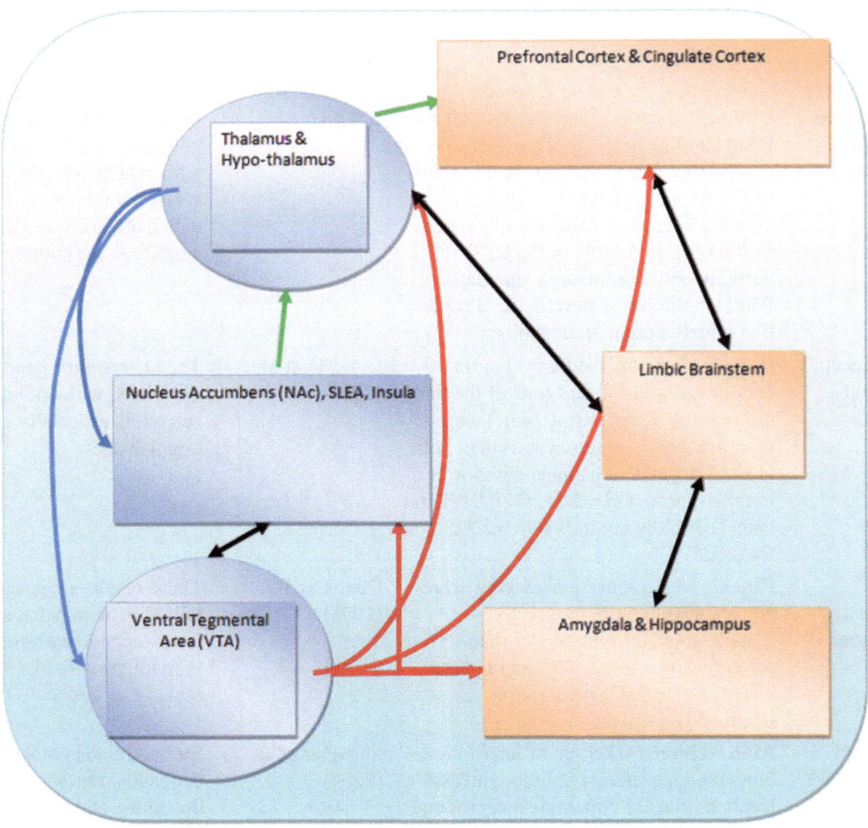

Fig. 18.5 Reward circuitry (*Source*: Blum et al. 2012f, with permission)

mine system: the ventral tegmental area and medial accumbens shell (Bossert et al. 2006). This process enhances dopamine net release in the NAc. In brief, glutamate reduction increases dopamine release; this fits well with Li's (2008) KARG addiction network map.

Many studies have focused on activation of D2-like dopamine receptors in the NAc of the mesolimbic dopamine system, as a crucial neural substrate where various stimuli produce relapse. Neural circuits from the prefrontal cortex, the hypothalamic–pituitary–adrenal axis, and the amyg-dala may also be activated by drug-associated stimuli and stress (Self and Nestler 1998). There are other dopamine-independent mechanisms in relapse like upregulation of the cAMP pathway in the NAc. This neuroadaptation represents a drug-opposite or opponent process that occurs after chronic drug exposure (Perez et al. 2011; Lintas et al. 2011). This system directly effects relapse to drug-seeking behavior. It is likely that the long-lasting risks for relapse are mediated by neuroadaptations via drug-induced changes in gene expression (mRNA) at both the NAc and prefrontal cortex–cingulate gyrus regions (Self and Nestler 1998; Fig. 18.5).

8 Neurogenetics of Clinical Issues in Pain Relief

Understanding the role of neurogenetics in pain relief including pharmacogenomic and nutrigenomic aspects will pave the way to better treatment to the millions suffering from both acute and chronic pain. We now know that dopaminergic tone is involved in pain sensitivity mechanisms (Chen et al. 2009).

Neurological loci for sensitivity to pain may reside in the mesolimbic system where a number of genes and polymorphisms associated with a predisposition to tolerance or intolerance to pain reside. The identification of certain gene polymorphisms as unique, therapeutic targets may assist in the treatment of pain. Pharmacogenetic testing for certain candidate genes, like mu receptors and PENK, is proposed as a means to improve clinical outcomes by provision of treatment personalized to each patient guided by their unique genetic makeup (Chen et al. 2009). The use of GARS, as described above, to identify clients with high addiction risk by providing valuable information about genetic predisposition to opioid addiction could become an important frontline approach on admission to pain clinics.

One notable paper evaluated the role of both mu-opioid receptors (MORs) and delta-opioid receptors (DORs), two genes expressed in the VTA that are thought to be involved in the addictive properties of opiates. Researchers David et al. (2008) found that knockout of the MOR gene abolished intra-VTA morphine self-administration at all doses tested, while male and female WT and DOR−/− mice exhibited self-administration similarly. Naloxone (4 mg/kg) disrupted this behavior in WT and DOR mutants, without triggering physical signs of withdrawal. Morphine ICSA was associated with an increase in FOS within the NAc, striatum, limbic cortices, amygdala, hippocampus, the lateral mammillary nucleus (LM), and the ventral posteromedial thalamus (VPM). This latter structure was found to express high levels of FOS exclusively in self-administering WT and DOR−/− mice. Abolition of morphine reward in MOR−/− mice was associated with a decrease in Fos-positive neurons in the mesocorticolimbic DA system, amygdala, hippocampus (CA1), and LM and a complete absence within the VPM. David et al. (2008) conclude that (1) VTA MORs, but not DORs, are critical for morphine reward and (2) the role of VTA-thalamic projections in opiate reward deserves to be further explored.

Moreover, clinical and laboratory studies indicate that the MOR gene contributes to inheritable vulnerability to the development of opiate addiction. Naturally occurring polymorphisms have been identified in the MOR gene. Substitutions occur at high allelic frequencies (10.5 and 6.6 %) in two coding regions of single-nucleotide polymorphisms (SNPs), the A118G and C17T, respectively, of the MOR gene. These SNPs cause amino acid changes in the receptor that impact on an individual's response to opioids and can influence (increases or decreases) vulnerability to opiate addiction (LaForge et al. 2000). Thus, the A118G substitution encodes a variant receptor with binding and signal transduction differences in response to beta-endorphin in cellular assays (LaForge et al. 2000).

Finally, to firmly establish the role of MOR in reward and response to buprenorphine, Ide et al. (2004) found that buprenorphine antinociception, assessed by hot-plate and tail-flick tests, was significantly reduced in heterozygous mu-opioid receptor knockout (MOR-KO) mice and abolished in homozygous MOR-KO mice. Buprenorphine, on the other hand, was able to establish a conditioned place preference (CPP) in homozygous MOR-KO, although as the number of copies of wild-type mu-opioid receptor genes was reduced, the magnitude of place preference was reduced. This study revealed that mu-opioid receptors mediate most of analgesic properties of buprenorphine (Ide et al. 2004).

9 NAAT Evidence in RDS

There is a series of published articles concerning the basic platform of utilizing Neuroadaptagen Amino Acid Therapy (NAAT) to treat various RDS behaviors (see Table 18.2). These results coupled with other qEEG studies of this compound suggest a putative anti-craving/anti-relapse role in drug addicts by direct or indirect dopaminergic interaction (Blum et al. 2010; Miller et al. 2010; Table 18.2).

Table 18.2 Phase 1 and 2 trials of Neuronutrient Amino Acid Therapy (NAAT)

Publications	Key findings
Blum et al. (1973) L-DOPA: effect on ethanol narcosis and brain biogenic amines in mice	Increased brain L-DOPA increases brain dopamine in mice and causes alcohol induced inebriated mice to sleep longer
	Dopamine, L-tryptophan, and alcohol work similarly in the brain
Blum et al. (1974) Ethanol narcosis in mice: serotonergic involvement	When mice were given alcohol and L-tryptophan compared to saline, the L-tryptophan plus alcohol group slept longer than saline plus alcohol group
	L-tryptophan and alcohol work similarly in the brain
Blum et al. (1987) Enkephalinase inhibition: Regulation of ethanol intake in mice	Mice genetically predisposed to like alcohol have a measured deficiency in enkephalin
	D-phenylalanine and hydrocinnamic acid are substances known to stop the breakdown of enkephalin in the brain – the amount of enkephalin available in the brain increases
	When the amount of enkephalin available in the brain increases, both voluntary and forced intake of alcohol decreases
	D-phenylalanine is one of the ingredients in NAAT
Blum et al. (1988) Improvement of inpatient treatment of the alcoholic as a function of neurotransmitter restoration: a pilot study	First small clinical trial of SAAVE (precursor amino acid loading and enkephalinase inhibition – earliest version of NAAT)
	Designed to elevate levels of enkephalin(s), serotonin, and catecholamines and regulate GABA, thought to be deficient in alcoholics
	Compared to controls those who took SAAVE had:
	Lower building up to drink score
	Required no PRN benzodiazepines
	Ceased having tremors 24 h earlier
	Less depression
Blum et al. (1988) Enkephalinase inhibition and precursor amino acid loading improves inpatient treatment of alcohol and polydrug abusers: double-blind placebo-controlled study of the nutritional adjunct SAAVE	Double-blind, placebo-controlled clinical trial of SAAVE of 62 people with substance use disorder (SUD)
	Results:
	Reduced stress as measured by skin conductance
	Improved physical and BESS (behavioral, emotional, social and spiritual) scores
	Sixfold decrease in leaving against medical advice (AMA) rates
Blum et al. (1988) Reduction of both drug hunger and withdrawal against advice rate of cocaine abusers in a 30 day inpatient treatment program by the neuronutrient Tropamine	Comparison of the effects of Tropamine [T] (amino acid and vitamin supplement), SAAVE [S] (a neuronutrient supplement), and no supplement [C] on a group of cocaine abusers in a 30-day hospital treatment program
	AMA rate:
	[C] 37.5 %
	[S] 26.6 %
	[T] 4.2 %
	Tropamine decreased the AMA rate by significant reduction of drug hunger
Brown et al. (1990) Neurodynamics of relapse prevention: a neuronutrient approach to outpatient DUI offenders	Relapse prevention using neuronutrients SAAVE and Tropamine in DUI offenders, either alcohol or cocaine
	Reduced relapse rates and enhanced recovery in 10-week outpatient setting
	After 10 months recovery rate was 73 % for SAAVE and 53 % for Tropamine
	These recovery rates are significantly better than the literature average for both alcoholism and cocaine dependence

(continued)

Table 18.2 (continued)

Publications	Key findings
Blum et al. (1990) Neuronutrient effects on weight loss in carbohydrate bingers; an open clinical trial	Examine the effects of PCAL-103 (NAAT) on compulsive eating and weight loss in 27 outpatients attending a supervised diet-controlled treatment program
	The PCAL-103 average weight loss was 26.96 vs. 10.2 lb in the control group. Relapse 18.2 % in the PCAL-103 group vs. 81.8 % in the control group
DeFrance et al. (1997) Enhancement of attention processing by Kantroll in healthy humans: a pilot study	Cognitive processing speeds in normal young adult volunteers were measured before and after 28–30 days of supplementation with a combination of amino acids–enkephalinase inhibition (NAAT), vitamins, and minerals
	Cognitive processing speeds were enhanced by statistically significant amplitude of the P300 component of the event-related potentials (ERPs). Focus improved
	These findings have relevance to relapse prevention because the resultant enhanced effect following NAAT as measured by the Contingent Continuous Performance Task (CCPT) and Spatial Orientation Task (SOT) and focus reflects better judgment and thus decision making
Blum et al. (1997) Clinical evidence for effectiveness of Phencal™ in maintaining weight loss in an open-label, controlled, 2-year study	Of 247 outpatients in a very-low-calorie fasting program, 130 who were having difficulty attaining their desired weight or maintaining their desired weight constituted the experimental group who took PhenCal™ plus Centrum vitamins, and the rest 117 who took only vitamins (Centrum) were the control group
	The PhenCal™ group compared to the control:
	Lost twice as much weight
	Regained 14.7 % of the weight, while the control group regained 41.7 %
	Decrease in food cravings (sugar) for females 70 % and males 63 %
	Decrease in binge eating for females 66 % and males 41 %
Ross (2001) Amino-acid precursor and enkephalinase inhibition therapy: evidence for effectiveness in treatment of Reward Deficiency Syndrome (RDS) with particular emphasis on eating disorders	Preliminary evaluation of six randomly selected former eating-disordered female clients (three were also chemically dependent) contacted at 9 months and 3 years of treatment with amino acid precursor and enkephalinase inhibition therapy
	All 6 reported initial benefit, one relapsed at 6 months, the other 5 all sustained and, in some cases, exceeded expectations.
	98 % of 100 patients similarly treated and evaluated reported significant improvement in both mood and reduced substance craving

(continued)

Table 18.2 (continued)

Publications	Key findings
Chen et al. (2004) Narcotic antagonists in drug dependence: pilot study showing enhancement of compliance with SYN-10, amino-acid precursors and enkephalinase inhibition therapy	A combination of Trexan (a narcotic antagonist) and amino acids was use to detoxify either methadone or heroin addicts
	Results were dramatic in terms of significantly enhancing compliance to continue taking Trexan
	Trexan alone for rapid detoxification, the average number of days of compliance calculated on 1,000 patients is 37 days
	12 subjects tested, receiving both the Trexan and amino acid therapy, continued to take the combination for an average of 262 days
	Suggests coupling amino acid therapy and enkephalinase inhibition while blocking the delta receptors with a pure narcotic antagonist as a novel method to induce rapid detox in chronic methadone patients and prevent relapse
	Testing this hypothesis with the sublingual combination of the partial opiate mu receptor agonist buprenorphine
Blum et al. (2006) Reward deficiency syndrome in obesity: a preliminary cross-sectional trial with a Genotrim variant	Consumption of large quantities of alcohol or carbohydrates (carbohydrate bingeing) stimulates production and usage of dopamine within the brain
	Obesity is due to the need to make up for inadequate dopaminergic activity in the reward center of the brain
	This has been called reward deficiency syndrome (RDS) used to categorize such genetic biologic influences on behavior
	RDS must be addressed at the same time as behavioral modifications are implemented to adequately treat obese patients
	In this small observational trial, 24 individuals completed a survey on which they documented 15 categories of benefit during their experience with a GenoTrim, a NAAT formulation customized to DNA
	Statistical analysis of the survey results demonstrated that stress reduction leads to:
	1. Improved sleep, enhanced energy, and improved focus and performance
	2. Reduced appetite, loss of unwanted weight, decreased body inches, and enhanced well-being
Chen et al. (2007a) Narcotic Attenuation Program attenuates substance use disorder, a clinical subtype of reward deficiency syndrome	1-year prospective study that evaluated the effects of taking Haveos (SynaptamineTM) on 61 compliant patients in a comprehensive outpatient clinical program
	Results after 12 weeks
	Results after 1 year
	Building up to relapse scores and ability to refrain from drug-seeking behavior both significantly improved
	Dropout rate: alcohol users 7 % and psychostimulant users 73 %
Blum et al. (2007) Synaptamine (SG8839), TM an amino-acid enkephalinase inhibition nutraceutical improves recovery of alcoholics, a subtype of Reward Deficiency Syndrome (RDS)	In an open clinical study, intravenous plus oral amino acid enkephalinase inhibition nutraceutical improved symptomatology of 600 recovering alcoholics
	Emotional and behavioral recovery scores significantly improved after administration of oral and intravenous Synaptamine
	Mean reductions for craving, depression, anxiety, anger, fatigue, lack of energy, and crisis were all significantly greater than 50 % ($P<0.001$)

(continued)

Table 18.2 (continued)

Publications	Key findings
Chen et al. (2007b) Chromium Picolinate (Crp) a putative anti-obesity nutrient induces changes in body composition: as function of the Taq1 dopamine D2 receptor gene	Chromium picolinate (CrP) was tested against placebo in groups of obese patients tested for the Taq1 dopamine D2 receptor gene
	In carriers of the DRD2 A2 genotype, weight loss and other changes in body composition were significant
	They were not significant for patients with the A1/A1 or A1/A2 allele
	These results suggest that the dopaminergic system, specifically the density of the D2 receptors, confers a significant differential therapeutic effect of CrP in terms of weight loss and change in body fat
	It is speculated that carriers of the DRD2 A1 allele had aberrant sugar cravings which masked the effects of CrP
Blum et al. (2008b) A short term pilot open label study to evaluate efficacy and safety of LG839, a customized DNA directed nutraceutical in obesity: exploring nutrigenomics	Preliminary investigational study to evaluate the impact of polymorphisms of five candidate genes on treatment for obesity with NAAT
	The formula for each patient was customized based on their genetic results
	Pre-NAAT compared to post-NAAT had significantly lower BMI
	Pre-NAAT compared to post-NAAT had significantly lower pounds
	Pre-NAAT compared to post-NAAT had trends for reduced late night snacking, carbohydrate craving reduction, reduction of stress, reduction of waist circumference
Blum et al. (2008a) LG839: anti-obesity effects and polymorphic gene correlates of reward deficiency syndrome	First DNA customized analysis of LG839 for weight management effects
	Out of 1,058 Dutch participants, a subset of 27 self-reported obese subjects were genotyped, and based on their polymorphisms, each subject utilized a customized LG839 variant assessed for pre and post for after 80 days of usage
	Significant results were observed for weight loss, sugar craving reduction, appetite suppression, snack reduction, reduction of late night eating (all $P<0.01$), increased perception of overeating, enhanced quality of sleep, increased happiness (all $P<0.05$), and increased energy ($P<0.001$)
	The study points to the importance of genotyping patients and providing DNA customized nutraceutical intervention to combat obesity
Blum et al. (2008c) Dopamine D2 Receptor Taq A1 allele predicts treatment compliance of LG839 in a subset analysis of pilot study in the Netherlands	This study evaluated the importance of carrying the dopamine D2 receptor A1 allele and treatment compliance of n in an obese Dutch population nutraceutical intervention to combat obesity
	Candidate genes to be associated with obesity include among others the dopamine D2 receptor (DRD2), methylenetetrahydrofolate reductase (MTHFR), serotonin receptor (5-HT2a), peroxisome proliferator-activated receptor-gamma (PPAR-γ), and leptin (OB) genes
	Compliance is two-fold better in carriers of DRD2 A1 allele compared to DRD2 A2 allele
	Suggests that if you need to enhance dopamine, compliance is better

(continued)

Table 18.2 (continued)

Publications	Key findings
Blum et al. (2009) Putative targeting of Dopamine D2 receptor function in Reward Deficiency Syndrome (RDS) by Synaptamine Complex™ Variant (KB220): clinical trial showing anti-anxiety effects	Double-blind, placebo-controlled study to determine antianxiety effects of KB220 (Synaptamine variant) in 62 alcoholic and polydrug abusers
	This was an objective test, not subjective, because antianxiety effect was evaluated by skin conductance level (SCL)
	Significant reduction of stress in the KB220 group compared to placebo including a time-by-treatment interaction. Positive antianxiety effect as monitored throughout a 28-day treatment period is most significant at the 7th day (a time with the most severe anxiety)
	These findings may be relevant to prevention of relapse
Braverman et al. (2010) Sustainable weight loss and muscle gain utilizing the Rainbow Diet™: targeting noradrenergic and dopaminergic mechanistic sites, hormonal deficiency repletion therapy and exercise: a case report	Case study of 58-year-old male identified as being obese utilized a special Rainbow Diet
	Patient received noradrenergic drug, NAAT (a natural D2 agonist), hormonal deficiency replacement therapy, and light exercise
	After 1 year BMI decreased, percent body fat decreased, improved cardiac function, fasting glucose level declined, prostate specific antigen (PSA) tripled, memory improved, testosterone levels increased
	There was sustainable weight loss and muscle gain
Miller et al. (2010) Acute intravenous synaptamine complex variant KB220™ "normalizes" neurological dysregulation in patients during protracted abstinence from alcohol and opiates as observed using quantitative electroencephalographic and genetic analysis for reward polymorphisms: part 1, pilot study with 2 case reports	Combination of both IV NAAT and oral forms
	Two case reports of an alcoholic and heroin addict
	Both patients were genotyped for a number of neurotransmitter reward genes to determine to what extent they carry putative dopaminergic risk alleles that may predispose them for alcohol or heroin dependence, respectively
	The genes tested included the dopamine transporter (DAT1, locus symbol SLC6A3), dopamine D4 receptor exon 3 VNTR (DRD4), DRD2 Taq1A (rs1800497), COMT val158 met SNP (rs4680), monoamine oxidase-A upstream VNTR (MAOA-uVNTR), and serotonin transporter-linked polymorphic region (5-HTTLPR, locus symbol SLC6A4)
	Both patients showed prevalence of at least one risk allele
	qEEG analysis revealed dysregulation in the PFC–cingulate gyrus in both addicts
	IV NAAT and oral produced a regulation of widespread theta activity
	These results have relevance for relapse prevention because of its effect on the part of brain involved in relapse (PFC–cingulate gyrus)
Blum et al. (2010) Overcoming qEEG abnormalities and reward gene deficits during protracted abstinence in male psychostimulant and polydrug abusers utilizing putative dopamine D2 agonist therapy: part 2	In a crossover study a total of 10 abstinent psychostimulant-dependent patients were randomized in a triple-blind, placebo-controlled study
	Each patient was genotyped for a number of reward genes for addiction risk assessment
	100 % of the patient carried a least on risk allele
	qEEG analysis was performed on each patient 1 h after administration of KB220Z powder
	KB220Z™ showed an increase of alpha waves and low-beta wave activity in the parietal brain region (relapse area)
	Authors propose that utilization of KB220Z may upregulate dopamine receptors in patients having moderate to high genetic addiction risk

(continued)

Table 18.2 (continued)

Publications	Key findings
Blum et al. (2011a) "Dopamine Resistance" in brain reward circuitry as a function of DRD2 gene receptor polymorphisms in RDS: Synaptamine complex variant (KB220) induced "Dopamine Sensitivity" and enhancement of happiness	In a crossover triple-blind, placebo-controlled study on ten Chinese abstinent (16 months) heroin-dependent patients, fMRI was utilized to assess dopaminergic
	BOLD activation in the brain
	KB220Z was administered to each patient and after 1 h it was found that the complex induced BOLD activation of caudate–accumbens dopaminergic activation
	The BOLD activation was significantly different compared to placebo
	In addition KB220Z also reduced the hyperexcitability of putamen
	The authors suggest that KB220Z induces "dopamine sensitivity" in genetically prone "dopamine-resistant" patients genotyped for various reward gene polymorphisms
Chen et al. (2012) Neurotransmitter-precursor-supplement Intervention for Detoxified Heroin Addicts	In the cluster-randomized, placebo-controlled trial, 83 detoxified heroin addicts were evaluated during withdrawal
	This study examined the effects of combined administration of tyrosine, lecithin, L-glutamine, and L-5-hydroxytryptophan (5-HTP) on heroin withdrawal syndromes and mental symptoms
	The experimental group compared to placebo had reduced insomnia and reduced withdrawal scores
	After 6 days of treatment, compared to placebo the experimental group had a significant reduction in tension–anxiety, depression–dejection, anger–hostility, fatigue–inertia, and total mood disturbance and a greater increase in their vigor–activity symptoms (all $P<0.05$)
Miller et al. (2012) Early intervention of intravenous KB220IV- Neuroadaptagen Amino-Acid Therapy (NAAT)™ improves behavioral outcomes in a residential addiction treatment program: a pilot study	In 129 patients the combination of IV and oral NAAT therapy (generic KB220) was assessed through Chronic Abstinence Symptom Severity (CASS) scale over a 30-day period
	Three scales were constructed based on this factor analysis: Emotion, Somatic, and Cognitive
	All three scales showed significant declines ($P=0.00001$) from pre- to posttreatment: $t=19.1$ for Emotion, $t=16.1$ for Somatic, and $t=14.9$ for Poor Cognitive
	A 2-year follow-up in a subset of 23 patients showed 21 (91 %) were sober at 6 months with 19 (82 %) having no relapse, 19 (82 %) were sober at 1 year with 18 (78 %) having no relapse, 21 (91 %) were sober at 2 years posttreatment with 16 (70 %) having no relapse
Blum et al. (2012) Neurogenetics and nutrigenomics of neuro-nutrient therapy for Reward Deficiency Syndrome: clinical ramifications and pitfalls	In one case report, a heroin addict targeted and treated using high amounts of GABA and/or L-glutamine showed serious complications as measured by SPEC
	A case report of an alcoholic utilizing appropriate and careful administration of balanced amounts of precursor amino acid and inhibitors of neurotransmitter catabolic enzymes showed significant improvement as measured by SPEC
	The authors caution clinicians to incorporate known appropriate neuro-therapy by adhering to basic neurogenetic and nutrigenomic principles

(continued)

Table 18.2 (continued)

Publications	Key findings
Blum et al. (2013) Withdrawal from buprenorphine/naloxone and maintenance with a natural dopaminergic agonist: a case study	A report of a worker's compensation case of a female who received pain medication and Suboxone
	Total monthly prescription costs including supplemental benzodiazepines, hypnotics, and stimulants exceeded $53,000
	The patient subsequently was placed on Suboxone for 2.5 years
	The authors carefully documented very severe withdrawal symptoms when she precipitously stopped taking buprenorphine/ naloxone and during follow-up while taking KB220Z daily. At 432 days post-Suboxone® withdrawal, the patient is being maintained on KB220Z, has been urine tested, and is opioid free

Source: Modified from Blum et al. (2012e), with permission

10 Identification of Compliance to Treatment Medications and Abstinence from Psychoactive Drug Abuse: Comprehensive Analysis of Reported Drugs (CARD)™

To evaluate both compliance to treatment medications and abstinence from psychoactive drug abuse, we utilized CARD™ (offered by Dominion Diagnostics, Inc., North Kingstown, RI) in a total of 24,000 specimens over a 2-year period (2010–2011) across six eastern states (unpublished). Specifically we accessed two important clinical issues: (1) compliance with prescription treatment medications during various types of level of care and (2) abstinence from all nonprescribed licit and illicit psychoactive drugs substances. Thus, by utilizing CARD, our group found significant evidence for both noncompliance and non-abstinence ($P<0.0001$) during different levels of care (e.g., inpatient/outpatient/intensive outpatient/ residential/opioid treatment programs) [these results will be published elsewhere]. Certainly this important outcome data suggests the need for better diagnosis and intelligently designed ways to enhance dopaminergic function in the mesolimbic brain reward circuit.

11 Conclusions

Our laboratory recently reviewed the role of dopaminergic activation and genetic stratification in process addictions, in a paper concerning "Sex, Drugs, and Rock 'N' Roll" (Blum et al. 2012b). We concluded that the reinforcing effects of drugs abuse and sex, food, and other addictions are mediated within the ventral striatum NAc. This structure also mandates motivated behaviors such as eating, drinking, and sexual activity, which are stimulated by natural rewards. We have discussed the mechanisms that are the basis of human motivation. Based on abundant scientific research, they share molecular-genetic antecedents with powerful biological drives for survival, which, if impaired, lead to aberrant compulsive, obsessive, and impulsive behaviors. We have hypothesized that neurotransmitter-related candidate genes and their polymorphisms can be used to predict predisposition to addiction and other RDS behaviors. The use and further development of dopaminergic agonistic agents to target specific gene polymorphisms has been proposed, and evidence that the proliferation of D2 receptors can attenuate human craving behavior was explored.

Innovative experiments have been developed during many years of scientific research into the nature of addiction. The challenge is to incorporate them in treatment programs for patients attending inpatient/outpatient addiction clinics. A go-forward model to assist in diagnosis and understanding treatment outcome and potential relapse prevention is proposed herein to include the following: (1) the Genetic Addiction Risk Score (GARS)™ for appropriate RDS diagnosis, (2) Comprehensive Analysis of Reported Drugs (CARD)™ to determine both compliance and abstinence during treatment, (3) natural D2 agonistic therapy (NAAT-KB220)™, and, eventually,

(4) mRNA testing (patent pending) to determine pre- and post-candidate gene expressions in reward deficiency syndrome (RDS). We are, therefore, proposing a paradigm shift we have called "Reward Deficiency Solutions System (RDSS)™."

References

Alia-Klein N, Parvaz MA, Woicik PA, Konova AB, Maloney T, Shumay E, Wang R, Telang F, Biegon A, Wang GJ, Fowler JS, Tomasi D, Volkow ND, Goldstein RZ (2011) Gene × disease interaction on orbitofrontal gray matter in cocaine addiction. Arch Gen Psychiatry 68(3):283–294

Baransel Isir AB, Oguzkan S, Nacak M, Gorucu S, Dulger HE, Arslan A (2008) The catechol-O-methyl transferase Val158Met polymorphism and susceptibility to cannabis dependence. Am J Forensic Med Pathol 29(4):320–322

Barr CL, Kidd KK (1993) Population frequencies of the A1 allele at the dopamine D2 receptor locus. Biol Psychiatry 34(4):204–209

Bart G, Kreek MJ, Ott J, LaForge KS, Proudnikov D, Pollak L, Heilig M (2005) Increased attributable risk related to a functional mu-opioid receptor gene polymorphism in association with alcohol dependence in central Sweden. Neuropsychopharmacology 30(2):417–422

Biederman J, Petty CR, Ten Haagen KS, Small J, Doyle AE, Spencer T, Mick E, Monuteaux MC, Smoller JW, Faraone SV (2009) Effect of candidate gene polymorphisms on the course of attention deficit hyperactivity disorder. Psychiatry Res 170(2–3):199–203

Blum K, Gold MS (2011) Neuro-chemical activation of brain reward mesolimbic circuitry is associated with relapse prevention and drug hunger: a hypothesis. Med Hypotheses 76(4):576–584

Blum K, Kozlowski GP (1990) Ethanol and neuromodulators interaction: a cascade model of reward. In: Ollat H, Parvez S, Parvez H (eds) Alcohol and behavior. VSP Press, Utrecht

Blum K, Wallace JE (1974) Effects of catecholamine synthesis inhibition on ethanol-induced withdrawal symptoms in mice. Br J Pharmacol 51(1):109–111

Blum K, Briggs AH, Trachtenberg MC, Delallo L, Wallace JE (1987) Enkephalinase inhibition: regulation of ethanol intake in genetically predisposed mice. Alcohol 4(6):449–456

Blum K, Calhoun W, Wallace JE, Merritt JH, Geller I (1973) Soporific action of ethanol in mice: possible role of biogenic amines. Pharmacol Biochem Behav 1(3):271–276

Blum K, Trachtenberg MC, Elliott CE, Dingler ML, Sexton RL, Samuels AI, Cataldie L (1988) Enkephalinase inhibition and precursor amino acid loading improves inpatient treatment of alcohol and polydrug abusers: double-blind placebo-controlled study of the nutritional adjunct SAAVE. Alcohol 5(6):481–493

Blum K, Noble EP, Sheridan PJ, Montgomery A, Ritchie T, Jagadeeswaran P, Nogami H, Briggs AH, Cohn JB (1990) Allelic association of human dopamine D2 receptor gene in alcoholism. JAMA 263(15): 2055–2060

Blum K, Sheridan PJ, Wood RC, Braverman ER, Chen TJ, Comings DE (1995a) Dopamine D2 receptor gene variants: association and linkage studies in impulsive-addictive-compulsive behavior. Pharmacogenetics 5(3):121–141

Blum K, Wood RC, Braverman ER, Chen TJ, Sheridan PJ (1995b) The D2 dopamine receptor gene as a predictor of compulsive disease: Bayes' theorem. Funct Neurol 10(1):37–44

Blum K, Braverman ER, Wood RC, Gill J, Li C, Chen TJ, Taub M, Montgomery AR, Sheridan PJ, Cull JG (1996a) Increased prevalence of the Taq I A1 allele of the dopamine receptor gene (DRD2) in obesity with comorbid substance use disorder: a preliminary report. Pharmacogenetics 6(4):297–305

Blum K, Sheridan PJ, Wood RC, Braverman ER, Chen TJ, Cull JG, Comings DE (1996b) The D2 dopamine receptor gene as a determinant of reward deficiency syndrome. J R Soc Med 89(7):396–400

Blum K, Braverman ER, Wu S, Cull JG, Chen TJ, Gill J, Wood R, Eisenberg A, Sherman M, Davis KR, Matthews D, Fischer L, Schnautz N, Walsh W, Pontius AA, Zedar M, Kaats G, Comings DE (1997) Association of polymorphisms of dopamine D2 receptor (DRD2), and dopamine transporter (DAT1) genes with schizoid/avoidant behaviors (SAB). Mol Psychiatry 2(3):239–246

Blum K, Chen TJ, Meshkin B, Downs BW, Gordon CA, Blum S, Mengucci JF, Braverman ER, Arcuri V, Varshavskiy M, Deutsch R, Martinez-Pons M (2006) Reward deficiency syndrome in obesity: a preliminary cross-sectional trial with a Genotrim variant. Adv Ther 23(6):1040–1051

Blum K, Chen TJH, Downs BW, Meshkin B, Blum SH et al (2007) Synaptamine (SG8839)™, an amino-acid enkephalinase inhibition nutraceutical improves recovery of alcoholics, a subtype of reward deficiency syndrome (RDS). Trends Appl Sci Res 2:132–138

Blum K, Chen AL, Chen TJ, Rhoades P, Prihoda TJ, Downs BW, Waite RL, Williams L, Braverman ER, Braverman D, Arcuri V, Kerner M, Blum SH, Palomo T (2008a) LG839: anti-obesity effects and polymorphic gene correlates of reward deficiency syndrome. Adv Ther 25(9):894–913

Blum K, Chen ALC, Chen TJH, Braverman ER, Reinking J, Blum SH, Cassel K, Downs BW, Waite L, Williams L, Prihoda TJ, Kerner MM, Palomo T, Comings DE, Tung H, Rhoades P, Oscar-Berman M (2008b) Activation instead of blocking mesolimbic dopaminergic reward circuitry is a preferred modality in the long term treatment of reward deficiency syndrome (RDS): a commentary. Theor Biol Med Model 5:24. doi:10.1186/1742-4682-5-24

Blum K, Chen TH, Chen ALC, Rhoades P, Prihoda TJ, Downs WB, Bagchi D, Bagchi M, Blum SH, Williams L, Braverman ER, Kerner M, Waite RL, Quirk B,

White L, Reinking J (2008c) Dopamine D2 receptor Taq A1 allele predicts treatment compliance of LG839 in a subset analysis of pilot study in the Netherlands. Gene Ther Mol Biol 12:129–140

Blum K, Chen TJ, Downs BW, Bowirrat A, Waite RL, Braverman ER, Madigan M, Oscar-Berman M, DiNubile N, Stice E, Giordano J, Morse S, Gold M (2009) Neurogenetics of dopaminergic receptor supersensitivity in activation of brain reward circuitry and relapse: proposing "deprivation-amplification relapse therapy" (DART). Postgrad Med 121(6):176–196

Blum K, Chen TJ, Morse S, Giordano J, Chen AL, Thompson J, Allen C, Smolen A, Lubar J, Stice E, Downs BW, Waite RL, Madigan MA, Kerner M, Fornari F, Braverman ER (2010) Overcoming qEEG abnormalities and reward gene deficits during protracted abstinence in male psychostimulant and polydrug abusers utilizing putative dopamine D2 agonist therapy: part 2. Postgrad Med 122(6):214–226

Blum K, Liu Y, Shriner R, Gold MS (2011a) Reward circuitry dopaminergic activation regulates food and drug craving behavior. Curr Pharm Des 17(12):1158–1167

Blum K et al (2011b) Understanding the high mind: humans are still evolving genetically. IIOAB J 2:1–22

Blum K, Oscar-Berman M, Giordano J, Downs B, Simpatico T, Han D, Femino J (2012) Neurogenetic impairments of brain reward circuitry links to reward deficiency syndrome (RDS): potential nutrigenomic induced dopaminergic activation. J Genet Syndr Gene Ther 3(4):1000e115

Blum K, Gardner E, Oscar-Berman M, Gold M (2012a) "Liking" and "wanting" linked to Reward Deficiency Syndrome (RDS): hypothesizing differential responsivity in brain reward circuitry. Curr Pharm Des 18(1):113–118

Blum K, Werner T, Carnes S, Carnes P, Bowirrat A, Giordano J, Oscar-Berman M, Gold M (2012b) Sex, drugs, and rock 'n' roll: hypothesizing common mesolimbic activation as a function of reward gene polymorphisms. J Psychoactive Drugs 44(1):38–55

Blum K, Chen AL, Giordano J, Borsten J, Chen TJ, Hauser M, Simpatico T, Femino J, Braverman ER, Barh D (2012c) The addictive brain: all roads lead to dopamine. J Psychoactive Drugs 44(2):134–143

Blum K, Giordano J, Hauser M, Saunders S, Campenella C, Han D, Smolen A, Borsten J, Downs BW, Waite RL, Braverman ER, Liu Y, Femino J, Simpatico T (2012d) Genetic addiction risk scores (GARS) coupled with the comprehensive analysis of reported drugs (CARD) provide diagnostic and outcome data supporting the need for Dopamine D2 agonist therapy: proposing a holistic – therapeutic model for reward deficiency syndrome (RDS). In: Omics addiction therapy conference, Las Vegas, 20 Aug 2012 (Keynote abstract)

Blum K, Berman MO, Stuller E, Miller D, Giordano J et al (2012e) Neurogenetics and nutrigenomics of neuronutrient therapy for Reward Deficiency Syndrome (RDS): clinical ramifications as a function of molecular neurobiological mechanisms. J Addict Res Ther 3:139. doi:10.4172/2155-6105.1000139

Blum et al. (2012f) Special issue proceedings of the meeting on addiction research and therapy, Journal Addiction Research and Therapy, Embassy Suites, Las Vegas, 20–22 August 2012

Blum K, Oscar-Berman M, Femino J, Waite RL, Benya L et al. (2013) Withdrawal from buprenorphine/naloxone and maintenance with a natural dopaminergic agonist: a cautionary note. J Addict Res Ther 4:146. doi:10.4172/2155-6105.1000146

Bossert JM, Gray SM, Lu L, Shaham Y (2006) Activation of group II metabotropic glutamate receptors in the nucleus accumbens shell attenuates context-induced relapse to heroin seeking. Neuropsychopharmacology 10:2197–2209

Boundy VA, Pacheco MA, Guan W, Molinoff PB (1995) Agonists and antagonists differentially regulate the high affinity state of the D2L receptor in human embryonic kidney 293 cells. Mol Pharmacol 48(5):956–964

Bowirrat A, Oscar-Berman M (2005) Relationship between dopaminergic neurotransmission, alcoholism, and Reward Deficiency Syndrome. Am J Med Genet B Neuropsychiatr Genet 132B(1):29–37

Bradaia A, Trube G, Stalder H, Norcross RD, Ozmen L, Wettstein JG, Pinard A, Buchy D, Gassmann M, Hoener MC, Bettler B (2009) The selective antagonist EPPTB reveals TAAR1-mediated regulatory mechanisms in dopaminergic neurons of the mesolimbic system. Proc Natl Acad Sci USA 106(47): 20081–20086

Braverman ER, Chen AL, Chen TJ, Schoolfield JD, Notaro A, Braverman D, Kerner M, Blum SH, Arcuri V, Varshavskiy M, Damle U, Downs BW, Waite RL, Oscar-Berman M, Giordano J, Blum K. (2010) Test of variables of attention (TOVA) as a predictor of early attention complaints, an antecedent to dementia. Neuropsychiatr Dis Treat 6:681–690. doi: 10.2147/NDT.S12243.

Brown RJ, Blum K, Trachtenberg MC (1990) Neurodynamics of relapse prevention: a neuronutrient approach to outpatient DUI offenders. J Psychoactive Drugs 22(2):173–187

Budygin EA, Park J, Bass CE, Grinevich VP, Bonin KD, Wightman RM (2012) Aversive stimulus differentially triggers subsecond dopamine release in reward regions. Neuroscience 201:331–337

Burger KS, Stice E (2012) Frequent ice cream consumption is associated with reduced striatal response to receipt of an ice cream-based milkshake. Am J Clin Nutr 95(4):810–817

Byerley W, Hoff M, Holik J, Caron MG, Giros B (1993) VNTR polymorphism for the human dopamine transporter gene (DAT1). Hum Mol Genet 2(3):335

Chen TJ, Blum K, Payte JT, Schoolfield J, Hopper D, Stanford M, Braverman ER (2004) Narcotic antagonists in drug dependence: pilot study showing enhancement of compliance with SYN-10, amino-acid precursors and enkephalinase inhibition therapy. Med Hypotheses 63(3):538–548

Chen TJ, Blum K, Waite RL, Meshkin B, Schoolfield J, Downs BW, Braverman EE, Arcuri V, Varshavskiy M, Blum SH, Mengucci J, Reuben C, Palomo T (2007a) Narcotic Attenuation Program attenuates substance use disorder, a clinical subtype of reward deficiency syndrome. Adv Ther 24(2):402–414

Chen TJH, Blum K, Kaats G, Braverman ER, Eisenberg A et al (2007b) Chromium Picolinate (Crp) A putative anti-obesity nutrient induces changes in body composition as function of the Taq1 Dopamine D2 receptor polymorphisms in a randomized double-blind placebo controlled study. Gene Ther Mol Biol 11:161–170

Chen AL, Chen TJ, Waite RL, Reinking J, Tung HL, Rhoades P, Downs BW, Braverman E, Braverman D, Kerner M, Blum SH, DiNubile N, Smith D, Oscar-Berman M, Prihoda TJ, Floyd JB, O'Brien D, Liu HH, Blum K (2009) Hypothesizing that brain reward circuitry genes are genetic antecedents of pain sensitivity and critical diagnostic and pharmacogenomic treatment targets for chronic pain conditions. Med Hypotheses 72(1):14–22

Chen TJ, Blum K, Chen AL, Bowirrat A, Downs WB, Madigan MA, Waite RL, Bailey JA, Kerner M, Yeldandi S, Majmundar N, Giordano J, Morse S, Miller D, Fornari F, Braverman ER (2011) Neurogenetics and clinical evidence for the putative activation of the brain reward circuitry by a neuroadaptagen: proposing an addiction candidate gene panel map. J Psychoactive Drugs 43(2):108–127

Chen D, Liu Y, He W, Wang H, Wang Z (2012) Neurotransmitter-precursor-supplement intervention for detoxified heroin addicts. J Huazhong Univ Sci Technolog Med Sci 32(3):422–427. doi:10.1007/s11596-012-0073-z

Contini V, Marques FZ, Garcia CE, Hutz MH, Bau CH (2006) MAOA-uVNTR polymorphism in a Brazilian sample: further support for the association with impulsive behaviors and alcohol dependence. Am J Med Genet B Neuropsychiatr Genet 141B(3):305–308

Cook EH Jr, Stein MA, Krasowski MD, Cox NJ, Olkon DM, Kieffer JE, Leventhal BL (1995) Association of attention-deficit disorder and the dopamine transporter gene. Am J Hum Genet 56(4):993–998

Cordeira J, Rios M (2011) Weighing in the role of BDNF in the central control of eating behavior. Mol Neurobiol 44(3):441–448

Dahlgren A, Wargelius HL, Berglund KJ, Fahlke C, Blennow K, Zetterberg H, Oreland L, Berggren U, Balldin J (2011) Do alcohol-dependent individuals with DRD2 A1 allele have an increased risk of relapse? A pilot study. Alcohol Alcohol 46(5):509–513

David V, Matifas A, Gavello-Baudy S, Decorte L, Kieffer BL, Cazala P (2008) Brain regional FOS expression elicited by the activation of mu- but not delta-opioid receptors of the ventral tegmental area: evidence for an implication of the ventral thalamus in opiate reward. Neuropsychopharmacology 33(7):1746–1759

DeFrance JF, Hymel C, Trachtenberg MC, Ginsberg LD, Schweitzer FC, Estes S, Chen TJ, Braverman ER, Cull JG, Blum K (1997) Enhancement of attention processing by Kantroll in healthy humans: a pilot study. Clin Electroencephalogr 28(2):68–75

Demetrovics Z, Varga G, Szekely A, Vereczkei A, Csorba J, Balazs H, Hoffman K, Sasvari-Szekely M, Barta C (2010) Association between Novelty Seeking of opiate-dependent patients and the catechol-O-methyltransferase Val(158)Met polymorphism. Compr Psychiatry 51(5):510–515

Duaux E, Gorwood P, Griffon N, Bourdel MC, Sautel F, Sokoloff P, Schwartz JC, Ades J, Lôo H, Poirier MF (1998) Homozygosity at the dopamine D3 receptor gene is associated with opiate dependence. Mol Psychiatry 3(4):333–336

Dunn JP, Kessler RM, Feurer ID, Volkow ND, Patterson BW, Ansari MS, Li R, Marks-Shulman P, Abumrad NN (2012) Relationship of dopamine type 2 receptor binding potential with fasting neuroendocrine hormones and insulin sensitivity in human obesity. Diabetes Care 35(5):1105–1111

Faraone SV, Doyle AE, Mick E, Biederman J (2001) Meta-analysis of the association between the 7-repeat allele of the dopamine D(4) receptor gene and attention deficit hyperactivity disorder. Am J Psychiatry 158(7):1052–1057

Feusner J, Ritchie T, Lawford B, Young RM, Kann B, Noble EP (2001) GABA(A) receptor beta 3 subunit gene and psychiatric morbidity in a post-traumatic stress disorder population. Psychiatry Res 104(2):109–117

Filip M, Zaniewska M, Frankowska M, Wydra K, Fuxe K (2012) The importance of the adenosine A(2A) receptor-dopamine D(2) receptor interaction in drug addiction. Curr Med Chem 19(3):317–355

Galeeva AR, Gareeva AE, Iur'ev EB, Khusnutdinova EK (2002) VNTR polymorphisms of the serotonin transporter and dopamine transporter genes in male opiate addicts. Mol Biol (Mosk) 36(4):593–598

Gokturk C, Schultze S, Nilsson KW, von Knorring L, Oreland L, Hallman J (2008) Serotonin transporter (5-HTTLPR) and monoamine oxidase (MAOA) promoter polymorphisms in women with severe alcoholism. Arch Womens Ment Health 11(5–6):347–355

Grzywacz A, Kucharska-Mazur J, Samochowiec J (2008) Association studies of dopamine D4 receptor gene exon 3 in patients with alcohol dependence. Psychiatr Pol 42(3):453–461

Hall FS, Sora I, Uhl GR (2001) Ethanol consumption and reward are decreased in mu-opiate receptor knockout mice. Psychopharmacology (Berl) 154(1):43–49

Hikida T, Kimura K, Wada N, Funabiki K, Nakanishi S (2010) Distinct roles of synaptic transmission in direct and indirect striatal pathways to reward and aversive behavior. Neuron 66(6):896–907

Hill SY, Hoffman EK, Zezza N, Thalamuthu A, Weeks DE, Matthews AG, Mukhopadhyay I (2008 Jun 5) Dopaminergic mutations: within-family association and linkage in multiplex alcohol dependence families. Am J Med Genet B Neuropsychiatr Genet 147B(4):517–526

Ide S, Minami M, Satoh M, Uhl GR, Sora I, Ikeda K (2004) Buprenorphine antinociception is abolished, but naloxone-sensitive reward is retained, in mu-opioid receptor knockout mice. Neuropsychopharmacology 29(9): 1656–1663

Ikemoto S (2010) Brain reward circuitry beyond the mesolimbic dopamine system: a neurobiological theory. Neurosci Biobehav Rev 35(2):129–150. doi:10.1016/j.neubiorev.2010.02.001

Johnson PM, Kenny PJ (2012) Dopamine D2 receptors in addiction-like reward dysfunction and compulsive eating in obese rats. Nat Neurosci 13(5):635–641

Kita JM, Kile BM, Parker LE, Wightman RM (2009) *In vivo* measurement of somatodendritic release of dopamine in the ventral tegmental area. Synapse 63(11):951–960. doi:10.1002/syn.20676

Kosek E, Jensen KB, Lonsdorf TB, Schalling M, Ingvar M (2009) Genetic variation in the serotonin transporter gene (5-HTTLPR, rs25531) influences the analgesic response to the short acting opioid Remifentanil in humans. Mol Pain 5:37

Kotler M, Cohen H, Segman R, Gritsenko I, Nemanov L, Lerer B, Kramer I, Zer-Zion M, Kletz I, Ebstein RP (1997) Excess dopamine D4 receptor (D4DR) exon III seven repeat allele in opioid-dependent subjects. Mol Psychiatry 2(3):251–254

Kraschewski A, Reese J, Anghelescu I, Winterer G, Schmidt LG, Gallinat J, Finckh U, Rommelspacher H, Wernicke C (2009) Association of the dopamine D2 receptor gene with alcohol dependence: haplotypes and subgroups of alcoholics as key factors for understanding receptor function. Pharmacogenet Genomics 19(7):513–527

LaForge KS, Shick V, Spangler R, Proudnikov D, Yuferov V, Lysov Y, Mirzabekov A, Kreek MJ (2000) Detection of single nucleotide polymorphisms of the human mu opioid receptor gene by hybridization or single nucleotide extension on custom oligonucleotide gelpad microchips: potential in studies of addiction. Am J Med Genet 96(5):604–615

Lai JH, Zhu YS, Huo ZH, Sun RF, Yu B, Wang YP, Chai ZQ, Li SB (2010) Association study of polymorphisms in the promoter region ofDRD4 with schizophrenia, depression, and heroin addiction. Brain Res 1359: 227–232

Lee SS, Lahey BB, Waldman I, Van Hulle CA, Rathouz P, Pelham WE, Loney J, Cook EH (2007) Association of dopamine transporter genotype with disruptive behavior disorders in an eight-year longitudinal study of children and adolescents. Am J Med Genet B Neuropsychiatr Genet 144B(3):310–317

Lee SY, Chen SL, Chen SH, Chu CH, Chang YH, Lin SH, Huang SY, Tzeng NS, Kuo PH, Lee IH, Yeh TL, Yang YK, Lu RB (2012) Interaction of the DRD3 and BDNF gene variants in subtyped bipolar disorder. Prog Neuropsychopharmacol Biol Psychiatry 39(2): 382–387

Li CY, Mao X, Wei L (2008) Genes and (common) pathways underlying drug addiction. PLoS Comput Biol 4(1):e2, Epub 2007 Nov 20

Li T, Hou Y, Cao W, Yan CX, Chen T, Li SB (2012) Role of dopamine D3 receptors in basal nociception regulation and in morphine-induced tolerance and withdrawal. Brain Res 1433:80–84

Limosin F, Romo L, Batel P, Adès J, Boni C, Gorwood P (2005) Association between dopamine receptor D3 gene BalI polymorphism and cognitive impulsiveness in alcohol-dependent men. Eur Psychiatry 20(3): 304–306

Lintas A, Chi N, Lauzon NM, Bishop SF, Gholizadeh S, Sun N, Tan H, Laviolette SR (2011) Identification of a dopamine receptor-mediated opiate reward memory switch in the basolateral amygdala-nucleus accumbens circuit. J Neurosci 31(31):11172–11183

Lobo MK, Covington HE 3rd, Chaudhury D, Friedman AK, Sun H, Damez-Werno D, Dietz DM, Zaman S, Koo JW, Kennedy PJ, Mouzon E, Mogri M, Neve RL, Deisseroth K, Han MH, Martinez D, Saccone PA, Liu F, Slifstein M, Orlowska D, Grassetti A, Cook S, Broft A, Van Heertum R, Comer SD (2012) Deficits in dopamine D(2) receptors and presynaptic dopamine in heroin dependence: commonalities and differences with other types of addiction. Boil Psychiatry 71(3):192–198

Martinez D, Saccone PA, Liu F, Slifstein M, Orlowska D, Grassetti A, Cook S, Broft A, Van Heertum R, Comer SD (2012) Deficits in dopamine D(2) receptors and presynaptic dopamine in heroin dependence: commonalities and differences with other types of addiction. Biol Psychiatry 71(3):192–198. doi:10.1016/j.biopsych.2011.08.024. Epub 19 Oct 2011

Merenäkk L, Mäestu J, Nordquist N, Parik J, Oreland L, Loit HM, Harro J (2011) Effects of the serotonin transporter (5-HTTLPR) and α2A-adrenoceptor (C-1291G) genotypes on substance use in children and adolescents: a longitudinal study. Psychopharmacology (Berl) 215(1):13–22

Mhatre M, Ticku MK (1994) Chronic ethanol treatment upregulates the GABA receptor beta subunitexpression. Brain Res Mol Brain Res 23(3):246–252

Miller DK, Bowirrat A, Manka M, Miller M, Stokes S, Manka D, Allen C, Gant C, Downs BW, Smolen A, Stevens E, Yeldandi S, Blum K (2010) Acute intravenous synaptamine complex variant KB220™ "normalizes" neurological dysregulation in patients during protracted abstinence from alcohol and opiates as observed using quantitative electroencephalographic and genetic analysis for reward polymorphisms: part 1, pilot study with 2 case reports. Postgrad Med 122(6):188–213

Miller M, Chen AL, Stokes SD, Silverman S, Bowirrat A, Manka M, Manka D, Miller DK, Perrine K, Chen TJ, Bailey JA, Downs W, Waite RL, Madigan MA, Braverman ER, Damle U, Kerner M, Giordano J, Morse S, Oscar-Berman M, Barh D, Blum K (2012) Early intervention of intravenous KB220IV- neuroadaptagen amino-acid therapy (NAAT) improves behavioral outcomes in a residential addiction treatment program: a pilot study. J Psychoactive Drugs 44(5):398–409

Mulert C, Juckel G, Giegling I, Pogarell O, Leicht G, Karch S, Mavrogiorgou P, Möller HJ, Hegerl U, Rujescu D (2006) A Ser9Gly polymorphism in the dopamine D3 receptor gene (DRD3) and event-related P300 potentials. Neuropsychopharmacology 31(6): 1335–1344

Myers RD, Robinson DE (1999) Mmu and D2 receptor antisense oligonucleotides injected in nucleus accumbens suppress high alcohol intake in genetic drinking HEP rats. Alcohol 18(2–3):225–233

Namkoong K, Cheon KA, Kim JW, Jun JY, Lee JY (2008) Association study of dopamine D2, D4 receptor gene, GABAA receptor beta subunit gene, serotonin transporter gene polymorphism with children of alcoholics in Korea: a preliminary study. Alcohol 42(2):77–81

Nedic G, Nikolac M, Sviglin KN, Muck-Seler D, Borovecki F, Pivac N (2011) Association study of a functional catechol-O-methyltransferase (COMT) Val108/158Met polymorphism and suicide attempts in patients with alcohol dependence. Int J Neuropsychopharmacol 4(3):377–388

Nestler EJ (2010) Cell type-specific loss of BDNF signaling mimics optogenetic control of cocaine reward. Science 330(6002):385–390

Neville MJ, Johnstone EC, Walton RT (2004) Identification and characterization of ANKK1: a novel kinase gene closely linked to DRD2 on chromosome band 11q23.1. Hum Mutat 23(6):540–545

Niehaus JL, Cruz-Bermúdez ND, Kauer JA (2009) Plasticity of addiction: a mesolimbic dopamine short-circuit? J Addict 18(4):259–271

Nikulina V, Widom CS, Brzustowicz LM (2012) Child abuse and neglect, MAOA, and mental health outcomes: a prospective examination. Biol Psychiatry 71(4):350–357

Nilsson KW, Comasco E, Åslund C, Nordquist N, Leppert J, Oreland L (2011) MAOA genotype, family relations and sexual abuse in relation to adolescent alcohol consumption. Addict Biol 16(2):347–355

Noble EP, Blum K, Ritchie T, Montgomery A, Sheridan PJ (1991) Allelic association of the D2 dopamine receptor gene with receptor-binding characteristics in alcoholism. Arch Gen Psychiatry 48(7):648–654

Noble EP, Zhang X, Ritchie T, Lawford BR, Grosser SC, Young RM, Sparkes RS (1998) D2 dopamine receptor and GABA(A) receptor beta3 subunit genes and alcoholism. Psychiatry Res 81(2):133–147

Palomo T, Archer T, Kostrzewa RM, Beninger RJ (2007) Comorbidity of substance abuse with other psychiatric disorders. Neurotox Res 12(1):17–27

Paloyelis Y, Asherson P, Mehta MA, Faraone SV, Kuntsi J (2010) DAT1 and COMT effects on delay discounting and trait impulsivity in male adolescents with attention deficit/hyperactivity disorder and healthy controls. Neuropsychopharmacology 35(12):2414–2426

Perez MF, Ford KA, Goussakov I, Stutzmann GE, Hu XT (2011) Repeated cocaine exposure decreases dopamine D-like receptor modulation of Ca(2+) homeostasis in rat nucleus accumbens neurons. Synapse 65(2):168–180. doi:10.1002/syn.20831

Piechota M, Korostynski M, Solecki W, Gieryk A, Slezak M, Bilecki W, Ziolkowska B, Kostrzewa E, Cymerman I, Swiech L, Jaworski J, Przewlocki R (2010) The dissection of transcriptional modules regulated by various drugs of abuse in the mouse striatum. Genome Biol 11(5):R48. doi:10.1186/gb-2010-11-5-r48

Pierce R, Kumaresan V (2006) The mesolimbic dopamine system: the final common pathway for the reinforcing effect of drugs of abuse? Neurosci Biobehav Rev 30(2):215–238

Ray R, Ruparel K, Newberg A, Wileyto EP, Loughead JW, Divgi C, Blendy JA, Logan J, Zubieta JK, Lerman C (2011) Human Mu Opioid Receptor (OPRM1 A118G) polymorphism is associated with brain mu-opioid receptor binding potential in smokers. Proc Natl Acad Sci USA 108(22):9268–9273

Reese J, Kraschewski A, Anghelescu I, Winterer G, Schmidt LG, Gallinat J, Rüschendorf F, Rommelspacher H, Wernicke C (2010) Haplotypes of dopamine and serotonin transporter genes are associated with antisocial personality disorder in alcoholics. Psychiatr Genet 20(4):140–152

Ross J (2001) Amino-acid precursor and enkephalinase inhibition therapy: evidence for effectiveness in treatment of "Reward Deficiency Syndrome (RDS)" with particular emphasis on eating disorders. Mol Psychiatry 6(1 Suppl 1):S1–S8

Rothman RB (1994) A review of the effects of dopaminergic agents in humans: implications for medication development. NIDA Res Monogr 145:67–87

Salomone JD, Correa M (2012) The mysterious motivational functions of mesolimbic dopamine. Neuron 76(3):470–485

Schellekens AF, Franke B, Ellenbroek B, Cools A, de Jong CA, Buitelaar JK, Verkes RJ (2012) Reduced dopamine receptor sensitivity as an intermediate phenotype in alcohol dependence and the role of the COMT Val158Met and DRD2 Taq1A genotypes. Arch Gen Psychiatry 69(4):339–348

Self DW, Nestler EJ (1998) Relapse to drug-seeking: neural and molecular mechanisms. Drug Alcohol Depend 51(1–2):49–60

Smith DE (2012) The process addictions and the new ASAM definition of addiction. J Psychoactive Drugs 44(1):1–4

Spangler R, Wittkowski KM, Goddard NL, Avena NM, Hoebel BG, Leibowitz SF (2004) Opiate-like effects of sugar on gene expression in reward areas of the rat brain. Brain Res Mol Brain Res 124(2):134–142

Stice E, Yokum S, Blum K, Bohon C (2010) Weight gain is associated with reduced striatal response to palatable food. J Neurosci 30(39):13105–13109

Szeto CY, Tang NL, Lee DT, Stadlin A (2001) Association between mu opioid receptor gene polymorphisms and Chinese heroin addicts. Neuroreport 12(6):1103–1106

Teh LK, Izuddin AF, FH MH, Zakaria ZA, SAlleh MZ (2012) Tridimensional personalities and polymorphism of dopamine D2 receptor among heroin addicts. Biol Res Nurs 14(2):188–196

Thanos PK, Volkow ND, Freimuth P, Umegaki H, Ikari H, Roth G, Ingram DK, Hitzemann R (2001) Overexpression of dopamine D2 receptors self-administration. J Neurochem 78(5):1094–1103

Thanos PK, Taintor NB, Rivera SN, Umegaki H, Ikari H, Roth G, Ingram DK, Hitzemann R, Fowler JS, Gatley SJ, Wang GJ, Volkow ND (2004) DRD2 gene transfer into the nucleus accumbens core of the alcohol preferring and nonpreferring rats attenuates alcohol drinking. Alcohol Clin Exp Res 28(5):720–728

Thanos PK, Rivera SN, Weaver K, Grandy DK, Rubinstein M, Umegaki H, Wang GJ, Hitzemann R, Volkow ND (2005) Dopamine D2R DNA transfer in dopamine D2 receptor-deficient mice: effects on ethanol drinking. Life Sci 77(2):130–139

Thanos PK, Michaelides M, Umegaki H, Volkow ND (2008) D2R DNA transfer into the nucleus accumbens attenuates cocaine self-administration in rats. Synapse 62(7):481–486

Tikkanen R, Auvinen-Lintunen L, Ducci F, Sjöberg RL, Goldman D, Tiihonen J, Ojansuu I, Virkkunen M (2011) Psychopathy, PCL-R, and MAOA genotype as predictors of violent reconvictions. Psychiatry Res 185(3):382–386

Tomasi D, Volkow ND (2012) Striatocortical pathway dysfunction in addiction and obesity: differences and similarities. Crit Rev Biochem Mol Biol 48(1):1–19

Treister R, Pud D, Ebstein RP, Laiba E, Gershon E, Haddad M, Eisenberg E (2009) Associations between polymorphisms in dopamine neurotransmitter pathway genes and pain response in healthy humans. Pain 147(1–3):187–193

van der Zwaluw CS, Engels RC, Vermulst AA, Rose RJ, Verkes RJ, Buitelaar J, Franke B, Scholte RH (2010) A serotonin transporter polymorphism (5-HTTLPR) predicts the development of adolescent alcohol use. Drug Alcohol Depend 112(1–2):134–139

Van Tol HH (1998) Structural and functional characteristics of the dopamine D4 receptor. Adv Pharmacol 42:486–489

Vengeliene V, Leonardi-Essmann F, Perreau-Lenz S, Gebicke-Haerter P, Drescher K, Gross G, Spanagel R (2006) The dopamine D3 receptor plays an essential role in alcohol-seeking and relapse. FASEB J 20(13):2223–2233

Wahlstrom D, White T, Luciana M (2010) Neurobehavioral evidence for changes in dopamine system activity during adolescence. Neurosci Biobehav Neurosci Biobehav Rev 34(5):631–648

Wang GJ, Smith L, Volkow ND, Telang F, Logan J, Tomasi D, Wong CT, Hoffman W, Jayne M, Alia-Klein N, Thanos P, Fowler JS (2011) Decreased dopamine activity predicts relapse in methamphetamine abusers. Mol Psychiatry 17(9):918–925. doi:10.1038/mp.2011.86, Epub 2011 Jul 12

Yacubian J, Sommer T, Schroeder K, Gläscher J, Kalisch R, Leuenberger B, Braus DF, Büchel C (2007) Gene-gene interaction associated with neural reward sensitivity. Proc Natl Acad Sci USA 104(19):8125–8130

Young RM, Lawford BR, Feeney GF, Ritchie T, Noble EP (2004) Alcohol-related expectancies are associated with the D2 dopamine receptor and GABAA receptor beta3 subunitgenes. Psychiatry Res 127(3):171–183

Zhang L, Dong Y, Doyon WM, Dani JA (2012) Withdrawal from chronic nicotine exposure alters dopamine sig-

Pharmacogenomics of Analgesics 19

Eugenia Ch. Yiannakopoulou

Abstract

Pharmacogenomics investigates interindividual genetic variability in DNA sequence of drug targets, drug-metabolising enzymes or disease genes, RNA expression or protein translation of genes affecting drug response and drug safety. Variation exists in patient response on analgesic treatment in terms of efficacy and safety. This variation may be in part explained by pharmacogenomics. This chapter overviews data on pharmacogenomics of opioids and nonsteroidal anti-inflammatory agents focusing on the effect of genetic variation on efficacy and safety of analgesic agents. However, there is limited level I evidence on this topic. Most data come from case-control studies and case reports. Basic research should focus on the identification of biologically meaningful polymorphisms enabling a hypothesis with biological plausibility driven research in the field of pharmacogenomics of analgesics.

1 Introduction

There is great variation in the response of individuals to standard doses of drug therapy that can lead to treatment failure or to life-threatening adverse drug reactions among patients with identical doses of the same drug. A drug's activity is the result of the interaction of the drug with proteins involved in absorption, distribution, metabolism, elimination (ADME proteins) and molecular drug targets or target pathways. Genetic variation in these proteins, i.e. single-nucleotide polymorphisms in genes coding for metabolising enzymes or drug transporters, might have a significant influence on the drug effect. Pharmacogenomics investigates interindividual genetic variability in DNA sequence of drug targets, drug-metabolising enzymes or disease genes, RNA expression or protein translation of genes affecting drug response and drug safety (Evans 1999).

Although clinical trials including thousands of patients may generally indicate efficacy and lack of toxicity of drugs, it is well known that it is not possible to predict a priori whether an individual patient will respond to a given medication without adverse effects. Individual variability in drug response can be explained by a number of parameters including

E.Ch. Yiannakopoulou (✉)
Department of Basic Medical Lessons, Faculty of Health and Caring Professions, Technological Educational Institute of Athens, Athens, Greece
e-mail: nyiannak@teiath.gr; egian@med.uoa.gr

D. Barh et al. (eds.), *Omics for Personalized Medicine*,
DOI 10.1007/978-81-322-1184-6_19, © Springer India 2013

age, gender, weight, pharmacokinetics, disease severity, concomitant diseases and environmental factors. Moreover, genetic differences among individuals in drug metabolism and/or cellular drug targets may explain a significant component of this variability (Siva et al. 2002). Especially interindividual variability in drug metabolism is a major cause of adverse drug effects. In many cases, such variability is linked to polymorphisms in genes coding for drug-metabolising enzymes. Individuals carrying enzyme-inactivating mutations display impaired drug metabolism. Thus, carriers of inactivating mutations when treated at standard doses have higher plasma drug concentrations and lower clearance rates, rendering them susceptible to adverse drug reactions (Martínez et al. 2004). There are two types of adverse drug reactions, Type A and Type B. Type A adverse drug reactions are dose dependent, referring to the augmentation of pharmacological action. Polymorphisms of pharmacokinetic genes may be important for narrow therapeutic window drugs with poor metabolisers having increased risk of Type A adverse drug reactions. Polymorphisms of drug targets may be also important for this type of adverse drug reactions. Type B idiosyncratic adverse drug reactions are not predicted by the pharmacological action. Idiosyncratic adverse drug reactions are thought to account for up to 20 % of all adverse drug reactions, although some researchers regard this as an overestimate, with 5 % being closer to reality (Pirmohamed 2010). Genetic variations of pharmacokinetic genes have been implicated in the aetiology of some idiosyncratic adverse drug reactions.

Acetaminophen, aspirin and other nonsteroidal anti-inflammatory agents are among the most commonly prescribed agents in the world. It is well known that interindividual variation exists in patient response on analgesic treatment in terms of efficacy and safety. This variation makes it difficult to rank these drugs by therapeutic efficacy. In clinical practice, if pain relief is not achieved with one agent, commonly the next step is rotation to other nonsteroidal anti-inflammatory agents and in case of failure, switch to narcotic analgesics ensues. Currently, the aetiology of this variability is not known; possible explanations could be disease variability (however it has not been verified in experimental trials), pharmacokinetic variation (although this has not been demonstrated), sex differences and the placebo effect. Pharmacogenomics might account for the differential response in over-the-counter analgesics. Thus, it has been suggested that pharmacogenomics of drug-metabolising enzymes and drug transporters should be incorporated in clinical trials of analgesics.

Table 19.1 Metabolising enzymes and transporters involved in the metabolism of opioids

CYP450 enzymes	Phase II enzymes	Drug transporters
CYP2D6	UGT1A1	ABCB1
CYP3A4	UGT2B7	
CYP3A5		

2 Pharmacogenomics and Narcotic Analgesics

Narcotic analgesics are widely used for the treatment of severe pain, especially cancer pain. Morphine and other mu-opioid agonists are among the most commonly prescribed narcotic analgesics for moderate to severe pain. There is great variation in human response to opioid analgesia. This variation could be explained by genetic variation in metabolising enzymes and transporters mediating opioid pharmacogenomics as well as by genetic variation in receptors and signal transduction elements mediating pharmacodynamics (Table 19.1).

2.1 Pharmacogenomics of Opioid Receptors and Opioid Analgesics

The pharmacologic actions of opioids are mediated through their interaction with the opioid receptors that are G-protein-coupled receptors located in the brain and spinal cord. Three subtypes have been cloned, mu-opioid receptors, kappa-opioid receptors and delta-opioid receptors. Delta receptors are the natural targets of encephalins. The mu-opioid receptor is the primary site of action of opioid analgesics including morphine, fentanyl and methadone. Although most of the currently prescribed opioids are mu-opioid receptor agonists, it has been well demonstrated that they exhibit overlapping affinity with kappa- and delta-opioid receptors.

More thoroughly investigated are polymorphisms of mu-opioid receptor gene that have been implicated in variation in opioid response. More than 100 polymorphisms have been identified in the human mu-opioid peptide receptor gene (OPMR1). A well-studied polymorphism of the mu-opioid receptor, the A118G polymorphism, has been described. Depending on the ethnicity, this polymorphism can be found in 2–48 % of the population. This polymorphism has been associated with both agonistic and antagonistic opioid effects. For example, it has been well demonstrated that the micro-opioid receptor single-nucleotide polymorphism (rs number 1799971) at nucleotide position 118 (OPRM1 118A>G), resulting in the substitution of one amino acid, has been associated with decrease in the analgesic effect of opioids. Data from clinical trials suggest that the A118G polymorphism has clinical implication in pain treatment. A cohort study of patients submitted to knee arthroplasty surgery and treated for postoperative pain with patient-controlled analgesia was genotyped for the A118G polymorphism in OPMR1. The study demonstrated that homozygous GG patients required higher doses of morphine during the first 24 h. This effect of A118G variant was also verified in a cohort of patients submitted to abdominal surgery. Another study of postoperative pain has shown that patients with the AA genotype need lower doses of morphine but have higher incidence of postoperative nausea and vomit. This finding indicates that OPMR1 pharmacogenomics might have relevance for opioid side effects. Other studies performed in cancer patients have also indicated the relevance of A118G polymorphism of OPMR1 in opioid efficacy in pain management. Available data imply that genotyping for this polymorphism might have clinical implications in pain management. However, this is a suggestion that needs to be further investigated. Defining OPMR1 haplotypes might be more clinically relevant as has been suggested by studies of postoperative pain after major abdominal surgery.

The A118G polymorphism has also been associated with substance dependence and susceptibility to other disorders such as epilepsy and schizophrenia. In addition, other common mu-opioid receptor polymorphisms have been associated with reduced pain relief to alfentanil or morphine. In addition, polymorphisms of the mu-opioid receptor have been identified in the third intracellular loop of the receptor that affect receptor signalling, but it is not known if these polymorphisms affect treatment efficacy.

Furthermore, variation in opioid sensitivity could be explained by polymorphisms in other molecules, such as endogenous opioid peptides, metabolising enzymes and second messenger systems.

2.2 Pharmacogenomics of Phase I Drug-Metabolising Enzymes and Opioid Analgesics

2.2.1 Codeine

The analgesic opioid drug codeine (3-methoxymorphine) is metabolised through several metabolic routes including *O*-demethylation into morphine through CYP2D6 and N-demethylation into norcodeine through CYP3A4. Norcodeine is not known to have any analgesic activity. Glucuronidation is also involved in the metabolism of codeine. Morphine is further glucuronidated to morphine-3-glucuronide, a metabolite not considered to possess analgesic properties and to morphine-6-glucuronide, known to have analgesic activity in humans.

Codeine is an important CYP2D6 substrate. It is activated to morphine exclusively by CYP2D6. Codeine has only mild opioid properties, while most of its analgesia and central nervous system depressant effects are based on its biotransformation to morphine, a reaction catalysed by CYP2D6 (Meyer 2000). Genetic polymorphism of this enzyme results in three separable phenotypes: poor metabolisers, extensive metabolisers and ultrarapid metabolisers. Poor metabolisers are homozygous for an inactive or deficient

CYP2D6 enzyme caused by mutations in the CYP2D6 gene, while ultrarapid metaboliser subjects have duplication of the gene, resulting in increased enzymatic activity (Weber 1999). Poor metabolisers have decreased activation of CYP2D6-dependent analgesic prodrugs such as codeine (Wilcox and Owen 2000). It is estimated that 7–10 % of the population does not express functional CYP2D6. According to a randomised, placebo-controlled, double-blind clinical trial that investigated the effect of CYP2D6 polymorphisms on codeine analgesia using an experimental pain model, codeine administration resulted in analgesia in extensive metabolisers but had no effect in poor metaboliser patients (Eckhardt et al. 1998). In addition, an equal incidence of side effects was observed in poor metabolisers in comparison with extensive metabolisers, although they did not receive any analgesic benefit (Eckhardt et al. 1998). In a small study of 11 patients treated with codeine for analgesia after hysterectomy, two patients had no analgesic effect from the codeine, one of whom was subsequently shown to be a poor metaboliser (Persson et al. 1995). However, other studies have shown no difference in the effect of codeine on postoperative pain intensity between extensive and poor metabolisers in both adults and children (Gaedigk et al. 2002; Poulsen et al. 1998).

CYP2D6 ultrarapid metabolisers (in an estimated 1 % of people in Finland and Denmark, 10 % of people in Greece and Portugal and 29 % of people in Ethiopia) have the potential for increased production of morphine from codeine and thus might be at greater risk for opioid-related adverse events and might benefit from a lower dose of opioids (Williams et al. 2002). The most common adverse reactions to codeine include drowsiness, light-headedness, dizziness, sedation, shortness of breath, nausea, vomiting and sweating. Serious adverse reactions include respiratory depression, circulatory depression, respiratory arrest, shock and cardiac arrest. Excessive activation of codeine in ultrarapid metabolisers with one additional copy of CYP2D6 has been reported in case reports (Table 19.2). The first described a patient prescribed a cough medicine containing codeine for bilateral pneumonia, who suffered from a life-threatening opioid intoxication. Upon genotyping, it was shown that the patient had at least three copies of CYP2D6, a finding consistent with ultrarapid metabolism of codeine (Gasche et al. 2004). The patient was also treated with a macrolide and an azole derivative, medicines that, being inhibitors of CYP3A4, might have further reduced the clearance of codeine and increased the risk of an opioid overdose associated with the CYP2D6 gene duplication. In addition, the patient was suffering from renal insufficiency that rendered him prone to the accumulation of codeine metabolites.

Another case report concerns the death of a 2-year-old boy that developed fever, wheezing and finally respiratory arrest 2 days after elective tonsillectomy. The boy had been prescribed codeine and acetaminophen for analgesia at recommended doses, i.e. 10–12.5 mg of codeine and 120 mg of acetaminophen orally every 4–6 h as needed. Codeine and morphine were detected at the peripheral blood, and CYP2D6 genotyping revealed functional duplication of the CYP2D6 allele, resulting in the ultrarapid metaboliser phenotype (Ciszkowski et al. 2009). In addition, the child had a history of recurrent episodes of hypoxemia due to bronchopneumonia that could possibly have led to alterations in the μ-opioid receptor and increased sensitivity to morphine. Voronov et al. 2007 have reported another case report of a 29-month-old previously healthy child who experienced apnoea resulting in brain injury following a dose of acetaminophen and codeine 2 days after tonsillectomy. A genetic polymorphism leading to ultrarapid metabolism of codeine into morphine was again confirmed in this child.

Another case report concerned the death of a breastfed baby 13 days after birth. His mother was prescribed codeine as an analgesic postdelivery. Post-mortem examination of stored breast milk samples showed morphine levels 4 times higher than expected. Upon genotyping, the mother was found to be heterozygous for a CYP2D6*2A allele and a CYP2D6*2X2 gene duplication. Thus, the mother had three functional CYP2D6 alleles and was classified as an ultrarapid metaboliser (Madadi et al. 2007). The extra CYP2D6 enzyme resulted in increased *O*-demethylation of

Table 19.2 Evidence on the impact of polymorphisms of metabolising enzymes on the safety of codeine

Type of evidence	Drug indication	Metabolising enzyme	Polymorphism	Adverse event	References
Case report	Codeine contained in a cough medicine	CYP2D6	CYP2D6*1x3, in a patient suffering from renal insufficiency and co-treated with CYP3A4 inhibitors	Life-threatening intoxication	Gasche et al. (2004)
Case report	Codeine for analgesia after tonsillectomy	CYP2D6	CYP2D6x2	Apnoea and brain injury	Voronov et al. (2007)
Case report	Codeine prescribed in mother as an analgesic postdelivery	CYP2D6	CYP2D6*2A and CYP2D6*2X2	Death of the breastfed 13 day boy	Madadi et al. (2007)
Case report	Codeine for analgesia after tonsillectomy	CYP2D6	CYP2D6*1xN	Death due to respiratory arrest	Ciszkowski et al. (2009)
Case report	Codeine for analgesia after tonsillectomy	CYP2D6	CYP2D6*1xN	Two deaths, one case of severe respiratory depression	Kelly et al. (2012)

codeine to morphine, and consequently, very high concentrations of morphine were found in both the breast milk and in the blood from the child. Following this fatal case, a case-control study investigated the characteristics of mothers and infants with or without signs and symptoms of central nervous system depression following codeine exposure during breast feeding. According to the study, breastfed infants of mothers who are CYP2D6 ultrarapid metabolisers (UMs) combined with the UGT2B7*2/*2 are at increased risk of potentially life-threatening CNS depression (Madadi et al. 2009).

Since 2007, the FDA requires the manufacturers of prescription codeine products to include information in the "Precautions" section of the label to inform prescribing doctors about these risks and to help prevent morphine overdose in breastfed infants.

In a very recent publication, Kelly et al. 2012 reported three cases of fatal or life-threatening events associated with codeine. In the 2 fatal cases, functional gene duplications encoding for CYP2D6 caused a significantly greater production of potent morphine from its parent drug, codeine, while a severe case of respiratory depression occurred in an extensive metaboliser. A very recent publication issued guidelines for codeine therapy in the context of CYP2D6 genotype (Crews et al. 2012). According to this publication, alternative analgesics should be prescribed in patients with CYP2D6 poor metaboliser phenotype or ultrarapid metaboliser phenotype. Opioids not metabolised by CYP2D6, including morphine, oxymorphone, buprenorphine, fentanyl, methadone and hydromorphone combined with non-opioid analgesics, have been proposed as alternative medications instead of codeine for patients prone to codeine toxicity or for those expected to be unresponsive in codeine treatment (Crews et al. 2012).

2.2.2 Oxycodone

Oxycodone is a semisynthetic opioid prescribed for moderate to severe pain. Oxycodone is similar in structure to codeine, but in contrast, it has an analgesic potency in humans similar to morphine. Both the parent compound and the metabolites of oxycodone are equally active in the opioid receptor. The principal metabolic pathway of oxycodone is the CYP3A accounting for approximately 50 % of the dose. Furthermore, the drug undergoes metabolism to oxymorphone via CYP2D6. The effect of polymorphisms of CYP enzymes on oxycodone safety and efficacy has been investigated, but results are discordant (Table 19.3) (Samer et al. 2010; Lemberg et al. 2010; Andreassen et al 2012).

Table 19.3 Evidence on the impact of polymorphisms of metabolising enzymes on the safety of hydrocodone and oxycodone

Type of evidence	Drug	Metabolising enzyme	Polymorphisms	Adverse event	References
Case report	Hydrocodone as antitussive agent	CYP2D6	CYP2D6*41	Death	Madadi et al. (2010)
Cross-over double-blind randomised controlled trial in healthy volunteers	Oxycodone	CYP2D6	32 alleles using the AmpliChip TM CYP450 DNA microarray	Volunteers who were CYP2D6 ultrarapid metabolisers experienced higher toxicity especially after CYP3A blockade with ketoconazole	Samer et al. (2010)
Cross-over randomised double-blind placebo controlled trial	Oxycodone	CYP2D6, CYP3A4, CYP3A5		No effect of genotype in patients treated for malignant and non-malignant chronic pain	Lemberg et al. (2010)
Cross-sectional study	Oxycodone	CYP2D6	CYP2D6*2x2	Genotype did not affect safety profile, i.e. the incidence of nausea, tiredness or cognitive failure in a cohort of cancer patients	Andreassen et al. (2012)

2.2.3 Hydrocodone

Hydrocodone is a hemisynthetic oral opioid structurally related to codeine. It is approximately 12 times more potent in the mu-opioid receptor than codeine. Hydrocodone is metabolised by oxidative metabolism via cytochrome P450 2D6 (CYP2D6) and cytochrome P450 3A4 (CYP3A4) into hydromorphone and norhydrocodone, respectively. These metabolites undergo conjugation by uridinediphosphateglucuronosyltransferases (UGTs) before excretion. Both the parent compounds and the metabolites of hydrocodone are equally active in opioid receptors. An effect of CYP2D6 polymorphism on hydrocodone metabolite production has been demonstrated. However, evidence suggests that analgesic efficacy of hydrocodone is not affected by metabolism via CYP2D6. Effect of CYP2D6 polymorphisms on hydrocodone toxicity has also been reported (Table 19.3). Madadi et al. (2010) reported a case of fatal opioid toxicity that occurred in a developmentally delayed child aged 5 years 9 months who was inadvertently administered high doses of hydrocodone for a respiratory tract infection. Genotyping was performed for *CYP2D6*4*, *CYP2D6*9*, *CYP2D6*10* and *CYP2D6*41* by using predesigned TaqMan allelic discrimination assays. Furthermore, the following assays were applied: (1) *CYP2D6* copy number attributable to the *CYP2D6*5* deletion allele, (2) gene duplication and (3) gene multiduplication. The analysis demonstrated that the child carried the allele *CYP2D6*41* allele and 1 fully functional wild-type allele. Therefore, the child genotype was *CYP2D6*2A/*41*, and the child was designated as a poor CYP2D6 metaboliser. Drug-drug interactions contributed to this fatal event, since the child was prescribed clarithromycin (a CYP3A4 inhibitor) for an ear infection and valproic acid, an inhibitor of UGT enzymes for seizures, since birth.

2.2.4 Tramadol

Tramadol is a synthetic, centrally acting analgesic for the treatment of moderate to severe pain. This analgesic is a racemic mixture containing 50 % (+) tramadol and 50 % (−) tramadol. Tramadol produces analgesia by the synergistic action of its two enantiomers and their metabolites. Its affinity for opioid receptors is 6,000 times lower than that of morphine. Tramadol is an example of analgesic drug that is metabolised by CYP2D6 to generate a pharmacologically

Table 19.4 Evidence on the impact of polymorphisms of phase I metabolising enzymes on the analgesic efficacy of tramadol

Type of evidence	Enzymes	Polymorphisms	Effects	References
Prospective study	CYP2D6	CYP2D6*1, CYP2D6*3, CYP2D6*4, CYP2D6*5, CYP2D6*9, CYP2D6*10, CYP2D6*17, CYP2D6*1xN	Poor metabolisers for CYP2D6 showed a lower response rate to tramadol analgesia for postoperative pain after abdominal surgery	Stamer et al. (2003)
Prospective study	CYP2D6	CYP2D6*10 C188T	This SNP reduced tramadol analgesic efficacy in a Chinese patient treated for postoperative pain after major abdominal surgery	Wang et al. (2006)
Prospective study in Malaysian patients	CYP2D6	CYP2D6*1, CYP2D6*3, CYP2D6*4, CYP2D6*5, CYP2D6*9, CYP2D6*10, CYP2D6*17, CYP2D6*1xN	The analgesic effects of tramadol were not measured adequately. Therefore, the effect of genotype on tramadol analgesic efficacy was not demonstrated	Gan et al. (2007)
Prospective study	CYP2D6	CYP2D6*3, CYP2D6*4, CYP2D6*5, CYP2D6*6, CYP2D6*1xN	CYP2D6 genotype did not affect tramadol analgesic efficacy in postoperative patients after knee arthroplasty	Slanar et al. (2012)

active product, the analgesic opioid receptor agonist *O*-desmethyltramadol (Galley et al. 2005). Genetic variations in *CYP2D6* have been shown to account for some of the variable pain response in the postoperative period because the *CYP2D6* activity has a clinically relevant impact on the level of analgesia mediated by the μ-opioid receptor (Table 19.4) (Stamer et al. 2003; Gan et al. 2007; Slanar et al. 2012). The effect of CYP2D6*10 polymorphism has been investigated in Chinese patients recovering from gastrectomy performed for gastric cancer (Wang et al. 2006). Patients were treated with self-administered tramadol via patient-controlled analgesia. Patients homozygous for CYP2D6*10 displayed higher consumption of tramadol in comparison with heterozygous group and with patient group without CYP2D6*10. Furthermore, tramadol is metabolised to N-demethyl tramadol by N-demethylation via CYP3A.

In addition, genetic variations in CYP2D6 affect the safety profile of tramadol (Table 19.5). Thus, case reports have indicated that the carriers of gene duplications, being CYP2D6 ultrarapid metabolisers (UM), are at high risk for toxic responses to tramadol treatment (Kirchheiner et al. 2008). A case report described a man with renal insufficiency and CYP2D6UM genotype who developed postoperative opioid-related respiratory insufficiency under intravenous patient-controlled tramadol analgesia (Stamer et al. 2008). Genotyping was performed by polymerase chain reaction and real-time polymerase chain reaction for the single-nucleotide polymorphisms *CYP2D6*3*, *CYP2D6*4*, *CYP2D6*5*, *CYP2D6*6*, *CYP2D6*7* and *CYP2D6*8* associated with poor metaboliser phenotype and for the SNP *CYP2D6*10* and *CYP2D6*41* associated with the intermediate metaboliser phenotype. The result of these tests was negative, while the duplication/multiduplication assay revealed a CYP2D6 gene duplication resulting in ultrarapid metabolism of tramadol to its active metabolite (+) *O*-desmethyltramadol. The opioid toxicity was enhanced by the decreased metabolic clearance due to renal insufficiency.

In another publication, Elkalioubie et al. (2011) have reported a near-fatal case of tramadol-induced cardiotoxicity in an ultrarapid CYP2D6 metaboliser. A 22-year-old girl was transferred to the ICU with cardiac arrest. The aetiology was suggested by the high concentrations of

Table 19.5 Evidence on the impact of polymorphisms of metabolising enzymes on the safety of tramadol

Type of evidence	Drug indications	Metabolising enzymes	Polymorphisms	Adverse events	References
Prospective study	Tramadol	CYP2D6	CYP2D6*1, *3, *4, *5, *9, *10, CYP2D6*17, CYP2D6*1xN	Intermediate metabolisers experienced more adverse events than extensive metabolisers and extensive metabolisers experienced more adverse effects than ultrarapid metabolisers	Gan et al. (2007)
Case report	Tramadol administered postoperatively as patient-controlled analgesia	CYP2D6	CYP2D6*1XN	Respiratory insufficiency in a patient suffering from renal impairment	Stamer et al. (2008)
Pharmacokinetic-pharmacodynamic study in the setting of experimental pain	Tramadol	CYP2D6	CYP2D6*1xN	Ultrarapid metabolisers experienced more adverse events than extensive metabolisers	Kirchheiner et al. (2008)
Case report	Tramadol as abuse	CYP2D6	CYP2D6*1xN	Near-fatal cardiac toxicity	Elkalioubie et al. (2011)

both tramadol and its main metabolite *O*-desmethyltramadol identified in the blood tests. Genotyping was performed, and the patient was found to be heterozygous for a wild-type allele duplication. The patient admitted occasional abuse of tramadol. Cardiotoxicity could be explained by high levels of norepinephrine due to tramadol-induced inhibition of norepinephrine reuptake.

Motivated by the abovementioned reports, Kirchheiner et al. (2008) systematically investigated the effect of CYP2D6 duplication on the pharmacodynamics and pharmacokinetics of tramadol after a single dose of 100 mg racemic tramadol in 11 carriers of a CYP2D6 gene duplication and compared it with 11 carriers of 2 active CYP2D6 genes. In addition, Gan et al. (2007) investigated the effect of CYP2D6 polymorphisms on tramadol pharmacokinetics and pharmacodynamics in Malaysian patients with different genotypes. Both studies demonstrated pharmacokinetic differences between the genotype groups as well as differences in the safety profile of tramadol. According to Gal et al., intermediate metabolisers experienced more adverse effects than extensive metabolisers, and extensive metabolisers experienced more adverse effects than ultrarapid metabolisers. However, in the study of Kirchheimer et al., ultrarapid metabolisers experienced more adverse events than extensive metabolisers. The discrepancies may be due to differences in the method of genotype determination, in the different polymorphisms investigated and in the type of assessed adverse events. Given that these results have not been verified in a large population study, at present, CYP2D6 genotyping cannot be recommended for tramadol dosing.

2.2.5 Methadone

Methadone metabolism is attributed primarily to cytochrome P450 enzymes CYP3A4, CYP2B6 and CYP2D6 (Wong et al. 2003; Levran et al. 2011). Interindividual differences in sensitivity to methadone have been observed. Fatal poisonings occur typically at concentrations between 0.4 and 1.8 mg/ml. However, in susceptible individuals, death may occur at much lower concentrations. In a study, 40 post-mortem cases have been reported, in which methadone had been implicated in the cause of death. The CYP2B6*6 allele that results in slow metabolism was associated with the highest post-mortem methadone concentration (Hamman et al. 1997).

2.3 Pharmacogenomics of Phase II Drug-Metabolising Enzymes and Opioid Analgesics

Phase II enzymes conjugate phase I metabolites, other intermediates or the parent compound for renal or biliary excretion. Phase II enzymes include glutathione S-transferases (GSTs), UDP glycurosyltransferases(UGT),N-acetyltransferases (NAT), NADH quinone oxidases and others. Knowledge on pharmacogenomics of phase II enzymes is not extensive. Thus, the relevance of the pharmacogenomics of the enzymes of the uridinediphosphateglucuronosyltransferase (UGT) superfamily in opioid disposition has been investigated, although it is not currently elucidated (Table 19.6) (Coffman et al. 1998; Holthe et al. 2002; Darbari et al. 2008; Rouguieg et al. 2010). Several polymorphisms of UGT1A1 have been described with the best studied, the one defined by a variable length "TA" tandem repeat in the regulatory TATA box of the UGT1A1 gene promoter (UGT1A1*28) that leads to reduced expression of the isozyme associated with Gilbert syndrome, the most common inherited cause of unconjugated hyperbilirubinemia. UGT1A1 catalyses the glucuronidation of opioids including morphine, buprenorphine and norbuprenorphine, hemisynthetic derivatives of the morphine alkaloid thebaine with partial agonist properties on opioid receptors. The effect of UGT1A1*28 polymorphism in the pharmacokinetics of morphine has been investigated in cancer patients, but no association has been demonstrated. Another enzyme involved in opioid metabolism is UGT2B7. The association of polymorphisms of this enzyme with morphine metabolism has been investigated. It has been shown that the UGT2B7 promoter variant −840G is associated with reduced glucuronidation of morphine, thus contributing in variability

Table 19.6 Evidence on the impact of polymorphisms of phase II metabolising enzymes on pharmacokinetics and pharmacodynamics of narcotic analgesics

Type of evidence	Drugs	Phase II enzymes	Polymorphisms	Effects	References
Experimental trial—in vitro model	Buprenor-phine	UGT2B7	rs7439366	The SNP alters UGT2B7 activity for buprenorphine as demonstrated in human kidney embryonic transfected cells	Coffman et al. (1998)
Pharmacokinetic study in cancer patients on chronic morphine therapy	Morphine	UGT2B7	UGT2B7	The investigated SNPs did not affect morphine glucuronide-to-morphine plasma rations	Holthe et al. (2002)
		UGT1A1	H268Y		
			UGT1A1*28		
Pharmacokinetic study	Morphine	UGT2B7	UGT2B7 −840G	This SNP was associated with reduced glucuronidation of morphine contributing thus in variability in hepatic clearance of morphine in patients with sickle cell anaemia	Darbari et al. (2008)
Experimental trial—in vitro model	Buprenorphine-norbupre-norphine	UGT1A1	UGT1A1*28	UGT1A1*28 resulted in a decrease of buprenorphine glucuronidation in pooled human liver microsomes without affecting glucuronidation rate in individual microsomes	Rouguieg et al. (2010)

in hepatic clearance of morphine in patients with sickle cell anaemia. Genetic modelling approaches integrating prior physiologic knowledge have shown that decreased UGT2B7 activity is associated with a decrease in active opioid exposure. The C802T exonic SNP of UGT2B7 has been shown to alter UGT2B7 affinity for buprenorphine.

2.4 Pharmacogenomics of Drug Transporters and Opioid Analgesics

Transport of drugs across cell membranes is determined by molecular weight, lipid solubility, ionisation and protein binding. Opioid analgesics are hydrophilic agents, and thus, the drug transporters mediate their entrance through cell membranes. Multidrug resistance proteins (MDR), multidrug resistance-associated proteins (MRP) and organic anion-transporting polypeptides (OATPs) contribute to opioid analgesic transport through cell membranes. Polymorphisms in these transporter genes may account for variation in analgesic efficacy of opioids. MDRs are ATP-dependent efflux pumps; *P*-glycoprotein is encoded by the multidrug resistance gene MDR1 (ABCB1). The clinical relevance of ABCB1 pharmacogenomics in the case of opioids has not been elucidated. Evidence suggests that the variations in the *P*-glycoprotein gene and ABCB1 polymorphisms influence opioid pharmacodynamics and dosage requirements. However, there are also studies that failed to demonstrate an effect of ABCB1 polymorphisms on morphine pharmacodynamics or pharmacokinetics. Fentanyl may be a substrate of ABCB1, and it has been shown that ABCB1 polymorphisms conferring decreased transporter function have been associated with increased respiratory depressive effects of fentanyl in Korean patients. A cohort study of patients submitted to colorectal surgery demonstrated that two SNPs of ABCB1, C3435T (in exon 24) and G2677T/A (in exon 21), were associated with opioid-induced side effects. In particular, homozygous patients (GG and CC, respectively) required less often ondansetron treatment, suggesting that the diplotype GG and CC was associated with fewer side effects of morphine. However, a very recent study in Turkish patients failed to demonstrate an effect of ABCB1 polymorphisms on respiratory depression caused by intravenous fentanyl.

Further research is needed in order to elucidate the clinical relevance of MRP and OATP polymorphisms in the context of opioid analgesics efficacy and safety.

2.5 Opioid-Induced Hyperalgesia and Pharmacogenomics

More and more scientific evidence accumulates on the opioid-induced analgesia, i.e. the pain enhancement induced by opioids. Opioid-induced hyperalgesia is defined as the state of increased sensitivity to pain or a decreased pain threshold in response to opioid therapy. The type of pain experienced may be the same as the underlying pain or may be different than the underlying pain. The mechanisms of opioid-induced analgesia are not elucidated. Proposed mechanisms include central glutaminergic system, spinal dynorphins, descending facilitation, decreased reuptake and enhanced nociceptive response and genetic mechanisms. Experimental trials performed in murine models have investigated the presence of genetic loci linked to opioid-induced hyperalgesia. Catechol-*O*-methyltransferase (COMT) is the most thoroughly investigated gene in the field of pain sensitivity modulation. COMT is an enzyme that inactivates biologically active catechols, including the neurotransmitters dopamine, noradrenaline and adrenaline. These neurotransmitters modulate multiple physiological processes including pain sensation. A relative polymorphism of catecholamine breakdown enzyme catechol-*O*-methyltransferase (COMT) has been described. This polymorphism leads to substitution of the amino acid valine for methionine (val(158)met). This polymorphism affects the breakdown of dopamine and noradrenaline being four times higher with valine allele in comparison with methionine. Thus, the polymorphism leads to reduced levels of dopamine/noradrenaline following noradrenaline release and has been associated with modulation of pain sensitivity. Three genotypes of this polymorphism have been described with the individuals that are homozygous for the met(158) allele having increased pain sensitivity. The effect of this polymorphism has been investigated in a cohort study of cancer patients, and it has been shown that homozygous val/val individuals required higher doses of morphine than heterozygous val/met or homozygous individuals met/met. Other investigators, in a study of healthy volunteers, combining the val(158)met polymorphism with three other SNPs located in the COMT promoter, have identified three haplotypes representing approximately 95 % of the population. These haplotypes conferred variant pain sensitivity and were characterised as low pain sensitivity, medium pain sensitivity and high pain sensitivity. The five most common combinations of these haplotypes were associated with variant sensitivity to experimental pain. The clinical implications of this finding have not further been elucidated. However, critical review of this finding suggests that investigating the effect of haplotypes might be more clinically meaningful than investigating the effect of single polymorphisms. The relevance of COMT SNPs on opioid-induced hyperalgesia is not proved but deserves further investigation.

2.6 Pharmacogenomics of Alternative Analgesics

Morphine has well-known side effects, and often its efficacy is limited at well-tolerated doses. It is known that a significant proportion of patients are only partially responsive at well-tolerated doses of morphine and other opiates. This has prompted research for alternative analgesics such as selective kappa receptor agonists, selective delta receptor agonists, a-adrenergic agonists, cannabinoids and nicotinic receptor agonists. Kappa-opioid receptor agonists are not promising analgesic drug targets, as both clinical studies and data from animal models have shown that such agents cause a psychotomimetic effect and dysphoria, water diuresis, salivation and emesis. Currently, peripherally acting selective kappa agonists are under development due to their lack of central side effects. Up to now, none of these agents has been approved as analgesic. Nalbuphine, pentazocine and butorphanol are analgesics available today with mixed kappa- and mu-opioid receptor activity. Delta receptor agonists are also not promising agents, as they have been demonstrated to cause convulsions in rodents.

Variable analgesic efficacy has also been observed with alternative analgesics, and thus, the pharmacogenetics of these alternative analgesics has also clinical implications. Due to pharmacologic interest of kappa- and delta-opioid receptors, polymorphisms of these receptors are of scientific interest. A total of 12 SNPs have been described in the kappa-opioid receptor human gene human OPRK1. Furthermore, 12 SNPs have been described for the human delta receptor OPRD1. However, the possible clinical relevance of these polymorphisms in the context of analgesia has not been elucidated.

3 Pharmacogenomics and Nonsteroidal Anti-inflammatory Agents

3.1 SNPs and Nonsteroidal Anti-inflammatory Agent Efficacy

Nonsteroidal anti-inflammatory agents (NSAIDs) are some of the most widely prescribed over-the-counter medications. At least 16 different registered nonsteroidal anti-inflammatory drugs are at least partially metabolised by CYP2C9. These include aceclofenac, acetylsalicylic acid, azapropazone, celecoxib, diclofenac, flurbiprofen, ibuprofen, indomethacin, lornoxicam, mefenamic acid, meloxicam, naproxen, phenylbutazone, piroxicam and tenoxicam (Hamman et al. 1997; Duggan et al. 1972; Zarza 2003; Brenner et al. 2003; Kirchheiner et al. 2003, 2004; Yasar et al. 2001; Miners et al. 1996; Zhao et al. 1992; Hutzler et al. 2002; Martínez et al. 2006; Kirchheiner and Seeringer 2007). Significant intergenotypic differences have been reported in the pharmacokinetics of celecoxib, flurbiprofen, ibuprofen and tenoxicam, which could be translated into dose recommendations based upon CYP2C9 genotype (Kirchheiner and Seeringer 2007). Celecoxib has been one of the first drugs for which the manufacturer's drug information recommends caution when administering celecoxib to poor metabolisers of CYP2C9 substrates, as they could have abnormally low levels (Kirchheiner and Seeringer 2007). However, the relative contribution of CYP2C9 on the pharmacokinetics of diclofenac has been found to be independent from CYP2C9 polymorphisms in several studies (Kirchheiner and Seeringer 2007). In addition, *CYP2C8* polymorphisms may influence interindividual variability in the pharmacokinetics of some NSAIDs, namely, ibuprofen and diclofenac (Tang 2003; Kumar et al. 2002; Martínez et al. 2005; García-Martín et al. 2004). In the case of ibuprofen, it has been shown that polymorphisms in cytochrome CYP2C8 significantly change the clearing capacity of individuals who are heterogenous or homogenous mutants of this metabolising enzyme.

3.2 SNPs and Nonsteroidal Anti-inflammatory Drug Safety

As already reported, several nonsteroidal anti-inflammatory drugs (NSAIDs) are metabolised by the cytochrome P450 2C9 (CYP2C9). Two common variants of the *CYP2C9* gene (*CYP2C9*2* and **3*) have been reported to significantly affect the activity of the CYP2C9 enzyme, and the presence of these CYP2C9 variant alleles has been associated with acute gastrointestinal haemorrhage due to NSAIDs (Table 19.7). A relevant case-control study included 26 patients with endoscopically documented NSAID-related gastroduodenal bleeding lesions and 52 controls with no lesions at endoscopy (Pilotto et al. 2007). The participants of both groups were *Helicobacter pylori* negative and acute users of a NSAID or cycloxygenase-2 inhibitor that undergoes CYP2C9 metabolism (i.e. celecoxib, diclofenac, ibuprofen, naproxen, or piroxicam). Two marker single-nucleotide polymorphisms in the *CYP2C9* gene, identifying the *CYP2C9 *2* and **3* allele, were

Table 19.7 Evidence on the impact of polymorphisms of metabolising enzymes on NSAIDs-induced gastrointestinal bleeding

Type of evidence	Drugs	Metabolising enzymes	Polymorphisms	Effects	References
Case report	Indomethacin	CYP2C9	CYP2C9*3	Acute gastrointestinal bleeding in a 71-year-old patient under long-term treatment with acenocoumarol, after treatment with indomethacin	Zarza (2003)
Case-control study	Aceclofenac, celecoxib, diclofenac, ibuprofen, indomethacin, lornoxicam, piroxicam and naproxen	CYP2C9	CYP2C9*2 CYP2C9*3	Carriers of CYP2C9 variant alleles were more prone to develop acute gastrointestinal bleeding when they received NSAIDs that were CYP2C9 substrates	Martínez et al. (2004)
Case-control study	Celecoxib, diclofenac, ibuprofen, naproxen, or piroxicam	CYP2C9	CYP2C9*2 CYP2C9*3	Significantly higher frequencies of CYP2C9*1/*3 and CYP2C9*1/*2 were identified in bleeding versus control patients	Pilotto et al. (2007)
Cross-sectional study	NSAIDs	CYP2C9 CYP2C8	CYP2C9*2 CYP2C9*3	The frequencies of the *CYP2C8*3* and *CYP2C9*2* alleles were higher in NSAID users who experienced a bleed versus those that did not experience a bleed	Blanco et al. (2008)

evaluated. Significantly higher frequencies of CYP2C9*1/*3 (34.6 % vs 5.8 %; $P<.001$; odds ratio [OR], 12.9; 95 % confidence interval [CI], 2.917–57.922) and CYP2C9*1/*2 (26.9 % vs 15.4 %; $P=.036$; OR, 3.8; 95 % CI, 1.090–13.190) were identified in bleeding versus control patients (Pilotto et al. 2007). In 2003, acute gastrointestinal bleeding was reported in a 71-year-old patient under long-term treatment with acenocoumarol, after treatment with indomethacin. Upon genotyping, the patient was found homozygous for the CYP2C9*3 allele, and the authors suggested that the interaction of genetically impaired metabolism of indomethacin and acenocoumarol, as well as a putative interaction in the metabolism of both CYP2C9 substrates, was the cause of bleeding (Zarza 2003). Following this case report, the first case-control study that investigated the effect of CYP2C9 polymorphisms in NSAID-induced gastrointestinal bleeding was published in 2004. This study demonstrated that the carriers of CYP2C9 variant alleles were more prone to develop acute gastrointestinal bleeding when they received NSAIDs that were CYP2C9 substrates (Martínez et al. 2004).

Apart from CYP2C9, a role for CYP2C8 polymorphisms on NSAIDs-induced gastrointestinal haemorrhage has been implicated. *CYP2C9*2* is in partial linkage disequilibrium with *CYP2C8*3*, and many NSAIDs are substrates for both enzymes. A cross-sectional study of NSAID users ($n=134$ bleeding cases, $n=177$ nonbleeding controls) investigated whether *CYP2C9* and *CYP2C8* polymorphisms were associated with gastrointestinal bleeding. The frequencies of the *CYP2C8*3* and *CYP2C9*2* alleles were higher in NSAID users who experienced a bleed versus those that did not experience a bleed (*CYP2C8*3*, odds ratio 3.4, $P<0.002$; *CYP2C9*2*, odds ratio 2.7, $P=0.013$). Further analysis of the data revealed that the highest bleeding risk was in patients who possessed both the variant *CYP2C8*3* and *CYP2C9*2* alleles. The authors hypothesised that *CYP2C8*3* and

Table 19.8 Evidence on the impact of polymorphisms of metabolising enzymes on diclofenac-induced hepatotoxicity

Enzymes	Effects	References
CYP2C9	The *CYP2C9* genotype was not associated with the risk of diclofenac-induced hepatitis	Aithal et al. (2000)
CYP2C8	*CYP2C8* polymorphisms could not predict the risk of hepatotoxicity	Daly et al. (2007)
UGT2B7	Associated with diclofenac hepatotoxicity	Daly et al. (2007)
ABCC2	Associated with diclofenac hepatotoxicity	Daly et al. (2007)

*CYP2C9*2* alleles conferred an increased risk of gastrointestinal bleeding due to decreased metabolic clearance and increased plasma concentrations of NSAIDs (Blanco et al. 2008).

Diclofenac has been associated with the rare but serious hepatotoxicity (de Abajo et al. 2004) (Table 19.8). Although the mechanism of diclofenac-induced hepatotoxicity is not elucidated, it is hypothesised that the diclofenac adducts are formed through either 5-hydroxydiclofenac metabolic pathway or the UGT2B7 glucuronidation pathway. These adducts seem to play a role in the pathogenesis of diclofenac-induced hepatotoxicity through covalent modification of proteins that results in alteration of protein function or in induction of immune response (Aithal et al. 2000). Relevant studies have demonstrated that the *CYP2C9* genotype is not associated with the risk of diclofenac-induced hepatitis (Aithal et al. 2000). In addition, a case-control study investigated the effect of *CYP2C8* polymorphisms on the risk of diclofenac hepatotoxicity (Daly et al. 2007). Cases were 24 patients who had experienced diclofenac hepatotoxicity and 160 control patients who had received diclofenac without developing hepatotoxicity. All patients were genotyped for the *CYP2C8*3* and *CYP2C8*4* alleles. *CYP2C8* polymorphisms could not predict the risk of hepatotoxicity. In contrast, genes involved in the metabolism and biliary excretion (i.e. *UGT2B7* and *ABCC2*, respectively) of diclofenac were associated with diclofenac hepatotoxicity(Daly et al. 2007).

4 Conclusions

Undoubtedly, the application of personalised medicine is anticipated to improve treatment efficacy and safety. Pharmacogenomics may be a critical pathway to personalised medicine. Concerning analgesics, current data suggest that pharmacogenomics contribute to variation in efficacy and safety of opioids and nonsteroidal anti-inflammatory agents. However, most data come from case-control studies and case reports. In addition, a recognised drawback in the field of pharmacogenomics is the common occurrence of false positive association between polymorphisms and the investigated outcome. Prospective studies are needed to further elucidate the clinical implications of available data as well as to define the guidelines for the clinical application of pharmacogenomic data. Furthermore, basic research should focus on the identification of biologically meaningful polymorphisms enabling a hypothesis with biological plausibility driven research in the field of pharmacogenomics of analgesics. Moreover, the publication of relevant negative results should be favoured.

References

Aithal GP, Day CP, Leathart JB, Daly AK (2000) Relationship of polymorphism in *CYP2C9* to genetic susceptibility to diclofenac-induced hepatitis. Pharmacogenetics 10:511–518

Andreassen TN, Eftedal I, Klepstad P, Davies A, Bjordal K, Lundström S, Kaasa S, Dale O (2012) Do CYP2D6 genotypes reflect oxycodone requirements for cancer patients treated for cancer pain? A cross-sectional multicentre study. Eur J Clin Pharmacol 68:55–64

Blanco G, Martinez C, Ladero JM, Garcia-Martin E, Taxonera C, Gamito FG et al (2008) Interaction of *CYP2C8* and *CYP2C9* genotypes modifies the risk for nonsteroidal anti-inflammatory drugs-related acute gastrointestinal bleeding. Pharmacogenet Genomics 18:37–43

Brenner SS, Herrlinger C, Dilger K, Mürdter TE, Hofmann U, Marx C et al (2003) Influence of age and cytochrome P450 2C9 genotype on the steady-state disposition of diclofenac and celecoxib. Clin Pharmacokinet 42:283–292

Ciszkowski C, Madadi P, Phillips MS, Lauwers AE, Koren G (2009) Codeine, ultrarapid-metabolism genotype, and postoperative death. N Engl J Med 361:827–828

Coffman BL, King CD, Rios GR, Tephly TR (1998) The glucuronidation of opioids, other xenobiotics, and androgens by human UGT2B7Y(268) and UGT2B7H(268). Drug Metab Dispos 26:73–77

Crews KR, Gaedigk A, Dunnenberger HM, Klein TE, Shen DD, Callaghan JT, Kharasch ED, Skaar TC (2012) Clinical Pharmacogenetics Implementation Consortium Clinical Pharmacogenetics Implementation Consortium (CPIC) guidelines for codeine therapy in the context of cytochrome P450 2D6 (CYP2D6) genotype. Clin Pharmacol Ther 91:321–326

Daly AK, Aithal GP, Leathart JB, Swainsbury RA, Dang TS, Day CP (2007) Genetic susceptibility to diclofenac-induced hepatotoxicity: contribution of *UGT2B7*, *CYP2C8*, and *ABCC2* genotypes. Gastroenterology 132:272–281

Darbari DS, van Schaik RH, Capparelli EV, Rana S, McCarter R, van den Anker J (2008) UGT2B7 promoter variant −840G>A contributes to the variability in hepatic clearance of morphine in patients with sickle cell disease. Am J Hematol 83:200–202

de Abajo FJ, Montero D, Madurga M, Garcia Rodriguez LA (2004) Acute and clinically relevant drug-induced liver injury: a population based case–control study. Br J Clin Pharmacol 58:71–80

Duggan DE, Hogans AF, Kwan KC, McMahon FG (1972) The metabolism of indomethacin in man. J Pharmacol Exp Ther 181:563–575

Eckhardt K, Li S, Ammon S, Schanzle G, Mikus G, Eichelbaum M (1998) Same incidence of adverse events after codeine administration irrespective of the genetically determined differences in morphine formation. Pain 76:27–33

Elkalioubie A, Allorge D, Robriquet L, Wiart JF, Garat A, Broly F, Fourrier F (2011) Near-fatal tramadol cardiotoxicity in a CYP2D6 ultrarapid metabolizer. Eur J Clin Pharmacol 67:855–858

Evans W (1999) Translating functional genomics into rational therapeutics. Science 286:487–491

Gaedigk A, Bradford LD, Marcucci KA, Leeder JS (2002) Unique CYP2D6 activity distribution and genotype-phenotype discordance in black Americans. Clin Pharmacol Ther 72:76–89

Galley HF, Mahdy A, Lowes DA (2005) Pharmacogenetics and anesthesiologists. Pharmacogenomics 6:849–856

Gan SH, Ismail R, Wan Adnan WA, Zulmi W (2007) Impact of CYP2D6 genetic polymorphism on tramadol pharmacokinetics and pharmacodynamics. Mol Diagn Ther 11:171–181

García-Martín E, Martínez C, Tabares B, Frías J, Agúndez JA (2004) Interindividual variability in ibuprofen pharmacokinetics is related to interaction of cytochrome P450 2C8 and 2C9 amino acid polymorphisms. Clin Pharmacol Ther 76:119–127

Gasche Y, Daali Y, Fathi M, Chiappe A, Cottini S, Dayer P et al (2004) Codeine intoxication associated with ultrarapid CYP2D6 metabolism. N Engl J Med 351:2827–2831

Hamman MA, Thompson GA, Hall SD (1997) Regioselective and stereoselective metabolism of ibuprofen by human cytochrome P450 2C. Biochem Pharmacol 54:33–41

Holthe M, Klepstad P, Zahlsen K, Borchgrevink PC, Hagen L, Dale O, Kaasa S, Krokan HE, Skorpen F (2002) Morphine glucuronide-to-morphine plasma ratios are unaffected by the UGT2B7 H268Y and UGT1A1*28 polymorphisms in cancer patients on chronic morphine therapy. Eur J Clin Pharmacol 58:353–356

Hutzler JM, Kolwankar D, Hummel MA, Tracy TS (2002) Activation of CYP2C9-mediated metabolism by a series of dapsone analogs: kinetics and structural requirements. Drug Metab Dispos 30:1194–1200

Kelly LE, Rieder M, van den Anker J, Malkin B, Ross C, Neely MN, Carleton B, Hayden MR, Madadi P, Koren G (2012) More codeine fatalities after tonsillectomy in North American children. Pediatrics 129:e1343–e1347

Kirchheiner J, Seeringer A (2007) Clinical implications of pharmacogenetics of P450 drug metabolising enzymes. Biochim Biophys Acta 1770:489–494

Kirchheiner J, Meineke I, Steinbach N, Meisel C, Roots I, Brockmoller J (2003) Pharmacokinetics of diclofenac and inhibition of cyclooxygenases 1 and 2: no relationship to the CYP2C9 genetic polymorphism in humans. Br J Clin Pharmacol 55:51–61

Kirchheiner J, Tsahuridu M, Jabrane W, Roots I, Brockmoller J (2004) The CYP2C9 polymorphism: from enzyme kinetics to clinical dose recommendations. Personalized Med 1:63–84

Kirchheiner J, Keulen JT, Bauer S, Roots I, Brockmöller J (2008) Effects of the CYP2D6 gene duplication on the pharmacokinetics and pharmacodynamics of tramadol. J Clin Psychopharmacol 28:78–83

Kumar S, Samuel K, Subramanian R, Braun MP, Stearns RA, Chiu SH et al (2002) Extrapolation of diclofenac clearance from in vitro microsomal metabolism data: role of acyl glucuronidation and sequential oxidative metabolism of the acyl glucuronide. J Pharmacol Exp Ther 303:969–978

Lemberg KK, Heiskanen TE, Neuvonen M, Kontinen VK, Neuvonen PJ, Dahl ML, Kalso EA (2010) Does co-administration of paroxetine change oxycodone analgesia: an interaction study in chronic pain patients. Scand J Pain 1:24–33

Levran O, Peles E, Hamon S, Randesi M, Adelson M, Kreek MJ (2011) CYP2B6 SNPs are associated with methadone dose required for effective treatment of opioid addiction. Addict Biol. doi:10.1111/j.1369-1600.2011.00349.x

Madadi P, Koren G, Cairns J et al (2007) Safety of codeine during breastfeeding: fatal morphine poisoning in the breastfed neonate of a mother prescribed codeine. Can Fam Physician 53:33–35

Madadi P, Ross CJ, Hayden MR, Carleton BC, Gaedigk A, Leeder JS et al (2009) Pharmacogenetics of neonatal opioid toxicity following maternal use of codeine during breastfeeding: a case-control study. Clin Pharmacol Ther 85(1):31–35

Madadi P, Hildebrandt D, Gong IY, Schwarz UI, Ciszkowski C, Ross CJ, Sistonen J, Carleton BC, Hayden MR, Lauwers AE, Koren G (2010) Fatal hydrocodone overdose in a child: pharmacogenetics and drug interactions. Pediatrics 126:e986–e989

Martínez C, Blanco G, Ladero JM, García-Martín E, Taxonera C, Gamito FG et al (2004) Genetic predisposition to acute gastrointestinal bleeding after NSAIDs use. Br J Pharmacol 141:205–208

Martínez C, García-Martín E, Blanco G, Gamito FJ, Ladero JM, Agúndez JA (2005) The effect of the cytochrome P450 CYP2C8 polymorphism on the disposition of (R)-ibuprofen enantiomer in healthy subjects. Br J Clin Pharmacol 59:62–69

Martínez C, Blanco G, García-Martín E, Agúndez JA (2006) Clinical pharmacogenomics for CYP2C8 and CYP2C9: general concepts and application to the use of NSAIDs. Farm Hosp 30:240–248

Meyer UA (2000) Pharmacogenetics of adverse drug reactions. Lancet 356:1667–1671

Miners JO, Coulter S, Tukey RH, Veronese ME, Birkett DJ (1996) Cytochromes P450, 1A2, and 2C9 are responsible for the human hepatic O-demethylation of R- and S-naproxen. Biochem Pharmacol 51:1003–1008

Persson K, Sjostrom S, Sigurdardottir I, Molnar V, Hammarlundudenaes M, Rane A (1995) Patient-controlled analgesia (PCA) with codeine for postoperative pain relief in ten extensive metabolizers and one poor metabolizer of dextromethorphan. Br J Clin Pharmacol 39:182–186

Pilotto A, Seripa D, Franceschi M, Scarcelli C, Colaizzo D, Grandone E et al (2007) Genetic susceptibility to nonsteroidal anti-inflammatory drug-related gastroduodenal bleeding: role of cytochrome P450 2C9 polymorphisms. Gastroenterology 133:465–471

Pirmohamed M (2010) Pharmacogenetics of idiosyncratic adverse drug reactions. In: Uetrecht J (ed) Adverse drug reactions, Handbook of experimental pharmacology. Springer, Hielderberg, pp 478–491

Poulsen L, Riishede L, Brosen K, Clemensen S, Sindrup SH (1998) Codeine in post-operative pain—study of the influence of sparteine phenotype and serum concentrations of morphine and morphine-6-glucuronide. Eur J Clin Pharmacol 54:451–454

Rouguieg K, Picard N, Sauvage FL, Gaulier JM, Marquet P (2010) Contribution of the different UDP-glucuronosyltransferase (UGT) isoforms to buprenorphine and norbuprenorphine metabolism and relationship with the main UGT polymorphisms in a bank of human liver microsomes. Drug Metab Dispos 38:40–45

Samer CF, Daali Y, Wagner M, Hopfgartner G, Eap CB, Rebsamen MC, Rossier MF, Hochstrasser D, Dayer P, Desmeules JA (2010) The effects of CYP2D6 and CYP3A activities on the pharmacokinetics of immediate release oxycodone. Br J Pharmacol 160:907–918

Siva C, Yokoyama WM, McLeod HL (2002) Pharmacogenetics in rheumatology: the prospects and limitations of an emerging field. Rheumatology 41:1273–1279

Slanar O, Dupal P, Matouskova O, Vondrackova H, Pafko P, Perlik F (2012) Tramadol efficacy in patients with postoperative pain in relation to CYP2D6 and MDR1 polymorphisms. Bratisl LekListy 113:152–155

Stamer UM, Kehnen K, Hothker F, Wolf S, Hoeft A (2003) Impact of CYP2D6 on postoperative tramadol analgesia. Pain 105:231–238

Stamer UM, Stüber F, Muders T, Musshoff F (2008) Respiratory depression with tramadol in a patient with renal impairment and CYP2D6 gene duplication. Anesth Analg 107:926–929

Tang W (2003) The metabolism of diclofenac-enzymology and toxicology perspectives. Curr Drug Metab 4:319–329

Voronov P, Przybylo HJ, Jagannathan N (2007) Apnea in a child after oral codeine: a genetic variant – an ultra-rapid metabolizer. Paediatr Anaesth 17:684–687

Wang G, Zhang H, He F, Fang X (2006) Effect of the CYP2D6*10 C188T polymorphism on postoperative tramadol analgesia in a Chinese population. Eur J Clin Pharmacol 62:927–931

Weber WW (1999) Populations and genetic polymorphisms. Mol Diagn 4:299–307

Wilcox R, Owen H (2000) Variable cytochrome P450 2D6 expression and metabolism of codeine and other opioid prodrugs: implications for the Australian anesthetist. Anaesth Intensiv Care 28:611–619

Williams DG, Patel A, Howard RF (2002) Pharmacogenetics of codeine metabolism in an urban population of children and its implications for analgesic reliability. Br J Anaesth 89:839–845

Wong SH, Wagner MA, Jentzen JM, Schur C, Bjerke J, Gock SB et al (2003) Pharmacogenomics as an aspect of molecular autopsy for forensic pathology/toxicology: does genotyping CYP 2D6 serve as an adjunct for certifying methadone toxicity? J Forensic Sci 48:1406–1415

Yasar U, Eliasson E, Forslund-Bergengren C, Tybring G, Gadd M, Sjöqvist F et al (2001) The role of CYP2C9 genotype in the metabolism of diclofenac in vivo and in vitro. Eur J Clin Pharmacol 57:729–735

Zarza J (2003) Major bleeding during combined treatment with indomethacin and low doses of acenocoumarol in a homozygous patient for 2C9*3 variant of cytochrome p-450 CYP2C9. Thromb Haemost 90:161–162

Zhao J, Leemann T, Dayer P (1992) In vitro oxidation of oxicam NSAIDS by a human liver cytochrome P450. Life Sci 51:575–581

Section IV

Personalized Medicine: Cardiovascular and Pulmonary Disorders

Cardiovascular Disease Pharmacogenomics

20

Bishwanath Chatterjee

Abstract

Cardiovascular disease (CVD), commonly known as heart disease, is the leading cause of death all over the world. Nearly one in four deaths worldwide is attributed to CVD. Generally, CVD constitutes a number of overlapping structural and functional irregularities in the heart and blood vessels, which together constitute the cardiovascular system. Given this, the corresponding drugs also either can treat or prevent these overlapping conditions. Often, more than one drug is prescribed to treat a medical condition, since one medication alone cannot remedy the problem. High mortality and morbidity due to CVD automatically requires physicians to treat CVD aggressively, leading to adverse drug reactions. Still, some patients do not respond positively even if the strongest recommended drug dosage is administered. Research in the CVD field has demonstrated linkage between gene variation and the severity of diseases such as blood cholesterol level or blood pressure. Correspondingly, research over more than three decades has established the linkage of genetic variation with the efficacy of CVD treatment in the cohorts of different CVD disease types. In this chapter, recent advances in the field of cardiovascular pharmacogenetics and pharmacogenomics are summarized.

1 Introduction

Cardiovascular disease (CVD), generally called heart disease, is a very broad term that encompasses a range of conditions that in someway affect the cardiovascular system and its normal function. Any problems with the heart and the blood vessels together constitute this category of disease. CVD is the leading cause of death specifically in developing countries (Gaziano 2005). According to the estimates, 17 million people worldwide die of CVD every year, particularly of heart attacks and strokes. Even in developed countries like the United States, in 2008, over 616,000 individuals died of heart disease. This amounted to nearly 25 % of the total deaths in the United States that year (Miniño et al. 2011). For prevention and treatment of CVD, cardiovascular drugs are prescribed abundantly, without consideration for

B. Chatterjee (✉)
Laboratory of Molecular Cancer Pathology,
National Institutes of Health, Building 10 Room 3B56,
10 Central Drive, Bethesda, MD 20892-1500, USA
e-mail: Chatterb@mail.nih.gov;
bishwachatterjee@gmail.com

D. Barh et al. (eds.), *Omics for Personalized Medicine*,
DOI 10.1007/978-81-322-1184-6_20, © Springer India 2013

the factors that alter drug response such as gender, age, diet, environmental factors, and genetic background. These variations, along with other concomitant drug use, may cause adverse drug reactions that lead to the hospitalization and death of many patients. Fighting this monster disease poses a very large economic burden worldwide. In 2010, coronary heart disease alone cost the United States an estimated $108.9 billion (Heidenreich et al. 2011), which included the cost of health-care services, medications, and lost productivity.

Technological advances in genomic research have enabled a deeper understanding of a variety of phenotypes relevant to clinical medicine. There are several approaches to identify and validate genetic influences on drug responses, and these approaches can be used individually or in combination with one another. For example, the candidate gene approach can be based on variable pharmacokinetics, pharmacodynamics, or other genes that are functional in the pathway(s) of the known candidate genes. On the other hand, unbiased approaches such as candidate genes selected from genome-wide association studies, or GWAS, can be successfully used to test hundreds of thousands of inherited genetic markers for association with medical traits. Another unbiased approach could be using animal models with manipulated genetic backgrounds. All such approaches used efficiently CVD drug response studies. These studies have generated a deeper understanding of variability in treatment outcomes of a disease, such as a particular drug's efficacy or toxicity.

2 Categories of Cardiovascular Disease and Treatment

Many of the heart disease problems are related to building of the plaque in the blood vessels, a process called as atherosclerosis. This leads to either clogging or narrowing of the arteries making slow blood flow, which can lead to heart attack, chest pain (angina), or stroke. Other kinds of heart disease may include heart failure, arrhythmias, or heart structural problems like muscle or valve problems. Problems with heart beating rhythm like fast, slow, or irregular beating also constitute other form of heart disease. Basically therefore a number of different overlapping conditions together constitute heart disease. Therefore heart disease is also treated using one of several medications together, or sometimes one medication can treat two overlapping conditions in conjunction with lifestyle changes and/or surgery. Table 20.1 describes the common cardiovascular disease categories and the group of medications that are prescribed to treat the disease.

Table 20.1 Common cardiovascular disease categories and their treatment plans

Cardiovascular disease category	Medication categories
Cardiovascular disease of clogged arteries	Treatments include blood pressure-lowering agents such as diuretics, angiotensin-converting enzyme (ACE) inhibitors, beta-blockers; blood-thinning medications; cholesterol-lowering medicines
Heart arrhythmias	Antiarrhythmic medicines
Cardiomyopathy	Treatments include medication that can improve the pumping ability of the heart such as ACE inhibitors or angiotensin II receptor blockers, beta-blockers
Valvular heart disease	Treatments include medications to open blood vessels (vasodilator), cholesterol-lowering medications, diuretics, and blood-thinning medication
Heart infection	Antibiotics and heart beat regulators such as angiotensin-converting enzyme inhibitors or beta-blockers

3 Pharmacogenetics and Pharmacogenomics of CVD Drugs

Since these diseases pose such large complications, a number of studies have been done to understand the efficacy and tolerance of cardiovascular drugs that are prescribed for these diseases. Population-based studies are a useful way to predict which drug may be the most effective in a particular ethnic group, even if the actual genetic determinant is not known for the efficacy of that drug. After the completion of the sequencing of the human genome, pharmacogenetics (single-gene study) has been transformed into pharmacogenomics (multiple-gene study). For each cardiovascular drug, a number of candidate genes have been studied along with pharmacokinetics. Therefore, the pharmacogenomics studies are divided into three broad categories: genes that are involved in drug targets, drug metabolism, and drug transport. These genes, along with therapeutic targets, environmental and physiological factors, and co-administrated drugs, make the therapeutic valuation of the administrated drug highly effective.

Given the complexity of cardiovascular disease, the effect of genetic variation was not well established. In most studies, there is very little data on how genetic variation affects drug therapy. Thus, large cohorts are needed for genome-wide studies for CVD drug therapy. Use of the approaches in CVD drug will be discussed later. Since the medications used in CVD are often overlapping with the spectrum of the disease, it will be useful to present pharmacogenomics of each medication separately.

3.1 Cholesterol Reducing Medications (Statins)

CVD risk increases with the increase of cholesterol in the body, specifically with decreasing HDL to LDL ratios. These traits are known to be highly heritable (Superko et al. 2012), and it has been suggested that more than 50 % of interindividual variations in LDL cholesterol-C levels are attributed to genetic factors (Heller et al. 1993). A number of studies correlated these findings (for review see Klos and Kullo 2007). Cholesterol-lowering drugs, or statins, are inhibitors of HMG-CoA reductase (3-hydroxy 3-methylglutaryl coenzyme A reductase), a key enzyme in cholesterol metabolism, and are prescribed widely to reduce the risk of cardiovascular disease. This class of drug works primarily to lower low-density lipoprotein cholesterol (LDLC). Although these medications are effective to a certain extent, certainly there is an individual-to-individual variability in lipid-lowering response, and some patients fail to meet lipid-lowering goals even after multiple dose adjustments (Mangravite et al. 2006).

Two different approaches were used to elucidate the pharmacogenomics of variable response to statins. A gene-specific candidate approach was used to find out the effect of gene sequence variation in the genes that may be directly involved in response to this class of drugs like HMGCR (3-hydroxy-3-methylglutaryl-CoA reductase) or LDLR (low-density lipoprotein receptor) (Hannuksela et al. 1994; Chasman et al. 2004; Thompson et al. 2005). Several GWAS studies have also been done for the role of statins in their cholesterol-lowering effect to understand the genetics beyond the candidate approach. Numbers of these studies are done with more than one statin, while other studies actually combined several previous studies. Though such efforts gave rise to some interesting findings, these findings need to be confirmed with larger cohorts and specific clinical guidance for the selection of patients and healthy controls. Table 20.2 summarizes a list of genes that may be involved in cholesterol-lowering effect of statins.

It is interesting to note that controversies still exist if a particular variation in a gene is responsible for the effect of statins. For example, two SNPs (rs17244841 and rs17238540) were found to be highly correlated with LDL-C reduction from statin therapy in the PRINCE study (Chasman et al. 2004); carriers of the minor allele of these SNPs were found to have smaller LDL-C reduction from Pravastatin treatment. In the CAP trial haplotype (H7) consisting of rs17244841, rs384662, and rs17238540, LDL-C reduction was associated from Simvastatin in African-American (but not Caucasian) patients (Krauss et al. 2008). A Chinese cross-sectional study showed rs12916 (a tightly lined SNP rs384662) to be associated with LDL-C reduction from statin therapy (Chien et al. 2010). However, in the Prospective Study of Pravastatin in the Elderly at Risk (PROSPER) or Atorvastatin Comparative Cholesterol Efficacy and Safety Study (ACCESS),

Table 20.2 Association of genes linked to cholesterol-lowering medications

Gene name	Gene function	Effective SNP/ splicing	References
ABCB1	Member of the superfamily of ATP-binding cassette, ATP-dependent drug efflux pump for xenobiotic compounds	C3435T (Ile1145Ile: rs1045642)	Kajinami et al. (2004), Fiegenbaum et al. (2005a), and Rodrigues et al. (2005)
ACE	Enzyme involved in catalyzing the conversion of angiotensin I into a physiologically active peptide angiotensin II, levels of circulating enzyme, or cardiovascular pathophysiology	Insertion/Deletion (I/D, rs4646994)	Thompson et al. (2005), Bray et al. (2001), Maitland-van der Zee et al. (2004, 2007), and Marian et al. (2003)
APOA1	Major protein component of high-density lipoprotein (HDL) in plasma, promotes cholesterol efflux from tissues to the liver	G(−75)A rs670	Thompson et al. (2005), Lahoz et al. (2003), and Sorkin et al. (2005)
APOB	Main apolipoprotein of chylomicrons and low-density lipoproteins	Xba I polymorphism rs693, rs115835	Chasman et al. (2004), Ojala et al. (1991), Guzman et al. (2000), Ye et al. (2003), and Vrablík et al. (2012)
APOC1	Member of the apolipoprotein C1 family, expressed primarily in the liver, activated when monocytes differentiate into macrophages	3′ region of the gene A to G rs4420638	Vrablík et al. (2012)
APOE	Essential for the normal catabolism of triglyceride-rich lipoprotein constituents	E2/E3/E4 rs429358, rs7699455, rs7412	Chasman et al. (2004), Thompson et al. (2005), Ojala et al. (1991), Ye et al. (2003), and Vrablík et al. (2012), Ordovas et al. (1995), Nestel et al. (1997), Sanllehy et al. (1998), Ballantyne (2003), Pedro-Botet et al. (2001), Pena et al. (2002), and Takane et al. (2006)
CETP	Transfers cholesteryl esters between lipoproteins. CETP may effect susceptibility to atherosclerosis	TaqIB rs708272	Kuivenhoven et al. (1998), Carlquist et al. (2003), Freeman et al. (2003), vanVenrooij et al. (2003), Winkelmann et al. (2003), Klerkx et al. (2003), Mohrschladt et al. (2005), and Marschang et al. (2006)
CYP7A1	Member of the cytochrome P450 superfamily of enzymes	−204A>C, rs3808607 Intron 1 A>C rs8192870	Thompson et al. (2005), Takane et al. (2006), Kajinami et al. (2005), and Jiang et al. (2012)
DNAJC5B	DnaJ homolog subfamily C member 5B	rs13279522	Shiffman et al. (2012)
HMGCR	Rate-limiting enzyme for cholesterol synthesis and is regulated via a negative feedback mechanism mediated by sterols and non-sterol metabolites	Alternate splicing and 33 SNPs including rs12654264, rs17244834, rs17238540, rs17244841, rs384662	Chasman et al. (2004), Thompson et al. (2005), Vrablík et al. (2012) Krauss et al. (2008), Mangravite et al. (2010), Medina & Krauss (2009), and Medina (2010)
LDLR	Low-density lipoprotein receptor	rs6511720, rs14158, rs1433099, rs7254521, rs5742911, rs273846, rs2569542, rs5925	Chasman et al. (2004), Vrablík et al. (2012), Mangravite et al. (2010), Salazar et al. (2000), Lahoz et al. (2005a), and Chasman et al. (2012)

(continued)

Table 20.2 (continued)

Gene name	Gene function	Effective SNP/ splicing	References
LIPC	Hepatic triglyceride lipase	C-514 T rs1800588 rs16940379	Thompson et al. (2005), Lahoz et al. (2005b), and Peters et al. (2011)
LPL	Lipoprotein lipase expressed in heart, muscle, and adipose tissue	rs328, rs249, rs7016529, rs1801177	Thompson et al. (2005), Sing et al. (1999), and Brautbar et al. (2012)
PON1	Arylesterase that mainly hydrolyzes paroxon to produce p-nitrophenol	R192Q rs662	Malin et al. (2001), Turban et al. (2001), Himbergen et al. (2005), and Fu et al. (2008)
PCSK9	Proprotein convertase subtilisin/kexin type 9	rs11206510; rs6235; rs10888896 and rs505151 (E670G)	Vrablík et al. (2012) Chasman et al. (2012), Peters et al. (2011), and Pisciotta et al. (2012)
SCAP	SREBF chaperone	A2386G rs12487736	Salek et al. (2002), Fan et al. (2001), and Fiegenbaum et al. (2005b)
SLCO1B1	Liver-specific member of the organic anion transporter family	c.T521C rs4149056	Tachibana-Iimori et al. (2004), Niemi et al. (2005), Igel et al. (2006), Santos et al. (2011, 2012), and Sortica et al. (2012)
SREBP1	Transcription factor that binds to the sterol regulatory element-1	−36del/G, rs9902941	Salek et al. (2002), Fiegenbaum et al. (2005b), Chien et al. (2010), and Berthold et al. (2008)
TLR4	Toll-like receptor (TLR) family	D299Q	Boekholdt et al. (2003) and Holloway et al. (2005)

rs17238540 was found not to be associated with LDL-C reduction from Pravastatin (Thompson et al. 2005). Similar results were obtained from the TNT trial (Thompson et al. 2009) using Atorvastatin. In the STRENGTH study (Voora et al. 2008), rs77244841 was found not to be associated with LDL-C reduction from statin therapy. Similarly, Assessment of Lescol in Renal Transplantation (ALERT) Study (Singer et al. 2007), neither rs17238540 nor rs77244841, was associated with LDL-C reduction from fluvastatin therapy. Similar results were obtained for most of the genes described in Table 20.2 (for recent review, see Superko et al. 2012). One recent study has indicated that there are no differences in the improved cholesterol level in African-Americans, Caucasian, or South Asians with standard dose of Atorvastatin (Chapman et al. 2011). Another analysis combining 22 other studies also found no differences in systemic exposure to Atorvastatin, indicating that dosing considerations in the current levels for Atorvastatin are similar for Asian subjects as compared with Caucasian subjects (Gandelman et al. 2012). Given this, it is important to have larger studies to pin down the effectiveness of statin therapy and various effectiveness of the outcome.

Besides the considerations of effectiveness of the statins, which in general are termed as safe and well tolerated (Baigent et al. 2005), these drugs also bring toxicity and adverse drug reaction. Statin-induced liver toxicity and muscle toxicity are the most common side effects within the limit of maximum permissible dose that may be prescribed (Armitage 2007). Present studies focus on one of the main carriers of statins, organic anion-transporting polypeptide 1B1 (OATP1B1), coded by *SLCO1B1* gene (Niemi 2007). Another important factor is the breast cancer-resistant protein (BCRP) coded by gene *ABCG2*, which modulates the absorption and elimination of statins and therefore alters the pharmacokinetics of statins. There are also several other genes that may be involved in the transport and kinetics of statins and some of these proteins may have some substrate specificity (Niemi 2010). Genetic polymorphism of ABCG2 c.421C>A (p.Gln141Lys; rs2231142) genotype shows effects on pharmacokinetics of fluvastatin and simvastatin lactone, but has no significant effect on pravastatin or active simvastatin acid (Keskitalo et al. 2009). Polymorphisms of ABCB1 gene (rs3789244 and rs1922242) were associated with the modification of the effectiveness of statins in the prevention of the clinical outcome, myocardial infarction (Peters et al. 2010). Pharmacokinetic and pharmacodynamic influences of SLCO1B1 polymorphism [c.521T>C (rs4149056), c.388A>G (rs2306283)] were found for many statins (for recent review, see Romaine et al. 2010; Niemi et al. 2011).

3.2 Anticoagulants

Anticoagulants are compounds that do not allow blood to clot. These include drugs such as heparin and coumarin. Anticoagulants are usually administered to patients with myocardial infarction, certain types of irregular heartbeat, and people with prosthetic (replacement or mechanical) heart valves among others. They have been used to prevent transient ischemic attacks and to reduce the risk of recurrent heart attacks. Coumarins include warfarin, phenprocoumon, and acenocoumarol. Coumarin dosage is problematic for a narrow window of therapeutic efficacy. Concomitant use of other drugs also brings variability of the window of therapeutically viable dosage. Warfarin is metabolized primarily via oxidation in the liver by CYP2C9 and exerts its anticoagulant effect by inhibiting the protein vitamin K epoxide reductase complex, subunit 1 (VKORC1). Genetic variation in these two genes was studied extensively, and 3 SNPs were found to be associated with warfarin metabolism and shed substantial insight into how a perfect dose could be determined for an individual (Higashi et al. 2002; Rieder et al. 2005; D'Andrea et al. 2005; Yuan et al. 2005; Sconce et al. 2005; Takahashi et al. 2006; Kimura et al. 2007; Gage et al. 2008; Takeuchi et al. 2009; Wadelius et al. 2009; Lubitz et al. 2010; Ma et al. 2012). In addition, a third gene CYP4F2 is suggested to play a role in a low percentage of patients (Takeuchi et al. 2009; Ma et al. 2012; Caldwell et al. 2008). Two variants of *CYP2C9* gene (CYP2C9*2 (C430T, p. R144Crs1799853) and CYP2C9*3 (A1075C, p. I359L, rs1057910)) were studied in more detail compared to wild-type CYP2C9*1. These two variants have very low enzyme activity compared to the wild type, leading to higher plasma levels of warfarin (Sanderson et al. 2005). A similar result was also observed for some other coumarins with the exception of phenprocoumaron (Visser et al. 2004; Schalekamp et al. 2004), though several other studies have found less a dramatic effect for the CYP2C9 genotype.

In the case of VKORC1 −1639 G/A (rs9923231), the promoter region, and for C1173T (rs9934438), the intronic region of the gene has been studied in respect to their role in dosing of warfarin. The T allele of C1173T is shown to be associated with reduced dose of warfarin and other coumarins (Reitsma et al. 2005; Wadelius and Pirmohamed 2007; El Din et al. 2012).

CYP4F2 1347 C/T at (rs2108622) is the third gene that is associated with warfarin dose selection in GWAS studies (Takeuchi et al. 2009). Among Caucasian patients treated with warfarin, CYP4F2 polymorphism had a measurable effect on warfarin responsiveness during induction; however, the observed differences failed to reach the level of statistical significance (Bejarano-Achache et al. 2012). Two other studies in Chinese populations have conflicting results of linkage of this genotype with outcome in patients (Ma et al. 2012; Cen et al. 2010). A small number of other studies also looked into the association of polymorphism in the vitamin K-dependent proteins such as clotting factor II and VII (D'Ambrosio et al. 2004; Shikata et al. 2004).

3.3 Platelet Aggregation Inhibitor

Platelet aggregation inhibitors are group of drugs that inhibits platelets from aggregating to form a plug. They are used to prevent clotting and alter the natural course of atherosclerosis. Platelet aggregation inhibitors work in different places of the clotting cascade. The commonly used drugs are aspirin and clopidogrel. Various genetic variants and polymorphisms of platelet surface receptors and CYP450 isozymes, which are substantially involved in the metabolism of antiplatelet drugs, have been postulated to influence efficacy of antiplatelet drug therapy. Aspirin irreversibly inhibits cyclooxygenase-1 (COX1), an enzyme that catalyzes the conversion of arachidonic acid to prostaglandin G2 (PGG2) which ultimately converts to thromboxane A2 (TAX2). There, polymorphisms in the COX genes are therefore studied to find their role in platelet aggregation inhibition. Five different SNPs exist for COX-1: 842A>G (promoter region), 22C>T (exon 2), 128G>A (exon 3), 644C>A (exon 6), and 714C>A (exon 7); these were studied in 144 patients and only −842A>G (rs10306114) has shown sensitivity to aspirin therapy (Maree et al. 2005). Another SNP 50C>T, rs3842787 was also investigated for its role with aspirin therapy (Gonzalez-Conejero et al. 2005). A recent study found statistically significant association between COX-1 gene polymorphisms (−842A>G and 50C>T) and bleeding risk (Motovska et al. 2010). Platelet surface glycoprotein IIIa (ITGB3) haplotype P1A1/P1A2 (807C>T, rs1126643) has been shown to have an effect on sensitivity to aspirin (Nurden 1997; Undas et al. 1999). Some of these results could not be confirmed for aspirin and clopidogrel in the later studies (Pamukcu et al. 2005; Lev et al. 2007).

Loss-of-function variants in the hepatic cytochrome 2C19 (mainly *2 allele) system have been found to be the predominant genetic mediators of clopidogrel response. There are four different alleles of the *2 allele, and 681G/A polymorphism in exon 5 (rs4244285) gives rise to a low-activity drug-metabolizing enzyme (Hulot et al. 2006). Other studies demonstrated the linkage between 2C19*2 allele with decreased responsiveness to clopidogrel (Brandt et al. 2007; Fontana et al. 2007; Geisler et al. 2008; Collet et al. 2009). Some recent reviews summarize the pharmacogenomics of platelets with respect to aspirin and clopidogrel (Zuern et al. 2010; Cuisset et al. 2012; Levitt et al. 2012).

3.4 Antihypertensive Drugs

This class of drugs is used to prevent high blood pressure, thereby preventing complications as a result of high blood pressure, such as stroke and myocardial infarction. Evidence suggests that reduction of blood pressure by 5 mmHg can decrease the risk of stroke by 34 % and of ischemic heart disease by 21 % and reduce the likelihood of dementia, heart failure, and mortality and other forms of cardiovascular diseases (Law et al. 2003). Mainly three broad classes of antihypertensive drugs are used to treat hypertension: diuretics, vasodilators (angiotensin-converting enzyme inhibitors, angiotensin receptor blockers), and cardioinhibitory drugs.

3.4.1 Diuretics

Three categories of the diuretics, thiazide diuretics, loop diuretics, and potassium-sparing diuretics, are used extensively to treat hypertension. Single-nucleotide polymorphism in the gene encoding alpha-adducin (ADD1 Gly460Trp-rs4961) has been studied and found to be associated with the response of thiazide (Cusi et al. 1997; Glorioso et al. 1999; Sciarrone et al. 2003). These studies led to the development of a novel hypertension drug, Rostafuroxin, that targets ADD1 and is now in the trial phase II (Lanzani et al. 2010). SNPs spanning ADD2 gene were also studied in detail, and SNPs in intronic region (A/T rs1541582) were found to be associated with differential response to diuretics and beta-blockers (Kardia et al. 2007). An insertion-deletion SNP (rs4646994) in angiotensin I-converting enzyme, which is involved in catalyzing the conversion of angiotensin I into a physiologically active peptide angiotensin II, is found to be in association with fosinopril in some studies (Stavroulakis et al. 2000). However, most of the later studies did not find a significant linkage (Arnett et al. 2005), even though some recent studies with Chinese populations found a significant linkage of the genotype which was associated with systolic blood pressure (SBP) response to hydrochlorothiazide (Zhou et al. 2007; Li et al. 2011). Other polymorphisms in the genes in the renin-angiotensin-aldosteron system includes AGT A(−6)G (rs5051) and M235T (rs699), and its receptor AGTR1 A1166C (rs5186) have been studied. A study with African-Americans and non-Hispanic whites indicated that A1166C of AGTR1 and angiotensinogen G-6A polymorphisms had a significant effect on systolic BP response to the diuretic in only African-American women (Frazier et al. 2004). Later studies showed both positive and negative correlations (Schelleman et al. 2006; Jiang et al. 2007). A synonymous SNP C825T (rs5443) of the G-protein beta (3)-subunit (GNB3) may help identify patients with essential hypertension who are more responsive to diuretic therapy (Turner et al. 2001). A recent study found that among GNB3 T allele carriers, the risk of diabetes due to thiazide use was less increased than among homozygous GNB3 CC subjects (Bozkurt et al. 2009). Another polymorphism (G to A rs4149601) in the gene NEDD4L (Manunta et al. 2008; Luo et al. 2009; Svensson-Färbom et al. 2011) is known to be associated with both diuretics and beta-blockers.

3.4.2 Beta-Blockers

Beta-blockers primarily block β1 and β2 receptors from binding to epinephrine and norepinephrine. By blocking the effect of norepinephrine and epinephrine, beta-blockers reduce heart rate, reduce blood pressure by dilating blood vessels, and may constrict air passages by stimulating the muscles that surround the air passages to contract. 60 % of the patients treated with beta-blockers do not show an adequate decrease in blood pressure (Materson et al. 1993). Given this, polymorphisms associated with beta-1 androgen receptor (ADRB1) were evaluated. Arg389Gly (rs1801253), located in the GTP-binding domain, was found to be associated with interindividual differences in pathophysiologic characteristics or in the response to therapeutic betaAR agonists and antagonists (Mason et al. 1999). Since then several studies have linked the genotype to the beta-blocker response (e.g., Liu et al. 2003, 2006; Sofowora et al. 2003; Petersen et al. 2012). Similarly, GNB3 C825T polymorphism was also investigated and found to be associated with blood pressure response to beta-blockers (Filigheddu et al. 2004, 2012; Dörr et al. 2010). A recent study in atrial fibrillation patients homozygous for Arg389 found a much smaller response to carvedilol than carriers of at least one Gly389 allele (Rau et al. 2012).

3.4.3 ACE Inhibitors, Angiotensin II Antagonists, and Calcium Channel Blockers

Angiotensin II causes the muscles surrounding blood vessels to contract, thereby narrowing the vessels. Angiotensin II is formed from angiotensin I in the blood by the enzyme angiotensin-converting enzyme (ACE). ACE inhibitors are medications that slow (inhibit) the activity of the enzyme ACE, which decreases the production of angiotensin II, resulting in blood vessels enlargement and reduced blood pressure. ACE insertion/deletion genotype has been studied for blood pressure in great detail in healthy patients, and a linkage to the genotype has been found (Harrap et al. 2003; Arnett et al. 2006; Brugts et al. 2011; Ned et al. 2012).

Angiotensinogen, AGT geneM235T, polymorphism has been found to be associated with modified response to ACE inhibitors (Hingorani et al. 1995). A recent study found significant association of M235T polymorphism with essential hypertension. Patients carrying TT genotype had higher blood pressure-lowering response when treated with the ACE inhibitor enalapril than those carrying MM (Srivastava et al. 2012). Angiotensin II antagonists are also governed by polymorphism in the renin-angiotensin-aldosteron system genes. Modified blood pressure response was shown to be associated with ACE insertion/deletion polymorphism and CYP11B2 C-344T (rs1799998) (Kurland et al. 2001, 2002, 2004).

The calcium channel blockers are drugs that work by slowing the movement of calcium into the cells of the heart and blood vessel walls, which makes it easier for the heart to pump and widens blood vessels. Studies are confined to calcium-signaling genes such as CACNA1C, CACNB2, and KCNMB19 (Beitelshees et al. 2007, 2009; Niu et al. 2010). In addition, nitric oxide synthase 3 (NOS3) catalyzes production of nitric oxide in the endothelium and may play a role in cardiovascular disease. A recent study have analyzed 3 SNPs in NOS3 gene polymorphisms −690 C>T (rs3918226), −922 A>G (rs1800779), and glu298asp G>T (rs1799983) with three antihypertensive drugs with CVD outcomes.

Table 20.3 FDA pharmacogenomic marker labels for cardiovascular drugs[a]

Drug	Biomarkers	Label sections
Carvedilol	CYP2D6	Drug interaction, clinical pharmacology
Clopidogrel	CYP2C19	Warning and precautions, drug interaction, clinical pharmacology
Isosorbide and hydralazine	NAT1, NAT2	Clinical pharmacology
Lenalidomide	Chromosome 5q	Boxed warning, indication and uses, clinical studies, patient counseling
Metoprolol	CYP2D6	Precautions, clinical pharmacology
Prasugrel	CYP2C19	Use in specific population, clinical pharmacology, clinical studies
Propaphenone	CYP2D6	Clinical pharmacology
Propranolol	CYP2D6	Clinical pharmacology, drug interaction, clinical pharmacology
Ticagrelor	CYP2C19	Clinical studies
Warfarin	CYP2C9, VKORC1	Dose and administration, precautions, and clinical pharmacology

[a]Data consolidated from FDA website as of August, 2012

This study reveals significant associations with NOS3 variants and CHD and heart failure and significant pharmacogenetic effects for stroke and all-cause mortality (Zhang et al. 2012). A recent study has reevaluated seven candidate genes linked to hypertension (Takeuchi et al. 2012). A comprehensive review of pharmacogenomics and hypertension management can also be found at (Johnson 2012).

4 Conclusions

Research in the area of pharmacogenomics of cardiovascular drugs has found hundreds of variations that may be linked to a particular treatment and particular ethnic group. Many of these studies gave rise to the urge to initiate new studies with larger population sizes and strict protocols that may dissect the results of some negative findings that may be associated with a particular variation. Combining certain results from different studies might give more power to understanding the role of the variation, but combination results must be looked with caution since different drugs from the same class and clinical variation in diagnosis and selection of the population are involved. These factors are important and should be taken into consideration. It is important to note that these studies have led to at least a few clinical actionable medicines in the market, and now a genetic test is required before a dose is prescribed for drugs like warfarin and clopidogrel. On top of this, the FDA has recommendation genotyping of patients for a number of drugs. Table 20.3 lists cardiovascular drugs that have FDA pharmacogenomic marker labels for genetic testing advisory.

In the genomic era, the introduction of next generation sequencing (NGS) has now taken up a leading role in genomic research, and it will slowly replace GWAS. While whole genome sequencing with deep coverage for less than $1,000 is still in development, effort has been in place to get an inexpensive testing of critical pharmacogenetic variants via genotyping arrays through which one can obtain the pharmacogenomics information for an individual at a lower cost (Johnson et al. 2012). Given the fact that several genes and hundreds of genotypes have been identified for CVD treatment efficacy, there are handful of tests that are commercially available to patients and healthy individuals in the United States. Companies like Ambry Genetics (http://www.ambrygen.com/), GeneMarkDx (http://www.genmarkdx.com/), 23andMe (https://www.23andme.com/health), and Genelex (http://www.genelex.com/) have tests specifically for warfarin, clopidogrel, and several cytochrome

P450 group of enzymes. However, this is a rapidly expanding field, and many university hospital clinics have their own genotyping facilities.

In the era of NGS, it is now easy to create data with low cost, but the real problem still exists in terms of storage and analysis of data. Generation of RNAseq data from each individual may be equally useful as this cannot only detect the coding variations but also would give indication of expression status of key genes that may be involved in drug absorption, drug elimination, drug transport, drug receptors, and more.

Besides the role of genetic variation in effectiveness of several CVD medications, there may be a role of epigenetic alteration in CVD drug metabolism. Epigenetics is known to affect the genes in drug metabolism and transport and may well affect the efficacy of interindividual variation of the CVD drugs. A comprehensive review of the epigenetic influence of genes required for absorption, distribution, metabolism, and excretion can be found elsewhere (Kacevska et al. 2012). Expression of several of these genes involved in the CVD pathway can also be influenced by miRNAs.

Given the complexity of the cardiovascular drugs, future studies must give strong emphasis in selecting a population with very strict clinical features for success of large cohorts that may be needed to deal with these in the future. The National Heart, Lung, and Blood Institute (NHLBI), USA, convened a meeting of the working group in January 2011 to provide recommendations to the NHLBI that would guide informed decisions on research directions and priorities in the field of cardiovascular pharmacogenomics. The conference brought together leaders from academia, industry, and government to (1) discuss personalized medicine's current and potential impact on cardiovascular patient outcomes, (2) review emerging technologies and applications that may shape the field in the future, (3) discuss the results of an American College of Cardiology survey examining personalized medicine adoption rates among US cardiologists, (4) identify the barriers to adoption of pharmacogenetics-based-personalized medicine in cardiovascular practice, and (5) develop recommendations for next steps with an emphasis on actions and evidence generation that are needed for adoption and improved quality of cardiovascular patient care (Musunuru et al. 2012). Guided by the proceedings of the *New Frontiers in Personalized Medicine: Cardiovascular Research and Clinical Care* conference (http://www.personalizedmedicinecoalition.org/events/highlights/2011-1-6), the NHLBI working group characterized and discussed challenges for cardiovascular pharmacogenomics in five domains as follows: clinical needs, clinical validation, information delivery, education and compliance, and cost-effectiveness (Musunuru et al. 2012). The team sought to identify and prioritize the most pressing clinical needs to focus research and translational efforts. In particular, three areas of emerging pharmacogenomic applications were reviewed: anticoagulation (warfarin), antiplatelet therapy (clopidogrel), and lipid-lowering therapy (statins). It was noted that the gold standard for pharmacogenomic applications (as with all clinical interventions) is a prospective trial with treatment determined by genotype and with a clinical endpoint as the primary outcome. One such example was a large warfarin pharmacogenomics study, the Medco-Mayo Warfarin Effectiveness Study with almost 4,000 individuals, which was designed to test whether the use of genotype information could reduce the incidence of hospitalizations from warfarin-related adverse effects (Epstein et al. 2010).

References

Armitage J (2007) The safety of statin in clinical practice. Lancet 370:1781–1790

Arnett DK, Davis BR, Ford CE, Boerwinkle E, Leiendecker-Foster C, Miller MB, Black H, Eckfeldt JH (2005) Pharmacogenetic association of the angiotensin converting enzyme insertion/deletion polymorphism on blood pressure and cardiovascular risk in relation to antihypertensive treatment: the Genetics of Hypertension-Associated Treatment (Gen-HAT) study. Circulation 111:3374–3383

Arnett DK, Claas SA, Glasser SP (2006) Pharmacogenetics of antihypertensive treatment. Vascul Pharmacol 44:107–118

Baigent C, Keech A, Kearney PM, Blackwell L, Buck G et al (2005) Cholesterol Treatment Trialists' (CTT) collaborators. Lancet 366:1267–1278

Ballantyne CM (2003) Current and future aims of lipid-lowering therapy: changing paradigms and lessons from the Heart Protection Study on standards of efficacy and safety. Am J Cardiol 92:3K–9K

Beitelshees AL, Gong Y, Wang D, Schork NJ, Cooper-Dehoff RM, INVEST Investigators et al (2007) KCNMB1 genotype influences response to verapamil SR and adverse outcomes in the INternational VErapamil SR/Trandolapril STudy (INVEST). Pharmacogenet Genomics 17:719–729

Beitelshees AL, Navare H, Wang D, Gong Y, Wessel J et al (2009) CACNA1C gene polymorphisms, cardiovascular disease outcomes, and treatment response. Circ Cardiovasc Genet 2(4):362–370

Bejarano-Achache I, Levy L, Mlynarsky L, Bialer M, Muszkat M, Caraco Y (2012) Effects of CYP4F2 polymorphism on response to warfarin during induction phase: a prospective, open-label, observational cohort study. Clin Ther 34:811–823

Berthold HK, Laaksonen R, Lehtimäki T, Gylling H, Krone W, Gouni-Berthold I (2008) SREBP-1c gene polymorphism is associated with increased inhibition of cholesterol-absorption in response to ezetimibe treatment. Exp Clin Endocrinol Diabetes 116:262–267

Boekholdt SM, Agema WR, Peters RJ, Zwinderman AH, van der Wall EE et al (2003) Variants of toll-like receptor 4 modify the efficacy of statin therapy and the risk of cardiovascular events. Circulation 107:2416–2421

Bozkurt O, de Boer A, Grobbee DE, de Leeuw PW, Kroon AA, Schiffers P, Klungel OH (2009 May) Variation in renin-angiotensin system and salt-sensitivity genes and the risk of diabetes mellitus associated with the use of thiazide diuretics. Am J Hypertens 22(5):545–551

Brandt JT, Close SL, Iturria SJ, Payne CD, Farid NA et al (2007) Common polymorphisms of CYP2C19 and CYP2C9 affect the pharmacokinetic and pharmacodynamic response to clopidogrel but not prasugrel. J Thromb Haemost 5:2429–2436

Brautbar A, Virani SS, Belmont J, Nambi V, Jones PH, Ballantyne CM (2012) LPL gene variants affect apoC-III response to combination therapy of statins and fenofibric acid in a randomized clinical trial of individuals with mixed dyslipidemia. J Lipid Res 53:556–560

Bray PF, Cannon CP, Goldschmidt-Clermont P, Moyé LA, Pfeffer MA, Sacks FM, Braunwald E (2001) The platelet Pl(A2) and angiotensin-converting enzyme (ACE) D allele polymorphisms and the risk of recurrent events after acute myocardial infarction. Am J Cardiol 88:347–352

Brugts JJ, Isaacs A, de Maat MP, Boersma E, van Duijn CM et al (2011) A pharmacogenetic analysis of determinants of hypertension and blood pressure response to angiotensin-converting enzyme inhibitor therapy in patients with vascular disease and healthy individuals. J Hypertens 29:509–519

Caldwell MD, Awad T, Johnson JA, Gage BF, Falkowski M et al (2008) CYP4F2 genetic variant alters required warfarin dose. Blood 111:4106–4112

Carlquist JF, Muhlestein JB, Horne BD, Hart NI, Bair TL, Molhuizen HO, Anderson JL (2003) The cholesteryl ester transfer protein Taq1B gene polymorphism predicts clinical benefit of statin therapy in patients with significant coronary artery disease. Am Heart J 146:1007–1014

Cen HJ, Zeng WT, Leng XY, Huang M, Chen X (2010) CYP4F2 rs2108622: a minor significant genetic factor of warfarin dose in Han Chinese patients with mechanical heart valve replacement. Br J Clin Pharmacol 70:234–240

Chapman N, Chang CL, Caulfield M, Dahlöf B, Feder G, Sever PS, Poulter NR (2011) Ethnic variations in lipid-lowering in response to a statin (EVIREST): a substudy of the Anglo-Scandinavian Cardiac Outcomes Trial (ASCOT). Ethn Dis 21:150–157

Chasman DI, Posada D, Subrahmanyan L et al (2004) Pharmacogenetic study of statin therapy and cholesterol reduction. JAMA 291:2821–2827

Chasman DI, Giulianini F, MacFadyen J, Barratt BJ, Nyberg F, Ridker PM (2012) Genetic determinants of statin-induced low-density lipoprotein cholesterol reduction: the Justification for the Use of Statins in Prevention: an Intervention Trial Evaluating Rosuvastatin (JUPITER) trial. Circ Cardiovasc Genet 5:257–264

Chien KL, Wang KC, Chen YC et al (2010) Common sequence variants in pharmacodynamics and pharmacokinetic pathway-related genes conferring LDL cholesterol response to statins. Pharmacogenomics 11:309–317

Collet JP, Hulot JS, Pena A, Villard E, Esteve JB et al (2009) Cytochrome P450 2C19 polymorphism in young patients treated with clopidogrel after myocardial infarction: a cohort study. Lancet 373:309–317

Cuisset T, Morange PE, Alessi MC (2012) Recent advances in the pharmacogenetics of clopidogrel. Hum Genet 131:653–664

Cusi D, Barlassina C, Azzani T, Casari G, Citterio L et al (1997) Polymorphisms of alpha-adducin and salt sensitivity in patients with essential hypertension. Lancet 349:1353–1357

D'Ambrosio RL, D'Andrea G, Cappucci F, Chetta M, Di Perna P, Brancaccio V, Grandone E, Margaglione M (2004) Polymorphisms in factor II and factor VII genes modulate oral anticoagulation with warfarin. Haematologica 89:1510–1516

D'Andrea G, D'Ambrosio RL, Di Perna P, Chetta M, Santacroce R, Brancaccio V, Grandone E, Margaglione M (2005) A polymorphism in the VKORC1 gene is associated with an interindividual variability in the dose-anticoagulant effect of warfarin. Blood 105:645–649

Dörr M, Schmidt CO, Spielhagen T, Bornhorst A, Hentschel K (2010) β-blocker therapy and heart rate control during exercise testing in the general population: role of a common G-protein β-3 subunit variant. Pharmacogenomics 11:1209–1221

El Din MS, Amin DG, Ragab SB, Ashour EE, Mohamed MH, Mohamed AM (2012) Frequency of VKORC1

(C1173T) and CYP2C9 genetic polymorphisms in Egyptians and their influence on warfarin maintenance dose: proposal for a new dosing regimen. Int J Lab Hematol 34:517–524

Epstein RS, Moyer TP, Aubert RE, O'Kane DJ, Xia F, Verbrugge RR, Gage BF, Teagarden JR (2010) Warfarin genotyping reduces hospitalization rates results from the MM-WES (Medco-Mayo Warfarin Effectiveness study). J Am Coll Cardiol 55:2804–2812

Fan YM, Laaksonen R, Janatuinen T, Vesalainen R, Nuutila P, Knuuti J, Lehtimäki T (2001) Effects of pravastatin therapy on serum lipids and coronary reactivity are not associated with SREBP cleavage-activating protein polymorphism in healthy young men. Clin Genet 60:319–321

Fiegenbaum M, da Silveira FR, Van der Sand CR, Van der Sand LC, Ferreira ME, Pires RC, Hutz MH (2005a) The role of common variants of ABCB1, CYP3A4, and CYP3A5 genes in lipid-lowering efficacy and safety of simvastatin treatment. Clin Pharmacol Ther 78(5):551–558

Fiegenbaum M, Silveira FR, Van der Sand CR, Van der Sand LC, Ferreira ME, Pires RC, Hutz MH (2005b) Determinants of variable response to simvastatin treatment: the role of common variants of SCAP, SREBF-1a and SREBF-2 genes. Pharmacogenomics J 5:359–364

Filigheddu F, Reid JE, Troffa C, PinnaParpaglia P, Argiolas G, Testa A, Skolnick M, Glorioso N (2004) Genetic polymorphisms of the beta-adrenergic system: association with essential hypertension and response to beta-blockade. Pharmacogenomics J 4:154–160

Filigheddu F, Argiolas G, Degortes S, Zaninello R, Frau F, Pitzoi S, Bulla E, Bulla P, Troffa C, Glorioso N (2012) Haplotypes of the adrenergic system predict the blood pressure response to beta-blockers in women with essential hypertension. Pharmacogenomics 11:319–325

Fontana P, Hulot JS, De Moerloose P, Gaussem P (2007) Influence of CYP2C19 and CYP3A4 gene polymorphisms on clopidogrel responsiveness in healthy subjects. J Thromb Haemost 5:2153–2155

Frazier L, Turner ST, Schwartz GL, Chapman AB, Boerwinkle E (2004) Multilocus effects of the renin-angiotensin-aldosterone system genes on blood pressure response to a thiazide diuretic. Pharmacogenomics J 4:17–23

Freeman DJ, Samani NJ, Wilson V, McMahon AD, Braund PS, Cheng S, Caslake MJ, Packard CJ, Gaffney D (2003) A polymorphism of the cholesteryl ester transfer protein gene predicts cardiovascular events in non-smokers in the West of Scotland Coronary Prevention Study. Eur Heart J 2420:1833–1842

Fu R, Sun YM, Su Y, Wu Y, Luan Y (2008) Effect of statin therapy on plasma high-density lipoprotein-cholesterol levels is modified by paraoxonase 1 in Chinese patients with coronary heart disease. Clin Exp Pharmacol Physiol 35:982–983

Gage BF, Eby C, Johnson JA, Deych E, Rieder MJ et al (2008) Use of pharmacogenetic and clinical factors to predict the therapeutic dose of warfarin. Clin Pharmacol Ther 84:326–331

Gandelman K, Fung GL, Messig M, Laskey R (2012) Systemic exposure to atorvastatin between Asian and Caucasian subjects: a combined analysis of 22 studies. Am J Ther 9:164–173

Gaziano TA (2005) Cardiovascular disease in the developing world and its cost-effective management. Circulation 112:3547

Geisler T, Schaeffeler E, Dippon J, Winter S, Buse V, Bischofs C, Zuern C, Moerike K, Gawaz M, Schwab M (2008) Cytochrome P450 2C19 and non-genetic factors predict poor responsiveness to clopidogrel loading dose after coronary stent implantation. Pharmacogenomics 9:1251–1259

Glorioso N, Manunta P, Filigheddu F, Troffa C, Stella P et al (1999) The role of alpha-adducin polymorphism in blood pressure and sodium handling regulation may not be excluded by a negative association study. Hypertension 34:649–654

Gonzalez-Conejero R, Rivera J, Corral J, Acuña C, Guerrero JA, Vicente V (2005) Biological assessment of aspirin efficacy on healthy individuals: heterogeneous response or aspirin failure? Stroke 36:276–280

Guzman EC, Hirata MH, Quintão EC, Hirata RD (2000) Association of the apolipoprotein B gene polymorphisms with cholesterol levels and response to fluvastatin in Brazilian individuals with high risk for coronary heart disease. Clin Chem Lab Med 38:731–736

Hannuksela ML, Liinamaa MJ, Kesäniemi YA, Savolainen MJ (1994) Relation of polymorphisms in the cholesteryl ester transfer protein gene to transfer protein activity and plasma lipoprotein levels in alcohol drinkers. Atherosclerosis 110:35–44

Harrap SB, Tzourio C, Cambien F, Poirier O, Raoux S, PROGRESS Collaborative Group et al (2003) The ACE gene I/D polymorphism is not associated with the blood pressure and cardiovascular benefits of ACE inhibition. Hypertension 42:297–303

Heidenreich PA, Trogdon JG, Khavjou OA et al (2011) Forecasting the future of cardiovascular disease in the United States: a policy statement from the American Heart Association. Circulation 123:933–944

Heller DA, de Faire U, Pedersen NL et al (1993) Genetic and environmental influences on serum lipid levels in twins. N Engl J Med 328:1150–1156

Higashi MK, Veenstra DL, Kondo LM, Wittkowsky AK, Srino-uanprachanh SL, Farin FM, Rettie AE (2002) Association between CYP2C9 genetic variants and anticoagulation-related outcomes during warfarin therapy. JAMA 287:1690–1698

Himbergen TM, van Tits LJ, Voorbij HA, de Graaf J, Stalenhoef AF, Roest M (2005) The effect of statin therapy on plasma high-density lipoprotein cholesterol levels is modified by paraoxonase-1 in patients with familial hypercholesterolaemia. J Intern Med 258:442–449

Hingorani AD, Jia H, Stevens PA, Hopper R, Dickerson JE, Brown MJ (1995) Renin-angiotensin system gene polymorphisms influence blood pressure and the response to angiotensin converting enzyme inhibition. J Hypertens 13:1602–1609

Holloway JW, Yang IA, Ye S (2005) Variation in the toll-like receptor 4gene and susceptibility to myocardial infarction. Pharmacogenet Genomics 15:15–21

Hulot JS, Bura A, Villard E, Azizi M, Remones V, Goyenvalle C, Aiach M, Lechat P, Gaussem P (2006) Cytochrome P450 2C19 loss-of-function polymorphism is a major determinant of clopidogrel responsiveness in healthy subjects. Blood 108:2244–2247

Igel M, Arnold KA, Niemi M, Hofmann U, Schwab M, Lütjohann D, von Bergmann K, Eichelbaum M, Kivistö KT (2006) Impact of the SLCO1B1 polymorphism on the pharmacokinetics and lipid lowering efficacy of multiple-dose pravastatin. Clin Pharmacol Ther 79:419–426

Jiang X, Sheng HH, Lin G, Li J, Lu XZ, Cheng YL, Huang J, Xiao HS, Zhan YY (2007) Effect of renin-angiotensin-aldosterone system gene polymorphisms on blood pressure response to antihypertensive treatment. Chin Med J (Engl) 120:782–786

Jiang XY, Zhang Q, Chen P, Li SY, Zhang NN, Chen XD, Wang GC, Wang HB, Zhuang MQ, Lu M (2012) CYP7A1 polymorphism influences the LDL cholesterol-lowering response to atorvastatin. J Clin Pharm Ther 37:719–723

Johnson JA (2012) Advancing management of hypertension through pharmacogenomics. Ann Med 44(Suppl 1):S17–S22

Johnson JA, Burkley BM, Langaee TY, Clare-Salzler MJ, Klein TE, Altman RB (2012) Implementing personalized medicine: development of a cost-effective customized pharmacogenetics genotyping array. Clin Pharmacol Ther 92:437–439

Kacevska M, Ivanov M, Ingelman-Sundberg M (2012) Epigenetic-dependent regulation of drug transport and metabolism: an update. Pharmacogenomics 13:1373–1385

Kajinami K, Brousseau ME, Ordovas JM, Schaefer EJ (2004) Polymorphisms in the multidrug resistance-1(MDR1) gene influence the response to atorvastatin treatment in a gender-specific manner. Am J Cardiol 93:1046–1050

Kajinami K, Brousseau ME, Ordovas JM, Schaefer EJ (2005) A promoter polymorphism in cholesterol 7alphahydroxylase interacts with apolipoprotein E genotype in the LDL lowering response to atorvastatin. Atherosclerosis 180:407–415

Kardia SL, Sun YV, Hamon SC, Barkley RA, Boerwinkle E, Turner ST (2007) Interactions between the adducin 2 gene and antihypertensive drug therapies in determining blood pressure in people with hypertension. BMC Med Genet 8:61

Keskitalo JE, Pasanen MK, Neuvonen PJ, Niemi M (2009) Different effects of the ABCG2 c.421C>A SNP on the pharmacokinetics of fluvastatin, pravastatin and simvastatin. Pharmacogenomics 10:1617–1624

Kimura R, Miyashita K, Kokubo Y, Akaiwa Y, Otsubo R et al (2007) Genotypes of vitamin K epoxide reductase, gamma-glutamyl carboxylase, and cytochrome P450 2C9 as determinants of daily warfarin dose in Japanese patients. Thromb Res 120:181–186

Klerkx AH, Tanck MW, Kastelein JJ, Molhuizen HO, Jukema JW, Zwinderman AH, Kuivenhoven JA (2003) Haplotype analysis of the CETP gene: not TaqIB, but the closely linked -629C–>A polymorphism and a novel promoter variant are independently associated with CETP concentration. Hum Mol Genet 12:111–123

Klos KL, Kullo IJ (2007) Genetic determinants of HDL: monogenic disorders and contributions to variation. Curr Opin Cardiol 22:344–351

Krauss RM, Mangravite LM, Smith JD et al (2008) Variation in the 3-hydroxyl-3-methylglutarylcoenzyme a reductase gene is associated with racial differences in low density lipoprotein cholesterol response to simvastatin treatment. Circulation 117:1537–1544

Kuivenhoven JA, Jukema JW, Zwinderman AH, de Knijff P, McPherson R, Bruschke AV, Lie KI, Kastelein JJ (1998) The role of a common variant of the cholesteryl ester transfer protein gene in the progression of coronary atherosclerosis. The Regression Growth Evaluation Statin Study Group. N Engl J Med 338:86–93

Kurland L, Melhus H, Karlsson J, Kahan T, Malmqvist K, Ohman KP, Nyström F, Hägg A, Lind L (2001) Angiotensin converting enzyme gene polymorphism predicts blood pressure response to angiotensin II receptor type 1 antagonist treatment in hypertensive patients. J Hypertens 19(10):1783–1787

Kurland L, Melhus H, Karlsson J, Kahan T, Malmqvist K, Ohman P, Nyström F, Hägg A, Lind L (2002) Polymorphisms in the angiotensinogen and angiotensin II type 1 receptor gene are related to change in left ventricular mass during antihypertensive treatment: results from the Swedish Irbesartan Left Ventricular Hypertrophy Investigation versus Atenolol (SILVHIA) trial. J Hypertens 20:657–663

Kurland L, Liljedahl U, Karlsson J, Kahan T, Malmqvist K, Melhus H, Syvänen AC, Lind L (2004) Angiotensinogen gene polymorphisms: relationship to blood pressure response to antihypertensive treatment. Results from the Swedish Irbesartan Left Ventricular Hypertrophy Investigation vs Atenolol (SILVHIA) trial. Am J Hypertens 17:8–13

Lahoz C, Peña R, Mostaza JM, Jiménez J, Subirats E, Pintó X, Taboada M, López-Pastor A, RAP Study Group (2003) Apo A-I promoter polymorphism influences basal HDL-cholesterol and its response to pravastatin therapy. Atherosclerosis 168:289–295

Lahoz C, Peña R, Mostaza JM, Laguna F, García-Iglesias MF, Taboada M, Pintó X (2005a) Baseline levels of low-density lipoprotein cholesterol and lipoprotein (a) and the AvaII polymorphism of the low-density lipoprotein receptor gene influence the response of low-density lipoprotein cholesterol to pravastatin treatment. Metabolism 54:741–747

Lahoz C, Peña R, Mostaza JM, Laguna F, García-Iglesias MF, Taboada M, Pintó X, RAP Study Group (2005b) The -514C/T polymorphism of the hepatic lipase gene significantly modulates the HDL-cholesterol response to statin treatment. Atherosclerosis 182(1):129–134, Epub 26 Feb 2005

Lanzani C, Citterio L, Glorioso N, Manunta P, Tripodi G et al (2010) Adducin- and ouabain-related gene variants predict the antihypertensive activity of rostafuroxin, part 2: clinical studies. Sci Transl Med 2(59):ra87

Law M, Wald N, Morris J (2003) Lowering blood pressure to prevent myocardial infarction and stroke: a new preventive strategy. Health Technol Assess 7:1–94

Lev EI, Patel RT, Guthikonda S, Lopez D, Bray PF, Kleiman NS (2007) Genetic polymorphisms of the platelet receptors P2Y(12), P2Y(1) and GP IIIa and response to aspirin and clopidogrel. Thromb Res 119:355–360

Levitt MR, Osbun JW, Kim LJ (2012) The pharmacogenomics of clopidogrel. World Neurosurg 77:406–407

Li Y, Zhou Y, Yang P, Niu JQ, Wu Y, Zhao DD, Wu SL (2011) Interaction of ACE and CYP11B2 genes on blood pressure response to hydrochlorothiazide in Han Chinese hypertensive patients. Clin Exp Hypertens 33:141–146

Liu J, Liu ZQ, Tan ZR, Chen XP, Wang LS, Zhou G, Zhou HH (2003) Gly389Arg polymorphism of beta1-adrenergic receptor is associated with the cardiovascular response to metoprolol. Clin Pharmacol Ther 74:372–379

Liu J, Liu ZQ, Yu BN, Xu FH, Mo W, Zhou G, Liu YZ, Li Q, Zhou HH (2006) Beta1-Adrenergic receptor polymorphisms influence the response to metoprolol monotherapy in patients with essential hypertension. Clin Pharmacol Ther 80:23–32

Lubitz SA, Scott SA, Rothlauf EB, Agarwal A, Peter I et al (2010) Comparative performance of gene-based warfarin dosing algorithms in a multiethnic population. J Thromb Haemost 8:1018–1026

Luo F, Wang Y, Wang X, Sun K, Zhou X, Hui R (2009) A functional variant of NEDD4L is associated with hypertension, antihypertensive response, and orthostatic hypotension. Hypertension 54:796–801

Ma C, Zhang Y, Xu Q, Yang J, Zhang Y, Gao L, Xu B, Wang H, Li Y, Lu C, Yin T (2012) Influence of warfarin dose-associated genotypes on the risk of hemorrhagic complications in Chinese patients on warfarin. Int J Hematol 96(6):719–728

Maitland-van der Zee AH, Klungel OH, Leufkens HG, de Boer A, Stricker BH, Hofman A, Witteman JC, van Duijn CM, Kastelein JJ (2004) Effectiveness of HMG-CoA reductase inhibitors is modified by the ACE insertion deletion polymorphism. Atherosclerosis 175:377–379

Maitland-van der Zee AH, Boerwinkle E, Arnett DK, Davis BR, Leiendecker-Foster C, Miller MB, Klungel OH, Ford CE, Eckfeldt JH (2007) Absence of an interaction between the angiotensin converting enzyme insertion-deletion polymorphism and pravastatin on cardiovascular disease in high-risk hypertensive patients: the Genetics of Hypertension-Associated Treatment (GenHAT) study. Am Heart J 153:54–58

Malin R et al (2001) Paraoxonase genotype modifies the effect of pravastatin on high-density lipoprotein cholesterol. Pharmacogenetics 11(7):625–633

Mangravite LM, Thorn CF, Krauss RM (2006) Clinical implications of pharmacogenomics of statin treatment. Pharmacogenomics J 6:360–374

Mangravite LM, Medina MW, Cui J, Pressman S, Smith JD, Rieder MJ, Guo X, Nickerson DA, Rotter JI, Krauss RM (2010) Combined influence of LDLR and HMGCR sequence variation on lipid-lowering response to simvastatin. Arterioscler Thromb Vasc Biol 30:1485–1492

Manunta P, Lavery G, Lanzani C, Braund PS, Simonini M, Bodycote C, Zagato L, Delli Carpini S, Tantardini C, Brioni E, Bianchi G, Samani NJ (2008) Physiological interaction between alpha-adducin and WNK1-NEDD4L pathways on sodium-related blood pressure regulation. Hypertension 52:366

Maree AO, Curtin RJ, Chubb A, Dolan C, Cox D, O'Brien J, Crean P, Shields DC, Fitzgerald DJ (2005) Cyclooxygenase-1 haplotype modulates platelet response to aspirin. J Thromb Haemost 3:2340–2345

Marian AJ, Safavi F, Ferlic L, Dunn JK, Gotto AM, Ballantyne CM (2003) Interactions between angiotensin-I converting enzyme insertion/deletion polymorphism and response of plasma lipids and coronary atherosclerosis to treatment with fluvastatin: the lipoprotein and coronary atherosclerosis study. J Am Coll Cardiol 35:89–95

Marschang P, Sandhofer A, Ritsch A, Fišer I, Kvas E, Patsch JR (2006) Plasma cholesteryl ester transfer protein concentrations predict cardiovascular events in patients with coronary artery disease treated with pravastatin. J Intern Med 260:151–159

Mason DA, Moore JD, Green SA, Liggett SB (1999) A gain-of-function polymorphism in a G-protein coupling domain of the human beta1-adrenergic receptor. J Biol Chem 274:12670–12674

Materson BJ, Reda DJ, Cushman WC, Massie BM, Freis ED et al (1993) Single-drug therapy for hypertension in men. A comparison of six antihypertensive agents with placebo. The Department of Veterans Affairs Cooperative Study Group on Antihypertensive Agents. N Engl J Med 328:914–921. Erratum in: N Engl J Med 330:1689, 1994

Medina MW (2010) The relationship between HMGCR genetic variation, alternative splicing, and statin efficacy. Discov Med 9:495–499

Medina MW, Krauss RM (2009) The role of HMGCR alternative splicing in statin efficacy. Trends Cardiovasc Med 19:173–177

Miniño AM, Murphy SL, Xu J, Kochanek KD (2011) Deaths: final data for 2008. National vital statistics reports, vol 59, no 10. National Center for Health Statistics, Hyattsville

Mohrschladt MF, van der Sman-de Beer F, Hofman MK, van der Krabben M, Westendorp RG, Smelt AH (2005) TaqIB polymorphism in CETP gene: the influence on incidence of cardiovascular disease in statin-treated patients with familial hypercholesterolemia. Eur J Hum Genet 13:877–882

Motovska Z, Kvasnicka J, Hajkova J, Kala P, Simek S et al (2010) Platelet gene polymorphisms and risk of bleeding in patients undergoing elective coronary angiography: a genetic substudy of the PRAGUE-8 trial. Atherosclerosis 212:548–552

Musunuru K, Roden DM, Boineau R, Bristow MR, McCaffrey TA et al (2012) Cardiovascular pharmacogenomics: current status and future directions-report of a national heart, lung, and blood institute working group. J Am Heart Assoc 1:e000554

Ned RM, Yesupriya A, Imperatore G, Smelser DT, Moonesinghe R, Chang MH, Dowling NF (2012) The ACE I/D polymorphism in US adults: limited evidence of association with hypertension-related traits and sex-specific effects by race/ethnicity. Am J Hypertens 25:209–215

Nestel P, Simons L, Barter P, Clifton P, Colquhoun D, Hamilton-Craig I, Sikaris K, Sullivan D (1997) A comparative study of the efficacy of simvastatin and gemfibrozil in combined hyperlipoproteinemia: prediction of response by baseline lipids, apo E genotype, lipoprotein(a) and insulin. Atherosclerosis 129:231–239

Niemi M (2007) Role of OATP transporters in disposition of drugs. Pharmacogenomics 8:787–802

Niemi M (2010) Transporter pharmacogenetics and statin toxicity. Clin Pharmacol Ther 87:130–133

Niemi M, Neuvonen PJ, Hofmann U, Backman JT, Schwab M, Lütjohann D, von Bergmann K, Eichelbaum M, Kivistö KT (2005) Acute effects of pravastatin on cholesterol synthesis are associated with SLCO1B1 (encoding OATP1B1) haplotype*17. Pharmacogenet Genomics 15:303–309

Niemi M, Pasanen MK, Neuvonen PJ (2011) Organic anion transporting polypeptide 1B1: a genetically polymorphic transporter of major importance for hepatic drug uptake. Pharmacol Rev 63:157–181

Niu Y, Gong Y, Langaee TY, Davis HM, Elewa H et al (2010) Genetic variation in the beta2 subunit of the voltage-gated calcium channel and pharmacogenetic association with adverse cardiovascular outcomes in the INternational VErapamil SR-Trandolapril STudy GENEtic Substudy (INVEST-GENES). Circ Cardiovasc Genet 3:548–555

Nurden AT (1997) Platelet glycoprotein IIIa polymorphism and coronary thrombosis. Lancet 350:1189–1191

Ojala JP, Helve E, Ehnholm C, Aalto-Setälä K, Kontula KK, Tikkanen MJ (1991) Effect of apolipoprotein E polymorphism and XbaI polymorphism of apolipoprotein B on response to lovastatin treatment in familial and non-familial hypercholesterolaemia. J Intern Med 230:397–405

Ordovas JM, Lopez-Miranda J, Perez-Jimenez F, Rodriguez C, Park JS, Cole T, Schaefer EJ (1995) Effect of apolipoprotein E and A-IV phenotypes on the low density lipoprotein response to HMG CoA reductase inhibitor therapy. Atherosclerosis 113:157–166

Pamukcu B, Oflaz H, Nisanci Y (2005) The role of platelet glycoprotein IIIa polymorphism in the high prevalence of in vitro aspirin resistance in patients with intracoronary stent restenosis. Am Heart J 149:675–680

Pedro-Botet J, Schaefer EJ, Bakker-Arkema RG, Black DM, Stein EM, Corella D, Ordovas JM (2001) Apolipoprotein E genotype affects plasma lipid response to atorvastatin in a gender specific manner. Atherosclerosis 158:183–193

Pena R, Lahoz C, Mostaza JM, Jiménez J, Subirats E, Pintó X, Taboada M, López-Pastor A, Rap Study Group (2002) Effect of apoE genotype on the hypolipidaemic response to pravastatin in an outpatient setting. J Intern Med 251:518–525

Peters BJ, Rodin AS, Klungel OH, van Duijn CM, Stricker BH, Van't Slot R, de Boer A, Maitland-van der Zee AH (2010) Pharmacogenetic interactions between ABCB1 and SLCO1B1 tagging SNPs and the effectiveness of statins in the prevention of myocardial infarction. Pharmacogenomics 11:1065–1076

Peters BJ, Pett H, Klungel OH, Stricker BH, Psaty BM, Glazer NL, Wiggins KL, Bis JC, de Boer A, Maitland-van der Zee AH (2011) Genetic variability within the cholesterol lowering pathway and the effectiveness of statins in reducing the risk of MI. Atherosclerosis 217:458–464

Petersen M, Andersen JT, Jimenez-Solem E, Broedbaek K, Hjelvang BR et al (2012) Effect of the Arg389Glyβ(1)-adrenoceptor polymorphism on plasma renin activity and heart rate, and the genotype-dependent response to metoprolol treatment. Clin Exp Pharmacol Physiol 39:779–785

Pisciotta L, Sallo R, Rabacchi C, Wunsch A, Calandra S, Bertolini S (2012) Leucine 10 allelic variant in signal peptide of PCSK9 increases the LDL cholesterol-lowering effect of statins in patients with familial hypercholesterolaemia. Nutr Metab Cardiovasc Dis 22:831–835

Rau T, Düngen HD, Edelmann F, Waagstein F, Lainščak M, Dimković S et al (2012) Impact of the β1-adrenoceptor Arg389Gly polymorphism on heart-rate responses to bisoprolol and carvedilol in heart-failure patients. Clin Pharmacol Ther 92:21–28. Erratum in: Clin Pharmacol Ther 92:397, 2012

Reitsma PH, van der Heijden JF, Groot AP, Rosendaal FR, Büller HR (2005) A C1173T dimorphism in the VKORC1 gene determines coumarin sensitivity and bleeding risk. PLoS Med 2:e312

Rieder MJ, Reiner AP, Gage BF, Nickerson DA, Eby CS, McLeod HL, Blough DK, Thummel KE, Veenstra DL, Rettie AE (2005) Effect of VKORC1 haplotypes on transcriptional regulation and warfarin dose. N Engl J Med 352:2285–2293

Rodrigues AC, Rebecchi IM, Bertolami MC, Faludi AA, Hirata MH, Hirata RD (2005) High baseline serum total and LDL cholesterol levels are associated with MDR1 haplotypes in Brazilian hypercholesterolemic individuals of European descent. Braz J Med Biol Res 38:1389–1397

Romaine SP, Bailey KM, Hall AS, Balmforth AJ (2010) The influence of SLCO1B1 (OATP1B1) gene polymorphisms on response to statin therapy. Pharmacogenomics J 10:1–11

Salazar LA, Hirata MH, Quintão EC, Hirata RD (2000) Lipid-lowering response of the HMG-CoA reductase inhibitor fluvastatin is influenced by polymorphisms in the low-density lipoprotein receptor gene in Brazilian patients with primary hypercholesterolemia. J Clin Lab Anal 14:125–131

Salek L, Lutucuta S, Ballantyne CM, Gotto AM Jr, Marian AJ (2002) Effects of SREBF-1a and SCAP polymorphisms on plasma levels of lipids, severity, progression and regression of coronary atherosclerosis and response to therapy with fluvastatin. J Mol Med (Berl) 80:737–744

Sanderson S, Emery J, Higgins J (2005) CYP2C9 gene variants, drug dose, and bleeding risk in warfarin-treated patients: a HuGEnet systematic review and meta-analysis. Genet Med 7:97–104

Sanllehy C, Casals E, Rodriguez-Villar C, Zambón D, Ojuel J, Ballesta AM, Ros E (1998) Lack of interaction of apolipoprotein E phenotype with the lipoprotein response to lovastatin or gemfibrozil in patients with primary hypercholesterolemia. Metabolism 47:560–565

Santos PC, Soares RA, Nascimento RM, Machado-Coelho GL, Mill JG, Krieger JE, Pereira AC (2011) SLCO1B1 rs4149056 polymorphism associated with statin-induced myopathy is differently distributed according to ethnicity in the Brazilian general population: Amerindians as a high risk ethnic group. BMC Med Genet 12:136

Santos PC, Gagliardi AC, Miname MH, Chacra AP, Santos RD, Krieger JE, Pereira AC (2012) SLCO1B1 haplotypes are not associated with atorvastatin-induced myalgia in Brazilian patients with familial hypercholesterolemia. Eur J Clin Pharmacol 68:273–279

Schalekamp T, van Geest-Daalderop JH, de Vries-Goldschmeding H, Conemans J, Bernsen Mj M, de Boer A (2004) Acenocoumarol stabilization is delayed in CYP2C93 carriers. Clin Pharmacol Ther 75:394–402

Schelleman H, Stricker BH, Verschuren WM, de Boer A, Kroon AA, de Leeuw PW, Kromhout D, Klungel OH (2006) Interactions between five candidate genes and antihypertensive drug therapy on blood pressure. Pharmacogenomics J 6:22–26

Sciarrone MT, Stella P, Barlassina C, Manunta P, Lanzani C, Bianchi G, Cusi D (2003) ACE and alpha-adducin polymorphism as markers of individual response to diuretic therapy. Hypertension 41:398–403

Sconce EA, Khan TI, Wynne HA, Avery P, Monkhouse L, King BP, Wood P, Kesteven P, Daly AK, Kamali F (2005) The impact of CYP2C9 and VKORC1 genetic polymorphism and patient characteristics upon warfarin dose requirements: proposal for a new dosing regimen. Blood 106:2329–2333

Shiffman D, Trompet S, Louie JZ, Rowland CM, Catanese JJ et al (2012) Genome-wide study of gene variants associated with differential cardiovascular event reduction by pravastatin therapy. PLoS One 7:e38240

Shikata E, Ieiri I, Ishiguro S, Aono H, Inoue K, Koide T, Ohgi S, Otsubo K (2004) Association of pharmacokinetic (CYP2C9) and pharmacodynamic (factors II, VII, IX, and X; proteins S and C; and gamma-glutamyl carboxylase) gene variants with warfarin sensitivity. Blood 103:2630–2635

Sing K, Ballantyne CM, Ferlic L, Brugada R, Cushman I, Dunn JK, Herd JA, Pownall HJ, Gotto AM, Marian AJ (1999) Lipoprotein lipase gene mutations, plasma lipid levels, progression/regression of coronary atherosclerosis, response to therapy, and future clinical events. Lipoproteins and Coronary Atherosclerosis Study. Atherosclerosis 144:435–442

Singer JB, Holdaas H, Jardine AG, Fellstrøm B, Os I et al (2007) Genetic analysis of fluvastatin response and dyslipidemia in renal transplant recipients. J Lipid Res 48:2072–2078

Sofowora GG, Dishy V, Muszkat M, Xie HG et al (2003) A common beta1-adrenergic receptor polymorphism (Arg389Gly) affects blood pressure response to beta-blockade. Clin Pharmacol Ther 73:366–371

Sorkin SC, Forestiero FJ, Hirata MH, Guzmán EC, Cavalli SA, Bertolami MC, Salazar LA, Hirata RD (2005) APOA1 polymorphisms are associated with variations in serum triglyceride concentrations in hypercholesterolemic individuals. Clin Chem Lab Med 43:1339–1345

Sortica VA, Fiegenbaum M, Lima LO, Van der Sand CR, Van der Sand LC, Ferreira ME, Pires RC, Hutz MH (2012) SLCO1B1 gene variability influences lipid-lowering efficacy on simvastatin therapy in Southern Brazilians. Clin Chem Lab Med 50:441–448

Srivastava K, Chandra S, Bhatia J, Narang R, Saluja D (2012) Association of angiotensinogen (M235T) gene polymorphism with blood pressure lowering response to angiotensin converting enzyme inhibitor (Enalapril). J Pharm Pharm Sci 15(3):399–406

Stavroulakis GA, Makris TK, Krespi PG, Hatzizacharias AN, Gialeraki AE, Anastasiadis G, Triposkiadis P, Kyriakidis M (2000) Predicting response to chronic antihypertensive treatment with fosinopril: the role of angiotensin-converting enzyme gene polymorphism. Cardiovasc Drugs Ther 14:427–432

Superko HR, Momary KM, Li Y (2012) Statins personalized. Med Clin North Am 96:123–139

Svensson-Färbom P, Wahlstrand B, Almgren P, Dahlberg J, Fava C, Kjeldsen S, Hedner T, Melander O (2011) A functional variant of the NEDD4L gene is associated with beneficial treatment response with β-blockers and diuretics in hypertensive patients. J Hypertens 29:388–395

Tachibana-Iimori R, Tabara Y, Kusuhara H, Kohara K, Kawamoto R, Nakura J, Tokunaga K, Kondo I, Sugiyama Y, Miki T (2004) Effect of genetic polymorphism of OATP-C (SLCO1B1) on lipid-lowering response to HMG-CoA reductase inhibitors. Drug Metab Pharmacokinet 19:375–380

Takahashi H, Wilkinson GR, Nutescu EA, Morita T, Ritchie MD et al (2006) Different contributions of polymorphisms in VKORC1 and CYP2C9 to intra- and inter-population differences in maintenance dose of warfarin in Japanese. Caucasians and African-Americans. Pharmacogenet Genomics 16:101–110

Takane H, Miyata M, Burioka N, Shigemasa C, Shimizu E, Otsubo K, Ieiri I (2006) Pharmacogenetic determinants of variability in lipid-lowering response to pravastatin therapy. J Hum Genet 51:822–826

Takeuchi F, McGinnis R, Bourgeois S, Barnes C, Eriksson N et al (2009) A genome-wide association study confirms VKORC1, CYP2C9, and CYP4F2 as principal genetic determinants of warfarin dose. PLoS Genet 5:e1000433

Takeuchi F, Yamamoto K, Katsuya T, Sugiyama T, Nabika T et al (2012) Reevaluation of the association of seven candidate genes with blood pressure and hypertension: a replication study and meta-analysis with a larger sample size. Hypertens Res 35:825–831

Thompson JF, Man M, Johnson KJ et al (2005) An association study of 43 SNPs in 16 candidate genes with atorvastatin response. Pharmacogenomics J 5:352–358

Thompson JF, Hyde CL, Wood LS et al (2009) Comprehensive whole-genome and candidate gene analysis for response to statin therapy in the treating to new targets (TNT) cohort. Circ Cardiovasc Genet 2:173–181

Turban S, Fuentes F, Ferlic L, Brugada R, Gotto AM, Ballantyne CM, Marian AJ (2001) A prospective study of paraoxonase gene Q/R192 polymorphism and severity, progression and regression of coronary atherosclerosis, plasma lipid levels, clinical events and response to fluvastatin. Atherosclerosis 154:633–640

Turner ST, Schwartz GL, Chapman AB, Boerwinkle E (2001) C825T polymorphism of the G protein beta(3)-subunit and antihypertensive response to a thiazide diuretic. Hypertension 37:739–743

Undas A, Sanak M, Musial J, Szczeklik A (1999) Platelet glycoprotein IIIa polymorphism, aspirin, and thrombin generation. Lancet 353:982–983

vanVenrooij FV, Stolk RP, Banga JD, Sijmonsma TP, van Tol A, Erkelens DW, Dallinga-Thie GM, DALI Study Group (2003) Common cholesteryl ester transfer protein gene polymorphisms and the effect of atorvastatin therapy in type2 diabetes. Diabetes Care 26:1216–1223

Visser LE, van Schaik RH, van Vliet M, Trienekens PH, De Smet PA, Vulto AG, Hofman A, van Duijn CM, Stricker BH (2004) The risk of bleeding complications in patients with cytochrome P450 CYP2C9*2 or CYP2C9*3 alleles on acenocoumarol or phenprocoumon. Thromb Haemost 92:61–66

Voora D, Shah SH, Reed CR, Zhai J, Crosslin DR et al (2008) Pharmacogenetic predictors of statin mediated low-density lipoprotein cholesterol reduction and dose response. Circ Cardiovasc Genet 1:100–106

Vrablík M, Hubáček JA, Dlouhá D, Lánská V, Rynekrová J, Zlatohlávek L, Prusíková M, Ceška R, Adámková V (2012) Impact of variants within seven candidate genes on statin treatment efficacy. Physiol Res 61(6):609–617

Wadelius M, Pirmohamed M (2007) Pharmacogenetics of warfarin: current status and future challenges. Pharmacogenomics J 7:99–111

Wadelius M, Chen LY, Lindh JD, Eriksson N, Ghori MJ, Bumpstead S, Holm L, McGinnis R, Rane A, Deloukas P (2009) The largest prospective warfarin-treated cohort supports genetic forecasting. Blood 113: 784–792

Winkelmann BR, Hoffmann MM, Nauck M, Kumar AM, Nandabalan K et al (2003) Haplotypes of the cholesteryl ester transfer protein gene predict lipid-modifying response to statin therapy. Pharmacogenomics J 3:284–296

Ye P, Shang Y, Ding X (2003) The influence of apolipoprotein B and E gene polymorphisms on the response to simvastatin therapy in patients with hyperlipidemia. Chin Med Sci J 18:9–13

Yuan HY, Chen JJ, Lee MT, Wung JC, Chen YF et al (2005) A novel functional VKORC1 promoter polymorphism is associated with inter-individual and interethnic differences in warfarin sensitivity. Hum Mol Genet 14:1745–1751

Zhang X, Lynch AI, Davis BR, Ford CE, Boerwinkle E, Eckfeldt JH, Leiendecker-Foster C, Arnett DK (2012) Pharmacogenetic association of NOS3 variants with cardiovascular disease in patients with hypertension: the GenHAT study. PLoS One 7:e34217

Zhou Y, Wu SL, Liu JQ, Liang WN, Liu GF (2007) Possible association of ACE gene I/D polymorphism with blood pressure–lowering response to hydrochlorothiazide. Biomed Environ Sci 20:351–356

Zuern CS, Schwab M, Gawaz M, Geisler T (2010) Platelet pharmacogenomics. J Thromb Haemost 8:1147–1158

Pharmacogenetics of Oral Anticoagulants

21

Enrique Jiménez-Varo, Marisa Cañadas-Garre, Margarita Aguilera, Desiree Gonzalez Callejas, Cristina Perez Ramirez, and Miguel A. Calleja Hernández

Abstract

Coumarin derivates have been considered an ideal target of personalized medicine because of their pharmacological characteristics as vitamin K antagonists. Thus, many pharmacogenetic investigators have focused in this field, developing numerous warfarin algorithms based on CYP2C9 and VKORC1 genotypes. These pharmacogenetic algorithms include both genetic and nongenetic factors.

Coumarin derivates (warfarin in the UK and USA, acenocoumarol and phenprocoumon in European countries) are the most prescribed oral anticoagulants in the prevention and treatment of thromboembolic events associated with atrial fibrillation, prosthetic valves, venous thromboembolism, and orthopedic surgery.

Vitamin K antagonists have demonstrated efficacy in anticoagulation although they have a narrow therapeutic window. Patients need frequent monitoring by measuring the prothrombin time. The main challenge for physicians is to introduce patients as soon as possible within the therapeutic range (INR 2–3) in the beginning of therapy as the risk of undercoagulation, INR <2 (risk of thrombus), or overanticoagulation, INR> 4 (risk of bleeding), is higher in this period.

The adjustment of the dose to achieve effective and stable anticoagulation depends on several environmental factors including age, gender, weight, concomitant medications and interactions with foods containing vitamin K, smoking status and alcohol intake, and genetic factors.

E. Jiménez-Varo • M. Cañadas-Garre (✉)
D.G. Callejas • C.P. Ramirez • M.A.C. Hernández
Pharmacogenetics Unit, Pharmacy Service,
Hospital Universitario Virgen de las Nieves,
Granada, Spain
e-mail: ejimenezvaro@gmail.com;
marisacgarre@gmail.com

M. Aguilera
Pharmacogenetics Unit, Pharmacy Service,
Hospital Universitario Virgen de las Nieves,
Granada, Spain

Instituto de Nutrición y Tecnología de los Alimentos
"José Mataix Verdú" de la Universidad de Granada
(INYTA); Centro de Investigación Biomédica (CIBM),
Granada, Spain
e-mail: maguiler@ugr.es

D. Barh et al. (eds.), *Omics for Personalized Medicine*,
DOI 10.1007/978-81-322-1184-6_21, © Springer India 2013

Allelic variants in CYP2C9*2 (Arg144Cys, rs1799853), CYP2C9*3 Ile359Leu (rs1057910), and VKORC1 (rs9923231, rs9934438) gene polymorphisms have demonstrated an influence of about 40 % on coumarin-required dose.

Despite of knowledge that pharmacogenetic models can improve dosing recommendations according to CYP2C9 and VKORC1 polymorphisms and thus would decrease the risk of thrombus and bleeding events during initial phases of anticoagulation, the translation of pharmacogenetic algorithms into clinical practice is a challenge to pursue in the advancement of personalized medicine.

On the other hand, a new wave of oral anticoagulants has been developed to improve the efficiency and safety of anticoagulant therapy and avoid the several drawbacks of vitamin K antagonists. This new generation of anticoagulants is expected to be the alternative in prevention of stroke and systemic embolism in non-valvular atrial fibrillation patients.

In this chapter, we will review the most relevant pharmacogenetic studies performed with coumarin derivates, the application of pharmacogenetics in predicting the optimal coumarin derivates' initial dose based on genetic and environmental factors, genetic tests available for determination of main gene polymorphisms associated to warfarin sensitivity, the development of new oral anticoagulants (dabigatran, rivaroxaban, apixaban), and comparison of different pharmacological therapies available to use in patients with non-valvular atrial fibrillation requiring long-term anticoagulation.

1 Introduction

1.1 Main Indications for Anticoagulation Therapy

Since the 1950s, coumarin derivates (warfarin, acenocoumarol, and phenprocoumon) have been the cornerstone for prevention of stroke and systemic embolism in patients with atrial fibrillation (AF) and for prevention of recurrent venous thrombosis in patients with venous thromboembolism (VTE) and in patients with prosthetic heart valves (Hirsh 1992; Hirsh et al. 2001; Daly and King 2003; Davis et al. 2011; Dahl 2012).

1.2 Management of Antithrombotic Therapy

1.2.1 Oral Anticoagulation with Coumarin Derivates in AF Patients

Starting long-term oral anticoagulation therapy depends on stroke recurrence risk factors present in the patient. The AF Investigators and Stroke Prevention in Atrial Fibrillation (SPAF) group developed a scheme for assessing the risk of stroke through the $CHADS_2$ (congestive heart failure, hypertension, age ≥ 65 years, diabetes mellitus, stroke) system (Camm et al. 2010; Furie et al. 2011; Skanes et al. 2012; You et al. 2012). This system assigned two points to a stroke antecedent and one point to other conditions (congestive heart failure, hypertension, age ≥ 65 years, diabetes mellitus). Risk assessment of thromboembolic and hemorrhagic events based on $CHADS_2$ classification is described in Table 21.1.

Other clinically relevant nonmajor risk factors include female sex, age 65–74 years, and vascular disease (complex aortic plaque and peripheral arterial disease). The $CHADS_2$ classification was modified including new parameters to improve the evaluation of the embolism risk, especially for patients classified with $CHADS_2$ score < 2, creating the CHA_2DS_2-VASc (congestive heart failure, hypertension, age ≥ 75 [doubled], diabetes mellitus, stroke [doubled], vascular disease,

Table 21.1 Risk assessment of thromboembolic and hemorrhagic events (Camm et al. 2010; Furie et al. 2011; Skanes et al. 2012; You et al. 2012)

$CHADS_2$	Description	Score
C (congestive heart failure)	Recent history of congestive heart failure	1
H (hypertension)	History of hypertension	1
A (age ≥ 65 years)	Age ≥ 65 years	1
D (diabetes mellitus)	History of diabetes mellitus	1
S_2 (stroke)	History of stroke	2
	Transitory ischemic accident	
Total score possible		6

age 65–74, and sex female). According to this new classification, the European Society of Cardiology (ESC) recommendations for vitamin K antagonists (AVKs) treatment are made in function of CHA2DS2-VASc scores (Camm et al. 2010):

- CHA2DS2-VASc ≥ 2 (one major risk factor or ≥2 risk factors not clinically relevant) treatment with oral anticoagulant (OAC) with a dose adjusted to INR 2–3
- CHA2DS2-VASc = 1 (1 risk factor for not clinically relevant) OAC treatment with a dose adjusted to INR 2–3 or aspirin 75–325 mg. OAC preferable to aspirin
- CHA2DS2-VASc = 0 (no risk factors) aspirin 75–325 mg daily or no antithrombotic therapy. Preferably aspirin to antithrombotic therapy

This modification of CHADS2 is not considered in other countries like USA and Canada (Furie et al. 2011; Skanes et al. 2012; You et al. 2012).

1.2.2 Oral Anticoagulation with Coumarin Derivates in VTE

Anticoagulant therapy is effective in prevention of recurrent venous thrombosis in patients with VTE. This treatment begins with low molecular weight heparin (LMWH) or unfractionated heparin (UFH) and oral anticoagulant therapy one day after. Double anticoagulation LMWH + OAC is maintained until INR range 2–3 is reached on two consecutive days, which happens approximately in the first 4–5 days (Duran Parrondo et al. 2003).

Anticoagulant therapy is advisable to be maintained for 6 weeks to 3 months in patients with symptomatic distal or proximal thrombosis after surgery and at least 6 months in patients with idiopathic proximal thrombosis. OAC is recommended indefinitely in situations of idiopathic recurrent thrombosis or inherited or acquired thrombophilia (Duran Parrondo et al. 2003).

1.2.3 Oral Anticoagulation with Coumarin Derivates in Patients with Prosthetic Heart Valve

The risk of thromboembolism in the first 6 months after implantation of a prosthetic valve is particularly high (Eitz et al. 2008). There are three approaches for anticoagulant therapy in patients after valve surgery, depending on the bleeding risk score:

- Subcutaneous UFH prophylactic dose + OAC started from the first postoperative day
- Intravenous UFH + OAC from the second postoperative day. HNF to reach therapeutic INR
- LMWH + OAC since the first postoperative day. LMWH to achieve therapeutic INR

After the first 6 months, the risk of bleeding is greatly reduced if the patient is regular and adherent with the treatment and complies with the drug dosage regimens; due to these, patients must take lifetime oral anticoagulation (Eitz et al. 2008).

1.3 Drugs Pharmacology and Pharmacokinetics

1.3.1 Vitamin K Antagonists

Warfarin

Warfarin is worldwide prescribed. This coumarin derivate is administered as a racemic mixture of R-warfarin and S-warfarin. The S-enantiomer has an anticoagulant activity of 3–5 times greater than the enantiomer R-warfarin (Choonara et al. 1986). S-warfarin is mainly metabolized by CYP2C9. Although R-warfarin is also metabolized through the CYP2C9 pathway, there are other minor isoenzymes involved in the metabolism of this enantiomer, such as CYP1A1, CYP1A2, CYP2C8, CYP2C18, and CYP2C19 (Rettie et al. 1992; Zhang et al. 1995; Wienkers et al. 1996; Ngui et al. 2001).

Acenocoumarol

Acenocoumarol is the main oral anticoagulant prescribed in many European countries (Beinema et al. 2008; Markatos et al. 2008; Spreafico et al. 2008; López-Parra et al. 2013). It is orally administered as a racemic mixture of enantiomers (R-) and (S-). Biotransformation is performed in the liver by Cytochrome P450. The most important metabolism enzyme of this drug is CYP2C9 isoenzyme, which performs a hydroxylation at position 6, 7, and 8 of both enantiomers. The S-enantiomer is rapidly metabolized ($t_{1/2}$<2 h); therefore, the anticoagulant activity depends on R-acenocoumarol enantiomer. Unlike S-acenocoumarol enantiomer, R-acenocoumarol enantiomer is also metabolized by CYP1A2, CYP3A5, and CYP2C19 isoenzymes (Thijssen et al. 2000).

Phenprocoumon

Phenprocoumon, like other coumarin derivates, is administered as a racemic mixture of the S- and R-enantiomers. Bioavailability is over 90 %, and it is stereoselectively metabolized by Cytochrome P-450 to inactive hydroxylated metabolites (Alberio 2003). Approximately 65 % of phenprocoumon dose is eliminated through the urinary tract and the remaining 35 % by the fecal via (Toon et al. 1985). CYP2C9 and CY3A4 are the main isoenzymes in the metabolism of phenprocoumon (Ufer 2005). The two enantiomers R- and S-phenprocoumon reach their half-life time around 110–130 h from the first-dose administration (Jähnchen et al. 1976).

1.3.2 Mechanism of Action of Vitamin K Antagonists

Vitamin K antagonists exert their effect by inhibiting vitamin K epoxide reductase complex 1 (VKORC1), decreasing vitamin K. Vitamin K is an essential cofactor in the carboxylation of glutamate residues at the N-terminal region of the vitamin K-dependent proteins, synthesis of vitamin K–dependent coagulation factors (factors II, VII, IX, and X), and fibrinolytic proteins, such as protein C and S (Fig. 21.1). These vitamin K–dependent proteins are inhibited by the action of coumarin derivates. When antagonizing these vitamin K dependent coagulation factors, no interaction between them and the calcium present in the vascular subendothelium occurs. The inhibition of coagulation factors causes low plasma levels of prothrombin, which generates the formation of fibrin required to trigger coagulation (DrugBank 2012b; The Pharmacogenomics Knowledge Base 2012).

1.3.3 Pharmacokinetic Differences Between AVKs

The differences between these molecules lie on their pharmacokinetics properties. Acenocoumarol presents a half-life time of 8–11 h, whereas warfarin half-life time is fourfold the acenocoumarol half-life, from 34 to 42 h, and reaches its maximum plasma concentration peak at 90 min after oral administration, and phenprocoumon has a half-life time of 110–130 h after initial oral administration (Shirolkar et al. 2010). Main differences in pharmacokinetics of vitamin K antagonists are shown in Table 21.2.

2 Pharmacogenetics of Vitamin K Antagonists

Although coumarin derivates are effective, the management of oral anticoagulation therapy is complicated. The pharmacokinetic and pharmacodynamic properties lead to a variable, unpredictable, and independent response in each individual treatment.

The anticoagulant activity is measured by a standardized laboratory test, the international normalized ratio (INR). The beginning of anticoagulation therapy is the most problematic period due to the risk of thrombi (associated to an INR<2), and the risk of bleeding (associated to INR>4) is higher in the first 3 months (Hylek et al. 2003; Ufer 2005). This risk is high because each patient requires individualized doses to achieve stability.

Currently, clinicians prescribe an initial dose of coumarin derivates based on clinical parameters, and dose adjustment is made by titration. The first INR monitoring is performed at day 3 or 4 after beginning anticoagulant therapy. Depending

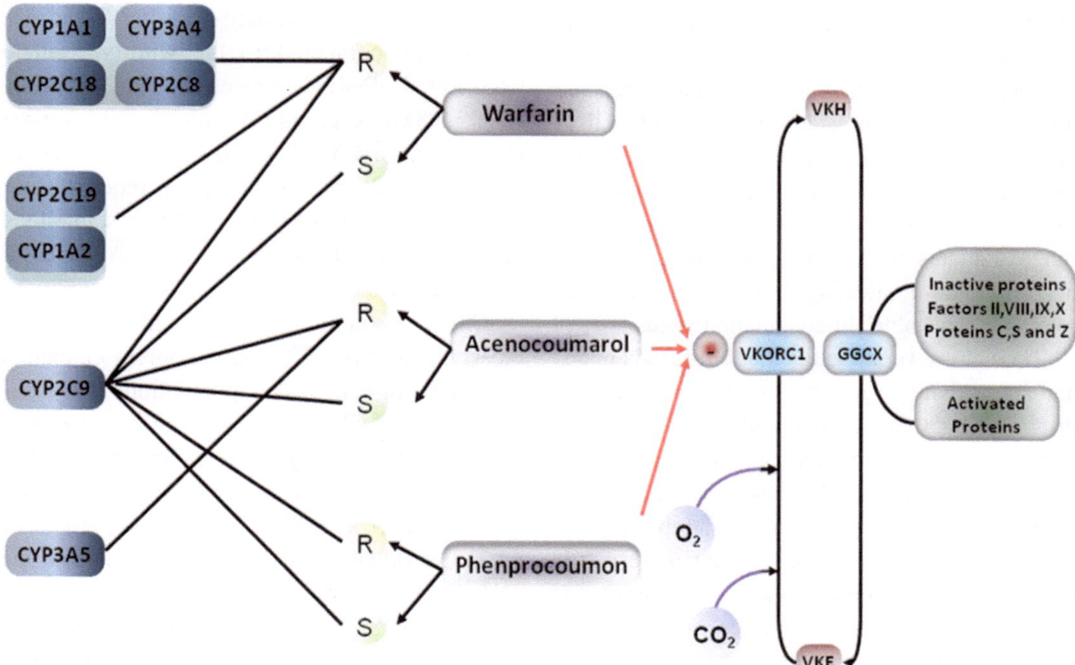

Fig. 21.1 Mechanism of action of coumarin derivates. *CYP1A1* Cytochrome P450 family 1, subfamily A, polypeptide 1; *CYP1A2* Cytochrome P450 family 1, subfamily A, polypeptide 2; *CYP3A4* Cytochrome P450 family 3, subfamily A, polypeptide 4; *CYP2C9* Cytochrome P450 family 2, subfamily C, polypeptide 9; *CYP2C8* Cytochrome P450 family 2, subfamily C, polypeptide 8; *CYP2C18* Cytochrome P450 family 2, subfamily C, polypeptide 18; *CYP2C19* Cytochrome P450 family 2 subfamily C, polypeptide19; *CYP2C3A5* Cytochrome P450 family 3, subfamily A, polypeptide 5; *VKORC1* vitamin K epoxide reductase complex, subunit 1; *R-W* enantiomer R-warfarin; *S-W* enantiomer S-warfarin; *R-A* enantiomer R-acenocoumarol; *S-A* enantiomer S-acenocoumarol; *R-P* enantiomer R-phenprocoumon; *S-P* enantiomer S-phenprocoumon; *inhibition VKE* vitamin K epoxide; *VKH* vitamin K hydroquinone; CO_2 carbon dioxide; O_2 oxygen

Table 21.2 Pharmacokinetics of vitamin K antagonists

Drug	$t_{1/2}$ (h)	Main isoenzyme metabolism R-enantiomer	S-enantiomer	References
Warfarin	34–42	CYP2C9		Choonara et al. (1986), Rettie et al. (1992), Zhang et al. (1995), Wienkers et al. (1996), Ngui et al. (2001)
		CYP1A1		
		CYP1A2		
		CYP2C18		
		CYP2C19		
		CYP3A4		
Acenocoumarol	8–11	CYP2C9		Beinema et al. (2008), Markatos et al. (2008), Spreafico et al. (2008), López-Parra et al. (2013)
		CYP1A2		
		CYP3A5		
		CYP2C19		
Phenprocoumon	110–130	CYP2C9		Jähnchen et al. (1976), Alberio (2003), Ufer (2005), Shirolkar et al. (2010)
		CYP3A4		

CYP1A1 Cytochrome P450 family 1, subfamily A, polypeptide 1; *CYP1A2* Cytochrome P450 family 1, subfamily A, polypeptide 2; *CYP2C18* Cytochrome P450 family 2, subfamily C, polypeptide 18; *CYP2C19* Cytochrome P450 family 2, subfamily C, polypeptide19; *CYP3A4* Cytochrome P450 family 3, subfamily A, polypeptide 4; *CYP2C9* Cytochrome P450 family 2, subfamily C, polypeptide 9; *CYP2C3A5* Cytochrome P450 family 3, subfamily A, polypeptide 5 on the INR value obtained, the initial dose is increased, decreased, or maintained until three consecutive INR values are obtained within the therapeutic range with a dose variation lower than 10 %, which is considered as a stable anticoagulation status. The main goal of oral anticoagulation therapy is to keep the patient in INR therapeutic range. Patients treated with oral anticoagulants need periodic monitoring of anticoagulant doses to maintain a stable anticoagulation status.

There are several factors that affect the anticoagulant activity. Age, sex, weight, concomitant drugs (CYP450 inducers or inhibitors), other associated diseases, alcohol and smoking status, vitamin K intake, and genetic factors are responsible for interindividual dose–response variability.

For all these reasons, coumarin derivates have been considered as an ideal target for personalized medicine. Studies have focused on finding genetic markers that can predict a better response to treatment in terms of efficacy and safety.

2.1 Main Gene Polymorphisms Involved in Oral Anticoagulant Therapy

The most studied genetic variations correspond to genes encoding CYP2C9 and vitamin K epoxide reductase complex 1 (VKORC1). Genetic variations in these two genes are responsible for approximately 30 % of the variability in the variability dose required (Rieder et al. 2005; Li et al. 2009). Table 21.3 shows the main gene polymorphisms involved in anticoagulant response and the associated effect.

2.1.1 CYP2C9

CYP2C9 isoenzyme is important in the liver. It is responsible of metabolizing approximately 15 % of drugs that undergo phase I metabolism (Evans and Relling 1999). Two common polymorphisms in CYP2C9 have been described. These SNPs (single nucleotide polymorphisms) are CYP2C9*2 (Arg144Cys, rs1799853) and CYP2C9*3 (Ile359-Leu; rs1057910). These allelic variants encode enzymes with reduced activity of 12% and 5% respectively compared to the wild-type allele (Rettie et al. 1994; Haining et al. 1996; Takanashi et al. 2000). Carriers of either allele variant CYP2C9*2 or CYP2C9*3 have more overanticoagulation risk (risk of bleeding), require lower doses, and need more time to achieve stable INR than wild-type patients for this gene (Aithal et al. 1999; Becquemont 2008).

Allelic variant frequencies differ between races, for example, CYP2C9*2 polymorphism is extremely rare in Asian population, the other allelic variant CYP2C9*3 was not found in this population (Rosemary and Adithan 2007), and in African Americans is around 6 % of the population (Limdi et al. 2008). However, 15–30 % of white individuals have any allelic variant for this gene (Jonas and McLeod 2009).

CYP2C9*8 (Arg150His, rs7900194) is manifested in 10–12 % of African American individuals (Scott et al. 2009; Perera et al. 2011). This allelic variant is responsible for a decrease of enzyme activity. Therefore, individuals carrying this minor allelic variant needed lower warfarin dose to achieve optimal levels of INR compared with wild-type individuals (Cavallari et al. 2010). These data have been corroborated by another study conducted with South African individuals with similar results (Mitchell et al. 2011).

2.1.2 VKORC1

Oral anticoagulants act by inhibiting vitamin K regeneration. This effect is performed through inhibition of vitamin K epoxide reductase. The vitamin K epoxide reductase complex 1 (VKORC1) gene encodes this protein and is located on the short arm of chromosome 16. It contains three exons and two introns and encodes a membrane protein of 163 amino acid residues (D'Andrea et al. 2005). It is located in the endoplasmic reticulum membrane (Cain et al. 1997). Several polymorphisms in the VKORC1 gene are associated to warfarin dose variability, mainly located in the noncoding region (Spreafico et al. 2008). Four haplotypes (VKORC1*1,*2,*3,*4) have been associated to variability in warfarin dose in Caucasian population (Osman et al. 2006).

Table 21.3 Main gene polymorphisms involved in anticoagulant response

Genes	Haplotypes	Reference SNPs	SNP changes	Protein changes	Effects	References
VKORC1	VKORC1*2	rs9923231	G1639A	–	Coumarin derivates sensitivity	Geisen et al. (2005), Rieder et al. (2005), Veenstra et al. (2005), Wadelius et al. (2005), Carlquist et al. (2006), Obayashi et al. (2006), Osman et al. (2006), Takahashi et al. (2006), Loebstein et al. (2007), Borgiani et al. (2007), Momary et al. (2007), Rieder et al. (2007), Schelleman et al. (2007), Wadelius et al. (2007), Stehle et al. (2008), Wang et al. (2008), Cini et al. (2012)
		rs9934438	C1137T	–		
		rs2359612	C2255T	–		
	VKORC1*3	rs7294	G9041A	–	Coumarin derivates resistance	
	VKORC1*4	rs17708472	C6009T	–		
CYP2C9	CYP2C9*2	rs1799853	C8633T	Arg144Cys	Reduced enzyme activity	Rettie et al. (1994), Haining et al. (1996), Takanashi et al. (2000)
	CYP2C9*3	rs1057910	A47639C	Ile359Leu	Reduced enzyme activity	
	CYP2C9*8	rs7900194	G474A	Arg150His	Reduced enzyme activity	Scott et al. (2009), Cavallari et al. (2010), Mitchell et al. (2011), Perera et al. (2011)
			G474T	Arg150 Leu		
CYP4F2	CYP4F2*3	rs2108622	G1347A	Val 433 Met	Low increased dose required	Fava et al. (2008), Fu et al. (2009), Scott et al. (2010)
GGCX	–	rs12714145	G6317A	Intron region	Low reduction of dose required	Wadelius et al. (2005), King et al. (2010)
CALU	–	rs339097	582+133A>G	Intron region	Inhibition of anticoagulant effect	Wajih et al. (2004), Shahin et al. (2011)

VKORC1 vitamin K epoxide reductase complex, subunit 1; *CYP2C9*2* Cytochrome P450 family 2, subfamily C, polypeptide 9, allele variant: 2; *CYP2C9*3* Cytochrome P450 family 2, subfamily C, polypeptide 9, allele variant: 3; *CYP4F2* Cytochrome P450, family 4, subfamily F, polypeptide 2; *GGCX* gamma-glutamyl carboxylase; *CALU* calumenin

VKORC1*1 haplotype corresponds to absence of polymorphism; VKORC1*2 haplotype with C1173T allelic variation (rs9934438), which is in linkage disequilibrium with the G1639A polymorphism (rs9923231) in the 5′-upstream region and with the C2255T polymorphism (rs2359612); VKORC1*3 haplotype with the G9041A polymorphism (rs7294); and VKORC1*4 haplotype with C6009T (rs17708472) polymorphism (Loebstein et al. 2007; Stehle et al. 2008). According to Geisen et al. (2005), these haplotypes cover about 99 % of the genetic variability in European population. The VKORC1*2 haplotype is associated with coumarin derivate sensitivity, and haplotypes *3 and *4 are associated with an increase in warfarin dose requirements to achieve the same level of anticoagulation than wild-type patients for this gene (Rieder et al. 2005, 2007; Wadelius et al. 2005; Cini et al. 2012).

VKORC1 *2 is the most prevalent haplotype among Caucasian and Asian population. These haplotype alters the binding site of the transcription factor, which leads to decreased expression of the protein (Wang et al. 2008). According to HapMap Project (dbSNP Short Genetic Variation, 2012), VKORC1*2 haplotype frequency population observed in Caucasians is 39.8 % for wild-type, 40.7 % for heterozygous mutant, and 19.5 % homozygous mutant. It has been described that this VKORC1 polymorphism has an influence of 11–30 % of warfarin dose variability in Caucasian and Asian populations (Veenstra et al. 2005; Carlquist et al. 2006; Obayashi et al. 2006; Borgiani et al. 2007; Wadelius et al. 2007). However, the influence of this polymorphism in warfarin dose variability in North American blacks is 4–10 % (Takahashi et al. 2006; Momary et al. 2007; Schelleman et al. 2007).

2.2 Other Gene Polymorphisms Involved in Anticoagulant Response

2.2.1 CYP4F2

CYP4F2 catalyzes vitamin K1 hydroxylation in the liver. It is reported an allelic variant (rs2108622; V433M) in CYP4F2 gene that has influence on required dose of coumarin derivates. The allelic variant CYP4F2*3 is responsible for a reduced enzyme activity. Therefore, CYP4F2 V433M polymorphism has been associated with warfarin resistance (Caldwell et al. 2008). Although there is an association between the presence of CYP4F2*3 polymorphism and increased dose required, the influence of this polymorphism on total variability warfarin dose is of 1–2 % (Scott et al. 2010). However, several studies have shown that the presence of the allelic variant CYP4F2 V433M is associated to ischemic stroke and myocardial infarction (Fava et al. 2008; Fu et al. 2009).

2.2.2 CYP2C19

Acenocoumarol anticoagulant activity lies mainly on R-acenocoumarol, which is metabolized mainly by CYP2C9, but other CYP450 isoenzymes, including CYP1A2 and CYP2C19, are also involved. Although polymorphisms in these isoenzymes have not shown influence on the required dose, taking concomitant drugs metabolized by CYP450, such as proton pump inhibitors, mainly metabolized by CYP2C19, may increase the risk of overanticoagulation in long-term oral anticoagulation therapy (Teichert et al. 2011b)

2.2.3 GGCX

The gene encoding gamma-glutamyl carboxylase has been studied as a potential candidate gene that affects the pharmacodynamics of warfarin. The gamma-glutamyl carboxylase enzyme catalyzes biosynthesis of vitamin K–dependent coagulation factor. GGCX is an essential cofactor in the reduction of vitamin K epoxide 2–3 to biologically active vitamin K hydroquinone, responsible for synthesizing coagulation factors that trigger the coagulation cascade (Stafford 2005). A Swedish study examined the effect of a polymorphism in the GGCX (rs12714145) resulting in a small but significant association with the dose of warfarin (Wadelius et al. 2005). Another variant allele in the GGCX gene (rs11676382) is associated with a reduction in the required dose of warfarin and a total implication of 2 % in the dose variability (King et al. 2010).

2.2.4 CALU

Calumenin, encoded by CALU, binds to GGCX and inhibits the binding site of vitamin K epoxide. It produces an inhibition of the anticoagulant effect. An allelic variant (rs339097) in CALU gene has been described to predict higher warfarin doses in African-American and in a Egyptian population (Wajih et al. 2004; Shahin et al. 2011).

2.3 Influence of Gene Polymorphisms in Oral Anticoagulant Response: Drug Efficacy, Prognosis, and Adverse Drug Effects

Early research focused on the influence of CYP2C9 polymorphisms on the variability of the anticoagulant response. A small study conducted in England compared 36 patients with doses of ≤1.5 mg/day warfarin with 100 individuals from a random control group who had started as standard anticoagulation therapy (Aithal et al. 1999). The goals were to analyze the effect of CYP2C9*2 and CYP2C9*3 polymorphisms in the dose and the incidence of bleeding complications. In the low-dose group, 89 % (29/36) had at least one variant allele for CYP2C9. The incidence of major bleeding was higher in the low-dose group compared with the random controls (ratio 3.68 [1-43-9.5] p-value = 0.007).

An Italian study was one of the first to manifest that patients carrying CYP2C9*2 and/or CYP2C9*3 polymorphisms require lower doses of warfarin to achieve therapeutic INR range and had a higher rate of bleeding complications than wild-type patients for this gene (Margaglione et al. 2000). The study recruited 180 Caucasian patients and concluded that patients with CYP2C9*1 (wild-type) required higher dose than patients with a variant allele CYP2C9*2 or CYP2C9*3 (6.7, 5.2, and 3.8 mg, respectively). Fifty-nine bleeding episodes occurred (10 major and 49 minor events) in 36 patients; the incidence of bleeding episodes in patients with a variant allele CYP2C9*2 and/or CYP2C9*3 haplotypes was 27.9 % while only 12.8 % of wild-type patients showed some hemorrhagic event.

Despite the evidence that some patients carrying *2 or *3 polymorphisms in CYP2C9 metabolism isoenzyme showed warfarin sensitivity and increased risk of overanticoagulation, this could only explain 10 % of the warfarin dose variability (Gage et al. 2004). For this reason, research efforts focused on polymorphisms in the target of coumarin derivates (Li et al. 2004; Rost et al. 2004). It has been described that the presence of polymorphisms in the VKORC1 associated to the required dose of coumarin derivates. The identification of polymorphisms in the VKORC1 led researchers to study the influence of genes encoding VKORC1 and variability in the warfarin-required dose variability.

Several studies showed association between the presence of VKORC1 haplotypes and dosage requirements and with low required dose (sensitivity) warfarin (D'Andrea et al. 2005; Rieder et al. 2005; Wadelius et al. 2005; Yuan et al. 2005) . Rieder et al. (2005) identified 10 SNPs and two haplotypes related to variability in the dose required. Common haplotypes (H1, H2, H7, H8, and H9) were clustered forming two distinct evolutionarily distant groups designated A (H1 and H2) haplotype and B haplotype (H7, H8, and H9). The A haplotype was associated to warfarin sensitivity and B haplotype was associated to warfarin resistance. Patients were classified according to the haplotypes, being the A/A haplotype low dose, A/B haplotype intermediate dose, and B/B haplotype high dose. The maintenance dose of warfarin varied from 2.7 ± 0.2 mg/day for low dose (A/A) to 4.9 ± 0.2 mg/day for intermediate dose (A/B) and 6.2 ± 0.3 mg/day for high dose (B/B) (p-value < 0.001).

Aquilante et al. (2006) tested the impact of CYP2C9 and VKORC1 polymorphisms on warfarin response in a cohort of 350 patients with stable INR and dose. Patients with the AA genotype for VKORC1 3673 SNP received 22.5 mg/week less than GG carriers for this gene (p-value < 0.001). On the other hand, patients carrying one allelic variant CYP2C9*2 or CYP2C9*3 (*1/*2 or *1/*3) required 9.5 mg/week less than wild-type patients (*1/*1). In addition, patients carrying two allelic variants (CYP2C9*2/*2, CYP2C9*2/*3, or CYP2C9*3/*3) needed a dose of 14.1 mg/week lower than wild-type patients.

The influence of GGCX gene polymorphisms in addition to VKORC1 polymorphisms was analyzed in 201 patients, mostly Caucasian, treated with long oral anticoagulation therapy (Wadelius et al. 2005). They found five common SNPs, four of them in linkage disequilibrium (rs9923231, rs9934438, rs2359612, and rs7294) and a fifth SNP (rs11150606) with lower allele frequency (4 %). These SNPs were responsible for approximately 30 % of the variability in the required dose of warfarin. The GGCX gene polymorphism (rs12714145) showed a small but significant association with the dose of warfarin.

Allelic variants in CYP2C9 and VKORC1 and clinical factors can explain 54 % of the dose variability. An allelic variant (rs2108622) in CYP4F2 gene was identified in a small sample (Caldwell et al. 2008) which contributes to increase the prediction of the required dose of warfarin to 56 %. CC carriers (wild-type) require lower doses than CT or TT patients for this gene. This allelic variant is associated to coumarin derivates resistance.

The main objectives when a patient starts oral anticoagulation therapy with coumarin derivates are the time necessary to reach the INR therapeutic range and time within therapeutic range (TTR). Pharmacogenetics aims to improve the achievement of these goals through genotyping patients. Schwarz et al. 2008 evaluated the influence of CYP2C9 and VKORC1 polymorphisms on warfarin anticoagulation response in Caucasian populations. Patients with A/A haplotype for VKORC1 gene reached INR therapeutic range and overanticoagulation (INR > 4) before the remaining patients. Allelic variants in CYP2C9 were not decisive in achieving INR therapeutic range. However, patients carrying a variant allele CYP2C9*2 or CYP2C9*3 reached INR > 4 before wild-type patients for this gene. These findings are consistent with those found by Millican et al. (2007) in which those VKORC1 allelic variant carriers showed initial sensitivity to warfarin in a Caucasian cohort.

Personalized medicine and genetic algorithms should be useful for multiracial population; influence of genetic polymorphisms has been also evaluated in other populations different from Caucasian. Limdi et al. (2009) studied the influence of genetic polymorphisms in CYP2C9 and VKORC1 in a cohort of 521 patients that included African Americans and European Americans. They observed the influence of CYP2C9 and VKORC1 gene polymorphisms in TTR and risk of overanticoagulation (INR > 4). African Americans showed other allelic variants in addition to CYP2C9*2 and CYP2C9*3 allelic variants, CYP2C9*5, CYP2C9*6, and CYP2C9*11, while American European allelic variants showed only CYP2C9*2 and CYP2C9*3. Furthermore, the genotype frequency of VKORC1 G1639A polymorphism was higher among the European population than in the African population (60.4 % vs. 20.1; p-value < 0.001). Carriers of an allelic variant in CYP2C9 and VKORC1 reached the therapeutic INR range sooner than wild-type patients for these genes. On the other hand, VKORC1 polymorphism is more decisive than CYP2C9 polymorphisms in reaching the INR therapeutic range. Patients carrying CYP2C9 and/or VKORC1 allelic variant showed more overanticoagulation incidence (p-value < 0.0001) than patients without allelic variants for these genes. Overanticoagulation frequency was lower in African American patients (60 episodes in 46 patients) than European American patients (98 episodes in 78 patients, p-value = 0.017), so genetic polymorphisms in CYP2C9 and VKORC1 influence on the variability of warfarin dose in both populations.

Prediction of the required warfarin dose before starting oral anticoagulation treatment with coumarin derivates is based on the determination of genetic polymorphisms that alter pharmacokinetics (CYP2C9), and pharmacodynamics (VKORC1) is more accurate than the dose prediction based on clinical parameters. The question is whether genetic polymorphisms keep their influence after the stable dose has been established for each patient. A recent study (Lund et al. 2012) tested the influence of VKORC1 G1639A CYP2C9*2 and CYP2C9*3 polymorphisms, during the first 3 months of anticoagulation therapy in 557 patients. The main goal was to assay the prevalence of INR > 5 (risk of bleeding), and the secondary aims were to quantify rate of bleeding events, time to reach the therapeutic range of warfarin stable dose, and time to achieve stable anticoagulation. Thirty-two percent of patients with genotype AA in VKORC1 had some INR > 5 in the first month, while patients with AG or GG genotype who showed INR > 5 were 12.4 and 5.7 %, respectively (p-value < 0.001). In the same analysis, repeated at the third month, VKORC1 polymorphisms had no influence in presenting INR > 5 (AA 2.9 %, AG 3.2 %, and GG 5.2 %). Patients with the AA genotype for VKORC1 presented more bleeding events than wild-type patients for this gene (4.9 % vs. 0.47 %, p-value < 0.009). Allelic variants in CYP2C9 showed no influence on INR > 5 or bleeding rate. VKORC1 and CYP2C9*2 polymorphisms also showed an influence on required stable doses between genotypes: (1) AA 2 mg, AG 4 mg, and GG 4.5 mg for VKORC1 genotype; (2) 4 mg for CYP2C9 wild-type; and (3) 3.6 mg for CYP2C9*2. According to these results, the determination of CYP2C9 and VKORC1 genotypes prior to therapy initiation is positive, but polymorphisms showed no influence after the first month.

In conclusion, polymorphisms in CYP2C9 and VKORC1 are responsible for approximately 40 % of the variability in dose (Bodin et al. 2005; Manolopoulos et al. 2010; Teichert et al. 2009, 2011a; Wadelius et al. 2009). Environmental factors together with genetic factors account for at least 50 % of the variability in the required dose of coumarin derivates (Cadamuro et al. 2010; D'Andrea et al. 2008; Hirsh et al. 2001; Limdi and Veenstra 2008; Schelleman et al. 2008). Individuals carrying CYP2C9 (*2,*3) and/or VKORC1 (*2) allelic variants need lower coumarin derivate dose than wild-type patients to achieve the same level of anticoagulation (Schalekamp et al. 2006). After CYP2C9*2 and CYP2C9*3 allelic variants, CYP2C9*5, CYP2C9*6, CYP2C9*11 allelic variants also show prominent prevalence in African Americans, compared to Caucasian population. The SNP G1639A in VKORC1 gene was higher among the European population than in the African population (Limdi et al. 2008). VKORC1 G1639A and CYP2C9*2 and CYP2C9*3 polymorphisms showed influence in the initial required warfarin dose (D'Andrea et al. 2005; Rieder et al. 2005; Lund et al. 2012).

2.4 Application of CYP2C9 and VKORC1 Gene Polymorphisms in the Management of Anticoagulant Therapy. Development of a Treatment Protocol: Pharmacogenetic Dosing Algorithms

Environmental factors together with genetic factors account for at least 50 % of the variability in the required dose of coumarin derivates (Cadamuro et al. 2010; D'Andrea et al. 2008; Hirsh et al. 2001; Limdi and Veenstra 2008; Schelleman et al. 2008). In 2007, the FDA (Food and Drug Administration) accepted that CYP2C9 and VKORC1 genotypes may be useful in recommending initial warfarin dosage (FDA News and Events 2012c). In 2010, the determination of polymorphisms CYP2C9*2 (rs1799853), CYP2C9*3 (rs1057910), and VKORC1 (rs9923231, rs9934438) was included in a table of recommended starting dose of warfarin (FDA. U.S. Food and Drug Administration 2012a).

Several dosing algorithms including demographic, clinical, and genetic variants (CYP2C9 and VKORC1) as predictors of the required dose of coumarin derivates have been proposed. The goals of these studies were improving efficiency and reducing adverse events (bleeding and thrombosis) that occur during initiation of oral anticoagulation therapy (Sconce et al. 2005; Gage et al. 2008; International Warfarin Pharmacogenetics Consortium et al. 2009; Wadelius et al. 2009; Lenzini et al. 2010; Markatos et al. 2008).

Studies conducted to predict warfarin, acenocoumarol, or phenprocoumon dose can be classified according to their particular objectives. Many pharmacogenetic studies have focused on searching a prediction of warfarin maintenance dose. These maintenance doses of warfarin include genetic factors (CYP2C9 and VKORC1) and also demographic and clinical parameters (age, sex, weight, amiodarone). These studies provided a basis for studies focused on developing algorithms that could be able to predict the initial required dose of warfarin before starting anticoagulation therapy with coumarin derivates. Other type of studies in the prediction of the required dose included dose adjustments depending on the INR values obtained in the first few days of therapy.

2.4.1 Pharmacogenetic Warfarin Dosing. Online Resources: Warfarin Dosing and IWPC Algorithms

Advances in pharmacogenetics and personalized medicine have encouraged researchers to conduct numerous investigations looking for warfarin pharmacogenetic dose algorithms, which include genetic and clinical factors in order to improve the management of oral anticoagulation therapy. Most of the developed algorithms are not powerful enough to recommend its use in a clinical setting. Two of the best-validated algorithms have been developed by Gage et al. and by the International Warfarin Pharmacogenetics Consortium (IWPC) (International Warfarin Pharmacogenetics Con-sortium (IWPC) Algorithm 2012) and are available online.

Gage et al. (2008) recruited a diverse multicentric cohort of 1,015 patients, 83 % Caucasian and 64 % women, to develop this algorithm. VKORC1 3673G>A was the most important factor in the prediction of warfarin initial dosing. CYP2C9*2 and CYP2C9*3 allele variants predicted 19 and 33 % of reductions in the therapeutic dose of warfarin, respectively. The derivation cohort was able to explain a 53–54 % of the variability in the stable therapeutic dose of warfarin.

IWPC algorithm was widely validated by several research groups that collected clinical and genetic data from more than 5,000 patients who were under warfarin treatment and with INR range of 2–3. This study classified patients into three ranges:

- Low dose: ≤21 mg/week
- High dose: ≥49 mg/week
- Intermediate dose: 21–49 mg/week

The pharmacogenetic algorithm obtained a prediction of 35 and 38 % variability in warfarin dosing in the low and high dose, respectively. In both cases, low-dose (≤21 mg/week) and high-dose (≥49 mg/week) pharmacogenetic algorithm was more capable to predict the variability dose than clinical algorithm dose (35 % vs. 24 %, $p<0.001$ and 32.8 % vs. 13.3 %, p-value <0.001, respectively). Both groups (low dose and high dose) represented 46 % of the total cohort. This algorithm resulted to be more useful for patients who are in one of these groups. Thus, the pharmacogenetic algorithm is not recommendable for patients who are within the middle group.

Both algorithms are available online for free use and simple application, by inserting patient parameters in a calculation table. The algorithm developed by Gage and colleagues is available on the website www.warfarindosing.org, and IWPC algorithm is in the pharmgkb website. These pharmacogenetic algorithms have proved to be more accurate in predicting warfarin dose compared to other pharmacogenetic algorithms (Sagreiya et al. 2010; Shaw et al. 2010; Bazan et al. 2012). Table 21.4 shows the different algorithms and required parameters to calculate the individualized dose warfarin.

2.4.2 CoumaGen-II: A Randomized Trial and Clinical Effectiveness Comparing Two Pharmacogenetic Algorithms and Standard Care for Individualizing Warfarin Dosing

A total of 504 patients were randomized into three arms in the CoumaGen-II trial (Anderson et al. 2012). Recruitment period was 2 years and the follow-up was 3 months. The population was divided in three groups. Two pharmacogenetic (PG) algorithm groups and a control standard clinical group were designed. Nomenclature PG-1 and PG-2 was used to designate the pharmacogenetic groups. The algorithm of PG-1 group was based on the algorithm developed by the IWPC group with slight modifications. PG-2 algorithm incorporated two modifications; the first modification was not including CYP2C9 allelic variants in the PG algorithm. Considering that this isoenzyme exerts its effect on the elimination route of

Table 21.4 Pharmacogenetic warfarin dosing

Algorithms	Drugs	Gene polymorphisms	Clinical variables	Races	Indications	Prediction of the doses	Websites	References
Warfarin dosing	Warfarin	CYP2C9*2, CYP2C9*3, CYP2C9*3, CYP2C9*5, CYP2C9*6, VKORC1-1639/3673, CYP4F2 V433M, GGCX rs11676382	Age, sex, weight, height, smoke status, liver disease, baseline INR, target INR, amiodarone/Cordarone, statin/HMG CoA reductase inhibitor, azole, sulfamethoxazole/Septra/Bactrim/Cotrim/Sulfatrim	African American or Black American Indian or Alaska native Asian or Indian subcontinent Native Hawaiian or other Pacific Islander White, Caucasian, or Middle Eastern Other	Atrial fibrillation, cardioembolic stroke, DVT, heart failure, cardiomyopathy, heart valve replacement, hip fracture, hip replacement, knee replacement, myocardial infarction, pulmonary embolism, pulmonary hypertension, others	53 % of the variability in a warfarin dose	Warfarindosing.org	Gage et al. (2008)
IWPC	Warfarin	VKORC1-A1639G, CYP2C9*2, CYP2C9*3, CYP2C9*3	Age, sex, weight, height, amiodarone, enzyme inducer	Asian Black or African American White or Caucasian Unknown or mixed race	DVT, pulmonary embolism, atrial fibrillation/flutter, heart valve, cardiomyopathy left ventricular dilation, stroke, post-orthopedic, others	35 % variability in warfarin dosing in the low and 38 % of high dose	http://www.pharmgkb.org/drug/PA451906#tabview=tab0&subtab=31	(IWPC Algorithm 2012)

Online resources: warfarin dosing and IWPC algorithms (International Warfarin Pharmacogenetics Consortium (IWPC) Algorithm 2012)
VKORC1 vitamin K epoxide reductase complex, subunit 1; *CYP2C9*2* Cytochrome P450 family 2, subfamily C, polypeptide 9, allele variant: 2; *CYP2C9*3* Cytochrome P450 family 2, subfamily C, polypeptide 9, allele variant: 3

Table 21.5 Results of TTR and OOR at month 1 and month 3 of CoumaGen-II trial

	PG-1 group ($n=245$)	PG-2 group ($n=232$)	p-value	Combined PG group ($n=477$)	Parallel controls ($n=1,866$)	p-value
TTR month 1, mean CI	70.2 (67.2–73.3)	67.5 (64.5–70.6)	Pni = 0.12 Ps = 0.22	68.9 (67.2–70.6)	58.4 (56.8–60.0)	<0.001
OOR INRs % month 1, mean CI	30.6 (27.6–33.7)	31.8 (28.8–34.8)	Pni = 0.0025 Ps = 0.59	31.2 (29.2–33.5)	41.5 (39.9–43.2)	<0.001
TTR month 3, mean CI	71.7 (68.7–74.6)	70.8 (67.9–73.6)	Pni = 0.019 Ps = 0.68	71.2 (69.2–73.3)	58.6 (57.0–60.2)	<0.001
OOR INRs % month 3, mean CI	30.3 (27.3–33.2)	30.3 (27.5–33.1)	Pni = 0.0068 Ps = 0.99	30.3 (28.5–32.6)	42.3 (40.8–43.8)	<0.001

Data from Anderson et al. (2012)
TTR indicates time into therapeutic range, *OOR* out of range, *INR* international normalized ratio, *PG* pharmacogenetics, *ni* non-inferiority, *s* superiority, *CI* indicates 95 % confidence interval

this molecule, i.e., after the third day ($t_{½}$: 15–42 h), they hypothesized that it should not affect the initial sensitivity to this drug (Schwarz et al. 2008). The second change was a reevaluation of the dose after 4 or 5 days (3 or 4 doses), based on INR values. They made two kinds of comparisons to evaluate clinical effectiveness between groups. The first comparison was blind, randomized comparison between groups using pharmacogenetic algorithm (PG-1 vs. PG-2) and the second comparison was between a combined algorithm-guided group (PG-1 + PG-2) vs. control standard clinical group. The main objectives were to calculate the percentage of "out of range" (OOR) time and the percentage of "in therapeutic range" (TTR) time at months one and three after initiation of oral anticoagulation therapy with warfarin. The groups PG-1 and PG-2 obtained similar results for both TTR as OOR (Table 21.5). However, there were differences between combined pharmacogenetic subgroups (PG-1 + PG-2) compared with the control group. Differences in these parameters were found at months one and three and are shown in Table 21.5.

Despite this study shows some limitations, as control group was not randomized and there were differences in baseline characteristics, it demonstrates that it is more accurate to adjust individualized dose using the pharmacogenetic algorithm rather than clinical dosage. This study constitutes an advance in personalized medicine and serves as contrast source to ongoing clinical trials.

2.4.3 Acenocoumarol Pharmacogenetic Algorithms

Many studies have been developed for warfarin pharmacogenetic algorithms. On the contrary, there are only three pharmacogenetic algorithms described in patients receiving acenocoumarol.

The first study included 193 Caucasian Spanish patients with atrial fibrillation, thromboembolic disease, valve replacement, and other conditions with stable anticoagulation dose adjusted to INR 3–4 for patients with prosthetic valves and INR 2–3 for the remaining patients (Verde et al. 2010). The aim of the study was to evaluate the influence of allelic variants of CYP2C9 (*2 and *3) and VKORC1 (G1639A, C497G, and C1173T) on the acenocoumarol dose to build an "acenocoumarol-dose genotype score" (AGS) in order to predict the dose required to achieve a stable and effective anticoagulation. Data from this study suggested that the AGS algorithm could be used to help in discriminating patients requiring high acenocoumarol doses to achieve stable anticoagulation. Particularly, the mean AGS was higher in the high-dose (>28 mg/week) compared with the low-dose (<7 mg/week) group. An AGS > 70 was associated with an increased odds ratio (OR) of requiring high acenocoumarol dosage (OR: 3.347; 95 % CI: 1.112–10.075; $p=0.032$).

The second pharmacogenetic study includes phenprocoumon-treated patients in addition to acenocoumarol-treated patients. They collected data from over 1,000 patients with atrial fibrillation, thromboembolic disease, valve replacement, and other conditions in the Netherlands. They developed genotype-guided and non-genotype-guided algorithms to predict the maintenance dose by using genetic information (CYP2C9 and VKORC1) and clinical parameters. Their pharmacogenetic algorithm was validated externally in 229 patients treated with phenprocoumon and 168 with acenocoumarol explaining 59 and 49 % of the variability in maintenance dose of phenprocoumon and acenocoumarol, respectively (Van Schie et al. 2011).

The third published algorithm is a Spanish study that enrolled 147 patients with thromboembolic disease who had stable anticoagulation status or within INR 2–3 range. They developed an algorithm that included clinical parameters (age, weight, and concomitant medications) and allelic variations in VKORC1, CYP2C9, CYP4F2, and APOE genes. The goal was to evaluate the influence of clinical and genetic variables in acenocoumarol maintenance dose. The pharmacogenetic algorithm was able to explain 60.6 % of acenocoumarol variability dose. The two different allelic variants introduced were CYP4F2 and APOE that explain 3.6 and 1.3 % in the variability dose, respectively (Borobia et al. 2012).

Despite the small sample size of these studies, which may lead to controversial results and not enough statistic power, they have reached similar results, demonstrating that acenocoumarol pharmacogenetic algorithm is useful to predict dose requirement and explains variability in anticoagulant dose response. Currently there are three clinical trials ongoing that are expected to be powerful enough to resolve the question of whether it is really useful to genotype patients prior to coumarin derivative therapy (ClinicalTrials.gov identifier NCT00839657; ClinicalTrials.gov identifier NCT01006733; ClinicalTrials.gov identifier NCT01119300 (Warfarin), NCT01119261 (Acenocoumarol), NCT01119274 (Phenprocoumon)).

2.4.4 Ongoing Clinical Trials for Coumarin Derivates

The COAG trial (Clarification of Optimal Anticoagulation through Genetics) aims to improve anticoagulation control for patients by using clinical plus genetic information to guide warfarin therapy initiation. This trial is intended to enroll 1,238 patients to evaluate time within therapeutic INR range and risk of presenting INR > 4. This study is currently recruiting participants diagnosed with stroke, venous thrombosis, atrial fibrillation, or atrial flutter. They are comparing two approaches (clinical vs. genetic guidance) in warfarin dosing to examine the utility of using genetic information for warfarin dosing; they have designed two different groups, a genotype-guided dosing algorithm warfarin group and a clinical-guided dosing algorithm warfarin group. Each study arm includes a baseline dose-initiation algorithm and a dose-revision algorithm applied over the first 4–5 doses of warfarin therapy. The first aim of efficacy is percentage of TTR and the second efficacy goal is occurrence of INR > 4 or serious clinical event. Outcomes are expected in 2013 (National Heart, Lung, and Blood Institute 2012).

The GIFT (Genetic Informatics Trial of Warfarin to Prevent DVT) plans to recruit 1,600 patients. They hypothesize that the use of genetics to guide warfarin therapy will reduce the risk of venous thromboembolism (VTE) postoperatively. They further hypothesize that using a target INR of 1.8 is non-inferior to using a target INR of 2.5 in thrombi prevention. The primary endpoint is to evaluate nonfatal thromboembolism, nonfatal major bleeding, or INR>4; and the secondary endpoint is TTR. Results are expected in 2015 (Washington University School of Medicine 2012).

EU-PACT (European Pharmacogenetics of Anticoagulant Therapy) trial aims to enroll 970 patients with venous thromboembolism or atrial fibrillation. The main objective of this trial is to demonstrate that a dosing algorithm containing genetic information is able to increase the time within therapeutic INR range during anticoagulation therapy with each of warfarin, acenocoumarol, and phenprocoumon compared to a dosing regimen that does not contain this information. Secondary outcomes of the study include cost-effectiveness, number of thromboembolic and bleeding events, time to reach stable dose, and number of supratherapeutic INR peaks. The design is a two-armed, single-blinded, randomized controlled trial. One arm is constituted by a genotype-guided dosing algorithm group starting anticoagulation therapy with warfarin, acenocoumarol, or phenprocoumon and the other group of patients is being dosed according to a non-genotype-guided dosing. Primary outcome measures are to assay TTR and INR>4. Secondary outcome measures include supratherapeutic or undertherapeutic INR range; time to achieve a stable dose; quantify number of dose adjustments and number of major bleeding events or thromboembolic events followed for three months (Utrecht Institute for Pharmaceutical Sciences 2012). This trial is expected to be powerful enough to answer questions regarding clinical implementation of genotype dosing, if genetic dosing algorithm increases the TTR-INR range during anticoagulation therapy, cost-effectiveness, number of thromboembolic and bleeding events, time to reach stable dose, and number of supratherapeutic INR peaks. Outcomes are expected in November 2012 (Johnson et al. 2011).

2.5 Pharmacogenetic Tests

The challenge of pharmacogenetic tests is to improve efficacy and safety of anticoagulant treatment with coumarin derivates. Several genetic tests have been developed, based on performing determination of CYP2C9 and VKORC1 polymorphisms. Implementation of these tests aims to be useful at predicting the individualized warfarin-required dose.

Currently, there are several available pharmacogenetic tests approved by the FDA, such as Verigene Warfarin Metabolism, eSensor Warfarin Sensitivity, INFINITI Warfarin Assay, eQ-PCR LC Warfarin Genotyping kit, and ParagonDx Genetic Warfarin Assay, and other genetic tests that have been developed but are not authorized by the FDA yet, such as Luminex, Invader, and SimpleProbe Warfarin Assay (King et al. 2008; Babic et al. 2009; Langley et al. 2009; Lefferts et al. 2010; Maurice et al. 2010, Infiniti Warfarin Assay 2012). Table 21.6 shows the main characteristics of a selection of the most important pharmacogenetic tests available for determination of warfarin sensitivity. Test, company, genotyping methodology, gene polymorphisms, time after DNA isolation, FDA status, advantages, and limitations are detailed, and differences between different genetic tests are also described in Table 21.6.

Genetic tests are intended to be useful at multiracial populations; some genetic tests include allelic variants more prevalent in black and Asian populations. For example, Infinity performs the determination of CYP2C9*3, *4, *5, *6, and *11 allelic variants, and eSensor genetic test determines CYP2C9*5, *6,*11,*14, *15, and *16 allelic variants, which are more common in non-Caucasian populations, in addition to the common CYP2C9*2 and CYP2C9*3 allelic variants.

One of the most important variables in the choice of the genetic test is its cost-effectiveness. Maurice et al. (2010) compared the performance of a variety of commercial genotyping assays (Verigene, eSensor, Invader, and Luminex), to detect variants in the CYP2C9 and VKORC1 genes, used in genotype-based warfarin dosing algorithms (Maurice et al. 2010). According to their results, Verigene Warfarin Metabolism Test offers an acceptable cost per genotyped sample when determining from 1 to 24 samples, without a great increase in the price. Based on that study, Verigene is also the easiest system to operate and the least complex to perform, based on pipetting steps. Furthermore, it has the shortest turnaround time, due to the avoiding of the nucleic acid amplification step. Unlike the other systems, Verigene does not require separate external controls for each assay run; internal controls are preloaded into the test cartridges, and external controls are only run to verify the performance of each new batch of cartridges. On the other hand, eSensor, Luminex, and Invader have added costs and hands-on time because of their requirement for 2–3 external controls per run, even for single sample analysis. This results in a significant cost advantage of Verigene for single sample analysis that is reduced as the sample run size increases, so it has the same price for 1, 8, or 24 samples. It is the least expensive system for single and eight sample analysis and second,

Table 21.6 Main characteristics of the most important pharmacogenetic tests available for determination of warfarin sensitivity

Tests	Companies	Methods	Genes	Time after DNA isolation (h)	FDA status	Advantages	Limitations	References
Verigene Warfarin Metabolism	Nanosphere, Northbrook, IL	Gold nanoparticle-based microarray assay system	CYP2C9*2,*3	1.5	Approved	No PCR	Required the most DNA by volume and concentration (10–80 ng/μL, 25 μL)	Lefferts et al. (2010), Maurice et al. (2010)
			VKORC1 (1173C>T)			Minimal manipulations needed		
						Random access		
						Automated calls		
						Modular expandability		
						No external controls required		
						The easiest to operate		
						The least complex to perform based on pipetting steps		
eSensor Warfarin Sensitivity	Osmetech, Pasedena, CA	Competitive DNA hybridization and electrochemical detection	CYP2C9*2,*3, *5, *6,*11,*14, *15, *16	3	Approved	Minimal manipulations are needed	External control are required	Maurice et al. (2010) and Babic et al. (2009)
			VKORC1-1639G>A			Modular expandability	Required a DNA amplification step	
						Random access		
						Required the least DNA/reaction		
						Extended CYP2C9 SNP panels		

(continued)

Table 21.6 (continued)

Tests	Companies	Methods	Genes	Time after DNA isolation (h)	FDA status	Advantages	Limitations	References
Luminex	Luminex-Corporation, Austin, TX	Multiplex PCR and allele-specific primer extension	CYP2C9: *2 *3, *4, *5, *6 VKORC1: −1639G>A, +85G>T, 121G>T, 134 T>C, 172A>G, 1331G>A, 3487 T>G	6	Approved	Have the capability of detecting some of other less common CYP2C9 and VKORC1 variants	No modular expandability External controls are required No random access Required a DNA amplification step A lot of pipetting steps and hands-on time	Maurice et al. (2010)
Invader	Third Wave, Madison, WI	Combination of PCR, signal amplification, and fluorescent resonance energy transfer	CYP2C9*2,*3 VKORC1-1639G>A	4	Approved	Low capital expense Established methodology Readily scalable to larger run sizes	No modular expandability External controls are required No random access Required a DNA amplification step A lot of pipetting steps and hands-on time Post-PCR manipulations	Lefferts et al. (2010), Maurice et al. (2010), Langley et al. (2009)
Infinity	AutoGenomics Carlsbad, CA	Allele-specific primer extension (ASPE) and quantification of generated fluorescent signal	CYP2C9 *2, *2 *3, *4, *5, *6,*11 VKORC1-1639G/A, 497 T/G, 698C/T, 1173C/T, 1542G/C, 2255C/T, 3462C/T, 3730G/A	8	Approved	Number of SNPs offered post-PCR automated Easily adaptable to include additional SNPs	Long turnaround time Large capital investment Less established methodology	Babic et al. (2009), Langley et al. (2009), and King et al. (2008)

eQ-PCR	TrimGen Corp. Sparks, MD	Combination of real-time PCR and melting curve analysis	CYP2C9*2,*3	1	Approved	Established methodology	The kit does not apply to detecting any other SNPs in CYP2C9 or VKORC1	Trimgen website
			VKORC1-1639G>A			Minimal manipulations needed		
ParagonDx	ParagonDx Morrisville, NC	Taqman methodology	CYP2C9 *2, CY2C9*3	4	Approved	Short turnaround time	Not commercially available	Babic et al. (2009) and Langley et al. (2009)
			VKORC1-1639G>A			Easy to use	Capital investment	
			6484 (1173C>T)			Established methodology	Small run size	
			9041 (3730G>A)			Low hands-on time		
SimpleProbe Warfarin Assay	The Idaho Technology LightCycler Salt Lake City, UT	Combination of PCR with real-time fluorescent monitoring and melting curve analysis	CYP2C9*2,*3	3 h	Approved	Short turnaround time	Capital investment	Langley et al. (2009)
			VKORC1-1639G>A			Easy to use		
						Established methodology		
						Low hands-on time		

behind Invader test, in cost-effectiveness for 24 samples. Despite this advantages, Verigene is only available for warfarin metabolism, F5/F2/ MTHFR, and CYP2C19 genetics test, so it does provide a broad panel of polymorphisms to study, unlike, for example, Luminex assay, which uses a method based on multiplex PCR, and allele-specific primer extension, which allows determination of more polymorphisms than the other systems. Verigene is therefore a closed system that cannot determine other polymorphisms outside its genotyping kit, while other genotyping tests offer the possibility of including gene polymorphism determinations chosen by the researchers. The versatility of the instrument is another factor to take into consideration in the cost-effectiveness of the selected genotyping platform.

The use of genetic testing for warfarin dosing may not be useful for the entire population that is susceptible of starting coumarin derivative treatment. A limitation of these genetic tests is that they are based on the identification of warfarin sensitivity, and therefore, allelic variants associated to coumarin derivate resistance cannot be determined. Furthermore, the use of closed genetic testing can significantly increase health costs.

3 New Oral Anticoagulant Drugs

Chronic treatment with VKAs (warfarin, acenocoumarol, and phenprocoumon) has been the treatment of choice for stroke prevention in atrial fibrillation patients and prevention and treatment of venous thromboembolism during the last 60 years (Dezee et al. 2006; Kamali and Pirmohamed 2006; Banerjee et al. 2011; Wartak and Bartholomew 2011). Because of several coumarin derivate drawbacks (Francis 2008), such as narrow therapeutic margin, delayed onset of action, many drugs and dietary interactions, the high interindividual variability in the anticoagulant response, and the necessity of periodic monitoring of anticoagulants doses, pharmaceutical development has focused on the search for new molecules that improve oral anticoagulation therapy.

The new anticoagulants can be classified in function of their mechanism of action into three groups (Douketis 2011):

1. Oral direct thrombin inhibitor. Dabigatran etexilate
2. Oral factor Xa inhibitor. Rivaroxaban, apixaban, betrixaban, edoxaban, and eribaxaban
3. Parenteral factor Xa inhibitor. Idrabiotaparinux

Ximelagatran, the first oral direct thrombin inhibitor which was evaluated in a clinical trial compared to warfarin in prophylaxis and treatment of VTE and stroke prevention in atrial fibrillation patients was not approved by FDA because it showed hepatotoxicity (Francis et al. 2003; Schulman et al. 2003; Colwell et al. 2005).

Dabigatran etexilate and rivaroxaban are currently available on the market. In 2008, dabigatran etexilate was authorized for the prevention of VTE after major orthopedic surgery by the European Medicines Agency (EMA). Based on the RE-LY study results, the FDA (Food and Drugs Administration) approved dabigatran etexilate in the prevention of stroke and systemic embolism in adult patients with non-valvular atrial fibrillation (FDA News and Events 2012a). After the results shown in the ROCKET AF study, in 2010, the FDA approved the use of rivaroxaban in the prevention of stroke and embolism in adult patients with non-valvular atrial fibrillation (FDA News and Events 2012b).

3.1 Oral Direct Thrombin Inhibitors

3.1.1 Dabigatran Etexilate

Dabigatran etexilate is rapidly absorbed and is hydrolyzed by plasma and liver esterases to active dabigatran (Baetz and Spinler 2008). It is absorbed in the stomach and intestine and depends on an acid environment. Its absorption is diminished by 30 % when dabigatran etexilate is coadministered with proton pump inhibitors (Francis 2008). To facilitate the absorption of dabigatran etexilate, capsules contain tartaric acid. The peak plasma concentration of dabigatran is reached within 2–3 h after oral administration. The half-life time of dabigatran is around 12–17 h, and steady state is attained within 3 days of treatment with this new oral anticoagulant (Liesenfeld et al. 2011).

The oral bioavailability of dabigatran etexilate in capsules is 6.5 % and can be increased by up to 75 % when the active substance is separated from the capsule, so it should not be separated. It has a low plasma protein binding (35 %). Dabigatran is not metabolized by the CYP450 isoenzymes; microsomal esterases are responsible of its metabolism. Pharmacologically active acyl glucuronides are formed through conjugation metabolism. It is mostly eliminated unmodified (80–85 % of the dose) and partially in the form of glucuronides through glomerular filtration (Eriksson et al. 2009).

There are no known interactions with food, and direct interaction with alcohol intake in animal models has not been described (Liesenfeld et al. 2011). The prodrug dabigatran etexilate, but not the active form dabigatran, is a P-glycoprotein (p-pg) substrate (Liesenfeld et al. 2011). Dabigatran interaction may occur with p-pg inhibitors, such as amiodarone (increased 50–60 % dabigatran area under the curve), quinidine, verapamil, or ketoconazole (contraindicated), and p-pg inducers such as rifampicin, carbamazepine, and St. John's wort (Eriksson et al. 2009; Schulman and Majeed 2012).

Plasma and liver esterases hydrolyze the biotransformation of dabigatran etexilate into dabigatran. Dabigatran binds to thrombin reversibly and competitively inhibiting its ability to convert fibrinogen to fibrin in the coagulation cascade. Since thrombin (serine protease) enables the conversion of fibrinogen to fibrin in the coagulation cascade, its inhibition prevents thrombi formation. Dabigatran also inhibits free thrombin, fibrin-bound thrombin, and thrombin-induced platelet aggregation (Brighton 2010).

RE-LY (Evaluation of Long-Term Anticoagulation Therapy)

Dabigatran etexilate was compared to warfarin in the RE-LY (Randomized Evaluation of Long-Term Anticoagulation Therapy) study in 18,113 patients with atrial fibrillation (AF) recruited during two years. Patients were randomized to dabigatran 110 mg twice a day, dabigatran 150 mg twice a day, or warfarin INR dose adjusted. The main objective of this study was to evaluate the effectiveness in the prevention of thromboembolic events, stroke, and hemorrhage of dabigatran, compared with standard warfarin dosing. The dose of 110 mg twice daily showed similar efficacy to warfarin in preventing stroke and systemic embolism. The incidence of major, minor, and intracranial bleeding was significantly lower in dabigatran 110 mg compared with warfarin (2.87 % vs. 3.57 % dabigatran warfarin, $p=0.003$). However, the incidence of myocardial infarction was higher with dabigatran 110 mg and 150 mg twice a day than with warfarin (0.82 %, 0.81 %, and 0.64 %, respectively) (Connolly et al. 2011). A meta-analysis published in January 2012 concluded that dabigatran was associated to 33 % of increased myocardial infarction or acute coronary syndrome risk (OR 1.33, 95 % CI 1.03–1.71, $p=0.03$) compared to warfarin, enoxaparin, and placebo (Uchino and Hernandez 2012).

3.2 Oral Factor Xa Inhibitors

3.2.1 Rivaroxaban

Rivaroxaban is orally administered. It has a nearly 100 % bioavailability. It has 92–95 % protein binding (FDA. U.S. Food and Drug Administration 2012b). Rivaroxaban is a p-glycoprotein substrate and is transported through the membrane. About 2/3 of the dose is metabolized; metabolism is conducted through CYP3A4, CYP2J2, and CYP450-independent mechanisms (Perzborn et al. 2011). Approximately one third of the administered dose is excreted unmodified by the kidneys, another third is metabolized to inactive metabolites and eliminated by the kidneys, and the last third of the dose metabolized to inactive metabolites and eliminated via fecal. The half-life time in young subjects is 5–9 h and in older individuals is 11–13 h (FDA. U.S. Food and Drug Administration 2012b). Rivaroxaban has a low renal clearance (3–4 l/h). Toxicity can be reversed using activated charcoal and supportive measures such as mechanical compression and hemodynamic support. If the bleeding does not stop, activated prothrombin complex concentrate, prothrombin complex concentrate, and recombinant factor VIIa can be administrated.

Rivaroxaban is an oral anticoagulant that inhibits directly factor Xa. Factor Xa is required to perform the conversion of prothrombin (FII) into thrombin (FIIa). Thrombin is a serine protease required for the conversion of fibrinogen to fibrin, which is responsible for triggering the coagulation cascade.

A molecule of factor Xa can generate more than 1,000 molecules of thrombin. Therefore, rivaroxaban is useful in blocking the coagulation cascade; rivaroxaban exerts its effect irreversibly (DrugBank 2012a).

ROCKET AF (Rivaroxaban Once-Daily Oral Direct Factor Xa Inhibition Compared with the Vitamin K Antagonism for Prevention of Stroke and Embolism Trial in Atrial Fibrillation)

Rivaroxaban, a factor Xa inhibitor, was studied in the prevention of stroke and systemic embolism in patients with non-valvular atrial fibrillation. The ROCKET AF study compared the efficacy of rivaroxaban versus warfarin in 14,264 patients with non-valvular atrial fibrillation with moderate to high risk of stroke (Patel et al. 2011). Patients were randomized to rivaroxaban 20 mg or warfarin dose adjustment. The results of this study showed that rivaroxaban was not inferior to warfarin in preventing stroke and systemic embolism. The incidence of stroke or systemic embolism was 1.7 % in the rivaroxaban group and 2.2 % in the warfarin group. Moreover, the incidence of bleeding was 14.9 and 14.5 % in warfarin and rivaroxaban groups, respectively. The incidence of intracranial hemorrhage in a year was lower in the rivaroxaban group (0.5 % vs. 0.7 %). The incidence of gastrointestinal bleeding was higher in the rivaroxaban group compared to warfarin (3.2 % vs. 2.2 %).

3.2.2 Apixaban

Apixaban is the latest highly selective and reversible inhibitor oral direct factor Xa. It is orally administered twice daily, and like the other new oral anticoagulants (dabigatran and rivaroxaban), it does not require dose monitoring routine laboratory. It has a high oral bioavailability, reaching peak plasma concentration at 3 h after oral administration, with a half-life time of 9–14 h (Raghavan et al. 2009; Weitz 2010). It is metabolized by hepatic metabolism via CYP3A4, CYP3A5, and sulfotransferase 1A1 and by kidney and intestinal metabolism. It is eliminated through multiple pathways, renal and intestinal routes. It has a renal elimination of 25 % and the remaining 75 % through the hepatobiliary system (Deedwania and Huang 2012; Eriksson et al. 2009).

Two phase III randomized trials have been performed to evaluate the efficacy and safety of apixaban in prevention of stroke and systemic embolism in AF patients. The first trial was "Apixaban Versus Acetylsalicylic Acid to Prevent Stroke in Atrial Fibrillation Patients who have Failed or are Unsuitable for Vitamin K Antagonist Treatment" (AVERROES trial) and the second was "the Apixaban for Reduction in Stroke and Other Events in Atrial Thromboembolic Fibrillation" (ARISTOTLE trial) (Diener et al. 2012; Granger et al. 2011).

AVERROES Trial

AVERROES trial was a randomized study which included 5599 AF patients with increased risk of stroke and intolerance to therapy with vitamin K antagonists. Patients were randomized into two groups: the first group to apixaban 5 mg twice daily or 2.5 mg twice daily if the patient was ≥80 years, ≤60 kg body weight, or serum creatinine of ≥1.5 mg/dl plus aspirin placebo and the other group to aspirin (81–324 mg daily) plus placebo apixaban. The aims were efficacy (preventing stroke or systemic embolism) and safety (major bleeding). Major bleeding was defined as the overbleeding with one or more of the following clinical parameters: drop in hemoglobin level of ≥2 g/dl over 24-h period or transfusion of ≥2 units packed cells and critical bleeding or fatal bleeding. Patients treated with apixaban obtained lower incidence of stroke and embolic system compared to the aspirin group (1.6 % vs. 3.7 % per year, p-value<0.001). Ischemic stroke was lower in apixaban group compared to aspirin (1.1 % vs. 3.0 % per year, p-value<0.001), and hemorrhages were also lower in the apixaban group compared to aspirin (6 vs. 9). No significant differences in the rate of death per year were found (3.5% apixaban group vs 4.4% aspirin group, p-value<0.07). This study concludes that apixaban is more effective than aspirin in preventing stroke and systemic embolism in patients who are not eligible for oral anticoagulation therapy with vitamin K antagonists (Eikelboom et al. 2010; Connolly et al. 2011).

ARISTOTLE Trial

ARISTOTLE trial was a double-blind, randomized study that included 18,201 patients with atrial fibrillation and at least one risk factor for stroke. Patients were randomized into two groups: apixaban group (5 mg daily or 5 mg twice daily if the patient was ≥80 years with body weight ≤60 kg or creatinine concentration ≥1.5 mg/dl) and warfarin group with dose adjusted to INR 2–3. The first efficacy objective was to evaluate the incidence of systemic embolism and stroke, and the first safety aim was bleeding and death from any cause. The result of the first efficacy endpoint was 1.7 % in the apixaban group vs. 1.60 % per year in the warfarin group ($p<0.001$ for non-inferiority and $p<0.01$ for superiority). Bleeding rate was lower in apixaban group compared with warfarin group in a year (2.13 % vs. 3.09 %, $p<0.001$), and the rates of death from any cause were 3.52 % in the apixaban group compared to 3.94 % in the warfarin group ($p=0.047$). ARISTOTLE trial concluded that apixaban was superior to warfarin in the prevention of stroke or systemic embolism results with less bleeding and death in both cases (Granger et al. 2011).

Apixaban has been approved in Europe for the prevention of venous thromboembolic events VTE in adults following a hip or knee replacement operation in 2011 (European Medicine Agency 2012a). Apixaban showed greater efficacy to warfarin in the prevention of stroke and systemic embolism in AF patients in the ARISTOTLE study. Apixaban is currently being evaluated by the FDA to obtain approval for the prevention of stroke and systemic embolism in AF patients.

3.3 Strengths and Limitations of the New Oral Anticoagulants

The new oral anticoagulants dabigatran, rivaroxaban (approved by the FDA for the prevention of stroke and systemic embolism in patients with AF), and apixaban (currently under review by the FDA for authorized for prevention of stroke and systemic embolism in AF patients) have showed to be not inferior in efficacy and safety compared to warfarin in preventing stroke and systemic embolism in AF patients in Phase III studies (Connolly et al. 2011; Diener et al. 2012; Granger et al. 2011; Patel et al. 2011). They are considered as the alternative to coumarin derivates in long-term oral anticoagulation therapy.

They have a rapid onset of action compared to coumarin derivates. This feature could allow avoiding bridge therapy with parenteral anticoagulants (UHF or LWMH). Figure 21.2 shows the comparison of action mechanisms of VKA vs. new oral anticoagulants.

One of the main limitations of VKA is that they present a high variability interindividual dose response. The new oral anticoagulants have a predictable anticoagulant response, and therefore a daily fixed dose can be given, and routine laboratory monitoring is not required in patients without severe renal impairment (Kubitza et al. 2005, 2008; Connolly et al. 2009; Schulman et al. 2009). No routine laboratory monitoring is a priori an advantage because it can increase the number of patients receiving long-term oral anticoagulation therapy. However, the daily fixed dose can cause a potential decrease of treatment adherence with consequent therapeutic failure.

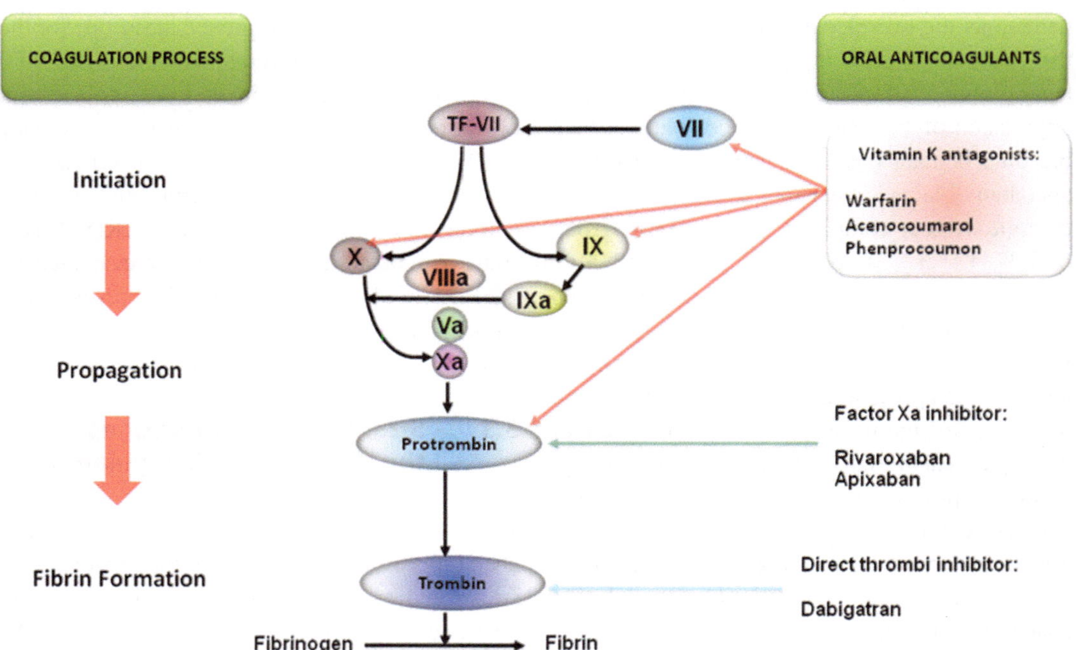

Fig. 21.2 Comparison of action mechanisms of VKA vs. new oral anticoagulants in the inhibition of coagulation cascade factors

To avoid this potential poor adherence, education is needed in patients who start anticoagulation therapy with the new oral anticoagulants. In the case of dabigatran and apixaban, they are administrated in a twice-daily dose, while daily dose of rivaroxaban can be administrated once after the first weeks (European Medicine Agency 2012b).

New oral anticoagulants are less dependent of CYP450 metabolism compared to VKA; therefore, drug and dietary interactions are reduced. The presence of food increases the time to reach peak plasma concentration (≈4 h) and decreases 30 % in plasma levels of rivaroxaban (Kubitza et al. 2006).

Dabigatran is metabolized by CYP450-independent mechanisms. The prodrug dabigatran etexilate and rivaroxaban are substrates of p-glycoprotein and transported through the membrane. Pg-p inductor drugs (rifampicin, carbamazepine, or St. John's wort) or Pg-p inhibitor drugs (ketoconazole, verapamil, amiodarone, or quinidine) may decrease or increase the anticoagulant response, respectively. Ketoconazole is contraindicated in patients taking dabigatran (Schulman and Majeed 2012). Factor Xa inhibitors (rivaroxaban and apixaban) are metabolized by CYP3A4, and therefore, concomitant drugs that are inhibitors (ketoconazole, ritonavir, erythromycin, or clarithromycin) or inducers (rifampicin) of this isoenzyme can influence the plasma drug concentration (Giorgi and Miguel 2012). Table 21.7 shows the molecular structures involved in pharmacodynamics and pharmacokinetics of new oral anticoagulants (dabigatran, rivaroxaban, and apixaban).

In the RE-LY study, dabigatran group showed a higher incidence of myocardial infarction compared to warfarin group. A recent analysis suggests that warfarin may be protective against myocardial infarction (Lip and Lane 2010). However, in the Phase III studies conducted with other new oral anticoagulants (rivaroxaban and apixaban), the rate of myocardial infarction was lower compared to warfarin group (Lip et al. 2010).

Although new oral anticoagulants have a shorter duration of the anticoagulant effect compared to VKA, there is no antidote available to reverse the anticoagulant effects if toxicity occurs (Schulman and Bijsterveld 2007; Wittkowsky 2010).

Table 21.7 Molecular structures involved in pharmacodynamics and pharmacokinetics of new oral anticoagulants

Genes	Drugs	Effects	References
ABCB1 (P-glycoprotein)	Dabigatran	Transport through the membrane	Eriksson et al. (2009), Liesenfeld et al. (2011), Schulman and Majeed (2012)
	Rivaroxaban	Interaction with p-gp inductors and inhibitors	
CYP3A4	Rivaroxaban Apixaban	Hepatic metabolism. Interactions with CYP3A4 inductors or inhibitors	Perzborn et al. (2011), Giorgi and Miguel (2012)
CYP2J2	Rivaroxaban	Hepatic metabolism of rivaroxaban	Perzborn et al. (2011)
CYP3A5 SULT1A1	Apixaban	Hepatic metabolism of apixaban	Eriksson et al. (2009), Deedwania and Huang (2012)

ABCB1 indicates ATP-binding cassette, subfamily B; *CYP3A4* Cytochrome P450 family 3, subfamily A, polypeptide 4; *CYP2J2* Cytochrome P450 family 2, subfamily J, polypeptide 2; *CYP3A5* Cytochrome P450 family 3, subfamily A, polypeptide 5; *SULT1A1* sulfotransferase 1A1

Because of the relevance of renal elimination in new anticoagulants (dabigatran and rivaroxaban), creatinine clearance (CL_{CR}) of each patient has to be analyzed before starting anticoagulant therapy. Patients with moderate renal clearance (CL CR<50 ml/min) or severe renal impairment (CL CR<30 ml/min) can exhibit prolonged time excretion and increased drug concentrations in plasma (Stangier et al. 2010, (European Medicine Agency 2012b). Dabigatran is contraindicated in patients with severe renal impairment (CL CR<30 ml/min) (European Medicines Agency 2012c)).

Dabigatran absorption is favored by an acidic environment, so the capsules contain tartaric acid. In the RE-LY study, dabigatran showed a higher rate of dyspepsia compared to warfarin patients, which contributed to a discontinuation in the dabigatran anticoagulation therapy in 2 % of patients (Connolly et al. 2009).

One of the most important limitations is the cost of the new oral anticoagulants, which is far superior to the known coumarin derivates. According to Freeman (AÑO), dabigatran may be cost-effective in patients with age ≥ 65 years with non-valvular atrial fibrillation and an increased risk of CHADS2 ≥ 1 (Reeman et al. 2011).

Although they are not ideal oral anticoagulants, dabigatran, rivaroxaban, and apixaban have suitable properties to become real alternatives to traditional anticoagulation therapy with coumarin derivates.

4 Future Perspectives

Prediction of the initial dose of coumarin derivates using CYP2C9 and VKORC1 polymorphisms increases efficacy and minimizes the toxicity of VKA in the first months of treatment. The inclusion of genotyping in the clinical setting reduces the risk of hospitalization for bleeding or thrombosis by approximately 30 % in outpatients treated with warfarin (Epstein et al. 2010). These results are consistent with the conclusions reached by Eckman et al. (2009), when evaluating the cost-effectiveness of genotype-guided dosing of warfarin versus standard induction therapy for non-valvular atrial fibrillation patients. The genotype-guided dosing resulted cost-effective in patients at high risk of bleeding but not for typical non-valvular atrial fibrillation patients, except for when the following conditions were complied: the genotyping was capable of reducing 32 % the rate of bleeding events, and it was available in 24 h and costs less than 200 dollars. Currently, PCR real-time techniques allow performing genotyping of CYP2C9 and VKORC1 within 24 h, and the cost of the genotyping method can be available for less than $40 per patient if it is performed "in-house," at the hospital laboratory facilities. If genotyping-guided dose is implemented in clinical anticoagulant therapy, it can reach a 30 % reduction of thromboembolism or bleeding events in typical non-valvular atrial fibrillation patients.

Furthermore, VKAs still remain a recommended option for anticoagulation therapy in patients who are already on anticoagulant therapy with VKAs and with good INR control, for new patients with non-valvular atrial fibrillation except if there are any criteria that justify initiating treatment with the new oral anticoagulants, and also for atrial fibrillation patients with valvular involvement. Thereby, VKAs will keep being a real possibility for anticoagulant therapy despite of the increasing trend of using new oral anticoagulants in prevention of stroke and systemic embolism for patients with atrial fibrillation. However, there are some situations in which the new oral anticoagulants may have a benefit compared with the VKAs, such as patients with known hypersensitivity or contraindication to the use of VKA, patients with antecedents of intracranial hemorrhage or ischemic stroke patients with high-risk clinical and neuroimaging criteria for intracranial hemorrhage, and finally in patients treated with VKAs which have been presented thromboembolic and/or major bleeding episodes despite a good control of INR (Spanish Agency of Medicines and Health Products 2012).

Dabigatran has demonstrated to be cost effective compared to warfarin in preventing stroke and systemic embolism in patients with non-valvular atrial fibrillation (Davidson et al. 2013; González-Juanatey et al. 2012). Although the new molecules are presented as the future in the prevention of thrombotic events in patients requiring long-term oral anticoagulation, the high cost of these molecules is a limitation in the clinical implementation, especially for some countries. In those countries, if new molecules cannot be introduced in their hospitals, routine genotyping of patients with a high bleeding risk starting long-term oral anticoagulation with coumarin derivates or poor INR control remains a real possibility to be considered to improve the efficiency and safety of treatment.

References

Aithal GP, Day CP, Kesteven PJ, Daly AK (1999) Association of polymorphisms in the cytochrome P450 CYP2C9 with warfarin dose requirement and risk of bleeding complications. Lancet 353:717–719

Alberio L (2003) Oral anticoagulation with vitamin K antagonists. Ther Umsch 60:5–9

Anderson JL, Horne BD, Stevens SM, Woller SC, Samuelson KM, Mansfield JW, Robinson M, Barton S, Brunisholz K, Mower CP, Huntinghouse JA, Rollo JS, Siler D, Bair TL, Knight S, Muhlestein JB, Carlquist JF (2012) A randomized and clinical effectiveness trial comparing two pharmacogenetic algorithms and standard care for individualizing warfarin dosing (CoumaGen-II). Circulation 125:1997–2005

Aquilante CL, Langaee TY, Lopez LM, Yarandi HN, Tromberg JS, Mohuczy D, Gaston KL, Waddell CD, Chirico MJ, Johnson JA (2006) Influence of coagulation factor, vitamin K epoxide reductase complex subunit 1, and cytochrome P450 2C9 gene polymorphisms on warfarin dose requirements. Clin Pharmacol Ther 79:291–302

Babic N, Haverfield EV, Burrus JA, Lozada A, Das S, Yeo KT (2009) Comparison of performance of three commercial platforms for warfarin sensitivity genotyping. Clin Chim Acta 406:143–147

Baetz BE, Spinler SA (2008) Dabigatran etexilate: an oral direct thrombin inhibitor for prophylaxis and treatment of thromboembolic diseases. Pharmacotherapy 28:1354–1373

Banerjee A, Marín F, Lip GY (2011) A new landscape for stroke prevention in atrial fibrillation: focus on new anticoagulants, antiarrhythmic drugs, and devices. Stroke 42(11):3316–3322

Bazan NS, Sabry NA, Rizk A, Mokhtar S, Badary O (2012) Validation of pharmacogenetic algorithms and warfarin dosing table in Egyptian patients. Int J Clin Pharm 34(6):837–844

Becquemont L (2008) Evidence for a pharmacogenetic adapted dose of oral anticoagulant in routine medical practice. Eur J Clin Pharmacol 64:953–960

Beinema M, Brouwers JR, Schalekamp T, Wilffert B (2008) Pharmacogenetic differences between warfarin, acenocoumarol and phenprocoumon. Thromb Haemost 100(6):1052–1057. Review

Bodin L, Verstuyft C, Tregouet DA, Robert A, Dubert L, Funck-Brentano C, Jaillon P, Beaune P, Laurent-Puig P, Becquemont L, Loriot MA (2005) Cytochrome P450 2C9 (CYP2C9) and vitamin K epoxide reductase (VKORC1) genotypes as determinants of acenocoumarol sensitivity. Blood 106(1):135–140

Borgiani P, Ciccacci C, Forte V, Romano S, Federici G, Novelli G (2007) Allelic variants in the CYP2C9 and VKORC1 loci and interindividual variability in the anticoagulant dose effect of warfarin in Italians. Pharmacogenomics 8:1545–1550

Borobia AM, Lubomirov R, Ramírez E, Lorenzo A, Campos A, Muñoz-Romo R, Fernández-Capitán C, Frías J, Carcas AJ (2012) An acenocoumarol dosing algorithm using clinical and pharmacogenetic data in spanish patients with thromboembolic disease. PLoS One 7:e41360

Brighton T (2010) Experimental and clinical pharmacology new oral anticoagulant drugs– mechanisms of action. Aust Prescr 33:38–41

Cadamuro J, Dieplinger B, Felder T, Kedenko I, Mueller T, Haltmayer M, Patsch W, Oberkofler H (2010) Genetic determinants of acenocoumarol and phenprocoumon maintenance dose requirements. Eur J Clin Pharmacol 66(3):253–260

Cain D, Hutson SM, Wallin R (1997) Assembly of the warfarin-sensitive vitamin K 2,3-epoxide reductase enzyme complex in the endoplasmic reticulum membrane. J Biol Chem 272:29068–29075

Caldwell MD, Awad T, Johnson JA, Gage BF, Falkowski M, Gardina P, Hubbard J, Turpaz Y, Langaee TY, Eby C, King CR, Brower A, Schmelzer JR, Glurich I, Vidaillet HJ, Yale SH, Qi Zhang K, Berg RL, Burmester JK (2008) CYP4F2 genetic variant alters required warfarin dose. Blood 111:4106–4112

Camm AJ et al (2010) Guías de práctica clínica para el manejo de la fibrilación auricular. Rev Esp Cardiol 63(12):1483.e1–1483.e83

Carlquist JF, Horne BD, Muhlestein JB, Lappé DL, Whiting BM, Kolek MJ, Clarke JL, James BC, Anderson JL (2006) Genotypes of the Cytochrome p450 isoform, CYP2C9, and the vitamin K epoxide reductase complex subunit 1 conjointly determine stable warfarin dose: a prospective study. J Thromb Thrombolysis 22:191–197

Cavallari LH, Langaee TY, Momary KM, Shapiro NL, Nutescu EA, Coty WA, Viana MA, Patel SR, Johnson JA (2010) Genetic and clinical predictors of warfarin dose requirements in African Americans. Clin Pharmacol Ther 87:459–464

Choonara IA, Haynes BP, Cholerton S, Breckenridge AM, Park BK (1986) Enantiomers of warfarin and vitamin K1 metabolism. Br J Clin Pharmacol 22:729–732

Cini M, Legnani C, Cosmi B, Guazzaloca G, Valdrè L, Frascaro M, Palareti G (2012) A new warfarin dosing algorithm including VKORC1 3730 G>A polymorphism: comparison with results obtained by other published algorithms. Eur J Clin Pharmacol 68:1167–1174

Colwell CW Jr, Berkowitz SD, Lieberman JR, Comp PC, Ginsberg JS, Paiement G, McElhattan J, Roth AW, Francis CW, EXULT B Study Group (2005) Oral direct thrombin inhibitor ximelagatran compared with warfarin for the prevention of venous thromboembolism after total knee arthroplasty. J Bone Joint Surg Am 87:2169–2177

Connolly SJ, Ezekowitz MD, Yusuf S, Eikelboom J, Oldgren J, Parekh A, Pogue J, Reilly PA, Themeles E, Varrone J, Wang S, Alings M, Xavier D, Zhu J, Diaz R, Lewis BS, Darius H, Diener HC, Joyner CD, Wallentin L, RE-LY Steering Committee and Investigators (2009) Dabigatran versus warfarin in patients with atrial fibrillation. N Engl J Med 361:1139–1151

Connolly SJ, Eikelboom J, Joyner C, Diener HC, Hart R, Golitsyn S, Flaker G, Avezum A, Hohnloser SH, Diaz R, Talajic M, Zhu J, Pais P, Budaj A, Parkhomenko A, Jansky P, Commerford P, Tan RS, Sim KH, Lewis BS, Van Mieghem W, Lip GY, Kim JH, Lanas-Zanetti F, Gonzalez-Hermosillo A, Dans AL, Munawar M, O'Donnell M, LawrenceJ LG, Afzal R, Yusuf S, AVERROES Steering Committee and Investigators (2011) Apixaban in patients with atrial fibrillation. N Engl J Med 364:806–817

D'Andrea G, D'Ambrosio RL, Di Perna P, Chetta M, Santacroce R, Brancaccio V, Grandone E, Margaglione M (2005) A polymorphism in the VKORC1 gene is associated with an interindividual variability in the dose-anticoagulant effect of warfarin. Blood 105:645–649

D'Andrea G, D'Ambrosio R, Margaglione M (2008) Oral anticoagulants: pharmacogenetics relationship between genetic and non-genetic factors. Blood Rev 22:127–140

Dahl OE (2012) New oral antithrombotics: focus on dabigatran, an oral, reversible direct thrombin inhibitor for the prevention and treatment of venous and arterial thromboembolic disorders. Vasc Health Risk Manag 8:45–57

Daly AK, King BP (2003) Pharmacogenetics of oral anticoagulants. Pharmacogenetics 13:247–252

Davidson T, Husberg M, Janzon M, Oldgren J, Levin LÅ (2013) Cost-effectiveness of dabigatran compared with warfarin for patients with atrial fibrillation in Sweden. Eur Heart J 34(3):177–183

Davis EM, Packard KA, Knezevich JT, Campbell JA (2011) New and emerging anticoagulant therapy for atrial fibrillation and acute coronary syndrome. Pharmacotherapy 31(10):975–1016

Deedwania P, Huang GW (2012) An evidence-based review of apixaban and its potential in the prevention of stroke in patients with atrial fibrillation. Core Evid 7:49–59

Dezee KJ, Shimeall WT, Douglas KM, Shumway NM, O'malley PG (2006) Treatment of excessive anticoagulation with phytonadione (vitamin K): a meta-analysis. Arch Intern Med 166(4):391–397

Diener HC, Eikelboom J, Connolly SJ, Joyner CD, Hart RG, Lip GY, O'Donnell M, Hohnloser SH, Hankey GJ, Shestakovska O, Yusuf S, AVERROES Steering Committee and Investigators (2012) Apixaban versus aspirin in patients with atrial fibrillation and previous stroke or transient ischaemic attack: a predefined subgroup analysis from AVERROES, a randomised trial. Lancet Neurol 1:225–231

Douketis JD (2011) Dabigatran as anticoagulant therapy for atrial fibrillation. Which patients should receive it, which patients may not need it, and other practical aspects of patient management. Pol Arch Med Wewn 121(3):73–80

DrugBank (2012a) Rivaroxaban http://www.drugbank.ca/drugs/DB06228. Accessed 7 May 2012

DrugBank (2012b) Acenocoumarol (DB01418). http://www.drugbank.ca/drugs/DB01418. Accessed 7 May 2012

Duran Parrondo C et al (2003) Anticoagulación oral. An Med Interna 20:377–384

Eckman MH, Rosand J, Greenberg SM, Gage BF (2009) Cost-effectiveness of using pharmacogenetic information in warfarin dosing for patients with nonvalvular atrial fibrillation. Ann Intern Med 150:73–83

Eikelboom JW, O'Donnell M, Yusuf S, Diaz R, Flaker G, Hart R, Hohnloser S, Joyner C, Lawrence J, Pais P, Pogue J, Synhorst D, Connolly SJ (2010) Rationale and design of AVERROES: apixaban versus acetylsalicylic acid to prevent stroke in atrial fibrillation patients who have failed or are unsuitable for vitamin K antagonist treatment. Am Heart J 159:348–353

Eitz T et al (2008) International normalized ratio self-management lowers the risk of thromboembolic events after prosthetic heart valve replacement. Ann Thorac Surg 85:949–955

Epstein RS, Moyer TP, Aubert RE, O'Kane DJ, Xia F, Verbrugge RR, Gage BF, Teagarden JR (2010) Warfarin genotyping reduces hospitalization rates results from the MM-WES (Medco-Mayo Warfarin effectiveness study). J Am Coll Cardiol 55: 2804–2812

Eriksson BI, Quinlan DJ, Weitz JI (2009) Comparative pharmacodynamics and pharmacokinetics of oral direct thrombin and factor xa inhibitors in development. Clin Pharmacokinet 48:1–22

European Medicine Agency (2012a) EPAR summary for the public. EMA/240842/2011 EMEA/H/C/002148. http://www.ema.europa.eu/docs/en_GB/document_library/EPAR_Summary_for_the_public/human/002148/WC500107773.pdf. Accessed 10 Sept 2012

European Medicine Agency (2012b) Summary of Rivaroxaban characteristics. http://www.ema.europa.eu/docs/en_GB/document_library/EPAR_-Product_Information/human/000944/WC500057108.pdf. Accessed 10 Sept 2012

European Medicines Agency (2012c) Updates on safety of Pradaxa http://www.ema.europa.eu/docs/en_GB/document_library/Press_release/2011/11/WC500117818.pdf. Accessed 10 Sept 2012

Evans WE, Relling MV (1999) Pharmacogenomics: translating functional genomics into rational therapeutics. Science 286:487–491

FDA News and Events (2012a) FDA approves Pradaxa to prevent stroke in people with atrial fibrillation. http://www.fda.gov/NewsEvents/Newsroom/PressAnnouncements/2010/ucm230241.htm?utm_campaign=Google2&utm_source=fdaSearch&utm_medium=website&utm_term=dabigatran%20atrial%20fibrillation&utm_content=3. Accessed 7 May 2012

FDA News and Events (2012b) FDA approves Xarelto to prevent stroke in people with common type of abnormal heart rhythm. http://www.fda.gov/NewsEvents/Newsroom/PressAnnouncements/ucm278646.htm?utm_campaign=Google2&utm_source=fdaSearch&utm_medium=website&utm_term=fda%20news%20and%20events%20rivaroxaban&utm_content=3. Accessed 7 May 2012

FDA News and Events (2012c) NEWS RELEASE. FDA clears genetic lab test for Warfarin sensitivity. http://www.fda.gov/newsevents/newsroom/pressannouncements/2007/ucm108984.htm. Accessed 10 Sept 2012

Fava C, Montagnana M, Almgren P, Rosberg L, Lippi G, Hedblad B, Engström G, Berglund G, Minuz P, Melander O (2008) The V433M variant of the CYP4F2 is associated with ischemic stroke in male Swedes beyond its effect on blood pressure. Hypertension 52:373–380

FDA. U.S. Food and Drug Administration (2012a) COUMADIN® TABLETS (Warfarin sodium Tablets, USP) Crystalline COUMADIN® FOR INJECTION (Warfarin Sodium for Injection, USP). http://www.accessdata.fda.gov/drugsatfda_docs/label/2010/009218s108lbl.pdf. Accessed 7 May 2012

FDA. U.S. Food and Drug Administration (2012b) XARELTO (rivaroxaban) tablets, for oral use. http://www.accessdata.fda.gov/drugsatfda_docs/label/2011/202439s000lbl.pdf. Accessed 7 May 2012

Francis CW (2008) New issues in oral anticoagulants. Hematol Am Soc Hematol Educ Program 2008(1):259–265. doi:10.1182/asheducation-2008.1.259

Francis CW, Berkowitz SD, Comp PC, Lieberman JR, Ginsberg JS, Paiement G, Peters GR, Roth AW, McElhattan J, Colwell CW Jr, EXULT A Study Group (2003) Comparison of ximelagatran with warfarin for the prevention of venous thromboembolism after total knee replacement. N Engl J Med 349:1703–1712

Fu Z, Nakayama T, Sato N, Izumi Y, Kasamaki Y, Shindo A, Ohta M, Soma M, Aoi N, Sato M, Ozawa Y, Ma Y, Matsumoto K, Doba N, Hinohara S (2009) A haplotype of the CYP4F2 gene associated with myocardial infarction in Japanese men. Mol Genet Metab 96:145–147

Furie KL, Kasner SE, Adams RJ, Albers GW, Bush RL, Fagan SC, Halperin JL, Johnston SC, Katzan I, Kernan WN, Mitchell PH, Ovbiagele B, Palesch YY, Sacco RL, Schwamm LH, Wassertheil-Smoller S, Turan TN, Wentworth D, American Heart Association Stroke Council, Council on Cardiovascular Nursing, Council on Clinical Cardiology, Interdisciplinary Council on Quality of Care and Outcomes Research (2011) Guidelines for the prevention of stroke in patients with stroke or transient ischemic attack: a guideline for healthcare professionals from the American Heart Association/American Stroke Association. Stroke 42(1):227–276

Gage BF, Eby C, Milligan PE, Banet GA, Duncan JR, McLeod HL (2004) Use of pharmacogenetics and clinical factors to predict the maintenance dose of warfarin. Thromb Haemost 91:87–94

Gage BF, Eby C, Johnson JA, Deych E, Rieder MJ, Ridker PM, Milligan PE, Grice G, Lenzini P, Rettie AE, Aquilante CL, Grosso L, Marsh S, Langaee T, Farnett LE, Voora D, Veenstra DL, Glynn RJ, Barrett A, McLeod HL (2008) Use of pharmacogenetic and clinical factors to predict the therapeutic dose of warfarin. Clin Pharmacol Ther 84:326–331

Geisen C, Watzka M, Sittinger K et al (2005) VKORC1 haplotypes and their impact on the inter-individual and inter-ethnical variability of oral anticoagulation. Thromb Haemost 94:773–779

Giorgi MA, Miguel LS (2012) Rivaroxaban in atrial fibrillation. Vasc Health Risk Manag 8:525–531

González-Juanatey JR, Alvarez-Sabin J, Lobos JM, Martínez-Rubio A, Reverter JC, Oyagüez I, González-Rojas N, Becerra V (2012) Cost-effectiveness of Dabigatran for stroke prevention in non-valvular atrial fibrillation in Spain. Rev Esp Cardiol 65(10):901–910

Granger CB, Alexander JH, McMurray JJ, Lopes RD, Hylek EM, Hanna M, Al-Khalidi HR, Ansell J, Atar D, Avezum A, Bahit MC, Diaz R, Easton JD, Ezekowitz JA, Flaker G, Garcia D, Geraldes M, Gersh BJ, Golitsyn S, Goto S, Hermosillo AG, Hohnloser SH, Horowitz J, Mohan P, Jansky P, Lewis BS, Lopez-Sendon JL, Pais P, Parkhomenko A, Verheugt FW, Zhu J, Wallentin L, ARISTOTLE Committees and Investigators (2011) Apixaban versus warfarin in patients with atrial fibrillation. N Engl J Med 365:981–992

Haining RL, Hunter AP, Veronese ME, Trager WF, Rettie AE (1996) Allelic variants of human cytochrome P450 2C9: baculovirus-mediated expression, purification, structural characterization, substrate stereoselectivity, and prochiral selectivity of the wild-type and I359L mutant forms. Arch Biochem Biophys 333:447–458

Hirsh J (1992) Antithrombotic therapy in deep vein thrombosis and pulmonary embolism. Am Heart J 123(4 Pt 2):1115–1122

Hirsh J, Dalen J, Anderson DR, Poller L, Bussey H, Ansell J, Deykin D (2001) Oral anticoagulants: mechanism of action, clinical effectiveness, and optimal therapeutic range. Chest 119(1 Suppl):8S–21S

Hylek EM, Go AS, Chang Y, Jensvold NG, Henault LE, Selby JV, Singer DE (2003) Effect of intensity of oral anticoagulation on stroke severity and mortality in atrial fibrillation. N Engl J Med 349:1019–1026

Infiniti Warfarin Assay (2012) Available from: http://www.autogenomics.com/pdf/EM-34004-D-Warfarin-Assay-Package-Insert.pdf. Accessed 10 Sept 2012

International Warfarin Pharmacogenetics Consortium (IWPC) Algorithm (2012) Available from: http://www.pharmgkb.org/do/serve?objId=PA162372936&objCls=Dataset#tabview=tabnull&subtab=. Accessed 7 May 2012

International Warfarin Pharmacogenetics Consortium, Klein TE, Altman RB, Eriksson N, Gage BF, Kimmel SE, Lee MT, Limdi NA, Page D, Roden DM, Wagner MJ, Caldwell MD, Johnson JA (2009) Estimation of the warfarin dose with clinical and pharmacogenetic data. N Engl J Med 360:753–764

Jähnchen E, Meinertz T, Gilfrich HJ, Groth U, Martini A (1976) The enantiomers of phenprocoumon: pharmacodynamic and pharmacokinetic studies. Clin Pharmacol Ther 20:342–349

Johnson EG, Horne BD, Carlquist JF, Anderson JL (2011) Genotype-based dosing algorithms for warfarin therapy: data review and recommendations. Mol Diagn Ther 15:255–264

Jonas DE, McLeod HL (2009) Genetic and clinical factors relating to warfarin dosing. Trends Pharmacol Sci 30:375–386

Kamali F, Pirmohamed M (2006) The future prospects of pharmacogenetics in oral anticoagulation therapy. Br J Clin Pharmacol 61(6):746–751

King CR, Porche-Sorbet RM, Gage BF, Ridker PM, Renaud Y, Phillips MS, Eby C (2008) Performance of commercial platforms for rapid genotyping of polymorphisms affecting warfarin dose. Am J Clin Pathol 129:876–883

King CR, Deych E, Milligan P, Eby C, Lenzini P, Grice G, Porche-Sorbet RM, Ridker PM, Gage BF (2010) Gamma-glutamyl carboxylase and its influence on warfarin dose. Thromb Haemost 104:750–754

Kubitza D, Becka M, Wensing G, Voith B, Zuehlsdorf M (2005) Safety, pharmacodynamics, and pharmacokinetics of BAY 59-7939 an oral, direct Factor Xa inhibitor – after multiple dosing in healthy male subjects. Eur J Clin Pharmacol 61(12):873–880

Kubitza D, Becka M, Zuehlsdorf M, Mueck W (2006) Effect of food, an antacid, and the H2 antagonist ranitidine on the absorption of BAY 59–7939 (rivaroxaban), an oral, direct factor Xa inhibitor, in healthy subjects. J Clin Pharmacol 46:549–558

Kubitza D, Becka M, Roth A, Mueck W (2008) Dose-escalation study of the pharmacokinetics and pharmacodynamics of rivaroxaban in healthy elderly subjects. Curr Med Res Opin 24:2757–2765

Langley MR, Booker JK, Evans JP, McLeod HL, Weck KE (2009) Validation of clinical testing for warfarin sensitivity: comparison of CYP2C9-VKORC1 genotyping assays and warfarin-dosing algorithms. J Mol Diagn 11:216–225

Lefferts JA, Schwab MC, Dandamudi UB, Lee HK, Lewis LD, Tsongalis GJ (2010) Warfarin genotyping using three different platforms. Am J Transl Res 2:441–446

Lenzini P, Wadelius M, Kimmel S, Anderson JL, Jorgensen AL, Pirmohamed M, Caldwell MD, Limdi N, Burmester JK, Dowd MB, Angchaisuksiri P, Bass AR, Chen J, Eriksson N, Rane A, Lindh JD, Carlquist JF, Horne BD, Grice G, Milligan PE, Eby C, Shin J, Kim H, Kurnik D, Stein CM, McMillin G, Pendleton RC, Berg RL, Deloukas P, Gage BF (2010) Integration of genetic, clinical, and INR data to refine warfarin dosing. Clin Pharmacol Ther 87:572–578

Li T, Chang CY, Jin DY, Lin PJ, Khvorova A, Stafford DW (2004) Identification of the gene for vitamin K epoxide reductase. Nature 427:541–544

Li C, Schwarz UI, Ritchie MD, Roden DM, Stein CM, Kurnik D (2009) Relative contribution of CYP2C9 and VKORC1 genotypes and early INR response to the prediction of warfarin sensitivity during initiation of therapy. Blood 113:3925–3930

Liesenfeld KH, Lehr T, Dansirikul C, Reilly PA, Connolly SJ, Ezekowitz MD, Yusuf S, Wallentin L, Haertter S, Staab A (2011) Population pharmacokinetic analysis of the oral thrombin inhibitor dabigatran etexilate in patients with non-valvular atrial fibrillation from the RE-LY trial. J Thromb Haemost 9:2168–2175

Limdi NA, Veenstra DL (2008) Warfarin pharmacogenetics. Pharmacotherapy 28(9):1084–1097

Limdi NA, Arnett DK, Goldstein JA, Beasley TM, McGwin G, Adler BK, Acton RT (2008) Influence of CYP2C9 and VKORC1 on warfarin dose, anticoagulation attainment and maintenance among European-Americans and African-Americans. Pharmacogenomics 9:511–526

Limdi NA, Wiener H, Goldstein JA, Acton RT, Beasley TM (2009) Influence of CYP2C9 and VKORC1 on warfarin response during initiation of therapy. Blood Cells Mol Dis 43:119–128

Lip GY, Lane DA (2010) Does warfarin for stroke thromboprophylaxis protect against MI in atrial fibrillation patients? Am J Med 123:785–789

Lip GY, Huber K, Andreotti F, Arnesen H, Airaksinen KJ, Cuisset T, Kirchhof P, Marín F, European Society of Cardiology Working Group on Thrombosis (2010) Management of antithrombotic therapy in atrial fibrillation patients presenting with acute coronary syndrome and/or undergoing percutaneous coronary intervention/stenting. Thromb Haemost 103:13–28

Loebstein R, Dvoskin I, Halkin H, Vecsler M, Lubetsky A, Rechavi G, Amariglio N, Cohen Y, Ken-Dror G, Almog S, Gak E (2007) A coding VKORC1 Asp36Tyr polymorphism predisposes to warfarin resistance. Blood 109:2477–2480

López-Parra AM, Borobia AM, Baeza C, Arroyo-Pardo E, Carcas AJ (2013) A multiplex assay to detect variations in the CYP2C9, VKORC1, CYP4F2 and APOE genes involved in acenocoumarol metabolism. Clin Biochem 46(1–2):167–169

Lund K, Gaffney D, Spooner R, Etherington AM, Tansey P, Tait RC (2012) Polymorphisms in VKORC1 have more impact than CYP2C9 polymorphisms on early warfarin International Normalized Ratio control and bleeding rates. Br J Haematol 158(2):256–261

Manolopoulos VG, Ragia G, Tavridou A (2010) Pharmacogenetics of coumarinic oral anticoagulants. Pharmacogenomics 4:493–496

Margaglione M, Colaizzo D, D'Andrea G, Brancaccio V, Ciampa A, Grandone E, Di Minno G (2000) Genetic modulation of oral anticoagulation with warfarin. Thromb Haemost 84:775–778

Markatos CN, Grouzi E, Politou M, Gialeraki A, Merkouri E, Panagou I, Spiliotopoulou I, Travlou A (2008) VKORC1 and CYP2C9 allelic variants influence acenocoumarol dose requirements in Greek patients. Pharmacogenomics 9:1631–1638

Maurice CB, Barua PK, Simses D, Smith P, Howe JG, Stack G (2010) Comparison of assay systems for warfarin-related CYP2C9 and VKORC1 genotyping. Clin Chim Acta 411:947–954

Millican EA, Lenzini PA, Milligan PE, Grosso L, Eby C, Deych E, Grice G, Clohisy JC, Barrack RL, Burnett RS, Voora D, Gatchel S, Tiemeier A, Gage BF (2007) Genetic-based dosing in orthopedic patients beginning warfarin therapy. Blood 110:1511–1515

Mitchell C, Gregersen N, Krause A (2011) Novel CYP2C9 and VKORC1 gene variants associated with warfarin dosage variability in the South African black population. Pharmacogenomics 12:953–963

Momary KM, Shapiro NL, Viana MA, Nutescu EA, Helgason CM, Cavallari LH (2007) Factors influencing warfarin dose requirements in African-Americans. Pharmacogenomics 8:1535–1544

National Heart, Lung, and Blood Institute (2012) Clarification of Optimal Anticoagulation Through Genetics (COAG) [ClinicalTrials.gov identifier NCT00839657]. US National Institutes of Health, ClinicalTrials.gov. Available from: http://www.clinicaltrials.gov/show/NCT00839657. Accessed 20 Feb 2012

Ngui JS, Chen Q, Shou M, Wang RW, Stearns RA, Baillie TA, Tang W (2001) In vitro stimulation of warfarin metabolism by quinidine: increases in the formation of 4′- and 10-hydroxywarfarin. Drug Metab Dispos 29:877–886

Obayashi K, Nakamura K, Kawana J, Ogata H, Hanada K, Kurabayashi M, Hasegawa A, Yamamoto K, Horiuchi R (2006) VKORC1 gene variations are the major contributors of variation in warfarin dose in Japanese patients. Clin Pharmacol Ther 80:169–178

Osman A, Enström C, Arbring K, Söderkvist P, Lindahl TL (2006) Main haplotypes and mutational analysis of vitamin K epoxide reductase (VKORC1) in a Swedish population: a retrospective analysis of case records. J Thromb Haemost 4:1723–1729

Patel MR, Mahaffey KW, Garg J, Pan G, Singer DE, Hacke W, Breithardt G, Halperin JL, Hankey GJ, Piccini JP, Becker RC, Nessel CC, Paolini JF, Berkowitz SD, Fox KA, Califf RM, Investigators ROCKETAF (2011) Rivaroxaban versus warfarin in nonvalvular atrial fibrillation. N Engl J Med 365:883–891

Perera MA, Gamazon E, Cavallari LH, Patel SR, Poindexter S, Kittles RA, Nicolae D, Cox NJ (2011) The missing association: sequencing-based discovery of novel SNPs in VKORC1 and CYP2C9 that affect warfarin dose in African Americans. Clin Pharmacol Ther 89:408–415

Perzborn E, Roehrig S, Straub A, Kubitza D, Misselwitz F (2011) The discovery and development of rivaroxaban, an oral, direct factor Xa inhibitor. Nat Rev Drug Discov 10(1):61–75

Raghavan N, Frost CE, Yu Z, He K, Zhang H, Humphreys WG, Pinto D, Chen S, Bonacorsi S, Wong PC, Zhang D (2009) Apixaban metabolism and pharmacokinetics after oral administration to humans. Drug Metab Dispos 37:74–81

Reeman JV, Zhu RP, Owens DK, Garber AM, Hutton DW, Go AS, Wang PJ, Turakhia MP (2011) Cost-effectiveness of dabigatran compared with warfarin for stroke prevention in atrial fibrillation. Ann Intern Med 154:1–11

Rettie AE, Korzekwa KR, Kunze KL, Lawrence RF, Eddy AC, Aoyama T, Gelboin HV, Gonzalez FJ, Trager WF (1992) Hydroxylation of warfarin by human cDNA-expressed Cytochrome P-450: a role for P-4502C9 in the etiology of (S)-warfarin-drug interactions. Chem Res Toxicol 5:54–59

Rettie AE, Wienkers LC, Gonzalez FJ, Trager WF, Korzekwa KR (1994) Impaired (S)-warfarin metabolism catalysed by the R144C allelic variant of CYP2C9. Pharmacogenetics 4:39–42

Rieder MJ, Reiner AP, Gage BF, Nickerson DA, Eby CS, McLeod HL, Blough DK, Thummel KE, Veenstra DL, Rettie AE (2005) Effect of VKORC1 haplotypes on transcriptional regulation and warfarin dose. N Engl J Med 352:2285–2293

Rieder MJ, Reiner AP, Rettie AE (2007) Gamma-glutamyl carboxylase (GGCX) tag SNPs have limited utility for predicting warfarin maintenance dose. J Thromb Haemost 5:2227–2234

Rosemary J, Adithan C (2007) The pharmacogenetics of CYP2C9 and CYP2C19: ethnic variation and clinical significance. Curr Clin Pharmacol 2:93–109

Rost S, Fregin A, Ivaskevicius V, Conzelmann E, Hörtnagel K, Pelz HJ, Lappegard K, Seifried E, Scharrer I, Tuddenham EG, Müller CR, Strom TM, Oldenburg J (2004) Mutations in VKORC1 cause warfarin resistance and multiple coagulation factor deficiency type 2. Nature 427:537–541

Sagreiya H, Berube C, Wen A, Ramakrishnan R, Mir A, Hamilton A, Altman RB (2010) Extending and evaluating a warfarin dosing algorithm that includes CYP4F2 and pooled rare variants of CYP2C9. Pharmacogenet Genomics 20:407–413

Schalekamp T, Brassé BP, Roijers JF, Chahid Y, van Geest-Daalderop JH, de Vries-Goldschmeding H, van Wijk EM, Egberts AC, de Boer A (2006) VKORC1 and CYP2C9 genotypes and acenocoumarol anticoagulation status: interaction between both genotypes affects overanticoagulation. Clin Pharmacol Ther 80:13–22

Schelleman H, Chen Z, Kealey C, Whitehead AS, Christie J, Price M, Brensinger CM, Newcomb CW, Thorn CF, Samaha FF, Kimmel SE (2007) Warfarin response and vitamin K epoxide reductase complex 1 in African Americans and Caucasians. Clin Pharmacol Ther 81(5):742–747

Schelleman H, Limdi NA, Kimmel SE (2008) Ethnic differences in warfarin maintenance dose requirement and its relationship with genetics. Pharmacogenomics 9(9):1331–1346

Schulman S, Bijsterveld NR (2007) Anticoagulants and their reversal. Transfus Med Rev 21:37–48

Schulman S, Majeed A (2012) The oral thrombin inhibitor dabigatran: strengths and weaknesses. Semin Thromb Hemost 38:7–15

Schulman S, Wåhlander K, Lundström T, Clason SB, Eriksson H, Investigators THRIVEIII (2003) Secondary prevention of venous thromboembolism with the oral direct thrombin inhibitor ximelagatran. N Engl J Med 349(18):1713–1721

Schulman S, Kearon C, Kakkar AK, Mismetti P, Schellong S, Eriksson H, Baanstra D, Schnee J, Goldhaber SZ, RE-COVER Study Group (2009) Dabigatran versus warfarin in the treatment of acute venous thromboembolism. N Engl J Med 361:2342–2352

Schwarz UI, Ritchie MD, Bradford Y, Li C, Dudek SM, Frye-Anderson A, Kim RB, Roden DM, Stein CM (2008) Genetic determinants of response to warfarin during initial anticoagulation. N Engl J Med 358:999–1008

Sconce EA, Khan TI, Wynne HA, Avery P, Monkhouse L, King BP, Wood P, Kesteven P, Daly AK, Kamali F (2005) The impact of CYP2C9 and VKORC1 genetic polymorphism and patient characteristics upon warfarin dose requirements: proposal for a new dosing regimen. Blood 106:2329–2333

Scott SA, Jaremko M, Lubitz SA, Kornreich R, Halperin JL, Desnick RJ (2009) CYP2C9*8 is prevalent among African-Americans: implications for pharmacogenetic dosing. Pharmacogenomics 10:1243–1255

Scott SA, Khasawneh R, Peter I, Kornreich R, Desnick RJ (2010) Combined CYP2C9, VKORC1 and CYP4F2 frequencies among racial and ethnic groups. Pharmacogenomics 11:781–791

Shahin MH, Khalifa SI, Gong Y, Hammad LN, Sallam MT, El Shafey M, Ali SS, Mohamed ME, Langaee T, Johnson JA (2011) Genetic and nongenetic factors associated with warfarin dose requirements in Egyptian patients. Pharmacogenet Genomics 21:130–135

Shaw PB, Donovan JL, Tran MT, Lemon SC, Burgwinkle P, Gore J (2010) Accuracy assessment of pharmacogenetically predictive warfarin dosing algorithms in patients of an academic medical center anticoagulation clinic. J Thromb Thrombolysis 30:220–225

Shirolkar SC, Fiuzat M, Becker RC (2010) Dronedarone and vitamin K antagonists: a review of drug-drug interactions. Am Heart J 160:577–582

Skanes AC, Healey JS, Cairns JA, Dorian P, Gillis AM, McMurtry MS, Mitchell LB, Verma A, Nattel S, Canadian Cardiovascular Society Atrial Fibrillation Guidelines Committee (2012) Focused 2012 update of the Canadian Cardiovascular Society atrial fibrillation guidelines: recommendations for stroke prevention and rate/rhythm control. Can J Cardiol 28(2):125–136

Spanish Agency of Medicines and Health Products (2012) Criteria and recommendations for the use of the new oral anticoagulants in the prevention of stroke and systemic embolism in patients with non-valvular atrial fibrillation. http://www.aemps.gob.es/medicamentosUsoHumano/informesPublicos/docs/criterios-aco-rev_05-09-12.pdf. Accessed Sept 2012

Spreafico M, Lodigiani C, van Leeuwen Y, Pizzotti D, Rota LL, Rosendaal F, Mannucci PM, Peyvandi F (2008) Effects of CYP2C9 and VKORC1 on INR variations and dose requirements during initial phase of anticoagulant therapy. Pharmacogenomics 9:1237–1250

Stafford DW (2005) The vitamin K cycle. J Thromb Haemost 3:1873–1878

Stangier J, Rathgen K, Stähle H, Mazur D (2010) Influence of renal impairment on the pharmacokinetics and pharmacodynamics of oral dabigatran etexilate: an open-label, parallel-group, single-centre study. Clin Pharmacokinet 49:259–268

Stehle S, Kirchheiner J, Lazar A, Fuhr U (2008) Pharmacogenetics of oral anticoagulants: a basis for dose individualization. Clin Pharmacokinet 47(9):565–594

Takahashi H, Wilkinson GR, Nutescu EA, Morita T, Ritchie MD, Scordo MG, Pengo V, Barban M, Padrini R, Ieiri I, Otsubo K, Kashima T, Kimura S, Kijima S, Echizen H (2006) Different contributions of polymorphisms in VKORC1 and CYP2C9 to intra- and inter-population differences in maintenance dose of warfarin in Japanese, Caucasians and African-Americans. Pharmacogenet Genomics 16:101–110

Takanashi K, Tainaka H, Kobayashi K, Yasumori T, Hosakawa M, Chiba K (2000) CYP2C9 Ile359 and Leu359 variants: enzyme kinetic study with seven substrates. Pharmacogenetics 10:95–104

Teichert M, Eijgelsheim M, Rivadeneira F, Uitterlinden AG, van Schaik RH, Hofman A, De Smet PA, van Gelder T, Visser LE, Stricker BH (2009) A genome-wide association study of acenocoumarol maintenance dosage. Hum Mol Genet 18(19):3758–3768

Teichert M, Eijgelsheim M, Uitterlinden AG, Buhre PN, Hofman A, De Smet PA, Visser LE, Stricker BH (2011a) Dependency of phenprocoumon dosage on polymorphisms in the VKORC1, CYP2C9, and CYP4F2 genes. Pharmacogenet Genomics 21(1):26–34

Teichert M, van Noord C, Uitterlinden AG, Hofman A, Buhre PN, De Smet PA, Straus S, Stricker BH, Visser LE (2011b) Proton pump inhibitors and the risk of overanticoagulation during acenocoumarol maintenance treatment. Br J Haematol 153:379–385

The Pharmacogenomics Knowledge Base (2012) Acenocoumarol. Available from: http://www.pharmgkb.org/drug/PA452632#tabview=tab3&subtab=31. Accessed May 2012

Thijssen HH, Flinois JP, Beaune PH (2000) Cytochrome P4502C9 is the principal catalyst of racemic acenocoumarol hydroxylation reactions in human liver microsomes. Drug Metab Dispos 28:1284–1290

Toon S, Heimark LD, Trager WF, O'Reilly RA (1985) Metabolic fate of phenprocoumon in humans. J Pharm Sci 74(10):1037–1040

Uchino K, Hernandez AV (2012) Dabigatran association with higher risk of acute coronary events: meta-analysis of noninferiority randomized controlled trials. Arch Intern Med 172:397–402

Ufer M (2005) Comparative pharmacokinetics of vitamin K antagonists: warfarin, phenprocoumon and acenocoumarol. Clin Pharmacokinet 44:1227–1246

Utrecht Institute for Pharmaceutical Sciences (2012). European Pharmacogenetics of Anticoagulant Therapy – Warfarin (EU-PACT) [ClinicalTrials.gov identifier NCT01119300]. US National Institutes of Health, ClinicalTrials.gov Available from: http://www.clinicaltrials.gov/show/NCT01119300. Accessed 20 Feb 2012

van Schie RM, Wessels JA, le Cessie S, de Boer A, Schalekamp T, van der Meer FJ, Verhoef TI, van Meegen E, Rosendaal FR, der Zee AH M-v, EU-PACT Study Group (2011) Loading and maintenance dose algorithms for phenprocoumon and acenocoumarol using patient characteristics and pharmacogenetic data. Eur Heart J 32:1909–1917

Veenstra DL, You JH, Rieder MJ, Farin FM, Wilkerson HW, Blough DK, Cheng G, Rettie AE (2005) Association of Vitamin K epoxide reductase complex 1 (VKORC1) variants with warfarin dose in a Hong Kong Chinese patient population. Pharmacogenet Genomics 15:687–691

Verde Z, Ruiz JR, Santiago C, Valle B, Bandrés F, Calvo E, Lucía A, Gómez Gallego F (2010) A novel, single algorithm approach to predict acenocoumarol dose based on CYP2C9 and VKORC1 allele variants. PLoS One 5:e11210

Wadelius M, Chen LY, Downes K, Ghori J, Hunt S, Eriksson N, Wallerman O, Melhus H, Wadelius C, Bentley D, Deloukas P (2005) Common VKORC1 and GGCX polymorphisms associated with warfarin dose. Pharmacogenomics J 5:262–270

Wadelius M, Chen LY, Eriksson N, Bumpstead S, Ghori J, Wadelius C, Bentley D, McGinnis R, Deloukas P (2007) Association of warfarin dose with genes involved in its action and metabolism. Hum Genet 121(1):23–34

Wadelius M, Chen LY, Lindh JD, Eriksson N, Ghori MJ, Bumpstead S, Holm L, McGinnis R, Rane A, Deloukas P (2009) The largest prospective warfarin-treated cohort supports genetic forecasting. Blood 113:784–792

Wajih N, Sane DC, Hutson SM, Wallin R (2004) The inhibitory effect of calumenin on the vitamin K-dependent gamma-carboxylation system. Characterization of the system in normal and warfarin-resistant rats. J Biol Chem 279:25276–25283

Wang D, Chen H, Momary KM, Cavallari LH, Johnson JA, Sadée W (2008) Regulatory polymorphism in vitamin K epoxide reductase complex subunit 1 (VKORC1) affects gene expression and warfarin dose requirement. Blood 112:1013–1021

Wartak SA, Bartholomew JR (2011) Dabigatran: will it change clinical practice? Cleve Clin J Med 78(10):657–664

Washington University School of Medicine (2012) Genetics Informatics Trial (GIFT) of warfarin to prevent DVT [ClinicalTrials.gov identifier NCT01006733]. US National Institutes of Health, ClinicalTrials.gov. Available from: http://www.clinicaltrials.gov/show/NCT01006733. Accessed 20 Feb 2012

Weitz JI (2010) New oral anticoagulants in development. Thromb Haemost 103:62–70

Wienkers LC, Wurden CJ, Storch E, Kunze KL, Rettie AE, Trager WF (1996) Formation of (R)-8-hydroxywarfarin in human liver microsomes. A new metabolic marker for the (S)-mephenytoin hydroxylase, P4502C19. Drug Metab Dispos 24:610–614

Wittkowsky AK (2010) New oral anticoagulants: a practical guide for clinicians. J Thromb Thrombolysis 29:182–191

You JJ, Singer DE, Howard PA, Lane DA, Eckman MH, Fang MC, Hylek EM, Schulman S, Go AS, Hughes M, Spencer FA, Manning WJ, Halperin JL, Lip GY, American College of Chest Physicians (2012) Antithrombotic therapy for atrial fibrillation: antithrombotic therapy and prevention of thrombosis, 9th ed: American College of Chest Physicians Evidence-Based Clinical Practice Guidelines. Chest 141(2 Suppl):e531S–e575S

Yuan HY, Chen JJ, Lee MT, Wung JC, Chen YF, Charng MJ, Lu MJ, Hung CR, Wei CY, Chen CH, Wu JY, Chen YT (2005) A novel functional VKORC1 promoter polymorphism is associated with inter-individual and inter-ethnic differences in warfarin sensitivity. Hum Mol Genet 14:1745–1751

Zhang Z, Fasco MJ, Huang Z, Guengerich FP, Kaminsky LS (1995) Human Cytochromes P4501A1 and P4501A2: R-warfarin metabolism as a probe. Drug Metab Dispos 23:1339–1346

Pharmacogenomics and Personalized Medicine of the Antiplatelet Drugs

22

Hong-Guang Xie and Ying-Dong Zhang

Abstract

Platelet activation and aggregation play a critical role in hemostasis and thrombosis. There is a fast-growing list of the antiplatelet drugs that are either marketed or under development, of which combination of aspirin and clopidogrel is now the standard of care for acute coronary syndromes or percutaneous coronary intervention for stenting. Overwhelming data have well demonstrated that aspirin monotherapy can greatly improve patient outcomes by irreversible suppression of the cyclooxygenase-1 enzyme responsible for the arachidonic acid pathway, that clopidogrel can exert its well-documented platelet inhibition through irreversible blockade of the platelet ADP receptor P2Y12, and that dual antiplatelet therapy (aspirin plus clopidogrel) is clinically more effective than either of the two for the secondary prevention of the recurrence of myocardial infarction, in-stent thrombosis, ischemic stroke, or even death. However, individuals may vary in their response to the drug, characterized by less or no response to either one or both in some patients when taking the same doses. It is well known that almost all genetic and nongenetic factors may contribute to that variation as summarized in this book chapter, and that DNA or pharmacogenomics is not the whole story about personalized medicine. The future landscape of optimal drug therapy would be much clearer over time and thus more attractive.

1 Introduction

Platelet activation and aggregation play a central role in the initiation and development of thrombus formation, to which the stimulation of platelet ADP receptor P2Y12 makes a greater contribution (Dorsam and Kunapuli 2004) than other components involved. In addition to ADP, thrombin is the most potent activator of platelets. Human platelets express two thrombin

H.-G. Xie (✉) • Y.-D. Zhang
General Clinical Research Center (H.-G.X.), and Department of Neurology (Y.-D.Z.), Nanjing Medical University Nanjing First Hospital, 68 Changle Road, Nanjing 210006, China

Department of Pharmacology (H.-G.X.), Nanjing Medical University School of Pharmacy, Nanjing 210029, China
e-mail: hongg.xie@gmail.com; zhangyingdong@aliyun.com

D. Barh et al. (eds.), *Omics for Personalized Medicine*, DOI 10.1007/978-81-322-1184-6_22, © Springer India 2013

receptors – protease-activated receptor (PAR)-1 and PAR4, both of which signal platelet activation (Kahn et al. 1998). Activation of the platelet P2Y12 receptors is only partially required for PAR1- and glycoprotein (GP) IIb/IIIa-mediated platelet activation, and thus the P2Y12 antagonists (such as clopidogrel) can partially inhibit PAR1- rather than PAR4-mediated platelet aggregation (Holinstat et al. 2006). Therefore, for patients treated with clopidogrel or aspirin, thrombin still maintains the ability to signal platelet aggregation, in particular at the site of a local thrombus or at the late (irreversible) phase of platelet activation and aggregation, which is mediated by PAR4 (Holinstat et al. 2006), and residual platelet reactivity after clopidogrel is, at least in part, the result of PAR-mediated response (Badr et al. 2012). Based on this concept, PAR1 antagonists represent a new class of oral antiplatelet agents (Capodanno et al. 2012).

Acetylsalicylic acid (also known as aspirin) acts through irreversible suppression of cyclooxygenase 1 (COX1) activity, thus inhibiting the synthesis of proaggregatory thromboxane A_2 (or TXA_2) and exerting its strong antiplatelet effect. Aspirin monotherapy can improve patient outcomes. The subsequently developed thienopyridines further enhance platelet inhibition and improve ischemic outcomes in patients with acute coronary syndromes (also known as ACS) or those undergoing percutaneous coronary intervention (also known as PCI) for coronary stenting through irreversible blockade of ADP binding to the platelet ADP receptor P2Y12. In clinical practice, dual antiplatelet therapy (DAT) with aspirin and clopidogrel has been thought to be the "good standard" antiplatelet regimen for their shared indications.

Platelet GP IIb/IIIa is the integrin receptor thought to play a crucial role in hemostasis and thus is the antithrombotic target (Armstrong and Peter 2012). The GPIIb/IIIa receptor antagonists suppress fibrinogen-mediated platelet aggregation (Rivera et al. 2009) as an adjunctive therapy concomitant for the triple antiplatelet therapy (or TAT) (Valgimigli and Minarelli 2011; Schneider 2011). In addition, other adjunct drugs (such as cilostazol and ketanserin) are being evaluated for triple antiplatelet therapy (Angiolillo 2012).

Table 22.1 Classification of the antiplatelet drugs

Class of drugs	Example
COX1 inhibitor	Aspirin
P2Y12 receptor antagonist	
Indirect acting or irreversible[a]	Ticlopidine, clopidogrel, prasugrel
Direct acting or reversible	Ticagrelor, cangrelor, elinogrel
Adjunctive or new drug	
PDE3 inhibitor	Cilostazol, K-134
GP IIb/IIIa blocker/ inhibitor	Abciximab, tirofiban, eptifibatide
GP Ib receptor antagonist	Agkistin
PAR1 receptor antagonist	Atopaxar, vorapaxar, SCH530348, E-5555
PAR4 receptor antagonist	YD-3
P2Y1 receptor antagonist	MRS2179, MRS2500
P2Y12 inverse agonist	AR-C78511
TXA_2 receptor antagonist	Ifetroban, ramatroban, S18886, SQ29548, BM-144, GR32191B, Z-335
TXA_2 synthase inhibitor	BM-573, BM-613
PI3Kβ inhibitor	AZD6482
Anti-vWF drug	ARC1779
5-HT receptor antagonist	Ketanserin
Miscellaneous	Dipyridamole, citalopram, NO donors

[a]Thienopyridine; PDE3, phosphodiesterase 3; 5-HT, serotonin. NO, nitric oxide, such as nitroglycerin and sodium nitroprusside

In this chapter, clinical pharmacogenomics and personalized medicine of the antiplatelet drugs are detailed with the major focus on aspirin and clopidogrel in patient care. In addition, other antiplatelet drugs that have been summarized in Table 22.1 are shown just for the whole landscape.

2 Aspirin

Aspirin, a highly lipophilic molecule, is the first clinically effective antiplatelet drug, acting through an irreversible modification of its main target COX1 and consequently blocking the generation of TXA_2. Up to date, low-dose aspirin, routinely combined with clopidogrel, has become the standard of care for secondary prevention of the recurrence of atherothrombotic events in ACS patients or those undergoing PCI for coronary stenting (Patrono et al. 2004a; Collet et al. 2009; Malek et al. 2008; Shuldiner et al. 2009; Sibbing et al. 2010; Simon et al. 2009; Harmsze et al. 2010; Wallentin et al. 2010; Anderson et al. 2007; Antman et al. 2008b; Smith et al. 2006a; Wurtz et al. 2012) and also for secondary prevention of ischemic stroke (Ansara et al. 2010). Although the dose of aspirin varies greatly worldwide (Peters et al. 2003), high-dose aspirin (>300 mg/day) does not confer a further reduced risk of ischemic events compared with low-dose aspirin (75–100 mg/day) but might be associated with an increased risk of major bleeding (Mehta et al. 2010; Topol et al. 2003; Peters et al. 2003).

Despite a wide range of daily effective doses (30–1,500 mg) (Patrono et al. 2005), aspirin completely abrogates the production of TXA_2 from platelets even at its low doses in humans (Patrono et al. 2004b, 2005), partially inhibiting TXA_2-mediated platelet aggregation (Patrono et al. 2004b).

Combination of aspirin and clopidogrel can result in simultaneous inhibition of two affected pathways involved in platelet aggregation (Shuldiner et al. 2009), producing greater antiplatelet effects than either of the two. On the other hand, the major bleeding complications (notably in the upper GI tract) are associated with the use of low-dose aspirin or clopidogrel (Patrono et al. 2004b; Pirmohamed et al. 2004), and thus concurrent use of aspirin and clopidogrel may have a higher risk of bleeding than each agent alone. Clinically, aspirin 75–325 mg once daily has been found to be optimal with no additional benefit from increasing doses (Jacobson 2004), since limiting aspirin dose to less than 100 mg daily could reduce the risk of major bleeding complications but could not compromise the efficacy of dual antiplatelet therapy (Peters et al. 2003).

In addition, platelets are also potent inflammatory cells, in turn, inducing inflammatory response in adjacent cells (such as leukocytes and endothelial cells) and leading to inflammatory processes in the vessel wall and consequently ACS (Hansson 2005) and atherogenesis (Aukrust et al. 2010). For instance, consistent with previous studies (Debrunner et al. 2008; Mockel et al. 2007), the levels of serum TNFα (a proinflammatory cytokine) are significantly higher in ACS patients than in age- and gender-matched healthy subjects (Bergandi et al. 2010), confirming that there is an increased inflammatory state in ACS patients, and that TNFα is indeed involved in platelet aggregation (Pignatelli et al. 2005). Furthermore, in a pilot study of 157 patients with coronary artery disease (CAD) undergoing PCI and receiving clopidogrel and aspirin, levels of interleukin-6 and C-reactive protein correlated well with ADP- and arachidonic acid-induced platelet aggregation (Muller et al. 2010). A retrospective study of a cohort of 903 PCI patients taking dual antiplatelet therapy revealed that recurrent ischemic events, such as death, myocardial infarction (MI), and stent thrombosis, were associated with a significantly high level of C-reactive protein and high on-treatment residual platelet aggregation (Muller et al. 2010). Moreover, in a cohort of 1,223 patients on aspirin and clopidogrel for a prior coronary stenting, greater ADP-induced platelet aggregation was associated with elevated levels of C-reactive protein and also with elevated white blood cell count and fibrinogen levels (Bernlochner et al. 2010), suggesting a cross-link between platelet activation and inflammatory pathways.

The term "aspirin resistance" is defined as less or no responsiveness to aspirin as measured by arachidonic acid-induced platelet aggregation in some individuals compared with others. The frequency of aspirin resistance varies by platelet function testing assay and patient profiling. An attenuated antiplatelet effect of aspirin can be explained by several mechanisms that are largely determined with unique clinical, pharmacodynamic (PD), biological, and genetic profiling (Wurtz et al. 2012). Isolated nonresponsiveness to aspirin or in combination with clopidogrel can strongly predict therapy failure of an antiplatelet agent, a need for dose adjustment or switching from one drug to another (Siller-Matula et al. 2013; Collet et al. 2012; Sambu et al. 2012). Some genetic polymorphisms identified in the genes encoding COX1 and thromboxane A synthase (which is responsible for the generation of TXA_2) may be associated with impaired response to aspirin or aspirin resistance (Chen et al. 2012; Li et al. 2012a, b; Verschuren et al. 2012).

Clopidogrel (SR-25990C)

2-Oxo-clopidogrel

Active metabolite (R-130964)

Inactive metabolite (SR-26334)

Fig. 22.1 The metabolites of clopidogrel in humans (Modified from Xie et al. 2011)

3 Clopidogrel

Clopidogrel (Plavix®), a prodrug, exerts its antiplatelet effect through its active metabolite generated in the liver, where two-step bioactivation of clopidogrel occurs as illustrated in Figs. 22.1 and 22.2. Accumulating evidence has demonstrated that clopidogrel active metabolite binds to platelet ADP receptor P2Y12 specifically and irreversibly, blocking ADP binding to that receptor and inhibiting platelet activation and aggregation (Savi et al. 1992, 1994a, b, 2000; Geiger et al. 1999; Sharis et al. 1998; Quinn and Fitzgerald 1999; Savi and Herbert 2005; Herbert and Savi 2003; Ding et al. 2003). Moreover, clopidogrel can suppress thrombin-induced TXA_2 generation (Shankar et al. 2006) and induce platelet apoptosis through increased Bak/Bax activation (Zhang et al. 2013), suggesting that in addition to irreversible blockade of platelet P2Y12 receptor, multiple other mechanisms are also involved.

Since approval in 1997 by the US FDA, clopidogrel has been widely used to prevent MI, stroke, or vascular death in patients with recent MI, stroke, or established peripheral arterial disease (Anonymous 2010c) according to its superiority over aspirin for reduction of such

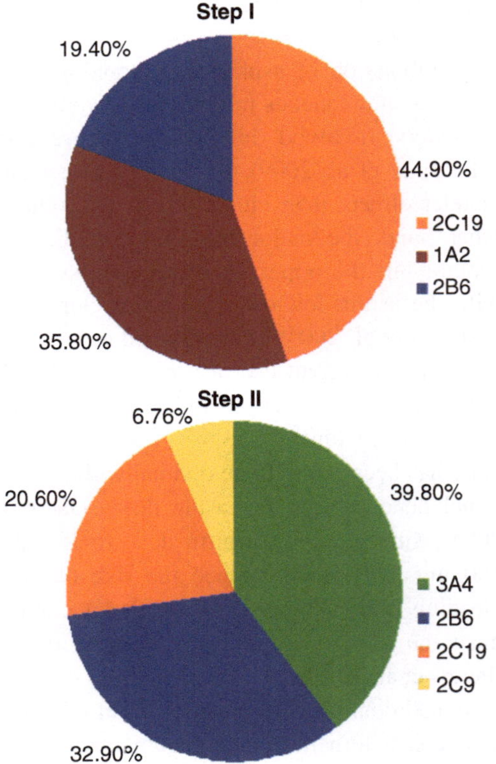

Fig. 22.2 The bioactivation of clopidogrel in the liver. The relative contribution of each human P450 isoform responsible for the two sequential P450-dependent oxidative steps in the bioactivation of clopidogrel (Data used are from the literature Kazui et al. 2010)

events (CAPRIE Steering Committee 1996). Furthermore, dual antiplatelet therapy (clopidogrel plus aspirin) was approved in 2002 by the US FDA for ACS patients or those undergoing PCI for stenting based on further clinical trial data (The CURE Trial Investigators 2001). In addition, pretreatment with a loading dose of clopidogrel alone or in combination with aspirin has become standard intervention in the setting of coronary stenting, because antiplatelet therapy can significantly improve clinical outcomes and prevent in-stent thrombosis in PCI patients (Bertrand et al. 2000; Mehta et al. 2001; Steinhubl et al. 2002; Petersen et al. 2003; Gurbel et al. 2003; Gurbel and Anderson 1997) or for patients with atrial fibrillation and without candidates for warfarin (Helton et al. 2007; Connolly et al. 2009).

Dual antiplatelet therapy inhibits platelet function, prevents recurrent ischemic events, and improves clinical outcomes following ACS or PCI (Antman et al. 2008a, b; King et al. 2008; Sabatine et al. 2005; Anderson et al. 2007). However, individuals may vary in their response to drugs. Fast-growing evidence has demonstrated that the pharmacological and clinical response to clopidogrel has substantial interindividual variability (Angiolillo et al. 2005a, 2007; Geisler et al. 2006, 2008b; Gurbel et al. 2003, 2005b; Gurbel and Tantry 2006; Matetzky et al. 2004; Mobley et al. 2004; Muller et al. 2003; Nguyen et al. 2005; O'Donoghue and Wiviott 2006; Schroeder et al. 2006; Snoep et al. 2007; Wang et al. 2006b) and that patients with impaired antiplatelet effect after receiving clopidogrel therapy would be at increased risk for subsequent adverse cardiovascular outcomes, such as death, recurrent MI, stroke, or stent thrombosis associated with coronary stenting (Collet et al. 2009; Gurbel et al. 2005c; Hochholzer et al. 2006; Matetzky et al. 2004; Mega et al. 2009a; Simon et al. 2009).

Similar to aspirin resistance, the individual response to clopidogrel is highly variable, and up to 20–50 % of patients would not achieve an optimal antiplatelet effect (as measured by low or no ex vivo inhibition of ADP-induced platelet aggregation) after receiving a loading dose of clopidogrel. Based on this trait, some patients can be categorized as nonresponders, poor responders, or resistant to clopidogrel (Angiolillo et al. 2005a, 2007; Gurbel et al. 2003, 2005b; Gurbel and Tantry 2006; Matetzky et al. 2004; Mobley et al. 2004; Muller et al. 2003; Snoep et al. 2007; Geisler et al. 2006, 2008b). The poor response to clopidogrel is associated with an increased risk for recurrent thrombotic events (such as death, MI, or stroke) during clopidogrel therapy (Gurbel et al. 2003, 2005a; Matetzky et al. 2004; Snoep et al. 2007; Cuisset et al. 2006; Geisler et al. 2008a), and is also an independent predictor of stent thrombosis in PCI patients (Buonamici et al. 2007; Price et al. 2008; Wenaweser et al. 2005; Zou et al. 2013).

Mechanisms underlying clopidogrel resistance are most likely multifactorial, although they are not fully elucidated (Anderson et al. 2007; Antman et al. 2008b; Nguyen et al. 2005; Wang et al. 2006b; Angiolillo et al. 2007; Schroeder et al. 2006). More importantly, the pharmacodynamic response of poor responders to clopidogrel would be greatly improved after ex vivo addition of an adequate amount of clopidogrel active metabolite (Wallentin et al. 2008). Such solid evidence suggests that impaired metabolic bioactivation of clopidogrel may be the major cause of poor response to that drug.

3.1 Absorption, Distribution, and Transport of Clopidogrel

The recently clarified metabolic bioactivation of clopidogrel has revealed the relative contribution of several cytochrome P450s (also known as CYP or P450) in the formation of clopidogrel active metabolite (Kazui et al. 2010; Hochholzer et al. 2010; Zahno et al. 2010). The data showed that clopidogrel exerted its antiplatelet effect only after the formation of its active metabolite (R-130964, see Fig. 22.1) in the liver (Mega et al. 2009b). Thus, the amount of circulating clopidogrel active metabolite is chemical basis for its antiplatelet effect.

The large interindividual difference in the formation of clopidogrel active metabolite and its antiplatelet effect may be the result of pronounced variability in clopidogrel absorption and bioavailability (Taubert et al. 2004). In humans, the plasma peak concentrations (C_{max}) of unchanged clopidogrel correlate well with the levels of the active metabolite (Taubert et al. 2004). Furthermore, the concentrations of clopidogrel and its active metabolite, respectively, correlate strongly with the degree of platelet inhibition induced by clopidogrel (Taubert et al. 2004; von Beckerath et al. 2005a; Ganesan et al. 2013). Similarly, there is a significant but negative correlation between clopidogrel bioavailability, as measured by the area under the plasma drug concentration-time curve (AUC) or C_{max}, and platelet inhibition, as measured by maximal antiplatelet effect or area under the antiplatelet effect-time curve in healthy subjects after taking a 300-mg loading dose of clopidogrel on day 1, followed by a 75-mg daily maintenance dose taken on days 2–7 (Kim et al. 2008a, b). Therefore, interindividual variation in clopidogrel intestinal absorption and bioavailability may contribute to variable response to clopidogrel.

As for the mechanisms underlying that variation, the absorptive apical-to-basolateral (A-to-B) flux of clopidogrel across intact Caco-2 monolayer was increased in a dose-dependent manner in the presence of well-characterized P-glycoprotein (P-gp) inhibitors (Taubert et al. 2006), suggesting that clopidogrel is a substrate of P-gp. However, relevant evidence demonstrated that the role of P-gp in limiting intestinal absorption of clopidogrel may not be the major cause of remarkable interindividual variation in the formation of clopidogrel active metabolite and its antiplatelet activity (Taubert et al. 2006).

In terms of the fact that intracellular clopidogrel accumulation was increased in the Caco-2 cells lacking P-gp, other efflux transporters – MRP2 and BCRP – may be involved in decreased clopidogrel absorption in the intestine. However, use of indomethacin (a subtype-unspecific MRP inhibitor) or MK-571 (a subtype 1- and 2-specific MRP inhibitor) did not increase clopidogrel absorption as anticipated (Taubert et al. 2006). Moreover, cyclosporine or elacridar – a dual inhibitor of P-gp and BCRP – was not more potent than verapamil (a P-gp inhibitor not inhibiting BCRP) to increase intracellular clopidogrel accumulation (Taubert et al. 2006), indicating that MRP2 and BCRP may not play an important role in active luminal secretion of clopidogrel.

More recently, in a cohort of 26 male CAD patients treated with clopidogrel (75 mg/day) for at least 5 days, increased expression of MRP3 (or ABCC3) was associated with impaired response to clopidogrel and vice versa, indicating that altered MRP3 expression and function may affect clopidogrel response (Luchessi et al. 2012).

According to the fact that clopidogrel and its active thiol metabolite and inactive carboxyl metabolite all are highly protein bound (Bhindi et al. 2008), high concentrations of inactive clopidogrel carboxylate in plasma could displace clopidogrel active metabolite from protein binding sites. In a cohort of female patients with stable CAD not taking clopidogrel, there was a significant correlation between VASP (a measure of inhibition of platelet aggregation) and both free and total concentrations of clopidogrel active metabolite and also between VASP and both free and total concentrations of clopidogrel inactive carboxylate (Ganesan et al. 2013). Moreover, free fractions of clopidogrel active metabolite could increase with protein binding of clopidogrel inactive carboxylate (Ganesan et al. 2013). Not surprisingly, clopidogrel inactive carboxylate might be a useful surrogate marker that could predict the effect of clopidogrel active metabolite in vivo.

3.2 Metabolism and Bioactivation of Clopidogrel

3.2.1 Carboxylesterase 1 (CES1) and Clopidogrel Carboxylate

Clopidogrel, an enantiopure carboxylic ester of *S*-configuration (Mullangi and Srinivas 2009; Savi et al. 2000; Reist et al. 2000), is a substrate of CES1 (Yang et al. 2007; Mao et al. 2011; Sato et al. 2012a, b). Oral clopidogrel is rapidly and predominantly hydrolyzed by hepatic CES1 to its inactive carboxylate (or SR26334, see Fig. 22.1), which represents some 85 % of the clopidogrel-related compounds in plasma (Caplain et al. 1999; Lins et al. 1999). More recently, in vitro evidence has indicated that clopidogrel carboxylate is not a P450 substrate (Maseneni et al. 2012; Zahno et al. 2010) and has no effect on platelet aggregation (Zahno et al. 2010). However, such an inactive metabolite can serve as a surrogate endpoint to obtain information on either absorption/elimination of that drug (Caplain et al. 1999) or compliance of patients to drug use (Mani et al. 2008). In addition, CES1-mediated clopidogrel inactive metabolite may increase the free concentrations of clopidogrel active metabolite, leading to amplified inhibition of platelet aggregation (Ganesan et al. 2013). However, the formation of further metabolites from clopidogrel carboxylate by myeloperoxidase (MPO) may contribute to myelotoxicity (Maseneni et al. 2012).

Clopidogrel active thiol metabolite (R-130964) contains a methylester function at carbon 10 as does its parent drug as shown in Fig. 22.1, whose hydrolysis by CES1 would lead to inactive carboxylic acid derivative (Reist et al. 2000). As mentioned above, in a competitive metabolic reaction, approximately 85 % of ingested clopidogrel is rapidly hydrolyzed to such an inactive carboxylic acid derivative by CES1 (Lins et al. 1999; Mega et al. 2009b; Sangkuhl et al. 2010; Simon et al. 2009; Reist et al. 2000). Thus, increased amount of clopidogrel would be switched preferentially into CES1-mediated metabolic pathway of clopidogrel (resulting in bioinactivation) in carriers with reduced-function P450s responsible for the first P450-mediated oxidative step of clopidogrel as compared with noncarriers (Mega et al. 2009b; Sangkuhl et al. 2010). Similar results would also occur when CES1 activity was increased due to its induction or increased transcription. Conversely, due to inhibition, drug interactions, or genetic polymorphism (Zhu et al. 2008; Sai et al. 2010), impaired activity of CES1 could lead to an increase in the proportion of clopidogrel that would be diverted to its metabolic bioactivation pathway.

CES1, a phase I drug-metabolizing enzyme, is independent of P450s but is expressed just in human liver microsomes and, to a less extent, cytosol, rather than in intestinal microsomes (Zhu et al. 2008). There is wide interindividual variation in the CES1 activity (Sato et al. 2012a). A nonsynonymous mutation in exon 4 of the human *CES1* gene (c. G428A) results in a catalytic site glycine (or G) to glutamic acid (or E) substitution at amino acid residual position 143, with the 143E variant leading to a 20–100 % loss of function (varying by the substrate) (Zhu et al. 2008). Because CES1 is responsible for the hydrolysis of clopidogrel, and its intermediate metabolite (2-oxo-clopidogrel) as well as the final bioactive thiol metabolite into the pharmacologically inactive carboxylate (Lewis et al. 2013; Bouman et al. 2011), there were significantly greater levels of plasma clopidogrel active metabolite and greater response to clopidogrel (as measured by ADP-induced platelet aggregation) in the *CES1* 143E carriers than noncarriers in a cohort of 566 apparently healthy white subjects or 350 clopidogrel-treated CAD patients, respectively (Lewis et al. 2013). Although the *CES1* 143E variant allele is relatively uncommon, it appears to exert a large effect on both clopidogrel metabolism and ADP-stimulated platelet aggregation as indicated with the absolute β value of 11.6 for the effect of *CES1* 143E on ADP-induced platelet aggregation (vs. 7.4 for the magnitude of effect of *CYP2C19**2 on the same trait). Moreover, carriers of the *CES1A2* -816C variant allele had markedly lower antiplatelet response to clopidogrel (as measured by VASP phosphorylation) compared with noncarriers (i.e., *CES1A2* -816A/A), and this polymorphism may explain 6.4 % of overall individual variation in response to clopidogrel in CAD patients (Ding et al. 2012).

3.2.2 Paraoxonase-1 (PON1)

PON1 was first identified as a rate-limiting enzyme responsible for the conversion of 2-oxo-clopidogrel to clopidogrel active metabolite, with *PON1* Q192R being an important genetic variant that could affect the formation of clopidogrel active metabolite and the antiplatelet effect in clinical practice (Bouman et al. 2011). Subsequent results were conflicting. However, convincing evidence is that the PON1 pathway is not a major bioactivation pathway of clopidogrel in vitro and in vivo (Gong et al. 2012; Ohmori et al. 2012;

Ancrenaz et al. 2012), and that the role of PON1 in the clopidogrel bioactivation now appears to be less important than we previously thought.

3.2.3 Cytochrome P450s

Via another series of metabolic pathways, some 15 % of absorbed clopidogrel is metabolized in the liver by several P450 enzymes to generate its pharmacologically active metabolite (R-130964, see Fig. 22.1), which irreversibly blocks the binding of ADP to platelet P2Y12 receptor, inhibiting platelet aggregation induced by exogenous and released ADP (Savi et al. 1994a, b, 2000; Pereillo et al. 2002). Attempts to identify certain a P450 responsible for clopidogrel metabolic activation in humans would help establish the pharmacological basis to further improve the prediction of interindividual variation in drug response and the potential drug interactions.

Although CYP1A was identified as the major P450 responsible for clopidogrel oxidation (as measured by the disappearance of the C^{14}-labeled clopidogrel in rat liver microsomes) in an earlier study (Savi et al. 1994a, b), subsequent evidence demonstrated that CYP1A2 was less important for clopidogrel metabolism than assumed, and that CYP3A activity was essential for the oxidative metabolism of clopidogrel (Clarke and Waskell 2003).

More recent data demonstrated that the formation rate of 2-oxo-clopidogrel (an intermediate metabolite of clopidogrel, see Fig. 22.1) in human liver microsomes (called the step I oxidative pathway as shown in the top panel of Fig. 22.2) exhibited the biphasic pattern in the Eadie-Hofstee plot, indicating that multiple enzymes – high-affinity and low-affinity components – were involved in that process (Kazui et al. 2010). When using monoclonal antibodies to specifically inhibit the production of 2-oxo-clopidogrel from clopidogrel in human liver microsomes, monoclonal antibody displayed their inhibitory effects as ranked in the descending order like CYP2C19 > CYP1A2 > CYP2B6 (Kazui et al. 2010), consistent with the results of the contribution of each P450 involved in the same step with the same descending order of CYP2C19 > CYP1A2 > CYP2B6 (Kazui et al. 2010). Furthermore, chemical inhibitors, such as furafylline (against CYP1A2) and omeprazole (against CYP2C19), suppressed the formation of 2-oxo-clopidogrel from clopidogrel by 35 and 27 %, respectively, consistent with the results of the anti-P450 antibody experiments (Kazui et al. 2010). Similar to the step I oxidation, the formation rate of clopidogrel active metabolite from 2-oxo-clopidogrel in human liver microsomes (called step II oxidative pathway) also exhibited the biphasic pattern in Eadie-Hofstee plots, suggesting the involvement of multiple enzymes with high-affinity and low-affinity components (Kazui et al. 2010). Similar inhibition studies (anti-P450 antibodies) were performed in human liver microsomes to determine their inhibitory effects on the formation of clopidogrel active metabolite from 2-oxo-clopidogrel in the step II process with a descending ranked order: CYP2B6 > CYP3A4 > CYP2C19 > CYP2C9, concordant with the contribution of each P450 responsible for clopidogrel active metabolite formation – CYP3A4 > CYP2B6 > CYP2C19 > CYP2C9 (Kazui et al. 2010). Moreover, ketoconazole (a nonselective, strong CYP3A inhibitor), sulfaphenazole (a CYP2C9 inhibitor), and omeprazole (a potent, mechanism-based CYP2C19 inhibitor) suppressed the generation of clopidogrel active metabolite by 38, 36, and 31 %, respectively (Kazui et al. 2010). Taken together, both CYP2C19 and CYP2B6 are involved in the whole process of clopidogrel active metabolite formation as shown in Fig. 22.2. Current data have well demonstrated that CYP2C19 contributes substantially to the whole process of clopidogrel oxidative metabolism (Kazui et al. 2010) as shown by several recent clinical studies (Brandt et al. 2007a, b; Mega et al. 2009a; Sugidachi et al. 2007; Umemura et al. 2008; Hochholzer et al. 2010; Zou et al. 2013).

CYP3A4 is involved in the only step II oxidative pathway of clopidogrel, contributing to the formation of the active metabolite from 2-oxo-clopidogrel (Kazui et al. 2010). When clopidogrel concentrations are greater than 10 μM, CYP2C19 is suppressed by clopidogrel itself, and CYP3A4 is primarily responsible for clopidogrel biotransformation in vitro (Zahno et al. 2010). This notion is strongly supported by the fact that either rifampin (a well-known CYP3A inducer) or St. John's wort (also a CYP3A4 inducer as determined by the erythromycin breath test) can further increase antiplatelet effect of clopidogrel, whereas concomitant use of clopidogrel and pravastatin (not a substrate of CYP3A) does not result in attenuated antiplatelet effect of clopidogrel (Lau et al. 2003, 2011), suggesting the involvement of CYP3A in the bioactivation of clopidogrel.

4 Platelet Pharmacogenomics

The platelets play a central role in the physiologic hemostasis and thrombotic disorders. It has long been well known that aspirin targets COX1, suppressing the generation of TXA_2 and arachidonic acid-induced platelet aggregation. By contrast, clopidogrel targets the platelet ADP receptor P2Y12, inhibiting platelet aggregation ex vivo induced by ADP, collagen, and low concentrations of thrombin (Feliste et al. 1987). There is great interindividual, rather than intraindividual, variation in response to platelet aggregation, implying strong heritability (Jones et al. 2009). The highly variable platelet function could be regulated by a large number of genes that encode platelet membrane receptors and intracellular signaling networks, each exerting a small effect.

In the case of clopidogrel, the ADP signaling pathway is of importance. Initial evidence has demonstrated that 44 % of the variation in platelet response to ADP is inherited (O'Donnell et al. 2001). A total of 1,327 SNPs were identified in 97 candidate genes from 500 healthy subjects (predominantly whites) recruited for the platelet phenotype-genotype correlation studies, of which 8 genes were highly specific for platelet ADP signaling pathways (Jones et al. 2009). These SNPs represent 13 and 16 % of variation in P2Y12-associated P-selectin expression and fibrinogen binding in response to ADP, and 30–36 % of the total hereditable variation in the ADP signaling pathway central to platelet function, respectively (Jones et al. 2009). For example, the variant rs1472122 in the *P2Y12* gene was associated with increased P-selectin expression in response to ADP, whereas another one rs10935839 had a decreased P-selectin expression (Jones et al. 2009). Moreover, a common *ITGA2* gene variant rs246406 was associated with P-selectin expression and two rare variants (rs41305896 and rs6450105) with fibrinogen binding in response to ADP (Jones et al. 2009). In addition, the other variants in the *PEAR1* gene (rs11264579), *VAV3* gene (rs17229705, rs12410842, and rs2769668), and *ITPR1* gene (rs17786144) were all associated with fibrinogen binding, increased P-selectin expression, or both in response to ADP (Jones et al. 2009).

Human platelet P2Y12 receptor, a Gi-coupled receptor whose activation can reduce the platelet cAMP level through the suppression of adenyl cyclase, has been identified as the target of clopidogrel (Hollopeter et al. 2001; Savi et al. 2000). Two nonsynonymous mutations – Agr256Gln and Arg265Trp – were identified in a patient with congenital bleeding, which led to impaired Gi signaling in response to P2Y12 stimulation (Cattaneo et al. 2003). A haplotype (termed H2) in the *P2Y12* gene was identified in healthy subjects to be associated with the maximum aggregation response to ADP and more marked downregulation of the intra-platelet cAMP levels by ADP compared with noncarriers (Fontana et al. 2003), but these variants were unlikely to have a marked effect on clopidogrel response (Angiolillo et al. 2005b; von Beckerath et al. 2005b; Cuisset et al. 2007; Lev et al. 2007; Smith et al. 2006a, b). In 2,208 consecutive acute MI patients receiving clopidogrel, three variants in the *P2Y12* gene – rs6785930, rs6809699, and rs16846673 – were not associated with the risk of the major adverse cardiovascular events (called MACE, such as death, MI, stent thrombosis, or stroke) in 12 months after taking the drug (Simon et al. 2009). Clearly, genetic variants in platelet P2Y12 receptors were neither consistently associated with altered platelet function and drug response nor linked to subsequent clinical outcomes after taking clopidogrel. Since cangrelor is a direct-acting P2Y12 antagonist without the need of metabolic bioactivation as required for clopidogrel, one could directly determine whether genetic polymorphisms in *P2Y12* receptor could affect cangrelor-induced inhibition of platelet aggregation ex vivo and clinical outcomes in cangrelor-treated patients.

Although P2Y1 platelet ADP receptor is not the target of clopidogrel, it may play an important role in platelet aggregation as ADP initiates platelet aggregation via P2Y1 receptor (Storey 2009) and its activation causes an increase in the intracellular Ca^{2+} level (Davi and Patrono 2007). Moreover, P2Y1-deficient mice and mice treated with the P2Y1 antagonist MRS2179 had significantly less arterial thrombosis than their respective control mice (Lenain et al. 2003). After blockade of P2Y12 receptor, P2Y1-deficient mice had a significant additive effect on the localized thrombosis in the mouse model (Lenain et al. 2003). Furthermore, over-expression of platelet P2Y1 receptor can induce platelet hyper-reactivity in transgenic mice (Hechler et al. 2003). This strongly demonstrates that platelet P2Y1 receptor is also a potential target for antithrombotic drugs. As anticipated, in healthy subjects, a common A1622G transition was associated with a significant platelet response to ADP in a dose-dependent manner, with the G-allele carriers conferring a greater response than noncarriers (Hetherington et al. 2005). These results suggest that polymorphism in the *P2Y1* gene or altered density of P2Y1 receptor on the platelet surface may partly explain why platelet response to ADP varies among individuals in the absence of clopidogrel. However, this variant was not associated with altered response to clopidogrel in patients (Lev et al. 2007; Harmsze et al. 2010; Sibbing et al. 2006).

Clopidogrel active metabolite directly blocks ADP binding to platelet P2Y12 receptor, inhibiting ADP-mediated activation of the GPIIb/IIIa (or called platelet integrin αIIbβ3) complex (Humbert et al. 1996), which is the main membrane receptor for platelet adhesion and aggregation (Kulkarni et al. 2000). The activation of GP IIb/IIIa is the final pathway for all known platelet agonists (Davi and Patrono 2007). The integrin αIIb can induce platelet granule secretion and TXA_2 production through PI3K/Akt activation (or outside-in signaling pathway) (Niu et al. 2012). The β3 subunit of GP IIb/IIIa receptor (also known as fibrinogen receptor) is encoded by the *ITGB3* gene, in which a nonsynonymous polymorphism (Leu33Pro) has been widely studied. For example, in 500 healthy subjects (predominantly whites) recruited for platelet functional testing, the *ITGB3* variants (including rs5918) showed no association with any of the markers of platelet response (Jones et al. 2009). In 38 stented patients receiving a 300-mg loading dose of clopidogrel, this variant had a modulatory effect on clopidogrel response as seen with a higher degree of GP IIb/IIIa activation and P-selectin expression in carriers of the Pro33 and a lower antiplatelet effect to a 300-mg loading dose of clopidogrel up to 24 h following intervention (Angiolillo et al. 2004a). However, in the 2,208 consecutive acute MI patients taking clopidogrel, this variant had no effect on the risk of the MACE in the first year (Simon et al. 2009). These data have demonstrated that *ITGB3* polymorphism may not contribute to variation in platelet response to clopidogrel.

Platelet integrin α2β1 (also known as GP Ia/IIa), a major platelet collagen receptor, plays a key role in thrombosis following collagen exposure. In the *ITGA2* gene (which encodes the α2 integrin or called GPIa), a C807T transition (rs1126643) may be of clinical relevance, with 807T carriers having increased receptor density and thrombotic risk (Kunicki et al. 1997; Ajzenberg et al. 2005). In 289 MI patients undergoing PCI and receiving dual antiplatelet therapy, carriers of the 807T/873A allele had significantly higher platelet aggregation than noncarriers (Giusti et al. 2008). Interestingly, clopidogrel-induced platelet inhibition in carriers of the 807T allele was not different from that in noncarriers in response to ADP stimuli among 44 stented patients receiving a 300-mg loading dose of clopidogrel, but clopidogrel response was seen with collagen stimuli only in noncarriers of the 807T allele (Angiolillo et al. 2004b), implying an increased platelet reactivity to collagen in carriers of the 807T allele due to increased exposure to collagen during coronary stenting. A more recent functional genomics study of 500 healthy subjects showed that *ITGA2* C807T polymorphism (rs1126643) was not in association with platelet function testing, and that other *ITGA2* variants (rs246406 and rs41305896) were linked to altered P-selectin expression and fibrinogen binding in response to ADP in addition to another variant (rs41315906) linked to fibrinogen binding in response to collagen (Jones et al. 2009).

GPIbα (also known as platelet adhesion receptor) mediates the initial adhesion of platelets to the blood vessel wall following injury, and thus its genetic polymorphisms may affect individual predisposition to platelet response in PCI patients. The *GPIbα* gene often contains several perfect tandem repeats of a mucin-like sequence, with four length variants differing in the number of tandem repeats of a 39-bp sequence encoding 13 amino acid residues (Ser399 to Thr411) corresponding to the heavily *O*-glycosylated region (Lopez et al. 1992; Ishida et al. 1995).

Platelet membrane GPIb is the primary platelet receptor for von Willebrand factor (vWF) that is responsible for shear-induced platelet activation. The addition of tandem repeats could increase the length of this elongated glycosylated region and extend the distance between the ligand-binding domain of GPIb and the platelet plasma membrane and may alter the susceptibility of platelets to shear-induced activation (Lopez et al. 1992). In addition, a C-to-T transition at the position 434 of the coding region, resulting in a Thr (ACG) to Met (ATG) dimorphism at residue 145, was in complete linkage disequilibrium with this repeats variant (Ishida et al. 1995; Kuijpers et al. 1992). Moreover, a T-to-C transition at position -5 upstream of the ATG start codon in the *GPIbα* gene was associated with increased expression of GPIbα protein (Afshar-Kharghan et al. 1999), suggesting a determinant of the surface levels of a cell adhesion receptor.

vWF mediates vWF-dependent platelet adhesion and aggregation at sites of vascular injury through binding to platelet membrane glycoprotein receptor GPIbα, since patients with vWF defects are at increased risk for excessive bleeding with invasive procedure. For example, a common vWF exon 28 variant (D1472H) was associated with decreased and delayed platelet aggregation induced by ristocetin, a withdrawn antibiotic that can cause profound thrombocytopenia by binding to vWF and promoting vWF-dependent platelet aggregation (Sadler 2010), as compared with its wild-type allele (Flood et al. 2010), conferring genetic risk for bleeding. Interestingly, 63 % of black Americans carry the *vWF* D1472H versus 17 % of whites (Flood et al. 2010).

It has been recognized that ADP plays a major role in platelet activation and is the final common agonist for platelet recruitment and thrombus formation. As a constitutively expressed potent inhibitor of platelet reactivity, ecto-ADPase (also known as CD39/ENTPD1) on the surface of endothelial cells efficiently hydrolyzes ATP and ADP released from red cells and activated platelets to AMP and drastically reduces or even abrogates platelet aggregation and recruitment (Marcus et al. 1997, 2003). Thus, platelets become unresponsive to most agonists (including ADP) when they are in motion and in proximity to endothelial cells (Marcus et al. 1991; Davi and Patrono 2007). Altered CD39 activity due to genetic polymorphisms (Friedman et al. 2009) may be one of the endogenous mechanisms (Jin et al. 2005), by which increased intrinsic platelet reactivity may occur persistently in some individuals rather than others, irrespective of the presence of clopidogrel therapy.

Endothelial nitric oxide synthase (eNOS) expressed in platelets is an upstream signaling protein of nitric oxide (NO)/cyclic GMP (cGMP) signaling pathway. Compared with age- and gender-matched healthy subjects, ACS patients had significantly high basal levels of cGMP and enhanced serine phosphorylation in basal eNOS and VASP (Bergandi et al. 2010), suggesting that increased platelet activation at baseline (as measured by increased levels of serum soluble P-selectin) in ACS patients may result in decreased platelet responsiveness to the antiaggregatory effects of endogenous NO and exogenous NO donors (such as nitroglycerin and sodium nitroprusside). Several SNPs have been identified in the *eNOS* gene, such as -786T/C in the promoter, 894G/T in exon 7 (causing Glu298Asp), 4a/4b in intron 4, and more. For example, carriers of *eNOS* -786C variant and 4a variant (the 27-bp variable number of tandem repeats in intron 4) had a significantly higher maximal platelet aggression value and increased residual platelet reaction induced by arachidonic acid (Fatini et al. 2009), whereas the variant 894T (or 298Asp) was an independent risk factor for MI after multiple logistic regression analysis (Shimasaki et al. 1998). In addition, the 4a variant was associated with a decreased risk for upper GI bleeding in patients taking low-dose aspirin for secondary prevention (Piazuelo et al. 2008).

5 Nongenetic Factors that Could Affect the Disposition of and Response to Aspirin or Clopidogrel

5.1 Age

It is well known that the elderly are underrepresented in the majority of clinical trials, resulting in much uncertainty and less optimal medical care. A prospective analysis of adverse drug reactions (ADRs) as cause of 18,820 admissions to hospitals in the UK has documented that patients admitted with ADRs were significantly older than those without ADRs (Pirmohamed et al. 2004). Older age is an important covariate that could be used to predict poor platelet response to clopidogrel, persistently high residual platelet reactivity, worsening clinical outcomes, and increased risk of bleeding complications in ACS patients. For example, poorer response to clopidogrel (as defined by ADP-induced platelet aggregation after 7 days of clopidogrel dosing) was associated with increased age (Shuldiner et al. 2009). Consistent with an observation of a cohort of 237 PCI patients loading with 600 mg of clopidogrel (Geisler et al. 2008b), age over 65 years was identified as a significant nongenetic factor that would better predict patients with persistently high residual platelet reactivity after coronary stenting, and dual antiplatelet therapy in another cohort of 1,092 consecutively enrolled patients (Geisler et al. 2008a), possibly because elderly subjects (aged >65 years) have decreased CYP2C19 activity as compared with young people (Kimura et al. 1999). Furthermore, in 2,208 consecutive acute MI patients receiving clopidogrel, 294 patients who had an outcome event were older than those without any outcome event (Simon et al. 2009). In 760 stented patients loading with 600 mg of clopidogrel, age was associated with high on-clopidogrel residual platelet aggregation, and effect of age on on-clopidogrel residual platelet aggregation was prominent in nondiabetic patients than diabetic patients (Hochholzer et al. 2010), indicating an interaction with diabetes mellitus. However, the relative contribution of age to the observed variation in on-clopidogrel residual platelet aggregation was estimated to be 1 % (Hochholzer et al. 2010), suggesting that age (>65 years) is a possible but not strong predictor for a low response to clopidogrel.

As a rare but serious thrombotic event, MI in young patients is routinely treated with PCI and stenting as well as chronic antiplatelet therapy. In 259 young (aged <45 years) survivors with acute MI who were initially treated with clopidogrel for at least 1 month, followed by the median drug exposure period of 1.07 (IQR 0.28–3.0) years with the maximum follow-up of 8 years, either a composite of a recurrent episode of ischemic events (death, recurrent MI, urgent PCI or CABG) or secondary stent thrombosis occurred more frequently in carriers versus noncarriers of *CYP2C19*2*, suggesting the *CYP2C19*2* variant is a major determinant of prognosis in this clinical setting (Collet et al. 2009). Moreover, the frequency of secondary stent thrombosis varied by the length of therapy and the type of stents used, with 67 % of stent thrombosis events occurring late (30 days to 1 year) or very late (>1 year) and 80 % of very late stent thrombosis events linked to drug-eluting stents (Collet et al. 2009). However, whether CYP2C19 activity has a more important role in the response to clopidogrel in young patients than older patients with ACS needs further studies (Storey 2009).

In addition, in a cohort of 1,524 PCI patients after pretreatment of 600-mg clopidogrel, age (per 10-year increment) was a significant risk factor for TIMI (Thrombolysis in Myocardial Infarction) major or minor bleeding complications as assessed by a multivariable logistic regression model (Sibbing et al. 2010). Similarly, patients with major bleeding are older than those with minor or no bleeding, and thus old age (in particular>80 years) is a strong predictor for major bleeding in PCI patients (Iijima et al. 2009; Kinnaird et al. 2003; Feit et al. 2007).

Compared with the middle-aged adults, the elderly have lower aspirin esterase activity and thus have increased response to aspirin when taking the recommended doses. In addition, the elderly may have a lower CES1 activity than young adults, because CES1 induction may decrease with the age (Xiao et al. 2012).

5.2 Gender

Women are also underrepresented in the majority of clinical trials. Compared with men, women undergo coronary intervention less frequently, and a lower proportion of women receive evidence-based drug therapy (Ostadal and Ostadal 2012). As for the baseline platelet reactivity (as measured by VerifyNow P2Y12 assay), female patients had a significantly higher mean value than males (Lin et al. 2012). Although there was no gender difference in arachidonic acid- or ADP-induced platelet aggregation, postmenopausal women had a significantly higher thrombin receptor-activating peptide (TRAP)-induced platelet reactivity (as measured by impedance aggregometry) than men and thus had a significantly increased maximal platelet reactivity (Bobbert et al. 2012). Multivariate linear regression further revealed female sex to be an independent risk factor for clopidogrel nonresponse (Lin et al. 2012) and a significant prognostic risk for increased TRAP-induced platelet aggregation (Bobbert et al. 2012). Consistent with the above findings, female gender was an independent predictor of higher on-treatment platelet reactivity in patients treated with dual antiplatelet agents alone or in combination with cilostazol (Park et al. 2011). In other words, patients with high on-treatment platelet reactivity are more likely to be women (Price et al. 2011). In addition, women may benefit less than men from therapy with fibrinolytic or glycoprotein IIb/IIIa antagonists (Ostadal and Ostadal 2012).

Opposite to the above, gender could affect CYP2C19 catalytic activity to a certain extent, with females having a higher activity than males (Ramsjo et al. 2010; Xie et al. 1997). In addition, females have higher CYP3A4 activity than males (Wolbold et al. 2003; Chen et al. 2006). The additive or synergic role of CYP2C19 and CYP3A4 in the generation of clopidogrel active metabolite may lead to higher levels of clopidogrel active metabolite in females than males and thus to greater clopidogrel-induced antiplatelet effects in women. As anticipated, in 2,208 consecutively recruited acute MI patients receiving clopidogrel, females had a significantly lower risk for an adverse cardiovascular event (death, recurrent MI, or stroke) than males (5.0 % vs. 8.3 %; $P=0.001$) during the period of the first-year follow-up (Simon et al. 2009). Similarly, in a cohort of 1,524 PCI patients pretreated with 600-mg clopidogrel, gender was a significant covariate determining ADP-induced platelet aggregation ($P=0.01$) as assessed by a multivariable linear regression model (Sibbing et al. 2010). Furthermore, female sex was also an independent predictor of increased risk of bleeding in PCI-treated patients (Iijima et al. 2009; Feit et al. 2007).

Other relevant evidence is given below. In a megakaryocytic DAMI cell line, testosterone could induce P2Y12 mRNA and protein expression and suppress VASP phosphorylation; these effects of testosterone on P2Y12 could be reversed by the androgen receptor antagonist bicalutamide (Lee et al. 2012). In addition, testosterone also induced TXA_2 synthase protein levels and increased TXA_2-mediated function in rats (Gonzales et al. 2005). In contrast, 17β-oestradiol had no effect on P2Y12 expression (Lee et al. 2012). Induction of hepatic MRP3 (which is involved in altered platelet response to clopidogrel) by ethynylestradiol was mediated by estrogen receptor (Ruiz et al. 2013). The net effects from the opposite effects of each other would be predicted on the basis of the complex model.

5.3 Body Weight

Body weight may affect the PK profile of clopidogrel and its active metabolite and subsequent antiplatelet effect. For instance, decreased

CYP3A4 activity could contribute to high on-clopidogrel platelet reactivity in overweight patients (Bonello-Palot et al. 2009; Kotlyar and Carson 1999). Human adipose tissue can upregulate CES1 expression and activity (Jernas et al. 2009), leading to faster bioinactivation of clopidogrel and aspirin. Moreover, high body weight or greater BMI (body mass index) has been identified as a risk factor for clopidogrel nonresponder (Lin et al. 2012) or failed dose adjustment (Bonello et al. 2012). A multivariate analysis indicated that *CYP2C19* loss-of-function polymorphisms and high body weight were two independent predictors of clopidogrel resistance, because the body weight was significantly greater in poor responders genotyped with *CYP2C19*1/*1* than in responders carrying the *CYP2C19*2* variant (Aleil et al. 2009). A similar effect of greater body weight or BMI on the response to clopidogrel was reported elsewhere (Shuldiner et al. 2009; Feher et al. 2007; Hochholzer et al. 2010). Furthermore, an increased risk of bleeding was seen in patients with low body weight when taking prasugrel (a new clopidogrel alternative less dependent on CYP2C19) (Wiviott et al. 2007), and persistent platelet activation was often seen in obese women who were otherwise healthy and relatively young (Davi et al. 2002). In addition, in a cohort of 1,524 PCI patients pretreated with 600-mg clopidogrel, BMI was a significant covariate determining ADP-induced platelet aggregation as assessed by a multivariable linear regression model (Sibbing et al. 2010). On the other hand, low body weight is an independent predictor of increased risk of bleeding and recurrent MI in 4,570 PCI patients receiving antithrombotic therapy (Iijima et al. 2009). Furthermore, in a prospective clinical study of 73 PCI patients, the only independent predictor of failed dose adjustment of clopidogrel in about 10 % patients with high on-clopidogrel platelet reactivity was a higher BMI, although ACS, type II diabetes mellitus, and *CYP2C19*2* were all associated with high on-clopidogrel platelet reactivity after a 600-mg loading dose. These studies strongly suggest that body weight may be an influential covariate that could affect platelet response to clopidogrel, and that higher dose requirements of clopidogrel would be needed for patients with high on-treatment platelet reactivity (Bonello-Palot et al. 2009). Therefore, the use of more than one weight-adjusted loading dose of clopidogrel tailored according to platelet reactivity monitoring would achieve an optimal level of platelet inhibition.

Obesity has a marked effect on the pharmacokinetics (PK) of lipophilic drugs, like aspirin (Hanley et al. 2010). For example, higher body weight is associated with lower biochemical responsiveness to aspirin as assessed by TXB_2 or platelet function testing assays (Bordeaux et al. 2010; Maree et al. 2005; Peace et al. 2010), suggesting a possible lower clinical efficacy (Fontana et al. 2010).

5.4 Ethnicity or Racial Background

Ethnicity of the patient may be one of the important demographic covariates that could affect the disposition of and response to some drugs across a wide spectrum of patients in clinical practice (Xie et al. 2001; Xie 2010; Yasuda et al. 2008; Ozdemir et al. 2008; Yang et al. 2012; Jang et al. 2012). Emerging data have well demonstrated that the median values of intrinsic clearance of CYP1A2 and CYP2C19 present in Chinese liver microsomes were observed to be only 54 % and 26 % of the corresponding values of the subjects of European descent, and that the median Km (Michaelis constant) value in the Chinese liver microsomes (76 μM) was significantly higher than that of the white subjects (32 μM), indicating that activity levels of CYP2C19, CYP2B6, and CYP1A2 (all of which catalyze the formation of clopidogrel active metabolite) are significantly lower in Chinese subjects than in white subjects (Yang et al. 2012). Consistent with the above findings, *CYP2C19*2* is reproducibly associated with a marked decrease in active metabolite bioavailability and antiplatelet effect and an increased risk for recurrent ischemic events or stent thrombosis. Although *CYP2C19*2* variant is common across many ethnic populations, its allelic and genotypic frequencies vary considerably by ethnicity as assessed by a series of meta-analysis (Xie 1997, 2000; Xie et al. 1999a, b). For instance, in healthy unrelated white subjects of European descent worldwide, the population frequency of the *CYP2C19*-genotyped PM (predominantly *2/*2) is 2.1 %, highly consistent with its PM frequency as determined by phenotyping (2.8 %) (Xie et al. 1999b). Similarly, in 922 healthy unrelated black subjects of African ancestry, the CYP2C19-phenotyped PM frequency is 3.9 %, consistent with its genotyped PM frequency of 3.7 % (Xie et al. 1999a). By contrast, in native and overseas healthy Chinese populations, the CYP2C19 PM frequency is 13.6 % by phenotyping and 13.8 % by genotyping (Xie 2000), showing that an estimated 182 million Chinese subjects could lack CYP2C19 activity due to loss-of-function genetic variations of *CYP2C19* based on a total of 1.3 billion Chinese individuals derived from the National Census of China in 2005. Similarly, the frequencies of *CYP2C19* PM alleles are common in the other Eastern Asian populations, such as Japanese (Jinnai et al. 2009) and Korean (Lee et al. 2009; Ramsjo et al. 2010). Therefore, the East Asian populations require special attention on clopidogrel resistance (Hasan et al. 2013). Clearly, genotyping of *CYP2C19* is clinically more interesting for patients of Eastern Asian origin than those of other ethnic backgrounds (Bhatt 2009; Geisler et al. 2008b), since the allele frequency of *CYP2C19*2* variant in the Chinese population (30 %) is twice that in whites (14.7 %) and blacks (17.3 %) (Xie et al. 2001). It is of interest to answer to what extent cardiovascular events in Asian patients receiving clopidogrel reflect the prevalence of *CYP2C19 *2* and **3* (Roden and Stein 2009). If just based on the ethnic diversity of *CYP2C19* nonfunctional polymorphism, less exposure to the active drug metabolite, less platelet inhibition, and less prevention from recurrent ischemic cardiovascular events would be expected for Eastern Asian patients than black and white patients when they all receive the same doses of clopidogrel. For Eastern Asians, such reduced benefits from clopidogrel in preventing recurrent ischemic events may be overcome actually by increasing the loading dose of clopidogrel to improve platelet inhibition or extending the duration they take the drug. On the other hand, genotyping of *CYP2C19* has proven helpful for clinicians to choose the most effective antiplatelet drug and doses for a given individual patient as suggested by the more recently released black box warning of clopidogrel by the US FDA (Anonymous 2010a; Holmes et al. 2010).

In addition, altered P-gp activity is responsible for the interindividual variation in intestinal absorption and bioavailability of clopidogrel, carriers of *P-gp* 3435T variant having significantly reduced bioavailability of that drug compared with noncarriers (Taubert et al. 2006). Furthermore, patients with the 3435T/T genotype had significantly higher rates of cardiovascular events (death, MI or stroke) at 1 year than those with the 3435C/C genotype (Simon et al. 2009). Similar to the case of CYP2C19, there are marked ethnic differences in P-gp transport activity (Kim et al. 2001) and at the allele frequency of the *P-gp* 3435T (Ameyaw et al. 2001; Schaeffeler et al. 2001; Marzolini et al. 2004; Balram et al. 2003).

The population frequencies of the loss-of-function variants *CYP2C9 *2* and **3* (Xie et al. 2002) and *CYP3A5*3* (Xie et al. 2004) vary by ethnicity. For CYP2C9, its altered enzymatic activity (due to the presence of *CYP2C9* nonfunctional variant) would play a less role in the formation of clopidogrel active metabolite since it is responsible for clopidogrel active metabolite formation from 2-oxo-clopidogrel with a relative contribution of 7 % (Kazui et al. 2010). On the other hand, *CYP3A5*3* allele (resulting in a truncated, nonfunctional enzyme) (Wang et al. 2006a; Xie et al. 2004) is common across all the major ethnic populations (Xie et al. 2004), but the contribution of CYP3A5 to the formation of clopidogrel active metabolite remains to be identified in the future.

Interestingly, a retrospective analysis of 7,236 consecutive patients (who underwent PCI of 13,135 lesions and received more than one drug-eluting stent) revealed that clopidogrel compliance at the time of the stent thrombosis event was higher in the black than non-black population, but that black race is an independent predictor of definite stent thrombosis after implantation of drug-eluting stent (Collins et al. 2010).

5.5 Cigarette Smoking

An earlier study indicated that CYP1A2 catalyzes biotransformation of clopidogrel in rats (Savi et al. 1994a, b). More recent in vitro study has well confirmed that CYP1A2 is one of the three identified P450s involved in the first oxidative step of clopidogrel to generate 2-oxo-clopidogrel (Kazui et al. 2010) as shown in Fig. 22.2. Previous studies have indicated that cigarette smoking can significantly induce the expression of CYP1A2 (Schrenk et al. 1998). More importantly, some patients who were receiving clopidogrel were also active or heavy cigarette smokers. Therefore, it is of interest to determine the PK/PD profile of clopidogrel in patients who are cigarette smoking. As expected, in 26 healthy subjects, there were lower levels of clopidogrel carboxyl metabolite in smokers than in nonsmokers, suggesting an increase in in vivo bioactivation of clopidogrel in the smoking cohort of subjects (Yousef et al. 2008). Moreover, in 259 consecutive stented patients (104 current smokers vs. 155 nonsmokers) and receiving 600-mg loading dose of clopidogrel or on chronic clopidogrel therapy, current smokers had significantly lower ADP-induced platelet aggregation and active GP IIb/IIIa expression than nonsmokers (Bliden et al. 2008). Moreover, the antiplatelet response of clopidogrel increased with the levels of serum cotinine (the major metabolite of nicotine) in a dose-dependent manner (Ueno et al. 2012), and enhanced clopidogrel response in smokers was reversed after smoking cessation (Park et al. 2012), further suggesting a cause-effect relationship. In a cohort of 722 stented Chinese, smokers had lower rate of poor response to clopidogrel than nonsmokers, with smokers homozygous for the *CYP2C19*1/*1* having the lowest rate of clopidogrel resistance and nonsmokers carrying the *CYP2C19*2* being at highest risk for clopidogrel resistance (Liu et al. 2010). Furthermore, in a total of 1,213 consecutive patients with acute ST-segment elevation MI (588 smokers vs. 625 nonsmokers), the rate of cardiac death was significantly lower in the smoke group than in the nonsmoke group (Sun et al. 2012). All above evidence suggests that smoking appears to increase clopidogrel response as compared with nonsmoking. However, a logistic regression analysis indicated that current smoking is not a major determinant of the platelet reactivity index (VASP assay) when clopidogrel is given (Aleil et al. 2009). In terms of reversal of enhanced clopidogrel response after smoking discontinuation, the recommendation for patients who are receiving clopidogrel to stop smoking (Wijns et al. 2010) should be reevaluated in the future clinical practice (Sibbing et al. 2012).

6 Clinical Pharmacogenomics of Clopidogrel

There is considerable interindividual variability in ADP-induced platelet aggregation before and after dosing of clopidogrel. The heritability of ADP-stimulated platelet aggregation at baseline and in response to clopidogrel was estimated to be 0.33 and 0.73, respectively, indicating a substantial genetic component (Shuldiner et al. 2009). The variation explained by all known covariates is less than 10 %, including increased age (accounting for 3.8 % of variance), BMI (2.3 %), triglyceride level (1.3 %), and decreased levels of HDL (high-density lipoprotein) cholesterol (1 %) as assessed by multivariate analysis of clinical study of clopidogrel response (Shuldiner et al. 2009). These results have well revealed that genetic factor is the major determinant of variable clopidogrel response.

A candidate gene testing approach (CGTA) has been widely used to identify and examine SNPs in the genes encoding the P450s responsible for the bioactivation of clopidogrel to an active metabolite, with the major focus on *CYP2C19* (Aleil et al. 2009; Brandt et al. 2007a; Collet et al. 2009; Giusti et al. 2007, 2009; Hulot et al. 2006; Kim et al. 2008a, b; Malek et al. 2008; Mega et al. 2009a; Sibbing et al. 2009, 2010; Trenk et al. 2008; Umemura et al. 2008). More recently, a genome-wide association study (GWAS) of ADP-induced platelet aggregation in response to clopidogrel was performed in 429 healthy, drug-naïve, related white subjects recruited from a relatively homozygous population (which minimized the potential for confounding factors), further confirming that *CYP2C19*2* variant is associated with impaired antiplatelet effect of clopidogrel. In addition to revealing a cluster of 13 SNPs (all of which were in strong linkage disequilibrium with each other) spanning 1.5 megabases on chromosome 10q24 that was most significantly associated with clopidogrel-induced platelet inhibition (Shuldiner et al. 2009), a well-designed GWAS study demonstrated that this cluster is localized within the *CYP2C18-CYP2C19-CYP2C9-CYP2C8* gene cluster, which encodes corresponding P450s that play an important role in the metabolism of drugs, including the bioactivation of clopidogrel. Further analysis would reveal that the loss-of-function *CYP2C19*2* accounts for almost the original 10q24 association signal (Shuldiner et al. 2009).

6.1 CYP2C19

CYP2C19 has a major effect on the formation of active metabolite from clopidogrel as it contributes substantially to the whole process of clopidogrel bioactivation (Kazui et al. 2010; Fontana et al. 2007). Therefore, the loss-of-function polymorphisms in the *CYP2C19* may contribute to decreased formation of clopidogrel active metabolite, reduced antiplatelet effects, and increased risk for recurrent ischemic events in patients receiving clopidogrel.

For the healthy subjects who were drug naïve, CYP2C19 catalytic activity (as assessed by omeprazole hydroxylation) correlated strongly with clopidogrel-induced inhibition of platelet aggregation, and the antiplatelet effect of clopidogrel was significantly lower in carriers of *CYP2C19*2/*2* or **2/*3* than carriers of *CYP2C19*1* (homozygotes or heterozygotes) (Kim et al. 2008a, b), suggesting that impaired CYP2C19 activity (due to its genetic polymorphisms) may be one of the causes that could result in "clopidogrel resistance." In 162 healthy subjects taking clopidogrel, carriers of at least one *CYP2C19* nonfunctional allele had a 32 % reduction in plasma exposure to clopidogrel active metabolite and also had a 25 % reduction in response to clopidogrel-induced inhibition of platelet aggregation as compared with noncarriers (Mega et al. 2009a). Furthermore, heterozygotes of the *CYP2C19*2* had intermediate value of platelet inhibition after taking clopidogrel compared with homozygotes of this variant and noncarriers, suggesting a gene-dose effect (Geisler et al. 2008b; Shuldiner et al. 2009). Moreover, carriers of *CYP2C19*2* did not have difference in baseline platelet aggregation but showed greater residual platelet aggregation after dosing of clopidogrel as compared with noncarriers (Shuldiner et al. 2009). In 74 healthy subjects (free from concomitant medications and predominantly whites) receiving a 300-mg loading dose of clopidogrel, the presence of the *CYP2C19*2* variant was significantly associated with lower exposure to clopidogrel active metabolite (as measured by AUC_{0-24} and C_{max}), lower inhibition of platelet aggregation at 4 h, and poor responder status (defined as less than 20 % of inhibition of platelet aggregation to 20 μM ADP at a given time t) than the absence of this variant allele (Brandt et al. 2007a), suggesting that an attenuated inhibition of platelet aggregation is associated with decreased formation of clopidogrel active metabolite. Moreover, in 47 healthy Japanese subjects, the mean AUC_{0-8} and C_{max} of clopidogrel active metabolite differed significantly between carriers with or without the loss-of-function *CYP2C19 *2* and **3*, and the lower response to clopidogrel was observed in homozygotes of *CYP2C19*2* or carriers of the compound genotype *CYP2C19*2/*3* than noncarriers (Umemura et al. 2008). More importantly, AUC_{0-8} and C_{max} of clopidogrel active metabolite correlated well with inhibition of platelet aggregation at 4 h postdose (Umemura et al. 2008), further confirming that decreased production of clopidogrel active metabolite would result in less or no responsiveness to clopidogrel.

In addition, a series of clopidogrel clinical trials have been performed in patients (Collet et al. 2009; Mega et al. 2009a; Simon et al. 2009; Trenk et al. 2008; Aleil et al. 2009; Giusti et al. 2007, 2009; Frere et al. 2008; Wallentin et al. 2010), further reproducibly and reliably confirming the previous findings obtained in healthy subjects (Brandt et al. 2007a; Hulot et al. 2006; Umemura et al. 2008; Fontana et al. 2007; Kim et al. 2008a, b; Angiolillo et al. 2006b; Mega et al. 2009a). For example, in 5,148 ACS patients receiving clopidogrel of 75 mg daily, the rate of MACE was higher in carriers of any *CYP2C19* loss-of-function variants than in those without during the first 30 days after start of clopidogrel treatment; however, that difference was not significant when the full follow-up period (up to 12 months) was considered (Wallentin et al. 2010). Moreover, in 2,485 stented patients pretreated with 600 mg of clopidogrel, the cumulative 30-day incidence of definite stent thrombosis following PCI was significantly higher in *CYP2C19*2* carriers than noncarriers, with *CYP2C19*2/*2* carriers having the highest risk of stent thrombosis (Sibbing et al. 2009). In a large cohort of consecutively recruited acute MI patients receiving clopidogrel, a worsening clinical outcome (such as death, MI or stroke) was frequently seen in patients harboring *CYP2C19* loss-of-function variant alleles (**2*, **3*, **4*, or **5*), particularly in the subgroup of patients undergoing PCI (Simon et al. 2009). Interestingly, a more significant association was found between high residual platelet aggregation and the combination of *CYP2C19*2* genotype and five known nongenetic risk factors, suggesting that addition of *CYP2C19*2* variant to nongenetic risk factors may improve the prediction of poor clopidogrel response in patients (Geisler et al. 2008b).

However, the existence of either poor responders not carrying nonfunctional variant *CYP2C19*2* or responders carrying this variant seen in healthy subjects (Hulot et al. 2006) and in patients (Aleil et al. 2009) indicates that other factors also affect the decreased response to clopidogrel, because the *CYP2C19*2* variant has accounted for just 10–12 % of the variation in clopidogrel response as estimated by two independent clinical studies (Shuldiner et al. 2009; Fontana et al. 2007). For carriers of *CYP2C19*2* variant, clopidogrel-induced inhibition of ADP-triggered platelet aggregation would be improved with high-dose clopidogrel therapy in healthy subjects (Anonymous 2010a), but this strategy is not well supported by currently available data (Pena et al. 2009) although it seems rational.

*CYP2C19**3 is another loss-of-function variant, relatively common in Chinese subjects at an allele frequency of 5 % but is very rare in blacks and whites (<0.5 %) (Xie et al. 2001). In Korean patients with CAD who underwent PCI with drug-eluting stents and received clopidogrel, *CYP2C19*3* variant was significantly more prevalent in the clopidogrel-resistant group (28.9 % of the patients tested) than the clopidogrel-responsive groups, and multiple logistic regression analysis revealed that *CYP2C19*3* was an independent predictor of clopidogrel resistance (Lee et al. 2009).

In contrast, the gain-of-function variant *CYP2C19*17* (due to an increased transcription of *CYP2C19*) is expected to lead to an enhanced response to antiplatelet treatment with clopidogrel and consequently improved prevention of thrombotic events. On the other hand, it may increase the risk of bleeding complications. Current results are conflicting to each other, and thus a meta-analysis was performed accordingly to solve these issues (Li et al. 2012a, b). The *CYP2C19*17* seems to be a double-edged sword in thrombosis and bleeding, and thus genotyping of *CYP2C19*17* may be clinically helpful in better predicting bleeding complications in clopidogrel-treated patients undergoing PCI (Li et al. 2012a, b).

6.2 CYP2B6

CYP2B6 is responsible, at least in part, for the production of clopidogrel active metabolite (Hagihara et al. 2008; Kazui et al. 2010; Nishiya et al. 2009; Richter et al. 2004; Turpeinen et al. 2005; Mega et al. 2009a). However, variable CYP2B6 activity (due to its genetic polymorphisms) may not be clinically relevant, because no significant associations between *CYP2B6* polymorphisms and the primary efficacy and safety outcomes (such as death, nonfatal MI, stroke, or stent thrombosis) at 15 months were observed in a cohort of 1,477 ACS patients treated with clopidogrel (Mega et al. 2009a).

6.3 CYP3A4/3A5

CYP3A4 is the major P450 responsible for the formation of clopidogrel active metabolite in vitro (Kazui et al. 2010; Zahno et al. 2010), in particular at clopidogrel concentrations >10 μM, which leads to the inhibition of CYP2C19 by clopidogrel itself (Zahno et al. 2010). So far, there are no or rare *CYP3A4* variant alleles identified that appear at high allele frequencies among diverse ethnic populations and that may alter CYP3A4 catalytic activity to a clinically relevant extent as determined in subjects taking clopidogrel (Mega et al. 2009a).

In PCI patients treated with clopidogrel, loss-of-function variant *CYP3A5*3* was associated with attenuated response to clopidogrel and an increased risk for thrombotic events when itraconazole (a selective CYP3A4 inhibitor) was taken concurrently (Suh et al. 2006), indicating that CYP3A5 may be involved in clopidogrel active metabolite formation. However, decreased function of CYP3A5 (due to the presence of *CYP3A5*3*) (Wang et al. 2006a; Xie et al. 2004) did not affect the formation of the active metabolite of and response to clopidogrel in healthy subjects (Brandt et al. 2007a; Kim et al. 2008a, b; Mega et al. 2009a), had no influence on residual platelet aggregation in patients taking clopidogrel (Geisler et al. 2008b; Smith et al. 2006a, b), and was not associated with a risk of subsequent cardiovascular events (such as death from any cause, nonfatal MI, or stroke) during 1-year follow-up in 2,208 acute MI patients (Simon et al. 2009), suggesting that the contribution of CYP3A5 to the bioactivation of clopidogrel may be less important than assumed.

6.4 CYP1A2

CYP1A2 is responsible for the formation of 2-oxo-clopidogrel from clopidogrel with a corresponding contribution of 36 % but does not catalyze the formation of clopidogrel active metabolite directly (Kazui et al. 2010). However, in subjects taking clopidogrel, carriers of reduced-function CYP1A2 allele variants were not associated with a consistent reduction of the PK and PD response to clopidogrel (Mega et al. 2009a).

6.5 P-Glycoprotein (P-gp)

Clopidogrel is a P-gp substrate (Taubert et al. 2006). After oral administration of a loading dose (300 or 600 mg), the C_{max} and AUC levels of clopidogrel and its active metabolite were lower in PCI patients carrying the *P-gp* 3435T/T genotype than in carriers of the 3435C (Taubert et al. 2006), suggesting that carriers of the genotype 3435T/T exhibit a greater P-gp activity as demonstrated elsewhere (Kim et al. 2001). When the loading dose was increased to 900 mg, the C_{max} and AUC levels of clopidogrel and its active metabolite were increased in carriers of the 3435T/T compared with the 3435C carriers due to the marked downregulation of P-gp expression and decreased function in the former (Taubert et al. 2006). Furthermore, the C_{max} and AUC levels of clopidogrel and its active metabolite were lower in the *MDR1*2* homozygotes (haplotype: 1236T-2677T-3435T) than in the *MDR1*1* homozygotes (haplotype: 1236C-2677G-3435C) for all three dosage groups (Taubert et al. 2006). In 321 healthy subjects treated with clopidogrel, as compared with *P-gp* 3435C carriage, *P-gp* 3435T/T had a significantly reduced platelet inhibition in response to a clopidogrel loading dose (300 mg) but not to maintenance dosing (Mega et al. 2010). Similarly, in 1,471 PCI patients treated with clopidogrel, *P-gp* 3435C/T polymorphism was significantly associated with the risk of cardiovascular death, MI, or stroke until 15 months, the 3435T/T homozygotes (approximately 27 % of the patient population) having a 72 % increased risk of the primary efficacy endpoints as compared with 3435C carriers (Mega et al. 2010). Inconsistent with a previous finding (Taubert et al. 2006), *P-gp* 3435T/T did not have reduced exposure to clopidogrel active metabolite, and the other two common P-gp polymorphisms – 2677G/T(A) or 1236C/T – did not add additional significance to the role of *P-gp* 3435C/T (Mega et al. 2010). Simon et al. observed that patients homozygous for *P-gp* 3435T had a higher rate of subsequent cardiovascular events at 1 year than homozygotes of 3435C (15.5 % vs. 10.7 %) in 2,208 acute MI patients receiving clopidogrel therapy (Simon et al. 2009), but a further subgroup analysis indicated that *P-gp* 3435T variant did not have a significantly independent influence on the adjusted risk of death, MI, or stroke in the 1,535 acute

MI patients undergoing PCI (Simon et al. 2009). However, these results will need to be confirmed in prospective, large-scale patient populations of diversely ethnic backgrounds.

6.6 CYP2C9

CYP2C9 is responsible for clopidogrel active metabolite formation from 2-oxo-clopidogrel with a relative contribution of 7 % (Kazui et al. 2010), indicating a less important role in the formation of clopidogrel active metabolite. For example, in 162 healthy subjects taking clopidogrel, carriers of reduced-function *CYP2C9* variants were not associated with a consistent reduction of the PK and PD response to clopidogrel (Mega et al. 2009a).

On the other hand, *CYP2C9* expressed in vascular endothelium is responsible for the generation of significant amounts of reactive oxygen species (e.g., O_2^-) (Fichtlscherer et al. 2004), indirectly enhancing platelet recruitment most likely by inactivating a platelet ecto-ADPase (thereby increasing the bioavailability of ADP) and also by scavenging nitric oxide (NO) (thereby impairing the antiplatelet activity of NO due to a decrease in NO bioavailability) (Davi and Patrono 2007). Therefore, impaired CYP2C9 activity (due to loss-of-function genetic polymorphisms) may exhibit two opposite effects: one is reduced formation of clopidogrel active metabolite (Kazui et al. 2010) and the other is reduced intrinsic platelet activation and recruitment due to a decrease in the CYP2C9-mediated formation of reactive oxygen species (Davi and Patrono 2007). The net effect during a dynamic process would be more complicated than we previously thought. Inconsistent results on the effects of *CYP2C9* polymorphisms on platelet response to clopidogrel are not surprising accordingly.

7 Comorbidities

The liver is the most important organ responsible for the metabolic conversion of clopidogrel to its active metabolite (Savi et al. 1992). In addition, both the parent drug and its active metabolite were found to be highly protein bound (>94 %) (Lins et al. 1999). Moreover, hepatic CYP2C19 is involved in clopidogrel bioactivation (Kazui et al. 2010), and there is a marked decrease in enzymatic activity of CYP 2C19, 3A4, and 1A2 in patients with severe liver disease (Kimura et al. 1999; Adedoyin et al. 1998; d'Esposito et al. 2010; George et al. 1995), and, thus, severe liver disease might impair the metabolism of and response to clopidogrel.

In addition to *CYP2C19*2*, type II diabetes mellitus, obesity, ACS, PCI, and stenting all can increase platelet aggregability (Bhatt 2008; Bonello-Palot et al. 2009). For example, diabetes mellitus was associated with low response to clopidogrel in patients after coronary intervention (Angiolillo et al. 2006a; Geisler et al. 2008a; Hochholzer et al. 2010), and patients with high on-treatment platelet reactivity after a clopidogrel loading dose were more often diabetics than good responders (44 % vs. 20 %) (Bonello-Palot et al. 2009). That may be because diabetic patients had a less generation of clopidogrel active metabolite (Erlinge et al. 2008), were more likely to have high on-clopidogrel residual platelet aggregation, and exhibited less prominent interaction with age and BMI than nondiabetic patients (Hochholzer et al. 2010). Moreover, in a large-scale study of 1,092 consecutively enrolled patients, type II diabetes mellitus, ACS, renal failure, and decreased left ventricular function were identified as significant nongenetic factors that would better predict patients with persistent high residual platelet reactivity after coronary stenting and dual antiplatelet therapy with clopidogrel and aspirin (Geisler et al. 2008a). Moreover, elevated residual platelet aggregation after PCI for symptomatic CAD is an independent predictor for adverse coronary events in acute MI patients (Geisler et al. 2008b). In addition, increased triglyceride levels and decreased levels of HDL cholesterol were used to explain 1.3 and 1 % of the variation of clopidogrel response, respectively (Shuldiner et al. 2009). In the 2,208 consecutively enrolled patients with acute MI and receiving clopidogrel, 319 patients who had an outcome event (such as death, nonfatal MI, or stroke) more frequently had a history of hypertension, diabetes, MI, PCI, stroke, or heart failure compared with those who did not (Simon et al. 2009).

In addition, impaired creatinine clearance is an independent baseline predictor of the major bleeding complications in 6,001 PCI patients (Feit et al. 2007). Patients with chronic kidney disease (CKD) undergoing hemodialysis have significantly higher baseline platelet aggregation than those with normal renal function, and CKD is an important predictor of low response to clopidogrel (Woo et al. 2011). Diabetic patients with impaired renal function are also characterized by attenuated antiplatelet response to clopidogrel. However, in nondiabetic patients with CAD, the presence of impaired renal function (defined as an estimated glomerular filtration rate or eGFR < 60 ml/min) is not associated with differences in clopidogrel-induced antiplatelet effects when compared with those with preserved renal function (Tello-Montoliu et al. 2013). However, eGFR < 15 ml/min is associated with significantly increased residual platelet reactivity, regardless of the presence of diabetes mellitus in patients who are receiving a maintenance dose of clopidogrel (Muller et al. 2012). As the most important predictor of CKD, diabetes mellitus may contribute predominantly to attenuated response to clopidogrel when impairment of renal function is not too severe.

8 Concurrent Medications and Drug-Drug Interactions

Concomitant drug use is frequently seen in patients who are required to take clopidogrel, and, thus, drug-drug interactions could lead to different clinical outcomes in a variety of clinical settings. For example, when co-prescribed with potent inhibitors of CYP2C19 (such as omeprazole, esomeprazole, cimetidine, ranitidine, and fluoxetine), clopidogrel-induced inhibition of platelet aggregation would be attenuated, especially in carriers of the loss-of-function *CYP2C19* variants (Gilard et al. 2006; Gurbel et al. 2008; Ho et al. 2009; Small et al. 2008) or when treated also with strong CYP3A4 inhibitors (Zahno et al. 2010). Moreover, in the 2,208 consecutively enrolled patients with acute MI and receiving clopidogrel, 319 patients who had an outcome event (such as death, recurrent MI, or stroke) were less likely to receive statins, beta-blockers, ACEI (angiotensin-converting enzyme inhibitors), GP IIb/IIIa inhibitors, and heparin as compared with those who did not (Simon et al. 2009). In addition, combined use of abciximab (a GP IIb/IIIa inhibitor) was a significant predictor of bleeding complications in 1,524 PCI patients receiving a loading dose of clopidogrel (600 mg) (Sibbing et al. 2010). In a multicenter, prospective, observational study of 11,823 consecutive hospital survivors of acute MI, concurrent administration of aspirin, clopidogrel, beta-blocker, ACEI/sartan, and statin was an effective guideline-adherent secondary prevention drug therapy strategy with an improved 1-year survival as compared with some cases with less combined drugs (Bauer et al. 2010).

Clinical relevance of the drug interactions may vary from patient to patient, from drug to drug, and even from disease to disease. Thus, the extent of the drug interactions, the site of the interaction, the mechanism of the interaction, and the therapeutic margin of the interacting drugs must be taken into account (Abernethy 1997), since clinical implications of drug-drug interactions may vary case by case.

Clopidogrel and statins (such as atorvastatin, cerivastatin, fluvastatin, lovastatin, pravastatin, and simvastatin) are frequently prescribed concomitantly in some patients (Michelson et al. 2007; Aleil et al. 2009; Collet et al. 2009; Giusti et al. 2009; Malek et al. 2008; Sibbing et al. 2010; Simon et al. 2009; Harmsze et al. 2010; Wallentin et al. 2010). Because the transport and biotransformation of clopidogrel in the body are mediated by P-gp and certain P450 isozymes that are also responsible for the disposition of statins, the potential drug-drug interactions may exist in various clinical settings. For example, drug interactions between clopidogrel and atorvastatin (a substrate of CYP3A4 and P-gp) or pravastatin (not a substrate of CYP3A4) were investigated in 44 stented patients and results indicated that clopidogrel was less effective in inhibition of platelet aggregation in patients who were concurrently taking atorvastatin rather than pravastatin (Lau et al. 2003). However, in a randomized clinical trial with 75 mg/day clopidogrel versus placebo in 15,603 patients, there was no interaction between randomized treatment and the type of statins used (CYP3A4-metabolized vs. non-CYP3A4-metabolized statins) for the rates of MACE (e.g., death, MI, or stroke) at the median follow-up of 28 months (Saw et al. 2007). Such evidence obtained from randomized clinical trials has well demonstrated no clinical interactions between concomitant use of clopidogrel and CYP3A4-metabolized statins (such as atorvastatin, lovastatin, and simvastatin). Most of published evidence has consistently demonstrated that the antiplatelet activity of clopidogrel is not affected by a statin whose metabolism is predominantly catalyzed by CYP3A4 or CYP2C9 or not, and that no dosage adjustment is required when a statin is co-administrated with clopidogrel.

Clopidogrel is a typical substrate of CYP3A (Clarke and Waskell 2003). The metabolism of an estimated half of all marketed drugs, including some statins, is catalyzed by CYP3A4. In clinical practice, CYP3A4 substrates, inducers, and inhibitors are often administrated with clopidogrel simultaneously, and it is of interest to determine whether there could be potential drug-drug interactions that should be avoided. For example, co-administration of ketoconazole (a potent noncompetitive CYP3A inhibitor) can result in a 30 % decrease in exposure to clopidogrel active metabolite and hence reduced inhibition of platelet aggregation (Farid et al. 2007). Similar results have been observed in the presence of erythromycin (a CYP3A4 substrate/suicide inhibitor) or troleandomycin (a CYP3A4 substrate/suicide inhibitor) (Lau et al. 2003). As expected, the presence of rifampin (a CYP3A4 inducer) can promote the formation of clopidogrel active metabolite and hence increase its antiplatelet activity (Lau et al. 2003), and increased CYP3A4 activity induced by St John's wort is associated with enhanced antiplatelet effect of clopidogrel in healthy subjects and patients who had a low platelet response to clopidogrel (Lau et al. 2011).

Warfarin is often associated with GI bleeding and high INR (Pirmohamed et al. 2004). In acute MI (with ST-elevation) patients, dual antiplatelet therapy and anticoagulant treatment regimens are routinely used to reduce the risk of thrombotic complications (Chen et al. 2005), since many patients on dual antiplatelet therapy still have thrombotic events. Therefore, combination of triple antithrombotic drugs (aspirin, clopidogrel, and warfarin) would be given carefully according to the balance on the benefits and risks patients might have (Hermosillo and Spinler 2008). If needed, the doses of aspirin should be reduced, since bleeding risk of aspirin is dose dependent (Hirschowitz and Hawkey 2001; Hirsh et al. 1995; Jacobson 2004). In addition to warfarin and fluvastatin, the sulfonylurea antidiabetics tolbutamide is also a substrate of CYP2C9 (Xie et al. 2002). In a cohort of 158 PCI patients with type II diabetes mellitus treated with clopidogrel and aspirin, 53 patients were on sulfonylurea antidiabetics, whereas the remaining 105 patients were taking other oral hypoglycemic drugs and insulin. Mean ADP-induced on-clopidogrel platelet reactivity was significantly higher in the users of sulfonylurea antidiabetics than nonusers. Such differences remained significant after the confounding factors (such as age, gender, body weight, use of insulin) were adjusted, suggesting that concurrent use of sulfonylurea antidiabetics would attenuate clopidogrel-induced platelet inhibition (Harmsze et al. 2009).

Bleeding is an extension of the antiplatelet effect and thus is more common with combined antiplatelet therapy than monotherapy alone. As anticipated, one of the major side effects of clopidogrel is GI bleeding (Pirmohamed et al. 2004), in particular when combined with aspirin. To avoid the bleeding, proton pump inhibitors (Aleil et al. 2009; Collet et al. 2009; Giusti et al. 2009; Sibbing et al. 2010; Simon et al. 2009; Harmsze et al. 2010; Wallentin et al. 2010) are frequently co-prescribed with clopidogrel and aspirin to prevent their GI bleeding complications (Ray et al. 2010; Bhatt et al. 2008). On the other hand, proton pump inhibitors (PPI) with a strong inhibition of CYP2C19 should best be avoided in patients treated with clopidogrel, in particular when treated with CYP3A4 inhibitors concomitantly due to a double hit of clopidogrel metabolic activation (Zahno et al. 2010).

Although clopidogrel is a substrate of CYP2C19 (Kazui et al. 2010; Zahno et al. 2010), it is also an inhibitor of CYP2C19 at higher concentrations (>10 μM) in vitro (Zahno et al. 2010). In healthy subjects, inhibition of CYP2C19 by clopidogrel (as measured by decreased formation of 5-hydroxyomeprazole) occurred only in carriers of *CYP2C19*1/*1*, not in carriers of *CYP2C19 *2/*2* or **2/*3* (Chen et al. 2009), indicating an attenuated inhibition of CYP2C19 when enzyme activity is impaired due to genetic polymorphism. Conversely, an obvious possibility raised by the above clopidogrel findings is that comedication of potent inhibitors of CYP2C19 (such as omeprazole, esomeprazole, cimetidine, ranitidine, and fluoxetine) may attenuate clopidogrel-induced inhibition of platelet aggregation, especially in carriers of the loss-of-function *CYP2C19* variants (Gilard et al. 2006; Gurbel et al. 2008; Ho et al. 2009; Small et al. 2008), strongly indicating a double impairment of CYP2C19 (Roden and Stein 2009).

ACEI is frequently administrated in most patients receiving clopidogrel (Aleil et al. 2009; Chen et al. 2005; Collet et al. 2009; Giusti et al. 2009; Malek et al. 2008; Sibbing et al. 2010; Simon et al. 2009; Harmsze et al. 2010). Some ACEI prodrug may be the substrate of CES1, such as trandolapril and temocapril (Zhu et al. 2008, 2009; Vistoli et al. 2009), and thus competitive drug-drug interactions may occur between such ACEI prodrugs and clopidogrel.

The metabolism of calcium channel blockers (CCB) of the non-dihydropyridine class (such as verapamil and diltiazem) is catalyzed by CYP3A4 (Liu et al. 2007; Chen et al. 2005; Collet et al. 2009; Giusti et al. 2009; Simon et al. 2009). Therefore, concurrent intake of clopidogrel and these CCBs could impair the formation of clopidogrel active metabolite due to a competitively metabolic inhibition and consequently reduce clopidogrel-induced platelet inhibition. For example, in 200 CAD patients undergoing PCI, ADP-stimulated platelet aggregation was 30 % higher in patients on concomitant intake of clopidogrel and CCB than in patients receiving clopidogrel alone, and decreased platelet inhibition by clopidogrel was seen in 40 % of patients on combined use of clopidogrel and CCB versus 20 % in patients taking clopidogrel alone (Siller-Matula et al. 2008). Furthermore, in 760 stented patients loading with 600 mg of clopidogrel, concomitant administration of a non-dihydropyridine CCB (verapamil or diltiazem) was associated with a statistically significant high on-clopidogrel residual platelet aggregation (Hochholzer et al. 2010), resulting in a low response to clopidogrel (Hochholzer et al. 2010; Siller-Matula et al. 2008). Therefore, it has been suggested that CCBs (at least non-dihydropyridines) should be avoided in patients required to take clopidogrel (Hochholzer et al. 2010).

The angiotensin II type 1-receptor blockers (also known as ARB or the sartans) are a class of antihypertensive drugs that are frequently co-administrated with clopidogrel in patients (Aleil et al. 2009; Sibbing et al. 2010; Simon et al. 2009), some of which are substrates of CYP2C9 (Xie et al. 2002). CYP2C9 plays a less important role in the metabolic bioactivation of clopidogrel (Kazui et al. 2010), and thus sartan-clopidogrel interactions could be less clinically significant. On the other hand, losartan and its active metabolite EXP3174, irbesartan, valsartan, and telmisartan all dose-depedently suppressed TXA_2-dependent human platelet activation (independent of angiotensin II involvement), whereas candesartan failed (Schwemmer et al. 2001; Monton et al. 2000), indicating that certain sartans may exert their antiplatelet effects through the blockade of TXA_2 receptor-dependent signaling rather than acting at the AT1 receptor itself and that concomitant use of DAT and a certain sartan may have an additive effect on platelet inhibition. However, for the ACS or PCI patients with impaired renal function, concurrent use of clopidogrel and a sartan should be avoided, because such a drug interaction could result in further impairment of renal function, in particular switching of clopidogrel to ticagrelor for the patients with eGFR < 30 ml/min (Dinicolantonio and Serebruany 2012).

9 Food, Drink, and Dietary Supplementation

Despite increased unchanged clopidogrel and slightly decreased exposure of clopidogrel active metabolite after administration of the drug, the numerical increase in maximum platelet aggregation in the fed versus fasted state was within the prespecified equipotency range of ±15 %, indicating that clopidogrel can be taken with or without food (Hurbin et al. 2012).

Grapefruit juice (GJ) is a popular drink, and thus GJ-drug interactions often occur (Anonymous 2010b), in particular its interaction with cardiovascular drugs (Bailey and Dresser 2004). GJ can irreversibly inactivate intestinal CYP3A4 (reducing presystemic metabolism of affected drugs), inhibit P-gp activity (reducing excretion of affected drugs to bile or intestinal lumen) and also inhibit OATP activity (reducing intestinal absorption of affected drugs) (Bailey and Dresser 2004). In terms of the fact that GJ decreases intestinal CYP3A protein expression (Lown et al. 1997) and that the formation of clopidogrel active metabolite occurs predominantly in the liver (Savi et al. 1992), the effect of GJ on clopidogrel metabolic activation seems to be limited, although an attenuated antiplatelet effect of clopidogrel was noted (Bailey and Dresser 2004). Of importance is GJ interacting with other concurrent drugs in patients given clopidogrel with concomitant GJ ingestion, such as statins, the CCB, the antidiabetic drugs, β-receptor antagonists, and angiotensin II type 1 receptor blockers, which are more frequently co-administrated with clopidogrel (Collet et al. 2009; Giusti et al. 2009; Malek et al. 2008; Sibbing et al. 2010; Simon et al. 2009). Thus, avoiding GJ entirely should be required in order to avoid GJ-drug interactions during the drug therapy.

A large number of clinical studies and case reports have well documented that *Hypericum perforatum* (also known as St. John's wort, or SJW) can induce the expression and function of CYP2C19, CYP3A4, CYP2B6, CYP1A2, CYP2C9, and P-gp (Xie and Kim 2005; Lei et al. 2010; Rahimi and Abdollahi 2012), which all are responsible for metabolic bioactivation (Kazui et al. 2010) and transport (Taubert et al. 2006) of clopidogrel. As expected, in 10 healthy subjects hyporesponsive to clopidogrel, SJW (300 mg, tid, 14 days) significantly induced hepatic CYP3A4 activity as measured by erythromycin breath test ($^{14}CO_2$ exhaled per hour) and decreased platelet aggregation after a 300-mg clopidogrel as compared with baseline testing in a prospective, open-labeled, control study with a 7-day washout period apart (Lau et al. 2011). Furthermore, SJW or placebo each was given to 10 post-coronary stent patients (less responsive to clopidogrel) in a randomized, double-blind manner; all patients had a lower clopidogrel response and received chronic clopidogrel therapy (75 mg daily maintenance dose) concomitantly for 14 days. SJW significantly decreased platelet reactivity (as measured by platelet receptor P2Y12 reaction units) and platelet aggregation (Lau et al. 2011), demonstrating an enhanced platelet inhibition in patients with a low platelet response to clopidogrel. On the other hand, the risk of bleeding would be increased in patients taking clopidogrel and SJW simultaneously. However, further studies are still needed to clarify such a preliminary finding in patient care.

10 Patient Adherence or Compliance

Poor or no compliance (or termed as nonadherence) to clopidogrel therapy may be one of the causes that could account for clopidogrel resistance (Serebruany et al. 2005; Latry et al. 2012), since noncompliance or under-dosing can result in poor bioavailability of clopidogrel. Although the main carboxylic metabolite of clopidogrel is pharmacologically inactive, determination of this metabolite can be used for the therapeutic drug monitoring to evaluate compliance of patients to clopidogrel treatment because it may exist in the body for 1–2 days after cessation of drug intake (Mani et al. 2008). In addition, adherence may be assessed by using the proportion of days covered by the treatment and persistence of the reimbursement form of medication from the health insurance reimbursement database as described elsewhere (Latry et al. 2012). The importance of adherence is well supported by a large, prospective, multicenter, observational study, in which guideline-adherent secondary prevention drug therapy resulted in an improved 1-year survival (Bauer et al. 2010). For the patients who received DAT, the treatment is often stopped within 1 year after discharge (Latry et al. 2012), and there were higher incidence rates of adverse cardiovascular events (e.g., MI, stent thrombosis, stroke, or death) observed in the initial 0–90 days after clopidogrel cessation in PCI patients compared with other time durations (Sachdeva et al. 2012).

11 Summary and Future Perspectives

Clopidogrel is one of the most extensively studied drugs in clinical practice. Based on currently available data and clinical concerns about the efficacy and safety of clopidogrel, the US FDA consecutively released the black box warnings for the three clopidogrel labels in 2010 (Anonymous 2010a; Holmes et al. 2010; Roden and Shuldiner 2010), alerting clinicians and patients of the role of *CYP2C19**2 and *3 variants in response to clopidogrel in various clinical settings, in particular ACS patients undergoing PCI for stenting (Collet et al. 2009; Mega et al. 2009a; Pare et al. 2010; Johnson et al. 2012; Scott et al. 2011) with a special concern about the first 30 days after start of clopidogrel treatment (Hulot et al. 2010; Mega et al. 2009a; Sofi et al. 2011; Wallentin et al. 2010). Thus, all relevant information summarized in the chapter and elsewhere would help physicians how to make a better choice of the drug and the dose on the basis of patient-specific profiling.

Individuals vary in their response to drugs due to huge patient heterogeneity. In addition to *CYP2C19* genetic polymorphisms involved, other unknown genetic factors and nongenetic factors may be contributory, such as ethnicity, gender, increased age and BMI, drug-drug interactions, coexisting diseases (such as diabetes, and inflammation), microRNA, gene promoter methylation, and other factors to be determined. In clinical practice, the recurrence of ischemic episodes or thrombotic events after DAT is a complex process, in which gene makeup and functional expression, platelets, procoagulant and fibrinolytic systems, ethnicity, gender, aging, disease heterogeneity, and environment factors all play a role. Thus, DNA or pharmacogenomics is not the whole story about personalized medicine,

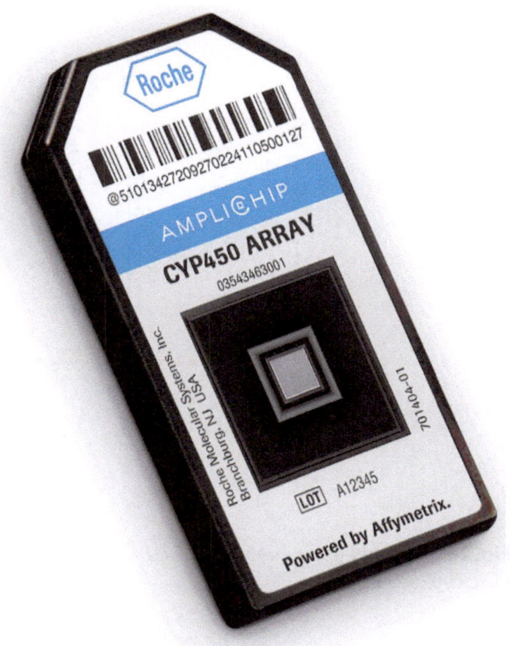

Fig. 22.3 The P450 assay chip for *CYP2C19* genotyping (Reproduced with permission from Roche)

and the simple view of directing drug therapy on the basis of genetic makeup is probably too simple (Xie et al. 2011; Longo 2012). However, identifying a subgroup of patients genetically unable to generate clopidogrel active thiol metabolite would be the first step to help identify those at high risk for therapy failure, and thus genotyping the pieces of the puzzle would help personalize clopidogrel therapy (Roberts et al. 2012; Shahabi et al. 2012; Guzauskas et al. 2012; Scott et al. 2011; Xie 2011; Pulley et al. 2012) as shown in Figs. 22.3 and 22.4 for *CYP2C19* genotyping. A more individualized approach may be to prescribe clopidogrel for patients without high-risk genotypes and to adjust the doses or duration of clopidogrel therapy, or to switch from clopidogrel to an alternative drug (such as prasugrel less dependent on CYP2C19) for patients with at-risk genotypes (such as *CYP2C19**2/*2), since genotyping rather than multiple repeated platelet function monitoring could be a cost-effective strategy to identify patients at high risk for atherothrombotic events (O'Donoghue and Wiviott 2006).

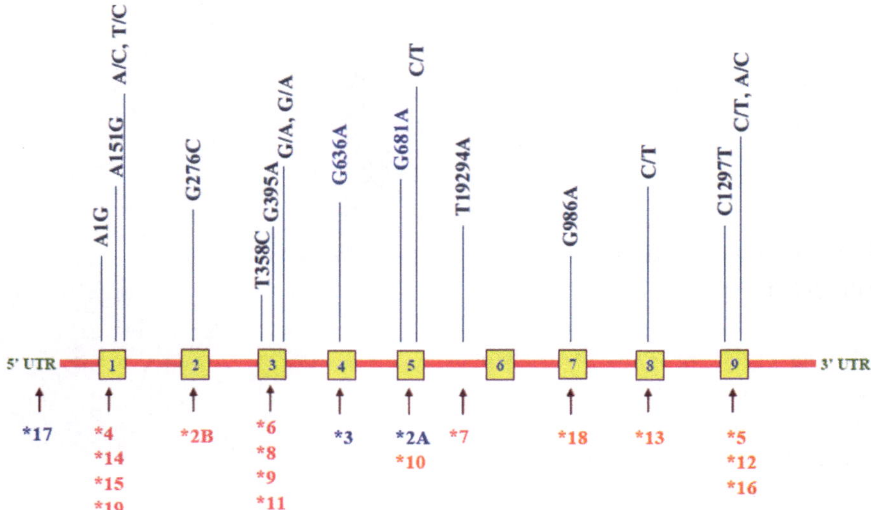

Fig. 22.4 Highly polymorphic *CYP2C19* gene sequences in humans (Data from the website at: http://www.cypalleles.ki.se/cyp2c19.htm)

In summary, multiple genetic and nongenetic factors contribute, to a varying extent, to platelet aggregation and clopidogrel resistance and thereby to wide interindividual variability in pharmacological response to clopidogrel and consequently clinical outcomes. Integration of all known factors into the decision-making would be helpful to maximize the likelihood of drug safety and efficacy after the drug and dose are tailored individually (Xie and Frueh 2005).

Acknowledgments This work was supported in part by a starting research grant No. 31010300010339, funded by the Nanjing First Hospital, Nanjing Medical University, China; a grant No. BK 2012525, funded by the Jiangsu Natural Science Foundation (JSNSF), China; and a starting research grant (No. 2012–258), funded by the Ministry of Human Resources and Social Security, China.

References

Abernethy DR (1997) Grapefruits and drugs: when is statistically significant clinically significant? J Clin Invest 99:2297–2298

Adedoyin A, Arns PA, Richards WO, Wilkinson GR, Branch RA (1998) Selective effect of liver disease on the activities of specific metabolizing enzymes: investigation of cytochromes P450 2C19 and 2D6. Clin Pharmacol Ther 64:8–17

Afshar-Kharghan V, Li CQ, Khoshnevis-Asl M, Lopez JA (1999) Kozak sequence polymorphism of the glycoprotein (GP) Ibalpha gene is a major determinant of the plasma membrane levels of the platelet GP Ib-IX-V complex. Blood 94:186–191

Ajzenberg N, Berroeta C, Philip I, Grandchamp B, Ducellier P, Huart V et al (2005) Association of the -92C/G and 807C/T polymorphisms of the alpha2 subunit gene with human platelets alpha2beta1 receptor density. Arterioscler Thromb Vasc Biol 25:1756–1760

Aleil B, Leon C, Cazenave JP, Gachet C (2009) CYP2C19*2 polymorphism is not the sole determinant of the response to clopidogrel: implications for its monitoring. J Thromb Haemost 7:1747–1749

Ameyaw MM, Regateiro F, Li T, Liu X, Tariq M, Mobarek A et al (2001) MDR1 pharmacogenetics: frequency of the C3435T mutation in exon 26 is significantly influenced by ethnicity. Pharmacogenetics 11:217–221

Ancrenaz V, Desmeules J, James R, Fontana P, Reny JL, Dayer P et al (2012) The paraoxonase-1 pathway is not a major bioactivation pathway of clopidogrel in vitro. Br J Pharmacol 166:2362–2370

Anderson JL, Adams CD, Antman EM, Bridges CR, Califf RM, Casey DE Jr et al (2007) ACC/AHA 2007 guidelines for the management of patients with unstable angina/non ST-elevation myocardial infarction: a report of the American College of Cardiology/American Heart Association Task Force on Practice Guidelines (Writing Committee to Revise the 2002 Guidelines for the Management of Patients With Unstable Angina/Non ST-Elevation Myocardial Infarction): developed in collaboration with the American College of Emergency Physicians, the Society for Cardiovascular Angiography and Interventions, and the Society of Thoracic Surgeons: endorsed by the American Association of

Cardiovascular and Pulmonary Rehabilitation and the Society for Academic Emergency Medicine. Circulation 116:e148–e304

Angiolillo DJ (2012) The evolution of antiplatelet therapy in the treatment of acute coronary syndromes: from aspirin to the present day. Drugs 72:2087–2116

Angiolillo DJ, Fernandez-Ortiz A, Bernardo E, Alfonso F, Sabate M, Fernandez C et al (2004a) PlA polymorphism and platelet reactivity following clopidogrel loading dose in patients undergoing coronary stent implantation. Blood Coagul Fibrinolysis 15:89–93

Angiolillo DJ, Fernandez-Ortiz A, Bernardo E, Ramirez C, Escaned J, Moreno R et al (2004b) 807 C/T Polymorphism of the glycoprotein Ia gene and pharmacogenetic modulation of platelet response to dual antiplatelet treatment. Blood Coagul Fibrinolysis 15:427–433

Angiolillo DJ, Fernandez-Ortiz A, Bernardo E, Ramirez C, Barrera-Ramirez C, Sabate M et al (2005a) Identification of low responders to a 300-mg clopidogrel loading dose in patients undergoing coronary stenting. Thromb Res 115:101–108

Angiolillo DJ, Fernandez-Ortiz A, Bernardo E, Ramirez C, Cavallari U, Trabetti E et al (2005b) Lack of association between the P2Y12 receptor gene polymorphism and platelet response to clopidogrel in patients with coronary artery disease. Thromb Res 116:491–497

Angiolillo DJ, Bernardo E, Ramirez C, Costa MA, Sabate M, Jimenez-Quevedo P et al (2006a) Insulin therapy is associated with platelet dysfunction in patients with type 2 diabetes mellitus on dual oral antiplatelet treatment. J Am Coll Cardiol 48:298–304

Angiolillo DJ, Fernandez-Ortiz A, Bernardo E, Ramirez C, Cavallari U, Trabetti E et al (2006b) Contribution of gene sequence variations of the hepatic cytochrome P450 3A4 enzyme to variability in individual responsiveness to clopidogrel. Arterioscler Thromb Vasc Biol 26:1895–1900

Angiolillo DJ, Fernandez-Ortiz A, Bernardo E, Alfonso F, Macaya C, Bass TA et al (2007) Variability in individual responsiveness to clopidogrel: clinical implications, management, and future perspectives. J Am Coll Cardiol 49:1505–1516

Anonymous (2010a) FDA drug safety communication: reduced effectiveness of Plavix (clopidogrel) in patients who are poor metabolizers of the drug. Available at: http://www.fda.gov/Drugs/DrugSafety/PostmarketDrugSafetyInformationforPatientsandProviders/ucm203888.htm

Anonymous (2010b) Grapefruit-drug interactions. Available at: http://www.powernetdesign.com/grapefruit/

Anonymous (2010c) Plavix prescribing information. Available at: http://products.sanofi-aventis.us/plavix/plavix.html

Ansara AJ, Nisly SA, Arif SA, Koehler JM, Nordmeyer ST (2010) Aspirin dosing for the prevention and treatment of ischemic stroke: an indication-specific review of the literature. Ann Pharmacother 44:851–862

Antman EM, Hand M, Armstrong PW, Bates ER, Green LA, Halasyamani LK et al (2008a) 2007 Focused Update of the ACC/AHA 2004 Guidelines for the Management of Patients With ST-Elevation Myocardial Infarction: a report of the American College of Cardiology/American Heart Association Task Force on Practice Guidelines: developed in collaboration With the Canadian Cardiovascular Society endorsed by the American Academy of Family Physicians: 2007 Writing Group to Review New Evidence and Update the ACC/AHA 2004 Guidelines for the Management of Patients With ST-Elevation Myocardial Infarction, Writing on Behalf of the 2004 Writing Committee. Circulation 117:296–329

Antman EM, Hand M, Armstrong PW, Bates ER, Green LA, Halasyamani LK et al (2008b) 2007 focused update of the ACC/AHA 2004 guidelines for the management of patients with ST-elevation myocardial infarction: a report of the American College of Cardiology/American Heart Association Task Force on Practice Guidelines. J Am Coll Cardiol 51:210–247

Armstrong PC, Peter K (2012) GPIIb/IIIa inhibitors: from bench to bedside and back to bench again. Thromb Haemost 107:808–814

Aukrust P, Halvorsen B, Ueland T, Michelsen AE, Skjelland M, Gullestad L et al (2010) Activated platelets and atherosclerosis. Expert Rev Cardiovasc Ther 8:1297–1307

Badr ER, Lang IM, Koppensteiner R, Calatzis A, Panzer S, Gremmel T (2012) Residual platelet activation through protease-activated receptors (PAR)-1 and −4 in patients on P2Y12 inhibitors. Int J Cardiol doi:10.1016/j.ijcard.2012.09.103

Bailey DG, Dresser GK (2004) Interactions between grapefruit juice and cardiovascular drugs. Am J Cardiovasc Drugs 4:281–297

Balram C, Sharma A, Sivathasan C, Lee EJ (2003) Frequency of C3435T single nucleotide MDR1 genetic polymorphism in an Asian population: phenotypic-genotypic correlates. Br J Clin Pharmacol 56:78–83

Bauer T, Gitt AK, Junger C, Zahn R, Koeth O, Towae F et al (2010) Guideline-recommended secondary prevention drug therapy after acute myocardial infarction: predictors and outcomes of nonadherence. Eur J Cardiovasc Prev Rehabil 17:576–581

Bergandi L, Cordero M, Anselmino M, Ferraro G, Ravera L, Dalmasso P et al (2010) Altered nitric oxide/cGMP platelet signaling pathway in platelets from patients with acute coronary syndromes. Clin Res Cardiol 99:557–564

Bernlochner I, Steinhubl S, Braun S, Morath T, Jaitner J, Stegherr J et al (2010) Association between inflammatory biomarkers and platelet aggregation in patients under chronic clopidogrel treatment. Thromb Haemost 104:1193–1200

Bertrand ME, Rupprecht HJ, Urban P, Gershlick AH (2000) Double-blind study of the safety of clopidogrel with and without a loading dose in combination with aspirin compared with ticlopidine in combination with aspirin after coronary stenting: the clopidogrel aspirin stent international cooperative study (CLASSICS). Circulation 102:624–629

Bhatt DL (2008) What makes platelets angry: diabetes, fibrinogen, obesity, and impaired response to antiplatelet therapy? J Am Coll Cardiol 52:1060–1061

Bhatt DL (2009) Tailoring antiplatelet therapy based on pharmacogenomics: how well do the data fit? JAMA 302:896–897

Bhatt DL, Scheiman J, Abraham NS, Antman EM, Chan FK, Furberg CD et al (2008) ACCF/ACG/AHA 2008 expert consensus document on reducing the gastrointestinal risks of antiplatelet therapy and NSAID use: a report of the American College of Cardiology Foundation Task Force on Clinical Expert Consensus Documents. Circulation 118:1894–1909

Bhindi R, Ormerod O, Newton J, Banning AP, Testa L (2008) Interaction between statins and clopidogrel: is there anything clinically relevant? QJM 101:915–925

Bliden KP, DiChiara J, Lawal L, Singla A, Antonino MJ, Baker BA et al (2008) The association of cigarette smoking with enhanced platelet inhibition by clopidogrel. J Am Coll Cardiol 52:531–533

Bobbert P, Stellbaum C, Steffens D, Schutte C, Bobbert T, Schultheiss HP et al (2012) Postmenopausal women have an increased maximal platelet reactivity compared to men despite dual antiplatelet therapy. Blood Coagul Fibrinolysis 23:723–728

Bonello L, Camoin-Jau L, Mancini J, Bessereau J, Grosdidier C, Alessi MC et al (2012) Factors associated with the failure of clopidogrel dose-adjustment according to platelet reactivity monitoring to optimize P2Y12-ADP receptor blockade. Thromb Res 130: 70–74

Bonello-Palot N, Armero S, Paganelli F, Mancini J, De Labriolle A, Bonello C et al (2009) Relation of body mass index to high on-treatment platelet reactivity and of failed clopidogrel dose adjustment according to platelet reactivity monitoring in patients undergoing percutaneous coronary intervention. Am J Cardiol 104:1511–1515

Bordeaux BC, Qayyum R, Yanek LR, Vaidya D, Becker LC, Faraday N et al (2010) Effect of obesity on platelet reactivity and response to low-dose aspirin. Prev Cardiol 13:56–62

Bouman HJ, Schomig E, van Werkum JW, Velder J, Hackeng CM, Hirschhauser C et al (2011) Paraoxonase-1 is a major determinant of clopidogrel efficacy. Nat Med 17:110–116

Brandt JT, Close SL, Iturria SJ, Payne CD, Farid NA, Ernest CS et al (2007a) Common polymorphisms of CYP2C19 and CYP2C9 affect the pharmacokinetic and pharmacodynamic response to clopidogrel but not prasugrel. J Thromb Haemost 5:2429–2436

Brandt JT, Payne CD, Wiviott SD, Weerakkody G, Farid NA, Small DS et al (2007b) A comparison of prasugrel and clopidogrel loading doses on platelet function: magnitude of platelet inhibition is related to active metabolite formation. Am Heart J 153:66–76

Buonamici P, Marcucci R, Migliorini A, Gensini GF, Santini A, Paniccia R et al (2007) Impact of platelet reactivity after clopidogrel administration on drug-eluting stent thrombosis. J Am Coll Cardiol 49:2312–2317

Caplain H, Donat F, Gaud C, Necciari J (1999) Pharmacokinetics of clopidogrel. Semin Thromb Hemost 25(Suppl 2):25–28

Capodanno D, Bhatt DL, Goto S, O'Donoghue ML, Moliterno DJ, Tamburino C et al (2012) Safety and efficacy of protease-activated receptor-1 antagonists in patients with coronary artery disease: a meta-analysis of randomized clinical trials. J Thromb Haemost 10:2006–2015

Cattaneo M, Zighetti ML, Lombardi R, Martinez C, Lecchi A, Conley PB et al (2003) Molecular bases of defective signal transduction in the platelet P2Y12 receptor of a patient with congenital bleeding. Proc Natl Acad Sci U S A 100:1978–1983

Chen ZM, Jiang LX, Chen YP, Xie JX, Pan HC, Peto R et al (2005) Addition of clopidogrel to aspirin in 45,852 patients with acute myocardial infarction: randomised placebo-controlled trial. Lancet 366:1607–1621

Chen M, Ma L, Drusano GL, Bertino JS Jr, Nafziger AN (2006) Sex differences in CYP3A activity using intravenous and oral midazolam. Clin Pharmacol Ther 80:531–538

Chen BL, Chen Y, Tu JH, Li YL, Zhang W, Li Q et al (2009) Clopidogrel inhibits CYP2C19-dependent hydroxylation of omeprazole related to CYP2C19 genetic polymorphisms. J Clin Pharmacol 49:574–581

Chen CY, Poole EM, Ulrich CM, Kulmacz RJ, Wang LH (2012) Functional analysis of human thromboxane synthase polymorphic variants. Pharmacogenet Genomics 22:653–658

Clarke TA, Waskell LA (2003) The metabolism of clopidogrel is catalyzed by human cytochrome P450 3A and is inhibited by atorvastatin. Drug Metab Dispos 31:53–59

Collet JP, Hulot JS, Pena A, Villard E, Esteve JB, Silvain J et al (2009) Cytochrome P450 2C19 polymorphism in young patients treated with clopidogrel after myocardial infarction: a cohort study. Lancet 373:309–317

Collet JP, Cuisset T, Range G, Cayla G, Elhadad S, Pouillot C et al (2012) Bedside monitoring to adjust antiplatelet therapy for coronary stenting. N Engl J Med 367:2100–2109

Collins SD, Torguson R, Gaglia MA Jr, Lemesle G, Syed AI, Ben Dor I et al (2010) Does black ethnicity influence the development of stent thrombosis in the drug-eluting stent era? Circulation 122:1085–1090

Connolly SJ, Pogue J, Hart RG, Hohnloser SH, Pfeffer M, Chrolavicius S et al (2009) Effect of clopidogrel added to aspirin in patients with atrial fibrillation. N Engl J Med 360:2066–2078

Cuisset T, Frere C, Quilici J, Barbou F, Morange PE, Hovasse T et al (2006) High post-treatment platelet reactivity identified low-responders to dual antiplatelet therapy at increased risk of recurrent cardiovascular events after stenting for acute coronary syndrome. J Thromb Haemost 4:542–549

Cuisset T, Frere C, Quilici J, Morange PE, Saut N, Lambert M et al (2007) Role of the T744C polymorphism of the P2Y12 gene on platelet response to a 600-mg loading dose of clopidogrel in 597 patients with non-ST-segment elevation acute coronary syndrome. Thromb Res 120:893–899

d'Esposito F, Nebot N, Edwards RJ, Murray M (2010) Impaired irinotecan biotransformation in hepatic microsomal fractions from patients with chronic liver disease. Br J Clin Pharmacol 70:400–408

Davi G, Patrono C (2007) Platelet activation and atherothrombosis. N Engl J Med 357:2482–2494

Davi G, Guagnano MT, Ciabattoni G, Basili S, Falco A, Marinopiccoli M et al (2002) Platelet activation in obese women: role of inflammation and oxidant stress. JAMA 288:2008–2014

Debrunner M, Schuiki E, Minder E, Straumann E, Naegeli B, Mury R et al (2008) Proinflammatory cytokines in acute myocardial infarction with and without cardiogenic shock. Clin Res Cardiol 97:298–305

Ding Z, Kim S, Dorsam RT, Jin J, Kunapuli SP (2003) Inactivation of the human P2Y12 receptor by thiol reagents requires interaction with both extracellular cysteine residues, Cys17 and Cys270. Blood 101: 3908–3914

Ding XL, Xie C, Miao LY (2012) Effects of CYP2C19 *2 and *3 as well as CES1A2 -816A/C genetic polymorphisms on clopidogrel-mediated platelet aggregation. In: The 13rd national meeting of Chinese Society for Clinical Pharmacology, Chengdu, October 26–28, pp 186 (abstract)

Dinicolantonio JJ, Serebruany VL (2012) Angiotensin receptor blockers worsen renal function and dyspnea on ticagrelor: a potential ticagrelor-angiotensin receptor blocker interaction? Clin Cardiol 36:647–648

Dorsam RT, Kunapuli SP (2004) Central role of the P2Y12 receptor in platelet activation. J Clin Invest 113:340–345

Erlinge D, Varenhorst C, Braun OO, James S, Winters KJ, Jakubowski JA et al (2008) Patients with poor responsiveness to thienopyridine treatment or with diabetes have lower levels of circulating active metabolite, but their platelets respond normally to active metabolite added ex vivo. J Am Coll Cardiol 52:1968–1977

Farid NA, Payne CD, Small DS, Winters KJ, Ernest CS, Brandt JT et al (2007) Cytochrome P450 3A inhibition by ketoconazole affects prasugrel and clopidogrel pharmacokinetics and pharmacodynamics differently. Clin Pharmacol Ther 81:735–741

Fatini C, Sticchi E, Bolli P, Marcucci R, Giusti B, Paniccia R et al (2009) eNOS gene influences platelet phenotype in acute coronary syndrome patients on dual antiplatelet treatment. Platelets 20:548–554

Feher G, Koltai K, Alkonyi B, Papp E, Keszthelyi Z, Kesmarky G et al (2007) Clopidogrel resistance: role of body mass and concomitant medications. Int J Cardiol 120:188–192

Feit F, Voeltz MD, Attubato MJ, Lincoff AM, Chew DP, Bittl JA et al (2007) Predictors and impact of major hemorrhage on mortality following percutaneous coronary intervention from the REPLACE-2 Trial. Am J Cardiol 100:1364–1369

Feliste R, Delebassee D, Simon MF, Chap H, Defreyn G, Vallee E et al (1987) Broad spectrum anti-platelet activity of ticlopidine and PCR 4099 involves the suppression of the effects of released ADP. Thromb Res 48:403–415

Fichtlscherer S, Dimmeler S, Breuer S, Busse R, Zeiher AM, Fleming I (2004) Inhibition of cytochrome P450 2C9 improves endothelium-dependent, nitric oxide-mediated vasodilatation in patients with coronary artery disease. Circulation 109:178–183

Flood VH, Gill JC, Morateck PA, Christopherson PA, Friedman KD, Haberichter SL et al (2010) Common VWF exon 28 polymorphisms in African Americans affecting the VWF activity assay by ristocetin cofactor. Blood 116:280–286

Fontana P, Dupont A, Gandrille S, Bachelot-Loza C, Reny JL, Aiach M et al (2003) Adenosine diphosphate-induced platelet aggregation is associated with P2Y12 gene sequence variations in healthy subjects. Circulation 108:989–995

Fontana P, Hulot JS, De Moerloose P, Gaussem P (2007) Influence of CYP2C19 and CYP3A4 gene polymorphisms on clopidogrel responsiveness in healthy subjects. J Thromb Haemost 5:2153–2155

Fontana P, Berdague P, Castelli C, Nolli S, Barazer I, Fabbro-Peray P et al (2010) Clinical predictors of dual aspirin and clopidogrel poor responsiveness in stable cardiovascular patients from the ADRIE study. J Thromb Haemost 8:2614–2623

Frere C, Cuisset T, Morange PE, Quilici J, Camoin-Jau L, Saut N et al (2008) Effect of cytochrome p450 polymorphisms on platelet reactivity after treatment with clopidogrel in acute coronary syndrome. Am J Cardiol 101:1088–1093

Friedman DJ, Talbert ME, Bowden DW, Freedman BI, Mukanya Y, Enjyoji K et al (2009) Functional ENTPD1 polymorphisms in African Americans with diabetes and end-stage renal disease. Diabetes 58: 999–1006

Ganesan S, Williams C, Maslen CL, Cherala G (2013) Clopidogrel variability: role of plasma protein binding alterations. Br J Clin Pharmacol 75:1468–1477

Geiger J, Brich J, Honig-Liedl P, Eigenthaler M, Schanzenbacher P, Herbert JM et al (1999) Specific impairment of human platelet P2Y(AC) ADP receptor-mediated signaling by the antiplatelet drug clopidogrel. Arterioscler Thromb Vasc Biol 19:2007–2011

Geisler T, Langer H, Wydymus M, Gohring K, Zurn C, Bigalke B et al (2006) Low response to clopidogrel is associated with cardiovascular outcome after coronary stent implantation. Eur Heart J 27:2420–2425

Geisler T, Grass D, Bigalke B, Stellos K, Drosch T, Dietz K et al (2008a) The residual platelet aggregation after deployment of intracoronary stent (PREDICT) score. J Thromb Haemost 6:54–61

Geisler T, Schaeffeler E, Dippon J, Winter S, Buse V, Bischofs C et al (2008b) CYP2C19 and nongenetic factors predict poor responsiveness to clopidogrel loading dose after coronary stent implantation. Pharmacogenomics 9:1251–1259

George J, Murray M, Byth K, Farrell GC (1995) Differential alterations of cytochrome P450 proteins in livers from patients with severe chronic liver disease. Hepatology 21:120–128

Gilard M, Arnaud B, Le Gal G, Abgrall JF, Boschat J (2006) Influence of omeprazol on the antiplatelet action of clopidogrel associated to aspirin. J Thromb Haemost 4:2508–2509

Giusti B, Gori AM, Marcucci R, Saracini C, Sestini I, Paniccia R et al (2007) Cytochrome P450 2C19 loss-of-function polymorphism, but not CYP3A4 IVS10+12G/A and P2Y12 T744C polymorphisms, is associated with response variability to dual antiplatelet treatment in high-risk vascular patients. Pharmacogenet Genomics 17:1057–1064

Giusti B, Gori AM, Marcucci R, Sestini I, Saracini C, Paniccia R et al (2008) Role of glycoprotein Ia gene polymorphisms in determining platelet function in myocardial infarction patients undergoing percutaneous coronary intervention on dual antiplatelet treatment. Atherosclerosis 196:341–348

Giusti B, Gori AM, Marcucci R, Saracini C, Sestini I, Paniccia R et al (2009) Relation of cytochrome P450 2C19 loss-of-function polymorphism to occurrence of drug-eluting coronary stent thrombosis. Am J Cardiol 103:806–811

Gong IY, Crown N, Suen CM, Schwarz UI, Dresser GK, Knauer MJ et al (2012) Clarifying the importance of CYP2C19 and PON1 in the mechanism of clopidogrel bioactivation and in vivo antiplatelet response. Eur Heart J 33:2856–2864

Gonzales RJ, Ghaffari AA, Duckles SP, Krause DN (2005) Testosterone treatment increases thromboxane function in rat cerebral arteries. Am J Physiol Heart Circ Physiol 289:H578–H585

Gurbel PA, Anderson RD (1997) New concept in coronary angioplasty: dilatation with a helical balloon that allows simultaneous autoperfusion. Cathet Cardiovasc Diagn 40:109–116

Gurbel PA, Tantry US (2006) Drug insight: clopidogrel nonresponsiveness. Nat Clin Pract Cardiovasc Med 3:387–395

Gurbel PA, Bliden KP, Hiatt BL, O'Connor CM (2003) Clopidogrel for coronary stenting: response variability, drug resistance, and the effect of pretreatment platelet reactivity. Circulation 107:2908–2913

Gurbel PA, Bliden KP, Guyer K, Cho PW, Zaman KA, Kreutz RP et al (2005a) Platelet reactivity in patients and recurrent events post-stenting: results of the PREPARE POST-STENTING Study. J Am Coll Cardiol 46:1820–1826

Gurbel PA, Bliden KP, Hayes KM, Yoho JA, Herzog WR, Tantry US (2005b) The relation of dosing to clopidogrel responsiveness and the incidence of high post-treatment platelet aggregation in patients undergoing coronary stenting. J Am Coll Cardiol 45:1392–1396

Gurbel PA, Bliden KP, Samara W, Yoho JA, Hayes K, Fissha MZ et al (2005c) Clopidogrel effect on platelet reactivity in patients with stent thrombosis: results of the CREST Study. J Am Coll Cardiol 46:1827–1832

Gurbel PA, Lau WC, Tantry US (2008) Omeprazole: a possible new candidate influencing the antiplatelet effect of clopidogrel. J Am Coll Cardiol 51:261–263

Guzauskas GF, Hughes DA, Bradley SM, Veenstra DL (2012) A risk-benefit assessment of prasugrel, clopidogrel, and genotype-guided therapy in patients undergoing percutaneous coronary intervention. Clin Pharmacol Ther 91:829–837

Hagihara K, Nishiya Y, Kurihara A, Kazui M, Farid NA, Ikeda T (2008) Comparison of human cytochrome P450 inhibition by the thienopyridines prasugrel, clopidogrel, and ticlopidine. Drug Metab Pharmacokinet 23:412–420

Hanley MJ, Abernethy DR, Greenblatt DJ (2010) Effect of obesity on the pharmacokinetics of drugs in humans. Clin Pharmacokinet 49:71–87

Hansson GK (2005) Inflammation, atherosclerosis, and coronary artery disease. N Engl J Med 352:1685–1695

Harmsze AM, van Werkum JW, ten Berg JM, Hackeng CM, Deneer VH (2009) Concomitant use of sulfonylurea antidiabetics attenuates on-clopidogrel platelet reactivity. Circulation 120:S501 (abstract)

Harmsze A, van Werkum JW, Bouman HJ, Ruven HJ, Breet NJ, ten Berg JM et al (2010) Besides CYP2C19*2, the variant allele CYP2C9*3 is associated with higher on-clopidogrel platelet reactivity in patients on dual antiplatelet therapy undergoing elective coronary stent implantation. Pharmacogenet Genomics 20:18–25

Hasan MS, Basri HB, Hin LP, Stanslas J (2013) Genetic polymorphisms and drug interactions leading to clopidogrel resistance: why the Asian population requires special attention. Int J Neurosci 123:143–154

Hechler B, Zhang Y, Eckly A, Cazenave JP, Gachet C, Ravid K (2003) Lineage-specific overexpression of the P2Y1 receptor induces platelet hyper-reactivity in transgenic mice. J Thromb Haemost 1:155–163

Helton TJ, Bavry AA, Kumbhani DJ, Duggal S, Roukoz H, Bhatt DL (2007) Incremental effect of clopidogrel on important outcomes in patients with cardiovascular disease: a meta-analysis of randomized trials. Am J Cardiovasc Drugs 7:289–297

Herbert JM, Savi P (2003) P2Y12, a new platelet ADP receptor, target of clopidogrel. Semin Vasc Med 3:113–122

Hermosillo AJ, Spinler SA (2008) Aspirin, clopidogrel, and warfarin: is the combination appropriate and effective or inappropriate and too dangerous? Ann Pharmacother 42:790–805

Hetherington SL, Singh RK, Lodwick D, Thompson JR, Goodall AH, Samani NJ (2005) Dimorphism in the P2Y1 ADP receptor gene is associated with increased platelet activation response to ADP. Arterioscler Thromb Vasc Biol 25:252–257

Hirschowitz BI, Hawkey CJ (2001) Questions regarding future research on aspirin and the gastrointestinal tract. Am J Med 110:74S–78S

Hirsh J, Dalen JE, Fuster V, Harker LB, Patrono C, Roth G (1995) Aspirin and other platelet-active drugs. The relationship among dose, effectiveness, and side effects. Chest 108:247S–257S

Ho PM, Maddox TM, Wang L, Fihn SD, Jesse RL, Peterson ED et al (2009) Risk of adverse outcomes associated with concomitant use of clopidogrel and proton pump inhibitors following acute coronary syndrome. JAMA 301:937–944

Hochholzer W, Trenk D, Bestehorn HP, Fischer B, Valina CM, Ferenc M et al (2006) Impact of the degree of peri-interventional platelet inhibition after loading with clopidogrel on early clinical outcome of elective coronary stent placement. J Am Coll Cardiol 48:1742–1750

Hochholzer W, Trenk D, Fromm MF, Valina CM, Stratz C, Bestehorn HP et al (2010) Impact of cytochrome P450 2C19 loss-of-function polymorphism and of major demographic characteristics on residual platelet function after loading and maintenance treatment with clopidogrel in patients undergoing elective coronary stent placement. J Am Coll Cardiol 55:2427–2434

Holinstat M, Voss B, Bilodeau ML, McLaughlin JN, Cleator J, Hamm HE (2006) PAR4, but not PAR1, signals human platelet aggregation via Ca2+ mobilization and synergistic P2Y12 receptor activation. J Biol Chem 281:26665–26674

Hollopeter G, Jantzen HM, Vincent D, Li G, England L, Ramakrishnan V et al (2001) Identification of the platelet ADP receptor targeted by antithrombotic drugs. Nature 409:202–207

Holmes DR Jr, Dehmer GJ, Kaul S, Leifer D, O'Gara PT, Stein CM (2010) ACCF/AHA Clopidogrel Clinical Alert: Approaches to the FDA "Boxed Warning". A Report of the American College of Cardiology Foundation Task Force on Clinical Expert Consensus Documents and the American Heart Association. Circulation 122:537–557

Hulot JS, Bura A, Villard E, Azizi M, Remones V, Goyenvalle C et al (2006) Cytochrome P450 2C19 loss-of-function polymorphism is a major determinant of clopidogrel responsiveness in healthy subjects. Blood 108:2244–2247

Hulot JS, Collet JP, Silvain J, Pena A, Bellemain-Appaix A, Barthelemy O et al (2010) Cardiovascular risk in clopidogrel-treated patients according to cytochrome P450 2C19*2 loss-of-function allele or proton pump inhibitor coadministration: a systematic meta-analysis. J Am Coll Cardiol 56:134–143

Humbert M, Nurden P, Bihour C, Pasquet JM, Winckler J, Heilmann E et al (1996) Ultrastructural studies of platelet aggregates from human subjects receiving clopidogrel and from a patient with an inherited defect of an ADP-dependent pathway of platelet activation. Arterioscler Thromb Vasc Biol 16:1532–1543

Hurbin F, Boulenc X, Daskalakis N, Farenc C, Taylor T, Bonneau D et al (2012) Clopidogrel pharmacodynamics and pharmacokinetics in the fed and fasted state: a randomized crossover study of healthy men. J Clin Pharmacol 52:1506–1515

Iijima R, Ndrepepa G, Mehilli J, Byrne RA, Schulz S, Neumann FJ et al (2009) Profile of bleeding and ischaemic complications with bivalirudin and unfractionated heparin after percutaneous coronary intervention. Eur Heart J 30:290–296

Ishida F, Furihata K, Ishida K, Yan J, Kitano K, Kiyosawa K et al (1995) The largest variant of platelet glycoprotein Ib alpha has four tandem repeats of 13 amino acids in the macroglycopeptide region and a genetic linkage with methionine145. Blood 86:1357–1360

Jacobson AK (2004) Platelet ADP receptor antagonists: ticlopidine and clopidogrel. Best Pract Res Clin Haematol 17:55–64

Jang JS, Cho KI, Jin HY, Seo JS, Yang TH, Kim DK et al (2012) Meta-analysis of cytochrome P450 2C19 polymorphism and risk of adverse clinical outcomes among coronary artery disease patients of different ethnic groups treated with clopidogrel. Am J Cardiol 110:502–508

Jernas M, Olsson B, Arner P, Jacobson P, Sjostrom L, Walley A et al (2009) Regulation of carboxylesterase 1 (CES1) in human adipose tissue. Biochem Biophys Res Commun 383:63–67

Jin RC, Voetsch B, Loscalzo J (2005) Endogenous mechanisms of inhibition of platelet function. Microcirculation 12:247–258

Jinnai T, Horiuchi H, Makiyama T, Tazaki J, Tada T, Akao M et al (2009) Impact of CYP2C19 polymorphisms on the antiplatelet effect of clopidogrel in an actual clinical setting in Japan. Circ J 73:1498–1503

Johnson JA, Roden DM, Lesko LJ, Ashley E, Klein TE, Shuldiner AR (2012) Clopidogrel: a case for indication-specific pharmacogenetics. Clin Pharmacol Ther 91:774–776

Jones CI, Bray S, Garner SF, Stephens J, de Bono B, Angenent WG et al (2009) A functional genomics approach reveals novel quantitative trait loci associated with platelet signaling pathways. Blood 114:1405–1416

Kahn ML, Zheng YW, Huang W, Bigornia V, Zeng D, Moff S et al (1998) A dual thrombin receptor system for platelet activation. Nature 394:690–694

Kazui M, Nishiya Y, Ishizuka T, Hagihara K, Farid NA, Okazaki O et al (2010) Identification of the human cytochrome P450 enzymes involved in the two oxidative steps in the bioactivation of clopidogrel to its pharmacologically active metabolite. Drug Metab Dispos 38:92–99

Kim RB, Leake BF, Choo EF, Dresser GK, Kubba SV, Schwarz UI et al (2001) Identification of functionally variant MDR1 alleles among European Americans and African Americans. Clin Pharmacol Ther 70: 189–199

Kim KA, Park PW, Hong SJ, Park JY (2008a) The effect of CYP2C19 polymorphism on the pharmacokinetics and pharmacodynamics of clopidogrel: a possible mechanism for clopidogrel resistance. Clin Pharmacol Ther 84:236–242

Kim KA, Park PW, Park JY (2008b) Effect of CYP3A5*3 genotype on the pharmacokinetics and antiplatelet effect of clopidogrel in healthy subjects. Eur J Clin Pharmacol 64:589–597

Kimura M, Ieiri I, Wada Y, Mamiya K, Urae A, Iimori E et al (1999) Reliability of the omeprazole hydroxylation index for CYP2C19 phenotyping: possible effect of age, liver disease and length of therapy. Br J Clin Pharmacol 47:115–119

King SB III, Smith SC Jr, Hirshfeld JW Jr, Jacobs AK, Morrison DA, Williams DO et al (2008) 2007 Focused Update of the ACC/AHA/SCAI 2005 Guideline Update for Percutaneous Coronary Intervention: a report of the American College of Cardiology/ American Heart Association Task Force on Practice Guidelines: 2007 Writing Group to Review New Evidence and Update the ACC/AHA/SCAI 2005 Guideline Update for Percutaneous Coronary Intervention, Writing on Behalf of the 2005 Writing Committee. Circulation 117:261–295

Kinnaird TD, Stabile E, Mintz GS, Lee CW, Canos DA, Gevorkian N et al (2003) Incidence, predictors, and prognostic implications of bleeding and blood transfusion following percutaneous coronary interventions. Am J Cardiol 92:930–935

Kotlyar M, Carson SW (1999) Effects of obesity on the cytochrome P450 enzyme system. Int J Clin Pharmacol Ther 37:8–19

Kuijpers RW, Faber NM, Cuypers HT, Ouwehand WH, dem Borne AE (1992) NH2-terminal globular domain of human platelet glycoprotein Ib alpha has a methionine 145/threonine145 amino acid polymorphism, which is associated with the HPA-2 (Ko) alloantigens. J Clin Invest 89:381–384

Kulkarni S, Dopheide SM, Yap CL, Ravanat C, Freund M, Mangin P et al (2000) A revised model of platelet aggregation. J Clin Invest 105:783–791

Kunicki TJ, Kritzik M, Annis DS, Nugent DJ (1997) Hereditary variation in platelet integrin alpha 2 beta 1 density is associated with two silent polymorphisms in the alpha 2 gene coding sequence. Blood 89: 1939–1943

Latry P, Martin-Latry K, Lafitte M, Peter C, Couffinhal T (2012) Dual antiplatelet therapy after myocardial infarction and percutaneous coronary intervention: analysis of patient adherence using a French health insurance reimbursement database. EuroIntervention 7:1413–1419

Lau WC, Waskell LA, Watkins PB, Neer CJ, Horowitz K, Hopp AS et al (2003) Atorvastatin reduces the ability of clopidogrel to inhibit platelet aggregation: a new drug-drug interaction. Circulation 107:32–37

Lau WC, Welch TD, Shields T, Rubenfire M, Tantry US, Gurbel PA (2011) The effect of St. John's Wort on the pharmacodynamic response of clopidogrel in hyporesponsive volunteers and patients: Increased platelet inhibition by enhancement of CYP3A4 metabolic activity. J Cardiovasc Pharmacol 57:86–93

Lee JM, Park S, Shin DJ, Choi D, Shim CY, Ko YG et al (2009) Relation of genetic polymorphisms in the cytochrome P450 gene with clopidogrel resistance after drug-eluting stent implantation in Koreans. Am J Cardiol 104:46–51

Lee SJ, Kwon JA, Cho SA, Jarrar YB, Shin JG (2012) Effects of testosterone and 17beta-oestradiol on expression of the G protein-coupled receptor P2Y12 in megakaryocytic DAMI cells. Platelets 23:579–585

Lei HP, Yu XY, Xie HT, Li HH, Fan L, Dai LL et al (2010) Effect of St. John's wort supplementation on the pharmacokinetics of bupropion in healthy male Chinese volunteers. Xenobiotica 40:275–281

Lenain N, Freund M, Leon C, Cazenave JP, Gachet C (2003) Inhibition of localized thrombosis in P2Y1-deficient mice and rodents treated with MRS2179, a P2Y1 receptor antagonist. J Thromb Haemost 1:1144–1149

Lev EI, Patel RT, Guthikonda S, Lopez D, Bray PF, Kleiman NS (2007) Genetic polymorphisms of the platelet receptors P2Y(12), P2Y(1) and GP IIIa and response to aspirin and clopidogrel. Thromb Res 119:355–360

Lewis JP, Horenstein RB, Ryan K, O'Connell JR, Gibson Q, Mitchell BD et al (2013) The functional G143E variant of carboxylesterase 1 is associated with increased clopidogrel active metabolite levels and greater clopidogrel response. Pharmacogenet Genomics 23:1–8

Li XL, Cao J, Fan L, Ye L, Wang Q, Cui CP et al (2012a) Correlation analysis of aspirin resistance and cyclooxygenase-1 haplotype in old Chinese patients with cardio-cerebrovascular diseases. Chin J Ying Yang Physiol 28:225–229 (in Chinese)

Li Y, Tang HL, Hu YF, Xie HG (2012b) The gain-of-function variant allele CYP2C19*17: a double-edged sword between thrombosis and bleeding in clopidogrel-treated patients. J Thromb Haemost 10:199–206

Lin SY, Cui HB, Chen XM, Wang SH, Zhou HL, Du WP et al (2012) Clinical application of VerifyNow-P2Y12 assay in evaluation of platelet inhibition efficacy of clopidogrel. Chin J Cardiovasc Dis 40:662–666 (in Chinese)

Lins R, Broekhuysen J, Necciari J, Deroubaix X (1999) Pharmacokinetic profile of 14C-labeled clopidogrel. Semin Thromb Hemost 25(Suppl 2):29–33

Liu YT, Hao HP, Liu CX, Wang GJ, Xie HG (2007) Drugs as CYP3A probes, inducers, and inhibitors. Drug Metab Rev 39:699–721

Liu XL, Wang ZJ, Yang Q, Ge HL, Gao F, Liu YY et al (2010) Impact of CYP2C19 polymorphism and smoking on response to clopidogrel in patients with stable coronary artery disease. Chin Med J 123:3178–3183

Longo DL (2012) Tumor heterogeneity and personalized medicine. N Engl J Med 366:956–957

Lopez JA, Ludwig EH, McCarthy BJ (1992) Polymorphism of human glycoprotein Ib alpha results from a variable number of tandem repeats of a 13-amino acid sequence in the mucin-like macroglycopeptide region. Structure/function implications. J Biol Chem 267:10055–10061

Lown KS, Bailey DG, Fontana RJ, Janardan SK, Adair CH, Fortlage LA et al (1997) Grapefruit juice increases felodipine oral availability in humans by decreasing intestinal CYP3A protein expression. J Clin Invest 99:2545–2553

Luchessi AD, Silbiger VN, Cerda A, Hirata RD, Carracedo A, Brion M et al (2012) Increased clopidogrel response is associated with ABCC3 expression: a pilot study. Clin Chim Acta 413:417–421

Malek LA, Kisiel B, Spiewak M, Grabowski M, Filipiak KJ, Kostrzewa G et al (2008) Coexisting polymorphisms of P2Y12 and CYP2C19 genes as a risk factor for persistent platelet activation with clopidogrel. Circ J 72:1165–1169

Mani H, Toennes SW, Linnemann B, Urbanek DA, Schwonberg J, Kauert GF et al (2008) Determination of clopidogrel main metabolite in plasma: a useful tool for monitoring therapy? Ther Drug Monit 30:84–89

Mao Z, Li Y, Peng Y, Luan X, Gui H, Feng X et al (2011) Lipopolysaccharide down-regulates carbolesterases 1 and 2 and reduces hydrolysis activity in vitro and in vivo via p38MAPK-NF-kappaB pathway. Toxicol Lett 201:213–220

Marcus AJ, Safier LB, Hajjar KA, Ullman HL, Islam N, Broekman MJ et al (1991) Inhibition of platelet function by an aspirin-insensitive endothelial cell ADPase. Thromboregulation by endothelial cells. J Clin Invest 88:1690–1696

Marcus AJ, Broekman MJ, Drosopoulos JH, Islam N, Alyonycheva TN, Safier LB et al (1997) The endothelial cell ecto-ADPase responsible for inhibition of platelet function is CD39. J Clin Invest 99:1351–1360

Marcus AJ, Broekman MJ, Drosopoulos JH, Islam N, Pinsky DJ, Sesti C et al (2003) Metabolic control of excessive extracellular nucleotide accumulation by CD39/ecto-nucleotidase-1: implications for ischemic vascular diseases. J Pharmacol Exp Ther 305:9–16

Maree AO, Curtin RJ, Dooley M, Conroy RM, Crean P, Cox D et al (2005) Platelet response to low-dose enteric-coated aspirin in patients with stable cardiovascular disease. J Am Coll Cardiol 46:1258–1263

Marzolini C, Paus E, Buclin T, Kim RB (2004) Polymorphisms in human MDR1 (P-glycoprotein): recent advances and clinical relevance. Clin Pharmacol Ther 75:13–33

Maseneni S, Donzelli M, Taegtmeyer AB, Brecht K, Krahenbuhl S (2012) Toxicology of clopidogrel and ticlopidine on human myeloid progenitor cells; Importance of metabolites. Toxicology 299:139–145

Matetzky S, Shenkman B, Guetta V, Shechter M, Bienart R, Goldenberg I et al (2004) Clopidogrel resistance is associated with increased risk of recurrent atherothrombotic events in patients with acute myocardial infarction. Circulation 109:3171–3175

Mega JL, Close SL, Wiviott SD, Shen L, Hockett RD, Brandt JT et al (2009a) Cytochrome p-450 polymorphisms and response to clopidogrel. N Engl J Med 360:354–362

Mega JL, Close SL, Wiviott SD, Shen L, Hockett RD, Brandt JT et al (2009b) Cytochrome P450 genetic polymorphisms and the response to prasugrel: relationship to pharmacokinetic, pharmacodynamic, and clinical outcomes. Circulation 119:2553–2560

Mega JL, Close SL, Wiviott SD, Shen L, Walker JR, Simon T et al (2010) Genetic variants in ABCB1 and CYP2C19 and cardiovascular outcomes after treatment with clopidogrel and prasugrel in the TRITON-TIMI 38 trial: a pharmacogenetic analysis. Lancet 376: 1312–1319

Mehta SR, Yusuf S, Peters RJ, Bertrand ME, Lewis BS, Natarajan MK et al (2001) Effects of pretreatment with clopidogrel and aspirin followed by long-term therapy in patients undergoing percutaneous coronary intervention: the PCI-CURE study. Lancet 358:527–533

Mehta SR, Tanguay JF, Eikelboom JW, Jolly SS, Joyner CD, Granger CB et al (2010) Double-dose versus standard-dose clopidogrel and high-dose versus low-dose aspirin in individuals undergoing percutaneous coronary intervention for acute coronary syndromes (CURRENT-OASIS 7): a randomised factorial trial. Lancet 376:1233–1243

Michelson AD, Linden MD, Furman MI, Li Y, Barnard MR, Fox ML et al (2007) Evidence that pre-existent variability in platelet response to ADP accounts for 'clopidogrel resistance'. J Thromb Haemost 5:75–81

Mobley JE, Bresee SJ, Wortham DC, Craft RM, Snider CC, Carroll RC (2004) Frequency of nonresponse antiplatelet activity of clopidogrel during pretreatment for cardiac catheterization. Am J Cardiol 93:456–458

Mockel M, Muller R, Vollert JO, Muller C, Danne O, Gareis R et al (2007) Lipoprotein-associated phospholipase A2 for early risk stratification in patients with suspected acute coronary syndrome: a multi-marker approach: the North Wuerttemberg and Berlin Infarction Study-II (NOBIS-II). Clin Res Cardiol 96:604–612

Monton M, Jimenez A, Nunez A, Lopez-Blaya A, Farre J, Gomez J et al (2000) Comparative effects of angiotensin II AT-1-type receptor antagonists in vivo on human platelet activation. J Cardiovasc Pharmacol 35:906–913

Mullangi R, Srinivas NR (2009) Clopidogrel: review of bioanalytical methods, pharmacokinetics/pharmacodynamics, and update on recent trends in drug-drug interaction studies. Biomed Chromatogr 23:26–41

Muller I, Besta F, Schulz C, Massberg S, Schonig A, Gawaz M (2003) Prevalence of clopidogrel non-responders among patients with stable angina pectoris scheduled for elective coronary stent placement. Thromb Haemost 89:783–787

Muller K, Aichele S, Herkommer M, Bigalke B, Stellos K, Htun P et al (2010) Impact of inflammatory markers on platelet inhibition and cardiovascular outcome including stent thrombosis in patients with symptomatic coronary artery disease. Atherosclerosis 213:256–262

Muller C, Caillard S, Jesel L, El Ghannudi S, Ohlmann P, Sauleau E et al (2012) Association of estimated GFR with platelet inhibition in patients treated with clopidogrel. Am J Kidney Dis 59:777–785

Nguyen TA, Diodati JG, Pharand C (2005) Resistance to clopidogrel: a review of the evidence. J Am Coll Cardiol 45:1157–1164

Nishiya Y, Hagihara K, Ito T, Tajima M, Miura S, Kurihara A et al (2009) Mechanism-based inhibition of human cytochrome P450 2B6 by ticlopidine, clopidogrel, and the thiolactone metabolite of prasugrel. Drug Metab Dispos 37:589–593

Niu H, Chen X, Gruppo RA, Li D, Wang Y, Zhang L et al (2012) Integrin αIIb-mediated PI3k/Akt activation in platelets. PLoS One 7:e47356

O'Donnell CJ, Larson MG, Feng D, Sutherland PA, Lindpaintner K, Myers RH et al (2001) Genetic and environmental contributions to platelet aggregation: the Framingham heart study. Circulation 103: 3051–3056

O'Donoghue M, Wiviott SD (2006) Clopidogrel response variability and future therapies: clopidogrel: does one size fit all? Circulation 114:e600–e606

Ohmori T, Yano Y, Sakata A, Ikemoto T, Shimpo M, Madoiwa S et al (2012) Lack of association between serum paraoxonase-1 activity and residual platelet aggregation during dual anti-platelet therapy. Thromb Res 129:e36–e40

Ostadal P, Ostadal B (2012) Women and the management of acute coronary syndrome. Can J Physiol Pharmacol 90:1151–1159

Ozdemir V, Graham JE, Godard B (2008) Race as a variable in pharmacogenomics science: from empirical ethics to publication standards. Pharmacogenet Genomics 18:837–841

Pare G, Mehta SR, Yusuf S, Anand SS, Connolly SJ, Hirsh J et al (2010) Effects of CYP2C19 genotype on outcomes of clopidogrel treatment. N Engl J Med 363:1704–1714

Park KW, Park JJ, Lee SP, Oh IY, Suh JW, Yang HM et al (2011) Cilostazol attenuates on-treatment platelet reactivity in patients with CYP2C19 loss of function alleles receiving dual antiplatelet therapy: a genetic substudy of the CILON-T randomised controlled trial. Heart 97:641–647

Park KW, Kang SH, Kang J, Jeon KH, Park JJ, Han JK et al (2012) Enhanced clopidogrel response in smokers is reversed after discontinuation as assessed by VerifyNow assay: additional evidence for the concept of 'smokers' paradox'. Heart 98:1000–1006

Patrono C, Bachmann F, Baigent C, Bode C, De Caterina R, Charbonnier B et al (2004a) Expert consensus document on the use of antiplatelet agents. The task force on the use of antiplatelet agents in patients with atherosclerotic cardiovascular disease of the European society of cardiology. Eur Heart J 25:166–181

Patrono C, Coller B, FitzGerald GA, Hirsh J, Roth G (2004b) Platelet-active drugs: the relationships among dose, effectiveness, and side effects: the Seventh ACCP Conference on Antithrombotic and Thrombolytic Therapy. Chest 126:234S–264S

Patrono C, Garcia Rodriguez LA, Landolfi R, Baigent C (2005) Low-dose aspirin for the prevention of atherothrombosis. N Engl J Med 353:2373–2383

Peace A, McCall M, Tedesco T, Kenny D, Conroy RM, Foley D et al (2010) The role of weight and enteric coating on aspirin response in cardiovascular patients. J Thromb Haemost 8:2323–2325

Pena A, Collet JP, Hulot JS, Silvain J, Barthelemy O, Beygui F et al (2009) Can we override clopidogrel resistance? Circulation 119:2854–2857

Pereillo JM, Maftouh M, Andrieu A, Uzabiaga MF, Fedeli O, Savi P et al (2002) Structure and stereochemistry of the active metabolite of clopidogrel. Drug Metab Dispos 30:1288–1295

Peters RJ, Mehta SR, Fox KA, Zhao F, Lewis BS, Kopecky SL et al (2003) Effects of aspirin dose when used alone or in combination with clopidogrel in patients with acute coronary syndromes: observations from the Clopidogrel in Unstable angina to prevent Recurrent Events (CURE) study. Circulation 108:1682–1687

Petersen JL, Dery JP, Fischi MC, Hernandez AF, Hranitzky PM, Rao SV et al (2003) Highlights from the American College of Cardiology Annual Scientific Sessions 2003: March 28 to April 2, 2003. Am Heart J 146:19–26

Piazuelo E, Fuentes J, Garcia-Gonzalez MA, Jimenez P, Lanas A (2008) A case–control study of the association between polymorphisms of the endothelial nitric oxide synthase and glycoprotein IIIa genes and upper gastrointestinal bleeding in users of low-dose aspirin. Clin Ther 30:121–130

Pignatelli P, De Biase L, Lenti L, Tocci G, Brunelli A, Cangemi R et al (2005) Tumor necrosis factor-alpha as trigger of platelet activation in patients with heart failure. Blood 106:1992–1994

Pirmohamed M, James S, Meakin S, Green C, Scott AK, Walley TJ et al (2004) Adverse drug reactions as cause of admission to hospital: prospective analysis of 18 820 patients. BMJ 329:15–19

Price MJ, Endemann S, Gollapudi RR, Valencia R, Stinis CT, Levisay JP et al (2008) Prognostic significance of post-clopidogrel platelet reactivity assessed by a point-of-care assay on thrombotic events after drug-eluting stent implantation. Eur Heart J 29:992–1000

Price MJ, Berger PB, Teirstein PS, Tanguay JF, Angiolillo DJ, Spriggs D et al (2011) Standard- vs high-dose clopidogrel based on platelet function testing after percutaneous coronary intervention: the GRAVITAS randomized trial. JAMA 305:1097–1105

Pulley JM, Denny JC, Peterson JF, Bernard GR, Vnencak-Jones CL, Ramirez AH et al (2012) Operational implementation of prospective genotyping for personalized medicine: the design of the Vanderbilt PREDICT project. Clin Pharmacol Ther 92:87–95

Quinn MJ, Fitzgerald DJ (1999) Ticlopidine and clopidogrel. Circulation 100:1667–1672

Rahimi R, Abdollahi M (2012) An update on the ability of St.John's wort to affect the metabolism of other drugs. Expert Opin Drug Metab Toxicol 8:691–708

Ramsjo M, Aklillu E, Bohman L, Ingelman-Sundberg M, Roh HK, Bertilsson L (2010) CYP2C19 activity comparison between Swedes and Koreans: effect of genotype, sex, oral contraceptive use, and smoking. Eur J Clin Pharmacol 66:871–877

Ray WA, Murray KT, Griffin MR, Chung CP, Smalley WE, Hall K et al (2010) Outcomes with concurrent use of clopidogrel and proton-pump inhibitors: a cohort study. Ann Intern Med 152:337–345

Reist M, Roy-de Vos M, Montseny JP, Mayer JM, Carrupt PA, Berger Y et al (2000) Very slow chiral inversion of clopidogrel in rats: a pharmacokinetic and mechanistic investigation. Drug Metab Dispos 28:1405–1410

Richter T, Murdter TE, Heinkele G, Pleiss J, Tatzel S, Schwab M et al (2004) Potent mechanism-based inhibition of human CYP2B6 by clopidogrel and ticlopidine. J Pharmacol Exp Ther 308:189–197

Rivera J, Lozano ML, Navarro-Nunez L, Vicente V (2009) Platelet receptor and signaling in the dynamics of thrombus formation. Haematologica 94:700–711

Roberts JD, Wells GA, Le May MR, Labinaz M, Glover C, Froeschl M et al (2012) Point-of-care genetic testing for personalisation of antiplatelet treatment (RAPID GENE): a prospective, randomised, proof-of-concept trial. Lancet 379:1705–1711

Roden DM, Shuldiner AR (2010) Responding to the Clopidogrel Warning by the US Food and Drug Administration. Real life is complicated. Circulation 122:445–448

Roden DM, Stein CM (2009) Clopidogrel and the concept of high-risk pharmacokinetics. Circulation 119: 2127–2130

Ruiz ML, Rigalli JP, Arias A, Villanueva S, Banchio C, Vore M et al (2013) Induction of hepatic multidrug resistance associated protein 3 by ethynylestradiol in independent of cholestasis and mediated by estrogen receptor. Drug Metab Dispos 41:275–280

Sabatine MS, Cannon CP, Gibson CM, Lopez-Sendon JL, Montalescot G, Theroux P et al (2005) Addition of clopidogrel to aspirin and fibrinolytic therapy for myocardial infarction with ST-segment elevation. N Engl J Med 352:1179–1189

Sachdeva A, Bavisetty S, Beckham G, Shen AY, Aharonian V, Mansulonani P et al (2012) Discontinuation of long-term clopidogrel therapy is associated with death and myocardial infarction after saphenous vein graft percutaneous coronary intervention. J Am Coll Cardiol 60:2357–2363

Sadler JE (2010) Redeeming ristocetin. Blood 116:155–156

Sai K, Saito Y, Tatewaki N, Hosokawa M, Kaniwa N, Nishimaki-Mogami T et al (2010) Association of carboxylesterase 1A genotypes with irinotecan pharmacokinetics in Japanese cancer patients. Br J Clin Pharmacol 70:222–233

Sambu N, Radhakrishnan A, Dent H, Calver AL, Corbett S, Gray H et al (2012) Personalised antiplatelet therapy in stent thrombosis: observations from the Clopidogrel Resistance in Stent Thrombosis (CREST) registry. Heart 98:706–711

Sangkuhl K, Klein TE, Altman RB (2010) Clopidogrel pathway. Pharmacogenet Genomics 20:463–465

Sato Y, Miyashita A, Iwatsubo T, Usui T (2012a) Conclusive identification of the oxybutynin-hydrolyzing enzyme in human liver. Drug Metab Dispos 40:902–906

Sato Y, Miyashita A, Iwatsubo T, Usui T (2012b) Simultaneous absolute protein quantification of carboxylesterases 1 and 2 in human liver tissue fractions using liquid chromatography-tandem mass spectrometry. Drug Metab Dispos 40:1389–1396

Savi P, Herbert JM (2005) Clopidogrel and ticlopidine: P2Y12 adenosine diphosphate-receptor antagonists for the prevention of atherothrombosis. Semin Thromb Hemost 31:174–183

Savi P, Herbert JM, Pflieger AM, Dol F, Delebassee D, Combalbert J et al (1992) Importance of hepatic metabolism in the antiaggregating activity of the thienopyridine clopidogrel. Biochem Pharmacol 44:527–532

Savi P, Combalbert J, Gaich C, Rouchon MC, Maffrand JP, Berger Y et al (1994a) The antiaggregating activity of clopidogrel is due to a metabolic activation by the hepatic cytochrome P450-1A. Thromb Haemost 72:313–317

Savi P, Laplace MC, Maffrand JP, Herbert JM (1994b) Binding of [3H]-2-methylthio ADP to rat platelets–effect of clopidogrel and ticlopidine. J Pharmacol Exp Ther 269:772–777

Savi P, Pereillo JM, Uzabiaga MF, Combalbert J, Picard C, Maffrand JP et al (2000) Identification and biological activity of the active metabolite of clopidogrel. Thromb Haemost 84:891–896

Saw J, Brennan DM, Steinhubl SR, Bhatt DL, Mak KH, Fox K et al (2007) Lack of evidence of a clopidogrel-statin interaction in the CHARISMA trial. J Am Coll Cardiol 50:291–295

Schaeffeler E, Eichelbaum M, Brinkmann U, Penger A, Asante-Poku S, Zanger UM et al (2001) Frequency of C3435T polymorphism of MDR1 gene in African people. Lancet 358:383–384

Schneider DJ (2011) Anti-platelet therapy: glycoprotein IIb-IIIa antagonists. Br J Clin Pharmacol 72:672–682

Schrenk D, Brockmeier D, Morike K, Bock KW, Eichelbaum M (1998) A distribution study of CYP1A2 phenotypes among smokers and non-smokers in a cohort of healthy Caucasian volunteers. Eur J Clin Pharmacol 53:361–367

Schroeder WS, Ghobrial L, Gandhi PJ (2006) Possible mechanisms of drug-induced aspirin and clopidogrel resistance. J Thromb Thrombolysis 22:139–150

Schwemmer M, Sommer O, Bassenge E (2001) Angiotensin receptor blocker losartan suppresses platelet activity by interfering with thromboxane signaling. Cardiovasc Drugs Ther 15:301–307

Scott SA, Sangkuhl K, Gardner EE, Stein CM, Hulot JS, Johnson JA et al (2011) Clinical pharmacogenomics of implementation consortium guidelines for cytochrome P450-2c19 (CYP2C19) genotype and clopidogrel therapy. Clin Pharmacol Ther 90:328–332

Serebruany VL, Oshrine BR, Malinin AI, Atar D, Michelson AD, Ferguson JJ III (2005) Noncompliance in cardiovascular clinical trials. Am Heart J 150:882–886

Shahabi P, Siest G, Visvikis-Siest S (2012) Clinical interest of point-of-care pharmacogenomic testing: clopidogrel behind warfarin. Pharmacogenomics 13:1215–1218

Shankar H, Garcia A, Prabhaker J, Kim S, Kunapuli SP (2006) P2Y12 receptor-mediated potentiation of thrombin-induced thromboxane A2 generation in platelets occurs through regulation of Erk1/2 activation. J Thromb Haemost 4:638–647

Sharis PJ, Cannon CP, Loscalzo J (1998) The antiplatelet effects of ticlopidine and clopidogrel. Ann Intern Med 129:394–405

Shimasaki Y, Yasue H, Yoshimura M, Nakayama M, Kugiyama K, Ogawa H et al (1998) Association of the missense Glu298Asp variant of the endothelial nitric oxide synthase gene with myocardial infarction. J Am Coll Cardiol 31:1506–1510

Shuldiner AR, O'Connell JR, Bliden KP, Gandhi A, Ryan K, Horenstein RB et al (2009) Association of cytochrome P450 2C19 genotype with the antiplatelet effect and clinical efficacy of clopidogrel therapy. JAMA 302:849–857

Sibbing D, von Beckerath O, Schomig A, Kastrati A, von Beckerath N (2006) P2Y1 gene A1622G dimorphism is not associated with adenosine diphosphate-induced platelet activation and aggregation after administration of a single high dose of clopidogrel. J Thromb Haemost 4:912–914

Sibbing D, Stegherr J, Latz W, Koch W, Mehilli J, Dorrler K et al (2009) Cytochrome P450 2C19 loss-of-function polymorphism and stent thrombosis following percutaneous coronary intervention. Eur Heart J 30:916–922

Sibbing D, Koch W, Gebhard D, Schuster T, Braun S, Stegherr J et al (2010) Cytochrome 2C19*17 allelic variant, platelet aggregation, bleeding events, and stent thrombosis in clopidogrel-treated patients with coronary stent placement. Circulation 121:512–518

Sibbing D, Bernlochner I, Schulz S, Massberg S, Schomig A, Mehilli J et al (2012) The impact of smoking on the antiplatelet action of clopidogrel in non-ST-elevation myocardial infarction patients: results from the ISAR-REACT 4 platelet substudy. J Thromb Haemost 10:2199–2202

Siller-Matula JM, Lang I, Christ G, Jilma B (2008) Calcium-channel blockers reduce the antiplatelet effect of clopidogrel. J Am Coll Cardiol 52:1557–1563

Siller-Matula JM, Delle-Karth G, Christ G, Neunteufl T, Maurer G, Huber K et al (2013) Dual non-responsiveness to antiplatelet treatment is a stronger predictor of cardiac adverse events than isolated non-responsiveness to clopidogrel or aspirin. Int J Cardiol 167:430–435

Simon T, Verstuyft C, Mary-Krause M, Quteineh L, Drouet E, Meneveau N et al (2009) Genetic determinants of response to clopidogrel and cardiovascular events. N Engl J Med 360:363–375

Small DS, Farid NA, Li YG, Ernest CS, Payne CD, Salazar DE et al (2008) Effect of ranitidine on the pharmacokinetics and pharmacodynamics of prasugrel and clopidogrel. Curr Med Res Opin 24:2251–2257

Smith SC Jr, Allen J, Blair SN, Bonow RO, Brass LM, Fonarow GC et al (2006a) AHA/ACC guidelines for secondary prevention for patients with coronary and other atherosclerotic vascular disease: 2006 update: endorsed by the National Heart, Lung, and Blood Institute. Circulation 113:2363–2372

Smith SM, Judge HM, Peters G, Armstrong M, Fontana P, Gaussem P et al (2006b) Common sequence variations in the P2Y12 and CYP3A5 genes do not explain the variability in the inhibitory effects of clopidogrel therapy. Platelets 17:250–258

Snoep JD, Hovens MM, Eikenboom JC, van der Bom JG, Jukema JW, Huisman MV (2007) Clopidogrel nonresponsiveness in patients undergoing percutaneous coronary intervention with stenting: a systematic review and meta-analysis. Am Heart J 154:221–231

Sofi F, Giusti B, Marcucci R, Gori AM, Abbate R, Gensini GF (2011) Cytochrome P450 2C19*2 polymorphism and cardiovascular recurrences in patients taking clopidogrel: a meta-analysis. Pharmacogenomics J 11:199–206

Steering Committee CAPRIE (1996) A randomized, blinded, trial of clopidogrel versus aspirin in patients at risk for ischemic events (CAPRIE). Lancet 348:1329–1339

Steinhubl SR, Berger PB, Mann JT III, Fry ET, DeLago A, Wilmer C et al (2002) Early and sustained dual oral antiplatelet therapy following percutaneous coronary intervention: a randomized controlled trial. JAMA 288:2411–2420

Storey RF (2009) Clopidogrel in acute coronary syndrome: to genotype or not? Lancet 373:276–278

Sugidachi A, Ogawa T, Kurihara A, Hagihara K, Jakubowski JA, Hashimoto M et al (2007) The greater in vivo antiplatelet effects of prasugrel as compared to clopidogrel reflect more efficient generation of its active metabolite with similar antiplatelet activity to that of clopidogrel's active metabolite. J Thromb Haemost 5:1545–1551

Suh JW, Koo BK, Zhang SY, Park KW, Cho JY, Jang IJ et al (2006) Increased risk of atherothrombotic events associated with cytochrome P450 3A5 polymorphism in patients taking clopidogrel. CMAJ 174:1715–1722

Sun YJ, Li YZ, Jiang DM, Zhang B, Gao Y, Yu HJ et al (2012) Effects of smoke on clinical prognosis of patients with acute ST-segment elevation myocardial infarction. Natl Med J China 92:1963–1966 (in Chinese)

Taubert D, Kastrati A, Harlfinger S, Gorchakova O, Lazar A, von Beckerath N et al (2004) Pharmacokinetics of clopidogrel after administration of a high loading dose. Thromb Haemost 92:311–316

Taubert D, von Beckerath N, Grimberg G, Lazar A, Jung N, Goeser T et al (2006) Impact of P-glycoprotein on clopidogrel absorption. Clin Pharmacol Ther 80:486–501

Tello-Montoliu A, Ferreiro JL, Kodali MK, Ueno M, Tomasello SD, Rollini F et al (2013) Impact of renal function on clopidogrel-induced antiplatelet effects in coronary artery disease patients without diabetes mellitus. J Thromb Thromb 36:14–17

The CURE Trial Investigators (2001) Effects of clopidogrel in addition to aspirin in patients with acute coronary syndrome without ST-elevation. N Engl J Med 345:494–502

Topol EJ, Easton D, Harrington RA, Amarenco P, Califf RM, Graffagnino C et al (2003) Randomized, double-blind, placebo-controlled, international trial of the oral IIb/IIIa antagonist lotrafiban in coronary and cerebrovascular disease. Circulation 108:399–406

Trenk D, Hochholzer W, Fromm MF, Chialda LE, Pahl A, Valina CM et al (2008) Cytochrome P450 2C19 681G>A polymorphism and high on-clopidogrel platelet reactivity associated with adverse 1-year clinical outcome of elective percutaneous coronary intervention with drug-eluting or bare-metal stents. J Am Coll Cardiol 51:1925–1934

Turpeinen M, Tolonen A, Uusitalo J, Jalonen J, Pelkonen O, Laine K (2005) Effect of clopidogrel and ticlopidine on cytochrome P450 2B6 activity as measured by bupropion hydroxylation. Clin Pharmacol Ther 77:553–559

Ueno M, Ferreiro JL, Desai B, Tomasello SD, Tello-Montoliu A, Capodanno D et al (2012) Cigarette smoking is associated with a dose–response effect in clopidogrel-treated patients with diabetes mellitus and coronary artery disease: results of a pharmacodynamic study. JACC Cardiovasc Interv 5:293–300

Umemura K, Furuta T, Kondo K (2008) The common gene variants of CYP2C19 affect pharmacokinetics and pharmacodynamics in an active metabolite of clopidogrel in healthy subjects. J Thromb Haemost 6:1439–1441

Valgimigli M, Minarelli M (2011) Triple antiplatelet therapy in acute coronary syndromes. Drugs 71:1703–1719

Verschuren JJ, Boden H, Wessels JA, van der Hoeven BL, Trompet S, Heijmans BT et al (2012) Value of platelet pharmacogenetics in common clinical practice of patients with ST-segment elevation myocardial infarction. Int J Cardiol (in press)

Vistoli G, Pedretti A, Mazzolari A, Bolchi C, Testa B (2009) Influence of ionization state on the activation of temocapril by hCES1: a molecular-dynamics study. Chem Biodivers 6:2092–2100

von Beckerath N, Taubert D, Pogatsa-Murray G, Schomig E, Kastrati A, Schomig A (2005a) Absorption, metabolization, and antiplatelet effects of 300-, 600-, and 900-mg loading doses of clopidogrel: results of the ISAR-CHOICE (Intracoronary Stenting and Antithrombotic Regimen: Choose Between 3 High Oral Doses for Immediate Clopidogrel Effect) Trial. Circulation 112:2946–2950

von Beckerath N, von Beckerath O, Koch W, Eichinger M, Schomig A, Kastrati A (2005b) P2Y12 gene H2 haplotype is not associated with increased adenosine diphosphate-induced platelet aggregation after initiation of clopidogrel therapy with a high loading dose. Blood Coagul Fibrinolysis 16:199–204

Wallentin L, Varenhorst C, James S, Erlinge D, Braun OO, Jakubowski JA et al (2008) Prasugrel achieves greater and faster P2Y12receptor-mediated platelet inhibition than clopidogrel due to more efficient generation of its active metabolite in aspirin-treated patients with coronary artery disease. Eur Heart J 29:21–30

Wallentin L, James S, Storey RF, Armstrong M, Barratt BJ, Horrow J et al (2010) Effect of CYP2C19 and ABCB1 single nucleotide polymorphisms on outcomes of treatment with ticagrelor versus clopidogrel for acute coronary syndromes: a genetic substudy of the PLATO trial. Lancet 376:1320–1328

Wang GJ, Yang P, Xie HG (2006a) Gene variants in noncoding regions and their possible consequences. Pharmacogenomics 7:203–209

Wang TH, Bhatt DL, Topol EJ (2006b) Aspirin and clopidogrel resistance: an emerging clinical entity. Eur Heart J 27:647–654

Wenaweser P, Dorffler-Melly J, Imboden K, Windecker S, Togni M, Meier B et al (2005) Stent thrombosis is associated with an impaired response to antiplatelet therapy. J Am Coll Cardiol 45:1748–1752

Wijns W, Kolh P, Danchin N, Di Mario C, Falk V, Folliguet T et al (2010) Guidelines on myocardial revascularization. Eur Heart J 31:2501–2555

Wiviott SD, Braunwald E, McCabe CH, Montalescot G, Ruzyllo W, Gottlieb S et al (2007) Prasugrel versus clopidogrel in patients with acute coronary syndromes. N Engl J Med 357:2001–2015

Wolbold R, Klein K, Burk O, Nussler AK, Neuhaus P, Eichelbaum M et al (2003) Sex is a major determinant of CYP3A4 expression in human liver. Hepatology 38:978–988

Woo JS, Kim W, Lee SR, Jung KH, Kim WS, Lew JH et al (2011) Platelet reactivity in patients with chronic kidney disease receiving adjunctive cilostazol compared with a high-maintenance dose of clopidogrel: results of the effect of platelet inhibition according to clopidogrel dose in patients with chronic kidney disease (PIANO-2 CKD) randomized study. Am Heart J 162:1018–1025

Wurtz M, Kristensen SD, Hvas AM, Grove EL (2012) Pharmacogenetics of the antiplatelet effect of aspirin. Curr Pharm Des 18:5294–5308

Xiao D, Chen YT, Yang D, Yan B (2012) Age-related inducibility of carboxylesterases by the antiepileptic agent phenobarbital and implications in drug metabolism and lipid accumulation. Biochem Pharmacol 84:232–239

Xie HG (1997) Direct evidence for the higher frequency of CYP2C19 allelic heterozygotes in Chinese subjects than in white subjects. Clin Pharmacol Ther 62:691–692

Xie HG (2000) Genetic variations of S-mephenytoin 4′-hydroxylase (CYP2C19) in the Chinese population. Life Sci 66:L175–L181

Xie HG (2010) Personalized immunosuppressive therapy in pediatric heart transplantation: progress, pitfalls and promise. Pharmacol Ther 126:146–158

Xie HG (2011) Chapter 52. Genetically polymorphic cytochrome P450s and transporters and personalized antimicrobial chemotherapy. In: Persing DH, Tenover FC, Tang YW, Nolte FS, Hayden RT, van Belkum A (eds) Molecular microbiology: diagnostic principles and practice, 2nd edn. ASM Press, Washington, DC, pp 803–832

Xie HG, Frueh FW (2005) Pharmacogenomics steps toward personalized medicine. Personalized Med 2:325–337

Xie HG, Kim RB (2005) St John's wort-associated drug interactions: short-term inhibition and long-term induction? Clin Pharmacol Ther 78:19–24

Xie HG, Huang SL, Xu ZH, Xiao ZS, He N, Zhou HH (1997) Evidence for the effect of gender on activity of (S)-mephenytoin 4′-hydroxylase (CYP2C19) in a Chinese population. Pharmacogenetics 7:115–119

Xie HG, Kim RB, Stein CM, Wilkinson GR, Wood AJ (1999a) Genetic polymorphism of (S)-mephenytoin 4′-hydroxylation in populations of African descent. Br J Clin Pharmacol 48:402–408

Xie HG, Stein CM, Kim RB, Wilkinson GR, Flockhart DA, Wood AJ (1999b) Allelic, genotypic and phenotypic distributions of S-mephenytoin 4′-hydroxylase (CYP2C19) in healthy Caucasian populations of European descent throughout the world. Pharmacogenetics 9:539–549

Xie HG, Kim RB, Wood AJ, Stein CM (2001) Molecular basis of ethnic differences in drug disposition and response. Annu Rev Pharmacol Toxicol 41:815–850

Xie HG, Prasad HC, Kim RB, Stein CM (2002) CYP2C9 allelic variants: ethnic distribution and functional significance. Adv Drug Deliv Rev 54:1257–1270

Xie HG, Wood AJ, Kim RB, Stein CM, Wilkinson GR (2004) Genetic variability in CYP3A5 and its possible consequences. Pharmacogenomics 5:243–272

Xie HG, Zou JJ, Hu ZY, Zhang JJ, Ye F, Chen SL (2011) Individual variability in the disposition of and response to clopidogrel: pharmacogenomics and beyond. Pharmacol Ther 129:267–289

Yang J, Shi D, Yang D, Song X, Yan B (2007) Interleukin-6 alters the cellular responsiveness to clopidogrel, irinotecan, and oseltamivir by suppressing the expression of carboxylesterases HCE1 and HCE2. Mol Pharmacol 72:686–694

Yang J, He MM, Niu W, Wrighton SA, Li L, Liu Y et al (2012) Metabolic capabilities of cytochrome P450 enzymes in Chinese liver microsomes compared with those in Caucasian liver microsomes. Br J Clin Pharmacol 73:268–284

Yasuda SU, Zhang L, Huang SM (2008) The role of ethnicity in variability in response to drugs: focus on clinical pharmacology studies. Clin Pharmacol Ther 84:417–423

Yousef AM, Arafat T, Bulatova NR, Al Zumyli R (2008) Smoking behaviour modulates pharmacokinetics of orally administered clopidogrel. J Clin Pharm Ther 33:439–449

Zahno A, Brecht K, Bodmer M, Bur D, Tsakiris DA, Krahenbuhl S (2010) Effects of drug interactions on biotransformation and antiplatelet effect of clopidogrel in vitro. Br J Pharmacol 161:393–404

Zhang S, Ye J, Zhang Y, Xu X, Liu J, Zhang SH (2013) P2Y(12) protects platelets from apoptosis via PI3K-depedent bak/bax inactivation. J Thromb Haemost 11:149–160

Zhu HJ, Patrick KS, Yuan HJ, Wang JS, Donovan JL, DeVane CL et al (2008) Two CES1 gene mutations lead to dysfunctional carboxylesterase 1 activity in man: clinical significance and molecular basis. Am J Hum Genet 82:1241–1248

Zhu HJ, Appel DI, Johnson JA, Chavin KD, Markowitz JS (2009) Role of carboxylesterase 1 and impact of natural genetic variants on the hydrolysis of trandolapril. Biochem Pharmacol 77:1266–1272

Zou JJ, Xie HG, Chen SL, Tan J, Lin L, Zhao YY et al (2013) Influence of CYP2C19 loss-of-function variants on the antiplatelet effects and cardiovascular events in clopidogrel-treated Chinese patients undergoing percutaneous coronary intervention. Eur J Clin Pharmacol 69:771–777

23 Pharmacogenomics of Pulmonary and Respiratory Diseases

Luciana B. Crotti, Fernanda Kehdy, Eduardo Tarazona-Santos, and Luis A. Espinoza

Abstract

The identification of variations or mutations in genes encoding proteins that are involved in drug processing or metabolism can provide key information relevant to differential responses to therapeutic agents in specific genetic populations groups. It is well accepted that genetic variability (functional polymorphisms) may explain the failure of therapies and/or serious adverse side effects during and after treatment. Therefore, there is enormous interest in identifying these variants and determining their clinical relevance. In this regard, the main focus of pharmacogenomics is the study of inherited variations in genes that modulate drug response and their influence in predicting patient response to a specific treatment. For this purpose, pharmacogenomics integrates genomic information and technologies driving drug discovery and developing large-scale genomic studies to identify genetic variations. These findings may provide benefits in designing therapies more targeted to specific diseases, maximizing therapeutic effects, decreasing adverse reactions, and developing better methods to determine effective drug dosages. In addition to the anticipated benefit of personalized medicine, the identification of genetic markers may also influence the lifestyle, environment, and diet of those individuals with high susceptibility to develop a particular disease(s) and to prevent or delay the development of diseases. This chapter will focus on mutations and the variety of polymorphisms that may be associated with therapy

L.B. Crotti (✉)
Department of Biochemistry, Uniformed Services University of the Health Sciences, 4301 Jones Bridge Rd, Bethesda, MD 20814, USA
e-mail: luciana.crotti-espinoza@usuhs.edu

F. Kehdy • E. Tarazona-Santos
Departamento de Biologia Geral, Universidade Federal de Minas Gerais, Av. Antonio Carlos 6627, Pampulha, Caixa Postal 486, 31270-910, Belo Horizonte, Brazil

L.A. Espinoza
Department of Biochemistry and Molecular Biology & Cell Biology, Georgetown University, 3900 Reservoir Rd, Washington, DC 20057, USA

D. Barh et al. (eds.), *Omics for Personalized Medicine*,
DOI 10.1007/978-81-322-1184-6_23, © Springer India 2013

resistance for people with different types of lung diseases such as chronic obstructive pulmonary disease (COPD), tuberculosis (TB), idiopathic pulmonary fibrosis (IPF), pulmonary arterial hypertension (PAH), and interstitial lung damage (ILD). In fact, insights from recent evidences strongly support the notion that pharmacogenomics will be essential in improving innovative genomic-based therapies based on the genomic profiles of patients.

1 Introduction

1.1 Pharmacogenomics

The completion of the Human Genome Project allowed the discovery of millions of variants (polymorphisms) in the human genome. Consequently, the principal goal of current human genetics and genomics studies is to identify the small portion of these variants that are relevant to the biomedical basis of cellular function and disease processes and how they may also have a clinical impact by measuring the impact of new and existing therapies. There is no doubt that the application of new technologies and genomic information has opened new possibilities for the design of innovative therapeutics and the development of large-scale genomic testing for genetic variations in some chronic diseases of the airway, such as chronic obstructive pulmonary disease (COPD). Most of these variants are single-nucleotide polymorphisms (SNPs) with more than 15 million reported in public databases (2010b). Copy number variations (CNVs) are another type of polymorphisms that are critically important in determining biomedical analyses, including pharmacogenetic traits (Redon et al. 2006). In the case of some genes, for example, the cytochrome P450 2D6 (*CYP2D6*), the pharmacogenetic community has recognized the functional importance of CNVs for this gene in a series of diseases, and, therefore, the genomics community has recognized that CNV polymorphism is probably a major contributor in determining the impact of gene variations in rare and common clinically relevant outcomes such as pulmonary disorders (He et al. 2011).

1.2 Genetic Variability in Lung Diseases

1.2.1 Chronic Obstructive Pulmonary Disease (COPD)

This heterogeneous pulmonary disease is characterized by shortness of breath, cough, and sputum production. There is multiple disease subtypes associated with COPD, which may comprise overlapping disorders such as emphysema or chronic bronchitis (Snider 1985a, b). The emphysema destroys the alveoli, which lose elasticity and enlarge, making difficult for the lungs to move air both in and out and, consequently, making hard the absorption of enough oxygen or expelling enough carbon dioxide (Hogg 2004; Hogg et al. 2004). Meanwhile, the symptoms of emphysema are the shortness of breath, which gradually becomes more severe over many years (Hogg and Timens 2009). The chronic bronchitis causes scarring on the walls of airways, swelling in the lining, partial blockage of the air passages by mucus, and spasm from time to time.

Although many factors have been indicated to cause emphysema and chronic bronchitis, different environmental factors such as respiratory infections (Metz and Kraft 2010), exposure to allergens (Kennedy et al. 2012) and air pollutants (Jacquemin et al. 2012), deficiency of the alpha-1-protease inhibitor (Bachmann and Laurell 1963), and active or passive cigarette smoke in different age groups (Polosa and Thomson 2013; Sadof and Kaslovsky 2011) are also important determinants in developing COPD. Most of these environment-related diseases are, however, not easily detected and may be acquired during childhood and manifested later in adulthood. In addition, more than half of the world's population live in cities, a trend that is rapidly accelerating, especially in developing countries, which generate large quantities of waste products and pollutants that result in unsustainable environment damage that potentially can cause the development of asthma (Balmes et al. 2009; Meng et al. 2007).

At the moment, the main problem in treating COPD is the lack of evidence on how all the different mechanisms interplay and lead to the heterogeneous clinical symptoms observed in patients suffering from this lung disorder. Accordingly, diagnostic markers are the main priority in order to achieve a more accurate diagnosis that sometimes is difficult to determine, essentially because COPD is a complex inflammatory disease that involves many different types of cells. Although environmental variability may also contribute to the interindividual response, the intraindividual variability is highly consistent. Even when cigarette smoking is the most important risk factor associated with these diseases and some predictions have been made on who will get COPD, still it is difficult to predict who is a higher risk. Therefore, genetic variability was pointed out as the major reason why patients with similar clinical phenotype respond differently to the same drug treatment (Drazen et al. 2000).

Thus, for example, tobacco smoking is the largest cause of premature morbidity and mortality throughout the world, including chronic respiratory diseases such as asthma and chronic obstructive pulmonary disease. Smoking cessation is the most effective way to reduce the risk of developing COPD, which has high morbidity and mortality worldwide. Furthermore, chronic exposure to tobacco has been identified as a heritable association (interindividual variation) with the cytochrome P450 2A6 (*CYP2A6*) activity (Minematsu et al. 2003; Mwenifumbo et al. 2007; Schoedel et al. 2004). This enzyme is responsible for over 90 % of the metabolic activation of nicotine to cotinine, which has been attributed to variations in the *CYP2A6* gene, which have fast metabolizer genotype being linked to an increased number of cigarettes smoked per day. Consequently, individuals with the impaired function of this gene appear to not only reduce significantly the consumption of cigarettes (Pianezza et al. 1998), but variants on the gene also seem to provide a protective effect against the development of pulmonary emphysema independent of the smoking habit (Minematsu et al. 2003). Large differences in the frequency of slow metabolizers are also evident among ethnic populations, which were associated with ethnic differences and may affect the rate of smoking and the incidence of tobacco-related diseases. For that reason, genotyping for *CYP2A6* could improve and individualize smoking cessation pharmacotherapy in the future to determine the most effective treatment for an individual smoker.

The utilization of expression array and linked analyses allowed the identification of serpin peptidase inhibitor, clade E (*SERPINE2*), a gene that is located within the region of chromosome 2q. This extracellular serine protease inhibitor toward trypsin, thrombin, plasmin, uPA, and other serine proteases is consistently expressed in the vascular adventitia. In family-based association with COPD pedigrees, it was observed that 18 SNPs in the gene *SERPINE2* showed significant association with spirometric phenotypes. Furthermore, a consistent association identified a group of SNPs (*rs1438831*, *rs7579646*, *rs840088*, *rs7562213*, *rs920251*, *rs3795877*, *rs6747096*, and *rs3795879*) after the replication of the data described above in a case–control association analysis that included individuals with severe smoking-related COPD (DeMeo et al. 2006). Noteworthy in this respect, the robust association was exclusive in the male population. These findings lead the authors to suggest that *SERPINE2* may define susceptibility phenotypes of COPD by regulating inflammatory responses through gene-smoking interactions. The association of variants in the gene *SERPINE2* with this condition was supported by another study that conducted a robust analysis in which the SNP *rs3795875* appears to contribute to the development of this lung condition in smokers (Young et al. 2006). However, using the same methods, a different study was unable to find an association between *SERPINE2* SNPs and this pulmonary disease in a large cohort (Chappell et al. 2006). The difference in the disease phenotype of the patients evaluated was pointed as the first explanation for this inconsistence. It was also suggested that the findings described by DeMeo et al. may represent false positives (Chappell et al. 2006). As a matter of fact, different associations indicated to SNPs appeared to be in linkage disequilibrium with one another.

In addition to *SERPINE2* variants, SNPs from other four candidate genes, such as epoxide hydrolase 1 (*EPHX1*), glutathione S-transferase pi (*GSTP1*), surfactant protein B (*SFTPB*), and the transformer growth factor beta 1 (*TGFB1*), evidenced strong association with bronchodilator responsiveness phenotypes in individuals from the National Emphysema Treatment Trial (NETT) (Kim et al. 2009a). A different study in the NETT cohort found that the fast variant (His139Arg) provides protection against emphysema (Hersh et al. 2006), while the slow variant (Tyr113His) has been associated with COPD. These findings supported a previous report in which SNPs in *EPHX1* were associated with COPD-related phenotypes. In this last study, it was also suggested that SNPs in *TGFB1*, latent transforming growth factor beta binding protein 2 (*LTBP*), and *SFTPB* may contribute in varying combinations to several of the phenotypes related to COPD (Hersh et al. 2006). The wild-type Ile105Val *GSTP1* variant has a higher efficiency in metabolizing aromatic epoxides of polycyclic aromatic hydrocarbons and has been associated with emphysema and upper lobe-predominant disease (DeMeo et al. 2007). The Ile105 variant is predominant in Japanese patients with COPD (Ishii et al. 1999) and it was also associated with the rapid decline of lung function in smokers (He et al. 2002). It has been proposed that genetic variants in *GSTP1* and *EPHX1* play relevant roles in the process of COPD susceptibility and they are determinant in for the distribution and severity of this disease (DeMeo et al. 2007).

There is no doubt that variations in 308 NETT subjects, the nonsynonymous Leu10-Pro *TGFB1* SNP rs1800470 showed association with those suffering from emphysema (Cho et al. 2010). This is in agreement with a previous report indicating that *TGFB1* is associated with the emphysema phenotype (Ito et al. 2008). The frequency of this polymorphism has an interesting functional effect by influencing the higher levels of TFGB1 protein that may facilitate its own transcription (Suthanthiran et al. 2000). The significant association of two SNPs in the promoter region of *TGFB1* (*rs2241712* and *rs1800469*) and another one in exon 1 of *TGFB1* (*rs1982073*) with former and current smokers suffering from the critical COPD phenotype, FEV_1, indicates that individuals carrying *TGFB1* variations are more likely to develop this pulmonary condition (Celedon et al. 2004). Indeed, two *TGFB1* SNPs (*rs1800469* and *rs1982073*) have shown strong association with the emphysema phenotype in Japanese subjects with COPD (Ito et al. 2008), and a new SNP (*rs6957*) was associated with development of this lung disease. However, these variants on *TGFB1* do not affect lung function decline in the general population (van Diemen et al. 2006). Also, some *TGFB1* variants appear to have no specific association with the pathogenesis of COPD. As a matter of fact, the *TGFB1* polymorphisms C-509T and T869C did not show any type of association in a Chinese population (Mak et al. 2009). Similar results were reported in a Korean population (Yoon et al. 2006) indicating the existence of significant ethnic differences in susceptibility to COPD.

The association between the corticotropin-releasing hormone receptor 1 (*CRHR1*) polymorphism and improved lung function in response to inhaled corticosteroid and β_2-agonist inhalers in patients suffering from COPD was only evident in those individuals carrying the SNP *rs242941* and at age over 60 years (Kim et al. 2009b). *CRHR1* encodes a G-protein receptor that binds neuropeptides of the corticotropin-releasing hormone family that is a critical factor in mediating corticotropin-releasing hormone expression into the hypothalamic-pituitary-adrenal axis (Kokkotou et al. 2006). The authors proposed that utilization of both corticosteroid and β_2-agonist may account for the improved effect than the monotherapy with corticosteroid alone.

Since environmental pollutants are associated with COPD, it is well accepted that increased oxidative stress occurs in the lungs of individuals suffering from this deadly disease. Several reports have proposed that the superoxide dismutase enzyme 3 (*SOD3*), which is exclusively expressed in the lung by alveolar type II cells and alveolar macrophages (Folz et al. 1997; Loenders et al. 1998), catalyzes the conversion of superoxide anions to hydrogen peroxide and oxygen and appears to be a critical component in regulating the presence of oxidants species that may promote COPD development. It was reported that the SNP R213G (*rs1799895*) that causes substitution of arginine for glycine at position 213 in the heparin domain of the enzyme might protect against the development of COPD in different populations of smokers (Juul et al. 2006; Siedlinski et al. 2009; Young et al. 2006). In addition, two other polymorphisms E1 (*rs8192287*) and I1 (*rs8192288*) were also identified in the conserved region of *SOD3* of individuals that have reduced forced vital capacity (FVC) %, and, therefore, they probably will have a higher risk in COPD exacerbation requiring hospitalization (Dahl et al. 2008). A more recent study reported that the two SNPs described by Dahl et al. as well as the SNP R213G (*rs1799895*) evidenced strong correlation with emphysema but were not associated with COPD susceptibility (Sorheim et al. 2010). These findings suggested that *SOD3* may have an explicit role in the emphysema subtype of COPD.

1.2.2 Genetic Variability in Tuberculosis (TB)

Tuberculosis is a major health problem worldwide and its therapy is in most of the cases based on a set of first-line medications that include isoniazid (INH), pyrazinamide, and rifampin (2010a). This therapy is usually cheap and effective (http://www.who.int). However, adverse drug reactions such as hepatotoxicity can occur, leading to decreased medication adherence and to drug resistance. Whether the incidence of hepatotoxicity differs among phenotypes known as slow and fast acetylators is a controversial topic, some studies have suggested that slow acetylator status is a risk factor for adverse drug reactions. Indeed, a series of epidemiological studies had determined that approximately 10 % of individuals infected with *Mycobacterium tuberculosis* (MTB) develop active TB disease, which manifest as pulmonary disease (Philips and Ernst 2012). In contrast, there are individuals that have non-contagious latent TB in which the bacteria remain inactive in their bodies. These differences suggest that genetic differences may strongly contribute to TB pathogenesis in different populations (Bellamy et al. 2000; Chaussabel et al. 2003).

Drug labels of isoniazid, pyrazinamide, and rifampin report that "slow acetylation may lead to higher blood levels of the drug, and thus, an increase in toxic reactions" (http://www.pharmgkb.org/drug/PA451182?tabType=tabDrugLabels#tabview=tab0&subtab=32). The NAT2 protein is encoded by the N-acetyltransferase 2 (*NAT2*) gene (OMIM 243400), located at 8p22. *NAT2* has two exons, the second of which includes a single open reading frame of 870 bp. In addition to the wild-type reference haplotype NAT2*4 (GenBank NM_000015), more than 40 variants have been described, most of which have been associated with impaired metabolic activity. The most common polymorphisms are 191G>A (Arg64Gln), ^341T>C (Ile114Thr), 590G>A (Arg197Gln), 803A>G (Lys268Arg), 857G>A (Lys286Glu), 282C>T, and 481C>T, which are unevenly distributed across ethnically diverse human populations (Fuselli et al. 2007).

Polymorphisms in the *NAT2* gene are responsible for variation in the acetylator phenotypes. Individuals can be classified as slow or fast acetylators (or metabolizers) on the basis of their *NAT2* genotype, although some authors have considered a third intermediate category for this gene (Kuznetsov et al. 2009). Slow, intermediate, and fast acetylators are defined as carriers of zero, one, or two full functioning *NAT2* haplotypes, respectively, although exceptions have been reported, in particular in tuberculosis HIV-infected patients (Fuselli et al. 2007). Subjects carrying the slow acetylator status of *NAT2* have significant susceptible risk factor for anti-tuberculosis drug-induced hepatotoxicity when taking drugs concomitantly with isoniazid. An unbalance in the drug metabolism process during the coadministration of drugs was suggested as a possible explanation for this phenomenon, although other events affecting this dysfunction may also be involved (Lee et al. 2010).

The solute carrier family 11 (proton-coupled divalent metal ion transporters), member 1 (*SLC11A1*), located on 2q35 has several polymorphisms and has been associated with the risk of TB in different ethnic groups (Ates et al. 2009; Jin et al. 2009; Taype et al. 2006). However, other reports have provided inconclusive results about the potential involvement of this gene in TB development (Png et al. 2012). A meta-analysis that tested ethnic and TB form differences identified four polymorphisms (D543N, 3′UTR TGTG ins/del, INT4, and [GT]n) that showed a strong association with increased risk of TB, particularly in individuals of Asian origin, but not in Europeans (Meilang et al. 2012).

Among 702 TB patients, the SNPs *rs3809849* in the MYB-binding protein (P160) 1a (*MYBBP1A*) and the *rs9061* in the SP110 nuclear body protein (*SP110*) genes were found to be significantly associated with TB. These two genes may be potentially gene candidates for determining the risk of developing pulmonary TB in Chinese subjects (Cai et al. 2013). These findings also suggested that the interaction between MYBBP1A and SP110 is critical for the regulation of the pathogenesis of TB. In support of this statement, others have shown that the SP110 protein is necessary to integrate intracellular signals to control innate immunity and cell death in infected cells with MTB (Pan et al. 2005). Additionally, the interaction of MYBBP1A with SP110 is critical for apoptosis induction in macrophages. Also, MYBBP1 itself is an important corepressor of NF-κB (Owen et al. 2007). In contrast, a systematic review and meta-analysis in a wide range of populations that included African, European, and Asian individuals found that the *SP110* variants *rs1135791*, *rs9061*, *rs11556887*, *rs3948464,* and *rs1346311* had no statistically significant association with TB. The combination of results after overall, sensitivity, and stratified analyses showed that these SNPs might not be significantly related with the development of pulmonary TB (Lei et al. 2012).

On the other side, the examination of human samples (alveolar macrophages) with different clinical types of TB by a combination of microarray and DNA polymorphisms approaches identified seven SNPs (*rs10491110*, *rs159319*, *rs3138031*, *rs159290*, *rs159291*, *rs159294*, and *rs210837*) in the chemokine (C-C motif) ligand 1 (*CCL1*) gene, an inflammatory mediator, that are significantly associated with pulmonary TB (Thuong et al. 2008). Of interest, its receptor, CCR8 plays critical roles modulating the development of Th2 type T cell response in vivo (Chensue et al. 2001), and it is essential for the migration of dendritic cells to lymph nodes as well as in the stimulation of peritoneal macrophages that may account for cell aggregation at a tissue damage (Qu et al. 2004). However, the study performed by Thuong et al. was not able to confirm a previous report (Mistry et al. 2007) in which a panel of genes that include Ras and Rab interactor 3 (*RIN3*), lymphocyte antigen 6 complex, locus G6D (*LY6G6D*), testis expressed 264 (*TEX264*), chromosome 14 open reading frame 2 (*C14orf2*), suppressor of cytokine signaling 3 (*SOCS3*), *KIAA2013,* arsA arsenite transporter, ATP-binding (*ASNA1*), ATP synthase, H+transporting, mitochondrial Fo complex, subunit C1 (*ATP5G1*), and NOP10 ribonucleoprotein homolog (yeast) (*NOLA3*) were able to discriminate between individuals with active, latent, and recurrent tuberculosis. Differences in the study design between the two studies have been considered as the major reason for this divergence.

Genetic polymorphisms were also identified in other chemokines such as the chemokine (C-C motif) ligand 2 (*CCL2*), the chemokine (C-C motif) ligand 5 (*CCL5*), the chemokine (C-C motif) receptor 2 (*CCR2*), and the chemokine (C-C motif) receptor 5 (*CCR5*) in the ethnic Sahariya population, a primitive tribe in the North Central India. These variants showed a strong association with either susceptibility or resistance to TB (Mishra et al. 2012). It has to be emphasized that there was a variation in the impact of each SNP. Indeed, the -403G/A variant for *CCL5* evidenced a major susceptibility to TB, while the V64I (*CCR2*) variant showed an increased protection against the disease.

1.2.3 Genetic Variability in Idiopathic Pulmonary Fibrosis (IPF)

IPF is a progressive diffuse interstitial lung disorder, characterized by epithelial cell hyperplasia, extracellular matrix deposition, and myofibroblast foci formation, which are indicators of poor prognosis. The effective treatments against IPF are extremely limited and in most of cases, IPF patients are unresponsive to corticosteroid and immunosuppressive therapy and even worse, they may acquire resistance to steroid therapy. In this regard, rituximab, a chimeric monoclonal antibody against the protein CD20 used to treat rheumatoid arthritis and B cell non-Hodgkin's lymphoma, has been utilized as an alternative approach in patients with the steroid-resistant nephrotic syndrome (Bagga et al. 2007). A report showing that administration of this monoclonal anti-CD20 antibody worsened the condition of a patient leading to the development of acute respiratory insufficiency with hypoxemia and death from diffuse pulmonary fibrosis appears to indicate that rituximab may lead to the development of a serious and potentially fatal toxic complication in this high-risk group of patients (Chaumais et al. 2009).

The enzyme thiopurine S-methyltransferase (*TPMT*) metabolizes thiopurine drugs via S-adenosyl-L-methionine as the S-methyl donor and S-adenosyl-L-homocysteine as a by-product (Perri et al. 2007). The activity of TPMT associated with genetic polymorphisms has been indicated as an efficient genotypic analysis before the initiation of azathioprine treatment in combination with N-acetylcysteine (NAC) and steroids in patients with IPF. Azathioprine is an immunosuppressant that if frequently limited due to its bone marrow toxicity and other side effects such as diarrhea, nausea, and vomiting (Launay et al. 2006). *TMPT* is located on chromosome 6 (6p22.3) and deficient, intermediate, or high activity of the enzyme has been associated with variation in the nine introns and ten exons as well as in the 5′-flanking promoter region of this gene (Gisbert et al. 2007; Krynetski et al. 1995; Yates et al. 1997).

Alterations at the *TPMT* locus have identified the *2, *3A, *3B, *3C, and *7 polymorphisms, which are the most common variants in TPMT-deficient patients (Yates et al. 1997). Although rare variants corresponding to the nonfunctional *TPMT* alleles involving *3D, *4, *5, and *6 have also been reported in this group of patients (Spire-Vayron de la Moureyre et al. 1998), these polymorphisms were omitted in this genotyping study. Polymorphisms were also identified in the promoter; however, they are usually not considered in the toxicity and therapeutic efficacy of thiopurine drugs because of their minor effect on the modulation of TPMT activity compared with SNPs in the open reading frame (Spire-Vayron de la Moureyre et al. 1999). Considering that polymorphisms in the TPMT gene vary among ethnic groups, the screening for *TPMT* polymorphisms would considerably reduce the negative treatment effects (http://clinicaltrials.gov/ct2/show/NCT00650091) of azathioprine (Hagaman et al. 2010).

The interleukin 1 receptor antagonist (*IL1RN*) gene encodes a protein that is a member of the interleukin 1 cytokine family and inhibits the activities of interleukin 1, alpha (*IL1A*), and interleukin 1, beta (*IL1B*), as well as modulates a variety of interleukin 1 related immune and inflammatory responses (Barlo et al. 2011). A predisposing of a genetic polymorphism (*rs2637988*) in *IL1RN* has been linked with susceptibility to IPF and reduced ratios of IL-1Ra/IL-1β ratios in bronchoalveolar lavage fluid of patients with IPF (Barlo et al. 2011). However, multiple reports have provided variable results about the association of the SNPs *rs408392*, *rs419598*, and *rs2637988* in *IL1RN* and risk of IPF. For example, increased risk of fibrosing alveolitis was associated with the variant *rs419598* in English subjects (Whyte et al. 2000) and the rs2637988 variant were seen as key contributors to proinflammatory and pro-fibrotic phenotypes in the lungs (Barlo et al. 2011). Also, no association was established between a variable number tandem repeat (VNTR), *rs408392*, or *rs2637988* and increased risk of developing this disease in Australian and Czech subjects with IPF (Hutyrova et al. 2002; Riha et al. 2004). In contrast, a meta-analysis that combined all the studies discussed above and also investigated different polymorphisms affecting *IL1RN* found a strong association between VNTR, *rs408392*, and *rs419598* with increased susceptibility to develop IPF (Korthagen et al. 2012).

The frequency of distribution of the major histocompatibility complex, class I, A and B (*HLA-A, B*), located at chromosome 6p21.31 (Volz et al. 1994), plays an important role in the generation of immune responses. Variations involving these genes may vary from one ethnic to another or even at members of the same ethnic group living in different geographic areas (Shen et al. 2010) and have been associated with susceptibility to several immune-related disorders (de Bakker et al. 2006; Voorter et al. 2005). The HLA polymorphisms (DRB1*1501) in patients with end-stage pulmonary disease showed a consistent involvement in IPF pathogenesis (Xue et al. 2011). This unique variant was observed Caucasian IPF subjects with highly variable disease severities and also appeared to be associated with greater magnitudes of gas exchange impairment.

An attempt to associate HLA-A, B gene polymorphisms with IPF, peripheral blood samples from Chinese individuals diagnosed with this pulmonary disease were evaluated using a PCR-SSP grouping method (Zhang et al. 2012). The comparison of IPF samples with the control group identified the haplotype frequency of HLA-A03, B15, B40, as well as the linked gene frequency of A2B15, A2B40, A11B15, and A24B58, which are significantly increased in patients suffering from IPF. These variants are in agreement with previous reports indicating the high prevalence of HLA-B15 in patients suffering from cryptogenic fibrosing alveolitis (Varpela et al. 1979) and also a strong association between the increased frequency of HLA-A03 and IPF in Caucasian patients (Libby et al. 1983). Furthermore, a small-scale study found that the variants *rs1801198* and *rs7286680* in the gene *TCN2* (transcobalamin II) showed significant association with IPF in Italian patients (Martinelli et al. 2013). Since these SNPs are part of the folate pathway, the authors suggested that these polymorphisms might interfere in the process of folate absorption in pulmonary macrophages. This assumption is based in evidence showing that activated macrophages into injured lung tissue expressed high levels of the folate receptor beta (FRβ), which appeared to increase the rate in folate uptake as observed recently in cells with accelerated proliferation (Feng et al. 2011).

Table 23.1 The clinical classification of pulmonary hypertension

Group 1	Pulmonary arterial hypertension (PAH)
	Associated with APAH
	Drug- and toxin-associated
	Heritable (HPAH)
	Idiopathic (IPAH)
	Persistent pulmonary hypertension of the newborn (PPHN)
	Pulmonary veno-occlusive disease (PVOD) and/or pulmonary capillary hemangiomatosis (PCH)
Group 2	Pulmonary hypertension due to left heart diseases
	Diastolic dysfunction
	Systolic dysfunction
	Valvular disease
Group 3	Pulmonary hypertension due to lung diseases and/or hypoxemia
	Alveolar hypoventilation disorders
	Chronic exposures to high altitude
	Chronic obstructive pulmonary disease (COPD)
	Developmental abnormalities
	Interstitial lung disease
	Other pulmonary diseases with mixed restrictive and obstructive partner
	Sleep-disorder breathing
Group 4	Chronic thromboembolic pulmonary hypertension (CTEPH)
Group 5	PH with unclear multifactorial mechanisms
	Hematological disorders
	Metabolic disorders
	Systemic disorders

1.2.4 Genetic Variability in Pulmonary Arterial Hypertension (PAH)

The PAH is an abnormal high pulmonary vascular resistance and mean pulmonary vascular resistance that lead to cardiac death or death from progressive right ventricular failure (Table 23.1). This disorder is much common in females than in males and can occur at any age (Loyd et al. 1984). In the familial (inherited) FPAH, there is an alteration in the gene bone morphogenetic protein receptor, type II (*BMPR2*). The BMPR2 protein helps regulate the growth of cells in the walls of the small arteries of the lungs. Other factors, probably genetic and/or environmental, are also necessary to initiate the disease because approximately 20 % of all individuals with a BMPR2 mutation ever develop IPAH (Machado et al. 2009). Therefore, a molecular classification of PAH, based upon the presence or absence of *BMPR2* mutations, has important implications for patient management and screening of relatives (Thomson et al. 2000). Some individuals in families with a different genetic condition called hereditary hemorrhagic telangiectasia (HHT), may also develop idiopathic PAH (IPAH), due to a mutation in the activin A receptor type II-like 1 (*ALK1*), a membrane-bound receptor member of the TGFβ signaling superfamily (Harrison et al. 2003; Shi and Massague 2003).

Although numerous mutations have been identified at the *BMPR2* locus in most families affected with PAH, no mutations were detected in approximately 30 % of the probands screened (Machado et al. 2006). Of interest, most of the mutations identified were deletions or insertions. In addition, different groups also reported point mutations and modifications of large fragments in *BMPR2* (Souza et al. 2008; Thomson et al. 2000). The characterization of the *BMPR2* 5′-untranslated region in subjects with PAH that are negative for mutations within the coding region and intron-exon junctions identified a GC>AT double-substitution mutation 944 bp upstream of the translation start site. This variation designed as c.*-944/5gc>AT was indicated as a pathogenic variant that may lead to an abnormal mRNA decay rate (Aldred et al. 2007). The utilization of additional experimental techniques also allowed the identification of deletions and duplications in different exons of *BMPR2* mutation-negative PPH families that were undetected by standard screening tests (Cogan et al. 2005).

In an earlier study, evidence was obtained, suggesting that individuals with a family history of PAH having *BMPR2* variations may develop the disease earlier and with more severe symptoms (Loyd et al. 1995); however, these data have not yet been confirmed in registry studies. Additional studies had shown that individuals carrying rare mutations on *BMPR2* and exposed to appetite suppressants aminorex such as fenfluramine derivatives evidenced an increased risk of developing severe PAH (Humbert et al. 2002). In addition, the utilization of new experimental approaches has demonstrated that *BMPR2* genetic variations lead to haploinsufficiency of its protein, which affects the stoichiometric imbalance of the receptor complex, a phenomenon that causes negative effects on the signaling pathway regulated by the receptor in PAH (Nasim et al. 2008). Knowledge about genes that cause IPAH is still growing, so it is possible that other genes may also contribute to the development of this lung disease and hopefully will be discovered in the near future. In fact, in a recent report, the identification of novel mutations in the *BMPR2* gene was indicated as key player in the initiation and progression stages of pathological lesions associated with primary pulmonary hypertension (PPH) (Machado et al. 2001). Altogether, this evidence suggested that haploinsufficiency appears to contribute to the molecular events underlying *BMPR2* susceptibility to the development of PAH.

Finally, a recent experimental in vivo study attempted to apply the concept of pharmacogenomics and clinical applications to PAH. This target-based approach examined two animal models of the disease in which the histologic features are similar: the vasoactive intestinal peptide knockout ($VIP^{-/-}$) and rats injected with monocrotaline (MCT). Treatment with vasoactive intestinal peptide was able to correct the phenotypic and genotypic alterations of mice $VIP^{-/-}$; however, this therapeutic approach was not effective in the rat MCT model, which develops a more severe PAH (Said and Hamidi 2011). These findings suggested that effective therapy against complex and mixed abnormalities affecting the airways may require multiple therapeutic approaches.

1.2.5 Genetic Variability in Interstitial Lung Damage (ILD)

Interstitial lung disease occurs when an injury to the lungs triggers an abnormal healing response, forming progressive scarring, which affects the ability to breathe. The common feature in this group of lung diseases is the appearance of diffuse alveolar damage manifested by inflammation and fibrosis, which may also affect airways and arteries (Crystal et al. 1981; Travis et al. 2008). Although different types of ILD have been identified on the basis of clinical signs and symptoms, here we will discuss the correlation between cocaine abuse and ILD development. This illicit drug acts as stimulant of the central nervous system and its addiction is a major health and social problem. Cocaine absorption through nasal tissues makes snorting cocaine nearly as fast-acting. Addiction to this drug causes different types of interstitial lung damage, including acute pulmonary hemorrhage, diffuse alveolar infiltrates, and eosinophilic infiltrates (Drent et al. 2012).

In the lung cells, cytochrome P450 (CYP) enzymes metabolize many drugs. Polymorphisms affecting these microsomal enzymes cause response variation among individuals (Weinshilboum 2003; Wilkinson 2005). The altered metabolic capacities associated to CYP polymorphisms can result in drugs inadequately metabolized. This can cause serious adverse respiratory reactions that in long term may lead to the development of drug-induced ILD. This was evident in samples of individuals diagnosed with a noninfectious or infectious disorder that has one or more *CYP* genetic variants and inadequately metabolized drugs, which appear to be at a substantially greater risk of developing a drug-induced ILD (DI-ILD) when exposed to multiple medications (Wijnen et al. 2008).

The development of ILD in individuals with cocaine abuse was associated with allelic variants in *CYP2C9*, *CYP2C19*, and vitamin K epoxide (*VKORC1*) genes. Considering that CYP enzymes are critical for the metabolization of drugs in the lung (Wijnen et al. 2010), the heterozygous subjects for either *CYP2C9*1/*2* or *CYP2C19*1/*3* were intermediate metabolizers of cocaine (Wijnen et al. 2011). These patients also displayed a variant allele for the *VKORC1* (-1639 GA and 113CT) enzyme. Since *VKORC1* encodes the enzyme that is responsible for reducing vitamin K 2,3-epoxide to the enzymatically activated form, variants affecting this gene may have a negative impact on the functional vitamin K-dependent proteins (Wallin et al. 2008). A previous report has shown that patients with the functionally defective *2 (430T) and *3 (1075C) allelic variants of the *CYP2C9* gene prescribed with the anticoagulant Coumadin required lower maintenance doses and took longer to achieve dose stabilization. These subjects are also at higher risk of serious and life-threatening bleeding in response to this anticoagulant when compared to patients without these variants. Therefore, vitamin K supplementation has been suggested as an alternative approach to prevent adverse effects associated with cocaine addiction (Wijnen et al. 2011).

1.3 Clinical Applications of Pharmacogenomics in Lung Diseases

Variations in drug response are critical in determining the response between different ethnic groups and in determining the most appropriate doses, so patients do not experience any treatment side effects. A series of studies have reported genetic polymorphisms in drug metabolism, particularly affecting CY2C9, CYP2D6, and CYP2C19 (Bauer et al. 2011; Lam and Cheung 2012; Mega et al. 2010). Thus, for example, approximately 15–30 % of Asians are poor metabolizers of CYP2C19, while only 3–6 % of Caucasians and 2–4 % of Africans have low to absent metabolic activity for this gene (Chen et al. 2008, 2012; Dandara et al. 2011). A similar pattern is observed in the CYP2D6 gene for which only 5–10 % of Caucasians and 1–2 % of Asians are poor metabolizers (Armstrong et al. 1994; Zhou et al. 2009), while polymorphisms for these genes are present in almost 50 % of Asian subjects. Because of these ethnic differences, new pharmacogenomic tests are now available for some of them.

The first test approved by the FDA for the analysis of variations on genes, the AmpliChip CYP450 from Roche, is narrow in scope (Table 23.2). It was designed to identify gene variations only for the CYP2D6 and CYP2C19 genes (http://molecular.roche.com/assays/Pages/AmpliChipCYP450Test.aspx). Considering that the number of CYP families is continuously

Table 23.2 AmpliChip CYP450 (Roche) test

Methodology	Markers	CYP alterations	Amount of sample	Benefits
Microarray	CYP2D6	33 CYP2D6 alleles 3 CYP2C19 alleles	3 mL of whole blood	There is no cross-contamination of samples
	CYP2C19	CYP2D6 gene duplications and deletions		Provide a customized genotype and predicted genotype information

Patent: CE-IVD; Japan-IVD; US-IVD

growing (Flockhart 2007), this test has serious limitations to be performed in individuals suffering from pulmonary disorders, basically due to the higher number of CYP enzymes that are expressed in human pulmonary tissues. On the other side, the DMET™ (drug metabolism enzymes and transporters) Plus Solution from Affymetrix (Table 23.3) can genotype up to 1,936 genes based in deletions, insertions, rare SNPs, trialleles, and copy number variants (Burmester et al. 2010). These genes are mainly involved in the absorption, distribution, metabolism, and drug disposition (http://www.affymetrix.com/estore/browse/products.jsp?productId=131412#1_3). Although this platform target genes that are impacted by environmental pollutants and may be important for the screening in patients suffering from lung diseases, the FDA has not yet approved it and because of that, it is basically utilized for standardization of exploratory studies and for the improvement of clinical pilot studies (Mijatovic et al. 2012).

Alpha-1 antitrypsin deficiency is the most commonly known genetic risk factor for COPD. Although no symptoms are associated with this deficiency for some individuals, in others the enzyme deficiency causes alveoli injury, which limits the expansion and contraction of the chest, increasing the development of pulmonary dysfunction associated with COPD (Larsson and Eriksson 1977). Tobacco consumption worsens this condition by increasing secretion of neutrophil elastase and inhibiting alpha-1 antitrypsin. Therefore, testing the activity of this enzyme in the blood is recommended to identify a rare liver disease (cirrhosis) in children and adults, a disorder that accelerates lung damage and development of COPD; however, approximately 3 % of individuals with COPD do not have alpha-1 antitrypsin deficiency and not all smokers develop COPD (Table 23.4). In addition, studies that focused on the association of the β_2-adrenergic receptor (*ADRB2*), which is extensively studied in asthma, and a series of other genes with COPD exacerbation have provided conflictive data (Hersh 2010; Hersh et al. 2005). Therefore, more studies are necessary to identify additional markers in COPD that allow a better association with the risk of COPD development. It is also important to determine the levels of markers, genetic functions, and clinical impact. In this regard, a genetic epidemiology study is being conducted at the Brigham and Women's Hospital in Boston, Massachusetts, to identify novel determinants of COPD in the US population. This study will analyze differences in nucleotides at over 500,000 selected positions across the entire human genome to identify the primary gene variations that make certain individuals more susceptible to developing COPD than others.

The use of corticosteroids and azathioprine for the treatment of IPF in patients with genetic polymorphisms of thiopurine methyltransferases (*TPMT*) may lead to severe azathioprine myelotoxicity (Perri et al. 2007). Therefore, a pretreatment screening for *TMPT* (Table 23.5) in patients with interstitial lung disease prior to treatment with mercaptopurines has provided safe prescription doses of this immunosuppressive drug in subjects with normal or deficient TMPT levels (Bakker et al. 2007). Nevertheless, the frequent use of azathioprine in combination with NAC and steroids (Table 23.6), which is one of the few treatment regimens with minimal adverse reactions, has, in certain way, restricted the testing for TPMT activity (Collard et al. 2007; Jang et al. 1996). A comparative study found that TPMT testing could reduce the risk of developing azathioprine

Table 23.3 DMET™ Plus Solution (Affymetrix) assay

Methodology	Markers	Amount of sample	Cost	Benefits
Microarray	CYP450 genes (478)	1 μg of genomic DNA	$2,450.00/7 Rx	Genotypes 1,936 markers in 225 high-value genes
	CYP1B1			High quality of positive assay controls
	CYP1A1			Detect signals from rare alleles
	CYP2C9			Compare alleles across multiple cohorts receiving the same drug(s)
	CYP2D6			Detect polymorphisms associated with PK/PD in phase I
	CYP2C19			
	Others			
	Drug transporter genes (64)			
	ABCB2			
	ABCG2			
	MDR1			
	SLCO1A2			
	SULT1A1			
	SULT4A1			
	Others			
	Phase II enzymes (408)			
	DPYD			
	GSTM1			
	GSTT1			
	NAT1			
	NAT2			
	UGT1A1			
	UGT1A9			
	UGT2B17			
	Others			
	Transcription regulators and other enzymes (413)			
	AHR			
	ARNT			
	NR1/2			
	PPARD			
	PPARG			
	RXRA			
	Others			

Patent: Products may be covered by one or more of the following patents and/or sold under license from Oxford Gene Technology: US patent Nos. 5,445,934; 5,700,637; 5,744,305; 5,945,334; 6,054,270; 6,140,044; 6,261,776; 6,291,183; 6,346,413; 6,399,365; 6,420,169; 6,551,817; 6,610,482; 6,733,977; and EP 619 321; 373 203 and other US or foreign patents

toxicity to almost 2 % (Hagaman et al. 2010), and determining the activity levels of this enzyme over the determination of allelic polymorphism may reduce the overall cost of treatment (Booth et al. 2010), supporting the hope that this test may be a common practice in patients with IPF. However, there are some gaps that preclude of the massive application of this test: first, the rate of abnormal TPMT activity differs significantly between ethnic groups (Oh et al. 2004), and second, clinical trials describing the impact of these differences are not yet available. Still, there is hope that this test may be a common practice in the clinic in the near future.

Table 23.4 Impact of variations on genes in therapies for COPD

Therapeutic	Type	Variations on genes
Symbicort	Monotherapy	AAT
Cilomilast	Monotherapy	PDE4
		Diarrhea
		Nausea
		Vomiting
Infliximab	Monotherapy	TNFA
		Back pain
		COPD exacerbation
		Diarrhea
		Headache
		Sinusitis
		Upper respiratory tract infection
N-acetylcysteine	Monotherapy	EPHX1, GSTP1, GSTM1, HMOX1, SOD3
		Constipation
		Diarrhea
		Epigastric pain
		Flushing
		Nausea
		Transient king rush
		Vomiting
Salmeterol + fluticasone	Polytherapy	HDAC2
		Risk of pneumonia
Other glucocorti-costeroids	Polytherapy	HDAC2
		Oral short-term use
		Increased blood sugar level
		Mood changes
		Weight gain and fluid retention
		Oral long-term use
		Osteoporosis
		Recurrent infection
		Cataracts
		Ulcers
		Sore throat
		Heavy growth of a fungus in the mouth, throat, or esophagus

Table 23.5 TMPT test

Methodology	Markers	TPMT activity	Amount of sample	Costs	Benefits
Enzymatic/high-performance liquid chromatography	TPMT	Normal	3 mL of whole blood	$300 per assay	Identifies risk of developing severe side effects
		Abnormal			Subjects with prescription to thiopurines
		High			

Patent: US patent numbers: 6576438; 7138250; 8188067

The current practice in the diagnosis of individuals affected with PAH shows that it is mostly diagnosed late in the course of the disease, with patients displaying severe functional and hemodynamic compromise (Badesch et al. 2010; Humbert et al. 2006). Therefore, the early diagnosis for the presence of PAH in high-risk groups definitely may improve diagnosis and therapeutic

Table 23.6 Impact of variations on metabolic genes in therapies for IPF

Therapeutic	Type	Variations on metabolic genes
NAC	Polytherapy	TPMT
Steroids		Bone marrow toxicity
Azathioprine		Fatigue
		Nausea
		Skin rashes
		Vomiting

Table 23.7 Impact of variations on metabolic genes in therapies for PAH

Therapeutic	Type	Variations on metabolic genes
Calcium channel blockers	Monotherapy	BMPR2
		Poor prognosis
Warfarin	Adjuvant	CYP2C9
		Reduced enzymatic of warfarin
		Low dose requirement for warfarin
		VKORC1
		Variable doses of warfarin

interventions (Benza et al. 2012). Familial PAH results from mutations on BMPR2 and hereditary transmission occurs in approximately 6–10 % of individuals with PAH (Lane et al. 2000). Genetic alteration on BMPR2 was also detected in approximately 25 % of patients with IPAH, in 15 % of subjects suffering of PAR related to fenfluramine use, and in some patients with other forms of associated PAH. Therefore, the American College of Chest Physicians has indicated a periodic screening for this gene in individuals affected by portal hypertension and scleroderma spectrum disorders (McLaughlin et al. 2009). However, the high cost of the genetic testing for BMPR2 mutations has limited the application of this test in the clinic. In addition, alterations in other genes are also involved in the heterogeneous response of subject with PAH to specific therapies (Table 23.7), and, hence, the genetic screening for this disease cannot be limited to only one gene (Smith et al. 2012).

1.4 The Future of Pharmacogenomics in Lung Diseases

Numerous evidences have demonstrated the critical role of basic science approaches in interpreting the mechanistic events to identify the most effective treatments. The recent reports are very encouraging; however, the identification of mutation(s) that drives a particular disease is still challenging. Although there is an enormous amount of evidence indicating that genetic polymorphism is associated with poor response to drug treatment, there are no clear explanations on how to translate this information into clinical application (Kirchheiner et al. 2005). In this regard, it is imperative to determine the impact of all the available pharmacogenomic data showing consistent and reproducible results for an increasing number of genetic markers that may have potential applicability in clinical practice (Table 23.8). Therefore, the close interaction between basic and clinical research is critical and is expected to enhance the fundamental genomic findings for the clinical settings.

Table 23.8 Impact of pharmacogenomics into clinical applications in pulmonary diseases

Improved diagnosis accuracy
Identification of the genetic risk factors associated with disease
Combination of family history and genomic data
Development of more efficient therapeutics
Integration of genomic data to support clinic application
Availability of genomic data to patients
Effectiveness on therapy using individual genomic data
Promotion of education and research in pharmacogenomics
Increased access to genomic medicine

The explosion of new methodologies and design approaches for analysis in pharmacogenomics is now able to integrate a large and complex amount of data produced by high-throughput assays in samples of patients with lung disorders. This information is contributing to our understanding of how the genetic differences in the expression and function of gene products affect drug responses of patients to medication and how those responses may be translated into new concepts for clinical application (Manolio and Green 2011). However, the whole-genome analysis still is not able to reveal the causes of rare genetic diseases, basically due to the many variants involved on different intracellular signaling pathways that modulate drug metabolism, drug receptors, and drug transporters as well as the different routes of excretion that may restrict the actions of drugs (Stingl et al. 2010). All this information is raising concerns regarding the selection of the most appropriate experimental approaches to determine the efficacy of a drug and the risk of adverse drug reactions.

Due to the gene and environmental interaction, in most cases it is difficult to correlate diagnosis of a disease with a specific polymorphism. Even though the current research programs in pharmacogenomics have increased substantially in the last decade, at the present time medications are not much safer and worse, many of the state-of-the-art medications are expensive. Taking into consideration that genetic variability can account for approximately 50–60 % of interindividual variability, pharmacogenomics holds great promise for elucidating the role of genetic variants responsible for the development of a particular pulmonary disease as well as for designing more effective therapies to minimize the chance of developing any adverse drug reactions.

The effect of single genes and their effects on specific individuals in the response to specific therapies may also be a consequence of the interaction of all genes in the genome in which the pharmacokinetics and/or pharmacodynamics may play key roles. In this regard, genetic testing may be an effective approach to reduce serious adverse reactions and events to current medications. However, still there are some remarkable challenges that need to be overcome such as the lack of consensus in selecting the appropriate controls, choosing the most suitable experimental approaches to validate the new target, and consistent evidence about the specific role of the target in the airways.

There is no doubt that pharmacogenomics can improve the efficacy of drugs that are currently used for the treatment of asthma symptoms. Furthermore, the identification of novel targets may be particular important to develop more effective strategies for therapeutic intervention against lung diseases. However, a few drug therapies available in accordance with genetic variations appear to indicate that most pharmaceutical companies are reluctant to invest in the development of this type of diagnostic and therapeutic agents. This is probably a big concern because of the great expectation with the promise of personalized medicine. To reach this goal, the identification of genes that drive disease development and progression in the lung is imperative. However, this is a challenging task, mainly due to the limited dataset used in the studies, which exhibit low power and have no consistency to determine whether genetic variants are associated with different respiratory disorders. Therefore, it was suggested that groups should participate in consortium working groups, so they can have access to different databases to study the pharmacogenomics of pulmonary diseases.

2 Conclusions

In this chapter we have summarized the evidence for a genetic contribution to asthma as well as the recent advances in pharmacogenomics for identifying the location and function of genes that cause complex diseases. There is no doubt that understanding the impact of genetic variation on predisposition to respiratory disorders could have major implications for diagnostic and therapeutic approaches. Giving the heterogeneous condition of the pulmonary diseases discussed in this chapter and the fact that pathological conditions vary greatly from patient to patient, several independent studies are required to validate the impact of SNPs in the susceptibility to the development of a particular disease. One of the main limitations in most of the studies is the sample size, which is widely recognized as not large enough to predict the relevance of

SNPs on outcomes in specific ethnic patient group. Understanding the effects of gene-gene and gene-environment interactions is critical to identify the events associated with the evolution of diseases in the lung. We have also reviewed how these techniques have been applied to the study of a series of lung disorders. It is likely that rapid additional advances will yield high-impact data and new avenues of investigation for novel therapeutic interventions that will have the potential to change the way patients are treated. In this regard, pharmacogenomics is critical for the development of personalized medicine to improve efficacy, tolerability, and drug safety.

References

(2010a) Global tuberculosis control: key findings from the December 2009 WHO report. Wkly Epidemiol Rec 85:69–80

1000 Genomes Project Consortium, Abecasis GR, Altshuler D, Auton A, Brooks LD, Durbin RM, Gibbs RA, Hurles ME, McVean GA (2010b) A map of human genome variation from population-scale sequencing. Nature 467:1061–1073

Aldred MA, Machado RD, James V, Morrell NW, Trembath RC (2007) Characterization of the BMPR2 5′-untranslated region and a novel mutation in pulmonary hypertension. Am J Respir Crit Care Med 176:819–824

Armstrong M, Fairbrother K, Idle JR, Daly AK (1994) The cytochrome P450 CYP2D6 allelic variant CYP2D6J and related polymorphisms in a European population. Pharmacogenetics 4:73–81

Ates O, Dalyan L, Musellim B, Hatemi G, Turker H, Ongen G, Hamuryudan V, Topal-Sarikaya A (2009) NRAMP1 (SLC11A1) gene polymorphisms that correlate with autoimmune versus infectious disease susceptibility in tuberculosis and rheumatoid arthritis. Int J Immunogenet 36:15–19

Bachmann R, Laurell CB (1963) Electrophoretic and immunologic classification of M-components in serum. Scand J Clin Lab Invest 15(Suppl 69):11–24

Badesch DB, Raskob GE, Elliott CG, Krichman AM, Farber HW, Frost AE, Barst RJ, Benza RL, Liou TG, Turner M et al (2010) Pulmonary arterial hypertension: baseline characteristics from the REVEAL Registry. Chest 137:376–387

Bagga A, Sinha A, Moudgil A (2007) Rituximab in patients with the steroid-resistant nephrotic syndrome. N Engl J Med 356:2751–2752

Bakker JA, Bierau J, Drent M (2007) Therapeutic regimens in interstitial lung disease guided by genetic screening: fact or fiction? Eur Respir J 30:821–822

Balmes JR, Earnest G, Katz PP, Yelin EH, Eisner MD, Chen H, Trupin L, Lurmann F, Blanc PD (2009) Exposure to traffic: lung function and health status in adults with asthma. J Allergy Clin Immunol 123:626–631

Barlo NP, van Moorsel CH, Korthagen NM, Heron M, Rijkers GT, Ruven HJ, van den Bosch JM, Grutters JC (2011) Genetic variability in the IL1RN gene and the balance between interleukin (IL)-1 receptor agonist and IL-1beta in idiopathic pulmonary fibrosis. Clin Exp Immunol 166:346–351

Bauer T, Bouman HJ, van Werkum JW, Ford NF, ten Berg JM, Taubert D (2011) Impact of CYP2C19 variant genotypes on clinical efficacy of antiplatelet treatment with clopidogrel: systematic review and meta-analysis. BMJ 343:d4588

Bellamy R, Beyers N, McAdam KP, Ruwende C, Gie R, Samaai P, Bester D, Meyer M, Corrah T, Collin M et al (2000) Genetic susceptibility to tuberculosis in Africans: a genome-wide scan. Proc Natl Acad Sci U S A 97:8005–8009

Benza RL, Gomberg-Maitland M, Miller DP, Frost A, Frantz RP, Foreman AJ, Badesch DB, McGoon MD (2012) The REVEAL Registry risk score calculator in patients newly diagnosed with pulmonary arterial hypertension. Chest 141:354–362

Booth RA, Ansari MT, Tricco AC, Loit E, Weeks L, Doucette S, Skidmore B, Hoch JS, Tsouros S, Sears M et al (2010) Assessment of thiopurine methyltransferase activity in patients prescribed azathioprine or other thiopurine-based drugs. Evid Rep Technol Assess (Full Rep) 196:1–282

Burmester JK, Sedova M, Shapero MH, Mansfield E (2010) DMET microarray technology for pharmacogenomics- based personalized medicine. Methods Mol Biol 632:99–124

Cai L, Deng SL, Liang L, Pan H, Zhou J, Wang MY, Yue J, Wan CL, He G, He L (2013) Identification of genetic associations of SP110/MYBBP1A/RELA with pulmonary tuberculosis in the Chinese Han population. Hum Genet 132(3):265–273

Celedon JC, Lange C, Raby BA, Litonjua AA, Palmer LJ, DeMeo DL, Reilly JJ, Kwiatkowski DJ, Chapman HA, Laird N et al (2004) The transforming growth factor-beta1 (TGFB1) gene is associated with chronic obstructive pulmonary disease (COPD). Hum Mol Genet 13:1649–1656

Chappell S, Daly L, Morgan K, Baranes TG, Roca J, Rabinovich R, Millar A, Donnelly SC, Keatings V, MacNee W et al (2006) The SERPINE2 gene and chronic obstructive pulmonary disease. Am J Hum Genet 79:184–186, author reply 186–187

Chaumais MC, Garnier A, Chalard F, Peuchmaur M, Dauger S, Jacqz-Agrain E, Deschenes G (2009) Fatal pulmonary fibrosis after rituximab administration. Pediatr Nephrol 24:1753–1755

Chaussabel D, Semnani RT, McDowell MA, Sacks D, Sher A, Nutman TB (2003) Unique gene expression profiles of human macrophages and dendritic cells to phylogenetically distinct parasites. Blood 102:672–681

Chen L, Qin S, Xie J, Tang J, Yang L, Shen W, Zhao X, Du J, He G, Feng G et al (2008) Genetic polymorphism analysis of CYP2C19 in Chinese Han populations from different geographic areas of mainland China. Pharmacogenomics 9:691–702

Chen M, Liu XJ, Yan SD, Peng Y, Chai H, Li Q, Wei JF, Xu YN, Huang DJ (2012) Association between cytochrome P450 2C19 polymorphism and clinical outcomes in Chinese patients with coronary artery disease. Atherosclerosis 220:168–171

Chensue SW, Lukacs NW, Yang TY, Shang X, Frait KA, Kunkel SL, Kung T, Wiekowski MT, Hedrick JA, Cook DN et al (2001) Aberrant in vivo T helper type 2 cell response and impaired eosinophil recruitment in CC chemokine receptor 8 knockout mice. J Exp Med 193:573–584

Cho MH, Washko GR, Hoffmann TJ, Criner GJ, Hoffman EA, Martinez FJ, Laird N, Reilly JJ, Silverman EK (2010) Cluster analysis in severe emphysema subjects using phenotype and genotype data: an exploratory investigation. Respir Res 11:30

Cogan JD, Vnencak-Jones CL, Phillips JA 3rd, Lane KB, Wheeler LA, Robbins IM, Garrison G, Hedges LK, Loyd JE (2005) Gross BMPR2 gene rearrangements constitute a new cause for primary pulmonary hypertension. Genet Med 7:169–174

Collard HR, Loyd JE, King TE Jr, Lancaster LH (2007) Current diagnosis and management of idiopathic pulmonary fibrosis: a survey of academic physicians. Respir Med 101:2011–2016

Crystal RG, Gadek JE, Ferrans VJ, Fulmer JD, Line BR, Hunninghake GW (1981) Interstitial lung disease: current concepts of pathogenesis, staging and therapy. Am J Med 70:542–568

Dahl M, Bowler RP, Juul K, Crapo JD, Levy S, Nordestgaard BG (2008) Superoxide dismutase 3 polymorphism associated with reduced lung function in two large populations. Am J Respir Crit Care Med 178:906–912

Dandara C, Lombard Z, Du Plooy I, McLellan T, Norris SA, Ramsay M (2011) Genetic variants in CYP (-1A2, -2C9, -2C19, -3A4 and -3A5), VKORC1 and ABCB1 genes in a black South African population: a window into diversity. Pharmacogenomics 12:1663–1670

de Bakker PI, McVean G, Sabeti PC, Miretti MM, Green T, Marchini J, Ke X, Monsuur AJ, Whittaker P, Delgado M et al (2006) A high-resolution HLA and SNP haplotype map for disease association studies in the extended human MHC. Nat Genet 38:1166–1172

DeMeo DL, Mariani TJ, Lange C, Srisuma S, Litonjua AA, Celedon JC, Lake SL, Reilly JJ, Chapman HA, Mecham BH et al (2006) The SERPINE2 gene is associated with chronic obstructive pulmonary disease. Am J Hum Genet 78:253–264

DeMeo DL, Hersh CP, Hoffman EA, Litonjua AA, Lazarus R, Sparrow D, Benditt JO, Criner G, Make B, Martinez FJ et al (2007) Genetic determinants of emphysema distribution in the national emphysema treatment trial. Am J Respir Crit Care Med 176:42–48

Drazen JM, Silverman EK, Lee TH (2000) Heterogeneity of therapeutic responses in asthma. Br Med Bull 56:1054–1070

Drent M, Wijnen P, Bast A (2012) Interstitial lung damage due to cocaine abuse: pathogenesis, pharmacogenomics and therapy. Curr Med Chem 19:5607–5611

Feng Y, Shen J, Streaker ED, Lockwood M, Zhu Z, Low PS, Dimitrov DS (2011) A folate receptor beta-specific human monoclonal antibody recognizes activated macrophage of rheumatoid patients and mediates antibody-dependent cell-mediated cytotoxicity. Arthritis Res Ther 13:R59

Flockhart DA (2007) Drug interactions: cytochrome P450 drug interaction table. Indiana University School of Medicine, http://www.medicine.edu/clinpharm/ddis/

Folz RJ, Guan J, Seldin MF, Oury TD, Enghild JJ, Crapo JD (1997) Mouse extracellular superoxide dismutase: primary structure, tissue-specific gene expression, chromosomal localization, and lung in situ hybridization. Am J Respir Cell Mol Biol 17:393–403

Fuselli S, Gilman RH, Chanock SJ, Bonatto SL, De Stefano G, Evans CA, Labuda D, Luiselli D, Salzano FM, Soto G et al (2007) Analysis of nucleotide diversity of NAT2 coding region reveals homogeneity across Native American populations and high intra-population diversity. Pharmacogenomics J 7:144–152

Gisbert JP, Gomollon F, Cara C, Luna M, Gonzalez-Lama Y, Pajares JM, Mate J, Guijarro LG (2007) Thiopurine methyltransferase activity in Spain: a study of 14,545 patients. Dig Dis Sci 52:1262–1269

Hagaman JT, Kinder BW, Eckman MH (2010) Thiopurine S- methyltransferase [corrected] testing in idiopathic pulmonary fibrosis: a pharmacogenetic cost-effectiveness analysis. Lung 188:125–132

Harrison RE, Flanagan JA, Sankelo M, Abdalla SA, Rowell J, Machado RD, Elliott CG, Robbins IM, Olschewski H, McLaughlin V et al (2003) Molecular and functional analysis identifies ALK-1 as the predominant cause of pulmonary hypertension related to hereditary haemorrhagic telangiectasia. J Med Genet 40:865–871

He JQ, Ruan J, Connett JE, Anthonisen NR, Pare PD, Sandford AJ (2002) Antioxidant gene polymorphisms and susceptibility to a rapid decline in lung function in smokers. Am J Respir Crit Care Med 166:323–328

He Y, Hoskins JM, McLeod HL (2011) Copy number variants in pharmacogenetic genes. Trends Mol Med 17:244–251

Hersh CP (2010) Pharmacogenetics of chronic obstructive pulmonary disease: challenges and opportunities. Pharmacogenomics 11:237–247

Hersh CP, Demeo DL, Lange C, Litonjua AA, Reilly JJ, Kwiatkowski D, Laird N, Sylvia JS, Sparrow D, Speizer FE et al (2005) Attempted replication of reported chronic obstructive pulmonary disease candidate gene associations. Am J Respir Cell Mol Biol 33:71–78

Hersh CP, Demeo DL, Lazarus R, Celedon JC, Raby BA, Benditt JO, Criner G, Make B, Martinez FJ, Scanlon

PD et al (2006) Genetic association analysis of functional impairment in chronic obstructive pulmonary disease. Am J Respir Crit Care Med 173:977–984

Hogg JC (2004) Pathophysiology of airflow limitation in chronic obstructive pulmonary disease. Lancet 364:709–721

Hogg JC, Timens W (2009) The pathology of chronic obstructive pulmonary disease. Annu Rev Pathol 4:435–459

Hogg JC, Chu F, Utokaparch S, Woods R, Elliott WM, Buzatu L, Cherniack RM, Rogers RM, Sciurba FC, Coxson HO et al (2004) The nature of small-airway obstruction in chronic obstructive pulmonary disease. N Engl J Med 350:2645–2653

Humbert M, Deng Z, Simonneau G, Barst RJ, Sitbon O, Wolf M, Cuervo N, Moore KJ, Hodge SE, Knowles JA et al (2002) BMPR2 germline mutations in pulmonary hypertension associated with fenfluramine derivatives. Eur Respir J 20:518–523

Humbert M, Sitbon O, Chaouat A, Bertocchi M, Habib G, Gressin V, Yaici A, Weitzenblum E, Cordier JF, Chabot F et al (2006) Pulmonary arterial hypertension in France: results from a national registry. Am J Respir Crit Care Med 173:1023–1030

Hutyrova B, Pantelidis P, Drabek J, Zurkova M, Kolek V, Lenhart K, Welsh KI, Du Bois RM, Petrek M (2002) Interleukin-1 gene cluster polymorphisms in sarcoidosis and idiopathic pulmonary fibrosis. Am J Respir Crit Care Med 165:148–151

Ishii T, Matsuse T, Teramoto S, Matsui H, Miyao M, Hosoi T, Takahashi H, Fukuchi Y, Ouchi Y (1999) Glutathione S-transferase P1 (GSTP1) polymorphism in patients with chronic obstructive pulmonary disease. Thorax 54:693–696

Ito M, Hanaoka M, Droma Y, Hatayama O, Sato E, Katsuyama Y, Fujimoto K, Ota M (2008) The association of transforming growth factor beta 1 gene polymorphisms with the emphysema phenotype of COPD in Japanese. Intern Med 47:1387–1394

Jacquemin B, Kauffmann F, Pin I, Le Moual N, Bousquet J, Gormand F, Just J, Nadif R, Pison C, Vervloet D et al (2012) Air pollution and asthma control in the Epidemiological study on the Genetics and Environment of Asthma. J Epidemiol Community Health 66:796–802

Jang IJ, Shin SG, Lee KH, Yim DS, Lee MS, Koo HH, Kim HK, Sohn DR (1996) Erythrocyte thiopurine methyltransferase activity in a Korean population. Br J Clin Pharmacol 42:638–641

Jin J, Sun L, Jiao W, Zhao S, Li H, Guan X, Jiao A, Jiang Z, Shen A (2009) SLC11A1 (Formerly NRAMP1) gene polymorphisms associated with pediatric tuberculosis in China. Clin Infect Dis 48:733–738

Juul K, Tybjaerg-Hansen A, Marklund S, Lange P, Nordestgaard BG (2006) Genetically increased antioxidative protection and decreased chronic obstructive pulmonary disease. Am J Respir Crit Care Med 173:858–864

Kennedy JL, Heymann PW, Platts-Mills TA (2012) The role of allergy in severe asthma. Clin Exp Allergy 42:659–669

Kim WJ, Hersh CP, DeMeo DL, Reilly JJ, Silverman EK (2009a) Genetic association analysis of COPD candidate genes with bronchodilator responsiveness. Respir Med 103:552–557

Kim WJ, Sheen SS, Kim TH, Huh JW, Lee JH, Kim EK, Lee SM, Lee S, Lim SY, Shin TR et al (2009b) Association between CRHR1 polymorphism and improved lung function in response to inhaled corticosteroid in patients with COPD. Respirology 14:260–263

Kirchheiner J, Fuhr U, Brockmoller J (2005) Pharmacogenetics-based therapeutic recommendations–ready for clinical practice? Nat Rev Drug Discov 4:639–647

Kokkotou E, Torres D, Moss AC, O'Brien M, Grigoriadis DE, Karalis K, Pothoulakis C (2006) Corticotropin-releasing hormone receptor 2-deficient mice have reduced intestinal inflammatory responses. J Immunol 177:3355–3361

Korthagen NM, van Moorsel CH, Kazemier KM, Ruven HJ, Grutters JC (2012) IL1RN genetic variations and risk of IPF: a meta-analysis and mRNA expression study. Immunogenetics 64:371–377

Krynetski EY, Schuetz JD, Galpin AJ, Pui CH, Relling MV, Evans WE (1995) A single point mutation leading to loss of catalytic activity in human thiopurine S-methyltransferase. Proc Natl Acad Sci U S A 92:949–953

Kuznetsov IB, McDuffie M, Moslehi R (2009) A web server for inferring the human N-acetyltransferase-2 (NAT2) enzymatic phenotype from NAT2 genotype. Bioinformatics 25:1185–1186

Lam MP, Cheung BM (2012) The pharmacogenetics of the response to warfarin in Chinese. Br J Clin Pharmacol 73:340–347

Lane KB, Machado RD, Pauciulo MW, Thomson JR, Phillips JA 3rd, Loyd JE, Nichols WC, Trembath RC (2000) Heterozygous germline mutations in BMPR2, encoding a TGF-beta receptor, cause familial primary pulmonary hypertension. Nat Genet 26:81–84

Larsson C, Eriksson S (1977) Liver function in asymptomatic adult individuals with severe alpha1-antitrypsin deficiency (Pi Z). Scand J Gastroenterol 12:543–546

Launay D, Remy-Jardin M, Michon-Pasturel U, Mastora I, Hachulla E, Lambert M, Delannoy V, Queyrel V, Duhamel A, Matran R et al (2006) High resolution computed tomography in fibrosing alveolitis associated with systemic sclerosis. J Rheumatol 33:1789–1801

Lee SW, Chung LS, Huang HH, Chuang TY, Liou YH, Wu LS (2010) NAT2 and CYP2E1 polymorphisms and susceptibility to first-line anti-tuberculosis drug-induced hepatitis. Int J Tuberc Lung Dis 14:622–626

Lei X, Zhu H, Zha L, Wang Y (2012) SP110 gene polymorphisms and tuberculosis susceptibility: a systematic review and meta-analysis based on 10 624 subjects. Infect Genet Evol 12:1473–1480

Libby DM, Gibofsky A, Fotino M, Waters SJ, Smith JP (1983) Immunogenetic and clinical findings in idiopathic pulmonary fibrosis. Association with the

B-cell alloantigen HLA-DR2. Am Rev Respir Dis 127:618–622

Loenders B, Van Mechelen E, Nicolai S, Buyssens N, Van Osselaer N, Jorens PG, Willems J, Herman AG, Slegers H (1998) Localization of extracellular superoxide dismutase in rat lung: neutrophils and macrophages as carriers of the enzyme. Free Radic Biol Med 24:1097–1106

Loyd JE, Primm RK, Newman JH (1984) Familial primary pulmonary hypertension: clinical patterns. Am Rev Respir Dis 129:194–197

Loyd JE, Butler MG, Foroud TM, Conneally PM, Phillips JA 3rd, Newman JH (1995) Genetic anticipation and abnormal gender ratio at birth in familial primary pulmonary hypertension. Am J Respir Crit Care Med 152:93–97

Machado RD, Pauciulo MW, Thomson JR, Lane KB, Morgan NV, Wheeler L, Phillips JA 3rd, Newman J, Williams D, Galie N et al (2001) BMPR2 haploinsufficiency as the inherited molecular mechanism for primary pulmonary hypertension. Am J Hum Genet 68:92–102

Machado RD, Aldred MA, James V, Harrison RE, Patel B, Schwalbe EC, Gruenig E, Janssen B, Koehler R, Seeger W et al (2006) Mutations of the TGF-beta type II receptor BMPR2 in pulmonary arterial hypertension. Hum Mutat 27:121–132

Machado RD, Eickelberg O, Elliott CG, Geraci MW, Hanaoka M, Loyd JE, Newman JH, Phillips JA 3rd, Soubrier F, Trembath RC et al (2009) Genetics and genomics of pulmonary arterial hypertension. J Am Coll Cardiol 54:S32–S42

Mak JC, Chan-Yeung MM, Ho SP, Chan KS, Choo K, Yee KS, Chau CH, Cheung AH, Ip MS (2009) Elevated plasma TGF-beta1 levels in patients with chronic obstructive pulmonary disease. Respir Med 103:1083–1089

Manolio TA, Green ED (2011) Genomics reaches the clinic: from basic discoveries to clinical impact. Cell 147:14–16

Martinelli M, Scapoli L, Carbonara P, Valentini I, Girardi A, Farinella F, Mattei G, Pacilli AM, Fasano L, Nava S et al (2013) Idiopathic pulmonary fibrosis and polymorphisms of the folate pathway genes. Clin Biochem 46(1–2):85–88

McLaughlin VV, Archer SL, Badesch DB, Barst RJ, Farber HW, Lindner JR, Mathier MA, McGoon MD, Park MH, Rosenson RS et al (2009) ACCF/AHA 2009 expert consensus document on pulmonary hypertension: a report of the American College of Cardiology Foundation Task Force on Expert Consensus Documents and the American Heart Association: developed in collaboration with the American College of Chest Physicians, American Thoracic Society, Inc., and the Pulmonary Hypertension Association. Circulation 119:2250–2294

Mega JL, Simon T, Collet JP, Anderson JL, Antman EM, Bliden K, Cannon CP, Danchin N, Giusti B, Gurbel P et al (2010) Reduced-function CYP2C19 genotype and risk of adverse clinical outcomes among patients treated with clopidogrel predominantly for PCI: a meta-analysis. JAMA 304:1821–1830

Meilang Q, Zhang Y, Zhang J, Zhao Y, Tian C, Huang J, Fan H (2012) Polymorphisms in the SLC11A1 gene and tuberculosis risk: a meta-analysis update. Int J Tuberc Lung Dis 16:437–446

Meng YY, Wilhelm M, Rull RP, English P, Ritz B (2007) Traffic and outdoor air pollution levels near residences and poorly controlled asthma in adults. Ann Allergy Asthma Immunol 98:455–463

Metz G, Kraft M (2010) Effects of atypical infections with Mycoplasma and Chlamydia on asthma. Immunol Allergy Clin North Am 30:575–585, vii–viii

Mijatovic V, Iacobucci I, Sazzini M, Xumerle L, Mori A, Pignatti PF, Martinelli G, Malerba G (2012) Imputation reliability on DNA biallelic markers for drug metabolism studies. BMC Bioinformatics 13(Suppl 14):S7

Minematsu N, Nakamura H, Iwata M, Tateno H, Nakajima T, Takahashi S, Fujishima S, Yamaguchi K (2003) Association of CYP2A6 deletion polymorphism with smoking habit and development of pulmonary emphysema. Thorax 58:623–628

Mishra G, Poojary SS, Raj P, Tiwari PK (2012) Genetic polymorphisms of CCL2, CCL5, CCR2 and CCR5 genes in Sahariya tribe of North Central India: an association study with pulmonary tuberculosis. Infect Genet Evol 12:1120–1127

Mistry R, Cliff JM, Clayton CL, Beyers N, Mohamed YS, Wilson PA, Dockrell HM, Wallace DM, van Helden PD, Duncan K et al (2007) Gene-expression patterns in whole blood identify subjects at risk for recurrent tuberculosis. J Infect Dis 195:357–365

Mwenifumbo JC, Sellers EM, Tyndale RF (2007) Nicotine metabolism and CYP2A6 activity in a population of black African descent: impact of gender and light smoking. Drug Alcohol Depend 89:24–33

Nasim MT, Ghouri A, Patel B, James V, Rudarakanchana N, Morrell NW, Trembath RC (2008) Stoichiometric imbalance in the receptor complex contributes to dysfunctional BMPR-II mediated signalling in pulmonary arterial hypertension. Hum Mol Genet 17:1683–1694

Oh KT, Anis AH, Bae SC (2004) Pharmacoeconomic analysis of thiopurine methyltransferase polymorphism screening by polymerase chain reaction for treatment with azathioprine in Korea. Rheumatology (Oxford) 43:156–163

Owen HR, Elser M, Cheung E, Gersbach M, Kraus WL, Hottiger MO (2007) MYBBP1a is a novel repressor of NF-kappaB. J Mol Biol 366:725–736

Pan H, Yan BS, Rojas M, Shebzukhov YV, Zhou H, Kobzik L, Higgins DE, Daly MJ, Bloom BR, Kramnik I (2005) Ipr1 gene mediates innate immunity to tuberculosis. Nature 434:767–772

Perri D, Cole DE, Friedman O, Piliotis E, Mintz S, Adhikari NK (2007) Azathioprine and diffuse alveolar haemorrhage: the pharmacogenetics of thiopurine methyltransferase. Eur Respir J 30:1014–1017

Philips JA, Ernst JD (2012) Tuberculosis pathogenesis and immunity. Annu Rev Pathol 7:353–384

Pianezza ML, Sellers EM, Tyndale RF (1998) Nicotine metabolism defect reduces smoking. Nature 393:750

Png E, Alisjahbana B, Sahiratmadja E, Marzuki S, Nelwan R, Adnan I, van de Vosse E, Hibberd M, van Crevel R,

Ottenhoff TH et al (2012) Polymorphisms in SP110 are not associated with pulmonary tuberculosis in Indonesians. Infect Genet Evol 12:1319–1323

Polosa R, Thomson NC (2013) Smoking and asthma: dangerous liaisons. Eur Respir J 41(3):716–728

Qu C, Edwards EW, Tacke F, Angeli V, Llodra J, Sanchez-Schmitz G, Garin A, Haque NS, Peters W, van Rooijen N et al (2004) Role of CCR8 and other chemokine pathways in the migration of monocyte-derived dendritic cells to lymph nodes. J Exp Med 200:1231–1241

Redon R, Ishikawa S, Fitch KR, Feuk L, Perry GH, Andrews TD, Fiegler H, Shapero MH, Carson AR, Chen W et al (2006) Global variation in copy number in the human genome. Nature 444:444–454

Riha RL, Yang IA, Rabnott GC, Tunnicliffe AM, Fong KM, Zimmerman PV (2004) Cytokine gene polymorphisms in idiopathic pulmonary fibrosis. Intern Med J 34:126–129

Sadof M, Kaslovsky R (2011) Adolescent asthma: a developmental approach. Curr Opin Pediatr 23:373–378

Said SI, Hamidi SA (2011) Pharmacogenomics in pulmonary arterial hypertension: toward a mechanistic, target-based approach to therapy. Pulm Circ 1:383–388

Schoedel KA, Hoffmann EB, Rao Y, Sellers EM, Tyndale RF (2004) Ethnic variation in CYP2A6 and association of genetically slow nicotine metabolism and smoking in adult Caucasians. Pharmacogenetics 14:615–626

Shen CM, Zhu BF, Deng YJ, Ye SH, Yan JW, Yang G, Wang HD, Qin HX, Huang QZ, Zhang JJ (2010) Allele polymorphism and haplotype diversity of HLA-A, -B and -DRB1 loci in sequence-based typing for Chinese Uyghur ethnic group. PLoS One 5:e13458

Shi Y, Massague J (2003) Mechanisms of TGF-beta signaling from cell membrane to the nucleus. Cell 113:685–700

Siedlinski M, van Diemen CC, Postma DS, Vonk JM, Boezen HM (2009) Superoxide dismutases, lung function and bronchial responsiveness in a general population. Eur Respir J 33:986–992

Smith BP, Best DH, Elliott CG (2012) Genetics and pharmacogenomics in pulmonary arterial hypertension. Heart Fail Clin 8:319–330

Snider GL (1985a) Distinguishing among asthma, chronic bronchitis, and emphysema. Chest 87:35S–39S

Snider GL (1985b) Interstitial pulmonary fibrosis environmental or idiopathic? Hosp Pract (Off Ed) 20:94C-94H, 94M-94P, 94T-94U passim

Sorheim IC, DeMeo DL, Washko G, Litonjua A, Sparrow D, Bowler R, Bakke P, Pillai SG, Coxson HO, Lomas DA et al (2010) Polymorphisms in the superoxide dismutase-3 gene are associated with emphysema in COPD. COPD 7:262–268

Souza R, Humbert M, Sztrymf B, Jais X, Yaici A, Le Pavec J, Parent F, Herve P, Soubrier F, Sitbon O et al (2008) Pulmonary arterial hypertension associated with fenfluramine exposure: report of 109 cases. Eur Respir J 31:343–348

Spire-Vayron de la Moureyre C, Debuysere H, Mastain B, Vinner E, Marez D, Lo Guidice JM, Chevalier D, Brique S, Motte K, Colombel JF et al (1998) Genotypic and phenotypic analysis of the polymorphic thiopurine S-methyltransferase gene (TPMT) in a European population. Br J Pharmacol 125:879–887

Spire-Vayron de la Moureyre C, Debuysere H, Fazio F, Sergent E, Bernard C, Sabbagh N, Marez D, Lo Guidice JM, D'Halluin JC, Broly F (1999) Characterization of a variable number tandem repeat region in the thiopurine S-methyltransferase gene promoter. Pharmacogenetics 9:189–198

Stingl JC, Parmar S, Huber-Wechselberger A, Kainz A, Renner W, Seeringer A, Brockmoller J, Langsenlehner U, Krippl P, Haschke-Becher E (2010) Impact of CYP2D6*4 genotype on progression free survival in tamoxifen breast cancer treatment. Curr Med Res Opin 26:2535–2542

Suthanthiran M, Li B, Song JO, Ding R, Sharma VK, Schwartz JE, August P (2000) Transforming growth factor-beta 1 hyperexpression in African-American hypertensives: a novel mediator of hypertension and/or target organ damage. Proc Natl Acad Sci U S A 97:3479–3484

Taype CA, Castro JC, Accinelli RA, Herrera-Velit P, Shaw MA, Espinoza JR (2006) Association between SLC11A1 polymorphisms and susceptibility to different clinical forms of tuberculosis in the Peruvian population. Infect Genet Evol 6:361–367

Thomson JR, Machado RD, Pauciulo MW, Morgan NV, Humbert M, Elliott GC, Ward K, Yacoub M, Mikhail G, Rogers P et al (2000) Sporadic primary pulmonary hypertension is associated with germline mutations of the gene encoding BMPR-II, a receptor member of the TGF-beta family. J Med Genet 37:741–745

Thuong NT, Dunstan SJ, Chau TT, Thorsson V, Simmons CP, Quyen NT, Thwaites GE, Thi Ngoc Lan N, Hibberd M, Teo YY et al (2008) Identification of tuberculosis susceptibility genes with human macrophage gene expression profiles. PLoS Pathog 4:e1000229

Travis WD, Hunninghake G, King TE Jr, Lynch DA, Colby TV, Galvin JR, Brown KK, Chung MP, Cordier JF, du Bois RM et al (2008) Idiopathic nonspecific interstitial pneumonia: report of an American Thoracic Society project. Am J Respir Crit Care Med 177:1338–1347

van Diemen CC, Postma DS, Vonk JM, Bruinenberg M, Nolte IM, Boezen HM (2006) Decorin and TGF-beta1 polymorphisms and development of COPD in a general population. Respir Res 7:89

Varpela E, Tiilikainen A, Varpela M, Tukiainen P (1979) High prevalences of HLA-B15 and HLA-Dw6 in patients with cryptogenic fibrosing alveolitis. Tissue Antigens 14:68–71

Volz A, Boyle JM, Cann HM, Cottingham RW, Orr HT, Ziegler A (1994) Report of the second international workshop on human chromosome 6. Genomics 21:464–472

Voorter CE, Drent M, van den Berg-Loonen EM (2005) Severe pulmonary sarcoidosis is strongly associated with the haplotype HLA-DQB1*0602-DRB1*150101. Hum Immunol 66:826–835

Wallin R, Wajih N, Hutson SM (2008) VKORC1: a warfarin-sensitive enzyme in vitamin K metabolism and biosynthesis of vitamin K-dependent blood coagulation factors. Vitam Horm 78:227–246

Weinshilboum R (2003) Inheritance and drug response. N Engl J Med 348:529–537

Whyte M, Hubbard R, Meliconi R, Whidborne M, Eaton V, Bingle C, Timms J, Duff G, Facchini A, Pacilli A et al (2000) Increased risk of fibrosing alveolitis associated with interleukin-1 receptor antagonist and tumor necrosis factor-alpha gene polymorphisms. Am J Respir Crit Care Med 162:755–758

Wijnen PA, Drent M, Nelemans PJ, Kuijpers PM, Koek GH, Neef C, Haenen GR, Bekers O (2008) Role of cytochrome P450 polymorphisms in the development of pulmonary drug toxicity: a case–control study in the Netherlands. Drug Saf 31:1125–1134

Wijnen PA, Bekers O, Drent M (2010) Relationship between drug-induced interstitial lung diseases and cytochrome P450 polymorphisms. Curr Opin Pulm Med 16:496–502

Wijnen PA, Bekers O, Drent M (2011) Development of cocaine-induced interstitial lung damage in two CYP2C and VKORC1 variant allele carriers. Mol Diagn Ther 15:177–180

Wilkinson GR (2005) Drug metabolism and variability among patients in drug response. N Engl J Med 352:2211–2221

Xue J, Gochuico BR, Alawad AS, Feghali-Bostwick CA, Noth I, Nathan SD, Rosen GD, Rosas IO, Dacic S, Ocak I et al (2011) The HLA class II Allele DRB1*1501 is over-represented in patients with idiopathic pulmonary fibrosis. PLoS One 6:e14715

Yates CR, Krynetski EY, Loennechen T, Fessing MY, Tai HL, Pui CH, Relling MV, Evans WE (1997) Molecular diagnosis of thiopurine S-methyltransferase deficiency: genetic basis for azathioprine and mercaptopurine intolerance. Ann Intern Med 126: 608–614

Yoon HI, Silverman EK, Lee HW, Yoo CG, Lee CT, Chung HS, Kim YW, Han SK, Shim YS, Yim JJ (2006) Lack of association between COPD and transforming growth factor-beta1 (TGFB1) genetic polymorphisms in Koreans. Int J Tuberc Lung Dis 10:504–509

Young RP, Hopkins R, Black PN, Eddy C, Wu L, Gamble GD, Mills GD, Garrett JE, Eaton TE, Rees MI (2006) Functional variants of antioxidant genes in smokers with COPD and in those with normal lung function. Thorax 61:394–399

Zhang J, Xu DJ, Xu KF, Wu B, Zheng MF, Chen JY, Huang JA (2012) HLA-A and HLA-B gene polymorphism and idiopathic pulmonary fibrosis in a Han Chinese population. Respir Med 106:1456–1462

Zhou SF, Liu JP, Chowbay B (2009) Polymorphism of human cytochrome P450 enzymes and its clinical impact. Drug Metab Rev 41:89–295

Pharmacogenomics of Allergy and Asthma

24

Anjana Munshi, Luciana B. Crotti, Vandana Sharma, Sanjeev Sharma, and Luis A. Espinoza

Abstract

Allergy and asthma are the diseases of multifactorial etiologies increasing dramatically all over the world. Numerous factors (e.g., aeroallergen, toxin, pathogen, food, chemical insect debris, and drugs) play a key role in the pathophysiology of allergy and asthma. Recent advances in the field of genetics have led to the identification of genetic factors which are involved not only in the pathogenesis of asthma and allergy but also significantly (60–70 %) contribute toward the interindividual variability to drug response and adverse drug reactions. There are several common categories of medications for treating asthma and allergy. Significant heterogeneity in the efficacy and adverse drug reactions of anti-allergic and anti-asthmatic drugs have been observed and efforts have been made to study the role of genetic determinants in the variable interindividual response to medication. Armed with the knowledge of a patient's pharmacogenomic information, a clinician could predict the response to certain drugs and adverse drug reactions to improve the efficacy and tolerability of the therapeutic interventions.

A. Munshi (✉)
Department of Molecular Biology, Institute of Genetics and Hospital for Genetic Diseases, Osmania University, Begumpet, Hyderabad 500016, Andhra Pradesh, India
e-mail: anjanadurani@yahoo.co.in

L.B. Crotti
Department of Biochemistry, Uniformed Services University of the Health Sciences, 4301 Jones Bridge Rd, Bethesda, MD 20814, USA

V. Sharma
Institute of Genetics and Hospital for Genetic Diseases, Osmania University, Begumpet, Hyderabad 500016, Andhra Pradesh, India

S. Sharma
Department of Pharmacology, Apollo Health City, Jubillee Hills, Hyderabad, Andhra Pradesh, India

L.A. Espinoza
Department of Biochemistry and Molecular Biology & Cell Biology, Georgetown University, 3900 Reservoir Rd, Washington, DC 20057, USA

D. Barh et al. (eds.), *Omics for Personalized Medicine*,
DOI 10.1007/978-81-322-1184-6_24,

Abbreviations

ABP Diamine oxidase
ADAM-33 A disintegrin and metalloproteinase domain-containing protein 33
ADRB2 Beta adrenergic receptors
ADRB-2 Beta-2 adrenergic receptor
ADRs Adverse Drug Reactions
ALOX5 Arachidonate 5 lipoxygenase
ATP Adenosine triphosphate
BALF Bronchoalveolar lavage fluid
cAMP Cyclic adenosine monophosphate
CCL11 Chemokine (C-C motif) ligand 11
CD-14 Cluster of differentiation 14 (CD14)
CFTR Cystic fibrosis transmembrane conductance regulator
CHRM2 Cholinergic receptor muscarinic 2
CLCA1 Calcium-activated chloride channel regulator 1
COX-2 Cycloxygenase-2
CRDCP15 Caspase recruitment domain-containing protein 15
CRHR1 Corticotrophin-releasing hormone receptor 1
CYF1P2 Cytoplasmic FMR1-interacting protein
CYP1A2 Cytochrome P450 family 1 subfamily A, polypeptide 2
CYP2E1 Cytochrome P450 family 2 subfamily E, polypeptide 1
CysLTR1 Cysteinyl leukotriene receptor
DEFB1 Defensin beta 1
DNA Deoxyribonucleic acid
DPP10 Dipeptidyl-peptidase 10
FCER1B Fc epsilon receptor I beta-chain
FDA Food and Drug Administration
FKB51 FK506-binding protein 51
FLG Flaggrin
GATA3 Trans-acting T-Cell-specific transcription factor
GC-R Glucocorticoid resistant
GC-S Glucocorticoid sensitive
GM-CSF Granulocyte-macrophage colony-stimulating factor
GMCSF Granulocyte-monocyte colony-stimulating factor
GPRA G protein-coupled receptor for asthma
GSNOR *S*-Nitrosoglutathione reductase
GST Glutathione S-transferase
GWAS Genome-wide association study
HDC Histidine decarboxylase
hGR Human glucocorticoid receptor
HLA Human leukocyte antigen
HNM Histamine *N*-methyl transferase
HRH Histamine receptor
IFNG Interferon γ
IFNGR1 Interferon gamma receptor-1
IgE Immunoglobulin E
IL Interleukin
IL-1β Interleukin-1β
ISSAC The International Study of Asthma and Allergies in Childhood
JAK1 Janus kinase 1
Km Michaelis-Menton Constant
LABA Long-acting beta-2 agonists
LT Leukotriene
LTC4S Leukotriene C4 synthase
mRNA Messenger ribonucleic acid
NAD$^+$ Nicotinamide adenine dinucleotide
NF-κB Nuclear factor kappa B
PARP-1 Poly (ADP-ribose) polymerase 1
PBMC Peripheral blood mononuclear cells
PDE4D phosphodiesterase-4D
PHF-11 Plant homeodomain finger protein 11
POSTN Periostin osteoblast-specific factor
RNA Ribonucleic acid
SABA Short-acting beta-2 agonists
Serpin B2 Serpin peptidase inhibitor clade B2
SLE Systemic Lupus Erythematous
SNPs Single nucleotide polymorphisms
SPINK5 Serine protease inhibitor kazal type 5
STAT6 Signal transducer and activator of transcription
TBX2 T-box transcription factor
TCγ T cell antigen receptor γ chain
TGF-β Transforming growth factor B
TH1 T helper 1
TH2 T helper 2
TLR Toll-like receptors
TNF-α Tumor necrosis factor-α
Vmax Maximum rate of the enzyme catalyzed reaction

1 Introduction

Allergy and asthma are the diseases of complex etiologies which are dramatically increasing all over the world. Allergic diseases are the most common causes of chronic illnesses in developed countries. The rates may appear to be lower in developing countries but there is an evidence of a steady increase over time in the proportion of the population suffering from allergies (ISSAC Steering Committee 1998; European Community Respiratory Health Survey 1996; Sheikh and Alves 2000; Garabrant and Schweitzer 2002; Demoly and Bousquet 2001; Bousquet 2000). Allergic reactions occur when an individual reacts abnormally to its surrounding environment. There is a worldwide epidemic of allergic diseases because of the changing environment and improved general health, superimposed on a range of genetic susceptibilities (Evans and McLeod 2003; Silverman et al. 2001). Allergic diseases include a wide range of immunological responses which may occur on exposure to aeroallergens, food, chemicals, microbes, parasites, insect debris, dust mites, smoking, and drugs. Up to 30 % of the general population suffers from various forms of allergies during a normal life span (Marshall 2004).

The common examples of allergies are food allergy, chemical energy, allergy due to poisons or environmental changes, and drug allergy (idiosyncratic reactions). Idiosyncratic reactions or adverse drug effects (ADRs) are the set of allergic reactions which occur when the usual therapeutic dose of a drug affects the patient having another underlying condition (Dolen 2004). ADRs are one the major causes of hospitalization and death in the United States and other developed countries. Approximately, 2,000,000 patients suffer from ADRs each year (Dolen 2004). There are four types of allergic/hypersensitivity reactions: type I, type II, type III, and type IV reactions. Type I reactions are mediated via immunoglobulin E (IgE) and are immediate anaphylactic shock reactions. The examples of type I allergic or hypersensitivity reactions are allergic rhinitis, food allergy, drug allergy, insect venom allergy, and allergic asthma. Type II reactions are cytotoxic type and involve IgG and IgM, e.g., pernicious anemia, acute rheumatic fever, and transplant rejections. Type III reactions are immune complex-mediated reactions where an antibody binds to an antigen leading to abnormal activation of complement system, e.g., glomerulonephritis and systemic lupus erythematous (SLE). Type IV are the delayed or cell-mediated reactions mediated via T cell lymphocytes, e.g., contact dermatitis such as poison ivy rashes (Gell and Coombs 1963).

Asthma is a multifactorial disease of the respiratory tract. It might develop due to exposure to an allergen, infection, exercise, obesity, smoking, and interaction of environment and genetic factors (Kumari and Rana 2012). It results on account of intermittent narrowing of the small airways of the lung with subsequent airflow obstruction, increased bronchial responsiveness, mucus production to a variety of stimuli accompanied by symptoms like wheeze, cough, and breathlessness (Szalai et al. 2006). Asthma is classified into two main types, intermittent or persistent. If persistent, it is assessed in terms of severity: mild, moderate, or severe. Atopic (allergic) asthma is the most common disease of childhood and has been reported to be strongly genetic in origin.

Evidence suggests that the prevalence of asthma increases globally by 50 % every decade (Masoli et al. 2004). Incidences of asthma have substantially increased in last two decades, but the reasons are unknown (Keller and Lowenstein 2002). In the United States, the prevalence of asthma is 7.7 % among adults and 9.6 % in children. It has been documented as 2–4 % among Asians, 15–20 % in the United Kingdom, Canada, Australia, New Zealand, and other developed countries (ISSAC 1998; Janson et al. 2001; Asher et al. 2006; Zock et al. 2006). A survey conducted in India showed that approximately 20–30 % of the population suffer from allergy and more than 15 % develop asthma (Singh and Kumar 2003).

2 Genetic Contribution to Asthma and Allergy

Genome-wide linkage screens reported in different populations investigating chromosomal regions linked to asthma and atopy or related phenotypes like elevated IgE levels, wheezing, and bronchial hyperresponsiveness have identified genes of biological relevance to these diseases on chromosome 5q [containing interleukin-3 (IL-3), IL-5, and granulocyte-macrophage colony-stimulating factor (GM-CSF)], 11q[FCER1B], 12q [IFNG (interferon γ) and STAT6], and 16p [IL-4R (the IL-4Rα chain, also part of the IL-13R)] (March et al. 2011).

Linkage studies followed by positional cloning approaches have identified novel susceptibility genes including cytoplasmic FMR1-interacting protein 2 (CYFIP2), dipeptidyl-peptidase 10 (DPP10), human leukocyte antigen (HLAG), plant homeodomain finger protein 11 (PHF11), G protein-coupled receptor for asthma (GPRA), and a disintegrin and metalloproteinase domain-containing protein 33 (ADAM33) suggesting their role in the pathophysiology of asthma and pulmonary allergic diseases (Van Eerdewegh et al. 2002; Allen et al. 2003; Zhang et al. 2003; Nicolae et al. 2005; Noguchi et al. 2005; Laitinen et al. 2004).

Candidate gene association studies have identified more than 100 genes in association with asthma and related phenotypes. These include receptors for detection of microbial products [toll-like receptors (TLR), cluster of differentiation 14 (CD14), caspase recruitment domain-containing protein 15 (CRDCP15)]; various cytokines and cytokine signaling proteins involved in T cell survival, proliferation, and differentiation; genes involved in lung function, development, and response to stimuli [beta-2 adrenergic receptor (ADRB-2), cystic fibrosis transmembrane conductance regulator (CFTR), serine protease inhibitor kazal-type 5 (SPINK5)]; genes involved in epithelial barrier function and innate immunity [Flaggrin (FLG) and Defensin beta 1 (DEFB1)]; and genes involved in the responses to environmental exposures, e.g., glutathione S-transferase (GSTM1, GSTP1, and GSTT1) (Halapi and Hakonarson 2004; Hoffjan et al. 2003; Kabesch 2005; Levy et al. 2005; Palmer et al. 2007; Vercelli 2008). However, these associations could not be established in subsequent replication studies. The genes which have been replicated extensively include ADRB2; the cytokines, receptors, signaling proteins, and transcription factors involved in T helper 1 (TH1) and T helper 2 (TH2) differentiation of T cells, like interleukin-4 and interleukin receptor 4R alpha (IL-4, IL-4RA), interferon G and interferon G receptor 1(IFNG, IFNGR1), signal transducer and activator of transcription(STAT6), trans-acting T-cell-specific transcription factor (GATA3), and T-box transcription factor (TBX21), and genes involved in cellular responses that characterize atopic disease, such as interleukin-13 (IL-13) and its receptor and the Fc epsilon receptor I beta-chain (FCER1B) gene (Liggett 1995; Martinez et al. 1997; Potter et al. 1993; Howard et al. 2002; Kabesch et al. 2006; Potaczek et al. 2009; Vladich et al. 2005; Wu et al. 2010; Basehore et al. 2004; Haller et al. 2009; Munthe-Kaas et al. 2008; Pykalainen et al. 2005; Randolph et al. 2004; Suttner et al. 2009; Zhou et al. 2009a).

The first genome-wide association study (GWAS) focusing on bronchial asthma was carried out in 2007 by Moffatt et al. Multiple markers on 17q21 were identified which reproducibly associated with childhood asthma. Independent replication studies in population of diverse ethnic backgrounds also confirmed the 17q21 association (March et al. 2011). Some of the genetic variants associated with asthma and allergy have been described in detail in the following subsections:

2.1 Interleukin-4 Receptor α-Chain (IL-4Rα)

The cytokines IL-4, IL-5, IL-9, and IL-13 are derived from T helper type 2 (Th2) (Renauld 2001) and induce structural changes in the airways and lung parenchyma, which are associated with asthma symptoms. These proinflammatory factors also contribute to a variety of clinically manifestations caused by reversible airway obstruction, involving difficulty in breathing with wheezing and coughing (Busse and Lemanske 2001). Multiple evidences support the concept that IL-4 and IL-13 are associated with the initiation and progression of asthma, through binding to a receptor complex in lung tissues and promoting many of the pathological features of asthma (Corry et al. 1998). Although epidemiological data and experimental evidence have firmly established a key role for the IL-4R axis in asthma pathogenesis, the relative dominance of IL-13 in asthma pathogenesis still remains unanswered. The new genomic information providing an understanding about the role of IL-4Rα in the lung in response to insults is very valuable; however, more information is required to define its role in those signaling pathways that are modulated in acute versus chronic pathophysiological changes in asthma. Of particular interest is the identification of polymorphisms in components of the IL-4 and IL-13 cytokine receptor that added important evidences about the requisite of these cytokines for asthma development. Also important is the relevant information of how IL-4 signaling is conferred to the effector cells by a series of phosphorylation events through binding to one of the variants of the IL-4Rα, a transmembrane receptor that recruits a second chain, the gamma common chain (yc), when activated by its ligand (Kondo et al. 1993). IL-4 can also activate a second form, the IL-4 type II receptor composed by the IL-4Rα and the IL-13 receptor alpha chain (IL-13Rα) (Cosman et al. 1990). Other important indicators of the events in which this cytokine is involved in modulating allergy responses are the IL-4 role in regulating immunoglobulin E (IgE) production, isotype switching from IgM to IgE production by B lymphocytes, stimulating B-cell proliferation, and modulating differentiation of T cells to the Th2 phenotype (Bloemen et al. 2007). The fact that these effects were not detected in IL-4 and IL-13 knockout mice (Kips et al. 1995; Leigh et al. 2004) and were also significantly inhibited in animals treated with IL-2 during initial antigen exposure (Kips et al. 1995) suggested that these IL-4 and IL-13 are participants in the multiple physiological processes that lead to asthmatic symptoms.

The identification of polymorphisms in IL-4Rα, which are more common in African Americans (Wenzel et al. 2007), has been associated with severe asthma exacerbations, lower lung function, and increased mast cell-related tissue inflammation. This was also evident in tissue biopsies of individuals with IL-4Rα polymorphisms in which increased mast cell numbers and a more specific increase in IgE+ were detected (Wenzel et al. 2007). These findings support the association of alleles at E375A with a gain in function of the receptor, which may have a causal role in promoting severe asthma exacerbations (Balzar et al. 2007; Busse et al. 2001). A meta-analyses of data from reports of case–control studies for associations of I50V and Q551R IL-4R variants and asthma proved that the R551 IL-4R variant had a highly significant association for a modest risk of atopic asthma, whereas the 150 V variants did not have any consistent involvement (Loza and Chang 2007). It was noted that the racial differences in the incidence of the Arg551 IL-4R (African Americans) and the Q551R (whites) impacted host defenses against parasites in tropical populations compared with temperate and cold regions, a strong environmental factor that appears to have strong influence on these mutations (Wenzel et al. 2007).

A significantly lower frequency of allele Arg551 of IL-4RA Q551R polymorphism was found in Slovenian children with nonatopic asthma, but it was absent in atopic asthmatics (Berce and Potocnik 2010). A similar approach was made in Asian populations in which the low prevalence of IleA 50/Ile50 was also identified in a Chinese group and Ile50/Q576 in Malay populations (Zhang et al. 2007). Moreover, the identification of polymorphisms of IL-4Rα associated with different ethnicities was indicated as a consequence of human population history and structure (Ober et al. 1998). Indeed, the increased eosinophil fraction and the enhanced bronchial hyperresponsiveness, markers of asthma pathogenesis Korean children carrying the IL-4Rα Arg551 allele, was linked to increased susceptibility to atopic asthma (Lee et al. 2004). This confirmed the findings in a Chinese population in which Arg551 was significantly higher in children with asthma (Zhang et al. 2006). In contrast, studies in Hawaiian, English, and Italian populations found no associations between Q551R and susceptibility to develop asthma (Bottini et al. 2002; Patuzzo et al. 2000; Tam et al. 2003). However, a recent phase 2b clinical trial data testing pitrakinra, a IL-4R receptor antagonist, in patients with moderate and severe asthma has reported a significant reduction in asthma exacerbations on those patients carrying the SNP res8832 (Slager et al. 2012). There is no doubt that these exciting findings indicate the existence of a subgroup of subjects with moderate to severe persistent asthma that showed, after treatment, an evident reduction in asthma exacerbations and symptoms.

2.2 Cluster of Differentiation 14

The cluster of differentiation 14 (CD14) is a multifunctional receptor found in two distinct forms, the membrane-bound (mCD14) and the soluble (sCD14). Of all forms, the CD14 has genetic variations that affect the structure of the CD14 molecule or the amount of CD14 available. Also, a functional C-159T polymorphism located in the promoter region of the gene has been associated with asthma and allergies (Martinez 2005), occupational lung disease (LeVan et al. 2005), and post lung transplantation response. CD14 was also associated with the recognition of pathogens by cells and with potential involvement in triggering the activation of defense responses. In fact, a clinical study determined that the soluble CD14 was increased in the bronchoalveolar lavage fluid (BALF) of smokers and COPD patients. Complementary in vitro studies using human bronchial epithelial cells lines treated either with interleukins or pathogens confirmed that CD14 levels were consistently increased in supernatants of stimulated cells (Regueiro et al. 2009). This research group also implied that CD14 might modulate a cellular effector mechanism to prevent overwhelming the inflammatory response caused by smoking or by the pathogen-associated molecular patterns.

With respect to the second type of response, CD14 may act by limiting the internalization of pathogens by airway cells. Later, it was also observed that smokers who had the TT homozygote genotype for the C-159T nucleotide polymorphism for the CD14 gene promoter have greater obstructive impairment of lung function and increased disease severity relative to those with the CC or CT genotypes. While the increased effect on lung dysfunction in the CC genotype was strongly correlated with the increased amount of cigarettes consumed for the subjects involved in the study, the lower lung function on the TT genotype was associated with a moderate smoking history (Zhou et al. 2009b). These findings suggested that C-159T SNP of the CD14 gene promoter might be a marker of susceptibility to tobacco smoke exposure as well as to the pathophysiology of COPD. The same group also proposed that tobacco smoke might be able to alter the interaction of genes with the environment; consequently, these events may induce in some way, changes in the innate immune pathways, promoting chronic lung disease.

The link between CD14 and asthma is still a conflictive issue. Indeed, this relationship was not present in Australian, Korean, and Polish children as well as in adults with moderate and severe persistent asthma (Hong et al. 2007; Kedda et al. 2005; Lis et al. 2001). Recently, in a more rigorous study using asthma patients separated by groups according to disease subphenotype (Perin et al. 2011), it was confirmed that CD14 gene polymorphism did not show any association with asthma in Slovenian children, but rather with bronchial obstruction and hyperreactivity in Slovenian patients with nonatopic asthma. These findings suggest the heterogeneity in genetic risk factors for asthma and asthma traits among different populations.

2.3 Poly (ADP-Ribose) Polymerase 1 (PARP-1)

PARP-1 is a nuclear protein that catalyzes the covalent long chain poly(ADP-ribosylation) of a variety of nuclear proteins utilizing nicotinamide adenine dinucleotide (NAD^+) as a substrate, with PARP-1 itself being the major target of modifications (Alkhatib et al. 1987; Cherney et al. 1987). This nuclear protein is only activated when binding to single- or double-stranded DNA ends via its two zinc fingers, which recognize DNA breaks independent of the DNA sequence (Cherney et al. 1987; Gradwohl et al. 1990; Menissier-de Murcia et al. 1989). Since PARP-1 activation is strictly proportional to the number of DNA breaks, its activity is strictly proportional to the number of DNA breaks in vivo or in vitro, and it is particularly inactive in the absence of DNA breaks (Cherney et al. 1987; Menissier-de Murcia et al. 1989; Schreiber et al. 2006). The covalent poly(ADP-ribosyl)ation of nuclear DNA-binding proteins in eukaryotes is a phenomenon that contributes to various physiologic and pathophysiologic events associated with DNA strand breakage, repair of DNA damage, and apoptosis (Bhatia et al. 1990; Hassa and Hottiger 2002; Kidwell and Mage 1976; Menissier-de Murcia et al. 1989; Simbulan-Rosenthal et al. 2003). Multiple lines of evidence have demonstrated that in addition to its accessory role in DNA repair, PARP-1 also plays regulatory roles in other nuclear processes, including DNA replication and the regulation of transcription, as a longevity assurance factor associated with genome stability, and in redox signaling (Schreiber et al. 2006).

PARP-1 gene knockout reduced in an asthma model allergen-induced IL-5 expression without affecting expression or activation of IL-4-associated kinases, but absence of PARP-1 doesn't alter the expression of janus kinase 1 (JAK1) and JAK3 and their catalytic activity. In contrast, signaling transducers and degradation of the activator of transcription 6 were observed in spleens in an allergen-dependent manner, a phenomenon that was linked to a severe reduction in GATA binding protein 3 transcripts (Datta et al. 2011). The authors suggested that mediation of STA-6 expression by PARP-1 is essential to IL-5 production and that modulation of NF-κB by this nuclear protein may also be a critical event. This alternative mechanism may provide an alternative therapeutic strategy to inhibit PARP-1 that strongly reduces the production of IL-5 as well as subsequent eosinophilia upon allergen exposure. Additional studies have found that PARP-1 inhibition prevented infiltration of inflammatory cells into the lungs upon allergen exposure (Boulares et al. 2003) and eosinophil recruitment as well as the production of IL-5 in a murine model of allergic airway inflammation (Naura et al. 2008; Oumouna et al. 2006). This suggested that PARP-1 depletion displays therapeutic potential for the treatment of disorders and diseases affecting the lung.

Genetic variants or SNPs of PARP-1 have been pointed out as the major contributions to dissect the events associated with its catalytic activity (Yadollahi-Farsani et al. 1999), modification of protein function during aging (Cottet et al. 2000), and understanding of the pathology of human breast cancers (Cao et al. 2007). Indeed, 20 rare variants of PARP-1 have been identified so far, suggesting that some of them may alter PARP-1 protein interaction through its BRCA1 C Terminus domain and contribute to an increased risk for some pathological conditions (Jiang et al. 2009; Lockett et al. 2004). Among these polymorphisms, the codon 762 variant (exon 17, 2444T to C, V to A) is present in about 5–33 % of the general population with an allelic frequency of 0.2, 0.4, and 0.6 in Caucasians, Asians, and African Americans. The 762 variant has also been indicated as the main cause of about a 40 % decrease of PARP-1 activity (Jiang et al. 2009).

In vitro and in vivo experiments have provided evidence that the PARP-1Val762Ala polymorphism reduces the enzymatic activity of PARP-1 (Lockett et al. 2004; Wang et al. 2007), depending on substrate condition by increasing Km (Wang et al. 2007). Indeed, an interesting report showed that the PARP-1 Val762Ala polymorphism was associated with a decreased risk of asthma in a Turkish population when compared with their wild-type homozygotes (Tezcan et al. 2009). This protective effect of PARP-1 Val762Ala polymorphism is the first one reported to have a protective effect against asthma. PARP-1 is also required for the transcriptional activity of transcription factors involved in inflammatory processes (Espinoza et al. 2007), and its depletion suppresses the expression of IL-12 and tumor necrosis factor-α (Boulares et al. 2003). These findings suggest that the protective effect of the PARP-1 Val762Ala polymorphism may be directly associated with decreased PARP-1 enzymatic activity, which in turn regulates the expression of genes involved in inflammatory responses in the lung tissue from stressors.

2.4 β_2-Adrenergic Receptor

The β_2-adrenergic receptor is an important regulator of airway smooth muscle tone. Agonists against this receptor are very efficient in the treatment of asthma. The β_2-adrenergic receptor contains at least 17 SNPs within a 3 kilobase region in which is also included its regulatory regions and coding region, which have been associated with altered expression of the receptor (Reihsaus et al. 1993; Scott et al. 1999). Five of the nine polymorphisms in the coding region are degenerated, while four of them result in amino acid substitutions within the protein (Reihsaus et al. 1993). These variations have been associated with altered expression, downregulation, or coupling of the receptor response to β2-adrenoeceptors agonists (Liggett 2000). The polymorphisms Arg16Gly and Glu27Gln have been related with reduced lung function (Green et al. 1994) and increased risk of asthma or COPD in African Americans (Matheson et al. 2006). These polymorphisms are also the main targets in modulating response to β-agonist treatment. However, some recent studies in large populations have indicated that the Ile164 variant seems to be associated with reduced lung function and increased risk of COPD, and that Ile164, Arg16, and Glu27 polymorphisms are not necessarily related to risk of asthma (Thomsen et al. 2012).

On the other hand, studies in Latin American populations with asthma found that those patients carrying the Gly16 variant are less likely to respond to a single dose of albuterol than carriers of the Arg16 (Martinez et al. 1997), suggesting the significant role of this polymorphism in modulating responses to β2-agonists. However, different responses are observed in two Latin American populations. As a matter of fact, Puerto Rican individuals with the Arg16had have lower bronchodilator response than Mexican subjects (Choudhry et al. 2005). These findings suggest the existence of ethnic-specific pharmacogenetic differences in the effect of Arg16 with asthma severity and bronchodilator response. Although the two polymorphisms at β_2-adrenergic receptor gene: one at codon 16 (arginine to glycine) and the other at the codon 27 (glutamine to glutamate) affect individual airway responses or response to acute or chronic β_2-agonist; they are not considered risk factors for asthma. Interestingly, an interactive effect on asthma between the Gly-16 homozygotes and cigarette smoking was observed in a group of smokers' individuals when compared with a never smoking population (Wang et al. 2001).

An interesting observation was the identification of the genetic variants of the enzyme S-nitrosoglutathione reductase (GSNOR), an enzyme that regulates the response to β2-agonists and β2-AR (Que et al. 2009). This variant has shown a strong correlation with airway hyperreactivity in bronchoalveolar fluid of asthmatic Latino subjects. Indeed, a consistent association was observed between altered GSNOR activity and increased risk of asthma in Puerto Rican individuals. Also of great interest was the predictive potential of GSNOR in the response to the bronchodilator albuterol in Mexicans and Puerto Rican individuals with asthma (Choudhry et al. 2010). Promising data has also indicated that the gain of function mutation of this enzyme is critical for the protection of lung tissues against allergen challenge and from β_2-adrenergic receptor desensitization under basal conditions (Whalen et al. 2007).

A study conducted to determine the effects of polymorphisms of the β_2-adrenergic receptor at codons 16 and 27 on agonist-induced vasodilatation and desensitization in the human vasculature found that individuals carrying β_2-adrenergic receptor polymorphisms homozygous for Arg16 and Gln27 have enhanced agonist-mediated desensitization (Dishy et al. 2001). In addition, subjects homozygous for Gly16 have resistance to agonist-mediated desensitization. These findings are in agreement with previous reports showing that variants for Gly16 had significantly greater bronchodilative desensitization compared to variants homozygous for Arg16 (Tan et al. 1997), or individuals with asthma who were homozygous for Arg16 had significantly lower response to an inhaled β_2-adrenergic agonist than individuals homozygous for Gly16 (Israel et al. 2000). Similarly, it was also reported that homozygous Arg16 patients had a higher incidence of exacerbation of asthma compared to individuals that are heterozygous or homozygous for Gly16 in response to β_2-adrenergic agonists (Taylor et al. 2000).

The evidence showing that albuterol-evoked FEV1 is higher and the response is more rapid in asthmatic individuals Arg16 homozygotes compared with the cohort of carriers of patients with Gly16 variant, indicated that genetic polymorphisms of β_2-adrenergic receptors influence bronchodilator responsiveness (Lima et al. 1999). In contrast, Arg16 homozygote patients who use albuterol for a long period of time had a decline in the morning peak expiratory flow, which increased during the following weeks out of treatment. Considering that individuals heterozygous and homozygous for Gly16 did not show that decline, the authors postulated that the Arg/Arg group might be at risk for adverse, or less salutary, effects, especially as they discontinue the high-dose therapy. It was also suggested that alternative schedules of therapies including earlier initiation of anti-inflammatory agents or both may produce better outcomes for patients that are homozygous for Arg16 (Evans and McLeod 2003).

The above-described findings were the basis for a regularly genotype-stratified, randomized placebo-controlled trial that included patients with mild asthma that were regularly treated with albuterol or placebo for 16 weeks or ipratropium bromide as needed. Patients with Arg/Arg phenotype had lower peak expiratory flow rate (PEFR) during treatment, while those Gly/Gly patients had an increase in morning PEFR during treatment albuterol. Different patterns were observed in responses from patients who discontinued albuterol and used ipratropium bromide. Indeed, Arg/Arg patients had an increase in morning PEFR, while no changes were detected in those individuals having the Gly/Gly genotype. These results led the authors to the interpretation that bronchodilators provide a long-term response to albuterol, while Arg/Arg patients should use a different bronchodilator.

Although the enhanced vasodilative response to isoproterenol in subjects who are homozygous for Glu27 was associated to the relative resistance to agonist-mediated desensitization (Cockcroft et al. 2006), others have indicated that it is the allele encoding Arg16 that determines vascular desensitization and that individuals homozygous for Gly16, whether they are homozygous for Gln27 or for Glu27, have a resistance to desensitization (Dishy et al. 2001). This group also highlighted that haplotypes and not a single polymorphism are probably the most valuable criteria for defining the functional significance in vivo. In support to this notion, another report showed that a polymorphism in the promoter region of the gene encoding the β_2-adrenergic receptor, which is associated with increased receptor expression under base-line conditions, was in strong linkage disequilibrium with the Glu27 polymorphism. Unique interactions of multiple polymorphisms within a haplotype ultimately can affect biologic and therapeutic phenotypes. Consequently, this individual phenotype may have poor predictive power as pharmacogenetic loci.

The predicting model to short-acting bronchodilators was built using Bayesian networks, which may allow the process of simultaneous associations and interactions among variables simultaneously in acute bronchodilator response. This analytical method is very significant because it will allow more prediction of drug responses and the characterization of the complex biological pathways involved. Even the bronchodilator response Bayesian network has the potential to provide accurate prediction models in pharmacogenetics; the authors recognized the limitations regarding its applicability in the clinical setting. The main two reasons discussed were the no inclusion of ADRB2 polymorphisms due to the contradictory evidences regarding associations with different phenotypes and populations (Israel et al. 2000) and the selection bias of restricted candidate genes (15 SNPs from 15 candidate genes), which may have low power to detect genetic effects as well as the genetic data. It is therefore necessary to point out that function of many of the genes within the human genome whose function remains unknown may have significant relevance for drug development and clinical medicine.

2.5 Glucocorticoid Receptor

Glucocorticoids (GCs) are routinely used by inhalation as an effective anti-inflammatory approach in patients with asthma. Although inhaled GCs (ICS) are very efficient in preventing the progression to asthma exacerbation and reducing asthma mortality, its chronic use has been associated with significant long-term side effects (Barnes and Woolcock 1998). These steroids inhibit expression of a series of proinflammatory mediators such as interleukins IL-3, IL-5, and IL-6, which modulate inflammatory cell proliferation and eosinophil regulation and enhance the production of the cytokines IL-10 and IL-12 (John et al. 1998; Naseer et al. 1997). These factors drive the suppression of macrophage activation and differentiation of T cells to the Th1 subtype, which are critical events in alleviating airway inflammation in asthma (Fu et al. 2013).

GR modulates in asthma many important proinflammatory genes by binding to and inhibiting key transcription factors, like the nuclear factor kappa B (NF-κB) or the activator protein 1 (AP-1), which are important regulators of the inflammation process associated with this disease (De Bosscher et al. 2003). Indeed, inhibition of the tumor necrosis factor α, IL-1β, and granulocyte-monocyte colony-stimulating factor (GMCSF) by GR is an efficient treatment against the pathophysiology of airway inflammation in asthma, including vasodilation, increased vascular permeability, and inflammatory cell recruitment (Boulares et al. 2003; Hamid and Tulic 2009). NF-κB also induces the expression of nitric oxide synthase (NOS), which generates the production of nitric oxide (NO), a molecule that contributes to many of the inflammatory manifestations of asthma, including vasodilation and inflammatory cell recruitment (Ashutosh 2000).

Asthmatic subjects that inhaled GC repress NF-κB with the concomitant gene expression reduction of both the inducible NOS and cyclooxygenase-2 (COX-2) with the consequent suppression of nitric oxide and peroxynitrite formation in neutrophils, eosinophils, and macrophages (Redington et al. 2001; Saleh et al. 1998). In addition, inhibition of AP-1 may target collagenase, stromelysin, and other metalloproteinases (Revollo and Cidlowski 2009), which are important players in the process of remodeling observed in the airways of asthmatic patients (Bergeron et al. 2009). The secondary signaling events associated with GCs action can also affect mRNA levels or promote changes in the composition of other signaling pathways. Indeed, GCs targeted the transcriptional activation function of Smad3, both in vivo and in vitro, which is downstream from the transforming growth factor β (TGF-β) receptor (Song et al. 1999).

An early study identified polymorphisms in the glucocorticoid receptor gene, which was located at position 1220 (AAT to AGT) and was associated with increased sensitivity to GCs in elderly individuals that exhibited heterozygous for this polymorphism (Huizenga et al. 1998). It was also noted that sensitivity to GCs may promote some severe side effects such as increased body mass index and reduced bone mineral density. These findings were later confirmed by others in two different groups of populations (Italians and Dutch) in which heterozygous carriers of the glucocorticoid receptor were more susceptible to becoming more obese with a slightly unfavorable cardiovascular profile (Di Blasio et al. 2003). Since the glucocorticoids receptor is part of a large heterocomplex of proteins that cooperatively regulate GC-mediated signaling, polymorphisms affect genes that encode heterocomplex components that may lead to a different type of pharmacogenetic interactions. The stress-induced-phosphoprotein 1 genetic variation was identified in asthmatic subjects with reduced lung function (Hawkins et al. 2009). This study also identified haplotypes and haplotype pairs that correlate with baseline function measures and improvement in chronic airflow limitation after corticosteroid therapy. In addition, variation in FCER2, the gene that encodes for the low-affinity IgE receptor, was associated with severe exacerbations in asthmatic children who inhaled corticosteroids (Tantisira et al. 2007). Although this association was valid across multiple ethnic groups, African Americans showed a greater frequency of the allele T2206C SNP, which in turn was associated with the higher asthma prevalence in this racial group. More recent studies confirmed the pharmacogenetic association of the rs28364072 SNP (T2206C) in the FCER2 gene (Koster et al. 2011). Interestingly, this report also found that homozygous carriers for this allele have a higher risk of developing severe exacerbations and asthma-related hospitalization as well as a need for the increased inhaled corticosteroids dose.

A region that includes the intron 1 and intron 5 of STIP1 appears to be highly relevant in controlling STIP1 expression (Hawkins et al. 2009). However, pharmacogenetic associations may influence the interaction of many genes, which may also have a considerable impact in the response to glucocorticoids therapy. Indeed, the TBX21 gene that encodes for the T-box expressed in T cells is a transcription factor that modulates airways responsiveness within asthmatic individuals treated with GCs (Tantisira et al. 2004a). A compared gene expression profile study found a high number of genes (923) that were modulated by treatment with IL-1β/TNF-α. These results obtained in an in vitro experimental approach were reversed by GCs in the GC-sensitive (GC-S) patients. Additional stringent analysis of the molecular signatures allowed the identification of a reduced number of genes (15) that were differentially expressed in GC-S from GC-insensitive (GC-R) patients. The authors indicated that these genes may be potential markers of GC responsiveness and therefore could be used as a predictor to distinguish GC responders from non-responders (Hakonarson et al. 2005).

The association of longitudinal changes in lung function in response to corticosteroids identified a genetic variation in CRHR1 (rs242941), which is directly associated with enhanced pulmonary function response to 8 weeks of inhaled corticosteroid in three different asthmatic populations, involving both pediatric and adult patients (Tantisira et al. 2004b). The screen of 131 SNPs in 14 genes found four SNPs (rs242941, rs1990975, rs889182, and rs6191) from three genes, corticotrophin-releasing hormone receptor (CRHR1), FCER2, and the nuclear receptor subfamily 3, group C, member 1 (NR3C1) that showed strong association with the response to inhaled corticosteroids. Additional information on the haplotypes affecting CRHR1 distinguished three haplotypes. The most common of these, the haplotype termed GAT, showed a significantly enhanced response to treatment. The estimated improvement in FEV1 for those subjects carrying homozygous GAT/GAT haplotype was more than twice that for those homozygous for non-GAT haplotypes after treatment. Similar results were also observed in a Korean population with COPD in which improved FEV1 following inhaled corticosteroid together with a long-acting β2-agonist was evident in individuals homozygous for the variant rs2429419 (Kim et al. 2007).

Another similar study, however, did not find any correlation between the three CRHR1 polymorphisms with long-term function decrease in treated and untreated individuals, suggesting that polymorphisms affecting CRHR1 genes are not necessarily associated with immediate or long-term improvement in FEV1 treated with GCs or with prevention of accelerated FEV1 decrease in adult (Dijkstra et al. 2008). The authors also ruled out the possibility that other SNPs in CRHR1 or neighboring genes may be involved with treatment responses on lung function improvement. Another possible explanation for these contradictory results was that a large inversion region on chromosome 17, with which CRHR1 is in strong linkage disequilibrium, may influence the enhanced corticosteroid response as well as the duration of protective response to CRHR1 to treatment (Tantisira et al. 2008).

3 Treatment Strategies in Allergy and Asthma

A comprehensive treatment plan for allergic diseases including asthma is to avoid exposure to allergens that provoke immediate hypersensitivity reaction. It has to be coupled with appropriate pharmacotherapy which involves drug therapy (antihistamines, anticholinergics, bronchodilators, anti-leukotriene and steroidal drugs) and allergen immunotherapy treatment strategies for allergic diseases including asthma (Marshall 2004). Pharmacological agents especially used to treat asthma have been classified as controllers and relievers. Controllers are the anti-inflammatory medications to control the asthma on long-term basis (Cho 2010). These are corticosteroids (for systemic and inhalation purpose), long-acting beta-2 agonists or (LABA), leukotriene antagonists, methylxanthines, and anti-immunoglobulin (IgE) antibody, e.g., omalizumab. Relievers are used on an as needed basis, which act quickly to reverse bronchoconstriction. A few examples of relievers are short-acting beta-2 agonists, systemic glucocorticoids, and short-acting anticholinergic drugs (Cho 2010).

Significant heterogeneity in the efficacy and adverse drug reactions of anti-allergic and anti-asthmatic drugs have been observed, and efforts have been made to study the role of genetic determinants in the variable interindividual response to medication (Weinshilboum 2003). Pharmacogenomics links the genomic background of individual patients suffering from complex diseases like asthma and allergy with efficacy and tolerability of the therapeutic interventions to develop optimal individualized therapies. There are many examples of heritable differences in drug distribution and metabolism (pharmacokinetics) in individuals resulting in varied clinical response to medication. Other mechanisms underlying the genetic response to drugs include changes in the drug target (pharmacodynamics) and unidentified side effects in predisposed individuals (idiosyncratic reactions) (Silverman et al. 2003; Weinshilboum 2003; Evans and McLeod 2003). It has been estimated that genetic variation can account for 20–95 % of variability in drug disposition and effect. After the publication of human genome project, the idea that genetic variation will allow prediction of treatment response has been seen as an important goal (Lander et al. 2001; Venter et al. 2001; Sherrid 2001; Elmer-Dewitt 2001; Begley 2001). The proper guidelines have been drafted by the Food and Drug Administration (FDA) for drug development companies to submit pharmacogenetic information when reporting clinical trial results (Weiss et al. 2006). This will allow tailoring of individualized therapy according to an individual's genetic makeup.

In this chapter we have highlighted genetic variation accounting for the altered therapeutic response of prescribed pharmacological agents in asthma and allergy with the help of few examples.

4 Pharmacogenomics of Asthma and Allergy

Several genetic variants and haplotypes have been recently reported in genes that play a role in pharmacokinetics and pharmacodynamics of asthma drugs. In asthma, pharmacogenetic studies have been focused on the study of beta-2 adrenergic receptor antagonists and anti-leukotrienes. However, the field of allergy has been less fortunate on this trend because of involvement of multiple genes and also potential for multiple gene environment interactions (Dolen 2004). Some of the examples have been discussed below, but it is likely that many, as yet undiscovered genetic variants exist, given the many genes involved in pharmacokinetics and pharmacodynamics of the drugs used in the treatment of allergic reactions including asthma.

4.1 Antihistaminic Drugs

Histamine is a bronchoconstrictor, biogenic amine which plays a key role in the pathogenesis of allergic diseases (Dolen 2004). Histamine receptor antagonists, i.e., antihistaminics are mainly prescribed and have been used clinically for more than 50 years. The patients' response to these receptor antagonists has been reported to be remarkably heterogenous in terms of efficacy and side effects (Dolen 2004). Histamine has four main receptors: HRH1, HRH2, HRH3, and HRH4. HRH1 receptor is involved in the production of pain, smooth muscle contraction, and reduced contractility of the heart via release of histamine (Dolen 2004). HRH2 receptor stimulation results in acid secretion in stomach, smooth muscle relaxation, and increased cardiac contractility. Both the receptors HRH1 and HRH2 affect the vascular permeability leading to vasodilation. HRH3 is involved in modulation of neurotransmission, and HRH4 has been found to be involved in chemotaxis and inflammatory mediator release (García-Martín 2009).

Histamine is synthesized from its precursor histidine by decarboxylation reaction, catalyzed by the enzyme L-histidine decarboxylase (HDC). Further it is metabolized by histamine N-methyl transferase (HNMT) to tele-methylhistamine and domain oxidase (ABP1) to imidazole acetaldehyde (García-Martín et al. 2009). The clinical heterogeneity in response to histamine receptor antagonists could be explained by the genetic variations of the synthetic and metabolic pathways and the receptors (Igaz et al. 2002). Multiple variants in these genes have been found among individuals of different ethnicities. The polymorphisms of the enzymes involved in the metabolic pathway of histamine could either enhance or delay degradation (Dolen 2004).

HDC is a homodimeric enzyme composed of two 55-kDa subunits, with 12 exons, and is located on chromosome 15q21-22 (Garcia-Martin 2009). HDC is decarboxylase pyridoxal phosphate dependent and highly specific for substrate histidine. HDC is stimulated by cAMP-dependent protein kinase and calcium calmodulin-dependent protein kinase II pathways (Garcia-Martin 2009). It has been found that increased HDC expression levels are associated with allergic skin disorders, e.g., chronic urticaria (Papadopoulou et al. 2005). The most relevant nonsynonymous polymorphisms which have been identified in the HDC gene are Met31Thr (rs17740607), Leu553Phe (rs16963486), and Asp644Glu (rs2073440) (Garcia-Martin 2009). However, it remains to be elucidated whether HDC expression levels correlate with local histamine concentration and influencing the metabolization of antihistaminic drugs (Garcia-Martin 2009).

Diamine oxidase or amiloride binding protein (ABP1) is located on chromosome 7q34-q36. The enzyme catalyzes the degradation of histamine and some other compounds like spermidine, putrescine, and spermine, substances involved in allergic and immune response (Garcia-Martin 2009). The determination of enzyme activity is a useful marker of gut mucosal integrity (Garcia-Martin 2009). The enzyme activity has been reported to be modified by certain disorders including severe burns, gut injury, divorce enteropathies, abdominal surgery, chemotherapy, and kidney injury (Peng et al. 2004; DiSilvestro et al. 1997; Stein et al. 1994).

Several nonsynonymous polymorphisms have been reported in ABP1 gene, and among these 3 common nonsynonymous SNPs which bring about the amino acid substitution, Thr16Met, Ser332Phe, and His645Asp have been detected in Caucasians (Ayuso et al. 2007; Garcia-Martin 2009). In addition to this, two more SNPs causing the amino acid changes, i.e., Ile479Met and His659Asn, have been reported to occur in low frequency among Asians and African, respectively (Ayuso et al. 2007).

The functional effect of common ABP1 gene variants has been studied, and it was found that the variants Thr16Met and His645Asp cause changes in enzyme kinetics by increasing the Km of the enzyme (Ayuso et al. 2007). His645Asp gene variant of ABP1 has also been studied in patients with asthma and allergic rhinitis (Garcia-Martin 2007). It was found that patients who were carriers for Asp 645 allele might influence the clinical presentation of asthma and allergic rhinitis. Individuals bearing 645Asp allele in heterozygous and homozygous condition have Vmax/Km values of 66 and 51 %, respectively, in comparison with the carriers of His645 allele. The effect of 16 Met variant allele is similar with Vmax/Km values of 79 and 42 % for heterozygous and homozygous individuals, respectively, in comparison with the noncarriers of the variant allele (Garcia-Martin 2009). A study analyzing the role of Thr16Met, Ser332Phe, and His645Asp polymorphisms of ABP1 gene in intestinal biopsy samples of patients with food allergy did not find any major differences in enzyme activity with these variants (Peterson et al. 2005). However, this study did not investigate the effect of these variants on Km of the enzyme or putative differences in intestinal enzyme activity between individuals with food allergy in comparison with healthy subjects; therefore, this topic deserves further investigation (Garcia-Martin 2009).

The individuals carrying 645Asp allele tend to develop clinical symptoms even though IgE levels did not alter severely. This suggests that individuals with an altered histamine predisposition may display atopic symptom even in the absence of high release of histamine (Garcia-Martin 2007). The number of studies on the effect of ABP1 variation is still low, but overall findings are promising and suggestive that the clinical course of allergic and other diseases influenced by high histamine levels may be altered in patients with genetically determined impaired histamine metabolism. Therefore, ABP1 genetic variants could be a determinant in predicting the clinical outcome in such diseases (Garcia-Martin 2009).

HNMT gene is located on chromosome 2q22.1. Studies have reported that the only product of the 50 kb human HNMT gene was a single mRNA of approximately 1.6 kb with six exons (Aksoy et al. 1996). However, recent reports have revealed a new 1.0 kb mRNA encoded by this gene. Only two nonsynonymous SNPs have been described in HNMT gene, i.e., Thr105Ile and Ile99Val. Of these Ile105 is the most common variant allele (Preuss et al. 1998). This allele has been associated with the variant enzyme having reduced thermal stability and decreased HNMT activity. In addition to these nonsynonymous SNPs, a number of mutations in the 5′ flanking region of the HNMT gene have been described, but none of these seem to have a relevant effect on HNMT enzyme activity (Wang et al. 2002).

Since histamine is an important mediator of asthma and HNMT is the primary enzyme for histamine metabolism in bronchial epithelium, the studies have been carried out to understand the role of HNMT gene polymorphisms in the pathogenesis of asthma. Initially promising findings indicating the association of Thr105Ile allele with asthma were reported (Yan et al. 2000).

However, further independent studies failed to identify such an association either for patients with asthma or with bronchial hyperresponsiveness (Garcia-Martin 2007; Yan et al. 2000; Deindl et al. 2005; Sharma et al. 2005). These findings suggest that this polymorphism is not a relevant factor per se in the development of allergic diseases (Garcia-Martin 2009). However, a recent study suggested that the variant allele of this polymorphism may confer increased risk of developing atopic dermatitis. This needs further confirmation.

Studies involving SNPs in histamine receptors HRH1, HRH2, HRH3, and HRH4 are scarce and controversial, although few genetic polymorphisms have been described. No relationship between these polymorphisms and receptor function has yet been found. The genetic variants reported in these receptors have been reviewed by Garcia-Martin (2009).

HRH1 and HRH2 polymorphisms were investigated in association with allergy. Two studies carried out independently on a Japanese and Korean population have reported a lack of association between these polymorphisms and salicylate-induced urticaria/angioedema (Chen et al. 2004).

However, the clinical significance of these genetic polymorphisms is underway and yet to be explored in prospective studies including large sample size. Therefore, in case of histamine molecule, there are multiple variants of potential clinical significance considering the complex pathogenesis of allergy, asthma, and other related phenotypes as well as all the mediators involved a pharmacogenomic approach for treating these diseases would require development of a staggeringly large number of specific therapeutic agents (Dolen 2004).

Although some relevant information on histamine pharmacogenomics has evolved in the last few years, the knowledge of the clinical implication of the pharmacogenomics is still in its infancy.

4.2 Beta-2 Agonists

Beta-2 agonists represent the most important class of bronchodilator drugs which is used in the treatment of asthma. Beta-2 agonists, e.g., albuterol and salmeterol, target the beta adrenergic receptors (ADRB2) expressed in the bronchial smooth muscles and cause bronchodilation. There are two types of beta-2 agonists, one with shorter duration (4–6 h) of action (SABA) and the other with longer duration of action (LABA). The potential safety concerns with this drug class has increased the interest in the pharmacogenomics of this drug class. With the introduction of higher dose formulations of SABA isogriaterenol, an increase in asthma mortality was observed in the UK in the 1960s (Bleecker 2008). Similar observations were made in New Zealand with high-dose formulation of phenoterol (Holgate 2008). These increases in mortality rate subsided with the withdrawal of these drugs (Pearce et al. 1995).

Beta-2 agonists, e.g., albuterol and salmeterol, target the ADRB2 expressed in the bronchial smooth muscles and cause bronchodilation. ADRB2 receptor is a 65 kDa glycoprotein comprised of 413 amino acids, and the gene encoding is located on chromosome 5q31-32. It is a member of the large family of G protein-coupled receptors for histamine, adenosine, platelet-activating factor, and bradykinin (Dolen 2004).

The main mechanism involved in the process of bronchodilation is activation of beta-2 adrenergic receptors with agonist drug which uncouples stimulatory G proteins, followed by generation of increased amount of cyclic adenosine monophosphate (cAMP) from adenosine triphosphate or ATP, and thus relaxing the bronchial smooth muscles during an asthmatic attack (Cho 2010). The receptor mutations have been associated with susceptibility to nocturnal asthma and variable response to beta-2 agonists as well as subsensitivity in some individuals. Around 13 mutations have been reported either in the receptor itself or in the gene regulating the transcription of receptor gene (Cho 2010; Lipworth et al. 2002). Out of these, rs1042713, rs1042714, and rs1800888 are relatively common and have some functional significance (Cho 2010). Theses SNPs result in substitution of arginine for glycine at position 16 (Arg16Gly), glutamine for glutamic acid at position 27 (Gln27Glu), and threonine for isoleucine at position 164 (Thr164Ile), respectively. Studies have revealed the positive association of some ADRB2 gene variants with adverse clinical outcome in asthmatic patients in beta-2 agonists. The frequency of these variants varies between populations of different racial groups (Maxwell et al. 2005). The frequency of SNPs, Arg16Gly, and Gln27Glu has been found to be high in Asian population. Thr164Ile rarely observed only in European Americans (Malmstorm et al. 1999). The estimated frequency of the Arg16 variant has been found to be 39.3 % in White Americans, 49.2 % in Black Americans, and 51.0 % among Chinese (Maxwell et al. 2005; Litonjua et al. 2010).

The Gly16 variants produce a receptor which exhibits enhanced downregulation on agonist exposure in comparison with Arg16 allele in an in vitro model. A greater concentration of beta-2 agonist was needed to down-regulate the Glu 27 than Gln27 (Green et al. 1994). In case of polymorphism at 16th position, the frequency of Arg16 has been found to be 39.3 % in White Americans, 49.2 % in Black Americans, and 51 % in Chinese population (Maxwell et al. 2005; Litonjua et al. 2010).

In a clinical study including 195 children, it was observed that bronchodilating response was significantly higher among subjects homozygous for Arg16 as compared to subjects bearing Gly16 genotype. It was also found that patients with homozygous Arg16Gln27 haplotype pairs and heterozygous for Arg16Gln27/Gly16Glu 27 were found to have highest bronchodilating response (62.8 and 61.6 %, respectively) as compared with patients having homozygous Gly16Gln27 haplotype pairs who showed lowest response (43.2 %) (Cho et al. 2005).

Similarly it has been shown that albuterol-evoked forced expiratory volume in one second was higher and the response was more rapid in Arg 16 homozygotes in comparison with carriers of Gly16 (Lima et al. 1999).

The response to chronic doses with beta-2 agonists was observed by Israel et al. In a retrospective study including 190 subjects with mild asthma, randomized to albuterol drug as needed or two puffs four times daily (Israel et al. 2000). It was found that asthmatic patients bearing Arg16 homozygous genotype and taking regular albuterol for 16 weeks had reduced morning peak expiratory flow rates from baseline than the patients bearing same polymorphism but on albuterol as and when needed doses ($p=0.012$) (Israel et al. 2000; Cho 2010). The patients having Gly16 polymorphism were found not to be benefitted even when on the regular treatment of albuterol.

Wechsler et al. showed that homozygous for Arg16 associated with the reduced response to LABA as well (Wechsler et al. 2006). This observation has led to the use of combined inhaled corticosteroids. LABA medication might result in a poor clinical outcome in patients with this polymorphism. However, another study suggested that this polymorphism does not predict the worst outcome in moderate to severe asthmatic patients treated with this combined medication (Bleecker et al. 2008).

Currently there is no evidence that routine genotyping of asthmatic patients would be clinically useful. However, in patients with more severe disease and a suboptimal response to beta-2 agonist therapy, genotyping might provide an impetus for the early use of other therapies. In a study by Yelensky et al. indacaterol-treated patients were analyzed in two large randomized phase III studies and were genotyped for the most commonly studied polymorphisms in the ADRB2 gene: Gly16Arg, Gln27Glu, Thr164Ile, and a variant in the 5′ untranslated region (rs1042711). The analysis showed a little evidence for the association between these ADRB2 variants and indacaterol response, suggesting that ADRB2 genetic variation is unlikely to have a major role in differential response to indacaterol treatment in COPD patients (Yelensky et al. 2012).

4.3 Anti-leukotrienes

Anti-leukotriene drugs or leukotriene antagonists are the pharmacological agents mainly used in the treatment of asthma. Anti-leukotriene therapies include cysteinyl leukotriene receptor (CysLTR) blockers (e.g., montelukast, zafirlukast, and pranlukast) and inhibitors of 5-lipoxygenase enzyme (e.g., zileuton) (Tse et al. 2011). Leukotrienes are released from mast cells, eosinophils, and from other inflammatory cells in patients with asthma and some allergic diseases (Joos and Sandford 2002).

The leukotriene pathway starts with the conversion of arachidonic acid (derived from phospholipids) into leukotriene A4 (LTA4). This reaction is catalyzed by the enzyme 5-lipoxygenase (ALOX-5), a rate limiting step in the synthesis of leukotrienes (Tse et al. 2011). LTA4 is rapidly converted to LTC4 which is transported extracellular and stepwise removal of amino acids results in the formation of other leukotrienes such as LTE4 and LTD4 (Klotsman et al. 2007).

The ALOX-5 gene is located on chromosome 10q11.12 (Funk et al. 1989). Mutations in the promoter region of ALOX gene have been found to be associated with heterogenous response to anti-leukotrienes (Dolen 2004). The activity of the promoter region of the ALOX5 gene has been reported to be influenced by tandem repeat polymorphisms (Sp1 transcription factor-binding motifs present), and this varies between individuals with three, four, and six repeats (mutant type) and five repeats (wild type) (In et al. 1997).

In a retrospective pharmacogenetic study, the association of ALOX-5 gene, Sp1/Egr1 polymorphism, was studied in patients who were resistant to ALOX-5 inhibitor, ABT-761. In this study the mean forced expiratory volume (FEV1) was found to be improved in homozygous normal (18.8 ± 3.6 %) and heterozygous individuals (23.3 ± 6 %) as compared to homozygous mutant (Drazen et al. 1999). It was found that about 6 % of patients had mutated genotype and showed poor response to therapy (Dolen 2004; Drazen et al. 1999).

In another study, significantly higher FEV1 response to anti-leukotriene drug, montelukast was observed in patients homozygous (GG) for ALOX5 rs2115819 in comparison with patients having AA and AG genotypes (Lima et al. 2006). In this study, asthma exacerbation rates in participants after 6 months of placebo, montelukast, and theophylline treatment were found to be 6.1, 3.7, and 5.2 events/person-year, respectively.

Another enzyme of leukotriene pathway leukotriene C4 synthase (LTC4S) is responsible for the conjugation of glutathione with LTA4 to form LTC4, a potent bronchoconstrictor. Single nucleotide polymorphism in LTC4S promoter A444C has been reported to be associated with severe asthma (Sayers et al. 2003). The 444C allele variant has been found to create an activator protein-2 binding sequence and appears to be associated with enhanced cysteinyl leukotriene production (Sayers et al. 2003). Sampson et al. found that asthmatic patients treated with zafirlukast (20 mg twice daily), and bearing "A" allele for A444C polymorphism had a lower FEV1 response as compared to those having CC and AC genotypes (Sampson et al. 2000). The above examples indicate that genetic variation in the enzymes of leukotriene synthesis pathway can significantly influence the efficacy of anti-leukotriene drugs.

4.4 Theophylline

Theophylline is a methylxanthine used in the treatment of asthma and chronic obstructive pulmonary disease. It has both anti-inflammatory and bronchodilating effect. The drug exerts bronchodilating effect via inhibiting phosphodiesterase (PDE4D) enzyme. PDE4D is involved in the hydrolysis of cyclic adenosine monophosphate (cAMP) in airway smooth muscle cells and is the main target of theophylline. Studies have found that polymorphisms in PDE4D can significantly influence the response to theophylline. Some of the genetic variants of PDE4D have been found to alter the response of the theophylline, although it is not clear that whether the in vitro inhibitory response of theophylline also occurs in vivo (Polson et al. 1978; Bergstrand 1980; Fenechand Ellul-Micallef 2005). However, the development of second-generation theophyllines specifically inhibiting PDE4D in vivo is underway in phase III clinical trials (Schmidt et al. 1999; Fenech and Ellul-Micallef 2005).

Theophylline is metabolized to 1, 3-dimethyluric acid, 3 methylxanthine, and 1 methylxanthine mainly by cytochrome P450 1A2 (CYP1A2) and partly by CYP2E1 enzymes. Asthmatic patients have been found to differ in terms of their rate of drug metabolism and its serum concentration. One of the genetic variant -2964G/A (rs2069514) of CYP1A2 has been studied and was found to be associated with reduced theophylline clearance. Due to its potential side effects such as toxicity, tachycardia, headache, nausea, and seizures, theophylline may need to be used with care in patients bearing the "A" allele at the site -2964(G/A) in the CYP1A2 gene, because theophylline metabolism levels have been found to be very low in such patients, especially in young asthmatic individuals (Obase et al. 2003). Therefore, patients bearing "A" allele for -2964G/A polymorphism in the CYP1A2 gene may require less dose in order to avoid drug toxicity (Pignatti 2004).

4.5 Steroids

Glucocorticoids are the potent anti-inflammatory agents commonly prescribed in the treatment of asthma (Weiss et al. 2004). Glucocorticoids have been found to work by activation of anti-inflammatory genes and they repress expression of proinflammatory genes (Weiss et al. 2004). Inhaled corticosteroids have been found to be effective in asthmatics, but in few patients receiving higher doses, side effects such as growth retardation in children, adrenal suppression, bone demineralization, skin changes, and cataract formation have been observed (Paltiel et al. 2001; Sharek and Bergman 2000; Patel et al. 2001; Casale et al. 2001; Banerjee et al. 2001; Isaksson 2001). Moreover, there is a subset of the population which shows glucocorticoid resistance and adverse drug effects (Szefler and Leung 1997).

Two types of steroid-resistant asthma have been identified: type I and type II. Type I (>95 % of cases) is induced by cytokine and is associated with increased expression of glucocorticoid β (a less reactive glucocorticoid receptor). Type II (<5 % of cases) is because of less number of glucocorticoid receptors (Morrow 2007). Clinically type I steroid-resistant, asthmatic patients show severe side effects such as adrenal gland suppression and cushingoid features. Type II steroid-resistant patients develop a generalized cortisol resistance and do not develop steroid-induced side effects. Some of the genetic variants of the human glucocorticoids receptor have been identified to affect the response of steroidal drugs in asthma. For instance, single nucleotide polymorphism involving substitution of Val641Asp in human glucocorticoid receptor gene (hGR) leads to reduced binding affinity for dexamethasone. Another variant Val729Ile in the same gene confers a fourfold decrease in dexamethasone activity (Hurley et al. 1991). In another study a novel, nucleotide (T to C) substitution at position 2209 (exon 9α) of the hGR gene, or hGRF737L, within helix 11 of the ligand-binding domain of the protein has been identified which results into glucocorticoid resistance due to decreased affinity binding (2-fold) to the ligand (Charmandari et al. 2007). It has been found that subjects harboring Asn363Ser polymorphism in hGR gene were found to have a higher sensitivity to dexamethasone leading to increased body mass index and decreased bone mineral density as compared to subjects having normal allele (Huizenga et al. 1998).

Another important candidate gene corticotrophin-releasing hormone receptor (CRHR1) gene, located on chromosome 17q21-22, has been investigated for the presence of several single nucleotide polymorphisms in association with heterogeneity in response to corticosteroids (Weiss et al. 2004). These studies indicated that there exists a rationale for predicting therapeutic response in asthma and other corticosteroid-treated diseases.

Therefore, the genetic variants of CRHR1 have strong influence on response to corticosteroids in asthmatic patients and may provide new insights in the development of more efficacious corticosteroids with a targeted approach.

Recently genomic studies have focused on differential expression of genes associated with inhaled corticosteroid treatment response, e.g., corticosteroid treatment downregulated expression of calcium-activated chloride channel regulator (CLCA1), periostin, and serpin B2 while upregulating FK506-binding protein 51 (FKBP51) expression of in epithelial cells of respiratory tract of patients with asthma (Woodruff et al. 2007; Tse et al. 2011). In a study involving 4 weeks of treatment with inhaled corticosteroids, high baseline expression of CLCA1, periostin osteoblast-specific factor (POSTN) and serine peptidase inhibitor clad B2 (serpin B2) genes were correlated with improvement in FEV1 in asthmatic patients, whereas high expression of FKBP51 was associated with poor response to treatment (Woodruff et al. 2007).

Hakonarson et al. (2005) examined gene expression profile in freshly isolated peripheral blood mononuclear cells (PBMC) from two independent cohorts (training and test sets) of glucocorticoid sensitivity (GC-S) and glucocorticoid-resistant (GC-R) asthma patients to search the gene that predicted responders and nonresponders to inhaled corticosteroids (Hakonarson et al. 2005). 11,812 genes were examined using high-density oligonucleotide microarrays in both resting PBMC and cells treated in vitro with interleukin-Iβ (IL-1β) and tumor necrosis factor-α (TNF-α) combined. 5,011 genes expressed at significant levels in PBMS and out of these 1,334 were notably up- or downregulated by IL-1β/TNF-α. 923 genes were reversed by glucocorticoid treatment in GC-S group. The expression pattern of 15 of these 923 genes accurately separated from the nonresponders. This study uncovered gene expression profile in PBMC that predicts clinical response to inhaled glucocorticoid therapy with meaningful accuracy. The results of this study suggest that a pharmacogenomic approach may lead to the development of novel therapeutic strategies and diagnostic tests in asthma patients.

A very recent study that tested the association of the highest-powered SNPs in four independent patients with asthma was able to identify 13 SNPs that showed strong association with changes in FEV1 after ICS treatment. Of all these genes, GLCCI1 SNP, rs37972 was present in those asthmatic patients with the same phenotype that showed attenuation in response to the treatment with steroids. Since this SNP is in complete linkage disequilibrium with the functional rs37973, it appears that the decremental response to ICS may occur by decreasing the expression of GLCCI1. In another report, the same research team described the strong association of the SNPs rs37973 in GLCCI1 and rs1876828 in CRHR1 with good response to ICS treatment from very poor responders. Although these findings used a quite large dataset from four existing clinical trials of asthmatics and the findings are exciting, the main limitation was the utilization of adult individuals from only one population (Caucasian), which restricts the generalization of the results to children or other ethnic groups. Finally, using similar approaches, others have identified the T gene as a novel pharmacogenetic locus for ICS response in a large proportion of asthmatic individuals. Two top SNPs showed correlation with three known variants in the T gene, which also evidenced association with T gene variant maps to a potentially functional region of the gene in which rs2305089 encodes for an amino acid change (GLy177Asp) within the T protein. The three T variants also showed strong association with ICS response and two- to threefold differences in FEV1 response for patients homozygous for the wild type when compared with mutant alleles for each T gene SNP.

4.6 Cross Talk Between Corticosteroids and Beta-2 Agonists

A cross talk between signaling of beta-2 agonists and corticosteroids has been suggested to influence the therapeutic efficacy of both classes of drugs. LABA, e.g., salmeterol and formoterol, have been found to increase the nuclear localization of glucocorticoid receptors and can also enhance the function of receptors (Barnes 2002). The cross talk between these two classes of drugs may be linked with beta-2 agonists phenotypic outcomes, and genetic variation in ADRB2 might lead to the modulation of responses to corticosteroids (Bleecker 2008). This interaction has formed the basis for the approval of combination of inhaled corticosteroids and beta-2 agonists and guideline recommendation that these drugs should be used to treat the asthmatics who are not controlled with low to medium doses of inhaled corticosteroids alone (Kavuru et al. 2000; Palmqvist et al. 2001; Zetterstrom et al. 2001; GINA 2006; NAEPP 2007).

4.7 Anticholinergics

Anticholinergic drugs are the one which block the actions of acetylcholine exerted through muscarinic receptors. Polymorphic variants in muscarinic receptors M1 and M2 have been found to alter the response of anticholinergic drugs e.g. ipratropium bromide (Fenech and Ellul-Micallef 2005). A variable tandem repeat (CA) in the M2 receptor gene (CHRM2) located on chromosome 7q35-36, has been found to influence the responsiveness to muscarinic receptor antagonists (Fenech and Ellul-Micallef 2005). Several polymorphisms (-708A/G, -627G/C, -513C/A, 492C/T, a CTTT repeat and a GT repeat) have been reported in the CHRM2 receptor gene but the role of these SNPs affecting the anticholinergic drug efficacy remains yet to be unresolved (Donfack et al. 2003).

4.8 Anti-IgE Drugs

Omalizumab is a recombinant anti-immunoglobulin E (anti-IgE) antibody therapy for asthma-targeted patients with high levels of IgE. Eotaxin, a chemokine (CCL11motif), is a potent eosinophil chemoattractant which plays an important role in the pathophysiology of asthma. Studies have indicated that genetic variation in CCL11 locus is an important determinant of serum IgE levels among asthmatic patients (Raby et al. 2006). The genetic variation in CCL11 might predict the response to omalizumab in asthma patients. Genetic variants affecting drug response in asthma and allergy have been summed up in Table 24.1.

5 Future Directions

The clinical phenotypes of patients and causes of asthma and allergy are heterogenous. Allergy and asthma are multifactorial complex diseases which develop due to a combined interaction of genetic and environmental interactions. The variation in the genes involved in the biological pathways which are the targets of many anti-asthmatic and anti-allergic drugs should be evaluated and considered carefully along with the consideration of environmental factors to predict the clinical outcome of these diseases. Several pharmacogenetic

Table 24.1 Genes and pharmacogenetic effects on current asthma therapy

Gene	Position	Polymorphism	Pharmacogenetic effect in asthma
β_2-Adrenergic receptor agonists			
ADRB2 (β_2-adrenergic receptor)	5q31.32	Arg16GLy	Associated with enhanced acute response for bronchoprotection (controversial results)
		Gln27Glu	Association with terbutaline nebulizer response
		PolyC (3′UTR)	Counteracts ADRB2 downregulation
ARG I (Arginase 1)	6q23	rs2781659 (5′region)	Increase ARG1 transcription, improved response to β_2AR agonist
GSNOR (alcohol dehydrogenase)	4q23	rs1154400 (5′region)	Differential response to β_2AR agonist
			Different haplotypes determine GSNOR transcription
			Decreased response to salbutamol
Leukotriene modifiers			
ALOX5 (5-lipoxygenase)	10q11.2	rs2115189 (intronic)	Differential response to zileuton and montelukast therapy (FEV_1 change)
		rs4987105 (Thr120Thr)	Differential response to montelukast therapy (FEV_1 change)
		rs4986832 (5′ region)	
LTC4S (leukotriene C_4 synthase)	5q35	A-444C (rs730012)	Associated with increased production and expression of LTC_4 by blood eosinophils
		rs272431 (intronic)	Causes a differential response to zileuton therapy (FEV_1 change)
CYSLTR2 (cysteinyl leukotriene receptor 2)	13q14	rs91227 (3′UTR)	Associated with morning PEF in subjects taking montelukast
		rs912278 (3′UTR)	
MRP1 (Multidrug resistance-associated protein-1)	16p13.1	rs119774 (intronic)	Causes a differential response to zileuton and montelukast therapy (FEV_1 change)
OATP2B1 (organic anion transporters SB1)	11q13	rs12422149 (Arg312Gln)	Reduced morning plasma concentrations of montelukast, resulting in lack of clinical benefit
Corticosteroids			
CRHR1 (corticotrophin-releasing hormone receptor 1)	17q12-q22	rs242941 (intronic)	Associated with corticosteroid response (FEV_1 change)
TBX21 (T-box 21)	17q21.32	rs224001 (His33Gln)	Improvements in BHR with corticosteroid treatment
NK2R (Neurokinin 2 receptor)	10q22.1	rs7703891 (Gly231Glu)	Improved asthma control with corticosteroid treatment
STIP1 (stress-induced phosphoprotein 1)	11q13	rs4980524 (intron 1)	Associated with corticosteroid response (FEV_1 change)
		rs6591838 (intron 1)	
		rs2236647 (intron 5)	
DUSP1 (dual-specificity phosphatase 1)	5q34	rs881152 (5′region)	Associated with bronchodilator response and asthma control in the presence of corticosteroids
FCER2 (Fc fragment of IgE, low affinity II, receptor for (CD23))	19p13.3	rs28364072 (intronic)	Associated with asthma exacerbations in the presence of corticosteroids
GLCCI1 (glucocorticoid induced transcript 1)	7p21.3	rs37972 (5′region)	Associated with corticosteroid response (FEV_1 change); potential transcriptional mechanism

FEV forced expiratory volume, *UTR* untranslated region

studies have already set the stage for this approach. However, current pharmacogenomic studies in the field of asthma and allergy are limited by their small sample size, population stratification issues, imprecise phenotypic definition, and lack of replication of results in prospective studies. The current approach in asthma pharmacogenomics is to include epigenetics and copy number variation. Epigenetics studies the changes in gene expression pattern that are caused by mechanisms other than changes in nucleotide sequences of the genetic code itself. These include DNA methylation, transcription regulation, small interfering RNAs or microRNA, and posttranslational histone modifications (Kabesch et al. 2010). Although the field of asthma epigenetics is still in its infancy, a study of human bronchial biopsies demonstrated that histone deacetylation activity was decreased in untreated asthmatics (Ito et al. 2002). Asthmatics on inhaled corticosteroids treatment had a higher level but still lower than control subjects. These results suggest a role for epigenetics in asthma therapy, focusing on histone deacetylates as a potential pharmacogenomic target.

In addition to this copy number variation has also been implicated in the pathogenesis of childhood asthma. In a genome-wide analysis, children with allergic asthma were found to have decreased copy number variations at TCRγ gene. TCRγ gene encodes for a T cell surface protein, mainly involved in cell-mediated immunity (Walsh et al. 2010). However, further studies are required to replicate these findings.

Inclusion of epigenetics and copy number variations in planning the pharmacogenomic studies of asthma and allergy could prove more beneficial in detecting the novel genomic variations predicting the response to drugs. Understanding the impact of genomic information to determine the response to medication in the form of personalized medicine has the potential to improve the health care, decrease the adverse drug effects, and to improve the clinical outcome. The translation of pharmacogenomic information into clinical practice would improve the quality of life of patients with asthma and allergy.

References

Aksoy S, Raftogianis R, Weinshilboum R (1996) Human histamine N-methyltransferase gene: structural characterization and chromosomal location. Biochem Biophys Res Commun 219(2):548–554

Alkhatib HM, Chen DF, Cherney B, Bhatia K, Notario V, Giri C, Stein G, Slattery E, Roeder RG, Smulson ME (1987) Cloning and expression of cDNA for human poly(ADP-ribose) polymerase. Proc Natl Acad Sci USA 84:1224–1228

Allen M, Heinzmann A, Noguchi E, Abecasis G, Broxholme J, Ponting CP, Bhattacharyya S, Tinsley J, Zhang Y, Holt R, Jones EY, Lench N, Carey A, Jones H, Dickens NJ, Dimon C, Nicholls R, Baker C, Xue L, Townsend E, Kabesch M, Weiland SK, Carr D, von Mutius E, Adcock IM, Barnes PJ, Lathrop GM, Edwards M, Moffatt MF, Cookson WO (2003) Positional cloning of a novel gene influencing asthma from chromosome 2q14. Nat Genet 35(3):258–263

Asher MI, Montefort S, Björkstén B, Lai CK, Strachan DP, Weiland SK, Williams H, ISAAC Phase Three Study Group (2006) Worldwide time trends in the prevalence of symptoms of asthma, allergic rhinoconjunctivitis, and eczema in childhood: ISAAC Phases One and Three repeat multicountry cross-sectional surveys. Lancet 368(9537):733–743

Ashutosh K (2000) Nitric oxide and asthma: a review. Curr Opin Pulm Med 6(1):21–25

Ayuso P, García-Martín E, Martínez C, Agúndez JA (2007) Genetic variability of human diamine oxidase: occurrence of three nonsynonymous polymorphisms and study of their effect on serum enzyme activity. Pharmacogenet Genomics 17(9):687–693

Balzar S, Strand M, Rhodes D, Wenzel SE (2007) IgE expression pattern in lung: relation to systemic IgE and asthma phenotypes. J Allergy Clin Immunol 119:855–862

Banerjee TR, Fitzmaurice GM, Kotlov TV, LaHive K, LeBoff MS, Israel E (2001) Effects of inhaled glucocorticoids on bone density in premenopausal women. N Engl J Med 345:941–947

Barnes PJ (2002) Scientific rationale for inhaled combination therapy with long-acting beta2-agonists and corticosteroids. EurRespir J 19:182–191

Barnes PJ, Woolcock AJ (1998) Difficult asthma. Eur Respir J 12(5):1209–1218

Basehore MJ, Howard TD, Lange LA, Moore WC, Hawkins GA, Marshik PL, Harkins MS, Meyers DA, Bleecker ER (2004) A comprehensive evaluation of IL4 variants in ethnically diverse populations: association of total serum IgE levels and asthma in white subjects. J Allergy Clin Immunol 114(1):80–87

Begley S (2001) Made to order medicine. Newsweek 137(26):64–66, 68–9

Berce V, Potocnik U (2010) Association of Q551R polymorphism in the interleukin 4 receptor gene with nonatopic asthma in Slovenian children. Wien Klin Wochenschr 122(2):11–18

Bergeron C, Al-Ramli W, Hamid Q (2009) Remodeling in asthma. Proc Am Thorac Soc 6(3):301–305

Bergstrand H (1980) Phosphodiesterase inhibition and theophylline. Eur J Respir Dis Suppl 109:37–44

Bhatia K, Kang VH, Stein GS, Bustin M, Cherney BW, Notario V, Haque SJ, Huppi K, Smulson ME (1990) Cell cycle regulation of an exogenous human poly(ADP-ribose) polymerase cDNA introduced into murine cells. J Cell Physiol 144:345–353

Bleecker ER, Lawrance RM, Ambrose HJ, Goldman M (2008) Beta2- adrenergic receptor gene polymorphisms: is Arg/Arg genotype associated with serious adverse events during treatment with budesonide and formoterol in one pressurized metered dose inhaler. Am J RespirCrit Care Med 177:A775

Bloemen K, Verstraelen S, Van Den Heuvel R, Witters H, Nelissen I, Schoeters G (2007) The allergic cascade: review of the most important molecules in the asthmatic lung. Immunol Lett 113:6–18

Bottini N, Borgiani P, Otsu A, Saccucci P, Stefanini L, Greco E, Fontana L, Hopkin JM, Mao XQ, Shirakawa T (2002) IL-4 receptor alpha chain genetic polymorphism and total IgE levels in the English population: two-locus haplotypes are more informative than individual SNPs. Clin Genet 61:288–292

Boulares AH, Zoltoski AJ, Sherif ZA, Jolly P, Massaro D, Smulson ME (2003) Gene knockout or pharmacological inhibition of poly(ADP-ribose) polymerase-1 prevents lung inflammation in a murine model of asthma. Am J Respir Cell Mol Biol 28:322–329

Bousquet J (2000) Allergy as a global problem: think globally act globally. Allergy 57(8):661–662

Busse WW, Lemanske RF (2001) Asthma. N Engl J Med 344:350–362

Busse W, Corren J, Lanier BQ, McAlary M, Fowler-Taylor A, Cioppa GD, van As A, Gupta N (2001) Omalizumab, anti-IgE recombinant humanized monoclonal antibody, for the treatment of severe allergic asthma. J Allergy Clin Immunol 108:184–190

Cao WH, Wang X, Frappart L, Rigal D, Wang ZQ, Shen Y, Tong WM (2007) Analysis of genetic variants of the poly(ADP-ribose) polymerase-1 gene in breast cancer in French patients. Mutat Res 632:20–28

Casale TB, Nelson HS, Stricker WE, Raff H, Newman KB (2001) Suppression of hypothalamic-pituitary-adrenal axis activity with inhaled flunisolide and fluticasone propionate in adult asthma patients (see comments). Ann Allergy Asthma Immunol 87:379–385

Charmandari E, Kino T, Ichijo T, Jubiz W, Mejia L, Zachman K, Chrousos GP (2007) A novel point mutation in helix 11 of the ligand-binding domain of the human glucocorticoid receptor gene causing generalized glucocorticoid resistance. J Clin Endocrinol Metab 92(10):3986–3990

Chen GL, Zhu B, Nie WP, Xu ZH, Tan ZR, Zhou G, Liu J, Wang W, Zhou HH (2004) Single nucleotide polymorphisms and haplotypes of histamine N-methyltransferase in patients with gastric ulcer. Inflamm Res 53:484–488

Cherney BW, McBride OW, Chen DF, Alkhatib H, Bhatia K, Hensley P, Smulson ME (1987) cDNA sequence, protein structure, and chromosomal location of the human gene for poly(ADP-ribose) polymerase. Proc Natl Acad Sci USA 84:8370–8374

Cho SH (2010) Pharmacogenomics approaches to asthma treatment. Allergy Asthma Immunol Res 2(3):177–182

Choudhry S, Ung N, Avila PC, Ziv E, Nazario S, Casal J, Torres A, Gorman JD, Salari K, Rodriguez-Santana JR et al (2005) Pharmacogenetic differences in response to albuterol between Puerto Ricans and Mexicans with asthma. Am J Respir Crit Care Med 171:563–570

Choudhry S, Que LG, Yang Z, Liu L, Eng C, Kim SO et al (2010) GSNO reductase and beta2-adrenergic receptor gene-gene interaction: bronchodilator responsiveness to albuterol. Pharmacogenet Genomics 20(6): 351–358

Cockcroft JR, Gazis AG, Cross DJ, Wheatley A, Dewar J, Hall IP, Noon JP (2006) Beta(2)-adrenoceptor polymorphism determines vascular reactivity in humans. Hypertension 36:371–375

Corry DB, Grunig G, Hadeiba H, Kurup VP, Warnock ML, Sheppard D, Rennick DM, Locksley RM (1998) Requirements for allergen-induced airway hyperreactivity in T and B cell-deficient mice. Mol Med 4:344–355

Cosman D, Lyman SD, Idzerda RL, Beckmann MP, Park LS, Goodwin RG, March CJ (1990) A new cytokine receptor superfamily. Trends Biochem Sci 15:265–270

Cottet F, Blanche H, Verasdonck P, Le Gall I, Schachter F, Burkle A, Muiras ML (2000) New polymorphisms in the human poly(ADP-ribose) polymerase-1 coding sequence: lack of association with longevity or with increased cellular poly(ADP-ribosyl)ation capacity. J Mol Med (Berl) 78:431–440

Datta R, Naura AS, Zerfaoui M, Errami Y, Oumouna M, Kim H, Ju J, Ronchi VP, Haas AL, Boulares AH (2011) PARP-1 deficiency blocks IL-5 expression through calpain-dependent degradation of STAT-6 in a murine asthma model. Allergy 66:853–861

De Bosscher K, Vanden Berghe W, Haegeman G (2003) The interplay between the glucocorticoid receptor and nuclear factor-kappaB or activator protein-1: molecular mechanisms for gene repression. Endocr Rev 24(4):488–522

Deindl P, Peri-Jerkan S, Deichmann K, Niggemann B, Lau S, Sommerfeld C, Sengler C, Müller S, Wahn U, Nickel R, Heinzmann A (2005) German Multicenter Atopy Study Group. No association of histamine- N-methyltransferase polymorphism with asthma or bronchial hyperresponsiveness in two German pediatric populations. Pediatr Allergy Immunol 16(1):40–42

Demoly P, Bousquet J (2001) Epidemiology of drug allergy. Curr Opin Allergy Clin Immunol 1:305–310

Di Blasio AM, van Rossum EF, Mastrini S et al (2003) The relation between two polymorphisms in the glucocorticoid receptor gene and body mass index, blood pressure and cholesterol in obese patients. Clin Endocrinol 59:68–74

Dishy V, Sofowora GG, Xie HG, Kim RB, Byrne DW, Stein CM, Wood AJ (2001) The effect of common polymorphisms of the beta2-adrenergic receptor on agonist-mediated vascular desensitization. N Engl J Med 345:1030–1035

DiSilvestro RA, Jones AA, Smith D, Wildman R (1997) Plasma diamine oxidase activities in renal dialysis patients, a human with spontaneous copper deficiency and marginally copper deficient rats. Clin Biochem 30(7):559–563

Dolen WK (2004) Pharmacogenomics in clinical allergy. Allergy Clin Immunol Int J World Allergy Org 16:231–236

Donfack J, Kogut P, Forsythe S (2003) Sequence variation in the promoter region of the cholinergic receptor muscranic 3 gene and asthma and atopy. J Allergy Clin Immunol 111:527–532

Drazen JM, Yandava CN, Dubé L, Szczerback N, Hippensteel R, Pillari A, Israel E, Schork N, Silverman ES, Katz DA, Drajesk J (1999) Pharmacogenetic association between ALOX5 promoter genotype and the response to anti-asthma treatment. Nat Genet 22(2):168–170

Elmer-Dewitt P (2001) The future of drugs. Time 157(2):56–57

Espinoza LA, Smulson ME, Chen Z (2007) Prolonged poly(ADP-ribose) polymerase-1 activity regulates JP-8-induced sustained cytokine expression in alveolar macrophages. Free Radic Biol Med 42:1430–1440

Europian Community Respiratory Health Survey (1996) Variations in the prevalence of respiratory symptoms, self-reported asthma attacks and use of asthma medication in the Europian Community Respiratory Health Survey. Eur Respir J 9:687–695

Evans WE, McLeod HL (2003) Pharmacogenomics – drug disposition, drug targets, and side effects. N Engl J Med 348:538–549

Fenech AG, Ellul-Micallef R (2005) Pharmacogenetics of asthma therapeutics. Br J Clin Pharmacol 9:24–28

Fu JJ, Baines KJ, Wood LG, Gibson PG (2013) Systemic inflammation is associated with differential gene expression and airway neutrophilia in asthma. OMICS 17(4):187–199

Funk CD, Hoshiko S, Matsumoto T, Radmark O, Samuelsson B (1989) Characterization of the human 5-lipoxygenase gene. Proc Natl Acad Sci USA 86:2587–2591

Garabrant DH, Schweitzer S (2002) Epidemiology of latex sensitisation and allergies in health care workers. J Allergy Clin Immunol 110:S82–S95

Garcia-Martin E, Garcia-Menaya J, Sanchez B, Martinez C, Rosendo R, Agundez JA (2007) Polymorphisms of histamine metabolizing enzymes and clinical manifestations of asthma and allergic rhinitis. Clin Exp Allergy 37(8):1175–1182

García-Martín E, Ayuso P, Martínez C, Blanca M, Agúndez JA (2009) Histone. Pharmacogenomics 10(5):867–883

Gell PGH, Coombs RRA (1963) Clinical aspects of immunology. Blackwell Ist ed, Oxford

Global Initiative for Asthma (GINA) (2006) Global strategy for asthma management and prevention. http://www.ginasthma.org

Gradwohl G, Menissier de Murcia JM, Molinete M, Simonin F, Koken M, Hoeijmakers JH, de Murcia G (1990) The second zinc-finger domain of poly(ADP-ribose) polymerase determines specificity for single-stranded breaks in DNA. Proc Natl Acad Sci USA 87:2990–2994

Green SA, Turki J, Innis M, Liggett SB (1994) Amino-terminal polymorphisms of the human beta 2-adrenergic receptor impart distinct agonist-promoted regulatory properties. Biochemistry 33:9414–9419

Hakonarson H, Bjornsdottir US, Halapi E, Bradfield J, Zink F, Mouy M, Helgadottir H, Gudmundsdottir AS, Andrason H, Adalsteinsdottir AE, Kristjansson K, Birkisson I, Arnason T, Andresdottir M, Gislason D, Gislason T, Gulcher JR, Stefansson K (2005) Profiling of genes expressed in peripheral blood mononuclear cells predicts glucocorticoid sensitivity in asthma patients. Proc Natl Acad Sci USA 102(41):14789–14794

Halapi E, Hakonarson H (2004) Recent development in genomic and proteomic research for asthma. Curr Opin Pulm Med 10(1):22–30

Haller G, Torgerson DG, Ober C, Thompson EE (2009) Sequencing the IL4 locus in African Americans implicates rare noncoding variants in asthma susceptibility. J Allergy Clin Immunol 124(6):1204–1209.e9

Hamid Q, Tulic M (2009) Immunobiology of asthma. Annu Rev Physiol 71:489–507

Hassa PO, Hottiger MO (2002) The functional role of poly(ADP-ribose)polymerase 1 as novel coactivator of NF-kappaB in inflammatory disorders. Cell Mol Life Sci 59:1534–1553

Hawkins GA, Lazarus R, Smith RS, Tantisira KG, Meyers DA, Peters SP, Weiss ST, Bleecker ER (2009) The glucocorticoid receptor heterocomplex gene STIP1 is associated with improved lung function in asthmatic subjects treated with inhaled corticosteroids. J Allergy Clin Immunol 123(6):1376–1383.e7

Hoffjan S, Nicolae D, Ober C (2003) Association studies for asthma and atopic diseases: a comprehensive review of the literature. Respir Res 4(1):14

Holgate ST (2008) Are long acting β2 agonists safe in the treatment of asthma. Pol Arch Med Wewn 118(9):460–461

Hong SJ, Kim HB, Kang MJ, Lee SY, Kim JH, Kim BS, Jang SO, Shin HD, Park CS (2007) TNF-alpha (−308 G/A) and CD14 (−159T/C) polymorphisms in the bronchial responsiveness of Korean children with asthma. J Allergy Clin Immunol 119:398–404

Howard TD, Koppelman GH, Xu J, Zheng SL, Postma DS, Meyers DA, Bleecker ER (2002) Gene-gene interaction in asthma: IL4RA and IL13 in a Dutch population with asthma. Am J Hum Genet 70(1):230–236

Huizenga NA, Koper JW, de Lange P, Pols HA, Stolk RP, Burger H, Grobbee DE, Brinkmann AO, de Jong FH, Lamberts SW (1998) A polymorphism in the glucocorticoid receptor gene may be associated with and increased sensitivity to glucocorticoids in vivo. J Clin Endocrinol Metab 83(1):144–151

Hurley DM, Accili D, Stratakis CA, Karl M, Vamvakopoulos N, Rorer E, Constantine K, Taylor SI, Chrousos GP (1991) Point mutation causing a single amino acid substitution in the hormone binding domain of the glucocorticoid receptor in familial glucocorticoid resistance. J Clin Invest 87(2):680–686

Igaz P, Fitzimons CP, Szalai C, Falus A (2002) Histamine genomics in silico. Am J Pharmacogenomics 2(1):67–72

In KH, Asano K, Beier D, Grobholz J, Finn PW, Silverman EK, Silverman ES, Collins T, Fischer AR, Keith TP, Serino K, Kim SW, De Sanctis GT, Yandava C, Pillari A, Rubin P, Kemp J, Israel E, Busse W, Ledford D, Murray JJ, Segal A, Tinkleman D, Drazen JM (1997) Naturally occurring mutations in the human 5-lipoxygenase gene promoter that modify transcription factor binding and reporter gene transcription. J Clin Invest 99(5):1130–1137

International Study of Asthma and Allergies in Childhood (ISAAC) Steering Committee (1998) Worldwide variation in prevalence of symptoms of asthma, allergic rhinoconjunctivitis, and atopic eczema: ISAAC. Lancet 351(9111):1225–1232

Isaksson M (2001) Skin reactions to inhaled corticosteroids. Drug Saf 24(5):369–373

Israel E, Drazen JM, Liggett SB, Boushey HA, Cherniack RM, Chinchilli VM, Cooper DM, Fahy JV, Fish JE, Ford JG, Kraft M, Kunselman S, Lazarus SC, Lemanske RF, Martin RJ, McLean DE, Peters SP, Silverman EK, Sorkness CA, Szefler SJ, Weiss ST, Yandava CN (2000) The effect of polymorphisms of the beta (2)-adrenergic receptor on the response to regular use of albuterol in asthma. Am J Respir Crit Care Med 162:75–80

Ito K, Caramori G, Lim S, Oates T, Chung KF, Barnes PJ, Adcock IM (2002) Expression and activity of histone deacetylases in human asthmatic airways. Am J Respir Crit Care Med 166(3):392–396

Janson C, Anto J, Burney P, Chinn S, de Marco R, Heinrich J, Jarvis D, Kuenzli N, Leynaert B, Luczynska C, Neukirch F, Svanes C, Sunyer J, Wjst M, European Community Respiratory Health Survey II (2001) The European Community Respiratory Health Survey: What are the main results so far? European Community Respiratory Health Survey II. Eur Respir J 18:598–611

Jiang J, Zhang X, Yang H, Wang W (2009) Polymorphisms of DNA repair genes: ADPRT, XRCC1, and XPD and cancer risk in genetic epidemiology. Methods Mol Biol 471:305–333

John M, Au BT, Jose PJ, Lim S, Saunders M, Barnes PJ, Mitchell JA, Belvisi MG, Chung KF (1998) Expression and release of interleukin-8 by human airway smooth muscle cells: inhibition by Th-2 cytokines and corticosteroids. Am J Respir Cell Mol Biol 18(1):84–90

Joos L, Sandford AJ (2002) Genotype predictors of response to asthma medications. Curr Opin Pulm Med 8(1):9–15

Kabesch M (2005) Candidate gene association studies and evidence for gene-by-gene interactions. Immunol Allergy Clin North Am 25(4):681–708

Kabesch M, Schedel M, Carr D, Woitsch B, Fritzsch C, Weiland SK, von Mutius E (2006) IL-4/IL-13 pathway genetics strongly influence serum IgE levels and childhood asthma. J Allergy Clin Immunol 117(2):269–274

Kabesch M, Michel S, Tost J (2010) Epigenetic mechanisms and the relationship to childhood asthma. Eur Respir J 36(4):950–961

Kavuru M, Melamed J, Gross G, Laforce C, House K, Prillaman B, Baitinger L, Woodring A, Shah T (2000) Salmeterol and fluticasone propionate combined in a new powder inhalation device for the treatment of asthma: a randomized, double-blind, placebo-controlled trial. J Allergy Clin Immunol 105:1108–1116

Kedda MA, Lose F, Duffy D, Bell E, ThompsonPJ UJ (2005) The CD14 C-159T polymorphism is not associated with asthma or asthma severity in an Australian adult population. Thorax 60:211–214

Keller MB, Lowenstein SR (2002) Epidemiology of asthma. Semin Respir Crit Care Med 23(4):317–329

Kidwell WR, Mage MG (1976) Changes in poly(adenosine diphosphate-ribose) and poly(adenosine diphosphate-ribose) polymerase in synchronous HeLa cells. Biochemistry 15:1213–1217

Kim SH, Ye YM, Hur GY, Lee SK, Sampson AP, Lee HY, Park HS (2007) CysLTR1 promoter polymorphism and requirement for leukotriene receptor antagonist in aspirin-intolerant asthma patients. Pharmacogenomics 8:1143–1150

Kips JC, Brusselle GG, Joos GF, Peleman RA, Devos RR, Tavernier JH, Pauwels RA (1995) Importance of interleukin-4 and interleukin-12 in allergen-induced airway changes in mice. Int Arch Allergy Immunol 107:115–118

Klotsman M, York TP, Pillai SG, Vargas-Irwin C, Sharma SS, van den Oord EJ, Anderson WH (2007) Pharmacogenetics of the 5-lipoxygenase biosynthetic pathway and variable clinical response to montelukast. Pharmacogenet Genomics 17(3):189–196

Kondo M, Takeshita T, Ishii N, Nakamura M, Watanabe S, Arai K, Sugamura K (1993) Sharing of the interleukin-2 (IL-2) receptor gamma chain between receptors for IL-2 and IL-4. Science 262:1874–1877

Koster ES, Maitland-van der Zee AH, Tavendale R, Mukhopadhyay S, Vijverberg SJ, Raaijmakers JA, Palmer CN (2011) FCER2 T2206C variant associated with chronic symptoms and exacerbations in steroid-treated asthmatic children. Allergy 66(12):1546–1552

Kumari K, Rana AC (2012) A review on epidemiology, pathophysiology and management of Asthma. J Appl Pharm Sci 2(7):55–64

Laitinen T, Polvi A, Rydman P, Vendelin J, Pulkkinen V, Salmikangas P, Mäkelä S, Rehn M, Pirskanen A, Rautanen A, Zucchelli M, Gullstén H, Leino M, Alenius H, Petays T, Haahtela T, Laitinen T, Polvi A, Rydman P, Vendelin J, Pulkkinen V, Salmikangas P, Mäkelä S, Rehn M, Pirskanen A, Rautanen A, Zucchelli M, Gullstén H, Leino M, Alenius H, Petäys T, Haahtela T, Laitinen A, Laprise C, Hudson TJ, Laitinen LA, Kere J (2004) Characterization of a common susceptibility locus for asthma-related traits. Science 304(5668):300–304

Lander ES, Linton LM, Birren B, Nusbaum C, Zody MC, Baldwin J et al (2001) Initial sequencing and analysis of the human genome. Nature 409:860–921

Lee SG, Kim BS, Kim JH, Lee SY, Choi SO, Shim JY, Hong TJ, Hong SJ (2004) Gene-gene interaction between interleukin-4 and interleukin-4 receptor alpha in Korean children with asthma. Clin Exp Allergy 34:1202–1208

Leigh R, Ellis R, Wattie JN, Hirota JA, Matthaei KI, Foster PS, O'Byrne PM, Inman MD (2004) Type 2 cytokines in the pathogenesis of sustained airway dysfunction and airway remodeling in mice. Am J Respir Crit Care Med 169:860–867

LeVan TD, Von Essen S, Romberger DJ, Lambert GP, Martinez FD, Vasquez MM, Merchant JA (2005) Polymorphisms in the CD14 gene associated with pulmonary function in farmers. Am J Respir Crit Care Med 171:773–779

Levy H, Raby BA, Lake S, Tantisira KG, Kwiatkowski D, Lazarus R, Silverman EK, Richter B, Klimecki WT, Vercelli D, Martinez FD, Weiss ST (2005) Association of defensin beta-1 gene polymorphisms with asthma. J Allergy Clin Immunol 115(2):252–258

Liggett SB (1995) Genetics of beta 2-adrenergic receptor variants in asthma. Clin Exp Allergy 25(2):89–94, discussion 95–86

Liggett SB (2000) The pharmacogenetics of beta2-adrenergic receptors: relevance to asthma. J Allergy Clin Immunol 105:S487–S492

Lima JJ, Thomason DB, Mohamed MH, Eberle LV, Self TH, Johnson JA (1999) Impact of genetic polymorphisms of the beta2-adrenergic receptor on albuterol bronchodilator pharmacodynamics. Clin Pharmacol Ther 65:519–525

Lima JJ, Zhang S, Grant A, Shao L, Tantisira KG, Allayee H, Wang J, Sylvester J, Holbrook J, Wise R, Weiss ST, Barnes K (2006) Influence of leukotriene pathway polymorphisms on response to montelukast in asthma. Am J Respir Crit Care Med 173(4):379–385

Lipworth B, Koppelman GH, Wheatley AP, KeJeune I, Coutie W, Meurs H, Kauffman HF, Postma DS, Hall IP (2002) Beta-2 adrenoceptor promoter polymorphisms: Extended and functional effects in peripheral blood mononuclear cells. Thorax 57(1): 61–66

Lis G, Kostyk E, Sanak M, Pietrzyk JJ (2001) Molecular studies in a population of children with bronchial asthma. Polymorphism in the promotor region of gene CD14. Pneumonol Alergol Pol 69:265–272

Litonjua AA, Gong L, Duan QL, Shin J, Moore MJ, Weiss ST, Johnson JA, Klein TE, Altman RB (2010) Very important pharmacogene summary ADRB2. Pharmacogenet Genomics 20(1):64–69

Lockett KL, Hall MC, Xu J, Zheng SL, Berwick M, Chuang SC, Clark PE, Cramer SD, Lohman K, Hu JJ (2004) The ADPRT V762A genetic variant contributes to prostate cancer susceptibility and deficient enzyme function. Cancer Res 64:6344–6348

Loza MJ, Chang BL (2007) Association between Q551R IL4R genetic variants and atopic asthma risk demonstrated by meta-analysis. J Allergy Clin Immunol 120:578–585

Malmstorm K, Rodriguez-Gomez G, Guerra J, Villaran C, Pineiro A, Wei LX, Seidenberg BC, Reiss TF (1999) Oral montelukast, inhaled beclomethasone and placebo for chronic asthma. A randomized, controlled trial. Montelukast/Beclomethasone Study Group. Ann Intern Med 130:487–495

March ME, Sleiman PM, Hakonarson H (2011) The genetics of asthma and allergic disorders. Discov Med 11(56):35–45

Marshall GD (2004) Internal and external environmental influences in allergic diseases. J Am Osteopath Assoc 104(5):S1–S6

Martinez FD (2005) Gene-environment interactions in asthma and allergies: a new paradigm to understand disease causation. Immunol Allergy Clin North Am 25:709–721

Martinez FD, Graves PE, Baldini M, Solomon S, Erickson R (1997) Association between genetic polymorphisms of the beta2-adrenoceptor and response to albuterol in children with and without a history of wheezing. J Clin Invest 100(12):3184–3188

Masoli M, Fabian D, Holt S, Beasley R (2004) Global Initiative for Asthma (GINA) Program2004The global burden of asthma: executive summary of the GINA Dissemination Committee report. Allergy 59(5):469–478

Matheson MC, Ellis JA, Raven J, Johns DP, Walters EH, Abramson MJ (2006) Beta2-adrenergic receptor polymorphisms are associated with asthma and COPD in adults. J Hum Genet 51:943–951

Maxwell TJ, Ameyaw MM, Pritchard S, Thornton N, Folayan G, Githang'a J, Indalo A, Tariq M, Mobarek A, Evans DA, Ofori-Adjei D, Templeton AR, McLeod HL (2005) Beta-2 adrenergic receptor genotypes and haplotypes in different ethnic groups. Int J Mol Med 16:573–580

Menissier-de Murcia J, Molinete M, Gradwohl G, Simonin F, de Murcia G (1989) Zinc-binding domain of poly(ADP-ribose) polymerase participates in the recognition of single strand breaks on DNA. J Mol Biol 210:229–233

Moffatt MF, Kabesch M, Liang L, Dixon AL, Strachan D, Heath S, Depner M, von Berg A, Bufe A, Rietschel E, Heinzmann A, Simma B, Frischer T, Willis-Owen SA, Wong KC, Illig T, Vogelberg C, Weiland SK, von Mutius E, Abecasis GR, Farrall M, Gut IG, Lathrop GM, Cookson WO (2007) Genetic variants regulating

ORMDL3 expression contribute to the risk of childhood asthma. Nature 448(7152):470–473
Morrow TJ (2007) Implications of pharmacogenomics in the current and future treatment of asthma. J Manag Care Pharm 13(6):497–505
Munthe-Kaas MC, Carlsen KH, Håland G, Devulapalli CS, Gervin K, Egeland T, Carlsen KL, Undlien D (2008) T cell-specific T-box transcription factor haplotype is associated with allergic asthma in children. J Allergy Clin Immunol 121(1):51–56
NAEPP (National Asthma Education and Prevention Program) (2007) Full report of the Expert Panel: guidelines for the diagnosis and management of asthma (EPR-3). National Heart, Lung, and Blood Institute
Naseer T, Minshall EM, Leung DY, Laberge S, Ernst P, Martin RJ, Hamid Q (1997) Expression of IL-12 and IL-13 mRNA in asthma and their modulation in response to steroid therapy. Am J Respir Crit Care Med 155(3):845–851
Naura AS, Hans CP, Zerfaoui M, You D, Cormier SA, Oumouna M, Boulares AH (2008) Post-allergen challenge inhibition of poly(ADP-ribose) polymerase harbors therapeutic potential for treatment of allergic airway inflammation. Clin Exp Allergy 38:839–846
Nicolae D, Cox NJ, Lester LA, Schneider D, Tan Z, Billstrand C, Kuldanek S, Donfack J, Kogut P, Patel NM, Goodenbour J, Howard T, Wolf R, Koppelman GH, White SR, Parry R, Postma DS, Meyers D, Bleecker ER, Hunt JS, Solway J, Ober C (2005) Fine mapping and positional candidate studies identify HLA-G as an asthma susceptibility gene on chromosome 6p21. Am J Hum Genet 76(2):349–357
Noguchi E, Yokouchi Y, Zhang J, Shibuya K, Shibuya A, Bannai M, Tokunaga K, Doi H, Tamari M, Shimizu M, Shirakawa T, Shibasaki M, Ichikawa K, Arinami T (2005) Positional identification of an asthma susceptibility gene on human chromosome 5q33. Am J Respir Crit Care Med 172(2):183–188
Obase Y, Shimoda T, Kawano T, Saeki S, Tomari SY, Mitsuta-Izaki K, Matsuse H, Kinoshita M, Kohno S (2003) Polymorphisms in the CYP1A2 gene and theophylline metabolism in patients with asthma. Clin Pharmacol Ther 73:468–474
Ober C, Cox NJ, Abney M, Di Rienzo A, Lander ES, Changyaleket B, Gidley H, Kurtz B, Lee J, Nance M et al (1998) Genome-wide search for asthma susceptibility loci in a founder population. The collaborative study on the genetics of asthma. Hum Mol Genet 7:1393–1398
Oumouna M, Datta R, Oumouna-Benachour K, Suzuki Y, Hans C, Matthews K, Fallon K, Boulares H (2006) Poly(ADP-ribose) polymerase-1 inhibition prevents eosinophil recruitment by modulating Th2 cytokines in a murine model of allergic airway inflammation: a potential specific effect on IL-5. J Immunol 177:6489–6496
Palmer CN, Ismail T, Lee SP, Terron-Kwiatkowski A, Zhao Y, Liao H, Smith FJ, McLean WH, Mukhopadhyay S (2007) Filaggrin null mutations are associated with increased asthma severity in children and young adults. J Allergy Clin Immunol 120(1):64–68
Palmqvist M, Arvidsson P, Beckman O, Peterson S, Lotvall J (2001) Onset of bronchodilation of budesonide/formoterol vs. salmeterol/fluticasone in single inhalers. Pulm Pharmacol Ther 14:29–34
Paltiel AD, Fuhlbrigge AL, Kitch BT, Liljas B, Weiss ST, Neumann PJ, Kuntz KM (2001) Cost-effectiveness of inhaled corticosteroids in adults with mild-to-moderate asthma: results from the asthma policy model. J Allergy Clin Immunol 108:39–49
Papadopoulou N, Kalogeromitros D, Staurianeas NG, Tiblalexi D, Theoharides TC (2005) Corticotropin-releasing hormone receptor-1 and histidine decarboxylase expression in chronic urticaria. J Invest Dermatol 125(5):952–955
Patel L, Wales JK, Kibirige MS, Massarano AA, Couriel JM, Clayton PE (2001) Symptomatic adrenal insufficiency during inhaled corticosteroid treatment. Arch Dis Child 85:330–334
Patuzzo C, Trabetti E, Malerba G, Martinati LC, Boner AL, Pescollderungg L, Zanoni G, Pignatti PF (2000) No linkage or association of the IL-4Ralpha gene Q576R mutation with atopic asthma in Italian families. J Med Genet 37:382–384
Pearce N, Beasley R, Crane J, Burgess C, Jackson R (1995) End of the New Zealand asthma mortality epidemic. Lancet 345:41–44
Peng X, Yan H, You Z, Wang P, Wang S (2004) Effects of enteral supplementation with glutamine granules on intestinal mucosal barrier function in severe burned patients. Burns 30(2):135–139
Perin P, Berce V, Potocnik U (2011) CD14 gene polymorphism is not associated with asthma but rather with bronchial obstruction and hyperactivity in Slovenian children with non-atopic asthma. Respir Med 105(1):S54–S59
Pignatti PF (2004) Trends in pharmacogenomics of drugs used in the treatment of asthma. Pharmacol Res 49:343–349
Polson JB, Krzanowski JJ, Goldman AL, Szentivanyi A (1978) Inhibition of human pulmonary phosphodiesterase activity by therapeutic levels of theophylline. Clin Exp Pharmacol Physiol 5(5):535–539
Potaczek DP, Nishiyama C, Sanak M, Szczeklik A, Okumura K (2009) Genetic variability of the high-affinity IgE receptor alpha-subunit (Fc epsilon RI alpha). Immunol Res 45(1):75–84
Potter PC, Van Wyk L, Martin M, Lentes KU, Dowdle EB (1993) Genetic polymorphism of the beta-2 adrenergic receptor in atopic and non-atopic subjects. Clin Exp Allergy 23(10):874–877
Preuss CV, Wood TC, Szumlanski CL, Raftogianis RB, Otterness DM, Girard B, Scott MC, Weinshilboum RM (1998) Human histamine N-methyltransferase pharmacogenetics: common genetic polymorphisms that alter activity. Mol Pharmacol 53(4):708–717

Pykalainen M, Kinos R, Valkonen S, Rydman P, Kilpeläinen M, Laitinen LA, Karjalainen J, Nieminen M, Hurme M, Kere J, Laitinen T, Lahesmaa R (2005) Association analysis of common variants of STAT6, GATA3, and STAT4 to asthma and high serum IgE phenotypes. J Allergy Clin Immunol 115(1):80–87

Que LG, Yang Z, Stamler JS, Lugogo NL, Kraft M (2009) S-nitrosoglutathione reductase: an important regulator in human asthma. Am J Respir Crit Care Med 180:226–231

Raby BA, Van Steen K, Lazarus R, Celedon JC, Silverman EK, Weiss ST (2006) Eotaxin polymorphisms and serum total IgE levels in children with asthma. J Allergy Clin Immunol 117:298–305

Randolph AG, Lange C, Silverman EK, Lazarus R, Silverman ES, Raby B, Brown A, Ozonoff A, Richter B, Weiss ST (2004) The IL12B gene is associated with asthma. Am J Hum Genet 75(4):709–715

Redington AE, Meng QH, Springall DR, Evans TJ, Créminon C, Maclouf J, Holgate ST, Howarth PH, Polak JM (2001) Increased expression of inducible nitric oxide synthase and cyclo-oxygenase-2 in the airway epithelium of asthmatic subjects and regulation by corticosteroid treatment. Thorax 56(5):351–357

Regueiro V, Campos MA, Morey P, Sauleda J, Agusti AG, Garmendia J, Bengoechea JA (2009) Lipopolysaccharide-binding protein and CD14 are increased in the bronchoalveolar lavage fluid of smokers. Eur Respir J 33:273–281

Reihsaus E, Innis M, MacIntyre N, Liggett SB (1993) Mutations in the gene encoding for the beta 2-adrenergic receptor in normal and asthmatic subjects. Am J Respir Cell Mol Biol 8:334–339

Renauld JC (2001) New insights into the role of cytokines in asthma. J Clin Pathol 54:577–589

Revollo JR, Cidlowski JA (2009) Mechanisms generating diversity in glucocorticoid receptor signaling. Ann N Y Acad Sci 1179:167–178

Saleh D, Ernst P, Lim S, Barnes PJ, Giaid A (1998) Increased formation of the potent oxidant peroxynitrite in the airways of asthmatic patients is associated with induction of nitric oxide synthase: effect of inhaled glucocorticoid. FASEB J 12(11):929–937

Sampson AP, Siddiqui S, Buchanan D, Howarth PH, Holgate ST, Holloway JW, Sayers I (2000) Variant LTC(4) synthase allele modifies cysteinyl leukotriene synthesis in eosinophils and predicts clinical response to zafirlukast. Thorax 55:S28–S31

Sayers I, Barton S, Rorke S, Beghé B, Hayward B, Van Eerdewegh P, Keith T, Clough JB, Ye S, Holloway JW, Sampson AP, Holgate ST (2003) Allelic association and functional studies of promoter polymorphism in the leukotriene C4 synthase gene (LTC4S) in asthma. Thorax 58:417–424

Schmidt D, Dent G, Rabe KF (1999) Selective phosphodiesterase inhibitors for the treatment of bronchial asthma and chronic obstructive pulmonary disease. Clin Exp Allergy 29(2):99–109

Schreiber V, Dantzer F, Ame JC, de Murcia G (2006) Poly(ADP-ribose): novel functions for an old molecule. Nat Rev Mol Cell Biol 7:517–528

Scott MG, Swan C, Wheatley AP, Hall IP (1999) Identification of novel polymorphisms within the promoter region of the human beta2 adrenergic receptor gene. Br J Pharmacol 126:841–844

Sharek PJ, Bergman DA (2000) The effect of inhaled steroids on the linear growth of children with asthma: a meta-analysis. Pediatrics 106:E8

Sharma S, Mann D, Singh TP, Ghosh B (2005) Lack of association of histamine-N-methyltransferase (HNMT) polymorphisms with asthma in the Indian population. J Hum Genet 50(12):611–617

Sheikh A, Alves B (2000) Hospital admissions for anaphylaxis: time trend study. BMJ 320:1441

Sherrid P (2001) Designer drugs: what's best for patients isn't always what's best for profits. US News World Rep 131:30–32

Silverman ES, Liggett SB, Gelfand EW, Rosenwasser LJ, Baron RM, Bolk S, Weiss ST, Drazen JM (2001) The pharmacogenetics of asthma: a candidate gene approach. Pharmacogenomics J 1(1):27–37

Silverman ES, Hjoberg J, Palmer LJ, Tantisira KG, Weiss ST, Drazen JM (2003) Application of pharmacogenetics to the therapeutics of asthma. In: Huston D (ed) Therapeutic targets of airway inflammation. Marcel Dekker, New York, pp 823–838

Simbulan-Rosenthal CM, Rosenthal DS, Luo R, Samara R, Espinoza LA, Hassa PO, Hottiger MO, Smulson ME (2003) PARP-1 binds E2F-1 independently of its DNA binding and catalytic domains, and acts as a novel coactivator of E2F-1-mediated transcription during re-entry of quiescent cells into S phase. Oncogene 22:8460–8471

Singh AB, Kumar P (2003) Aeroallergens in clinical practice of allergy in India. An overview. Ann Agric Environ Med 10:131–136

Slager RE, Otulana BA, Hawkins GA, Yen YP, Peters SP, Wenzel SE, Meyers DA, Bleecker ER (2012) IL-4 receptor polymorphisms predict reduction in asthma exacerbations during response to an anti-IL-4 receptor alpha antagonist. J Allergy Clin Immunol 130(516–522):e514

Song CH, Tian X, Gelehrter TD (1999) Glucocorticoid receptor inhibits transforming growth factor- beta signaling by directly targeting the transcriptional activation function of Smad3. Proc Natl Acad Sci U S A 96(21):11776–11781

Stein J, Scheuermann EH, Yazdi R, Lembcke B, Caspary WF (1994) Reduced post heparin plasma diamine oxidase activity in patients with chronic renal failure. Z Gastroenterol 32(4):236–239

Suttner K, Rosenstiel P, Depner M, Schedel M, Pinto LA, Ruether A, Adamski J, Klopp N, Illig T, Vogelberg C, Schreiber S, von Mutius E, Kabesch M (2009) TBX21 gene variants increase childhood asthma risk in combination with HLX1 variants. J Allergy Clin Immunol 123(5):1062–1068, 1068.e1-8

Szalai C, Buzas EI, Fulop AK, Kohidai L, Falus A (2006) Genetic strategies in pharmacology of asthma and autoimmunity. Anti-Inflamm Anti-Allergic Agents Med Chem 5:383–399

Szefler SJ, Leung DY (1997) Glucocorticoid-resistant asthma: pathogenesis and clinical implications for management. Eur Respir J 10(7):1640–1647

Tam EK, Jourdan-LeSaux C, Stauder S, Bollt O, Reber B, Yamamoto F, Haymer D (2003) Polymorphisms in the interleukin-4 receptor alpha chain: association with traits of allergy and asthma in an admixed population in Hawaii. Cell Mol Biol (Noisy-le-Grand) 49:1345–1349

Tan S, Hall IP, Dewar J, Dow E, Lipworth B (1997) Association between beta 2-adrenoceptor polymorphism and susceptibility to bronchodilator desensitisation in moderately severe stable asthmatics. Lancet 350:995–999

Tantisira KG, Hwang ES, Raby BA (2004a) TBX21: a functional variant predicts improvement in asthma with the use of inhaled corticosteroids. Proc Natl Acad Sci 101:18099–18104

Tantisira KG, Lake S, Silverman E (2004b) Corticosteroid pharmacogenetics: association of sequence variants in CRHR1 with improved lung function in asthmatics treated with inhaled corticosteroids. Hum Mol Genet 13:1353–1359

Tantisira KG, Silverman ES, Mariani TJ, Xu J, Richter BG, Klanderman BJ (2007) FCER2: a pharmacogenetic basis for severe exacerbations in children with asthma. J Allergy Clin Immunol 120:1285–1291

Tantisira KG, Lazarus R, Litonjua AA, Klanderman B, Weiss ST (2008) Chromosome 17: association of a large inversion polymorphism with corticosteroid response in asthma. Pharmacogenet Genomics 18(8): 733–737

Taylor DR, Drazen JM, Herbison GP, Yandava CN, Hancox RJ, Town GI (2000) Asthma exacerbations during long term beta agonist use: influence of beta(2) adrenoceptor polymorphism. Thorax 55:762–767

Tezcan G, Gurel CB, Tutluoglu B, Onaran I, Kanigur-Sultuybek G (2009) The Ala allele at Val762Ala polymorphism in poly(ADP-ribose) polymerase-1 (PARP-1) gene is associated with a decreased risk of asthma in a Turkish population. J Asthma 46:371–374

Thomsen M, Nordestgaard BG, Sethi AA, Tybjaerg-Hansen A, Dahl M (2012) Beta2-adrenergic receptor polymorphisms, asthma and COPD: two large population-based studies. Eur Respir J 39:558–566

Tse SM, Tantisira K, Weiss ST (2011) The pharmacogenetics and pharmacogenomics of asthma therapy. Pharmacogenomics J 11(6):383–392

Van Eerdewegh P, Little RD, Dupuis J, Del Mastro RG, Falls K, Simon J, Torrey D, Pandit S, McKenny J, Braunschweiger K, Walsh A, Liu Z, Hayward B, Folz C, Manning SP, Bawa A, Saracino L, Thackston M, Benchekroun Y, Capparell N, Wang M, Adair R, Feng Y, Dubois J, FitzGerald MG, Huang H, Gibson R, Allen KM, Pedan A, Danzig MR, Umland SP, Egan RW, Cuss FM, Rorke S, Clough JB, Holloway JW, Holgate ST, Keith TP (2002) Association of the ADAM33 gene with asthma and bronchial hyperresponsiveness. Nature 418(6896):426–430

Venter JC, Adams MD, Myers EW, Li PW, Mural RJ, Sutton GG et al (2001) Thesequence of the human genome. Science 291:1304–1351

Vercelli D (2008) Discovering susceptibility genes for asthma and allergy. Nat Rev Immunol 8(3):169–182

Vladich FD, Brazille SM, Stern D, Peck ML, Ghittoni R, Vercelli D (2005) IL-13 R130Q, a common variant associated with allergy and asthma, enhances effector mechanisms essential for human allergic inflammation. J Clin Invest 115(3):747–754

Walsh KM, Bracken MB, Murk WK, Hoh J, Dewan AT (2010) Association between reduced copy-number at T-cell receptor gamma (TCRgamma) and childhood allergic asthma: a possible role for somatic mosaicism. Mutat Res 690(1–2):89–94

Wang Z, Chen C, Niu T, Wu D, Yang J, Wang B, Fang Z, Yandava CN, Drazen JM, Weiss ST et al (2001) Association of asthma with beta(2)-adrenergic receptor gene polymorphism and cigarette smoking. Am J Respir Crit Care Med 163:1404–1409

Wang L, Thomae B, Eckloff B, Wieben E, Weinshilboum R (2002) Human histamine N-methyltransferase pharmacogenetics: gene resequencing, promoter characterization, and functional studies of a common 5′-flanking region single nucleotide polymorphism (SNP). Biochem Pharmacol 64(4):699–710

Wang XG, Wang ZQ, Tong WM, Shen Y (2007) PARP1 Val762Ala polymorphism reduces enzymatic activity. Biochem Biophys Res Commun 354:122–126

Wechsler ME, Lehman E, Lazarus SC, Lemanske RF Jr, Boushey HA, Deykin A, Fahy JV, Sorkness CA, Chinchilli VM, Craig TJ, DiMango E, Kraft M, Leone F, Martin RJ, Peters SP, Szefler SJ, Liu W, Israel E, National Heart, Lung, and Blood Institute's Asthma Clinical Research Network (2006) beta-Adrenergic receptor polymorphisms and response to salmeterol. Am J Respir Crit Care Med 173:519–526

Weinshilboum R (2003) Inheritance and drug response. N Engl J Med 348(6):529–537

Weiss ST, Lake SL, Silverman ES et al (2004) Asthma steroid pharmacogenetics: a study strategy to identify replicated treatment responses. Proc Am Thorac Soc 1:364–367

Weiss ST, Litonjua AA, Lange C, Lazarus R, Ligett SB, Bleecker ER, Tantisira KG (2006) Overview of pharmacogenetics of asthma. Pharmacogenomics J 6:311–326

Wenzel SE, Balzar S, Ampleford E, Hawkins GA, Busse WW, Calhoun WJ, Castro M, Chung KF, Erzurum S, Gaston B et al (2007) IL4R alpha mutations are associated with asthma exacerbations and mast cell/IgE expression. Am J Respir Crit Care Med 175:570–576

Whalen EJ, Foster MW, Matsumoto A, Ozawa K, Violin JD, Que LG, Nelson CD, Benhar M, Keys JR, Rockman HA et al (2007) Regulation of beta-adrenergic receptor signaling by S-nitrosylation of G-protein-coupled receptor kinase 2. Cell 129:511–522

Woodruff PG, Boushey HA, Dolganov GM, Barker CS, Yang YH, Donnelly S, Ellwanger A, Sidhu SS, Dao-Pick TP, Pantoja C, Erle DJ, Yamamoto KR, Fahy JV (2007) Genome-wide profiling identifies epithelial cell genes associated with asthma and with treatment response to corticosteroids. Proc Natl Acad Sci USA 104(40):15858–15863

Wu X, Li Y, Chen Q, Chen F, Cai P, Wang L, Hu L (2010) Association and gene-gene interactions of eight common single-nucleotide polymorphisms with pediatric asthma in middle china. J Asthma 47(3):238–244

Yadollahi-Farsani M, Kefalas P, Saxty BA, MacDermot J (1999) Polymorphic forms of human ADP-ribosyltransferase-1 differences in their catalytic activities revealed by labeling of membrane-associated substrates. Eur J Biochem 262:342–348

Yan L, Galinsky RE, Bernstein JA, Liggett SB, Weinshilboum RM (2000) Histamine N-methyltransferase pharmacogenetics: association of a common functional polymorphism with asthma. Pharmacogenetics 10(3):261–266

Yelensky R, Li Y, Lewitzky S, Leroy E, Hurwitz C, Rodman D, Trifilieff A, Paulding CA (2012) A pharmacogenetic study of ADRB2 polymorphisms and indacaterol response in COPD patients. Pharmacogenomics J 12(6):484–488

Zetterstrom O, Buhl R, Mellem H, Perpiñá M, Hedman J, O'Neill S, Ekström T (2001) Improved asthma control with budesonide/formoterol in a single inhaler, compared with budesonide alone. Eur Respir Dis 18(2):262–268

Zhang Y, Leaves NI, Anderson GG, Ponting CP, Broxholme J, Holt R, Edser P, Bhattacharyya S, Dunham A, Adcock IM, Pulleyn L, Barnes PJ, Harper JI, Abecasis G, Cardon L, White M, Burton J, Matthews L, Mott R, Ross M, Cox R, Moffatt MF, Cookson WO (2003) Positional cloning of a quantitative trait locus on chromosome 13q14 that influences immunoglobulin E levels and asthma. Nat Genet 34(2):181–186

Zhang AM, Li HL, Hao P, Chen YH, Li JH, Mo YX, Dai M (2006) Association of Q576R polymorphism in the interleukin-4 receptor gene with serum IgE levels in children with asthma. Zhongguo Dang Dai Er Ke Za Zhi 8:109–112

Zhang W, Zhang X, Qiu D, Sandford A, Tan WC (2007) IL-4 receptor genetic polymorphisms and asthma in Asian populations. Respir Med 101:186–190

Zhou H, Hong X, Jiang S, Dong H, Xu X, Xu X (2009a) Analyses of associations between three positionally cloned asthma candidate genes and asthma or asthma-related phenotypes in a Chinese population. BMC Med Genet 10:123

Zhou H, Alexis NE, Almond M, Donohue J, LaForce C, Bromberg PA, Peden DB (2009b) Influence of C-159T SNP of the CD14 gene promoter on lung function in smokers. Respir Med 103:1358–1365

Zock JP, Heinrich J, Jarvis D, Verlato G, Norbäck D, Plana E, Sunyer J, Chinn S, Olivieri M, Soon A, Villani S, Ponzio M, Dahlman-Hoglund A, Svanes C, Luczynska C, Indoor Working Group of the European Community Respiratory Health Survey II (2006) Distribution and determinants of house dust mite allergens in Europe: the European Community Respiratory Health Survey II. J Allergy Clin Immunol 118:682–690

Section V

Personalized Medicine: Metabolic Disorders

Pharmacogenomics in Type 2 Diabetes Mellitus and Metabolic Disorders

25

Sandhiya Selvarajan, Melvin George, and Suresh Kumar Srinivasamurthy

Abstract

Pharmacogenomics, a newly growing branch of medicine, makes use of an individual's genetic information to guide therapy and has become an important tool in achieving 'personalised medicine'. The discovery of novel genetic polymorphisms in drug transporters, targets and metabolising enzymes has given an insight into the biological phenomena of drug efficacy and toxicity. Type 2 diabetes mellitus has been on the rise in both developed and developing countries owing to increase in life span as well as change in lifestyle. The advances in the field of pharmacogenomics has revealed diverse issues associated with onset of type 2 diabetes and genetic variants associated with varied responses to commonly prescribed antidiabetic drugs. Genetic polymorphisms like *A1369* variant increasing MgATPase activity of K_{ATP} channel have been found to provide a plausible molecular mechanism by which the K23/A1369 haplotype increases susceptibility to type 2 diabetes in humans homozygous for these variants. The oral hypoglycaemic drugs sulfonylureas act through K_{ATP} channel blockade resulting in stimulation of insulin release from pancreatic β-cells and have been a basis of type 2 diabetes pharmacotherapy since long time. However, sulfonylureas have been found to have interindividual variability in drug response and adverse effects. Interindividual variations in efficacy and adverse events are also recognised with metformin, a biguanide which improves insulin resistance in type 2 diabetes. Genetic polymorphisms in *OCT1* and *OCT2,* two organic cation transporters, have been found to be associated with changes in responses to metformin. Similarly, the regulatory and glucose

S. Selvarajan (✉) • S.K. Srinivasamurthy
Division of Clinical Pharmacology, Jawaharlal Institute of Postgraduate Medical Education and Research (JIPMER), Pondicherry, India
e-mail: sandhiyaselvarajan@yahoo.com; sandhiyaselvarajan@gmail.com

M. George
Department of Cardiology, SRM Medical College and Hospital Research Centre, Kattankulathur, Kancheepuram District, Tamil Nadu, India
e-mail: melvingeorge2003@gmail.com

D. Barh et al. (eds.), *Omics for Personalized Medicine*,
DOI 10.1007/978-81-322-1184-6_25, © Springer India 2013

homeostasis-related single nucleotide polymorphisms (SNPs) in peroxisome proliferator-activated receptor (PPAR) agonist-modulated genes can be used to explain the interindividual variability in response to thiazolidinediones. In addition, the identification of genetic defects responsible for inherited metabolic disorders like phenylketonuria, urea cycle disorder and Niemann–Pick disease has opened a promising approach to promote drug development in enzyme-deficient diseases. The detection of association of Niemann–Pick disease, type C1, gene-like 1 (NPC1L1) variation with response to ezetimibe, a cholesterol uptake inhibitor, has confirmed the role of pharmacogenomics in drug therapy of metabolic disorders. Hence, this chapter will discuss in detail the application of pharmacogenomics in drug response pertaining to type 2 diabetes mellitus and other metabolic disorders.

1 Introduction

Pharmacogenomics (PGx) is the study of variations in an individual's drug response related to DNA and RNA characteristics. The advances in molecular biology and bioinformatics have brought out mounting evidence towards the application of pharmacogenomics in assessing the efficacy as well as adverse drug reactions of a drug. Thus, pharmacogenomics plays a vital role in paving the way for personalised medicine as well as drug development (del Barrio 2010; Kalow 2002, 2004). The identification of disease-related genes by genome-wide association studies has led to the better perception of the underlying pathogenesis of many common diseases. Moreover these findings have paved the way for novel pharmacogenomic targets like genes encoding receptors, transporters, ion channels, growth factors and transcription factors (Wang et al. 2013). The spreading of genome-based epidemiological research in health care for early detection of illness and individualisation of therapy will help in the discovery of a promising drugs with better response (Schork and Topol 2010). Furthermore, the clinical test for evaluating interindividual pharmacogenomic variants to drug response has become available to the physicians with the introduction of US Food and Drug Administration-mediated product insert revisions that include pharmacogenetic information for selected drugs. Thus, the progress in genomic research has a major role in drug discovery, drug development, drug approval and clinical application, thus enabling earlier launch of the drug with lesser cost (Nebert et al. 2008).

The commonly occurring diseases like diabetes mellitus, hypertension and coronary artery disease are considered as complex disorders with multiple aetiological factors (Ezzati and Riboli 2012). The association between genetic variations and diseases like diabetes, cardiovascular diseases and obesity can be used to caution high-risk population to follow lifestyle modifications. Among these disorders, diabetes mellitus, a metabolic disease is considered as coronary artery disease risk equivalent (Whiteley et al. 2005). The incidence of diabetes has been increasing all over the world owing to the sedentary lifestyle, dietary habits and prolonged life span. Currently the disease is being treated with drugs like insulin, sulfonylureas, biguanides, thiazolidinediones, glucagon-like peptide (GLP)-1 analogues, dipeptidyl peptidase-4 inhibitors, glinides, amylinomimetics and α-glucosidase inhibitors. The selection of drugs in the treatment of diabetes is based on the efficacy, safety, tolerability, cost-effectiveness of drug and patient's disease status. With multifactorial and polygenic aetiologies as well as with the availability of many drugs, personalised therapy for metabolic disorders like diabetes mellitus could be achieved only with the help of personalised medicine through tools like pharmacogenomics (Aquilante 2010). Further, the identification of genetic defects responsible for inherited metabolic disorders like phenylketonuria, urea cycle disorder and Niemann–Pick disease has opened a promising approach to promote drug development in such enzyme-deficient diseases. The credentials of association of Niemann–Pick disease, type C1, gene-like 1 (NPC1L1) variation with response to ezetimibe, a cholesterol uptake inhibitor, have further confirmed the role of pharmacogenomics in drug therapy of metabolic disorders. Hence, this chapter explains the application of pharmacogenomics in drug response pertaining to diabetes mellitus and other metabolic disorders.

2 Metabolic Disorders

Metabolism refers to the biochemical processes involved in the breakdown of substrates like carbohydrates, proteins and fat in the body. Since enzymes play a major role in metabolic processes, any alteration in their function due to genetic polymorphisms can result in altered metabolism leading to metabolic disorders. These metabolic disorders may be either acquired or congenital and can be classified as shown in Table 25.1. The incidence of metabolic disorders can vary in different populations from either very rare to relatively more common. The inherited metabolic diseases are usually treated either by induction of enzyme activity, diet, drug or combination therapy (Pearce 2011). The treatment of metabolic disorders mainly focuses on supportive care as well as suppression and elimination of toxic products. The newer advances in clinical research including recombinant DNA technology have led to further advances in the treatment of metabolic disorders through replacement of deficient enzymes (Trefz et al. 2005; Hjiej et al. 2008).

Table 25.1 Classification of congenital metabolic disorders

Group	Metabolic disorders	Examples
I	Disorders that disrupt the synthesis or catabolism of complex molecules resulting in permanent and progressive symptoms, independent of food intake	Lysosomal disorders, peroxisomal disorders, disorders of intracellular transport and processing
II	Disorders that lead to an acute or progressive accumulation of toxic compounds due to blockade in metabolic pathway	Blockade in amino acid metabolism, e.g. phenylketonuria, homocystinuria, maple syrup urine disease, organic blockade acidurias, congenital urea cycle defects and carbohydrate intolerances, e.g. galactosaemia
III	Disorders due to deficiency of energy production or utilisation of substrates in the liver, myocardium muscle or brain	Congenital lactic acidaemias, fatty acid oxidation defects, gluconeogenesis defects, mitochondrial respiratory chain disorders

3 Diabetes Mellitus

Diabetes mellitus is a complex disease influenced by both environmental and genetic factors. It may either be due to insulin deficiency as in type 1 diabetes or insulin resistance as in type 2 diabetes mellitus. However, type 2 diabetes mellitus (T2DM) is one of the major public health concerns of the world as the prevalence is increasing over the years owing to sedentary lifestyle. Nearly 250 million people worldwide are suffering from type 2 diabetes mellitus and are usually diagnosed only after the patient becomes symptomatic. In addition, microvascular complications occurring in type 2 diabetes can result in nephropathy, retinopathy, neuropathy and gangrene. The available therapies for type 2 diabetes mellitus include lifestyle modifications with dietary restrictions and exercise in the early stage. This is followed by treatment with oral antidiabetic drugs and injectable insulin depending on the patient's glycaemic status. Maintenance of near-normal glycaemic status has been proven to reduce the risk for the occurrence and progression of microvascular complications (Holman et al. 2008). Majority of the patients with T2DM respond to antihyperglycaemic drugs during the initial stages. Later as the duration of disease progresses, monotherapy fails, and either addition of a second antidiabetic agent or transition to insulin becomes necessary to reinstate acceptable glycaemic control. However, approximately 40 % of the patients with type 2 diabetes do not achieve the desired glycated haemoglobin (HbA1c) level of less than 7 % following treatment (Hoerger et al. 2008). A study done on long-term glycaemic control in type 2 diabetes found a cumulative incidence of monotherapy failure of 15 % with rosiglitazone, 21 % with metformin and 34 % with glyburide at the end of 5 years (Kahn et al. 2006). Hence, the treatment of diabetes mellitus favours combination therapy to control blood glucose levels. As the treatment of diabetes has a huge impact on both health-care system and the individual patient's quality of life, better understanding of genetic basis of the disease would help in both prevention and therapy.

Diabetes being a complex disease with multiple genetic variants, the predictive test based on genetic variations coupled with the possibility of occurrence of type 2 diabetes could potentially be used for prediction and early treatment of the disease (Li et al. 2012). The genome-wide association (GWA) studies have lead to the rapid replication of the common genetic variants associated with type 2 diabetes (Mooyaart et al. 2011).

Insulin resistance (IR) in diabetes mellitus may be a complex phenomenon due to interaction between genetic and environmental factors. Meta-Analysis of Glucose and Insulin-related traits Consortium (MAGIC) has documented 16 genetic variants associated with fasting glucose in nondiabetic Europeans. However, some of these SNPs were associated with type 2 diabetes while other variants were just associated with physiological changes in glycaemic status (Dupuis et al. 2010). A meta-analysis done in 1,028 individuals showed evidence for significant interaction between body mass index (BMI) and rs1800795 polymorphism of the interleukin-6 gene on influencing both insulin resistance and onset of type 2 diabetes mellitus in obese individuals homozygous for the C allele. Thus, rs1800795 polymorphism of the interleukin-6 gene may be used as a marker for insulin resistance in obese people (Underwood et al. 2012).

Common variants of *MTNR1B, G6PC2* and *GCK* were found to be associated with elevated fasting plasma glucose and impaired insulin secretion, both individually and in combination, suggesting the role of these alleles on the precipitation of hyperglycaemia in individuals predisposed to the risk of developing diabetes mellitus (Tam et al. 2010). Several SNPs in genes associated with risk for diabetes were found to be associated with reduction in glucose and incretin-induced insulin secretion. In a study, the SNP rs7903146 in TCF7L2 (transcription factor 7-like 2) was shown to have a significant interaction between baseline glucose tolerance and change in insulin secretion. This study showed that higher glucose levels at baseline predicted an increase in insulin secretion following glycaemic improvement in those carrying the risk alleles following lifestyle changes. In addition to TCF7L2 the other diabetic risk gene WFS1 has also been found to be associated with impaired incretin signalling. Such findings show the relevance of these SNPs in the progression of prediabetic stage towards clinically apparent type 2 diabetes (Heni et al. 2010). A study done to evaluate the hepatic Wnt activity in TOPGAL transgenic mouse found that both Wnt signalling and TCF7L2 are negative regulators of hepatic gluconeogenesis, while TCF7L2 is involved in the downstream effector pathway of insulin in hepatocytes (Ip et al. 2012). Similarly in another study done by the Japanese, the clusterin (CLU) gene was found to be associated with diabetes, through an increase in insulin resistance as well as impairment of insulin secretion (Daimon et al. 2011). A study done to investigate the association of nicotinamide phosphoribosyltransferase (NAMPT)-3186 C/T and (NAMPT)-948G/T polymorphisms with the risk of type 2 diabetes mellitus (T2DM) and their impact on the efficacy of repaglinide in 170 Chinese Han T2DM patients showed significant association between plasma levels of postprandial serum insulin and cholesterol following repaglinide monotherapy for 8 weeks (Sheng et al. 2011).

A large meta-analysis of 36 studies examining the association of type 2 diabetes mellitus (T2DM) with polymorphisms in the *TCF7L2* gene in various ethnicities to evaluate the size of gene effect and the possible genetic mode of action showed an association between four variants of *TCF7L2* gene and T2DM. This indicates that the TCF7L2 gene may be involved in gene–gene as well as gene–environmental interactions in influencing the risk of T2DM (Tong et al. 2009). Thus, pharmacogenomic studies, assessing the role of genetic determinants of drug response, promise useful information for improving personalised treatment strategies and help to reduce the risk of adverse drug reactions in susceptible individuals.

3.1 Genetic Polymorphisms in Type 2 Diabetes Mellitus (T2DM)

The genetic variations play vital role in any disease as they are either responsible for the occurrence of the disease or for the individual variations in drug response. The genetic polymorphisms may alter the individual response to antidiabetic drugs either through variations in genes involved in the drug transport, metabolism and drug targets. Genetic variations in organic anion-transporting polypeptide [OATP], organic cation transporter [OCT], peroxisome proliferator-activated receptor gamma [PPARG], adenosine triphosphate [ATP]-sensitive potassium channel [K(ATP)], incretin receptors, adipokines, transcription factor 7-like 2 [TCF7L2], insulin receptor substrate 1 [IRS1], nitric oxide synthase 1 adaptor protein [NOS1AP] and solute carrier family 30 member 8 [SLC30A8] CDK5 regulatory subunit associated protein 1-like 1 [CDKAL1], insulin-like growth factor 2 mRNA binding protein 2 [IGF2BP2], potassium voltage-gated channel, KQT-like subfamily, member 1 [KCNQ1], paired box 4 [PAX4] and neuronal differentiation 1 [NEUROD1] transcription factors, ataxia telangiectasia mutated [ATM] and serine racemase [SRR] have been found to be associated with the occurrence of diabetes (Mannino and Sesti 2012). It has been shown that type 2 diabetes mellitus is associated with the variants in genes like HHEX, KIF11, IDE, CDKN2A/B and IFG2BP2 (Grarup et al. 2007). Moreover variations in DGKB/TMEM195, ADRA2A, GLIS3 and C2CD4B loci were found to be associated with decreased glucose-stimulated insulin secretion explaining the significance of pancreatic beta-cell dysfunction on genetic predisposition to type 2 diabetes (Boesgaard et al. 2010). The defect in guanine nucleotide binding 3 (GNB3) gene, coding for G-protein has been associated with the occurrence of type 2 diabetes along with other diseases like hypertension and obesity (Kiani et al. 2005). Studies in Chinese Hans, with common variants in *CDKAL1, CDKN2A/B, IGF2BP2* and *SLC30A8* loci, have shown the association of these variations in causing type 2 diabetes either independently or in combination with other risk factors (Wu et al. 2008). The genetic variations associated with diabetes mellitus are shown in Table 25.2.

Table 25.2 Various genes associated with the pathogenesis of type 2 diabetes mellitus

SL no	Genes associated with type 2 diabetes mellitus	Outcome of the genetic variations
1	*ADCY5, PROX1, GCK, GCKR, DGKB, TMEM195* (Dupuis et al. 2010)	Modest increase in glucose levels
2	*DGKB, TMEM195, ADRA2A, GLIS3, C2CD4B* (Boesgaard et al. 2010)	Decreased glucose-stimulated insulin response
3	*HHEX, KIF11, IDE* and *CDKN2A/B l* (Grarup et al. 2007)	Pancreatic beta-cell dysfunction
4	*CDKAL1, CDKN2A/B, IGF2BP2 and SLC30A8*	Pancreatic beta-cell dysfunction

3.2 Genetic Polymorphisms in Maturity-Onset Diabetes Mellitus (MODY)

A rare, autosomal dominant form of diabetes mellitus known as maturity-onset diabetes of the young (MODY) has six primary forms, each occurring as a result of mutations in six different genes (Capuano et al. 2012). Currently MODY is also classified as other specific types of diabetes mellitus. It is characterised by β-cell dysfunction in the absence of insulin resistance or obesity as well as the onset of diabetes before 25 years of age. Among the various forms of MODY, those with MODY3 due to mutations in the hepatocyte nuclear factor 1 homeobox A gene (*HNF1A*) has been shown to be more sensitive to the hypoglycaemic effects of sulfonylureas. In a study, three MODY3 patients with (*HNF1A*) mutations were found to have dramatic change in HbA1c levels with severe hypoglycaemia following cessation and reintroduction of sulfonylureas, respectively. In another study MODY3 patients showed a 5.2-fold higher response to gliclazide compared to metformin. Similarly compared to patients with type 2 diabetes, gliclazide showed 3.9-fold greater response in MODY3 (Pearson et al. 2000; Ellard 2000). In 30–58 % of cases with diabetes diagnosed in less than 6 months of age, a rare heterozygous mutation was detected in the potassium inwardly rectifying channel, subfamily J, member 11 (*KCNJ11*). These mutations result in continuous activation of ATP-dependent K^+ channel and prevention of insulin secretion by pancreatic β-cells leading to the misdiagnosis of type 1 diabetes. Study by Pearson et al. demonstrated that patients with mutations in *KCNJ11* could be successfully treated with sulfonylureas, instead of insulin (Pearson et al. 2000). Such mutations mislead the physician resulting in improper treatment with insulin resulting in poor glycaemic response. Similarly studies have identified mutations in the ATP-binding cassette, subfamily C, member 8 gene (*ABCC8*), have resulted in neonatal diabetes. In this disease, only those patients carrying *F132V* mutation have shown positive response to sulfonylureas apart from having to be maintained on insulin therapy (Oztekin et al. 2012). Such findings reveal that the genetic aetiology of hyperglycaemia may modulate response to hypoglycaemic agents and may have strong implications on patient management. Moreover, these finding pave the way for explaining additional genetic factors that might influence drug response in the treatment of T2DM.

3.3 Genetic Variations and Response to Sulfonylureas

Sulfonylureas like gliclazide, glibenclamide and glimepiride are the most commonly used oral hypoglycaemic agents in patients with type 2 diabetes mellitus. However, studies have found that 10–20 % of treated individuals do not achieve adequate glycaemic control even while on the highest recommended dose and are known as 'primary sulfonylurea failure' cases. On the contrary, 5–10 % of diabetic patients initially responding to sulfonylurea treatment are subsequently unable to maintain normal glycaemic levels with the same drug and are known as 'secondary sulfonylurea failure'. Diabetic patients with *TCF7L2* genetic polymorphism have been found to develop secondary failure at an earlier stage and need a higher dose of sulfonylurea or combination therapy with other antidiabetic drugs (Holstein et al. 2011).

3.4 Genetic Variations and Response to Metformin

Metformin, a biguanide used in the treatment of type 2 diabetes mellitus, acts by suppressing hepatic glucose production through activation of AMP-activated protein kinase (AMPK). Metformin apart from reducing blood glucose levels has an added advantage of reduction in weight as well as insulin resistance with little or no risk of inducing hypoglycaemia. Metformin, being a hydrophilic organic cation, serves as a substrate for organic cation transporters (OCTs), including OCT1 in the liver and OCT2, expressed abundantly in the kidney. Both OCT1 and OCT2 transport metformin into the hepatocytes and renal epithelium, respectively. The elimination of metformin from liver and kidney into bile and urine, respectively, is taken care of by the multidrug and toxin extrusion 1 protein (MATE1). OCT1 and OCT2 belonging to the SLC22A family of solute carriers are encoded by *SLC22A1* and *SLC22A2* genes, respectively, while MATE1 is encoded by the *SLC47A1* gene. Gene polymorphisms in drug transporters could be a cause of individual variations in drug response, and studies have focused on the genes encoding OCTs (Tarasova et al. 2012; Becker et al. 2009, 2010).

Metformin uptake into hepatocytes through OCT1 is a critical step for achieving its effects, and variations in *SLC22A1* may result in altered glycaemic response to the drug (Tkac et al. 2013). Following metformin administration, volunteers carrying *SLC22A1* risk alleles were found to have significantly higher plasma glucose levels and greater area under the curve (AUC) for most of the sampling time points compared to those of wild-type alleles. Insulin levels in individuals carrying risk alleles were found to be higher even 2 h following glucose administration compared to those with wild-type alleles (Shu et al. 2008). These findings have been further supported by in vitro characterisation of the variant alleles, especially with 420del allele, showing reduced activity for metformin.

Wang et al. showed the role OCT1 in the transport of metformin by demonstrating lesser levels of metformin in the liver and intestine of OCT1 −/− mice (Wang et al. 2002). Metformin concentration in the blood was found to be higher in OCT1 −/− mice, although 24-h plasma concentration-time profiles were similar between knockout and control animals. In hepatocytes isolated from OCT1−/− mice, metformin effects on AMPK activation and gluconeogenesis were found to be less compared to cells cultured from OCT1+/+ mice. In contrary, in a study done by Shikata et al. to evaluate the role of OCT1 on patients receiving metformin for more than 3 months, no remarkable variations were observed between responder and nonresponders in the frequency of *OCT1* polymorphisms (Shikata et al. 2007). However, the result could be attributed to lesser sample, and a study in large population is needed to confirm the results.

OCT2 expressed in the basolateral membrane of the renal epithelium and involved in the transportation of metformin is critical for the regulation of metformin accumulation in the kidney. Hence, OCT2 plays an extensive role in the pharmacokinetic profile of metformin than OCT1. A study done in 15 healthy Chinese volunteers showed association between variant T allele in *SLC22A2* and reduced renal clearance of metformin in the presence of the OCT2 inhibitor cimetidine. Similarly, a study done in Chinese Hans with T2DM demonstrated 808G>T variance in the *SLC22A2* gene to be associated with an increased incidence of hyperlactacidaemia in patients receiving metformin (Li et al. 2010). Another study conducted by Song et al. found that three *SLCA22A2* variants (596C>T, 602C>T and 808G>T) were associated with reduced renal excretion and enhanced plasma concentration of metformin (Song et al. 2008). In spite of the changes in volume of distribution and clearance of metformin in OCT1-/OCT2-knockout mice, there was no significant change in pharmacodynamic effects. These findings may challenge the hypothesis that inhibition of OCT affects efficacy of metformin (Higgins et al. 2012). Similarly, in a study done in 23 healthy volunteers of Caucasian and African American ethnicities, the G/T genotype at the *SLC22A2* 808G>T locus showed lower renal clearance compared to carriers of the wild-type G/G genotype. However, a study done by Tzvetkov et al. in 103 healthy participants did not report statistically significant evidence for association between 14 *SLC22A2* markers, including 808G>T and renal clearance of metformin (Tzvetkov et al. 2009). Hence, it may be possible that *SLC22A2* markers are essential for the renal elimination of metformin in Asians alone. Regardless of these differences, findings have supported the clinical relevance *of SLC22A2* polymorphisms. However, a larger prospective study in type 2 diabetic patients is required before the clinical application of these markers. *SLC47A1* gene encoding the *MATE1* protein located in the bile canalicular membrane in the hepatocyte and the brush border of the renal epithelium serves in the clearance of metformin through bile and urine. *MATE1* is colocalised with OCT1 and OCT2 in the hepatocyte and renal epithelium, respectively. Due to its role in metformin excretion, the variations in *MATE1* gene may contribute to individual variations in drug response. However, the association between *SLC47A1* genetic variants with metformin response has not been established so far.

3.5 Genetic Variations and Response to Thiazolidinediones

Thiazolidinediones (TZD), like troglitazone, rosiglitazone and pioglitazone, are peroxisome proliferator-activated receptor (PPAR)-gamma agonists and are used as insulin sensitisers in type 2 diabetes mellitus. These drugs are banned due to member-specific, PPAR gamma-independent toxicity like troglitazone-induced hepatotoxicity, rosiglitazone-induced QT prolongation and pioglitazone for bladder carcinoma as well as cardiovascular complications. Thiazolidinediones may be influencing the expression of multiple genes involved in carbohydrate and lipid metabolism to enhance insulin sensitivity. Although many genes and metabolic pathways are likely to be involved in the insulin-sensitising action of PPARγ ligands, it has been proposed that the effect of thiazolidinediones on genes regulating free fatty acid (FFA) transport and metabolism may be playing a major role in the antidiabetic effects of these drugs. Studies have suggested that thiazolidinediones may attenuate insulin resistance by increasing the metabolic clearance of free fatty acids (FFA). The fatty acid transporter CD36, a target for PPARγ ligands, is supposed to mediate the uptake of FFA through adipocytes and myocytes. In both human and animal studies, mutations in *Cd36* have been found to be associated with impaired carbohydrate and lipid metabolism. Studies done to evaluate the metabolic effects of pioglitazone in spontaneously hypertensive rats (SHR) with wild-type Cd36 have shown significantly lower fatty acids, triglycerides, insulin and epididymal fat pad weights compared to SHR carrying deletion mutation in Cd36. This study demonstrates that Cd36 may play a key determinant role in insulin-sensitising actions of a thiazolidinedione (Qi et al. 2002). A study done in 121 type 2 diabetes mellitus patients with 15–30 mg/day of pioglitazone for 12 weeks in addition to other medications showed that G/G genotype was correlated with a reduction in fasting plasma glucose (FPG) and homeostasis model assessment of insulin resistance (HOMA-IR) compared to C/C. Moreover, G/G genotype was found to be an independent factor for the reduction of FPG and HOMA-IR after adjusting for age, gender and BMI (Makino et al. 2009). Since all these drugs are banned the prospect of these drugs remains doubtful.

3.6 Genetic Variations in Drug Metabolising Enzymes

Among the cytochrome P450 enzymes, CYP2C8 comprises 7 % of the total hepatic CYP content and is responsible for the oxidative metabolism of many antidiabetic drugs including thiazolidinediones and meglitinides. In a pharmacokinetic study done to evaluate the effect of single dose of 15 mg pioglitazone in healthy volunteers, the weight-adjusted area under the curve (AUC) was found to be 34 % less in *CYP2C8*3* homozygous individuals and 26 % less in heterozygous subjects compared to wild-type homozygotes. Apart from thiazolidinediones, the *CYP2C8*3* allele has been found to be associated with increased metabolism and decreased plasma concentrations of repaglinide (Kirchheiner et al. 2005). Hence, patients with genetic polymorphisms in *CYP2C8* may show poor response to thiazolidinediones and repaglinide when used alone or in combination.

3.7 Genetic Polymorphisms in Indian Diabetic Patients

An epidemiology study conducted in 1,038 normal subjects and 1,031 type 2 diabetic subjects from Chennai showed that the frequency of the 'T' allele of both rs12255372(G/T) and rs7903146(C/T) polymorphisms of *TCF7L2* was significantly higher in diabetic subjects (23 and 33 %) compared to normal subjects (19 and 28 %). This study has demonstrated the association between *TCF7L2* gene polymorphisms and susceptibility to type 2 diabetes mellitus in Asian Indians (Bodhini et al. 2007). Another study done in Punjabi ancestry found an association between SNPs in *GLIS3* and *ADCY5* and risk for developing type 2 diabetes (Rees et al. 2011). As far as *OCT1* polymorphism is concerned, South Indian Tamil population have an allele frequency of 8.9,

80.3 and 24.5 % of rs2282143, rs628031 and rs622342 polymorphisms, respectively. This frequency was found to be similar to that of African Americans but different from that of Caucasians (Umamaheswaran et al. 2011). The role of this polymorphism on drug response in South Indians needs to be confirmed.

4 Genetic Polymorphisms and Congenital Metabolic Disorders

4.1 Phenylketonuria

Phenylketonuria (PKU), a common inborn metabolic disorder with an incidence of 1 in 8,000 in the United States, occurs due to deficiency of phenylalanine hydroxylase (PAH). As a result the normal conversion of phenylalanine to tyrosine is affected resulting in high concentration of phenylalanine in blood leading to mental retardation (Eisensmith and Woo 1991). Currently phenylketonuria is treated by dietary restriction of phenylalanine-containing foods. Care needs to be taken to avoid the artificial sweetener aspartame, also known as L-aspartyl-phenylalanine as an alternative to sucrose in these patients. The untreated disease may result in the occurrence of infantile spasms, developmental delay, disturbed behaviour, hyperactivity, etc. The use of tetrahydrobiopterin has been found to be promising as a new, alternative treatment method in some phenylketonuria patients. However, a study found that the mutations causing phenylketonuria determine the susceptibility of the patient to tetrahydrobiopterin (Quirk et al. 2012). Another study found that response to tetrahydrobiopterin is more in patients with mild hyperphenylalaninaemia phenotypes and that responsiveness may not be predicted on the basis of genotype, especially in mixed heterozygotes (Muntau et al. 2002). Yet another study done in Japanese patients with PKU demonstrated that genotypes would be useful for predicting clinical phenotype and the information may be used in decision making for treatment with tetrahydrobiopterin in East Asia (Okano et al. 2011).

The molecular mechanism underlying the occurrence of mental retardation is unclear and may not be explained by the primary enzyme deficiency alone. The lower levels of tyrosine caused by high concentration of phenylalanine in blood has been shown to decrease catecholamine synthesis in the brain of mouse model of PKU (Pascucci et al. 2002). In mouse model, the mutation of PAH has been shown to alter the expression of orexin gene, resulting in an increase in the levels of mRNA and the levels of orexin. The gene expression of the precursor for orexins, namely, prepro-orexin, and the mRNA is found to be upregulated along with increased orexin A levels in the PKU mouse brain (Surendran et al. 2003). Phenylketonuria is the only metabolic disorder showing higher levels of brain orexin A, and since orexin is associated with wakefulness and increased activity, patients with PKU are usually hyperactive. However, orexins have not been studied in PKU so far, and this may be an area for novel research in the field of metabolic disorders.

4.2 Galactosaemia

Galactosaemia is an autosomal recessive disorder characterised by the deficiency of the enzyme galactose-1-phosphate uridyltransferase (GALT). As a result patients with this inborn metabolic disorder are unable to metabolise galactose and untreated patients, especially infants present with vomiting, diarrhoea, failure to thrive, jaundice, liver dysfunction, hepatomegaly, hypoglycaemia, altered coagulation, mental retardation, cataract, etc. Treatment of severe galactosaemia includes total elimination of dietary galactose including milk and milk-based infant formulae. The GALT enzyme is comprised of 16 distinct mutations of which N314D has been found to be the most common mutation with a frequency of 40 % followed by Q188R at 2.7 %. The novel GALT gene mutations include 6 missense mutations, namely, *Y89H, Q103R, P166A, S181F, K285R* and *R333L*. GALT also may have one nonsense mutation, S307X and 3 silent mutations, namely, Q103Q, K210K and H319H. In an Indian study, the

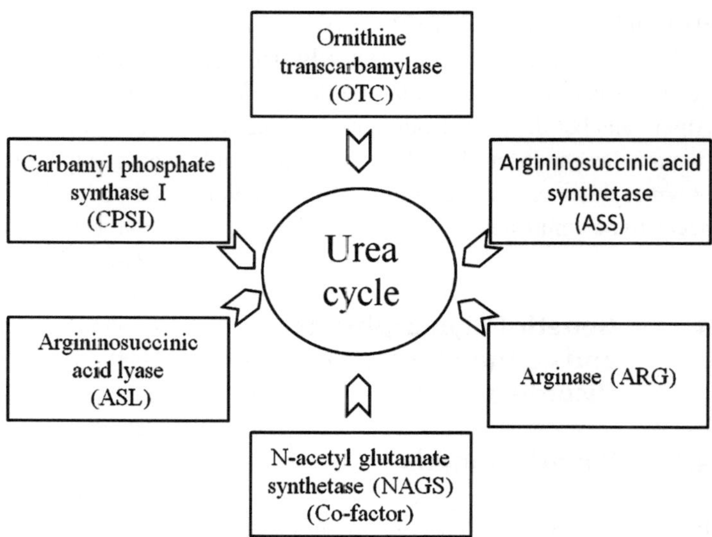

Fig. 25.1 Enzymes involved in urea cycle

functional significance of the novel GALT missense mutations was investigated using SNPs&GO and SIFT tools. The study highlighted the heterogeneity of classical galactosaemia in the Indian population and emphasised the importance of GALT gene analysis in diagnosis of galactosaemia. It also revealed that the mutational profile of Indian GALT is significantly different from the other populations (Singh et al. 2012).

4.3 Medium-Chain Acyl-CoA Dehydrogenase (MCAD) Deficiency

Medium-chain acyl-CoA dehydrogenase (MCAD) deficiency with an estimated incidence of 1:10,000 is the most common inborn error of metabolism. It is the most common fatty acid oxidation disorder that uses tandem mass spectrometry for its detection. Affected individuals appear normal until an episode of illness is provoked by an excessive period of fasting. Mortality is reported to be up to 25 % in symptomatic individuals. The first presentation is usually during 3 months to 2 years of age. Moreover, accumulated medium-chain acyl-CoA esters bind to carnitine and get excreted in urine. This results in secondary carnitine deficiency with muscle weakness or hypotonia. Treatment includes administration of oral or intravenous glucose, depending on the severity of the disease to provide energy and thus removing the need to utilise fat. Carnitine can also be administered during a crisis to treat secondary carnitine deficiency and to promote excretion of excess esters. The mainstay of treatment is to avoid prolonged periods of fasting and treat signs of hypoglycaemia at an earlier stage (Huidekoper et al. 2012).

4.4 Urea Cycle Disorders (UCD)

The urea cycle disorders (UCD) arise due to the defects in the metabolism of the excess nitrogen produced from the breakdown of protein and other nitrogen-containing compounds. The excess nitrogen is converted into ammonia (NH_4) and transported to the liver to enter the urea cycle. This cycle composed of five enzymes in the direct pathway and one enzyme as a necessary cofactor (Fig. 25.1). Defects in the genes CPSI, *OTC*, *ASS*, *ASL* and *ARG1* can cause carbamyl phosphate synthase I deficiency, ornithine transcarboxylase deficiency, citrullinaemia, arginosuccinicaciduria, argininaemia, etc. Molecular tests are already available for the detection of *CPSI*, *OTC*, *ASS*, etc., while the test for ASL and ARG1 is still in research (Wang et al. 2012). Inherited defect in urea cycle is characterised by hyperammonaemia and results in non-specific symptoms like lethargy, hypothermia, apnoea, convulsions and encephalopathy. Patients with urea cycle disorder may present at any age but are more likely to develop symptoms following exposure to metabolic stress as in infectious states with enhanced protein catabolism. In between the episodes, the patients are relatively well though some children may have developmental delay. During treatment of urea cycle disorder, frequent monitoring of plasma levels of ammonia is needed to ensure treatment response. In addition, these patients may also be treated with protein-restricted diet, supplemented with arginine and high-calorie feeds except in arginase deficiency.

5 Conclusions

Personalised medicine makes it possible to prescribe the right drug for the right patient at the right time. In due course personalised medicine will provide the basis for cost-effectiveness in health-care setup. This will help to improve the drug's efficacy, prevent side effects and make health care more efficient and cost effective in the future. Though studies imply that pharmacogenetic testing could improve the safety or efficacy of the treatment, this knowledge is not being currently used in clinical practice. The clinical implementation of pharmacogenomics is challenging as ample facilities are required to carry out genomic studies and experts are needed to interpret the results. Moreover, the clinical utility of pharmacogenomic tests will depend on the relative cost–benefit ratio pertaining to the disease as well as therapy. In addition, one more factor crucial for the clinical application of pharmacogenomic is the availability of valid rapid reporting genetic assays. The genetic tests used in clinical practice needs to be ethical, legal and socially acceptable. However, since genetic markers of drug response may not be always the clinically useful predictors of adverse effect, genotype–phenotype association should be always done before opting for clinical utility. Currently, the genomic attempt to bridge the gap between genetic variations and response to drug seems to be very slow and underutilised, and in the coming years, there is a promising future for application of pharmacogenomics in personalised medicine.

References

Aquilante CL (2010) Sulfonylurea pharmacogenomics in Type 2 diabetes: the influence of drug target and diabetes risk polymorphisms. Expert Rev Cardiovasc Ther 8(3):359–372

Becker ML, Visser LE, van Schaik RH, Hofman A, Uitterlinden AG, Stricker BH (2009) Genetic variation in the organic cation transporter 1 is associated with metformin response in patients with diabetes mellitus. Pharmacogenomics J 9(4):242–247

Becker ML, Visser LE, van Schaik RH, Hofman A, Uitterlinden AG, Stricker BH (2010) Interaction between polymorphisms in the OCT1 and MATE1 transporter and metformin response. Pharmacogenet Genomics 20(1):38–44

Bodhini D, Radha V, Dhar M, Narayani N, Mohan V (2007) The rs12255372(G/T) and rs7903146(C/T) polymorphisms of the TCF7L2 gene are associated with type 2 diabetes mellitus in Asian Indians. Metabolism 56(9):1174–1178

Boesgaard TW, Grarup N, Jorgensen T, Borch-Johnsen K, Hansen T, Pedersen O (2010) Variants at DGKB/TMEM195, ADRA2A, GLIS3 and C2CD4B loci are associated with reduced glucose-stimulated beta cell function in middle-aged Danish people. Diabetologia 53(8):1647–1655

Capuano M, Garcia-Herrero CM, Tinto N, Carluccio C, Capobianco V, Coto I et al (2012) Glucokinase (GCK) mutations and their characterization in MODY2 children of southern Italy. PLoS One 7(6):e38906

Daimon M, Oizumi T, Karasawa S, Kaino W, Takase K, Tada K et al (2011) Association of the clusterin gene polymorphisms with type 2 diabetes mellitus. Metabolism 60(6):815–822

del Barrio SJ (2010) Pharmacogenomics in clinical trials. Rev Derecho Genoma Hum 33:197–216

Dupuis J, Langenberg C, Prokopenko I, Saxena R, Soranzo N, Jackson AU et al (2010) New genetic loci implicated in fasting glucose homeostasis and their impact on type 2 diabetes risk. Nat Genet 42(2):105–116

Eisensmith RC, Woo SL (1991) Phenylketonuria and the phenylalanine hydroxylase gene. Mol Biol Med 8(1):3–18

Ellard S (2000) Hepatocyte nuclear factor 1 alpha (HNF-1 alpha) mutations in maturity-onset diabetes of the young. Hum Mutat 16(5):377–385

Ezzati M, Riboli E (2012) Can noncommunicable diseases be prevented? Lessons from studies of populations and individuals. Science 337(6101):1482–1487

Grarup N, Rose CS, Andersson EA, Andersen G, Nielsen AL, Albrechtsen A et al (2007) Studies of association of variants near the HHEX, CDKN2A/B, and IGF2BP2 genes with type 2 diabetes and impaired insulin release in 10,705 Danish subjects: validation and extension of genome-wide association studies. Diabetes 56(12):3105–3111

Heni M, Ketterer C, Thamer C, Herzberg-Schafer SA, Guthoff M, Stefan N et al (2010) Glycemia determines the effect of type 2 diabetes risk genes on insulin secretion. Diabetes 59(12):3247–3252

Higgins JW, Bedwell DW, Zamek-Gliszczynski MJ (2012) Ablation of both organic cation transporter (OCT)1 and OCT2 alters metformin pharmacokinetics but has no effect on tissue drug exposure and pharmacodynamics. Drug Metab Dispos 40(6):1170–1177

Hjiej H, Doyen C, Couprie C, Kaye K, Contejean Y (2008) Substitutive and dietetic approaches in childhood autistic disorder: interests and limits. Encéphale 34(5):496–503

Hoerger TJ, Segel JE, Gregg EW, Saaddine JB (2008) Is glycemic control improving in U.S. adults? Diabetes Care 31(1):81–86

Holman RR, Paul SK, Bethel MA, Matthews DR, Neil HA (2008) 10-year follow-up of intensive glucose control in type 2 diabetes. N Engl J Med 359(15):1577–1589

Holstein A, Hahn M, Korner A, Stumvoll M, Kovacs P (2011) TCF7L2 and therapeutic response to sulfonylureas in patients with type 2 diabetes. BMC Med Genet 12:30

Huidekoper HH, Ackermans MT, Koopman R, van Loon LJ, Sauerwein HP, Wijburg FA (2012) Normal rates of whole-body fat oxidation and gluconeogenesis after overnight fasting and moderate-intensity exercise in patients with medium-chain acyl-CoA dehydrogenase deficiency. J Inherit Metab Dis Sep 14. [Epub ahead of print]

Ip W, Shao W, Chiang YT, Jin T (2012) The Wnt signaling pathway effector TCF7L2 is up-regulated by insulin and represses hepatic gluconeogenesis. Am J Physiol Endocrinol Metab 303(9):E1166–E1176

Kahn SE, Haffner SM, Heise MA, Herman WH, Holman RR, Jones NP et al (2006) Glycemic durability of rosiglitazone, metformin, or glyburide monotherapy. N Engl J Med 355(23):2427–2443

Kalow W (2002) Pharmacogenetics and personalised medicine. Fundam Clin Pharmacol 16(5):337–342

Kalow W (2004) Human pharmacogenomics: the development of a science. Hum Genomics 1(5):375–380

Kiani JG, Saeed M, Parvez SH, Frossard PM (2005) Association of G-protein beta-3 subunit gene (GNB3) T825 allele with Type II diabetes. Neuro Endocrinol Lett 26(2):87–88

Kirchheiner J, Roots I, Goldammer M, Rosenkranz B, Brockmoller J (2005) Effect of genetic polymorphisms in cytochrome p450 (CYP) 2C9 and CYP2C8 on the pharmacokinetics of oral antidiabetic drugs: clinical relevance. Clin Pharmacokinet 44(12):1209–1225

Li Q, Liu F, Zheng TS, Tang JL, Lu HJ, Jia WP (2010) SLC22A2 gene 808 G/T variant is related to plasma lactate concentration in Chinese type 2 diabetics treated with metformin. Acta Pharmacol Sin 31(2):184–190

Li C, Han J, Shang D, Li J, Wang Y, Wang Y et al (2012) Identifying disease related sub-pathways for analysis of genome-wide association studies. Gene 503(1):101–109

Makino H, Shimizu I, Murao S, Kondo S, Tabara Y, Fujiyama M et al (2009) A pilot study suggests that the G/G genotype of resistin single nucleotide polymorphism at −420 may be an independent predictor of a reduction in fasting plasma glucose and insulin resistance by pioglitazone in type 2 diabetes. Endocr J 56(9):1049–1058

Mannino GC, Sesti G (2012) Individualized therapy for type 2 diabetes: clinical implications of pharmacogenetic data. Mol Diagn Ther 16(5):285–302

Mooyaart AL, Valk EJ, van Es LA, Bruijn JA, de Heer E, Freedman BI et al (2011) Genetic associations in diabetic nephropathy: a meta-analysis. Diabetologia 54(3):544–553

Muntau AC, Roschinger W, Habich M, Demmelmair H, Hoffmann B, Sommerhoff CP et al (2002) Tetrahydrobiopterin as an alternative treatment for mild phenylketonuria. N Engl J Med 347(26):2122–2132

Nebert DW, Zhang G, Vesell ES (2008) From human genetics and genomics to pharmacogenetics and pharmacogenomics: past lessons, future directions. Drug Metab Rev 40(2):187–224

Okano Y, Kudo S, Nishi Y, Sakaguchi T, Aso K (2011) Molecular characterization of phenylketonuria and tetrahydrobiopterin-responsive phenylalanine hydroxylase deficiency in Japan. J Hum Genet 56(4):306–312

Oztekin O, Durmaz E, Kalay S, Flanagan SE, Ellard S, Bircan I (2012) Successful sulfonylurea treatment of a neonate with neonatal diabetes mellitus due to a novel missense mutation, p.P1199L, in the ABCC8 gene. J Perinatol 32(8):645–647

Pascucci T, Ventura R, Puglisi-Allegra S, Cabib S (2002) Deficits in brain serotonin synthesis in a genetic mouse model of phenylketonuria. Neuroreport 13(18):2561–2564

Pearce N (2011) Epidemiology in a changing world: variation, causation and ubiquitous risk factors. Int J Epidemiol 40(2):503–512

Pearson ER, Liddell WG, Shepherd M, Corrall RJ, Hattersley AT (2000) Sensitivity to sulphonylureas in patients with hepatocyte nuclear factor-1alpha gene mutations: evidence for pharmacogenetics in diabetes. Diabet Med 17(7):543–545

Qi N, Kazdova L, Zidek V, Landa V, Kren V, Pershadsingh HA et al (2002) Pharmacogenetic evidence that cd36 is a key determinant of the metabolic effects of pioglitazone. J Biol Chem 277(50):48501–48507

Quirk ME, Dobrowolski SF, Nelson BE, Coffee B, Singh RH (2012) Utility of phenylalanine hydroxylase genotype for tetrahydrobiopterin responsiveness classification in patients with phenylketonuria. Mol Genet Metab 107(1–2):31–36

Rees SD, Hydrie MZ, O'Hare JP, Kumar S, Shera AS, Basit A et al (2011) Effects of 16 genetic variants on fasting glucose and type 2 diabetes in South Asians: ADCY5 and GLIS3 variants may predispose to type 2 diabetes. PLoS One 6(9):e24710

Schork NJ, Topol EJ (2010) Genotype-based risk and pharmacogenetic sampling in clinical trials. J Biopharm Stat 20(2):315–333

Sheng FF, Dai XP, Qu J, Lei GH, Lu HB, Wu J et al (2011) NAMPT -3186C/T polymorphism affects repaglinide response in Chinese patients with Type 2 diabetes mellitus. Clin Exp Pharmacol Physiol 38(8):550–554

Shikata E, Yamamoto R, Takane H, Shigemasa C, Ikeda T, Otsubo K et al (2007) Human organic cation transporter (OCT1 and OCT2) gene polymorphisms and therapeutic effects of metformin. J Hum Genet 52(2):117–122

Shu Y, Brown C, Castro RA, Shi RJ, Lin ET, Owen RP et al (2008) Effect of genetic variation in the organic cation transporter 1, OCT1, on metformin pharmacokinetics. Clin Pharmacol Ther 83(2):273–280

Singh R, Thapa BR, Kaur G, Prasad R (2012) Biochemical and molecular characterization of GALT gene from Indian galactosemia patients: identification of 10 novel mutations and their structural and functional implications. Clin Chim Acta 12:10

Song IS, Shin HJ, Shim EJ, Jung IS, Kim WY, Shon JH et al (2008) Genetic variants of the organic cation transporter 2 influence the disposition of metformin. Clin Pharmacol Ther 84(5):559–562

Surendran S, Campbell GA, Tyring SK, Matalon K, McDonald JD, Matalon R (2003) High levels of orexin A in the brain of the mouse model for phenylketonuria: possible role of orexin A in hyperactivity seen in children with PKU. Neurochem Res 28(12):1891–1894

Tam CH, Ho JS, Wang Y, Lee HM, Lam VK, Germer S et al (2010) Common polymorphisms in MTNR1B, G6PC2 and GCK are associated with increased fasting plasma glucose and impaired beta-cell function in Chinese subjects. PLoS One 5(7):e11428

Tarasova L, Kalnina I, Geldnere K, Bumbure A, Ritenberga R, Nikitina-Zake L et al (2012) Association of genetic variation in the organic cation transporters OCT1, OCT2 and multidrug and toxin extrusion 1 transporter protein genes with the gastrointestinal side effects and lower BMI in metformin-treated type 2 diabetes patients. Pharmacogenet Genomics 22(9):659–666

Tkac I, Klimcakova L, Javorsky M, Fabianova M, Schroner Z, Hermanova H et al (2013) Pharmacogenomic association between a variant in SLC47A1 gene and therapeutic response to metformin in type 2 diabetes. Diabetes Obes Metab 15(2):189–191

Tong Y, Lin Y, Zhang Y, Yang J, Zhang Y, Liu H et al (2009) Association between TCF7L2 gene polymorphisms and susceptibility to type 2 diabetes mellitus: a large Human Genome Epidemiology (HuGE) review and meta-analysis. BMC Med Genet 10:15

Trefz FK, Scheible D, Frauendienst-Egger G, Korall H, Blau N (2005) Long-term treatment of patients with mild and classical phenylketonuria by tetrahydrobiopterin. Mol Genet Metab 86(Suppl 1):S75–S80, Epub 2005 Oct 20

Tzvetkov MV, Vormfelde SV, Balen D, Meineke I, Schmidt T, Sehrt D et al (2009) The effects of genetic polymorphisms in the organic cation transporters OCT1, OCT2, and OCT3 on the renal clearance of metformin. Clin Pharmacol Ther 86(3):299–306

Umamaheswaran G, Praveen RG, Arunkumar AS, Das AK, Shewade DG, Adithan C (2011) Genetic analysis of OCT1 gene polymorphisms in an Indian population. Indian J Hum Genet 17(3):164–168

Underwood PC, Chamarthi B, Williams JS, Sun B, Vaidya A, Raby BA et al (2012) Replication and meta-analysis of the gene-environment interaction between body mass index and the interleukin-6 promoter polymorphism with higher insulin resistance. Metabolism 61(5):667–671

Wang DS, Jonker JW, Kato Y, Kusuhara H, Schinkel AH, Sugiyama Y (2002) Involvement of organic cation transporter 1 in hepatic and intestinal distribution of metformin. J Pharmacol Exp Ther 302(2):510–515

Wang IM, Stone DJ, Nickle D, Loboda A, Puig O, Roberts C (2013) Systems biology approach for new target and biomarker identification. Curr Top Microbiol Immunol 363:169–99

Wang J, Zhan H, Li FY, Pursley AN, Schmitt ES, Wong LJ (2012) Targeted array CGH as a valuable molecular diagnostic approach: experience in the diagnosis of mitochondrial and metabolic disorders. Mol Genet Metab 106(2):221–230

Whiteley L, Padmanabhan S, Hole D, Isles C (2005) Should diabetes be considered a coronary heart disease risk equivalent?: results from 25 years of follow-up in the Renfrew and Paisley survey. Diabetes Care 28(7):1588–1593

Wu Y, Li H, Loos RJ, Yu Z, Ye X, Chen L et al (2008) Common variants in CDKAL1, CDKN2A/B, IGF2BP2, SLC30A8, and HHEX/IDE genes are associated with type 2 diabetes and impaired fasting glucose in a Chinese Han population. Diabetes 57(10):2834–2842

Pharmacogenomics and Gut Microbiota Biomarkers in Obesity

26

Margarita Aguilera, María Luján Jiménez-Pranteda, Barbara Jenko, Verónica Jiménez, and Marisa Cañadas Garre

Abstract

Obesity development is a dependent multifactor disease due to imbalance between energy intake and energy expenditure. This balance represents a combination of numerous variables such as behavior, diet, environment, genetics, and social structures. Complex interactions among these variables contribute to the individual differences for development of obesity which leads and allows individualizing of treatment response.

Physicians are exploring the causes and the main approaches and strategies to tailor individualized treatments based on importance of host genetic variability itself and the microbiota variability.

The chapter will be focus on omics application impact involved in clinical biomarkers differentially detected nowadays with a demonstrated impact on obese individuals, mainly those that detect host biomarkers and also gut microbiota biomarkers.

Disease biomarkers for obesity are multiple, such us immunological (inflammatory status, defined by elevated C-reactive protein and IL-6), biochemical (glucose, lipids, satiety-related hormones), microbiological (the fecal microbial composition), and genetic (Tiihonen K, Ouwehand AC, Rautonen N, Br J Nutr 103(7):1070–1078, 2010).

Genetic markers for prognosis of weight loss response are genes coding adrenergic receptors (ADBRs), uncoupling proteins (UCPs), leptin (LEP),

M. Aguilera (✉)
INYTA-CIBM, University of Granada, Granada, Spain

Pharmacogenetics Unit, Pharmacy Service, University Hospital Virgen de las Nieves, Granada, Spain
e-mail: maguiler@ugr.es

M.L. Jiménez-Pranteda • V. Jiménez
INYTA-CIBM, University of Granada, Granada, Spain

B. Jenko
INYTA-CIBM, University of Granada, Granada, Spain

Pharmacogenetics Unit, Pharmacy Service, University Hospital Virgen de las Nieves, Granada, Spain

Institute of Biochemistry, Faculty of Medicine, University of Ljubljana, Ljubljana, Slovenia

M.C. Garre
Pharmacogenetics Unit, Pharmacy Service, University Hospital Virgen de las Nieves, Granada, Spain
e-mail: marisacgarre@gmail.com

D. Barh et al. (eds.), *Omics for Personalized Medicine*,
DOI 10.1007/978-81-322-1184-6_26, © Springer India 2013

leptin receptor (LEPR), melanocortin pathways genes (MC3R, POMC), serotonin receptor, peroxisome proliferator-activated receptor PPAR-γ-2, and genes related to cytokines (Martinez AJ, Parra DM, Santos JL et al. Asia Pac J Clin Nutr 17:119–122, 2008).

Moreover, through current literature, it is clear that variable individual composition of intestinal microbiota plays a role in energy extraction from non-digested carbohydrates in the form of production of short-chain fatty acids. The microbiota also plays a role in host metabolism by influencing and modulating host gene expression in various tissues. For this reason, a specific part of the chapter is deserved for molecular and genetics microbiota variability explanations.

Furthermore, it is suggested by several researchers that the host genetic may influence the composition of gut microbiota. The proposed genes which affect the microbiota composition are related to the immune system function and metabolism (Mathes WF, Kelly SA, Pomp D Br J Nutr 106(Suppl 1):S1–10,2011). Several studies indicate that lean and obese microbiome differ primarily in their interface with the host and in the way they interact with host metabolism, suggest the importance of host genetic control in shaping individual microbiome diversity in mammals (Benson AK, Kelly SA, Legge R, Ma F, Low SJ, Kim J, Zhang M, Oh PL, Nehrenberg D, Hua K, Kachman SD, Moriyama EN, Walter J, Peterson DA, Pomp D, Proc Natl Acad Sci U S A 107:18933–18938, 2010).

1 Introduction

Obesity results from the accumulation of excess adipose tissue. However, it is not a single disorder but a heterogeneous group of conditions with multiple causes. Major causes of the increasing prevalence of obesity include behavioral and environmental factors, such as excessive consumption of energy-dense foods and a sedentary lifestyle (Fig. 26.1). Still, it is now recognized that a series of underexplored physiological and environmental predispositions underlies the traditional risk factors for obesity and its associated metabolic disorders (Harris et al. 2012). Obesity is generally defined as an excess of body fat. However, there is no clear delineation between how much fat is normal and how much fat is abnormal. Furthermore, body fat is difficult and expensive to measure directly in large samples. Therefore, obesity is often defined as excess weight after adjusting for height. Body mass index (BMI, calculated as weight in kg/height in m^2) is used in adults to ascertain obesity status, and a BMI $\geq$ 30 is considered obese (Table 26.1).

Obesity has dramatically increased during the past decades and has now reached epidemic proportions in both developed and developing countries. In the United States it is of grave public health significance. In 2003–2004, approximately 66 % of population was overweight (BMI $\geq$ 25 kg/m^2) and 32 % was obese (BMI $\geq$ 30 kg/m^2) (Agurs-collins et al. 2009). Even in Japan where the self-reported prevalence of obesity has remained consistently low over the last 30 years, obesity is now increasing in middle-aged adults and children partly due to a Western-style change in diet (Lakhan and Kirchgessner 2011). According to the World Health Organization and to the International Obesity Task Force, more than 155 million children and adolescents worldwide are overweight and 40 million are clearly obese.

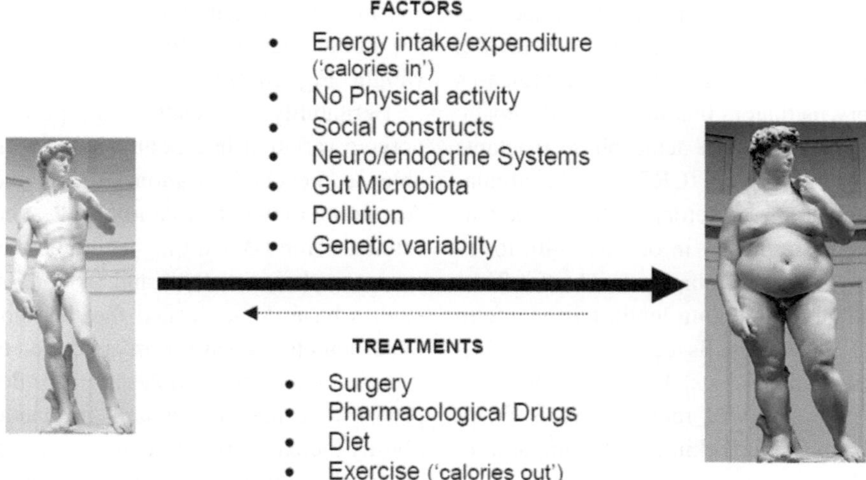

Fig. 26.1 Factors capable of impacting the obese phenotype and current treatments of obesity

Table 26.1 Obesity grades

Phenotype	Normal	Overweight	Obesity	Severe obesity	Morbid obesity
Body mass index BMI = weight/height	18.5–24.9	25–29.9	30–34.9	35–39.9	>40

It is also associated with a low-grade inflammation, which contributes to the development of insulin resistance, the metabolic syndrome, type 2 diabetes mellitus, and, in the long term, cardiovascular diseases. Several factors appear to be implicated in this inflammatory process. The plasma fatty acid (FA) composition, for example, which is influenced by the amount and quality of the dietary fat, has been associated with the low-grade inflammation of overweight and obese individuals. In particular, the arachidonic acid/eicosapentaenoic acid (ARA/EPA) ratio has been proposed as a biomarker of inflammation and in the development of inactivity-associated insulin resistance. Another participating factor is the increasing mass of adipose tissue, which is infiltrated by activated macrophages and releases inflammatory mediators that increase the hepatic expression and release of acute phase reactants such as C-reactive protein (CRP) into the circulation. In addition, the secretory profile of the adipocytes changes radically in obesity, with lower levels of adiponectin and higher levels of inflammatory mediators including leptin (Brignardello et al. 2010). The adipose tissue of obese individuals, including adolescents, has been shown to produce higher levels of tumor necrosis factor (TNF)-alpha and interleukin (IL)-6 compared to lean individuals.

The analyses and correlation between gastrointestinal, immunological, and blood biomarkers will make the comprehension of this complex disease, obesity, much easier (Tiihonen et al. 2010; Brignardello et al. 2010).

Thus, big efforts will be required toward the integration of multiple data obtained in high-throughput studies. This chapter will be focused on host biomarkers and microbiota biomarkers due to the high impact on energy metabolism.

2 Genetics in Obesity

2.1 Hereditability

The epidemic of obesity is attributed to recent environmental changes. The main two reasons for the global increase in the prevalence of early-onset obesity are easy access to high-energy palatable food combined with decreased physical activity level. However, other putative environmental contributors to the recent obesity epidemic have been proposed, such as an obesity-prone intrauterine environment, assortative mating among obese individuals, decreasing sleep duration, infections, and low-grade inflammation or the increasingly controlled ambient temperature. However, if these factors are responsible for the global shift in BMI distribution, genetic factors must explain most of the interindividual differences in obesity risk observed in populations. In fact, the risk of obesity in a child is ten times higher if both parents are obese than if both parents are of normal weight (Fig. 26.2).

Heritability represents the proportion of phenotypic variation in a population that is attributable to genetic variation among individuals. According to twin and family studies, heritability estimates for BMI during childhood or adolescence are between 0.20 and 0.86. Longitudinal studies have demonstrated that heritability estimates tend to increase from childhood to preadolescence and from preadolescence to adolescence, probably because individuals at genetic risk for obesity increasingly select environments that promote gaining weight, correlated with their genetic propensities. In addition, physical activity reduces the influence of genetic factors on BMI in young adults, and it is likely that these results are transposable to children or adolescents.

Studies performed in twins have given light to contribute to elucidation of the causes and consequences of obesity (Naukkarinen et al. 2012). There is an indication that conscious efforts to prevent weight gain in the lean and the more unhealthy lifestyle in the obese co-twins explains the weight disparities, which appeared at 18 years of age, at a time when environmental influences may also start to diverge as the young adults move out of their parental home and establish more individualistic lifestyles. Thus, recent studies have shed some interesting light on the interplay of physical exercise and its modulatory effects on an individual's genetic predisposition to weight gain and obesity (Li et al. 2010).

Moreover, several other studies have investigated the effect of environmental factors on the apparent heritability of other obesity-associated phenotypes such as insulin sensitivity and lipid profiles and blood pressure, strengthening the view that genetic predisposition can indeed be modified by environmental factors such as physical exercise and dietary habits.

2.2 Obesity Monogenic Forms

Several single-gene disorders result in severe and early-onset obesity. These monogenic forms of

Fig. 26.2 Hereditability of obese phenotype: transmission of genetic and environmental factors

early-onset obesity show the biological importance of the mutant gene in body-weight control. The main genes affected in these monogenic disorders are leptin (LEP), leptin receptor (LEPR), pro-opiomelanocortin (POMC), prohormone convertase 1 (PCSK1), melanocortin 4 receptor (MC4R), brain-derived neurotrophic factor (BDNF), and neurotrophic tyrosine kinase receptor type 2 (NTRK2). They encode hormones or neurotransmitters and their hypothalamic receptors of the highly conserved leptin-melanocortin pathway, which is critical for the regulation of food intake and body weight. Monogenic obesity forms are frequently accompanied by additional clinical features (e.g., severe hyperphagia, intestinal troubles) and normal development, except for BDNF, SIM1, and NTRK2 deficiencies, which are associated with cognitive impairment, behavioral problems, or syndrome features. The focus on the leptin-melanocortin pathway as a target for pharmacological intervention in patients with severe obesity turned out to be effective.

3 Omics Applications Toward "Obesomics"

The measurement of the simultaneous expression values of thousands of genes or proteins from high-throughput omics platforms creates a large amount of data whose interpretation by inspection can be a daunting task. A major challenge of using such data is to translate these lists of genes/proteins into a better understanding of the underlying biological phenomena (Kumar 2011).

The huge amount of information available makes it necessary for the reconstruction and prediction of pathways, establishing connections between genes and disease, finding the relationships between genes and specific biological functions, and integrating these data with biochemical pathways in order to facilitate the interpretation.

Several databases that provide one-stop access to comprehensive information on genes or gene products are widely available. The NCBI Entrez browser (http://www.ncbi.nlm.nih.gov/Entrez) is one of the cross database portal to sequence data, conserved domains, gene location, protein structure, published microarray expression data (Gene Expression Omnibus, http://www.ncbi.nlm.nih.gov/geo), and function (OMIM, Online Mendelian Inheritance in Man, http://www.ncbi.nlm.nih.gov/omim).

MeSH concepts http://www.ncbi.nlm.nih.gov/mesh is generating hypotheses about which biological properties are shared within a set of genes or proteins based on their associations with data extracted from published literature and then use those relationships to cluster both the genes and the concepts. The database uses a combination of descriptors, qualifier, and gene identifiers giving a final result as a "bucket" collection for phenomenon and process concept hierarchies (Agarwal and Searls 2008).

Omics data approaches show the ability to generate hypotheses about which biological properties are shared within a set of genes or proteins based on their associations with MeSH concepts from the literature and those relationships between genes and concepts. In a specific "Diet-Induced Obesity Model Study," the most representative buckets were involved in energy metabolism, lipid metabolism, and adipose tissue metabolism, and there were also buckets of genes overexpressed that were related to glycolysis, gluconeogenesis, pentose phosphate pathway, and citric acid cycle. The same analysis showed a xenobiotic metabolism genes bucket downregulated (Kumar 2011).

Theses tools will evolve according to assisting the interpretation of high-throughput data from multiple omics platforms and will improve as the literature expands and becomes increasingly available as electronic full text and as computational tools for processing language become more robust.

Recent and important studies about human obesity and microbiota-associated studies are summarized in Table 26.2: "Obesomics," whose contents describe in a linked manner: Omics methodology applications, key short definitions, techniques, biomarkers type, database, and main references. It is mainly highlighting the results on human obese against to murine models studies, which even if they have given valuable information, it would be recommendable avoid the multiple barriers derived from escalation results between different studies.

Recent studies are approaching promised strategies on quantitative genetics for better understanding the overall host genetic architecture that influences the microbiota (Spor et al. 2011). In general, the evolution of knowledge of human genetic population studies are aimed to detect the importance of gene variability involved in differential responses to the putative structure that interacts with the individual (Benson et al. 2010). Previous experience in other areas such as nutrigenomics and nutrigenetics could be extended and applied to gut interactions (Dimitrov 2011; Fenech et al. 2011; Ferguson et al. 2007) and study the effects probiotics exert when they are employed as functional foods or nutraceuticals.

The detection of specific polymorphisms or single nucleotide polymorphisms (SNPs), as determinants or biomarkers of clinical genomics, is important for those that occur with high frequency in the general population and those able to modify or regulate important proteins in metabolic pathways, generating changes in the activity of these enzymes. Benson et al. have performed the first study taking a quantitative trait loci (QTL)-based approach to relate genomic variation in the host to microbial diversity in the gut. Testing of the CMM abundances for co-segregation with 530 SNPs identifies 18 host-associated QTLs that show a linkage with the abundances of specific microbial taxa (Benson et al. 2010).

4 Pharmacogenomics in Obesity

4.1 Applications in Disease Management and Development of Treatment Protocol

There are currently three main treatment modalities for obesity: lifestyle modifications, pharmacotherapy, and bariatric surgery. The cornerstone of lifestyle modifications includes changes to dietary and exercise habits. However, less than 5 % of the obese people who follow these recommendations effectively lose weight and maintain that weight loss.

Bariatric surgery is a new treatment for morbid obesity, but the relevance of an invasive surgery procedure mainly in childhood or adolescence is still under debate. It is leading to major improvements in glucose and lipid homeostasis and to the amelioration of some systemic inflammatory markers (Clément 2011). A recent study reported the first case of laparoscopic sleeve gastrectomy successfully performed in a 6-year-old morbidly obese child. Because of the lack of efficiency of the cur-

Table 26.2 "Obesomics": omics methodologies potentially used in human obesity and gut microbiota studies

OMICS methodology applications	Key short definitions	Techniques	Biomarkers type	Databases	Reference
Genetic variability	Interindividual variability gene profiles	Affymetrix/Luminex	SNP, CNV	HAPMAP	Mathes et al. (2011)
		PCR-RFLP		SNP	Li-Wan-Po and Farndon (2011)
				PHARMAGKB	Chen et al. (2007)
Epigenenetics	Methylation pattern independent of primary DNA sequence	Immunoblot for histone hyperacetylation	SCFA, HATs, HDACi	EPIGENIE_***Epigenetics Databases*** and Tools	Canani et al. (2011)
		ChIP-Seq (specific epigenetic modifications including acetylation)			
Metagenomics	Genome-wide determinations	Sequencing	16S rDNA; 18S rDNA	MEDLINE	Greenblum et al. (2012)
	Microbiome data	Pyrosequencing	Functional rDNA	Metagenomics of the human intestinal tract consortium	Maccaferri et al. (2011)
					Liu et al. (2010)
				MetaHIT Consortium website	Ferrer et al. (2013)
Transcriptomics	Gene expression profile	RNA microarrays; RT-PCR; Qpcr	Transcripts of RNA	H-DBAS – H-Invitational *database*	Ji et al. (2012)
			cDNA		Birzele et al. (2012)
Proteomics	Determinations of protein differentially expressed	2-DE Analysis, multidimensional, chromatography, LC-MS/MS, MALDI-TOF	Proteins, peptides, oligopeptides	UNIPROt	Pérez-Pérez et al. (2012)
				Proteomics Databases | EBI	Peinado et al. (2012)
					Lehr et al. (2011)
Metabolomics	Quantitative measurement of multiparametric metabolic responses	NRM analysis, MS, GC, HPLC	Metabotype: lipids, sugars, vitamins, antioxidants	The Human *Metabolome Database* (HMDB)	Xie et al. (2012a)
		CE/MS, FTRI arrays			Ament et al. (2012)
Gut Bacterial Phenomics	Bacterial phenotype manifestation of genotype	Base on genotype and environmental factors	Bacterial phyla, genus, clusters	NA	Serino et al. (2012)
Gut Host Phenomics	Cellular phenotype manifestation of genotype	Base on genotype and environmental factors	Epithelial proliferation	NA	
			Epithelial barrier integrity		

(continued)

Table 26.2 (continued)

OMICS methodology applications	Key short definitions	Techniques	Biomarkers type	Databases	Reference
Nutrigenomics	Correlated data between genomic and diet collections	Programmed diets/sequencing	Metabolic enzymes; key diet parameters	*european nutrigenomics organization*	Rudkowska and Pérusse (2012)
Integromics	Data filtering global evaluation	Molecular data bioinformatic and interpretation softwares	Bioinformatic Software	MeSH	Katoh et al. (2007)
				Global *Database* on Body Mass Index	
Interactomics	Molecular communication between probiotic and host cells	Hypothesis on bacteria Phenotype and host phenotype	Likely ligand and receptor	FunCoup	Williams et al. (2012)
			MAMPs/PRRs		Alexeyenko et al. (2012)

rent approaches to reverse childhood obesity, prevention was proposed as the first line of treatment in 2003 by the American Academy of Pediatrics. In its policy statement, the Academy promoted breastfeeding, healthy eating habits and physical activity, and encouraged limitation of television viewing. However, current prevention programs have had little success and have proven ineffective in reversing the rising rates of childhood obesity. These disappointing observations reveal the urgent need to better understand the complex molecular and physiologic mechanisms involved in human obesity in order to propose better disease prevention and care (Choquet and Meyre 2010).

4.2 Drugs and Genetic Variability: SNPs and Drug Efficacy vs. Adverse Drug Effects with Molecular Mechanisms

The long-term safety and efficacy of the antiobesity drugs (orlistat, sibutramine, evodiamine, berberine, phentermine, exenatide, topiramate, metformin) have not been determined in specific populations as children or adolescents, mainly because pharmacotherapy is not routinely proposed as a treatment for specific population obesity.

Orlistat is a drug designed to treat obesity. Its primary function is preventing the absorption of fats from the human diet, thereby reducing caloric intake. Orlistat works by inhibiting pancreatic lipase, an enzyme that breaks down triglycerides in the intestine. Without this enzyme, triglycerides from the diet are prevented from being hydrolyzed into absorbable free fatty acids and are excreted undigested.

Sibutramine is an orally administered agent for the treatment of obesity. It is a centrally acting stimulant chemically related to amphetamines. Sibutramine is classified as a Schedule IV controlled substance in the United States. In October 2010, sibutramine was withdrawn from Canadian and US markets due to concerns that the drug increases the risk of heart attack and stroke in patients with a history of heart disease.

Evodiamine, the botanic alkaloid, has also been shown to inhibit adipocyte differentiation, improve diet-induced obesity, and decrease rate of food intake. Although, the mechanistic action of evodiamine in these processes has not yet been fully elucidated, it has been strongly suggested that the pathways involved include the serotonin transporter system.

Phentermine is a central nervous system stimulant and sympathomimetic with actions and uses similar to those of dextroamphetamine. It has been used most frequently in the treatment of obesity.

Exenatide, derived from a compound found in the saliva of the Gila monster, a large lizard native to the southwestern USA, is a functional analog of glucagon-like peptide-1 (GLP-1), a naturally occurring peptide.

Metformin is a biguanide antihyperglycemic agent used for treating non-insulin-dependent diabetes mellitus (NIDDM). It improves glycemic control by decreasing hepatic glucose production, decreasing glucose absorption, and increasing insulin-mediated glucose uptake. Metformin is the only oral antihyperglycemic agent that is not associated with weight gain. *Metformin may induce weight loss and is the drug of choice for obese NIDDM patients***.** When used alone, metformin does not cause hypoglycemia; however, it may potentiate the hypoglycemic effects of sulfonylureas and insulin. Its main side effects are dyspepsia, nausea, and diarrhea. Dose titration and/or use of smaller divided doses may decrease side effects.

Topiramate is an anticonvulsant drug produced by Ortho-McNeil Neurologics, a division of Johnson & Johnson. It is used to treat epilepsy in both children and adults. In children it is also indicated for the treatment of Lennox-Gastaut syndrome (a disorder that causes seizures and developmental delays). It is also Food and Drug Administration (FDA) approved for, and now most frequently prescribed for, the prevention of migraines.

Table 26.3 is a compilation of previously described main drugs used on obese populations

Table 26.3 Drugs and substances used in obesity treatments and target genetic variability

Drugs	Chemical structure	Gene	Variants localization	SNPs	References
Orlistat FDA drug label at DailyMed: 5bbdc95b-82a1-4ba5-8185-6504ff68cc06	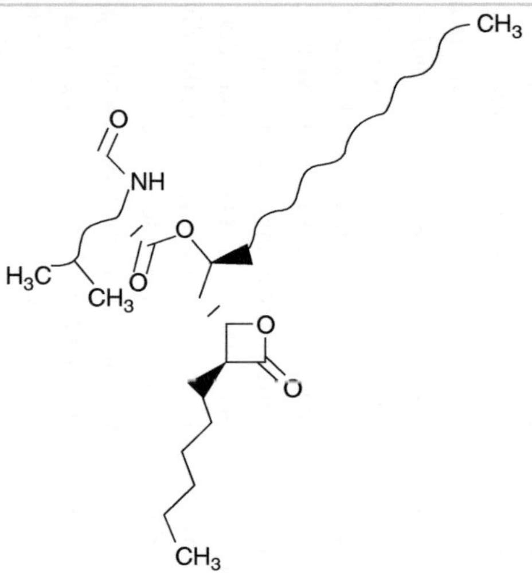	CD36	chr7: 80244939	rs1761667	Pepino et al. (2012)
Sibutramine FDA drug label at DailyMed: c0333e62-bb00-4722-27b6-dca6ab4d1eb1		GNAQ	chr9 : 80335188	Promotor haplotype	Klenke and Siffert (2011)
				G(−173)A	
				G(−168)A	
		GNAS	chr20:57478806	rs7121	Hahn et al. (2006)
		GNB3	chr12: 6954875	rs5443 (C825T)	Hsiao et al. (2009)
		CYP2B6	chr19:41496460	CYP2B6 *4/*6	Zhang et al. (2012)
		CYP3A5	chr 7:99270538	CYP3A5*3	Zhang et al. (2012)
		CYP2C19	chr 10:96541615	CYP2C19 *2/*3	Hsiao et al. (2010)
		UCP2	chr 11: 73694754	rs659366	Grudell et al. (2008)
Evodiamine Not FDA drug label	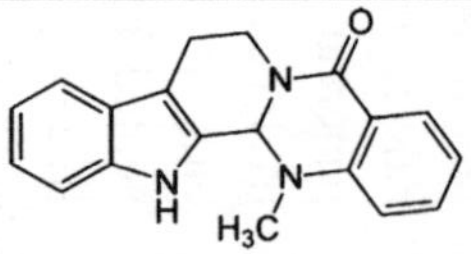	SLC6A4	chr17:28564358	rs4795541	Hu et al. (2012)

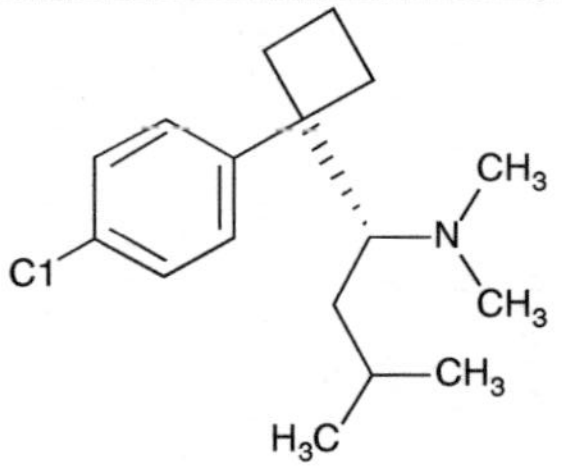

Beberine Not FDA drug label	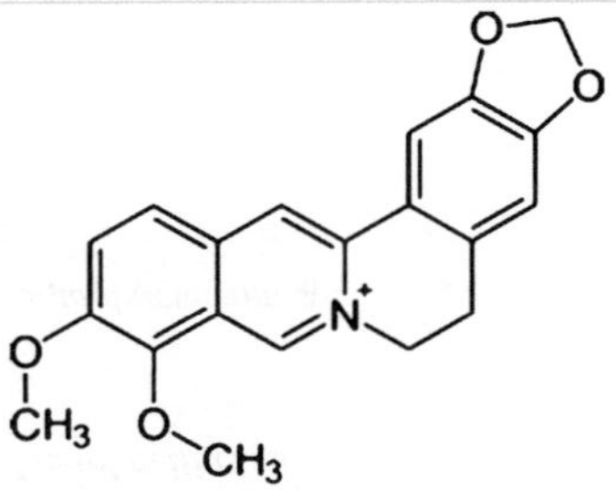	SLC6A4	chr17: 28564358	rs4795541	Hu et al. (2012)
Phentermine FDA drug label at DailyMed: adf965bd-86 cc-4313-9253-c151585f3f94	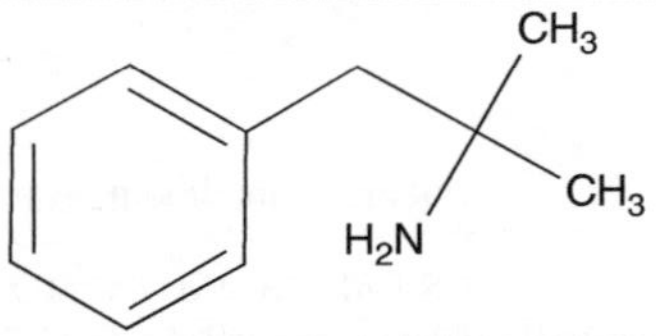	MTHFR	chr1 : 11845786	NA	Grasso (2011)
		CYP2D6	chr 22 : 42522500	NA	Roth et al. (2008)
Topiramate FDA drug label at DailyMed: 61f36cba-e314-4ec9-8c84-5259153049b5	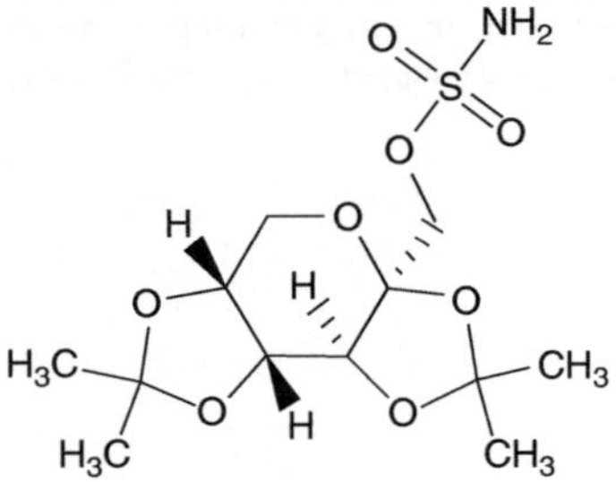	SCN2A	chr2:166168502	rs2304016	Baretić (2012)
		CA12	chr 15:63633238	rs2306719	
			chr15:63623463	rs4984241	

(continued)

Table 26.3 (continued)

Drugs	Chemical structure	Gene	Variants localization	SNPs	References
Metformin FDA drug label at DailyMed: e455ec79-08cd-4d75-9cb2-625d3912ee88	H_3C–C(=NH)–NH–C(=NH)–N(CH_3)–CH_3	ATM	chr11:108283161	rs11212617	Florez and Castillo-Florez (2012)
		SLC47A2	chr17:19619998	rs12943590	Choi et al. (2011)
		PPARG	chr3:12393125	rs1801282	Delahanty et al. (2011)
			chr1:72812440	rs2815752	
		SRR	chr17:2216258	rs391300	
		KCNJ11	chr11:17409572	rs5219	
		BDNF	chr11:27679915	rs6265	
		SLC22A1	chr6:160575837	rs34059508	
			chr 6:160560844	rs628031	
			chr6:160560845	rs72552763	
			chr6: 160560883	rs622342	
			chr6: 160572866	rs6265	
			chr6:160543148	rs12208357	
		SLC22A3	chr6: 160858154	rs8187725	
Exenatide (GLP-1-mimetic)	His-Gly-Glu-Gly-Thr-Phe-Thr-Ser-Asp-Leu-Ser-Lys-Gln-Met-Glu-Glu-Glu-Ala-Val-Arg-Leu-Phe-Ile-Glu-Trp-Leu-Lys-Asp-Gly-Gly-Pro-Ser-Ser-Gly-Ala-Pro-Pro-Pro-Ser	NA	NA	NA	Colagiuri (2010)
FDA drug label at DailyMed: 751747da-7c1f-41ad-b1a6-a6d920f70599					Russell-Jones and Gough (2012)
					Kelly et al. (2012)
V06A: Diet formulations for treatment of obesity	ATC therapeutic category	APOE	chr19:45411941	rs429358	Petkeviciene et al. (2012)
		DRD2	chr19:45412079	rs7412	Winkler et al. (2012)
			chr11:113270828	rs1800497	
Nutraceuticals	*Lactobacillus* sp.	NA	NA	NA	Angelakis et al. (2012)
Probiotics					Yoda et al. (2012)
Prebiotics	*Bifidobacterium* sp.				Delzenne et al. (2011)
					Parnell and Reimer (2012)

and concomitant diseases. There is a huge variability of chemical structures and diverse pathway targets that are affecting. However, very few studies have been performed on human genetic variability and the effect of obesity drugs; however, this review has summarized studies about genes mutated and their reference SNPs that are involved and demonstrated difference on safety, adverse drugs events, or effective responses.

4.3 Pharmacogenomics in Risk Assessment, Diagnosis, and Prognosis

The objective of individual genotyping for drug SNP has the potential to help to redefine how pharmacologic agents undergo clinical development, because, by identifying carriers of known genomic variants that contribute to susceptibility, a high risk population can be stratified as well as individuals with potential for a better response to a drug. Specific analyses of pharmacogenetic studies have shown that the increase in incidence of outcomes in trials restricted to individuals with specific genotypic profiles can result in substantial reductions in requisite sample sizes for such trials. In addition, they have also derived realistic bounds for samples sizes for clinical trials investigating pharmacogenetic effects that leverage genetic variations are identified in recent association studies (Schork and Topol 2010).

4.3.1 Environmental Factors in Obesity

Environmental factors, such as diet, physical activity, age, gender, socioeconomic status, and ethnicity, among others, have been shown to modulate the risk for obesity. As obesity genetics makes further progress, considerable interest has recently been turned to the potential interactions between obesity-predisposing gene variants and specific environmental situations. A significant interaction between the effects of high-fat diet and FTO genotype on BMI has been reported recently, the observed increase in BMI across FTO genotypes being restricted to people who reported a high-fat diet. An interaction between the apolipoprotein A-II (APOA2) -265T>C SNP and high-saturated fat in relation to BMI and obesity has been observed. An interaction between the FTO obesity-risk genotype and physical activity on BMI variation or obesity risk has been consistently reported in nine independent studies including adults and adolescents. These results strongly suggest that the increased risk of obesity provided by FTO variants can be attenuated through physical activity.

4.3.2 Genome-Wide Association Studies in Obesity

Recent advancements in genetics research have employed the use of genome-wide association studies (GWAS) to determine the genetic underpinnings of energy balance and obesity in human populations. Several SNP have been implicated in obesity, yet many have failed to replicate across studies. Moreover some limitations are linked to GWAS, for example, hundreds of thousands to millions of SNP may be evaluated in a single sample, reducing the statistical power to identify significant associations. Moreover, heterogeneity requires extremely large sample sizes to reveal moderate to small genetic effects, limiting the discovery of contributions of rare alleles. Table 26.4 summarizes the most important GWAS performed on obese populations, and they are recovered on http://www.genome.gov/gwastudies/.

Genes and the SPNs with significant association values to obesity and BMI disorders have been recovered. Moreover, some of them have been also associated to concomitant diseases as cardiovascular alterations, diabetes, or metabolic syndrome (Table 26.4). These genes are FAAH (fatty acid amide hydrolase), CCDC93 (coiled-coil domain containing 93), ROBO1 (roundabout, axon guidance receptor, homolog 1), GHSR (growth hormone secretagogue receptor), PCSK1(proprotein convertase subtilisin/kexin type 1), NPC1(Niemann-Pick disease, type C1), ENPP1 (ectonucleotide pyrophosphatase/phosphodiesterase 1), ANKK1 (ankyrin repeat and kinase domain containing 1), FTO (fat mass and obesity associated),

Table 26.4 Obesity and concomitant diseases host variability genetic marker studies

GWAS studies http://www.genome.gov/gwastudies/	Genes	Variants localization	Reference SNP	Phenotype association	References
Study size: 1644 individuals from 261 families of northern European ancestry	FAAH	chr1:46870760	rs324420	Increased BMI, ↑TG, ↓HLD	Zhang et al. (2009)
Framingham Heart Study cohort	CCDC93	chr2:118763068	rs11684454	Association with obesity	Herbert et al. (2006)
		chr2:118836025	rs7566605		Dina et al. 2007a, b
					Loos et al. (2007)
					Rosskopf et al. (2007)
Framingham Heart Study cohort	ROBO1	chr3:79137230	rs1455824	Association with BMI	Lasky-Su et al. (2008)
		chr3:79176116	rs1455832		
		chr3:79197923	rs2311350		
A study in 507 middle-aged overweight persons	GHSR	chr3:172163449	rs509035	Association with obesity	Baessler et al. (2005)
		chr3:172170104	rs9819506		
		chr3:172175074	rs490683		
Study at 13,659 individuals of European	PCSK1	chr5:95728898	rs6235	Association with obesity	Benzinou et al. (2008)
		chr5:95728974	rs6234		
		chr5:95751785	rs6232		
1,380 Europeans with early-onset and morbid adult obesity		chr6:22078615	rs4712652	Risk of pooled childhood and adult severe obesity	Hofker and Wijmenga (2009)
		chr10:16299951	rs10508503		
	NPC1	chr18:21140432	rs1805081		Hofker and Wijmenga (2009)
A study of 809 children, 409 obese and 400 lean controls	ENPP1	chr6:132167977	rs997509	Association with obesity and insulin resistance	Santoro et al. (2009)
	ANKK1	chr11:113270828	rs1800497	Association with obesity	Stice et al. (2008)
	FTO	chr16:53800954	rs1421085	Association with several obesity	Do et al. (2008)
					Hofker and Wijmenga (2009)
		chr16:53813367	rs17817449	Association with BMI and obesity	Do et al. (2008)
A total of 17,508 Danes from five different study groups		chr16:53820527	rs9939609	Association with BMI and obesity	Frayling et al. (2007)
			rs6499640		Grau et al. (2009)
			rs3751812		Andreasen et al. (2008)
			rs7202116		Razquin et al. (2011)
					Yang et al. (2012)
		chr16:79682751	rs1424233	Association with diabetes mellitus type II	Andreasen et al. (2008)

	NPC1	chr18:21140432	rs1805081		Hofker and Wijmenga (2009)
	MCR4	chr18:57851097	rs17782313	Association with BMI and obesity	Hofker and Wijmenga (2009)
		chr18:58039276	rs2229616		Heid et al. (2005)
					Brönner et al. (2006)
					Young et al. (2007)
					Schmid et al. (2012)
4 Finnish cohorts consisting of 2637 MetS cases and 7927 controls	APOA1/C3/A4/A5	chr11: 116648917	rs964184	Association with metabolic syndrome linked to dyslipidemia and obesity	Kristiansson et al. (2012)
Other candidate genes	*NEGR1*	chr1:72765116	rs2568958	Association with obesity	Ng et al. (2010)
	SEC16B	chr1:177913519	rs10913469		
	TMEM18	chr2:644953	rs7561317		
	ETV5/DGKG	chr3: 85834290	rs7647305		
	GNPDA2	chr4:45182527	rs10938397		
	LIN7C/BDNF	chr11:27656910	rs4923461		
		chr11:27667202	rs925946		
	MTCH2	chr11:47663049	rs10838738		
	BCDIN3D/FAIM2	chr12: 50247468	rs7138803		
	SH2B1	chr16: 28883241	rs7498665		
	KCTD15	chr19: 34309532	rs29941		
	ADRB1	chr 115804035	Trp64Arg		
	G6PD	chr10:153764216	rs1050828		
	HTR2C	chr10:113818281	rs3813928		
	LEP	chr7:127882069	rs4731426		
		chr7:127878782	rs7799039		
	LEPR	chr1:6605851	rs1137101		
	PPARγ	chr3:12393124	rs1801282 (Pro12Ala)	Association with obesity and morbid obesity	Zuo et al. (2011)

MCR4 (melanocortin receptor), and APOA1/C3/A4/A5 (Apoenzima).

Other very recent studies, a joint analysis of GWAS for early-onset extreme obesity study groups identified gene variants in or near FTO, MC4R, transmembrane protein 18 (TMEM18), serologically defined colon cancer antigen 8 (SDCCAG8), and TRF1-interacting ankyrin-related ADP-ribose polymerase/methionine sulfoxide reductase A (the TNKS/MSRA gene cluster). The TMEM18 locus was previously identified as associated with adult BMI in the international GIANT consortium, and the TNKS/MSRA gene cluster had previously been linked to waist circumference in adults. Interestingly, some of the new loci associated with BMI in adults and identified by GWAS approaches have also been associated with childhood extreme obesity (TMEM18, glucosamine-6-phosphate deaminase 2 (GNPDA2) and neuronal growth regulator 1 (NEGR1)) or BMI in children (insulin-induced gene 2 (INSIG2), FTO, MC4R, TMEM18, GNPDA2, NEGR1, BDNF, and potassium channel tetramerization domain containing 15 (KCTD15). Several of the likely causal obesity-predisposing genes are highly expressed or known to act in the central nervous system, emphasizing, as with Mendelian forms of obesity, a key role for central regulation of food intake in predisposition to obesity.

The most important associations have been described for MCR4 and specifically for FTO (Mathes et al. 2011; Yang et al. 2012). These authors have found that the difference in variance for BMI among individuals with opposite homozygous genotypes (rs7202116) at the FTO locus is approximately 7 %, corresponding to a difference of ~0.5 kg in the standard deviation of weight. These results indicate that genetic variants can be discovered that are associated with variability and that between-person variability in obesity can partly be explained by the genotype at the FTO locus.

Apart from these SNPs, several polymorphisms have been reported in immune related genes that are involved in interaction with various microorganisms and associated with a number of metabolic disorders and gut associated cancers. A numbers of such SNP markers are listed in Table 26.5.

4.3.3 Obesity Candidate Gene Studies

In spite of the numerous genes assigned to be involved in obesity (Table 26.4), the most important association variability gene with obesity has been with FTO and different SNPs. The function of FTO is mostly unknown. Mice with an FTO syntenic fused toes (Ft) mutation present developmental defects possibly owing to a programmed cell death anomaly. However, the Ft mutation involves several genes. In humans, individuals with a small chromosomal duplication on 16q12.2 including FTO present with mental retardation, dysmorphic faces, digital anomalies, and obesity. FTO is highly expressed in the human hypothalamus, pituitary, and adrenal glands, suggesting a potential role in the hypothalamic-pituitary-adrenal axis (HPA) implicated in body weight regulation. Moreover, the protein has no identified structural domain and no informed network link to any other proteins, although the SNP rs1421085 and rs1781744 were potentially functional (El-Sayed et al. 2012).

On the other hand, MCR4 has been also associated to obesity and BMI in different population studies. In a total of 19 studies (comprising 34,195 cases and 89,178 controls) of the rs17782313 polymorphism (or its proxy rs12970134) included in the meta-analysis demonstrated that association between this SNP remained significant even after adjustment for body mass index (BMI) (Xie et al. 2012b).

5 Gut Microbiota Biomarkers and Obesity

As previously it has been described, obesity is one of the major public health problems because of its increasing prevalence and association with important chronic disorders such as type 2 diabetes mellitus, atherosclerosis, nonalcoholic fatty liver disease, and cancer. Obesity is becoming a social problem with a large health expenditure associated; hence, the study of its causality is gaining more and more attention nowadays in order to offer a more effective therapy in conjunction with exercise and a healthy diet.

Table 26.5 Variability of host molecular markers involved in the interaction with microorganisms and their impact in health/disease

Gene	SNP/CNV	Variants Localization	Phenotype/clinical impact	References
MUC2	rs10902073	Chr11: 1060934	Associated with gastric cancer	Marín et al. (2012)
	rs10794281	Chr11:1063149		
	rs2071174	Chr11: 1073712		
	rs7944723	chr11:1093710		
	rs10794293	chr11: 1098939		
	rs3924453	chr11: 1105806		
	rs4077759	chr11:1105976		
	rs11825977	chr11:1075920	Associated with Crohn's disease	Moehle et al. (2006)
	rs2856111	chr11:1075747	Associated with gel forming	
	Minisatellites			
	MUC2-MS6		Associated with gastric cancer	Jeong et al. (2007)
TLR2	rs121917864	chr4:154626088	With probable pathogenic allele – germ line cell studies	
	rs137853177	chr4:154625965	Clinical source – germ line cell studies	
	rs137853176	chr4:154624324	Clinical source – germ line cell studies	
	rs7656411	chr4:154627655	Colon cancer and NSAID	Slattery et al. (2012)
	rs5743704	chr4:154625951	Survival in colon and rectal cancer	
	rs5743708	chr4:154626317	Survival in colon and rectal cancer	
TLR4	rs148638298	chr9:120475463	Clinical source – germ line cell studies	
	rs4986790	chr9:120475302	Clinical source – Associated with metabolic syndrome	Cuda et al. (2011) Steinhardt et al. (2010)
			Host-microbial interactions	Penders et al. (2010)
	rs5030728	chr9:120474282	Associated with metabolic syndrome	Cuda et al. (2011)
	rs4986791	chr9:120475602	Clinical source – germ line cell studies;	Penders et al. (2010)
			Host-microbial interactions	Kolz et al. (2008)
			Association with type 2 diabetes	
	rs7873784	chr9:120478936	Associated with colorectal cancer	Tsilidis et al. (2009)
	rs11536891	chr9:120479337	Associated with colorectal cancer	Tsilidis et al. (2009)
TLR5	rs5744168	chr1:223285200	With probable pathogenic allele – germ line cell studies	Hawn et al. (2003)
	rs2072493	chr1:223284599	With probable pathogenic allele – germ line cell studies	Michelsen and Arditi (2007)
MYD88	rs137853065	chr3:38180469		Michelsen and Arditi (2007)
	rs137853064	chr3: 38182001		
	rs74315309	chromosome 1		
	rs6853	chromosome 3	Association with vaccine immunogenetics	Dhiman et al. (2008)

Obesity is the result of a long-term positive imbalance between energy intake and expenditure and is regulated by multiple pathways involving metabolites, hormones, and neuropeptides (Sanz et al. 2010). Both genetic and environmental factors, such as a sedentary lifestyle or poor diet, are involved in the development of obesity. However, these factors do not fully explain the excessive prevalence of obesity. That is why it has recently been proposed that gut bacteria may play an important role in the pathophysiology of obesity as well as contribute to differences in body weight, insulin sensitivity, glucose metabolism, and other cardiovascular risk factors among individuals (Diamant et al. 2011).

The human gastrointestinal tract contains approximately 1×10^{14} microorganisms, with a large bacterial diversity and methanogenic archaea

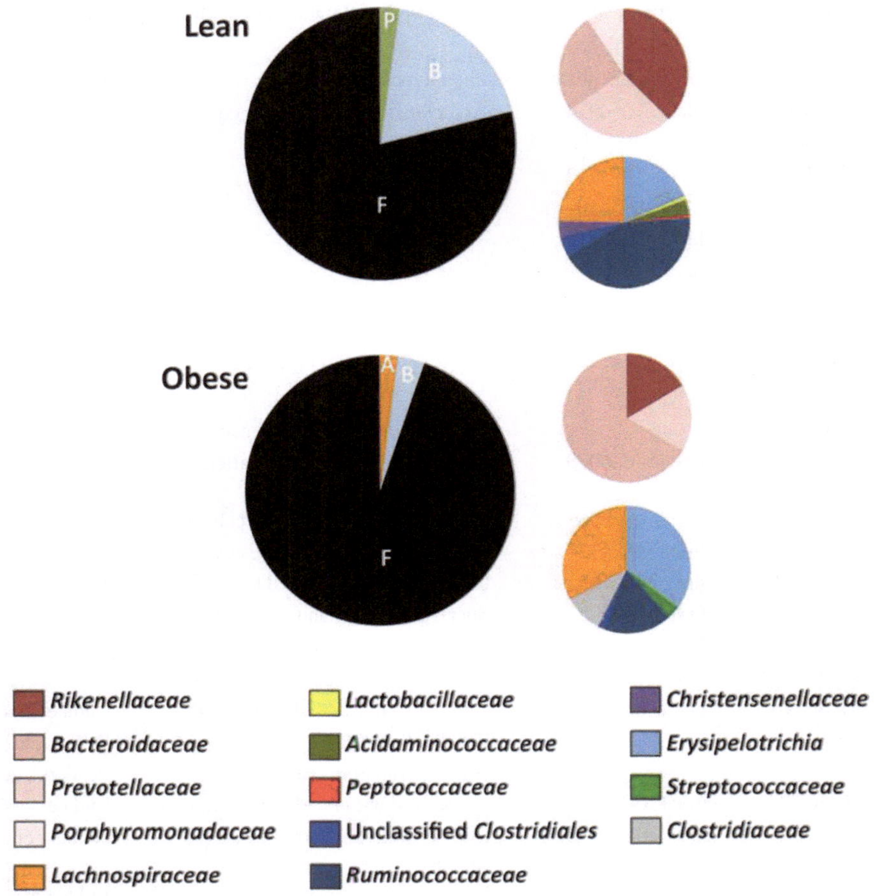

Fig. 26.3 Phylogenetic compositions of the clone libraries retrieved from "Obese" and "Lean" individuals. Labels are as follows: *A* Actinobacteria, *B* Bacteroidetes, *F* Firmicutes, *P* Proteobacteria (Ferrer et al. 2013)

included. In fact, thanks to methods recently developed in molecular biology, such as pyrosequencing and microarrays, it is estimated for the intestinal microbiota to be constituted by over 5,000 different bacterial phylotypes (Diamant et al. 2011). Altogether, the genome of the gastrointestinal tract microbiota forms the commonly called *microbiome*. We could therefore say that the human *metagenome* consists of a mixture of genes embedded in the human genome (Homo sapiens) and in the genomes of our microbial partners. The gut microbiota can be considered as a "microbial organ" in the gut, which will contribute in multiple processes in the host, such as defense against pathogens at the gut level, immunity, the development of the intestinal microvilli, and the synthesis of several vitamins (Diamant et al. 2011).

Gut microbial composition in healthy humans is influenced by the host genotype, diet, age, and sex, as well as by drugs and organic diseases that can modulate its composition and metabolic activity (Diamant et al. 2011). Moreover, the composition of this microbiota is altered in obese individuals. In this sense, comparative studies of the gut microbiota of both genetically obese (leptin-deficient) and their corresponding lean littermates (Ley et al. 2005) and between obese and lean humans (Ley et al. 2006) have shown that obese individuals have a significantly greater proportion of Firmicutes and a significantly lower proportion of Bacteroidetes than the lean ones (Fig. 26.3). Furthermore, according to a study performed by Ley et al. (2006), obese humans consumption of an hypo-energy diet (low in carbohydrates or fats) has shown a significant increase in the fecal Bacteroidetes proportions that went together with weight loss, showing a direct relationship between the population of Bacteroidetes and the tendency of losing weight.

Over the last 5 years, an increasing effort has been made to understand the role of intestinal microbiota in health and disease, resulting in regarding to it as a new organ actively involved in the control of host metabolism, both in humans and mice. Among hundreds (up to thousands) of germ species inhabiting the intestine, few of them are cultivable. Nevertheless, next-generation sequencing-based molecular technologies have been developed, allowing to overcome this problem and shed light on the way the gut microbiota undergoes dramatic changes during (patho)-physiological modifications of the host. Hence, the study of the overall gut germ genome (metagenome) and transcriptome (microbiome) has been launched. Thus, genomics and transcriptomics have begun to be increasingly used, opening the so-called omics era, including proteomic and metabolomic techniques as well. Taken together, the "omics" allow the study of gut microbiota impact on whole host metabolism, resulting in the definition of new metabolic profiles (i.e., the presence of metabolites within the blood defines a metabolomic profile), other than those based on nucleic acid analyses only. Once demonstrated the involvement of the gut microbiota within metabolic diseases, "omics" analyses has allowed the identification of the obesity-induced gut microbiota imbalance, characterized by increased Firmicutes to Bacteroidetes ratio (metagenomics) and of the so-called core microbiome, focusing on the gut microbiota at a gene – rather than, solely, at a taxonomic level. In addition, metabolomic studies revealed, for instance, the implication of gut microbiota to nonalcoholic fatty liver disease in insulin-resistant mice. Additionally, the use of germ-free (axenic) mice has made possible the microflora transfer to investigate the mechanisms through which gut microbes modulate host metabolism, albeit the molecular actors of the host gut microbiota interplay remain to be fully determined. Here, we report the role of "omics" in the multiple analyses of gut microbiota-driven metabolic modifications of the host, proposing also to focus on lipopolysaccharides (LPS), the Gram-negative proinflammatory molecules we already showed to be the initiators of metabolic diseases.

These changes in the microbiota are associated with alterations in the metagenome of the intestinal microbiota, particularly with an enrichment of those genes involved in energy harvest (Turnbaugh et al. 2006). Thus, gut microbiota is not only related to the lipid metabolism by variations of the gut microbial populations in terms of obesity or thinness of the individual but is also much more complex. In fact, comparisons between germ-free mice (GF) and mice colonized by the conventional distal gut microbiota showed that the microbiota, as a whole, increases the host's ability to extract energy from the diet and store energy in the adipocytes, contributing to body-weight gain (Bäckhed et al. 2004).

The gut microbiota may influence the energy absorption in the gut, by indirect mechanisms (increased intestinal transit) or direct mechanisms (Bäckhed and Crawford 2010), including:

1. Increased absorption of monosaccharides such as glucose: according to a study performed by Bäckhed et al. (2004), the colonization of GF mice led to a two-fold increase in intestinal glucose absorption.
2. Fermentation of short-chain fatty acids (SCFAs): the commensal microbiota is able to use dietary complex polysaccharides, otherwise not accessible to humans, producing SCFAs, mainly acetate, propionate, and butyrate. These SCFAs are employed as a substrate for lipogenesis in the liver (human), and the liver and adipose tissue (rodents), in case of acetate; as energy substrates and for the synthesis of membrane lipids in the colonocytes, both propionate and butyrate; and as substrate for gluconeogenesis in the liver, the propionate (Bäckhed and Crawford 2010). Furthermore, according to a study carried out by Samuel et al. (2008), these SCFAs may act as ligands of G protein-coupled receptors (Gpr41) whose activation stimulates the expression of peptide hormones (e.g., leptin and peptide tyrosine-tyrosine) involved in appetite and energy metabolism.

3. Modulation of lipid absorption: the bile salts facilitate the emulsification and absorption of dietary lipids; however, some bacterial enzymes may catalyze the deconjugation and dehydroxylation of these bile salts, which would affect its hydrophobicity and alter the solubility and absorption of lipids along the gut (Ridlon et al. 2006).

In addition to these effects on energy absorption, the gut microbiota may affect the lipid composition and metabolism in the serum, liver, and peripheral organs. According to a study performed by Velagapudi et al. (2010) on mice metabolomics and lipidomics, the gut microbiota has an important effect on the different species of triglycerides in the serum, adipose tissue, and liver, as well as in the phosphatidylcholine species analyzed, so it would be interesting to study not only the behavior of different type of lipids but also its particular species in detail.

Some of the effects of the gut microbiota on the lipid composition and metabolism in the serum, liver, and peripheral organs are detailed below:

- *Serum*: the gut microbiota decreases the expression of an inhibitor (angiopoietin-like protein 4) of lipoprotein lipase (LPL), increasing its activity. This affects both the lipid storage and the serum triglycerides (TGs) levels by increasing the clearance of the last ones (Bäckhed and Crawford 2010). Hence, TGs are removed in higher quantities from the serum, increasing therefore their storage in the adipose tissue and liver (Velagapudi et al. 2010).
- *Liver*: increased intestinal glucose absorption in colonized mice can stimulate lipogenesis either directly or through an increase in the insulin levels. The increase in de novo synthesis is reflected by increased levels of the liver TGs in colonized mice (Bäckhed et al. 2004).
- *Adipose tissue*: obesity and the increased adipocytes size are associated with macrophage infiltration and elevated levels of inflammatory markers, i.e., a low-grade inflammation related with changes in the gut microbiota composition and increased plasma lipopolysaccharide (LPS) levels. According to a study performed by Cani et al. (2007) LPS, which is derived from the outer membrane of a Gram-negative bacteria, may contribute to increase adipose mass and elevate expression of pro-inflammatory markers in the adipose tissue, thus suggesting a connection between gut microbiota, lipid accumulation, and inflammation. This inflammation is also associated with other diseases such as atherosclerosis (Caesar et al. 2010).

Obesity is characterized by altered gut microbiota, low-grade inflammation, and increased endocannabinoid (eCB) system tone; however, a clear connection between gut microbiota and eCB signaling has yet to be confirmed. We have reported that the gut microbiota modulates the intestinal eCB system tone, which in turn regulates gut permeability and plasma lipopolysaccharide (LPS) levels. The impact of the increased plasma LPS levels and eCB system tone found in obesity on adipose tissue metabolism (e.g., differentiation and lipogenesis) remains unknown. By interfering with the eCB system using a CB1 agonist and antagonist in lean and obese mouse models, we found that the eCB system controls gut permeability and adipogenesis. We also show both in vivo and ex vivo that LPS and eCB system control adipose tissue metabolism. In conclusion, our data indicate that the gut microbiota could determine adipose tissue physiology through LPS-eCB system regulatory loops and may play critical roles in adipose tissue plasticity during obesity.

An important concern in drug metabolism is also that the ADME process may be affected by many of these microbial-mediated pathways, no studies have evaluated how the ADME of obesogenic chemicals are affected by variations in the human microbiome (Snedeker and Hay 2012). The extent of biotransformation variation by the gut microbes, the so-called *presystemic metabolism,* of different individuals is also not known. Despite phylogenetic diversity, the implied functional metabolic redundancy observed in the gut metagenomes of individual twins (Turnbaugh et al. 2010) raises the question as to whether or not important differences exist between the enzymatic capacities of individuals. This has been indirectly established in the area of colon cancer where variation in fecal enzyme activities has been found to correlate with cancer risk (Rowland 2009).

The pharmacology literature provides valuable evidence demonstrating how chemical fate can be affected by variability in the host microbiome (Wallace et al. 2011). For example, a review identified 30 drugs (including chloramphenicol) that can be metabolized by the gut microbiota whose metabolism shows considerable interindividual variability depending on the presence or absence of specific bacteria genera (Sousa et al. 2008).

6 Obesity Test Providers and Cost/Effectiveness

Nutrigenetic/nutrigenomic tests are available in different forms depending on the country; however, the customer should be very well advised about the cost-effective test and the value of the data (see list below).

How does nutrigenomics work? First DNA is collected, usually by swabbing the inside of the mouth, and the sample is sent off to the test kit's laboratory. After a week or more, a report is completed which outlines the individual's genetic risk for certain diseases based upon measuring SNPs (single nucleotide polymorphisms). SNPs are slight variations in an individual's genetic makeup which influence an individual's susceptibility to certain factors that can increase their risk for disease. The DNA assessment kit usually costs between $100 and 500+ depending on the type and number of tests.

There are specific pharmacogenetic global tests that analyze more than 90 SNP belonging to phase I and II drug metabolic enzymes, transporters, receptors, and specific drug targets (Pharmachip®, DruginCode).

However, the cost of these tests constitutes one of the bottleneck to translate the validated and in progress pharmacogenetic biomarkers to the health system and the clinical applications, because of its current dubitably cost-effectiveness, economic incentives and reimbursement issues, and the difficulty of real clinical evaluations. Nowadays, there are few studies evaluating the economical aspect on pharmacogenetics, and because of that, it remains mainly as an abstract concept, which delay the demonstration of effectiveness to the health and governmental organisms in order to overcome the first step of implementation of pharmacogenetics as a useful tool that is able to earn in a global hospital context, more in costly therapies that are based on error-assay determination.

Concretely, around the world, for countries where the cost of obesity has been estimated, the burden ranges from 2 to 10 % of total healthcare costs and is on the rise. Using a prevalence and population attributable risk-based approach to estimate the cost of obesity in the United States, estimated its direct and indirect costs at 100 billion dollars per year. Using the same approach but with the new World Health Organization (WHO) criteria for obesity (Body Mass Index or BMI 30 kgr/m2), the same authors have subsequently estimated the direct cost of obesity at 70 billion dollars in 1995 or 7% of total health care cost. In addition, another 50 billion dollars must be added for the indirect costs associated to a loss of productivity of persons disabled by obesity. Clearly, preventing and treating obesity would have a significant effect on the cost of health care in the United States and in other countries.

Clearly, the estimates of the costs of obesity in developed countries around the world are strongly determined by the prevalence of the disease in these countries.

In view of the magnitude of the problem, there is, however, an urgent need for more funding for major research effort into the etiology of the disease including genetic factors, as well as on new strategies to treat, or better still, to prevent it. The prevention of obesity should be among the top priorities of the public health agenda in industrialized countries and help should be provided to developing countries.

7 Conclusions/Future Directions

- The investigation in obesity and its pharmacological treatment is currently in development; the use of new technologies and omics applications constitutes a huge area of comprehension for clinicians and researchers due to its increasing impact on health and consequently economic areas.

- There are large amount of polymorphism studies regarding obesity; however, only specific SNPs that belonging to FTO and MCR4 genes have been significantly associated with obesity in genome-wide association and replicated in gene candidates studies.
- There exist few drugs for treatment of obesity; thus, the design of new drugs and clinical trials in obese individuals that include pharmacogenomic assays will contribute to improve the therapeutic arsenal.
- The role of gut microbiota is being considered a determinant on modifications of energy intake and lipid metabolism. Researchers include the possibility to modify microbiota in order to exert a systemic effect on obese individuals.

The future directions on obesity pharmacologic treatments will be oriented and leaded by new discovery regarding human genome and its variability, even plasticity. Personalized treatments will be pursued during the next years taking into account the idiosyncrasy of individuals, since the chemical design is until the phase IV of clinical trials.

8 Web Resources

www.decodeme.com/obesity
www.athenadiagnostics.com/…/early-onset-obseit
http://www.athenadiagnostics.com/content/diagnostic-ed/endocrinology/early-onset-obseity
http://www.decodeme.com/obesity
http://www.cdc.gov/genomics/hugenet/CaseStudy/obesity/obesity.html
https://www.23andme.com/health/Obesity/ (FDA)
http://www.ibbiotech.com/en/3-5/servicios-medicos/genetica/obesidad/
http://www.ibdna.com/regions/UK/EN/?page=genetic-predisposition-obesity
http://www.ehow.com/about_6082694_genetic-testing-obesity.html
http://www.health24.com/dietnfood/DietDocs_articles/15-1871,57658.asp
http://www.easy-dna.com/health-dna-testing/obesity.html

References

Agarwal P, Searls DB (2008) Literature mining in support of drug discovery. Brief Bioinform 9(6):479–492, Epub 2008 Sep 27. Review. PubMed PMID: 18820304

Agurs-collins T, Khoury MJ, Simon-morton D, Olster DH, Harris JR, Milner JA (2009) NIH public access. Obesity 16(Suppl 3):1–18

Alexeyenko A, Schmitt T, Tjärnberg A, Guala D, Frings O, Sonnhammer EL (2011) Comparative interactomics with Funcoup 2.0. Nucleic Acids Res 40(Database issue):D821–D828, PMID: 22110034

Ament Z, Masoodi M, Griffin JL (2012) Applications of metabolomics for understanding the action of peroxisome proliferator-activated receptors (PPARs) in diabetes, obesity and cancer. Genome Med 4(4):32, PMID: 22546357

Andreasen CH, Stender-Petersen KL, Mogensen MS, Torekov SS, Wegner L, Andersen G, Nielsen AL, Albrechtsen A, Borch-Johnsen K, Rasmussen SS, Clausen JO, Sandbaek A, Lauritzen T, Hansen L, Jørgensen T, Pedersen O, Hansen T (2008) Low physical activity accentuates the effect of the FTO rs9939609 polymorphism on body fat accumulation. Diabetes 57(1):95–101, PMID:17942823

Angelakis E, Bastelica D, Ben Amara A, El Filali A, Dutour A, Mege JL, Alessi MC, Raoult D (2012) An evaluation of the effects of Lactobacillus ingluviei on body weight, the intestinal microbiome and metabolism in mice. Microb Pathog 52(1):61–68, PMID: 22020311

Bäckhed F, Crawford PA (2010) Coordinated regulation of the metabolome and lipidome at the host-microbial interface. Biochim Biophys Acta – Mol Cell Biol Lipids 1801:240–245

Bäckhed F, Ding H, Wang T, Hooper LV, Koh GY, Nagy A, Semenkovich CF, Gordon JI (2004) The gut microbiota as an environmental factor that regulates fat storage. Proc Natl Acad Sci U S A 101: 15718–15723

Baessler A, Hasinoff MJ, Fischer M, Reinhard W, Sonnenberg GE, Olivier M, Erdmann J, Schunkert H, Doering A, Jacob HJ, Comuzzie AG, Kissebah AH, Kwitek AE (2005) Genetic linkage and association of the growth hormone secretagogue receptor(ghrelin receptor) gene in human obesity. Diabetes 54(1): 259–267, PMID: 15616037

Baretić M (2012) Targets for medical therapy in obesity. Dig Dis 30(2):168–172, PMID: 22722433

Benson AK, Kelly SA, Legge R, Ma F, Low SJ, Kim J, Zhang M, Oh PL, Nehrenberg D, Hua K, Kachman SD, Moriyama EN, Walter J, Peterson DA, Pomp D (2010) Individuality in gut microbiota composition is a complex polygenic trait shaped by multiple environmental and host genetic factors. Proc Natl Acad Sci U S A 107:18933–18938

Benzinou M, Creemers JW, Choquet H, Lobbens S, Dina C, Durand E, Guerardel A, Boutin P, Jouret B, Heude

B, Balkau B, Tichet J, Marre M, Potoczna N, Horber F, Le Stunff C, Czernichow S, Sandbaek A, Lauritzen T, Borch-Johnsen K, Andersen G, Kiess W, Körner A, Kovacs P, Jacobson P, Carlsson LM, Walley AJ, Jørgensen T, Hansen T, Pedersen O, Meyre D, Froguel P (2008) Common nonsynonymous variants in PCSK1 confer risk of obesity. Nat Genet 40(8):943–945, PMID: 18604207

Birzele F, Fässler S, Neubauer H, Hildebrandt T, Hamilton BS (2012) Analysis of the transcriptome of differentiating and non-differentiating preadipocytes from rats and humans by next generation sequencing. Mol Cell Biochem 369(1–2):175–181, PMID: 22773307

Brignardello J, Morales P, Diaz E, Romero J, Brunser O, Gotteland M (2010) Pilot study: alterations of intestinal microbiota in obese humans are not associated with colonic inflammation or disturbances of barrier function. Aliment Pharmacol Ther 32(11–12):1307–1314

Brönner G, Sattler AM, Hinney A, Soufi M, Geller F, Schäfer H, Maisch B, Hebebrand J, Schaefer JR (2006) The 103I variant of the melanocortin 4 receptor is associated with low serum triglyceride levels. J Clin Endocrinol Metab 91(2):535–538, PMID:16278267

Caesar R, Fåk F, Bäckhed F (2010) Effects of gut microbiota on obesity and atherosclerosis via modulation of inflammation and lipid metabolism. J Intern Med 268:320–328

Canani RB, Costanzo MD, Leone L, Bedogni G, Brambilla P, Cianfarani S, Nobili V, Pietrobelli A, Agostoni C (2011) Epigenetic mechanisms elicited by nutrition in early life. Nutr Res Rev 24(2):198–205, PMID: 22008232

Cani PD, Amar J, Iglesias MA, Poggi M, Knauf C, Bastelica D, Neyrinck AM, Fava F, Tuohy KM, Chabo C, Waget A, Delmée E, Cousin B, Sulpice T, Chamontin B, Ferrières J, Tanti JF, Gibson GR, Casteilla L, Delzenne NM, Alessi MC, Burcelin R (2007) Metabolic endotoxemia initiates obesity and insulin resistance. Diabetes 56:1761–1772

Clément K (2011 Sep) Bariatric surgery, adipose tissue and gut microbiota. Int J Obes (Lond) 35(Suppl 3):S7–S15. doi:10.1038/ijo.2011.141, Review. PubMed PMID: 21912389

Chen HH, Lee WJ, Wang W, Huang MT, Lee YC, Pan WH (2007) Ala55Val polymorphism on UCP2 gene predicts greater weight loss in morbidly obese patients undergoing gastric banding. Obes Surg 17(7):926–933, PMID: 17894153

Choi SH, Kwak SH, Lee Y, Moon MK, Lim S, Park YJ, Jang HC, Kim MS (2011) Plasma vaspin concentrations are elevated in metabolic syndrome in men and are correlated with coronary atherosclerosis in women. Clin Endocrinol (Oxf) 75(5):628–635. doi:10.1111/j.1365-2265.2011.04095.x, PubMed PMID: 21545478

Choquet H, Meyre D (2010) Genomic insights into early-onset obesity. Genome Med 2(6):36

Colagiuri S (2010) Diabesity: therapeutic options. Diabetes Obes Metab 12(6):463–473, PMID: 20518802

Cuda C, Badawi A, Karmali M, El-Sohemy A (2011) Polymorphisms in Toll-like receptor 4 are associated with factors of the metabolic syndrome and modify the association between dietary saturated fat and fasting high-density lipoprotein cholesterol. Metabolism 60(8):1131–1135, PMID: 21306745

Delahanty RJ, Beeghly-Fadiel A, Xiang YB, Long J, Cai Q, Wen W, Xu WH, Cai H, He J, Gao YT, Zheng W, Shu XO (2011) Association of obesity-related genetic variants with endometrial cancer risk: a report from the Shanghai Endometrial Cancer Genetics Study. Am J Epidemiol 174(10):1115–1126. doi 10.1093/aje/kwr233. Epub 5 Oct 2011. PubMed PMID: 21976109; PubMed Central PMCID: PMC3246689

Delzenne NM, Neyrinck AM, Cani PD (2011) Modulation of the gut microbiota by nutrients with prebiotic properties: consequences for host health in the context of obesity and metabolic syndrome. Microb Cell Fact 10(Suppl 1):S10, PMID: 21995448

Dhiman N, Ovsyannikova IG, Vierkant RA, Ryan JE, Pankratz VS, Jacobson RM, Poland GA (2008) Associations between SNPs in toll-like receptors and related intracellular signaling molecules and immune responses to measles vaccine: preliminary results. Vaccine 26(14):1731–1736, PMID: 18325643

Diamant M, Blaak EE, de Vos WM (2011) Do nutrient-gut-microbiota interactions play a role in human obesity, insulin resistance and type 2 diabetes? Obes Rev 12(4):272–281.doi:10.1111/j.1467-789X.2010.00797.x. Epub 26 Aug 2010. Review. PubMed PMID: 20804522

Dimitrov DV (2011) The human gutome: nutrigenomics of the host-microbiome interactions. OMICS. 15(7–8):419–430. doi:10.1089/omi.2010.0109. Epub 1 Dec 2010. Review. PubMed PMID: 21121704

Dina C, Meyre D, Samson C, Tichet J, Marre M, Jouret B, Charles MA, Balkau B, Froguel P (2007a) Comment on "A common genetic variant is associated with adult and childhood obesity". Science 315(5809):187, PMID:17218508

Dina C, Meyre D, Gallina S, Durand E, Körner A, Jacobson P et al (2007b) Variation in FTO contributes to childhood obesity and severe adult obesity. Nat Genet 39(6):724–726

Do R, Bailey SD, Desbiens K, Belisle A, Montpetit A, Bouchard C, Pérusse L, Vohl MC, Engert JC (2008) Genetic variants of FTO influence adiposity, insulin sensitivity, leptin levels, and resting metabolic rate in the Quebec Family Study. Diabetes 57(4):1147–1150, PMID: 18316358

El-Sayed Moustafa JS, Eleftherohorinou H, de Smith AJ, Andersson-Assarsson JC, Alves AC, Hadjigeorgiou E, Walters RG, Asher JE, Bottolo L, Buxton JL, Sladek R, Meyre D, Dina C, Visvikis-Siest S, Jacobson P, Sjöström L, Carlsson LM, Walley A, Falchi M, Froguel P, Blakemore AI, Coin LJ (2012) Novel association approach for variable number tandem repeats (VNTRs) identifies DOCK5 as a susceptibility gene for severe obesity. Hum Mol Genet 21(16):3727–2738. doi:10.1093/hmg/dds187. Epub 16 May 2012. PubMed PMID: 22595969; PubMed Central PMCID: PMC3406755

Fenech M, El-Sohemy A, Cahill L, Ferguson L, French TA, Tai ES, Milner J, Koh WP, Xie L, Zucker M, Buckley M, Cosgrove L, Lockett T, Fung KY, Head R (2011) Nutrigenetics and nutrigenomics: viewpoints on the current status and applications in nutrition research and practice. J Nutrigenet Nutrigenomics 28(4):69–89

Ferguson LR, Shelling AN, Lauren D, Heyes JA, McNabb WC (2007) Nutrigenomics and gut health. Mutat Res 622:1–6

Ferrer M, Ruiz A, Lanza F, Haange SB, Oberbach A, Till H, Bargiela R, Campoy C, Segura MT, Richter M, von Bergen M, Seifert J, Suarez A (2013) Microbiota from the distal guts of lean and obese adolescents exhibit partial functional redundancy besides clear differences in community structure. Environ Microbiol 15(1):211–226. doi:10.1111/j.1462-2920.2012.02845.x, PubMed PMID: 22891823

Florez H, Castillo-Florez S (2012) Beyond the obesity paradox in diabetes: fitness, fatness, and mortality. JAMA 308(6):619–620. doi:10.1001/jama.2012.9776, PubMed PMID: 22871873

Frayling TM, Timpson NJ, Weedon MN, Zeggini E, Freathy RM, Lindgren CM, Perry JR, Elliott KS, Lango H, Rayner NW, Shields B, Harries LW, Barrett JC, Ellard S, Groves CJ, Knight B, Patch AM, Ness AR, Ebrahim S, Lawlor DA, Ring SM, Ben-Shlomo Y, Jarvelin MR, Sovio U, Bennett AJ, Melzer D, Ferrucci L, Loos RJ, Barroso I, Wareham NJ, Karpe F, Owen KR, Cardon LR, Walker M, Hitman GA, Palmer CN, Doney AS, Morris AD, Smith GD, Hattersley AT, McCarthy MI (2007) A common variant in the FTO gene is associated with body mass index and predisposes to childhood and adult obesity. Science 316(5826):889–894, PMID: 17434869

Grasso P (2011) Novel approaches to the treatment of obesity and type 2 diabetes mellitus: bioactive leptin-related synthetic peptide analogs. Recent Pat Endocr Metab Immune Drug Discov 5(3):163–175, PMID: 21913885

Grau K, Hansen T, Holst C, Astrup A, Saris WH, Arner P, Rössner S, Macdonald I, Polak J, Oppert JM, Langin D, Martinez JA, Pedersen O, Sørensen TI (2009) Macronutrient-specific effect of FTO rs9939609 in response to a 10-week randomized hypo-energetic diet among obese Europeans. Int J Obes (Lond) 33(11):1227–1234, PMID: 19687793

Greenblum S, Turnbaugh PJ, Borenstein E (2012) Metagenomic systems biology of the human gut microbiome reveals topological shifts associated with obesity and inflammatory bowel disease. Proc Natl Acad Sci U S A 109(2):594–599, PMID: 22184244

Grudell AB, Sweetser S, Camilleri M, Eckert DJ, Vazquez-Roque MI, Carlson PJ, Burton DD, Braddock AE, Clark MM, Graszer KM, Kalsy SA, Zinsmeister AR (2008) A controlled pharmacogenetic trial of sibutramine on weight loss and body composition in obese or overweight adults. Gastroenterology 135(4):1142–1154, PMID: 18725220

Hahn S, Frey UH, Siffert W, Tan S, Mann K, Janssen OE (2006) The CC genotype of the GNAS T393C polymorphism is associated with obesity and insulin resistance in women with polycystic ovary syndrome. Eur J Endocrinol 155(5):763–770, PMID: 17062894

Harris K, Kassis A, Major G, Chou CJ (2012) Is the gut microbiota a new factor contributing to obesity and its metabolic disorders? J Obes 879151

Hawn TR, Verbon A, Lettinga KD, Zhao LP, Li SS, Laws RJ, Skerrett SJ, Beutler B, Schroeder L, Nachman A, Ozinsky A, Smith KD, Aderem A (2003) A common dominant TLR5 stop codon polymorphism abolishes flagellin signaling and is associated with susceptibility to legionnaires' disease. J Exp Med 198(10):1563–1572, PMID: 14623910

Heid IM, Vollmert C, Hinney A, Döring A, Geller F, Löwel H, Wichmann HE, Illig T, Hebebrand J, Kronenberg F, KORA Group (2005) Association of the 103I MC4R allele with decreased body mass in 7937 participants of two population based surveys. J Med Genet 42(4):e21, PMID: 15805150

Herbert A, Gerry NP, McQueen MB, Heid IM, Pfeufer A, Illig T, Wichmann HE, Meitinger T, Hunter D, Hu FB, Colditz G, Hinney A, Hebebrand J, Koberwitz K, Zhu X, Cooper R, Ardlie K, Lyon H, Hirschhorn JN, Laird NM, Lenburg ME, Lange C, Christman MF (2006) A common genetic variant is associated with adult and childhood obesity. Science 312(5771):279–283, PMID: 16614226

Hofker M, Wijmenga C (2009) A supersized list of obesity genes. Nat Genet 41(2):139–140, PMID: 19174833

Hsiao DJ, Wu LS, Huang SY, Lin E (2009) Weight loss and body fat reduction under sibutramine therapy in obesity with the C825T polymorphism in the GNB3 gene. Pharmacogenet Genomics 19(9):730–733, PMID: 19687782

Hsiao TJ, Wu LS, Hwang Y, Huang SY, Lin E (2010) Effect of the common -866G/A polymorphism of the uncoupling protein 2 gene on weight loss and body composition under sibutramine therapy in an obese Taiwanese population. Mol Diagn Ther 14(2):101–106, PMID: 20359253

Hu Y, Ehli EA, Hudziak JJ, Davies GE (2012) Berberine and evodiamine influence serotonin transporter (5-HTT) expression via the 5-HTT-linked polymorphic region. Pharmacogenomics J 12(5):372–378. doi:10.1038/tpj.2011.24. Epub 7 June 2011. PubMed PMID: 21647174

Jeong YH, Kim MC, Ahn EK, Seol SY, Do EJ, Choi HJ, Chu IS, Kim WJ, Kim WJ, Sunwoo Y, Leem SH (2007) Rare exonic minisatellite alleles in MUC2 influence susceptibility to gastric carcinoma. PLoS One 2(11):e1163, PMID: 18000536

Ji B, Ernest B, Gooding JR, Das S, Saxton AM, Simon J, Dupont J, Métayer-Coustard S, Campagna SR, Voy BH (2012) Transcriptomic and metabolomic profiling of chicken adipose tissue in response to insulin neutralization and fasting. BMC Genomics 13(1):441, PMID: 22938590

Katoh M, Katoh M (2007) AP1- and NF-kappaB-binding sites conserved among mammalian WNT10B orthologs elucidate the TNFalpha-WNT10B signaling loop implicated in carcinogenesis and adipogenesis. Int J Mol Med 19(4):699–703, PubMed PMID: 17334647

Kelly AS, Bergenstal RM, Gonzalez-Campoy JM, Katz H, Bank AJ (2012) Effects of exenatide vs. metformin on endothelial function in obese patients with pre-diabetes: a randomized trial. Cardiovasc Diabetol 11:64, PMID: 22681705

Klenke S, Siffert W (2011) SNPs in genes encoding G proteins in pharmacogenetics. Pharmacogenomics 12(5):633–654, PMID: 21619427

Kolz M, Baumert J, Müller M, Khuseyinova N, Klopp N, Thorand B, Meisinger C, Herder C, Koenig W, Illig T (2008) Association between variations in the TLR4 gene and incident type 2 diabetes is modified by the ratio of total cholesterol to HDL-cholesterol. BMC Med Genet 9:9, PMID: 18298826

Kristiansson K, Perola M, Tikkanen E, Kettunen J, Surakka I, Havulinna AS, Stancáková A, Barnes C, Widen E, Kajantie E, Eriksson JG, Viikari J, Kähönen M, Lehtimäki T, Raitakari OT, Hartikainen AL, Ruokonen A, Pouta A, Jula A, Kangas AJ, Soininen P, Ala-Korpela M, Männistö S, Jousilahti P, Bonnycastle LL, Järvelin MR, Kuusisto J, Collins FS, Laakso M, Hurles ME, Palotie A, Peltonen L, Ripatti S, Salomaa V (2012) Genome-wide screen for metabolic syndrome susceptibility Loci reveals strong lipid gene contribution but no evidence for common genetic basis for clustering of metabolic syndrome traits. Circ Cardiovasc Genet 5(2):242–249, PMID: 22399527

Kumar V (2011) Omics and literature mining. Methods Mol Biol 719:457–477, PMID: 21370097

Lakhan SE, Kirchgessner A (2011) Gut microbiota and sirtuins in obesity-related inflammation and bowel dysfunction. J Transl Med 9(1):202

Lasky-Su J, Lyon HN, Emilsson V, Heid IM, Molony C, Raby BA, Lazarus R, Klanderman B, Soto-Quiros ME, Avila L, Silverman EK, Thorleifsson G, Thorsteinsdottir U, Kronenberg F, Vollmert C, Illig T, Fox CS, Levy D, Laird N, Ding X, McQueen MB, Butler J, Ardlie K, Papoutsakis C, Dedoussis G, O'Donnell CJ, Wichmann HE, Celedón JC, Schadt E, Hirschhorn J, Weiss ST, Stefansson K, Lange C (2008) On the replication of genetic associations: timing can be everything! Am J Hum Genet 82(4):849–858, PMID: 18387595

Lehr S, Hartwig S, Sell H (2011) Adipokines: a treasure trove for the discovery of biomarkers for metabolic disorders. Proteomics Clin Appl 6(1–2):91–101, PMID: 22213627

Ley RE, Bäckhed F, Turnbaugh P, Lozupone CA, Knight RD, Gordon JI (2005) Obesity alters gut microbial ecology. Proc Natl Acad Sci U S A 102:11070–11075

Ley RE, Turnbaugh PJ, Klein S, Gordon JI (2006) Microbial ecology: human gut microbes associated with obesity. Nature 444:1022–1023

Li S, Zhao JH, Luan J, Ekelund U, Luben RN, Khaw KT et al (2010) Physical activity attenuates the genetic predisposition to obesity in 20,000 men and women from EPIC-Norfolk prospective population study. PLoS Med 7:e1000332

Liu Y, Zhang C, Zhao L, Nardini C (2010) Adapting functional genomic tools to metagenomic analyses: investigating the role of gut bacteria in relation to obesity. Brief Funct Genomics 9(5–6):355–361, PMID: 21266343

Li-Wan-Po A, Farndon P (2011) Barking up the wrong genome-we are not alone. J Clin Pharm Ther 36(2):125–127, PMID: 21366639

Loos RJ, Barroso I, O'rahilly S, Wareham NJ (2007) Comment on "A common genetic variant is associated with adult and childhood obesity". Science 315(5809):187, PMID: 17218509

Maccaferri S, Biagi E, Brigidi P (2011) Metagenomics: key to human gut microbiota. Dig Dis 29(6):525–530, PMID: 22179207

Marín F, Bonet C, Muñoz X, García N, Pardo ML, Ruiz-Liso JM, Alonso P, Capellà G, Sanz-Anquela JM, González CA, Sala N (2012) Genetic variation in MUC1, MUC2 and MUC6 genes and evolution of gastric cancer precursor lesions in a long-term follow-up in a high-risk area in Spain. Carcinogenesis 33(5):1072–1080, PMID: 22402132

Martinez AJ, Parra DM, Santos JL et al (2008) Genotype-dependent response to energy-restricted diets in obese subjects: towards personalized nutrition. Asia Pac J Clin Nutr 17:119–122

Mathes WF, Kelly SA, Pomp D (2011) Advances in comparative genetics: influence of genetics on obesity. Br J Nutr 106(Suppl 1):S1–S10, PMID: 22005399

Michelsen KS, Arditi M (2007) Toll-like receptors and innate immunity in gut homeostasis and pathology. Curr Opin Hematol 14(1):48–54, PMID: 17133100

Moehle C, Ackermann N, Langmann T, Aslanidis C, Kel A, Kel-Margoulis O, Schmitz-Madry A, Zahn A, Stremmel W, Schmitz G (2006) Aberrant intestinal expression and allelic variants of mucin genes associated with inflammatory bowel disease. J Mol Med 84(12):1055–1066, PMID: 17058067

Naukkarinen J, Rissanen A, Kaprio J, Pietiläinen KH (2012) Causes and consequences of obesity: the contribution of recent twin studies. Int J Obes (Lond) 36(8):1017–1024. doi:10.1038/ijo.2011.192. Epub 11 Oct 2011. Review. PubMed PMID: 21986704

Ng MC, Tam CH, So WY, Ho JS, Chan AW, Lee HM, Wang Y, Lam VK, Chan JC, Ma RC (2010) Implication of genetic variants near NEGR1, SEC16B, TMEM18, ETV5/DGKG, GNPDA2, LIN7C/BDNF, MTCH2, BCDIN3D/FAIM2, SH2B1, FTO, MC4R, and KCTD15 with obesity and type 2 diabetes in 7705 Chinese. J Clin Endocrinol Metab 95(5):2418–2425, PMID: 20215397

Parnell JA, Reimer RA (2012) Prebiotic fiber modulation of the gut microbiota improves risk factors for obesity and the metabolic syndrome. Gut Microb 3(1):29–34, PMID: 22555633

Peinado JR, Pardo M, de la Rosa O, Malagón MM (2012) Proteomic characterization of adipose tissue constituents, a necessary step for understanding adipose tissue complexity. Proteomics 12(4–5):607–620, PMID: 22246603

Penders J, Thijs C, Mommers M, Stobberingh EE, Dompeling E, Reijmerink NE, van den Brandt PA, Kerkhof M, Koppelman GH, Postma DS (2010) Host-microbial interactions in childhood atopy: toll-like receptor 4 (TLR4), CD14, and fecal Escherichia coli. J Allergy Clin Immunol 125(1):231–236, e1–5. PMID: 20109750

Pepino MY, Love-Gregory L, Klein S, Abumrad NA (2012) The fatty acid translocase gene CD36 and lingual lipase influence oral sensitivity to fat in obese subjects. J Lipid Res 53(3):561–566, PMID: 22210925

Pérez-Pérez R, López JA, García-Santos E, Camafeita E, Gómez-Serrano M, Ortega-Delgado FJ, Ricart W, Fernández-Real JM, Peral B (2012) Uncovering suitable reference proteins for expression studies in human adipose tissue with relevance to obesity. PLoS One 7(1), PMID: 22272336

Petkeviciene J, Smalinskiene A, Luksiene DI, Jureniene K, Ramazauskiene V, Klumbiene J, Lesauskaite V (2012) Associations between apolipoprotein E genotype, diet, body mass index, and serum lipids in Lithuanian adult population. PLoS One 7(7):e41525, PMID: 22844488

Razquin C, Marti A, Martinez JA (2011) Evidences on three relevant obesogenes: MC4R, FTO and PPARγ. Approaches for personalized nutrition. Mol Nutr Food Res 55(1):136–149, PMID:21207518

Ridlon JM, Kang DJ, Hylemon PB (2006) Bile salt biotransformations by human intestinal bacteria. J Lipid Res 47:241–259

Rosskopf D, Bornhorst A, Rimmbach C, Schwahn C, Kayser A, Krüger A, Tessmann G, Geissler I, Kroemer HK, Völzke H (2007) Comment on "A common genetic variant is associated with adult and childhood obesity". Science 315(5809):187, PMID: 17218510

Roth JD, Trevaskis JL, Wilson J, Lei C, Athanacio J, Mack C, Kesty NC, Coffey T, Weyer C, Parkes DG (2008) Antiobesity effects of the beta-cell hormone amylin in combination with phentermine or sibutramine in diet-induced obese rats. Int J Obes 32(8):1201–1210, PMID: 18560368

Rowland IR (2009) The role of the gastrointestinal microbiota in colorectal cancer. Curr Pharm Des 15(13):1524–1527

Rudkowska I, Pérusse L (2012) Individualized weight management: what can be learned from nutrigenomics and nutrigenetics? Prog Mol Biol Transl Sci 108: 347–382, PMID: 22656384

Russell-Jones D, Gough S (2012) Recent advances in incretin-based therapies. Clin Endocrinol 77(4): 489–499, PMID: 22804841

Samuel BS, Shaito A, Motoike T, Rey FE, Backhed F, Manchester JK, Hammer RE, Williams SC, Crowley J, Yanagisawa M, Gordon JI (2008) Effects of the gut microbiota on host adiposity are modulated by the short-chain fatty-acid binding G protein-coupled receptor, Gpr41. Proc Natl Acad Sci U S A 105:16767–16772

Santoro N, Cirillo G, Lepore MG, Palma A, Amato A, Savarese P, Marzuillo P, Grandone A, Perrone L, Del Giudice EM (2009) Effect of the rs997509 polymorphism on the association between ectonucleotide pyrophosphatase phosphodiesterase 1 and metabolic syndrome and impaired glucose tolerance in childhood obesity. J Clin Endocrinol Metab 94(1):300–305, PMID: 18940878

Sanz Y, Santacruz A, Gauffin P (2010) Gut microbiota in obesity and metabolic disorders. Proc Nutr Soc 69:434–441

Schmid PM, Heid I, Buechler C, Steege A, Resch M, Birner C, Endemann DH, Riegger GA, Luchner A (2012) Expression of fourteen novel obesity-related genes in Zucker diabetic fatty rats. Cardiovasc Diabetol 11:48, PMID: 22553958

Schork NJ, Topol EJ (2010) Genotype-based risk and pharmacogenetic sampling in clinical trials. J Biopharm Stat 20(2):315–333

Serino M, Chabo C, Burcelin R (2012) Intestinal MicrobiOMICS to define health and disease in human and mice. Curr Pharm Biotechnol 13(5):746–758, PMID: 22122483

Slattery ML, Lundgreen A, Welbourn B, Corcoran C, Wolff RK (2012) Genetic variation in selenoprotein genes, lifestyle, and risk of colon and rectal cancer. PLoS One 7(5):e37312, PMID: 22615972

Snedeker SM, Hay AG (2012) Do interactions between gut ecology and environmental chemicals contribute to obesity and diabetes? Environ Health Perspect 120(3):332–339

Sousa T, Paterson R, Moore V, Carlsson A, Abrahamsson B, Basit AW (2008) The gastrointestinal microbiota as a site for the biotransformation of drugs. Int J Pharm 363(1–2):1–25

Spor A, Koren O, Ley R (2011) Unravelling the effects of the environment and host genotype on the gut microbiome. Nat Rev Microbiol 9:279–290

Steinhardt AP, Aranguren F, Tellechea ML, Gómez Rosso LA, Brites FD, Martínez-Larrad MT, Serrano-Ríos M, Frechtel GD, Taverna MJ (2010) A functional nonsynonymous toll-like receptor 4 gene polymorphism is associated with metabolic syndrome, surrogates of insulin resistance, and syndromes of lipid accumulation. Metabolism 59(5):711–717, PMID: 19922963

Stice E, Spoor S, Bohon C, Small DM (2008) Relation between obesity and blunted striatal response to food is moderated by TaqIA A1 allele. Science 322(5900):449–452, PMID: 18927395

Tsilidis KK, Helzlsouer KJ, Smith MW, Grinberg V, Hoffman-Bolton J, Clipp SL, Visvanathan K, Platz EA (2009) Association of common polymorphisms in IL10, and in other genes related to inflammatory response and obesity with colorectal cancer. Cancer Causes Control 20(9):1739–1751. doi:10.1007/s10552-009-9427-7, PubMed PMID: 19760027

Tiihonen K, Ouwehand AC, Rautonen N (2010) Effect of overweight on gastrointestinal microbiology and immunology: correlation with blood biomarkers. Br J Nutr 103(7):1070–1078, Epub 2009; PubMed PMID: 19930761

Turnbaugh PJ, Ley RE, Mahowald MA, Magrini V, Mardis ER, Gordon JI (2006) An obesity-associated gut microbiome with increased capacity for energy harvest. Nature 444:1027–1031

Turnbaugh PJ, Quince C, Faith JJ, McHardy AC, Yatsunenko T, Niazi F et al (2010) Organismal, genetic, and transcriptional variation in the deeply sequenced gut microbiomes of identical twins. Proc Natl Acad Sci U S A 107(16):7503–7508

Velagapudi VR, Hezaveh R, Reigstad CS, Gopalacharyulu P, Yetukuri L, Islam S, Felin J, Perkins R, Borén J, Orešič M, Bäckhed F (2010) The gut microbiota modulates host energy and lipid metabolism in mice. J Lipid Res 51:1101–1112

Wallace TC, Guarner F, Madsen K, Cabana MD, Gibson G, Hentges E, Sanders ME (2011) Human gut microbiota and its relationship to health and disease. Nutr Rev 69(7):392–403. doi:10.1111/j.1753-4887.2011.00402.x. Epub 30 June 2011. Review. PubMed PMID: 21729093

Williams MJ, Almén MS, Fredriksson R, Schiöth HB (2012) What model organisms and interactomics can reveal about the genetics of human obesity. Cell Mol Life Sci 69(22):3819–3834. doi:10.1007/s00018-012-1022-5. Epub 23 May 2012. Review. PubMed PMID: 22618246

Winkler JK, Woehning A, Schultz JH, Brune M, Beaton N, Challa TD, Minkova S, Roeder E, Nawroth PP, Friederich HC, Wolfrum C, Rudofsky G (2012) TaqIA polymorphism in dopamine D2 receptor gene complicates weight maintenance in younger obese patients. Nutrition 28(10):996–1001, PMID: 22541053

Xie B, Waters MJ, Schirra HJ (2012a) Investigating potential mechanisms of obesity by metabolomics. J Biomed Biotechnol 2012:805683, PMID: 22665992

Xie B, Takeuchi F, Chandak GR, Kato N, Pan HW, AGEN-T2D Consortium, Zhou DH, Pan HY, Mi J (2012b) Common polymorphism near the MC4R gene is associated with type 2 diabetes: data from a meta-analysis of 123,373 individuals. Diabetologia 55(10):2660–6

Yang RY, Havel PJ, Liu FT (2012) Galectin-12: a protein associated with lipid droplets that regulates lipid metabolism and energy balance. Adipocyte 1(2):96–100. PubMed PMID: 23700518; PubMed Central PMCID: PMC3609087

Yoda K, He F, Miyazawa K, Kawase M, Kubota A, Hiramatsu M (2012) Orally administered heat-killed Lactobacillus gasseri TMC0356 alters respiratory immune responses and intestinal microbiota of diet-induced obese mice. J Appl Microbiol 113(1): 155–162, PMID: 22519947

Young EH, Wareham NJ, Farooqi S, Hinney A, Hebebrand J, Scherag A, O'rahilly S, Barroso I, Sandhu MS (2007) The V103I polymorphism of the MC4R gene and obesity: population based studies and meta-analysis of 29 563 individuals. Int J Obes (Lond) 31(9):1437–1441, PMID:17356525

Zhang Y, Sonnenberg GE, Baye TM, Littrell J, Gunnell J, DeLaForest A, MacKinney E, Hillard CJ, Kissebah AH, Olivier M, Wilke RA (2009) Obesity-related dyslipidemia associated with FAAH, independent of insulin response, in multigenerational families of Northern European descent. Pharmacogenomics 10(12):1929–1939, PMID: 19958092

Zhang W, Roederer MW, Chen WQ, Fan L, Zhou HH (2012) Pharmacogenetics of drugs withdrawn from the market. Pharmacogenomics 13(2):223–231, PMID: 22256871

Zuo HJ, Xie ZM, Zhang WW, Li YR, Wang W, Ding XB, Pei XF (2011) Gut bacteria alteration in obese people and its relationship with gene polymorphism. World J Gastroenterol 17(8):1076–1081, PMID: 21448362

Pharmacogenomics and Personalized Medicine for Infectious Diseases

27

Nirmal Kumar Ganguly and Gautam Kumar Saha

Abstract

Humans have been plagued by the scourge of invasion by pathogens leading to infectious diseases from the time in memoriam and are still the cause of morbidity and mortality among millions of individuals. Trying to understand the disease mechanisms and finding the remedial measures have been the quest of humankind. The susceptibility to disease of an individual in a given population is determined by ones genetic buildup. Response to treatment and the disease prognosis also depends upon individual's genetic predisposition. The environmental stress induces mutations and is leading to the emergence of ever-increasing more dreaded infectious pathogens, and now we are in the era of increasing antibiotic resistance that has thrown up a challenge to find new treatment regimes. Discoveries in the science of high-throughput sequencing and array technologies have shown new hope and are bringing a revolution in human health. The information gained from sequencing of both human and pathogen genomes is a way forward in deciphering host-pathogen interactions. Deciphering the pathogen virulence factors, host susceptibility genes, and the molecular programs involved in the pathogenesis of disease has paved the way for discovery of new molecular targets for drugs, diagnostic markers, and vaccines. The genomic diversity in the human population leads to differences in host responses to drugs and vaccines and is the cause of poor response to treatment as well as adverse reactions. The study of pharmacogenomics of infectious diseases is still at an early stage of development, and many intricacies of the host-pathogen interaction are yet to be understood in full measure. However, progress has been made over the decades of research in some of the important infectious diseases revealing

N.K. Ganguly (✉) • G.K. Saha
Policy Centre for Biomedical Research, Translational Health Science and Technology Institute
(Department of Biotechnology Institute, Government of India), Office@National Institute of Immunology, New Delhi, India
e-mail: nkganguly@nii.ac.in; gautamkumarsaha@gmail.com

D. Barh et al. (eds.), *Omics for Personalized Medicine*,
DOI 10.1007/978-81-322-1184-6_27,

how the host genetic polymorphisms of drug-metabolizing enzymes and transporters affect the bioavailability of the drugs which further determine the efficacy and toxicology of the drugs used for treatment. Further, the field of structural biology and chemistry has intertwined to give rise to medical structural genomics leading the way to the discovery of new drug targets against infectious diseases. This chapter explores how the advent of "omics" technologies is making a beginning in bringing about a change in the prevention, diagnosis, and treatments of the infectious diseases and hence paving way for personalized medicine.

1 Introduction

Over the millennia with the progression of human civilization, the condition of human health has changed considerably. The lifespan of the human has increased considerably with the advent of vaccines against several diseases which has eradicated small pox, and now we are embarking on the global campaign to eradicate poliomyelitis and have controlled the disease in most parts of the world. The treatment of infectious diseases got a boost with the discovery of penicillin by Alexander Fleming in the earlier part of the twentieth century. As a range of antibiotics were later discovered, infectious diseases such as meningitis, bacterial pneumonia, sepsis, and other life-threatening bacterial infection were treatable. Also the survivability of patients undergoing operative procedures and aggressive chemotherapy was feasible and their recovery high. The last 50 years of the twentieth century have been eventful with the discovery of antimicrobials that had given us the hope that we shall eradicate all infectious diseases. Despite all the efforts and progress we have made in medical science, infectious disease remained a major health problem.

But the golden era of the antibiotics will not be long if we go on unregulated administering and promoting rampant use of antibiotics, as is evident from the rise of antibiotic resistance (Caramia and Ruffini 2012), and the new emerging diseases have posed major challenge (Table 27.1). In the last few decades with the advent of field of genomics, there is a new hope in prevention, diagnostics, and treatment of infectious diseases. We will explore in this chapter

Table 27.1 Timeline of pathogen discovery

Year	Microbe
1973	Rotavirus
1975	Parvovirus B-19
1976	*Cryptosporidium parvum*
1977	Ebola virus
1977	*Legionella pneumophila*
1977	Hantaan virus
1977	*Campylobacter jejuni*
1980	Human T-lymphotropic virus I (HTLV-I)
1981	Toxin producing strains of *Staphylococcus aureus*
1982	*Escherichia coli* O157:H7
1982	HTLV-II
1982	*Borrelia burgdorferi*
1983	Human immunodeficiency virus
1983	*Helicobacter pylori*
1985	*Enterocytozoon bieneusi*
1986	*Cyclospora cayetanensis*
1988	Hepatitis E virus
1989	*Ehrlichia chaffeensis*
1989	Hepatitis C
1991	Guanarito virus
1991	*Encephalitozoon hellem*
1991	New species of *Babesia*
1992	*Vibrio cholerae* O139
1992	*Bartonella henselae*
1993	Sin nombre virus
1993	Encephalitozoon cuniculi
1994	Sabia virus
1995	HHV-8
1999	Nipah virus
2002	SARS virus

The table highlights the organisms that are of public health importance and their year of discovery
Source: World Health Organization: Newly discovered organisms of public health importance: Page 6. From WHO Regional Office East-Asia: Combating Emerging Infectious Diseases in South East Asia Region (2005)

how the advent of new technologies is bringing about a change in medical treatment of the infectious diseases. For the use of "omics" technology to be successful requires considerable information of pathogen genome as well as genome information of the host. The pathogen genomic and proteomic information helps to identify antigens that can give us information necessary for making a diagnostic tool and vaccine design. The pathogen genome on one hand gives us the information about the important genes conferring disease pathogenesis as well as drug resistance, while the genome of the host on the other hand will reveal the susceptibility genes, and the further knowledge of polymorphisms in genes of the host metabolic and immune system will lead to the new vaccine strategies, drugs targets, and also their treatment outcomes.

Rapid advances of biotechnological and informatics tools in the past few decades mainly in fields of genetics, genomics, and proteomics are leading the way in identifying treating and thus improving the health of human beings. The effective treatment in a patient can only be achieved by first rapid diagnosis of the disease and also identifying its causative agent that is particularly important in the cases of infectious diseases. New insights gained by the analysis of genome and structural feature of pathogen macromolecules have brought about new hope in the treatment of the dreaded diseases. The knowledge of system biology in respect to the microbial infections is still in development, and data available is mostly for few human infections. The rapid development of new generation sequencing technologies have led to generation of new knowledge base and with more advancement of such technologies in the coming years has brought in a hope that all diseases will conquered. In the near future, we will have complete sequences of the total transcriptomes, like genome sequences a decade earlier, and proteomic technologies will attain the throughput and sensitivity of microarrays. Other technologies like metabolomics, glycomics, lipidomics, and phosphoproteomics when referred to in the context of infectious diseases are still in various stages of development, but we are taking the right steps in the direction of development of such technologies (Antony et al. 2012).

2 Host-Pathogen Interactions: Technologies Enhancing Our Understanding

The technologies of transcriptomics and functional genomics are transforming our understanding of microbial infections and helping us decipher the reason of infections susceptibility in the humans.

2.1 Transcriptomics in Infectious Disease

Transcriptomics have been developed and used by scientists to broaden our understanding of infectious diseases. To elucidate the host-pathogen interaction, cDNA microarrays have been widely used. The studies have focused on how the pathogens effect the host cell gene expression. The wild-type virulent strains and isogenic mutants have been used to gauge the responses of the host cell. The major findings of these studies have shown how the pathogen virulence factors modify host cell factor expression (Roy and Mocarski 2007). The role of pathogen recognition receptors (PRR) affecting host-pathogen interaction has been studied. These studies have shown that the host cell responses have an alarm signal (Jenner and Young 2005). Studies have also shown that gene cluster signals are responsible for generating the alarm signals that are the target of attack by the invading pathogens (Hamon and Cossart 2008). Array technologies are being used with molecular probes of host human (also animals/plants) and microbial genes to monitor and at the same time point the expression of genes from host cells and those of the pathogens to better understand the full complexity of host-pathogen interaction.

2.2 Functional Genomics in Infectious Diseases

Functional genomics have also led to the development of tools to decipher infectious diseases by manipulating the cellular mechanisms. The technology of the RNA interference (RNAi) has undergone tremendous development in the last decade which has led to the large-scale reverse genetic screens in human cells and model organisms (Boutros and Ahringer 2008). RNAi technology uses double-stranded ribonucleic acid (dsRNA) with having complementary sequence to the target mRNA sequence and is used to silence or downregulate the gene expression of its target. Long dsRNA induces interferons or other unspecific responses in mammalian cells which is avoided by the use of small interfering dsRNA (siRNA) directly or small hairpin RNA (shRNA). RNAi screening using RNA probes, which induces loss of function of host genes, leads to discovery of host resistance factors (HRF). It is made possible when silencing this restrictive factors leads to invading pathogen replication enhancement and may also identify host susceptibility factors (HSF) and also identify permissive factors, that when silenced will decrease the pathogen replication. The RNAi screens still have some limitations due to off-target siRNA effect (Echeverri et al. 2006). Thus, the primary screening validation is made by using additional siRNA screens. From the several RNAi screenings in human and in fruit fly cells, only 300 host factors were validated from about 10,000 and 20,000 targets identified in the initial screen (Agaisse et al. 2005; Brass et al. 2008; Konig et al. 2008; Krishnan et al. 2008; Zhou et al. 2008).

To make the system biological tools like RNAi more effective in finding mechanisms in host-pathogen interaction and thus finding cure for the microbial infections, there is a need for integrations of all data obtained from the omics technologies. In addition several rounds of biological experimentations are required by using mutant pathogens, cellular RNAi knockdowns, or humanized animal models using mice or primate infection model. The resulting inferences from the validated data would help us build predictive models which could lead us to the better understanding of pathogen interactions with the host.

2.3 Susceptibility to Infection Is Determined by Host Genes

It is evident from human history of infectious diseases that not everybody in a given population is affected by an infective disease. For an infective organism to cause an infection, both the virulence of the pathogen and the host susceptibility are important. The identification of genetic factors of host innate and adaptive immunity that determine the protection from pathogen is an important endeavor of scientists. Animal models of infectious diseases especially mouse models have been used to find the genetic factors and biochemical mechanism of disease susceptibility (Marquet and Schurr 2001). The identification of candidate genes responsible for disease susceptibility or resistance and the occurrence of genetic polymorphism in them give us the best possible biological scenario of the disease. Researchers have found that in bacterial diseases, tuberculosis and leprosy seem to have similar genetic susceptibility determinants in the host as exemplified from the finding that higher incidence of these diseases was found in the monozygotic twins than in dizygotic twins and siblings (Abel et al. 1995; Vidal et al. 1995). Mouse model of infections has revealed that gene encoding the natural resistance-associated macrophage protein 1 (*Nramp1*) confers natural resistance to infections caused by *Mycobacterium*, *Salmonella*, and *Leishmania*. *Nramp1* is found in the membrane of the phagosome of the macrophages where it seems to be probably affecting the replication of the infecting intracellular bacterium (Gruenheid et al. 1997). The genomic analysis in humans has found a similar gene to that of mice which is also having similar pattern of expression. Hence, it has been inferred that humans too carry similar susceptibility gene. There have been further studies which have shown that polymorphisms found in the *Nramp1* gene are related to the infectivity of leprosy and tuberculosis (Abel et al. 1998). In one of the studies, it was found that persons who carry an *Nramp1* heterozygous variant will be four times more likely to be infected by tuberculosis than persons who are carrying more common variants of the *Nramp1* (Bellamy et al. 1998).

Cell-mediated immunity plays an important role in the context of tuberculosis and is well studied. Research has further gone ahead to find links between tuberculosis susceptibility and polymorphism in the gene coding for receptors of interferon-γ and interleukin-12, which are cytokines belonging to T-helper cell type 1 (Th 1). The absence of functional copies of either of these genes in families and of isolated patients leads to high susceptibility to *M. tuberculosis* infection (Jouanguy et al. 1996, 1997; Newport et al. 1996; Altare et al. 1998).

The highly polymorphic human leukocyte antigen (HLA) system is the name of the major histocompatibility complex (MHC) in humans. The HLA class I glycoproteins are highly expressed on the surface of every nucleated human cell, and they present endogenous peptides derived from the cell to the cytotoxic T cells. HLA class I glycoprotein plays a major role during the viral infection as it presents intracellular viral peptides on the surface which leads to cell-mediated immune response, which further leads to the destruction of virus-infected cells. HLA class II glycoproteins which are present on the surface of antigen-presenting cells (APCs) on the other hand present 9–14 amino acids long peptides that are derived from the engulfed pathogen and displayed on the surface which then are recognized by T cell as foreign antigens, and it will elicit an immune response to the antigen. The length of antigen as well as composition are important in deciding if the antigenic peptide will bind to the antigenic peptide-binding cleft. Polymorphisms occur almost solely in the peptide-binding cleft of HLA class I and II in the glycoproteins. The diversity of HLA-binding region ensures that some pathogenic peptides will be preferentially presented compared to the others. Thus, in a given population, the HLA diversity ensures the advantage that some of the HLA glycoprotein peptide-binding clefts will be able to bind and present the pathogenic antigen peptide which will lead to an immune response to any invading pathogen. This ensures the survivability of the species against an infection.

Thus, genomic studies have focused on the identification of susceptibility genes which would lead to the better management of infectious diseases in the population. The tuberculosis susceptibility has been associated with HLA class II genetics. The association is evident from the studies that have been found between pulmonary tuberculosis and class II HLA antigens in several populations (Marquet and Schurr 2001).

As is now clear, the knowledge of the mechanism of action of the pathogen and the identification of the susceptibility genes goes a long way in the management of the disease in context of a public health perspective to prevent, diagnose, identify, and target the vulnerable populations against a given infectious disease.

3 Pharmacogenomics in Infectious Disease Management

The requirement of the genomic information of both the host and pathogen is important to fully carry out infectious disease management. The first sequence map of human genome being completed in June 2000 (Lander et al. 2001; Venter et al. 2001), followed by discovery of genome-wide single-nucleotide polymorphisms (Sachidanandam et al. 2001) and further genomic sequencing of several pathogens by institutes like the J. Craig Venter Institute (www.jcvi.org) gradually opened the pathway to create new treatment and disease management methods. The ultimate aim of genomic technologies to bring personalized medicine for every infectious disease scenario is still decades away, but here we will focus only on important breakthrough which has shown us the way forward in respect to infectious disease identification and treatment.

3.1 The Use of Pathogen Genome for Antigen Identification

Neisseria meningitidis is a bacterium that can cause meningitis and other forms of meningococcal disease such as meningococcemia, a life-threaten-ing sepsis (www.cdc.gov/meningococcal/).

N. meningitidis is a major cause of morbidity and mortality during childhood in industrialized countries and has been responsible for epidemics in Africa and in Asia. *Neisseria meningitidis* serogroup B is responsible for causing about one third of infection. The genome sequence of *N. meningitidis* has opened a new way for disease management (Pizza et al. 2000; Tettelin et al. 2000). The vaccine that was used only contained capsular polysaccharides from the serogroups A, C, Y, and W135 only. Serogroup B polysaccharide contained elements that resemble human polysialic acid and hence is poorly immunogenic and might generate autoantibodies (Hayrinen et al. 1995). In this scenario the *N. meningitidis* serogroup B was sequenced, and 350 potential antigens from the serogroup B were expressed in *Escherichia coli* to find the potential vaccine candidate. The expressed proteins are injected into mice to find immunogenic antigens that can developed into a vaccine. Similar strategies are being used to find vaccine candidates for other serotypes of Neisseria species and for other pathogenic organisms.

Pathogen genomic information is also being used to find the immunologically important peptides for cytotoxic T lymphocytes (CTLs) epitopes. The response of CTLs is to seek out virally infected cells by recognizing the peptides presented by human leukocyte antigen (HLA) glycoproteins on the cell surface and killing the infected cells. CTL epitopes are the viral peptide that is presented by the HLA and recognized by CTLs. The peptides are of the length of 10 amino acids, and genome sequence is used to find out and synthesize these peptides for immunogenic evaluation. Amino acids are divided into segments of peptide, measuring 10 amino acids in length and overlapping the previous peptides 9 amino acids, for example, West Nile virus genome translates into 3,433 amino acids which can be segmented into 3,424 peptides that are 10 amino acids in length.

Immunoinformatics, a field of bioinformatics, is speeding up the finding of CTL epitopes for the scientist working in the field. Algorithms on computer softwares are being used to match the viral peptides with the HLA glycoproteins in silico for binding based on previously known results and are being tested (De Groot et al. 2001). Informatics based algorithms help eliminate 99 % of the peptides that would not be used in the experimental screens. Thus, the time and effort to screen for the CTL epitopes have been reduced drastically. CTL epitopes may be used for making subunit-based vaccines and diagnostic tests. Virus-specific antibody found using CTL epitopes can be used in enzyme-linked immunosorbent assay. It may be even possible to use CTL epitopes to test for the antigen itself.

3.2 Genomic Information for Identifying Infectious Pathogens

Basically as of now only four types of molecular diagnostic tests are carried out to detect infection in laboratory setup. First is by direct detection where the pathogens can be detected directly by imaging technologies of microscopy and cell culture. Second method of diagnosis is by the detection of proteins produced by pathogens by the use of specific antibodies, like that used in enzyme-linked immunosorbent assay (ELISA). Third method is by detection of the specific antibodies IgA, IgM, and IgG against the pathogens and the changes in their titers using antibody capture assay. Fourth method uses detection of nucleic acid of the pathogens and amplifying their signal using techniques like polymerase chain reaction. Latest diagnostic technologies have been developed on these basic four biotechnological technologies (Speers 2006).

Pathogen genome can also be used to identify the infecting organism itself. Microbial DNA in the clinical specimen can be used to identify the disease-causing pathogens. Human immunodeficiency virus (HIV), hepatitis virus, *Borrelia burgdorferi* (causative agent of Lyme disease), and mycobacteria are few examples of pathogens that can be identified by their genomic sequences. *Mycobacterium tuberculosis* antimicrobial resistance strain-caused infections are becoming quite common, and genomic information has deciphered few potential candidates like *katG* gene mutations in resistant strains (Siqueira et al. 2009a, b; Marahatta et al. 2011). The traditional culture test for mycobacterium which is time-consuming and less sensitive is giving way to restriction fragment length polymorphism (RFLP), a specific technique used in DNA fingerprinting (Van Soolingen 2001). The technology uses restriction enzymes that cut DNA at the places having certain particular nucleotide sequences. The nucleotide pattern that is obtained is then compared to the previously identified specific nucleotide pattern of the genome of the pathogen DNA. DNA patterns can be separated on the basis of length, and the pattern of DNA fragments in the DNA fingerprint is characteristic of particular isolate, and each particular pathogen has a unique pattern. DNA fingerprint technology is faster and reliable than culturing of the mycobacteria, ideal for discovering new drug-resistant strains from unique genomic sequences of each mycobacterium. The technology is very useful for identification of strains during the time of outbreaks and further epidemiological studies.

The knowledge of viral load in patients is also important for dosage determination in drug therapy; hence, detection of the viral pathogenic DNA and RNA in clinical specimens is of paramount importance. Treatment of viral diseases like HIV, chronic hepatitis B, and hepatitis C often depends on the knowledge of viral load (Revets et al. 1996). For example, HIV viral loads are detected by enzymatic amplification of the viral nucleic acid and detection of the signal from the labeled probes that hybridizes with them. The signal usually is either a color signal conjugated to the probe or a chemiluminescent probe, and the intensity of the signal corresponds to the number of copies of the nucleic acid RNA. Capillary electrophoresis detects hybridized probes at a very high sensitivity with detecting as low as 2,000 copies of HIV RNA in milliliters of plasma (Kolesar et al. 1997).

3.3 Pathogen Genomic Information Determines Antimicrobial Resistance

The rampant uncontrolled use of antimicrobials has led to increased number of antibiotic bacterial strains. Genomic mutations allow the certain bacterial strains to overcome the effects of antimicrobials and are able to propagate in spite of the presence of antibiotics. The pathogenic bacteria have started showing resistance and have become a major problem for human health. *Staphylococcus aureus*, *Streptococcus pneumoniae*, *Pseudomonas aeruginosa,* and *E. coli* are a few examples of such bacteria.

Mechanisms of gaining resistance have been elucidated by means of omics technologies. Here are few examples. Fluoroquinolones are drugs that act on the bacterial DNA replication by binding to bacterial enzymes involved in bacterial DNA replication, that is, DNA gyrase and topoisomerase. The bacterial resistance to quinolone occurs due to mutation in the quinolone-binding site in the enzymes mentioned above. The mutation leads to change in the amino acid at the site of binding of fluoroquinolones to the enzymes. If both the bacterial enzymes are mutated, then high-level resistance occurs to the quinolone drug affecting the treatment of infection as compared to when either of the enzymes is mutated (Hooper 2001).

Now genetic test is available to detect antimicrobial resistance in the infecting pathogens. The information is important because it would lead to a better treatment management of the infection. The methicillin-resistant *Staphylococcus aureus* phenotype is detected when cultured in the presence of oxacillin after a period of 24 h. Before the era of omics technology, the only means of resistance detection was by culture test which is a very time-consuming test. Methicillin resistance in *S. aureus* is controlled by alternations of penicillin-binding protein PBP2a. Gene *mecA* controls the production of PBP2a. Polymerase chain reaction test is used to detect the presence of *mecA* in reference laboratories, while commercially developed kit can detect the same using a fluorescein-labeled *mecA* probe. Both DNA probe and PCR technology when used for analysis can detect *mecA*-resistant gene in a given sample in less than 3 h. The rapid detection of antimicrobial resistance in pathogens helps patients in providing adequate treatment opportunities (Louie et al. 2000).

3.4 Genomic Factors Determine Response to Therapy in Infectious Diseases

The study of the host genome becomes important to fully understand the drug effects and as such design more effective methods of treatments. Although the ultimate goal is to decipher the system biological effect, the trend of single gene effects is also very important.

Cytokines play a very important role in human immunity (Paul and Seder 1994). In hepatitis C infection, interferon alpha is used to stimulate cell-mediated immunity against the viral infection and is the primary treatment. However, studies have shown that response to interferon-alpha treatment is only 50 % in some cases even when combined with other antiviral treatment (Manns et al. 2001). Further studies have shown that if chronic hepatitis C patients have IL-10 polymorphism variant, it leads to the reduction in expression of IL-10 itself, and they will have five times more chance of effective treatment with interferon alpha than those who do not carry the polymorphism (Edwards-Smith et al. 1999).

Interleukin-10 (IL-10) is a polymorphic cytokine and is a T-helper cell type II (Th2) cytokine that is associated with the induction of the production of large amount of antibodies in body's immune response. Th1 cytokines which promote cell-mediated immunity inhibit Th2 response and vice versa. Thus, people with high-expressing IL-10 genotype if infected and suffering with chronic hepatitis C infection are less likely to respond to interferon-alpha treatment. New treatment regimes have to be developed for patients suffering from chronic hepatitis C infection and carrying IL-10 polymorphism associated with high cytokine expression.

Vaccine responses can be used as a system of gauging the state of immune system (Poland 1999a, b). Vaccines are administered to large number of population as an integral part of public health system. Vaccines are used to mimic the infective disease conditions that induce immunological memory to protect the individual against subsequent exposure to the pathogen and lead to prevention of disease. The phenomenon to gain protective immunity against a pathogen upon being vaccinated for the particular pathogen depends on individual genetic build. As studies have shown, not all healthy individuals are able to generate a protective immune response upon vaccination. It has been observed in the case of measles vaccination that only 10 % of the population was seronegative and clustered in family (Poland 1999a, b; Poland et al. 1999c). Both HLA I and HLA II class alleles have been responsible for the measles vaccine response, while HLA-B7, HLA-B51, HLA-DRB1*13, and HLA-DQA1*01 are associated with positive measles vaccine response, and HLA-B, HLA-DR, and HLA-DQA1 have been responsible for the vaccine being noneffective (Hayney et al. 1996, 1998; Poland et al. 1998).

4 Infection Treatment: Response to Drug Treatment Determined by Host Genomics

Drugs used for targeting any pathogenic infection can only be successful if we are aware how it is affecting the host and pathogen at genomic level and hence are able to explain the host efficacy and toxicity. We look at few important infectious diseases where pharmacogenomic research has been bringing a landscape change in the disease treatment.

4.1 Leishmania: Pharmacogenomics in Disease Management

4.1.1 The Disease

Leishmaniasis is a very complex major tropical infection transmitted by the vector Sand fly is all right. The infection is caused by intracellular protozoan parasites of Leishmania genus. There are more than 20 species of *Leishmania*. The type of infective species, virulence factors, and host immune responses and depending on the clinical symptoms, the disease is categorized into cutaneous leishmaniasis (CL) and visceral leishmaniasis (VL). VL is also known as kala-azar; the origin of the name is from the eastern and

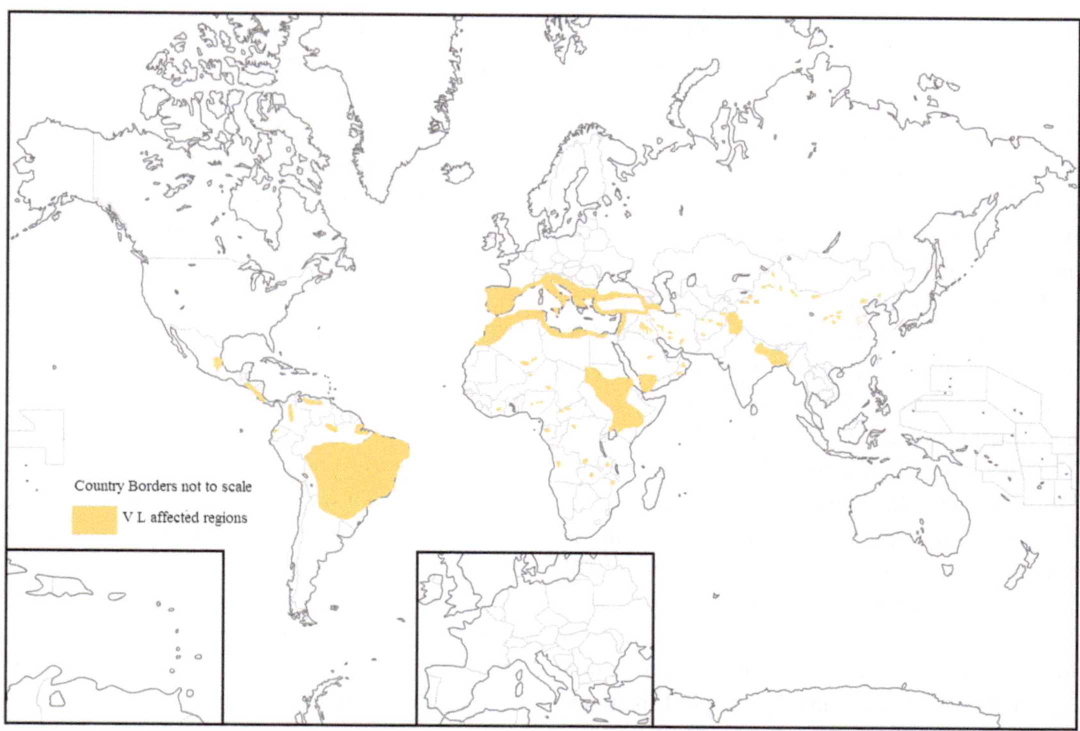

Fig. 27.1 Geographical distribution of visceral leishmaniasis in the Old and New World (*Source: World Health Organization:* URL: http://www.who.int/leishmaniasis/leishmaniasis_maps/en/, Data source and Map production with permission of WHO headqauters: WHO/NTD/IDM HIV/AIDS, Tuberculosis and Malaria (HTM) World Health Organization, October 2010, Accessed February 2013)

northeastern part of the Indian subcontinent where the disease is endemic. Depending upon the place where one has acquired the infection, CL is further categorized into "New World" from Central America and South America and "Old World" if from Asia, Middle East, Africa, or southern Europe. More than 1–1.5 million cases of leishmaniasis occur worldwide (about 80 countries are affected) with major countries being the developing nations of Asia, Africa, and Latin America (www.who.int/topics/leishmaniasis/en/).

Species of Leishmania are several causing different clinical manifestations of the infectious disease. *L. donovani* produces primary cutaneous disease as well as gives rise to visceral leishmaniasis (VL) and also post-kala-azar dermal leishmaniasis (PKDL) that is manifest after the treatment of the initial visceral disease. Visceral leishmaniasis main causative pathogen is the *L. donovani* complex with Old World VL disease being caused by the species *L. donovani and L. infantum,* and New World disease is mainly caused by different species of *L. chagasi* (Fig. 27.1).

Old World CL causative Leishmania species are *L. tropica* which are mainly found in urban areas and *L. major* being prevalent in the desert areas, while the New World CL disease is caused by *L. mexicana complex* (includes *L. mexicana, L. amazonensis,* and *L. venezuelensis*) and *Leishmania (Viannia subgenus) braziliensis complex (L.(V.) braziliensis, L.(V.)colombiensis, L.(V.) panamensis, L.(V.) guyanensis, and L.(V.) peruviana*). In the cases of infection caused by *L. braziliensis complex,* there is always a chance that the infection dissemination to mucosal region can occur to give rise to mucocutaneous leishmaniasis (MCL) (Herwaldt 1999).

The complex disease is manifested due to multiple factors ranging from environmental factors such as time and number of exposure with infected vector sand flies, species of the infecting Leishmania pathogen, to host genetic factors that include immune status of both innate and adoptive immune systems that determine the clinical outcome of the disease. Other reasons for *Leishmania* disease susceptibility are malnutrition, immunodeficiency with HIV coinfection, and young age. The protection against invading pathogenic Leishmania protozoa and even the curative resolution of the disease is provided by Th1 cytokine response involving cytokine interferon gamma (IFN-γ), interleukin (IL)-12, and tumor necrosis factor alpha (TNF-α), whereas Th2 response cytokines IL-10, transforming (TGF)-β, and IL-4 have been implicated in increasing susceptibility to the disease in the experimental animals (Reed and Scott 1993; Sacks and Noben-Trauth 2002). Nonhealing lesions and diffused lesions in CL have been implicated to Th2 response, while self-healing lesion has been associated with Th1 response (Melby et al. 1994). However, in some situation IL-4 (a Th2 cytokine) has been implicated to induce IL-12 production and lead to Th1 cytokine response, and it has also been found in some cases that Th2 response occurs independent of IL-4 (Alexander and Bryson 2005; Mansueto et al. 2007). Leishmania infection is a complex infection depending on host factors as well as strain polymorphism. *Leishmania mexicana* cysteine proteases which target IL-12 that prevents Th1 protective response (Buxbaum et al. 2003), while the Leishmania analogue of activated C kinase (LACK) from the *Leishmania major* induces Th2 response that leads to host parasitization (Kelly et al. 2003). Polymorphism of *L. braziliensis* also affects disease outcomes (Cupolillo et al. 2003; Schriefer et al. 2004). PKDL is a complication arising after treatment of VL, affecting 50 % of VL patients in Sudan (study carried out in United Sudan) and also 5–10 % patients in India. PKDL has been found to be associated with increased levels of IL-10 (Zijlstra et al. 2003; Ganguly et al. 2008).

4.1.2 Disease Treatment

The major treatment regime of CL which has propensity of dissemination towards VL and MCL is with parenteral antimonials like sodium stibogluconate or meglumine antimoniate, pentamidine, and oral miltefosine (Olliaro et al. 2005; Ameen 2007; Amato et al. 2008), whereas CL with low risk of spread is treated with local and physical therapies such as intralesional antimonials, topical paromomycin, cryotherapy, and thermotherapy or by oral azoles. However, when the disease progresses to MCL, treatment is prolonged, and toxicity from such long-duration drug use is a common occurrence (Marsden 1986; Amato et al. 2008).

Vaccine is still elusive in the case of Leishmania. Some trials with DNA vaccines do have shown a way forward. These vaccines have shown the promise to be effective as they have been able to induce 1L-12 production, which was in response from the persistent antigen exposure from the DNA vaccine (Requena et al. 2004). In Venezuela killed *Leishmania* promastigotes along with bacillus Calmette-Guerin (BCG) used as immunotherapy have shown results with a high cure rate in clinical trials by inducing Th1 response (Convit et al. 2003). *L. major* vaccine trial with BCG and parenteral antimony combined have been successfully used for treatment of PKDL (Mansueto et al. 2007). The search for effective vaccine for Leishmania had got a boost with knowledge from genome sequence data of several Leishmania strains. More vaccine candidate genes will be evaluated in the future (Stober et al. 2006).

4.1.3 Genetic Susceptibility and Pharmacogenetic Implications

In the absence of an effective vaccine with recurring infection such as PKDL, dissemination infection to mucosa leads to aggravating of the disease. Prolonged treatment with parenteral antimonials that give rise to high-level risk of toxicity with high morbidity and mortality from the disease is a problem of concern (Convit et al. 2003; Muse et al. 2008). The new technologies are trying to address these very problems. Epidemiological studies in different ethnic populations in several countries have shown the variation susceptibility to the disease be it CL, MCL, or VL which caused different strains of Leishmania. Variation in disease presentation and progression, familial clustering, in a population indicates the underlying genetic reason of susceptibility. Recent studies have shown that *L. donovani* cause of CL has been found to affect exclusively a particular ethnic group in Sri Lanka (Samaranayake et al. 2008). *L. donovani* though normally associated with causing VL is shown in few places, Kenya, Yemen, Cyprus, and the Himalayan region of North India, and is the main causative pathogen of CL (Mebrahtu et al. 1993; Pratlong et al. 1995; Sharma et al. 2005; Antoniou et al. 2008).

To deduce the genetic susceptibility of the Leishmania disease, experimental murine animal models along with clonal parasite line (to control environmental variable) have been used to find the genes responsible for disease progression along with their human homologues of disease susceptibility (Handman et al. 2005). First genes that were used to deduce from such analysis in murine model were *NRAMP 1* and the H-2 locus had been implicated in *L. donovani* infection (Blackwell et al. 1980). HLA class II antigen HLA-PQ3 is found to be associated with CL in Venezuela (Lara et al. 1991) and MCL in Brazil caused by *L. braziliensis* (Petzl-Erler et al. 1991). PCR genotyping studies in Mexico on Leishmania patients has found an association with HLA class II genes with Cutaneous Leishmaniasis (CL) (Olivo-Diaz et al. 2004). High blood TNF has been found to be associated with MCL (Castes et al. 1993) and acute VL (Barral-Netto et al. 1991a, b). A Venezuelan study has implicated that allele 2 of TNF-β polymorphism with high risk of developing MCL caused *L. braziliensis* and higher frequency of allele 2 of TNF-α polymorphism was also associated with MCL (Cabrera et al. 1995). In Brazil by using family-based disequilibrium test analysis (TDT), investigation has shown that TNF polymorphism has been linked to *L. chagasi* infection (Karplus et al. 2002). In asymptomatic patients having positive skin test, *L. chagasi* has been associated with TNF-1 allele of TNF-α gene, while in case of symptomatic *L. chagasi* VL patients, TNF-2 allele is implicated. Due to parasite heterogeneity, this TNF polymorphism association has not been correlated to infection by other Leishmania species such as in *L. infantum* VL (Meddeb-Garnaoui et al. 2001) and *L. major* CL (Kamali-Sarvestani et al. 2006). Variation in promoter of *IL-4* and IFN-γ gene polymorphism has been found to be linked to *L. major* CL disease susceptibility and progression respectively in an Iranian study, while in a Sudanese VL patient study, *IL-4* polymorphism has been shown to increase disease susceptibility (Mohamed et al. 2003). Polymorphism in promoter region of *IL-10* gene leads to higher IL-10 production which has been shown to increase the risk of having skin lesions during an infection of *L. braziliensis* (Salhi et al. 2008). IL-6 can diminish the high Th1 proinflammatory response that occurs when *L. braziliensis* CL progresses to MCL (Hatzigeorgiou et al. 1993; Bacellar et al. 2002). IL-6 polymorphism plays an important role in the progression of *L. braziliensis* CL to MCL, and this finding is important since their genetic markers have high prognostic value (Castellucci et al. 2006).

Genome-wide linkage have been performed for *L. donovani*-infected VL patients in Artinga ethnic group in Sudan to help identify loci on chromosome 22q12 and is associated with disease susceptibility genes (Bucheton et al. 2003a, b). IL-2 receptor β chain (IL2RB) gene is present in the highly susceptible loci on chromosome 22q12 that was identified from this study. IL-2 receptor has been detected in high levels during VL infection and plays a critical role in T cell genetic responses (Barral-Netto 1991b). Further studies have shown *IL2RB* polymorphism in association with *L. donovani* VL (Bucheton et al. 2007).

Another candidate gene found is *SLC11A1* (formerly NRAMPI) on chromosome 2q35, an innate resistance gene that regulates macrophage activation and contributes to increased VL risk in Sudanese population (Bucheton et al. 2003a, b; Mohamed et al. 2004) as well as increased susceptibility to several intracellular pathogens (Blackwell et al. 2001). Other studies have shown that genotypes having significantly high level of mannan binding lecithin occur more prominently in patients with clinical VL. An opsonin, mannan-binding protein, is known to enhance pathogen infection. Polymorphism in mannan-binding gene has been shown to increase risk of developing *L. chagasi* VL in Brazilian study population (Alonso et al. 2007). In PKDL there is elevated level of IFN-γ. Polymorphism of

IFN-γ receptor 1 from study in Sudan has been implicated in PKDL (Salih et al. 2007). The IFN receptor expression is important for the activation of macrophages via IFN-γ.

Drug treatments are not very efficient in the treatment of leishmaniasis disease; more effective treatment regimes can be developed by thoroughly knowing the genetic factors that lead to disease progression. Thus, unnecessary drug use and adverse reaction can be avoided. As various genetic susceptibility studies have shown, cytokine response determines the disease progression in leishmaniasis. Role of IL-10 in pathogenesis of leishmaniasis is known and is well established, and IL-10 polymorphisms have shown to increase risk of lesions in *L. braziliensis* infection. In a study with *L. guyanensis* infected CL patients from French Guiana, high level of mRNA IL-10 within lesions leads to poor chemotherapy response and treatment failure (Bourreau et al. 2001). It is hypothesized that IL-10 might be regulating the response to chemotherapy by blocking the Th1 response. The increased level of IL-10 has been linked to the active VL (Nylen and Sacks 2007) and PKDL (Saha et al. 2007) and also associated with persistent CL infection occurring from *L. major* (Melby et al. 1996) and *L. mexicana* (Louzir et al. 1998). Success of VL treatment with amphotericin B and the complete elimination of IL-10 are associated with one another (Saha et al. 2007). On the other hand, MCL is associated with low IL-10 receptor expression and low IL-10 secretion that decrease the ability for modulation of proinflammatory response (Faria et al. 2005).

Progress has been made to find the susceptibility genes and will provide further insight into disease pathogenesis and will lead to progress in the field of diagnostic markers, drug targets, and vaccine development to control, treat, and eradicate leishmaniasis.

4.2 Improving the Treatment of Malaria by Pharmacogenomics

4.2.1 Disease Burden

Malaria is vector-borne (mosquito) disease that has been one of the top causes of mortality in the world for generations especially in tropical countries of Asia and Africa. Even after renewed global efforts, still there is high infectivity and mortality. Three billion people are at risk with 1–2 million deaths attributed to malaria each year (www.who.int/topics/malaria/en/). Four species of protozoan parasite are involved from genus *Plasmodium,* i.e., *P. falciparum*, *P. vivax*, *P. malariae,* and *P. ovale*. These malaria-causing combination parasitic species occur in human population and occur in infected individuals (Gurarie et al. 2006). In respect to prevalence, virulence, and multidrug resistance, *P. falciparum* has been a major cause of mortality and morbidity. *P. falciparum* accounts for about 80 % cases of malaria in Africa (Roca-Feltrer et al. 2008). Next to it is *P. vivax* which causes 100–300 million cases annually (Price et al. 2007). The most commonly used drugs are chloroquine (CQ) and sulfadoxine-pyrimethamine (S-P Fansidar®) that are becoming less effective due to the development of resistance in malaria parasite by *P. falciparum,* and the species has become predominant and become a threat to travelers and people alike (Schlagenhauf and Petersen 2008). In the absence of vaccine and in addition, development of resistance even in the mosquito vector control against chemical methods using insecticides has thrown new challenges for the researchers (Greenwood et al. 2008) (Fig. 27.4a).

4.2.2 Malarial Therapy

Some of the recent developments in malarial treatment using pharmacogenomics are bringing about improvements in the efficacy of treatment regime of malaria. Current treatment regimes have recommended artemisinin combination treatments (ACTs) in cases of uncomplicated falciparum malaria in nonpregnant adults (Lin et al. 2010). The drug regime is highly efficacious and has reduced development to resistance. In cases of uncomplicated malaria, the ACT is being used in 88 countries by 2009. In the coming years, a number of patients including women and children will be brought under ACT therapy regime as per World Health Organization.

Like the treatment of HIV and tuberculosis, combination therapy is now being used for malaria treatment too, which reduces resistance among the highly efficacious drug the artemisinins which rapidly eliminate the parasite from blood and thus limit the number of parasites so that the other more bioavailable drugs given in combination act on the parasite. Unrelated mode of action of two or more combination drugs also reduces the chances of resistance (Yeung et al. 2004).

4.2.3 Pharmacogenomics Way Forward to Effective Treatment

Among many other factors which contribute to drug effectiveness, malarial drug bioavailability and tolerability are depended upon the host metabolic mechanisms. The severe drug reaction to primaquine in the 1950s used in antimalarial treatment was instrumental in the discovery of glucose-6-phosphate dehydrogenase (G6PD) deficiency in 1956; thus, importance of the use of pharmacogenetics in malarial treatment was realized (Alving et al. 1956). The polymorphism leading to variation G6PD or even its deficiency is a grave problem in designing the effective drugs. Even now during malarial terminal prophylaxis to decrease transmission, primaquines are administered. Thus, the G6PD status of patient becomes quite important (Luzzatto 2010).

Knowledge of both the host and parasitic genetics is necessary to designing drugs and dosage for effective treatment regimes. Parasitic genetics helps us in deciphering the modes of resistance, and host genetics help us in giving the information about host drug bioavailability and explain adverse reaction to the drugs. G6PD polymorphisms and genetic variation in CYP2C8 can play pivotal role in point of care diagnostics, but these genetic testings will have to be incorporated into the laboratories and national health programs. The knowledge of this important genetic variations in population would ultimately reduce cost and make the treatment regime more effective and with lesser adverse reaction and ultimately reduce the suffering of the patients.

The pharmacogenetic drug policy in context of malaria is slowly becoming a reality as per efforts of the WHO and other agencies. Genetics is becoming a guide to new drug policy. Amodiaquine was generally known to be tolerated in malarial treatment, but later when it was found in the Caucasian population during the decades of 1980 and 1990 to be responsible to cause agranulocytosis with fatalities and also cause hepatotoxicity (Hatton et al. 1986; Raymond et al. 1989; Phillips-Howard and West 1990), the drug was first removed from the list of essential drugs against malaria but then had to be added back to the list as alternate drugs started showing resistance. Amodiaquine induced adverse reaction in individuals was attributed to genetic make up of the individual. The genotypes of individuals harboring CYP2C8, CYPIA1, and CYP1B1 have been reported in studies to show immunogenic adverse reaction to amodiaquine (Li et al. 2002; Kerb et al. 2009). Some population in Africa has shown hepatotoxicity and leucopenia with only two doses given 3 weeks apart (Orrell et al. 2008).

Amodiaquine when administered to an individual with reduced CYP2C8 activity impairs the metabolism of the drug and hence leads to the cause of hepatotoxicity and agranulocytosis. Other common variants of the enzymes CYP2C8*2 and CYP2C8*3 have been associated with decrease in the metabolizing activity of CYP2C8 enzyme as is evident from studies with anticancer drugs (Dai et al. 2001). Individuals with CYP2C8*3 genotype have no CYP2C8 enzymatic activity in vitro (Parikh et al. 2007). In a study from Burkina Faso, patients carrying CYP2C8*2 genotype showed common adverse effects to Amodiaquine and in addition patients have also reported to experience more abdominal pain when compared to healthy individuals. The study from Burkina Faso and Ghana could not clearly establish the relation between drug efficacy and CYP2C8 genotype (Adjei et al. 2008). Though the inactivated gene of CYP2C8 is not very high in population, estimates have shown that CYP2C8*2 and CYP2C8*3 occur in about 2.1 % of the population in Zanzibar, United Republic of Tanzania, which was about 30,000 patients of the total malarial patients ~100,000 (Cavaco et al. 2005). In Ghana it was found that 1.5 % of the population has been estimated to have metabolic variants of CYP2C8. Hence, due to high disease burden, the study of pharmacogenomics for drug metabolism was carried out in large patient samples from the population to get a clear correlation between genotype and efficacy of drug treatment as well as adverse reaction.

Fig. 27.2 World map showing frequencies of the CYP2C8*2 allele in different populations. x=allele frequency in reference population (US Caucasians), y=allele frequency in country with data analysis refer (*Source:* Roederer et al. (2011), Map by: Pharmacogenetics for every nation initiative: Accessed Feb 2013)

Fig. 27.3 World map showing frequencies of the CYP2C8*3 allele in different populations. x=allele frequency in reference population (U S Caucasians), y=allele frequency in country with data analysis refer (*Source:* Roederer et al. (2011), Map by: Pharmacogenetics for every nation initiative: Accessed Feb 2013)

These pharmacovigilance studies further reduced the effective treatment cost incurred on public health both monetary and from the point of view suffering of the patients (Figs. 27.2 and 27.3).

4.2.4 Pharmacogenomics of ACTs

Major active antimalaria metabolite of artemisinin is dihydroartemisinin (DHA) (Ilett et al. 2002). Artesunate is rapidly converted to its active metabolite catalyzed via CYP2A6 which is a major enzyme; conversion to DHA also includes minor enzymes CYP2B6, CYP1A1, and CYP1A2 (Li et al. 2003). CYP2A6 has about 40 variant forms of which at least 13 have been implicated as slow metabolizing enzymes, and 5 have been reported to show no activity in vitro (Di et al. 2009). Hence, lower level of CYPBA6 enzymes in patients will have reduced bioavailability of DHA the major antimalarial metabolite and hence have lower antimalarial activity.

Major endemic areas of malaria like sub-Saharan Africa, Ghana, Sabah region of Malaysia have been evaluated for the presence of CYP2A6 genotype. Among these population of Ghana has high wild-type CYP2A6 along with 80 % alleles being CYP2A6*1A (Gyamfi et al. 2005), whereas Malaysian population has an allele CYP2A6*1A frequency of 32 % with only 8 % wild-type enzyme (Yusof and Gan 2009). Other Asian populations have been reported to carry several other alleles of CYP2A6 with even alleles that do not show any CYP2A6 enzyme activity at all. No activity variant of CYP2A6 is found about 11.5–20.1 % in Japanese, Chinese, and Thai populations (Gyamfi et al. 2005). Hence, artesunate is expected to be more effective in population of Ghana. In some parts of Thailand, about 10 % of patients have shown resistance to artemisinins (White 2008). Though it has been found by study that about 14 % frequency of CYPZ86 alleles have no activity in the Thai population and the antibiotic resistance is indicative to be related to CYP2A6 activity and ability to convert artesunate to DHA (Noedl et al. 2009), more studies require to be done to clearly establish the relation between the genotype and resistance to artemisinin-based therapy.

4.2.5 Malarial Parasite Resistance Genes

Several mutations in gene targeted by antimalarial drug have been identified which led to the resistance in vivo of ACT drug partners such as mefloquine, lumefantrine, amodiaquine, and chlorproguanil (Kerb et al. 2009; Mehlotra et al. 2009). Identification of genes and mechanism is important for controlling the infection. Research has yielded the information regarding the gene responsible and underlying mechanism of action resistance of some drugs against *P. falciparum* and *P. vivax*.

Chloroquine resistance (CQR) in *P. falciparum* has been linked to point mutation CQ resistance transporter gene (Pfcrt chromosome 7). The mutation *Pfcrt*-K76T is a reliable marker for CQR. While CQ-sensitive strain carries wild-type allele *CVMNK*, the variant *CQR* alleles are S_{agt}*VMNT* (Asia, South America, Africa), S_{tct}*VMNT* (South America), *CVMNT* (South America, Philippines), *CVIET* (Southeast Asia, Africa), and *CVMET* (Colombia). Another multidrug resistance gene (*pfmdr1* chromosome 5) is a parasite transporter gene. Polymorphism, point mutation, and copy number variation have been implicated in multidrug resistance. In different geographic regions, the *pfmdr1* two mutant alleles have been reported, namely, 86Y_184Y_1034S_104N_1246D found mostly in Asia and Africa and 86N_184F_1034C_1042D_1246Y predominantly from South America (Valderramos and Fidock 2006). The *pfcrt-76* and *pfmdr1-86Y* mutations have been related jointly to contribute in giving rise to CQR phenotype in addition to other likely parasite genes (Hayton and Su 2004).

P. falciparum DHPS enzyme (*pf-dhps*, chromosome 8) has been linked to resistance to the sulfa class of antimalarial drugs, while mutations in DHFR (*pf-dhfr*, chromosome 8) domain have been linked to high level of pyrimethamine resistance. Combination of sulfadoxine-pyrimethamine (S-P) treatment failure has been found to be associated with *pf-dhps* double mutant (437G with either 540E or 581G), combined with the *pf-dhfr* triple mutant (108N_511_59R) (Hayton and Su 2004; Hyde 2007).

Point mutations in *P. vivax* ortholog of *pfcrt (pcvg10)* are associated with clinical CQR. *pfmdr1 P. vivax* ortholog that is *pvmdr1* has been proven and has also been identified. Y97CF point mutation of *pvmdr1* has been linked to CQR. *pv-dhs* and *pv-dhr* gene point mutations have been identified and are suspected to link to clinical resistance in antimalarial S-P treatment (Hayton and Su 2004).

More data is required for new mutations in the parasite genes, and in addition more data is needed for therapy of other ACT drug partners like sulfadoxine-pyrimethamine and lumefantrine. A new rejuvenation is taking place in pharmacology and pharmacokinetics development of databases of antimalarial pharmacogenetics. Worldwide Antimalarial Resistance Network (http://www.wwarn.org/) has set up a module together with high-quality pharmacological research data for optimum drug dosage in light drug resistance information and adverse event reporting. The aim to achieve global cooperation will go a long way to personalized malarial treatment as per population needs.

4.3 Pharmacogenomics in Tuberculosis Treatment

4.3.1 Disease Burden

Infectious diseases are still a major challenge to our society; however, newer technologies have brought in new hope for control to this dreaded disease. Tuberculosis is caused by pathogenic bacterium *Mycobacterium* tuberculosis (MTB) and still infects about one third of the population of the world (www.who.int/topics/tuberculosis/en/). A person with active tuberculosis will infect about 10–15 persons in a year. For decades now we have very efficacious treatment regimes, but still we have not been able to eradicate the disease from the population. Now with the rise of human immunodeficiency (HIV) infection in the last few decades, people infected with AIDS are more at risk due to diminished immunity. The year 2010 saw as per estimates about ~8–8 million cases of tuberculosis, of which ~1.1 million deaths were reported among HIV negative patients, while about ~0.35 million deaths in HIV-related TB were also reported.

During the last half a century (for about 50 years), the most effective treatment regimes have been the combination therapies of drugs that was because a single drug treatment was found to be in invariantly leading to resistance for the drug, leading to much more severity and complications (Crofton 1994). Due to rampant and unregulated use of tuberculosis drugs, however, this has led to emergence of multidrug resistant tuberculosis (MDR) (Fig. 27.4b).

4.3.2 Treatment Regimes

Now the treatment course is usually for 6 months with the combination of isoniazid, rifampicin, pyrazinamide, and ethambutol for the first 2 months. This has to be followed up by the next 4 months with isoniazid and rifampicin treatment. If the treatment is taken up with diligence and patient completes the whole drug course, then it has been reported that efficiency of the treatment is very high with more than >95 % patients getting cured and relapse is in less than 5 % of patients (Menzies et al. 2009). Another advantage of multidrug treatment is that the treatment regime helps in treating different population of tubercle bacilli. For the last 20 years, knowledge from the field of genetic molecular basis of drug treatment outcomes has helped us in the better management of and understanding of treatment efficiency and of drug. The difference in drug response is found among different individuals of the population. The individual person tends to show similar type of response to tuberculosis drugs that do not change over time. Thus, in light of above observations, we say that there is a huge variation in drug response among individuals due to variation in genes involved in drug metabolism, drug transporters, and drug targets compared to minimal within-subject variation as found from studies. Further studies on drug response revealed that 20–95 % of variation in drug pharmacokinetics is due to genetic factors (Kalow et al. 1998). The sequence variation in drug-metabolizing enzymes, drug transporters, or drug targets leads to the variation in drug response among individuals (Evans and Relling 1999; Evans and Johnson 2001).

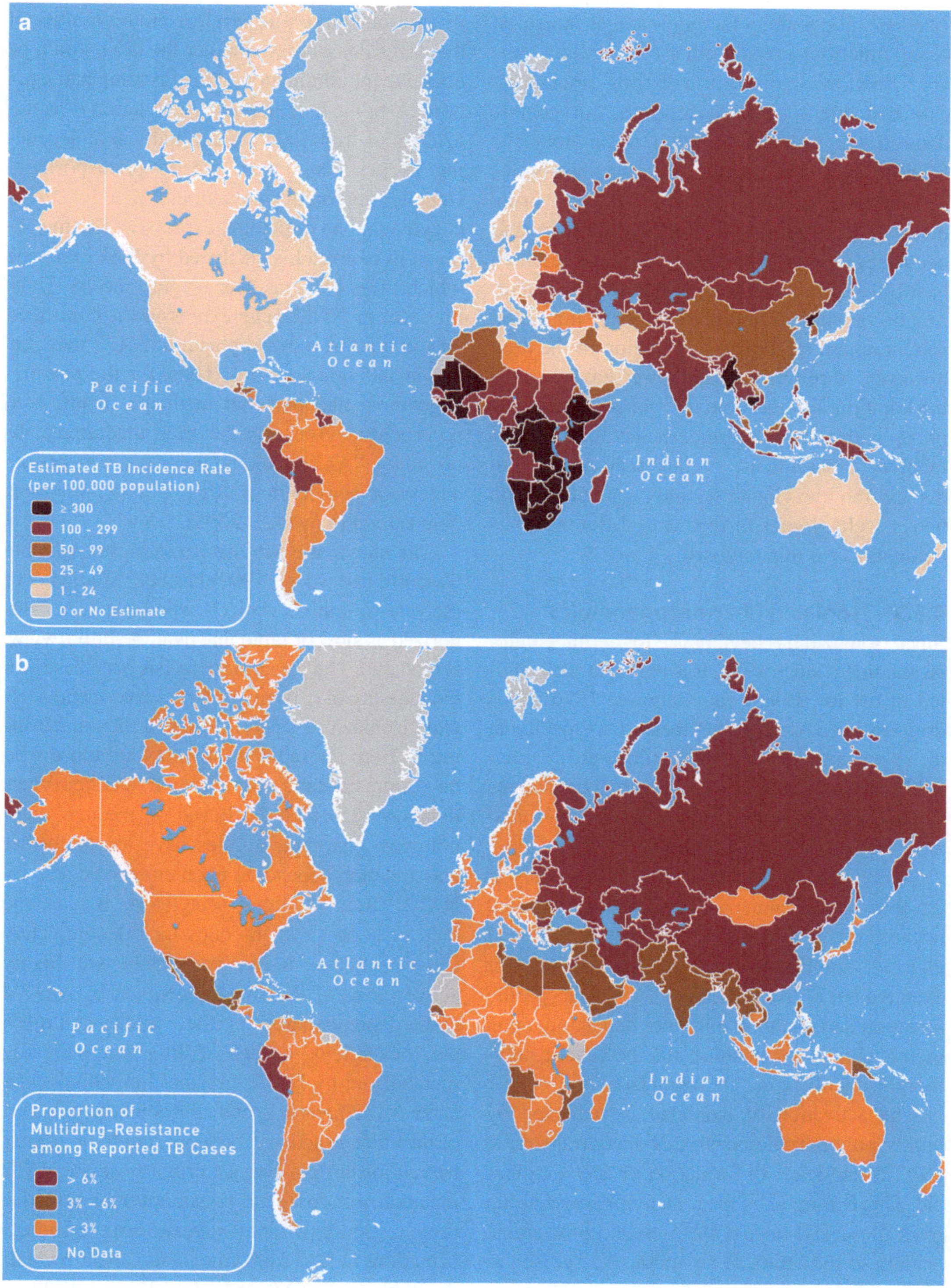

Fig. 27.4 (**a**) Estimated incidences of tuberculosis 2009. (**b**) Proportion of MDR TB among new TB cases 2009 (Map Source: Centre for Disease Control, USA- Health Information of International Travel 2012)

Some nongenetic factors such as nutrition organ function, age and other concomitant therapies, nature of disease, and drug interaction can also effect in drug response, but genetic determinant remains constant throughout the lifetime of the individual. Pharmacogenomics have played an important role in deciphering therapeutic efficacy of drug metabolism and occurrence of adverse events. Though research is still being pursued to decipher the intricacies of how genetic differences play an important role in regard to clinical application of the drug however, through research we have gained information on the role of genetic polymorphism with respect to drug efficacy for the treatment of tuberculosis. In this section we will discuss the knowledge we have gained through newer technologies in regard to different drugs being used for tuberculosis.

4.3.3 Isoniazid Pharmacogenomics

Since isoniazid has been in use for antituberculosis treatment since 1952, it is the most well studied of the lot (Ellard and Gammon 1976). This drug has been found to be tuberculosis specific in its action against tubercle bacilli and has relatively minimal toxicity. Now pharmacogenomics is playing a very important role in making isoniazid the first-line treatment drug.

Acetylation of isoniazid takes place mainly in the liver and gut mucosa. For any drug ingested in the body, it is absorbed and metabolized and then its soluble by-products are released or excreted out of the body. The drugs have specific retention and metabolizing rates depending upon their chemical composition and the genetic polymorphisms of the metabolizing enzymes.

The activation of isoniazid is catalyzed by highly polymorphic enzyme N-acetyltransferase (NAT2) and leads to formation of acetyl isoniazid. This is formed by the transfer of acetyl group from the acetyl coenzyme A to acceptor amine leading to formation of an amide. Acetyl isoniazid combines with several other cellular compounds to give a variety of metabolites which do not have any antituberculosis activity. The level of acetylating isoniazid that will be subjected to during metabolism in the body determines the disease outcome. The level of bioavailability of the drugs determines whether the drug would be effective for elimination of the invading pathogen or toxic to the human body. Acetylation of isoniazid varies from individual to individual depending as per his or her genetic predisposition. Genetics determines the amount of active NAT2 enzyme that an individual expresses. The metabolism of isoniazid is catalyzed by NAT2 enzyme which takes place in liver or gut mucosa. Thus, the level of NAT2 gene expression is controlled by the type of polymorphism in *NAT2* that particular individual carries. Thus, for the pharmacogenomic and personal medicine in effect to succeed, the dosage for the drugs that are metabolized by NAT2 should be tailor made as per the enzymatic activity depending upon the polymorphic variant (Roy et al. 2008).

The enzymatic activity being highly variable Cascorbi and Roots (1999) has been studied over the years in human subjects who have been categorized as slow or rapid inactivators (http://www.brti.co.zw). The categorization has been based on the measure of capacity of NAT2 enzyme to acetylate isoniazid to acetyl isoniazid thus inactivating it. Here the rapid inactivators are those who have more concentration of active NAT2 enzymes than slow inactivators. Based on the new technologies, genotypic studies have led to further classification depending upon enzymatic activity NAT2 variant as rapid acetylators, that is, the wild type gene which codes for the completely active enzyme. Rapid acetylators are highly active forms of the enzyme denoted by *NAT2*4* allele. Patients harboring these alleles can tolerate conventional dosage of drug that is rapidly metabolized by NAT2 enzyme. Individuals who carry *NAT2* heterozygous alleles where only one of the allele is active/functional should be administered lower than average drug (those are NAT2 metabolized) dosage to get an optimum effective drug response without adverse drug response. Mutations in NAT2 enzyme in human individuals designated as *NAT2*5A*, *NAT2*5B*, *NAT2*6A*, and several others which lead to rendering the NAT2 gene activity are termed as slow acetylators which can lead to diminished drug clearance and toxicity.

The variation of frequency of slow acetylator gene is depended on the race, population type, and the ethnicity from one country to the next. It is found in a study that 90 % of Middle Eastern, 60 % in South Indian population, Caucasian and Negroid, and 72 % of the US population harbor slow acetylator gene. In mongoloid populations like the Eskimos, Japanese, and the Chinese, slow acetylators are found in only 10 % of study subjects. In another study carried out in a population of 18 healthy Caucasian, there is variability in isoniazid clearance. While isoniazid preparation is responsible for only 2 % variation and body weight accounted for only 3 % variation in isoniazid clearance, the majority variation of 88 % in isoniazid clearance was due to *NAT2* genotypes (Kinzig-Schippers et al. 2005). High-activity *NAT2* allele-carrying individuals have higher isoniazid clearance. Other studies have shown that 4–6 times more isoniazid concentration in individual is carrying slow acetylator NAT2 genotype (Parkin et al. 1997). A study estimating the comparison of urinary isoniazid excretion in Japanese patients to normal, healthy individuals showed that persons with higher number of active *NAT2* alleles had higher level of isoniazid acetylation (Kita et al. 2001).

The relation between isoniazid concentration in blood with drug efficacy and toxicity knowledge is important. Peak isoniazid concentration to minimum inhibitory concentration ratio has been proposed to serve as a means to outcome tuberculosis treatment (Mitchison 1984). Mean early bactericidal activity of isoniazid depends on its level in the plasma which in turn depends upon the variant *NAT2* genotype carried by an individual. Comparatively the mean bacterial activity is lower in rapid acetylators than in the slow acetylators (Donald et al. 2004). Therapeutic failure or relapse of infection is thus attributed to the lower plasma level due to rapid metabolism of isoniazid in rapid acetylator genotypes, while on other hand high level of isoniazid in slow acetylators may lead to the high level of toxicity (Weiner et al. 2003). NAT2 allele genotyping of tuberculosis patients prior to the treatment with isoniazid is the way forward. Dosage adjustment of isoniazid could be carried out depending upon if the patient is harboring none, one, or two alleles NAT2 rapid acetylators. Thus, isoniazid would be more pharmaceutically viable for treatment of tuberculosis.

In pulmonary tuberculosis patients with known acetylator state, the response to isoniazid treatment analysis was carried out when it is administered alone or in combination with p-aminosalicylic acid. The study compared isoniazid response between *NAT2* slow and fast acetylators, and the study revealed that there is association with treatment response and bacteriological negativity (Selkon et al. 1961). Tuberculosis treatment trials used dosage regimes of daily, twice weekly (Tuberculosis Research Centre Madras, study 1970, 1973), or three times weekly drug regimes (Ellard and Gammon 1976). By means of controlled clinical trials, it was observed that using once a week uptake of isoniazid showed better clinical response to treatment compared to rapid acetylators, with cure rate of 20–35 %. It was postulated that metabolic status of isoniazid may have lesser clinical significance for daily isoniazid treatment regime as compared to thrice weekly or twice weekly treatments. In slow acetylator individuals, the peak concentration of isoniazid was higher than rapid acetylators, and the level of isoniazid decreased more gradually. The effectiveness of a drug in tuberculosis treatment is determined in terms of coverage and the exposure. Coverage has been defined as the number of hours for which bacteriostatic concentration of isoniazid is (0.2 μg/ml) maintained in the blood, while exposure is defined as area under concentration time curve. Both parameters have been found to be significantly greater in slow acetylators. Hence, in rapid acetylator individuals, there is a suboptimal concentration of isoniazid which leads to failure of once-weekly regime of isoniazid (Sarma et al. 1975). Other studies using once-weekly isoniazid-rifapentine were compared with twice-weekly isoniazid-rifampicin; treatment also showed that in case of once-weekly treatment regimes, treatment outcome was poor and was associated with isoniazid acetylator status of the patients (Weiner et al. 2003).

The clinical studies have shown that rapid acetylators having infected from combined tuberculosis and HIV infection are at a further disadvantage since it has been found that antituberculosis drug bioavailability becomes suboptimal in those individuals (Gurumurthy et al. 2004). Tuberculosis patients having chronic renal failure are also at a risk from adverse drug reactions if they also happen to harbor slow acetylator genotypes of *NAT2*. Studies have shown that slow acetylators have higher peak isoniazid concentration, exposure, and half-life compared to rapid acetylators and healthy subjects (Gurumurthy et al. 1992). Hence, in the case of pulmonary tuberculosis patient also suffering from chronic renal failure, the isoniazid dosage should be administered based on their *NAT2* genotypes status. In adult pulmonary patients, studies were carried out to determine correlation between isoniazid dosage and *NAT2* genotypic and phenotypic status. Determination of isoniazid therapeutic dosage has shown that the fast acetylators need higher drug dosage to have an optimum positive treatment response. Fast acetylators tuberculosis patients when administered with 6 mg/kg isoniazid had similar exposure level as 3 mg/kg isoniazid administered to slow acetylators does (Donald et al. 2007).

In a further study in a population of South African tuberculosis patients, it was found that current treatment regimes were causing suboptimal exposure of isoniazid in patients having rapid acetylator status (Wilkins et al. 2011). Several field studies have further suggested that there is a need for calibration of isoniazid dosage as per the individual tuberculosis patient's age, acetylator status, and disease process for an effective antimicrobial outcome of drug treatment (Jeena et al. 2011). In children affected with tuberculosis, it was shown through several studies that the exposure of isoniazid was reduced in the rapid acetylators when compared to the slow acetylators and thus likely to affect the outcome of the treatment of tuberculosis (Cranswick and Mulholland 2005; Schaaf et al. 2005; McIlleron et al. 2009).

Though isoniazid has been found to be nontoxic during conventional regimes, two types of adverse reactions to isoniazid have been reported. The most common isoniazid toxicity reported is hepatotoxicity which affects 2–28 % of the patients (Tostmann et al. 2008). Another isoniazid-associated adverse event is peripheral neuropathy. Neuropathy usually occurs in slow acetylators due to administration of high doses of isoniazid (Devadatta et al. 1960). Hepatotoxicity is the major adverse reaction of isoniazid, and the factors that are responsible are NAT2 acetylation, oxidation by cytochrome P450s oxidation (CYP) 2EI, and detoxification by glutathione S-transferase (GST) enzyme activity (Roy et al. 2008).

Accumulation of acetyl hydrazine, a toxic metabolite of isoniazid, has been implicated in peripheral neuropathy, and the condition in humans is reversible by concomitant administration of pyridoxine (Zilber et al. 1963). Further, it has also been deduced that hepatotoxicity occurs due to hydrazine metabolites of isoniazid. Rifampicin also causes induction of isoniazid metabolism and inducing isoniazid hydrolase to produce isonicotinic acid and hydrazine. The rifampicin induction is more pronounced in slow acetylators compared to rapid acetylators (Sarma et al. 1986).

In some populations studies have established association with *NAT2* acetylator status and isoniazid-induced hepatotoxicity, while in other studies it has not. Studies in Japanese and Taiwanese populations have shown that the acetylator status of *NAT2* increased the risk factor for hepatotoxicity by 28-fold isoniazid-induced hepatotoxicity (Ohno et al. 2000; Huang et al. 2002). In another study on the Korean population, the *NAT2* slow acetylator status has been implicated to increase isoniazid-induced hepatotoxicity by two- to eightfold, and hence the *NAT2* acetylator genotype could serve as predictor of hepatotoxicity (Cho et al. 2007). *NAT2*53/5B*, *NAT2*6A/6A*, *481T/T*, and *590A/A* diplotypes have been indicated and could be used as biomarkers for prediction of antituberculosis drug-induced toxicity (Ben Mahmoud et al. 2012). Slow acetylator *NAT2* alleles have been attributed to increase 3–8-fold to 28-fold higher risk in isoniazid-induced hepatotoxicity. But other studies in tuberculosis patients have not been able to find any association of *NAT2* acetylator status and the drug-induced hepatotoxicity. Case studies of Caucasian origin patients with tuberculosis (Leiro-Fernandez et al. 2011), genotyping in an Indian population (Roy et al. 2001), and study on heterogeneous population of Hispanics, Africans, Caucasian, South American, and Asian have not reported any linkage between *NAT2* acetylator status and isoniazid-induced hepatotoxicity polymorphisms (Vuilleumier et al. 2006).

Cytochrome P450 2E1 is one of the enzymes of the hepatic microsomal enzyme system. *CYP2E1* gene encodes a member of the cytochrome P450 superfamily of enzymes. The cytochrome P450 proteins are monooxygenases which catalyze many reactions involved in drug metabolism and synthesis of cholesterol, steroids, and other lipids. CYP2E1 is an enzyme which brings about conversion of acetyl hydrazine to hepatotoxins, such as acetyl diazone and ketene, and brings about conversion of acetylonium ion (Nelson et al. 1976).

Polymorphism in CYP2E1 has been linked with increasing the risk factor associated with isoniazid-induced liver injury (Lee et al. 2010). The enzyme relocates to the endoplasmic reticulum and can be induced by isoniazid or its metabolite hydrazine. In animal model studies using rat, it has been found that CY2PE1 activity is linked to blood isoniazid levels (Yue et al. 2004). In the presence of variant genotype of *CYP2E1,* isoniazid could on the other hand inhibit the activity of the cytochrome P450 2E1 enzyme. Enhanced cytochrome P450 2E1 activity leads to the increased production of hepatotoxins and hence causing hepatotoxicity. Both NAT2 and CYP2E1 polymorphisms have been shown to be associated with susceptibility of first-line drug-induced hepatitis. CYP2E1 polymorphisms have been related to increase in risk of antituberculosis drug-induced liver toxicity. The common **1A/*1A* genotype of CYP2E1, in tuberculosis patients from Taiwan, has been linked to increase in the liver damage risk by 2.5 times. Presence of both slow *NAT2* acetylator status and the **1A/*1A* genotype further increases risk of hepatotoxicity when compared to presence of either of the single polymorphism (Huang et al. 2003).

The *CYPLE1 *6* and **1A*6*1D* haplotypes in Indian pediatric patients have been shown in separate study and have shown to increase the liver toxicity. Further, the common **1A* allele at *CYP2E1* has been implicated to hepatotoxicity in various heterogeneous population comprising of Asians, Africans, Caucasians, Hispanics, and South Americans (Leiro-Fernandez et al. 2011), but study done on a Korean population on the other hand could not find any association between *CYP2E1* polymorphism and liver toxicity (Huang et al. 2002).

The glutathione S-transferases are class of two distinct supergene families of proteins located in cytosolic and membrane-bound forms. Glutathione S-transferases are a class of enzymes that are responsible for detoxification of therapeutic medication, carcinogens, therapeutic medication, and toxic chemicals that are mostly electrophilic in nature. GSTs are present both in eukaryotes and prokaryotes. At present, eight distinct classes of the soluble and cytoplasmic mammalian glutathione S-transferases have been identified: alpha, kappa, mu, omega, pi, sigma, theta, and zeta. The cytosolic GST enzymes are encoded by at least five different loci coding for GST enzymes, distantly related gene families (designated class alpha, mu, pi, sigma, and theta GST), whereas the membrane-bound enzymes, microsomal GST, and leukotriene C4 synthetase are encoded by single genes and both have arisen separately from the soluble GST (Simon et al. 2000; Strange et al. 2000).

Glutathione S-transferase catalyzed elimination of toxic chemicals from the human body is carried out by making the toxic chemical soluble by conjugation with glutathione. In context of isoniazid-related hepatotoxicity, studies have indicated that deletions of GST mu 1 (*GSTM1*) and GST theta 1(*GSTT1*) are associated with liver damage (Cho et al. 2007; Huang et al. 2007). GST enzymes play an important role in removing the harmful metabolites of isoniazid from the body. The toxic metabolites generated by isoniazid metabolisms are from intracellular free radicals that are scavenged by conjugation with glutathione in reactions catalyzed by GST enzymes. Now, studies in Indian patients suffering from tuberculosis show that those harboring homozygous *GSTM1* mutations have higher risk of hepatotoxicity. It was also found in a study on Taiwanese tuberculosis that patients have twice the risk of isoniazid-induced hepatotoxicity if they have homozygous *GSTM1* deletion. Thus, it can be inferred from similar studies that identification of GSTM1 deletion in patients will lead to the better management of isoniazid-induced hepatotoxicity.

4.3.4 Manganese Superoxide Dismutase Role in Drug-Induced Hepatotoxicity

Reactive oxygen species as we have been aware is a causative agent for damage to hepatic tissue. It has been deduced that level of mitochondrial oxygen species is reduced by the action of manganese superoxide dismutase (MSD). As the name suggests, MSD catalyzes the dismutation of superoxide into oxygen and hydrogen peroxide and is the first line of defense against reactive oxygen species. Polymorphism in the MSD enzyme has been found in a study in the population in Taiwan, where genotypes having T>C polymorphism in codon 47 lead to variant amino acid valine in place of alanine which increases in risk associated with antituberculosis drug-induced hepatotoxicity (Huang et al. 2007). The presence of valine at codon 47 causes the increased activity in the enzyme manganese superoxide dismutase which leads to the accumulation of the toxic byproduct hydrogen peroxide which can cause hepatotoxicity.

4.3.5 Rifampicin in Tuberculosis Treatment

Rifampicin has been proven to show concentration-dependent activity against *M. tuberculosis* and is a very important first-line drug against tuberculosis (Ji et al. 1993; Jayaram et al. 2003). Drug transporters P-glycoprotein and OATP1B transporters uptake rifampicin as a substrate and hence play an important role in distributing the drug throughout the body. The drug transporters are transcriptionally regulated by the nuclear receptors, i.e., pregnane X receptor and constitutive androstane receptor.

The phenomenon variation in bioavailability of rifampicin among individuals in a population on administration of standard dosage has been subjected to investigation by the scientists. The pharmacokinetics of rifampicin depends upon the uptake of machinery of the cells in the body. It has been found that there is a relation between pharmacokinetics and polymorphisms of genes that is responsible for drug efflux and influx. A study group of individuals suffering from tuberculosis who were categorize as per place of origin Africans versus non-Africans, it was observed that in drug transporter gene, *SLCO1B1* 463 C>A polymorphism leads to reduced rifampicin exposure and bioavailability (Weiner et al. 2010). The people of African origin (black subjects) that carry *SLCO1B1* 463 C>A gene polymorphisms have been associated with more pronounced reduced rifampicin exposure compared to people from other races. The study thus showed for the very first time that marked interindividual variation in rifampicin exposure can be attributed to *SLCO1B1* polymorphism. Another study in South Africa has also highlighted that the variant allele of *SLCO1B1* rs 4149032 polymorphism reduced the bioavailability of rifampicin in the body of the patients when present both in the homozygous and heterozygous states (Chigutsa et al. 2011). This finding has been attributed to the observation that there is about 21 % variability in drug clearance among the patients. Polymorphisms in the *ABCB1*, PXR, and CAR genes have not been found to affect the pharmacokinetics of rifampicin in the patients in any significant manner. Researchers have further predicted by means of stimulation that increasing the dosage of rifampicin in patients carrying *SLCO1B1* rs 4149032 would increase the plasma availability of the drug and thus would have a positive impact on the treatment outcome. However, more studies needed to be carried out to know the exact association of *SLCO1B1* gene polymorphisms between rifampicin bioavailability to provide an effective treatment regime.

As has been already mentioned, the variation in human leukocyte antigens also is known cause of disease susceptibility and the response to treatment has also in Indian patients the lack of human allele *HLA DQAI* 01201,* while the presence of DQB1*0207 has been reported to be associated with antituberculosis-induced hepatotoxicity (Sharma et al. 2002).

4.3.6 Aminoglycosides: Genetics of Adverse Reaction in Tuberculosis Treatment

Aminoglycosides are antibiotics which are molecules that consist of amino-modified sugars, and some of the drugs of this class have been used for treatment of tuberculosis. Aminoglycosides such as kanamycin, streptomycin, and amikacin have been used to treat tuberculosis. Aminoglycosides have known to cause ototoxicity. Ototoxicity is term used when there is damage to the ear (oto-), specifically the cochlea or auditory nerve and sometimes the vestibular system due to toxins. The association between aminoglycoside-induced ototoxicity and mitochondrial mutations has been found in a study in Chinese family. The deafness phenotype was found to be associated with C>T 1494 12S rRNA gene polymorphisms which could be induced with the administration of aminoglycosides or even get more aggravated (Zhao et al. 2004).

There is still no clear relationship of ethnicity and genetic background and response to antituberculosis treatment, and no single variant of *NAT2* and *CYP2E1* genes is associated with significant liver damage (Yamada et al. 2009). More extensive pharmacogenomic research is still needed for realization of robust personalized medicine for tuberculosis.

5 Case of Amphotericin B Toxicity

The antifungal medicine amphotericin B has been found to be effective and well quite toxic. Further investigations revealed that the immunomodulatory role of amphotericin B also involves the induction of production of proinflammatory cytokine. In human cell the amphotericin-induced higher mRNA expression and cytokine production have been detected in studies (Rogers et al. 2000). The discovery of induction of proinflammatory cytokine production was able to explain the infusion-related toxicity effect like nausea, fever, chills, and hypotension that are characteristics of this cytokine release. It was also able to explain the mechanism of action of amphotericin B since the proinflammatory cytokines are responsible for the activation of monocytes, macrophages, and promote chemotaxis that led to enhanced immune response to the infection.

6 Omics on the Path of Drug Discovery

Since the advent of the era of omics technology, the number of drugs that have been discovered have not delivered as was first predicted especially in the case of infectious disease. Some of the shortcoming and the remedial measures have already been discovered in the previous sections. It has been found that even with high-throughput screening of number of drug, candidates have not been successful always (Payne et al. 2007). The field of scientific research which has become all encompassing and interdisciplinary has added strength along the way and opened new avenues. The field of biology is intertwined with structural biology and chemistry has given rise to the field of medical structural genomics. The exact causes of failure of high-throughput screens have not been well defined. The field of structure-based drug discovery has tried to overcome these limitations in the availability of chemical libraries and absence of structural information of many of the targets. The field of structure-based drug discovery has its origin from the field of X-ray crystallography and nuclear magnetic resonance (NMR) technology. With the availability of human genome sequences and pathogen genome sequence databases, the field of structural genomics has gained importance, and hence over the past decade, more than 20 such projects have been taken up. The field of structural biology has got a boost with the coming together of robotics and informatics in the biological research sphere (Haquin et al. 2008).

For the synthesis of an effective drug, by means of medical structural genomics, the protein which drug will affect should be well defined experimentally in both structure and functional aspects as the potential target. The protein should not only be well characterized structurally but also should be well defined as essential for the survival of the pathogen. Once drug and its protein target in a microorganism is identified, the field of medical structural genomics provides rapid mechanisms using high-throughput X-ray crystallography and NMR assay system to find the ligand-bound structures. To identify such drug targets, it is very essential to know the complex host and pathogen interaction. The mode the pathogen uses to cause the infections is very difficult to elucidate and is a long process. The technologies of RNA interference and other gene knockout techniques should be complimented with experimental chemical biology approaches as microorganisms adopt multiple mechanisms for survival.

This has been emphasized for the fact examples of efforts of scientist for targeting the fatty acid biosynthesis pathways of bacteria. At first drugs were found to have high bioavailability and are potent against the bacterial replication in vitro. These compounds were subjected to be tested in animals and have been found to be not effective, the reason being that the bacterium utilizes the fatty acids present in their host vertebrates (Brinster et al. 2009). Hence, this study proves that there is need for more effective screening using the services of scientist from several spectra of biology like microbiologist, biochemists along with structural biologist, and chemical biologists to find effective molecules and compounds which can eliminate the pathogen under proper infective conditions (Hoon et al. 2008).

In pharmaceutical research scenario, it might also be possible that the drug target for a cell active compound is not known and then medical structural genomics provides a number of purified protein targets which can be assayed for binding interaction with bioactive compound by means of number of biophysical techniques like thermal stability (Ericsson et al. 2006). Such efforts have already been carried out in the field of protozoan pathogens. The program of Medical Structural Genomics of Protozoan Pathogens (http://msgpp.org/description.shtml) has been initiated to screen for drugs for ten protozoan diseases. The initiative has screening of thousands of potential antimalarial drugs against about 67 putative *Plasmodium falciparum* protein targets by expressing them in bacterial expression system in the laboratory and deciphering their 3D structures. Further, the compounds are assayed for their effectiveness in live organisms and further validated in appropriate disease model. The terms chemical validation and drugability are often used in conjunction in such cases. Drugability is meant to be used how tractable a given drug target is for the development of a drug candidate, while chemical validation means that drugs have been found to be active in live organism. Drugs which fulfill the abovementioned criteria are worth the effort, time, and resources. In the future more collaborative efforts between medical structural genomic centers and the chemical biology institutes would be possible with the availability of collection of phenotypically defined compound that would have proven anti-pathogen activity resulting in the synergistic target validation and hit to lead development using structure-based drug design.

Pharmaceutical industry has now taken fragment-based drug discovery methodology as an alternatively less expensive and at times more effective than high throughput screening. Variety of methods like X-ray crystallography, NMR, surface fluorescence polarization, plasmon resonance yield, and differential thermal denaturation have been used to obtain macromolecular structure to screen libraries of small fragment that are obtained from compounds that are building blocks of drugs and hence can be more drug like. The fragment-based drug discovery is based on the screening libraries of small molecules on the rule of three which has the molecular weight <300 Da, the calculated log of octanol/water coefficient (Clog P) <3, and $3 \leq$ rotatable bonds and hydrogen bonds (Rees et al. 2004; Congreve et al. 2008).

Protein-protein interactions are important for all biological processes. Metabolic activities in the biological system are catalyzed by protein-based enzyme where in certain cases their activities are regulated by modulation of an equilibrium of an alternate, nonadditive, functionally distinct oligomeric assemblies (morpheeins) that have now been described as mode allostery. The oligomerization from the protein-protein interaction need not lead to gain in free energy, and it has been found that small molecules can block or disrupt any protein-protein interaction that is necessary for biological systems, for example, being in the development of potent peptidomimetic inhibitor of HSV ribonucleotide reductase with antiviral activity (Liuzzi et al. 1994). The discoveries have opened avenues where structure-based information can be used to develop small novel antimicrobial molecules that can be made which can target protein-protein interfaces (Wells and McClendon 2007).

An example of this technology has been used to find small-molecule species-specific allosteric drugs for porphobilinogen synthase (PBGS). The oligomeric equilibrium for porphobilinogen synthase (PBGS) consists of high-activity octamers, low-activity hexamers, and two dimer conformations. In silico docking analysis from a small molecule library helped in selecting suitable compounds and molecules that had more affinity for docking PBGS allosteric site and thus were subjected to testing in vitro. In one compound whose inhibition mechanism is species specific, conversion of PBGS octamers to hexamers was thus identified (Lawrence et al. 2008). The above findings have led the way of targeting of oligomeric enzymes in pathogenic organism bacteria. Prime example is bacterial inorganic pyrophosphatases, which function as hexamer (Kankare et al. 1996). On the other hand, the eukaryotic, cytosolic, and mitochondrial pyrophosphatases function as homodimers (Oksanen et al. 2007) and hence have different interfaces than its bacterial counterparts as evident from the study of evolutionary aspects of inorganic phosphatases. In this context the strategy has been to target the oligomeric state of the bacterial inorganic pyrophosphatase enzymes to inhibit their activity rather than their conserved active site (Sivula et al. 1999). The technology has opened a novel pathway where more antibiotics can be developed.

The amount of knowledge of protein structures being generated is enormous; the need of the hour is dissemination of the knowledge databases among scientists and academic researchers on a worldwide scale. The protein structural know-how should be in the public domain without any constrains and copyright restrictions; also in addition the databases should be available free of any monetary charge. Structural genomic projects the world over have solved the structures of many proteins and have made the knowledge available for world community by submitting the structures to Protein Data Bank {PDB; http://www.wwpdb.org/}. Worldwide Protein Data Bank is the site whose mission is to maintain a single protein structural public database which can be accessed by the global community (Berman et al. 2007).

There is lot of structural data of protein-ligand complexes that is in private pharmaceutical industries not in the public domain. The economic incentives of drug discovery are driving force for this secrecy, but in this process there are a lot of valuable data that are duplicated and lots of valuable resources and energy efforts. The learning process from failures and successes in pharmaceutical corporate sectors is never known to the scientific community, and a major loss is of most valuable time. Hence, as we can see, the drug discovery resources are not being adequately utilized across the academia and industry, so there is suggestion to have open-access industry-academia partnerships as possible mechanisms to overcome the problem. A frame work is need where both financial and intellectual properties of the innovators are safeguarded when there structural data are deposited in the databases like PDB. A simple proposition would delay the release date of such structural data so that protection of intellectual property is feasible. Policies which can bring into the public domain structural data from the corporate world could only be possible by the concerted efforts of all stakeholders from industry, national, and international research funding agencies from all nations (Edwards et al. 2009).

Apart from easier dissemination of structural information related to infectious diseases and collaboration of structural biologist with medical chemist and molecular biologist, there is need for development of automation in several technologies to bring about unprecedented growth in the new drug discovery. Fragment-based drug design needs the support of high throughput technologies such that along with structural genomics, there will be more success in the determining protein-ligand structure determination.

7 Conclusion

Decades of experience have shown that the infectious diseases would emerge with more vigor and virulence. When the diseases are not controlled, then it would take a considerable toll of human health both in terms of mortality and morbidity. The life would be affected by emerging microbial disease-causing pathogen whatever the region, ethnicity, lifestyle, socioeconomic status, and ethnic background. Hence, the threat from infectious diseases is real and the situation is overtly challenging. The great advances in the genetics, genomics, and proteomics have the potential to take up the challenge in the coming decades. It is evident that these technologies have the potential to change the field of diagnostics, treatment, and discovery of drugs and vaccines. The need of the hour is to strengthen the public health programs at both the national and the international levels to strengthen research in the field of omics to fully realize the potential of scientific technologies that would usher in the era of pharmacogenomic-based personalized medicine.

References

Abel L, Vu DL, Oberti J, Nguyen VT, Van VC, Guilloud-Bataille M, Schurr E, Lagrange PH (1995) Complex segregation analysis of leprosy in southern Vietnam. Genet Epidemiol 12(1):63–82

Abel L, Sanchez FO, Oberti J, Thuc NV, Hoa LV, Lap VD, Skamene E, Lagrange PH, Schurr E (1998) Susceptibility to leprosy is linked to the human Nrampl gene. J Infect Dis 177(1):133–145

Adjei GO, Kristensen K, Goka BQ, Hoegberg LC, Alifrangis M, Rodrigues OP, Kurtzhals JA (2008) Effect of concomitant artesunate administration and cytochrome P4502c8 polymorphisms on the pharmacokinetics of amodiaquine in Ghanaian children with uncomplicated malaria. Antimicrob Agents Chemother 52(12):4400–4406

Agaisse H, Burrack LS, Philips JA, Rubin EJ, Perrimon N, Higgins DE (2005) Genome-wide RNAi screen for host factors required for intracellular bacterial infection. Science 309(5738):1248–1251

Alexander J, Bryson K (2005) T helper (H)1/Th2 and leishmania: paradox rather than paradigm. Immunol Lett 99(1):17–23

Alonso DP, Ferreira AF, Ribolla PE, de Miranda Santos IK, do Socorro Pires e Cruz M, de Carvalho FA, Abatepaulo AR et al (2007) Genotypes of the mannan-binding lectin gene and susceptibility to visceral leishmaniasis and clinical complications. J Infect Dis 195(8):1212–1217

Altare F, Durandy A, Lammas D, Emile JF, Lamhamedi S, Le Deist F, Drysdale P et al (1998) Impairment of mycobacterial immunity in human interleukin-12 receptor deficiency. Science 280(5368):1432–1435

Alving AS, Carson PE, Flanagan CL, Ickes CE (1956) Enzymatic deficiency in primaquine-sensitive erythrocytes. Science 124(3220):484–485

Amato VS, Tuon FF, Bacha HA, Neto VA, Nicodemo AC (2008) Mucosal leishmaniasis: current scenario and prospects for treatment. Acta Trop 105(1):1–9

Ameen M (2007) Cutaneous leishmaniasis: therapeutic strategies and future directions. Expert Opin Pharmacother 8(16):2689–2699

Antoniou M, Haralambous C, Mazeris A, Pratlong F, Dedet JP, Soteriadou K (2008) Leishmania donovani leishmaniasis in Cyprus. Lancet Infect Dis 8(1):6–7

Antony PM, Balling R, Vlassis N (2012) From systems biology to systems biomedicine. Curr Opin Biotechnol 23(4):604–608

Bacellar O, Lessa H, Schriefer A, Machado P, Ribeiro de Jesus A, Dutra WO, Gollob KJ, Carvalho EM (2002) Up-regulation of Th1-type responses in mucosal leishmaniasis patients. Infect Immun 70(12):6734–6740

Barral-Netto M, Badaro R, Barral A, Almeida RP, Santos SB, Badaro F, Pedral-Sampaio D et al (1991a) Tumor necrosis factor (cachectin) in human visceral leishmaniasis. J Infect Dis 163(4):853–857

Barral-Netto M, Barral A, Santos SB, Carvalho EM, Badaro R, Rocha H, Reed SG, Johnson WD Jr (1991b) Soluble Il-2 receptor as an agent of serum-mediated suppression in human visceral leishmaniasis. J Immunol 147(1):281–284

Bellamy R, Ruwende C, Corrah T, McAdam KP, Whittle HC, Hill AV (1998) Variations in the Nrampl gene and susceptibility to tuberculosis in West Africans. N Engl J Med 338(10):640–644

Ben Mahmoud L, Ghozzi H, Kamoun A, Hakim A, Hachicha H, Hammami S, Sahnoun Z et al (2012) Polymorphism of the N-acetyltransferase 2 gene as a susceptibility risk factor for antituberculosis drug-induced hepatotoxicity in Tunisian patients with tuberculosis. Pathol Biol (Paris) 60(5):324–330

Berman H, Henrick K, Nakamura H, Markley JL (2007) The worldwide protein data bank (wwPDB): ensuring a single, uniform archive of Pdb data. Nucleic Acids Res 35(Database issue):D301–D303

Blackwell J, Freeman J, Bradley D (1980) Influence of H-2 complex on acquired resistance to Leishmania donovani infection in mice. Nature 283(5742):72–74

Blackwell JM, Goswami T, Evans CA, Sibthorpe D, Papo N, White JK, Searle S et al (2001) SLC11A1 (formerly Nrampl) and disease resistance. Cell Microbiol 3(12):773–784

Bourreau E, Prevot G, Gardon J, Pradinaud R, Launois P (2001) High intralesional interleukin-10 messenger RNA expression in localized cutaneous leishmaniasis is associated with unresponsiveness to treatment. J Infect Dis 184(12):1628–1630

Boutros M, Ahringer J (2008) The art and design of genetic screens: RNA interference. Nat Rev Genet 9(7):554–566

Brass AL, Dykxhoorn DM, Benita Y, Yan N, Engelman A, Xavier RJ, Lieberman J, Elledge SJ (2008) Identification of host proteins required for HIV infection through a functional genomic screen. Science 319(5865):921–926

Brinster S, Lamberet G, Staels B, Trieu-Cuot P, Gruss A, Poyart C (2009) Type II fatty acid synthesis is not a suitable antibiotic target for gram-positive pathogens. Nature 458(7234):83–86

Bucheton B, Abel L, El-Safi S, Kheir MM, Pavek S, Lemainque A, Dessein AJ (2003a) A major susceptibility locus on chromosome 22q12 plays a critical role in the control of kala-azar. Am J Hum Genet 73(5):1052–1060

Bucheton B, Abel L, Kheir MM, Mirgani A, El-Safi SH, Chevillard C, Dessein A (2003b) Genetic control of visceral leishmaniasis in a Sudanese population: candidate gene testing indicates a linkage to the Nramp1 region. Genes Immun 4(2):104–109

Bucheton B, Argiro L, Chevillard C, Marquet S, Kheir MM, Mergani A, El-Safi SH, Dessein AJ (2007) Identification of a novel G245R polymorphism in the IL-2 receptor beta membrane proximal domain associated with human visceral leishmaniasis. Genes Immun 8(1):79–83

Buxbaum LU, Denise H, Coombs GH, Alexander J, Mottram JC, Scott P (2003) Cysteine protease B of Leishmania mexicana inhibits host Th1 responses and protective immunity. J Immunol 171(7):3711–3717

Cabrera M, Shaw MA, Sharples C, Williams H, Castes M, Convit J, Blackwell JM (1995) Polymorphism in tumor necrosis factor genes associated with mucocutaneous leishmaniasis. J Exp Med 182(5):1259–1264

Caramia G, Ruffini E (2012) Proper antibiotic therapy. From penicillin to pharmacogenomic. Minerva Pediatr 64(2):225–237

Cascorbi I, Roots I (1999) Pitfalls in N-acetyltransferase 2 genotyping. Pharmacogenetics 9(1):123–127

Castellucci L, Menezes E, Oliveira J, Magalhaes A, Guimaraes LH, Lessa M, Ribeiro S et al (2006) Il6–174 G/C promoter polymorphism influences susceptibility to mucosal but not localized cutaneous leishmaniasis in Brazil. J Infect Dis 194(4):519–527

Castes M, Trujillo D, Rojas ME, Fernandez CT, Araya L, Cabrera M, Blackwell J, Convit J (1993) Serum levels of tumor necrosis factor in patients with American cutaneous leishmaniasis. Biol Res 26(1–2):233–238

Cavaco I, Stromberg-Norklit J, Kaneko A, Msellem MI, Dahoma M, Ribeiro VL, Bjorkman A, Gil JP (2005) CYP2C8 polymorphism frequencies among malaria patients in Zanzibar. Eur J Clin Pharmacol 61(1):15–18

Center of Disease Control researchers, The Yellow Book, CDC Health Information for International Travel (2012), Chapter 3, Infectious diseases related to travel. http://wwwnc.cdc.gov/travel/yellowbook/2012/chapter-3-infectious-diseases-related-to-travel/tuberculosis.htm. Accessed Feb 2013

Chigutsa E, Visser ME, Swart EC, Denti P, Pushpakom S, Egan D, Holford NH et al (2011) The SLCO1B1 rs4149032 polymorphism is highly prevalent in South Africans and is associated with reduced rifampin concentrations: dosing implications. Antimicrob Agents Chemother 55(9):4122–4127

Cho HJ, Koh WJ, Ryu YJ, Ki CS, Nam MH, Kim JW, Lee SY (2007) Genetic polymorphisms of NAT2 and CYP2E1 associated with antituberculosis drug-induced hepatotoxicity in Korean patients with pulmonary tuberculosis. Tuberculosis (Edinb) 87(6):551–556

Congreve M, Chessari G, Tisi D, Woodhead AJ (2008) Recent developments in fragment-based drug discovery. J Med Chem 51(13):3661–3680

Convit J, Ulrich M, Zerpa O, Borges R, Aranzazu N, Valera M, Villarroel H, Zapata Z, Tomedes I (2003) Immunotherapy of American cutaneous leishmaniasis in Venezuela during the period 1990–99. Trans R Soc Trop Med Hyg 97(4):469–472

Cranswick N, Mulholland K (2005) Isoniazid treatment of children: can genetics help guide treatment? Arch Dis Child 90(6):551–553

Crofton J (1994) Global challenge of tuberculosis. Lancet 344(8922):609

Cupolillo E, Brahim LR, Toaldo CB, de Oliveira-Neto MP, de Brito ME, Falqueto A, de Farias Naiff M, Grimaldi G Jr (2003) Genetic polymorphism and molecular epidemiology of Leishmania (Viannia) braziliensis from different hosts and geographic areas in Brazil. J Clin Microbiol 41(7):3126–3132

Dai D, Zeldin DC, Blaisdell JA, Chanas B, Coulter SJ, Ghanayem BI, Goldstein JA (2001) Polymorphisms in human CYP2C8 decrease metabolism of the anticancer drug paclitaxel and arachidonic acid. Pharmacogenetics 11(7):597–607

De Groot AS, Saint-Aubin C, Bosma A, Sbai H, Rayner J, Martin W (2001) Rapid determination of HLA B*07 ligands from the West Nile virus NY99 genome. Emerg Infect Dis 7(4):706–713

Devadatta S, Gangadharam PR, Andrews RH, Fox W, Ramakrishnan CV, Selkon JB, Velu S (1960) Peripheral neuritis due to isoniazid. Bull World Health Organ 23:587–598

Di YM, Chow VD, Yang LP, Zhou SF (2009) Structure, function, regulation and polymorphism of human cytochrome P450 2A6. Curr Drug Metab 10(7):754–780

Donald PR, Sirgel FA, Venter A, Parkin DP, Seifart HI, van de Wal BW, Werely C, van Helden PD, Maritz JS (2004) The influence of human N-acetyltransferase genotype on the early bactericidal activity of isoniazid. Clin Infect Dis 39(10):1425–1430

Donald PR, Parkin DP, Seifart HI, Schaaf HS, van Helden PD, Werely CJ, Sirgel FA, Venter A, Maritz JS (2007) The influence of dose and N-acetyltransferase-2

(NAT2) genotype and phenotype on the pharmacokinetics and pharmacodynamics of isoniazid. Eur J Clin Pharmacol 63(7):633–639

Echeverri CJ, Beachy PA, Baum B, Boutros M, Buchholz F, Chanda SK, Downward J et al (2006) Minimizing the risk of reporting false positives in large-scale RNAi screens. Nat Methods 3(10):777–779

Edwards AM, Bountra C, Kerr DJ, Willson TM (2009) Open access chemical and clinical probes to support drug discovery. Nat Chem Biol 5(7):436–440

Edwards-Smith CJ, Jonsson JR, Purdie DM, Bansal A, Shorthouse C, Powell EE (1999) Interleukin-10 promoter polymorphism predicts initial response of chronic hepatitis C to interferon alfa. Hepatology 30(2):526–530

Ellard GA, Gammon PT (1976) Pharmacokinetics of isoniazid metabolism in man. J Pharmacokinet Biopharm 4(2):83–113

Ericsson UB, Hallberg BM, Detitta GT, Dekker N, Nordlund P (2006) Thermofluor-based high-throughput stability optimization of proteins for structural studies. Anal Biochem 357(2):289–298

Evans WE, Johnson JA (2001) Pharmacogenomics: the inherited basis for interindividual differences in drug response. Annu Rev Genomics Hum Genet 2:9–39

Evans WE, Relling MV (1999) Pharmacogenomics: translating functional genomics into rational therapeutics. Science 286(5439):487–491

Faria DR, Gollob KJ, Barbosa J Jr, Schriefer A, Machado PR, Lessa H, Carvalho LP et al (2005) Decreased in situ expression of interleukin-10 receptor is correlated with the exacerbated inflammatory and cytotoxic responses observed in mucosal leishmaniasis. Infect Immun 73(12):7853–7859

Ganguly S, Das NK, Panja M, Pal S, Modak D, Rahaman M, Mallik S et al (2008) Increased levels of interleukin-10 and IgG3 are hallmarks of Indian post-kala-azar dermal leishmaniasis. J Infect Dis 197(12):1762–1771

Greenwood BM, Fidock DA, Kyle DE, Kappe SH, Alonso PL, Collins FH, Duffy PE (2008) Malaria: progress, perils, and prospects for eradication. J Clin Invest 118(4):1266–1276

Gruenheid S, Pinner E, Desjardins M, Gros P (1997) Natural resistance to infection with intracellular pathogens: the Nramp1 protein is recruited to the membrane of the phagosome. J Exp Med 185(4):717–730

Gurarie D, Zimmerman PA, King CH (2006) Dynamic regulation of single- and mixed-species malaria infection: insights to specific and non-specific mechanisms of control. J Theor Biol 240(2):185–199

Gurumurthy P, Raghupati Sarma G, Jayasankar K et al (1992) Single-dose pharmacokinetics of isoniazid and rifampicin in patients with chronic renal failure. Indian J Tuberc 39:221–228

Gurumurthy P, Ramachandran G, Hemanth Kumar AK, Rajasekaran S, Padmapriyadarsini C, Swaminathan S, Bhagavathy S et al (2004) Decreased bioavailability of rifampin and other antituberculosis drugs in patients with advanced human immunodeficiency virus disease. Antimicrob Agents Chemother 48(11):4473–4475

Gyamfi MA, Fujieda M, Kiyotani K, Yamazaki H, Kamataki T (2005) High prevalence of cytochrome P450 2A6*1A alleles in a black African population of Ghana. Eur J Clin Pharmacol 60(12):855–857

Hamon MA, Cossart P (2008) Histone modifications and chromatin remodeling during bacterial infections. Cell Host Microbe 4(2):100–109

Handman E, Elso C, Foote S (2005) Genes and susceptibility to leishmaniasis. Adv Parasitol 59:1–75

Haquin S, Oeuillet E, Pajon A, Harris M, Jones AT, van Tilbeurgh H, Markley JL, Zolnai Z, Poupon A (2008) Data management in structural genomics: an overview. Methods Mol Biol 426:49–79

Hatton CS, Peto TE, Bunch C, Pasvol G, Russell SJ, Singer CR, Edwards G, Winstanley P (1986) Frequency of severe neutropenia associated with amodiaquine prophylaxis against malaria. Lancet 1(8478):411–414

Hatzigeorgiou DE, He S, Sobel J, Grabstein KH, Hafner A, Ho JL (1993) Il-6 down-modulates the cytokine-enhanced antileishmanial activity in human macrophages. J Immunol 151(7):3682–3692

Hayney MS, Poland GA, Jacobson RM, Schaid DJ, Lipsky JJ (1996) The influence of the HLA-DRB1*13 allele on measles vaccine response. J Investig Med 44(5):261–263

Hayney MS, Poland GA, Jacobson RM, Rabe D, Schaid DJ, Jacobsen SJ, Lipsky JJ (1998) Relationship of HLA-DQA1 alleles and humoral antibody following measles vaccination. Int J Infect Dis 2(3):143–146

Hayrinen J, Jennings H, Raff HV, Rougon G, Hanai N, Gerardy-Schahn R, Finne J (1995) Antibodies to polysialic acid and its N-propyl derivative: binding properties and interaction with human embryonal brain glycopeptides. J Infect Dis 171(6):1481–1490

Hayton K, Su XZ (2004) Genetic and biochemical aspects of drug resistance in malaria parasites. Curr Drug Targets Infect Disord 4(1):1–10

Herwaldt BL (1999) Leishmaniasis. Lancet 354(9185):1191–1199

Hoon S, Smith AM, Wallace IM, Suresh S, Miranda M, Fung E, Proctor M et al (2008) An integrated platform of genomic assays reveals small-molecule bioactivities. Nat Chem Biol 4(8):498–506

Hooper DC (2001) Mechanisms of action of antimicrobials: focus on fluoroquinolones. Clin Infect Dis 32(Suppl 1):S9

Huang YS, Chern HD, Su WJ, Wu JC, Lai SL, Yang SY, Chang FY, Lee SD (2002) Polymorphism of the N-acetyltransferase 2 gene as a susceptibility risk factor for antituberculosis drug-induced hepatitis. Hepatology 35(4):883–889

Huang YS, Chern HD, Su WJ, Wu JC, Chang SC, Chiang CH, Chang FY, Lee SD (2003) Cytochrome P450 2E1 genotype and the susceptibility to antituberculosis drug-induced hepatitis. Hepatology 37(4):924–930

Huang YS, Su WJ, Huang YH, Chen CY, Chang FY, Lin HC, Lee SD (2007) Genetic polymorphisms of manganese superoxide dismutase, NAD(P)H: quinone oxidoreductase, glutathione S-transferase M1 and T1, and the susceptibility to drug-induced liver injury. J Hepatol 47(1):128–134

Hyde JE (2007) Drug-resistant malaria – an insight. FEBS J 274(18):4688–4698

Ilett KF, Ethell BT, Maggs JL, Davis TM, Batty KT, Burchell B, Binh TQ et al (2002) Glucuronidation of dihydroartemisinin in vivo and by human liver microsomes and expressed UDP-glucuronosyltransferases. Drug Metab Dispos 30(9):1005–1012

Jayaram R, Gaonkar S, Kaur P, Suresh BL, Mahesh BN, Jayashree R, Nandi V et al (2003) Pharmacokinetics-pharmacodynamics of rifampin in an aerosol infection model of tuberculosis. Antimicrob Agents Chemother 47(7):2118–2124

Jeena PM, Bishai WR, Pasipanodya JG, Gumbo T (2011) In silico children and the glass mouse model: clinical trial simulations to identify and individualize optimal isoniazid doses in children with tuberculosis. Antimicrob Agents Chemother 55(2):539–545

Jenner RG, Young RA (2005) Insights into host responses against pathogens from transcriptional profiling. Nat Rev Microbiol 3(4):281–294

Ji B, Truffot-Pernot C, Lacroix C, Raviglione MC, O'Brien RJ, Olliaro P, Roscigno G, Grosset J (1993) Effectiveness of rifampin, rifabutin, and rifapentine for preventive therapy of tuberculosis in mice. Am Rev Respir Dis 148(6 Pt 1):1541–1546

Jouanguy E, Altare F, Lamhamedi S, Revy P, Emile JF, Newport M, Levin M et al (1996) Interferon-gamma-receptor deficiency in an infant with fatal bacille calmette-guerin infection. N Engl J Med 335(26):1956–1961

Jouanguy E, Lamhamedi-Cherradi S, Altare F, Fondaneche MC, Tuerlinckx D, Blanche S, Emile JF et al (1997) Partial interferon-gamma receptor 1 deficiency in a child with tuberculoid bacillus calmette-guerin infection and a sibling with clinical tuberculosis. J Clin Invest 100(11):2658–2664

Kalow W, Tang BK, Endrenyi L (1998) Hypothesis: comparisons of inter- and intra-individual variations can substitute for twin studies in drug research. Pharmacogenetics 8(4):283–289

Kamali-Sarvestani E, Rasouli M, Mortazavi H, Gharesi-Fard B (2006) Cytokine gene polymorphisms and susceptibility to cutaneous leishmaniasis in Iranian patients. Cytokine 35(3–4):159–165

Kankare J, Salminen T, Lahti R, Cooperman BS, Baykov AA, Goldman A (1996) Structure of *Escherichia coli* inorganic pyrophosphatase at 2.2 A resolution. Acta Crystallogr D Biol Crystallogr 52(Pt 3):551–563

Karplus TM, Jeronimo SM, Chang H, Helms BK, Burns TL, Murray JC, Mitchell AA et al (2002) Association between the tumor necrosis factor locus and the clinical outcome of Leishmania chagasi infection. Infect Immun 70(12):6919–6925

Kelly BL, Stetson DB, Locksley RM (2003) Leishmania major lack antigen is required for efficient vertebrate parasitization. J Exp Med 198(11):1689–1698

Kerb R, Fux R, Morike K, Kremsner PG, Gil JP, Gleiter CH, Schwab M (2009) Pharmacogenetics of antimalarial drugs: effect on metabolism and transport. Lancet Infect Dis 9(12):760–774

Kinzig-Schippers M, Tomalik-Scharte D, Jetter A, Scheidel B, Jakob V, Rodamer M, Cascorbi I et al (2005) Should we use N-acetyltransferase type 2 genotyping to personalize isoniazid doses? Antimicrob Agents Chemother 49(5):1733–1738

Kita T, Tanigawara Y, Chikazawa S, Hatanaka H, Sakaeda T, Komada F, Iwakawa S, Okumura K (2001) N-Acetyltransferase2 genotype correlated with isoniazid acetylation in Japanese tuberculous patients. Biol Pharm Bull 24(5):544–549

Kolesar JM, Allen PG, Doran CM (1997) Direct quantification of HIV-1 RNA by capillary electrophoresis with laser-induced fluorescence. J Chromatogr B Biomed Sci Appl 697(1–2):189–194

Konig R, Zhou Y, Elleder D, Diamond TL, Bonamy GM, Irelan JT, Chiang CY et al (2008) Global analysis of host-pathogen interactions that regulate early-stage HIV-1 replication. Cell 135(1):49–60

Krishnan MN, Ng A, Sukumaran B, Gilfoy FD, Uchil PD, Sultana H, Brass AL et al (2008) RNA interference screen for human genes associated with West Nile virus infection. Nature 455(7210):242–245

Lander ES, Linton LM, Birren B, Nusbaum C, Zody MC, Baldwin J, Devon K et al (2001) Initial sequencing and analysis of the human genome. Nature 409(6822):860–921

Lara ML, Layrisse Z, Scorza JV, Garcia E, Stoikow Z, Granados J, Bias W (1991) Immunogenetics of human American cutaneous leishmaniasis: study of HLA haplotypes in 24 families from Venezuela. Hum Immunol 30(2):129–135

Lawrence SH, Ramirez UD, Tang L, Fazliyez F, Kundrat L, Markham GD, Jaffe EK (2008) Shape shifting leads to small-molecule allosteric drug discovery. Chem Biol 15(6):586–596

Lee SW, Chung LS, Huang HH, Chuang TY, Liou YH, Wu LS (2010) NAT2 and CYP2E1 polymorphisms and susceptibility to first-line anti-tuberculosis drug-induced hepatitis. Int J Tuberc Lung Dis 14(5):622–626

Leiro-Fernandez V, Valverde D, Vazquez-Gallardo R, Botana-Rial M, Constenla L, Agundez JA, Fernandez-Villar A (2011) N-acetyltransferase 2 polymorphisms and risk of anti-tuberculosis drug-induced hepatotoxicity in Caucasians. Int J Tuberc Lung Dis 15(10):1403–1408

Li XQ, Bjorkman A, Andersson TB, Ridderstrom M, Masimirembwa CM (2002) Amodiaquine clearance and its metabolism to N-desethylamodiaquine is mediated by CYP2C8: a new high affinity and turnover enzyme-specific probe substrate. J Pharmacol Exp Ther 300(2):399–407

Li XQ, Bjorkman A, Andersson TB, Gustafsson LL, Masimirembwa CM (2003) Identification of human cytochrome P(450)S that metabolise anti-parasitic drugs and predictions of in vivo drug hepatic clearance from in vitro data. Eur J Clin Pharmacol 59(5–6):429–442

Lin JT, Juliano JJ, Wongsrichanalai C (2010) Drug-resistant malaria: the Era of Act. Curr Infect Dis Rep 12(3):165–173

Liuzzi M, Deziel R, Moss N, Beaulieu P, Bonneau AM, Bousquet C, Chafouleas JG et al (1994) A potent peptidomimetic inhibitor of HSV ribonucleotide reductase with antiviral activity in vivo. Nature 372(6507):695–698

Louie L, Matsumura SO, Choi E, Louie M, Simor AE (2000) Evaluation of three rapid methods for detection of methicillin resistance in staphylococcus aureus. J Clin Microbiol 38(6):2170–2173

Louzir H, Melby PC, Ben Salah A, Marrakchi H, Aoun K, Ben Ismail R, Dellagi K (1998) Immunologic determinants of disease evolution in localized cutaneous leishmaniasis due to Leishmania major. J Infect Dis 177(6):1687–1695

Luzzatto L (2010) The rise and fall of the antimalarial Lapdap: a lesson in pharmacogenetics. Lancet 376(9742):739–741

Manns MP, McHutchison JG, Gordon SC, Rustgi VK, Shiffman M, Reindollar R, Goodman ZD et al (2001) Peginterferon alfa-2b plus ribavirin compared with interferon alfa-2b plus ribavirin for initial treatment of chronic hepatitis C: a randomised trial. Lancet 358(9286):958–965

Mansueto P, Vitale G, Di Lorenzo G, Rini GB, Mansueto S, Cillari E (2007) Immunopathology of leishmaniasis: an update. Int J Immunopathol Pharmacol 20(3):435–445

Marahatta SB, Gautam S, Dhital S, Pote N, Jha AK, Mahato R, Mishra S et al (2011) Katg (Ser 315 Thr) gene mutation in isoniazid resistant *Mycobacterium tuberculosis*. Kathmandu Univ Med J (KUMJ) 9(33):19–23

Marquet S, Schurr E (2001) Genetics of susceptibility to infectious diseases: tuberculosis and leprosy as examples. Drug Metab Dispos 29(4 Pt 2):479–483

Marsden PD (1986) Mucosal leishmaniasis (Espundia Escomel, 1911). Trans R Soc Trop Med Hyg 80(6):859–876

McIlleron H, Willemse M, Werely CJ, Hussey GD, Schaaf HS, Smith PJ, Donald PR (2009) Isoniazid plasma concentrations in a cohort of South African children with tuberculosis: implications for international pediatric dosing guidelines. Clin Infect Dis 48(11):1547–1553

Mebrahtu YB, Van Eys G, Guizani I, Lawyer PG, Pamba H, Koech D, Roberts C et al (1993) Human cutaneous Leishmaniasis caused by Leishmania donovani s.l. in Kenya. Trans R Soc Trop Med Hyg 87(5):598–601

Meddeb-Garnaoui A, Gritli S, Garbouj S, Ben Fadhel M, El Kares R, Mansour L, Kaabi B et al (2001) Association analysis of HLA-class II and class III gene polymorphisms in the susceptibility to mediterranean visceral leishmaniasis. Hum Immunol 62(5):509–517

Mehlotra RK, Henry-Halldin CN, Zimmerman PA (2009) Application of pharmacogenomics to malaria: a holistic approach for successful chemotherapy. Pharmacogenomics 10(3):435–449

Melby PC, Andrade-Narvaez FJ, Darnell BJ, Valencia-Pacheco G, Tryon VV, Palomo-Cetina A (1994) Increased expression of proinflammatory cytokines in chronic lesions of human cutaneous leishmaniasis. Infect Immun 62(3):837–842

Melby PC, Andrade-Narvaez F, Darnell BJ, Valencia-Pacheco G (1996) In situ expression of interleukin-10 and interleukin-12 in active human cutaneous leishmaniasis. FEMS Immunol Med Microbiol 15(2–3):101–107

Menzies D, Benedetti A, Paydar A, Martin I, Royce S, Pai M, Vernon A, Lienhardt C, Burman W (2009) Effect of duration and intermittency of rifampin on tuberculosis treatment outcomes: a systematic review and meta-analysis. PLoS Med 6(9):e1000146

Mitchison DA (1984) Drug resistance in mycobacteria. Br Med Bull 40(1):84–90

Mohamed HS, Ibrahim ME, Miller EN, Peacock CS, Khalil EA, Cordell HJ, Howson JM et al (2003) Genetic susceptibility to visceral leishmaniasis in the Sudan: linkage and association with IL4 and IFNGR1. Genes Immun 4(5):351–355

Mohamed HS, Ibrahim ME, Miller EN, White JK, Cordell HJ, Howson JM, Peacock CS et al (2004) SLC11A1 (formerly Nramp1) and susceptibility to visceral leishmaniasis in the Sudan. Eur J Hum Genet 12(1):66–74

Musa AM, Khalil EA, Mahgoub FA, Elgawi SH, Modabber F, Elkadaru AE, Aboud MH et al (2008) Immunochemotherapy of persistent post-kala-azar dermal leishmaniasis: a novel approach to treatment. Trans R Soc Trop Med Hyg 102(1):58–63

Nelson SD, Mitchell JR, Timbrell JA, Snodgrass WR, Corcoran GB 3rd (1976) Isoniazid and iproniazid: activation of metabolites to toxic intermediates in man and rat. Science 193(4256):901–903

Newport MJ, Huxley CM, Huston S, Hawrylowicz CM, Oostra BA, Williamson R, Levin M (1996) A mutation in the interferon-gamma-receptor gene and susceptibility to mycobacterial infection. N Engl J Med 335(26):1941–1949

Noedl H, Socheat D, Satimai W (2009) Artemisinin-resistant malaria in Asia. N Engl J Med 361(5):540–541

Nylen S, Sacks D (2007) Interleukin-10 and the pathogenesis of human visceral leishmaniasis. Trends Immunol 28(9):378–384

Ohno M, Yamaguchi I, Yamamoto I, Fukuda T, Yokota S, Maekura R, Ito M et al (2000) Slow N-acetyltransferase 2 genotype affects the incidence of isoniazid and rifampicin-induced hepatotoxicity. Int J Tuberc Lung Dis 4(3):256–261

Oksanen E, Ahonen AK, Tuominen H, Tuominen V, Lahti R, Goldman A, Heikinheimo P (2007) A complete structural description of the catalytic cycle of yeast pyrophosphatase. Biochemistry 46(5):1228–1239

Olivo-Diaz A, Debaz H, Alaez C, Islas VJ, Perez-Perez H, Hobart O, Gorodezky C (2004) Role of HLA class II alleles in susceptibility to and protection from localized cutaneous leishmaniasis. Hum Immunol 65(3):255–261

Olliaro PL, Guerin PJ, Gerstl S, Haaskjold AA, Rottingen JA, Sundar S (2005) Treatment options for visceral leishmaniasis: a systematic review of clinical studies done in India, 1980–2004. Lancet Infect Dis 5(12):763–774

Orrell C, Little F, Smith P, Folb P, Taylor W, Olliaro P, Barnes KI (2008) Pharmacokinetics and tolerability of artesunate and amodiaquine alone and in combination in healthy volunteers. Eur J Clin Pharmacol 64(7):683–690

Parikh S, Ouedraogo JB, Goldstein JA, Rosenthal PJ, Kroetz DL (2007) Amodiaquine metabolism is impaired by common polymorphisms in CYP2C8: implications for malaria treatment in Africa. Clin Pharmacol Ther 82(2):197–203

Parkin DP, Vandenplas S, Botha FJ, Vandenplas ML, Seifart HI, van Helden PD, van der Walt BJ, Donald PR, van Jaarsveld PP (1997) Trimodality of isoniazid elimination: phenotype and genotype in patients with tuberculosis. Am J Respir Crit Care Med 155(5):1717–1722

Paul WE, Seder RA (1994) Lymphocyte responses and cytokines. Cell 76(2):241–251

Payne DJ, Gwynn MN, Holmes DJ, Pompliano DL (2007) Drugs for bad bugs: confronting the challenges of antibacterial discovery. Nat Rev Drug Discov 6(1):29–40

Petzl-Erler ML, Belich MP, Queiroz-Telles F (1991) Association of mucosal leishmaniasis with HLA. Hum Immunol 32(4):254–260

Phillips-Howard PA, West LJ (1990) Serious adverse drug reactions to pyrimethamine-sulphadoxine, pyrimethamine-dapsone and to amodiaquine in Britain. J R Soc Med 83(2):82–85

Pizza M, Scarlato V, Masignani V, Giuliani MM, Arico B, Comanducci M, Jennings GT et al (2000) Identification of vaccine candidates against serogroup B meningococcus by whole-genome sequencing. Science 287(5459):1816–1820

Poland GA (1999a) Current paradoxes and changing paradigms in vaccinology. Vaccine 17(13–14):1605–1611

Poland GA (1999b) Immunogenetic mechanisms of antibody response to measles vaccine: the role of the HLA genes. Vaccine 17(13–14):1719–1725

Poland GA, Jacobson RM, Schaid D, Moore SB, Jacobsen SJ (1998) The association between HLA class I alleles and measles vaccine-induced antibody response: evidence of a significant association. Vaccine 16(19):1869–1871

Poland GA, Jacobson RM, Colbourne SA, Thampy AM, Lipsky JJ, Wollan PC, Roberts P, Jacobsen SJ (1999c) Measles antibody seroprevalence rates among immunized Inuit, Innu and Caucasian subjects. Vaccine 17(11–12):1525–1531

Pratlong F, Bastien P, Perello R, Lami P, Dedet JP (1995) Human cutaneous leishmaniasis caused by Leishmania donovani sensu stricto in Yemen. Trans R Soc Trop Med Hyg 89(4):398–399

Price RR, Viscount HB, Stanley MC, Leung KP (2007) Targeted profiling of oral bacteria in human saliva and in vitro biofilms with quantitative real-time PCR. Biofouling 23(3–4):203–213

Raymond JM, Dumas F, Baldit C, Couzigou P, Beraud C, Amouretti M (1989) Fatal acute hepatitis due to amodiaquine. J Clin Gastroenterol 11(5):602–603

Reed SG, Scott P (1993) T-cell and cytokine responses in leishmaniasis. Curr Opin Immunol 5(4):524–531

Rees DC, Congreve M, Murray CW, Carr R (2004) Fragment-based lead discovery. Nat Rev Drug Discov 3(8):660–672

Requena JM, Iborra S, Carrion J, Alonso C, Soto M (2004) Recent advances in vaccines for leishmaniasis. Expert Opin Biol Ther 4(9):1505–1517

Revets H, Marissens D, de Wit S, Lacor P, Clumeck N, Lauwers S, Zissis G (1996) Comparative evaluation of NASBA HIV-1 RNA Qt, AMPLICOR-HIV monitor, and QUANTIPLEX HIV RNA assay, three methods for quantification of human immunodeficiency virus type 1 RNA in plasma. J Clin Microbiol 34(5):1058–1064

Roca-Feltrer A, Carneiro I, Armstrong Schellenberg JR (2008) Estimates of the burden of malaria morbidity in Africa in children under the age of 5 years. Trop Med Int Health 13(6):771–783

Roederer MW, McLeoda H, Juliano JJ (2011) WHO Bulletin of World Health Organization: can pharmacogenomics improve malaria drug policy? http://www.who.int/bulletin/volumes/89/11/11-087320/en/index.html

Rogers PD, Stiles JK, Chapman SW, Cleary JD (2000) Amphotericin B induces expression of genes encoding chemokines and cell adhesion molecules in the human monocytic cell line THP-1. J Infect Dis 182(4):1280–1283

Roy CR, Mocarski ES (2007) Pathogen subversion of cell-intrinsic innate immunity. Nat Immunol 8(11):1179–1187

Roy B, Chowdhury A, Kundu S, Santra A, Dey B, Chakraborty M, Majumder PP (2001) Increased risk of antituberculosis drug-induced hepatotoxicity in individuals with glutathione S-transferase M1 'null' mutation. J Gastroenterol Hepatol 16(9):1033–1037

Roy PD, Majumder M, Roy B (2008) Pharmacogenomics of anti-TB drugs-related hepatotoxicity. Pharmacogenomics 9(3):311–321

Sachidanandam R, Weissman D, Schmidt SC, Kakol JM, Stein LD, Marth G, Sherry S et al (2001) A map of human genome sequence variation containing 1.42 million single nucleotide polymorphisms. Nature 409(6822):928–933

Sacks D, Noben-Trauth N (2002) The immunology of susceptibility and resistance to Leishmania major in mice. Nat Rev Immunol 2(11):845–858

Saha S, Mondal S, Ravindran R, Bhowmick S, Modak D, Mallick S, Rahman M et al (2007) IL-10- and TGF-beta-mediated susceptibility in kala-azar and post-kala-azar dermal leishmaniasis: the significance of amphotericin B in the control of Leishmania donovani infection in India. J Immunol 179(8):5592–5603

Salhi A, Rodrigues V Jr, Santoro F, Dessein H, Romano A, Castellano LR, Sertorio M et al (2008) Immunological and genetic evidence for a crucial role of IL-10 in cutaneous lesions in humans infected with Leishmania braziliensis. J Immunol 180(9):6139–6148

Salih MA, Ibrahim ME, Blackwell JM, Miller EN, Khalil EA, ElHassan AM, Musa AM, Mohamed HS (2007) IFNG and IFNGR1 gene polymorphisms and susceptibility to post-kala-azar dermal leishmaniasis in Sudan. Genes Immun 8(1):75–78

Samaranayake TN, Dissanayake VH, Fernando SD (2008) Clinical manifestations of cutaneous leishmaniasis in Sri Lanka – possible evidence for genetic susceptibility among the Sinhalese. Ann Trop Med Parasitol 102(5):383–390

Sarma GR, Kailasam S, Mitchison DA, Nair NG, Radhakrishna S, Tripathy SP (1975) Studies of serial plasma isoniazid concentrations with different doses of a slow-release preparation of isoniazid. Tubercle 56(4):314–323

Sarma GR, Immanuel C, Kailasam S, Narayana AS, Venkatesan P (1986) Rifampin-induced release of hydrazine from isoniazid. A possible cause of hepatitis during treatment of tuberculosis with regimens containing isoniazid and rifampin. Am Rev Respir Dis 133(6):1072–1075

Schaaf HS, Parkin DP, Seifart HI, Werely CJ, Hesseling PB, van Helden PD, Maritz JS, Donald PR (2005) Isoniazid pharmacokinetics in children treated for respiratory tuberculosis. Arch Dis Child 90(6):614–618

Schlagenhauf P, Petersen E (2008) Malaria chemoprophylaxis: strategies for risk groups. Clin Microbiol Rev 21(3):466–472

Schriefer A, Schriefer AL, Goes-Neto A, Guimaraes LH, Carvalho LP, Almeida RP, Machado PR et al (2004) Multiclonal Leishmania braziliensis population structure and its clinical implication in a region of endemicity for American tegumentary leishmaniasis. Infect Immun 72(1):508–514

Selkon JB, Fox W, Gangadharam PR, Ramachandran K, Ramakrishnan CV, Velu S (1961) Rate of inactivation of isoniazid in south Indian patients with pulmonary tuberculosis. 2. Clinical implications in the treatment of pulmonary tuberculosis with isoniazid either alone or in combination with PAS. Bull World Health Organ 25:779–792

Sharma SK, Balamurugan A, Saha PK, Pandey RM, Mehra NK (2002) Evaluation of clinical and immunogenetic risk factors for the development of hepatotoxicity during antituberculosis treatment. Am J Respir Crit Care Med 166(7):916–919

Sharma NL, Mahajan VK, Kanga A, Sood A, Katoch VM, Mauricio I, Singh CD et al (2005) Localized cutaneous leishmaniasis due to Leishmania donovani and Leishmania tropica: preliminary findings of the study of 161 new cases from a new endemic focus in Himachal Pradesh, India. Am J Trop Med Hyg 72(6):819–824

Simon T, Becquemont L, Mary-Krause M, de Waziers I, Beaune P, Funck-Brentano C, Jaillon P (2000) Combined glutathione-S-transferase M1 and T1 genetic polymorphism and tacrine hepatotoxicity. Clin Pharmacol Ther 67(4):432–437

Siqueira HR, Freitas FA, Oliveira DN, Barreto AM, Dalcolmo MP, Albano RM (2009a) Clinical evolution of a group of patients with multidrug-resistant TB treated at a referral center in the city of Rio De Janeiro, Brazil. J Bras Pneumol 35(1):54–62

Siqueira HR, Freitas FA, Oliveira DN, Barreto AM, Dalcolmo MP, Albano RM (2009b) Isoniazid-resistant mycobacterium tuberculosis strains arising from mutations in two different regions of the katG gene. J Bras Pneumol 35(8):773–779

Sivula T, Salminen A, Parfenyev AN, Pohjanjoki P, Goldman A, Cooperman BS, Baykov AA, Lahti R (1999) Evolutionary aspects of inorganic pyrophosphatase. FEBS Lett 454(1–2):75–80

Speers DJ (2006) Clinical applications of molecular biology for infectious diseases. Clin Biochem Rev 27(1):39–51

Stober CB, Lange UG, Roberts MT, Gilmartin B, Francis R, Almeida R, Peacock CS, McCann S, Blackwell JM (2006) From genome to vaccines for leishmaniasis: screening 100 novel vaccine candidates against murine Leishmania major infection. Vaccine 24(14):2602–2616

Strange RC, Jones PW, Fryer AA (2000) Glutathione S-transferase: genetics and role in toxicology. Toxicol Lett 112–113:357–363

Tettelin H, Saunders NJ, Heidelberg J, Jeffries AC, Nelson KE, Eisen JA, Ketchum KA et al (2000) Complete genome sequence of *Neisseria meningitidis* serogroup B strain MC58. Science 287(5459):1809–1815

Tostmann A, Boeree MJ, Aarnoutse RE, de Lange WC, van der Ven AJ, Dekhuijzen R (2008) Antituberculosis drug-induced hepatotoxicity: concise up-to-date review. J Gastroenterol Hepatol 23(2):192–202

Tuberculosis Research Centre Madras, study (1970) A controlled comparison of a twice-weekly and three once-weekly regimens in the initial treatment of pulmonary tuberculosis. Bull World Health Organ 43(1):143–206

Tuberculosis Research Centre Madras, study (1973) Controlled comparison of oral twice-weekly and oral daily isoniazid plus PAS in newly diagnosed pulmonary tuberculosis. Br Med J 2(5857):7–11

Valderramos SG, Fidock DA (2006) Transporters involved in resistance to antimalarial drugs. Trends Pharmacol Sci 27(11):594–601

Van Soolingen D (2001) Molecular epidemiology of tuberculosis and other mycobacterial infections: main methodologies and achievements. J Intern Med 249(1):1–26

Venter JC, Adams MD, Myers EW, Li PW, Mural RJ, Sutton GG, Smith HO et al (2001) The sequence of the human genome. Science 291(5507):1304–51

Vidal S, Tremblay ML, Govoni G, Gauthier S, Sebastiani G, Malo D, Skamene E et al (1995) The Ity/Lsh/Bcg locus: natural resistance to infection with intracellular parasites is abrogated by disruption of the Nramp1 gene. J Exp Med 182(3):655–666

Vuilleumier N, Rossier MF, Chiappe A, Degoumois F, Dayer P, Mermillod B, Nicod L, Desmeules J, Hochstrasser D (2006) CYP2E1 genotype and isoniazid-induced hepatotoxicity in patients treated for latent tuberculosis. Eur J Clin Pharmacol 62(6):423–429

Weiner M, Burman W, Vernon A, Benator D, Peloquin CA, Khan A, Weis S et al (2003) Low isoniazid concentrations and outcome of tuberculosis treatment with once-weekly isoniazid and rifapentine. Am J Respir Crit Care Med 167(10):1341–1347

Weiner M, Peloquin C, Burman W, Luo CC, Engle M, Prihoda TJ, Mac Kenzie WR et al (2010) Effects of tuberculosis, race, and human gene SLCO1B1 polymorphisms on rifampin concentrations. Antimicrob Agents Chemother 54(10):4192–4200

Wells JA, McClendon CL (2007) Reaching for high-hanging fruit in drug discovery at protein-protein interfaces. Nature 450(7172):1001–1009

White NJ (2008) Qinghaosu (artemisinin): the price of success. Science 320(5874):330–334

Wilkins JJ, Langdon G, McIlleron H, Pillai G, Smith PJ, Simonsson US (2011) Variability in the population pharmacokinetics of isoniazid in South African tuberculosis patients. Br J Clin Pharmacol 72(1):51–62

Yamada S, Tang M, Richardson K, Halaschek-Wiener J, Chan M, Cook VJ, Fitzgerald JM et al (2009) Genetic variations of NAT2 and CYP2E1 and isoniazid hepatotoxicity in a diverse population. Pharmacogenomics 10(9):1433–1445

Yeung S, Pongtavornpinyo W, Hastings IM, Mills AJ, White NJ (2004) Antimalarial drug resistance, artemisinin-based combination therapy, and the contribution of modeling to elucidating policy choices. Am J Trop Med Hyg 71(2 Suppl):179–186

Yue J, Peng RX, Yang J, Kong R, Liu J (2004) CYP2E1 mediated isoniazid-induced hepatotoxicity in rats. Acta Pharmacol Sin 25(5):699–704

Yusof W, Gan SH (2009) High prevalence of CYP2A6*4 and CYP2A6*9 alleles detected among a Malaysian population. Clin Chim Acta 403(1–2):105–109

Zhao H, Li R, Wang Q, Yan Q, Deng JH, Han D, Bai Y, Young WY, Guan MX (2004) Maternally inherited aminoglycoside-induced and nonsyndromic deafness is associated with the novel C1494T mutation in the mitochondrial 12s rRNA gene in a large Chinese family. Am J Hum Genet 74(1):139–152

Zhou H, Xu M, Huang Q, Gates AT, Zhang XD, Castle JC, Stec E et al (2008) Genome-scale RNAi screen for host factors required for HIV replication. Cell Host Microbe 4(5):495–504

Zijlstra E, Zijlstra EE, Musa AM, Khalil EA, el-Hassan IM, el-Hassan AM (2003) Post-kala-azar dermal leishmaniasis. Lancet Infect Dis 3(2):87–98

Zilber LA, Bajdakova ZL, Gardasjan AN, Konovalov NV, Bunina TL, Barabadze EM (1963) The prevention and treatment of isoniazid toxicity in the therapy of pulmonary tuberculosis. 2. An assessment of the prophylactic effect of pyridoxine in low dosage. Bull World Health Organ 29:457–481

Pharmacogenomics of Viral Diseases

28

Shailendra Dwivedi, Suraj Singh Yadav, Manish Kumar Singh, Shailja Shukla, Sanjay Khattri, and Kamlesh Kumar Pant

Abstract

Viral diseases are leading cause of deaths worldwide as WHO report suggests that hepatitis A virus (HAV) infects more than 80 % of the population of many developing countries. Viral hepatitis B (HBV) affects an estimated 360 million people, whereas hepatitis C affects 123 million people worldwide, and last but not least, at current, India has an HIV/AIDS population of approximately 2.4 million people and more than 30 million in whole world and now it has become a reason for 1.8 million death globally; thus, millions of people still struggle for their lives.

The progress in medical science has made it possible in overcoming the various fatal diseases such as small pox, chicken pox, dengue, etc., but human immunodeficiency viruses, influenza, and hepatitis virus have renewed challenge surprisingly. The obstacles and challenges in therapy include existence of antibiotic resistance strains of common organisms due to overuse of antibiotics, lack of vaccines, adverse drug reaction, and last but not least the susceptibility concerns. Emergence of pharmacogenomics and pharmacogenetics has shown some promises to take challenges. The discovery of human genome project has opened new vistas to understand the behaviors of genetic makeup in development and progression of diseases and treatment in various viral diseases. Current and previous decade have been engaged in making repositories of polymorphisms (SNPs) of various genes including drug-metabolizing enzymes, receptors, inflammatory cells related with immunity, and antigen-presenting cells, along with the prediction of risks. The genetic makeup alone is most likely an adequate way to handle the therapeutic decision-making process for

S. Dwivedi • S.S. Yadav • M.K. Singh • S. Shukla
S. Khattri (✉) • K.K. Pant
Department of Pharmacology and Therapeutics, CSM Medical University, Erstwhile-King George Medical College, Lucknow, Uttar Pradesh, India
e-mail: pharmacsmmu@gmail.com; tarang2016@gmail.com

D. Barh et al. (eds.), *Omics for Personalized Medicine*,
DOI 10.1007/978-81-322-1184-6_28, © Springer India 2013

previous regimen failure. With the introduction of new antiviral therapeutic agents, a significant improvement in progression and overall survival has been achieved, but these drugs have shown several adverse responses in some individuals, so the success is not up to the expectations. Research and acquisition of new knowledge of pharmacogenomics may help in overcoming the prevailing burden of viral diseases. So it will definitely help in selecting the most effective therapeutic agents, effective doses, and drug response for the individuals. Thus, it will be able to transform the laboratory research into the clinical bench side and will also help in understanding the pathogenesis of viral diseases with drug action, so the patients will be managed more properly and finally become able to fulfill the promise of the future.

1 Introduction

Infectious diseases like viral are the number one cause of premature death in the world. A large number of antiviral drugs are used clinically, but their effectiveness is being eroded by the development of resistance and concerns over safety. Research in this area has led to fundamental discoveries, which have helped our understanding of the reasons why individuals differ in the way they handle drugs and ultimately in the way they respond to drugs, either in terms of efficacy or toxicity. However, not much of this knowledge has been translated into clinical practice, most drug–gene associations that have some evidence of clinical validity have not progressed to clinical settings. Advances in genomics since 2000, after exploration of the human genome, have provided us with unprecedented opportunities to understand variability in drug responses and the opportunity to incorporate this into patient cure. The needs for newer and safer antiviral drugs remain continued. Nowadays an explosion in genome sequencing of both viral pathogens and their human host is helping us to understand the complex interactions involved in the infection process. The pharmaceutical industry is exploiting this information to identify better targets for treating viral diseases and to improve understanding of patient responses to a drug so-called pharmacogenomics. Common human diseases caused by viruses include the common cold, the flu, chickenpox, and cold sores. Serious diseases such as Ebola, hepatitis, and AIDS are also caused by viruses. Many viruses cause little or no disease and are said to be "benign." The more harmful viruses are described as virulent. Viruses cause different diseases depending on the types of cell that they infect. Some viruses can cause lifelong or chronic infections where the viruses continue to reproduce in the body despite the host's defense mechanisms, as it is common in hepatitis B virus, hepatitis C virus, and HIV infections.

1.1 Overview of Viral Diseases Burden on Health

Hepatitis A occurs sporadically and epidemically worldwide, with a tendency to cyclic recurrences (Lemon 1994). Epidemics are uncommon in developing countries where adults are generally immune. Poor sanitation and hygiene conditions in different parts of the world leave large segments of the population susceptible to infection, and outbreaks may result whenever the virus is introduced (Lemon 1994; Melnick 1995; Shapiro and Margolis 1993). Worldwide, HAV infections account for 1.4 million cases annually (Viral Hepatitis Prevention Board 1997).

Hepatitis B virus (HBV) infection is a serious global health problem, with 2 billion people infected worldwide and 350 million suffering from chronic HBV infection. The tenth leading cause of death worldwide, HBV infections result in 500,000–1.2 million deaths per year caused by chronic hepatitis, cirrhosis, and hepatocellular carcinoma; the last accounts for 320,000 deaths per year (WHO Report.3 1997; WHO fact sheet 2000). Hepatitis B virus (HBV) infection is a major global public health problem. Of the approximately 2 billion people who have been infected worldwide, more than 350 million act as chronic carriers of HBV (WHO fact sheet 2000). Approximately 15–40 % of infected patients will develop cirrhosis, liver failure, or hepatocellular carcinoma (Lok 2002). HBV infection accounts for 500,000–1.2 million deaths each year (Mahoney 1999; Lee 1997) and is the tenth leading cause of death worldwide.

Since its discovery in 1989, hepatitis C virus (HCV) has been recognized as a major cause of chronic liver disease worldwide. The most recent WHO estimate of the prevalence of HCV infection is 2 %, representing 123 million people (Perz et al. 2004). HCV is the leading cause of liver transplantation in developed countries and the most common chronic blood-borne infection in the USA.

There is a wide range of prevalence estimates among developing countries and generally less data available to validate assumptions about the burden of disease than in the developed world. This range in prevalence is reflected in reviewing the estimates from developing countries that are among the world's most populous nations (Population Reference Bureau 2004). China, whose citizens account for one-fifth of the world's population, has a reported seroprevalence of 3.2 % (Xia et al. 1996). In India, which holds an additional one-fifth of the world's population, one community-based survey reported an overall rate of 0.9 % (Chowdhury et al. 2003).

Countries with the highest reported prevalence rates are located in Africa and Asia; areas with lower prevalence include the industrialized nations in North America, Northern and Western Europe, and Australia. Populous nations in the developed world with relatively low rates of HCV seroprevalence include Germany (0.6 %) (Palitzsch et al. 1999), Canada (0.8 %) (Zou et al. 2000), France (1.1 %) (Desenclos 2000), and Australia (1.1 %) (Law et al. 2003; Australian Census 2001). Low, but slightly higher, seroprevalence rates have been reported in the USA (1.8 %) (Alter et al. 1999), Japan (1.5–2.3 %) (Ohshima et al. 2000), and Italy (2.2 %).

Human papillomavirus (HPV) causes cervical cancer, the second biggest cause of female cancer mortality worldwide. Estimates of the number of cervical cancer deaths are around 250,000 per year. The prevalence of genital HPV infection in the world is around 440 million. There are over 100 genotypes of HPV, 40 of which infect human mucosal areas of the upper digestive tract and genital tract. The majority of adenocarcinomas of the cervix and of squamous cell cancers (SCC) of the vulva, vagina, penis, and anus are caused by HPV-16 and HPV-18 (together accounting for about 70 % of cases globally), the remaining 30 % being due to other high-risk HPV types (such as HPV-31, HPV-33, HPV-35, HPV-39, HPV-45, HPV-51, HPV-66). The relative importance of different high-risk types varies between countries and regions, but type 16 has the greatest contribution to cervical cancer in all regions. HPV is also associated with other cancers of the anus, head and neck, and rarely, recurrent respiratory papillomatosis in children.

About 500,000 cases of cervical cancer are estimated to occur each year, over 80 % of which occur in developing countries, where neither population-based routine screening (e.g., Papanicolaou smear test) nor optimal treatment is available. The highest estimated incidence rates of cervical cancer occur in Africa, Central and South America, and Asia. Epidemiological studies in the USA have reported that 75 % of the 15–50-year-old population is infected with genital HPV over their lifetime, 60 % with transient infection, 10 % with persistent infection (confirmed by detection of HPV DNA in genital samples), 4 % with mild cytological signs, and 1 % with clinical lesions.

HPV belongs to the family Papovaviridae. These are small nonenveloped icosahedral viruses with an 8 kbp long double-stranded circular DNA genome. The papillomavirus genome comprises early and late genes that encode early proteins E1–E7 and late proteins L1–L2. The early proteins are nonstructural proteins involved in replication and transcription of the genome (E1–E5) or in host cell tumoral transformation (E6 and E7), whereas L1 and L2 are the structural capsid proteins of the virion. The low-grade cervical dysplasias correspond to productively infected cells that actively shed virus, whereas high-grade dysplasias and cancers do not produce virions: viral gene expression in these cells is limited to the E6 and E7 oncogenes that are transcribed from randomly integrated viral DNA. The E7 protein is thought to induce cell proliferation and disrupt the cell cycle regulation by inactivation of the Rb family proteins, whereas E6 blocks cell apoptosis by directing the p53 tumor suppressor protein to the proteasome.

Prophylactic HPV vaccine candidates are based on recombinant capsid protein L1 and aim to elicit neutralizing antiviral antibodies to protect against infection, while therapeutic vaccine candidates are based on viral oncogenic proteins E6 and E7, with or without L1, and aim to induce cell-mediated immune responses to eliminate the transformed tumor cells. The most advanced and promising approach for a prophylactic vaccine involves the use of noninfectious recombinant virus-like particles (VLPs) which self-assemble spontaneously from pentamers of the L1 capsid protein. Two prophylactic vaccine candidates are at the level of phase III clinical evaluation and the companies have filed for licensure. GSK is focusing on a bivalent HPV-16, HPV-18 VLP vaccine candidate and Merck is developing a tetravalent vaccine based on VLPs from HPV-6, HPV-11, HPV-16, and HPV-18. Both showed very high efficacy in proof-of-principle studies and the manufacturers have announced results showing almost 100 % protection against high-grade cervical cancer precursors caused by HPV types 16 and 18 in women aged 16–25 years.

Most HPV infections in young females are temporary and have little long-term significance. Seventy percent of infections are gone in 1 year and 90 % in 2 years. However, when the infection persists – in 5–10 % of infected women – there is high risk of developing precancerous lesions of the cervix, which can progress to invasive cervical cancer. This process usually takes 10–15 years, providing many opportunities for detection and treatment of the precancerous lesion. Progression to invasive cancer can be almost always prevented when standard prevention strategies are applied, but the lesions still cause considerable burden necessitating preventive surgeries, which do in many cases involve loss of fertility.

In more developed countries, cervical screening using a Papanicolaou (Pap) test or liquid-based cytology is used to detect abnormal cells that may develop into cancer. If abnormal cells are found, women are invited to have a colposcopy. During a colposcopic inspection, biopsies can be taken and abnormal areas can be removed with a simple procedure, typically with a cauterizing loop or, more commonly in the developing world, by freezing (cryotherapy). Treating abnormal cells in this way can prevent them from developing into cervical cancer.

Pap smears have reduced the incidence and fatalities of cervical cancer in the developed world, but even so, there were 11,000 cases and 3,900 deaths in the USA in 2008. Cervical cancer has substantial mortality in resource-poor areas; worldwide, there are an estimated 490,000 cases and 270,000 deaths each year (Kahn 2009).

Several influenza epidemics in the twentieth century caused millions of deaths worldwide, including the worst epidemic in American history, the Spanish influenza outbreak that killed more than 500,000 in 1918. Seasonal influenza is an acute viral infection caused by an influenza virus. There are three types of seasonal influenza – A, B, and C. Type A influenza viruses are further typed into subtypes according to different kinds and combinations of virus surface proteins. Among many subtypes of influenza A viruses, currently influenza A (H1N1) and A (H3N2) subtypes are circulating among humans. Influenza viruses circulate in every part of the world. Type C influenza cases occur much less frequently than A and B. That is why only influenza A and B viruses are included in seasonal influenza vaccines.

Avian influenza in humans as H5N1 avian influenza is an infectious disease of birds that can be spread to people but is difficult to transmit from person to person. Almost all people with H5N1 infection have had close contact with infected birds or H5N1-contaminated environments.

When people do become infected, the mortality rates get up to 60 %.

Most swine influenza viruses (SIVs) do not cause disease in humans. However, some countries have reported cases of human infection with SIVs. Most of these human infections have been mild and the viruses have not spread further to other people. The H1N1 virus that caused the influenza pandemic in 2009–2010, thought to have originated in swine, is an example of SIV that was able to spread easily among people and also cause disease.

HIV is the human immunodeficiency virus. It is the virus that can lead to acquired immune deficiency syndrome, or AIDS. CDC estimates that about 56,000 people in the United States contracted HIV in 2006. There are two types of HIV, HIV-1 and HIV-2. In the USA, unless otherwise noted, the term "HIV" primarily refers to HIV-1. Both types of HIV damages a person's body by destroying specific blood cells, called CD4+ T cells, which are crucial to helping the body immune defense. AIDS is the late stage of HIV infection, when a person's immune system gets severely damaged and feels difficulty in fighting against diseases and certain cancers. Before the development of certain medications, people with HIV could progress to AIDS in just a few years. At this time, there is no cure for HIV infection. Despite major advances in diagnosing and treating HIV infection, in 2007, 35,962 cases of AIDS were diagnosed and 14,110 deaths among people living with HIV were reported in the USA. The above-discussed diseases are continuously threatening the whole world and insisting to search and explore new medicine concept so that management would be possible and people may live long disease-free and free from economical burden.

2 Genome and Proteome Complexity of HIV

Understanding the incredibly complex biology of the HIV virus is essential for building effective diagnostics and drugs. HIV belongs to a class of viruses known as retroviruses because it contains ribonucleic acid (RNA) as its genetic material. After the HIV virus infects a cell, it uses an enzyme called reverse transcriptase to convert its RNA into DNA so that it can replicate itself using the host cell's own replication machinery. The HIV replication cycle begins with fusion of the virus to the host cell surface, which begins the influx of viral proteins into the host cell. After viral DNA is formed by reverse transcription and integrated into the host DNA, new viral RNA is used as genomic RNA to make viral proteins, which travel to the cell surface to form a new HIV virus. This vicious cycle rapidly produces several billion new viruses every day in persons infected with HIV and is further complicated by the ability of reverse transcriptase to mutate, causing new strains of HIV to develop in infected individuals. HIV targets the immune system directly by infecting CD4+ lymphocytes, which also leads to the systematic degradation of the immune system because CD4+ cells are pivotal in helping immune responses (Blum 2005). The constant process of evolution and replication in the HIV virus creates incredible stress for the immune system and has been one of the reasons why HIV has been especially difficult for medical researchers to combat. Drugs against HIV are called antiretroviral drugs because HIV is considered a retrovirus as previously described and the drugs fall into three classes consistent with the underlying biology of the virus: (1) nucleoside reverse transcriptase inhibitors (NRTIs), (2) non-nucleoside reverse transcriptase inhibitors (NNRTIs), and (3) protease inhibitors (PIs) (Pirmohamed 2001). Reverse transcriptase inhibitors directly inhibit the reproductive capacity of the HIV virus because of the mandatory role the reverse transcriptase enzyme plays in viral reproduction. NRTIs contain faulty versions of the nucleotides used by reverse transcriptase to convert RNA to DNA, causing improper build of the new DNA so that HIV's genetic material cannot be incorporated into the healthy genetic material of the host cell. On the other hand, NNRTIs work by attaching themselves to reverse transcriptase and prevent the enzyme from converting RNA to DNA. After viral RNA is translated into a polypeptide sequence, the sequence is assembled into a long chain that includes the proteins like reverse transcriptase and protease.

3 Obstacles in Drug Resigning or Therapy Against Viral Diseases

Viral diseases affect hundreds of millions of people worldwide, resulting in a devastating toll on human health and socioeconomic development. Along with the emergence of newly recognized human pathogens (the SARS corona virus, the recent influenza viruses H5N1 and H1N1), the ever-increasing incidence of chronic viral infections caused by HIV and hepatitis B and C viruses continues to increase the global burden of infectious diseases (Haagmans et al. 2009; De Clercq 2004). Vaccines have been developed for some of the most important viral pathogens. Although vaccines against HIV (Rerks-Ngarm et al. 2009) and hepatitis C virus (Wedemeyer et al. 2009) are in clinical phases III and II, respectively, there is still little prospect of effective vaccines against these agents. Enormous burden and death due to these viral diseases compel the scientists/ virologists to think forcefully to overcome such mortality globally.

There are around 40 antiviral compounds in clinical use targeting various viral diseases (over half of these drugs are being used in the treatment of patients with HIV infection), while there is no treatment for most acute infections, such as the ones that cause severe illnesses, including hemorrhagic fever, encephalitis, and even cancer. Most of the available drugs are of limited efficacy and cause severe side effects (Dykxhoorn and Lieberman 2006). HIV protease, which cleaves Gag and Gag–Pro–Pol polyproteins at ten varied sites necessary for the maturation of virus (Kohl et al. 1988), is a major therapeutic target for antiviral drugs. In the last 20 years, structure-based drug discovery efforts have led to the development of nine approved competitive active site protease inhibitors (PIs). These inhibitors are the most potent anti-HIV drugs and essential components of the highly active antiretroviral therapy (HAART) (Bartlett et al. 2001). The development of drug resistance is a major reason for the failure of protease inhibitor therapy. The virus accumulates many mutations within the protease that prevent PIs from binding to the protease. More than half the residues within the protease mutate in different combinations and lead to drug resistance (Wu et al. 2003).

Drug resistance is defined, in a clinical setting, as the point at which administration of the drug can no longer safely treat the disease state due to an induced change in the drug target or an inability of the drug to reach the target. With an antimicrobial agent, clinical resistance occurs when the minimum inhibitory concentration (MIC) of the drug, for a given microbial strain, exceeds that concentration of drug that can safely be administered. Resistance to a drug can arise by (1) mutation of the gene (or gene cluster); (2) by acquisition of extrachromosomal DNA, or a transposable plasmid, that carries the resistance gene or genes; (3) upregulation of the target; or (4) upregulation of an efflux mechanism. One study has recently suggested (Peet 2010) an approach to combating drug resistance that involves the selection of resilient drug targets (Lefebvre and Schiffer 2008) that are evolutionally constrained and the development of robust drugs (Nalam and Schiffer 2008) that are less susceptible to the development of resistance. Importantly, antiviral chemotherapy is plagued by the rapid development of drug resistance strains, resulting from the high rate of replication of viruses combined with the low fidelity with which they replicate their genomes.

The biggest challenge is to deal with the variations among HIV-1 clades. HIV-1 is capable of establishing latent infection in the early phases of infection (Mehandru et al. 2004). HIV-1 can also escape from CTL sequestration of infected cells in the central nervous system owing to infrequent access of T cells in the CNS (McMichael and Rowland-Jones 2001). Increased susceptibility of vaccinated individuals might be owing to vector-specific immune responses, which make T cells more prone to HIV infection. However, recent reports demonstrate no role of vector-specific CD4+ T cells in an increasing susceptibility to HIV infection (Hutnick et al. 2010). The mechanism of T-cell exhaustion and antiviral control would be important to assess in vaccine studies to ensure the generation and maintenance of memory T cells (Trautmann et al. 2006). It has been reported that

HIV-specific CD4+ T cells are induced during acute HIV infection, but their helper function gets compromised (Malhotra et al. 2001). The large sequence diversity of envelope glycoprotein is another barrier in the development of an HIV-1 vaccine.

Another major hurdle is to achieve the balanced immune response against vaccine candidates. It is likely that epitopes that are less well conserved between clades but conserved within a clade are capable of eliciting clade-specific, rather than cross-reactive, NAbs. HIV-1 clade C viruses seem to be more sensitive to neutralization than HIV-1 variants from other clades (Van Gils et al. 2010). HIV-1 clade C infection was most prevalent among elite neutralizers (Simek et al. 2009). Vaccine candidates using Env as the immunogen have not been able to produce bNAbs. bNAbs with long HCDR3 regions seem to be polyreactive for non-HIV antigens. Lack of bNAb generation has been attributed to autoreactivity of bNAbs directed toward Env. Autoreactivity of anti-Env bNAbs induces central tolerance mechanisms and results in removal of B cells producing antibody with the same epitope specificity. The heavy chain of 2F5 bNAb was cloned in B cells in mice models and resulted in the loss of B cells expressing mature 2F5 IgM in mice (Verkoczy et al. 2010). Another hypothesis emphasizes that bNAbs are not immunogenic enough to induce proliferation in B cells followed by fine-tuning of the B-cell receptor (antibody coding) through somatic hypermutation. The fine-tuning of the B-cell receptor gene is necessary to maximize the specificity of antibody to its antigen (Harro et al. 2009). This raises concerns about the approach of using conserved Env epitopes as HIV-1 vaccine candidates for bNAb generation.

Natural infection with HIV does not result in virus clearance by the host immune system and the development of natural immunity to reinfection. In spite of intense and sustained immune responses by both the humoral and cell-mediated defenses, HIV is able to resist eradication and continues depleting CD4+ T cells, which eventually leads to clinical progression to AIDS. There even is evidence that superinfection with a second HIV isolate can readily occur in HIV-infected persons, leading to the emergence of recombinant virus variants and generating increased virus diversity (Mc Cutchan et al. 2005). HIV-1 integrates itself as latent proviral DNA into the genome of long-lived memory CD4+ T cells, which provide a persistent reservoir of the virus that escapes immune surveillance (Peterlin and Trono 2003). It has been calculated that it would take up to 60 years to eradicate a reservoir of as few as 1×10E5 latently infected cells. The window of opportunity for an HIV vaccine is therefore narrowly limited to the very early stages of infection, before the virus can seed the lymphoid organs in mucosal tissues (Gallo 2005). HIV also has developed multiple mechanisms to circumvent the host immune responses including its ability to downregulate the major histocompatibility complex (MHC) class I molecules and by doing so to minimize its recognition by CTL, as well as its high genetic evolution rates, which allows it to evade immune responses through the emergence of viral CTL (Feeney et al. 2005), and neutralizing antibody escape variants. Another difficulty with the development of an effective HIV vaccine stems from the fact that the virus envelope glycoprotein conceals its conserved receptor- and co-receptor-binding sites in crypts that are masked by the hypervariable loops of the molecule and by glycan residues (Wei et al. 2003). Neutralizing antibodies induced in response to gp120 are primarily targeted to the hypervariable loops of the molecule, and only rarely do they recognize the receptor-binding sites, which makes it hard to generate broadly cross-reactive neutralizing antibodies against primary virus isolates from patients (Yang et al. 2004). Lack in efficacy of antibody responses raised by monomeric gp120 vaccines in protection against HIV infection has been proven beyond any doubt in the world's first two phase III clinical trials of AIDS vaccines (Mascola et al. 1996), although neutralizing antibodies administered passively to nonhuman primates can provide protection against experimental SHIV infection (Ferrantelli et al. 2002). However, contrary to laboratory-adapted virus strains, which use CXCR-4 as a co-receptor ("X4" strains), and against which protection in chimpanzees could readily be achieved by inducing neutralizing antibodies targeted to the hypervariable V3 loop (Girard et al. 1995), primary virus isolates, which use CCR-5 as a co-receptor ("R5" strains), are difficult to neutralize, which casts doubt on the possibility for a vaccine to elicit protection against infection by the induction of neutralizing antibodies alone. In view of all these problems, recent vaccine approaches have focused on the induction of cellular immune responses (McMichael and Hanke 2002). The induction of a cellular immune response against HIV, especially a CD8+ CTL response, although not being able to provide sterilizing immunity and protection from infection, should hopefully enable vaccines to control virus replication following infection, reduce their virus load, slow down their progression toward disease, and reduce the probability of secondary transmission of the virus. However, some viruses, for some reasons, are not fully amenable to this approach, such as influenza, retroviruses, herpes viruses, the slow viruses, rhinoviruses, and arboviruses. Obstacles to the use of vaccines include (1) multiplicity of serotypes, e.g., rhinoviruses and togaviruses; (2) antigenic change, e.g., influenza and retroviruses; and (3) latent infections.

Drug resistance is also a major obstacle in the treatment of hepatitis C virus (HCV). The essential HCV NS3/4A protease is an attractive therapeutic target responsible for cleaving at least four sites along the viral polyprotein. Many protease inhibitors are currently in clinical trials; however, multidrug resistance is widespread and arises very quickly. Certain studies clearly compared the co-crystal structures of substrate with co-crystal structures of inhibitor complexes and show that, as in the case of HIV-1 protease (Chellappan et al. 2007; Altman et al. 2008), primary drug resistance occurs in HCV NS3/4A where the inhibitors protrude away from the substrate envelope. Similarly prolonged therapy against HBV with lamivudine is associated with an increased incidence of viral resistance. The low efficacy, undesirable side effects, and occurrence of resistance to HBV mutations remain the major obstacles in their clinical application in treating HBV infection (Lau et al. 1997; Hoofnagle and di Bisceglie 1997). The need for alternative therapeutic approaches has provided the impetus to develop novel therapeutic reagents for inhibiting HBV replication.

3.1 Antiretroviral Resistance Assays

The purpose of resistance testing is to make available the information to assist in the selection of the antiretroviral regimen(s) and more likely achieve and maintain viral suppression. All guidelines come to the result that HIV drug resistance testing should be performed when a HIV-infected person enters into clinical care, whether he is treated immediately or not. The aim of this strategy is to detect chances of transmitted resistance as maximum as possible. In HIV-infected individuals receiving antiretroviral therapy, resistance testing should be performed in the presence of virological failure. To ensure adequate performance of resistance testing, HIV-1 RNA levels should be at least 1,000 copies/ml at the time of testing, although guidelines agree that resistance testing could be also attempted in individuals with HIV-1 RNA levels between 500 and 1,000 copies/ml. However, in this last group of patient, the chances of amplifying HIV-1 sequences are markedly lower. Drug resistance testing might also be helpful when managing suboptimal viral load reduction. However, this is less clear because the addition of, or switch to, new antiretroviral drugs could be helpful to achieve viral suppression. Importantly, given that drug resistance mutations wane after treatment interruption, drug resistance testing in the setting of virological failure should be performed while the patient is taking his/her antiretroviral drugs or within 4 weeks after discontinuing therapy.

Two types of antiretroviral resistance assays are currently available to assist the clinician in assessing HIV resistance: genotypic assays and phenotypic assays (Hanna and D'Aquila 2001; Hirsch et al. 2008). But these two tests are also not up to mark as not useful for samples with HIV-1 RNA levels <500–1,000 copies/ml. Table 28.1 shows the most common assays along with their advantage and disadvantages.

Table 28.1 The most common assays along with their advantage and disadvantages

	Advantages	Disadvantages
Genotypic assays	Rapid turnaround time (days versus week)	Measurement of drug resistance is indirect – identified mutations are subjectively translated into conclusions on viral resistance
	Reduced cost	Only identifies mutations in predominant viral quasispecies (<25 % of viral population)
	Proven utility in the predicting short-term virologic outcome	Unclear relevance of certain mutations
	Allow detection of emerging mutations before onset of resistance	Drug resistance does not correlate with mutations in all cases
	Detect contamination between specimens	Interpretation can be difficult when multiple mutations are present
	Simple technology, available in regular hospital laboratories	Unable to detect mutation linkage
Phenotypic assays	Directly measure drug susceptibility: determine IC50 or IC90 (contamination of drug required to inhibit viral replication by 50 or 90 %)	Nonstandarized cutoff values
	Measure susceptibility to any drug, determine the presence of cross-resistance and multidrug resistance	Only identifies mutations in predominant viral quasispecies (<25 % of viral population)
	Measure overall effect of mutations, can assess non-B-clade HIV-1 strains and potential to measure viral tropism and replication capacity in parallel	Long-term drug response cannot be determined, fails to account for interaction between drugs in combination therapy
	Straightforward interpretation	Highly complex testing platform with longer turnaround time
		Increased cost
Virtual phenotype	Pattern-matching algorithm that uses a large genotypic–phenotypic correlative database to infer phenotypic properties based on sequence data	It is not known which mutations are used to match a new sequence to those sequences that are already in the database
	Includes a tabulation of the number of matches in the database for each drug and the distribution of phenotypes (fold increase in IC50) for the matching sample	Its predictive power depends on the number of matched data sets available. Thus, variation is frequently higher in smaller data sets, as well as for newer drugs or complex resistant patterns
	Provides a quantitative prediction of drug resistance	Matches are based on preselected codons, not on the entire nucleotide sequence
	Showed good correlation for most drugs	

These tests have limited sensitivity for the detection of minority variants in the viral population, so detection of non-B subtypes may be limited for some tests, and last but not least these results require expert interpretation. In addition, a genotype test can be used to generate a predicted phenotype, referred to as a virtual phenotype. With the virtual phenotype, the viral sequence (genotype) is entered into a database consisting of paired genotypes and phenotypes in order to derive an estimated phenotype. More recently, some investigators have employed newer techniques, such as allele-specific PCR, single-genome, and ultra-deep sequencing, to assess the role of minority HIV variants that harbor drug resistance, but are not detectable by current standard genotypic or phenotypic assays (Hirsch et al. 2008). Table 28.2 shows the recent advances in resistances testing at minority levels.

Table 28.2 The recent advances in resistances testing at minority levels

	Allele-specific PCR	Single genome sequencing	Ultra-deep sequencing
Principle	Differential amplification of mutants versus wild type in real-time PCR	Massive sequencing of single-genome molecules	Massively parallel microfluidic solid surface sequencing of single molecules
Sensitivity	0.003–0.4 %	2 %	0.5–1 %
No. of mutations	1	Multiple	300–400 base pairs
Linked mutations	No	Yes	Yes
Goodness	Sensitivity, positive predictive value, and negative predictive value	Linkage mutation detection	Linkage, sensitivity, accuracy, negative predictive value, rapidity of results
Limitations	Only one allele per reactions detectable, limited in polymorphisms	Cost, time, and labor investment	STRING Bioinformatics interpretation required

Thus, current emerging techniques like sequencing, microarray, and real-time PCR revolutionized the modern concept of medicine, i.e., well-known as personalized medicine concept where treatment will be based on individuals' genome composition as well as genetic makeup of infective agents (strains). One important aspect of personalized medicine is patient-to-patient variation in drug response. Pharmacogenomics addresses this issue by seeking to identify genetic contributors to human variation in drug efficacy and toxicity in viral diseases. Here, we are going to discuss current updates, success, and challenges of this field, which has evolved from studies of single candidate genes to comprehensive genome-wide analyses, and thus, this new field will ultimately open new vistas to understand the better management against the fatal viral diseases, so more effective personalized clinical treatment strategies may be developed.

4 Personalized Medicine: Promising a Solution for Viral Disease Management

The ultimate goal of personalized medicine is to take advantage of a molecular understanding of disease both to optimize drug development and direct preventive resources and therapeutic agents at the right population of people while they are still well. The goals of personalized medicine in drug development include (1) the selection of optimal drug targets; (2) the selection of optimal drug dosage; (3) the selection and monitoring of patients for shorter, less-expensive advanced clinical trials; (4) the ability to predict which individuals will respond to drugs at high rates and who will be less likely to suffer toxic side effects; (5) reducing the overall cost of drug development and increasing drug value; and (6) to ultimately improve and provide more effective healthcare for all individuals whether they are well or suffering from the early or late stages of illness. Genetic variants can be used to predict the predisposition of an individual for future disease development. By applying the principles of personalized medicine, it is possible to significantly enhance the productivity of drug discovery and development, so that the identification of the target gene and the appropriate pathways of the suitable drug can be developed which will definitely help in combating the viral diseases.

4.1 Target Selection

Recent emergence and progress in molecular biological techniques make the route easier to select the right target by various traditional and modern genomic approaches nowadays in practices. Thus, one of the major challenges is to identify and characterize the target which is essential for the virus to survive, but which is absent, or significantly divergent, in their mammalian host may sort out in drug discovery. For viral diseases, the small genome and relatively few viral proteins make this process fairly straightforward. The modern molecular techniques like real-time PCR, microarray, and next-gen sequencing approach are currently giving the most precise and specific results; thus they are helping in progressing the newer therapeutic approach. Table 28.3 shows various molecular techniques for target selections.

4.2 Viral Load Assay

The quantitative analysis of HIV-1 is based on viral load assay. Viral load assay is expressed by the copy number in a given unit of plasma, usually using copies/ml. An alternative expression is the international unit suggested by WHO (IU/ml). Based on the interlaboratory standardization using the three FDA-approved HIV-1 viral load kits, the exchange relationship between the copies/ml and the IU/ml is between 1:0.92 and 1:1 (www.fda.gov/cber/pmalabel/p050069LB). Viral load is a dynamic and relative parameter. Its value changes depending on methods used and it can fluctuate even on a daily basis. The variances between repeated tests using the same kit are usually smaller than those determined with different methods. Thus, a dynamic comparison of viral load values using the same assay method over time is strongly suggested. HIV viral load has multiple clinical implications (Report on the global AIDS epidemic 2007; Torti et al. 2007). Firstly, it currently serves as a complementary diagnosis of HIV infection and may become one of the diagnosis standards in the near future. For qualitative determination of HIV infection, the higher the viral load value is measured, the more confident a diagnosis of HIV infection can be drawn. For example, if the viral load is higher than 3,000 copies/ml, the probability of HIV infection of the individual from whom the sample is taken is significantly high, particularly if high viral load is repeatable with another sample taken from the same individual at a different time. Secondly, viral load is helpful for early diagnosis of HIV infection. It is reported that there is a viral burst in blood in the early stage of HIV infection. The viral load in the early stage is sometimes even higher than that in the disease stage of AIDS. The viral load assay can also be used in the complementary diagnosis of HIV infection for neonates from HIV-infected mothers. Although the antibody-based HIV assay is the diagnosis standard in clinical practice, it is useless during its window period in the early stage of infection and has no diagnostic value for infants from HIV-infected mothers as discussed earlier. Thirdly, the analysis of the viral load is an efficient parameter in assessing antiviral therapy. The effective standard is the viral load decreased at least by 0.5 log after 4-week or 1 log after an 8-week antiviral therapy or the viral load reduced to 1,000 copies/ml after a 16–24-week therapy. Finally, the viral load has some values in predicting the progress of AIDS. For example, the viral load can predict the probability (p value) of becoming AIDS in 6 years for patients with normal CD4+ counts. The p value is 0.054 when the viral load is less than 500 copies/ml and the p value dramatically increased to 0.8 when the viral load is more than 30,000 copies/ml. When CD4+ counts are less than 200/μl, the viral load can be used to predict a shorter progress of AIDS. The probability of turning to AIDS within 3–6 months from viral carrier is proportionally associated with the viral load values. A similar conclusion was drawn from a recent clinical trial including 751 HIV-infected patients with b200 CD4+/mm^3 before HAART (Torti et al. 2007). Patients with higher CD4+ T-cell counts following the treatment appeared to have survived after month 3, whereas those with increasing HIV RNA N400 copies/ml did not. The three methods currently employed in commercial kits for HIV-1 viral load assays are rt-PCR-, b DNA-, and NASBA-based assays. Table 28.4 shows the comparison of three FDA-approved HIV-1 viral load assays commonly used in assessment with their advantages and disadvantages.

The major difference between these assays is the requirement of specific equipment rather than their sensitivity or specificity. Therefore, the resources of the various settings remain the

Table 28.3 The molecular techniques for target selections

Technology	Use	Advantages	Disadvantages
Signature-tagged mutagenesis	Identification of genes required for pathogen survival in animal model	Identifies essential genes in vivo	Limited to genes that are not required in vitro, library construction
In vivo expression technology	Identification of pathogen induced in vivo	Identifies genes induced in vivo	Increased expression does not necessarily mean the gene is essential in vivo, further validation required, construct initial library
Microarrays	Understanding host response to pathogens, correlating gene expression with pathogenicity, identifying molecular targets of antimicrobial compounds, inferring function of unknown genes	Low resource, no library construction required, looks at all genes simultaneously	Increased expression does not necessarily mean the gene is essential, further validation required, technically difficult to use for pathogens grown in vivo
Comparative genomics	Identification of pathogenicity-related genes, identification of antigens for vaccine development, selecting targets conserved across multiple pathogens, selecting targets with lowest homology to human proteins	No experimental work required, total in silico work	Spectrum and selectivity analysis based on linear sequencing, analysis could be misleading
Structural genomics	Select conserved target across multiple pathogen, selecting targets with lowest homology to human protein	Spectrum and selectivity analysis based on 3D analysis of the active site	High resource required to solve the crystal structure
Next-gen sequencing	Whole genome sequencing, investigation of genome diversity, metagenomics, epigenetics, discovery of uncoding RNAs, protein-binding sites	Higher sensitivity and the potential to detect the full spectrum of virus including unknown and unexpected viruses	Sequencing in bulk is error-prone

Table 28.4 Comparison of three FDA-approved HIV-1 viral load assays commonly used in assessment

Product	Amplicor	Versant	NucliSENS
Principle (company)	Rt-PCR (Roche)	b DNA (Bayer)	NASBA (bioMerieux)
Method of amplification	Exponential amplification of the target molecules	Linear amplification of the signal	Method of amplification
Genomic region targeted	Gag	Pol	Gag and pol
Technical platform	Microplate can be upgraded to real-time analysis	Microplate	Microplate can be upgraded to real-time analysis
Advantage /Disadvantage	Less false-positive than b DNA	Wide range/time consuming	For many types of sample
Dynamic range	50 (V1.0) or 400 (V1.5) to 750,000 copies/ml	b DNA V3.0: 75–500,000 copies/ml	NucliSENS HIV-1 QT: 176-3, 500,000 copies/ml
Sample amount required	0.2–0.5 ml	1 ml	10 μl–2 ml

determinant factor for choosing the PCR-based, hybridization-based, or isothermal amplification-based HIV detection assays. Polymerase chain reaction (PCR)- or rt-PCR-based assay is the most frequently used technology in molecular biology and molecular diagnostics. PCR-based assays can be used for either the qualitative or quantitative analysis of HIV. As HIV is a retrovirus, both PCR and rt-PCR are used as qualitative assays of viral infection targeting the integrated form of provirus or the free HIV themselves, respectively. The rt-PCR is however specifically required in HIV viral load assay. The amplified products from PCR or rt-PCR can be visualized by a variety of methods, such as agarose gel electrophoresis, real-time visualization, and enzymatic reaction. One commercial kit, Amplicor, uses enzymatic reaction for product visualization with microplate platform (Murphy et al. 2000). Considering the complicated procedure for the Amplicor assay as the post-amplification hybridization and enzymatic reaction for visualization involves multiple steps and many types of reagents, real-time visualization is advantageous in the development of new rt-PCR-based assays for HIV viral load analysis (Stevens et al. 2007). The simplicity and shorter time required for finishing the test make the rt-PCR-based HIV viral load assay to be more competitive among all available kits and possibly to be one of the focus in future kits under development.

The amplified products from this nucleic acid-based method can be electrophoretically analyzed and fluorescently visualized in real-time pattern. For single-molecule identification, nonspecific fluorescent dye such as SYBR Green can be chosen; currently our clinical pharmacology laboratory uses this approach. High-fidelity DNA polymerase is different from Taq polymerase in their exonuclease activities: the former has 3′–5′ exonuclease activity or proofreading function; the latter has 5′–3′ exonuclease activity. These differences are differentially employed in real-time visualization of PCR-amplified products. The Taq polymerase is used in TaqMan technology, whereas real-time visualization of assays using the mutation-sensitive on/off switch mediated by the high-fidelity DNA polymerase depends on molecular beacon or FRET. The molecular beacon structure can have target-independent sequences and be designed at the 5′ end of primers. The molecular beacon can have a mix of target-specific sequences in its loop and target-independent sequences in its hairpin. This mixed beacon adds one short fragment to be identified and thus helps to enhance the specificity of the assay. Similarly, the combination of the mutation-sensitive on/off switch and the FRET technology also adds additional sequence fragments under discrimination test. The mutation-sensitive on/off switch offers high sensitivity and specificity in nucleic acid identification/mutation detection. Its combination with fluorescent real-time visualization makes it a highly competitive technology in developing both qualitative and quantitative assays for HIV infection.

5 Database of Mutation or SNPs and Their Effects on Drug Response

Genetic variation is likely to contribute substantially to the variation in drug response observed across human populations. The field of pharmacogenomics, which relates genetic variability to variability in human drug response, has evolved considerably from candidate gene studies to studies of variation across whole genomes of human populations containing individuals who exhibit a range of responses to different drugs. The initial successes in the field were often the identification of genetic variants within drug-metabolizing genes that had large effects on sensitivity to a given drug. The field has since broadened in scope to encompass regulatory mutations, and refined techniques have made us able to identify the mutations with smaller effect sizes. Whereas early pharmacogenomics studies sought primarily to identify associations between common genetic variation and drug response, more recent approaches have begun to identify mRNAs, miRNAs, and other downstream events that are influenced by genetic variation and may underlie variation in pharmacologic responses. The primary aim of pharmacogenomics has been to uncover novel human genetic variants that affect therapeutic response phenotypes and to identify the genes responsible for those phenotypic differences. The ultimate goal of the field has been to use an understanding of these relations to devise novel personalized pharmacological treatment strategies that maximize the potential for therapeutic benefit and minimize the risk of adverse effects for any given medication. Advances in DNA sequencing and polymorphism characterization technologies have enabled the field to evolve from the sole reliance on hypothesis-driven approaches to the use of discovery-oriented, genome-wide approach that requires fewer a priori assumptions regarding genetic variants. Candidate gene approaches resulted primarily in the identification of genetic variants in drug-metabolizing genes with large effects on toxicity or response (Weinshilboum and Sladek 1980); however, many genome-wide association studies (GWAS) have identified novel associations between drug response and genetic variants with unknown functional relevance and often with relatively small effect sizes (Daly 2010). The recent development of high-throughput sequencing techniques has enabled researchers to begin to examine the contribution of rare variants to drug sensitivity (Ramsey et al. 2012). Genetic screening of HIV for prediction of resistance is recommended by experts and is being increasingly used in the clinical setting (Hirsch et al. 2000). Testing for mutations in the HIV genome that are predictive of resistance can be done before therapy or after failure of an initial regimen. Genotypic resistance testing has been judged to be cost-effective, even with the relatively high current costs of the tests (Weinstein et al. 2001). At present, the majority of genetic resistance testing for HIV drugs involves sequence-based approaches; however, kit-based tests are expected to become common and testing costs should fall automatically. These tests are also likely to expand beyond the viral genome, as additional polymorphisms in the host genome are linked to HIV treatment outcome.

Before exploring the applications of pharmacogenomics on HIV, it is important to understand the underlying biological principles. Genetic variation in the human genome occurs predominantly as single-nucleotide polymorphisms (SNPs) where single nucleotides differ between humans. An individual's DNA contains as many as ten million SNPs, which are responsible for the diversity of human phenotypes, as well as susceptibility to drugs and diseases (Twyman 2004). Scientists approach the problem of identifying, cataloging, and characterizing SNPs from two different angles – a genomic approach and a functional approach. The genomic approach involves scientists comparing the genomes of numerous individuals to study differences and recording their results in forums such as the dbSNP short genetic variation database hosted by the National Center for Biotechnology Information (NCBI). This method of SNP characterization requires a large amount of computer-powered data analysis unlike the functional approach where scientists focus on select genes known to be associated with a particular process or disease, and then they examine them across a population (Perelson et al. 1996). In pharmacogenomics, such genetic variation with respect to drug response is studied and used to guide patient care. The primary bioinformatics resource for pharmacogenomics is the PharmGKB database developed by Stanford University that is studying the impact of genetic variation on drug response and focuses on clinical interpretation of variants, drug dosing guidelines, genetic tests, and other information that is practically applicable for actors in the health sector. Even without expanding the numbers of drugs available, however, pharmacogenomics is still incredibly impactful because doses and preferences for existing drugs can be employed for patient care. In fact, some major innovations have already occurred in resistance testing technologies and in HIV treatment selection based on genomics. Genotyping and phenotyping are two established methods for identifying antiretroviral resistance in patients on therapy. In HIV genotyping, a discrete sequence of reverse transcriptase and protease genes in extracted RNA specimens is amplified to generate cDNA amplicon, which then undergoes sequencing. A software system called OpenGene then aligns the sequences, reports mutations, and produces an interpretive report. The interpretation does not provide any insight though with regard to the degree of resistance to a drug because the output has either (1) no evidence of resistance, (2) possible resistance, (3) resistance, or (4) insufficient evidence. HIV drug resistance databases also play a key role by providing advanced information to clinicians. Important databases often used in HIV genetic research are the Los Alamos HIV Drug Resistance database and the Stanford HIV RT and Protease Sequence Database. The former collects all sequences and focuses on annotation and data analysis. A recent study prepared the database that contains a compilation of nearly all published HIV reverse transcriptase and protease sequences, which are linked to data about the source of the sequence sample and the antiretroviral drug treatment history of the individual from whom the isolate was obtained (Shafer 2003). Another group developed the HIVbase software solution that helps researchers effectively manage DNA/amino acid sequences and related genetic/clinical data using storage and query capabilities (Salemi 2004). Storing genotypic resistance data and linking to other clinical information is an important tool for successful disease management. Examples of such systems that identify mutation patterns associated with resistance are Virodec (Roche Diagnostics) and ViroScore (ABL) published on their diagnostic website. Virodec HIV is an online application engineered to upload, analyze, interpret, deliver, and store genetic sequence data from genotyping assays. ViroScore is an HIV resistance sequence management system with a sequence database that is used by analysis tools/algorithms for resistance interpretation. Despite these advanced technologies, interpretation is still the limitation with all resistance testing, and the systematic approaches developed to predict phenotypes based on mutational patterns are complicated by the complex mutation patterns for resistance.

Mutations and SNPs in HBV also studied in details as the most frequently observed precore mutation is a G to A transversion at nucleotide 1896. This substitution introduces a translation stop codon (TAG) in the distal precore gene and prevents expression of the preC/C fusion protein that functions as a precursor of HBeAg (Carman et al. 1989). Less common precore mutations resulting in HBeAg negativity include initiation codon mutations (at positions 1814 or 1815), a nonsense mutation at 1874, a missense mutation at 1862, and frame shift mutations (Kramvis et al. 1997). The 1896 stop codon mutant is often present in patients with chronic active or inactive hepatitis and in asymptomatic carriers in Mediterranean and Oriental countries (Lee et al. 1996). A mutation at 1899, which may occur in association with the 1896 mutation or other mutations that are associated with HBeAg negativity, is another mutation that improves the stability of by providing an additional A–T(U) base pair (Kramvis et al. 1997). The 1896 stop codon mutation is also present in a high proportion of patients with fulminant hepatitis B, an observation that initially suggested a causal role for the mutant (Maruyama et al. 1998).

The core gene contains both humoral and cytotoxic T-cell epitopes (Bertoletti et al. 1991). Mutations within immunodominant cytotoxic T-cell epitopes may be exploited by viruses to evade protective immune responses critical for viral clearance. Deletions of the core gene have been reported in immunocompromised and Oriental patients with chronic hepatitis B or hepatocellular carcinoma (Yuan et al. 1998). These deletions almost always involve loss of B- and T-cell epitopes and may confer a selective advantage on the virus by evading immune surveillance.

Point substitutions in the S gene are of particular interest because they affect the immunogenicity of HBsAg, especially the determinant (against which neutralizing antibodies are raised). In an effort to explain the effects that mutations in one region exert, both locally and on linearly distant epitopes, the original two-loop model of the *a* determinant (positions 124–147) (Carman et al. 1990), with disulfide bridges between amino acids 124 and 137, has recently been replaced by the cysteine web model of the MHR (positions 100–160 or 169) of the S protein (Carman et al. 1990). The current model still takes account of potential disulfide bridges but additionally supposes cysteines 107, 137, 138, 139, and 149 to be located in a webbed structure in the viral envelope. Two loops (107–137 and 139–147) are external to the virion and probably in opposition, and there is another tight loop between amino acids 121 and 124. The whole MHR is divided into five antigenic regions, named HBs1 (up to position 120), HBs2 (120–123), HBs3 (124–137), HBs4 (139–147), and HBsS (148–169). There are indications that the loops formed by HBs2 and HBs4, respectively, are spatially close. This mutant was found in studies in Singapore, Italy, Japan, Taiwan, Indonesia, and Brunei (Hsu et al. 1997).

Mutations of the polymerase gene may be associated with resistance to the therapeutic effects of nucleoside analogues and with viral persistence (Ono-Nita et al. 1999). Lamivudine (2,3-dideoxy-3-thiacytidine) is a potent inhibitor of RNA-dependent DNA polymerase of HBV, irreversibly blocking reverse transcription and inhibiting viral replication. It thus effectively reduces viral burden in chronic HBV carriers. Long-term treatment with lamivudine may, however, lead to resistance as the result of the generation of mutations that disrupt the YMDD (tyrosine, methionine, aspartate, and aspartate) locus in the C domain of the polymerase gene (Ling et al. 1996). The mutation consists of either a methionine to valine (M552V) or a methionine to isoleucine (M552I) substitution. Both mutations result in amino acid substitutions in codons 195 and 196 in the overlapping S gene. Lamivudine-resistant variants may also have a leucine to methionine (L528M) change in the B domain, occurring often in association with the M552V mutation and rarely with the M552I mutation (Chayama et al. 1998). The L528M substitution has no effect on the amino acid sequence of the S gene. Replication efficiency of the YMDD mutant is less than that of wild-type virus and, after cessation of treatment, the wild-type virus re-overtakes the mutant. Lamivudine-resistant viruses remain functional and pathogenic. Famciclovir is the prodrug of penciclovir, an acyclic deoxyguanosine analogue. Penciclovir inhibits DNA-dependent as well as RNA-dependent DNA polymerase activity. It has similar therapeutic effects to lamivudine and is responsible for the emergence of mutants usually involving the B domain of the polymerase gene. However, mutants resistant to famciclovir appear to be less common than those induced by lamivudine (Pichoud et al. 1999).

The X gene protein exhibits numerous activities affecting intracellular signal transmission, gene transcription, cell proliferation, DNA repair, and apoptosis (Arbuthnot et al. 2000). An eight-nucleotide deletion at the 3′ end of the gene and within the core promoter/enhancer II (CP/ENII) region (positions 1770–1777) (Fukuda et al. 1996) and a 20-nucleotide deletion at 1752–1772 (Okamoto et al. 1994) study have been described in HBsAg- and HBeAg-negative patients. These deletions have been shown to downregulate the preC promoter, and this may be the reason for the suppression of HBV protein secretion.

The core promoter plays a central role in HBV replication and morphogenesis, directing the transcription of both pregenomic RNA and precore mRNA. It overlays the 3′ end of the X gene and the 5′ end of the preC/C gene. Sequence variation in the core promoter is limited because of its pivotal role in viral replication. The double mutation, A to T transversion at 1762 and G to A transition at 1764, is often present in patients with chronic hepatitis, hepatocellular carcinoma, and fulminant hepatitis and less often in asymptomatic carriers, in immunosuppressed patients, and in carriers without HBV markers (Kramvis and Kew 1999). Mutants are inextricably bound to an evolution to chronicity and may be important in hepatocarcinogenesis, development of fulminant hepatitis, or an asymptomatic course. The clinical importance of HBV surface antigen variants has been the subject of several reviews (Carman 1996).

An individual's response to a drug is the complex interaction of both genetic and nongenetic factors. Genetic variants in the drug target itself, disease pathway genes, or drug-metabolizing enzymes can all be used as predictors of drug efficacy or toxicity. More than one million SNPs are now available for genotyping and phenotyping studies (Durbin et al. 2010). Novel genotyping strategies are emerging on a regular basis using a variety of techniques designed to increase the rate of data generation and analysis. A high-resolution SNP map recently developed by the SNP Consortium (http://snp.cshl.org/) could expedite the identification of genes for complex viral diseases such as HIV/AIDS and hepatitis. In virology, the SNP technology has focused on detecting the predisposition for predicting toxic responses to drugs and selecting the best individual and combinations of antiretroviral drugs.

5.1 Prediction of Drug Toxicity

SNP detection has been used to predict adverse events in antiretroviral therapy in patients with HIV infections (Abo et al. 2012). The potential clinical value of the pharmacogenetics approach for predicting drug toxicity will be uncovered as more candidate polymorphisms are discovered. The application of genotyping strategies to predict antiretroviral drug efficacy has recently emerged in a variety of clinical settings. Genotype resistance testing of HIV isolates has demonstrable clinical use and provides a way to assist therapeutic decision making in patients whose HIV RNA levels are rising (Klein et al. 2009). Moreover, HIV viral load testing has served as the major guide to the selection and maintenance of antiretroviral therapy (Sagreiya et al. 2010).

5.2 Transcriptional Profiling and Genomic Microarrays

The development of printed and spotted genomic microarrays has enabled the rapid accumulation of new information concerning gene mutation and expression in human. Microarrays can be used to gene mutations and SNPs as well as provide rapid screening information regarding mRNA expression. Transcriptional profiling has the ability to generate hundreds of thousands of data points requiring sophisticated and complex information systems necessary for accurate and useful data analysis. This technique has generated a wealth of new information in the drug and biomarker target discovery and pharmacogenomic drug efficacy testing.

6 Susceptibility of Various Genotypes and Therapy in Viral Diseases

Highly active antiretroviral therapy for HIV-1, although resulting in dramatic suppression of viral replication, has also furnished a strong selective force for the emergence of drug-resistant variants. Here, the distribution of polymorphisms can be extreme. For example, examination of the HIV-1 reverse transcriptase (RT) and HIV-1 protease structural variant databases (Variome™ modules) established by Structural Bioinformatics (SBI; http://www.strubix.com) and Quest Diagnostics (http://www.questdaignostics.com) has revealed that no two patients have exactly the same sequence – each individual patient exhibits a unique sequence or structural variant for these drug targets because a principal component

Table 28.5 Significance levels (p values) for gene regions with $p \leq 0.005$, for either an association with (i) cervical precancer/cancer, (ii) progression to cervical precancer/cancer, or (iii) HPV persistence

Gene	Chromosome	SNPs	(i) CIN3+ (n=415) vs. RC (n=425)[a]	(ii) CIN3+ (n=415) vs. HPV persistence (n=356)	(iii) HPV persistence (n=356) vs. RC (n=425)
*PRDX3**	10q25–q26	11	**0.00015**	0.0744	0.16314
*RPS19**	19q13.2	4	**0.00045**	0.00585	0.45823
DDX1	2p24	13	**0.0006**	0.69087	0.21284
TELO2	16p13.3	20	**0.0009**	0.17849	0.65932
C1RL	12p13.31	20	**0.00165**	0.14029	0.00875
ILDR1	3q13.33	11	**0.00285**	0.05830	0.88461
THRAP4		6	**0.00370**	0.80776	0.05615
GDF10	10q11.22	20	**0.00400**	0.13639	0.04230
GDF2	10q11.22	19	**0.00400**	0.12639	0.04080
TYMS	18p11.32	20	0.0055	0.30233	**0.0015**
EVPL	17q25	10	0.02055	0.08985	**0.00180**
GC	4q12–q13	19	0.0343	**0.0004**	0.00965
IL2RA	10p15–p14	44	0.24394	**0.00115**	0.44318
PIK3CA	3q26.3	11	0.91060	0.05145	**0.00490**

*FDR for both *PRDX3* and *RPS19*=0.19
[a]RC=random controls
The values in bold represents the $p<0.005$

determining drug efficacy is the distribution of drug-target structural variants within the patient population (Chander et al. 2002).

6.1 Host Genetics Factor and HPV and Its Association with Pathogenesis of Cervical Cancer

While it is well known that carcinogenic human papillomaviruses (HPVs) are the causal agents of cervical cancer, HPV infections are extremely common relative to rare cancer incidence, indicating that many infections spontaneously resolve (Schiffman et al. 2007) or persist without progression. Host genetic factors may play a role in cervical carcinogenesis and are thought to influence who develops persistent HPV infection and perhaps who further progresses to cancer (Czene et al. 2002; Hemminki and Chen 1999; Hemminki et al. 2006; Hussain et al. 2008; Hildesheim and Wang 2002; Carrington et al. 2005). The roles of host genetic factors and other cofactors associated with cervical cancer are summarized in Tables 28.5 and 28.6. As discussed by Safaeian et al. (2012), it is particularly very interesting because the stepwise pathogenesis of the disease has been extensively studied. From its initiation through HPV infection at the cervical transformation zone, and subsequent steps related to viral persistence, progression to precancer, and invasion (Schiffman et al. 2007), the same or different factors can be associated with each step toward pathogenesis. The role of nongenetic cofactors in persistence and progression has been well studied, but there are fewer studies on the host genetics role on the pathogenesis of cervical cancer.

6.2 Interferon Therapy of Chronic Viral Hepatitis

Current therapies for chronic viral hepatitis include two regimens:

- IFN or antiviral nucleoside/nucleotide analogues such as lamivudine, adefovir dipivoxil, and ribavirin (Marcellin et al. 2002; Chander et al. 2002) and IFN or lamivudine alone can control hepatitis B in about one-third of patients (Marcellin et al. 2002; Liaw 2002).

Table 28.6 Association of SNPs with (i) cervical precancer/cancer, (ii) progression to cervical precancer/cancer, or (iii) HPV persistence (all models adjusted for age)

					(i)	(ii)	(iii)
		Random control	HPV persistence	CIN3/cancer	CIN3/cancer vs. RC	CIN3/cancer vs. HPV persistence	HPV persistence vs. RC
Gene	rs#	*N* (%)	*N* (%)	*N* (%)	OR (95 %)	OR (95 %)	OR (95 %)
PRDX3	RS7082598						
	CC	288 (68)	266 (75)	346 (83)	1.0 (ref)	1.0 (ref)	1.0 (ref)
	CT	127 (30)	87 (24)	64 (15)	0.41 (0.29–0.58)	0.55 (0.38–0.80)	0.74 (0.53–1.01)
	TT	10 (2)	3 (1)	5 (1)	0.45 (0.15–1.35)	1.29 (0.29–5.65)	0.33 (0.09–1.20)
					p-trend < 0.00001	*p*-trend = 0.008	*p*-trend = 0.02
	CT/TT	137 (32)	90 (25)	69 (17)	0.41 (0.30–0.58)	0.58 (0.40–0.83)	0.71 (0.52–0.97)
RPS9	RS230589						
	CC	100 (24)	102 (29)	141 (34)	1.0 (ref)	1.0 (ref)	1.0 (ref)
	CT	228 (54)	172 (48)	213 (51)	0.66 (0.48–0.91)	0.83 (0.59–1.15)	0.74 (0.53–1.05)
	TT	97 (23)	82 (23)	60 (14)	0.43 (0.28–0.66)	0.51 (0.33–0.78)	0.83 (0.56–1.25)
					p-trend = 0.0007	*p*-trend = 0.003	*p*-trend = 0.34
	CT/TT	325 (76)	254 (71)	273 (66)	0.59 (0.43–0.80)	0.72 (0.53–0.99)	0.77 (0.56–1.06)
Il2RA	*RS247641*						
	AA	262 (62)	196 (55)	289 (70)	1.0 (ref)	1.0 (ref)	1.0 (ref)
	AT	147 (35)	135 (38)	114 (27)	0.69 (0.51–0.93)	0.56 (0.41–0.77)	1.22 (0.91–1.65)
	TT	15 (4)	25 (7)	12 (3)	0.68 (0.31–1.5)	0.34 (0.16–0.70)	2.24 (1.15–4.37)
					p-trend = 0.02	*p*-trend = 0.00002	*p*-trend = 0.02
	AT/TT	162 (38)	160 (45)	126 (30)	0.69 (0.51–0.92)	0.53 (0.39–0.71)	1.32 (0.99–1.76)
TELO2	RS4786772						
	AA	185 (44)	151 (42)	134 (32)	1.0 (ref)	1.0 (ref)	1.0 (ref)
	AG	197 (46)	147 (41)	211 (51)	1.54 (1.14–2.09)	1.65 (1.19–2.27)	0.92 (0.68–1.24)
	GG	42 (10)	58 (16)	69 (17)	2.51 (1.59–3.94)	1.40 (0.91–2.15)	1.68 (1.06–2.65)
					p-trend = 0.0002	*p*-trend = 0.03	*p*-trend = 0.13
	AG/GG	239 (56)	205 (58)	280 (68)	1.71 (1.28–2.28)	1.57 (1.16–2.13)	1.05 (0.79–1.39)
C1RL	RS12227050						
	GG	402 (95)	310 (87)	357 (86)	1.0 (ref)	1.0 (ref)	1.0 (ref)
	AG	22 (5)	45 (13)	57 (14)	2.91 (1.73–4.89)	1.06 (0.69–1.62)	2.73 (1.60–4.65)
	AA	1 (0.2)	1 (0.3)	1 (0.2)	1.37 (0.08–22.18)	0.77 (0.05–12.91)	1.36 (0.08–21.92)
					p-trend = 0.0001	*p*-trend = 0.85	*p*-trend = 0.0004
	AG/AA	23 (5)	46 (13)	58 (14)	2.85 (1.71–4.74)	1.05 (0.69–1.61)	2.67 (1.58–4.51)
TYMS	RS2342700						
	CC	227 (53)	145 (41)	169 (41)	1.0 (ref)	1.0 (ref)	1.0 (ref)
	CG	169 (40)	164 (46)	199 (48)	1.56 (1.17–2.09)	1.06 (0.78–1.44)	1.53 (1.13–2.06)
	GG	29 (7)	47 (13)	46 (11)	2.18 (1.30–3.64)	0.87 (0.54–1.40)	2.57 (1.54–4.27)
					p-trend = 0.0002	*p*-trend = 0.78	*p*-trend = 0.00005
	CG/GG	198 (47)	211 (59)	245 (59)	1.65 (1.25–2.18)	1.02 (0.76–1.37)	1.68 (1.26–2.23)

Odds ratios (95 % confidence intervals); for top-ranked SNPs ($p \leq 0.0001$)

- A combination of IFN with ribavirin is standard therapy for hepatitis C and has resulted in eradication of HCV in about 50 % of patients (Chander et al. 2002; McHutchison et al. 1998).

The outcome of viral therapy is influenced by viral load and viral genomic variations (especially viral genotypes and certain specific genomic variants). Various therapeutic agents showed variation in their efficacy during treatment;

Table 28.7 The susceptibility of various genotypes in viral disease hepatitis

Genotype	Disease	Result/association	Researcher/references
Promoter gene polymorphism at −238 of TNF-α	Chronic hepatitis	Positive associations	Höhler et al. (1998a, b)
CC genotype than TT/TC SNP(rs12979860) OF IL-28 B	Hepatitis	Sustained virologic Response rates of HCV	Ge et al. (2009)

Table 28.8 Therapeutic agents and their effect (drug efficacy) on various genotype/strain of hepatitis

Therapeutic agents	Researchers/references	Genotypes and disease	Association and results
IFN	Kao et al. (2002, 2000)	C and D of HBV vs. B and A of HBV	Liver diseases severity increases in C and D
IFN and ribavirin	Farci and Purcell (2000)	Genotype 1 and genotype 2 of HCV	Poorer response (40 vs. 80 %)
IFN	King et al. (2002)	A/G heterozygous eIf-2α of HBV	Significant (OR = 12.28; 95 % CI, 1.5–107.8; $p = 0.009$)
peg-IFN-alpha and RBV	Suppiah et al. (2009)	IL-28B with HCV	Significant ($p = 7.06 \times 10^{-8}$; OR = 3.36; 95 % CI, 2.15–5.35)
peg-IFN-alpha and RBV	Tanaka et al. (2009)	Genotype 1 of HCV two SNPs (rs8099917) and (rs12980275) of IL-28B	Strong association with null virologic response (NVR) ($p = 3.11 \times 10^{-15}$) and ($p = 1.93 \times 10^{-13}$)
peg-IFN-alpha and RBV	Rauch et al. (2010)	(rs8099917)T/G SNP with HCV genotype 1:G allele fail to respond of IL-28B	Null virologic response (NVR) (OR = 2.31; 95 % CI, 1.74–3.06; $p = 6.07 \times 10^{-9}$) vs. sustained vs. sustained viral response (SVR) (OR = 5.19; 95 % CI, 2.90–9.30; $p = 3.11 \times 10^{-8}$)

Table 28.7 shows the susceptibility of various genotypes in viral disease hepatitis, and Table 28.8 shows the therapeutic agents and their effect (drug efficacy) on various genotype/strain of hepatitis. In addition, major histocompatibility complex (MHC) class I and MHC class II polymorphisms (Thursz and Thomas 1997), interleukin 10 polymorphism (Yee et al. 2001), MxA promoter single-nucleotide polymorphisms (SNPs), and mannose-binding protein SNPs (Hijikata et al. 2001) all have been supported to affect host immune and antiviral responses and thus are associated with disease progression and treatment response.

In the human body, humoral IFNs serve as the first lines of cellular defense in the control of viral infection. These IFN-induced molecules all lead to effective control of viral expansion, either by inhibiting viral replication or by promoting infected cells to undergo apoptosis. Supposedly, a combination of these molecules in each person will determine the individual's varying degree of response to IFN treatment. However, the genes encoding these molecules are conserved in humans with only certain genetic polymorphisms.

6.3 IL-28B Expression and Viral Diseases

The recent GWAS articles linking the IL28B genotype to IFN-alpha therapeutic response have triggered intensive research to establish underlying mechanisms for the association

Table 28.9 The most significant genetic predictors of drug response

Organ or system involved phenotype	Associated gene/allele	Drug/drug response
HIV-1 infection	*CCR5*	Maraviroc efficacy
Hepatitis C infection	*IL28B*	Interferon-alpha efficacy

(Liapakis and Jacobson 2010). At this point, there are very few possible explanations and few strong, mechanistic clues (Thio and Thomas 2010). It is speculated that these variants correlate with the regulation of the cytokine IL28B transcription, because these SNPs are located upstream of the IL28B gene (Thio and Thomas 2010). The observations cited earlier by Suppiah and colleagues and Tanaka and colleagues strongly suggest that the identified SNPs do indeed alter the expression of the IL28B and, perhaps, IL28A genes (Suppiah et al. 2009; Tanaka et al. 2009). These two studies found that those who carried the G risk allele at the SNP rs8099917 had lower mRNA expression of the cytokine IL28B in peripheral blood mononuclear cells (Suppiah et al. 2009; Tanaka et al. 2009). On the other hand, Ge and colleagues reported no difference in cytokine IL28B expression in peripheral blood mononuclear cells from 80 HCV-uninfected persons homozygous for a proxy allele for the SNP rs12979860, using the SNPExpress database (Ge et al. 2009). Thus, various SNPs showed their variation during response; Table 28.9 shows the most significant predictors of drug response.

6.4 Type III Interferon-Based Therapy

As presented earlier, at least four independent GWAS studies provide significant evidence for the role of the IL28B gene in the pathogenesis of HCV infection (Ge et al. 2009; Suppiah et al. 2009). However, still more pieces of data are needed to complete the mechanistic picture (Thio and Thomas 2010). The IL28B gene encodes IFN-lambda-3, which is one of the recently discovered type III interferons (or lambda interferons), and belongs to the IL10 superfamily (Dellgren et al. 2009). Other type III interferons are the cytokines IL28A and IL29. Some murine studies have revealed the preferential expression of IFN-lambda receptors on epithelial surfaces and have suggested that type III interferons may allow the host to rapidly eliminate viruses at the major portals of entry into the body before infection is established, without activating other arms of the immune system (Ank et al. 2006, 2008; Ank and Paludan 2009). Nonetheless, one important difference established between the murine and human systems is that the interleukin (Liapakis and Jacobson 2010); receptor alpha chain (receptor complex induced by IFN-lambda) is expressed in human hepatocytes, whereas the murine liver seems unlikely to respond to IFN-lambda (Ank et al. 2006, 2008; Ank and Paludan 2009). This finding suggests that IFN-lambda contributes to host defense against hepatotropic viruses, such as HCV in humans. Like type I interferons, lambda interferons have activity against HCV and other viral infections in vitro and in vivo (Marcello et al. 2006). However, in vitro exogenous IFN-lambda induces a slower, more sustained abundance of IFN-stimulated genes than IFN-alpha (Marcello et al. 2006). Although these findings answer why IFN-lambda might play a key role in HCV recovery, they do not explain why treatment-associated resolution of HCV infection is associated with certain base sequences located upstream of the start codon for the cytokine IL28B (Thio and Thomas 2010).

7 Transition: From Lab to Bench Site

In addition to the mechanistic questions of profound importance, the recent GWAS findings on genetic variation as a predictor of outcome in patients treated for HCV infection will have major practical implications. Several recent GWAS studies have demonstrated remarkable associations between SNPs near or within the region of the IL28B gene, which codes for IFN-lambda-3 (Ge et al. 2009; Suppiah et al. 2009; Tanaka et al. 2009; Rauch et al. 2010).

These results promise to lead to important mechanistic findings related to IFN responsiveness in this disease and will probably have major contributions for individualized medicine and therapeutic decision making. Ge and colleagues suggest that, in the near future, advance knowledge of the genotype of patients infected with HCV could become an important component of the clinical decision to initiate treatment. These findings may also lead to more individualized treatment regimens with regard to both the chosen medicines and the duration of therapy (Ge et al. 2009; Suppiah et al. 2009; Tanaka et al. 2009; Rauch et al. 2010). Moreover, the critical importance of the IL28B gene region in mediating response has emphasized interest in the development of IFN-lambda as a therapeutic agent for patients with CHC (Doyle et al. 2006; Pagliaccetti et al. 2008). For instance, the early studies of a pegylated form of IL29 (IFN-lambda-1) have demonstrated promising antiviral activity with the potential benefit of reduced hematologic toxicity, owing to the hepatocyte-specific receptor profile for IFN-lambda (versus IFN-alpha) (Doyle et al. 2006; Pagliaccetti et al. 2008).

7.1 IFN-Alpha Pathway

In a recent candidate gene study, Welzel and colleagues investigated the association between genetics and HCV treatment response with a focus on the IFN-alpha pathway (Welzel et al. 2009). They genotyped 56 SNPs along the IFN-alpha pathway in 1,051 patients in the Hepatitis C Long-term Treatment Against Cirrhosis (HALT-C) trial, utilized TaqMan® assays (Applied Biosystems, Carlsbad, CA, USA) with analysis on the ABI 7900HT platform (Applied Biosystems), and focused on European Americans ($n = 581$) for purposes of statistical power (Welzel et al. 2009). Participants with fibrosis score 3 had not previously responded to IFN treatment with or without RBV, had a Child–Turcotte–Pugh score of less than seven, and were treated with peg-IFN and RBV for 24 weeks and then up to 48 weeks if undetectable at week 20. Recent study by Welzel and colleagues, they examined the SNPs in genes encoding IFN-alpha, the IFN-alpha receptor, and JAK/tyrosine kinase/STAT 1 and 2, all part of the signal transduction via the JAK–STAT pathway; IFN-alpha-induced genes with antiviral properties, including adenosine deaminase/eukaryotic translation initiation factor 2A-alpha kinase 2/NFKB1/myxovirus resistance 1/2′5′-oligoadenylate synthetase 1; as well as interferon regulatory factor (Welzel et al. 2009). In a recent study, Ke and colleagues (2010) extended the previous research and applied both ANN algorithms and logistic regression with feature selection to predict IFN-alpha and RBV treatment outcomes using genetic factors. The cohort of 523 CHC patients was original to the previous study by Lin (2006).

In addition to rapid patient-specific drug-resistance phenotyping for the management of antiretroviral therapy, structural pharmacogenomics can be used for the prediction of clinical trial outcomes. By computationally analyzing the interaction of a putative drug with a sampling of the polymorphism structural repertoire that a new drug will encounter in the clinic, it is possible to predict the binding effectiveness of a new drug before the initiation of expensive clinical evaluation in patients. This is a straightforward application and extension of the recently reported successful computational prediction of drug resistance phenotypes for HIV-1 protease on a patient-by-patient basis (Shenderovich et al. 2003). The interaction and movements of side chains in individual polymorphism structures in response to drug binding can be computed, for example, comparison of the computed structural changes upon drug binding with the observed changes in the corresponding X-ray crystallographic structure (Hong et al. 2000) of the saquinavir–HIV protease complex.

7.2 Drug Efficacy and Adverse Effects of Viral Drugs

The sensitivity and specificity for predicting drug resistance phenotypes on a patient-by-patient basis for various Food and Drug Administration (FDA; http://www.fda.gov)-approved drugs,

Table 28.10 Prediction of drug efficacy

Protease inhibitor	Sensitive below	Resistant above	Sensitivity (%)	Specificity (%)	Kappa²	*p* value
Amprenavir	0.7	1.4	86.7	100	0.907	<0.0001
Indinavir	0.6	1.5	94.1	100	0.958	<0.0001
Nelfinavir	0.7	1.0	60.6	96.8	0.567	<0.0001
Ritonavir	0.7	1.4	100	84.1	0.754	<0.0001
Saquinavir	0.6	1.1	68.4	100	0.752	<0.0001
Lopinavir	0.3	0.7	100	83	0.755	<0.0001

Comparison of computational drug resistance phenotype with laboratory phenotyping ΔE bind cuttoff (kcal/mol)
ViroLogic PhenoSense (ViroLogic, http:/www.virologic.com)
Kappa is measure of inter-assay agreement: kappa > 0.75, excellent agreement; 0.4 < kappa < 0.75, good agreement; kappa < 0.4, poor agreement
Nevirapine and tipranavir (Boehringer Ingelheim, http://www.boehringer-ingelheim.com)
Saquinavir (Roche, http//www.roche.com)
Amprenavir (Glaxo Wellcome, http://www.gsk.com)
Indinavir (Merck, http://www.merck.com)
Nelfinavir (Pfizer, http://www.pfizer.com) (Lopinavir)
Ritonavir (Abbot, http://www.abbot.com)

based upon the computed relative energy (ΔE *bind*) of interaction of each drug with each patient's unique drug-target polymorphism structure. The numbers appear to be quite good in comparison to what is achievable in the laboratory at present. However, perhaps what is most important, with respect to increased efficiency and speed in drug design, is that answers can be generated overnight rather than in 3–6 weeks (the typical timeframe for laboratory drug resistance phenotyping). Another way of looking at the interaction of putative drugs or drug leads with a set of drug-target polymorphisms is to dock the molecules to the individual polymorphisms and measure the distance of the drug to each residue within the protein. Such an analysis for five different protease inhibitors is shown in Table 28.10: as prediction of drug efficacy.

The mutation frequency profile is superimposed at the bottom of the graph, and this analysis clearly identifies those aminoacyl side chains that mutate with a high frequency and are in close proximity to the bound drug, those that are proximal and stable, those that are distal and mutate, and those that are distal and are stable. The Variome™ structural pharmacogenomics technology has broad applications in the rational design of highly effective infectious disease therapies – bacteria as well as viruses, drugs as well as vaccines – that offer the prospect of stable efficacy in the face of drug selection pressure. In addition to naturally occurring infectious agents, it is clear that there is significant value in biodefense-related applications in rational drug and vaccine design and in threat assessment and prediction. Understanding and managing the adverse effects of antiretroviral therapy highly active antiretroviral therapy (HAART) has changed the landscape of HIV disease in a way that seemed unthinkable a decade ago. The first HAART regimens worked in suppressing virus but were encumbered by a variety of short-term and long-term side effects. More recent regimens became simpler, easier to take, and with fewer adverse events. Knowledge of both the short- and long-term adverse events associated with HAART is essential for providers and for patients. For new drugs to be acceptable in the current field, they will have to pass a litmus test of tolerability. Since adverse events are often remarkably idiosyncratic, pharmacogenomics may offer a way of predicting side effects and their severity from a particular drug or drug class in individual patients.

Adverse events (AEs) play a major role in determining adherence to highly active antiretroviral therapy (HAART), and adherence is perhaps the most significant determinant of a

Table 28.11 Adverse effects associated with different classes of antiretroviral

Class	Drug	Adverse effects
NRTIs	Zidovudine	Anemia, nausea, rash, myopathy, dyslipidemia
	Stavudine, didanosine	Nausea, lipoatrophy, dyslipidemia, pancreatitis, lactic acidosis, hepatic steatosis
	Abacavir	HSR, hepatotoxicity
	Tenofovir	Renal insufficiency, bone loss
NNRTIs	Efavirenz	CNS adverse effects, rash, hepatotoxicity, teratogenicity, hypertriglyceridemia
	Nevirapine	HSR, rash, hepatotoxicity,
	Etravirine	Rash, hepatotoxicity
PIs	All PIs	Nausea, rash, hepatotoxicity, dyslipidemia, diarrhea, insulin resistance
	Atazanavir	Jaundice, sclera icterus, nephrolithiasis
	Lopinavir, fosamprenavir	Heart disease
Integrase inhibitors	Raltegravir	Headache, insomnia, dizziness, fatigue

regimen's success (d'Arminio Monforte et al. 2000). While randomized, controlled clinical trials are the gold standard for evaluating the efficacy of drugs, they may underestimate short-term toxicity in the general clinic population because of the desire of the subjects to stay in the trial and the support from the trial staff in enabling them to do so. Long-term toxicities may be missed because of the often younger age of the subjects in clinical trials and because the relatively short-term duration of these trials may not detect toxicities with a low prevalence rate; Table 28.11 shows the adverse effects associated with different classes of antiretroviral. AEs of antiretroviral drugs can be usefully divided into short- and long-term toxicities and also by the class of agent used.

8 Mechanisms of Resistance

Considerable progress has been made in identifying mutations associated with drug resistance (Table 28.12) and in understanding the mechanisms through which they confer resistance (Table 28.13). A variety of mechanisms have been identified that differ both for different classes of drugs and for drugs of a given class.

8.1 Resistance to Nucleoside and Nucleotide Analogues

Nucleoside analogues and nucleotide analogues (Table 28.13) block the synthesis of viral DNA by reverse transcriptase enzyme. After phosphorylation by cellular kinases, these compounds are incorporated by reverse transcriptase into the nascent chain of viral DNA. Because these drugs lack a 3′ hydroxyl group, no additional nucleotides can be attached to them, and the synthesis of viral DNA is arrested. Two distinct mechanisms are involved in HIV resistance to these drugs: impairment of the incorporation of the analogue into DNA and removal of the analogue from the prematurely terminated DNA chain.

8.2 Impairment of Analogue Incorporation

Several mutations or groups of mutations in reverse transcriptase enzyme can promote resistance by selectively impairing the ability of reverse transcriptase to incorporate an analogue into DNA. They essentially include the M184V mutation, the Q151M complex of mutations, and the K65R mutation (Table 28.12). The M184V

Table 28.12 Mutation involved in resistance of HIV to nucleoside analogues, nonnucleoside reverse transcriptase inhibitors (NNRTIs), and protease inhibitors

Mutation	
Reverse transcriptase	Comments
Mutations conferring resistance to nucleoside analogues	Family of mutations known as thymidine analogue mutations. Associated with resistance to most nucleoside analogue except lamivudine
M41L, D67N	In vitro cause high-level resistance to zidovudine and low-level resistance to stavudine, didanosine, and abacavir
K70R, L210W	Segregate in two pathways, one comprising T215Y and L210W and the other T215F and K219Q
T215Y, T215F, K219Q, K219E	Pathway comprising T215Y and L210W associated with decreased responsiveness to tenofovir
M184V	Observed in most viruses resistant to treatment with lamivudine
	Confers high-level resistance to lamivudine in vitro
	Can interfere with resistance to zidovudine and stavudine when the number of thymidine analogue mutations is small
	Increase the level of resistance to didanosine and abacavir owing to thymidine analogue mutations
Q151M	Rare pathway for resistance of HIV-1 to nucleoside analogues
F116Y, F77L, V75I, A62V	In vitro, cause high-level resistance to most nucleoside analogues except lamivudine and tenofovir
69 Insertional mutations	Insertion of two or more amino acids (usually serines) next to codon 69
	Emerge only in viruses that already have several thymidine analogue mutations
	Confer high-level resistance to all nucleoside analogues
K65R	Selected for by zalcitabine, abacavir, and tenofovir therapy
Y115F	Selected for by abacavir therapy
L74V	Selected for by didanosine therapy, usually when didanosine is the only nucleoside analogue
Mutations conferring resistance to NNRTIs	
K103N	Mutation most frequently selected for by efavirenz therapy
	Occasionally selected for by nevirapine therapy
	Confers high-level resistance to all available NNRTIs
Y181C	Mutations most frequently selected by nevirapine
Y188C	Confers high-level resistance to nevirapine but lower-level resistance to efavirenz
Y108I	Y188L, unlike Y188C, seen mostly with efavirenz therapy
L100I, V106A, G190A, G190S	Mutations that accumulate during prolonged ineffective therapy with most NNRTIs
Protease and gag mutations	
L90M	Frequent resistance mutation, observed during failure of therapy with most protease inhibitors
	Mutation most frequently selected for by saquinavir therapy
V82A, V82T, V82F	Common resistance mutations
	Can emerge early during failure of therapy with most protease inhibitors
	Mutations most frequently selected for by ritonavir and indinavir therapy
D30N	Mutations most frequently selected for by nelfinavir therapy
N88D, N88S	D30N always first
L101, L10F, K20R,K20M, M36I, M46I, M46L, I54V, I54L, A71V, A71T, G73S, V77I, M93L	Mutations that can accumulate during failure of therapy with most protease inhibitors, causing gradual increases in the level of resistance

(continued)

Table 28.12 (continued)

Mutation Reverse transcriptase	Comments
I84V	Frequently found after prolonged ineffective therapy with protease inhibitors Associated with high-level resistance to most protease inhibitors
G48V	Exclusively selected for by saquinavir therapy
	Associated with high-level resistance to saquinavir
L24I	Emerges occasionally during failure of indinavir therapy
	Also found with lopinavir therapy
I47V, I50V	Most often selected for by amprenavir therapy
	Also found with lopinavir therapy
V32I, F53L	Rare mutations
	Confer high-level resistance to most protease inhibitors
A431V	Mutations in gag, the main viral substrate of the protease
L449F	Increases resistance and partially compensates for resistance-associated loss of viral replicative capacity

Mutations are designated according to the letter of the wild-type amino acid that is subject to substitution, followed in turn by the position of that amino acid in the reverse transcriptase or protease sequence and by the letter of the mutant amino acid. For example, M184V indicates that methionine at position 184 is replaced by a valine. A complete description of HIV drug-resistance mutations and the latest information on their interpretation can be found at http://hivdb.stanford.edu and http://www.iasusa.org/resistance_mutations/index.html
30 (International AIDS Society-USA 2002)

Table 28.13 Antiretroviral agents used in the treatment of HIV infection

Drugs	Mechanisms of action	Mechanisms of resistance
Nucleoside analogues	Analogues of normal nucleosides	Thymidine analogue mutations promote ATP-mediated and pyrophosphate-mediated excision of the incorporated terminator M184V or Q151M complex mutations and impair incorporation of nucleoside analogues
Zidovudine	Active as triphosphate derivatives	
Stavudine	Incorporated into nascent viral DNA	
Lamivudine	Prematurely terminate HIV DNA synthesis	
Didanosine		
Zalcitabine		
Abacavir		
Nucleotide analogues	Same as nucleoside analogues	K65 R impairs incorporation of tenofovir into DNA thymidine analogue mutations often associated with cross-resistance to tenofovir
Tenofovir		
Nucleoside reverse transcriptase inhibitors	Bind a hydrophobic pocket of HIV type 1 reverse transcriptase	Mutations reduce affinity of the inhibitors for the enzyme
Nevirapine	Block polymerization of viral DNA	Single mutations generally sufficient to induce high level of resistance
Efavirenz	Inactive against HIV type 2	
Delavirdine		
Protease inhibitors	Structure derived from natural peptidic substrates of HIV type 1 protease	Mutations reduce affinity of the inhibitors for the enzyme
Saquinavir		
Ritonovir	Bind the active site of the protease	High-level resistance requires accumulation of mutations
Indinavir		
Nelfinavir		
Amprenavir		
Lopinavir		
Fusion inhibitors	Thirty-six amino acid peptide derived from the HR2 domain of glycoprotein 41	Mutations affect HR1, a domain of glycoprotein 41 whose interaction with HR-2 promotes membrane fusion
Enfuvirtide	Interferes with glycoprotein 41-dependent membrane fusion	

mutation involves the replacement of methionine by valine at position 184 of the reverse transcriptase and is the main mutation that confers resistance to lamivudine (Boucher et al. 1993). Methionine 184 is located at the heart of the catalytic site of reverse transcriptase, and its replacement by a valine, which has a different side chain, interferes with the proper positioning of lamivudine triphosphate within the catalytic site (Sarafianos et al. 1999). The M184V mutation induces very high levels of resistance to lamivudine. When lamivudine is used as a single agent, resistant strains overtake wild-type virus in a few weeks (Schuurman et al. 1995), and when lamivudine is used as part of a failing regimen of HAART, the M184V mutation is almost always the first mutation to emerge (Havlir et al. 2000). The group of mutations referred to as the Q151M complex (Iversen et al. 1996) (Table 28.12) is most often selected for in the course of the failure of regimens containing stavudine and didanosine. This pathway always starts with the Q151M substitution, a residue located in the immediate vicinity of the nucleotide-binding site of reverse transcriptase, and is followed by the gradual accumulation of secondary mutations that enhance resistance and increase the activity of the enzyme (Kosalaraksa et al. 1999). The Q151M complex is relatively rare in HIV-1 (fewer than 5 % of all HIV strains with resistance to nucleoside analogues) but can confer high-level resistance to most but not all (e.g., lamivudine and tenofovir) analogues (Iversen et al. 1996). Interestingly, the Q151M complex is markedly more frequent in HIV-2 than in HIV-1. The K65R mutation is seen with increasing frequency in patients in whom therapy with nucleoside or nucleotide analogues fails, especially when the regimen includes tenofovir or abacavir. This mutation appears to confer resistance to most analogues, with the exception of zidovudine.

8.3 Removal of the Analogue from the Terminated DNA Chain

Removal of the nucleoside analogue from the terminated DNA chain is associated with a group of mutations commonly termed "thymidine analogue mutations" (Table 28.12). Mutations from this group are most frequently selected for after the failure of drug combinations that include thymidine analogues, such as zidovudine and stavudine, but they can promote resistance to almost all nucleoside and nucleotide analogues, including tenofovir (Larder and Kemp 1989; Picard et al. 2001). These mutations occur gradually, and their order of emergence can vary. Thymidine analogue mutations promote resistance by fostering ATP- or pyrophosphate-mediated removal of nucleoside analogues from the 3′ end of the terminated DNA strand (Meyer et al. 1999). ATP and pyrophosphate, which are abundant in normal lymphocytes, do not participate in the DNA-polymerization reaction, but the structure of a reverse transcriptase expressing thymidine analogue mutations facilitates their entry into a site adjacent to the incorporated analogue (Chamberlain et al. 2002). In this position, ATP or pyrophosphate can attack the phosphodiester bond that links the analogue to DNA, resulting in removal of the analogue. Interestingly, the efficiency of this process, also known as "primer rescue," can be significantly decreased by the presence of other mutations in reverse transcriptase, a phenomenon that has been best described in the case of the M184V mutation (Larder et al. 1995). As a consequence; M184V slows the selection of thymidine analogue mutations by thymidine analogues (Picard et al. 2001) and may slightly increase the residual antiviral activity of some nucleoside analogues in spite of the presence of thymidine analogue mutations.

8.4 Resistance to Nonnucleoside Reverse Transcriptase Inhibitors

Nonnucleoside reverse transcriptase inhibitors are small molecules that have a strong affinity for a hydrophobic pocket located close to the catalytic domain of the reverse transcriptase. The binding of the inhibitors affects the flexibility of the enzyme, thereby blocking its ability to synthesize DNA. The mutations that are selected for after the failure of treatment with nonnucleoside reverse transcriptase inhibitors are all located in the pocket targeted by these compounds, and they reduce the affinity of the drug (Esnouf et al. 1997; Hsiou et al. 2001). Because of subtle differences in the interaction between various nonnucleoside reverse transcriptase inhibitors and the hydrophobic pocket, however, the mutations that emerge most frequently are somewhat drug dependent (Table 28.12). Resistance to nevirapine is often associated with the Y181C mutation, but other mutations, such as Y188C, K103N, G190A, and V106A, also occur. Initial resistance to efavirenz is generally characterized by the K103N mutation, but the Y188L mutation is also seen.

8.5 Resistance to Protease Inhibitors

The HIV protease cleaves large polyprotein precursors at specific sites, releasing the structural proteins and enzymes necessary for the assembly of infectious viral particles. In the absence of a functional protease, viral particles are produced, but they are immature and are not infectious. The protease of HIV is a symmetrically assembled homodimer with a central, symmetric, substrate-binding cavity. Detailed knowledge of the structure of this domain and of the structure of the natural protein substrates of the enzyme has led to the design of specific inhibitors whose chemical structure mimics that of the viral peptides that are normally recognized and cleaved by the protease (Roberts et al. 1990).

These compounds display a strong affinity for the active site of the HIV protease and inhibit the catalytic activity of the enzyme in a highly selective manner. Resistance to protease inhibitors is the consequence of amino acid substitutions that emerge either inside the substrate-binding domain of the enzyme or at distant sites (Molla et al. 1996) (Table 28.13). Directly or indirectly, these amino acid changes modify the number and the nature of the points of contact between the inhibitors and the protease, thereby reducing their affinity for the enzyme (Hong et al. 2000; Prabu-Jeyabalan et al. 2002). As an example, the common resistance mutation V82A reduces the size of an amino acid residue in the protease that is more important for binding most inhibitors than for binding the natural viral protein substrate (Prabu-Jeyabalan et al. 2002).

Protease inhibitors have been designed to bind the protease with maximal affinity and tend to occupy more space inside the active site cavity than do natural substrates. Unlike the inhibitors, the natural substrates of the protease have a variable, but generally less tight, interaction with the catalytic site, a phenomenon that promotes the ordered sequential cleavage of the polyproteins required for proper assembly of the viral particle. Resistance mutations in the protease, which result in an overall enlargement of the catalytic site of the enzyme, would thus be predicted to have a greater effect on the binding of inhibitors than the natural templates. Some mutations are selected for only by certain protease inhibitors (Table 28.12), reflecting particularities in the chemical structure of the inhibitors that influence their interaction with the substrate-binding domain of the enzyme. However, there is considerable overlap between the combinations of mutations in HIV strains that develop resistance to protease inhibitors. This overlap explains the wide cross-resistance that is generally observed within this drug class (Schapiro et al. 1999).

Remarkably, resistance to protease inhibitors can also be promoted by mutations in some of the natural viral substrates of the protease (Mammano et al. 1998). Characteristic substitutions of amino acids near cleavage sites of the gag polyprotein have been identified that can increase the level of resistance and the replicative capacity of the virus by facilitating cleavage under conditions in which the amount of active enzyme is suboptimal or improving the ability of proteases containing resistance mutations to interact with the substrate.

8.6 Resistance to Fusion Inhibitors

HIV-1 enters target cells through an intricate sequence of interactions between the HIV envelope glycoprotein (gp) complex (gp120–gp41) and specific cell-surface receptors (Kilby and Eron 2003). The early steps in this process allow gp41, the fusogenic component of the complex, to interact with the cell membrane, thereby tethering the virus to its target. The membranes of the virus and target cell are then brought into close proximity, fostering their fusion, by further rearrangement of gp41. In this step, a distal hydrophobic region of gp41, HR2, folds onto a more proximal hydrophobic region, HR1, effectively shortening the molecule. Enfuvirtide, a 36-aminoacid peptide derived from HR2, destabilizes this process by binding to HR1 and blocks the infectivity of HIV-1. Viral resistance to enfuvirtide usually results from mutations located in a stretch of ten amino acids within HR1 (Rimsky et al. 1998). Interestingly, changes in amino acids in gp41 outside HR and even changes in gp120 appear to be associated with significant differences in the susceptibility of the virus to enfuvirtide. These mutations or polymorphisms probably explain the remarkably wide range of natural susceptibility to enfuvirtide among HIV-1 strains (Derdeyn et al. 2001; Reeves et al. 2002) and could participate in the evolution of acquired resistance to enfuvirtide.

8.7 Cross-Resistance

Cross-resistance, defined as resistance to drugs to which a virus has never been exposed, results from mutations that have been selected for by the use of another drug. Cross-resistance is always restricted to drugs within a given class of antiretroviral agents but all three classes of antiretroviral drugs are affected. Early in the evolution of resistance to nucleoside analogues or protease inhibitors, viruses may have only a low level of cross-resistance to alternative agents within each of these two classes of drugs (Richman 1990). Nevertheless, these strains may need to add only one or a few additional mutations to this preexisting scaffolding for high-level cross-resistance to develop. Therefore, in patients infected with strains that have low levels of cross-resistance, the switch to apparently active alternative drugs can be accompanied by rapid selection for highly resistant variants, at the expense of minimal evolutionary changes (Dulioust et al. 1999). Resistance is not an all-or-nothing phenomenon and generally increases over time (Richman 1990; Barbour et al. 2002; Ross et al. 2000).

Single mutations rarely produce complete resistance to antiretroviral drugs, although the M184V mutation in reverse transcriptase, which results in complete resistance to lamivudine, is an exception to this rule. Single mutations in the hydrophobic pocket of reverse transcriptase also provide strong resistance to nonnucleoside reverse transcriptase inhibitors, but the viral strains in patients in whom regimens using these drugs are failing often incur additional mutations, suggesting that the level of resistance provided by single mutations is not optimal (Hanna et al. 2000). Resistance to reverse transcriptase inhibitors through the accumulation of thymidine analogue mutations or Q151M complex mutations and resistance to protease inhibitors are always gradual processes, leading to progressive increases in the level of resistance (Mammano et al. 2000). Nonetheless, even after several mutations have accumulated in the HIV protease and reverse transcriptase sufficient to produce patent treatment failure, resistance may not have reached maximal levels, and additional mutations, associated with increases in resistance, can occur even in the absence of any changes in treatment (Barbour et al. 2002). Indeed, the development of complete resistance may represent an exception. Because resistance mutations can impair viral replicative capacity, the solution adopted by the dominant viral population may entail making concessions in terms of resistance. Efforts are being made to evaluate the antiviral activity retained by individual drugs in the face of resistance mutations. Such knowledge may help in the treatment of patients infected with viruses that express resistance to multiple classes of drugs, since keeping HIV under some pharmacologic pressure may reduce its pathogenic potential.

8.8 Effect of Resistance on Viral Replicative and Pathogenic Capacity

Many resistance mutations impair viral replication. Because these mutations modify key viral proteins, they have deleterious effects of variable extent on protein function. Although some of these deficits can be partially corrected by compensatory mutations (Nijhuis et al. 1999), viruses forced to develop higher levels of resistance under intense and continuous pressure by antiretroviral drugs often have a substantial impairment in their replicative capacity (Barbour et al. 2002). The degree of replicative impairment conferred by resistance mutations is highly variable. The most crippling mutations appear to be those associated with resistance to protease inhibitors (Zennou et al. 1998), but significant replicative defects can be related to mutations in reverse transcriptase that confer resistance to nucleoside analogues or nonnucleoside reverse transcriptase inhibitors (Back et al. 1996; Bleiber et al. 2001) as well as envelope mutations associated with resistance to enfuvirtide. The clinical effect of resistance-associated loss of viral replicative capacity is the subject of intense investigation. Several observations suggest that resistant viruses have lost some of their virulence. When antiretroviral treatment is interrupted in patients infected with HIV that is resistant to multiple drugs, the resistant strain is more or less rapidly replaced by wild-type virus. This change is accompanied by a drop in CD4 T-cell counts, suggesting that wild-type virus has a greater replicative and pathogenic potential (Deeks et al. 2001). In a significant proportion of patients in whom HAART is failing, the CD4 T-cell counts remain significantly above pretreatment levels, despite the poor control of replication (Kaufmann et al. 1998). Whether this apparent immunologic benefit of HAART in spite of patent virologic failure is the direct consequence of reduced viral pathogenicity, a sign of persistent residual activity of some drugs in the regimen, or both, remains to be elucidated. Consequently, therapeutic strategies that take advantage of resistance-associated loss of HIV replicative capacity have yet to be identified.

However, recent data of CATCH and the US-based studies suggested that transmitted drug-resistant HIV can remain the dominant population in peripheral blood for an extended period of time. Furthermore, a study analyzed the baseline nucleotide sequences among 11 subjects with primary HIV infection who showed variation against antiretroviral therapy (Little et al. 2004). The patients had been infected for approximately 65 days and had at least one major drug resistance mutation. Longitudinal samples were collected for a median of 225 days after infection and analyzed for persistence of transmitted drug-resistant variants. Seven patients had evidence of NNRTI resistance, two patients had resistance to both NRTIs and PIs, and one patient had triple-class resistance. The average time to reversion of the K103N variants to mixed 103 N/K populations in the seven patients with NNRTI resistance was 196 days following the estimated date of infection. In the four patients with mutations conferring resistance to protease inhibitors, no reversion to wild type was detected at 64, 191, 327, and 342 days after infection. Complete reversion of genotypic resistance was observed in only one patient at 1,019 days after infection. Dr. Kuritzkes concluded the data presented at the 11th CROI by a team of investigators at Gilead Sciences (Borroto-Esoda et al. 2004). Study FTC-301A compared once-daily emtricitabine (Emtriva) to once-daily stavudine (Zerit), both combined with didanosine (Videx) and efavirenz (Sustiva), in 571 treatment-naïve patients from North America, Latin America, and Europe. Overall, 90/546 (16 %) evaluable patients entered the study with HIV mutations at positions associated with resistance to NNRTIs. There were no differences in the prevalence or type of mutations between the groups. For subjects with wild-type virus at baseline, the incidence of virologic failure was 12 % in the stavudine group and 5 % in the emtricitabine group. Among patients with drug-resistant virus at baseline, virologic failure occurred in 32 % of the stavudine group and 13 % of the emtricitabine group.

8.9 Mutations

8.9.1 The K65R Mutation

Perhaps the most widely cited case in point regarding the dangers of K65R is GlaxoSmithKline's ESS3000009 trial (Gallant et al. 2003). In this study, 345 patients were randomized to receive either tenofovir or efavirenz, combined with a fixed-dose combination tablet containing 600 mg abacavir and 300 mg lamivudine, all to be taken once a day. Patients were naïve to antiretroviral therapy prior to starting the study and had an average baseline viral load of 4.63 log10 copies/ml and a baseline CD4+ cell count of 260 cells/mm^3. Because of a glaringly high number of treatment failures being documented in the triple-NRTI group, the study investigators conducted an unplanned analysis involving the first 194 patients who completed 8 weeks of follow-up. Approximately 49 % of patients in the triple-NRTI group met the definition of virologic failure, compared to only 5.4 % patients in the efavirenz-based arm. Tenofovir, abacavir, and lamivudine can all select for the K65R mutation. Among 21 patients who had viral loads that were high enough to test for drug resistance, 10 (48 %) had the K65R mutation. Tenofovir, abacavir, and lamivudine all exert concerted pressure on the 65 locus, and, with the emergence of that single mutation, it is able to abrogate the efficacy of the regimen with a single stroke.

8.9.2 TAMs and K65R

NRTIs exert a blocking effect by plugging a non-extendable nucleoside analogue monophosphate to the 3′ end of the growing proviral DNA chain. This effectively terminates chain extension and, ultimately, inhibits replication of the virus. However, this process can be reversed by a reverse transcriptase reaction that removes the chain-terminating residue and reinstates an extendable primer. This reverse reaction of DNA polymerization, termed pyrophosphorolysis, enables reverse transcription and DNA synthesis to resume. Pyrophosphorolysis can be enhanced by key mutations, often referred to as thymidine analogue mutations (TAMs). While pyrophosphorolysis is believed to be the primary mechanism of resistance to zidovudine and stavudine, the process is not drug specific in the way that discriminatory mutations tend to be. Consequently, these pyrophosphorolysis-enhancing mutations can confer reduced susceptibility to all of the NRTIs.

In essence, NRTI resistance is the sum of two mechanisms: (1) their ability to alter drug binding or incorporation, through the mechanism described in the beginning of this section, plus (2) their ability to alter the excision of drug from the nascent DNA chain. The 65R mutation impedes both drug binding and primer excision.

In effect, variants harboring both K65R and TAMs still considered to be a rare occurrence can have surprising effects. Many TAMs are in the "fingers" domain of the polymerase, which makes up the deoxynucleotide triphosphate binding site. In turn, TAMs interfere with proper function of the enzyme. Because K65R impedes primer excision, it is believed to sensitize HIV to NRTIs affected by TAMs, most notably zidovudine. It has also been noted that TAMs can reverse some of the effects of K65R on drug binding and incorporation; there are data suggesting that concurrent zidovudine use can prevent the emergence of the K65R mutation though the mechanism responsible for this has not yet been elucidated.

Of interest are data from six recent clinical trials evaluating all reverse transcriptase inhibitor regimens containing tenofovir, either with or without zidovudine. Among patients initiating therapy for the first time, tenofovir-inclusive regimens not involving zidovudine were much more likely to fail in association with development of the K65R mutation. Conversely, tenofovir- and zidovudine-inclusive regimens were much more likely to fail in association with the development of TAMs.

8.9.3 The Role of Minor Variants

Moving on to resistance issues in heavily pretreated HIV-positive patients, Dr. Kuritzkes reviewed data stemming from ACTG 398, a clinical trial designed to determine whether the addition of a second protease inhibitor to a regimen containing amprenavir (Agenerase) improved the 24-week response to salvage therapy (Mellors et al. 2003). The study enrolled 481 heavily pretreated HIV-positive patients: 21 % had been on one protease inhibitor in the past, 53 % had been on two prior protease inhibitors in the past, and 26 % had been on three protease inhibitors in the past. Approximately 44 % of the patients had also been on an NNRTI in the past.

All of the patients received efavirenz, adefovir dipivoxil, abacavir, and amprenavir. In addition, subjects were randomized to receive either placebo, nelfinavir (Viracept), saquinavir (Fortovase), or indinavir (Crixivan). There were no statistically significant differences between the three active drug arms. There was, however, a significant difference between the combined active arms and the placebo arm, demonstrating that dual protease inhibitor regimens are superior to single protease inhibitor regimens in heavily pretreated patients.

Not surprisingly, the ACTG 398 investigators reported that NNRTI-naive patients had significantly better responses than NNRTI-experienced patients: 84 % of NNRTI-experienced patients, compared to 57 % of the NNRTI-naive patients, failed to achieve an undetectable viral load after 24 weeks of follow-up. Of interest, though, is an analysis of the NNRTI-experienced patients. After 24 weeks, however, the difference in response rates between these two groups became insignificant and eventually converged.

Reviewing some recent work in the area of thymidine analogue mutations, a review study results explained that certain TAMs tend to occur together but that others are rarely found together in the same virus. For example, the M41L/L210W/T215Y pathway is a common TAM pattern and is associated with high-level resistance to zidovudine and with cross-resistance to other NRTIs, including tenofovir and abacavir. In contrast, the D67N/K70R/K219Q pathway isolate is a less common TAM pattern and is associated with a lower fold resistance than the M41L/L210W/T215Y cluster. It was also showed that T215F mutation rarely appears in the reverse transcriptase of viruses also harboring the L210W mutation or the M41L mutation (Kuritzkes 2004) and the divergent TAM pathways (Hu et al. 2004). In a nutshell, these studies found that isolates carrying the T215Y mutation replicated more efficiently than isolates harboring the 215F mutation. With incorporation of the L210W mutation into isolates harboring the T215Y mutation, viral fitness was substantially reduced. Conversely, with the incorporation of the K70R mutation into isolates harboring the T215Y mutation, there was a substantial growth advantage in the presence of zidovudine (Hu et al. 2004). Thus, it was concluded that T215Y pathway is more common and confers higher-level resistance to zidovudine and other NRTIs, whereas viruses carrying T215F have lower replication capacity and are poorly fit. With respect to the K70R mutation, it confers a significant advantage to HIV in the presence of zidovudine and that this mutation plays a much larger role in the development of zidovudine resistance than is usually considered (Kuritzkes 2004).

Perhaps the best-known reverse transcriptase mutation is M184V, which is known to cause high-level resistance to lamivudine and emtricitabine. However, there have been studies suggesting that high-level resistance to lamivudine does not necessarily mean that the drug is rendered worthless. An example of this phenomenon can be found in a clinical trial of partial treatment interruptions conducted by Dr. Steven Deeks and his colleagues (Deeks et al. 2003). This study focused on a cohort of HIV-positive individuals who had a history of excellent treatment adherence, had drug-resistant viremia (greater than 400 copies/ml), and were experiencing a documented treatment-mediated benefit (e.g., a viral load below and CD4+ cell count above pretreatment levels). The patients either stopped their PI(s) or their NRTIs to determine the selective effects of these two drug classes in terms of maintaining the less-fit virus. One more study discontinued lamivudine in four highly treatment-experienced patients with no viable alternative treatment options and evidence of the M184V mutation while on a regimen consisting of at least three antiretrovirals (Campbell et al. 2003). Six weeks after stopping lamivudine, HIV RNA levels increased an average of 0.6 log10 copies/ml above baseline, even though the M184V mutation remained detectable in virus from all patients. This basically showed us that, even in the presence of the M184V mutation, lamivudine was still contributing to HIV RNA suppression. Reversion of M184V to wild type occurred in all four patients between 6 and 14 weeks, accompanied by an additional average increase in viral load of 0.3 log10 copies/ml. Upon resuming lamivudine therapy, the M184V mutation reappeared within 8 weeks in all four patients. The COLATE trial was an open-label trial involving 131 patients experiencing virologic failure while on a lamivudine-containing regimen (Dragsted et al. 2004). In switching to another regimen, approximately half of the patients were randomized to continue lamivudine therapy, with the remaining patients randomized to discontinue lamivudine treatment. Forty-eight weeks later, there were no significant differences between the two groups.

9 Genetic Markers for the Toxicity of Antiretroviral Drugs

Pharmacogenomics examines the influence of genetic variability on drug efficacy or toxicity by correlating gene expression or single-nucleotide polymorphisms with patient outcomes. The PREDICT study (Mallal et al. 2008) showed that the presence of the allele HLA-B5701 is highly predictive of an HSR to ABC. Positive and negative predictive value was 60 and 100 %, respectively. The used of a genetic tests for HLA-B5701 led to a reduction in ABC-related HSR to <1 % in the ARIES trial (Squires et al. 2008), compared to a rate of between 4 and 8 % seen in other trials not using HLA screening. This is the first example of a clinical use of genetic screening in HIV disease management to get widespread approval. Other potential mutation at position 516G>T in CYP2B6 signals a longer half-life and increased levels of efavirenz (Haas et al. 2004). Other haplotypes within CYP2B6 have been associated with a longer half-life for nevirapine (Chantarangsu 2009). The genetic variant associated with Gilbert's disease, UGT 1A1*28, has been identified as part of a haplotype of four UGT1A variants spanning three genes at the UGT1A gene locus (Lankisch et al. 2006). Initial assessment performed in a small cohort of patients with TDF-associated renal dysfunction identified a single substitution at position of 1,249 of MRP2 gene as being potentially associated with the side effect (Izzedine et al. 2006).

An example is the nucleotide variation at position 3,435 of the human P-glycoprotein (MDR-1) gene; the number of patients in each genotype group (3,435 C/C, C/T, and T/T) still experiencing virological suppression at 24 months was 55/112, 72/125, and 74/133, respectively ($p=0.07$) (Brumme et al. 2003). Greater than 80 % of these patients were on a regimen of 2 NRTIs plus a PI. This was a small effect, though significant, for a slower time to immunological and virological failure for the C/C genotype.

10 Conclusions

Pharmacogenomics is providing a miraculous opportunity to bring change in clinical practices, medicine, and health due to upcoming researches in to this field. It also helps in minimizing toxicity and adverse effects by correlating newly found resistance mutations with historical antiretroviral use, which could be of interest to clinicians in planning subsequent antiretroviral regimens for patients who have undergone a range of treatments. If the molecular diagnostic tests can be made less expensive, it will make able to give more imperative therapy against viral disease as AIDS/HIV and hepatitis which are currently giving an enormous economical burden. However, for this to happen, there will need to be major technological improvements in our ability to sequence or screen genomic DNA to assess an individual's genome in the context of population data. Extensive population-based studies will be required to evaluate the phenotypic significance both acute and long term of genomic variation, at the level of individual variants and combinations of variants. Clinical trials must be carried out with the goal of identifying the outcomes of various combinations of genomic, expression-based, metabolomic, and proteomic data to arrive at complete clinico-genomic profiles predictive of disease risk and drug response. All these efforts jointly open a new ray of hope for a miraculous change in the history of medicine for people suffering from viral diseases.

Acknowledgments Authors are thankful to Prof. Kamlesh K Pant, head of the department Pharmacology and Therapeutics for their proper guidance and Capt. Sudhanshu Pandey for the language editing and Mrs Shashi Dwivedi, Miss Vinita Gupta, Mr. Shobhit Srivastava, Mahendra Dwivedi, Narendra Dwivedi, Dr. Sadashiv, Virendra Saini, Durgesh, and Ram Swarup for their hearty support.

References

Abo R et al (2012) Merging pharmacometabolomics with pharmacogenomics using '1000 Genomes' single-nucleotide polymorphism imputation: selective serotonin reuptake inhibitor response pharmacogenomics. Pharmacogenet Genomics 22:247–253

Alter MJ, Kruszon-Moran D, Nainan OV et al (1999) The prevalence of hepatitis C virus infection in the United States, 1988 through 1994. N Engl J Med 341:556–562

Altman MD et al (2008) HIV-1 protease inhibitors from inverse design in the substrate envelope exhibit subnanomolar binding to drug-resistant variants. J Am Chem Soc 130(19):6099–6113

Ank N, Paludan SR (2009) Type III IFNs: new layers of complexity in innate antiviral immunity. Biofactors 35(1):82–87

Ank N, West H, Bartholdy C et al (2006) Lambda interferon (IFN-lambda), a type III IFN, is induced by viruses and IFNs and displays potent antiviral activity against select virus infections in vivo. J Virol 80(9):4501–4509

Ank N, Iversen MB, Bartholdy C et al (2008) An important role for type III interferon (IFN-lambda/IL-28) in TLR-induced antiviral activity. J Immunol 180(4): 2474–2485

Arbuthnot P, Capovilla A, Kew M (2000) Putative role of hepatitis B virus X protein in hepatocarcinogenesis: effects on apoptosis, DNA repair, mitogen-activated protein kinase and JAK/STAT pathways. J Gastroenterol Hepatol 15:357–368

Australian Census (2001) Total population. http://www.abs.gov.au/ausstats/abs%40census.nsf/ddc9b4f92657325cca256c3e000bdbaf/7dd97c937216e32fca256bbe008371f0!OpenDocument#Census Counts. Accessed 25 July 25 2005

Back NK, Nijhuis M, Keulen W et al (1996) Reduced replication of 3TC-resistant HIV-1variants in primary cells due to a processivity defect of the reverse transcriptase enzyme. EMBO J 15:4040–4049

Barbour JD, Wrin T, Grant RM et al (2002) Evolution of phenotypic drug susceptibility and viral replication capacity during long term virologic failure of protease inhibitor therapy in human immunodeficiency virus infected adults. J Virol 76:11104–11112

Bartlett JA et al (2001) Overview of the effectiveness of triple combination therapy in antiretroviral-naive HIV-1 infected adults. AIDS 15(11):1369–1377

Bertoletti A, Ferrari C, Fiaccadori F et al (1991) HLA class I-restricted human cytotoxic T cells recognize endogenously synthesized hepatitis B virus nucleocapsid antigen. Proc Natl Acad Sci USA 88:10445–10449

Bleiber G, Munoz M, Ciuffi A, Meylan P, Telenti A (2001) Individual contributions of mutant protease and reverse transcriptase to viral infectivity, replication, and protein maturation of antiretroviral drug-resistant human immunodeficiency virus type 1. J Virol 75:3291–3300

Blum RA (2005) HIV resistance testing in the USA – a model for the application of pharmacogenomics in the clinical setting. Pharmacogenomics 06: future medicine. Web. http://www.futuremedicine.com/doi/full/10.1517/14622416.6.2.169

Borroto-Esoda K, Harris J, Waters J et al (2004) Baseline genotype as a predictor of virological failure in patients receiving emtricitabine once daily or stavudine twice daily in combination with didanosine and efavirenz [Abstract 672]. In: 11th conference on retroviruses and opportunistic infections, San Francisco

Boucher CA, Cammack N, Schipper P et al (1993) High-level resistance to (¡) enantiomeric2′-deoxy-3′-thiacytidine in vitro is due to one amino acid substitution in the catalytic site of human immunodeficiency virus type 1 reverse transcriptase. Antimicrob Agents Chemother 37:2231–2234

Brumme Z, Dong W, Chan K, Hogg R et al (2003) Influence of polymorphisms within the CX3CR1 and MDR-1 genes on initial antiretroviral therapy response. AIDS 17:201–208

Campbell TB, Young RK, Johnson SC et al (2003) Antiviral activity of lamivudine in persons infected with HIV-1 that has M184V and multiple thymidine analogue mutations [Abstract 140]. In: 12th international HIV drug resistance workshop, Los Cabos

Carman WF (1996) Molecular variants of hepatitis B virus. Clin Lab Med 16:407–428

Carman WF, Jacyna MR, Hadziyannis S et al (1989) Mutation preventing formation of hepatitis B e antigen in patients with chronic hepatitis B infection. Lancet 2:588–591

Carman WF, Zanetti AR, Karayiannis P et al (1990) Vaccine-induced escape mutant of hepatitis B virus. Lancet 336:325–329

Carrington M, Wang S, Martin MP, Gao X, Schiffman M et al (2005) Hierarchy of resistance to cervical neoplasia mediated by combinations of killer immunoglobulin-like receptor and human leukocyte antigen loci. J Exp Med 201:1069–1075

Chamberlain PP, Ren J, Nichols CE et al (2002) Crystal structures of zidovudine- or lamivudine-resistant human immunodeficiency virus type 1 reverse transcriptases containing mutations at codons 41, 184, and 215. J Virol 76:10015–10019

Chander G, Sulkowski MS, Jenckes MW et al (2002) Treatment of chronic hepatitis C: a systematic review. Hepatology 36(Suppl):S135–S144

Chantarangsu S (2009) A CYP2B6 haplotype influences NVP plasma concentrations postpartum following a single intrapartum dose for the prevention of mother to child transmission of HIV in Thai women [Abstract O-02]. In: 10th international workshop on HIV clinical pharmacology, Amsterdam, 15–17 Apr 2009

Chayama K, Suzuki Y, Kobayashi M et al (1998) Emergence and takeover of YMDD motif mutant hepatitis B virus during long-term lamivudine therapy and re-takeover by wild type after cessation of therapy. Hepatology 27(6):1711–1716

Chellappan S et al (2007) Design of mutation-resistant HIV protease inhibitors with the substrate envelope hypothesis. Chem Biol Drug Des 69(5):298–313

Chowdhury A, Santra A, Chaudhuri S et al (2003) Hepatitis C virus infection in the general population: a community-based study in West Bengal, India. Hepatology 37:802–809

Czene K, Lichtenstein P, Hemminki K (2002) Environmental and heritable causes of cancer among 9.6 million individuals in the Swedish Family-Cancer Database. Int J Cancer 99:260–266

D'Arminio Monforte A, Lepri AC, Rezza G (2000) Insights into the reasons for discontinuation of the first highly active antiretroviral therapy (HAART) regimen in a cohort of antiretroviral naïve patients. I.CO.N.A. Study Group. Italian Cohort of Antiretroviral-Naïve Patients. AIDS 14:499–507

Daly AK (2010) Genome-wide association studies in pharmacogenomics. Nat Rev Genet 11:241–246

De Clercq E (2004) Antivirals and antiviral strategies. Nat Rev Microbiol 2:704–720

Deeks SG, Wrin T, Liegler T et al (2001) Virologic and immunologic consequences of discontinuing combination antiretroviral-drug therapy in HIV-infected patients with detectable viremia. N Engl J Med 344:472–480

Deeks SG, Martin JN, Hoh R et al (2003) Continued reverse transcriptase inhibitor therapy is sufficient to maintain short-term partial suppression of multi-drug resistant viremia [Abstract 640]. In: 10th conference on retroviruses and opportunistic infections, Boston

Dellgren C, Gad HH, Hamming OJ, Melchjorsen J, Hartmann R (2009) Human interferon-lambda3 is a potent member of the type III interferon family. Genes Immun 10(2):125–131

Derdeyn CA, Decker JM, Sfakianos JN et al (2001) Sensitivity of human immunodeficiency virus type 1 to fusion inhibitors targeted to the gp41 first heptad repeat involves distinct regions of gp41 and is consistently modulated by gp120 interactions with the coreceptor. J Virol 75:8605–8614

Desenclos JC (2000) Epidemiology of hepatitis C. Revue du Praticien 50:1066–1070

Doyle SE, Schreckhise H, Khuu-Duong K et al (2006) Interleukin-29 uses a type 1 interferon- like program to promote antiviral responses in human hepatocytes. Hepatology 44:896–906

Dragsted U, Fox Z, Mathiesen L et al (2004) Final week 48 analysis of a phase 4, randomised, open-label, multi-center trial to evaluate safety and efficacy of continued lamivudine twice daily versus discontinuation of lamivudine in HIV-1-infected adults with virological failure on ongoing combination treatments containing lamivudine: the COLATE trial [Abstract 549]. In: 11th conference on retroviruses and opportunistic infections, San Francisco

Dulioust A, Paulous S, Guillemot L, Delavalle AM, Boue F, Clavel F (1999) Constrained evolution of human immunodeficiency virus type 1 protease during sequential therapy with two distinct protease inhibitors. J Virol 73:850–854

Durbin RM et al (2010) A map of human genome variation from population-scale sequencing. Nature 467:1061–1073

Dykxhoorn DM, Lieberman J (2006) Silencing viral infection. PLoS Med 3:1000–1004

Esnouf RM, Ren J, Hopkins AL et al (1997) Unique features in the structure of the complex between HIV-1 reverse transcriptase and the bis(heteroaryl)piperazine (BHAP)U-90152 explain resistance mutations for this nonnucleoside inhibitor. Proc Natl Acad Sci USA 94:3984–3989

Farci P, Purcell RH (2000) Clinical significance of hepatitis C virus genotypes and quasispecies. Semin Liver Dis 20:103–126

Feeney ME, Tang Y, Pfafferott K et al (2005) HIV-1 viral escape in infancy followed by emergence of a variant-specific CTL response. J Immunol 174(12):7524–7530

Ferrantelli F, Rasmussen RA, Hofmann-Lehmann R et al (2002) Do not underestimate the power of antibodies—lessons from adoptive transfer of antibodies against HIV. Vaccine 20(suppl 4):A61–A65

Fukuda R, Ishimura N, Kushiyama Y et al (1996) Hepatitis B virus with X gene mutation is associated with the majority of serologically silent non-B, non-C chronic hepatitis. Microbiol Immunol 40:481–488

Gallant JE, Rodriguez AE, Weinberg W et al (2003) Early non-response to tenofovir DF (TDF) + abacavir (ABC) and lamivudine (3TC) in a randomized trial compared to efavirenz (EFV) + ABC and 3TC: ESS30009[Abstract 1722a]. In: 43rd interscience conference on antimicrobial agents and chemotherapy, Chicago

Gallo RC (2005) The end or the beginning of the drive to an HIV preventive vaccine: a view from over 20 years. Lancet 366(9500):1894–1898

Ge D, Fellay J, Thompson AJ et al (2009) Genetic variation in IL28B predicts hepatitis C treatment-induced viral clearance. Nature 461:399–401

Girard M, Meignier B, Barre-Sinoussi F et al (1995) Vaccine-induced protection of chimpanzees against infection by a heterologous human immunodeficiency virus type 1. J Virol 69(10):6239–6248

Haagmans BL, Andeweg AC, Osterhaus A (2009) The application of genomics to emerging zoonotic viral diseases. PLoS Pathog 5(10):e1000557

Haas D, Ribaudo H, Kim R et al (2004) Pharmacogenetics of efavirenz and central nervous system side effects. AIDS 18:2391–2400

Hanna GJ, D'Aquila RT (2001) Clinical use of genotypic and phenotypic drug resistance testing to monitor antiretroviral chemotherapy. Clin Infect Dis 32:774–782

Hanna GJ, Johnson VA, Kuritzkes DR et al (2000) Patterns of resistance mutations selected by treatment of human immunodeficiency virus type 1 infection with zidovudine, didanosine, and nevirapine. J Infect Dis 181:904–911

Harro C et al (2009) Safety and immunogenicity of the Merck adenovirus serotype 5 (MRKAd5) and MRKAd6 human immunodeficiency virus type 1 trigene vaccines alone and in combination in healthy adults. Clin Vaccine Immunol 16:1285–1292

Havlir DV, Hellmann NS, Petropoulos CJ et al (2000) Drug susceptibility in HIV infection after viral rebound in patients receiving indinavir-containing regimens. JAMA 283:229–234

Hemminki K, Chen B (1999) Familial risks for cervical tumors in full and half siblings: etiologic apportioning. Cancer Epidemiol Biomarkers Prev 15:1413–1414

Hemminki K, Dong C, Vaittinen P (2006) Familial risks in cervical cancer: is there a hereditary component? Int J Cancer 82:775–781

Hijikata M, Mishiro S, Miyamoto C et al (2001) Genetic polymorphism of the MxA gene promoter and interferon responsiveness of hepatitis C patients: revisited by analyzing two SNP sites (−123 and −88) in vivo and in vitro. Intervirology 44(6):379–382

Hildesheim A, Wang SS (2002) Host and viral genetics and risk of cervical cancer: a review. Virus Res 89:229–240

Hirsch MS et al (2000) Antiretroviral drug resistance testing in adult HIV-1 infection: recommendations of an International AIDS Society-USA panel. J Am Med Assoc 283:2417–2426

Hirsch MS, Günthard HF, Schapiro JM et al (2008) Antiretroviral drug resistance testing in adult HIV-1 infection: 2008 recommendations of an International AIDS Society-USA panel. Clin Infect Dis 47:266–285

Höhler T, Kruger A, Gerken G et al (1998a) Tumor necrosis factor alpha promoter polymorphism at position −238 is associated with chronic active hepatitis C infection. J Med Virol 54:173–177

Höhler T, Kruger A, Gerken G et al (1998b) A tumor necrosis factor-alpha (TNF-alpha) promoter polymorphism is associated with chronic hepatitis B infection. Clin Exp Immunol 111(3):579–582

Hong L et al (2000) Crystal structure of an *in vivo* HIV-1 protease mutant in complex with saquinavir: insights into the mechanisms of drug resistance. Protein Sci 9:1898–1904

Hoofnagle JH, di Bisceglie AM (1997) The treatment of chronic viral hepatitis. N Engl J Med 336(5): 347–356

Hsiou Y, Ding J, Das K et al (2001) The Lys103Asn mutation of HIV-1 RT: a novel mechanism of drug resistance. J Mol Biol 309:437–445

Hsu HY, Chang MH, Ni Y-H, Lin HH, Wang SM, Chen DS (1997) Surface gene mutants of hepatitis B virus in infants who develop acute or chronic infections despite immunoprophylaxis. Hepatology 26:786–791

Hu ZX et al (2004) Fitness of T215Y vs T215F mutants in HIV-1 RT: comparison of specific thymidine analogue-resistance mutation patterns [Abstract 638]. In: 11th conference on retroviruses and opportunistic infections, San Francisco

Hussain SK, Sundquist J, Hemminki K (2008) Familial clustering of cancer at human papillomavirus-associated sites according to the Swedish Family-Cancer Database. Int J Cancer 122:1873–1878

Hutnick NA et al (2010) Adenovirus-specific human T cells are pervasive, polyfunctional, and cross-reactive. Vaccine 28:1932–1941

International AIDS Society-USA (2002) HIV drug resistance mutations, San Francisco. http://www.iasusa.org/resistance_mutations/index.html. Accessed 9 Feb 2004

Iversen AK, Shafer RW, Wehrly K et al (1996) Multidrug-resistant human immunodeficiency virus type 1 strains resulting from combination antiretroviral therapy. J Virol 70:1086–1090

Izzedine H, Hulot J, Villard E (2006) Association between ABCC2 gene haplotypes and tenofovir-induced proximal tubulopathy. J Infect Dis 194:1481–1491

Kahn JA (2009) HPV vaccination for the prevention of cervical intraepithelial neoplasia. N Engl J Med 361(3):271–278

Kao JH, Wu NH, Chen PJ et al (2000) Hepatitis B genotypes and the response to interferon therapy. J Hepatol 33:998–1002

Kao JH, Chen PJ, Lai MY et al (2002) Genotypes and clinical phenotypes of hepatitis B virus in patients with chronic hepatitis B virus infection. J Clin Microbiol 40:1207–1209

Kaufmann D, Pantaleo G, Sudre P, Telenti A (1998) CD4-cell count in HIV-1-infected individuals remaining viraemic with highly active antiretroviral therapy (HAART): Swiss HIV Cohort Study. Lancet 351:723–724

Kilby JM, Eron JJ (2003) Novel therapies based on mechanisms of HIV-1 cell entry. N Engl J Med 348:2228–2238

King JK, Yeh SH, Lin MW et al (2002) Genetic polymorphisms in interferon pathway and response to interferon treatment in hepatitis B patients: a pilot study. Hepatology 36:1416–1424

Klein TE et al (2009) Estimation of the warfarin dose with clinical and pharmacogenetic data. N Engl J Med 360:753–764

Kohl NE et al (1988) Active human immunodeficiency virus protease is required for viral infectivity. Proc Natl Acad Sci USA 85(13):4686–4690

Kosalaraksa P, Kavlick MF, Maroun V, Le R, Mitsuya H (1999) Comparative fitness of multididеoxynucleoside-resistant human immunodeficiency virus type 1 (HIV-1) in an in vitro competitive HIV-1 replication assay. J Virol 73:5356–5363

Kramvis A, Kew MC (1999) The core promoter of hepatitis B virus. J Viral Hepat 6:415–427

Kramvis A, Bukofzer S, Kew MC, Song E (1997) Nucleic acid sequence analysis of the precore region of the hepatitis B virus from sera of southern African black adult carriers of the virus. Hepatology 25:235–240

Kuritzkes DR (2004) Quantification of human immunodeficiency virus type 1 by reverse transcriptase-coupled polymerase chain reaction. J Infect Dis 190(11): 2047–2054

Lankisch T, Moebius UM, Wehmeier M (2006) Gilbert's disease and atazanavir: from phenotype to UDP glucuronosyltransferase haplotype. Hepatology 44:1324–1332

Larder BA, Kemp SD (1989) Multiple mutations in HIV-1 reverse transcriptase confer high-level resistance to zidovudine (AZT). Science 246:1155–1158

Larder BA, Kemp SD, Harrigan PR (1995) Potential mechanism for sustained antiretroviral efficacy of AZT-3TC combination therapy. Science 269:696–699

Lau DT et al (1997) Long-term follow-up of patients with chronic hepatitis B treated with interferon alpha. Gastroenterology 113(5):1660–1667

Law MG, Dore GJ, Bath N et al (2003) Modeling hepatitis C virus incidence, prevalence, and long-term sequelae in Australia. Int J Epidemiol 32:717–724

Lee WM (1997) Hepatitis B virus infection. N Engl J Med 337(24):1733–1745

Lee YI, Hur GM, Suh DJ, Kim SH (1996) Novel pre-C/C gene mutants of hepatitis B virus in chronic active hepatitis: naturally occurring escape mutants. J Gen Virol 77:1129–1138

Lefebvre E, Schiffer CA (2008) Resilience to resistance of HIV-1 protease inhibitors: profile of darunavir. AIDS Rev 10(3):131–142

Lemon SM (1994) Hepatitis A virus. In: Webster RG, Granoff A (eds) Encyclopedia of virology. Academic Press Ltd., London, pp 546–554

Liapakis A, Jacobson I (2010) Pharmacogenetics of hepatitis C therapy. Pharmacogenomics 11(2):135–139

Liaw YF (2002) Therapy of chronic hepatitis B: current challenges and opportunities. J Viral Hepat 9:393–399

Lin C (2006) HCV NS3-4A Serine Protease: Hepatitis C Viruses: Genomes and Molecular Biology. Norfolk (UK): Horizon Biosci. PMID: 21250386

Ling R, Mutimer D, Ahmed M et al (1996) Selection of mutations in the hepatitis B virus polymerase during therapy of transplant recipients with lamivudine. Hepatology 24:711–713

Little SJ, KK Koelsch, Ignacio CC et al (2004) Persistence of transmitted drug-resistant virus among patients with primary HIV infection deferring antiretroviral therapy [Abstract 36LB]. In: 11th conference on retroviruses and opportunistic infections, San Francisco

Lok AS (2002) Chronic hepatitis B. N Engl J Med 346(22):1682–1683

Mahoney FJ (1999) Update on diagnosis, management, and prevention of hepatitis B virus infection. Clin Microbiol Rev 12(2):351–366

Malhotra U et al (2001) Role for HLA class II molecules in HIV-1 suppression and cellular immunity following antiretroviral treatment. J Clin Invest 107:505–517

Mallal S, Phillips E, Carosi G, Workman C (2008) PREDICT-1 Study Team. HLA-B*5701 screening for hypersensitivity to abacavir. N Engl J Med 358:568–657

Mammano F, Petit C, Clavel F (1998) Resistance-associated loss of viral fitness in human immunodeficiency virus type 1: phenotypic analysis of protease and gag coevolution in protease inhibitor-treated patients. J Virol 72:7632–7637

Mammano F, Trouplin V, Zennou V, Clavel F (2000) Retracing the evolutionary pathways of human immunodeficiency virus type 1 resistance to protease inhibitors: virus fitness in the absence and in the presence of drug. J Virol 74:8524–8531

Marcellin P, Castelnau C, Boyer N (2002) Treatment of hepatitis B. Rev Med Interne 23(S5):535s–536s

Marcello T, Grakoui A, Barba-Spaeth G et al (2006) Interferons alpha and lambda inhibit hepatitis C virus replication with distinct signal transduction and gene regulation kinetics. Gastroenterology 131:1887–1898

Maruyama T, Kuwata S, Koike K et al (1998) Precore wild-type DNA and immune complexes persist in chronic hepatitis B after seroconversion: no association between genome conversion and seroconversion. Hepatology 27:245–253

Mascola JR, Snyder SW, Weislow OS et al (1996) Immunization with envelope subunit vaccine products elicits neutralizing antibodies against laboratory-adapted but not primary isolates of human immunodeficiency virus type 1. The National Institute of Allergy and Infectious Diseases AIDS Vaccine Evaluation Group. J Infect Dis 173(2):340–348

Mc Cutchan FE, Hoelscher M, Tovanabutra S et al (2005) In-depth analysis of a heterosexually acquired human immunodeficiency virus type 1 superinfection: evolution, temporal fluctuation, and intercompartment dynamics from the seronegative window period through 30 months post infection. J Virol 79(18):11693–11704

McHutchison JG, Gordon SC, Schiff ER et al (1998) Interferon alfa-2b alone or in combination with ribavirin as initial treatment for chronic hepatitis C. Hepatitis Interventional Therapy Group. N Engl J Med 339:1485–1492

McMichael A, Hanke T (2002) The quest for an AIDS vaccine: is the CD8+ T-cell approach feasible? Nat Rev Immunol 2(4):283–291

McMichael AJ, Rowland-Jones SL (2001) Cellular immune responses to HIV. Nature 410:980–987

Mehandru S et al (2004) Primary HIV-1 infection is associated with preferential depletion of CD4+ T lymphocytes from effector sites in the gastrointestinal tract. J Exp Med 200:761–777

Mellors J, Palmer S, Nissley D et al (2003) Low frequency non-nucleoside reverse transcriptase inhibitor (NNRTI)-resistant variants contribute to failure of efavirenz-containing regimens in NNRTI-experienced patients with negative standard genotypes for NNRTI mutations [Abstract 134]. In: 12th international HIV drug resistance workshop, Los Cabos

Melnick JL (1995) History and epidemiology of hepatitis A virus. J Infect Dis 171(Suppl 1):S2–S8

Meyer PR, Matsuura SE, Mian AM, So AG, Scott WA (1999) A mechanism of AZT resistance: an increase in nucleotide-dependent primer unblocking by mutant HIV-1 reverse transcriptase. Mol Cell 4:35–43

Molla A, Korneyeva M, Gao Q et al (1996) Ordered accumulation of mutations in HIV protease confers resistance to ritonavir. Nat Med 2:760–766

Murphy DG, Cote L, Fauvel M, Rene P, Vincelette J (2000) Multicenter comparison of Roche COBAS AMPLICOR MONITOR version 1.5, Organon Teknika NucliSens QT with Extractor and Bayer Quantiplex version 3.0 for quantification of human Immunodeficiency virus type 1 RNA in plasma. J Clin Microbiol 38:4034–4041

Nalam MN, Schiffer CA (2008) New approaches to HIV protease inhibitor drug design II: testing the substrate envelope hypothesis to avoid drug resistance and discover robust inhibitors. Curr Opin HIV AIDS 3(6):642–646

Nijhuis M, Schuurman R, de Jong D et al (1999) Increased fitness of drug resistant HIV-1 protease as a result of acquisition of compensatory mutations during suboptimal therapy. AIDS 13:2349–2359

Ohshima S, Komatsu M, Nakane K et al (2000) Iatrogenic GB virus C/hepatitis G virus infection in an area endemic for hepatitis C virus. J Hosp Infect 44:179–185

Okamoto H, Tsuda F, Akahane Y et al (1994) Hepatitis B virus with mutations in the core of the promoter for an e antigen-negative phenotype in carriers with antibody to e antigen. J Virol 68:8102–8110

Ono-Nita SK, Kato N, Shiratori Y et al (1999) Susceptibility of lamivudine-resistant hepatitis B virus to other reverse transcriptase inhibitors. J Clin Invest 103:1635–1640

Pagliaccetti NE, Eduardo R, Kleinstein SH et al (2008) Interleukin-29 functions cooperatively with interferon to induce antiviral gene expression and inhibit hepatitis C virus replication. J Biol Chem 283(44):30079–30089

Palitzsch KD, Hottentrager B, Schlottmann K et al (1999) Prevalence of antibodies against hepatitis C virus in the adult German population. Eur J Gastroenterol Hepatol 11:1215–1220

Peet NP (2010) Drug resistance: a growing problem. Drug Discov Today 15(15–16):583–586

Perelson AS, Neumann AU, Markowitz M, Leonard JM, Ho DD (1996) HIV-1 dynamics in vivo: virion clearance rate, infected cell life-span, and viral generation time. Science 271(5255):1582–1586

Perz JF, Farrington LA, Pecoraro C, Hutin YJF, Armstrong GL (2004) Estimated global prevalence of hepatitis C virus infection. In: 42nd annual meeting of the Infectious Diseases Society of America, Boston, 30 Sept–3 Oct 2004

Peterlin BM, Trono D (2003) Hide, shield and strike back: how HIV-infected cells avoid immune eradication. Nat Rev Immunol 23(2):97–107

Picard V, Angelini E, Maillard A et al (2001) Comparison of genotypic and phenotypic resistance patterns of human immunodeficiency virus type 1 isolates from patients treated with stavudine and didanosine or zidovudine and lamivudine. J Infect Dis 184:781–784

Pichoud C, Seigne`res B, Wang Z, Trepo C, Zoulim F (1999) Transient selection of a hepatitis B virus polymerase gene mutant associated with a decreased replication capacity and famciclovir resistance. Hepatology 29:230–237

Pirmohamed M (2001) The pharmacogenomics of HIV therapy. Pharmacogenomics J 1:243–253. Web. http://www.nature.com/tpj/journal/v1/n4/full/6500069a.html

Population Reference Bureau (2004) World population datasheethttp://www.prb.org/pdf04/04WorldDataSheet_Eng.pdf. Accessed 25 July 2005

Prabu-Jeyabalan M, King N, Nalivaika E, Scott W, Schiffer C (2002) Drug resistance and substrate recognition in HIV-1 protease. Antivir Ther 7(Suppl 1):S36, abstract

Ramsey LB et al (2012) Rare versus common variants in pharmacogenetics: SLCO1B1 variation and methotrexate disposition. Genome Res 22:1–8

Rauch A, Kutalik Z, Descombes P et al (2010) Genetic variation in IL28B is associated with chronic hepatitis C and treatment failure: a genomewide association study. Gastroenterology 138:1338–1345

Reeves JD, Gallo SA, Ahmad N et al (2002) Sensitivity of HIV-1 to entry inhibitors correlates with envelope/coreceptor affinity, receptor density, and fusion kinetics. Proc Natl Acad Sci USA 99:16249–16254

Report on the Global AIDS Epidemic (2007) http://data.unaids.org/pub/EPISlides/2007/071118_epi_regional%20factsheet_en.pdf

Rerks-Ngarm S, Pitisuttithum P, Nitayaphan S (2009) Vaccination with ALVAC and AIDSVAX to prevent HIV-1 infection in Thailand. N Engl J Med 361:2209–2220

Richman DD (1990) Susceptibility to nucleoside analogues of zidovudine-resistant isolates of human immunodeficiency virus. Am J Med 88:8S–10S

Rimsky LT, Shugars DC, Matthews TJ (1998) Determinants of human immunodeficiency virus type 1 resistance to gp41-derived inhibitory peptides. J Virol 72:986–993

Roberts NA, Martin JA, Kinchington D et al (1990) Rational design of peptide-based HIV proteinase inhibitors. Science 248:358–361

Ross L, Johnson M, DeMasi R et al (2000) Viral genetic heterogeneity in HIV-1-infected individuals is associated with increasing use of HAART and higher viremia. AIDS 14:813–819

Safaeian M, Hildesheim A, Gonzalez P, Yu K, Porras C et al (2012) Single nucleotide polymorphisms in the PRDX3 and RPS19 and risk of HPV persistence and cervical precancer/cancer. PLoS One 7(4):e33619. doi:10.1371/journal.pone.0033619

Sagreiya H et al (2010) Extending and evaluating a warfarin dosing algorithm that includes CYP4F2 and pooled rare variants of CYP2C9. Pharmacogenet Genomics 20:407–413

Salemi, M (2004) HIVbase: a PC/Windows-based software offering storage and querying power for locally held HIV-1 genetic, experimental and clinical data. Oxford J Bioinform 20(3): 436–438. Web. http://bioinformatics.oxfordjournals.org/content/20/3/436.short

Sarafianos SG, Das K, Clark AD Jr et al (1999) Lamivudine (3TC) resistance in HIV-1 reverse transcriptase involves steric hindrance with beta-branched amino acids. Proc Natl Acad Sci USA 96:10027–10032

Schapiro JM, Winters MA, Lawrence J, Merigan TC (1999) Clinical cross-resistance between the HIV-1 protease inhibitors saquinavir and indinavir and correlations with genotypic mutations. AIDS 13:359–365

Schiffman M, Castle PE, Jeronimo J, Rodriguez AC, Wacholder S (2007) Human papillomavirus and cervical cancer. Lancet 370:890–907

Schuurman R, Nijhuis M, van Leeuwen R et al (1995) Rapid changes in human immunodeficiency virus type 1 RNA load and appearance of drug-resistant virus populations in persons treated with lamivudine (3TC). J Infect Dis 171:1411–1419

Shafer, R (2003) Human immunodeficiency virus reverse transcriptase and protease sequence database. Nucleic Acids Res 31:1298–1303. Web. 08 Dec 2011. http://nar.oxfordjournals.org/content/31/1/298.long

Shapiro CN, Margolis HS (1993) Worldwide epidemiology of hepatitis A virus infection. J Hepatol 18(Suppl 2): S11–S14

Shenderovich MD, Kagan RM, Heseltine PN, Ramnarayan K (2003) Structure-based phenotyping predicts HIV-1 protease inhibitor resistance. Protein Sci 12(8):1706–1718

Simek MD et al (2009) Human immunodeficiency virus type 1 elite neutralizers: individuals with broad and potent neutralizing activity identified by using a high through put neutralization assay together with an analytical selection algorithm. J Virol 83:7337–7348

Squires K, Young B, Patel P et al (2008) First large, multicenter, open label study utilizing HLA-B*5701 screening for abacavir hypersensitivity in North America. AIDS 22:1673–1675

Stevens W, Horsfield P, Scott LE (2007) Evaluation of the performance of the automated NucliSENS EasyMAG and EasyQ systems versus the Roche AmpliPrep-AMPLICOR combination for high-throughput monitoring of human immunodeficiency virus Load. J Clin Microbiol 45:1244–1249

Suppiah V, Moldovan M, Ahlenstiel G et al (2009) IL28B is associated with response to chronic hepatitis C interferon-alpha and ribavirin therapy. Nat Genet 41(10):1100–1104

Tanaka Y, Nishida N, Sugiyama M et al (2009) Genome-wide association of IL28B with response to pegylated interferon-a and ribavirin therapy for chronic hepatitis C. Nat Genet 10:1105–1111

Thio CL, Thomas DL (2010) Interleukin-28b: a key piece of the hepatitis C virus recovery puzzle. Gastroenterology 138(4):1240–1243

Thursz MR, Thomas HC (1997) Host factors in chronic viral hepatitis. Semin Liver Dis 17:345–350

Torti C, Lapadula G, Maggiolo F et al (2007) Predictors of AIDS defining events among advanced naïve patients after HAART. HIV Clin Trials 8:112–120

Trautmann L et al (2006) Upregulation of PD-1 expression on HIV-specific CD8+T cells leads to reversible immune dysfunction. Nat Med 12:1198–1202

Twyman RM (2004) SNP discovery and typing technologies for pharmacogenomics. Curr Top Med Chem 4(13):1423–1431

Van Gils MJ et al (2010) High prevalence of neutralizing activity against multiple unrelated human immunodeficiency virus type 1 (HIV-1) subtype B variants in sera from HIV-1 subtype B-infected individuals: evidence for subtype specific rather than strain-specific neutralizing activity. J Gen Virol 91(Part 1):250–258

Verkoczy L et al (2010) Autoreactivity in an HIV-1 broadly reactive neutralizing antibody variable region heavy chain induces immunologic tolerance. Proc Natl Acad Sci USA 107:181–186

Viral Hepatitis Prevention Board (1997) News from the VHPB meeting in St. Julians, Malta. Viral Hepatitis 6. http://hgins.uia.ac.be/esoc/VHPB/maltatxt.html

Wedemeyer H, Schuller E, Schlaphoff V, Stauber RE, Wiegand J, Schiefke I, Firbas C, Jilma B, Thursz M, Zeuzem S, Hofmann WP, Hinrichsen H, Tauber E, Manns MP, Klade CS (2009) Therapeutic vaccine IC41 as late add-on to standard treatment in patients with chronic hepatitis C. Vaccine 27(37):5142–5151. doi:10.1016/j.vaccine.2009.06.027

Wei X, Decker JM, Wang S et al (2003) Antibody neutralization and escape by HIV-1. Nature 422(6929):307–312

Weinshilboum RM, Sladek SL (1980) Mercaptopurine pharmacogenetics: monogenic inheritance of erythrocyte thiopurine methyltransferase activity. Am J Hum Genet 32:651–662

Weinstein C et al (2001) Use of genotypic resistance testing to guide hiv therapy: clinical impact and cost-effectiveness. Ann Intern Med 134:440–450

Welzel TM, Morgan TR, Bonkovsky HL et al (2009) Variants in interferon alpha pathway genes and response to pegylated interferon-Alpha2a plus ribavirin for treatment of chronic hepatitis C virus infection in the hepatitis C antiviral long-term treatment against cirrhosis trial. Hepatology 49(6):1847–1858

World Health Organization (1997) The world health report.3. World Health Organization, Geneva

World Health Organization (2000) Hepatitis B. World Health Organization fact sheet 204. WHO 4 Web. http://who.int/inf-fs/en/fact204.html

Wu TD et al (2003) Mutation patterns and structural correlates in human immunodeficiency virus type 1 protease following different protease inhibitor treatments. J Virol 77(8):4836–4847

Xia GL, Liu CB, Cao HL et al (1996) Prevalence of hepatitis B and C virus infections in the general Chinese population: results from a nationwide cross-sectional seroepidemiologic study of hepatitis A, B, C, D, and E virus infections in China, 1992. Int Hepatol Commun 5:62–73

Yang X, Tomov V, Kurteva S et al (2004) Characterization of the outer domain of the gp120 glycoprotein from human immunodeficiency virus type 1. J Virol 78(23): 12975–12986

Yee LJ, Tang J, Gibson AW et al (2001) Interleukin 10 polymorphisms as predictors of sustained response in antiviral therapy for chronic hepatitis C infection. Hepatology 33:708–712

Yuan TT, Lin MH, Qiu SM, Shih C (1998) Functional characterization of naturally occurring variants of human hepatitis B virus containing the core internal deletion mutation. J Virol 72:2168–2176

Zennou V, Mammano F, Paulous S, Mathez D, Clavel F (1998) Loss of viral fitness associated with multiple Gag and Gag-Pol processing defects in human immunodeficiency virus type 1 variants selected for resistance to protease inhibitors in vivo. J Virol 72:3300–3306

Zou S, Tepper M, El Saadany S (2000) Prediction of hepatitis C burden in Canada. Can J Gastroenterol 14:575–580

Section VI

Personalized Medicine: Ortho-Gyneco-Andro-Opthalmology

Pharmacogenomics of Osteoporosis-Related Bone Fractures

29

Karen Rojo Venegas, Margarita Aguilera, Marisa Cañadas Garre, and Miguel A. Calleja Hernández

Abstract

Osteoporosis is a systemic skeletal disease characterised by low bone mass and deterioration in the microstructure of bone tissue, which causes bone fragility and consequent increase in fracture risk. The lifetime fracture risk of a patient with osteoporosis is as high as 40 %, and fractures most commonly occur in the hip, spine or wrist. From a patient's perspective, a fracture and the subsequent loss of mobility and autonomy often represent a major drop in the quality of life. Additionally, osteoporotic fractures of the hip and spine carry a 12-month excess mortality of up to 20 %, because they require hospitalisation and they subsequently present enhanced risk of other complications, such as pneumonia or thromboembolic disease due to chronic immobilisation.

On the other hand, osteoporosis is a multifactorial and polygenic disease, so the individual likelihood of having a fracture depends on the combination of several risk factors such as low bone mineral density (BMD) and genetic predisposition.

This chapter will present the most interesting concerns related to pharmacogenetics of the pathologies most prevalent in the elderly population. It will identify fracture risk and estimation of the probability of major osteoporotic fracture by assessment tools. The genetics of osteoporosis will be approached by showing general pathways involved in bone homeostasis and polymorphisms of genes related with BMD and osteoporotic fractures. Genetic or genomic screening approaches that have been used and also potential benefits from other existing or emerging technologies/

K.R. Venegas • M.C. Garre • M.A.C. Hernández
Pharmacogenetics Unit. Pharmacy Service,
University Hospital Virgen de las Nieves,
Granada, Spain
e-mail: marisacgarre@gmail.com

M. Aguilera (✉)
Pharmacogenetics Unit, Pharmacy Service,
Hospital Universitario Virgen de las Nieves,
Granada, Spain

Instituto de Nutrición y Tecnología de los Alimentos
"José Mataix Verdú" de la Universidad de Granada
(INYTA), Centro de Investigación Biomédica (CIBM),
Granada, Spain
e-mail: maguiler@ugr.es

D. Barh et al. (eds.), *Omics for Personalized Medicine*,
DOI 10.1007/978-81-322-1184-6_29, © Springer India 2013

approaches will be discussed. In addition, the pharmacogenetics of osteoporosis will be discussed, in particular antiresorptive medications, reviewing efficacy limitations and adverse drug reactions associated with current osteoporosis treatments and how these could be attributed to genetic factors. Additionally, it will analyse the therapeutic targets in osteoporosis (bone anabolic and bone resorptive) and the development of new antiresorptive therapies. Furthermore, gene–environment, age and gender differences interactions known to affect response to osteoporosis treatment will be considered. Finally, the translation of pharmacogenetic findings to the clinical practice will be discussed.

1 Introduction

1.1 Background in Osteoporosis

Osteoporosis is a systemic skeletal disease characterised by low bone mass and deterioration in the microstructure of bone tissue which causes bone fragility and consequent increase in fracture risk (NIH Consensus Development Panel on Osteoporosis Prevention, Diagnosis, and Therapy 2001). The lifetime fracture risk of a patient with osteoporosis is as high as 40 %, and fractures most commonly occur in the hip, spine or wrist (Rachner et al. 2011). From a patient's perspective a fracture and the subsequent loss of mobility and autonomy often represent a major drop in the quality of life. Additionally, osteoporotic fractures of the hip and spine carry a 12-month excess mortality of up to 20 % because they require hospitalisation and they subsequently present enhanced risk of other complications, such as pneumonia or thromboembolic disease due to chronic immobilisation (Center et al. 1999).

On the other hand, osteoporosis is a multifactorial and polygenic disease (Ralston 2002), so the individual likelihood of having a fracture depends on the combination of several risk factors such as low bone mineral density (BMD) and genetic predisposition (Cummings et al. 1995; Rojo et al. 2010).

This will present the most interesting concerns related to pharmacogenetics of the pathologies most prevalent in the elderly population. It will identify fracture risk and estimate the probability of major osteoporotic fracture by assessment tools. The genetics of osteoporosis will be approached by showing general pathways involved in bone homeostasis and polymorphisms of genes related with BMD and osteoporotic fractures. Genetic, or genomic, screening approaches that have been used and also potential benefits from other existing or emerging technologies will be discussed. In addition, the pharmacogenetics of osteoporosis will be discussed. In particular antiresorptive medications, reviewing efficacy limitations and adverse drug reactions associated with current osteoporosis treatments and how these could be attributed to genetic factors. Moreover, it will analyse the therapeutic targets in osteoporosis (bone anabolic and bone resorptive) and the development of new antiresorptive therapies. Furthermore, gene–environment, age and gender interactions known to affect response to osteoporosis treatment will be considered. Finally, the translation of pharmacogenetic findings to the clinical practice will be discussed.

1.2 Identification and Estimation of Major Fracture Risks by Assessment Tools

One of the best and most widely used clinical determinants of bone status of an individual is the measurement of BMD (Kanis 2002) which is considered a valid parameter for diagnosing osteoporosis and predicting fracture risk (Cummings et al. 2002). BMD can be assessed with dual-energy X-ray absorptiometry (DEXA), and osteoporosis is defined by a T-score of less than 2.5. Although many external factors play fundamental roles in determining BMD, it has been estimated that over 50 % of women and 70 % of men who have suffered fractures have not previously had osteoporotic BMD values determined (Nguyen et al. 2007). Furthermore, in studies of osteoporosis therapy, increases in BMD were not linearly proportional to fracture risk reductions. The change in BMD induced by antiresorptive drugs explains only approximately 15 % of the reduction in fracture risk (Delmas et al. 2004). The International Osteoporosis Foundation projects that measurements of BMD using DEXA are believed to overestimate BMD by 20–50 % (Johnell and Kanis 2006); they are therefore poor predictors of fracture in individuals.

Although DEXA is widely available and has been commonly used for clinical phase-3 studies, it has some limitations. As an area-based measure of bone mineral, DEXA does not allow assessment of bone geometry; it also does not distinguish between cortical bone, the outer shell and trabecular bone (the spongy inner part) which are important determinants of bone strength and bone loss at different rates. Advances in imaging techniques with high-resolution peripheral CT that yield volumetric bone density data might allow better prediction of bone strength and thus fracture risk if indication such as intracortical porosity is taken into account (Zebaze et al. 2010). Whether these novel techniques will be useful in daily practice remains to be seen.

New decision-making methods, such as the fracture risk assessment tool (FRAX), have integrated clinical risk factors with DEXA-based BMD to predict an individual's 10-year risk of sustaining a hip fracture as well as the 10-year probability of having a major osteoporotic fracture. This defined as clinical spine, forearm, hip or shoulder fracture (Unnanuntana et al. 2010). The algorithm is adjusted for geographic variation in risk. Version 3.0 of the FRAX has been shown to give lower probability estimates for major osteoporotic fracture and hip fracture than version 2.0 but has little impact on rank order of risk. This tool bridges the chasm between the parameters for diagnosing osteoporosis and identifying individuals with structurally compromised bone that causes an increased risk of fracture. However, to be a new scale has limitations due to that these do not contain important risk factors associated with falls and fractures, such as drugs inhibitors of the proton pump (Azagra et al. 2011).

National and international guidelines have been established to address the challenge of screening for osteoporosis in an evidence-based and cost-effective manner (Hodgson et al. 2003; Compston et al. 2009; Brown and Josse 2002). Several risk factors, such as age, low body mass index, previous fragility fractures, a family history of fractures, use of glucocorticoids, alcohol intake and active cigarette smoking, have to be taken into account (Kanis 2002).

Evidently, osteoporosis is a complex multifactorial disease, and it is now understood that genetic factors play a central role in its pathogenesis. It is therefore necessary to consider that the genetics of osteoporosis comprise two main areas: genetics of disease susceptibility and pharmacogenetics of drug response. This chapter will present both of these independent yet interrelated areas and their implications in clinical practice after an adequate validation process.

2 Genetics of Osteoporosis

2.1 Genetic and Genomic Screening Advances

A variety of different approaches have been used to assess osteoporosis-related genes. The main genetic and genomic screening advances include linkage analysis of affected families and relevant animal models, candidate gene association studies, genome-wide sequencing, genome-wide association studies (GWAS) and meta-analyses. Each of these approaches will be explained.

Linkage Analyses: In this type of study, panels of single nucleotide polymorphism (SNP) markers are analysed for evidence of co-segregation with the disease within a family (Sawcer et al. 2004). Although several bone-related phenotype traits have been investigated (femoral neck geometry (Peacock et al. 2005), bone ultrasound properties (Wilson et al. 2004), ADRB1, bone loss), genome-wide linkage scans have only been able to identify one candidate gene for osteoporosis, namely, a non-synonymous coding change in BMP2 (Ser37Ala) in an Icelandic patient cohort (Styrkarsdottir et al. 2003). Relatively small sample sizes, restricted ethnic groups analysed and inaccurate measurement of some structural variables have contributed to study limitations in other cases (Wilson et al. 2004; Peacock et al. 2005).

Animal Studies: Linkage studies in animals provide another way of identifying genes that regulate BMD and other phenotypes relevant to osteoporosis. These involve crossing inbred laboratory strains of mice with low and high bone density, resulting in a second generation of mice whose levels of BMD vary in function of allele segregation. This is analysed by undergoing a genome-wide search in the second generation. The main advantages of these studies are the great statistical power achieved by the large number of individuals that can be analysed and that the confounding factors can be easily reduced due to the fine control of the environment. Linkage studies in mice and other animal models have led to the identification of the arachidonate 15-lipoxygenase gene (ALOX15) which regulates bone mass in mice (Klein et al. 2004) and whose human homologue (ALOX12) has been associated with spine BMD in Caucasian population (Ichikawa et al. 2006).

Candidate Gene Association Studies: Candidate gene association studies evaluate the potential association of known polymorphisms in candidate genes with osteoporosis traits. These studies include case–control approaches and transmission disequilibrium tests (TDT) which are family-based association tests that detect genetic linkage when there is genetic association between a genetic marker and a trait. The analysis of the association between candidate gene polymorphisms and relevant osteoporotic features has yielded several interesting markers to be considered in further studies but with conflicting results up to date. Some of the genes depicted so far include the following: collagen type 1 α1 (COL1A1), oestrogen receptor α (ER-α), lipoprotein receptor-related protein 5 and 6 (LRP-5, LRP-6), sclerostin (SOST), integrin beta-3 (ITGB-3) and vitamin D receptor (VDR) (Ioannidis et al. 2002, 2004; Mann and Ralston 2003; ; Garnero et al. 2005; Ralston et al. 2006; Tofteng et al. 2007; van Meurs et al. 2008; Mencej-Bedrac et al. 2009a, b; Liu et al. 2010).

Genome-Wide Association Studies: GWAS involve the comparison of a vast quantity of gene polymorphisms (usually single nucleotide polymorphisms, SNPs) in hundreds or thousands of individuals, normally divided into two groups of participants, people with the disease (cases) and similar people without it (controls). The density of genetic markers and the extent of linkage disequilibrium in these studies allow the identification of a large proportion of the common variation in the human genome in the population under study. The number of specimens genotyped provides sufficient power to detect variants exerted through modest effect. GWAS have led to the discovery of associations of particular genes and molecular pathways with diseases such as osteoporosis. GWAS have identified and confirmed a large number of osteoporosis-related SNPs, mainly BMD and/or fractures, and susceptibility to osteoporosis as such: ADAMTS18, COL1A1, CRHR1, CTNNB1, DCDC5, ESR1, FLJ42280, FOXC2 and FOXL1, GPR177, HDAC5, LRP4, LRP5, MARK3, MEF2C, MEPE, MHC Locus, SOST, SOX6, SP7, SPTNB1, STARD3NL, TGFB1, TGFB3, TNFRSF11B, TNFSF11, VDR and ZBTB40. In this type of analysis, some genes that had been found interesting through previous candidate gene association studies have been confirmed by GWAS (COL1A1, LRP-4, LRP-5, LRP-6, MARK-3, SOST, OPG, RANK, RANKL, VDR) (Cho et al. 2009; Xiong et al. 2009; Richards et al. 2008; Ralston 2010; Uitterlinden and Ralston 2010).

Genome-Wide Sequencing: Genome-wide sequencing involves the sequencing of whole genomes, with the aim of identifying variations and mutations that can be associated with a phenotype. Sequencing the genomes of a large number of people will provide a comprehensive resource on human genetic variation (1000 Genomes Project Consortium 2010). Once the disease-associated regions or determinate candidate loci are identified by GWAS and other approaches, sequencing technology can be applied to analyse more exhaustively the variants in those regions. This will provide a more extensive knowledge of those gene variants in different individuals, not only single markers, creating the whole catalogue of all variants contained in a determined DNA sequence. In the near future, these advances will allow the examination of more sequence variations and their involvement in the disease (1000 Genomes Project Consortium 2010; Zaghloul and Katsanis 2010). These studies do not identify which variants cause the increased risk for the disease but can provide the set of candidate molecular structures, and further experimental studies will be required to determine which genes, genetic elements and variants functionally cause the increased risk of disease (Zaghloul and Katsanis 2010).

Meta-analyses: Meta-analyses are retrospective or prospective analyses of combined data from several published studies or unpublished datasets that allow the enhancing of sample size and statistical power. They can be applied to association studies, linkage approaches and GWAS. This strategy is increasingly providing confirmation of osteoporosis-related gene associations (Ioannidis et al. 2007; Rivadeneira et al. 2009).

The following section will describe the major genes involved in bone homeostasis.

2.2 General Pathways/Genes Involved in Bone Homeostasis

The genetics of osteoporosis have been studied extensively, and numerous associations between hundreds of polymorphisms in candidate genes and bone traits, both quantitative and qualitative, have been published in the literature (Richards et al. 2008). Currently, many candidate genes have been investigated as valuable tools for their association with BMD and osteoporotic fracture risk (Rojo Venegas et al. 2010). However, the excitement surrounding early studies of allelic variation has often continued into controversy owing to the failure of independent replication, possibly due to insufficient statistical power and false-positive results (Huang et al. 2003; Ioannidis 2005). Genes involved in common pathways have been described as being related to the risk of osteoporosis, risk of hip and vertebral fractures and BMD values (Cheung et al. 2010). These gene variants could affect homeostasis and bone structure and therefore BMD values.

The main genetic factors described in potential candidate genes, GWAS and meta-analyses related with osteoporosis, BMD and osteoporotic fractures are mentioned in Table 29.1, which summarises the main genes with their respective signalling pathways and genetic or genomic screening technologies/approaches.

Due to the available studies from different reviews and the clinical importance of osteoporosis, we will focus on the mostly studied pathways and genes in relation to physiological function of osteoclast (bone resorption) and osteoblast (bone formation), with particular effect on BMD and osteoporosis fracture risk. Nevertheless, the results of these studies are controversial, and no convincing conclusions have emerged yet.

GWAS have allowed the association of large numbers of SNPs with osteoporosis and the identification of novel pathways involved. A large-scale meta-analysis of GWAS data analysed 150 genes and found SNPs from 9 gene loci associated with BMD (*ESRα*, *LRP4*, *ITGA1*, *LRP5*, *SOST*, *SPP1*, *RANK*, *OPG* and *RANKL*) and SNPs from the *LRP5*, *SOST*, *SPP1* and *RANK* loci associated with fracture risk (Richards et al. 2009).

Moreover, the GEFOS meta-analysis identified 20 BMD loci that reached genome-wide significance. Thirteen of these loci were mapped to regions that had not been previously associated with this trait: 1p31.3 (GPR177), 2p21 (SPTBN1), 3p22 (CTNNB1), 4q21.1 (MEPE), 5q14 (MEF2C), 7p14 (STARD3NL), 7q21.3 (FLJ42280), 11p11.2 (ARHGAP1), 11p14.1 (DCDC5), 11p15 (SOX6), 16q24 (FOXL1), 17q21 (HDAC5) and 17q12 (CRHR1) (Rivadeneira et al. 2009). This meta-analysis also confirmed seven known BMD loci on

Table 29.1 Signalling pathways and some candidate genes involved in osteoporosis and osteoporotic fracture

Name of pathways	Name genes (Alternate symbols)	Genetic and genomic screening advances	References
Osteoclastogenesis	Osteoprotegerin (OPG), receptor activator of NF-$_k$B ligand (RANKL), receptor activator of NF-$_k$B (RANK)	GWAS candidate gene meta-analysis	Hsu et al. (2010), Richards et al. (2009), Rivadeneira et al. (2009), Styrkarsdottir et al. (2009), and Ralston (2010)
Wingless-type MMTV integration site (Wnt) signalling	Low-density lipoprotein receptor-related protein 5 and 6 (LRP-5, LRP-6), integrin-β (ITGB), integrin-α, sclerostin (SOST), secreted frizzled-related protein (Sfrp)	GWAS candidate gene meta-analysis	Richards et al. (2009), Rivadeneira et al. (2009), Styrkarsdottir et al. (2009), Ralston (2010), Mencej-Bedrac et al. (2009a, b), and Reppe et al. (2010)
Other main signalling proliferation/inhibition growth cellular: Canonical Wnt TGF-B signalling Beta-catenin phosphorylation MAPK pathway	Secreted phosphoprotein 1 (SPP1), spectrin beta non-erythrocytic 1 (SPTBN1), G protein-coupled receptor 177 (GPR177), catenin (cadherin-associated protein) beta 1 (CTNNB1), myocyte enhancer factor 2C (MEF2C), SRY (sex determining region Y)-box 6 (SOX6), histone deacetylase 5 (HDAC5), corticotropin-releasing hormone receptor 1 (CRHR1), zinc finger and BTB domain-containing protein 40 (ZBTB40), forkhead box protein L1 (FOXL1)	GWAS candidate gene meta-analysis	Hsu et al. (2010), Richards et al. (2009), Rivadeneira et al. (2009), Ralston (2010), Mencej-Bedrac et al. (2009a, b), and Uitterlinden et al. (2006)
Vitamin D	Vitamin D receptor (VDR), vitamin D receptor binding protein (DBP)	GWAS candidate gene meta-analysis	Ralston (2010), Uitterlinden et al. (2006), Ji et al. (2010), Fang et al. (2006), and Macdonald et al. (2006)
Oestrogens	Oestrogen receptor-α (ESR-α), oestrogen receptor-β (ESR-β)	GWAS candidate gene	Richards et al. (2009), and Ralston (2010)
Collagen	Collagen type 1 A (COL1A1)	GWAS candidate gene meta-analysis	Mann and Ralston (2003), Yazdanpanah et al. (2007), and Ralston (2010)
Homeostasis calcium phosphorus	Calcium-sensing receptor (CaSR)	GWAS candidate gene meta-analysis	Hsu et al. (2010), Kapur et al. (2010), Tsukamoto et al. (2000), and Harding et al. (2006)
Mevalonate	Farnesyl-diphosphate synthase (FDPS), geranylgeranyl diphosphate synthase 1 (GGPS1)	GWAS candidate gene	Marini et al. (2008), Choi et al. (2010), and Levy et al. (2007)

the GWAS level: 1p36 (ZBTB40), 6q25 (ESR1), 8q24 (OPG), 11q13.4 (LRP5), 12q13 (SP7), 13q14 (TNFSF11) and 18q21 (RANKL). The deCODE GWAS research group identified four new genome-wide significant loci near the SOST gene at 17q21, the MARK3 gene at 14q32, the SP7 gene at 12q13 and the TNFRSF11A (RANK) gene at 18q21 (Styrkarsdottir et al. 2009).

The analysis of the association between candidate gene polymorphisms and relevant osteoporotic features has yielded several interesting markers to be taken into consideration. One of those relevant genes is the vitamin D receptor, VDR. However, controversial results have been found in different polymorphisms in the VDR gene (*Bsm*I, *Fok*I, *Cdx2*, *Taq*I, *Apa*I) by candidate gene association studies. Three different meta-analysis studies performed on the elderly (fractured, osteoporotic and healthy) have summarised association studies between VDR SNPs and BMD and/or risk of hip fracture. Data from two of these studies, including the GENOMOS study, show that these polymorphisms are not associated with BMD or fractures (Uitterlinden et al. 2006; Ji et al. 2010), though the Cdx2 polymorphism may be associated with risk of vertebral fractures (Uitterlinden et al. 2006). The third meta-analysis found out a modest significance between risk of HF and GG *Bsm*I genotypes (Fang et al. 2006). In another large-scale population-based study performed in early postmenopausal Caucasian women, it was found that *FokI* and *BsmI* SNPs do not influence BMD in this population if calcium intake is adequate (Macdonald et al. 2006). The GEFOS meta-analysis also failed to find an association between VDR polymorphisms and BMD or fracture (Richards et al. 2009).

COL1A1 has also been reported to be related with low BMD and the risk of OF (Ralston et al. 2006). In a meta-analysis study, a significant decrease was found in BMD values in patients with GT genotype of the *Sp1* SNP and especially in those individuals homozygous for TT (Mann and Ralston 2003). Moreover, in a large population-based cohort study of elderly Caucasians, an increased risk of fragility fracture and lower BMD were observed in female carriers of the T allele in the *Sp*1 SNP (Yazdanpanah et al. 2007). No influence of the rs1107946 (G296T) polymorphism on fracture or BMD in postmenopausal women was observed, although power limitations cannot be excluded.

The conclusion results compiled for each pathway remain conflicting, possibly owing to the complexity of the osteoporosis phenotype itself, added to limitations in the molecular available tools. These problems should be approached and solved by common effort for developing and improving screening, risk assessment, diagnosis and treatment initiation (MacLaughlin 2010). In this regard, an important contribution would be the complementary use of validated pharmacogenetic tests in clinical practice (Lamberts and Uitterlinden 2009; van Straaten and Van Schaik 2010).

3 Pharmacogenetics of Osteoporosis

3.1 Pharmacology Treatment Used for Osteoporosis

The therapeutic breakthroughs that have emerged for the treatment of osteoporosis may improve the quality (term referred to the constellation of bone architecture, bone turnover and damage accumulation and mineralisation) and quantity (integration of bone mass, estimated by BMD) of bone among a range of pharmacological alternatives, all of which are used for prevention of osteoporosis fracture (Delmas et al. 2005).

The treatments used in this disease fall into two classes (Gates et al. 2009): antiresorptive drugs and anabolic drugs. Antiresorptive drugs as bisphosphonates (BP) slow down bone resorption, whereas anabolic drugs stimulate bone formation, including teriparatide (parathyroid hormone) and possibly strontium ranelate, which has been suggested to induce a combination of modest effects on bone formation and resorption (Gates et al. 2009). Prestigious clinical guidelines have concluded that all these drugs have been shown to reduce the risk of osteoporosis fracture to a greater or lesser extent along with concomitant increases in bone density and decreases in high bone turnover (NICE 2010a, b).

Due to advances in genomic and proteomic revolutionised drug discoveries and target validation processes, new prospects have emerged for the identification of novel therapeutics against skeletal diseases (Cho and Nuttall 2002). According to new pharmacology treatment, there are drugs that have been developed based on monoclonal antibody actions and small molecules mimicking similar designs established for anticancer molecular-directed therapy: denosumab, odanacatib, saracatinib and antibodies against the proteins sclerostin and dickkopf-1 (two endogenous inhibitors of bone formation) (Lewiecki 2011; Rachner et al. 2011). These new anti-osteoporotic drugs aim to modulate bone metabolism through regulation of osteoclastic proliferation, apoptosis, cell activation, osteoclastic protein folding and aggregation (Zhu et al. 2010).

3.2 Pharmacogenetics in the Osteoporosis Treatment

Pharmacogenetics studies the impact of variability of individual genomes on the response to treatment. Although extensive efforts are focusing on harmonising the various definitions of pharmacogenetics (including those of the European Agency for the Evaluation of Medicinal Products), in practice, this term is used loosely and interchangeably in the literature (EMA 2002). Pharmacogenetics became widely acknowledged in the 1990s and is based on the concept that inherited DNA, RNA and protein-level differences influence metabolism and thereby individual patients' responses to drugs. Those studying pharmacogenetics test the hypothesis in vitro and in vivo that the reason why certain patients respond well to a particular drug or dosage, while others do not, has a genetic basis (Gervasini et al. 2010).

Regarding genetic achievements in osteoporosis, the use of high-throughput whole genome expression microarrays application (a technology that allows researchers to study the expression of many genes at once) has pursued findings of specific upregulated or downregulated genes and the discovery of putative clinical biomarkers. A global microarray expression analysis demonstrated that eight genes are highly associated with BMD variation in postmenopausal Caucasian women. Members of the Wnt signalling pathway, SOST and DKK1, were differentially expressed, and they can be effective targets for osteoporosis drug actions (Reppe et al. 2010). This work describes the functional studies of expression whose results are controversial. The results of osteoporosis-related genes based on microarray identification have been obtained at cellular level in primary cultures of human osteoblasts (Trost et al. 2010). Also, specific genes belonging to osteoporosis pathways have been described with a significant association that have been depicted in Table 29.1, such as pathways of hormone replacement, vitamin D metabolism, Ca/P metabolism-related molecule structures and antiresorptive and anabolic agents.

For above reason, the emerging field of pharmacogenetics is very useful for refining and optimising osteoporosis drug treatment. It could potentially allow the identification of the most effective drug and dose for each patient in terms of beneficial and adverse effects, based on the single genotype. In order to develop this area, the study of pharmacogenetics of osteoporosis should include the understanding of molecular mechanisms of drug action, the identification of drug response candidate genes and their variants and the expansion of clinical trials to include patients' genetic profiling. All these approaches could provide useful tools to tailor decisions about osteoporosis drug treatments in order to maximise the health and well-being of osteoporotic patients.

Curiously enough, pharmacogenetics of osteoporosis has not been widely valued up to date. The available studies have investigated and demonstrated modest effects of some major osteoporosis candidate genes, such as those encoding the VDR (Marc et al. 1999; Palomba et al. 2003, 2005) ER-α/ER-β (Arko et al. 2002; Kurabayashi et al. 2004; Heilberg et al. 2005) and COL1A1 (Qureshi et al. 2002) which have demonstrated an interesting effect in the responses of the antiresorptive drugs. In the next subsection of the chapter, the pharmacogenetics of treatment particularity of antiresorptive drugs will be explained. Furthermore, the studies performed in the last 3 years will be reviewed and analysed.

3.2.1 Pharmacogenetics of Antiresorptive Treatments

The great majority of association studies have investigated only genes affecting BMD, bone turnover marker variation and fracture risk; they might be independent from genes which affect drug responses (adverse affect and efficacy limitations).

The variability in drug response is much more complicated than simple variability in BMD or bone turnover markers. Thus, it will be very important in future works to define the phenotypes of antiresorptive drug response and to enlarge pharmacogenetic studies to include genes involved in drug-specific pharmacokinetics and pharmacodynamics.

In the last 3 years, four novel studies have been published in the field of genetics of osteoporosis with specific relation to response osteoporosis treatment. One study performed in 2008 analysed the effects of the rs1800012 (G>T) SNP of the COL1A1 gene on BMD response to at least 3 years of low-dose hormone replacement therapy in 111 postmenopausal Turkish women (Simsek et al. 2008) and demonstrated that the increase in lumbar BMD and femoral BMD were higher in women with the GG genotype compared to those with the GT genotype.

In the same year, another study associated the rs2297480 (A>C) SNP of FDPS, the molecular target of amino-bisphosphonates in osteoclasts, with the response to a 2-year amino-bisphosphonate treatment in 234 osteoporotic Danish women (Marini et al. 2008). They found that subjects with the homozygous CC genotype showed a decreased response by urinary Crosslaps after 2 years but not after 1 year of amino-bisphosphonate therapy when compared to the heterozygous AC and to the homozygous AA genotypes.

In 2009, in a cohort of 249 osteoporotic or osteopenic men (Kruk et al. 2009), the rs3736228 (C>T) polymorphism of the LRP5 gene was associated with hip BMD in osteoporotic men. However, there was no association between this and other SNPs of this gene (rs4988321 (G>A)), and BMD and bone turnover response after 2 years with risedronate treatment was observed.

The last pharmacogenetics study on osteoporosis was performed in 144 osteoporotic Korean women (Choi et al. 2010). It analysed the role of rs2297480 and rs11264361 SNPs of the FDPS gene and the rs3840452 and rs3841735 SNPs of the GGPS1 on the response to 1-year bisphosphonate treatment (alendronate or risedronate). It was found that women with two deletion alleles (-8188A del) of the rs3840452 polymorphism of the GGPS1 gene presented a significantly lower improvement in BMD as opposed to those with only one deletion allele or no deletion alleles. Women with two deletion alleles had a sevenfold higher risk of nonresponse to bisphosphonates compared with women with the other two genotypes, after adjusting for baseline BMD.

Results from these studies in the pharmacogenetics of osteoporosis suggest that patient genotyping could be useful to target osteoporosis drug treatments to subjects most likely to respond in terms of BMD and bone turnover marker variation. However, association studies can have some limitations, such as inadequate sample size or sampling errors, genetic differences between different ethnic groups, the presence of gene–gene and/or gene–environment interactions acting as confounding factors, the complexity of genome and gene regulation (epigenetic factors, somatic mutations, microRNAs and so on) and frequent accidental statistical association not due to a real association between genotype and phenotype. For all of these reasons, currently, no definite gene variations have conclusively shown to be responsible for the regulation of any anti-osteoporosis drug response.

3.2.2 Adverse Drug Reactions Associated to Osteoporosis Treatment

It is well known that drugs do not always reach their therapeutic target because they are not equally effective in all patients and often, they may lead to a variety of unwanted adverse effects.

Although the osteoporosis drugs are effective in the majority of cases, most are associated with adverse effects that render long-term administration and adherence problem (Siris et al. 2009). From the genetic perspective, several genes that could be related to these unwanted adverse effects of osteoporosis treatments have been evaluated. Table 29.2 shows the main side effects of treatments for osteoporosis.

Table 29.2 Side effect of established treatments for osteoporosis

Type of therapy	Drugs	Doses/interval/route	Side effects
Antiresorptive	Bisphosphonate		Osteonecrosis of the jaw, subtrochanteric fractures. Possible risk of atrial fibrillation
	Alendronate	70 mg/weekly/oral	Oesophageal irritation
	Risedronate	35 mg/weekly/oral 150 mg/monthly/oral	Oesophageal irritation
	Ibandronate	150 mg/monthly/oral	Oesophageal irritation
	Zoledronate	5 mg/yearly/IV	Hypocalcaemia, potential renal toxic effects
	Raloxifene	60 mg/daily/oral	Thromboembolic disease
Anabolic	Strontium[a] Ranelate	2 g/daily/oral	Thromboembolic disease, drugs rash with eosinophilia systemic syndrome, abdominal discomfort
	Teriparatide	20 μg/daily/SC	Hypercalcaemia, nauseas, diarrhoea
	PTH (1–84)[b]	100 μg/daily/SC	Hypercalcaemia, nauseas, diarrhoea

IV intravenous, *SC* subcutaneous
[a]Approved in more 70 countries
[b]Approved in Europe but not in the USA

Several adverse effects, like oesophageal irritation and thromboembolic disease, have been described for antiresorptive treatments, particularly bisphosphonates (Table 29.2). The most threatening side effect of bisphosphonates is the development of osteonecrosis of the jaw (ONJ), which is also the only one associated to gene factors (Kennel and Drake 2009). The incidence of ONJ in patients treated for osteoporosis is low (Reid 2009), while the incidence in cancer patients (due to being treated with high doses of intravenous drugs) is higher (Franken et al. 2011). During the last 4 years, six genes have been proposed to be involved in the risk of developing ONJ (Sarasquete et al. 2008, 2009; Lehrer et al. 2009; Katz et al. 2011; Marini et al. 2011). A short review of the most relevant of these studies is presented in the present work.

One of these studies, by GWAS, determined the involvement of genes in the pathogenesis of ONJ after 2 years of intravenous BP treatment (pamidronate or zoledronate) through the study of two groups of patients with multiple myeloma (22 with ONJ and 65 without) (Sarasquete et al. 2008). Homozygosity for the T allele of the CYP2C8rs1934951 (C>T) polymorphism was associated with risk of ONJ with increased likelihood of developing ONJ. The authors concluded that among genetic factors, CYP2C8 polymorphisms arise as a promising risk factor and that the bisphosphonate-related ONJ can be predicted with a conjunction of genetic and environmental risk factors (Sarasquete et al. 2008, 2009). However, this study has not been confirmed by studies on independent series of patients, so it remains to be seen if that association is a consequence of the genetics background and environmental factors unique to the population studied. Despite the interesting contribution of this preliminary study through a GWAS, variability in gene encoding CYP2C8 would not play a role in the metabolism of BP, due to these drugs that do not undergo any physical–chemical modification.

Another study (Lehrer et al. 2009) performed in seven patients with ONJ and abnormalities in serum bone markers (two metastatic breast cancer, three osteoporosis, one prostate cancer and one Gaucher's disease) treated with intravenous or oral BPs (pamidronate, zoledronate or alendronate) analysed the effect of matrix metalloproteinase 2 (MMP2) protein and the development of the ONJ. This protein was selected on the basis of its potential involvement in the breakdown of the extracellular matrix in normal physiological processes, such as embryonic development, reproduction and tissue remodelling, as well as in disease processes, such as arthritis and metastasis. In this study, they concluded that MMP2 is a candidate gene for bisphosphonate-induced ONJ for three reasons: (1) MMP2 is associated with bone abnormalities which could be related to ONJ, (2) MMP2 is the only gene known to be associated with bone abnormalities and atrial fibrillation and (3) a network of disorders and diseases, genes linked by known disorder-gene associations, indicates that cardiovascular diseases and bone diseases are related. This suggests that a single drug such as bisphosphonate, acting on a single gene, MMP2, could have both bone and cardiovascular side effects different from the osteoclast inhibition that is characteristic of bisphosphonate. Nevertheless, further studies on patients without ONJ are needed.

A third work (Katz et al. 2011) studied the main genes associated with BPs (pamidronate or zoledronate)—related to ONJ in a cohort of 78 patients with multiple myeloma (12 with ONJ) receiving intravenous BP treatment for over 1 year. This study showed that the risk of developing ONJ was 4 times higher in patients with smoking history and in patients treated with pamidronate versus those treated with zoledronate. A trend towards higher odds ratio for developing ONJ in multiple myeloma patients undergoing intravenous BP therapy was observed for the combined genotype of five gene SNPs: COL1A1 (rs1800012), RANK (rs12458117), MMP2 (rs243865), OPG (rs2073618) and OPN (rs11730582).

Finally, a recent study (Marini et al. 2011) makes an interesting analysis of the influence of rs2297480 (A/C) SNP in the target FDPS gene (an enzyme involved in the mechanism of action of the BP) in a cohort of 68 Caucasian patients with multiple myeloma, metastatic mammary and prostate cancer treated with an intravenous BP (zoledronic acid). The results of this case–control study demonstrated significant differences in the allele and genotype frequencies of this SNP between ONJ cases and controls. These results are in agreement with the previously described higher responsiveness of the AC and AA genotypes to oral treatment with amino-BP when compared to CC genotype. It could be assumed that the A allele segregates with the ONJ complications through a positive modulation of the response to a potent amino-BP, as zoledronic acid. However, the authors determined a no direct causative relationship between ONJ and FDPS gene polymorphisms. Also, they propose to confirm this study in other patients' cohorts and the validation of this genetic marker for ONJ, as well.

Therefore, if the BPs are causal to the development of osteonecrosis is something that remains to be determined, suggesting that environmental and/or genetic variation between individuals may confer susceptibility or resistance to developing ONJ. Pharmacogenetic studies on any other side effects have not been reported so far.

3.2.3 Other Genetic and Nongenetic Factors Associated to Osteoporosis Treatment Responses

Many risk factors contribute to the pathogenesis of osteoporosis, and ageing affects bone density variations that are also mainly determined by genetic factors. Nevertheless, lifestyle factors such as bone-loading physical activities, nutrition (insufficient calcium and vitamin D intake and alcohol consumption), nicotine abuse, illnesses or the intake of medications with a negative impact on bone metabolism are additional risk factors for osteoporosis and can also affect drug responses (Weinshilboum 2003).

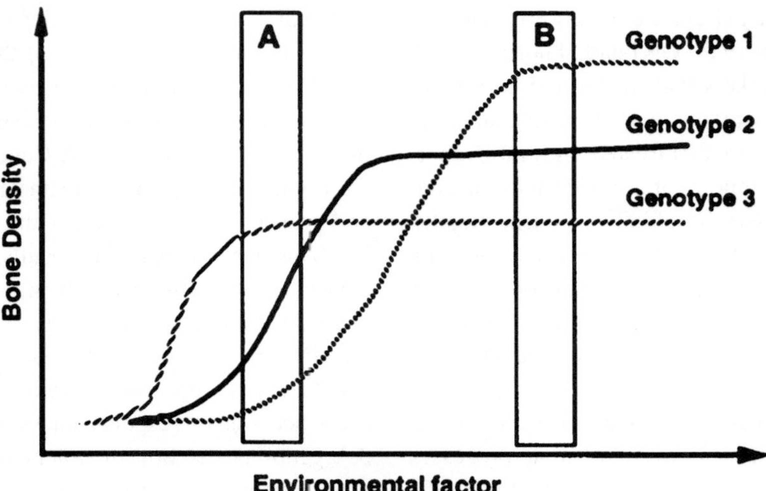

Fig. 29.1 Gene–environment interactions. The concept that a gene variant could be an advantage under one set of conditions and a disadvantage under another is depicted. Under conditions A, individuals with genotype 1 would be worse off compared with those with genotype 2 or 3. However, under conditions B, the reverse order would apply (Eisman 1999)

Genetic variation in a population creates an impressive spectrum of phenotypic diversity, particularly when changes in diet or in environment are imposed to the population. GWAS have become powerful tools for linking sequence variants with overlying systems-level phenotypes, but they do not provide insight into the mechanisms through which genetic variation drives phenotypic variation. Systems genetics is an emerging discipline that provides means to fill this knowledge gap by assembling the hierarchy of interactions among genes, proteins and other intermediate phenotypes that manifest themselves as phenotypic variations (Voy 2011).

Figure 29.1 shows a graphic which illustrates the interaction between gene and environment data. The concept for which a gene variant could be an advantage under one set of conditions and a disadvantage under another one is depicted.

So far, no conclusive work has been found to determine whether there is a difference in the response to the treatments used for osteoporosis, taking into account age, gender and environmental factors (disease, calcium intake, exercises, among others) or none of these factors are associated with or influenced by genetics. However, it is necessary to consider some aspects that will be described below.

Osteoporosis is an age-related disease with several gender-specific differences. Its prevalence is higher in women than men. Due to their larger bones, men have greater bending strength than women; however, once a hip fracture has occurred, mortality is higher in men (Center et al. 1999; NIH Consensus Development Panel on Osteoporosis Prevention, Diagnosis, and Therapy 2001; Rojo et al. 2010).

According to the International Osteoporosis Foundation, an estimated 75 million men and women in Europe, the USA and Japan are affected by osteoporosis. Osteoporosis does not affect only women; men experience both osteoporosis and its clinical consequences. Differences in sex hormone production 'especially the abrupt decline of oestrogen in women' are responsible for inter-gender differences in the pathophysiology of osteoporosis. One in 3 women and 1 in 5 men over the age of 50 have experienced osteoporotic fractures. Of the 9 million osteoporotic fractures estimated in 2000, 1.6 million were of the hip (Johnell and Kanis 2006). It is estimated that in 2050, the number of hip fractures will increase from the current 4.5 million to 8.2 million (Gullberg et al. 1997). Thirty-nine percent of these osteoporotic fractures and 30 % of hip fractures specifically occur in men. Hasserius and colleagues showed that men generally have higher rates of fracture-related mortality (Hasserius et al. 2003). The osteoporosis resulting disability in Europe is greater than the combined disability of all non-lung cancers (Cummings-Vaughn and Gammack 2010).

On the other hand, about 40 % of white postmenopausal women are affected by osteoporosis, and, with an ageing population, this number is expected to steadily rise in the near future (Burge et al. 2007).

Thus, the treatment of osteoporosis also differs between genders; therapy options have been studied only in women (Pietschmann et al. 2009) and in men that were not protected from the disease (Patsch et al. 2011). Therefore, the treatment in men is understudied, and the magnitude of the impact of male osteoporosis has also been underestimated (Haney and Bliziotes 2008).

On the other hand, some studies have shown the influence of certain diseases in the response to treatment for osteoporosis. For example, the influence of vitamin D insufficiency at initial status on the response to alendronate and raloxifene was tested in patients with low BMD (Antoniucci et al. 2005, 2009), showing that basal status does not affect BMD response to BPs when they are co-administered with vitamin D and calcium. On the contrary, a study performed in women with postmenopausal osteoporosis showed that optimal vitamin D repletion seems to be necessary to maximise the response to antiresorptive drugs (alendronate, risedronate and raloxifene) in terms of both BMD changes and anti-fracture efficacy (Adami et al. 2009).

In addition, the decrease in calcium absorption could contribute to the pathogenesis of osteoporosis. A recently randomised, double-blind, placebo-controlled and multicenter clinical trial (Shapses et al. 2011) evaluated whether the alendronate (ALN) increased the fractional calcium absorption (FCA) in postmenopausal women with low BMD and vitamin D$\leq$25 ng/ml. Patients who were treated with five weekly doses of ALN 70 mg+vitamin D (3) 2800IU (ALN+D) showed an increase in FCA compared with those who received placebo.

In an interesting retrospective study (case–control), the efficacy of ALN was evaluated in postmenopausal osteoporotic women with type 2 diabetes mellitus (DM) to genetics level by measuring the efficacy with the BMD. Patients with type 2 DM proved resistant to long-term bisphosphonate treatment, especially in hip, femoral neck and forearm regions of the BMD (Dagdelen et al. 2007). Metabolic syndrome diseases as DM are in genetic linkage to a missense mutation in LRP6 (Wnt signalling), one of the genes related with osteoporosis (Mani et al. 2007). In general, modifiable external factors such as age and gender, as well as modifiable factors such as the intake of calcium and vitamin D, not only have implications as a risk factor for osteoporosis but also probably in response to treatment.

4 Molecular Structures as Osteoporosis Therapeutic Targets: Pharmacogenomics

To understand the development and creation of new drugs that are being used or are in early phases of use in the treatment or prevention of osteoporosis, it is necessary to consider a new concept in this chapter: pharmacogenomics. However, this important and necessary term is closely related to the area of pharmacogenetics.

In contrast to pharmacogenetics, which study the effect of an individual's genetics variability in their response to certain drugs, pharmacogenomics study the molecular and genetic basis (including DNA, RNA and proteins) of disease to develop new ways of treatment (Wang 2010).

Moreover, whereas pharmacogenomics refer to approaches that take into account the characteristics of the entire genome, using an integrated view, pharmacogenetics are restricted to the determinants of each individual as far as the efficacy or toxicity of a drug only at different levels (Weinshilboum and Wang 2006).

A detailed knowledge of bone biology with molecular insights into the communication

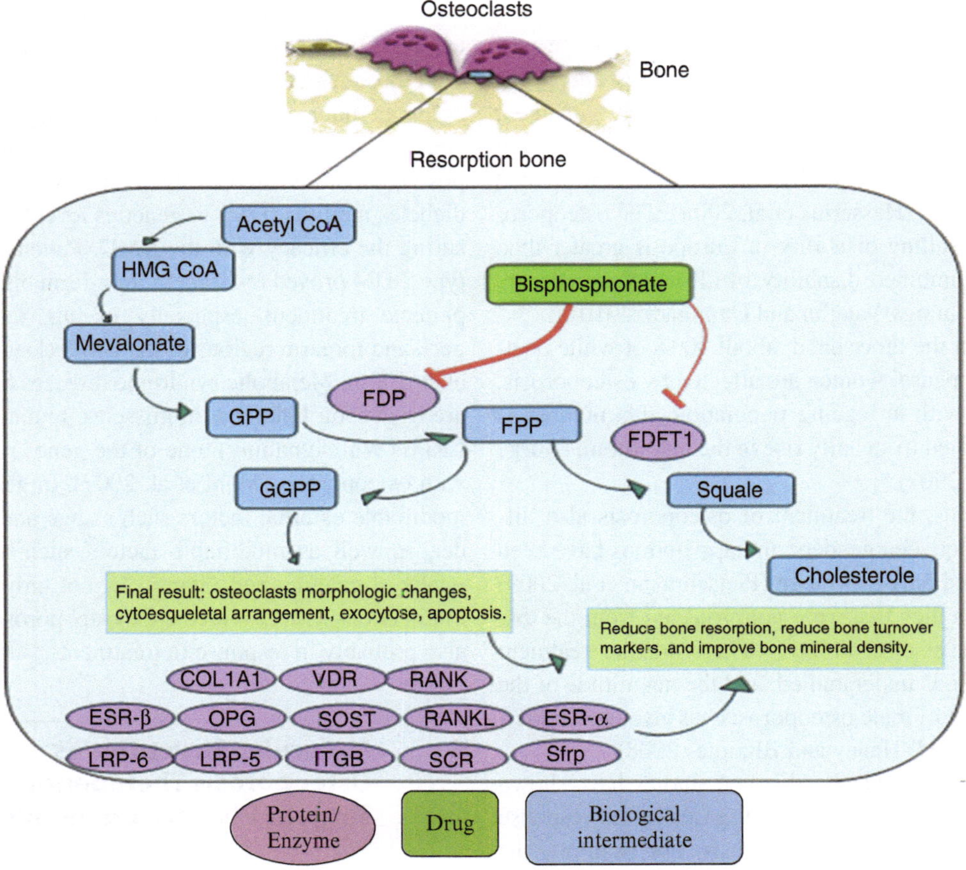

Fig. 29.2 Representation of molecular structures coded by genes involved in bisphosphonates' effects inside osteoclasts. *FPP* farnesyl pyrophosphate, *FDFT1* farnesyl-diphosphate farnesyltransferase, *FDPS* farnesyl-diphosphate synthase, *FPP* farnesyl, *GPP* geranyl pyrophosphate, *GGPP* geranylgeranyl pyrophosphate, *COL1A1* collagen type 1 A, *ESR-α* oestrogen receptor-α; *ESR-β* oestrogen receptor-β, *ITGB* integrin-β, *LRP-5* low-density lipoprotein receptor-related protein 5, *LRP-6* low-density lipoprotein receptor-related protein 6, *OPG* osteoprotegerin, *RANK* receptor activator of NF-$_{k}$B, *RANKL* receptor activator of NF-$_{k}$B ligand, *SCR* scrapper kinase, *Sfrp* secreted frizzled-related protein, *SOST* sclerostin, *VDR* vitamin D receptor

between bone-forming osteoblasts and bone-resorptive osteoclasts and the orchestrating signalling network has led to the identification of novel therapeutic targets. Bisphosphonate's therapy inhibits specifically genes involved on resorption bone pathway (Fig. 29.2). However, the most promising novel treatments include new molecules affecting systemic osteoclasts pathways: denosumab (a monoclonal antibody for receptor activator of NF-κB ligand, a key osteoclast cytokine), odanacatib (a specific inhibitor of the osteoclast protease cathepsin K) and antibodies against the proteins sclerostin and dickkopf-1 (two endogenous inhibitors of bone formation). This chapter examines these novel therapies and explains their underlying physiology.

4.1 Targets for Bone Anabolic Drugs

During the last years, the role of the wingless-type MMTV integration site family (Wnt signalling pathways) has been to be involved in bone formation where it appears to be an important regulator of bone accrual during growth

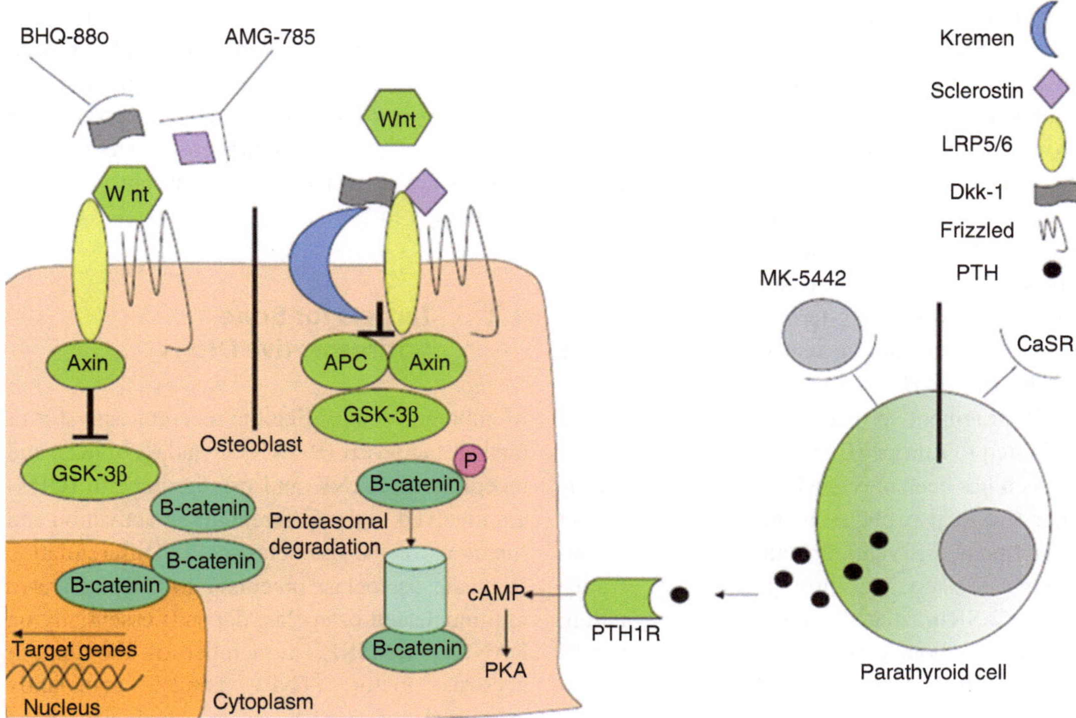

Fig. 29.3 Osteoblast physiology and potential therapeutic targets

(Harada and Rodan 2003). The discovery of this crucial role, the canonical Wnt signalling, in skeletal biology opens a new avenue for bone anabolic therapies (Rawadi and Roman-Roman 2005; Hoeppner et al. 2009). Pharmacological manipulation of the Wnt cascade can affect bone formation rates in adults; positive modulators of this pathway in vivo are expected to act as bone anabolic drugs. Thus, many elements of the Wnt cascade constitute potential targets for pharmacological intervention. However, it must be noted that most of the approaches presented below target protein–protein interactions that are relatively difficult to manipulate pharmacologically. Wnt/β-catenin signalling affects proliferation, commitment, differentiation/maturation, function and lifespan of osteoblasts. There are clinical trials with molecules such as BH-880 and AMG-785 that stimulate the Wnt pathway when Wnts bind to Frizzled (Fzd) receptors and LRP 5/6 (Fig. 29.3). This activation inhibits a cytoplasmic complex composed of glycogen synthase kinase (GSK-3β), Axin and adenomatous polyposis coli (APC). Cytoplasmic β-catenin levels rise and some β-catenins translocate to the nucleus where it associates transcription T-cell factors (Tcf) and transcription lymphoid enhancer-binding factors (Lef) to regulate gene expression.

On the other hand, some new molecular designs, like calcilytic drugs (MK-5442), antagonise CaSR and trigger short bursts of PTH secretion. Binding of PTH to its receptor enhances osteoblast functions and bone formation. The presence of Wnt antagonists Dickkopf-1 (Dkk-1) and sclerostin inhibits Wnt signalling. Dkk-1 needs to form a complex with Kremen to bind LRP5/6, whereas sclerostin binds LRP5/6 directly. BHQ-880 and AMG-785 are antibodies for Dkk-1 and sclerostin, respectively. After neutralising Dkk-1 and sclerostin, Wnt can bind to LRP5/6, which results in the degradation of GSK-3β. As a consequence, β-catenin is stabilised, accumulates and translocates into the nucleus where it regulates transcription of osteoblastic genes.

Findings regarding the small molecule inhibitors of GSK3β increase bone mass and lower adiposity and reduce fracture risk. Neutralising antibodies to Dkk-1, secreted frizzled-related protein 1 and sclerostin, which bind and block the Wnt receptor LRP-5/6, produce similar outcomes in animal models. These drugs are exciting breakthroughs, but they are not free of risks. The challenges include tissue-specific targeting and consequently, long-term safety.

Anabolic agents currently approved are the full-length PTH (1–84) and their N-terminal (1–34) fragment, teriparatide, though the first one is limited to Europe (Canalis et al. 2007).

As it has been previously mentioned, extracellular CaSR also plays an important role in the recruitment, differentiation and survival of osteoblasts and osteoclasts through activation of multiple CaSR-mediated intracellular signalling pathways in bone cells. CaSR on the parathyroid gland controls PTH release to maintain serum calcium concentrations within a narrow physiological range. Drugs that mimic high concentrations of calcium at the CaSR and suppress PTH secretion are termed calcimimetic drugs, such as cinacalcet.

Although the physiological role of the CaSR expressed on osteoblasts and osteoclasts is not fully understood, in a recent study (Pierre 2010) it has demonstrated that the CaSR is a target for the anabolic strontium ranelate (SR) in bone cells. The study noted that SR has been found to increase osteoblastic cell replication through activation of the CaSR and also that SR increases the expression of osteoclast proteins, such as OPG, and decreases expression of RANKL.

Calcilytic drugs represent a new class of bone-forming drugs. They act as antagonists of the CaSR and mimic hypocalcaemia, thus evoking a short pulse of PTH secretion. Calcilytics are given orally and obviate the need for injections, as opposed to PTH treatment. A major practical obstacle for these drugs has been their narrow therapeutic index. Several programmes involving calcilytic drugs have been discontinued because of unfavourable pharmacokinetics (Gowen et al. 2000) and lack of efficacy (NCT00471237). Currently, newer calcilytic drugs with an improved pharmacological profile are being assessed (Balan et al. 2009; Kumar et al. 2010). The most advanced compound of this class is MK-5442, which is currently in phase-2 trials for postmenopausal osteoporosis. Results are expected to be ready and available for therapy in 2012.

4.2 Targets for Bone Antiresorptive Drugs

Membrane proteins like the receptor activator of nuclear factor-κB (RANK) ligand (RANKL) and receptors as RANK and osteoprotegerin (OPG) are involved in the differentiation, activation and functions of the osteoclast cells. Differentiation from the osteoclast precursor to fully activated multinucleated osteoclast depends essentially on RANKL. RANKL, a member of the tumour necrosis factor (TNF) family, abundantly expressed by bone-forming osteoblasts, bone marrow stromal cells and T and B lymphocytes, activates its receptor, RANK, expressed on osteoclasts. After RANKL induced RANK stimulation, several key regulatory transcription factors and enzymes are induced to promote the differentiation, proliferation, multinucleation, activation and survival of osteoclasts (Fig. 29.4).

The Src kinase is highly expressed in osteoclasts and mediates multiple pathways regulating osteoclast activity. Interestingly, the absence of Src does not alter the number of osteoclasts (Boyce et al. 1992), and it is associated with an enhanced rate of osteoblastic bone formation (Marzia et al. 2000). With their jellyfish-like shape, motile cytoskeleton and adhesion molecules such as integrins (Fig. 29.5), osteoclasts attach to bone and create a sealing zone on the bone surface which provides a highly enriched acidic microenvironment that is essential for catalytic activity of osteoclastic enzymes such as cathepsin K.

On the other hand, cathepsin K is a key cystein proteinase of the mature osteoclast that degrades collagen and breaks down bone (Wilson et al. 2009). Cathepsin K is a crucial determinant of resorptive activity by osteoclasts; bone of poor

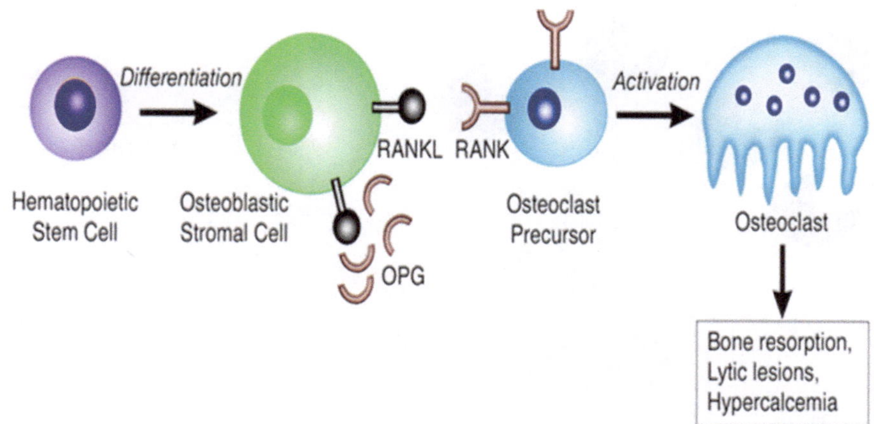

Fig. 29.4 Role of receptor activator of nuclear factor-κB (RANK) ligand (RANKL) and its two receptors (RANK and OPG) in controlling bone metabolism (Image modified from Younes and Kadin 2003)

quality in which microcracks accumulate is removed and hole-like lacunae appear (Fig. 29.5).

Recent and more targeted therapies are being designed and used based on approaches similar to cancer therapy, for example, monoclonal antibody and small molecules such as denosumab, odanacatib and saracatinib.

4.2.1 Denosumab

The prominent role of RANKL in osteoclastogenesis has made it a prime target in diseases characterised by excessive bone loss. Initially, a chimerical OPG (decoy) fusion protein was used to antagonise RANKL (Bekker et al. 2001). However, the formation of neutralising antibodies against OPG after administration of the fusion protein and its potential cross-reactivity with tumour necrosis factor-related apoptosis-inducing ligand (Emery et al. 1998) led to a new strategy, the development of denosumab, a fully human monoclonal antibody against RANKL (Fig. 29.5), with pre-authorisation evaluation of medicines for human use in Europe (EMA 2009).

In several clinical trials, the positive impact of denosumab has been demonstrated in patients with low BMD and/or osteoporotic patients. A phase-II controlled and randomised study in postmenopausal US women with a T-score between −1 and −2.5 at the lumbar spine showed that 60 mg denosumab every 6 months significantly increased BMD at lumbar spine and total hip compared to placebo and significantly reduced bone markers (BM) (Bone et al. 2008). Similarly, it was shown that among women with a T-score between −1.8 and −4 in lumbar spine and −1.8 and −3.5 at the proximal femur, a dose of 60 mg of denosumab every 6 months increased BMD and decreased the BM compared with placebo (McClung et al. 2006). Another phase-II study in postmenopausal women with osteoporosis who received 60 mg of denosumab every 6 months showed a significant decrease in the incidence of vertebral and hip fractures compared with placebo (Cummings et al. 2009).

A multicenter phase-III study compared the efficacy of denosumab (60 mg/6 months) with alendronate (70 mg/day) on BMD and BM in postmenopausal women with low BMD, showing significantly greater BMD gains and a further reduction of BM in patients treated with denosumab (Brown et al. 2009).

A multicenter randomised study demonstrated that switching the alendronate to denosumab for at least 6 months in postmenopausal women caused an increase in BMD at the lumbar spine, total hip, femoral neck and distal radius as well as a significant decrease in bone remodelling compared to women who continued treatment with alendronate (Kendler et al. 2010).

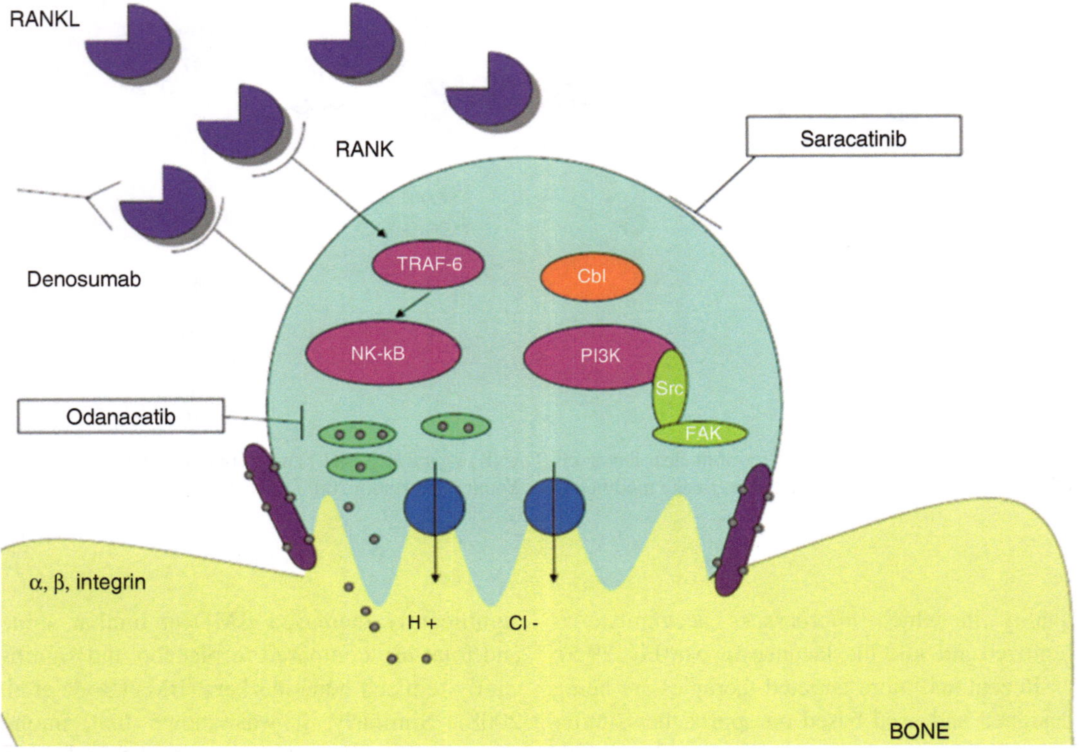

Fig. 29.5 Summary of osteoclast physiology and potential therapeutic targets

4.2.2 Odanacatib

Based on the concept that the protease cathepsin K has an important role in enzymatic bone degradation, the use of cathepsin-K inhibitors has emerged as a novel therapeutic approach (Fig. 29.5). A high specificity and affinity for cathepsin K over other cathepsins (B, L and S) that are widely expressed, particularly in the skin, was crucial for this class of compounds (Gauthier et al. 2008).

Odanacatib is currently the most advanced inhibitor of cathepsin K under clinical investigation. Programmes with less specific cathepsin-K inhibitors were stopped due to cutaneous side effects, such as a scleroderma-like skin thickening and rash (Gauthier et al. 2008).

In phase-I study, odanacatib at oral doses of 50–100 mg once a week reduced serum concentrations of the bone resorption marker CTX by 62 %. Daily administration of odanacatib (10 mg) reduced serum concentrations of CTX by 81 % (Stoch et al. 2009).

In a phase-II study (Bone et al. 2010), the effects of weekly oral doses of odanacatib were assessed in 399 women with postmenopausal osteoporosis. After 24 months, odanacatib (50 mg) increased BMD in the lumbar spine by 5.7 % and total hip by 4.1 % compared with placebo. In fact, a subset of 32 women undergoing bone biopsies showed that treatment with odanacatib resulted in a modest and transient reduction of bone formation markers with no suppression of bone formation rate. Adverse reactions with odanacatib were close to those with placebo, and scleroderma-like cutaneous lesions were not seen.

A current phase-III study is assessing the anti-fracture efficacy of odanacatib (NCT00529373) by undergoing a trial with over 16,000 postmenopausal women and is expected to be completed in 2012. Another cathepsin-K inhibitor, ONO-5334, is currently being assessed in a phase-II trial in women with postmenopausal osteoporosis (NCT00532337).

4.2.3 Saracatinib

The Src kinase is another important protein; it is highly expressed in osteoclasts and mediates multiple pathways regulating osteoclast activity (Fig. 29.5). The effect of impaired osteoclastic functions in Src-deficient mice provided the rationale to explore the skeletal effects of an inhibitor of Src kinase.

A phase-I trial (Hannon et al. 2010) assessed saracatinib (AZD0530) in 59 healthy young men and demonstrated decreased serum concentrations of CTX by 88 % and urinary excretion of NTX by 67 % (250 mg) after 25 days. Concentrations of bone formation markers in the administered saracatinib group were similar to those in the placebo group. Although there were no significant differences between these two groups, papular rash (30 % vs. 6 %) and loose stools (24 % vs. 0 %) were more frequent in the men given saracatinib than in those given placebo (Hannon et al. 2010).

Saracatinib is currently being explored in phase-II studies for osteosarcoma (NCT00752206) and bone metastases (NCT00558272), but not for osteoporosis.

5 Translations of Pharmacogenetic Findings to the Clinical Practice Environment

The translation of pharmacogenetic findings of osteoporosis-related bone fractures to the clinical practice should be addressed in a general context together with the parallel advancements in other pharmacogenetic treatment approaches.

In order to accelerate the translation and transition based on scientific results found in screening drug effects over a genetically different population, the findings must first be approved by the appropriate public and private investors. They should allow specific and reliable tests to be conducted for those valid biomarkers demonstrating association with the safety and efficacy of therapy.

There are three different types of pharmacogenetic biomarkers: validated pharmacogenetic biomarkers, probably validated biomarkers and in exploration biomarkers. These depend on the type of clinical activity for which they are to be developed; the first biomarker would be used in routine clinical practice and the other two in recently applied discoveries and basic research area.

The most specific valid biomarker definition is one that is widely accepted by the scientific community or preclinical predictor of clinical outcome. The probably valid biomarker seems to have predictive value, but the results have not yet been replicated in all populations and are not widely accepted while the exploratory biomarker is in hypothesis generation phase.

However, government expert agencies are the latest groups responsible for regulating and determining the status of a specific biomarker and also the initiatives pursued until their mandatory use in hospitals or clinics (Fig. 29.6). There are consortiums of experts such as Joint Voluntary Genomic Data and Submission (VGDS) from the FDA/EMA that process this information (US FDA, EMA 2009).

Genetic biomarkers may permit the subsetting of patient populations into groups who respond differently to various therapies. The analytical performance characteristics should be established before using the biomarker in stratification, identification of responders or as tests to avoid prescribing to either biomarker-positive or biomarker-negative subjects (Tesch et al. 2010).

Concretely, classical therapeutic approaches in osteoporosis have been established for controlling of the marker of turnover, bone formation and bone resorption that are now commercially available for decreasing fracture rate as a clinical end point. Then, all polymorphisms in genes that directly or indirectly affect these markers could be considered genomic biomarkers. However, no evidence of clinical use of multiple pharmacogenetic biomarkers has been shown due to the controversial result of association studies which are described throughout this chapter.

Bone pathway profiles of whole genes of individuals could solve the contradictory findings from isolated studies which can give erroneous conclusions. It is also important to verify and replicate the use of a valid biomarker in several populations. These important steps could become a realisation through the use of high-throughput technologies. If it is considered that fractures related to osteoporosis will continue to be a substantial and growing public health problem, important investments and coordinated macro studies will be necessary to increase the quality of life of the patient population.

Moreover, another complementary item is the elaboration of guidelines of anti-osteoporosis drugs that are necessary for pharmaceutical care. Also, these guidelines can include recommendations of a specific pharmacogenetics test.

Considering that fractures related to osteoporosis continue to be a substantial and growing public health problem, important investments and coordinated macro studies are necessary to increase the quality of life of this patient population.

Furthermore, other general important considerations must be taken into account before the translation of pharmacogenetic achievements, and these should be used in routine clinical practice. Thus, general terms that cover areas of molecular biology–pharmacogenetics formation, ethical–legal issues, quality management in genetic test laboratories, ISO norms and cost-effective successful strategies will be briefly disclosed in this chapter.

The first step before the transfer of effective pharmacogenetic discovery to the clinic and hospital environments will pass through the assessment that professionals involved in this circuit are properly formed, informed and coordinated for the management and application of requirements of pharmacogenetic tests. Moreover, they must have the proper infrastructure for collecting and analysing sample specimens.

Currently, international diffusion about how to develop and the level of consensus on specific formation programmes on pharmacogenetics are scarce and limited to a number of pharmaceutical and physician schools (Murphy et al. 2010; Maize et al. 2010). Also, minimal data have probably been shared about pilot assays performed for every school or university about efforts regarding the implementation of basic pharmacogenetic knowledge, such as the lack of eligible PhD courses and specific training. In any case, this general lack of knowledge is the first barrier to overcome in order to implement pharmacogenetics test into routine medical practice. In this sense, previous experiences shared across literature can help to establish models for effective progress in pharmacogenetics applications (Gurwitz 2010). There are also some specific centre investigators that study racially and ethnically diverse populations and that are pioneers in the education of PharmD, MD and PhD students in pharmacogenomics and have led the establishment of unique graduate and postdoctoral training programmes focused on pharmacogenomics (Kroetz et al. 2009). Moreover, deeper on globalisation knowledge Web, there are some offered pharmacogenetic course online and many online bioinformatic tools that facilitate working with pharmacogenetics and pharmacogenomics (PharmGKB: www.pharmagkb.com). Basic knowledge on general advance methodologies are revised in other items of this chapter; however, strategic designs of pharmacogenetic studies are the key for successful results in this area of research where many factors converge, and they also should be well integrated and interpreted.

Ethical-privacy protection and legal issues in pharmacogenetic translational research compile concepts such as individual specimen origin (anonymisation, decoding, banking, traceability, identifiable health data, etc.), private authorisation, informed consent and specific regulations establishing bases about genetic privacy (UNESCO 1997). There are international organisms that carry out the emission, correction and control of these norms of implementation (FDA, EMA), and subsequently there are national bioethics advisory commissions that warrant the application of regulations in hospital environment clinical research.

On the other hand, 'quality' as the degree to which health services for individuals or populations

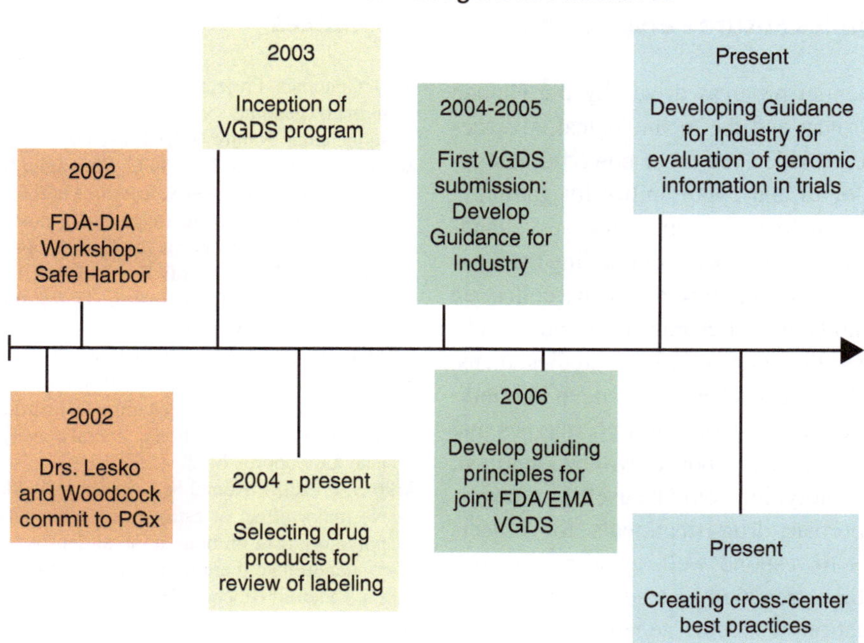

Fig. 29.6 FDA pharmacogenomics initiatives. Committees and processing information

increase the likelihood of desired outcomes has been defined at the health care sector, the definition being consistent with current professional knowledge. However, there is no clear mechanism for measuring the quality dimension of genetic services. The lack of international acceptance and user-driven, rigorously developed service quality indicators for clinical genetics makes difficult to follow an established and unique criteria for quality management method (Zellerino et al. 2009). Moreover, general aspects of health care should base on continuous improvements that comprise whole structure (tools, resources), process (activities) and outcome (results) that are focused in setting care, what takes place during care delivery and determination if the aims have been achieved, respectively (Lacalamita et al. 2008).

Following the global tendency, the first mandatory step prior to accreditation is the implementation of a quality management system. Thus, currently, hospitals worldwide are implementing general norms for acquiring quality assurance of their health care programmes (Guzel and Guner 2009; Lorenzo et al. 2010).

Further limitations to the translation of validated and in progress pharmacogenetics biomarkers in clinical applications include cost-effectiveness, economic incentives, reimbursement issues and the difficulty of making real evaluations of these limitations. Today, there are few studies evaluating the economical aspect on pharmacogenetics, and because of that, it remains as an abstract concept, which delays the demonstration of effectiveness to the governmental organisations. In order to overcome the first step of implementation of pharmacogenetics as a useful tool in a global hospital context, more costly therapies based on error-assay determination have to be conducted.

All these complex issues together with the idiosyncrasy of osteoporotic fractured patient care in hospital and controversial SNP-specific description of clinical impact evaluation constitute a huge challenge for the future advances. Finally, in this complex scenario, it will be absolutely necessary to harmonise the pharmacogenetic implementation, interpretation of results and the manner of reporting pharmacogenetic data

6 Omics Future Perspectives

The exponential progress driven by the Human Genome Project and the technological advances continue to provide even more powerful analytical technologies and opportunities for gaining a better understanding of complex diseases, including osteoporosis. The success in finding osteoporosis genes should clearly be based on a collection of large cohorts of well-characterised individuals. It is expected that with large-scale studies, many, if not all, of the genes that contribute to interindividual variation in osteoporosis phenotypes and influence therapeutic responses will be identified.

Patient genotyping could be useful for targeting osteoporosis drug treatments to subjects most likely to respond well, avoiding suboptimal long-term treatments or adverse reactions. The application of specific genetic tests to identify those subjects before the beginning of drug treatment is important mostly for diseases, such as osteoporosis, whose effective available therapies are so numerous, and therefore, the selection of the optimal therapy is foreseeable. Moreover, the pharmacogenetics could help to map novel molecular drug targets, with an impact on drug discovery, moving from 'one drug fits all' to personalised therapy. Certainly, the genes to be evaluated should always encompass those encoding drug targets, drug metabolising enzymes and drug transporters. Towards this goal pharmacogenomics does and will increasingly benefit from the novel technologies and experimental strategies, such as genome-wide scan association studies, microarray analysis, next generation sequencing and others.

Moreover, pharmacogenomics association studies need to be extended and confirmed in large cohorts, in different ethnic groups and/or in multicentric studies. All gene variants positively correlated with drug response in association studies will have to be validated by functional in vitro, in vivo and ex vivo studies.

Personalised medicine is expected to benefit from combining genomic information with regular monitoring of physiological states by multiple high-throughput methods.

References

1000 Genomes Project Consortium (2010) A map of human genome variation from population-scale sequencing. Nature 467:1061–1073

Adami S, Giannini S, Bianchi G, Sinigaglia L, Di Munno O, Fiore CE, Minisola S, Rossini M (2009) Vitamin D status and response to treatment in post-menopausal osteoporosis. Osteoporos Int 20:239–244

Antoniucci DM, Vittinghoff E, Blackwell T, Black DM, Sellmeyer DE (2005) Vitamin D insufficiency does not affect bone mineral density response to raloxifene. J Clin Endocrinol Metab 90:4566–4572

Antoniucci DM, Vittinghoff E, Palermo L, Black DM, Sellmeyer DE (2009) Vitamin D insufficiency does not affect response of bone mineral density to alendronate. Osteoporos Int 20:1259–1266

Arko B, Prezelj J, Komel R, Kocijancic A, Marc J (2002) No major effect of estrogen receptor beta gene RsaI polymorphism on bone mineral density and response to alendronate therapy in postmenopausal osteoporosis. J Steroid Biochem Mol Biol 81:147–152

Azagra R, Roca G, Encabo G, Prieto D, Aguyé A, Zwart M, Güell S, Puchol N, Gene E, Casado E, Sancho P, Solà S, Torán P, Iglesias M, Sabaté V, López-Expósito F, Ortiz S, Fernandez Y, Diez-Perez A (2011) Prediction of absolute risk of fragility fracture at 10 years in a Spanish population: validation of the WHO FRAX ™ tool in Spain. BMC Musculoskelet Disord 12:30

Balan G, Bauman J, Bhattacharya S, Castrodad M, Healy DR, Herr M, Humphries P, Jennings S, Kalgutkar AS, Kapinos B, Khot V, Lazarra K, Li M, Li Y, Neagu C, Oliver R, Piotrowski DW, Price D, Qi H, Simmons HA, Southers J, Wei L, Zhang Y, Paralkar VM (2009) The discovery of novel calcium sensing receptor negative allosteric modulators. Bioorg Med Chem Lett 19:3328–3332

Bekker PJ, Holloway D, Nakanishi A, Arrighi M, Leese PT, Dunstan CR (2001) The effect of a single dose of osteoprotegerin in postmenopausal women. J Bone Miner Res 16:348–360

Bone HG, Bolognese MA, Yuen CK, Kendler DL, Wang H, Liu Y, San Martin J (2008) Effect of Denosumab on bone mineral density and bone turnover in postmenopausal women. J Clin Endocrinol Metab 93:2149–2157

Bone HG, McClung MR, Roux C, Recker RR, Eisman JA, Verbruggen N, Hustad CM, DaSilva C, Santora AC, Ince BA (2010) Odanacatib, a cathepsin-K inhibitor for osteoporosis: a two-year study in postmenopausal women with low bone density. J Bone Miner Res 25:937–947

Boyce BF, Yoneda T, Lowe C, Soriano P, Mundy GR (1992) Requirement of pp 60c-src expression for osteoclasts to form ruffled borders and resorb bone in mice. J Clin Invest 90:1622–1627

Brown JP, Josse RG (2002) Clinical practice guidelines for the diagnosis and management of osteoporosis in Canada. CMAJ 167:S1–S34

Brown JP, Prince RL, Deal C, Recker RR, Kiel DP, de Gregorio LH, Hadji P, Hofbauer LC, Alvaro-Gracia JM, Wang H, Austin M, Wagman RB, Newmark R, Libanati C, San Martin J, Bone HG (2009) Comparison of the effect of denosumab and alendronate on BMD and biochemical markers of bone turnover in postmenopausal women with low bone mass: a randomized, blinded, phase 3 trail. J Bone Miner Res 24:153–161

Burge R, Dawson-Hughes B, Solomon DH, Wong JB, King A, Tosteson A (2007) Incidence and economic burden of osteoporosis-related fractures in the United States, 2005–2025. J Bone Miner Res 22:465–475

Canalis E, Giustina A, Bilezikian JP (2007) Mechanisms of anabolic therapies for osteoporosis. N Engl J Med 357:905–916

Center JR, Nguyen TV, Schneider D, Sambrook PN, Eisman JA (1999) Mortality after all major types of osteoporotic fracture in men and women: an observational study. Lancet 353:878–882

Cheung CL, Xiao SM, Kung AW (2010) Genetic epidemiology of age-related osteoporosis and its clinical applications. Nat Rev Rheumatol 6:507–517

Cho C, Nuttall M (2002) Emerging techniques for the discovery and validation of therapeutic targets for skeletal diseases. Expert Opin Ther Targets 6:679–689

Cho YS, Go MJ, Kim YJ, Heo JY, Oh JH, Ban HJ, Yoon D, Lee MH, Kim DJ, Park M, Cha SH, Kim JW, Han BG, Min H, Ahn Y, Park MS, Han HR, Jang HY, Cho EY, Lee JE, Cho NH, Shin C, Park T, Park JW, Lee JK, Cardon L, Clarke G, McCarthy MI, Lee JY, Lee JK, Oh B, Kim HL (2009) A large-scale genome-wide association study of Asian populations uncovers genetic factors influencing eight quantitative traits. Nat Genet 41:527–534

Choi HJ, Choi JY, Cho SW, Kang D, Han KO, Kim SA, Kim SY, Chung YS, Shin CS (2010) Genetic polymorphism of geranylgeranyl diphosphate synthase (GGSP1) predicts bone density response to bisphosphonate therapy in Korean women. Yonsei Med J 51:231–238

Compston J, Cooper A, Cooper C, Francis R, Kanis JA, Marsh D, McCloskey EV, Reid DM, Selby P, Wilkins M, National Osteoporosis Guideline Group (NOGG) (2009) Guidelines for the diagnosis and management of osteoporosis in postmenopausal women and men from the age of 50 years in the UK. Maturitas 62:105–108

Cummings SR, Nevitt MC, Browner WS, Stone K, Fox KM, Ensrud KE, Cauley J, Black D, Vogt TM (1995) Risk factors for hip fracture in white women. Study of Osteoporotic Fractures Research Group. N Engl J Med 332:767–773

Cummings SR, Bates D, Black DM (2002) Clinical use of bone densitometry: scientific review. JAMA 288:1889–1897

Cummings SR, San Martin J, McClung MR, Siris ES, Eastell R, Reid IR, Delmas P, Zoog HB, Austin M, Wang A, Kutilek S, Adami S, Zanchetta J, Libanati C, Siddhanti S, Christiansen C, Trial FREEDOM (2009) Denosumab for prevention of fractures in postmenopausal women with osteoporosis. N Engl J Med 361:756–765

Cummings-Vaughn LA, Gammack JK (2010) Falls, osteoporosis and hip fractures. Med Clin North Am 95:495–506

Dagdelen S, Sener D, Bayraktar M (2007) Influence of type 2 diabetes mellitus on bone mineral density response to bisphosphonates in late postmenopausal osteoporosis. Adv Ther 24:1314–1320

Delmas PD, Li Z, Cooper C (2004) Relationship between changes in bone mineral density and fracture risk reduction with antiresorptive drugs: some issues with meta analyses. J Bone Miner Res 19:330–337

Delmas PD, Rizzoli R, Cooper C, Reginster JY (2005) Treatment of patients with postmenopausal osteoporosis is worthwhile. The position of the International Osteoporosis Foundation. Osteoporos Int 16:1–5

Eisman JA (1999) Genetics of osteoporosis. Endocr Rev 20(6):788–804

Emery JG, McDonnell P, Burke MB, Deen KC, Lyn S, Silverman C, Dul E, Appelbaum ER, Eichman C, DiPrinzio R, Dodds RA, James IE, Rosenberg M, Lee JC, Young PR (1998) Osteoprotegerin is a receptor for the cytotoxic ligand TRAIL. J Biol Chem 273:14363–14367

European Agency for the Evaluation of Medicinal Products-Committee for Proprietary Medicinal Products (2002) Position paper on terminology in pharmacogenetics EMEA/CPMP/3070/01. Available at: http://www.eurogentest.org/uploads/1165308837041_Pharmacogenetics.pdf

European Medicines Agency (2009) Pre authorisation evaluation of medicines for human use: denosumab. Available at: www.ema.europa.eu/pdfs/human/opinion/Prolia_77616809en.pdf

Fang Y, Rivadeneira F, van Meurs JB, Pols HA, Ioannidis JP, Uitterlinden AG (2006) Vitamin D receptor gene BsmI and TaqI polymorphisms and fracture risk: a meta-analysis. Bone 39:938–945

Franken AA, van Blijderveen NJ, Witjes MJ, Netelenbos CJ (2011) Bisphosphonate-related osteonecrosis of the jaw. Ned Tijdschr Geneeskd 155:A3077

Garnero P, Munoz F, Borel O, Sornay-Rendu E, Delmas PD (2005) Vitamin D receptor gene polymorphisms are associated with the risk of fractures in postmenopausal women, independently of bone mineral density. J Clin Endocrinol Metab 90:4829–4835

Gates BJ, Sonnett TE, Duvall CA, Dobbins EK (2009) Review of osteoporosis pharmacotherapy for geriatric patients. Am J Geriatr Pharmacother 7:293–323

Gauthier JY, Chauret N, Cromlish W, Desmarais S, le Duong T, Falgueyret JP, Kimmel DB, Lamontagne S, Léger S, LeRiche T, Li CS, Massé F, McKay DJ, Nicoll-Griffith DA, Oballa RM, Palmer JT, Percival MD, Riendeau D, Robichaud J, Rodan GA, Rodan SB, Seto C, Thérien M, Truong VL, Venuti MC, Wesolowski G, Young RN, Zamboni R, Black WC (2008) The discovery of odanacatib (MK-0822), a selective inhibitor of cathepsin K. Bioorg Med Chem Lett 18:923–928

Gervasini G, Benítez J, Carrillo JA (2010) Pharmacogenetic testing and therapeutic drug monitoring are complementary tools for optimal individualization of drug therapy. Eur J Clin Pharmacol 66:755–774

Gowen M, Stroup GB, Dodds RA, James IE, Votta BJ, Smith BR, Bhatnagar PK, Lago AM, Callahan JF, DelMar EG, Miller MA, Nemeth EF, Fox J (2000) Antagonizing the parathyroid calcium receptor stimulates parathyroid hormone secretion and bone formation in osteopenic rats. J Clin Invest 105:1595–1604

Gullberg B, Johnell O, Kanis JA (1997) World-wide projections for hip fracture. Osteoporos Int 7:407–413

Gurwitz D (2010) Pharmacogenetics education: 10 years of experience at Tel Aviv University. Pharmacogenomics 11:647–649

Guzel O, Guner E (2009) ISO 15189 Accreditation: requirements for quality and competence of medical laboratories, experience of a laboratory. Clin Biochem 42:274–278

Haney EM, Bliziotes M (2008) Male osteoporosis: new insights in an understudied disease. Curr Opin Rheumatol 20:423–428

Hannon RA, Clack G, Rimmer M, Swaisland A, Lockton JA, Finkelman RD, Eastell R (2010) Effects of the Src kinase inhibitor saracatinib (AZD0530) on bone turnover in healthy men: a randomized, double-blind, placebo-controlled, multiple-ascending-dose phase I trial. J Bone Miner Res 25:463–471

Harada S, Rodan G (2003) Control of osteoblast function and regulation of bone mass. Nature 423:349–355

Harding B, Curley AJ, Hannan FM, Christie PT, Bowl MR, Turner JJ, Barber M, Gillham-Nasenya I, Hampson G, Spector TD, Thakker RV (2006) Functional characterization of calcium sensing receptor polymorphisms and absence of association with indices of calcium homeostasis and bone mineral density. Clin Endocrinol (Oxf) 65(5):598–605

Hasserius R, Karlsson MK, Nilsson BE, Redlund-Johnell I, Johnell O, European Vertebral Osteoporosis Study (2003) Prevalent vertebral deformities predict increased mortality and increased fracture rate in both men and women: a 10-year population-based study of 598 individuals from the Swedish cohort in the European Vertebral Osteoporosis Study. Osteoporos Int 14:61–68

Heilberg IP, Hernandez E, Alonzo E, Valera R, Ferreira LG, Gomes SA, Bellorin-Font E, Weisinger JR (2005) Estrogen receptor (ER) gene polymorphism may predict the bone mineral density response to raloxifene in postmenopausal women on chronic hemodialysis. Ren Fail 27:155–161

Hodgson SF, Watts NB, Bilezikian JP, Clarke BL, Gray TK, Harris DW, Johnston CC Jr, Kleerekoper M, Lindsay R, Luckey MM, McClung MR, Nankin HR, Petak SM, Recker RR, Osteoporosis Task Force AACE (2003) American Association of Clinical Endocrinologists medical guidelines for clinical practice for the prevention and treatment of postmenopausal osteoporosis: 2001 edition, with selected updates for 2003. Endocr Pract 9:544–564

Hoeppner L, Secreto F, Westendorf J (2009) Wnt signaling as a therapeutic target for bone diseases. Expert Opin Ther Targets 13:485–496

Hsu YH, Zillikens MC, Wilson SG, Farber CR, Demissie S, Soranzo N, Bianchi EN, Grundberg E, Liang L, Richards JB, Estrada K, Zhou Y, van Nas A, Moffatt MF, Zhai G, Hofman A, van Meurs JB, Pols HA, Price RI, Nilsson O, Pastinen T, Cupples LA, Lusis AJ, Schadt EE, Ferrari S, Uitterlinden AG, Rivadeneira F, Spector TD, Karasik D, Kiel DP (2010) An integration of genome-wide association study and gene expression profiling to prioritize the discovery of novel susceptibility Loci for osteoporosis-related traits. PLoS Genet 6(6):e1000977

Huang QY, Recker RR, Deng HW (2003) Searching for osteoporosis genes in the post-genome era: progress and challenges. Osteoporos Int 14:701–715

Ichikawa S, Koller DL, Johnson ML, Lai D, Xuei X, Edenberg HJ, Klein RF, Orwoll ES, Hui SL, Foroud TM, Peacock M, Econs MJ (2006) Human ALOX12, but not ALOX15, is associated with BMD in white men and women. J Bone Miner Res 21:556–564

Ioannidis JP (2005) Why most published research findings are false? PLoS Med 2:696–701

Ioannidis JP, Stavrou I, Trikalinos TA, Zois C, Brandi ML, Gennari L, Albagha O, Ralston SH, Tsatsoulis A, ER-alpha Genetics Meta-Analysis (2002) Association of polymorphisms of the estrogen receptor alpha gene with bone mineral density and fracture risk in women: a meta-analysis. J Bone Miner Res 17:2048–2060

Ioannidis JP, Ralston SH, Bennett ST, Brandi ML, Grinberg D, Karassa FB, Langdahl B, van Meurs JB, Mosekilde L, Scollen S, Albagha OM, Bustamante M, Carey AH, Dunning AM, Enjuanes A, van Leeuwen JP, Mavilia C, Masi L, McGuigan FE, Nogues X, Pols HA, Reid DM, Schuit SC, Sherlock RE, Uitterlinden AG, GENOMOS Study (2004) Differential genetic effects of ESR1 gene polymorphisms on osteoporosis outcomes. JAMA 292:2105–2114

Ioannidis JP, Ng MY, Sham PC, Zintzaras E, Lewis CM, Deng HW, Econs MJ, Karasik D, Devoto M, Kammerer CM, Spector T, Andrew T, Cupples LA, Duncan EL, Foroud T, Kiel DP, Koller D, Langdahl B, Mitchell BD, Peacock M, Recker R, Shen H, Sol-Church K, Spotila LD, Uitterlinden AG, Wilson SG, Kung AW, Ralston SH (2007) Meta-analysis of genome-wide scans provides evidence for sex- and site-specific regulation of bone mass. J Bone Miner Res 22:173–183

Ji GR, Yao M, Sun CY, Li ZH, Han Z (2010) BsmI, TaqI, ApaI and FokI polymorphisms in the vitamin D receptor (VDR) gene and risk of fracture in Caucasians: a meta-analysis. Bone 47:681–686

Johnell O, Kanis JA (2006) A estimate of the worldwide prevalence and disability associated with osteoporotic fractures. Osteoporos Int 17:1726–1733

Kanis JA (2002) Diagnosis of osteoporosis and assessment of fracture risk. Lancet 359:1929–1936

Kapur K, Johnson T, Beckmann ND, Sehmi J, Tanaka T, Kutalik Z, Styrkarsdottir U, Zhang W, Marek D, Gudbjartsson DF, Milaneschi Y, Holm H, Diiorio A, Waterworth D, Li Y, Singleton AB, Bjornsdottir US, Sigurdsson G, Hernandez DG, Desilva R, Elliott P,

Eyjolfsson GI, Guralnik JM, Scott J, Thorsteinsdottir U, Bandinelli S, Chambers J, Stefansson K, Waeber G, Ferrucci L, Kooner JS, Mooser V, Vollenweider P, Beckmann JS, Bochud M, Bergmann S (2010) Genome-wide meta-analysis for serum calcium identifies significantly associated SNPs near the calcium-sensing receptor (CASR) gene. PLoS Genet 6(7):e1001035. doi:10.1371/journal.pgen.1001035

Katz J, Gong Y, Salmasinia D, Hou W, Burkley B, Ferreira P, Casanova O, Langaee TY, Moreb JS (2011) Genetic polymorphisms and other risk factors associated with bisphosphonate induced osteonecrosis of the jaw. Int J Oral Maxillofac Surg 40:605–611

Kendler DL, Roux C, Benhamou CL, Brown JP, Lillestol M, Siddhanti S, Man HS, San Martin J, Bone HG (2010) Effects of denosumab on bone mineral density and bone turnover in postmenopausal women transitioning from alendronate therapy. J Bone Miner Res 25:72–81

Kennel KA, Drake MT (2009) Adverse effects of bisphosphonates: implications for osteoporosis management. Mayo Clin Proc 84:632–637

Klein RF, Allard J, Avnur Z, Nikolcheva T, Rotstein D, Carlos AS, Shea M, Waters RV, Belknap JK, Peltz G, Orwoll ES (2004) Regulation of bone mass in mice by the lipoxygenase gene Alox15. Science 303:229–232

Kroetz DL, Ahituv N, Burchard EG, Guo S, Sali A, Giacomini KM (2009) Institutional profile: The University of California Pharmacogenomics Center: at the interface of genomics, biological mechanisms and drug therapy. Pharmacogenomics 10:1569–1576

Kruk M, Ralston SH, Albagha OM (2009) LRP5 polymorphisms and response to risedronate treatment in osteoporotic men. Calcif Tissue Int 84:171–179

Kumar S, Matheny CJ, Hoffman SJ, Marquis RW, Schultz M, Liang X, Vasko JA, Stroup GB, Vaden VR, Haley H, Fox J, DelMar EG, Nemeth EF, Lago AM, Callahan JF, Bhatnagar P, Huffman WF, Gowen M, Yi B, Danoff TM, Fitzpatrick LA (2010) An orally active calcium-sensing receptor antagonist that transiently increases plasma concentrations of PTH and stimulates bone formation. Bone 46:534–542

Kurabayashi T, Matsushita H, Tomita M, Kato N, Kikuchi M, Nagata H, Honda A, Yahata T, Tanaka K (2004) Association of vitamin D and estrogen receptor gene polymorphism with the effects of long term hormone replacement therapy on bone mineral density. J Bone Miner Metab 22:241–247

Lacalamita R, Schirone M, Paradiso A (2008) ISO 9001:2000 applied to a research oncology laboratory: which problems? The experience of National Cancer Institute-Bari. Ann Oncol 19:1207–1208

Lamberts S, Uitterlinden A (2009) Genetic testing in clinical practice. Annu Rev Med 60:431–442

Lehrer S, Montazem A, Ramanathan L, Pessin-Minsley M, Pfail J, Stock RG, Kogan R (2009) Bisphosphonate-induced osteonecrosis of the jaws, bone markers, and a hypothesized candidate gene. J Oral Maxillofac Surg 67:159–161

Levy ME, Parker RA, Ferrell RE, Zmuda JM, Greenspan SL (2007) Farnesyl diphosphate synthase: a novel genotype association with bone mineral density in elderly women. Maturitas 7(3):247–252. Epub 26 Mar 2007

Lewiecki EM (2011) New targets for intervention in the treatment of postmenopausal osteoporosis. Nat Rev Rheumatol 7:631–638

Liu JM, Zhang MJ, Zhao L, Cui B, Li ZB, Zhao HY, Sun LH, Tao B, Li M, Ning G (2010) Analysis of recently identified osteoporosis susceptibility genes in Han Chinese women. J Clin Endocrinol Metab 95:E112–E120

Lorenzo TR, Sánchez PM, Santana CL, Cobian MJL, García del Rosario C (2010) [Management of quality in an Intensive Care Unit: implementation of ISO 9001:2008 international standard]. Gestión de la calidad en una unidad de cuidados intensivos: implementación de la norma ISO9001:2008. Med Intensiva 34:476–482

Macdonald HM, McGuigan FE, Stewart A, Black AJ, Fraser WD, Ralston S, Reid DM (2006) Large-scale population-based study shows no evidence of association between common polymorphism of the VDR gene and BMD in British women. J Bone Miner Res 21:151–162

MacLaughlin E (2010) Improving osteoporosis screening, risk assessment, diagnosis, and treatment initiation: role of the health system pharmacist in closing the gap. Am J Health Syst Pharm 67:4–8

Maize DF, Fuller SH, Hritcko PM, Matsumoto RR, Soltis DA, Taheri RR, Duncan W (2010) A review of remediation programs in pharmacy and other health professions. Am J Pharm Educ 74:25

Mani A, Radhakrishnan J, Wang H, Mani A, Mani MA, Nelson-Williams C, Carew KS, Mane S, Najmabadi H, Wu D, Lifton RP (2007) LRP6 mutation in a family with early coronary disease and metabolic risk factors. Science 315:1278–1282

Mann V, Ralston SH (2003) Meta analysis of COL1A1 Sp1 polymorphism in relation to bone mineral density and osteoporotic fracture. Bone 32:711–717

Marc J, Prezelj J, Komel R, Kocijancic A (1999) VDR genotype and response to etidronate therapy in late postmenopausal women. Osteoporos Int 10:303–306

Marini F, Falchetti A, Silvestri S, Bagger Y, Luzi E, Tanini A, Christiansen C, Brandi ML (2008) Modulatory effect of farnesyl pyrophosphate synthase (FDPS) rs2297480 polymorphism on the response to long-term amino-bisphosphonate treatment in postmenopausal osteoporosis. Curr Med Res Opin 24:2609–2615

Marini F, Tonelli P, Cavalli L, Cavalli T, Masi L, Falchetti A, Brandi ML (2011) Pharmacogenetics of bisphosphonate-associated osteonecrosis of the jaw. Front Biosci (Elite Ed) 3:364–370

Marzia M, Sims NA, Voit S, Migliaccio S, Taranta A, Bernardini S, Faraggiana T, Yoneda T, Mundy GR, Boyce BF, Baron R, Teti A (2000) Decreased c-Src expression enhances osteoblast differentiation and bone formation. J Cell Biol 151:311–320

McClung MR, Lewiecki EM, Cohen SB, Bolognese MA, Woodson GC, Moffett AH, Peacock M, Miller PD, Lederman SN, Chesnut CH, Lain D, Kivitz AJ, Holloway DL, Zhang C, Peterson MC, Bekker PJ, AMG 162 Bone Loss Study Group (2006) Denosumab in postmenopausal women with low bone mineral density. N Engl J Med 354:821–831

Mencej-Bedrac S, Prezelj J, Kocjan T, Komadina R, Marc J (2009a) Analysis of association of LRP5, LRP6, SOST, DKK1, and CTNNB1 genes with bone mineral density in a Slovenian population. Calcif Tissue Int 85:501–506

Mencej-Bedrac S, Prezelj J, Kocjan T, Teskac K, Ostanek B, Smelcer M, Marc J (2009b) The combinations of polymorphisms in vitamin D receptor, osteoprotegerin and tumor necrosis factor superfamily member 11 genes are associated with bone mineral density. J Mol Endocrinol 42:239–247

Murphy JE, Green JS, Adams LA, Squire RB, Kuo GM, McKay A (2010) Pharmacogenomics in the curricula of colleges and schools of pharmacy in the United States. Am J Pharm Educ 74:7

National Institute for Health and Clinical Excellence (NICE) (2010a) Technology appraisal guidance 160 (amended January 2010): Alendronate, etidronate, risedronate, raloxifene and strontium ranelate for the primary prevention of osteoporotic fragility fractures in postmenopausal women. http://www.nice.org.uk/nicemedia/pdf/TA160guidance.pdf

National Institute for Health and Clinical Excellence (NICE) (2010b) Technology appraisal guidance 161 (amended January 2010): Alendronate, etidronate, risedronate, raloxifene and strontium ranelate for the primary prevention of osteoporotic fragility fractures in postmenopausal women. http://www.nice.org.uk/nicemedia/live/11748/47177/47177.pdf

Nguyen ND, Eisman JA, Center JR, Nguyen TV (2007) Risk factors for fracture in nonosteoporotic men and women. J Clin Endocrinol Metab 92:955–962

NIH Consensus Development Panel on Osteoporosis Prevention, Diagnosis, and Therapy (2001) Osteoporosis prevention, diagnosis, and therapy. JAMA 285: 785–795

Palomba S, Numis FG, Mossetti G, Rendina D, Vuotto P, Russo T, Zullo F, Nappi C, Nunziata V (2003) Raloxifene administration in postmenopausal women with osteoporosis: effect of different BsmI vitamin D receptor genotypes. Hum Reprod 18:192–198

Palomba S, Orio F Jr, Russo T, Falbo A, Tolino A, Manguso F, Nunziata V, Mastrantonio P, Lombardi G, Zullo F (2005) BsmI vitamin D receptor genotypes influence the efficacy of antiresorptive treatments in postmenopausal osteoporotic women. A 1-year multicenter, randomized and controlled trial. Osteoporos Int 16:943–952

Patsch JM, Deutschmann J, Pietschmann P (2011) Gender aspects of osteoporosis and bone strength. Wien Med Wochenschr 161:117–123

Peacock M, Koller DL, Lai D, Hui S, Foroud T, Econs MJ (2005) Sex-specific quantitative trait loci contribute to normal variation in bone structure at the proximal femur in men. Bone 37:467–473

Pharmacogenomic knowledge base (Pharmgkb) Gene, variants, pathway. Available at: http://www.pharmgkb.org/. Jan 2012

Pierre JM (2010) The calcium-sensing receptor in bone cells: a potential therapeutic target in osteoporosis. Bone 46:571–576

Pietschmann P, Rauner M, Sipos W, Kerschan-Schindl K (2009) Osteoporosis: an age-related and gender-specific disease – a mini-review. Gerontology 55:3–12

Qureshi AM, Herd RJ, Blake GM, Fogelman I, Ralston SH (2002) COLIA1 Sp1 polymorphism predicts response of femoral neck bone density to cyclical etidronate therapy. Calcif Tissue Int 70:158–163

Rachner TD, Khosla S, Hofbauer LC (2011) Osteoporosis: now and the future. Lancet 377:1276–1287

Ralston SH (2002) Genetic control of susceptibility to osteoporosis. J Clin Endocrinol Metab 87: 2460–2466

Ralston SH (2010) Genetics of osteoporosis. Ann N Y Acad Sci 1192:181–189

Ralston SH, Uitterlinden AG, Brandi ML, Balcells S, Langdahl BL, Lips P, Lorenc R, Obermayer-Pietsch B, Scollen S, Bustamante M, Husted LB, Carey AH, Diez-Perez A, Dunning AM, Falchetti A, Karczmarewicz E, Kruk M, van Leeuwen JP, van Meurs JB, Mangion J, McGuigan FE, Mellibovsky L, del Monte F, Pols HA, Reeve J, Reid DM, Renner W, Rivadeneira F, van Schoor NM, Sherlock RE, Ioannidis JP, Investigators GENOMOS (2006) Large-scale evidence for the effect of the COLIA1 Sp1 polymorphism on osteoporosis outcomes: the GENOMOS study. PLoS Med 3:e90

Rawadi G, Roman-Roman S (2005) Wnt signalling pathway: a new target for the treatment of Osteoporosis. Expert Opin Ther Targets 9:1063–1077

Reid IR (2009) Osteonecrosis of the jaw: who gets it, and why? Bone 44:4–10

Reppe S, Refvem H, Gautvik VT, Olstad OK, Høvring PI, Reinholt FP, Holden M, Frigessi A, Jemtland R, Gautvik KM (2010) Eight genes are highly associated with BMD variation in postmenopausal Caucasian women. Bone 46:604–612

Richards JB, Rivadeneira F, Inouye M, Pastinen TM, Soranzo N, Wilson SG, Andrew T, Falchi M, Gwilliam R, Ahmadi KR, Valdes AM, Arp P, Whittaker P, Verlaan DJ, Jhamai M, Kumanduri V, Moorhouse M, van Meurs JB, Hofman A, Pols HA, Hart D, Zhai G, Kato BS, Mullin BH, Zhang F, Deloukas P, Uitterlinden AG, Spector TD (2008) Bone mineral density, osteoporosis, and osteoporotic fractures: a genome-wide association study. Lancet 371:1505–1512

Richards JB, Kavvoura FK, Rivadeneira F, Styrkársdóttir U, Estrada K, Halldórsson BV, Hsu YH, Zillikens MC, Wilson SG, Mullin BH, Amin N, Aulchenko YS, Cupples LA, Deloukas P, Demissie S, Hofman A, Kong A, Karasik D, van Meurs JB, Oostra BA, Pols HA, Sigurdsson G, Thorsteinsdottir U, Soranzo N, Williams FM, Zhou Y, Ralston SH, Thorleifsson G, van Duijn CM, Kiel DP, Stefansson K, Uitterlinden AG, Ioannidis JP, Spector TD, Genetic Factors for Osteoporosis Consortium (2009) Collaborative meta-analysis: associations of 150 candidate genes with osteoporosis and osteoporotic fracture. Ann Intern Med 151:528–537

Rivadeneira F, Styrkársdottir U, Estrada K, Halldórsson BV, Hsu YH, Richards JB, Zillikens MC, Kavvoura FK, Amin N, Aulchenko YS, Cupples LA, Deloukas P, Demissie S, Grundberg E, Hofman A, Kong A, Karasik D, van Meurs JB, Oostra B, Pastinen T, Pols HA, Sigurdsson G, Soranzo N, Thorleifsson G, Thorsteinsdottir U, Williams FM, Wilson SG, Zhou Y, Ralston SH, van Duijn CM, Spector T, Kiel DP, Stefansson K, Ioannidis JP, Uitterlinden AG, Genetic Factors for Osteoporosis (GEFOS) Consortium (2009) Twenty bone-mineral-density loci identified by large-scale meta-analysis of genome-wide association studies. Nat Genet 41:1199–1206

Rojo Venegas K, Aguilera Gómez M, Eisman JA, García Sánchez A, Faus Dader MJ, Calleja Hernández MA (2010) Pharmacogenetics of osteoporosis-related bone fractures: moving towards the harmonization and validation of polymorphism diagnostic tools. Pharmacogenomics 11:1287–1303

Rojo K, Aznarte P, Calleja MA, Contreras O, Martínez JL, López-Mezquita B, Faus-Dader MJ (2010) Factors of risk in an elderly population: evaluation for the prevention of hip fractures. Rev Cir Ortop Traumatol 54:167–173

Sarasquete ME, García-Sanz R, Marín L, Alcoceba M, Chillón MC, Balanzategui A, Santamaria C, Rosiñol L, de la Rubia J, Hernandez MT, Garcia-Navarro I, Lahuerta JJ, González M, San Miguel JF (2008) Bisphosphonate-related osteonecrosis of the jaw is associated with polymorphisms of the cytochrome P450 CYP2C8 in multiple myeloma: a genome-wide single nucleotide polymorphism analysis. Blood 112:2709–2712

Sarasquete ME, González M, San Miguel JF, García-Sanz R (2009) Bisphosphonate-related osteonecrosis: genetic and acquired risk factors. Oral Dis 15:382–387

Sawcer SJ, Maranian M, Singlehurst S, Yeo T, Compston A, Daly MJ, De Jager PL, Gabriel S, Hafler DA, Ivinson AJ, Lander ES, Rioux JD, Walsh E, Gregory SG, Schmidt S, Pericak-Vance MA, Barcellos L, Hauser SL, Oksenberg JR, Kenealy SJ, Haines JL (2004) Enhancing linkage analysis of complex disorders: an evaluation of high-density genotyping. Hum Mol Genet 13:1943–1949

Shapses SA, Kendler DL, Robson R, Hansen KE, Sherrell RM, Field MP, Woolf E, Berd Y, Mantz AM, Santora AC 2nd (2011) Effect of alendronate and vitamin D(3) on fractional calcium absorption in a double-blind, randomized, placebo-controlled trial in postmenopausal osteoporotic women. J Bone Miner Res 26:1836–1844

Simsek M, Cetin Z, Bilgen T, Taskin O, Luleci G, Keser I (2008) Effects of hormone replacement therapy on bone mineral density in Turkish patients with or without COL1A1 Sp1 binding site polymorphism. J Obstet Gynaecol Res 34:73–77

Siris ES, Selby PL, Saag KG, Borgström F, Herings RM, Silverman SL (2009) Impact of osteoporosis treatment adherence on fracture rates in North America and Europe. Am J Med 122(2 Suppl):S3–S13

Stoch SA, Zajic S, Stone J, Miller DL, Van Dyck K, Gutierrez MJ, De Decker M, Liu L, Liu Q, Scott BB, Panebianco D, Jin B, Duong LT, Gottesdiener K, Wagner JA (2009) Effect of the cathepsin K inhibitor odanacatib on bone resorption biomarkers in healthy postmenopausal women: two double-blind, randomized, placebo-controlled phase I studies. Clin Pharmacol Ther 86:175–182

Styrkarsdottir U, Cazier JB, Kong A, Rolfsson O, Larsen H, Bjarnadottir E, Johannsdottir VD, Sigurdardottir MS, Bagger Y, Christiansen C, Reynisdottir I, Grant SF, Jonasson K, Frigge ML, Gulcher JR, Sigurdsson G, Stefansson K (2003) Linkage of osteoporosis to chromosome 20p12 and association to BMP2. PLoS Biol 1(3):E69

Styrkarsdottir U, Halldorsson BV, Gretarsdottir S, Gudbjartsson DF, Walters GB, Ingvarsson T, Jonsdottir T, Saemundsdottir J, Snorradóttir S, Center JR, Nguyen TV, Alexandersen P, Gulcher JR, Eisman JA, Christiansen C, Sigurdsson G, Kong A, Thorsteinsdottir U, Stefansson K (2009) New sequence variants associated with bone mineral density. Nat Genet 4:15–17

Tesch G, Amur S, Schousboe JT, Siegel JN, Lesko LJ, Bai JP (2010) Successes achieved and challenges ahead in translating biomarkers into clinical applications. AAPS J 12:243–253

Tofteng CL, Bach-Mortensen P, Bojesen SE, Tybjaerg-Hansen A, Hyldstrup L, Nordestgaard BG (2007) Integrin beta3 Leu33Pro polymorphism and risk of hip fracture: 25 years follow-up of 9233 adults from the general population. Pharmacogenet Genomics 17: 85–91

Trost Z, Trebse R, Prezelj J, Komadina R, Logar DB, Marc J (2010) A microarray based identification of osteoporosis-related genes in primary culture of human osteoblasts. Bone 46:72–80

Tsukamoto K, Orimo H, Hosoi T, Miyao M, Ota N, Nakajima T, Yoshida H, Watanabe S, Suzuki T, Emi M (2000 Mar) Association of bone mineral density with polymorphism of the human calcium-sensing receptor locus. Calcif Tissue Int 66(3):181–183

U.S. Food and Drug Administration, European Commission, European Medicines Agency (2009) General principles processing joint FDA EMEA Voluntary Genomic Data Submissions (VGDSs) within the framework of the Confidentiality Arrangement. http://www.ema.europa.eu/docs/en_GB/document_library/Scientific_guideline/2009/09/WC500003887.pdf

Uitterlinden AG, Ralston SH (2010) Genetics of osteoporosis. Endocr Rev 31:629–662

Uitterlinden AG, Ralston SH, Brandi ML, Carey AH, Grinberg D, Langdahl BL, Lips P, Lorenc R, Obermayer-Pietsch B, Reeve J, Reid DM, Amedei A, Bassiti A, Bustamante M, Husted LB, Diez-Perez A,

Dobnig H, Dunning AM, Enjuanes A, Fahrleitner-Pammer A, Fang Y, Karczmarewicz E, Kruk M, van Leeuwen JP, Mavilia C, van Meurs JB, Mangion J, McGuigan FE, Pols HA, Renner W, Rivadeneira F, van Schoor NM, Scollen S, Sherlock RE, Ioannidis JP, APOSS Investigators, EPOS Investigators, EPOLOS Investigators, FAMOS Investigators, LASA Investigators, Rotterdam Study Investigators, GENOMOS Study (2006) The association between common vitamin D receptor gene variations and osteoporosis: a participant-level meta-analysis. Ann Intern Med 145:255–264

UNESCO (1997) Universal declaration on the human rights. 11th Nov 1997. http://unesdoc.unesco.org/images/0011/001102/110220e.pdf#page=47

Unnanuntana A, Gladnick BP, Donnelly E, Lane JM (2010) The assessment of fracture risk. J Bone Joint Surg Am 92:743–753

van Meurs JB, Trikalinos TA, Ralston SH, Balcells S, Brandi ML, Brixen K, Kiel DP, Langdahl BL, Lips P, Ljunggren O, Lorenc R, Obermayer-Pietsch B, Ohlsson C, Pettersson U, Reid DM, Rousseau F, Scollen S, Van Hul W, Agueda L, Akesson K, Benevolenskaya LI, Ferrari SL, Hallmans G, Hofman A, Husted LB, Kruk M, Kaptoge S, Karasik D, Karlsson MK, Lorentzon M, Masi L, McGuigan FE, Mellström D, Mosekilde L, Nogues X, Pols HA, Reeve J, Renner W, Rivadeneira F, van Schoor NM, Weber K, Ioannidis JP, Uitterlinden AG, GENOMOS Study (2008) Large-scale analysis of association between LRP5 and LRP6 variants and osteoporosis. JAMA 299:1277–1290

van Straaten T, Van Schaik R (2010) Genetic techniques for pharmacogenetic analyses. Curr Pharm Des 16:231–237

Voy BH (2011) Systems genetics: a powerful approach for gene-environment interactions. J Nutr 141:515–519

Wang L (2010) Pharmacogenomics: a systems approach. Wiley Interdiscip Rev Syst Biol Med 2:3–22

Weinshilboum R (2003) Inheritance and drug response. N Engl J Med 348:529–537

Weinshilboum RM, Wang L (2006) Pharmacogenetics and pharmacogenomics: development, science, and translation. Annu Rev Genomics Hum Genet 7:223–245

Wilson SG, Reed PW, Andrew T, Barber MJ, Lindersson M, Langdown M, Thompson D, Thompson E, Bailey M, Chiano M, Kleyn PW, Spector TD (2004) A genome-screen of a large twin cohort reveals linkage for quantitative ultrasound of the calcaneus to 2q33-37 and 4q12-21. J Bone Miner Res 19:270–277

Wilson SR, Peters C, Saftig P, Brömme D (2009) Cathepsin K activity-dependent regulation of osteoclast actin ring formation and bone resorption. J Biol Chem 284:2584–2592

Xiong DH, Liu XG, Guo YF, Tan LJ, Wang L, Sha BY, Tang ZH, Pan F, Yang TL, Chen XD, Lei SF, Yerges LM, Zhu XZ, Wheeler VW, Patrick AL, Bunker CH, Guo Y, Yan H, Pei YF, Zhang YP, Levy S, Papasian CJ, Xiao P, Lundberg YW, Recker RR, Liu YZ, Liu YJ, Zmuda JM, Deng HW (2009) Genome-wide association and follow-up replication studies identified ADAMTS18 and TGFBR3 as bone mass candidate genes in different ethnic groups. Am J Hum Genet 84(3):388–398. doi:10.1016/j.ajhg.2009.01.025

Yazdanpanah N, Rivadeneira F, van Meurs JB, Zillikens MC, Arp P, Hofman A, van Duijn CM, Pols HA, Uitterlinden AG (2007) The 1997 G/T and Sp1 polymorphisms in the collagen type I a1 (COLIA1) gene in relation to changes in femoral neck bone mineral density and the risk of fracture in the elderly: the Rotterdam study. Calcif Tissue Int 81:18–25

Younes A, Kadin ME (2003) Emerging applications of the tumor necrosis factor family of ligands and receptors in cancer therapy. J Clin Oncol 21(18):3526–3534

Zaghloul NA, Katsanis N (2010) Functional modules, mutational load and human genetic disease. Trends Genet 26:168–176

Zebaze RM, Ghasem-Zadeh A, Bohte A, Iuliano-Burns S, Mirams M, Price RI, Mackie EJ, Seeman E (2010) Intracortical remodelling and porosity in the distal radius and post-mortem femurs of women: a cross-sectional study. Lancet 375:1729–1736

Zellerino BC, Milligan SA, Gray JR, Williams MS, Brooks R (2009) Identification and prioritization of quality indicators in clinical genetics: an international survey. Am J Med Genet C Semin Med Genet 151C(3):179–190

Zhu Z, Xue LM, Han T, Jiao L, Qin LP, Li YS, Zheng HC, Zhang QY (2010) Anti-osteoporotic effects and proteomic characterization of the target and mechanism of an Er-Xian Decoction on osteoblastic UMR-106 and osteoclasts induced from RAW264.7. Molecules 15:4695–4710

30 Pharmacogenomics of Gynecological Disorders

Jaspreet Kaur, Beenish Rahat, and Jyotdeep Kaur

Abstract

Pharmacogenomics aims to identify the genome-based variability in drug disposition (metabolism and transport), which influences its efficacy and toxicity, and also to develop diagnostics and therapeutics on the basis of genetic variants so as individualize pharmacotherapy. Progress made in the last two decades in pharmacogenomics study of gynecological diseases has enabled us to predict the therapy outcome on the basis of inherited or acquired genetic mutations. The present chapter focuses on the clinical relevance of genetic variation in terms of diagnosis, prognosis, and therapeutics in context to this group of diseases with special reference to gynecological cancers. Various disease susceptibility genes and their polymorphisms of gynecological disorders have been identified, and their molecular diagnostics and targeted therapies developed. To name a few well studied are BRCA1/2, HER2, GSTT1, GSTM1, CYP1B1, PPAR-γ, VEGF, etc. Detailed account of the genetic components and their variants associated with each disease and how it influences the discourse of the disease and individualization of the therapy is given here.

1 Introduction

The aim of carrying out pharmacogenomics study is to identify the inherited variations (from person to person or within population) responsible for differences observed in drug metabolism, its efficacy and toxicity. In literature the terms pharmacogenetics and pharmacogenomics are generally used synonymously, but in true sense pharmacogenetics means the study of a gene or a fixed number of genes, whereas pharmacogenomics is the study of whole genome to understand its effects on drug response in the individuals. Here in this chapter the focus will be on the pharmacogenetics of gynecological diseases in the order of breast cancer, cervical cancer, ovarian cancer, endometrial cancer, gestational diabetes mellitus, and preeclampsia.

The search for genetic components of the diseases is one of the important goals of pharmacogenomics,

J. Kaur
University Institute of Engineering and Technology, Panjab University, Chandigarh, India

B. Rahat • J. Kaur (✉)
Biochemistry, Postgraduate Institute of Medical Education and Research (PGIMER), Chandigarh, India
e-mail: jyotdeep2001@yahoo.co.in

D. Barh et al. (eds.), *Omics for Personalized Medicine*,
DOI 10.1007/978-81-322-1184-6_30, © Springer India 2013

and focus of research had been to identify the genes or mutations in genome associated with human diseases (Williams-Jones and Corrigan 2003). For diagnosis of rare congenital human diseases such as Duchenne muscular dystrophy, chromosomal analysis or other clinical biochemistry tests have been developed for more than three decades. More recently, high-throughput genomic technologies, the results of Human Genome Project (HGP), and the advances in genotyping have made possible to perform the molecular/genetic tests that can predict the precise disposition of a few adult-onset human diseases such as Huntington disease, Alzheimer disease, and certain familial forms of breast cancer and other gynecological disorders. While genetic tests are quite reliable and can predict the risk of disease in the future, yet the clinical utility of predictive genetic testing for genetic diseases varies considerably and adds to the burden of health care services (Reviewed by Evans et al. 2001).

Currently, it is well established that inherited variations in drug-metabolizing genes and drug target genes lead to the observed differences in the chemotherapy response (Yan and Beckman 2005; Freedman et al. 2010). If these genetic variants of a disease are present at a rate of more than 1 % in a population, these are called polymorphisms. These can be insertions, deletions, or substitutions of nucleotides in the genome. Most common are single-nucleotide substitutions called single-nucleotide polymorphisms abbreviated as SNPs. More than 1.4 million SNPs have been reported so far in the human genome, and the number is likely to increase as new findings are reported with time and with molecular biology advances. An SNP may occur after 300–1,000 bases in a genome; there is possibility of several million SNPs in a genome. SNPs could be used to identify population-wide predisposition or susceptibility to the disease and may contribute to polygenic diseases. They can occur in noncoding regions and in coding regions. The latter often generate polymorphic variation in expressed proteins, affecting their functional properties such as drug transport and metabolism as well as cellular targets, signaling pathways, and cellular responses to treatment. If the SNP occurs in the promoter region of a gene, it may change the binding site for transcription factors and thus alter the regulation of expression (Gradhand and Kim 2008). The goal is thus to identify polymorphisms in the genes that contribute to disease, which will be targets for new therapies, and may predict the outcome and side effects of a given therapy. Many of these SNPs have actually led to the development of molecular diagnostics used now by the clinicians. Kim and Misra (2007) have reviewed the applications of genotyping in biomedical applications.

There are a very large number of gynecological disorders and these contribute significantly to the global burden of diseases. Much emphasis has been given to the research studies and to find better diagnostics and therapeutics for these plethora of diseases. Using proteomics technologies, several thousands of proteins in a pathological sample or a biological fluid can be compared simultaneously with the respective healthy tissue or the fluid to find a unique disease specific protein, called a biomarker. Biomarker studies have utilized the "omics" techniques to identify patterns or changes in thousands of proteins and can globally analyze almost all small molecular weight proteins. A few of these techniques are electrospray ionization, mass spectrometry, matrix-assisted laser desorption ionization time-of-flight mass spectrometry (MALDI-TOFMS), and surface-enhanced laser desorption ionization time-of-flight mass spectrometry (SELDI-TOFMS). More recently a review on the role of proteomics studies to identify biomarkers for gynecological diseases has been reported by Meehan et al. (2010). The unique biomarker of a disease will lead to development of highly specific diagnostics and can also be the drug target for therapeutics. Petricoin et al. (2002) have reported the analysis of the serum samples of patients with neoplastic ovarian carcinoma and women with nonmalignant ovary disease by MS-SELDI (mass spectroscopy-surface-enhanced laser desorption and ionization) and the used bioinformatics algorithm to interpret the proteome spectra generated. This results had a sensitivity of 100 and positive predictive value of 94 %. Authors concluded that such studies of proteomic technology can be applied as screening procedures in high-risk group and normal population for identification of malignancies.

The hormone-responsive cancers such as most of gynecological cancers have been reported to have a different mechanism of carcinogenesis as compared to chemical carcinogenesis. As early as in 1982, Handerson et al. on the basis of data of epidemiological and laboratory studies had reported that some cancers can be grouped together as hormone-related cancers such as breast, endometrium, testis, ovary, osteosarcoma, thyroid, and prostate. Hormones induce faster cell proliferation in the tissue, and during this process, random genetic mutations accumulate, leading to malignancy (Henderson et al. 1982, 1988). However, the genes involved in induction of such cancers were not known. Of the various hormones, the role of estrogen is more pronounced and well established (Henderson et al. 1988; Henderson and Feigelson 2000). The main reason for this is thought to be estrogen stimulation of cell proliferation in the target organs. As further research reported by Henderson and Feigelson (2000), breast cancer model was found to have involvement of the genes such as the cytochrome P459c17α (CYP17) gene, the cytochrome P450, subfamily 19 (CYP19) gene or aromatase (responsible for synthesis of estrogens from androgens also reported by Simpson et al. 1994.), 17β-hydroxysteroid dehydrogenase 1 (HSD17B1) gene, and the estrogen receptor alpha (*ER*) gene. Estrogens are metabolized by several enzymes, including cytochrome P450 1A1 and 1A2 (CYP1A1 and CYP1A2), through hydroxylation resulting in reactive metabolites (Badawi et al. 2001; Yamazaki et al. 1998; Xu et al. 1999). Further degradation of these metabolites via detoxification pathways involves participation of sulfotransferase1A1 (SULT1A1) (Raftogianis et al. 2000), forming products which are eliminated via urine. Much work has been done on these genes variants of the above enzymes in the population, and estrogen metabolism may get disturbed through altered activities of the enzymes produced (Eaton et al. 1995; Raftogianis et al. 1997; Kiyohara et al. 1998; Goodman et al. 2001). Thus, the above gene polymorphisms of CYP, aromatase, and SULT1A1 have been found to be linked with estrogen-induced cancers. Mikhailova et al. (2006) reported high risk of endometrial cancer and ovarian cancers with gene variants of hormone-metabolizing enzymes in the Caucasian Russian females. High correlation of several SNPs of above genes was found with estrogen-responsive cancers, whereas no association of SNPs of aromatase (CYP19) and CYP1A1 was found for hormone-dependent cancers.

Treatment plan of a cancer depends upon the tumor site, size, and type (pathology); generally grading is done depending upon one or more of the factors such as the mitotic index, number of sites involved, and the size of tumor. Efficacy and toxicity of the drugs also depend upon the genetic makeup of the host. The benefit from specific treatment and the adverse events experienced vary considerably. Effective treatments of cancer in general and for gynecological cancers in particular have increased with time specifically the development of monoclonal antibodies, hormone, and growth factor receptor agonists and antagonists, etc. as will be discussed in this chapter with reference to each disease.

2 Breast Cancer

Breast cancer is a global health problem and remains the most prevalent cancer diagnosed in women in the world. Although breast cancer mortality has declined in the last two decades, breast cancer continues to represent a major threat to the lives and productivity of women. There is an estimated age-standardized incidence rate of 37.4 per 100,000 and an estimated age-standardized mortality rate of 13.2 per 100,000 women (http://www-dep.iarc.fr, GLOBOCAN database). Both personal and family histories influence a woman's risk of developing breast cancer. The known risk factors for breast cancer are exposure to ionizing radiation, breast cancer in a first-degree relative, reproductive and hormonal factors, and alcohol consumption, but these factors explain only a portion of the variability in breast cancer risk (Ambrosone 2007). Of these risk factors, a family history of breast cancer is responsible for the greatest increase in risk, with women with a first-degree relative with breast cancer having twice the risk of those who do not (Ambrosone 2007). In addition, twin studies indicated that up to 30 % of breast cancer cases may be due to genetic factors (Lichtenstein et al. 2000). Genetics of breast cancer and prevalence of susceptibility genes will be discussed below.

2.1 Genetics of Breast Cancer and Disease Susceptibility Genes

So far several candidate genes for dominant breast cancer susceptibility have been discovered (Pitot and Loeb 2002). It is well established by now that many cases of hereditary breast cancer are because of accumulation of germ line mutations in a gene *BRCA1* located on chromosome 17q, shown for the first time by genetic linkage analysis (Hall and Lee 1990). BRCA1 was cloned in 1994 by Miki et al. (1994), and the normal cellular functions of the BRCA 1 gene have been well studied. Some of the reported multiple functions of the gene, expressed as BRCA 1 protein (1,863 amino acid), are tumor suppressor activity, control over cell cycle progression (regulated by kinases through phosphorylations), and many types of specialized DNA repair enzyme mechanisms in response to DNA damage, and if the damage is not repaired, then it leads to cell destruction and apoptosis, etc. Clinically significant, or deleterious, mutations of *BRCA1* and *BRCA2* genes are associated with increased susceptibility for breast and ovarian cancer.

Much research has gone into identify the deleterious *BRCA1* and *BRCA2* mutations and SNPs, enabling much specific treatment regimens to be administered in breast and ovarian cancer women with inherited predisposition to these cancers. The innovative feature of these clinical changes has been the genetic approach to identification of high-risk women. In the last few years, several genome-wide association studies (GWASs) investigating common genetic variants have successfully identified single-nucleotide polymorphisms (SNPs) in a number of independent loci to be associated with breast cancer risk. Each SNP confers only a small increase in breast cancer risk, but the higher number of SNPs present in the same locus gives a higher risk in individuals carrying multiple susceptibility SNPs. Genomics and pharmacogenomics of breast cancer has been reviewed recently by Ayoub et al. (2011).

A growth factor receptor gene, human epidermal growth factor receptor (*HER2*) (Coussens et al. 1985; King et al. 1985; Akiyama et al. 1986), is amplified in 25–30 % of breast cancers leading to the expression of HER2 proteins at abnormally high levels in malignant cells (Slamon et al. 1987, 1989). Women with breast cancers that overexpress HER2 have an aggressive form of the disease with significantly shortened disease-free survival and overall survival (Slamon et al. 1987, 1989; Seshadri et al. 1993; Press et al. 1993; Ravdin and Chamness 1995). Treatment with trastuzumab is successful only in breast cancers that overexpress *HER2* receptor.

Genotype analysis of the 2,999 samples (1,042 case probands, 508 control probands, and 1,449 relatives of case probands, of whom 111 had breast cancer diagnoses) for nine SNPs verified by the Breast Cancer Association Consortium (BCAC) for the study, namely, rs2981582 (FGFR2), rs3803662 (TNRC9), rs3817198 (LSP1), rs889312 (MAP3K), rs13281615 (8q24), rs2107425 (H19), rs17468277 (CASP8), rs13387042 (2q35), and rs10941679 (5p12), was done by scientists in Australia. Families with disease-causing BRCA1/2 mutations were excluded. There was no difference in all three SNP scores between the relatives (with SNP data generated) as defined by their breast cancer status, and the authors have questioned the utility of SNP-based risk prediction, even in the familial setting, although the addition of many more such SNPs may improve their utility. Because metastasis is an important prognostic determinant in patients of breast cancers, it is important to understand the role of germ line mutations in progression of disease and breast cancer metastasis. In fact there are not much reports either from microarray studies or from other experimentations to elucidate their contribution to metastasis.

Initial studies using rats have identified that the susceptibility to cancer metastasis is heritable. Quantitative analysis of genetic mapping identified a metastasis efficiency locus (*Mtes1*) on mouse chromosome 19 in a 10 Mb region which is analogous to human chromosome 11q12-13; in the same analysis a metastasis suppressor gene *Brms1* has also been located in the same locus, but genetic variants or polymorphisms have not been found even after extensive research (Seraj et al. 2000; Hunter et al. 2001). Another gene studied in this gene is *Sipa1* (signal-induced proliferation-associated gene 1), which gets induced in presence of mitogens, and its expressed protein is a GTPase activator. This protein has the function in controlling cell adhesion. SNPs have been reported in this gene. Park et al. (2005) have reported that *Sipa1* genes of the highly metastatic mouse strains AKR and FVB contain A739T SNP. This SNP changes an alanine to threonine in the PDZ protein–protein interaction domain. This evidence is consistent with *Sipa1* being a metastasis susceptibility gene in Mtes1 locus, and it can be inferred that the *Sipa1* polymorphism can have significant influence on tumor metastasis (Park et al. 2005).

2.2 Pharmacogenetics of Breast Cancer Therapy

Research in pharmacogenetics has led to elucidation of pathways of drug metabolism and the role of biomarkers to optimize chemotherapeutic strategies for each individual patient (Olopade et al. 2008; Dowsett and Dunbier 2008). Although breast cancer therapies are improving rapidly, but multidrug resistance (MDR) and treatment-related toxicities add to morbidity and mortality. Genome-wide studies have revealed genetic polymorphisms of deleterious consequences within population and ethnic groups of a population. Breast cancer is a heterogeneous disease, manifesting as different molecular subtypes, each of which responds to chemotherapeutic agents and radiation differently (O'Brien et al. 2008). Advanced or metastatic breast cancer therapy includes single or multiple combinations of hormonal therapy; monoclonal antibodies for overexpressed receptors such as trastuzumab, lapatinib, and bevacizumab; and chemotherapy regimens such as inclusion of tamoxifen, aromatase inhibitors, fulvestrant, anthracyclines, taxanes, cyclophosphamide, vinorelbine, platinum drugs, and vinca alkaloids. Here in this chapter these therapeutics have been classified as novel agents or monoclonal antibodies, hormone intervention therapy, and chemotherapy. Pharmacogenetics of these therapeutics is explained below.

2.2.1 Novel Agents and Monoclonal Antibody Trastuzumab (Herceptin)

Amplification of *HER2* has a direct role in the pathogenesis of these cancers, and several murine monoclonal antibodies generated against the *HER2* protein were found to inhibit the proliferation of human cancer cells that overexpressed *HER2* (Pietras et al. 1994, 1995; Shepard et al. 1991). To minimize immunogenicity, a human antibody called trastuzumab (Herceptin) was developed and tested against breast cancer cells that overexpressed *HER2* in vitro and in vivo. This antibody inhibited tumor growth when used alone but had synergistic effects when used in combination with cisplatin and carboplatin (Pietras et al. 1994; Pegram et al. 1999), docetaxel (Konecny et al. 1999), and ionizing radiation (Pietras et al. 1999) and additive effects when used with doxorubicin, cyclophosphamide, methotrexate, and paclitaxel (Pietras et al. 1998, 1999 Baselga et al. 1998; Pegram et al. 1999; Konecny et al. 1999). Later studies with trastuzumab therapy in patients of HER2-overexpressing cancers showed a high efficacy and a better prognosis with this monoclonal antibody (adjuvant to chemotherapy) but most of the highly aggressive cancers resistant to the treatment. A small percentage (15–25 %) of patients responding initially to monoclonal antibody were found to have relapse within 1 year of the monotherapy (Valabrega et al. 2007; Vogel et al. 2002).

Of the identified angiogenic factors, vascular endothelial growth factor (VEGF) is the most potent and specific regulator of both normal and pathologic angiogenesis (Ferrara and Vis-Smyth 1997). In preclinical studies, VEGF was found to be upregulated in human breast cancer cells. Hurvitz et al. (2009) have presented a phase II study data of 50 patients of breast cancer metastasis with overexpression of HER2, who were treated with trastuzumab and bevacizumab, and showed a clear benefit in terms of overall survival with this combination therapy.

A small new molecule produced by GlaxoSmithKline called lapatinib (Tykerb® or Tyverb®) is a reversible tyrosine kinase inhibitor (TKI) of both HER2 and epidermal growth factor receptor (EGFR). Recently Cameron et al. (2010) have reported that in patients with HER2 positive metastatic breast cancer and trastuzumab resistance disease, there has been improved response and prolonged time to progression (TTP) when lapatinib plus capecitabine was given and final survival analysis in phase III randomized trial was compared with capecitabine alone (Cameron et al. 2010).

Most of the monoclonal antibodies and targeted therapies though having much needed specificities and high efficacy are associated with a range of adverse events such as cardiotoxicity, thromboembolism, and hypertension. Recent data by Criscitiello et al. (2012) demonstrates that targeted therapies against breast cancer such as anti-HER2 monoclonal antibodies trastuzumab and lapatinib cause cardiovascular dysfunction or even congestive heart failure. Antiangiogenic monoclonal antibody bevacizumab may lead to cardiovascular inefficiency and hypertension. Monoclonal antibodies can cause necrosis of the cancer cells by complement activation, but cancer cells are protected by membrane-bound complement regulatory proteins (mCRP) that are overexpressed in cancer cells. Gelderman et al. (2004) have reported that the agents which can block the mCRP in cancer cells will make monoclonal antibody more effective. Beta-glucan used along with the antibody has shown better results.

2.2.2 Hormonal and Growth Factor Intervention Therapy

As already explained in Sect. 2.1.1, the focus in selection of therapy for women with breast cancer has been almost exclusively on the presence of estrogen receptor (ER) and human epidermal growth factor receptor (HER2), with minimal emphasis on the genetic makeup of the patient (Rebbeck et al. 2007). Endocrine antagonist therapy has been the treatment of choice for those women whose tumors are potentially endocrine sensitive, as indicated by expression of the ER and/or progesterone receptor.

Tamoxifen

The selective ER modulator tamoxifen has been the most important therapeutic agent in breast cancer for the past three decades (Goetz et al. 2008; Schroth et al. 2007, 2009). Tamoxifen is a nonsteroidal parent drug, US Food and Drug administration approved, having mixed agonist and antagonist activity. The drug exerts its effects by binding to the estrogen receptors and modulating estrogen-induced transcription. It is found to be effective in a metastatic disease as well as in patients of higher-risk category or predisposition for breast cancer. Tamoxifen has been used for last 35 years and extensively studied drug for blocking estrogen receptors in breast cancers (Schroth et al. 2007, 2009). In addition, tamoxifen has received approval for treatment of men with metastatic breast cancer and is commonly used in the adjuvant setting.

Tamoxifen undergoes extensive hepatic oxidation by the CYP450 isoforms to several primary and secondary metabolites with variable potencies toward the estrogen receptor. Several lines of evidence indicated that most of the tamoxifen therapeutic effects in breast cancer are mediated by its antiestrogenic metabolites 4-hydroxy-tamoxifen and 4-hydroxy-*N-desmethyl*-tamoxifen (endoxifen) (Schroth et al. 2009).

The metabolites exhibit significantly greater affinity for the estrogen receptor and greater potency in suppressing cell proliferation compared to tamoxifen. It is not known whether a specific metabolite or a relative ratio of several tamoxifen metabolites exerts its estrogenic effect in some tissues. Tamoxifen metabolites and the enzymes catalyzing the metabolism have been studied; one CYP2D6 enzyme is rate limiting for tamoxifen metabolism and forms active metabolites (Goetz et al. 2008). The polymorphisms present in this gene are responsible for a high, low, or diminished activity of the enzyme. Sixty-three different alleles of this gene have been identified and studied, forming the categories of nil or poor, intermediate, and ultrametabolizers (Goetz et al. 2008). Ultrametabolizers carry gene duplications and multiduplications of functional alleles, which lead to higher CYP2D6 expression and enzyme activity (Goetz et al. 2008). Recently, genetic variation in the metabolizing enzyme CYP2D6 has emerged as an important contributor to the interindividual variability in response after the administration of tamoxifen. For summary of tamoxifen metabolism polymorphisms and their effects refer to Table 30.1.

Genetic polymorphisms in ER may also influence tamoxifen-related toxicity or other benefits associated with the drug. Rehman et al. (2004) have reported that 4 months of tamoxifen treatment was associated with reduction in plasma lipid concentrations in women with wild-type ER but not in those with ER germ line variants. In other studies, tamoxifen treatment was associated with a significantly higher increase in bone mineral density in women who carried a 21 CA repeats allele in intron 5 of ERb compared to noncarriers (Yoneda et al. 2002).

Aromatase Inhibitors

During the past decade the third-generation aromatase inhibitors (AIs) anastrozole, exemestane, and letrozole have emerged as important agents against breast cancer and appear to be more efficacious than tamoxifen in the advanced disease setting (Ingle and Suman 2005). There efficacy is clearly established in the adjuvant setting (Ingle et al. 2006) and in the prevention in post-menopausal women (Ingle and Suman 2005). Aromatase inhibitors block the conversion of androgens to estrogens, exerting antitumor activity by reducing the circulating estrogens. Tamoxifen and the AIs have recently become foci of intense pharmacogenetic/pharmacogenomic research (Ingle et al. 2008 and reviewed by Sarah et al. 2010). Whether patients are likely to benefit or suffer toxicity from aromatase inhibitors has been studied by variants in aromatase (CYP19). The presence of an SNP in exon 10 of aromatase (Arg264Cys) is associated with differential RNA levels and a possible increased risk of developing

Table 30.1 Hormonal therapy agents and gene variants (polymorphisms) that influence the efficacy and toxicity of breast cancer treatment

Drug	Gene with its polymorphisms	Role of genes	Possible effects due to polymorphism
Tamoxifen	ER	Estrogen receptor (target)	Primary and acquired
			Resistance
	CYP2D6	Phase I DME	Diminished concentration of effective metabolite
			Endoxifen
	SULT1A1	Phase II DME	Slow metabolism of drug and high toxicity
		Elimination of drug	
Aromatase inhibitors	CYP19 (aromatase)	Target	Primary resistance to drug
			High toxicity
	CYP1A2	DME	Unknown
	CYP2C9	DME	
	CYP3A	DME	

SULT1A1 sulfotransferase, *DME* drug-metabolizing enzyme

breast cancer and prostate cancer (Kristensen et al. 2000; Modugno et al. 2001). Other polymorphisms in aromatase include a number of TTTA repeats in intron 4. A rare polymorphic allele of CYP19 (repeat (TTTA)12) has been reported to be significantly more frequent in breast cancer patients than in controls (Suspitsin et al. 2002), but a correlation between this polymorphism and response to treatment has not been examined. Finally, polymorphism in enzymes that metabolize aromatase inhibitors may also be associated with resistance. CYP1A2, CYP2C9, and CYP3A have been associated with anastrozole metabolism (Grimm and Dyroff 1997). Pharmacogenetics of breast cancer has been reviewed by Stearns et al. (2004). The use of genomic profiles to assess the risk of tumor recurrence, drug resistance, and drug toxicity could be tremendously useful in breast cancer management (Table 30.1).

2.2.3 Chemotherapy

The advances in pharmacogenomics study of cancer have the advantage of improved therapeutic index, improved dose regimen, and selection of optimal types of drug for an individual or set of individuals. Breast cancer is a heterogeneous disease, manifesting as different molecular subtypes, each of which responds to chemotherapeutic agents and radiation differently (O'Brien et al. 2008). Advanced or metastatic breast cancer therapy includes single or multiple combinations of chemotherapy regimens anthracyclines, taxanes, cyclophosphamide, vinorelbine, platinum drugs, and vinca alkaloids along with hormonal therapy such as tamoxifen, aromatase inhibitors, fulvestrant and monoclonal antibodies, trastuzumab, lapatinib, and bevacizumab. Several tumor characteristics, such as poor grade and lack of hormone receptors, have been associated with improved response to chemotherapy. Chemotherapeutics commonly used in breast cancer are described in Table 30.2 along with gene variants that may affect drug efficacy or safety. Relatively small retrospective analyses have been conducted, and germ line DNA samples have been collected in trials to examine pharmacogenetic effects on response or toxicities to chemotherapy in breast cancer.

3 Cervical Cancer

Cervical cancer is the second most lethal cancer among women worldwide with 300,000 deaths annually. Efforts to find the causative agents of cervical cancer have resulted many in the list, but the largest single agent found is human papillomavirus (HPV) 4, and around 99.7 % of the cervical cancers were found to have sequences of HPV. Persistent infection with HPV during the lifetime of women is found to be necessary for cervical cancer development. However, not all women who had infection of HPV will have cancer, and other risk factors are found to play significant role in the development of malignancy (Zur Hausen 2000; Martin et al. 2007). The risk factors that further increase the chance of cancer formation are use of oral contraceptive, smoking, inherited genetic components, and certain environmental factors. Human leukocyte antigens (HLA) have been implicated in HPV infection and persistence. It was initially thought that HPV testing would help in screening the population and identifying the high-risk group for the disease and could also be used for diagnosing and monitoring the carcinoma, but it was not the case because of occurrence of transient HPV infections. Removing the epithelial cells from cervix (call Pap smear) and cell cytology analysis has been found to be useful in screening the population and diagnosing cervical cancer at very initial stages (Papanicolaou 1942). Pap test is a histology study to identify abnormal cellular alterations manually by visual detection and is a tedious and time-consuming method but quite reliable (Broomall et al. 2010). But there is an urgent need of a more sensitive screening method for cervical dysplasia.

3.1 Genetics of Cervical Cancer Disease Susceptibility Genes

HPV sequences have been found in as many as 99.7 % of the cases cervical tumors analyzed, but presence of viral DNA has been shown to have poor diagnostic value for cervical cancer. Since adaptive immunity of the host plays a

Table 30.2 Chemotherapeutic agents and gene variants (polymorphisms) that influence the efficacy and toxicity of breast cancer treatment

Drug	Gene with its polymorphisms	Role of genes	Possible effects due to polymorphism
Doxorubicin	GST	Phase II enzyme	Slow metabolism
		Detoxification	Greater drug availability
	MDR	Drug transporter	Drug resistance
Paclitaxel	CYP2C8	Phase I DME	Slow metabolism of drug and high toxicity
Methotrexate	MTHFR	Maintains pool of one carbon moieties for NA and protein synthesis	Folate retention and bone marrow sensitivity to drug
5-Fluorouracil	DPD	DME	Neurotoxicity

GST glutathione S-transferase, *MDR* multidrug resistance, *DME* drug-metabolizing enzyme, *MTHFR* methyltetrahydrofolate reductase, *DPD* dihydropyrimidine dehydrogenase

major role in establishment of viral infection and its persistence, association between cervical cancer and MHC class II polymorphism has been studied. Various such immunogenic loci have been identified. Calhoun et al. (2002) studied the tumor necrosis factor (TNF) genes and their polymorphisms present in the population, as well as interleukin 4 (IL-4) receptor polymorphisms and found that many of these polymorphic loci influence the course of HPV infection. Several polymorphisms within the DNA damage-response gene, *p53*, have also been studied. The tumor necrosis factor _238 polymorphism was significantly underrepresented in cervical cancer patients. *NRAMP1* 3′-untranslated region STP_86 polymorphism also appeared to be inversely associated with cervical cancer, but this result did not reach statistical significance (Calhoun et al. 2002).

Chatterjee et al. (2009) genotyped two polymorphisms in Fas gene (FasR-1377G/A, FasR-670A/G) and one in FasL gene (FasL844T/C). None of the polymorphisms, or the Fas haplotypes, showed a significant association with cervical cancer in African population. This is the first study on the role of Fas and FasL polymorphisms in cervical cancer in African populations. Nunobiki et al. (2011) have reported that in HPV infected women, polymorphisms and deletions in the genes of glutathione S-transferase (GST) isoforms GSTM1 and GSTT1 are associated with development of cervical cancer. GST is an important enzyme for protection against oxidative stress and detoxification. Other loci studied and linked to carcinogenesis were for p53, MDM2, and FAS gene promoter.

3.2 Pharmacogenetics of Cervical Cancer Therapy

Various novel agents (monoclonal antibodies) and chemotherapy regimens are used along with other modalities such as surgery and radiotherapy. Chemoradiation has been shown to be more effective when cisplatin and/or 5-fluorouracil is given once weekly along with radiation but is associated with more side effects and toxicity. Chemotherapy for cervical cancer is done by combination of cisplatin, paclitaxel, ifosfamide, carboplatin, 5-FU, and cyclophosphamide.

5-fluorouracil has been most commonly used to treat metastatic and disseminated cervical cancers. It is a pyrimidine analog and gets incorporated into DNA and halts replication and thus has antitumor activity. Approximately 5 % gets used up for this function and rest undergoes degradation into metabolites which get excreted through urine. One important rate-limiting enzyme of this pathway is dihydropyrimidine dehydrogenase (DPD), which has been found to have several polymorphisms in its gene (DPYD), resulting in partial or complete loss of activity, thus is leading to acute toxicity (Milano et al. 1999). DPYD polymorphisms should be examined and identified in patients to avoid its fatal toxicity. Milano and Etienne (1994) have reviewed in detail its pharmacological effects.

The genes of glutathione S-transferase (GST) family encode enzymes that appear to be critical in cellular protection against the cytotoxic effects. GSTs play an important role in conjugating glutathione thus protecting the cell from the deleterious effects of oxidative stress (Rebbeck 1997; Ueda et al. 2003). GST isoforms GSTM1 and GSTT1 gene deletions may promote the development of cervical dysplasia by moderating the activation and detoxification of polycyclic hydrocarbons and other compounds that influence oxidative stress and DNA adduct formation (Goodman et al. 2001).

Recently, the cervical cancer vaccine Gardasil (effective against four types of human papillomavirus) was commercialized. While clearly of major benefit, Gardasil does not prevent all types of cervical cancer. Since the cancer is virus induced, several protective and therapeutic vaccine formulations have been tried. Vaccination with tumor antigens has been shown to be an effective treatment for cancer in various animal models. Dileo et al. (2003) have used the antigen from the HPV16 protein to formulate lipid-protamine-DNA (LPD) particles to be injected in mice. These particles when administered intravenously or subcutaneously, in tumor-bearing mice, were found to induce CTL responses specifically against the antigen E7. The tumors were regressed completely and in 100 % of the mice. It also prevented the formation of HPV-mediated tumors. Similar results have been presented by Cui and Huang (2005) using HPV 16 antigen E7 in liposomal polycationic DNA preparations and administered in murine models.

4 Endometrial Cancer

Endometrial cancer is the common cancer of female genital tract involving the cancer of the uterine corpus. It is more common in 60 and 70 years of age group, but it can occur in some conditions at 40 years of age. The estimated deaths due to endometrial cancer is around 50,000 with 200,000 cases diagnosed each year. Although the exact cause of endometrial cancer is unknown, increased levels of estrogen appear to play a role. The various risk factors associated with endometrial cancer are diabetes, estrogen replacement therapy without the use of progesterone, history of endometrial polyps, infertility, polycystic ovary syndrome, obesity, treatment with tamoxifen, etc. Although several different histologic subtypes of endometrial cancer are recognized, these are commonly explained by a dualistic model that categorizes cancers into two major types, type I and type II carcinomas. Type I tumors (endometrioid epithelial carcinoma) comprise 80 % of all new cases of endometrial cancer and are histologically well or moderately differentiated and estrogen dependent and typically have a favorable prognosis. Type II tumors (nonendometrioid) include other subtypes, often with serous papillary or clear cell histology. These tend to be poorly differentiated and are associated with a much more aggressive phenotype. Although type II tumors make up only 10–15 % of all endometrial cancer cases, they are responsible for 50 % of all relapses. The 2010 FIGO staging system for the carcinoma of endometrium is as follows

- IA Tumor confined to the uterus, no or < ½ myometrial invasion
- IB Tumor confined to the uterus, > ½ myometrial invasion
- II Cervical stromal invasion, but not beyond uterus
- IIIA Tumor invades serosa or adnexa
- IIIB Vaginal and/or parametrial involvement
- IIIC1 Pelvic lymph node involvement
- IIIC2 Para-aortic lymph node involvement, with or without pelvic node involvement
- IVA Tumor invasion bladder mucosa and/or bowel mucosa
- IVB Distant metastases including abdominal metastases and/or inguinal lymph nodes

The surgical therapy is the choice of treatment for patients with endometrial cancer who present with an early-stage disease. Women with stage 1 disease who are at increased risk for recurrence and those with stage 2 disease are often offered surgery in combination with radiation therapy.

Chemotherapy may be considered in some cases, especially for those with stage 3 and 4 disease, while hormonal therapy with progestins and antiestrogens has been used for the treatment of endometrial stromal sarcomas. The antibody herceptin which is used to treat breast cancers that overexpress the HER2/neu protein, has been tried with some success in a phase II trial in women with uterine papillary serous carcinomas that overexpress HER2/neu.

The reported response rates to systemic hormonal and cytotoxic therapy for advanced and recurrent endometrial cancer vary dramatically, but the overall median survival is generally extremely poor. Advanced or recurrent endometrial cancer is commonly treated with different combinations of cytotoxic chemotherapy including doxorubicin (DOX), cisplatin, and paclitaxel with or without hormonal agents. A systematic review of systemic therapy for advanced or recurrent endometrial cancer was carried out, and the findings suggested that combination chemotherapy with cisplatin and doxorubicin results in higher response rates than single-agent doxorubicin chemotherapy. Paclitaxel in combination with cisplatin and doxorubicin chemotherapy improves both response rate and median survival; however, the use of this three-drug combination is associated with increased toxicity.

The need of the hour is improvement in management of the disease by optimizing the use of existing drugs or developing new agents. Individualizing treatments by identifying patients who will, and will not, respond to specific agents will potentially increase the response rates as well as reduce the toxicities. In this regard, a study utilized microarray expression technology to identify genes and gene pathways associated with endometrial cancer cell line sensitivity and resistance to the most commonly used cytotoxic agents, DOX. The authors have identified Src tyrosine kinase pathway which is activated in DOX-resistant endometrial cancer. By using a selective Src inhibitor, they could demonstrate the ability to increase DOX sensitivity in both resistant and more sensitive endometrial cell line (Indermaur et al. 2010).

4.1 CYP1B1 and Endometrial Cancer

CYP1B1 has an important role in the development of cancers related to uterus. This is based on the observations that highest levels of CYP1B1 are found in the endometrium (Hayes et al. 1996). There is significantly higher levels of 4-OHE2 in endometrial myometrium as compared to surrounding normal myometrium. Further inhibition of CYP1B1 has been shown to remove this effect (Liehr et al. 1995). Furthermore, 4-OHE2 production was shown to be associated with the development of endometrial carcinoma in mice. Moreover, CYP1B1 may also be involved in the tamoxifen-induced endometrial cancers (Colacurci et al. 2000; Czernobilsky and Lifschitz-Mercer 1997; Sharma et al. 2003). CYP1B1 catalyzes isomerization of *trans-4-hydroxytamoxifen* to *cis*-4-hydroxytamoxifen (Crewe et al. 2002). *Cis*-4-hydroxytamoxifen, being a weak estrogen agonist, acts as a weak promoter of estrogen signaling. Indeed, clinical resistance to tamoxifen therapy has been associated with the increased formation of *cis-hydroxytamoxifen* (Osborne et al. 1992) for which CYP1B1-mediated metabolism may be directly responsible. In cell lines treated with tamoxifen in the absence of estrogen, upregulation of CYP1B1 was observed which possibly was due to binding of *cis*-4-hydroxytamoxifen to the estrogen-responsive element on the CYP1B1 promoter. Also, CYP1B1 promoter methylation, which results in decreased CYP1B1 expression, resulted in increased overall survival after tamoxifen therapy (Widschwendter et al. 2004). Hence, CYP1B1 expression might be involved in clinical resistance to tamoxifen therapy within tumor tissue as well in causation of tamoxifen-induced endometrial tumors.

4.1.1 CYP1B1 SNP and Endometrial Cancer Risk

There are two known genotypes in CYP1B1 that act as a risk factor for endometrial cancer development, namely, 432 V/V and 119S/S (Rylander-Rudqvist et al. 2004). 432 V/V genotype has been shown as a more relevant risk factor in premenopausal women for assessing endometrial cancer, while 453S/S genotype acts as a protective factor for endometrial cancer development (Doherty et al. 2005). The 119S/S allele was strongly correlated with positive ERa and ERh status, whereas the 432 V/V allele was weakly correlated. Another recent study found an association between *CYP1B1*3 and CYP1A1/2 genotypes* (145). CYP1B1*1 alone acts as a moderate risk (OR = 1.34), while if it is studied in combination of CYP1A1/2 low-risk alleles, the risk of endometrial cancer is significantly reduced (OR = 0.29) (Doherty et al. 2005). However, no association was recently found in a study between polymorphisms in the CYP1B1 gene and cancer risk (Rylander-Rudqvist et al. 2004). The contradictory results between the two sample populations may be because of environmental, genetic, or dietary differences. The population-specific results obtained on the risk of endometrial cancer and CYP1B1 genotypes are listed in Table 30.3. Thus, CYP1B1-induced carcinogenesis could contribute to the disease etiology of endometrial cancer, wherein the formation of genotoxic catechol estrogens, positive ER status, and prolonged tamoxifen therapy cause tumor formation. However, this has not been shown in endometrial cancer.

4.2 PPARγ and Endometrial Cancer

Endogenous estrogens play a crucial role in endometrial cancer pathogenesis, with most endometrial cancer risk factors causing an increase in estrogens. Adipose tissue, where androgens are converted to estrogens by the enzyme aromatase, is an important source of endogenous estrogen production in the postmenopausal woman. The peroxisome proliferator-activated receptor-gamma (PPARγ), a key transcriptional regulator of adipogenesis, may also play a role in the regulation of aromatase expression in adipose tissue. *PPARγ* Pro12Ala polymorphism known to alter aromatase expression ultimately might affect endometrial cancer susceptibility. In a study, the Pro12Ala polymorphism was genotyped in an invasive endometrial cancer cases ($n = 222$) and matched controls ($n = 666$) nested within the Nurses' Health Study Cohort. There was no evidence of an association between the Ala allele of the *PPARγ* codon 12 polymorphism and endometrial cancer risk (adjusted odds ratio = 1.18, 95 % confidence interval = 0.80–1.76) (Paynter et al. 2004).

Table 30.3 Population-specific results obtained on the risk of endometrial cancer and CYP1B1 genotypes

Polymorphism	Ethnicity	Genotype	Association	References
A119S	Caucasian		No	Rylander-Rudqvist et al. (2004)
	Japanese	S/S	Yes(increased risk)	Sasaki et al. (2003)
L432V	Caucasian/African-American	V/V	Yes (moderate increase)	Doherty et al. (2005)
	Caucasian	V/V	Yes(increased risk)	Doherty et al. (2005)
	Japanese	V/V	Yes(increased risk)	Sasaki et al. (2003)
N453S	Caucasian		No	Sissung et al. (2006)
	Japanese		No	Tanaka et al. (2002)
	–	N/S, S/S	Decreased risk	McGrath et al. (2004)

4.3 MMP (Matrix Metalloproteinase) and TIMP (Tissue Inhibitor of Metalloproteinase) Expression in Endometrial Cancer

Expression studies done in endometrial cancer have generally observed an increased production of MMPs in cancerous tissues. The mRNA or protein levels of MMP1 (Takemura et al. 1992), MMP2 (Aglund et al. 2004; Di Nezza et al. 2002; Graesslin et al. 2006a, b; Karahan et al. 2007), MMP7 (Graesslin et al. 2006a; Misugi et al. 2005; Ueno et al. 1999), MMP9 (Graesslin et al. 2006a, b; Karahan et al. 2007; Di Nezza et al. 2002; Aglund et al. 2004, Bogusiewicz et al. 2007; Laird et al. 1999; Soini et al. 1997), and MMP12 (Yang et al. 2007) were reported to be increased in endometrial cancer and/or associated with poor prognostic features. In case of MMM14, which is known to be a key enzyme in tumor invasion and metastasis, increased expression in endometrial cancer has been associated with more myometrial and lymphatic invasion (Di Nezza et al. 2002). However, contradictory reports are there where decreased levels of MMP14 mRNA have been reported in endometrial cancer. Therefore, further studies on a larger set of cohorts are required to be conclusive. Conflicting results have also been reported for MMP26 expression. In one study, endometrial tumor tissue showed increased MMP26 protein expression as compared to adjacent normal tissue in postmenopausal but not premenopausal women, and this increased expression in postmenopausal women was associated with increased grade, stage, and invasiveness of tumors (Tunuguntla et al. 2003). In another study, a decreased MMP26 mRNA and protein was observed in endometrial cancer compared with normal tissue, but they did not report the menopausal status of subjects (Isaka et al. 2003) which may be cause of conflicting results obtained.

Microarray and immunohistochemical studies have highlighted the differences of MMP expression in endometrial cancer subtypes and have revealed that increased MMP2, MMP9, and MMP11 protein/mRNA expression in less aggressive type I endometrial cancers compared with more aggressive type II (Monaghan et al. 2007; Moreno-Bueno et al. 2003). While a generally consistent expression of MMPs in endometrial cancer has been observed in various studies, results are more variable for TIMPs, TIMP1 (Ueno et al. 1999) and TIMP3 (Tunuguntla et al. 2003). mRNA and protein have been reported to be increased in endometrial cancer compared with normal tissue. However, both TIMP2 and TIMP4 have conflicting reports (Tunuguntla et al. 2003). TIMPs have roles as agonists to MMPs and later are upregulated in endometrial cancers. So it has been suggested that ratio of MMPs to TIMPs might be of clinical importance.

The 2G allele (MMP1 −16071G/2G) was observed to be more frequently present in Japanese endometrial cancer cases compared to controls (Nishioka et al. 2000). The 2G allele had significantly higher transcriptional activity as it results in the creation of an Ets binding site (Rutter et al. 1998). Also, the 2G allele affected the MMP1 protein expression level as endometrial tumors with 2G allele expressed more MMP1 protein. In case of MMP9 −1562C>T (rs3918242), T allele results in a loss of a repressor protein binding site and was found to be less frequently present in endometrial cancer cases (10.7 %) than controls (16.7 %) and was associated with the risk of endometrial carcinoma (O'Mara et al. 2009; Sugimoto et al. 2006). Yet another study suggests that individuals with the MMP-7 -181 G/G and A/G genotype may have an increased risk of developing endometrial cancer (O'Mara et al. 2009).

4.4 Progesterone and Endometrial Cancer

An adequate progesterone response is required in the endometrium to control normal cell proliferation. In support of this is the observation that the increased risk of endometrial cancer associated with estrogen replacement therapy can be abolished by addition of progesterone to hormone replacement therapy (Ito 2007). Progesterone receptor (PGR), responsible for physiological effects of progesterone, exists as two isoforms, progesterone receptor A (PRA) and progesterone receptor B (PRB) (Kastner et al. 1990). Progesterone therapy has been observed to be more successful for PGR-positive tumors than for PGR-negative tumors. Also, in case of endometrial cancers, PGR positivity was found to be an independent prognostic factor for disease-free survival. PGR single-nucleotide polymorphisms (SNPs) and endometrial cancer risk have been investigated in some studies. A case-control study of 187 cases and 397 controls nested within the Nurses' Health Study reported the rs10895068 SNP (331G/A) to be associated with an increased risk of endometrial cancer [OR = 1.90; 95 % CI, 1.10–3.29]. This particular SNP was found to increase the expression of PRB. However, a population-based case-control study of 275 cases and 314 controls conducted in Sweden found no association with risk. Another candidate polymorphism investigated has been the PROGINS polymorphism, a 306 bp Alu insertion in intron 7 of the PGR gene, which is in complete linkage disequilibrium with a missense SNP in exon 4 (rs1042838; Val660Leu) and a silent SNP in exon 5 (rs1042839; His770His) and has been associated with affecting the stability of PGR isoforms (Tong et al. 2001). Higher incidence of homozygote insertion carriers of the PROGINS polymorphism was found in a Brazilian case-control study in endometrial cancer subjects compared with control subjects. Studies based on SNP tagging approach using information from the International HapMap Project carried out in Chinese population and US Caucasian suggested that a 3′-untranslated region (UTR) SNP rs608995 is associated with increased risk of endometrial cancer. Xu et al. (2009) identified two SNPs located in the 3′-flanking region of the gene reported to be associated with endometrial cancer risk (Xu et al. 2009; Lee et al. 2010). Recently a large-scale study with a maximum of 2,888 endometrial cancer cases and 4,483 female control subjects from up to three studies were genotyped for four PGR polymorphisms (rs1042838, rs10895068, rs11224561, and rs471767) (O'Mara et al. 2011). Logistic regression with adjustment for age, study, ethnicity, and body mass index was performed to calculate odds ratios (ORs) and associated 95 % confidence intervals (CIs) and P-values. Out of these four SNPs in PGR gene, only rs11224561 was found significantly associated with endometrial cancer. Increased risk of endometrial cancer was found to be associated with the A allele of the rs11224561 (OR per allele 1.31, $P<0.001$, adjusted for age and study) (O'Mara et al. 2009).

4.5 Obesity and Endometrial Cancer

Obese persons have a 4.5–6.25 times higher risk of endometrial cancer compared with nonobese persons (Calle et al. 2003; Schouten et al. 2004). This suggests that the increase in obesity prevalence worldwide over the past few decades might have contributed to the increased incidence of endometrial cancer. In addition to the environmental factors, genetic susceptibility is also well recognized as the major determinants of obesity risk. Recent genome-wide association studies (GWAS) have established a link between the genetic architecture of BMI and obesity, while there exists a strong link between BMI and endometrial cancer. Hence, a study investigated the association of BMI-related GWAS markers with endometrial cancer. In this study obesity-related SNPs were related to endometrial cancer using GWAS data from the Shanghai endometrial cancer study (1996–2005). It was found that the BMI-associated risk variants were present at higher frequency in endometrial cancer patients than in control population ($P=0.0003$), for 22 unique loci. Out of 35 BMI-associated variants studied, 9 (corresponding to 7 loci) were significantly associated with the risk of endometrial cancer. The allelic OR ranges from 1.15 to 1.29 for consistent SNPs. These seven loci were in SEC16B/RASAL, MTCH2, FTO, MSRA, MC4R, SOX6, and TMEM18 genes. Even after adjusting for BMI, this association persisted. This suggests that these obesity markers also act as markers for determining endometrial cancer risk (Delahanty et al. 2011).

5 Ovarian Cancer

Ovarian cancer, the sixth most frequently occurring cancer in women, is a tumor with a low prevalence but high mortality. Incidence rates differ according to geographical location. Developed countries have witnessed higher than 10/100,000 incidence rates, whereas in Africa and India the incidence rate is 3–4/100,000. Out of the 25,000 diagnosed new cases of ovarian cancer annually, at least 14,000 patients succumb to the disease. The high rate of mortality for ovarian cancer can be ascribed to the difficulty of making an early diagnosis which in turn is due to lack of well-defined symptoms of this cancer. The interval between the beginning of symptoms and histological diagnosis is very short, usually 2–3 months, and may be less than a month in some cases. The risk factors for ovarian cancers include a family history for ovarian cancer and the length of the ovulatory period (early menarche and late menopause), while breast feeding and long-term use of progestogens contraceptive are the protective factors.

The first step of treatment includes surgery which may be conservative or radical, debulking, palliative, or only exploratory depending upon the spread of the disease. High-energy radiation therapy has been used in the past which has been replaced by chemotherapy (Vella et al. 2004).

First-Line Therapy: The most frequently used alkylating agents are anthracycline and melphalan. Cyclophosphamide, thioTEPA, and chlorambucil are some other commonly used drugs. Lately, platinum and taxanes have been shown to be the most active drugs. Cisplatin has numerous adverse reactions and is highly toxic. Carboplatin, a platinum derivative, shows similar activity but is less toxic. Taxanes are also considered to be effective drugs. The combination therapy with platinum and paclitaxel has improved the disease-free and overall survival rate of ovarian cancer.

Second-Line Chemotherapy: There is a high percentage of relapse patients after they have undergone first line of chemotherapy. In this group of patients, three categories can be identified, namely, those (1) with complete response in first-line therapy with platinum or platinum-containing regimens relapsing at different times, (2) with partial response to first-line chemotherapy, and (3) with no response to first-line drugs.

Although available therapy has improved complete clinical remission and progression-free survival, yet ovarian cancer is a disease with a high risk of relapse. Hence, the efficacy of newer drugs or anticancer agents needs to study.

Polymorphisms in a number of genes have been implicated in differential drug response to platinum drugs or taxanes. The genes studied in this regard are involved in drug metabolism, drug transport, and DNA repair. Characterization of the drug-metabolizing genetic profile of individuals is of great interest in clinical oncology as this can help to determine optimal chemotherapy for each patient with improved efficacy and at the same time reduce the incidence of drug toxicity and poor drug response (Lee et al. 2005). Some of the studied genes are as follows:

5.1 GSTM1-T1 Polymorphisms in Ovarian Cancer

GSTs are a family of detoxification enzymes. The general reaction catalyzed by all isoforms is the conjugation of glutathione (GSH) to a molecule containing an electrophilic group. The conjugated form is more soluble and can be easily eliminated by excretion. A recent study revealed the correlation between the polymorphisms of GSTM1-T1 enzymes and disease outcome in advanced ovarian carcinoma patients, following platinum-/paclitaxel-based chemotherapy (Medeiros et al. 2003).

5.2 MDR1 and CYP2C8 Polymorphism and Paclitaxel

The MDR1 gene encodes for P-glycoprotein that acts as an ATP-dependent drug export pump for taxanes and other cytotoxic drugs through the cell membranes. A high expression of P-glycoprotein on tumor cells leads to chemoresistance (Gottesman and Pastan 1993) and appears to be correlated with a poor response to paclitaxel treatment (Kamazawa et al. 2002; Penson et al. 2004). C3435T polymorphism of the MDR1 gene was found to be associated with a lower P-glycoprotein expression (Hoffmeyer et al. 2000). Out of approximately 25 SNPs for the MDR1 gene, G2677T/A and C3435T have been shown to be involved in P-glycoprotein expression and phenotype (Marzolini et al. 2004; Tanabe et al. 2001) among Caucasians. Gréen et al. found that patients with ovarian cancer and homozygous for the G2677T/A MDR1 SNP responded better to paclitaxel treatment, whereas the C3435T SNP appears not to have any impact on the treatment outcome (Green et al. 2006). These authors also studied the relationship between neurotoxicity in paclitaxel therapy and genetic alterations of the MDR1 gene and the genotypes of CYP2C8. A higher clearance of paclitaxel was demonstrated in patients heterozygous for G2677T/A MDR1 and a lower clearance of paclitaxel in patients heterozygous for CYP2C8*3. Thus, MDR1 and CYP2C8 genotypes may be used to predict inter-patient variability in paclitaxel pharmacokinetics and might be informative for tailored therapy (Green et al. 2009).

5.3 ERCC1 and MRP 2 Polymorphism and Platinum/Paclitaxel

ERCC1 is a protein encoded by the excision repair cross-complementation group 1 gene. The ERCC1 protein belongs to the nucleotide excision repair (NER) pathway and is activated when DNA is damaged by platinum-based chemotherapeutic agents such as cisplatin and carboplatin (Khrunin et al. 2012).

An elevated expression of ERCC1 in ovarian cancer has also been correlated with the development of resistance to platinum-based therapy (Dabholkar et al. 1992, 1994).

The *C118T* polymorphism of the ERCC1 gene can be used as a marker of ERCC1 expression as presence of T allele is associated with lower levels of ERCC1 mRNA and consequently a reduced DNA repair capability (Yu et al. 2000).

Also the association between the C/C genotype and increased resistance to platinum-based chemotherapy, resulting in poor treatment response, has been observed (Kang et al. 2006). Ovarian cancer patients harboring C/C genotypes were less responsive to chemotherapy and maintained a higher risk for disease progression and death in contrast to those harboring C/T or T/T genotypes or patients with a high ERCC1 expression or the C/C genotype at codon 118 may benefit from a platinum/paclitaxel combination, while the treatment with platinum without paclitaxel may be beneficial for the patients with low ERCC1 expression as well as for the patients with C/T or T/T genotype.

Very recently, Yan et al. have reported another SNP of *ERCC1* 19007C/T and found that T/T genotype may be a useful marker for predicting worse survival in EOC patients as it is associated with decreased response to platinum-based chemotherapy and increased risk of disease progression and death (Yan et al. 2012) in Chinese women with epithelial ovarian cancer.

MRP2, an organic anion transporter, is involved in methotrexate, cisplatin, and irinotecan active metabolite glucuronide transport (Green et al. 2009). In some studies carried out on the cell lines of clear cell carcinomas of ovary, MRP2 mRNA expression was associated with the development of pharmaco-resistance to anticancer agents, such as cisplatin (Itamochi et al. 2002). Another study reported that MRP2 expression may be a potential predictor of the response to standard chemotherapy in ovarian cancer. However, other in vivo studies failed to confirm this association in a population of epithelial ovarian carcinoma patients treated with combination therapies, including cisplatin (Arts et al. 1999).

Cytochrome P450 *CYP1A1* is involved in estrogen metabolism, and polymorphisms have been associated with functional changes and risk for ovarian cancer. Heubner et al. found a statistically significant association between the 462Val allele and the development of platinum resistance in ovarian tumors (Heubner et al. 2010).

5.4 IL-8 Polymorphism and Cyclophosphamide and Bevacizumab

In case of IL-8 gene (-251T/A) polymorphism patients with at least one A allele have an increased IL-8 production and a significant lower response to cyclophosphamide and bevacizumab chemotherapy rate than those patients who were homozygous for the wild-type (T allele) (Steffensen et al. 2010).

Findings of a Russian study showed the relationship between the GSTP1 Ile105Val polymorphism and cisplatin efficacy in ovarian cancer patients. These authors observed a better progression-free survival in patients with the homozygous genotype (Ile/Ile) as compared to patients with one or two Val alleles. On the other hand, overall survival did not present significant difference, although ovarian cancer patients carrying the Ile/Ile genotype exhibited a longer lifespan than those with the Ile/Val genotype (Khrunin et al. 2012).

5.5 Xeroderma Pigmentosum Gene Polymorphisms and Carboplatin/Paclitaxel

Members of xeroderma pigmentosum gene family (*XPA-XPG*) are important in nucleotide excision repair and hence have an important role in platinum therapy. Saldivar et al. (2007) have studied polymorphisms in many genes of this family and demonstrated an increased risk of recurrence in patients harboring XPA G allele and an increased risk of death in XPG C allele carriers. XPA G allele was associated with increased overall survival too. Moreover, two SNPs studied in case of XPD gene were found to be reducing the risk of death by fivefold in ovarian cancer patients undergoing platinum chemotherapy.

Xeroderma pigmentosum group D (XPD) is a member of the NER family and plays an important role in the repair of DNA damage caused by alkylating drugs, platinum analogs, and radiation. On the other hand, protein levels were demonstrated to correlate with resistance to these agents in human tumor cell lines (Chen et al. 2002). Notably, Khrunin et al. studied the relationship between XPD Asp312Asn and Lys751Gln polymorphisms (both responsible for lower DNA repair capacity) and survival in ovarian cancer patients treated with carboplatin/paclitaxel. These authors found that patients carrying at least one variant allele (312 Asn or 751 Gln) had a significant reduction of death risk compared to patients with the Asp/Asp or Lys/Lys genotype (Saldivar et al. 2007).

Certain other genes whose SNPs are studied with respect to response to various drug therapy in ovarian cancer patients are listed in Table 30.4.

Table 30.4 Gene polymorphism associated with ovarian cancer therapies

Gene	SNP	Drug	Effect	References
GSTP1	lle105val	Platinum	Progression-free Survival worsens	Stephens et al. (2001)
XPD	ASP312 Asn Lys 751Gln	Carboplatin/Pactitaxel	Decreased risk of death	Chen et al. (2002) and Saldivar et al. (2007)
MTHFR	C677T	Methotrexate	Increased toxicity, reduced response	Ueland et al. (2001)
ABCB1	2677G>T/A	Pactitaxel	Increased response	Grenetier et al. (2006)
ABCB1	1236C>T	Docetaxel	Decreased clearance of drug	Bosch et al. (2006)
ABCB1	2168G/A, 2677 T/A	Carboplatin/Pactitaxel	Increased response to drug	Obata et al. (2006) and Sampath et al. (2006)
Cytochrome P450	CyP3A4*1B	Docetaxel	Increased clearance of drug	Tran et al. (2006)
P53	R72P exon 4	Platinum	Reduced overall survival	Gadducci et al. (2006) and Galic et al. (2007)

5.6 Cisplatin-Based Chemotherapy in Ovarian Cancer Patients of Different Ethnic Origins

Upon comparative evaluation of pharmacogenomic relevance of the most common polymorphisms in glutathione *S*-transferases and DNA repair genes, the *p53* gene, and the *CYP2E1* in Yakut and Russian patients with ovarian cancer treated with the same cisplatin-based chemotherapy, 10 out of 19 studied SNPs showed differences. In spite of marked ethnic differences in the frequency of polymorphic variants of important cisplatin chemotherapy-related genes, none emerged as a clear casual candidate (Khrunin et al. 2012).

The information compiled above suggests that polymorphisms in several genes may be correlated with response or toxicity to chemotherapy or outcome in ovarian cancer. However, such studies are generally small and extrapolatory and have not been validated in independent population. Therefore, reliable validation is required before utilizing such data clinically. In this regard, a large study was carried out by Dabholkar et al. (1994) and Finucane et al. (2011). They examined the pharmacogenomics of ovarian cancer in 914 DNA samples collected within Scottish Randomised Trial in Ovarian Cancer (SCOTROC1) phase III trial. The patients receiving either carboplatin-paclitaxel or carboplatin-docetaxel were screened for 27 SNPs in a set of 16 genes which were previously shown to have associations with therapy. The genes included *ABCB1, CYP2C8, CYP3A4, GSTP1, ERCC1, XPD, and p53.* No significant reproducible association between the genotypes and the outcome or toxicity of the treatment was observed. This study stresses the importance of large, well-defined clinical sample sets for validating pharmacogenomic markers.

5.7 BRCA1 and BRCA2 Mutations in Ovarian Cancer and PARP1 Inhibitor Therapy

In normal cells, double-strand breaks are repaired via homologous recombination, which relies upon functional BRCA1 and BRCA2 enzymes (Ashworth 2008). Cancer with BRCA mutations has difficulty repairing double-strand DNA damage. In tumors with BRCA-associated mutations, therefore, PARP-1 inhibition leads to increased tumor cell death (Edwards et al. 2008). The presence of BRAC1 and BRAC2 mutations in around 40 % of sporadic serous ovarian cancers has made these cancers as good targets for PARP-1 inhibitors. PARP inhibition results in unrepaired DNA single-strand and double-strand breaks and eventual cell death. Cancer cell sensitivity and resistance to both PARP inhibition and platinum have been associated with loss and restoration of homologous recombination DNA repair, respectively, indicating similar mechanisms of anticancer activity and resistance. Use of PARP inhibitors for treatment of these cancers has given good results. Ovarian cancers with PTEN mutations can also serve as good candidates for PARP inhibitor-mediated treatment (Sakai et al. 2008, 2009).

Olaparib (AZD2281; KU-0059436) is a potent oral PARP inhibitor that is selectively cytotoxic to cells lacking either *BRCA1* or *BRCA2* function, while sparing repair-proficient cells expressing *BRCA1* or *BRCA2*, as in heterozygous mutant or wild-type cells. Preliminary results of a phase I clinical trial of olaparib in patients with a range of advanced solid tumors, including *BRCA* mutation carriers, suggested that Olaparib was well tolerated and had durable single-agent anticancer activity in *BRCA1* and *BRCA2* mutation carriers. This phase I study was further extended to evaluate olaparib in a cohort of *BRCA1* and *BRCA2* mutation carriers, including patients with advanced ovarian cancer with differing platinum-free intervals. It was observed that responses to PARP inhibition do correlate with platinum sensitivity, with a response rate of 61.5 % in patients with platinum-sensitive disease. However, patients with platinum-resistant and even platinum-refractory disease still have the potential to benefit (41.7 and 15.4 % response rate, respectively). However, association with prior platinum response requires confirmation in larger studies (Fong et al. 2010). Randomized trials are ongoing to study olaparib and other PARP inhibitors in both *BRCA* mutation-associated and sporadic ovarian cancers. Strategies being

Table 30.5 Novel drugs and their targets under study in ovarian cancer

Drugs	Targets
Bevacizumab, cediranib, sorafenib, sunitinib, aflibercept, pazopanib, BIBF-1120	Angiogenesis
Erlotinib	Epidermal growth factor receptor
BIBF-1120	Fibroblast growth factor receptor
Temsirolimus, everolimus, deforolimus	Mammalian target of rapamycin
Perifosine, PBI-05204, GSK 2141795	Phosphatase and tensin homolog, AKT protein kinase family
Cediranib, pzapanib	Platelet-derived growth factor receptor
ABT-888 (veliparib), AZD 2281(olaparib), AG014699, BSI-201(iniparib)	Poly (ADP-ribose) polymerase

explored include different regimes in which PARP inhibitors are combined with chemotherapy (Table 30.5).

6 Preeclampsia

Preeclampsia (PE) is identified clinically by maternal hypertension and proteinuria occurring after the 20th week of gestation (≥140 mmHg for systolic blood pressure, ≥90 mmHg for diastolic blood pressure and on at least two occasions, and more than 300 mg/day of protein in the maternal urine). It is also known as gestational hypertension. Hypertensive disorders concern up to 8 % of all pregnancies. PE is a major cause of maternal, fetal, and neonatal morbidity and mortality (Duley 2009), with an estimate of >60,000 maternal deaths per year (Goldenberg et al. 2008). Preeclamptic pregnancies are terminated early by cesarean delivery; such early deliveries may result in premature fetus or fetus with low birth weight, respiratory distress syndrome, and other complications of early delivery. Further preeclamptic pregnancy increases the risk of cardiovascular and renal disease in future years (Garovic and Hayman 2007; Smith et al. 2001). In PE trophoblast cells fail to invade uterine spiral arteries (Meekins et al. 1994), leading to a decreased perfusion with consequent placental hypoxia. PE is caused as a result of endothelial cells dysfunction with the decrease of vasodilators (prostacyclin, nitric oxide) and increase of vasoconstrictors (thromboxane, endothelin-1) levels, which results in systemic hypertension and proteinuria (Baumwell and Karumanchi 2007; Myatt and Webster 2009). These clinical features can be attributed to the renin–angiotensin system, endothelial dysfunction (Dekker and Sibai 1998), activity of coagulation cascade elements, influence of oxidative stress, and improper inflammatory reactions (Kupferminc 2005; LaMarca et al. 2007; Middeldorp 2007).

6.1 Preeclampsia and Gene Polymorphism

Genes that can act as good candidates for the role of polymorphism in PE development are those which are involved in regulating fluid homeostasis, blood pressure, and development of placental vasculature. There are few genes with known gene polymorphism associated with the risk of preeclampsia development like AGTR1 (rs5186), AGT (*rs4762*), NOS3(rs1799983), and TGFB1(rs1982073). Some of the genes studied in detail in this regard are as follows.

6.1.1 Endothelial Nitric Oxide Synthase (eNOS)

The eNOS gene product has a role in blood pressure regulation, vascular dilation, vascular smooth muscle proliferation and inhibition of platelet aggregation, making it an attractive candidate for predisposition to PE (Arngrimsson et al. 1997). eNOS synthesizes nitric oxide (NO) in endothelial cells from L-arginine. In normal pregnancy, nitric oxide (NO) level increases during the time of gestation contrary to PE in which very low NO molecule production discriminate vasodilatation (Napolitano et al. 2000). It is a potent vasodilator, which accounts for the biologic activity of endothelium-derived relaxing factor. It is important in regulation of vasomotor tone and blood flow by inhibiting smooth muscle contraction and platelet aggregation (de Pace et al. 2007; Myatt and Webster 2009). eNOS activity generates around 70 % of plasma nitrite, and inhibition of its activity is associated with corresponding decreases in circulation nitrite concentrations (Kleinbongard et al. 2003). Nitrite is a product of NO metabolism that can be recycled in vivo to form NO (Lundberg et al. 2008).

There are three known polymorphisms in the eNOS gene having role in PE, the T-786C (rs2070744), which is a single-nucleotide polymorphism (SNP) in the promoter region (Metzger et al. 2005), the G-894T (rs1799983) in exon 7 which causes the conversion of a glutamic acid to an aspartic acid (Glu298Asp) residue(Metzger et al. 2007; Yoshimura et al. 1998), and a 27 bp variable number of tandem repeats (VNTR) in intron 4 (Sandrim et al. 2007). eNOS polymorphisms affect endogenous NO formation in normal pregnancy. The "CGlub" haplotype acts as a protective haplotype against the development of PE by increasing endogenous NO formation. The "CGlub" haplotype is more frequent in the healthy pregnant women than in the PE women ($P=0.0044$). The healthy pregnant women with CC genotype instead of TT genotype for T-786C polymorphism have a higher nitrite levels ($P<0.05$), while a lower nitrite level is found in healthy pregnant women with 4a4a genotype compared with the 4b4b genotype for the VNTR polymorphism in intron 4 ($P<0.05$) (Sandrim et al. 2010).

G894T variant is known as a risk factor for the development of PE (Hakli et al. 2003). The combined frequency of combined frequency of the variant in heterozygous (GT) and homozygous (TT) of G894T polymorphism significantly increases the risk of abruption placentae development in PE women (OR 3.51), which is premature separation of placenta from uterine wall. Further eNOST allele is a major risk for development of abruption placentae ($P<0.0001$) (Hillermann et al. 2005). In general CC genotype of T786C polymorphism increases the risk of PE development 1.62-fold and mutated C allele as a risk allele with OR = 1.36. Moreover CC genotype increased the risk of severe preeclampsia development by 2.37-fold. CT and CC genotypes are overrepresented in severe preeclampsia women. Values of systolic and diastolic blood pressure vary according to the genotype of T786C polymorphism, TT:TC:CC¼171.9:173.2:176.3 mmHg and 110.9:109.2:112.4 mmHg, respectively; likewise proteinuria value also varies according to genotype (215.5:210.7:230.3 mg/dl, respectively) (Seremak-Mrozikiewicz et al. 2011).

6.1.2 Angiotensinogen (AGT)

AGT is a key component of renin–angiotensin system, encoding for angiotensinogen, which acts as a precursor of angiotensin II. It is a potent vasoconstrictor, modulating blood pressure regulation, body-fluid homeostasis, and vascular remodeling in pregnancy.

There are two known SNPs of AGT that has been studied in context of PE (M235T and T174M) both in AGT exon 2. M235T variant rs699 is a transversion involving a T-C change at nucleotide position 704 of the AGT gene, encoding threonine instead of methionine at residue 235. The T174M variant rs4762 encodes methionine instead of threonine at residue 174. There are controversial studies regarding the role of these two SNPs in PE. A recent meta-analysis study has shown that the TT genotype of M235T polymorphism increases the risk of PE compared to MM genotype by 1.61-fold ($P=0.001$). TT genotype acts as a risk factor in Caucasians and Mongolians but not in Africans. TT genotype compared to the MM genotype is also associated with an increased risk of hypertension in Asians (Sethi et al. 2003) and in Han Chinese (Ji et al. 2010). The T174M variant was not found significantly associated with PE development (Lin et al. 2012). In nulliparous French-Canadian pregnant women (Levesque et al. 2004), Thr allele of the polymorphism Met235Thr is a significant risk factor predisposing pregnant women to PE development (OR = 2.1, $P=0.015$); further the Met allele of Thr174Met is also associated with preeclampsia ($P<0.0033$) increasing 2-fold risk for PE development. G1035A (rs3889728) genotype is also involved in PE development via haplotype A-Met-Thr (G1035A-Thr174Met-Met235Thr) and increases a 2.1-fold increased risk of preeclampsia. The C allele of polymorphism A-20C (rs5050, SNP in the promoter region) is associated with an increase in plasma angiotensinogen levels. Thus, the women with A-Met-Thr haplotype have higher concentration of angiotensinogen in their decidual cells which leads to higher risk of placental hypoxemia by increasing vasoconstriction of spiral arteries. This further makes trophoblasts unable to invade spiral arteries there by leaving arteries responsive to angiotensin II, thus predisposing women to PE (Ishigami et al. 1999).

6.1.3 Angiotensin II Receptor Type 1 (AGTR1)

The maternal renin–angiotensin–aldosterone system (RAAS) is a hormone system involved pathogenesis of PE (Shah 2006; Vitoratos et al. 2012). In normal pregnancy, plasma levels of renin, angiotensin II, and aldosterone are upregulated via RAAS (Irani and Xia 2008; Lindheimer et al. 2010). The angiotensin-converting enzyme (ACE) is a key RAAS component and plays an important role in blood pressure homeostasis by generating the vasoconstrictor peptide angiotensin II and by inactivating the vasodilator peptides bradykinin and Ang-(Velloso et al. 2007). Angiotensin II is the major signaling molecule of the RAAS, and its major cardiovascular effects are mediated by AGTR1. This balance is disturbed in PE and the plasma level of these proteins decrease to normal nongravid range.

Seven SNPs are known in AGTR1, but the most well studied is the transversion in the 3′-UTR (+1166A>C) of this gene. In a meta-analysis based on 10 populations from eight studies, it was found to have no significant association with PE development (OR = 1.19, $P=0.11$) (Zhao et al. 2012), while in Western Iranian women the frequency of AC + CC of A1166C polymorphism predisposes women for mild preeclampsia ($P>0.05$) (Rahimi et al. 2013).

A deletion polymorphism (I/D) of the ACE gene is known to be involved in PE development, where the D allele acts as a risk factor by increasing ACE activity. The presence of D allele of ACE increases the risk of PE development by 1.8-fold ($P=0.002$) ID genotype decreases the level of total antioxidant capacity (TCA) significantly in mild preeclamptic women compared to those with II genotype (Rahimi et al. 2013). DD genotype has been found to be present in higher frequency in Turkish preeclamptic women than normal pregnant women ($P=0.006$) (Bereketoglu et al. 2012).

6.1.4 Inflammatory Mediators

Preeclampsia involves an excessive maternal systemic inflammatory response with cytokine-mediated endothelial damage seems which makes a number of proinflammatory mediators possible candidates involved in PE development. Some studies have shown abnormal production of interleukin 1 (IL-1) cytokine in preeclamptic patients. Inflammatory cytokines like IL-12 and IL-18 induce apoptosis of trophoblast and may inhibit angiogenesis. Therefore, both IL-12 and IL-18 have been assessed in PE patients. Toll-like receptors (TLRs) expressed at the fetal–maternal interface play an important role in normal placental development; since pathogenesis of PE involves an intense innate immune response, with a central role for TLRs, there is a possible association between abnormal expression of TLR and PE.

A recent study has shown that genotypic and allelic distributions for six polymorphisms in inflammatory mediators were similar between the study and control groups, *IL-1R1 (PstI, rs2234650)*, *IL-12 (+1188, rs3212227)*, *IL-18 (−137,rs187238)*, *IL-18 (−607, rs1946519)*, *TLR-2 (+2258, rs5743708)*, and *TLR-4 (+896, rs4986790)*, showing no association with PE (Franchim et al. 2011). However, allelic variants of the *TLR-2* polymorphism *TLR-2* SNP (Arg753Gln) and *TLR-4* SNPs (Asp299Gly andThr399Ile) aggregate with a lower threshold for the development of early-onset (<34 + 0 weeks'), but not late-onset (≥34 + 0 weeks') preeclampsia in British women (Xie et al. 2010). Proinflammatory cytokine, tumor necrosis factor (TNF), shows only a modest association between the *TNF* (G308A) polymorphism and PE development (Bombell and McGuire 2008).

6.1.5 Vascular Endothelial Growth Factor (VEGF)

VEGF is a major angiogenic factor and a prime regulator of endothelial cell proliferation. It plays a crucial role in physiological vasculogenesis and vascular permeability (Ferrara et al. 1992). During pregnancy it plays its role in vascular development of embryo, trophoblast proliferation, and growth of fetal as well as maternal blood cells. In PE there is deficiency in trophoblast invasion of the spiral arteries of the placental bed that leads to a poorly perfused fetoplacental unit resulting in poor placental growth and inadequate physiological changes in the vasculature of the placental bed.

A known polymorphism of *VEGF* 936C/T polymorphism increases the risk of severe PE development by 2.7- to 2-fold ($P=0.019–0.001$) (Papazoglou et al. 2004; Shim et al. 2007), with T allele acting as a major risk for severe PE, while in a study by Kim et al, there has been found no association in *VEGF* 936C/T polymorphism and PE development (Kim et al. 2008). T allele was associated with significantly lower VEGF plasma levels than subjects carrying the C allele (Renner et al. 2000).

7 Gestational Diabetes Mellitus (GDM)

Gestational diabetes mellitus (GDM) is considered to be a heterogeneous metabolic disorder with mixed genetic etiology and phenotypes (Lapolla et al. 2009). In GDM pregnant women exhibit various degrees of glucose intolerance. Reported rates of GDM range from 2 to 10 % of pregnancies. Approximately 5–10 % of women with GDM develop type 2 diabetes (T2D) shortly after pregnancy. Based on current data it is estimated that women who have had GDM have a 35–60 % chance of developing diabetes within the next 10–20 years. GDM is more common in women with ethnic background like African-Americans, Afro-Caribbeans, Native Americans, Hispanics, Pacific Islanders, and South Asian.

Increased maternal estrogen and progesterone in early pregnancy promotes pancreatic β-cell hyperplasia resulting in an increase in insulin level (Rieck and Kaestner 2010). In the second and third trimester, there is decrease maternal insulin sensitivity caused by the continuous increase in the fetoplacental factors (Catalano et al. 1992). During pregnancy, generally there is threefold increase of insulin resistance in whole body than in nonpregnant women, but some women have higher insulin resistance that they cannot compensate with increased production in the β-cells of the pancreas and consequently become hyperglycemic, developing gestational diabetes (Kuhl 1991). Placental hormones especially cortisol and progesterone and to some extent human placental lactogen, prolactin, and estradiol contribute to increased insulin resistance. GDM also increases the risk of hypertensive disorders in pregnant women with the increasing chance of cesarean delivery. GDM further increases the risk of fetal disorders like intrauterine fetal death due to fasting hyperglycemia (105 mg/dl or 5.8 mmol/l) during the last 4–8 weeks (Freathy et al. 2010).

7.1 GDM Gene Polymorphisms

Understanding the genetic underpinnings of GDM may lead to new markers to predict risk for earlier diagnosis and prevention in order to minimize pregnancy-related complications as well as risk to the fetus during pregnancy and after delivery. The strategy that has been used to find the genetic cause of GDM is based on the fact that both GDM and type 2 diabetes (T2D) share the same pathophysiological changes involving insulin resistance accompanied with the defect in insulin secretion (Buchanan 2001; Kautzky-Willer et al. 1997). Therefore, the candidate genes that underlie the pathogenesis of T2D are potential candidates of GDM study. Functional studies have shown that these genes are involved in many processes like insulin resistance (PPARG), abnormal beta-cell functioning (SLC308, CDKAL1, CDKN2A/B, MTNR1B, IGF2BP2, KCNQ1), and abnormal utilization of glucose (*GCK*) (Groenewoud et al. 2008; Lauenborg et al. 2009; Ubeda et al. 2006) (Table 30.6).

Among the T2D-related genes, those that are involved in pancreatic β-cell function are associated with GDM (*CDKAL1, MTNR1B, IGF2BP2, CDKN2A/2B, SLC30A8, IDE, KCNQ1, and CENTD2*), while as those genes which are involved in insulin resistance (*FTO, PPARG, IRS1, KLF14, and GCKR*) are not significantly associated with GDM. Proving that, these

Table 30.6 Summary of gene polymorphism related to GDM

Gene	SNP	Effect	Risk allele	OR value	References
CDKAL1	rs7754840	Decreases AUC	C	1.11	Wang et al. (2011)
	rs10440833		A	1.72	Kwak et al. (2012)
	rs7756992	Decreases FIC,HOMA-B, AUC		1.55, 1.22	Cho et al. (2009) and Lauenborg et al. (2009)
IGF2BP2	rs4402960	Decreases insulin secretion, HOMA-B	T	1.207, 1.18	Wang et al. (2011) and Cho et al. (2009)
	rs1470579		C	1.332	Kwak et al. (2012)
CDKN2A–CDKN2B	rs2383208	Decreases HOMA-B, increases FPG	A		
	rs1081166			1.49	Cho et al. (2009)
	rs10965250		G	1.31	Kwak et al. (2012)
SRR	rs391300	Increases FPG	T	1.2	Wang et al. (2011)
MTNR1B	rs10830962	Decreases FIC	G	1.45	Kwak et al. (2012)
	rs10830963	Decreases HOMA-B, increases FPG	G	1.46	Kim et al. (2011)
	rs1387153		T	1.44	Kim et al. (2011)
TCF7L2	rs7903146	Increases glucose during OGTT	T	2.04, 1.90	Pappa et al. (2011), Freathy et al. (2010), and Papadopoulou et al. (2011)
	rs12255372		T	1.58	Papadopoulou et al. (2011)
	rs7901695		C	1.87	Papadopoulou et al. (2011)

FIC fasting insulin concentration, *FPG* fasting plasma glucose, *HOMA-B* homeostasis model assessment of beta-cell function, *AUC* area under the curve, *OGTT* oral glucose tolerance test

mutations increase GDM risk through decreased ability of β-cells to compensate for the insulin resistance associated with pregnancy (Kwak et al. 2012) (Table 30.6).

7.2 CDK5 Regulatory Subunit-Associated Protein 1-Like 1 (CDKAL1: rs7754840, rs7756992)

Genome-wide association studies have linked single-nucleotide polymorphisms in an intron of CDKAL1 with the risk type 2 diabetes development (Steinthorsdottir et al. 2007). There are two SNPs (rs7756992 and rs7754840) *in CDKAL1* which are associated with GDM risk at a genome-wide significance level (Cho et al. 2009; Lauenborg et al. 2009). rs7754840 C allele of *CDKAL1*, which is the risk variant of GDM, alters the function of CDK5 in β-cell compensation during pregnancy and is significantly associated with decreased fasting insulin concentration and decreased homeostasis model assessment of beta-cell function (HOMA-B) in women with Korean GDM subjects. This shows a compromised β-cell compensation (Kwak et al. 2012). rs7754840 is significantly correlated with area under the curve (AUC) of insulin during a 100 g OGTT which is performed at the time of GDM diagnosis with insulin AUC ($b = -0.080$ mU $l^{-1} \times$ h, $P = 0.007$) in Chinese population (Wang et al. 2011). The variants in *CDKAL1* are associated with decreased birth weight, which could be explained by reduced fetal insulin secretion (Freathy et al. 2009).

7.3 Insulin-Like Growth Factor 2 mRNA-Binding Protein 2 (IGF2BP2: rs4402960, rs1470579)

IGF2BP2 binds to the 5′-UTR of the insulin-like growth factor 2 (IGF2) mRNA and regulates IGF2 translation. There are two known SNPs that play a role in GDM development: rs4402960 and rs1470579. In Chinese population rs4402960 has been found to be associated with GDM (OR = 1.207, 95 % CI = 1.029–1.417, $P=0.021$) (Wang et al. 2011); a similar association is also found with GDM in Korean population for this SNP with an OR value 1.18, 95 % CI = 1.01–1.38, $P=0.034$ (Cho et al. 2009). However, in Danish population a nonsignificant association was found for rs4402960 (OR = 1.18, 95 % CI = 0.97–1.42, $P=0.096$) (Lauenborg et al. 2009). rs1470579 in *IGF2BP2* shows a positive association with GDM (Kwak et al. 2012).

IGF2BP2 plays its role in GDM since it affects first-phase insulin secretion and the disposition index. The presence of T allele for rs4402960 shows a negative association with HOMA-B ($b=20.057$, $P=0.046$), thus showing its effect in beta-cell dysfunction (Wang et al. 2011).

7.4 Cyclin-Dependent Kinase Inhibitors 2A and 2B (CDKN2A–CDKN2B: rs2383208, rs10811661)

CDKN2B lies adjacent to the tumor suppressor gene CDKN2A; it forms a complex with *CDK4* or *CDK6*. This complex performs its cell growth regulatory function by preventing cyclin D-mediated activation of CDK kinase, thereby inhibiting cell cycle G1 progression. Two known polymorphisms of *CDKN2A–CDKN2B* are rs2383208 and rs10811661. rs10811661 increases the risk of GDM development (OR 1.49, $P=1.05\times10^{-7}$) (Cho et al. 2009). rs2383208, a variant at the same LD block with rs1081166, acts as a risk factor for GDM development in Chinese population (OR = 1.24, $P=0.003$). Compared with wild-type carriers, homozygous harboring the risk alleles of rs2383208 had a 1.5-fold increased risk of GDM. The carriers of rs2383208 A allele show a lower level of HOMA-B and increased fasting plasma glucose (Wang et al. 2011). Thus, rs2383208 plays a role in GDM development by regulating pancreatic beta-cell and fasting plasma glucose.

7.5 Serine Racemase (SRR: rs391300)

Serine racemase eliminates water from L-serine, thereby forming pyruvate and ammonia. It is also involved in the generation of D-serine from L-serine (De Miranda et al. 2002). rs391300 is associated with GDM risk (OR = 1.202, 95 % CI = 1.020–1.416, $P=0.028$). Further homozygous TT genotype increases the risk of GDM by 1.856-fold (95%CI = 1.236–2.789, $P=0.003$). rs391300 is associated with increased fasting plasma glucose indicating that this variant affects the incidence of GDM by modulating the secretion of insulin and/or glucagon (Wang et al. 2011).

7.6 Melatonin Receptor 1B (MTNR1B: rs10830963, rs1387153, rs10830962)

MTNR1B encodes the MT2 protein, one of two high-affinity forms of a receptor for melatonin, which is an integral membrane protein that is a G-protein-coupled, 7-transmembrane receptor. rs10830963 shows nonsignificant association with GDM in Chinese population although showing the role of risk allele G of rs10830963 in a lower value of HOMA-B and higher level of fasting plasma glucose with HOMA-B ($b=20.096$, $P=0.001$) and fasting plasma glucose ($b=0.062$, $P=0.019$) (Wang et al. 2011). Kim et al. has reported that two SNPs in MTNR1B (rs1387153 and rs10830963) are strongly associated with GDM in Korean women (Kim et al. 2011). A variant near MTNR1B, rs10830962, is also significantly associated with the risk of GDM (OR 1.454; $P=2.49\times10^{-13}$). The rs10830962 variant is in strong LD ($r^2=0.98$) with rs10830963, which is known to be associated with fasting glucose concentrations, insulin concentrations, and type 2 diabetes. The rs10830962 G allele near MTNR1B is associated with decreased fasting insulin concentrations ($b=20.018$; $P=0.029$) in women with GDM (Kwak et al. 2012).

7.7 Transcription Factor 7-Like 2 (TCF7L2 rs7903146, rs12255372, rs7901695)

T2D risk T allele of rs7903146 is found to be significantly associated with an increased risk of GDM (Pappa et al. 2011). The HAPO study has also found similar association between rs7903146 and GDM (Freathy et al. 2010). GDM subjects with the T allele of rs7903146 have higher glucose levels during an OGTT than in noncarriers. These findings suggest that defects in insulin secretion, associated with rs7903146 in TCF7L2, modestly increase glucose levels during pregnancy as well as risk of GDM (Konig and Shuldiner 2012). The heterozygous genotypes CT, GT, and TC of the rs7903146 (T is risk for type 2 diabetes), rs12255372 (T is risk for type 2 diabetes), and rs7901695 (C is risk for type 2 diabetes), respectively, as well as the homozygous genotypes TT, TT, and CC of the rs7903146, rs12255372, and rs7901695, respectively, were strongly associated with gestational diabetes ($P<0.0001$) (Papadopoulou et al. 2011). The *TCF7L2* rs7903146 SNP is asso ciated with an increased risk of GDM also in Scandinavian women (Shaat et al. 2007), while the rs12255372 SNP is associated with susceptibility for gestational diabetes and affected the insulin response to oral glucose through interaction with the percentage of body fat in probands with gestational diabetes (Watanabe et al. 2007). These three *TCF7L2* SNPs show the risk to type 2 diabetes and gestational diabetes in the order rs7903146 C>T, rs12255372 G>T, and rs7901695 T>C (Tong et al. 2009); these are also located within a well-defined linkage disequilibrium block (Grant et al. 2006; Helgason et al. 2007; Scott et al. 2006).

7.8 Peroxisome Proliferator-Activated Receptor-γ (PPARγ)

PPARγ is a nuclear receptor that plays a critical role in insulin signaling and adipogenesis and is a target for the insulin sensitizing thiazolidinediones. PPARγs regulates the number and size of peroxisomes produced in the cells by binding to peroxisome proliferators. It mediates the development of several chronic diseases, including diabetes. PPARγ2 isoform (Pro12Ala) is a known polymorphic form of PPARγ2. GDM patients carrying the Ala12 allele gained significantly more weight during their pregnancy (Konig and Shuldiner 2012).

7.9 Additive Effect of SNPs

Different studies have shown the additive effects of risk variants on GDM development. In Lauenborg et al. study, it increases the risk of GDM development by 1.18-fold per allele ($P=3.2\times10^{-6}$) while studying the association of 11 examined T2D risk variants. There is 3.30-fold increased risk (95%CI 1.69–6.39) of GDM for women carrying 15 or more risk alleles compared with women with nine or fewer risk alleles ($P=2.8\times10^{-4}$). Similarly additive effect was observed for of the risk alleles (T allele of rs4402960, A allele of rs2383208, and G allele of rs10830963) on HOMA-B level ($b=-0.108$ unit per risk allele, $P=1.86\times10^{-4}$) in Chinese population. In this study proportion of women with GDM increases in the subgroups with each additional risk allele increased GDM risk by 1.196-fold (95%CI=1.092–1.309, $P=1.08\times10^{-4}$). Moreover, the subjects having 4, 5, and 6 risk alleles has 2.008-fold ($P=0.011$), 5.576-fold ($P=3.31\times10^{-4}$), and 9.717-fold ($P=0.047$) increased risk of developing, respectively (Wang et al. 2011).

8 Conclusions

Emerging data suggest that in case of gynecological diseases, susceptibility to disease, responses to treatment, and toxicity to treatment vary according to genetic makeup of individual as well as genetic variations among various populations. Single-nucleotide variations in drug-metabolizing enzymes, DNA repair enzymes, various signaling proteins, tumor suppressor, and oncogenes contribute to these interindividual variations. Sufficient information has been gathered in some disease but data need to be strengthened in others to identify subgroup of patients based on genomic patterns. Appropriate drug trials should be then designed for the identified subgroups in order to aim toward individualized therapy. Government and industrial participation is required to encourage women and their clinicians to participate in such academic trials.

References

Aglund K, Rauvala M, Puistola U, Angstrom T, Turpeenniemi-Hujanen T, Zackrisson B, Stendahl U (2004) Gelatinases A and B (MMP-2 and MMP-9) in endometrial cancer-MMP-9 correlates to the grade and the stage. Gynecol Oncol 94:699–704

Akiyama T, Sudo C, Ogawara H, Toyoshima K, Yamamoto T (1986) The product of the human c-erbB-2 gene: a 185-kilodalton glycoprotein with tyrosine kinase activity. Science 232:1644–1646

Ambrosone CB (2007) The promise and limitations of genome-wide association studies to elucidate the causes of breast cancer. Breast Cancer Res 9:114

Arngrimsson R, Hayward C, Nadaud S, Baldursdottir A, Walker JJ, Liston WA, Bjarnadottir RI, Brock DJ, Geirsson RT, Connor JM, Soubrier F (1997) Evidence for a familial pregnancy-induced hypertension locus in the eNOS-gene region. Am J Hum Genet 61:354–362

Arts HJ, Katsaros D, de Vries EG, Massobrio M, Genta F, Danese S, Arisio R, Scheper RJ, Kool M, Scheffer GL, Willemse PH, van der Zee AG, Suurmeijer AJ (1999) Drug resistance-associated markers P-glycoprotein, multidrug resistance-associated protein 1, multidrug resistance-associated protein 2, and lung resistance protein as prognostic factors in ovarian carcinoma. Clin Cancer Res 5:2798–2805

Ashworth A (2008) A synthetic lethal therapeutic approach: poly(ADP) ribose polymerase inhibitors for the treatment of cancers deficient in DNA double-strand break repair. J Clin Oncol 26:3785–3790

Ayoub N, Lucas C, Kaddoumi A (2011) Genomics and pharmacogenomics of breast cancer: current knowledge and trends. Asian Pac J Cancer Prev 12:1127–1140

Badawi AF, Cavalieri EL, Rogan EG (2001) Role of human cytochrome P450 1A1, 1A2, 1B1, and 3A4 in the 2-, 4-, and 16alpha-hydroxylation of 17beta-estradiol. Metabolism 50:1001–1003

Baselga J, Norton L, Albanell J, Kim YM, Mendelsohn J (1998) Recombinant humanized anti-HER2 antibody (Herceptin) enhances the antitumor activity of paclitaxel and doxorubicin against HER2/neu overexpressing human breast cancer xenografts. Cancer Res 58:2825–2831

Baumwell S, Karumanchi SA (2007) Pre-eclampsia: clinical manifestations and molecular mechanisms. Nephron Clin Pract 106:c72–c81

Bereketoglu C, Kasap M, Pazarbasi A (2012) Studies on Angiotensin-converting enzyme insertion/deletion polymorphism and genotype distributions in Turkish preeclampsia patients. J Pregnancy 2012:108206

Bogusiewicz M, Stryjecka-Zimmer M, Postawski K, Jakimiuk AJ, Rechberger T (2007) Activity of matrix metalloproteinase-2 and -9 and contents of their tissue inhibitors in uterine leiomyoma and corresponding myometrium. Gynecol Endocrinol 23:541–546

Bombell S, McGuire W (2008) Tumour necrosis factor (-308A) polymorphism in pre-eclampsia: meta-analysis of 16 case-control studies. Aust N Z J Obstet Gynaecol 48:547–551

Bosch TM, Huitema AD, Doodeman VD, Jansen R, Witteveen E, Smit WM, Jansen RL, van Herpen CM, Soesan M, Beijnen JH, Schellens JH (2006) Pharmacogenetic screening of CYP3A and ABCB1 in relation to population pharmacokinetics of docetaxel. Clin Cancer Res 12:5786–5793

Broomall EM, Reynolds SM, Jacobson RM (2010) Epidemiology, clinical manifestations, and recent advances in vaccination against human papillomavirus. Postgrad Med 122:121–129

Buchanan TA (2001) Pancreatic B-cell defects in gestational diabetes: implications for the pathogenesis and prevention of type 2 diabetes. J Clin Endocrinol Metab 86:989–993

Calhoun ES, McGovern RM, Janney CA, Cerhan JR, Iturria SJ, Smith DI, Gostout BS, Persing DH (2002) Host genetic polymorphism analysis in cervical cancer. Clin Chem 48:1218–1224

Calle EE, Rodriguez C, Walker-Thurmond K, Thun MJ (2003) Overweight, obesity, and mortality from cancer in a prospectively studied cohort of U.S. adults. N Engl J Med 348:1625–1638

Cameron D, Casey M, Oliva C, Newstat B, Imwalle B, Geyer CE (2010) Lapatinib plus capecitabine in women with HER-2-positive advanced breast cancer: final survival analysis of a phase III randomized trial. Oncologist 15:924–934

Catalano PM, Tyzbir ED, Wolfe RR, Roman NM, Amini SB, Sims EA (1992) Longitudinal changes in basal hepatic glucose production and suppression during insulin infusion in normal pregnant women. Am J Obstet Gynecol 167:913–919

Chatterjee K, Engelmark M, Gyllensten U, Dandara C, van der Merwe L, Galal U, Hoffman M, Williamson AL (2009) Fas and FasL gene polymorphisms are not associated with cervical cancer but differ among Black and Mixed-ancestry South Africans. BMC Res Notes 2:238

Chen ZP, Malapetsa A, Monks A, Myers TG, Mohr G, Sausville EA, Scudiero DA, Panasci LC (2002) Nucleotide excision repair protein levels vis-a-vis anticancer drug resistance in 60 human tumor cell lines. Ai Zheng 21:233–239

Cho YM, Kim TH, Lim S, Choi SH, Shin HD, Lee HK, Park KS, Jang HC (2009) Type 2 diabetes-associated genetic variants discovered in the recent genome-wide association studies are related to gestational diabetes mellitus in the Korean population. Diabetologia 52:253–261

Colacurci N, De Seta L, De Franciscis P, Mele D, Fortunato N, Cassese S (2000) Tamoxifen effects on endometrium. Panminerva Med 42:45–47

Coussens L, Yang-Feng TL, Liao YC, Chen E, Gray A, McGrath J, Seeburg PH, Libermann TA, Schlessinger J, Francke U et al (1985) Tyrosine kinase receptor with extensive homology to EGF receptor shares chromosomal location with neu oncogene. Science 230:1132–1139

Crewe HK, Notley LM, Wunsch RM, Lennard MS, Gillam EM (2002) Metabolism of tamoxifen by recombinant human cytochrome P450 enzymes: formation of the 4-hydroxy, 4'-hydroxy and N-desmethyl metabolites and isomerization of trans-4-hydroxytamoxifen. Drug Metab Dispos 30:869–874

Criscitiello C, Metzger-Filho O, Saini KS et al (2012) Targeted therapies in breast cancer: are heart and vessels also being targeted? Breast Cancer Res 14:209

Cui Z, Huang L (2005) Liposome-polycation-DNA (LPD) particle as a carrier and adjuvant for protein-based vaccines: therapeutic effect against cervical cancer. Cancer Immunol Immunother 54:1180–1190

Czernobilsky B, Lifschitz-Mercer B (1997) Endometrial pathology. Curr Opin Obstet Gynecol 9:52–56

Dabholkar M, Bostick-Bruton F, Weber C, Bohr VA, Egwuagu C, Reed E (1992) ERCC1 and ERCC2 expression in malignant tissues from ovarian cancer patients. J Natl Cancer Inst 84:1512–1517

Dabholkar M, Vionnet J, Bostick-Bruton F, Yu JJ, Reed E (1994) Messenger RNA levels of XPAC and ERCC1 in ovarian cancer tissue correlate with response to platinum-based chemotherapy. J Clin Invest 94:703–708

De Miranda J, Panizzutti R, Foltyn VN, Wolosker H (2002) Cofactors of serine racemase that physiologically stimulate the synthesis of the N-methyl-D-aspartate (NMDA) receptor coagonist D-serine. Proc Natl Acad Sci U S A 99:14542–14547

de Pace V, Chiossi G, Facchinetti F (2007) Clinical use of nitric oxide donors and L-arginine in obstetrics. J Matern Fetal Neonatal Med 20:569–579

Dekker GA, Sibai BM (1998) Etiology and pathogenesis of preeclampsia: current concepts. Am J Obstet Gynecol 179:1359–1375

Delahanty RJ, Beeghly-Fadiel A, Xiang YB, Long J, Cai Q, Wen W, Xu WH, Cai H, He J, Gao YT, Zheng W, Shu XO (2011) Association of obesity-related genetic variants with endometrial cancer risk: a report from the Shanghai Endometrial Cancer Genetics Study. Am J Epidemiol 174:1115–1126

Di Nezza LA, Misajon A, Zhang J, Jobling T, Quinn MA, Ostor AG, Nie G, Lopata A, Salamonsen LA (2002) Presence of active gelatinases in endometrial carcinoma and correlation of matrix metalloproteinase expression with increasing tumor grade and invasion. Cancer 94:1466–1475

Dileo J, Banerjee R, Whitmore M, Nayak JV, Falo LD Jr, Huang L (2003) Lipid-protamine-DNA-mediated antigen delivery to antigen-presenting cells results in enhanced anti-tumor immune responses. Mol Ther 7:640–648

Doherty JA, Weiss NS, Freeman RJ, Dightman DA, Thornton PJ, Houck JR, Voigt LF, Rossing MA, Schwartz SM, Chen C (2005) Genetic factors in catechol estrogen metabolism in relation to the risk of endometrial cancer. Cancer Epidemiol Biomarkers Prev 14:357–366

Dowsett M, Dunbier AK (2008) Emerging biomarkers and new understanding of traditional markers in personalized therapy for breast cancer. Clin Cancer Res 14:8019–8026

Duley L (2009) The global impact of pre-eclampsia and eclampsia. Semin Perinatol 33:130–137

Eaton DL, Gallagher EP, Bammler TK, Kunze KL (1995) Role of cytochrome P4501A2 in chemical carcinogenesis: implications for human variability in expression and enzyme activity. Pharmacogenetics 5:259–274

Edwards SL, Brough R, Lord CJ, Natrajan R, Vatcheva R, Levine DA, Boyd J, Reis-Filho JS, Ashworth A (2008) Resistance to therapy caused by intragenic deletion in BRCA2. Nature 451:1111–1115

Evans JP, Skrzynia C, Burke W (2001) The complexities of predictive genetic testing. BMJ 322:1052–1056

Ferrara N, Vis-Smyth T (1997) The biology of vascular endothelial growth factor. Endocr Rev 18:4–25

Ferrara N, Houck K, Jakeman L, Leung DW (1992) Molecular and biological properties of the vascular endothelial growth factor family of proteins. Endocr Rev 13:18–32

Finucane MM, Stevens GA, Cowan MJ, Danaei G, Lin JK, Paciorek CJ, Singh GM, Gutierrez HR, Lu Y, Bahalim AN, Farzadfar F, Riley LM, Ezzati M (2011) National, regional, and global trends in body-mass index since 1980: systematic analysis of health examination surveys and epidemiological studies with 960 country-years and 9.1 million participants. Lancet 377:557–567

Fong PC, Yap TA, Boss DS, Carden CP, Mergui-Roelvink M, Gourley C, De Greve J, Lubinski J, Shanley S, Messiou C, A'Hern R, Tutt A, Ashworth A, Stone J, Carmichael J, Schellens JH, de Bono JS, Kaye SB (2010) Poly(ADP)-ribose polymerase inhibition: frequent durable responses in BRCA carrier ovarian cancer correlating with platinum-free interval. J Clin Oncol 28:2512–2519

Franchim CS, Sass N, Mattar R, Pendeloski KP, Lin LH, Torloni MR, Daher S (2011) Inflammatory mediators gene polymorphisms in preeclampsia. Hypertens Pregnancy 30:338–346

Freathy RM, Bennett AJ, Ring SM, Shields B, Groves CJ, Timpson NJ, Weedon MN, Zeggini E, Lindgren CM, Lango H, Perry JR, Pouta A, Ruokonen A, Hypponen E, Power C, Elliott P, Strachan DP, Jarvelin MR, Smith GD, McCarthy MI, Frayling TM, Hattersley AT (2009) Type 2 diabetes risk alleles are associated with reduced size at birth. Diabetes 58:1428–1433

Freathy RM, Hayes MG, Urbanek M, Lowe LP, Lee H, Ackerman C, Frayling TM, Cox NJ, Dunger DB, Dyer AR, Hattersley AT, Metzger BE, Lowe WL Jr (2010) Hyperglycemia and Adverse Pregnancy Outcome (HAPO) study: common genetic variants in GCK and TCF7L2 are associated with fasting and postchallenge glucose levels in pregnancy and with the new consensus definition of gestational diabetes mellitus from the International Association of Diabetes and Pregnancy Study Groups. Diabetes 59:2682–2689

Freedman AN, Sansbury LB, Figg WD, Potosky AL, Weiss Smith SR, Khoury MJ, Nelson SA, Weinshilboum RM, Ratain MJ, McLeod HL, Epstein RS, Ginsburg GS, Schilsky RL, Liu G, Flockhart DA, Ulrich CM, Davis RL, Lesko LJ, Zineh I, Randhawa G, Ambrosone CB, Relling MV, Rothman N, Xie H, Spitz MR, Ballard-Barbash R, Doroshow JH, Minasian LM (2010) Cancer pharmacogenomics and pharmacoepidemiology: setting a research agenda to accelerate translation. J Natl Cancer Inst 102:1698–1705

Gadducci A, Di Cristofano C, Zavaglia M, Giusti L, Menicagli M, Cosio S, Naccarato AG, Genazzani AR, Bevilacqua G, Cavazzana AO (2006) P53 gene status in patients with advanced serous epithelial ovarian cancer in relation to response to paclitaxel- plus platinum-based chemotherapy and long-term clinical outcome. Anticancer Res 26:687–693

Galic V, Willner J, Wollan M, Garg R, Garcia R, Goff BA, Gray HJ, Swisher EM (2007) Common polymorphisms in TP53 and MDM2 and the relationship to TP53 mutations and clinical outcomes in women with ovarian and peritoneal carcinomas. Genes Chromosomes Cancer 46:239–247

Garovic VD, Hayman SR (2007) Hypertension in pregnancy: an emerging risk factor for cardiovascular disease. Nat Clin Pract Nephrol 3:613–622

Gelderman KA, Tomlinson S, Ross GD, Gorter A (2004) Complement function in mAb- mediated cancer immunotherapy. Trends Immunol 25:158–164

Goetz MP, Suman VJ, Couch FJ, Ames MM, Rae JM, Erlander MG, Ma XJ, Sgroi DC, Reynolds CA, Lingle WL, Weinshilboum RM, Flockhart DA, Desta Z, Perez EA, Ingle JN (2008) Cytochrome P450 2D6 and homeobox 13/interleukin-17B receptor: combining inherited and tumor gene markers for prediction of tamoxifen resistance. Clin Cancer Res 14:5864–5868

Goldenberg RL, Culhane JF, Iams JD, Romero R (2008) Epidemiology and causes of preterm birth. Lancet 371:75–84

Goodman MT, McDuffie K, Guo C, Terada K, Donlon TA (2001) CYP17 genotype and ovarian cancer: a null case-control study. Cancer Epidemiol Biomarkers Prev 10:563–564

Gottesman MM, Pastan I (1993) Biochemistry of multidrug resistance mediated by the multidrug transporter. Annu Rev Biochem 62:385–427

Gradhand U, Kim RB (2008) Pharmacogenomics of MRP transporters (ABCC1-5) and BCRP (ABCG2). Drug Metab Rev 40:317–354

Graesslin O, Cortez A, Fauvet R, Lorenzato M, Birembaut P, Darai E (2006a) Metalloproteinase-2, -7 and -9 and tissue inhibitor of metalloproteinase-1 and -2 expression in normal, hyperplastic and neoplastic endometrium: a clinical-pathological correlation study. Ann Oncol 17:637–645

Graesslin O, Cortez A, Uzan C, Birembaut P, Quereux C, Darai E (2006b) Endometrial tumor invasiveness is related to metalloproteinase 2 and tissue inhibitor of metalloproteinase 2 expressions. Int J Gynecol Cancer 16:1911–1917

Grant SF, Thorleifsson G, Reynisdottir I, Benediktsson R, Manolescu A, Sainz J, Helgason A, Stefansson H, Emilsson V, Helgadottir A, Styrkarsdottir U, Magnusson KP, Walters GB, Palsdottir E, Jonsdottir T, Gudmundsdottir T, Gylfason A, Saemundsdottir J, Wilensky RL, Reilly MP, Rader DJ, Bagger Y, Christiansen C, Gudnason V, Sigurdsson G, Thorsteinsdottir U, Gulcher JR, Kong A, Stefansson K (2006) Variant of transcription factor 7-like 2 (TCF7L2) gene confers risk of type 2 diabetes. Nat Genet 38:320–323

Green H, Soderkvist P, Rosenberg P, Horvath G, Peterson C (2006) mdr-1 single nucleotide polymorphisms in ovarian cancer tissue: G2677T/A correlates with response to paclitaxel chemotherapy. Clin Cancer Res 12:854–859

Green H, Soderkvist P, Rosenberg P, Mirghani RA, Rymark P, Lundqvist EA, Peterson C (2009) Pharmacogenetic studies of Paclitaxel in the treatment of ovarian cancer. Basic Clin Pharmacol Toxicol 104:130–137

Grenetier S, Bouchoux C, Goguel V (2006) CTD kinase I is required for the integrity of the rDNA tandem array. Nucleic Acids Res 34:4996–5006

Grimm SW, Dyroff MC (1997) Inhibition of human drug metabolizing cytochromes P450 by anastrozole, a potent and selective inhibitor of aromatase. Drug Metab Dispos 25:598–602

Groenewoud MJ, Dekker JM, Fritsche A, Reiling E, Nijpels G, Heine RJ, Maassen JA, Machicao F, Schafer SA, Haring HU, t Hart LM, van Haeften TW (2008) Variants of CDKAL1 and IGF2BP2 affect first-phase insulin secretion during hyperglycaemic clamps. Diabetologia 51:1659–1663

Hakli T, Romppanen EL, Hiltunen M, Helisalmi S, Punnonen K, Heinonen S (2003) Endothelial nitric oxide synthase polymorphism in preeclampsia. J Soc Gynecol Investig 10:154–157

Hall JM, Lee MK (1990) Linkage of early onset familial breast cancer to chromosome 17q21. Science 250:1684–1689

Hayes CL, Spink DC, Spink BC, Cao JQ, Walker NJ, Sutter TR (1996) 17 beta-estradiol hydroxylation catalyzed by human cytochrome P450 1B1. Proc Natl Acad Sci U S A 93:9776–9781

Helgason A, Palsson S, Thorleifsson G, Grant SF, Emilsson V, Gunnarsdottir S, Adeyemo A, Chen Y, Chen G, Reynisdottir I, Benediktsson R, Hinney A, Hansen T, Andersen G, Borch-Johnsen K et al (2007) Refining the impact of TCF7L2 gene variants on type 2 diabetes and adaptive evolution. Nat Genet 39:218–225

Henderson BE, Feigelson HS (2000) Hormonal carcinogenesis. Carcinogenesis 21:427–433

Henderson BE, Ross RK, Pike MC, Cassagrande JT (1982) Endogenous hormones as a major factor in human cancer. Cancer Res 24:3232–3239

Henderson BE, Ross RK, Bernstein L (1988) Estrogens as a cause of human cancer: the Richard and Hinda Rosenthal Foundation Award Lecture. Cancer Res 48:246

Heubner M, Wimberger P, Riemann K, Kasimir-Bauer S, Otterbach F, Kimmig R, Siffert W (2010) The CYP1A1 Ile462Val polymorphism and platinum resistance of epithelial ovarian neoplasms. Oncol Res 18:343–347

Hillermann R, Carelse K, Gebhardt GS (2005) The Glu298Asp variant of the endothelial nitric oxide synthase gene is associated with an increased risk for abruptio placentae in pre-eclampsia. J Hum Genet 50:415–419

Hoffmeyer S, Burk O, von Richter O, Arnold HP, Brockmoller J, Johne A, Cascorbi I, Gerloff T, Roots I, Eichelbaum M, Brinkmann U (2000) Functional polymorphisms of the human multidrug-resistance gene: multiple sequence variations and correlation of one allele with P-glycoprotein expression and activity in vivo. Proc Natl Acad Sci U S A 97:3473–3478

Hunter KW et al (2001) Predisposition to efficient mammary tumor metastatic progression is linked to the breast cancer metastasis suppressor gene Brms1. Cancer Res 61:8866–8872

Hurvitz SA, Pegram MD, Lin L-S, Chan DS, Allen HJ, Dichmann RA, Hagenstad CT, Barstis J, Hermann RC, Hu EH, Moroose RL, Thomas SP, Vogel CL, Ryba N, Elashoff D (2009) Final results of a phase II trial evaluating trastuzumab and bevacizumab as first line treatment of HER2-amplified advanced breast cancer. Cancer Res 69:854

Indermaur MD, Xiong Y, Kamath SG, Boren T, Hakam A, Wenham RM, Apte SM, Lancaster JM (2010) Genomic-directed targeted therapy increases endometrial cancer cell sensitivity to doxorubicin. Am J Obstet Gynecol 203:158.e1–158.e40

Ingle JN, Suman VJ (2005) Aromatase inhibitors for therapy of advanced breast cancer. J Steroid Biochem Mol Biol 95:113–119

Ingle JN, Suman VJ, Rowland KM, Mirchandani D, Bernath AM, Camoriano JK, Fishkin PA, Nikcevich DA, Perez EA (2006) Fulvestrant in women with advanced breast cancer after progression on prior aromatase inhibitor therapy: North Central Cancer Treatment Group Trial N0032. J Clin Oncol 24:1052–1056

Ingle JN, Tu D, Pater JL, Muss HB, Martino S, Robert NJ, Piccart MJ, Castiglione M, Shepherd LE, Pritchard KI, Livingston RB, Davidson NE, Norton L, Perez EA, Abrams JS, Cameron DA, Palmer MJ, Goss PE (2008) Intent-to-treat analysis of the placebo-controlled trial of letrozole for extended adjuvant therapy in early breast cancer: NCIC CTG MA.17. Ann Oncol 19:877–882

Irani RA, Xia Y (2008) The functional role of the renin-angiotensin system in pregnancy and preeclampsia. Placenta 29:763–771

Isaka K, Nishi H, Nakai H, Nakada T, Feng Li Y, Ebihara Y, Takayama M (2003) Matrix metalloproteinase-26 is expressed in human endometrium but not in endometrial carcinoma. Cancer 97:79–89

Ishigami T, Tamura K, Fujita T, Kobayashi I, Hibi K, Kihara M, Toya Y, Ochiai H, Umemura S (1999) Angiotensinogen gene polymorphism near transcription start site and blood pressure: role of a T-to-C transition at intron I. Hypertension 34:430–434

Itamochi H, Kigawa J, Sultana H, Iba T, Akeshima R, Kamazawa S, Kanamori Y, Terakawa N (2002) Sensitivity to anticancer agents and resistance mechanisms in clear cell carcinoma of the ovary. Jpn J Cancer Res 93:723–728

Ito K (2007) Hormone replacement therapy and cancers: the biological roles of estrogen and progestin in tumorigenesis are different between the endometrium and breast. Tohoku J Exp Med 212:1–12

Ji LD, Zhang LN, Shen P, Wang P, Zhang YM, Xing WH, Xu J (2010) Association of angiotensinogen gene M235T and angiotensin-converting enzyme gene I/D polymorphisms with essential hypertension in Han Chinese population: a meta-analysis. J Hypertens 28:419–428

Kamazawa S, Kigawa J, Kanamori Y, Itamochi H, Sato S, Iba T, Terakawa N (2002) Multidrug resistance gene-1 is a useful predictor of Paclitaxel-based chemotherapy for patients with ovarian cancer. Gynecol Oncol 86:171–176

Kang S, Ju W, Kim JW, Park NH, Song YS, Kim SC, Park SY, Kang SB, Lee HP (2006) Association between excision repair cross-complementation group 1 polymorphism and clinical outcome of platinum-based chemotherapy in patients with epithelial ovarian cancer. Exp Mol Med 38:320–324

Karahan N, Guney M, Baspinar S, Oral B, Kapucuoglu N, Mungan T (2007) Expression of gelatinase (MMP-2 and MMP-9) and cyclooxygenase-2 (COX-2) in endometrial carcinoma. Eur J Gynaecol Oncol 28:184–188

Kastner P, Krust A, Turcotte B, Stropp U, Tora L, Gronemeyer H, Chambon P (1990) Two distinct estrogen-regulated promoters generate transcripts encoding the two functionally different human progesterone receptor forms A and B. EMBO J 9:1603–1614

Kautzky-Willer A, Prager R, Waldhausl W, Pacini G, Thomaseth K, Wagner OF, Ulm M, Streli C, Ludvik B (1997) Pronounced insulin resistance and inadequate beta-cell secretion characterize lean gestational diabetes during and after pregnancy. Diabetes Care 20:1717–1723

Khrunin A, Ivanova F, Moisseev A, Khokhrin D, Sleptsova Y, Gorbunova V, Limborska S (2012) Pharmacogenomics of cisplatin-based chemotherapy in ovarian cancer patients of different ethnic origins. Pharmacogenomics 13:171–178

Kim S, Misra A (2007) SNP genotyping: technologies and biomedical applications. Annu Rev Biomed Eng 9:289–320

Kim YJ, Park BH, Park H, Jung SC, Pang MG, Ryu HM, Lee KS, Eom SM, Park HY (2008) No association of the genetic polymorphisms of endothelial nitric oxide synthase, dimethylarginine dimethylaminohydrolase, and vascular endothelial growth factor with preeclampsia in Korean populations. Twin Res Hum Genet 11:77–83

Kim JY, Cheong HS, Park BL, Baik SH, Park S, Lee SW, Kim MH, Chung JH, Choi JS, Kim MY, Yang JH, Cho DH, Shin HD, Kim SH (2011) Melatonin receptor 1 B polymorphisms associated with the risk of gestational diabetes mellitus. BMC Med Genet 12:82

King CR, Kraus MH, Aaronson SA (1985) Amplification of a novel v-erbB-related gene in a human mammary carcinoma. Science 229:974–976

Kiyohara C, Nakanishi Y, Inutsuka S, Takayama K, Hara N, Motohiro A, Tanaka K, Kono S, Hirohata T (1998) The relationship between CYP1A1 aryl hydrocarbon hydroxylase activity and lung cancer in a Japanese population. Pharmacogenetics 8:315–323

Kleinbongard P, Dejam A, Lauer T, Rassaf T, Schindler A, Picker O, Scheeren T, Godecke A, Schrader J, Schulz R, Heusch G, Schaub GA, Bryan NS, Feelisch M, Kelm M (2003) Plasma nitrite reflects constitutive nitric oxide synthase activity in mammals. Free Radic Biol Med 35:790–796

Konecny G, Untch M, Pegram M (1999) Herceptin in the treatment of metastatic breast cancer. Gynäkolog 32:624–631

Konig M, Shuldiner AR (2012) The genetic interface between gestational diabetes and type 2 diabetes. J Matern Fetal Neonatal Med 25:36–40

Kristensen VN, Harada N, Yoshimura N, Haraldsen E, Lonning PE, Erikstein B, Karesen R, Kristensen T, Borresen-Dale AL (2000) Genetic variants of CYP19 (aromatase) and breast cancer risk. Oncogene 19:1329–1333

Kuhl C (1991) Aetiology of gestational diabetes. Baillieres Clin Obstet Gynaecol 5:279–292

Kupferminc MJ (2005) Thrombophilia and pregnancy. Curr Pharm Des 11:735–748

Kwak SH, Kim SH, Cho YM, Go MJ, Cho YS, Choi SH, Moon MK, Jung HS, Shin HD, Kang HM, Cho NH, Lee IK, Kim SY, Han BG, Jang HC, Park KS (2012) A genome-wide association study of gestational diabetes mellitus in Korean women. Diabetes 61:531–541

Laird SM, Dalton CF, Okon MA, Bunning RA, Marshall R, Li TC (1999) Metalloproteinases and tissue inhibitor of metalloproteinase 1 (TIMP-1) in endometrial flushings from pre- and post-menopausal women and from women with endometrial adenocarcinoma. J Reprod Fertil 115:225–232

LaMarca BD, Ryan MJ, Gilbert JS, Murphy SR, Granger JP (2007) Inflammatory cytokines in the pathophysiology of hypertension during preeclampsia. Curr Hypertens Rep 9:480–485

Lapolla A, Dalfra MG, Fedele D (2009) Diabetes related autoimmunity in gestational diabetes mellitus: is it important? Nutr Metab Cardiovasc Dis 19:674–682

Lauenborg J, Grarup N, Damm P, Borch-Johnsen K, Jorgensen T, Pedersen O, Hansen T (2009) Common type 2 diabetes risk gene variants associate with gestational diabetes. J Clin Endocrinol Metab 94:145–150

Lee W, Lockhart AC, Kim RB, Rothenberg ML (2005) Cancer pharmacogenomics: powerful tools in cancer chemotherapy and drug development. Oncologist 10:104–111

Lee E, Hsu C, Haiman CA, Razavi P, Horn-Ross PL, Van Den Berg D, Bernstein L, Le Marchand L, Henderson BE, Setiawan VW, Ursin G (2010) Genetic variation in the progesterone receptor gene and risk of endometrial cancer: a haplotype-based approach. Carcinogenesis 31:1392–1399

Levesque S, Moutquin JM, Lindsay C, Roy MC, Rousseau F (2004) Implication of an AGT haplotype in a multigene association study with pregnancy hypertension. Hypertension 43:71–78

Lichtenstein P, Holm NV, Verkasalo PK, Iliadou A, Kaprio J, Koskenvuo M, Pukkala E, Skytthe A, Hemminki K (2000) Environmental and heritable factors in the causation of cancer–analyses of cohorts of twins from Sweden, Denmark, and Finland. N Engl J Med 343:78–85

Liehr JG, Ricci MJ, Jefcoate CR, Hannigan EV, Hokanson JA, Zhu BT (1995) 4-Hydroxylation of estradiol by human uterine myometrium and myoma microsomes: implications for the mechanism of uterine tumorigenesis. Proc Natl Acad Sci U S A 92:9220–9224

Lin R, Lei Y, Yuan Z, Ju H, Li D (2012) Angiotensinogen gene M235T and T174M polymorphisms and susceptibility of pre-eclampsia: a meta-analysis. Ann Hum Genet 76:377–386

Lindheimer MD, Taler SJ, Cunningham FG (2010) Hypertension in pregnancy. J Am Soc Hypertens 4:68–78

Lundberg JO, Weitzberg E, Gladwin MT (2008) The nitrate-nitrite-nitric oxide pathway in physiology and therapeutics. Nat Rev Drug Discov 7:156–167

Martin CM, Kehoe L, Spillane CO, O'Leary JJ (2007) Gene discovery in cervical cancer: towards diagnostic and therapeutic biomarkers. Mol Diagn Ther 11:277–290

Marzolini C, Paus E, Buclin T, Kim RB (2004) Polymorphisms in human MDR1 (P-glycoprotein): recent advances and clinical relevance. Clin Pharmacol Ther 75:13–33

McGrath M, Hankinson SE, Arbeitman L, Colditz GA, Hunter DJ, De Vivo I (2004) Cytochrome P450 1B1 and catechol-O-methyltransferase polymorphisms and endometrial cancer susceptibility. Carcinogenesis 25:559–565

Medeiros R, Pereira D, Afonso N, Palmeira C, Faleiro C, Afonso-Lopes C, Freitas-Silva M, Vasconcelos A, Costa S, Osorio T, Lopes C (2003) Platinum/paclitaxel-based chemotherapy in advanced ovarian carcinoma: glutathione S-transferase genetic polymorphisms as predictive biomarkers of disease outcome. Int J Clin Oncol 8:156–161

Meehan KL, Rainczuk A, Salamonsen LA, Stephens AN (2010) Proteomics and the search for biomarkers of female reproductive diseases. Reproduction 140:505–519

Meekins JW, Pijnenborg R, Hanssens M, McFadyen IR, van Asshe A (1994) A study of placental bed spiral arteries and trophoblast invasion in normal and severe pre-eclamptic pregnancies. Br J Obstet Gynaecol 101:669–674

Metzger IF, Souza-Costa DC, Marroni AS, Nagassaki S, Desta Z, Flockhart DA, Tanus-Santos JE (2005) Endothelial nitric oxide synthase gene haplotypes associated with circulating concentrations of nitric oxide products in healthy men. Pharmacogenet Genomics 15:565–570

Metzger IF, Sertorio JT, Tanus-Santos JE (2007) Modulation of nitric oxide formation by endothelial nitric oxide synthase gene haplotypes. Free Radic Biol Med 43:987–992

Middeldorp S (2007) Pregnancy failure and heritable thrombophilia. Semin Hematol 44:93–97

Mikhailova ON, Gulyaeva LF, Prudnikov AV, Gerasimov AV, Krasilnikov SE (2006) Estrogen-metabolizing gene polymorphisms in the assessment of female hormone-dependent cancer risk. Pharmacogenomics J 6:189–193

Miki Y, Swensen J, Shattuck-Eidens D, Futreal PA, Harshman K, Tavtigian S, Liu Q, Cichran C, Bennett LM, Ding W et al (1994) A strong candidate for the breast and ovarian cancer susceptibility gene *BRCA1*. Science 266:66–71

Milano G, Etienne MC (1994) Dihydropyrimidine dehydrogenase (DPD) and clinical pharmacology of 5-fluorouracil (review). Anticancer Res 14: 2295–2297

Milano G, Etienne MC, Pierrefite V et al (1999) Dihydropyrimidine dehydrogenase deficiency and fluorouracil-related toxicity. Br J Cancer 79:627–630

Misugi F, Sumi T, Okamoto E, Nobeyama H, Hattori K, Yoshida H, Matsumoto Y, Yasui T, Honda K, Ishiko O (2005) Expression of matrix metalloproteinases and tissue inhibitors of metalloproteinase in uterine endometrial carcinoma and a correlation between expression of matrix metalloproteinase-7 and prognosis. Int J Mol Med 16:541–546

Modugno F, Weissfeld JL, Trump DL, Zmuda JM, Shea P, Cauley JA, Ferrell RE (2001) Allelic variants of aromatase and the androgen and estrogen receptors: toward a multigenic model of prostate cancer risk. Clin Cancer Res 7:3092–3096

Monaghan H, MacWhinnie N, Williams AR (2007) The role of matrix metalloproteinases-2, -7 and -9 and beta-catenin in high grade endometrial carcinoma. Histopathology 50:348–357

Moreno-Bueno G, Sanchez-Estevez C, Cassia R, Rodriguez-Perales S, Diaz-Uriarte R, Dominguez O, Hardisson D, Andujar M, Prat J, Matias-Guiu X, Cigudosa JC, Palacios J (2003) Differential gene expression profile in endometrioid and nonendometrioid endometrial carcinoma: STK15 is frequently overexpressed and amplified in nonendometrioid carcinomas. Cancer Res 63:5697–5702

Myatt L, Webster RP (2009) Vascular biology of preeclampsia. J Thromb Haemost 7:375–384

Napolitano M, Miceli F, Calce A, Vacca A, Gulino A, Apa R, Lanzone A (2000) Expression and relationship between endothelin-1 messenger ribonucleic acid (mRNA) and inducible/endothelial nitric oxide synthase mRNA isoforms from normal and preeclamptic placentas. J Clin Endocrinol Metab 85: 2318–2323

Nishioka Y, Kobayashi K, Sagae S, Ishioka S, Nishikawa A, Matsushima M, Kanamori Y, Minaguchi T, Nakamura Y, Tokino T, Kudo R (2000) A single nucleotide polymorphism in the matrix metalloproteinase-1 promoter in endometrial carcinomas. Japanese J Cancer Res 91:612–615

Nunobiki O, Ueda M, Toji E, Yamamoto M, Akashi K, Sato N et al (2011) Genetic polymorphism of cancer susceptibility genes and HPV infection in cervical carcinogenesis. Pathol Res Int 2011:364069

Obata H, Yahata T, Quan J, Sekine M, Tanaka K (2006) Association between single nucleotide polymorphisms of drug resistance-associated genes and response to chemotherapy in advanced ovarian cancer. Anticancer Res 26:2227–2232

O'Brien C, Cavet G, Pandita A, Hu X, Haydu L, Mohan S, Toy K, Rivers CS, Modrusan Z, Amler LC, Lackner MR (2008) Functional genomics identifies ABCC3 as a mediator of taxane resistance in HER2-amplified breast cancer. Cancer Res 68:5380–5389

Olopade OI, Grushko TA, Nanda R, Huo D (2008) Advances in breast cancer: pathways to personalized medicine. Clin Cancer Res 14:7988–7999

O'Mara TA, Clements JA, Spurdle AB (2009) The use of predictive or prognostic genetic biomarkers in endometrial and other hormone-related cancers: justification for extensive candidate gene single nucleotide polymorphism studies of the matrix metalloproteinase family and their inhibitors. Cancer Epidemiol Biomarkers Prev 18:2352–2365

O'Mara TA, Fahey P, Ferguson K, Marquart L, Lambrechts D, Despierre E, Vergote I, Amant F, Hall P, Liu J, Czene K, Rebbeck TR, Ahmed S, Dunning AM, Gregory CS, Shah M, Webb PM, Spurdle AB (2011) Progesterone receptor gene variants and risk of endometrial cancer. Carcinogenesis 32:331–335

Osborne CK, Wiebe VJ, McGuire WL, Ciocca DR, DeGregorio MW (1992) Tamoxifen and the isomers of 4-hydroxytamoxifen in tamoxifen-resistant tumors from breast cancer patients. J Clin Oncol 10:304–310

Papadopoulou A, Lynch KF, Shaat N, Hakansson R, Ivarsson SA, Berntorp K, Agardh CD, Lernmark A (2011) Gestational diabetes mellitus is associated with TCF7L2 gene polymorphisms independent of HLA-DQB1*0602 genotypes and islet cell autoantibodies. Diabet Med 28:1018–1027

Papanicolaou GN (1942) A new procedure for staining vaginal smears. Science 95:438–439

Papazoglou D, Galazios G, Koukourakis MI, Panagopoulos I, Kontomanolis EN, Papatheodorou K, Maltezos E (2004) Vascular endothelial growth factor gene polymorphisms and pre-eclampsia. Mol Hum Reprod 10:321–324

Pappa KI, Gazouli M, Economou K, Daskalakis G, Anastasiou E, Anagnou NP, Antsaklis A (2011) Gestational diabetes mellitus shares polymorphisms of genes associated with insulin resistance and type 2 diabetes in the Greek population. Gynecol Endocrinol 27:267–272

Park YG, Zhao X, Lesueur F, Lowy DR, Lancaster M, Pharoah P, Qian X, Hunter KW (2005) Sipa1 is a candidate for underlying the metastasis efficiency modifier locus Mtes1. Nat Genet 37:1055–1062

Paynter RA, Hankinson SE, Colditz GA, Hunter DJ, De Vivo I (2004) No evidence of a role for PPAR gamma Pro12Ala polymorphism in endometrial cancer susceptibility. Pharmacogenetics 14:851–856

Pegram M, Hsu S, Lewis G, Pietras R, Beryt M, Sliwkowski M, Coombs D, Baly D, Kabbinavar F, Slamon D (1999) Inhibitory effects of combinations of HER-2/neu antibody and chemotherapeutic agents used for treatment of human breast cancers. Oncogene 18:2241–2251

Penson RT, Oliva E, Skates SJ, Glyptis T, Fuller AF Jr, Goodman A, Seiden MV (2004) Expression of multidrug resistance-1 protein inversely correlates with paclitaxel response and survival in ovarian cancer patients: a study in serial samples. Gynecol Oncol 93:98–106

Petricoin EF, Ardekani AM, Hitt BA, Levine PJ, Fusaro VA, Steinberg SM, Mills GB, Simone C, Fishman DA, Kohn EC, Liotta LA (2002) Use of proteomic patterns in serum to identify ovarian cancer. Lancet 359:572–577

Pietras RJ, Fendly BM, Chazin VR, Pegram MD, Howell SB, Slamon DJ (1994) Antibody to HER-2/neu receptor blocks DNA repair after cisplatin in human breast and ovarian cancer cells. Oncogene 9:1829–1838

Pietras RJ, Arboleda J, Reese DM, Wongvipat N, Pegram MD, Ramos L, Gorman CM, Parker MG, Sliwkowski MX, Slamon DJ (1995) HER-2 tyrosine kinase pathway targets estrogen receptor and promotes hormone-independent growth in human breast cancer cells. Oncogene 10:2435–2446

Pietras RJ, Pegram MD, Finn RS, Maneval DA, Slamon DJ (1998) Remission of human breast cancer xenografts on therapy with humanized monoclonal antibody to HER-2 receptor and DNA-reactive drugs. Oncogene 17:2235–2249

Pietras RJ, Poen JC, Gallardo D, Wongvipat PN, Lee HJ, Slamon DJ (1999) Monoclonal antibody to HER-2/neureceptor modulates repair of radiation-induced DNA damage and enhances radiosensitivity of human breast cancer cells overexpressing this oncogene. Cancer Res 59:1347–1355

Pitot HC, Loeb DD (2002) Fundamentals of oncology, 4th edn. Marcel Dekker Publisher/Informa Healthcare, New York

Press MF, Pike MC, Chazin VR, Hung G, Udove JA, Markowicz M, Danyluk J, Godolphin W, Sliwkowski M, Akita R et al (1993) Her-2/neu expression in node-negative breast cancer: direct tissue quantitation by computerized image analysis and association of over-expression with increased risk of recurrent disease. Cancer Res 53:4960–4970

Raftogianis RB, Wood TC, Otterness DM, Van Loon JA, Weinshilboum RM (1997) Phenol sulfotransferase pharmacogenetics in humans: association of common SULT1A1 alleles with TS PST phenotype. Biochem Biophys Res Commun 239:298–304

Raftogianis R, Creveling C, Weinshilboum R, Weisz J (2000) Estrogen metabolism by conjugation. J Natl Cancer Inst Monogr 27:113–124

Rahimi Z, Mozafari H, Parsian A (2013) Preeclampsia and angiotensin converting enzyme (ACE) I/D and angiotensin II type-1 receptor (AT1R) A1166C polymorphisms: association with ACE I/D polymorphism. J Renin Angiotensin Aldosterone Syst 14:174–180

Ravdin PM, Chamness GC (1995) The c-erbB-2 proto-oncogene as a prognostic and predictive marker in breast cancer: a paradigm for the development of other macromolecular markers–a review. Gene 159:19–27

Rebbeck TR (1997) Molecular epidemiology of the human glutathione S-transferase genotypes GSTM1 and GSTT1 in cancer susceptibility. Cancer Epidemiol Biomarkers Prev 6:733–743

Rebbeck TR, Troxel AB, Norman S, Bunin G, DeMichele A, Schinnar R, Berlin JA, Strom BL (2007) Pharmacogenetic modulation of combined hormone replacement therapy by progesterone-metabolism genotypes in postmenopausal breast cancer risk. Am J Epidemiol 166:1392–1399

Rehman I, Cross SS, Azzouzi AR, Catto JW, Deloulme JC, Larre S, Champigneuille J, Fromont G, Cussenot O, Hamdy FC (2004) S100A6 (Calcyclin) is a prostate basal cell marker absent in prostate cancer and its precursors. Br J Cancer 91:739–744

Renner W, Kotschan S, Hoffmann C, Obermayer-Pietsch B, Pilger E (2000) A common 936 C/T mutation in the gene for vascular endothelial growth factor is associated with vascular endothelial growth factor plasma levels. J Vasc Res 37:443–448

Rieck S, Kaestner KH (2010) Expansion of beta-cell mass in response to pregnancy. Trends Endocrinol Metab 21:151–158

Rutter JL, Mitchell TI, Buttice G, Meyers J, Gusella JF, Ozelius LJ, Brinckerhoff CE (1998) A single nucleotide polymorphism in the matrix metalloproteinase-1 promoter creates an Ets binding site and augments transcription. Cancer Res 58:5321–5325

Rylander-Rudqvist T, Wedren S, Jonasdottir G, Ahlberg S, Weiderpass E, Persson I, Ingelman-Sundberg M (2004) Cytochrome P450 1B1 gene polymorphisms and postmenopausal endometrial cancer risk. Cancer Epidemiol Biomarkers Prev 13:1515–1520

Sakai W, Swisher EM, Karlan BY, Agarwal MK, Higgins J, Friedman C, Villegas E, Jacquemont C, Farrugia DJ, Couch FJ, Urban N, Taniguchi T (2008) Secondary mutations as a mechanism of cisplatin resistance in BRCA2-mutated cancers. Nature 451:1116–1120

Sakai W, Swisher EM, Jacquemont C, Chandramohan KV, Couch FJ, Langdon SP, Wurz K, Higgins J, Villegas E, Taniguchi T (2009) Functional restoration of BRCA2 protein by secondary BRCA2 mutations in BRCA2-mutated ovarian carcinoma. Cancer Res 69:6381–6386

Saldivar JS, Lu KH, Liang D, Gu J, Huang M, Vlastos AT, Follen M, Wu X (2007) Moving toward individualized therapy based on NER polymorphisms that predict platinum sensitivity in ovarian cancer patients. Gynecol Oncol 107:S223–S229

Sampath D, Greenberger LM, Beyer C, Hari M, Liu H, Baxter M, Yang S, Rios C, Discafani C (2006) Preclinical pharmacologic evaluation of MST-997, an orally active taxane with superior in vitro and in vivo efficacy in paclitaxel- and docetaxel-resistant tumor models. Clin Cancer Res 12:3459–3469

Sandrim VC, de Syllos RW, Lisboa HR, Tres GS, Tanus-Santos JE (2007) Influence of eNOS haplotypes on the plasma nitric oxide products concentrations in hypertensive and type 2 diabetes mellitus patients. Nitric Oxide 16:348–355

Sandrim VC, Palei AC, Sertorio JT, Cavalli RC, Duarte G, Tanus-Santos JE (2010) Effects of eNOS polymorphisms on nitric oxide formation in healthy pregnancy and in pre-eclampsia. Mol Hum Reprod 16:506–510

Sarah R, Esther NM, Shaker A (2010) Mousa Pharmacogenetics in breast cancer: steps toward personalized medicine in breast cancer management. Pharmacogenomics Pers Med 3:129–143

Sasaki M, Tanaka Y, Kaneuchi M, Sakuragi N, Dahiya R (2003) CYP1B1 gene polymorphisms have higher risk for endometrial cancer, and positive correlations with estrogen receptor alpha and estrogen receptor beta expressions. Cancer Res 63:3913–3918

Schouten LJ, Goldbohm RA, van den Brandt PA (2004) Anthropometry, physical activity, and endometrial cancer risk: results from the Netherlands Cohort Study. J Natl Cancer Inst 96:1635–1638

Schroth W, Antoniadou L, Fritz P, Schwab M, Muerdter T, Zanger UM, Simon W, Eichelbaum M, Brauch H (2007) Breast cancer treatment outcome with adjuvant tamoxifen relative to patient CYP2D6 and CYP2C19 genotypes. J Clin Oncol 25:5187–5519

Schroth W, Goetz MP, Hamann U, Fasching PA, Schmidt M, Winter S, Fritz P, Simon W, Suman VJ, Ames MM, Safgren SL, Kuffel MJ, Ulmer HU, Bolander J, Strick R, Beckmann MW, Koelbl H, Weinshilboum RM, Ingle JN, Eichelbaum M, Schwab M, Brauch H (2009) Association between CYP2D6 polymorphisms and outcomes among women with early stage breast cancer treated with tamoxifen. JAMA 302:1429–1436

Scott LJ, Bonnycastle LL, Willer CJ, Sprau AG, Jackson AU, Narisu N, Duren WL, Chines PS, Stringham HM, Erdos MR et al (2006) Association of transcription factor 7-like 2 (TCF7L2) variants with type 2 diabetes in a Finnish sample. Diabetes 55:2649–2653

Seraj MJ, Samant RS, Verderame MF, Welch DR (2000) Functional evidence for a novel human breast carcinoma metastasis suppressor, BRMS1, encoded at chromosome 11q13. Cancer Res 60:2764–2769

Seremak-Mrozikiewicz A, Drews K, Barlik M, Sieroszewski P, Grzeskowiak E, Mrozikiewicz P (2011) The significance of -786T>C polymorphism of endothelial NO synthase (eNOS) gene in severe preeclampsia. J Matern Fetal Neonatal Med 24:432–436

Seshadri R, Firgaira FA, Horsfall DJ, McCaul K, Setlur V, Kitchen P (1993) Clinical significance of HER-2/neu oncogene amplification in primary breast cancer. The South Australian Breast Cancer Study Group. J Clin Oncol 11:1936–1942

Sethi AA, Nordestgaard BG, Tybjaerg-Hansen A (2003) Angiotensinogen gene polymorphism, plasma angiotensinogen, and risk of hypertension and ischemic heart disease: a meta-analysis. Arterioscler Thromb Vasc Biol 23:1269–1275

Shaat N, Lernmark A, Karlsson E, Ivarsson S, Parikh H, Berntorp K, Groop L (2007) A variant in the transcription factor 7-like 2 (TCF7L2) gene is associated with an increased risk of gestational diabetes mellitus. Diabetologia 50:972–979

Shah DM (2006) The role of RAS in the pathogenesis of preeclampsia. Curr Hypertens Rep 8:144–152

Sharma M, Shubert DE, Lewis J, McGarrigle BP, Bofinger DP, Olson JR (2003) Biotransformation of tamoxifen in a human endometrial explant culture model. Chem Biol Interact 146:237–249

Shepard HM, Lewis GD, Sarup JC, Fendly BM, Maneval D, Mordenti J, Figari I, Kotts CE, Palladino MA Jr, Ullrich A et al (1991) Monoclonal antibody therapy of human cancer: taking the HER2 protooncogene to the clinic. J Clin Immunol 11:117–127

Shim JY, Jun JK, Jung BK, Kim SH, Won HS, Lee PR, Kim A (2007) Vascular endothelial growth factor gene +936 C/T polymorphism is associated with preeclampsia in Korean women. Am J Obstet Gynecol 197(271):e271–e274

Simpson ER, Mahendroo MS, Means GD, Kilgore MW, Hinshelwood MM, Graham-Lorence S, Amarneh B, Ito Y, Fisher CR, Michael MD et al (1994) Aromatase cytochrome P450, the enzyme responsible for estrogen biosynthesis. Endocr Rev 15:342–355

Sissung TM, Price DK, Sparreboom A, Figg WD (2006) Pharmacogenetics and regulation of human cytochrome P450 1B1: implications in hormone-mediated tumor metabolism and a novel target for therapeutic intervention. Mol Cancer Res 4:135–150

Slamon DJ, Clark GM, Wong SG, Levin WJ, Ullrich A, McGuire WL (1987) Human breast cancer: correlation of relapse and survival with amplification of the HER-2/neu oncogene. Science 235:177–182

Slamon DJ, Godolphin W, Jones LA, Holt JA, Wong SG, Keith DE, Levin WJ, Stuart SG, Udove J, Ullrich A et al (1989) Studies of the HER-2/neu proto-oncogene in human breast and ovarian cancer. Science 244:707–712

Smith GC, Pell JP, Walsh D (2001) Pregnancy complications and maternal risk of ischaemic heart disease: a retrospective cohort study of 129,290 births. Lancet 357:2002–2006

Soini Y, Alarakkola E, Autio-Harmainen H (1997) Expression of messenger RNAs for metalloproteinases 2 and 9, type IV collagen, and laminin in nonneoplastic and neoplastic endometrium. Hum Pathol 28:220–226

Stearns V, Davidson NE, Flockhart DA (2004) Pharmacogenetics in the treatment of breast cancer. Pharmacogenomics J 4:143–153

Steffensen KD, Waldstrom M, Brandslund I, Jakobsen A (2010) The relationship of VEGF polymorphisms with serum VEGF levels and progression-free survival in patients with epithelial ovarian cancer. Gynecol Oncol 117:109–116

Steinthorsdottir V, Thorleifsson G, Reynisdottir I, Benediktsson R, Jonsdottir T, Walters GB, Styrkarsdottir U, Gretarsdottir S, Emilsson V, Ghosh S, Baker A, Snorradottir S, Bjarnason H, Ng MC, Hansen T, Bagger Y, Wilensky RL, Reilly MP, Adeyemo A, Chen Y, Zhou J, Gudnason V, Chen G, Huang H, Lashley K, Doumatey A, So WY, Ma RC, Andersen G, Borch-Johnsen K, Jorgensen T, van Vliet-Ostaptchouk JV, Hofker MH, Wijmenga C, Christiansen C, Rader DJ, Rotimi C, Gurney M, Chan JC, Pedersen O, Sigurdsson G, Gulcher JR, Thorsteinsdottir U, Kong A, Stefansson K (2007) A variant in CDKAL1 influences insulin response and risk of type 2 diabetes. Nat Genet 39:770–775

Stephens M, Smith NJ, Donnelly P (2001) A new statistical method for haplotype reconstruction from population data. Am J Hum Genet 68:978–989

Sugimoto M, Yoshida S, Kennedy S, Deguchi M, Ohara N, Maruo T (2006) Matrix metalloproteinase-1 and -9 promoter polymorphisms and endometrial carcinoma risk in a Japanese population. J Soc Gynecol Investig 13:523–529

Suspitsin EN, Grigoriev MY, Togo AV, Kuligina ES, Belogubova EV, Pozharisski KM, Chagunava OL, Sokolov EP, Theillet C, Berstein LM, Hanson KP, Imyanitov EN (2002) Distinct prevalence of the CYP19 Delta3(TTTA)(7) allele in premenopausal versus postmenopausal breast cancer patients, but not in control individuals. Eur J Cancer 38:1911–1916

Takemura M, Azuma C, Kimura T, Tokugawa Y, Miki M, Ono M, Saji F, Tanizawa O (1992) Malignant cell-specific gelatinase activity in human endometrial carcinoma. Cancer 70:147–151

Tanabe M, Ieiri I, Nagata N, Inoue K, Ito S, Kanamori Y, Takahashi M, Kurata Y, Kigawa J, Higuchi S, Terakawa N, Otsubo K (2001) Expression of P-glycoprotein in human placenta: relation to genetic polymorphism of the multidrug resistance (MDR)-1 gene. J Pharmacol Exp Ther 297:1137–1143

Tanaka Y, Sasaki M, Kaneuchi M, Shiina H, Igawa M, Dahiya R (2002) Polymorphisms of the CYP1B1 gene have higher risk for prostate cancer. Biochem Biophys Res Commun 296:820–826

Tong D, Fabjani G, Heinze G, Obermair A, Leodolter S, Zeillinger R (2001) Analysis of the human progesterone receptor gene polymorphism progins in Austrian ovarian carcinoma patients. Int J Cancer 95:394–397

Tong Y, Lin Y, Zhang Y, Yang J, Liu H, Zhang B (2009) Association between TCF7L2 gene polymorphisms and susceptibility to type 2 diabetes mellitus: a large Human Genome Epidemiology (HuGE) review and meta-analysis. BMC Med Genet 10:15

Tran A, Jullien V, Alexandre J, Rey E, Rabillon F, Girre V, Dieras V, Pons G, Goldwasser F, Treluyer JM (2006) Pharmacokinetics and toxicity of docetaxel: role of CYP3A, MDR1, and GST polymorphisms. Clin Pharmacol Ther 79:570–580

Tunuguntla R, Ripley D, Sang QX, Chegini N (2003) Expression of matrix metalloproteinase-26 and tissue inhibitors of metalloproteinases TIMP-3 and -4 in benign endometrium and endometrial cancer. Gynecol Oncol 89:453–459

Ubeda M, Rukstalis JM, Habener JF (2006) Inhibition of cyclin-dependent kinase 5 activity protects pancreatic beta cells from glucotoxicity. J Biol Chem 281:28858–28864

Ueda M, Hung YC, Terai Y, Kanda K, Takehara M, Yamashita H, Yamaguchi H, Akise D, Yasuda M, Nishiyama K, Ueki M (2003) Glutathione S-transferase GSTM1, GSTT1 and p53 codon 72 polymorphisms in human tumor cells. Hum Cell 16:241–251

Ueland PM, Hustad S, Schneede J, Refsum H, Vollset SE (2001) Biological and clinical implications of the MTHFR C677T polymorphism. Trends Pharmacol Sci 22:195–201

Ueno H, Yamashita K, Azumano I, Inoue M, Okada Y (1999) Enhanced production and activation of matrix metalloproteinase-7 (matrilysin) in human endometrial carcinomas. Int J Cancer 84:470–477

Valabrega G, Montemurro F, Aglietta M (2007) Trastuzumab: mechanism of action, resistance and future perspectives in HER2-overexpressing breast cancer. Ann Oncol 18:977–984

Vella N, Aiello M, Russo AE, Scalisi A, Spandidos DA, Toffoli G, Sorio R, Libra M, Stivala F (2004) 'Genetic profiling' and ovarian cancer therapy. Mol Med Rep 4:771–777

Velloso EP, Vieira R, Cabral AC, Kalapothakis E, Santos RA (2007) Reduced plasma levels of angiotensin-(1-7) and renin activity in preeclamptic patients are associated with the angiotensin I- converting enzyme deletion/deletion genotype. Braz J Med Biol Res 40:583–590

Vitoratos N, Hassiakos D, Iavazzo C (2012) Molecular mechanisms of preeclampsia. J Pregnancy 2012: 298343

Vogel CL, Cobleigh MA, Tripathy D, Gutheil JC, Harris LN, Fehrenbacher L, Slamon DJ, Murphy M, Novotny WF, Burchmore M, Shak S, Stewart SJ, Press M (2002) Efficacy and safety of trastuzumab as a single agent in first-line treatment of HER2-overexpressing metastatic breast cancer. J Clin Oncol 20:719–726

Wang Y, Nie M, Li W, Ping F, Hu Y, Ma L, Gao J, Liu J (2011) Association of six single nucleotide polymorphisms with gestational diabetes mellitus in a Chinese population. PLoS One 6:e26953

Watanabe RM, Allayee H, Xiang AH, Trigo E, Hartiala J, Lawrence JM, Buchanan TA (2007) Transcription factor 7-like 2 (TCF7L2) is associated with gestational diabetes mellitus and interacts with adiposity to alter insulin secretion in Mexican Americans. Diabetes 56:1481–1485

Widschwendter M, Siegmund KD, Muller HM, Fiegl H, Marth C, Muller-Holzner E, Jones PA, Laird PW (2004) Association of breast cancer DNA methylation profiles with hormone receptor status and response to tamoxifen. Cancer Res 64:3807–3813

Williams-Jones B, Corrigan OP (2003) Rhetoric and hype: where's the 'ethics' in pharmacogenomics? Am J Pharmacogenomics 3:375–383

Xie F, Hu Y, Speert DP, Turvey SE, Peng G, Money DM, Magee LA, von Dadelszen P (2010) Toll-like receptor gene polymorphisms and preeclampsia risk: a case-control study and data synthesis. Hypertens Pregnancy 29:390–398

Xu X, Duncan AM, Merz-Demlow BE, Phipps WR, Kurzer MS (1999) Menstrual cycle effects on urinary estrogen metabolites. J Clin Endocrinol Metab 84:3914–3918

Xu WH, Long JR, Zheng W, Ruan ZX, Cai Q, Cheng JR, Xiang YB, Shu XO (2009) Association of the progesterone receptor gene with endometrial cancer risk in a Chinese population. Cancer 115:2693–2700

Yamazaki H, Shaw PM, Guengerich FP, Shimada T (1998) Roles of cytochromes P450 1A2 and 3A4 in the oxidation of estradiol and estrone in human liver microsomes. Chem Res Toxicol 11:659–665

Yan L, Beckman RA (2005) Pharmacogenetics and pharmacogenomics in oncology therapeutic antibody development. Biotechniques 39:565–568

Yan L, Shu-Ying Y, Shan K, Yip BH, Rong-Miao Z, Na W, Hai-Yan S (2012) Association between polymorphisms of ERCC1 and survival in epithelial ovarian cancer patients with chemotherapy. Pharmacogenomics 13:419–427

Yang X, Dong Y, Zhao J, Sun H, Deng Y, Fan J, Yan Q (2007) Increased expression of human macrophage metalloelastase (MMP-12) is associated with the invasion of endometrial adenocarcinoma. Pathol Res Pract 203:499–505

Yoneda K, Tanji Y, Ikeda N, Miyoshi Y, Taguchi T, Tamaki Y, Noguchi S (2002) Influence of adjuvant tamoxifen treatment on bone mineral density and bone turnover markers in postmenopausal breast cancer patients in Japan. Cancer Lett 186:223–230

Yoshimura M, Yasue H, Nakayama M, Shimasaki Y, Sumida H, Sugiyama S, Kugiyama K, Ogawa H, Ogawa Y, Saito Y, Miyamoto Y, Nakao K (1998) A missense Glu298Asp variant in the endothelial nitric oxide synthase gene is associated with coronary spasm in the Japanese. Hum Genet 103:65–69

Yu JJ, Lee KB, Mu C, Li Q, Abernathy TV, Bostick-Bruton F, Reed E (2000) Comparison of two human ovarian carcinoma cell lines (A2780/CP70 and MCAS) that are equally resistant to platinum, but differ at codon 118 of the ERCC1 gene. Int J Oncol 16:555–560

Zhao L, Dewan AT, Bracken MB (2012) Association of maternal AGTR1 polymorphisms and preeclampsia: a systematic review and meta-analysis. J Matern Fetal Neonatal Med 25:2676–2680

zur Hausen H (2000) Papillomaviruses causing cancer: evasion from host-cell control in early events in carcinogenesis. J Natl Cancer Inst 92:690–698

Pharmacogenomics and Personalized Medicine in Infertility

31

Subeer S. Majumdar, Indrashis Bhattacharya, and Meraj Khan

Abstract

Around 60–80 million couples are infertile around the globe, and in a large proportion of western world, constant decline in fertility is a major cause of concern. Although infertility per se may not threaten physical health, it is a source of distress, as societal norms tend to equate infertility with failure. Given the complexity of the process of gametogenesis and the large number of genes involved, it is likely that a significant proportion of infertility phenotypes are genetic in origin. Pharmacogenomics has aggressively been studied and addressed for life-threatening illnesses like cancer, cardiovascular disorders, depression, HIV, tuberculosis, asthma, and diabetes. However, this field has not advanced sufficiently to address issues related to personalized treatment for infertility, mainly because infertility fails to fit into the definition of a major disease and is lower in order of priority than other ailments. Here, we will discuss the present status of knowledge which may be a starting point for expansion into pharmacogenomics of male and female infertility. Polymorphisms in several genes are studied in relation to controlled ovarian hyperstimulation (COH) outcome in infertile women undergoing IVF treatment in clinics, and therapies based on genetic background has just begun to emerge. Except this, no major advancement has yet occurred in this field, but a huge scope of pharmacogenomics and personalized medicine for treatment of infertility exists because several genes and their SNPs have already been proven to be associated with infertility, and they may provide strong basis for personalized therapies.

S.S. Majumdar (✉)
Division of Cellular Endocrinology,
National Institute of Immunology, Delhi, India
e-mail: subeer@nii.ac.in

I. Bhattacharya
Department of Zoology, Hindu College,
University of Delhi, Delhi, India

M. Khan
Department of Physiology, AIIMS, Delhi, India

1 Overview

Infertility is a condition of reproductive system which affects a couple with almost equal frequency. Infertility is defined as an inability to conceive a child by a couple in a stable relationship during the year of regular intercourse without the use of

D. Barh et al. (eds.), *Omics for Personalized Medicine*,
DOI 10.1007/978-81-322-1184-6_31, © Springer India 2013

contraceptive methods. Around 60–80 million couples are infertile around the globe, and it contributes a great proportion to the overall reproductive disability (Baird et al. 2002; Altmäe et al. 2011). Although infertility per se may not threaten physical health, it is a source of distress, as societal norms and perceived religious dictums may equate infertility with failure. About 61 countries are facing fertility-related problems at or below the replacement level (United Nations Population Fund. Report of the International Conference on State of the World Population. United Nations Annual Conference 1999. UNFPA: state of world population, 1999). Although constant effort to identify the causes of infertility in both the sexes is being made, about half of the cases are classified as idiopathic in nature, and normalcy cannot be restored after hormonal supplementation (Matzuk and Lamb 2002, 2008). Given the complexity of the process of gametogenesis and the large number of genes involved, it is likely that a significant proportion of infertility phenotypes are genetic in origin.

1.1 Pharmacogenetics to Pharmacogenomics to Personalized Medicine

The development of genomic tools has led to a number of important discoveries in reproductive physiology and the etiology of infertility. Knockout mouse models have also proven to be a crucial tool to understand the functional significance of changes in gene expression (Matzuk and Lamb 2002, 2008; Hwang et al. 2011). With the use of new deep sequencing methods, the discovery of small RNAs, and the increased sensitivity of molecular methods, it is expected that mechanisms regulating the function of gonad and reproductive tract will become evident. Defining the etiology and progression of these abnormalities and pathologies is important for the identification of novel and more reliable therapeutics. Moreover, the development of new molecular tools in the fields of genomics, proteomics, and pharmacogenomics is providing new windows on personalized physiology and their impact on our knowledge of molecular orchestration to better understand the infertility. The biological system is not essentially linear; there are thousands of genes and proteins interacting in networks with time and space contexts along with redundancy, thus resulting in complex emerging properties. In the present time, there are available tools and techniques of generating high-throughput genomics and proteomics data. Such data mining and their analyses towards pattern discovery and prediction provide a promise for improvising novel approach to address complex multifactorial health condition like infertility. It is known that biological functions and transcriptomics are linked by translation of proteins and their complex interaction. Therefore high-throughput proteomics approach may compliment in the process of molecular classification. With the availability of interactome bank and powerful computational and mathematical modeling tools, it is now possible to predict the function of newly discovered proteins also (Chengalvala et al. 2006). Preemptive medicine is an emerging concept of predicting, anticipating, and preventing diseases and disease sequels before the irreversible tissue damage takes place. Analysis of data from genomic and proteomic profiles, as discussed above, may help in such predictions and may provide a lead time for designing preventive strategies in infertility and associated disorders. If a gene variant is associated with a particular drug response in a patient, there is the chance for making clinical decisions based on genetics by adjusting the dosage or selecting a specific drug. Modern approaches include multigene analysis or whole-genome *single-nucleotide polymorphism* (*SNP*) profiles, and these approaches are just coming into clinical use for drug discovery and development.

Therapeutic and preventive approaches in large populations can be markedly improved by a personalized evaluation of their genetic background. Many new methods are available now, and others are steadily improving, that can be used to rapidly genotype large numbers of genetic variants. These methods will lead to the creation of a complete SNP database that will prove to be a crucial tool for defining the genetic mechanisms underlying a given disease. This will lead to identify loci that influence the individual patient's response to therapy and, consequently, new pharmacological targets for the development of novel drugs (Roberto et al. 2004). Pharmacogenomics has aggressively been studied and addressed for life-threatening illnesses like cancer, cardiovascular disorders, depression, HIV, tuberculosis, asthma, and diabetes. However, this field has not advanced sufficiently to address issues related to personalized treatment for infertility, mainly because infertility fails to fit into the definition of a disease and is lower in order of priority than other ailments. Here, we will discuss the present status of knowledge which may be a starting point for expansion into pharmacogenomics of male and female infertility.

2 Male Infertility

About 15 % of couple globally are childless, where 50 % contribution stems from the "male factor" (Matzuk and Lamb 2002, 2008; Hwang et al. 2011). With the complexity of the spermatogenic process where a large number of genes are involved, there is a high likelihood that a significant proportion of male infertility phenotypes are of genetic origin (Matzuk and Lamb 2002, 2008; Hwang et al. 2011). However, till date the majority of all the clinical cases remain idiopathic in nature (Nuti and Krausz 2008), and no drug targets are yet been developed. Figure 31.1 describes the prevalence of male infertility.

2.1 Amenable Targets

Target amenable for use in pharmacogenomics of male infertility are discussed in this section. In case of male infertility, the numerical and structural chromosomal abnormalities interfere with normal meiosis and can cause spermatogenic failure, for example, the Klinefelter's syndrome (*47, XXY*) is the most commonly found abnormality. Structural abnormalities that are often encountered include reciprocal translocations, Robertsonian translocations, and pericentric inversions. Secondly, five microdeletions in the long arm of the Y-chromosome, e.g., *AZFa*, *P5/proximal P1* (*AZFb*), *P5/distal P1*, *b2/b4* (*AZFc*), and *gr/gr* deletions, are reported to be associated with spermatogenic failure (Matzuk and Lamb 2002, 2008; Hwang et al. 2011). The AZFc locus contains *DAZ1–4*, paralogs to the autosomal *DAZL*. Interestingly, 4 novel missense mutations in *DAZL* with one homozygous *DAZL*-null male patient showed infertility. It is noteworthy that 71.4 % of patients with an AZFc deletion had sperm in their testes and their partners achieved pregnancy via assisted reproductive techniques. Interestingly, new AZFc haplotypes named "gr/gr" or b2/b3, which both result in loss of *DAZ3* and *DAZ4*, have been identified to be linked with infertility but also been found in normozoospermic individuals (Matzuk and Lamb 2002, 2008; Hwang et al. 2011). The link between autosomal *DAZL* function and severity of male infertility with AZFc deletion in the population is not properly

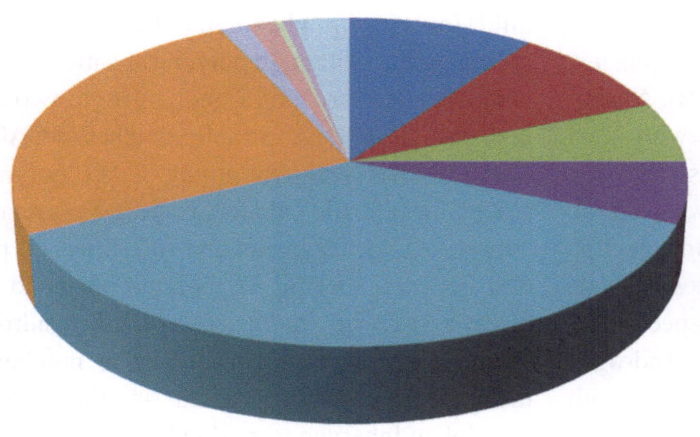

Fig. 31.1 Prevalence of male infertility. *UT* untreatable, *T* treatable

Table 31.1 Genetic and endocrine basis of male infertility

Defect	Phenotype	Frequency (%)
Chromosomal aberrations	Azoospermia to normospermia	2–10
Klinefelter's syndrome 47, XXY	Azoospermia to severe oligospermia	5–10
Other sex chromosome alterations	Azoospermia to normospermia	0.1–0.2
Robertsonian translocations	Azoospermia to severe oligospermia	0.5–1
Reciprocal translocation	do	do
Y chromosomal microdeletions	do	5–10
AZFa	Azoospermia to Sertoli cell syndrome	0.5–1
AZFb	Azoospermia to spermatogenic arrest	do
AZFc	Azoospermia to severe oligospermia	3–7
AZFb+c	Sertoli cell syndrome/ spermatogenic arrest	0.5–1
Partial AZFc deletion	Azoospermia to normospermia	3–5
Gene mutation		
CFTR	Obstructive azoospermia	60–70
AR	Azoospermia to oligospermia	2–3
INSL3–LGR8	Cryptorchidism	4–5
Hypogonadotropic hypogonadism		
Congenital: Kallmann syndrome	Azoospermia	0.5
Acquired: tumor, infection, autoimmune, pituitary infarction, and drug use	Azoospermia to oligospermia	0.25
Hypergonadotropic hypogonadism		
Classic: Klinefelter's syndrome	Azoospermia to severe oligospermia	5–10
Defective androgen synthesis response		
5-α-Reductase deficiency	Oligospermia	0.25
Complete androgen insensitivity	Azoospermia	
Partial androgen resistance	Oligozoospermia	

Azoospermia = complete absence of sperm; aspermia = ejaculation does not emit any semen; oligozoospermia = <10 million sperm/ml of semen; asthenozoospermia = >40 % of sperm have low motility; teratozoospermia = >40 % of sperm with abnormal morphology; necrospermia = nonviable/dead sperm; oligoasthenozoospermia = motile density <8 million sperm/ml; hematospermia = red blood cells present in semen; pyospermia = white blood cells present in semen; polyzoospermia = excessively high sperm concentration

established yet. Table 31.1 summarizes the genetic basis of male infertility along with the frequency of such disorder, globally.

It is important to note that infertility that is not related to chromosomal alterations may occur due to single-gene mutations or polymorphisms, single-nucleotide polymorphisms (SNPs). A complete list of human genes associated with male infertility with known SNPs is described by (Matzuk and Lamb 2002, 2008; Hwang et al. 2011) and others (see reviews in Tüttelmann et al. 2007; Nuti and Krausz 2008), and these can be used to find appropriate compounds specifically acting to overcome respective defects. Androgen receptor is estimated to be associated with infertility in 2–3 % of all cases of azoospermia/oligozoospermia. The mutated *AR* gene results into *mild to complete* androgen insensitivity. In men, complete androgen insensitivity observed in 46,XY patients is presented as phenotypic females with undeveloped testes, whereas in the mild forms with phenotypic males show low sperm counts. Although polymorphisms that have been examined in the *AR* gene relate to two sites in exon 1 that exhibit variable stretches of CAG or GGC repeats and examining CAG polymorphisms in infertile subjects reveals a correlation between CAG expansion and infertility, no relation between specific *AR* polymorphism to this disease severity (*mild to complete* androgen insensitivity) is reported. Finally, a number of monogenic disorders, listed in the "Online Mendelian Inheritance in Man" database, are also associated with spermatogenic failure, e.g., the classic *Kallmann syndrome* and *Noonan syndrome*, where spermatogenic failure results from *hypogonadotropic hypogonadism*, *cryptorchidism*, *delayed puberty* or *male pseudohermaphroditism*, and *ambiguous genitalia* (Hardelin and Dode 2008). In cystic fibrosis (CF), spermatogenesis is unaffected, but congenital bilateral aplasia of the vas deferens (CBAVD) causes ejaculatory azoospermia. Some genetic disorders may have marked phenotypic variation, e.g., in Yq microdeletions the testicular phenotype varies from Sertoli cell-only syndrome to germ cell arrest or hypospermatogenesis. Hence, treatments for them may also require different pharmacological compounds suiting to the need of the individual.

2.2 Diagnosis and Therapy

The common diagnosis for male infertility is routine semen analysis where sperm concentration, motility, morphology, and the presence of other cells (spermatogenic and white blood cells) are assessed along with the indicators of patency and function of the genital tract (volume, liquefaction time, pH, and fructose presence or absence). This analysis, however, is not a test of fertility potential and indicates infertility only in the case of azoospermia (absence of sperm in the ejaculate: nonobstructive and obstructive). The majority of endocrine disorders are relatively simple to diagnose by serum hormonal profile (FSH, LH, and testosterone). Although major endocrine cause behind male infertility is either *congenital* or *acquired hypogonadotropic hypogonadism* which can be restored during puberty or adulthood by hormones (http://www.endotext.org/male/male5/maleframe5.htm, and Salenave et al. 2012), it is important to note that in patients suffering from such male infertility, hormonal therapy sometimes fails to initiate spermatogenesis (Schaison et al. 1993; Nieschlag and Leifke 1997). However, due to lack of sufficient knowledge regarding the biology of the individual cell types present inside the seminiferous epithelium and regarding genes expressed by these cells affecting pathophysiology of the onset and progression of infertility, the idiopathic nature of certain forms of male infertility prevails.

2.3 Approaches to Understand Genes Regulating Spermatogenesis

There are major three independent approaches budding to understand the molecular basis of the pathophysiology of male infertility to develop potential therapeutics for treatment.

The functional genomics approach, where somatic and germinal cell-specific genes are being identified, which may have a probable role in the regulation of spermatogenesis. Till date, a number of genes related to male fertility have been identified with the use of targeted gene disruptions in mouse models, in which over 400 mutant mouse lines displaying abnormal reproductive function have been developed (for details see Matzuk and Lamb 2002, 2008; Hwang et al. 2011). Microarray analysis of transcripts expressed in the testes of neonatal to adult mice identified 351 genes expressed exclusively in the male germ line and 1,652 genes upregulated at the time of or after the onset of meiosis. Overexpression models by transgenesis or loss of expression either by knockout or knockdown mice models can reveal the function of these genes further (for details see Majumdar et al. 2009). MicroRNA (miRNA)-mediated posttranscriptional gene regulation has emerged as a fundamental mechanism controlling normal tissue development and function. Using a mouse mutant with a hypomorphic Dicer1 allele, Otsuka and group (2008) demonstrated that Dicer1 deficiency resulted into infertility in females. MiRNA has also been shown to be involved in the regulation of male infertility (Papaioannou and Nef 2010). Our current knowledge on miRNA regulation and posttranslational modification in each component of the healthy reproductive systems and associated infertility is limited and superficial. However, a very recent study by Ogorevc et al. (2011) has listed SNPs of miRNA target sites in some of the genes associated with male fertility, and this miRNA-related database may have future relevance for the development of novel male infertility-associated genetic markers.

The second line of research includes *transcriptome profiling* of the genes differentially expressed using testicular biopsy samples from infertile and fertile human males. The gene list made from the data of high-throughput profiling facilitates the identification of candidate genes involved in the pathophysiology of infertility and is the first step for practicing the pharmacogenetics of male infertility. Table 31.2 summarizes some of the pioneering studies carried out in this area in various populations across the globe.

In the final approach, the contribution of genetic polymorphisms to male infertility and, more importantly, as markers for a pharmacogenomic method to the treatment of infertile males (for details see reviews in Matzuk and Lamb 2008). Gene-targeted sequencing studies have reported a number of SNPs significantly associated with male infertility. For example, KIT tyrosine kinase receptor and its ligand KITLG are involved in the survival and proliferation of germ cells. Animal models and functional studies in humans suggest that this signalling pathway plays a role in male infertility (Matzuk and Lamb 2008). In a study, total of 167 idiopathic infertile men (sperm counts <5 million sperm/ml) and 465 unrelated healthy controls from the same geographical region were genotyped for SNPs in *KIT* gene (Galan et al. 2006). They found statistically significant association of the rs3819392 polymorphism, which is located within the *KIT* gene, with idiopathic male infertility. In addition, a deviation from the Hardy–Weinberg equilibrium (HWE) law was observed for rs10506957 polymorphism within the *KITLG* gene only in the infertile group indicating that the KIT/KITLG system may be involved in a low sperm count trait in humans. Similarly, SNPs of other genes associated with male infertility have been identified and replicated by multiple studies including aryl hydrocarbon receptor repressor (*AHRR*; Watanabe et al. 2004; Merisalu et al. 2007) and methylenetetrahydrofolate reductase (*MTHFR*; Park et al. 2005; Lee et al. 2006). Table 31.3 summarizes some other pioneering studies carried out in this area.

3 Female Infertility

Out of 15 % of couples globally who are childless, 50 % contribution stems from the "female factor" (http://www.americanpregnancy.org/infertility/femaleinfertility.html). With the complexity of the process of oogenesis where multiple genes are involved (Matzuk and Lamb 2002, 2008; Hwang et al. 2011), it is difficult to figure out the relative contribution of a single gene in this process. Figure 31.2 describes the types and prevalence of female infertility.

3.1 Ovarian Failure

Ovulatory disorders are one of the most common reasons why women are unable to conceive and account for about 30 % of women's infertility. Premature ovarian failure (POF) is a disorder affecting the ability of a woman's ovaries to function correctly. It is also known as primary ovarian insufficiency and is characterized by absence of menstrual bleeding, low estrogen levels, and possible onset of autoimmune diseases in women younger than 40 years. A proportion of women with spontaneous POF have mutations in the FMR1 gene (Verkerk et al. 1991). This X-linked disorder is caused by the absence of a protein coded by the fragile X mental retardation 1 gene (*FMR1*). The genetic determinants of *FMR1* are complex and are related to the length of a polymorphic expansion of CGG trinucleotides in the 5′ untranslated region of the gene. (This is the gene mutation responsible for fragile X syndrome – the most common cause of hereditary mental retardation). The risk of a woman having this mutation is higher if she has a family history of premature ovarian failure. Around 24 % of women with permutation in this gene show POF (Shwartz et al. 1994; Wittenberger et al. 2007).

3.1.1 SNPs Associated with Female Infertility with Respect to Ovulation

Unlike the males, a large number of studies have been done in the recent past to indentify the association of SNPs in controlling female fertility. For example, polymorphisms in genes including FSH-R, LH-R, LH-β, Cyp11A1, Cyp19A1, ESR1, ESR2, PGR, AMH/AMHR2, GDF9, BM15, SOD2, SHBG, studied in relation to controlled ovarian hyperstimulation (COH) outcome in infertile women undergoing IVF treatment, are

Table 31.2 Differential transcriptomics in testicular biopsies obtained from infertile subjects vs. fertile controls in different human population across the globe

Year and country	Experiment	Outcome	References
2004, China	Microarray analysis of testicular biopsy of two azoospermic males Vs healthy controls	56 up regulated and 72 down regulated genes identified in azoospermic males. Rapl A expression was significantly elevated in azoospermic males	Yang et al. (2004)
2006, Taiwan	Microarray analysis of testicular biopsy of 15 infertile males with maturational arrest and Sertoli cell only syndrome vs. 9 healthy controls	300 genes down regulated in infertile patients. 10 novel sterility related genes were indentified	Lin et al. (2006)
2007, UK	Microarray analysis of testicular biopsy of 33 infertile males with different spermatogenic phenotype vs. healthy controls	Multiple germ cell specific genes involved in elevated inflammatory response were observed in infertile testes	Ellis et al. (2007)
2007, Germany	Molecular signature of different developmental stages of 28 samples with homogeneous testicular pathology out of 289 azoospermic subjects	348 premeiotic, 81 postmeiotic and 38 terminal differentiation genes were identified	Feig et al. (2007)
2007, France	Global expression pattern of testicular biopsies of 69 infertile males with different Sertoli cell only phenotype vs. healthy controls	188 genes (like steroid metabolism, innate immunity, focal adhesion and MAPK pathway) are found to be upregulated in the infertile subjects	Spiess et al. (2007)
2008, UK	Microarray of testicular tissue of normal male vs. GnRH antagonist treated males	Down regulation of genes involved in steroidogenesis, cholesterol metabolism, Leydig cell specific INSL3, junctional protein of Sertoli-germ cells	Bayne et al. (2008)
2008, Japan	Genome wide transcription analysis of testicular tissue obtained from 47 non-obstructive azoospermic, 11 obstructive azoospermic subjects	Five different SNPs of ADP-ribosyltransferase identified, individual having common haplotype having more serum androgen probable protective for spermatogenesis	Okada et al. (2008)
2009, China	Differential expression of miRNA in non-obstructive azoospermia vs. normal control	154 down regulated (mir-17-19 clusters, mir-371,2,3 clusters) and 19 upregulated (miR-302a, miR-491-3p, miR-520 d-3p, miR-383) in azoospermic males	Lian et al. (2009)
2010, Italy	Differential transcriptome of testicular tissue obtained from 16 idiopathic infertile male patients with AZFc microdeletions vs. normal controls	Lack of DAZ, AZFc gene expression in Yq deletion essential for fertility	Gatta et al. (2010)
2010, Germany	Differential transcriptome of testicular tissue obtained from 26 idiopathic infertile male patients with AZFc or AZPb+c deletions vs. normal controls	Differential gene expression in hypospermatogenesis, meiotic arrest, Sertoli cell only syndrome	Cappallo-Obermann et al. (2010)
2011, Italy	Differential transcriptome of testicular tissue obtained from infertile male patients with AZFc deletions vs. normal controls	331 genes related to spermatogenesis were down regulated in the subjects; lack of DAZ gene expression was evident in subjects	Ferlin A. (2011)
2011, Chile	Expression pattern of DAX-1 (orphan nuclear receptor known for transcriptional repression in gonadal development) and its alternate splice variant DAX-1A in 65 subjects with primary testicular failure normalized against 33 healthy control	Protein and mRNA expression of both the isoforms are detected in Sertoli cells and in germ cells (spermatogonia to spermatids), thereby alternate splicing of DAX-1 not involved in primary testicular failure	Lardone et al. (2011)

(continued)

Table 31.2 (continued)

Year and country	Experiment	Outcome	References
2012, Portugal	Differential transcriptome of testicular tissue obtained from 2 infertile male patients with varicocele vs. normal controls	Up regulation of CCIN and PRM2 and down regulation of MTIM, PHLDA1, INSL3	Oliveira et al. (2012)
2012, France	Differential transcriptome of sperm obtained from 8 infertile male patients vs. 3 normal controls	33-fold down regulation of genes involved in sperm motility, DNA repair and oxidative stress regulation	Montjean et al. (2012)
2012, Switzerland	Global transcriptome profiling and protein network analysis of testicular tissue obtained from 47 prepubertal boys with undescended testes, vs. infertile adults patients with spermatogenic arrest vs. fertile controls	Total 3,580 differentially expressed genes reported in somatic and germ cells, out of which 933 were undetectable in 45 embryonic and adult non-testicular tissue	Chalmel et al. (2012)
2012, China	Presence of two selected miRNAs (miR-19b and let-7a) in seminal plasma from fertile men and idiopathic infertile patients with oligozoospermia and non-obstructive azoospermia (NOA) and expression patterns of two selected miRNAs in 96 idiopathic infertile males (48 oligozoospermia and 48 NOA) and 48 fertile controls	Expression of miR-19b and let-7a in the seminal plasma are reproducible and stable. Aberrant over-expression levels of miR-19b and let-7a may be an indicator of spermatogenic failure	Wu et al. (2012)

Table 31.3 Association of SNPs with male infertility in different human population across the globe

Year and country	Experiment	Outcome	References
2007, Germany	Genetic polymorphism and patho-physiology of infertility	Association of polymorphism in deletion of AZF gr/gr and MTHER, but not for POLG, DAZL, USP26, FSHR, and CAG repeat length in AR, PRM1/2, TNP1/2, ER1, TAF7L remains controversial	Tüttelmann et al. (2007)
2008, Germany	Nonsynonymous polymorphisms of LHCGR genes with 278 maldescended testes, 277 infertile subjects without maldescended testes and 271 healthy controls	No particular LHCGR genotype linked with LH resistance, S312N polymorphism in exon 10 of LHCGR correlated with spermatogenic damage rather than maldescensus	Simoni et al. (2008)
2008, Italy	Genetic polymorphism and patho-physiology of infertility	Lack of association of polymorphism of POLG, DAZL, USP26, FSHR with abnormal spermatogenesis	Nuti and Krausz (2008)
2009, Utah, USA	Genome wide SNP association study targeting male infertility	Several potential SNPs identified associated with infertility	Aston and Carrell (2009)
2009, Turkey	Relation between genetic polymorphism of xenobiotic metabolizing enzymes (glutathione S-transferase [GST M1] and CYP1A1*2C) and infertility in 110 infertile subjects vs. 105 controls	CYPIAI (val/val) or CYP1A1 (Ile/val) in association with GSTM1 null genotype has 7 times higher risk to be infertile rather than CYP1A1 (Ile/Ile) in association with GSTM1 wild type genotype	Aydos et al. (2009)
2009, China	Relation between genetic polymorphism of xenobiotic metabolizing enzymes (glutathione S-transferase T1 [GST T1] gene polymorphism) on 63 infertile subjects with varicocele vs. 54 controls	GSTT1 null genotype predisposes to over oxidative damage of sperm in infertile subjects	Wu et al. (2009)
2010, Iran	Association between functionally important polymorphism in xenobiotic metabolizing enzymes (glutathione S-transferase [GST] genes) on 166 infertile subjects with idiopathic oligoasthenoteratozoospermia	GSTM1 or GSTT1 null genotypes for developing infertility but non deleted GSTM1 or GSTT1 genotypes emerged as protective factors against infertility	Safarinejad et al. (2010)

(continued)

Table 31.3 (continued)

Year and country	Experiment	Outcome	References
2011, India	Role of AZFc haplotypes gr/gr or b2/b3 and partial deletion of AZFc (loss of DAZ3, DAZ4) in 236 azoospermic, 182 oligozoospermic and 240 healthy normozoospermic males	18 (gr/gr), 11 (b1/b3), 2 (b2/b3) in azoospermic subjects; 12 (gr/gr), 5 (b1/b3), 4 (b2/b3) in oligozoospermic; 7 (gr/gr) in normozoospermic males. Association between occurrence of subdeletion and male infertility with spermatogenic failure	Shahid et al. (2011)
2012, Republic of Macedonia	Association of 9 SNPs in 8 different genes (FASLG, JMJDIA, LOC203413, TEX15, BRDT, OR2W3, INSR and TAS2R38) with 136 idiopathic subjects (60 azoospermic; 76 oligozoospermic and 161 control)	SNPs rs5911500 (Albanian population) in LOC203413, rs3088232 in BRDT (Macedonian population), rs11204546 in OR2W3 associated with abnormal spermatogenesis	Plaseski et al. (2012)
2012, USA	Genome wide association study (250,000 autosomal SNPs) of two fertility traits (family size and birth rate) in 269 married men of European origin	9 SNPs were associated with reduced fertility USP8 (deubiquitinating enzyme important in acrosome assembly), UBD and EPSTI1 (innate immunity) LRRC32 (codes for TGF-β-receptor on T_{reg})	Kosova et al. (2012)
2010, Portugal	DNA methylation pattern of the promoter CpG island of two germ line regulator gene DAZL and DAZ in sperm of oligoasthenoteratozoospermia (OAT) and normozoospermic (NZ) subjects	OAT subjects displayed increased methylation defects in DAZL promoter CpG island compared to NZ	Navarro-Costa et al. (2010)

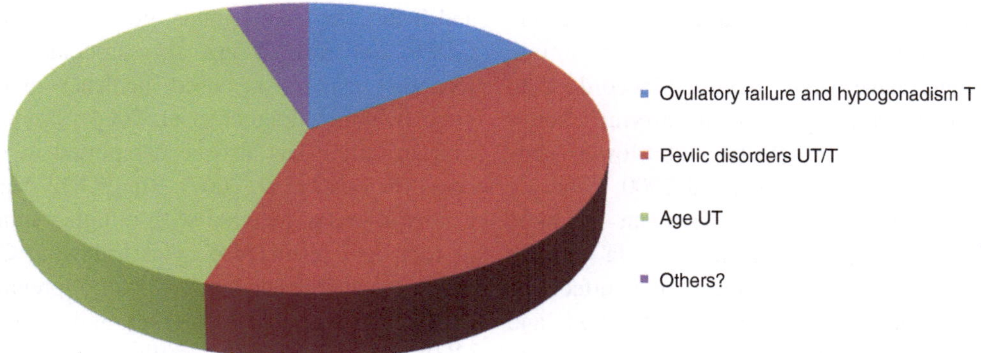

Fig. 31.2 Prevalence of female infertility. *UT* untreatable, *T* treatable

methodically studied by several investigators and compiled elegantly by Altmäe et al. (2011). Following excerpt on various genes is taken from this review. There is sufficient evidence to state that genetic variation in FSH-R has a role in COH outcome. In short, the FSH-R 680Ser variant was originally associated with elevated basal levels of FSH (a key marker of ovarian reserve and the best-known predictor of COH response) and elevated gonadotrophin requirements during COH (Perez Mayorga et al. 2000). Important proof of principle studies have been performed on the p.N680S polymorphism of the FSH-receptor gene. A large body of evidence from a study in healthy volunteers and from clinical studies suggests that granulosa cells of homozygous Ser680/Ser680-type women are more resistant to FSH action. Genotyping of women scheduled for ovarian stimulation for assisted reproduction could be an attractive test to tailor the gonadotrophin dose based on the individual patient's susceptibility to administered FSH (Haller-Kikkatalo et al. 2012). Several polymorphisms in the LH-β gene located at chromosome region 11p13 gene have been identified (179 SNPs; www.snpper.chip.org); three polymorphisms in the coding area have been related to decreased LH activity (Haavisto et al. 1995): polymorphisms Trp8Arg and Ile15Thr were associated with suppressed fertility (Berger et al. 2005), repeated loss of pregnancy, and menstrual irregularities leading to infertility (Furui et al. 1994). The Gly102Ser SNP has been associated with infertility and menstrual disorders (Ramanujam et al. 1999). The variant 8Arg-15Thr has been found to be more frequent among hyporesponders to rFSH in IVF clinics (Alviggi et al. 2009). Studies in women with higher rFSH consumption and low COH responses suggested that ovarian outcome was improved by treatment with exogenous LH and which also reduced consumption of rFSH (De Placido et al. 2004). Several polymorphisms have been identified (over 520 SNPs according to www.snpper.chip.org database) in LHR gene which is located at chromosome region 2p21, with 11 exons. In iatrogenic ovarian hyperstimulation syndrome (OHSS) patients, 18insLeuGln polymorphisms in the coding region of LHR showed no association between these variants and the development of OHSS (Kerkela et al. 2007). The cytochrome P450, family 11, subfamily A, polypeptide 1 (CYP11A1) gene is located at chromosome region 15q23–q24, with 9 exons, and several SNPs have been identified (241 SNPs, www.snpper.chip.org). A polymorphism in the promoter region of the gene (TTTTAn) was strongly related to hyperandrogenemia and polycystic ovary syndrome in COH-treated women, including those at a high risk of OHSS (Gharani et al. 1997).

In ovarian steroidogenesis, aromatase is one of the key enzymes which catalyzes the final stage of conversion of the androgens testosterone and androstenedione to estradiol and estrone, respectively. The aromatase gene (CYP19A1) is another candidate which may be targeted for pharmacogenetic approach of female infertility. The CYP19A1 gene is located at chromosome region 15q21.1 and comprises 10 exons, where the last 9 (II–X) are coding exons and the first exon, one of nine alternative untranslated first exons, regulates tissue-specific expression (Sebastian and Bulun 2001). Over 1,080 SNPs in the aromatase gene have been identified according to www.snpper.chip.org. Women carrying shorter CYP19A1 (TTTA)n repeats exhibit lower estrogen concentrations (Haiman et al. 2000; Tworoger et al. 2004). Undoubtedly, estrogen receptors (ESRs) are important candidates in ovarian responsiveness to FSH, since direct effects of estrogens on follicle growth, maturation, and oocyte release are well established (Goldenberg et al. 1972). Estrogens play an important role in endometrial preparation for implantation in addition to folliculogenesis (Speroff and Fritz 2005). SNP studies in estrogen receptors suggested that estrogen receptor signalling pathways have a role in COH outcome (Altmäe et al. 2011). Similarly, progesterone receptor (PGR) is important; several polymorphic variants in the PGR gene have been described, and over 910 SNPs have been identified (www.snpper.chip.org). During early follicle development, folic acid modulates growth of immature follicles (Twigt et al. 2011). In animal models, insufficient folate intake is known to impair fertility (Mooij et al. 1992). The MTHFR 677C/T polymorphism is the most influential and prevalent genetic variation affecting folate metabolism. MTHFR 677 CT heterozygotes have a higher proportion of good quality embryos and an increased chance of pregnancy (Laanpere et al. 2011). In addition to the MTHFR 677C/T polymorphism, the other studied folate pathway SNPs are known to affect outcome of IVF. MTHFR 1793 GA, SLC19A1 80 GA, and CTH 1208 GT heterozygotes showed a lower probability of previously failed IVF treatments, while women heterozygous for FOLR1 1816 C/delC and linked 1841 G/A showed an increased risk of pregnancy loss (Laanpere et al. 2011). Bone morphogenetic protein 15 (BMP15) and growth differentiation factor 9 (GDF9) are expressed in oocytes from early-stage follicles (Aaltonen et al. 1999). Both proteins play crucial roles in determining follicle growth and ovulation rates. The gene for GDF9 is located at chromosome region 5q31.1, and that for BMP15 is at Xp11.2. Naturally occurring mutations in sheep have been identified in both genes, affecting ovulation rate and fecundity (Gemmell and Slate 2006). In humans, polymorphisms in GDF9 and BMP15 are also associated with fecundity and an increased incidence of dizygotic twinning (Palmer et al. 2006); premature ovarian failure has also been reported in some cases (Kovanci et al. 2007). The SOD2 Ala/Ala genotype was associated with a higher number of total and fertilized oocytes following COH, and it was an independent predictor of pregnancy (Ruiz-Sanz et al. 2010). A study by Legro et al. (2008) is a substudy of a multicentric, double-blinded, prospective study on pregnancy in polycystic ovary syndrome (PCOS) and is reviewed by Goldenberg and Glueck (2008), as stated here. In this study, 626 infertile women with polycystic ovary syndrome were randomly assigned to receive 50 mg clomiphene citrate plus placebo ($n=209$), 2 g extended-release metformin plus placebo ($n=208$), or a combination of metformin and clomiphene ($n=209$) for up to six cycles. Of 626 patients in the original study, 312 women participated in the pharmacogenetic substudy; 98 received metformin XR (2 g/day), 102 clomiphene, and 112 combined clomiphene–metformin XR treatment. This study was designed "to identify predictive genetic polymorphism and other determinants of ovulatory response" in prospective fashion. Candidate genes tested included estrogen receptor 1 (*ESR1*), CYP genes (*CYP2C9* and *CYP2D6*), and *STK11*. *STK11*, formerly known as *LKB1*, is a serine–threonine kinase gene expressed in the liver, which phosphorylates and activates AMP-activated protein kinase. It was shown to be a site of metformin action. The C allele of a SNP in the *STK11* gene was associated with a significantly decreased chance of ovulation in polycystic ovary syndrome women treated with metformin. In analysis of ovulation per cycle, the adjusted odds ratio for CC versus GG (wild-type normal) was 0.30 (95 % CI: 0.14–0.66), and the odds ratio for CG versus GG was 0.30 (95 % CI: 0.14–0.66). This elegant study is of great importance because despite treatment, many women with polycystic ovary syndrome fail to ovulate, 24.9 % in the clomiphene group, 44.7 % in the metformin group, and 16.7 % in the clomiphene–metformin group. For further details also see Table 31.4.

Table 31.4 Association of SNPs with female infertility in different human population across the globe

Year and country	Experiment	Outcome	References
2008, USA	Association of single nucleotide polymorphisms (SNPs) of selected candidate genes [estrogen receptor 1 (*ESR1*, formerly known as *ERα*), two cytochrome P450 genes (*CYP2C9* and *CYP2D6*) and a serine-threonine kinase gene expressed in the liver, *STK11* (formerly known as*LKB1*)] with response to treatment with clomiphene and/or metformin in 312 subjects undergoing controlled ovarian hyperstimulation (CHO) having polycystic ovary syndrome (PCOS)	Substitution of G instead of C in the *STK11* gene was associated with a significantly decreased chance of ovulation in PCOS women treated with metformin	Legro (2008)
2009, Netherlands	Association of SNPs of selected candidate genes [FSH-R, cytochrome P450 gene (*CYP2D6*) and a serine-threonine kinase gene expressed in the liver, *STK11* (formerly known as*LKB1*), OCT1, an organic cation transporter critical for elimination of foreign substances] with response to treatment with clomiphene and/or metformin in 312 subjects undergoing controlled ovarian hyperstimulation (CHO) having polycystic ovary syndrome (PCOS)	Subjects harboring the least sensitive variant (Ser/Ser) of FSH-R need higher doses of exogenous FSH for ovarian stimulation in ART OCT1 was found to be highly polymorphic, contributing to variation in response to the drug in diabetic patients with PCOS treated with metformin	Overbeek and Lambalk (2009)
2011, USA	Association of SNPs in the coding region of FSH-R, which result in amino acid changes (p.307Thr/Ala, p.680Asn/Ser) with response to ovarian stimulation with FSH during infertility treatment	FSH response varies with presence of different SNPs, alternate splice variants of FSH-R affecting the extracellular domain without causing a frameshift have been found in women undergoing ovarian stimulation	Lalioti (2011)
2011, Sweden	Association of SNPs in subjects undergoing controlled ovarian hyperstimulation (CHO) to induce folliculogenesis	Polymorphisms in genes involved in FSH signalling (FSH-R Asn680Ser), estrogen biosynthesis (ESR1 PvuII T/C), influence the response to exogenous gonadotrophin administration, resulting in differences in COH and IVF outcomes	Altmäe et al. (2011)
2011, India	Association of FSH-R polymorphism at position −29 was studied in 100 subjects undergoing IVF treatment	Almost 72 % of subjects with the AA genotype (instead of GG genotype) at position −29 of *FSH-R* gene were poor ovarian responders due to lower expression of FSHR mRNA and protein	Desai et al. (2011)
2012, Greece	Association of SNPs in subjects undergoing controlled ovarian hyperstimulation (CHO) to induce ovulation	The Asn/Ser variant of FSH-R functions more efficiently, while the Ser/Ser and Asn/Asn variants have a tendency to resist to FSH stimulation, whereas estrogen receptor 1 (ESR1), the Pvull and the Xbal polymorphisms seem to be associated with differences in the response to ovarian stimulation, while the Rsal polymorphism in estrogen receptor 2 (ESR2) is currently under investigation	Loutradis et al. (2012)

3.2 Implantation and Endometrial Failure

Endometrium is a dynamic tissue that responds on a cyclic basis to circulating levels of the ovarian-derived steroid hormones, estradiol and progesterone. The orchestrated events in endometrium help and permit the successful blastocyst implantation and the associated events begin from endometrial receptivity to embryonic implantation in a particular time frame (Hess et al. 2006). The endometrium is receptive to embryonic implantation during a defined window that is temporally and spatially restricted, and if there is any perturbance, it may cause infertility in women. Kao et al. (2002) has applied paralleled gene expression profiling using high-density oligonucleotide microarrays to investigate the diagnostic tests for endometriosis and targeted drug discovery for treatment of endometriosis-based implantation failure. The current lack of a robust test for endometrial receptivity impairs clinician's ability to diagnose infertility of endometrial origin, confounding decision-making in the application of artificial reproductive technologies, for example, whether or not to transfer an embryo in any particular cycle. Gene array analyses have identified genes that are regulated across the menstrual cycle (Giudice 2004; Ponnampalam et al. 2006) and in women undergoing IVF (Horcajadas et al. 2007), but no clear genetic markers of the receptive phase have emerged. New proteomics technologies provide a clear opportunity for biomarker discovery aimed at improving patient outcomes from IVF and embryo transfer. Proteomic studies have demonstrated the complexity of the human uterine secretome during the period in which endometrial receptivity is attained and that secretome is altered in women with unexplained infertility. Many of the regulatory proteins anticipated to be present in uterine fluid were not detected. The existing studies have identified large numbers of potential biomarkers for a number of clinically important diseases of the female reproductive tract and infertility. While existing studies have identified large numbers of potential biomarkers for a number of clinically important diseases of the female reproductive tract and infertility, there remains ample scope to apply both existing and emerging proteomics technologies in the context of reproductive disease and unexplained infertility. Detection of crucial disturbances in the concentration of metabolites by metabolomic profiling of key biomarkers can be beneficial in the management of various medical conditions, including female factor infertility. Metabolomics can reflect various phenotypes at a functional level with close relations to the reproductive disorders and infertility (Goldsmith et al. 2010).

Despite of such studies and identification of SNPs involved in female infertility, the pharmacogenomics of ovulation induction facilitating decisions on who, when, and how to treat is in its infancy, and a lot is needed to be achieved (Overbeek and Lambalk 2009). However, genome-wide association studies (GWAS) are powerful tools used to identify genetic factors that influence drug responses, with no a priori assumptions (Altmäe et al. 2011). From the effect of individual variation on drug response, i.e., pharmacogenetics, GWAS has given rise to pharmacogenomics: whole-genome impact on drug responses. Ideally, pharmacogenetic/genomic studies will eventually lead to an era in which an individual's DNA sequence is regarded as an integral determinant of drug therapy (Giacomini et al. 2007). The most predictive evaluation of fertility including ovulation, pregnancy, and sperm production will involve a combination of phenotypic (patient's age, weight, gonadal morphology, basal gonadotrophin levels, and other blood serum markers) and genetic markers, which will be applied in routine diagnostic tests before a therapy is planned.

Although most commonly used drugs for regulation of male and female fertility and ovulation are antiestrogenic compounds for the release of gonadotrophins and other hormone agonists and antagonists, viz., Clomid, Serophene, Femara, Follistim, Gonal-F, Bravelle, Fertinex, Ovidrel, Proctinal, Parlodel, Novarel, Pregnyl, Menogon, Repronex, Menopur, Nugon, Antagon, Cetrotide, Lupron, Synarel, Zoladex, Ladogal, Nolvadex, Cetrorelix, Antagon, Buserelin, and Lucrin (for details see http://infertility.about.com/od/

infertilitytreatments/a/fertility_drugs.htm); none has yet been investigated in true sense with respect to genetic makeup of an individual. For the treatment of infertility, it will take some time before the choice of drug could be truly individualized.

4 Conclusion

Pharmacogenomics aims to develop rational means to optimize drug therapy based on gene expression or single-nucleotide polymorphisms, with respect to the patients' genotype for maximum efficacy with minimal adverse effects (Becquemont 2009; Squassina et al. 2010). In the field of molecular-targeted therapy for infertility and associated reproductive disorders, the application of array-based methodologies will be required for the identification of molecular targets with "key" roles in the diagnosis and pathogenesis of infertility. Further since there is no information, an emphasis is required on deciphering the molecular mechanisms of drug action, identifying novel therapeutic targets and suitable agents to target them with, and discovering molecular markers/signatures that predict response to therapy against multifactorial infertility conditions in each affected men and women. The way a person responds to a drug is a complex trait that is influenced by many different genes and factors. Without knowing all of the genes involved in drug response, scientists have found it difficult to develop genetic tests that could predict a person's response to a particular drug. With increasing incidents of infertility, pharmacogenomics is expected to bring many benefits to patients and the health-care industry, increase the number of new drugs, and reduce the costs associated with drug development (Grady and Ritchie 2011). It is clear that a lot of correlative studies with respect to mutation and the drug of choice are to be done before pharmacogenomics, and personalized medicine is successfully implemented in diagnosis and treatment of male and female infertility. In order to introduce personalized medicine into infertility management, the initiative should be taken that includes the following: (i) the development of large available omics data sets across various laboratories and countries in order to define and isolate distinct subpopulations, (ii) construction of a global biological model of infertility and associated disorders in order to define therapies accordingly, and (iii) shifting the focus on the treatment outcome from patient behavior to enhancing the clarity of results and treatment efficacy.

References

Aaltonen J, Laitinen MP, Vuojolainen K, Jaatinen R, Horelli-Kuitunen N, Seppa L, Louhio H, Tuuri T, Sjoberg J, Butzow R et al (1999) Human growth differentiation factor 9 (GDF-9) and its novel homolog GDF-9B are expressed in oocytes during early folliculogenesis. J Clin Endocrinol Metab 84:2744–2750

Altmäe S, Hovatta O, Stavreus-Evers A, Salumets A (2011) Genetic predictors of controlled ovarian hyperstimulation: where do we stand today? Hum Reprod Update 17(6):813–828. doi:10.1093/humupd/dmr034

Alviggi C, Clarizia R, Coppola M, De Rosa G, De Biasio K, Pettersson K, De Placido G, Umaidan P (2009) A common LH polymorphism is associated with higher FSH consumption during ovarian stimulation for IVF/CSI cycles. Int J Gynecol Obstet 107S2:S105

Aston KI, Carrell DT (2009) Genome-wide study of single-nucleotide polymorphisms associated with azoospermia and severe oligozoospermia. J Androl 30(6):711–725. doi:10.2164/jandrol.109.007971

Aydos SE, Taspinar M, Sunguroglu A, Aydos K (2009) Association of CYP1A1 and glutathione S-transferase polymorphisms with male factor infertility. Fertil Steril 92(2):541–547. doi:10.1016/j.fertnstert.2008.07.017

Baird DT, Benagiano G, Cohen J (2002) Physiopathological determinants of human infertility. Hum Reprod Update 8(5):435–447

Bayne RA, Forster T, Burgess ST, Craigon M, Walton MJ, Baird DT, Ghazal P, Anderson RA (2008) Molecular profiling of the human testis reveals stringent pathway-specific regulation of RNA expression following gonadotropin suppression and progestogen treatment. J Androl 29(4):389–403

Becquemont L (2009) Pharmacogenomics of adverse drug reactions: practical applications and perspectives. Pharmacogenomics 10(6):961–969

Berger K, Billerbeck AE, Costa EM, Carvalho LS, Arnhold IJ, Mendonca BB (2005) Frequency of the allelic variant (Trp8Arg/Ile15Thr) of the luteinizing hormone gene in a Brazilian cohort of healthy subjects and in patients with hypogonadotropic hypogonadism. Clinics (Sao Paulo) 60:461–464

Cappallo-Obermann H, von Kopylow K, Schulze W, Spiess AN (2010) A biopsy sample reduction approach to identify significant alterations of the testicular transcriptome in the presence of Y-chromosomal microdeletions that are independent of germ cell composition. Hum Genet 128(4):421–431. doi:10.1007/s00439-010-0865-9

Chalmel F, Lardenois A, Evrard B, Mathieu R, Feig C, Demougin P, Gattiker A, Schulze W, Jégou B, Kirchhoff C, Primig M (2012) Global human tissue profiling and protein network analysis reveals distinct levels of transcriptional germline-specificity and identifies target genes for male infertility. Hum Reprod 27(11):3233–3248. doi:10.1093/humrep/des301

Chengalvala MV, Meade EH Jr, Cottom JE, Hoffman WH, Shanno LK, Wu MM, Kopf GS, Shen ES (2006) Regulation of female fertility and identification of future contraceptive targets. Curr Pharm Des 12(30):3915–3928

Desai SS, Achrekar SK, Pathak BR, Desai SK, Mangoli VS, Mangoli RV, Mahale SD (2011) Follicle-stimulating hormone receptor polymorphism (G-29A) is associated with altered level of receptor expression in Granulosa cells. J Clin Endocrinol Metab 96(9):2805–2812. doi:10.1210/jc.2011-1064

De Placido G, Alviggi C, Mollo A, Strina I, Ranieri A, Alviggi E, Wilding M, Varricchio MT, Borrelli AL, Conforti S (2004) Effects of recombinant LH (rLH) supplementation during controlled ovarian hyperstimulation (COH) in normogonadotrophic women with an initial inadequate response to recombinant FSH (rFSH) after pituitary downregulation. Clin Endocrinol 60:637–643

Ellis PJ, Furlong RA, Conner SJ, Kirkman-Brown J, Afnan M, Barratt C, Griffin DK, Affara NA (2007) Coordinated transcriptional regulation patterns associated with infertility phenotypes in men. J Med Genet 44(8):498–508

Feig C, Kirchhoff C, Ivell R, Naether O, Schulze W, Spiess AN (2007) A new paradigm for profiling testicular gene expression during normal and disturbed human spermatogenesis. Mol Hum Reprod 13(1):33–43

Ferlin A (2011) Update on genetics in male infertility. Endocr Abstr 26:S11.1. http://www.endocrine-abstracts.org/ea/0026/ea0026s11.1.htm

Furui K, Suganuma N, Tsukahara S, Asada Y, Kikkawa F, Tanaka M, Ozawa T, Tomoda Y (1994) Identification of two point mutations in the gene coding luteinizing hormone (LH) beta-subunit, associated with immunologically anomalous LH variants. J Clin Endocrinol Metab 78:107–113

Galan JJ, De Felici M, Buch B, Rivero MC, Segura A, Royo JL, Cruz N, Real LM, Ruiz A (2006) Association of genetic markers within the KIT and KITLG genes with human male infertility. Hum Reprod 21(12):3185–3192

Gatta V, Raicu F, Ferlin A, Antonucci I, Scioletti AP, Garolla A, Palka G, Foresta C, Stuppia L (2010) Testis transcriptome analysis in male infertility: new insight on the pathogenesis of oligo-azoospermia in cases with and without AZFc microdeletion. BMC Genomics 11:401. doi:10.1186/1471-2164-11-401

Gemmell NJ, Slate J (2006) Heterozygote advantage for fecundity. PLoS One 1:e125

Gharani N, Waterworth DM, Batty S, White D, Gilling-Smith C, Conway GS, McCarthy M, Franks S, Williamson R (1997) Association of the steroid synthesis gene CYP11a with polycystic ovary syndrome and hyperandrogenism. Hum Mol Genet 6:397–402

Giacomini KM, Brett CM, Altman RB, Benowitz NL, Dolan ME, Flockhart DA, Johnson JA, Hayes DF, Klein T, Krauss RM, Kroetz DL, McLeod HL, Nguyen AT, Ratain MJ, Relling MV, Reus V, Roden DM, Schaefer CA, Shuldiner AR, Skaar T, Tantisira K, Tyndale RF, Wang L, Weinshilboum RM, Weiss ST, Zineh I (2007) The pharmacogenetics research network: from SNP discovery to clinical drug response. Clin Pharmacol Ther 81:328–345

Giudice LC (2004) Microarray expression profiling reveals candidate genes for human uterine receptivity. Am J Pharmacogenomics 4(5):299–312

Goldenberg N, Glueck CJ (2008) Is pharmacogenomics our future? Metformin, ovulation and polymorphism of the STK11 gene in polycystic ovary syndrome. Pharmacogenomics 9:1163–1165

Goldenberg RL, Vaitukaitis JL, Ross GT (1972) Estrogen and follicle stimulation hormone interactions on follicle growth in rats. Endocrinology 90:1492–1498

Goldsmith P, Fenton H, Morris-Stiff G, Ahmad N, Fisher J, Prasad KR (2010) Metabonomics: a useful tool for the future surgeon. J Surg Res 160:122–132

Grady BJ, Ritchie MD (2011) Statistical optimization of pharmacogenomics association studies: key considerations from study design to analysis. Curr Pharmacogenomics Personal Med 9(1):41–66

Haavisto AM, Pettersson K, Bergendahl M, Virkamaki A, Huhtaniemi I (1995) Occurrence and biological properties of a common genetic variant of luteinizing hormone. J Clin Endocrinol Metab 80:1257–1263

Haiman CA, Hankinson SE, Spiegelman D, De Vivo I, Colditz GA, Willett WC, Speizer FE, Hunter DJ (2000) A tetranucleotide repeat polymorphism in CYP19 and breast cancer risk. Int J Cancer 87:204–210

Haller-Kikkatalo K, Salumets A, Raivo U (2012) Review on autoimmune reactions in female infertility: antibodies to follicle stimulating hormone. Clin Dev Immunol 2012:1–15, Article ID 762541

Hardelin JP, Dode C (2008) The complex genetics of Kallmann syndrome: KAL1, FGFR1, FGF8, PROKR2, PROK2, et al. Sex Dev 2:181–193

Hess AP, Nayak NR, Giudice LC (2006) Oviduct and endometrium: cyclic changes in the primate oviduct and endometrium. In: Neill JD (ed) The physiology of reproduction, 3rd edn. Elsevier, St. Louis, pp 359–403

Horcajadas JA, Pellicer A, Simón C (2007) Wide genomic analysis of human endometrial receptivity: new times, new opportunities. Hum Reprod Update 13(1):77–86

Hwang K, Yatsenko AN, Jorgez CJ, Mukherjee S, Nalam RL, Matzuk MM, Lamb DJ (2011) Mendelian genetics of male infertility. Ann N Y Acad Sci 1214:E1–E17

Kao LC, Tulac S, Lobo S, Imani B, Yang JP, Germeyer A, Osteen K, Taylor RN, Lessey BA, Giudice LC (2002) Global gene profiling in human endometrium during the window of implantation. Endocrinology 143(6):2119–2138

Kerkela E, Skottman H, Friden B, Bjuresten K, Kere J, Hovatta O (2007) Exclusion of coding-region mutations in luteinizing hormone and follicle-stimulating hormone receptor genes as the cause of ovarian hyperstimulation syndrome. Fertil Steril 87:603–606

Kosova G, Scott NM, Niederberger C, Prins GS, Ober C (2012) Genome-wide association study identifies candidate genes for male fertility traits in humans. Am J Hum Genet 90(6):950–961. doi:10.1016/j.ajhg.2012.04.016

Kovanci E, Rohozinski J, Simpson JL, Heard MJ, Bishop CE, Carson SA (2007) Growth differentiating factor-9 mutations may be associated with premature ovarian failure. Fertil Steril 87:143–146

Laanpere M, Altma¨e S, Kaart T, Stavreus-Evers A, Nilsson TK, Salumets A (2011) Folate-metabolizing gene variants and pregnancy outcome of IVF. Reprod Biomed Online 22:603–614

Lalioti MD (2011) Impact of follicle stimulating hormone receptor variants in fertility. Curr Opin Obstet Gynecol 23(3):158–167.doi:10.1097/GCO.0b013e3283455288

Lardone MC, Parada-Bustamante A, Ebensperger M, Valdevenito R, Kakarieka E, Martínez D, Pommer R, Piottante A, Castro A (2011) DAX-1 and DAX-1A expression in human testicular tissues with primary spermatogenic failure. Mol Hum Reprod 17(12):739–746. doi:10.1093/molehr/gar051

Lee HC, Jeong YM, Lee SH, Cha KY, Song SH, Kim NK, Lee KW, Lee S (2006) Association study of four polymorphisms in three folate-related enzyme genes with non-obstructive male infertility. Hum Reprod 21:3162–3170

Legro RS (2008) Individualizing infertility therapy with pharmacogenomics: vanity or vanguard? Pharmacogenomics 9 (9):1179–1181. doi:10.2217/14622416.9.9.1179

Legro RS, Barnhart HX, Schlaff WD, Carr BR, Diamond MP, Carson SA, Steinkampf MP, Coutifaris C, McGovern PG, Cataldo NA, Gosman GG, Nestler JE, Giudice LC, Ewens KG, Spielman RS, Leppert PC, Myers ER (2008) Evaluation of ovulatory response to treatment of polycystic ovary syndrome is associated with a polymorphism in the *STK11* gene. J Clin Endocrinol Metab 93(3):792–800

Lian J, Zhang X, Tian H, Liang N, Wang Y, Liang C, Li X, Sun F (2009) Altered microRNA expression in patients with non-obstructive azoospermia. Reprod Biol Endocrinol 7:13. doi:10.1186/1477-7827-7-13

Lin YH, Lin YM, Teng YN, Hsieh TY, Lin YS, Kuo PL (2006) Identification of ten novel genes involved in human spermatogenesis by microarray analysis of testicular tissue. Fertil Steril 86(6):1650–1658

Loutradis D, Theofanakis C, Anagnostou E, Mavrogianni D, Partsinevelos GA (2012) Genetic profile of SNP(s) and ovulation induction. Curr Pharm Biotechnol 13(3):417–425

Majumdar SS, Usmani A, Bhattacharya I, Sarda K, Gautam M, Sharma D, Basu S, Dhup S (2009) A method for rapid generation of transgenic animals to evaluate testis genes during sexual maturation. J Reprod Immunol 83(1–2):36–39

Matzuk MM, Lamb DJ (2002) Genetic dissection of mammalian fertility pathways. Nat Cell Biol 4(Suppl): S41–S49

Matzuk MM, Lamb DJ (2008) The biology of infertility: research advances and clinical challenges. Nat Med 14(11):1197–1213

Merisalu A, Punab M, Altmae S, Haller K, Tiido T, Peters M, Salumets A (2007) The contribution of genetic variations of aryl hydrocarbon receptor pathway genes to male factor infertility. Fertil Steril 88:854–859

Montjean D, De La Grange P, Gentien D, Rapinat A, Belloc S, Cohen-Bacrie P, Menezo Y, Benkhalifa M (2012) Sperm transcriptome profiling in oligozoospermia. J Assist Reprod Genet 29(1):3–10. doi:10.1007/s10815-011-9644-3

Mooij PN, Wouters MG, Thomas CM, Doesburg WH, Eskes TK (1992) Disturbed reproductive performance in extreme folic acid deficient golden hamsters. Eur J Obstet Gynecol Reprod Biol 43:71–75

Navarro-Costa P, Nogueira P, Carvalho M, Leal F, Cordeiro I, Calhaz-Jorge C, Gonçalves J, Plancha CE (2010) Incorrect DNA methylation of the DAZL promoter CpG island associates with defective human sperm. Hum Reprod 25(10):2647–2654. doi:10.1093/humrep/deq200

Nieschlag E, Leifke E (1997) Empirical therapies for idiopathic infertility. In: Nieschlag E, Behre HM (eds) Andrology: male reproductive health and dysfunction. Springer, Heidelberg, pp 313–319

Nuti F, Krausz C (2008) Gene polymorphisms/mutations relevant to abnormal spermatogenesis. Reprod Biomed Online 16(4):504–513

Ogorevc J, Dovc P, Kunej T (2011) Polymorphisms in microRNA targets: a source of new molecular markers for male reproduction. Asian J Androl 13:505–508

Okada H, Tajima A, Shichiri K, Tanaka A, Tanaka K, Inoue I (2008) Genome-wide expression of azoospermia testes demonstrates a specific profile and implicates ART3 in genetic susceptibility. PLoS Genet 4(2):e26. doi:10.1371/journal.pgen.0040026

Oliveira A, Neto A, Almeida C, Silva-Ramos M, Versos R, Barros A, Sousa M, Carvalho F (2012) Comparative study of gene expression in patients with varicocele by microarray technology. Andrologia 44(Suppl 1):260–265. doi:10.1111/j.1439-0272.2011.01173.x

Otsuka M, Zheng M, Hayashi M, Lee JD, Yoshino O, Lin S, Han J (2008) Impaired micro RNA processing causes corpus luteum insufficiency and infertility in mice. J Clin Invest 118(5):1944–1954

Overbeek A, Lambalk N (2009) Pharmacogenomics of ovulation induction: facilitating decisions on who, when and how to treat. Pharmacogenomics 10(9):1377–1379. doi:10.2217/pgs.09.110

Palmer JS, Zhao ZZ, Hoekstra C, Hayward NK, Webb PM, Whiteman DC, Martin NG, Boomsma DI, Duffy DL, Montgomery GW (2006) Novel variants in growth differentiation factor 9 in mothers of dizygotic twins. J Clin Endocrinol Metab 91:4713–4716

Papaioannou MD, Nef S (2010) MicroRNAs in the testis: building up male fertility. J Androl 31(1):26–33

Park JH, Lee HC, Jeong YM, Chung TG, Kim HJ, Kim NK, Lee SH, Lee S (2005) MTHFR C677T polymorphism associates with unexplained infertile male factors. J Assist Reprod Genet 22:361–368

Perez Mayorga M, Gromoll J, Behre HM, Gassner C, Nieschlag E, Simoni M (2000) Ovarian response to follicle-stimulating hormone (FSH) stimulation depends on the FSH receptor genotype. J Clin Endocrinol Metab 85:3365–3369

Plaseski T, Noveski P, Popeska Z, Efremov GD, Plaseska-Karanfilska D (2012) Association study of single-nucleotide polymorphisms in FASLG, JMJDIA, LOC203413, TEX15, BRDT, OR2W3, INSR, and TAS2R38 genes with male infertility. J Androl 33(4):675–683. doi:10.2164/jandrol.111.013995

Ponnampalam AP, Weston GC, Susil B, Rogers PA (2006) Molecular profiling of human endometrium during the menstrual cycle. Aust N Z J Obstet Gynaecol 46(2):154–158

Ramanujam LN, Liao WX, Roy AC, Loganath A, Goh HH, Ng SC (1999) Association of molecular variants of luteinizing hormone with menstrual disorders. Clin Endocrinol 51:243–246

Roberto T, Donati MB, Iacoviello L (2004) Trends in pharmacogenomics of drugs acting on hypertension. Pharmacol Res 49:351–356

Ruiz-Sanz JI, Aurrekoetxea I, Matorras R, Ruiz-Larrea MB (2010) Ala16Val SOD2 polymorphism is associated with higher pregnancy rates in in vitro fertilization cycles. Fertil Steril 95:1601–1605

Safarinejad MR, Shafiei N, Safarinejad S (2010) The association of glutathione-S-transferase gene polymorphisms (GSTM1, GSTT1, GSTP1) with idiopathic male infertility. J Hum Genet 55(9):565–570. doi:10.1038/jhg.2010.59

Salenave S, Trabado S, Maione L, Brailly-Tabard S, Young J (2012) Male acquired hypogonadotropic hypogonadism: diagnosis and treatment. Ann Endocrinol 73(2):141–146

Schaison G, Young J, Pholsena M, Nahoul K, Couzinet B (1993) Failure of combined follicle stimulating hormone-testosterone administration to initiate and/or maintain spermatogenesis in men with hypogonadotropic hypogonadism. J Clin Endocrinol Metab 77(6):1545–1549

Schwartz CE, Dean J, Howard-Peebles PN et al (1994) Obstetrical and gynecological complications in fragile X carriers: a multicenter study. Am J Med Genet 51:400–402

Sebastian S, Bulun SE (2001) A highly complex organization of the regulatory region of the human CYP19 (aromatase) gene revealed by the Human Genome Project. J Clin Endocrinol Metab 86:4600–4602

Shahid M, Dhillon VS, Khalil HS, Sexana A, Husain SA (2011) Associations of Y-chromosome subdeletion gr/gr with the prevalence of Y-chromosome haplogroups in infertile patients. Eur J Hum Genet 19(1):23–29. doi:10.1038/ejhg.2010.151

Speroff L, Fritz MA (2005) Clinical gynecologic endocrinology and infertility. Lippincott Williams and Wilkins, Philadelphia

Spiess AN, Feig C, Schulze W, Chalmel F, Cappallo-Obermann H, Primig M, Kirchhoff C (2007 Nov) Cross-platform gene expression signature of human spermatogenic failure reveals inflammatory-like response. Hum Reprod 22(11):2936–2946

Squassina A, Manchia M, Manolopoulos VG, Artac M, Lappa-Manakou C, Karkabouna S, Mitropoulos K, Del Zompo M, Patrinos GP (2010) Realities and expectations of pharmacogenomics and personalized medicine: impact of translating genetic knowledge into clinical practice. Pharmacogenomics 11(8):1149–1167

Simoni M, Tüttelmann F, Michel C, Böckenfeld Y, Nieschlag E, Gromoll J (2008) Polymorphisms of the luteinizing hormone/chorionic gonadotropin receptor gene: association with maldescended testes and male infertility. Pharmacogenet Genomics 18(3):193–200. doi:10.1097/FPC.0b013e3282f4e98c

Tüttelmann F, Rajpert-De Meyts E, Nieschlag E, Simoni M (2007 Dec) Gene polymorphisms and male infertility–a meta-analysis and literature review. Reprod Biomed Online 15(6):643–658

Twigt JM, Hammiche F, Sinclair KD, Beckers NG, Visser JA, Lindemans J, de Jong FH, Laven JS, Steegers-Theunissen RP (2011) Preconception folic acid use modulates estradiol and follicular responses to ovarian stimulation. J Clin Endocrinol Metab 96:E322–E329

Tworoger SS, Chubak J, Aiello EJ, Ulrich CM, Atkinson C, Potter JD, Yasui Y, Stapleton PL, Lampe JW, Farin FM et al (2004) Association of CYP17, CYP19, CYP1B1, and COMT polymorphisms with serum and urinary sex hormone concentrations in postmenopausal women. Cancer Epidemiol Biomarkers Prev 13:94–101

Verkerk AJ, Pieretti M, Sutcliffe JS, Fu YH, Kuhl DP, Pizzuti A, Reiner O, Richards S, Victoria MF, Zhang FP (1991) Identification of a gene (FMR-1) containing a CGG repeat coincident with a breakpoint cluster region exhibiting length variation in fragile X syndrome. Cell 65:905–914

Watanabe M, Sueoka K, Sasagawa I, Nakabayashi A, Yoshimura Y, Ogata T (2004) Association of male infertility with Pro185Ala polymorphism in the aryl hydrocarbon receptor repressor gene: implication for the susceptibility to dioxins. Fertil Steril 82(suppl 3): 1067–1071

Wittenberger MD, Hagerman RJ, Sherman SL, McConkie-Rosell A, Welt CK, Rebar RW, Corrigan EC, Simpson JL, Nelson LM (2007) The FMR1 premutation and reproduction. Fertil Steril 87:456–465

Wu Wu Q, Xing J, Xue W, Sun J, Wang X, Jin X (2009) Influence of polymorphism of glutathione S-transferase T1 on Chinese infertile patients with varicocele. Fertil Steril 91(3):960–962. doi:10.1016/j.fertnstert.2007.08.061

Wu W, Hu Z, Qin Y, Dong J, Dai J, Lu C, Zhang W, Shen H, Xia Y, Wang X (2012) Seminal plasma microRNAs: potential biomarkers for spermatogenesis status. Mol Hum Reprod 18(10):489–497

Yang B, Wang H, Gao XK, Chen BQ, Zhang YQ, Liu HL, Wang Y, Qin WJ, Qin RL, Shao GX, Shao C (2004) Expression and significance of Rap1A in testes of azoospermic subjects. Asian J Androl 6(1):35–40

Pharmacogenomics in Ophthalmology

32

Stephen G. Schwartz, Tomomi Higashide, and Milam A. Brantley Jr.

Abstract

Pharmacogenomics is an evolving research discipline within ophthalmology, but genetic data are not currently used to guide daily clinical decisions. Ophthalmic pharmacogenomic research has thus far focused on open-angle glaucoma (OAG) and age-related macular degeneration (AMD), two common and worldwide causes of visual loss. In the treatment of OAG and allied disorders, there are reported associations between various polymorphisms in adrenergic receptor genes and topical β-antagonists as well as between the prostaglandin receptor gene and a topical prostaglandin analogue. In the treatment of exudative AMD, there are reported associations between AMD-associated genes, such as *complement factor H* (*CFH*) and *age-related maculopathy susceptibility 2* (*ARMS2*), and the efficacy of different treatment modalities including photodynamic therapy and intravitreal vascular endothelial growth factor (VEGF) antagonists. The steroid response associated with ophthalmic corticosteroids has been investigated, but no definite genetic associations have been reported. As additional pharmacogenomic trials are reported, the precise relationship between genotype and treatment response may become clearer.

S.G. Schwartz (✉)
Department of Ophthalmology, Bascom Palmer Eye Institute, University of Miami Miller School of Medicine, 311 9th Street North, #100, 34102 Naples, FL, USA
e-mail: sschwartz2@med.miami.edu

T. Higashide
Department of Ophthalmology and Visual Science, Kanazawa University Graduate School of Medical Sciences, 13-1 Takara-Machi, 920-8641 Kanazawa, Ishikawa, Japan
e-mail: eyetomo@med.kanazawa-u.ac.jp

M.A. Brantley Jr.
Department of Ophthalmology, Vanderbilt University, 2213 Garland Avenue MRB4 Room 11425, 37232 Nashville, TN, USA
e-mail: milam.brantley@vanderbilt.edu

D. Barh et al. (eds.), *Omics for Personalized Medicine*,
DOI 10.1007/978-81-322-1184-6_32, © Springer India 2013

1 Introduction

Pharmacogenomics is an evolving research discipline within ophthalmology. Open-angle glaucoma (OAG) and age-related macular degeneration (AMD) are common, worldwide causes of visual loss that are largely treated with pharmacologic therapies. Pharmacogenetic associations have been reported for both OAG (Schwartz et al. 2008; Moroi et al. 2009) and exudative AMD (Schwartz and Brantley 2011). At this time, genetic data are not typically used to make routine clinical decisions. However, as additional clinical trial results are collected, the potential benefits of pharmacogenomics knowledge in the care of patients with ophthalmic diseases will be better understood.

There are at least two potential roles for applying pharmacogenomics in the treatment of ophthalmic diseases. First, more therapies specifically targeted for individuals may lead to improved treatment outcomes and reduce patient exposure to inefficacious medications. Second, pharmacogenomics may lead to the development of novel therapies for these diseases.

2 Open-Angle Glaucoma

Control of intraocular pressure (IOP) is generally effective in delaying the progression of optic neuropathy and visual loss in patients with OAG, normal tension glaucoma (NTG), ocular hypertension, and related disorders (Costagliola et al. 2009a, b). Two of the major categories of medications used to lower IOP are β-adrenergic antagonists and prostaglandin analogues, both of which have a considerable rate of nonresponse. Pooled data from multiple randomized clinical trials reported a nonresponse rate of 28 % with the β-blocker timolol maleate and 18 % with the prostaglandin analogue latanoprost (Camras and Hedman 2003). There is currently no way to identify these nonresponders prior to initiation of therapy. This "trial and error" strategy may lead to extra office visits and exposure to additional medications for some patients.

2.1 β-Adrenergic Antagonists

The β-adrenergic antagonists include several nonselective agents (β_1- and β_2-blockers), such as timolol, and one β_1-selective agent, betaxolol hydrochloride. The nonselective agents are generally more effective in IOP reduction than betaxolol (Allen et al. 1986). The high nonresponse rate associated with topical betaxolol is similar to the high nonresponse rate associated with systemic β_1-blockers used to treat systemic hypertension and other diseases (Materson et al. 1993).

Most pharmacogenomic studies of β-adrenergic antagonists have focused on the polymorphisms in adrenergic receptor genes (Table 32.1). The β_1-adrenergic receptor (β_1-AR) gene contains two well-characterized single nucleotide polymorphisms (SNPs): Ser49Gly and Arg389Gly (Maqbool et al. 1999; Mason et al. 1999). The β_2-adrenergic receptor (β_2-AR) contains four common SNPs: Gly16Arg, Cys19Arg, Gln27Glu, and Thr164Ile (Green et al. 1993, 1994; Parola and Kobilka 1994; Liggett 2000). The α-adrenergic receptor (α-AR) contains multiple subtypes, many of which have well-characterized polymorphisms, including Del 301–303 in α_{2B}-*AR* and Del 322–325 in α_{2C}-*AR* (Flordellis et al. 2004). Furthermore, timolol is metabolized by cytochrome P40 2D6 (CYP2D6), and polymorphisms in *CYP2D6* are associated with the efficacy of oral timolol in the treatment of systemic hypertension (McGourty et al. 1985).

The relationship between adrenergic receptor genotype and presence of OAG or NTG has been evaluated in several studies. In a Japanese study of 211 OAG patients, 294 NTG patients, and 240 controls, there was a significant association between NTG and the β_1-*AR* Arg389Gly polymorphism (Inagaki et al. 2006). A US series of 299 OAG patients and 284 controls found no associations between β_2-*AR* polymorphisms and clinical status (McLaren et al. 2007). In a Japanese series of 92 untreated NTG patients, associations were reported between diurnal IOP measurements and β_1-*AR* Ser49Gly, α_{2B}-*AR* Del 301–303, and α_{2C}-*AR* Del 322–325 (Gao et al. 2010).

Table 32.1 Pharmacogenomics of β-adrenergic antagonists

Gene(s)	Results	References
β₁-AR	β₁-*AR* Arg389Gly associated with normal tension glaucoma	Inagaki et al. (2006)
β₂-AR	No association between β₂-*AR* and open-angle glaucoma	McLaren et al. (2007)
Multiple	Association between diurnal IOP in untreated NTG patients and β₁-*AR* Ser49Gly, α₂B-*AR* Del 301–303, and α₂C-*AR* Del 322–325	Gao et al. (2010)
β₁-AR	β₁-*AR* Arg389 homozygotes had higher baseline IOP and greater response to betaxolol	Schwartz et al. (2005)
β₂-AR	No association between β₂-*AR* and response to timolol	Fuchsjager-Mayrl et al. (2005)
β₂-AR	β₂-*AR* Gln27 homozygotes more likely to experience a 20 % or greater decrease in IOP following treatment with β-blockers	McCarty et al. (2008)
Multiple	Upon treatment with timolol, β₁-*AR* Ser49 homozygotes had higher systolic and diastolic blood pressure using head-up tilt test; poor *CYP2D6* metabolizers had less favorable pharmacokinetic and pharmacodynamic responses to aqueous timolol; *GNAS1* CC homozygotes of T393C polymorphism associated with lower change in diastolic blood pressure from rest to maximum during exercise	Nieminen et al. (2005)
CYP2D6	*CYP2D6* Arg296Cys associated with greater risk of systemic bradycardia when exposed to timolol	Yang et al. (2009)

In a multiracial US series of 48 normal volunteers, the *β₁-AR* Arg389 homozygote genotype was associated with higher baseline IOP and a greater magnitude of response to treatment with betaxolol. There were no associations with *β₁-AR* Ser49Gly (Schwartz et al. 2005). An Austrian series of 89 normal volunteers treated with timolol reported no association between IOP response and the homozygous haplotypes of *β₂-AR* polymorphisms Arg16/Gln27, Gly16/Gln27, and Gly16/Glu27 (Fuchsjager-Mayrl et al. 2005). In contrast, in a US series of 210 patients with OAG or suspected glaucoma, *β₂-AR* Gln27 homozygotes were more likely to have a 20 % or greater decrease in IOP following treatment with topical β-blockers (McCarty et al. 2008).

At least two series have studied associations between genotype and systemic toxicity to ophthalmic timolol. In a Finnish group of 19 OAG patients and 18 normal volunteers treated with timolol, *β₁-AR* Ser49 homozygotes had higher systolic blood pressure and higher diastolic blood pressure than Gly49 carriers following a head-up tilt test. In this series, the investigators also studied polymorphisms of *CYP2D6* and the α-subunit of G protein (*GNAS1*). Subjects determined to be poor *CYP2D6* metabolizers (zero functional alleles) had less favorable pharmacokinetic and pharmacodynamic parameters than other subjects with aqueous timolol, but not with hydrogel timolol. The CC allele of the thymine-by-cytosine replacement at the base 393 (T393C) in *GNAS1* was associated with a lower change in diastolic blood pressure from rest to maximum during exercise (Nieminen et al. 2005). Similarly, in a Chinese series of 133 OAG patients, systemic bradycardia associated with topical timolol was more common with the *CYP2D6* Arg296Cys polymorphism (Yang et al. 2009).

2.2 Prostaglandin Analogues

Latanoprost is a highly selective agonist of the prostaglandin $F_{2\alpha}$ (FP) receptor (Stjernschantz et al. 1995). A Japanese series of 100 normal volunteers treated with latanoprost reported that the SNPs rs3753380 and rs3766355 in the

FP receptor gene were associated with the magnitude of response to therapy (Sakurai et al. 2007). Using pathway analysis, the following related polymorphisms were studied and reported to have no relationship with IOP reduction: *prostaglandin transporter* T396A (van der Zwaag et al. 2002); *fatty acid amide hydrolase* P129T (Sipe et al. 2002); *matrix metalloproteinase-1* promoter-1607 insG (Rutter et al. 1998); *matrix metalloproteinase-2* promoter C-1306T (Price et al. 2001); *matrix metalloproteinase-3* promoter-1171 delA (Ye et al. 1995); and *matrix metalloproteinase-9* promoter C-1562T and CA repeats (Zhang et al. 1999; St. Jean et al. 1995).

3 Age-Related Macular Degeneration (AMD)

Age-related macular degeneration (AMD) is the leading cause of irreversible visual loss among the elderly in Western nations (Resnikoff et al. 2004). Patients with AMD are typically classified as having exudative (neovascular) disease if there is evidence of choroidal neovascularization (CNV) and non-exudative (non-neovascular) if there is not (Ambati et al. 2003).

3.1 Non-exudative AMD

At this time, there is no proven effective therapy to reduce visual loss due to non-exudative AMD. The Age-Related Eye Disease Study (AREDS) reported that a specific combination of antioxidants and zinc reduces progression to advanced disease and visual loss in certain patients with non-exudative AMD (Age-Related Eye Disease Study Research Group 2001).

A polymorphism in the complement factor H gene (*CFH*) is strongly associated with AMD presence (Edwards et al. 2005; Haines et al. 2005; Klein et al. 2005; Hageman et al. 2005) and progression (Seddon et al. 2007). A second major risk locus includes the *ARMS2* (*age-related maculopathy susceptibility 2*, also known as *LOC387715*) and *HTRA1* (*HtrA serine peptidase 1*) genes and is also strongly associated with AMD (Jakobsdottir et al. 2005; Rivera et al. 2005; Yang et al. 2006; Dewan et al. 2006). Because the two genes are in strong linkage disequilibrium and their effects are statistically indistinguishable, it has yet to be determined if it is *ARMS2* or *HTRA1* that is responsible for the association with AMD.

A subset of patients in the AREDS trials was evaluated with respect to polymorphisms in *CFH* and *ARMS2*. In this group from the AREDS, 264 of 876 AREDS category 3 and 4 patients (30.1 %) progressed to advanced AMD over 5 years. In these patients, nutritional supplementation with antioxidants and zinc was associated with a greater reduction in disease progression in patients with the *CFH* Y402H TT genotype (68 %) than in patients with the CC genotype (11 %). There were no significant differences in disease progression with respect to genotype at *ARMS2* A69S (Klein et al. 2008).

3.2 Exudative AMD

A variety of pharmacologic treatments have demonstrated efficacy in the treatment of CNV secondary to exudative AMD. Photodynamic therapy (PDT) with verteporfin (Visudyne, Novartis, Basel, Switzerland) was reported to reduce the risk of visual loss in patients with predominantly classic CNV (Treatment of Age-related Macular Degeneration With Photodynamic Therapy (TAP) Study Group 1999). Intravitreal injection of VEGF inhibitors (Fig. 32.1) has now become standard treatment for exudative AMD (Kovach et al. 2012). At this time, four anti-VEGF agents are in clinical use. Pegaptanib (Macugen, Eyetech, Palm Beach Gardens, FL) (Gragoudas et al. 2004), ranibizumab (Lucentis, Genentech, South San Francisco, CA) (Rosenfeld et al. 2006; Brown et al. 2006), and aflibercept (Eylea, Regeneron, Tarrytown, NY) (Heier et al. 2011) are US FDA-approved for the treatment of exudative AMD. Bevacizumab (Avastin, Genentech, South San Francisco, CA) is FDA-approved for the systemic treatment of metastatic colorectal and other cancers (Yang et al. 2003), but is used extensively as an off-label intravitreal

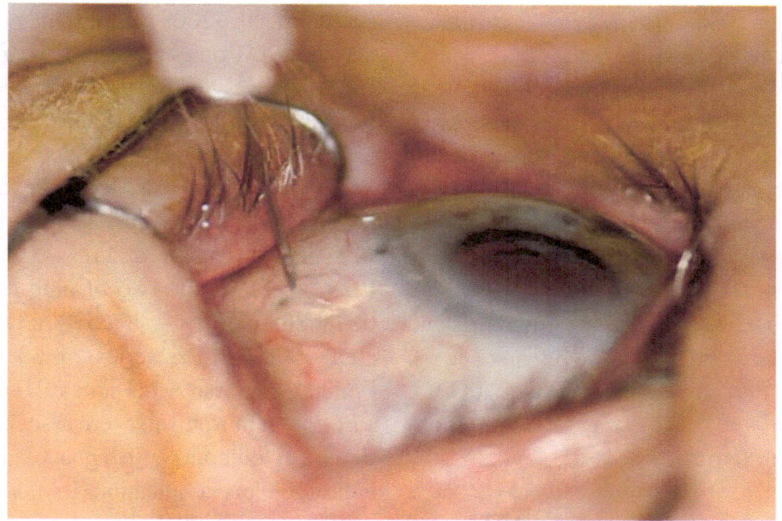

Fig. 32.1 Photograph demonstrating the technique of intravitreal injection, left eye

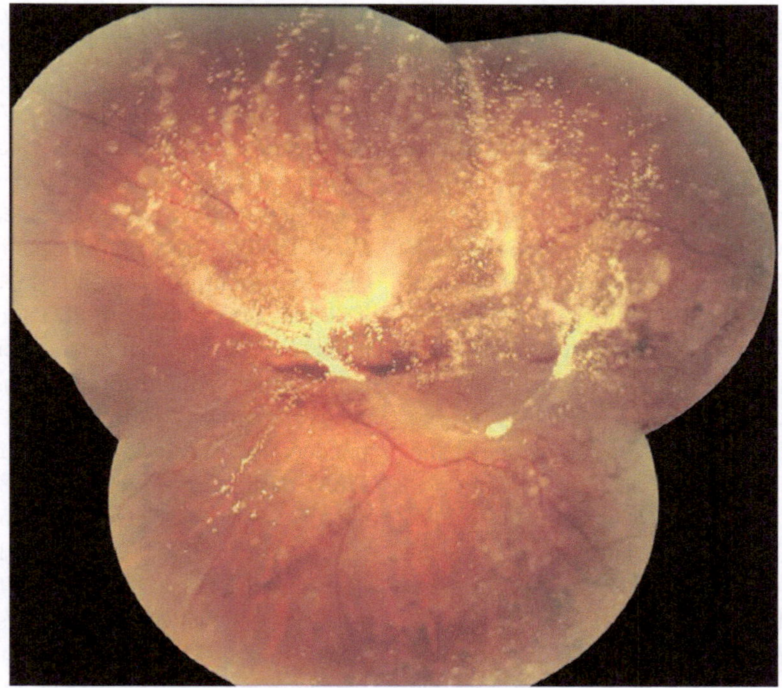

Fig. 32.2 Fundus photograph, left eye, demonstrating the typical appearance of triamcinolone acetonide in the vitreous cavity following injection (Image courtesy of Harry W. Flynn, Jr., M.D.)

treatment of exudative AMD (Rosenfeld et al. 2005). Intravitreal triamcinolone acetonide (Fig. 32.2) has proven somewhat efficacious as an adjunctive therapy, especially when combined with PDT (Chan et al. 2009) or bevacizumab (Ahmadieh et al. 2011).

Each of these agents has demonstrated efficacy, but there remains a persistent and unexplained variability among patients' individual treatment response, especially with the anti-VEGF agents (Menghini et al. 2010). In addition, triamcinolone is associated with several adverse events, especially elevated IOP, which remains poorly understood (Smithen et al. 2004). PDT is delivered as an intravenous infusion combined with the use of a photoactivator. The anti-VEGF agents are delivered as intravitreal injections, which are associated with a low risk of endophthalmitis and other serious complications (Schwartz et al. 2009). Similar to medications used to treat OAG, there is currently no reliable way to identify nonresponders prior to treatment.

3.2.1 Photodynamic Therapy (PDT)

Most pharmacogenetic studies of PDT have focused on the AMD-associated variants *CFH* Y402H and *ARMS2* A69S, although other genes have been studied, including *C-reactive protein* (*CRP*), *VEGF* (or *VEGFA*), *pigment epithelium-derived factor* (*PEDF*), and *apolipoprotein E* (*APOE*) (Table 32.2).

In a series of 27 patients from England treated with PDT, those with the *CFH* Y402H CC genotype lost, on average, more letters of visual acuity than did patients with the CT genotype (Goverdhan et al. 2008). A subsequent series of 69 patients from the USA reported that mean visual acuity following PDT was worse in patients with *CFH* Y402H TT than with CT or CC, in patients with classic CNV but not occult CNV. There was no association seen between visual outcome and *ARMS2* A69S (Brantley et al. 2009).

In a series of 110 patients from Japan, *HTRA1* rs11200638 GG was associated with improved visual acuity outcomes and less risk of disease recurrence following PDT. In this study, the combination of *CFH* rs1410996 and rs2274700 was associated with a reduction in the time interval until disease recurrence following PDT. There was no association between PDT outcomes and *CFH* rs1061170 (Y402H) and rs800292; *VEGF* rs699947, rs1570360, and rs2010963; and *PEDF* rs12150053, rs12948385, rs9913583, and rs1136287 (Tsuchihashi et al. 2011).

Other series from Finland (Seitsonen et al. 2007), Israel (Chowers et al. 2008a), and Australia (Feng et al. 2009) reported no association between PDT outcomes and *CFH* Y402H. A subsequent series form Israel reported no associations between PDT and *ARMS2* A69S or *HTRA1* rs11200638 (Chowers et al. 2008b).

Polymorphisms in other genes have been linked to PDT outcomes. An Australian series of 273 patients reported an association between PDT outcomes and two of nine polymorphisms (rs2808635 GG and rs876538 AA) in *CRP* (Feng et al. 2009). In a series of 86 patients from Finland, two (rs699947 and rs2146323) of three *VEGF* polymorphisms were associated with PDT treatment outcomes, using a binary responder/nonresponder classification (Immonen et al. 2010).

A series of 90 Italian patients treated with PDT for classic CNV was screened for polymorphisms in genes related to coagulation, including *factor V* G1691A, *prothrombin* G20210A, *factor XIII-A* G185T, *methylenetetrahydrofolate reductase (MTHFR)* C677T, *methionine synthase* A2756G, and *methionine synthase reductase* A66G. Using a binary responder/nonresponder classification, responders were associated with *prothrombin* G20210A and *MTHFR* 677T carriers, and nonresponders were associated with *factor XIII-A* 185T carriers (Parmeggiani et al. 2007). The same group subsequently reported a series of 84 patients treated with PDT for occult CNV; in these patients, responders were associated with the combination carriers of *factor V* 1691A and *prothrombin* 20210A alleles, while nonresponders were associated with f*actor XIII-A* G185T (Parmeggiani et al. 2008). The same investigators also reported a series of 234 patients treated with PDT for CNV secondary to pathologic myopia, rather than AMD. They reported an association between responders and carriers of *MTHFR* 677T allele and between nonresponders and *factor XIII-A* 185 GT/TT genotypes (Parmeggiani et al. 2010).

3.2.2 Anti-VEGF Agents

At this time, all reported pharmacogenetic studies of anti-VEGF agents have involved bevacizumab or ranibizumab. Again, most studies have focused on *CFH* Y402H and *ARMS2* A69S (Table 32.3).

Table 32.2 Pharmacogenomics of photodynamic therapy

Gene(s)	Results	References
CFH	*CFH* Y402H CC associated with worse visual outcomes	Goverdhan et al. (2008)
CFH and *ARMS2*	*CFH* Y402H TT associated with worse visual outcomes in classic CNV; no association with *ARMS2*	Brantley et al. (2009)
CFH, HTRA1, VEGF, and *PEDF*	*HTRA1* GG associated with more favorable treatment outcomes; combination of 2 *CFH* genotypes associated with reduced time interval until disease recurrence; no other associations	Tsuchihashi et al. (2011)
CFH	No association	Seitsonen et al. (2007)
CFH	No association	Chowers et al. (2008a)
CFH and *CRP*	No association with *CFH*; 2 of 9 *CRP* polymorphisms associated with more favorable treatment response	Feng et al. (2009)
ARMS2 and *HTRA1*	No associations	Chowers et al. (2008b)
VEGF	2 polymorphisms associated with response to treatment	Immonen et al. (2010)
Multiple	Classic CNV: *prothrombin* G20210A and *MTFHR* C677T associated with treatment response; *factor XIII-A* G185T associated with nonresponse; no other associations	Parmeggiani et al. (2007)
Multiple	Occult CNV: combination of *factor V* 1691A and *prothrombin* 20210A alleles associated with response; *factor XIII-A*G185T associated with nonresponse; no other associations	Parmeggiani et al. (2008)
Multiple	Pathologic myopia: *MTHFR*677T allele associated with response; *factor XIII-A*185 GT/TT genotypes associated with nonresponse	Parmeggiani et al. (2010)

Table 32.3 Pharmacogenomics of anti-vascular endothelial growth factor therapy

Treatment	Genes	Result	References
Bevacizumab	*CFH* and *ARMS2*	*CFH* Y402H CC associated with worse visual outcomes; no association with *ARMS2*	Brantley et al. (2007)
Bevacizumab	*CFH*	*CFH* Y402H CC associated with worse visual outcomes	Nischler et al. (2011)
Ranibizumab	*CFH*	*CFH* Y402H CC associated with more injections required	Lee et al. (2009)
Ranibizumab	*CFH* and *ARMS2*	*ARMS2* A69S TT and *CFH* Y402H CC associated with worse visual outcomes	Teper et al. (2010)
Ranibizumab	*CFH*	*CFH* Y402H CC associated with poor treatment response; combination heterozygotes at *CFH* Y402H and *FZD4* rs10896563 associated with more favorable outcomes; no other associations	Kloeckener-Gruissem et al. (2011)
Ranibizumab	*CFH, HTRA1*, and *VEGF*	*CFH* Y402H TC associated with better visual outcomes; no association with number of injections with any gene	McKibbin et al. (2012)
Bevacizumab and/or ranibizumab	*APOE*	*APOE* ε4 associated with better treatment outcomes	Wickremasinghe et al. (2011)
Bevacizumab or ranibizumab	*CFH, ARMS2, HTRA1*, and *C3*	No statistically significant differences for any gene	Hagstrom et al. (2013)

In a series of 86 US patients treated with bevacizumab, patients with the *CFH* Y402H CC genotype experienced less favorable visual results than other two genotypes (TC and TT), while there were no associations with *ARMS2/LOC387715* (Brantley et al. 2007). Similar results were reported in a series of 197 patients from Austria, in which patients with *CFH* Y402H CC on average lost visual acuity, while patients with CC or TC on average gained visual acuity following treatment with bevacizumab (Nischler et al. 2011).

In a series of 156 US patients treated using an as-needed protocol with ranibizumab, the *CFH* Y402H CC polymorphism correlated with an increased number of injections performed (Lee et al. 2009). Similar results were reported in a series of 90 patients from Poland, in which *CFH* Y402H CC and *ARMS2* A69S TT were associated with relatively less favorable visual outcomes following treatment with ranibizumab (Teper et al. 2010). In a series of 243 Swiss patients treated with ranibizumab and classified as poor responders vs. good responders, *CFH* Y402H CC was associated with poor responders and the combination of heterozygous genotypes at *CFH* Y402H and *frizzled homolog 4* (*FZD4*) rs10896563 was associated with good responders; *ARMS2*, *HTRA1*, *VEGFA*, *complement factor B* (*CFB*), *kinase insert domain receptor* (*KDR*), and *low-density lipoprotein receptor-related protein 5* (*LRP5*) were not associated with treatment outcomes (Kloeckener-Gruissem et al. 2011). A UK series of 104 patients treated with ranibizumab reported an association between visual acuity improvement of five letters or more and *CFH* Y402H CT compared with *CFH* Y402H TT; the investigators also reported nonsignificant trends towards more favorable outcomes associated with *HTRA1* and *VEGF* polymorphisms (McKibbin et al. 2012).

In a series of 172 Australian patients treated with bevacizumab, ranibizumab, or a combination of the two medications, the *APOE* ε4 allele associated with better visual acuity outcomes compared with the *APOE* ε2 allele (Wickremasinghe et al. 2011).

The Comparison of AMD Treatments Trials (CATT) was a prospective randomized clinical trial comparing ranibizumab to bevacizumab. A subset of 834 patients enrolled in CATT underwent pharmacogenetic testing. No statistically significant differences were found with respect to variants at *CFH, ARMS2, HTRA1*, and *C3* (Hagstrom et al. 2013).

3.2.3 Triamcinolone Acetonide

In a pilot study of 52 US patients treated with intravitreal triamcinolone acetonide for a variety of retinal diseases, including wet AMD, there were no significant associations between visual outcome and six common polymorphisms of the glucocorticoid receptor gene (*GR*) (Gerzenstein et al. 2008). In another study of 102 Hungarian patients, the *GR* N363S polymorphism was associated with steroid-induced ocular hypertension upon treatment with topical corticosteroids (prednisolone acetate 0.5 %, flurometholone 0.1 %, or combined) after photorefractive keratectomy (Szabo et al. 2007).

Other genes have been investigated with respect to steroid-induced IOP elevation, including spliceosome proteins (Xu et al. 2003; Yan et al. 2010) and immunophilins (Zhang et al. 2008). In a US series of 197 OAG patients, 107 steroid responders, and 400 normal volunteers, there were no associations with polymorphisms in *GR*, the immunophilin *FKBP4*, or the spliceosomes *SFRS3*, *SFRS5*, or *SFRS9* (Fingert et al. 2010).

4 Summary

At this time, pharmacogenomics is primarily useful as a research tool within ophthalmology. Pharmacogenomic testing of ophthalmic medications is not generally performed in clinical settings.

The Pharmacogenomics Knowledgebase (PharmGKB) states: "Timolol is a non-selective beta-adrenergic agonist applied to the eye to reduce intraocular pressure. It is metabolized via CYP2D6. The FDA recommends, but does not require, genetic testing prior to initiating or reinitiating treatment with Timolol (Istalol)" (http://www.pharmgkb.org/drug/PA451690, accessed 9/12/12). However, the FDA Table of Pharmacogenomic Biomarkers in Drug Labels, which lists medications that include pharmacogenomic information on their labels, includes the systemic beta-blockers carvedilol, metoprolol, and propanolol, but not timolol or any other ophthalmic medication (http://www.fda.gov/Drugs/ScienceResearch/ResearchAreas/Pharmacogenetics/ucm083378.htm, accessed 9/12/12).

Although several studies of glaucoma and AMD therapeutics have reported statistically significant associations between genotype and treatment response, none of these associations has been confirmed in large-scale clinical trials. In many instances, the data appear inconsistent among studies. This may be due to important underlying differences in baseline genetic characteristics between studies, particularly between studies from different continents. Different studies also used different enrollment criteria (normal volunteers vs. glaucoma patients, classic CNV vs. occult), different study endpoints (visual acuity, anatomic response, number of re-treatments required), and statistics (continuous outcomes vs. binary "good responder"/"poor responder" outcomes).

As data from clinical trials continues to be collected, various pharmacogenomic relationships may become clearer.

Acknowledgment Partially supported by NIH Center Core Grant P30EY014801, an Unrestricted Grant from Research to Prevent Blindness (New York, NY), and the Department of Defense (DOD Grant #W81XWH-09-1-0675).

Dr. Schwartz has performed consulting activities for Alimera, Bausch+Lomb, Eyetech, and ThromboGenics, has received lecture fees from Regeneron, and has received royalties from IC Labs related to the use of genetics to detect steroid responders.

References

Age-Related Eye Disease Study Research Group (2001) A randomized, placebo-controlled, clinical trial of high-dose supplementation with vitamins C and E, beta carotene, and zinc for age-related macular degeneration and vision loss: AREDS report no. 8. Arch Ophthalmol 119:1417–1436

Ahmadieh H, Taei R, Riazi-Esfahani M, Piri N, Homayouni M, Daftarian N, Yaseri M (2011) Intravitreal bevacizumab versus combined intravitreal bevacizumab and triamcinolone for neovascular age-related macular degeneration: six-month results of a randomized clinical trial. Retina 31:1819–1826

Allen RC, Hertzmark E, Walker AM, Epstein DL (1986) A double-masked comparison of betaxolol vs. timolol in the treatment of open-angle glaucoma. Am J Ophthalmol 101:535–541

Ambati J, Ambati BK, Yoo SH, Ianchulev S, Adamis AP (2003) Age-related macular degeneration: etiology, pathogenesis, and therapeutic strategies. Surv Ophthalmol 48:257–293

Brantley MA Jr, Fang AM, King JM, Tewari A, Kymes SM, Shiels A (2007) Association of complement factor H and LOC387715 genotypes with response of exudative age-related macular degeneration to intravitreal bevacizumab. Ophthalmology 114:2168–2173

Brantley MA Jr, Edelstein SL, King JM, Plotzke MR, Apte RS, Kymes SM, Shiels A (2009) Association of complement factor H and LOC387715 genotypes with response of exudative age-related macular degeneration to photodynamic therapy. Eye 23:626–631

Brown DM, Kaiser PK, Michels M, Soubrane G, Heier JS, Kim RY, Sy JP, Schneider S, ANCHOR Study Group (2006) Ranibizumab versus verteporfin for neovascular age-related macular degeneration. N Engl J Med 355:1432–1444

Camras CB, Hedman K (2003) Rate of response to latanoprost or timolol in patients with ocular hypertension or glaucoma. J Glaucoma 12:466–469

Chan A, Blumenkranz MS, Wu KH, Wang G, Berker N, Parast LM, Sanislo SR (2009) Photodynamic therapy with and without adjunctive intravitreal triamcinolone acetonide: a retrospective comparative study. Ophthalmic Surg Lasers Imaging 40:561–569

Chowers I, Cohen Y, Goldenberg-Cohen G, Vicuna-Kojchen J, Lichtinger A, Weinstein O, Pollack A, Axer-Siegel R, Hemo I, Averbukh E, Banin E, Meir T, Lederman M (2008a) Association of complement factor H Y402H polymorphism with phenotype of neovascular age related macular degeneration in Israel. Mol Vis 14:1829–1834

Chowers I, Meir T, Lederman M, Goldenberg-Cohen N, Cohen Y, Banin E, Averbukh E, Hemo I, Pollack A, Axer-Siegel R, Weinstein O, Hoh J, Zack DJ, Galbinur T (2008b) Sequence variants in *HTRA1* and *LOC387715/ARMS2* and phenotype and response to photodynamic therapy in neovascular age-related macular degeneration in populations from Israel. Mol Vis 14:2263–2271

Costagliola C, dell'Omo R, Romano MR, Rinaldi M, Zeppa L, Parmeggiani F (2009a) Pharmacotherapy of intraocular pressure: part I. Parasympathomimetic, sympathomimetic and sympatholytics. Expert Opin Pharmacother 10:2663–2677

Costagliola C, dell'Omo R, Romano MR, Rinaldi M, Zeppa L, Parmeggiani F (2009b) Pharmacotherapy of intraocular pressure: part II. Carbonic anhydrase inhibitors, prostaglandin analogues and prostamides. Expert Opin Pharmacother 10:2859–2870

Dewan A, Liu M, Hartman S, Zhang SS, Liu DT, Zhao C, Tam PO, Chan WM, Lam DS, Snyder M, Barnstable C, Pang CP, Hoh J (2006) HTRA1 promoter polymorphism in wet age-related macular degeneration. Science 314:989–992

Edwards AO, Ritter R III, Abel KJ, Manning A, Panjuysen C, Farrer LA (2005) Complement factor H polymorphism and age-related macular degeneration. Science 308:421–424

Feng X, Xiao J, Longville B, Tan AX, Wu XN, Cooper MN, McAllister IL, Isaacs T, Palmer LJ, Constable IJ (2009) Complement factor H Y402H and c-reactive protein polymorphism and photodynamic therapy response in age-related macular degeneration. Ophthalmology 116:1908–1912

Fingert JH, Alward WL, Wang K, Yorio T, Clark AF (2010) Assessment of SNPs associated with the human glucocorticoid receptor in primary open-angle glaucoma and steroid responders. Mol Vis 16:596–601

Flordellis C, Paris H, Karabinis A, Lymperopoulos A (2004) Pharmacogenomics of adrenoceptors. Pharmacogenomics 5:803–817

Fuchsjager-Mayrl G, Markovic O, Losert D, Lucas T, Wachek V, Muller M, Schmetterer L (2005) Polymorphism of the beta-2 adrenoceptor and IOP lowering potency of topical timolol in healthy subjects. Mol Vis 23:811–815

Gao Y, Sakurai M, Takeda H, Higashide T, Kawase K, Sugiyama K (2010) Association between genetic polymorphisms of adrenergic receptor and diurnal intraocular pressure in Japanese normal-tension glaucoma. Ophthalmology 117:2359–2364

Gerzenstein SM, Pletcher MT, Cervino AC, Tsinoremas NF, Young B, Puliafito CA, Fini ME, Schwartz SG (2008) Glucocorticoid receptor polymorphisms and intraocular pressure response to intravitreal triamcinolone acetonide. Ophthalmic Genet 29:166–170

Goverdhan SV, Hannan S, Newsom RB, Luff AJ, Griffiths H, Lotery AJ (2008) An analysis of the CFH Y402H genotype in AMD patients and controls from the UK, and response to PDT treatment. Eye 22:849–854

Gragoudas ES, Adamis AP, Cunningham ET Jr, Feinsod M, Guyer DR, VEGF Inhibition Study in Ocular Neovascularization Clinical Trial Group (2004) Pegaptanib for neovascular age-related macular degeneration. N Engl J Med 351:2805–2816

Green SA, Cole G, Jacinto M, Innis M, Liggett SB (1993) A polymorphism of the human beta 2-adrenergic receptor within the fourth transmembrane domain alters ligand binding and functional properties of the receptor. J Biol Chem 268:23116–23121

Green SA, Turki J, Innis M, Liggett SB (1994) Amino-terminal polymorphisms of the human beta 2-adrenergic receptor impart distinct agonist-promoted regulatory properties. Biochemistry 33:9414–9419

Hageman GS, Anderson DH, Johnson LV, Hancox LS, Taiber AJ, Hardisty LI, Hageman JL, Stockman HA, Borchardt JD, Gehrs KM, Smith RJ, Silvestri G, Russell SR, Klaver CC, Barbaetto I, Chang S, Yannuzzi LA, Barile GR, Merriam JC, Smith RT, Olsh AK, Bergeron J, Zernant J, Merriam JE, Gold B, Dean M, Allikmets R (2005) A common haplotype in the complement regulator gene factor H (HFI/CFH) predisposes individuals to age-related macular degeneration. Proc Natl Acad Sci 102:7227–7232

Hagstrom SA, Ying GS, Pauer GJ, Sturgill-Short GM, Huang J, Callanan DG, Kim IK, Klein ML, Maguire MG, Martin DF, Comparison of AMD Treatments Trials Research Group (2013) Pharmacogenetics for genes associated with age-related macular degeneration in the Comparison of AMD Treatments Trials (CATT). Ophthalmology 120:593–599

Haines JL, Hauser MA, Schmidt S, Scott WK, Olson LM, Gallins P, Spencer KL, Kwan SY, Noureddine M, Gilbert JR, Schnetz-Boutaud N, Agarwal A, Postel EA, Pericak-Vance MA (2005) Complement factor H variant increases the risk of age-related macular degeneration. Science 308:419–421

Heier JS, Boyer D, Nguyen QD, Marcus D, Roth DB, Yancopoulos G, Stahl N, Ingerman A, Vitti R, Berliner AJ, Yang K, Brown DM, CLEAR-IT 2 Investigators (2011) The 1-year results of CLEAR-IT 2, a phase 2 study of vascular endothelial growth factor trap-eye dosed as-needed after 12-week fixed dosing. Ophthalmology 118:1098–1106

Immonen I, Seitsonen S, Tommila P, Kangas-Kontio T, Kakko S, Savolainen ER, Liinamaa MJ (2010) Vascular endothelial growth factor gene variation and the response to photodynamic therapy in age-related macular degeneration. Ophthalmology 117:103–108

Inagaki Y, Mashima Y, Fuse N, Funayama T, Ohtake Y, Yasuda N, Murakami A, Hotta Y, Fukuchi T, Tsubota K (2006) Polymorphism of beta-adrenergic receptors and susceptibility to open-angle glaucoma. Mol Vis 12:673–680

Jakobsdottir J, Conley YP, Weeks DE, Mah TS, Ferrell RE, Gorin MB (2005) Susceptibility genes for age-related maculopathy on chromosome 10q26. Am J Hum Genet 77:389–407

Klein RJ, Zeiss C, Chew EY, Tsai JY, Sackler RS, Haynes C, Henning AK, SanGiovanni JP, Mane SM, Mayne ST, Bracken MB, Ferris FL, Ott J, Barnstable C, Hoh J (2005) Complement factor H polymorphism in age-related macular degeneration. Science 308:385–389

Klein ML, Francis PJ, Rosner B, Reynolds R, Hamon SC, Schultz DW, Ott J, Seddon JM (2008) *CFH* and *LOC387715/ARMS2* genotypes and treatment with antioxidants and zinc for age-related macular degeneration. Ophthalmology 115:1019–1025

Kloeckener-Gruissem B, Barthelmes D, Labs S, Schindler C, Kurz-Levin M, Michels S, Fleischhauer J, Berger W, Sutter F, Menghini M (2011) Genetic association with response to intravitreal ranibizumab in patients with neovascular AMD. Invest Ophthalmol Vis Sci 52:4694–4702

Kovach JL, Schwartz SG, Flynn HW Jr, Scott IU (2012) Anti-VEGF treatment strategies for wet AMD. J Ophthalmol 2012:786870

Lee AY, Raya AK, Kyrnes SM, Shiels A, Brantley MA Jr (2009) Pharmacogenetics of complement factor H

(Y402H) and treatment of exudative age-related macular degeneration with ranibizumab. Br J Ophthalmol 93:610–613

Liggett SB (2000) Pharmacogenomics of beta-1 and beta-2 adrenergic receptors. Pharmacology 61:167–173

Maqbool A, Hall AS, Ball SG, Balmforth AJ (1999) Common polymorphisms of β_1-adrenoceptor identification and rapid screening assay. Lancet 353:897

Mason DA, Moore JD, Green SA, Liggett SB (1999) A gain-of-function polymorphism in a G-protein coupling domain of the human β_1-adrenergic receptor. J Biol Chem 274:12670–12674

Materson BJ, Reda DJ, Cushman WC, Massie BM, Freis ED, Kochar MS, Hamburger RJ, Frye C, Lakshman R, Gottdiener J, Ramirez EA, Henderson WG, for the Department of Veterans Affairs Cooperative Study Group on Antihypertensive Agents (1993) Single-drug therapy for hypertension in men: a comparison of six antihypertensive agents with placebo. N Engl J Med 328:914–921

McCarty CA, Burmester JK, Mukesh BN, Patchett RB, Wilke RA (2008) Intraocular pressure response to topical beta-blockers associated with an ADRB2 single-nucleotide polymorphism. Arch Ophthalmol 126:959–963

McGourty JC, Silas JH, Fleming JJ, McBurney A, Ward JW (1985) Pharmacokinetics and beta-blocking effects of timolol in poor and extensive metabolizers of debrisoquine. Clin Pharmacol Ther 38:409–413

McKibbin M, Ali M, Bansal S, Baxter PD, West K, Williams G, Cassidy F, Inglehearn CF (2012) CFH, VEGF, and HTRA1 promoter genotype may influence the response to intravitreal ranibizumab therapy for neovascular age-related macular degeneration. Br J Ophthalmol 96:208–212

McLaren N, Reed DM, Musch DC, Downs CA, Higashi ME, Santiago C, Radenbaugh PA, Allingham RR, Richards JE, Moroi SE (2007) Evaluation of the β2-adrenergic receptor gene as a candidate glaucoma gene in 2 ancestral populations. Arch Ophthalmol 125:105–111

Menghini M, Kurz-Levin MM, Amstutz C, Michels S, Windisch R, Barthelmes D, Sutter FK (2010) Response to ranibizumab therapy in neovascular AMD: an evaluation of good and bad responders. Klin Monbl Augenheilkd 227:244–248

Moroi SE, Raoof DA, Reed DM, Zollner S, Qin Z, Richards JE (2009) Progress toward personalized medicine for glaucoma. Expert Rev Ophthalmol 4:146–161

Nieminen T, Uusitalo H, Maenpaa J, Turjanmaa V, Rane A, Lundgren S, Ropo A, Rontu R, Lehtimaki T, Kahonen M (2005) Polymorphisms of genes CY2D6, ADRB1, and GNAS1 in pharmacokinetics and systemic effects of ophthalmic timolol: a pilot study. Eur J Clin Pharmacol 61:811–819

Nischler C, Oberkofler H, Ortner C et al (2011) Complement factor H Y402H gene polymorphism and response to intravitreal bevacizumab in exudative age-related macular degeneration. Acta Ophthalmol 89:344–349

Parmeggiani F, Costagliola C, Gemmati D, D'Angelo S, Perri P, Scapoli GL, Catozzi L, Federici F, Sebastiani A, Incorvaia C (2007) Predictive role of coagulation-balance gene polymorphisms in the efficacy of photodynamic therapy with verteporfin for classic choroidal neovascularization secondary to age-related macular degeneration. Pharmacogenet Genomics 17:1039–1046

Parmeggiani F, Costagiola C, Gemmati D, D'Angelo S, Perri P, Campa C, Catozzi L, Federici F, Sebastiani A, Incorvaia C (2008) Coagulation gene predictors of photodynamic therapy for occult choroidal neovascularization in age-related macular degeneration. Invest Ophthalmol Vis Sci 49:3100–3106

Parmeggiani F, Gemmati D, Costagliola C, Semeraro F, D'Angelo S, Perri P, Sebastiani A, Incoravaia C (2010) Impact of coagulation-balance gene predictors on efficacy of photodynamic therapy for choroidal neovascularization in pathologic myopia. Ophthalmology 117:517–523

Parola AL, Kobilka BK (1994) The peptide product of a 5′ leader cistron in the beta 2 adrenergic receptor mRNA inhibits receptor synthesis. J Biol Chem 269:4497–4505

Price SJ, Greaves DR, Watkins H (2001) Identification of novel, functional genetic variants in the human matrix metalloproteinase-2 gene: role of Sp1 in allele-specific transcriptional regulation. J Biol Chem 276:7549–7558

Resnikoff S, Pascolini D, Etya'ale D, Kocur I, Pararajasegaram R, Pokharel GP, Mariotti SP (2004) Global data on visual impairment in the year 2002. Bull World Health Organ 82:844–851

Rivera A, Fisher SA, Fritsche LG, Keilhauer CN, Lichtner P, Meitinger T, Weber BH (2005) Hypothetical LOC387715 is a second major susceptibility gene for age-related macular degeneration, contributing independently of complement factor H to disease risk. Hum Mol Genet 14:3227–3236

Rosenfeld PJ, Moshfeghi AA, Puliafito CA (2005) Optical coherence tomography findings after an intravitreal injection of bevacizumab (Avastin) for neovascular age-related macular degeneration. Ophthalmic Surg Lasers Imaging 36:331–335

Rosenfeld PJ, Brown DM, Heier JS, Boyer DS, Kaiser PK, Chung CY, Kim RY, MARINA Study Group (2006) Ranibizumab for neovascular age-related macular degeneration. N Engl J Med 355:1419–1431

Rutter JL, Mitchell TI, Buttice G, Meyers J, Gusella JF, Ozelius LJ, Brinckerhoff CE (1998) A single nucleotide polymorphism in the matrix metalloproteinase-1 promoter creates an Ets binding site and augments transcription. Cancer Res 58:5321–5325

Sakurai M, Higashide T, Takahashi M, Sugiyama K (2007) Association between genetic polymorphisms of the prostaglandin $F_{2\alpha}$ receptor gene and response to latanoprost. Ophthalmology 114:1039–1045

Schwartz SG, Brantley MA Jr (2011) Pharmacogenetics and age-related macular degeneration. J Ophthalmol 2011:252549

Schwartz SG, Puckett BJ, Allen RC, Castillo IG, Leffler CT (2005) β_1-adrenergic receptor polymorphisms and clinical efficacy of betaxolol hydrochloride in normal volunteers. Ophthalmology 112:2131–2136

Schwartz SG, Ayala-Haedo JA, Kishor KS, Fini ME (2008) Pharmacogenomics of open-angle glaucoma. Curr Pharmacogenomics Personal Med 6:121–125

Schwartz SG, Flynn HW, Scott IU (2009) Endophthalmitis after intravitreal injections. Expert Opin Pharmacother 10:2119–2126

Seddon JM, Francis PJ, George S, Schultz DW, Rosner B, Klein ML (2007) Association of CFH Y402H and LOC387715 A69S with progression of age-related macular degeneration. JAMA 297:1793–1800

Seitsonen SP, Jarvela IE, Meri S, Tommila PV, Ranta PH, Immonen IJ (2007) The effect of complement factor H Y402H polymorphism on the outcome of photodynamic therapy in age-related macular degeneration. Eur J Ophthalmol 17:943–949

Sipe JC, Chiang K, Gerber AL, Beutler E, Cravatt BF (2002) A missense mutation in human fatty acid amide hydrolase associated with problem drug use. Proc Natl Acad Sci USA 99:8394–8399

Smithen LM, Ober MD, Maranan L, Spaide RF (2004) Intravitreal triamcinolone acetonide and intraocular pressure. Am J Ophthalmol 138:740–743

St. Jean PL, Zhang XC, Hart BK, Lamlum H, Webster MW, Steed DL, Henney AM, Ferrell RE (1995) Characterization of a dinucleotide repeat in the 92 kDa type IV collagenase gene (CLG4b), localization of CLG4B to chromosome 20 and the role of CLG4B in aortic aneurysmal disease. Ann Hum Genet 59:17–24

Stjernschantz J, Selen G, Sjoquist B, Resul B (1995) Preclinical pharmacology of latanoprost, a phenyl-substituted PGF2 alpha analogue. Adv Prostaglandin Thromboxane Leukot Res 23:513–518

Szabo V, Borgulya G, Filkorn T, Majnik J, Banyasz I, Nagy ZZ (2007) The variant N363S of glucocorticoid receptor in steroid-induced ocular hypertension in Hungarian patients treated with photorefractive keratectomy. Mol Vis 13:659–666

Teper SJ, Nowinska A, Pilat J, Palucha A, Wylegala E (2010) Involvement of genetic factors in the response to a variable-dosing ranibizumab treatment regimen for age-related macular degeneration. Mol Vis 16:2598–2604

Treatment of Age-Related Macular Degeneration with Photodynamic Therapy (TAP) Study Group (1999) Photodynamic therapy of subfoveal choroidal neovascularization in age-related macular degeneration with verteporfin: one-year results of 2 randomized clinical trials – TAP report. Arch Ophthalmol 117:1329–1345

Tsuchihashi T, Mori K, Horie-Inoue K, Gehlbach PL, Kabasawa S, Takita H, Ueyama K, Okazaki Y, Inoue S, Awata T, Katayama S, Yoneya S (2011) Complement factor H and high-temperature requirement A-1 genotypes and treatment response of age-related macular degeneration. Ophthalmology 118:93–100

van der Zwaag B, Verzijl HT, Beltran-Valero de Bernabe D, Schuster VL, van Bokhoven H, Kremer H, van Reen M, Wichers GH, Brunner HG, Padberg GW (2002) Mutation analysis in the candidate Mobius syndrome genes PGT and GATA2 on chromosome 3 and EGR2 on chromosome 10. J Med Genet 39:E30

Wickremasinghe SS, Xie J, Lim J, Chauhan DS, Robman L, Richardson AJ, Hageman G, Baird PN, Guymer R (2011) Variants in the APOE gene are associated with improved treatment outcome following anti-VEGF therapy for neovascular AMD. Invest Ophthalmol Vis Sci 52:4072–4079

Xu Q, Leung DY, Kisich KO (2003) Serine-arginine-rich protein p30 directs alternative splicing of glucocorticoid receptor pre-mRNA to glucocorticoid receptor beta in neutrophils. J Biol Chem 278:27112–27118

Yan XB, Tang CH, Huang Y, Fang H, Yu ZQ, Wu LM, Liu RY (2010) Alternative splicing in exon 9 of glucocorticoid receptor pre-mRNA is regulated by SRp40. Mol Biol Rep 37:1427–1433

Yang JC, Haworth L, Sherry RM, Hwu P, Schwartzentruber DJ, Topalian SL, Steinberg SM, Chen HX, Rosenberg SA (2003) A randomized trial of bevacizumab, an anti-vascular endothelial growth factor antibody, for metastatic renal cancer. N Engl J Med 349:427–434

Yang Z, Camp NJ, Sun H, Tong Z, Gibbs D, Cameron DJ, Chen H, Zhao Y, Pearson E, Li X, Chien J, Dewan A, Harmon J, Bernstein PS, Shridhar V, Zabriskie NA, Hoh J, Howes K, Zhang K (2006) A variant of the HTRA1 gene increases susceptibility to age-related macular degeneration. Science 314:992–993

Yang Y, Wu K, Yuan H, Yu M (2009) Cytochrome oxidase 2D6 gene polymorphism in primary open-angle glaucoma with various effects to ophthalmic timolol. J Ocul Pharmacol Ther 25:163–171

Ye S, Watts GF, Mandalia S, Humphries SE, Henney AM (1995) Preliminary report: genetic variation in the human stromelysin promoter is associated with progression of coronary atherosclerosis. Br Heart J 73:209–215

Zhang B, Ye S, Herrmann SM, Eriksson P, de Maat M, Evans A, Arvelier D, Luc G, Cambien F, Hamsten A, Watkins H, Henney AM (1999) Functional polymorphism in the regulatory region of gelatinase B gene in relation to severity of coronary atherosclerosis. Circulation 99:1788–1794

Zhang X, Clark AF, Yorio T (2008) FK506-binding protein 51 regulates nuclear transport of the glucocorticoid receptor beta and glucocorticoid responsiveness. Investig Ophtalmol Vis Sci 49:1037–1047

Section VII

Personalized Medicine: Cell Therapy and Transplantation

Personalized Medicine in Cell Therapy and Transplantation

33

Madhusudana Girija Sanal

Abstract

Developments in cell therapy and organ transplantation would transform personalized medicine as perceived today. Adult stem cells and induced pluripotent stem cell (iPSC) derivatives can be used to repair, rejuvenate, or replace damaged organs. Transplanted cells act by secreted factors, cell-cell interaction, and immune modulation or by repopulating the damaged tissue. Repopulating organs with genetically corrected progenitor cells or differentiated cells could correct several rare genetic diseases at a very personal level. Cell therapy has applications in curing infections as well from life-threatening fungal infections to HIV. Today tools such as zinc finger nucleases and TALENS make it possible to manipulate human genome precisely. In the future transplantable personalized whole organs will be generated using iPSC and tetraploid complementation, and techniques which are still in infancy. Cells and organs will be engineered and standardized to be compatible with a wide range of drugs and environmental conditions or become more personalized for special needs. Three dimensional printing technologies can now generate simple organs like urinary bladder in a personalized way, but in the future, it might be possible to "print" more complex organs. Genetically engineered cells would play a major role in the future of fighting cancer in a personalized manner. Stem cell biology, genetic engineering, and regenerative medicine catapulted by latest developments in basic sciences would revolutionize human life, and we need to prepare and sensitize our society well ahead.

1 End of the Beginning of a Great Revolution in Medicine

We are at the verge of a scientific revolution which is going to change how we human beings look, live, think, and interact. So far the revolutions have been in the shape of machines, which enabled us to fly faster than birds, swim faster than fish, run faster

M.G. Sanal (✉)
Institute of Liver and Biliary Sciences,
New Delhi, India
e-mail: sanalmg@gmail.com

D. Barh et al. (eds.), *Omics for Personalized Medicine*,
DOI 10.1007/978-81-322-1184-6_33, © Springer India 2013

than a cheetah, peer through the cosmic clouds, and communicate with people far away in space and so on. But the ongoing revolution in biotechnology is capable of changing the very nature of humankind – not only by conquering diseases but also enhancing the longevity, physical and intellectual power and even our physical appearance as human beings. The relentless scientific developments in this direction could result in several social issues. It is necessary to alert and prepare society early enough to accept these revolutionary consequences. Personalized medicine and stem cell medicine are just a part of the beginning events of this great revolution. It will not be surprising if a reader was to interpert this chapter to be a mosaic of facts and fiction. The fast-moving technology is realizing yesterday's fictions.

2 Cell Therapy Is Not New: Beginnings in Blood Transfusion

Cell therapy and other forms of regenerative medicine such as tissue engineering form a highly evolving branch of medicine. It may appear as a recent breakthrough but that is not the case. The cell therapy has its origins in blood transfusion, bone marrow transplantation, and reproductive in in vitro fertilization (freezing down semen and zygotes). Earlier (meaningful and successful) attempts at blood transfusion were carried out by the British obstetrician James Blundell. In 1818 Blundell performed the first successful blood transfusion of human blood, saving the life of a woman, bleeding during child birth. He extracted four ounces of blood from her husband's arm to transfuse into his bleeding wife (Welck et al. 2010). Modern cell-based therapies have progressed quite far from the first recorded human cell therapy – human blood transfusion to the advanced cellular therapies of today (see Table 33.1).

Probably the oldest, commonest, and successful form of "personalized cell therapy" is blood transfusion. The early history of immune cell therapy began 20 years ago for the treatment of blood cancer. Barnes and Loutit in 1957 performed an experiment on a mouse model of leukemia (cancer of the white blood cells) by injecting bone marrow from another mouse after giving a high dose of radiation to the leukemic mouse to destroy its own bone marrow cells. Following the success of this experiment, in 1965, Mathé and colleagues infused pooled white blood cells into patients suffering from leukemia. They observed a correlation between the dose and the incidence of graft-versus-host disease (GVHD): a disease

Table 33.1 Personalized cell therapy

Applications	Mechanisms	Examples	Major organ/cell type involved
Correction of genetic diseases	Repopulation by genetically corrected cells	Urea cycle disorder Primary oxaluria α1-anti trypsin deficiency Sickle cell anemia	Liver Liver, kidney Liver, lungs Red blood cells
Degenerative disorders	Repopulation by healthy cells	Alzheimer's disease Parkinson's disease Osteoarthritis Myocardial ischemic injury Cirrhosis	Brain Brain Joints Heart Liver
Auto immune	Repopulation by healthy cells	Type 1 diabetes	Pancreas/β cells
Radiation injury	Repopulation by healthy cells/immunomodulation or secreted factors	Accidental exposure to lethal radiation/radiotherapy	Hematological/gastrointestinal/lungs
Reconstruction	Repopulation/replacement or augmentation	Burns/cosmetic procedures Congenital esophageal atresia	Skin/soft tissue Food piipe

which follows transplantation when the immune system of the transplant recipient recognizes the transplant as foreign and starts attacking it resulting in the destruction and rejection of the transplant. It was later reported by Weiden and colleagues (1981) that there is an inverse correlation between relapse of the cancer and GVHD. Another landmark in blood cell therapy, in the late nineteenth century, was the use of blood mononuclear cells to treat severe immunodeficiency in people suffering from chronic fungal infections. The fungal infections were immediately cleared, and this proved the immunological benefits of mononuclear cell therapy (Shehata et al. 2007).

The use of adult stem cells in disease management is a relatively old practice (bone marrow stem cell, infusion in treatment of blood cancers and certain childhood immunological diseases are classic examples of stem cell therapy). There are mainly two kinds of bone marrow transplants: autologous bone marrow transplant, in which stem cells are removed from the patient himself, as when he gets a high dose of radiation or chemotherapy and preserved at ultralow temperatures (cryopreservation). The same cell would be infused back to the patient after the "destructive" therapy.

Allogenic bone marrow transplant is when bone marrow stem cells are collected from a donor preferably genetically identical like a sibling or from an immunologically matched but genetically normal (for treatment of certain inherited diseases). Allogenic transplantation, as noted before, has benefits in certain immunological diseases and cancers compared (immunotherapy) to autologous transplants.

Umbilical cord blood transplant is a recent addition. Stem cells are harvested from umbilical cord soon after birth. The stem cells are frozen down in cord blood banks till used. Umbilical cord stem cell transplant can be allogenic or autologous. However, umbilical cord blood cells are relatively immature but multipotent and are suitable for allogenic transplantations without significant immunological complications. Many parents are now storing the umbilical cord blood in "banks" because it might be lifesaving in an unlikely event that their child requires a transplant sometime in the future. In fact not only the child but the parents and siblings could be benifited as well. Umbilical cord cells possess better repopulating capacity, when transplanted, as well as increased proliferative capacity, good engraftment, and less GVHD compared to bone marrow (Rosler 2000). It may be noted that umbilical cord blood cells also act by modulating the stem cell niche of the host by secreting cell signaling molecules.

3 Stem Cells in Cell Therapy

Stem cells are cells which have the remarkable capacity to develop into one or many different cell types in our body. In adults they are important in maintaining the body cell mass and function by replenishing the cells we lose every day. They are also important in many tissues they serve as a sort of internal repair system, dividing essentially without limit to replace other cells as long as the person or animal is alive (Ref. http://stemcells.nih.gov/info/basics/basics1.asp). As noted by Till and McCulloch in 1963, they have two properties: (1) "unlimited" self-renewal and (2) differentiation. All stem cells do not have the capacity to give all other cell types found in the human body. Embryonic stem cells are special because they have the potential to give rise to all types of cells and tissues in our body. This quality of embryonic stem cells is known as "pluripotence." Embryonic stem cells differentiate to give rise to various adult-type stem cells which are committed stem cells which can also divide "unlimited" but can only differentiate to a limited types of cells. Theoretically embryonic stem cells can make any cell type/tissue/organ for transplantation, and unlike mature cells, they can divide without any limit. This means, a single embryonic stem cell has the potential to give rise to an entire adult. However the problem is that embryonic stems cells normally occur only once in a person's life time: when the individual was an embryo inside his or her mother's womb. This makes it impractical to relay on embryonic stem cells for personalized regenerative medical applications. It may be possible to isolate few embryonic stem cells from an early embryo, expand them, and freeze for future use of the developing individual. However this is a risky and expensive path which also involves ethical issues.

Adult stem cells offer another alternative for personalized regenerative medicine. Adult stem cells and mature cell types are used for treatment. Adult stem cells do not occur in large numbers; they are very sensitive and live in very special environments/niches and are difficult to isolate. So, getting sufficient number of cells for therapy is often difficult. It may be noted that several organs have huge regenerative capacities. Liver and bone are great examples. Even if two thirds of a liver is removed, the remaining mass is sufficient to grow back the entire liver. A single liver cell has the capacity to divide about 100 times, while an ordinary cell undergoes senescence after about 32 doublings. An interesting feature of a liver is that new liver cells come from adult liver cells rather than from "liver stem cells." It is even interesting that proliferating hepatocytes, for example, after a partial liver resection, can give rise to not only hepatocytes but also differentiate into cholangiocytes, epithelial cells lining the biliary tract (Michalopoulos and DeFrances 1997; Overturf et al. 1997; Kisseleva et al. 2010). After all it is relatively easy to isolate and transplant liver cells. Despite all these favorable properties, liver cell therapy has not become a main therapeutic option because it is still difficult to get healthy liver cells for transplantation. For example, in a patient with a failing liver, most of the patient's own liver cells are damaged. So we have to depend on his children, parents, or siblings or any genetically suitable individual for donating a part of their liver or liver cells. However, in many occasions, it is difficult to find a genetically matched voluntary donor. As noted earlier in this chapter, even after transplantation engraftment of transplanted cells is poor – not to mention the engraftment into a failing liver were the "environment" is even hostile – which is true in several causes of liver diseases such as long-time (chronic) hepatitis C viral infection and fibrosis of the liver from alcoholism. On the other hand, cell therapy may be of great use in liver failure from certain types of poisoning, inherited genetic diseases, etc.

3.1 Stem Cell Technology Might Eliminate the Need for Personalized Medicine

Stem cell technology would complement (or overtake and obliviate) the fields of genomics and proteomics in providing the "currently perceived" benefits of personalized medicine. It is believed that genomics and proteomics would facilitate medicine to become more personalized. However, during the next growth phase, we would be able to generate, replace, and even mass produce human organs which will not be rejected by the patient's immune system. These organs would match the HLA types (human leukocyte antigen: a cell surface determinant of acceptance or rejection of a transplanted organ by the organ recipient). In the early phase of this technological evolution, these "manufactured organs" would be generated from the patient's own cells or from an induced pluripotent stem cell (iPSC) bank, a bank much like a blood bank or semen bank which stores and retrieves iPSC for public or private use. iPSCs are cells which are derived usually from mature cells, which have the potential to differentiate into almost any cell type in an adult (more details are provided later in this chapter) (Table 33.2). It is estimated that a tissue bank from 150 selected homozygous HLA-typed volunteers could match 93 % of the UK population with a minimal requirement for immunosuppression (Taylor et al. 2012). The later generation organs would be more personalized and customized (by the patient himself or his doctor) having better genetic profiles than patients own original organs. For example, these later generation organs would be devoid of inherited/genetic defects and would have optimized, standardized performance profile. This would probably obliviate the need for medicine as organs would become more standardized and uniform. We will come back to this topic later in this chapter.

Table 33.2 Stages of evolution of personalized medicine

Stages	Features	Technologies
1	Limited areas of medicine: drug metabolism, transplantation, psychiatry, neurology, cardiology, diabetes, oncology	SNP-based : PCR, microarray
2	Every field of medicine, social structure and planning (family stucture, education, employment), finance (health care policies, insurance, banking)	Whole genome deep sequencing
3	Patient, specific/custom organs, drugs, and treatments	Whole genome and epigenome deep sequencing, iPSC
4	Standardized organs, tissues, standardized genomes	Future technologies propelled by major advances in basic sciences
5	Engineered and designer/custom genomes/ "metahomininans"	Future technologies, change from reverse engineering and copying nature to independent engineering

4 The Promise of Cell Therapy: Evolving Role of Cells as Medicine

Currently transplantation is the only hope for several terminal diseases, but the need for organs far exceeds the availability. For example, a person with liver failure can be saved by transplanting liver cells which were cultured and expanded in plates and stored frozen in the laboratory (when an appropriate donor is available for organ transplantation). Currently there are several practical hurdles. In the case of liver cells, for example, it is difficult to grow freshly isolated hepatocytes (which are the most abundant liver cells) over a week in vitro in the laboratory without losing their important functions. Even if they are grown, they will not engraft upon transplantation. The reasons are poorly understood.

Another example to illustrate the usefulness of cell therapy: a patient with fatal myocardial infarction ("heart attack") is unlikely to survive without the help of artificial life support. In the future cell therapy could give a patient a new life. Stem cell therapy would regenerate the damaged cardiac muscle. A phase II, multicenter, randomized, double-blind, placebo-controlled study to evaluate the safety and efficacy of ex vivo cultured adult human mesenchymal stem cells by intravenous infusion following acute myocardial infarction is being conducted by Osiris Therapeutics, a US-based company (ClinicalTrials.gov Identifier: NCT00877903). Mesenchymal stem cells (MSC) are stem cells that can preferentially mature into cells/tissue of "mesodermal" origin such as blood vessels, bone, blood, lymph nodes, heart, muscle, fat tissue, and fibrous tissue.

Diabetes (especially type 1 diabetes which is usually caused by autoimmune reaction in which one's own cells are destroyed by altered immune system) can be cured by repopulating the pancreas with functional insulin-secreting beta cells, which produce the insulin. A worn cartilage inside a knee joint may be repaired or replaced by repopulating the cartilage tissue with cartilage-forming cells called chondrocytes. It is hypothesized that type 2 diabetes could be mitigated by adipocyte (fat cell) transplantation and repopulation of the old fat-ridden adipocytes because it is recently found that adipocytes secrete hormones, namely, adipokines. Adiponectin, an adipokine, is important in sensitizing muscles and liver to insulin and glucose. Another adipokine, leptin, decreases appetite (Sanal 2009).

Similarly, a thalassemia (an inherited disease in which red blood cells are produced with defective hemoglobin resulting in anemia and weakness) patient could get a permanent cure by transplantation of blood-producing stem cells which are corrected for the genetic defect. A person suffering from Parkinson's disease would recover following cell therapy intended to replenish the "dopaminergic" neurons which are the special type of nerve cells selectively damaged in this disease.

Another major application of cell therapy is in the treatment of blood cancers (Shehata et al. 2007). Multiple myeloma is a type of blood cancer that is difficult to treat with anti-cancer drugs (chemotherapy) or radiation alone. More intense chemotherapy results in very low blood cell counts resulting in opportunistic infections, anemia, bleeding, and death. Using the patient's own stem cells (autologous) to help increase the blood counts after high doses of chemotherapy is called autologous stem cell transplantation (ASCT). During an ASCT, a patient's stem cells are given back to the patient all at once on a single day or in multiple small divisions. It has been shown that giving back stem cells over a period of days helps to increase bone marrow activity and decrease the time it takes for blood counts to recover after ASCT, lowering a patient's risk of side effects and infections and overall treatment costs.

Few clinical trials have shown the beneficial effects of cell therapy in blindness resulting from corneal opacity (cornea is the transparent layer through which the light enters the pupil, the aperture infront of the lens). Limbus insufficiency syndrome, for example, which is charecterized by corneal surface neovascularization (sprouting of new blood vessels in cornea making it progressively opaque), loss of corneal transparency, epithelial irregularities, or repetitive ulcers of cornea, can be mitigated if not cured by cell therapy that uses cells from bone marrow mesenchymal stem cells or limbal stem cells in amniotic membrane transplant (ClinicalTrials.gov Identifier: NCT01562002). Another approach is to make bioengineered corneas. The damaged, opaque parts of the cornea can be replaced by biosynthetic mimics of corneal extracellular matrix. Fagerholm, Griffith of Linköping University, Sweden and others, in a population of ten patients, replaced the opaque anterior cornea without the use of any human donor tissue. They used synthetically cross-linked collagen, a fibrous, transparent protein which is amenable to modification. It was reported that even after a 24-month follow-up, the biosynthetic implants remained without blood vessels invading and loss of transparency and without the need for long-term use of steroids (usually administered after transplantation to suppress the patient's own (host) white blood cells from destroying the graft) (Fagerholm et al. 2010). Yoshiki Sasai et al. from Kobe, Japan, have grown the precursor of a human eye in the lab. This was an eye-opening win for stem cell biologists. They successfully recapitulated the embryonic development of an embryonic eye, a structure called an optic cup which is 550 μm in diameter that contains multiple layers of retinal cells including light receptors. In future, it is possible that permanently damaged eyes can be repaired or replaced (Nakano et al. 2012).

Most of the therapies mentioned are highly experimental and have a high failure rate. These miraculous cures are highly possible, but it may be a long way before we develop and optimize all the necessary technology. However it may be noted that scientists are already successful in making artificial trachea (wind pipe) and urinary bladder by using cells. Paolo Macchiarini, a Professor at the Karolinska University Hospital, replaced the cancerous windpipe of a 36-year-old man with late-stage tracheal cancer, with one made in a laboratory and seeded with the man's cells (Macchiarini 2011). Anthony Atala, a Urologist at Wake Forest University's Institute for Regenerative Medicine, was the first to build a functioning urinary bladder made cell by cell, and it was transplanted into the patient (Figs. 33.1 and 33.2), a child who was born with a deformed bladder (Pincock 2011).

5 Mechanisms of Cell Therapy

Regeneration is widespread throughout the animal kingdom. The capacity for regeneration decreases as we go "up" along the ladder of evolution. Worms like planarian can regenerate their entire body from any tiny bit of their body. A salamander can regenerate an entire limb after amputation. Regeneration of a lizard's tail is a good example from reptiles.

Use of MSC therapy has increased tremendously in recent times – the MSC hype. MSC currently are being used for healing several types of injuries such as nerve cell injury, skin burns (ectodermal), gastric, colonic ulcers, pancreatitis

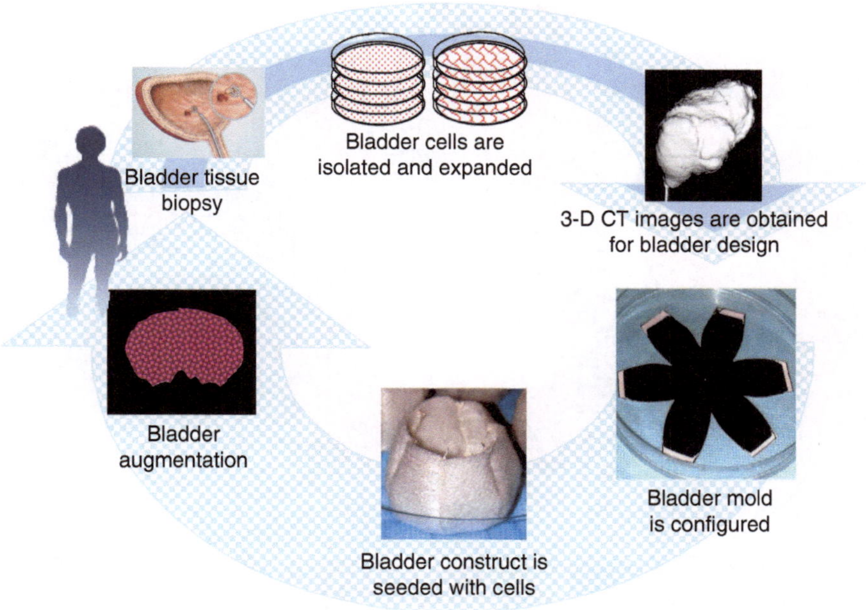

Fig. 33.1 Schemata for engineering a personalized urinary bladder for a patient. First a catheter is inserted through the external urinary orifice into the bladder. A small piece of the bladder tissue is removed (biopsy). This is grown in the laboratory in cell culture plates. The cells are allowed to proliferate to the required quantity. A three-dimensional CT/MRI scan of the patient's pelvis is done to decipher the dimensions of the bladder to be engineered. A bladder mold is configured based on this. Cells would be harvested from the culture and would be seeded on to the bladder mold and cultured further till cells proliferate and populate the bladder scaffold optimally. The bladder would be evaluated for quality and if it passes the tests would be implanted into the patient (The figure is adapted from the material provided by Dr. Antony Atala, W.H. Boyce Professor and Director of the Wake Forest Institute of Regenerative Medicine, Wake Forest Baptist Medical Center, Medical Center Boulevard, Winston-Salem, NC 27157)

(endodermal), bone, and kidney injury (mesenchymal). It is possible that during natural evolution, mesenchymal cells were selected for protection of body against injuries – a mobile protection force against injuries and facilitating regeneration, the same way white blood cells evolved as the defense force. MSC are attracted to an injury site, and they would get involved in the repair and regeneration process at the injury site by undergoing differentiation into the injured cell type taking signals from the environment. It is proposed that they undergo a niche-dependent differentiation to help regeneration or they aid regeneration by stimulating the surrounding cells through cell signaling molecules. As noted before, the MSC may have their ancestors in hydra, planarian, and lizard tail, but their regenerative capacity diminished progressively as life forms became "more complicated and evolved." It is hypothesized that regenerative capacity is a double-edged sword. The same genes which bestow regenerative and proliferative capacity are linked with formation and evolution of cancer as well. In long-lived multicellular organisms, cancer is a bigger threat compared to diminished ability to regenerate.

One direct mechanism is that stem cells or progenitor cells can engraft, differentiate, and replace damaged tissue. In this scenario the stem cells are directly transplanted into the damaged tissue or are differentiated into a specific cell type in vitro and then injected. In the former case, it will differentiate into the desired cell type after reaching the site of injury. These cells then integrate into the site of injury, repopulate the damaged tissue/organ, and thus improve the function of the organ/tissue, for example, improvement of liver function following injection of healthy hepatocytes in a child suffering from a life-threatening liver disease like a urea cycle disorder (Horslen et al. 2003).

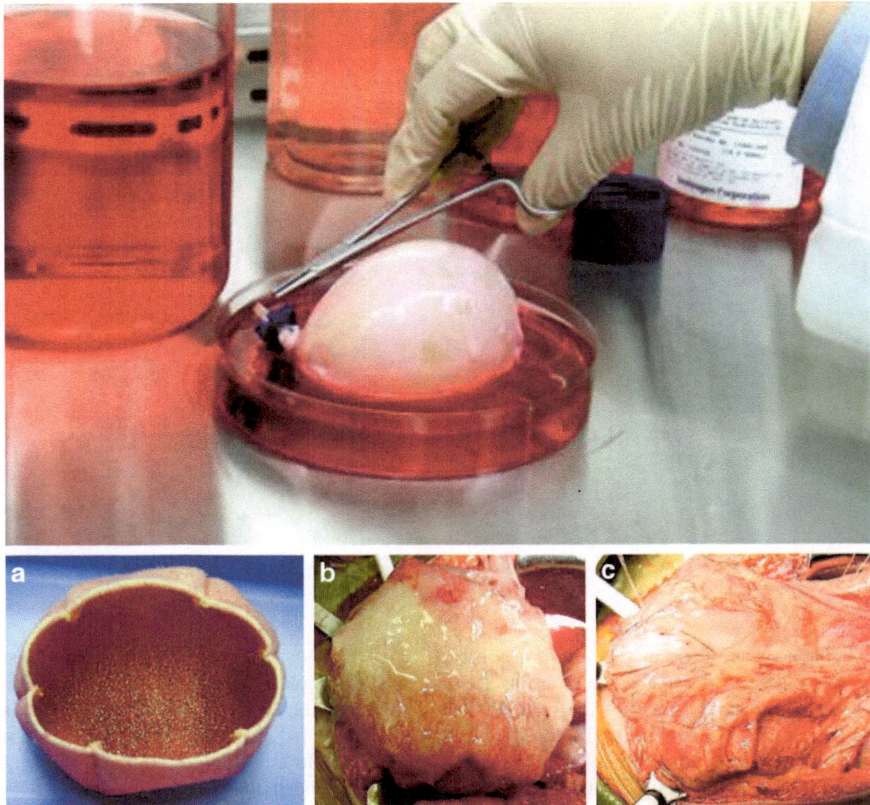

Fig. 33.2 A urinary bladder engineered in the laboratory. Printing of organs is still in experimental stage. Currently organs with relatively simple structure, such as a urinary bladder (shown in this picture), can be printed in the laboratory for transplantation into human patients. v. Scaffold b. Scaffold populated by patient's own cells c. Getting ready for implantation (Photographs A, B, and C are from Dr. Antony Atala, W.H. Boyce Professor and Director of the Wake Forest Institute of Regenerative Medicine, Wake Forest Baptist Medical Center, Medical Center Boulevard, Winston-Salem, NC 27157)

Certain forms of cell therapy rely more on the paracrine effects of transplanted cells than on transdifferentiation and functional replacement of damaged cells. To take advantage of this fact, scientists have tried to encapsulate naïve/genetically modified mature cells/stem cells with semipermeable/selectively permeable membranes and to implant them at or near the site of injury to enhance a supportive healing microenvironment. The cells are administered locally (e.g., a "Band-Aid" containing live mesenchymal progenitor cells) or systemically (e.g., intracoronary administration of endothelial cells engineered to secrete antiplatelet factors) and will remain viable only for a relatively short period (hours to weeks). However, this much time is sufficient for them to secrete the natural "healing factors" or the relevant "engineered" therapeutic factors or to induce epigenetic changes in the microenvironment. Cells that secrete factors that facilitate angiogenesis, anti-inflammation, and anti-apoptosis are prime examples of this. It has been observed that bone marrow transplantation mitigates hematopoietic syndrome (a disease condition caused by mass destruction of bone marrow cells resulting in decreased immunity and bleeding) and gastrointestinal syndrome (due to destruction of gut cells). Mitigation of radiation injury that occurs after transplantation of mesenchymal stem cells may be the result of paracrine effects of the released growth factors (Saha et al. 2011). There need to be several factors and cell signaling mechanisms

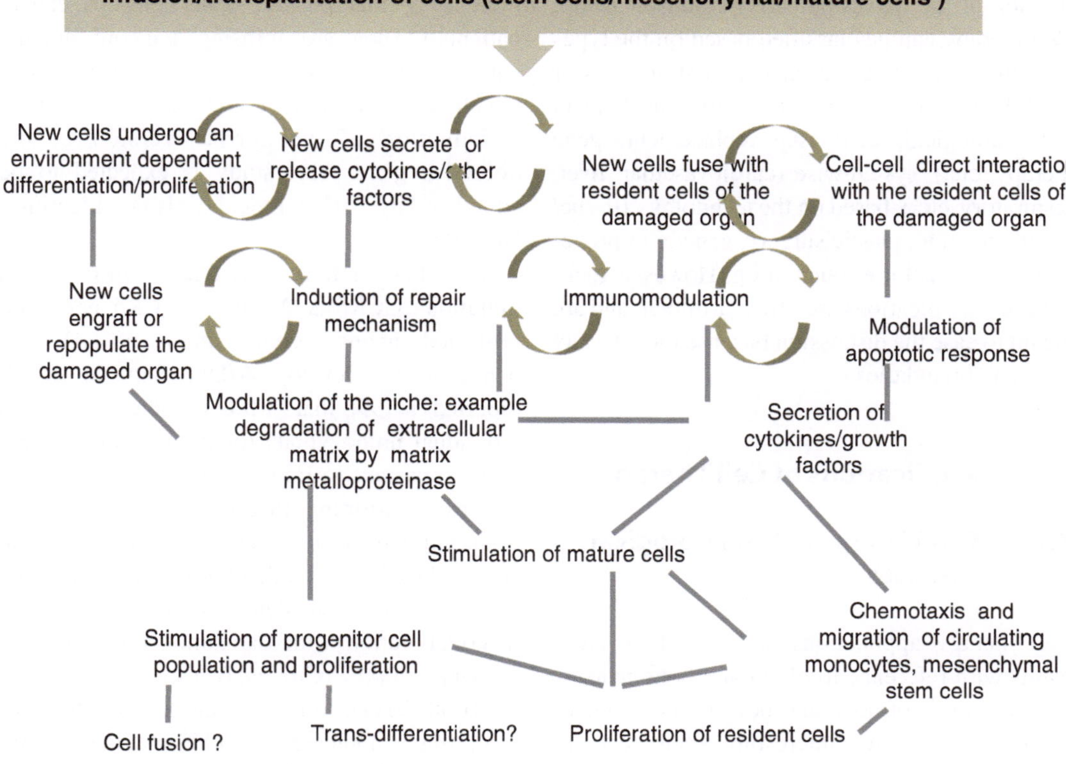

Fig. 33.3 Mechanisms involved in the beneficial effects of cell therapy. One or more mechanisms, often closely interacting and forming a complex web are involved depending on the type of cell therapy

to explain these observed beneficial effects, but a detailed discussion is beyond the scope of this book. Encapsulating and implanting pancreatic beta cells in a diabetic patient is an example of using the endocrine effects of the secreted protein, insulin.

The beneficial effects of cell therapy (especially cell therapy using cells other than induced pluripotent stem cells (iPSCs) or embryonic stem cells) are more complex and dependent on the type of cell therapy and the reason for the therapy (Fig. 33.3). The possible mechanisms are listed below:

1. Immunological/immune modulation
2. Release of cytokines/growth factors/vesicles/microRNA
3. Induction/prevention of apoptosis
4. Cell-cell fusion
5. Repopulation by the transplanted cells
6. Induction of repair and regeneration by attracting circulating mesenchymal/other cell types
7. Induction of epigenetic changes in the local niche

6 Current State of Cell Therapy

As of today (August 30, 2012) 4,176 clinical trials involving stem cells directly or indirectly are listed in ClinicalTrials.gov, a service of US National institute of Health. The bulk of cell therapy (excluding blood/blood products) is "autologous," meaning cells are harvested from the subject and then transplanted back to the same individual with or without processing. Processing can involve separation/chemical or biological stimulation/activation/in vitro proliferation/

tagging/genetic manipulation of the relevant fraction of the harvested cells (Trounson 2009). Cell therapy can be classified based on the types of cells used – like mesenchymal stem cells or cord blood cells – based on the mechanism (immunological, cell/tissue replacement, gene therapy, etc.), systemwise (cardiovascular, liver, respiratory, etc.), based on the technology (tissue/bioengineering, plastic surgery, genetic improvement, organ culture), and so on. However, these kind of classifications are often 'artificial' and are meant to ease the discussion because there hardly exist any boundaries.

7 Applications of Cell Therapy

7.1 Cell Therapy in Hematological Diseases

Cell therapy applications and recent developments with reference to blood and bone marrow diseases are enormous and beyond the scope of this book. Only a few interesting recent developments which are more relevant to personalized medicine would be discussed.

Technology has enabled us to generate iPSC from adult human cells with reasonable efficiency. This actualized the modeling of several diseases as well as personalized correction of genetic defects in several inherited diseases. In 2009, Belmonte's lab in Spain reported correction of Fanconi-anemia-specific iPSC that can differentiate into hematopoietic progenitor cells of the myeloid and erythroid lineages which are disease-free. This was a proof of concept that iPSC technology could revolutionize inherited hematological diseases (Raya et al. 2009). A clinical trial in progress is launched in 2012 by the University of Washington Cancer Consortium, USA, to "assess the toxicity and efficacy of infusion of genetically modified cells, as well as the feasibility of mobilization of peripheral blood stem cells with filgrastim and plerixafor for patients with Fanconi anemia." (Filgrastim is a drug that is a granulocyte colony-stimulating factor (G-CSF) analog used to stimulate the proliferation and differentiation of granulocytes, and plerixafor is an immunostimulant drug used to multiply hematopoietic stem cells in cancer patients.) They are infusing Fanconi anemia patients with autologous patient blood stem cells which are genetically corrected in the lab by introduction of the correct version of the defective gene. This study is expected to be completed by 2017 (ClinicalTrials.gov Identifier: NCT01331018).

It is known that individuals, who carry a mutation "CCR5-Δ32" in the CCR5 gene, are protected against certain strains of HIV and hence won't develop AIDS. Samson et al. reported in 1996 that "white blood cells from an individual homozygous for the null allele were highly resistant to infection by M-tropic HIV-1 viruses, confirming that CCR-5 is the major co-receptor for primary HIV-1 strains" (Samson et al. 1996). Based on this finding, a clinical trial was launched by the University of Pennsylvania in collaboration with Sangamo Biosciences in January 2009 (ClinicalTrials.gov Identifier: NCT00842634), to investigate the usefulness of "zinc finger nuclease-" or ZFN-mediated mutation of CCR5 gene. ZFNs are proteins that can be engineered to bind and cut specific DNA sequences, for example, the sequence of CCR5 gene in the current example (Fig. 33.4). Mutation of CCR5 gene will make CD4 T cell, the specific type of white blood cell involved in AIDS, resistant to HIV infection. This trial is expected to be completed by January 2013. This is a unique example in the history of gene therapy, where an otherwise normal gene is mutated and "improved" to make it disease resistant.

Natural killer (NK) cells are large, granular cells, a subtype of lymphocytes which in turn are a type of white blood cells. NK cells can kill cancer cells and virus-infected cells. Major histocompatibility complex (MHC) is a class of protein on cell surface which tags "self" from "nonself." NK-92 is an immortalized NK cell line of high potency isolated (92 stands for 1992, the year in the pre-Y2K notation). The immortalization procedure bestows NK cells with a capacity to proliferate inside cell culture plates unlimitedly (unlike normal healthy cells). These cells are harvested in the laboratory and irradiated, to

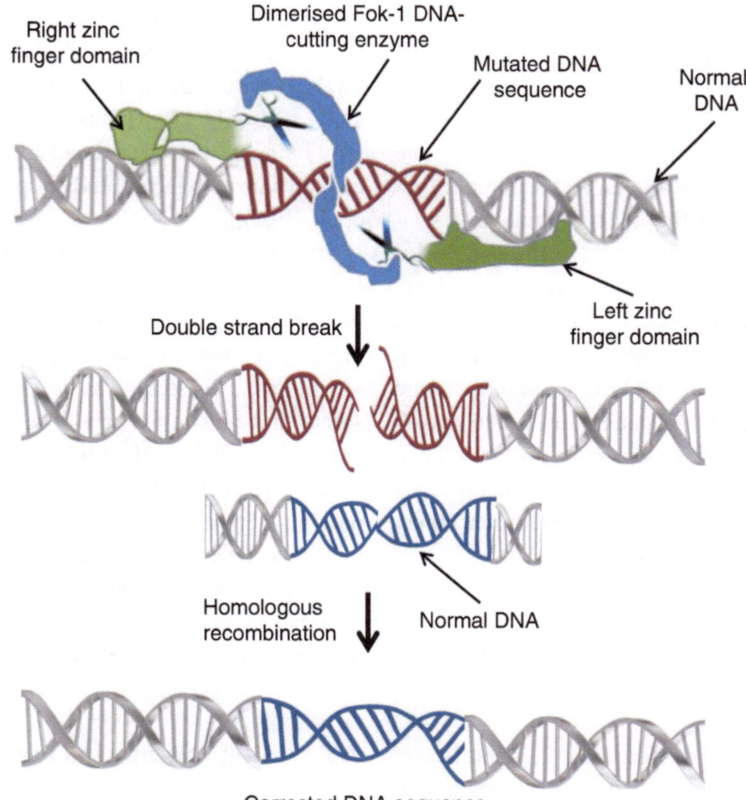

Fig. 33.4 Simplified mechanism of zinc finger nucleases correcting a genetic defect. Zinc-finger nucleases (ZFNs) are DNA binding proteins which can identify and cut a given sequence very precisely. Zinc-finger nucleases can be targeted to unique sequences within complex genomes. The endogenous DNA repair machinery is used to correct the genetic defect through homologous recombination in the presence of the right DNA sequence which is provided along with the ZNFs (Normal DNA means the 'wild type' DNA or the DNA sequence which is seen in individuals who are 'disease free' or 'normal'.)

prevent in vivo proliferation and engraftment. The processed cells are injected into cancer patients and the allogenic infused NK-92 cells are expected to destroy the cancer cells in the recipient. ZRx-101 cells are modified NK-92 cells with superior antitumor activity against hematologic and solid tumor targets in vitro and in vivo. A clinical trial was launched in 2009 at the University of Pittsburg to determine the safety and efficacy of the ZRx-101 cell line in patients with refractory or relapsed acute myeloid leukemia (Clinical Trials.gov identifier: NCT00900809).

The classic methods of gene therapy involves the use of viral vectors to transfer the correct version of the gene to the patient's cells. The disadvantage is that viruses such as retrovirus or lentivirus permanently integrates to the patient's genome. Integration of retrovirus and lentivirus is somewhat 'random' and alters the patient's genome. These changes in genome have the potential to initiate a cancer. After all, the original defective copy is retained, and this may be a problem in certain dominant disorders like a1-antitrypsin deficiency where the defective gene produces a protein that is toxic to cells (Ding et al. 2011). We will come back to this example later in this chapter. A rather successful example of gene therapy using virus-mediated insertion of corrected genes is severe combined immune deficiency (SCID) – a disease of the white blood cells. There are many causes for this disease like inherited deficiency of the enzyme adenosine deaminase (ADA). Children with ADA-deficient SCID often die because of infections in infancy or early childhood. Effective treatments are bone marrow transplantation (only if a suitable donor is available) or expensive lifelong treatment injections of PEG-ADA (marketed as Adagen) enzyme replacement therapy. The University of California launched a clinical trial in 2008 to evaluate the efficacy and safety of gene therapy using retrovirus as the gene carrier into stem cells from the bone marrow of patients with ADA-deficient SCID. The patient's own bone marrow cells are suppressed using busulfan, a chemotherapy agent that kills some of the bone marrow stem cells in the patient, to "make space" for the gene-modified stem cells to grow once they are infused back to the patients. This study is expected to be completed by 2014 (ClinicalTrials.gov Identifier: NCT00794508).

7.2 Cell Therapy in Myocardial Diseases

As noted elsewhere in this chapter, scientists as well as the public have huge expectations on the benefits of cell therapy in "incurable" ailments like myocardial infarction. A few studies have demonstrated some benefit in left ventricular function after intracoronary injection of patient's own bone marrow preparations. The serial cardiac magnetic resonance imaging substudy of the randomized multicenter REPAIR-AMI trial (Reinfusion of Enriched Progenitor Cells and Infarct Remodeling in Acute Myocardial Infarction study) undertaken at Kerckhoff Heart Center, Germany, reported a significant improvement (6.6 %, $p=0.01$) in global ejection fraction in patients treated with bone marrow-derived progenitor cells (Dill et al. 2009).

A 5-year follow-up of the local autologous transplantation of bone marrow cells enriched for stem cell fraction (selecting for CD133+ a marker of "stemness") in patients with myocardial infarction, completed in Tehran Heart Center, Tehran University of Medical Sciences, Iran, showed no clear benefit for patients treated by implantation of a CD133+ bone marrow cell population into an ischemic myocardium. However, it seems from this study and similar small studies that long-term result of transplantation of CD133+ is a safe and feasible procedure. This study had only 13 patients and three controls. It may be noted that there is a lack of blind, multicenter trials with larger study population (Ahmadi et al. 2012).

A similar study from Harvard Medical School, Boston, reported (2011) a decrease in the degree of ischemic myocardium, which was accompanied by a trend towards reduction in anginal symptoms. This study was designed to assess the safety and feasibility of a dose-escalating intracoronary infusion of autologous bone marrow (BM)-derived CD133+ stem cell therapy to the patients with chronic total occlusion (CTO) and ischemia. They injected CD133+ cells into the epicardial vessels supplying the collateral flow to areas of viable ischemic myocardium in the distribution of the CTO (Adler et al. 2011).

7.3 Cell Therapy in Liver Diseases

There are various approaches of "cell therapy" which can be allologous or autologous. The autologous "cell therapy" can be performed in different ways using different types of cells – hepatic administration of autologous CD34-positive cells (a marker of hematopoietic progenitor cells; see below) induced by granulocyte colony-stimulating factor, portal vein administration of CD133+ mononuclear cells, administration of autologous bone marrow-derived or adipose tissue-derived mesenchymal stem cells, administration of cord blood cells (if stored frozen at the time of birth of the patient), and transplantation of patient's own corrected hepatocytes. Induced pluripotent stem cell-derived "hepatocytes" or "mesenchymal cells" and hepatocytes derived through transdifferentiation of the patient's own fibroblasts or mesenchymal cells are still in the horizon. Effectiveness of CD34+cells has been shown by some groups but not generally agreed upon. Scientists have reported decrease in liver fibrosis and function following infusion of autologous bone marrow cells in mice, and on the basis of those results started autologous bone marrow cell infusion therapy for liver cirrhosis. The results so far are not convincing.

CD34 is a cell surface glycoprotein and functions as a cell-cell adhesion factor present in several types of cells especially stem cells in blood and bone marrow. It is known that certain subpopulation of CD34+ cells have the potential for regenerating damaged tissue. Granulocyte colony-stimulating factor (G-CSF or GCSF) is a potent inducer of bone marrow stem cell (HSC) mobilization from the bone marrow into the bloodstream. Gordon et al. from Imperial College London reported an improvement in serum bilirubin and serum albumin in four of five patients suffering from chronic liver disease when treated with G-CSF. They isolated $CD34^+$ cell from peripheral blood of these patients by a technique called leukapheresis and injected these cells back into the portal vein (Gordon et al. 2006). However, it may be noted that the study period was too short; the study was done on just five patients, and there is a lack of more convincing studies in this line after 6 years of this pilot study. A group from India led by Sarin reported short-term benefits in acute-on-chronic liver failure (ACLF) patients following treatment with G-CSF (Garg et al. 2012). The authors hypothesize that the benefits could be the result of mobilization of stem cells from bone marrow which helped in liver regeneration or mitigation of the symptoms. It may be argued that the observed benefits in this study following G-CSF administration could be attributed to a boost in the patient's immunity as a result of increased production of white blood cells and decreased bleeding tendency from increased platelet synthesis and function. Bleeding and infections are major causes of mortality in liver failure patients. This highlights the importance of investigator-blind, multicenter trials.

The most successful of the liver cell therapies is currently allologous or autologous transplantation of mature liver cells. Hepatocyte transplantation is considered as an alternative to whole-organ transplantation to support many forms of liver diseases. When it is difficult to find a living donor/brain dead for whole-organ transplantation, the alternative is to harvest hepatocytes for cell transplantation from fresh cadavers or parts of the whole organ removed for other purpose. As noted earlier in this chapter, these harvested hepatocytes can be frozen for later use. These hepatocytes are then injected into a patient when he is in need, especially in a hepatic emergency like an acute (sudden) liver failure (Guha et al. 2001; Puppi et al. 2012). Another application of hepatocyte transplantation is in gene therapy. Certain inherited liver diseases like urea cycle disorder, which requires very stringent medical management, can be corrected by transplanting immunologically compatible healthy hepatocytes in case a healthy donor is not available. Since hepatocyte transplantation does not preclude future organ transplantation, multiple hepatocyte transplantation and repopulation can be attempted. On the other hand it is difficult to perform multiple liver transplantations (Dhawan et al. 2006). A more ingenious and personalized approach would be to correct the genetic defect using biotechnological tools (such as zinc finger nucleases, TALEN proteins, CRISPR/Cas systems which enhance site directed recombination) (Papaioannou et al. 2012; Jinek et al. 2012). The idea is to take the patient's own cells, correct the gene defects using molecular scissors like TALEN proteins or zinc finger nucleases, and transplant the corrected cells back to the patient. There are two major road blocks. The first block is that the hepatocytes do not grow well and proliferate outside the body. The second and perhaps the biggest road block in liver cell therapy is poor engraftment or repopulation of the transplanted cells. The transplanted liver cells engraft well, proliferate, and repopulate only when there is significant advantage for the new comers over the old residents. The biggest challenge, perhaps, for the acceptance of liver cell therapy in clinics lies in the creation and maintenance of this selective advantage for the transplanted liver cells over host liver cells (Grompe 2002). Irradiating the host liver with appropriate radiation intensity will incapacitate the host liver cells from dividing and competing with the transplanted cells. Treating host liver with mitotic inhibitors is an alternative approach in this line, but neither radiation nor mitotic inhibiters are quite safe as they are potential mutagens and hence considered as carcinogens. So far a safe, reliable, and practical method of conditioning host liver for the proliferation of transplanted donor cells has not yet been established.

7.4 Cell Therapy in Lung Diseases

We have already mentioned in the previous section an example of the advances in bioengineering: how doctors and bioengineers worked together to create a bioengineered trachea with a man's own cells.

Lung cancer, pulmonary hypertension, chronic obstructive pulmonary disease (COPD), and asthma are the common difficult to cure if not incurable diseases in the respiratory medicine. It is well known that cigarette smoking is one of the most common risk factors in COPD. Another major cause is α1-antitrypsin deficiency, which is an inherited disease. α1-Antitrypsin is an enzyme produced in the liver, and one of its functions is to protect tissues, such as lung tissue from damage by protecting from certain protein-digesting enzymes released by neutrophils. Mutations in the α1-antitrypsin gene in some forms of the disease result in the production of a defective enzyme which is not only nonfunctional but also toxic as they aggregate inside liver cells over time. Recently Nagy's lab in Canada was able to "repair" the gene mutation in vitro using mutation-specific zinc finger nucleases, as noted in the previous section, protein scissors which can be engineered to recognize and cut highly specific DNA sequences. They corrected the gene defect in the iPSC derived from a human patient suffering from this disease. They were able to further differentiate these iPSC cells into liver cells, which were functional in terms of a1-antitrypsin activity and transplanted them into a mouse liver to show engraftment (Yusa et al. 2011).

Some animal models have demonstrated the beneficial effects of mesenchymal stem cell transplantation, especially in moderating inflammatory response (Li et al. 2012; Hansmann et al. 2012). However, there is a lack of more convincing studies – well controlled and double blind – published in reputed journals. However, based on the beneficial effects of animal studies, few clinical trials were undertaken. Safety Study of Cell Therapy to Treat Chronic Obstructive Pulmonary Disease (COPD-01) was carried out in Brazil to determine whether the cell therapy with bone marrow mononuclear cells is safe in the treatment of chronic obstructive pulmonary disease, specifically the pulmonary emphysema (ClinicalTrials.gov Identifier: NCT01110252). According to the undertakers, "twelve month follow-up showed a significant improvement in the quality of life, as well as a clinical stable condition, which suggest a change in the natural process of the disease." The study is insufficient because of the very small study population and short follow-up period highlighting the need of large multicenter well-controlled long-term blinded studies in this area.

Administration of MSC is believed to have a "soothing" effect on irradiated tissues and organs treated with cytotoxic agents. Several drugs used in chemotherapy as well as radiotherapy are cytotoxic and preferentially stop tumor cells from dividing, but normal cells are also affected, especially fast-dividing cells in the alveoli, intestine, and bone marrow. Therefore combining chemotherapy with bone marrow stem cell or mesenchymal stem cell transplantation may allow the physician to administer higher doses of radiation or chemotherapy drugs or to mitigate the damage on normal tissue (ClinicalTrials.gov Identifier: NCT00003284). Pulmonary (lung) fibrosis is another disease in which stem cell therapy may be of use. The lung tissue becomes thick, stiff, and scarred, diminishing the gas exchange over time. Certain forms of this disease have an immunological component and the immunomodulatory and cytoregulatory functions of transplanted cell may retard the progress of the disease. A clinical trial was initiated in 2010 at The Prince Charles Hospital, Australia, to evaluate the safety and feasibility of placental MSC treatment for subjects diagnosed with idiopathic pulmonary fibrosis (ClinicalTrials.gov Identifier: NCT01385644).

8 Induced Pluripotence: A Breakthrough Technology Facilitating Personalized Cell Therapy

Induced pluripotent stem cells (iPSCs) are pluripotent stem cells generated in vitro from non-embryonic stem cells, like skin cells or blood cells, but they have very similar if not the same properties of embryonic stem cells. They were first generated by Yamanaka's team by "forced" expression of specific genes now known as the Yamanaka's factors: Sox2, Oct3/4, Klf4, and Myc (Fig. 33.5a, b). This happened in 2006 and the experiment was in mouse fibroblast-like cells. This was followed by generation of human iPSC by the same group as well as by another group led by James Thomson in 2007 (Stadtfeld and Hochedlinger 2010). The iPSCs have advantages over embryonic stem cells. iPSCs can be derived from any individual of any age, and generation of iPSCs do not involve "killing" or manipulation of an embryo and hence not associated with those ethical issues which surround embryonic stem cells. It is possible to generate iPSC from any person, using his own cell, and from iPSC it is possible to generate any cell/tissue present in that individual. However, generation of a whole organ such as liver or kidney is theoretically possible, but there are several practical and ethical issues which are expected to be resolved in the near future.

Investigators have realized the potential of iPSCs in regenerative and transplantation medicine and disease modeling and as a tool for drug development. iPSCs can be generated from any individual to a great level irrespective of age and health condition. This has many advantages such as patient's own tissues could provide him/ her almost limitless, immune-matched supply of pluripotent cells which could be used to regenerate/ repair his body parts. Therapeutic cloning indeed is a possibility that is unacceptable due to ethical issues which could be solved in the future. Current road blocks for iPSC to clinics are the possibility of undesirable genetic and epigenetic changes in iPSC which could result in tumor formation or malignancy (Martins-Taylor et al. 2011; Ji et al. 2012). However the high-throughput DNA sequencing and epigenetic screening techniques are growing very fast. So it would be possible in the near future to select for human applications the best iPSC colonies, genetically and epigenetically sound, from hundreds of colonies in a cost-effective manner.

iPSC could be a lifesaver not only for humans but also for laboratory animals providing a non-exhausting source for primary cells. This means several studies (like drug testing) which use animals can be done on iPSC-derived cell types (e.g., iPSC-derived hepatocytes) or tissues. Every individual is different, genetically and epigenetically. Scientists currently use stem cells to test/screen drugs in a more personalized way or identify molecules or genes implicated to a disease, drug response, or a genomic or phenomic trait in highly personalized manner. This would be revolution in personalized medicine. Immunologic tests, genotyping, or quantifying drug response on a patient by patient basis using iPSC/iPSC-derived cells would provide the doctor the information needed to select and fine-tune drugs or drug combinations or to plan a cell therapy or transplantation. Techniques to derive more and more iPSC-derived functional cell types from hepatocytes, neuronal types, to cardiomyocytes are improving at great speed (Cayo et al. 2012; Swistowski and Zeng 2012; David and Franz 2012).

It is uncontroversial that iPSC-based technologies have the potential to improve medical practice by circumventing immunorejection, curing several genetic diseases, replacing/ rejuvenating aging tissues, and taking personalized medicine to the next level. iPSC combined with embryo manipulation technologies have huge potential for de novo organogenesis and permanent cure for degenerative diseases and several age-associated diseases.

World over governments are taking efforts to plan, formulate, and fund iPSC research. For example, in 2009 the American Recovery and Reinvestment Act (ARRA) NINDS funded three consortium "to develop and characterize iPSC cell lines for familial forms of Amyotrophic Lateral Sclerosis (ALS), Huntington's Disease (HD) and Parkinson's Disease (PD)"

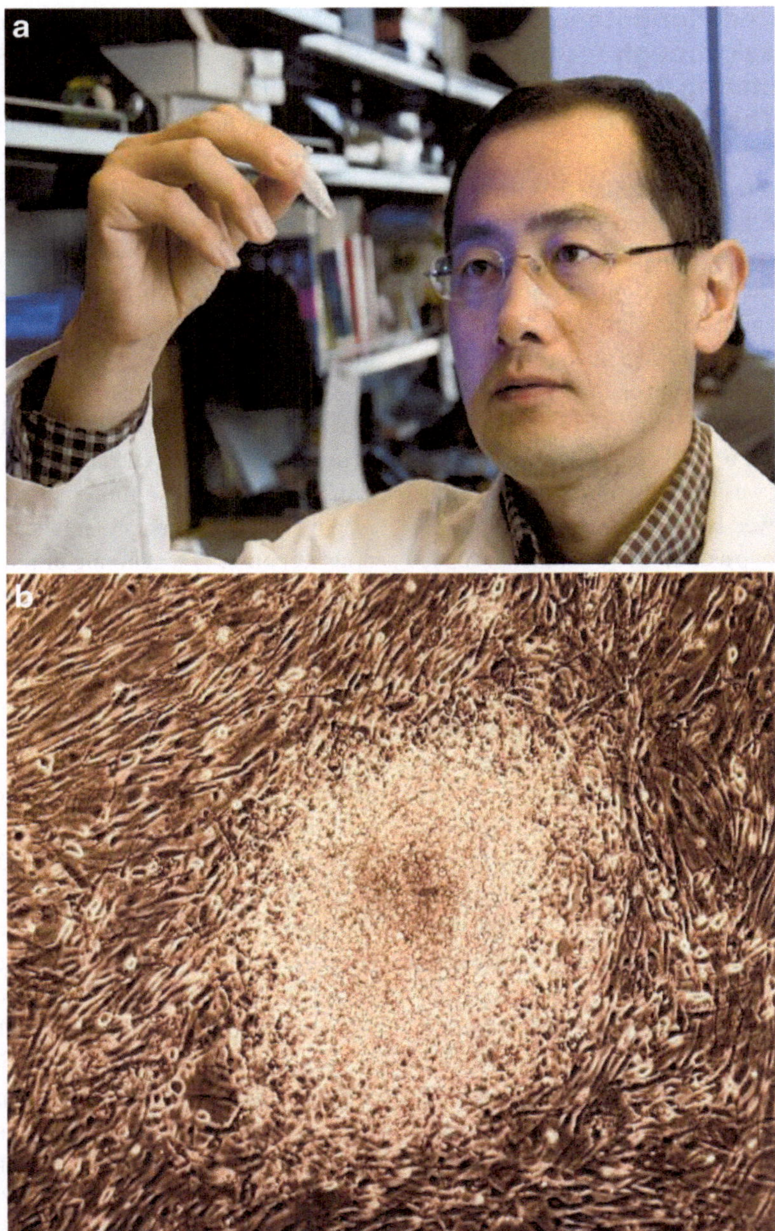

Fig. 33.5 Prof. Shinya Yamanaka, MD, PhD (**a**). Prof. Yamanaka was trained as an orthopedic surgeon but moved to full time research. He won the 2012 Nobel Prize in Physiology or Medicine for his discovery of how to transform ordinary adult skin cells into induced pluripotent stem cells which are capable of developing into any cell in the human body – this has revolutionary consequences in transforming personalized medicine. Induced pluripotent stem cell (iPSC) colony (cluster of cells in the center) growing on a layer of fibroblasts (elongated cell in the background). These iPSC are produced by putting DNA sequences inside fibroblasts (a process called transfection) which code for Yamanaka's transcription factors Oct3/4, Sox2, Klf4, and c-Myc (**b**)

(http://www.ninds.nih.gov/funding/areas/neurodegeneration/2nd_iPSC_Executive_Summary.htm).

Reprogramming of adult cells offers the potential to treat many inherited diseases through gene therapy such as Parkinson's disease, Alzheimer's disease, amyotrophic lateral sclerosis (ALS/Lou Gehrig's disease), certain inherited forms of cardiovascular disease, and diabetes. The genetic and epigenetic bases of more and more diseases are getting revealed day by day. A couple of recent research reports demonstrated the proof of principle. iPSCs generated from a patient with ALS were differentiated into motor neurons, the type of cells which are specifically destroyed in ALS (Dimos et al. 2008). Successful transplantation and integration (which is yet to be achieved) would permanently cure such diseases. This was to illustrate the usefulness of iPSCs in cell-replacement therapies. Few groups have demonstrated "derivation and differentiation of functional dopaminergic neurons from hESCs and iPSCs under xeno-free defined conditions" in a scalable manner following good manufacturing practice (GMP)-compliant protocols (Swistowski and Zeng 2012). Similarly several short steps are being made around the world to make mass production of iPSC a reality which is required in several disease conditions, for example, to mass produce patient-specific hepatocytes for transplantation. iPSCs are typically derived and are maintained in adherent culture. There is a recent report of an efficient protocol for derivation of mouse iPSCs in stirred suspension bioreactors, with and without the use of c-Myc (an avoidable transcription factor because of its bad reputation as an oncogene). The suspension-reprogrammed cells derived using this protocol expressed pluripotency markers and showed the potential to differentiate into multiple lineages both in vitro and in vivo and contributed to the germline in chimeric mice. In another report, Shafa M et al. demonstrated a yield of 58-fold expansion of iPSCs derived from human patients over 4 days in stirred suspension bioreactors. The resulting cells were karyotypically normal, expressed pluripotency markers. These experiments demonstrated the removal of another road block in bringing iPSC-derived cells from bench to bedside (Shafa et al. 2012a, b).

9 Personalized Cell Therapy to Organ Therapy

So far, we are rather limiting our discussion to cell therapy. Generation of whole personalized organs is the next big leap. Several technologies are at the horizon. Scanning and printing of patient's entire organ is an evolving reality rather than science fiction. Growth of basic sciences, namely, mathematics, physics, and chemistry, has realized the dreams of biologists. Another more promising and natural yet controversial method is to grow an entire organ in an engineered womb which is perfused with oxygenated defined nutrient medium. Growing larger mammals like chimpanzees or 'lesser animals' like pigs or sheep, with patient-specific human organs growing inside would be perhaps more acceptable for the society. This is because the general public would be more comfortable in killing pigs or sheep than more human-like animals like apes. Similarly the public or even scientists may not accept growing human bodies devoid of a cerebral cortex or sensory system solely for organ harvesting despite the fact that it is more ethical than killing intact animals (even after anesthesia) for biological experiments, drug testing, or human consumption.

10 Tissue on Demand: Progress in Organ Printing

The last three years have witnessed a huge revolution in 3D printing technologies. Now practically any complex shape can be printed, sometimes with the help of temporary scaffolding. Printer "inks" can range from plastics, gels, and cellular matrix to matrix containing human cells. Beating myocardium, blood vessels, and urinary bladder, the list of tissues/organs successfully printed are long. However, printing of complex organs like brain, kidney, or liver has not yet achieved and perhaps we have miles to go before!

Like stem cells and cell therapy, 3D tissue printing has also indeed attracted venture capitalists, and we find companies like "Organovo," a company based in San Diego, California, which has developed a printer to "print out" tissue structures and perhaps organs "on demand"! Currently the application is limited to drug testing by pharmaceutical companies and for laboratory use (http://www.organovo.com). Organovo's 3D printer can deposit a variety of live cells in any pattern desired, starting with small building blocks of any shape. These "small bricks" are further loaded into a print cartridge that is then attached to the tissue printer which "prints" the cell aggregates layer by layer, producing multiple stacked 2D patterns, this ultimately result in the 3D tissue or organ. The cell aggregates are held in place by a noninvasive gel and will retain sufficient rigidity and flexibility.

However the concept of "cell printing" is at least 15 years old, perhaps even more. In 1997, Mrksich M et al. from Harvard University reported "using microcontact printing to pattern the attachment of mammalian cells to self-assembled monolayers" (Mrksich et al. 1997). Currently research is going on to develop fully functional organs using scaffoldless, print-based engineering techniques described before that use self-assembling multicellular units and tissue matrix as "ink" and employ early developmental morphogenetic principles, such as cell sorting and tissue fusion (Jakab et al. 2010).

11 Future Possibilities: Growing Human Organs in "Anencephalic" Systems or Animals

The brain is the center of human senses, consciousness, and personality. Engineering a tissue mass without a developed nervous system cannot perceive any senses including pain. Therefore, it would be more ethical to work on this kind of systems than using live animals for experiments or organ harvesting. These bio-artificial systems ("vegetable systems") can be developed to model murine, porcine, or human systems for drug testing to organ harvesting (Sanal 2011). The path towards this kind of systems is complex but more practical and realizable in near future compared to "printing" of fully functional organs especially when it comes to complex organs like kidney and liver which are also the most commonly transplanted organs. We will discuss briefly about this possibility.

As noted before the recent developments in induced pluripotent stem cell, biology has huge potential applications. Combining the iPSC technology with the modern embryo manipulation techniques would allow us to develop human organs in other animals or within "vegetable" human bodies. Most of the normal cells are diploid meaning they have two sets of chromosomes. Gametes, eggs, and sperms are haploid and will fuse together to form a diploid zygote which would form an embryo which is capable of growing into a fetus. When two diploid cells fuse, it will give rise to a tetraploid cell. Cells can be forced to fuse by application of electric shock (Fig. 33.6). Artificially generated tetraploid cells from most of the cell types won't proliferate for long. Tetraploid cells are rare in human body except in organs such as liver or placenta. Tetraploid embryo can be produced by inducing two zygotes to fuse. A tetraploid embryo when implanted inside a uterus is incapable of growing into a living animal. However a tetraploid embryo can give rise to extraembryonic tissues such as placenta. On the other hand, embryonic stem (ES) cells or iPSCs cannot form a fetus when implanted into the uterus because ES or iPSC cells are incapable of generating extraembryonic tissue normally. If embryonic stem cells or iPSCs are injected into a tetraploid embryo, they will complement each other: a phenomenon called tetraploid complementation. The tetraploid embryo will give rise to extraembryonic tissues, while iPSC/ES will give rise to a viable fetus (Fig. 33.7a). The fetus thus generated is solely from ES/iPSC. This technique has wonderful applications. Injecting a patient-derived iPSC into a tetraploid embryo could give rise to a fetus genetically identical to the patient. This is therapeutic cloning. It may be noted that this is a very different method with reference to the method used to clone Dolly the sheep (the first mammal ever cloned) by Sir Ian Vimut in 1996 (cloning of Dolly made use of an inefficient technique called somatic cell nuclear transfer (SCNT) which involves transfer of the nucleus of an adult cell

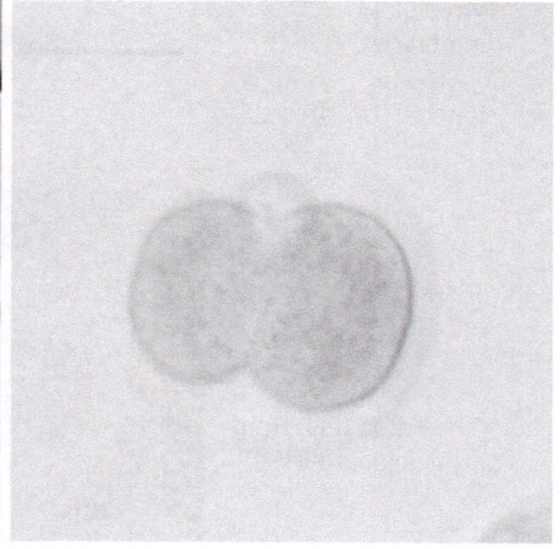

Fig. 33.6 Actual setup for giving electric shock to embryos (*right*) and a microphotograph of tetraploid embryo (*left*). Electric shock induces two diploid embryonic cells to fuse to form a tetraploid cell (the instrumentation set up). Cell membranes and cytoplasm of two cells fused as one can be seen in this picture (*left*). The nucleus would fuse after few hours (The pictures were provided by Ronald Naumann, Head of the Transgenic Core Facility, MPI of Molecular Cell Biology and Genetics, Pfotenhauerstrasse 108, 01307 Dresden, Germany)

into an oocyte) (Caulfield et al. 2007). However, it is very much unethical to grow a human fetus for therapeutic cloning, even if it is for the benefit of his genetically identical twin. A practical solution is to generate an iPSC line which is genetically edited out of the genes required for the formation of nervous system and reproductive system. If we are interested only in liver and heart geneses, several other organs such as limbs, facial structures, ears, and intestines can be edited out (knocked out). Many organs may not have single critical genes like pancreas, or knocking out certain genes could have effects on multiple organs. Currently, we know only a limited number of "organ critical genes." But in the future, the list will become longer and there will be more ways to circumvent these issues.

A milestone achievement in this direction is reported by Kobayashi et al. in Cell, a reputed journal in 2010. They injected wild-type mouse iPSCs into Pdx1 –/– rat blastocysts (an early stage of developing embryo). Pdx1 is the critical gene for genesis of the pancreas. Pdx1 –/– rats are genetically modified by editing out this gene (Kobayashi et al. 2010). These animals therefore do not develop pancreas – they are pancreatogenesis disabled. Upon injection of mouse iPSC cells which express Pdx1 gene, they will compensate the vacancy of the pancreatic "developmental niche," generating almost an entirely iPSC-derived mouse pancreas inside the rat. Following this path it could be possible to generate a human pancreas (or another organ) in animals – monkey, pig, or sheep, which may be genetically modified to enhance the implantation and survival of an embryo containing a human organ which is genetically identical to a patient from whom the iPSC was derived (Fig. 33.7b). Instead of using animals to "grow" human organs, it is possible to use "container bodies" derived from human cells but devoid of a developed nervous system. These will be the ultimate personalized organs because the organs derived by this method will be genetically identical to the patient in need. The only organ which cannot be generated and transplanted by this method would be the brain because the brain of an individual and the information which it stores defines and decides an individual. It is meaningless to transplant a new brain, despite being fully functional, without memories and all other information the individual accumulated over his life.

The British-Indian scientist J. B. S. Haldane (Fig. 33.8) as early as 1924 coined the term "ectogenesis" to describe how human pregnancy would 1 day give way to artificial wombs (Salako 2008). Haldane wrote, "Now that the technique is fully developed, we can take an ovary from a

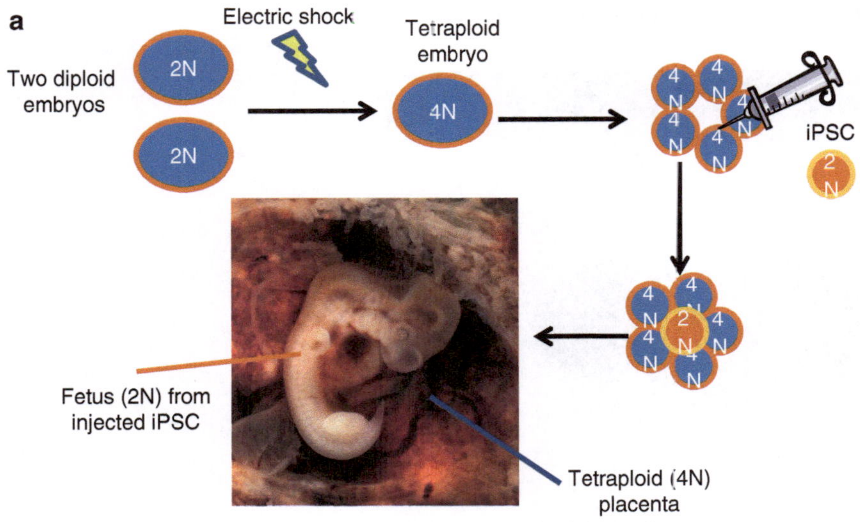

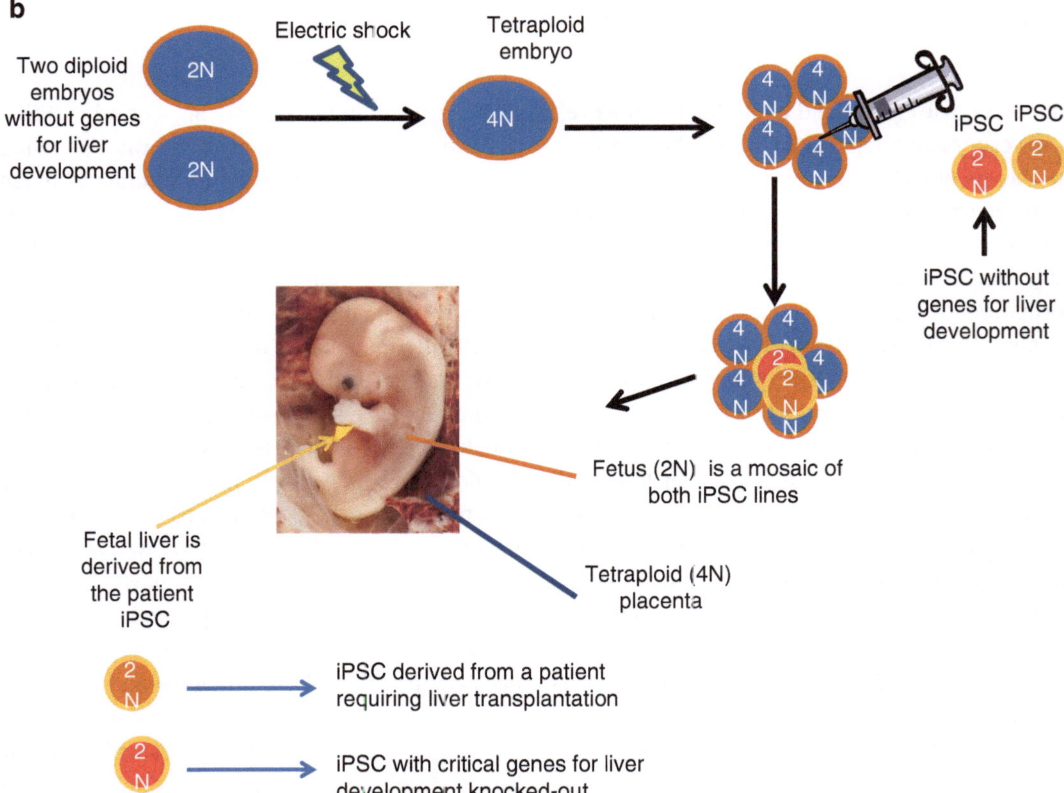

Fig. 33.7 (**a**) Tetraploid complementation. Normal animal cells are diploid: each chromosome is present in duplicate while a tetraploid has four copies of each chromosome. A tetraploid is made fusing two embryos at the two-cell stage by applying an electrical current. The resulting tetraploid cell will continue to divide, and can give rise to placenta, etc., but not the embryo upon implantation inside uterus. If normal iPSC are injected at appropriate time point into tetraploid embryo, a normal embryo can result from iPSC. (**b**) Tetraploid complementation can be used to generate patient specific organs. Two of more iPSC lines which mutually complement for certain critical genes specific for the organ of interest are mixed and allowed to form a chimera which would eventually result in the generation of patient specific organs for transplantation. Here in this example the liver bud is formed almost entirely from the patient because only patient iPSC derived cells will respond to the natural signals for liver budding because only these cells have the necessary genes for liver development. In the cell-cell competition they will therefore win over the other cells lacking these critical developmental genes

Fig. 33.8 *John Burdon Sanderson Haldane* FRS (5 November 1892–1 December 1964), a scientist, visionary, original thinker and philosopher. He was also one of the founders of population genetics. He left Britain and became an Indian citizen. His essay, Daedalus; or, Science and the Future (1924), predicted many scientific advances but has been criticized heavily. His writing was treated as "shocking" and "science fiction" at the time, being the first book about ectogenesis (the development of fetuses in artificial wombs)

woman, and keep it growing in a suitable fluid for as long as 20 years, producing a fresh ovum each month, of which 90 % can be fertilized, and the embryos grown successfully for 9 months, and then brought out into the air." According to him by the year 2074, ectogenesis will become a popular technique – with "less than 30 % of children… now born of woman" (Haldane 1923). In the real world, there is a slow but reasonable progress in embryo-endometrial coculture (Spandorfer et al. 2002) and "artificial uterus" research. As noted before in this chapter, there are quite few methods to grow human organs: (1) inside transgenic, "immunologically humanized" large animals (Sanal 2011; Jung et al. 2011) like sheep, pig, nonhuman primates, etc. and (2) genetically human "container bodies" which are modified "vegetable" human bodies (such as a "body" without nervous system, reproductive system, limbs, digestive system, sensory organs, skeleton, or voluntary muscles which may be grown inside a genetically modified animal, surrogate human uterus, or artificial uterus/womb). Scientists have engineered endometrial tissue by populating a biodegradable scaffolding shaped like a human uterus. When an embryo was introduced to the artificial uterine lining, it attached and successfully implanted. The embryo was allowed to grow for only 6 days due to ethical concerns.

iPSCs also offer enormous possibilities for genome editing: correcting existing disease-causing mutations, converting existing allele to a better disease-resistant allele (e.g., editing CCR5 allele to confer HIV resistance) or physiologically better allele or even inserting "new" genes conferring superior human or "superhuman" qualities – physical, psychological, or intellectual. Standardization of tissues or organs is a more socially "acceptable" application. Editing HLA genes would offer "invisibility" or improve host acceptability of tissue/organs produced for transplantation making mass production practical. At this level it is no longer required to produce tissue/organ from a patient's own iPSC. On the same line new organs, mass produced, can be genetically modified for better function, for example, cytochrome P450 superfamily (officially abbreviated as CYP) of enzymes may be optimized in mass-produced liver. These are liver enzymes involved in a variety of activities such as detoxification and drug metabolism (Sorich and McKinnon 2012). At this point "personalization" of medicine would meet an end (personalized medicine as defined or explained in the present time). Personalization would then mean, perhaps, personalizing organs (the way we personalize a computer while making a "custom" order) or personalizing the very "existence" of oneself.

12 Ethical Issues in Personalized Medicine, Evolving iPSC-Based Therapies, and Biotechnology

Development of science and technology often out phases the speed of dissemination of the new knowledge to the larger public and its assimilation and acceptance by the society. Advances in science often bring unimaginable social ethical, racial, political, and evolutionary consequences, and the society won't be prepared to absorb the shock. Recent advances in genetics, bioinformatics, and stem cell biology are going to revolutionize personalized medicine.

There exists a worldwide ban on reproductive human cloning which was proposed in 2001 by France and Germany to the UN, and it became effective since September 2006 (Arsanjani 2006). This followed a landmark in reproductive cloning which was reported by Zavos and Illmensee in Kentucky, USA, who extracted and injected a skin fibroblast nucleus from an infertile man into an oocyte isolated from his wife. According to a report one out of three SCNT attempts was successful – which is remarkable. However the four-celled embryo failed to implant and develop inside the uterus. However this is considered as the "first evidence of the creation and transfer of a human cloned embryo for reproductive purposes." The developmental arrest of SCNT embryo was independently verified recently in 2011 by two groups, one from New York and the other from Harvard. They found that "when the zygotic genome was replaced with that of a somatic cell, development progressed normally throughout the cleavage stages, but then arrested before the morula stage. This arrest was associated with a failure to activate transcription in the transferred somatic genome." However, mouse zygotes reprogrammed the somatic cell genome to a pluripotent state within hours after transfer in contrast to human zygotes (Egli et al. 2011). It may be noted that in the above experiment if human iPSC nucleus were used, probably Egli et al. would have been successful in taking the zygote into further divisions and development.

It is very unlikely that society is ready to embrace these new developments for several reasons. These developments can affect, for example, aging and longevity of humans in an unprecedented way, and it is quite clear that no nation has enough economic resources to support the resulting population explosion. It may be remembered that over 80 % of the whole world's wealth is concentrated in less than 10 % of the world population, and indeed they would be the first and prime beneficiaries of these advanced and expensive technologies. This would aggravate the economic and political disparities. In many societies around the world, more than half of the total wealth is controlled by the richest 10 %. A study by the World Institute for Development Economics Research at the United Nations University published a report that the richest 1 % of adults alone owned 40 % of global assets in the year 2000. The bottom half of the world adult population owns only 1 % of the global wealth (Davies et al. 2007). The ecological importance of death is recycling of biomass and energy. On the same line, the economic importance of death is that it gives the society an opportunity for redistribution and recycling of the wealth. Imagine a situation when rich people survive for hundreds of years accumulating more and more wealth compared to the poor. The increased lifespan has an impact on family structure, reproduction, parenthood, and parental behavior. This is just the economic side of the problem which is just a tip of an iceberg of problems on the horizon. The real effects of advances in technologies like therapeutic or reproductive cloning would be overwhelming involving all imaginable aspects of human life.

13 Conclusion

The concept of personalized medicine would undergo a revolutionary change hand in hand with new developments in stem cell biology, genetic engineering, and basic sciences. Cell therapy combined with gene manipulation techniques can evolve potential cures for inherited, infectious, and chronic degenerative diseases. Emerging technologies for induction of pluripotence and embryo manipulation have the potential to transform medicine and society to an unprecedented level by facilitating production of patient-specific cells, tissue, and organs in small to industrial scales and even cloning of individuals. There is an urgent need to prepare the society to accept the consequences, moderate the "unwanted effects" to acceptable limits, and minimize negative impact.

Acknowledgments The author acknowledges Prof. Anna Velcich, Ph.D., Albert Einstein College of Medicine, New York, for her constructive criticism and comments. The author is thankful to Prof. Mitradas Panicker, National Center for Biological Sciences, Bangalore, for the thoughts and inspiration imparted. The author is thankful to Megha Majumdar and friends for proof reading and corrections.

References

Adler DS, Lazarus H, Nair R, Goldberg JL, Greco NJ, Lassar T, Laughlin MJ, Das H, Pompili VJ (2011) Safety and efficacy of bone marrow-derived autologous CD133+ stem cell therapy. Front Biosci (Elite Ed) 3:506–514

Ahmadi H, Farahani MM, Kouhkan A, Moazzami K, Fazeli R, Sadeghian H, Namiri M, Madani-Civi M, Baharvand H, Aghdami N (2012) Five-year follow-up of the local autologous transplantation of CD133+ enriched bone marrow cells in patients with myocardial infarction. Arch Iran Med 15(1):32–35

Arsanjani MH (2006) Negotiating the UN declaration on human cloning. Am J Int Law 100(1):164, ISSN: 00029300

Caulfield T, Ogbogu U, Nelson E, Einsiedel E, Knoppers B, McDonald M, Brunger F, Downey R, Fernando K, Galipeau J, Geransar R, Griener G, Hyun I, Isasi R, Kardel M, Knowles L, Kucic T, Lotjonen S, Lyall D, Magnus D, Mathews DJ, Nisbet M, Nisker J, Pare G, Pattinson S, Pullman D, Rudnicki M, Williams-Jones B, Zimmerman S (2007) Stem cell research ethics: consensus statement on emerging issues. J Obstet Gynaecol Can 29(10):843–848. Erratum in: J Obstet Gynaecol Can 29(12):971. Grenier, Glenn [corrected to Griener, Glenn]

Cayo MA, Cai J, Delaforest A, Noto FK, Nagaoka M, Clark BS, Collery RF, Si-Tayeb K, Duncan SA (2012) JD induced pluripotent stem cell-derived hepatocytes faithfully recapitulate the pathophysiology of familial hypercholesterolemia. Hepatology 56(6):2163–2171

David R, Franz WM (2012) From pluripotency to distinct cardiomyocyte subtypes. Physiology (Bethesda) 27(3):119–129

Davies JB, Sandstrom S, Shorrocks A, Wolff, EN (2007) Estimating the level and distribution of global household wealth. Working Papers UNU-WIDER Research Paper, World Institute for Development Economic Research (UNU-WIDER)

Dhawan A, Mitry RR, Hughes RD (2006) Hepatocyte transplantation for liver-based metabolic disorders. J Inherit Metab Dis 29(2–3):431–435

Dill T, Schächinger V, Rolf A, Möllmann S, Thiele H, Tillmanns H, Assmus B, Dimmeler S, Zeiher AM, Hamm C (2009) Intracoronary administration of bone marrow-derived progenitor cells improves left ventricular function in patients at risk for adverse remodeling after acute ST-segment elevation myocardial infarction: results of the Reinfusion of Enriched Progenitor cells And Infarct Remodeling in Acute Myocardial Infarction study (REPAIR-AMI) cardiac magnetic resonance imaging substudy. Am Heart J 157(3):541–547

Dimos JT, Rodolfa KT, Niakan KK, Weisenthal LM, Mitsumoto H, Chung W, Croft GF, Saphier G, Leibel R, Goland R, Wichterle H, Henderson CE, Eggan K (2008) Induced pluripotent stem cells generated from patients with ALS can be differentiated into motor neurons. Science 321(5893):1218–1221, Epub 2008 Jul 31

Ding J, Yannam GR, Roy-Chowdhury N, Hidvegi T, Basma H, Rennard SI, Wong RJ, Avsar Y, Guha C, Perlmutter DH, Fox IJ, Roy-Chowdhury J (2011) Spontaneous hepatic repopulation in transgenic mice expressing mutant human α1-antitrypsin by wild-type donor hepatocytes. J Clin Invest 121(5):1930–1934. doi:10.1172/JCI45260

Egli D, Chen AE, Saphier G, Ichida J, Fitzgerald C, Go KJ, Acevedo N, Patel J, Baetscher M, Kearns WG, Goland R, Leibel RL, Melton DA, Eggan K (2011) Reprogramming within hours following nuclear transfer into mouse but not human zygotes. Nat Commun. 2011 Oct 4; 2:488. doi:10.1038/ncomms1503

Fagerholm P, Lagali NS, Merrett K et al (2010) A biosynthetic alternative to human donor tissue for inducing corneal regeneration: 24-month follow-up of a phase 1 clinical study. Sci Transl Med 2(46):46–61

Garg V, Garg H, Khan A, Trehanpati N, Kumar A, Sharma BC, Sakhuja P, Sarin SK (2012) Granulocyte colony-stimulating factor mobilizes CD34(+) cells and improves survival of patients with acute-on-chronic liver failure. Gastroenterology 142(3):505–512, e1

Gordon MY, Levicar N, Pai M, Bachellier P, Dimarakis I, Al-Allaf F, M'Hamdi H, Thalji T, Welsh JP, Marley SB, Davies J, Dazzi F, Marelli-Berg F, Tait P, Playford R, Jiao L, Jensen S, Nicholls JP, Ayav A, Nohandani M, Farzaneh F, Gaken J, Dodge R, Alison M, Apperley JF, Lechler R, Habib NA (2006) Characterization clinical application of human CD34+ stem/progenitor cell populations mobilized into the blood by granulocyte colony-stimulating factor. Stem Cells 24(7):1822–1830, Epub 2006 Mar 23

Grompe M (2002) Complete hepatic regeneration after somatic deletion of an albumin-plasminogen activator transgene. 1991. J Hepatol 37(4):422–424

Guha C, Roy-Chowdhury N, Jauregui H, Roy-Chowdhury J (2001) Hepatocyte-based gene therapy. J Hepatobiliary Pancreat Surg 8(1):51–57

Haldane JBS (1923) Science and the future. A paper read to the Heretics, Cambridge, on February 4th, 1923 Transcribed by: Cosma Rohilla Shalizi Berkeley, California 10 April 1993. Available online: http://masi.cscs.lsa.umich.edu/~crshalizi/Daedalus.html

Hansmann G, Fernandez-Gonzalez A, Aslam M, Vitali SH, Martin T, Mitsialis SA, Kourembanas S (2012) Mesenchymal stem cell-mediated reversal of bronchopulmonary dysplasia and associated pulmonary hypertension. Pulm Circ 2(2):170–181

Horslen SP, McCowan TC, Goertzen TC, Warkentin PI, Cai HB, Strom SC, Fox IJ (2003) Isolated hepatocyte transplantation in an infant with a severe urea cycle disorder. Pediatrics 111(6 Pt 1):1262–1267

Jakab K, Norotte C, Marga F, Murphy K, Vunjak-Novakovic G, Forgacs G (2010) Tissue engineering by self-assembly and bio-printing of living cells. Biofabrication 2(2):022001. doi:10.1088/1758-5082/2/2/022001, Epub 2010 Jun 2

Ji J, Ng SH, Sharma V, Neculai D, Hussein S, Sam M, Trinh Q, Church GM, McPherson JD, Nagy A, Batada NN (2012) Elevated coding mutation rate during the reprogramming of human somatic cells into induced pluripotent stem cells. Stem Cells 30(3):435–440. doi:10.1002/stem.1011

Jinek M, Chylinski K, Fonfara I, Hauer M, Doudna JA, Charpentier E (2012) A programmable dual-RNA-guided DNA endonuclease in adaptive bacterial immunity. Science 337(6096):816–821

Jung KC, Park CG, Jeon YK, Park HJ, Ban YL, Min HS, Kim EJ, Kim JH, Kang BH, Park SP, Bae Y, Yoon IH, Kim YH, Lee JI, Kim JS, Shin JS, Yang J, Kim SJ, Rostlund E, Muller WA, Park SH (2011) In situ induction of dendritic cell-based T cell tolerance in humanized mice and nonhuman primates. J Exp Med 208(12):2477–2488

Kisseleva T, Gigante E, Brenner DA (2010) Recent advances in liver stem cell therapy. Curr Opin Gastroenterol 26(4):395–402

Kobayashi T, Yamaguchi T, Hamanaka S, Kato-Itoh M, Yamazaki Y, Ibata M, Sato H, Lee YS, Usui J, Knisely AS, Hirabayashi M, Nakauchi H (2010) Generation of rat pancreas in mouse by interspecific blastocyst injection of pluripotent stem cells. Cell 142(5):787–799

Li J, Li D, Liu X, Tang S, Wei F (2012) Human umbilical cord mesenchymal stem cells reduce systemic inflammation and attenuate LPS-induced acute lung injury in rats. J Inflamm (Lond) 9(1):33

Macchiarini P (2011) Bioartificial tracheobronchial transplantation. Interview with Paolo Macchiarini. Regen Med 6(6 Suppl):14–15

Mathé G, Schwarzenberg L, Méry AM, Cattan A, Schneider M, Amiel JL, Schlumberger JR, Poisson J, Wajener G (1965) Extensive histological and cytological study of the so-called "complete" remission of acute leukemia. Bull Mem Soc Med Hop Paris 116(16):1691–1698

Martins-Taylor K, Nisler BS, Taapken SM, Compton T, Crandall L, Montgomery KD, Lalande M, Xu R (2011) Recurrent copy number variations in human induced pluripotent stem cells. Nat Biotechnol 29(6):488–491. doi:10.1038/nbt.1890

Michalopoulos GK, DeFrances MC (1997) Liver regeneration. Science 276(5309):60–66

Mrksich M, Dike LE, Tien J, Ingber DE, Whitesides GM (1997) Using microcontact printing to pattern the attachment of mammalian cells to self-assembled monolayers of alkanethiolates on transparent films of gold and silver. Exp Cell Res 235(2):305–313

Nakano T, Ando S, Takata N et al (2012) Self-formation of optic cups and storable stratified neural retina from human ESCs. Cell Stem Cell 10(6):771–785

Overturf K, al-Dhalimy M, Ou CN, Finegold M, Grompe M (1997) Serial transplantation reveals the stem-cell-like regenerative potential of adult mouse hepatocytes. Am J Pathol 151(5):1273–1280

Papaioannou I, Simons JP, Owen JS (2012) Oligonucleotide-directed gene-editing technology: mechanisms and future prospects. Expert Opin Biol Ther 12(3):329–342

Pincock S (2011) Anthony Atala: at the cutting edge of regenerative surgery. Lancet 378(9800):1371

Puppi J, Strom SC, Hughes RD, Bansal S, Castell JV, Dagher I, Ellis EC, Nowak G, Ericzon BG, Fox IJ, Gómez-Lechón MJ, Guha C, Gupta S, Mitry RR, Ohashi K, Ott M, Reid LM, Roy-Chowdhury J, Sokal E, Weber A, Dhawan A (2012) Improving the techniques for human hepatocyte transplantation: report from a consensus meeting in London. Cell Transplant 21(1):1–10

Raya A, Rodríguez-Pizà I, Guenechea G, Vassena R, Navarro S, Barrero MJ, Consiglio A, Castellà M, Río P, Sleep E, González F, Tiscornia G, Garreta E, Aasen T, Veiga A, Verma IM, Surrallés J, Bueren J, Izpisúa Belmonte JC (2009) Disease-corrected haematopoietic progenitors from Fanconi anaemia induced pluripotent stem cells. Nature 460(7251): 53–59

Rosler ES, Brandt JE, Chute J, Hoffman R (2000) An in vivo competitive repopulation assay for various sources of human hematopoietic stem cells. Blood 96(10):3414–3421

Saha S, Bhanja P, Kabarriti R, Liu L, Alfieri AA, Guha C (2011) Bone marrow stromal cell transplantation mitigates radiation-induced gastrointestinal syndrome in mice. PLoS One 6(9):e24072

Salako SE (2008) The UNESCO Universal Declaration on Bioethics and Human Rights: protecting future generations and the quest for a global consensus. Med Law 27(4):805–823

Samson M, Libert F, Doranz BJ, Rucker J, Liesnard C, Farber CM, Saragosti S, Lapoumeroulie C, Cognaux J, Forceille C, Muyldermans G, Verhofstede C, Burtonboy G, Georges M, Imai T, Rana S, Yi Y, Smyth RJ, Collman RG, Doms RW, Vassart G, Parmentier M (1996) Resistance to HIV-1 infection in caucasian individuals bearing mutant alleles of the CCR-5 chemokine receptor gene. Nature 382(6593): 722–725

Sanal MG (2009) Adipose tissue transplantation may be a potential treatment for diabetes, atherosclerosis and nonalcoholic steatohepatitis. Med Hypotheses 72(3): 247–249, Epub 2008 Nov 28

Sanal MG (2011) Future of liver transplantation: non-human primates for patient-specific organs from induced pluripotent stem cells. World J Gastroenterol 17(32):3684–3690

Shafa M, Sjonnesen K, Yamashita A, Liu S, Michalak M, Kallos MS, Rancourt DE (2012a) Expansion and long-term maintenance of induced pluripotent stem cells in stirred suspension bioreactors. J Tissue Eng Regen Med 6(6):462–472. doi:10.1002/term.450

Shafa M, Day B, Yamashita A, Meng G, Liu S, Krawetz R, Rancourt DE (2012b) Derivation of iPSCs in stirred suspension bioreactors. Nat Methods 9(5):465–466. doi:10.1038/nmeth.1973

Shehata N, Lin Y, Pendergrast J, Branch DR (2007) Cellular therapies: a Canadian blood services research and development symposium. Transfus Med Rev 21(4):317–336

Sorich MJ, McKinnon RA (2012) Personalized medicine: potential, barriers and contemporary issues. Curr Drug Metab 13(7):1000–1006

Spandorfer SD, Barmat L, Navarro J, Burmeister L, Veeck L, Clarke R, Liu HC, Rosenwaks Z (2002) Autologous endometrial coculture in patients with a previous history of poor quality embryos. J Assist Reprod Genet 19(7):309–312

Stadtfeld M, Hochedlinger K (2010) Induced pluripotency: history, mechanisms, and applications. Genes Dev 24(20):2239–2263

Swistowski A, Zeng X (2012) Scalable production of transplantable dopaminergic neurons from hESCs and iPSCs in xeno-free defined conditions. Curr Protoc Stem Cell Biol, Chapter 2:Unit2D.12. doi:10.1002/9780470151808.sc02d12s22, PMID: 22872425

Taylor CJ, Peacock S, Chaudhry AN, Bradley JA, Bolton EM (2012) Generating an iPSC bank for HLA-matched tissue transplantation based on known donor and recipient HLA types. Cell Stem Cell 11(2): 147–152

Till JE, McCulloch EA (1963) Early repair processes in marrow cells irradiated and proliferating in vivo. Radiat Res 18:96–105

Trounson A (2009) New perspectives in human stem cell therapeutic research. BMC Med 7:29, Review

Weiden PL, Sullivan KM, Flournoy N, Storb R, Thomas ED (1981) Antileukemic effect of chronic graft-versus-host disease: contribution to improved survival after allogeneic marrow transplantation. N Engl J Med 304(25):1529–1533

Welck M, Borg P, Ellis H (2010) James Blundell MD Edin FRCP (1790–1877): pioneer of blood transfusion. J Med Biogr 18(4):194–197

Yusa K, Rashid ST, Strick-Marchand H, Varela I, Liu PQ, Paschon DE, Miranda E, Ordóñez A, Hannan NR, Rouhani FJ, Darche S, Alexander G, Marciniak SJ, Fusaki N, Hasegawa M, Holmes MC, Di Santo JP, Lomas DA, Bradley A, Vallier L (2011) Targeted gene correction of α1-antitrypsin deficiency in induced pluripotent stem cells. Nature 478(7369):391–394. doi:10.1038/nature10424

Pharmacogenetics and Pharmacogenomics of Chronic Kidney Disease Comorbidities and Kidney Transplantation

34

Clarice Chemello, Margarita Aguilera, Marisa Cañadas Garre, and Miguel A. Calleja Hernández

Abstract

Chronic kidney disease (CKD) is a worldwide major public health problem associated with increased risk of mortality and rate of hospitalization and decreased life expectancy. Progression from early to late stages of CKD generally results in the onset of new symptoms and concomitant complications. Frequent complications and comorbidities of CKD include fluid and electrolyte abnormalities, secondary hyperparathyroidism and renal osteodystrophy (known as Chronic Kidney Disease-Bone Mineral Disorder – CKD-BMD), hypertension and hyperlipidemia, anemia, metabolic acidosis, and several other comorbidities involving malnutrition, pruritus, and uremic bleeding. CKD patients are at increased risk of cardiovascular disease (CVD), which includes coronary heart disease (CHD), cerebrovascular disease, peripheral vascular disease, and heart failure. The management and prevention of these comorbidities, as well as the kidney transplant complications, are complex.

Pharmacogenetics and pharmacogenomics have been applied to the management of CKD patients in both conservative and renal replacement treatments (dialysis and transplantation) trying to avoid the occurrence of drug-related problems and appearance of comorbidities. This chapter will discuss important findings in CKD pharmacogenetic and pharmacogenomic studies conducted to date and future research directions in this field. The focus will be on the CKD comorbidities (CKD-BMD and CVD) and calcineurin inhibitors (cyclosporine and tacrolimus) as immunosuppressive therapy.

C. Chemello (✉)
Department of Pharmaceutical Sciences, Center of Health Sciences, Federal University of Santa Catarina, Florianopolis, Brazil

Pharmacogenetics Unit. Pharmacy Service, University Hospital Virgen de las Nieves, Granada, Spain
e-mail: clachemello@yahoo.com.br

M. Aguilera • M.C. Garre • M.A.C. Hernández
Pharmacogenetics Unit. Pharmacy Service, University Hospital Virgen de las Nieves, Granada, Spain

D. Barh et al. (eds.), *Omics for Personalized Medicine*,
DOI 10.1007/978-81-322-1184-6_34, © Springer India 2013

Although many studies are limited by small sample sizes and replication of the findings is needed, several candidate genes have been identified and are discussed here: CYP3A5, CYP3A4, ABCB1, CASR, VDR, GC, MTHFR, and RFC1. Thus, the future is promising for a personalized treatment of CKD, which will improve therapeutic outcomes, minimize side effects, and lead to a more cost-effective care.

1 Chronic Kidney Disease

Chronic kidney disease (CKD; ICD 10-International Classification of Diseases: N18) is a pathophysiological process with multiple causes, whose consequence is inexorable loss in the number and function of nephrons, which often results in the end stage of renal disease (ESRD; ICD-10: N18-0) (WHO 2007), which in turn is a clinical situation in which there has been an irreversible loss of endogenous renal function of sufficient magnitude leaving the patient permanently dependent of renal replacement therapy (dialysis or transplant). Given the ability of the kidneys to recover its function after acute injury, the vast majority of patients with ESRD (over 90 %) reach this situation as a result of CKD (Kasper et al. 2005).

According to the *National Kidney Foundation/ Kidney Disease Outcome Quality Initiative* (NKF/KDOQI) practical guideline to Chronic Kidney Disease: Evaluation, Classification, and Stratification (KDOQI 2002), CKD is an important worldwide health problem, defined as either glomerular filtration rate (GFR) < 60 ml/min/1.73 m^2 for more than 3 months, with or without kidney damage, or kidney damage for more than 3 months as defined by structural or functional abnormalities of the kidney, with or without decreased GFR.

1.1 Pathophysiology

The pathophysiology of CKD involves initial specific mechanisms of cause and a progressive series of mechanisms that are a common consequence of reduced renal mass, independent of etiology. This reduction in renal mass causes structural and functional hypertrophy of surviving nephrons, which is mediated by vasoactive molecules, cytokines, and growth factors and is due to an initially adapted hyperfiltration which in turn is mediated by an increase in pressure and glomerular capillary flow. Finally, these short-term adaptations are revealed unfavorable because they predispose to sclerosis of residual population of viable nephrons (Kasper et al. 2005).

The earliest phase common to all types of CKD is a loss of renal reserve. Whereas renal function is completely normal, the GFR can increase about 20–30 % in response to stimulation of protein overload. During the initial phase of decreased renal reserve, GFR can be normal or high (hyperfiltration); however, the expected increase in response to a protein overload is mitigated. This early phase is particularly well documented in diabetic nephropathy. In this phase, the only clue could be laboratory measurements that determine GFR; the parameters most used are the concentrations of serum urea and creatinine. When urea and creatinine are slightly higher, there is already a significant chronic injury in the nephrons (KDOQI 2002).

When the GFR drops around 30 % the normal level, patients could remain asymptomatic with the rise in serum urea and creatinine. However, a careful analysis often reveals additional CKD clinical and laboratory manifestations. Among them are nocturia, slightly anemia and loss of energy, early alterations in appetite and nutrition, as well as alterations on calcium-phosphorus metabolism (moderate renal insufficiency). When the GFR drops below 30 % the normal level, clinical and biochemical manifestations of uremia grow in number and intensity (severe renal impairment). In the stages of mild and moderate renal insufficiency, intercurrent clinical situations of stress may further affect the renal function, with signs and symptoms of uremia.

Some processes that CKD patients may be especially sensitive are infections (urinary, respiratory, or digestive), poorly controlled hypertension, hypervolemia or hypovolemia, and nefrotoxicidad by radiocontrast. When the GFR drops below 5–10 % of normal, the survival turns impossible without renal replacement therapy (KDOQI 2002).

1.2 Etiology and Predictor Factors

There has been a dramatic increase in the incidence of ESRD and relative displacement of incidence of the causes of CKD during the past two decades. Although glomerulonephritis was the leading cause of CKD in the past, diabetic neuropathy and hypertension are now much more common causes. This may be a consequence of the prevention and more effective treatment of glomerulonephritis and decreased mortality from other causes in diabetic and hypertensive. The overall longevity and decreasing premature cardiovascular mortality also increased the average age of patients with ESRD. An especially common cause of CKD in the elderly is hypertension; in these patients, chronic renal ischemia caused by renovascular disease may be an additional factor that contributes to unidentified pathophysiological process described. Many patients present an advanced stage of CKD, which impedes to determine conclusively its cause (Kasper et al. 2005).

Among the major risk factors for CKD are older age, family history of CKD, hypertension, diabetes, renal mass reduction, low birth weight, systemic autoimmune diseases, urinary tract infections, urolithiasis, obstructive diseases of the urinary low tract by use of nephrotoxic drugs, African-American race, and low educational or social level (Cabrera 2004).

1.3 Prevalence and Incidence

The incidence rates of ESRD have increased steadily internationally since 1989. The USA has the highest incident rate of ESRD, followed by Japan. Japan has the highest prevalence per million population, with the USA taking second place (Bakris and Ritz 2009; Stenvinkel 2010).

The frequency of CKD continues to increase worldwide as does the prevalence of ESRD. The most common, but not only, causes of CKD are hypertension and diabetes. The presence of CKD is associated with a large increase in cardiovascular (CV) risk. Moreover, CV risk increases proportionally as GFR drops below 60 ml/min. Lastly, death from CV causes is eightfold higher in CKD, much higher than death from cancer. Consequently, the identification and reduction of CKD have become a vital public health priority (Bakris and Ritz 2009; Stenvinkel 2010).

The reported prevalence of CKD stages 1–4 in the most recent NHANES (National Health and Nutrition Examination Survey) between 1999 and 2006 was 26 million (13 %) out of approximately 200 million US residents aged 20 and older. Of these, 65.3 % had CKD stage 3 or 4.

The most recent report of the United States Renal Data System estimates that nearly one-half million patients in the USA were treated for ESRD in the year 2004, and by 2010, this figure is expected to increase by approximately 40 %. The elderly are a growing segment of the population and at increased risk for renal disease. Additionally, males and African-Americans with preexisting hypertension or diabetes and CKD are also at much higher risk for ESRD. These observations have also been confirmed throughout the developed world: Europe, Asia, Australia, as well as in developing regions such as China, India, and Africa (Bakris and Ritz 2009; Stenvinkel 2010).

2 Chronic Kidney Disease-Bone Mineral Disorder Mechanism and Pharmacogenetics/Pharmacogenomics

2.1 Chronic Kidney Disease-Bone Mineral Disorder Mechanism: Calcium-Phosphorus Metabolism and Related Treatment

As a result of the loss of excretory functions, endocrine regulatory, and kidneys, CKD patients suffer multiple medical complications, such as electrolyte abnormalities, anemia, secondary

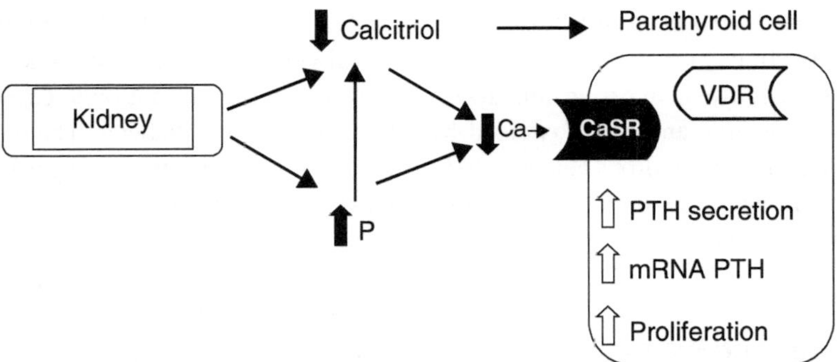

Fig. 34.1 Mechanism of the pathogenesis of secondary hyperparathyroidism

hyperparathyroidism, and renal dystrophy. Furthermore, these patients may also have other serious complications such as calcification of soft tissue and bone, cardiovascular disease, infection, and malnutrition, showing high rates of mortality and morbidity (U.S. Pharmacist 2007).

With the decrease in renal function, the reduction in Ca absorption is perceived by the calcium-sensing receptor (CASR) expressed on the surface of cells of the parathyroid glands causing PTH synthesis and secretion increasing. This increase leads to (1) increase Ca reabsorption by the kidneys, (2) increased activation of vitamin D in the renal tubules, and (3) increased release of Ca from bone. The CASR is also expressed in the kidney and perceives Ca reduction, resulting in increased resorption of Ca. This positive feedback mechanism acts to increase the levels of Ca (Waller 2011).

Furthermore, there is a decrease in the elimination of phosphorus resulting hypophosphatemia and in a reciprocal decrease in serum calcium level. Hypocalcemia is the first stimulus for PTH release by the parathyroid glands; such effects are mediated by the interaction between ionized calcium and the CASR in the cell membrane of the parathyroid gland. Hyperphosphatemia also increases the synthesis and release of PTH by acting directly on the parathyroid glands and the production of messenger ribonucleic acid (mRNA)-encoding PTH. Aiming to regulate the ionized calcium, PTH reduces P reabsorption and increases Ca reabsorption by the kidney proximal tubules (at least until the GFR is reduced to less than 30 ml/min), also increasing the mobilization of calcium from bone. This results in the correction in the levels of calcium and phosphorus, at least in the early stages of CKD; meanwhile, this occurs at the expense of higher rates of PTH. Increased PTH is most noticeable when the GFR is lower than 60 ml/min/1.73 m^2 (stage 3) and worsens with declining renal function (DiPiro et al. 2005).

This imbalance in the calcium-phosphorus metabolism that generates SHPT is defined by the KDOQI Clinical Guide as osteodystrophy and was recently defined by the working KDIGO (*Kidney Disease Improving Global Outcomes*) as *Chronic Kidney Disease-Bone Mineral Disorder* (CKD-BMD) that is characterized by one or more of the following clinical parameters: abnormal levels of Ca, P, PTH, and vitamin D; abnormalities in bone turnover, mineralization, volume, etc.; and vascular calcifications or soft tissue (KDOQI 2002; KDIGO 2009).

To clearly understand the pathogenesis of SHPT, see Fig. 34.1.

KDOQI clinical guideline (KDOQI 2002) for mineral-bone disease in CKD recommends that serum Ca, P, and PTH should be measured in all patients with CKD presenting GFR <60 ml/min/1.73 m^2. The frequency of such determinations must be based on the stage of the CKD. Table 34.1 contains the values recommended by the KDOQI Clinical Guide for levels of PTH,

Table 34.1 PTH, Ca, P, and CaxP levels recommended by KDOQI clinical guideline for stage 5 CKD patients

Parameters	Values
PTH	150–300 pg/ml
Ca	8.4–9.5 mg/dl
P	3.5–5.5 mg/dl
CaxP	<55 mg^2/dl^2

Ca, P, and Ca x P, for patients with stage 5 CKD (Frase 2009). Secondary hyperparathyroidism is one of the most important CKD morbidities due to the impairment in calcium-phosphorus metabolism which has an impact in bone disorders, called as Chronic Kidney Disease-Bone Mineral Disorder (CKD-BMD) (KDIGO 2009). CKD-BMD is a clinical syndrome that develops as a systemic disorder with a critical role in the pathogenesis of soft issue and vascular calcifications and fracture risk, thus increasing CKD patients' mortality. CKD-BMD manifests through one or a combination of the following: (1) abnormalities of bone turnover, mineralization, volume, linear growth, or strength; (2) abnormalities of calcium, phosphate, parathyroid hormone, or vitamin D metabolism; and/or (3) vascular or other soft tissue calcification (Zaza et al. 2009).

2.1.1 Calcium-Sensing Receptor (CASR), Vitamin D Receptor (VDR), and Vitamin D-Binding Protein (DBP) Polymorphisms Related to Calcium-Phosphorus Metabolism

The production and secretion of PTH are regulated by the concentrations of Ca ions and metabolites of vitamin D, having as target molecules CASR and VDR, respectively (Grzela et al. 2006). CASR and VDR play key roles in calcium homeostasis and, consequently, in the pathogenesis of SHPT. In addition to their individual role, important interactions occur between them, which contribute to the regulation of serum calcium (McCann and Beto 2010; Laaksonen et al. 2009).

CASR is a member of the G protein-coupled receptors superfamily and is present primarily in the parathyroid glands, kidneys, bones, and intestines. Its discovery in 1993 has allowed a better understanding of the regulation of calcium homeostasis. Its main function is to control parathyroid levels of extracellular Ca, maintaining balanced blood concentration and consequently the production and release of PTH controlling SHPT in CKD. It is the therapeutic target of cinacalcet, the first calcimimetic type II approved for use in humans to treat SHPT (Drüeke 2004).

The CASR gene (located on chromosome 3, position q13) encodes a membrane G protein-coupled receptor, which has a key role in the regulation of extracellular Ca level exerting its action on the parathyroid gland and kidney. CASR activation triggered by increasing the concentration of extracellular Ca inhibits PTH secretion and promotes urinary calcium excretion (Yun et al. 2007). In CKD, CASR expression is reduced by inactivating mutations, this being a key factor for the development of SHPT. The three main CASR single-nucleotide polymorphisms (SNP) with clinical impact related to SHPT pathophysiology reported are those in exon 7: G986T, A990G, and C1011G (Erturk 2006).

CASR, in response to alterations in extracellular Ca, is the key regulator of PTH secretion and, to a lesser degree, PTH synthesis. CASR is responsible for maintaining serum calcium concentrations within a narrow physiological range by sensing extracellular calcium concentrations and by mediating alterations in PTH secretion and renal calcium reabsorption (Laaksonen et al. 2009). CASR and vitamin D receptor (VDR) expression are decreased in patients with SHPT; thus, PTH secretion is downregulated, showing CASR role in the pathophysiology of SHPT (Riccardi and Martin 2008; Rodríguez et al. 2005; Goodman and Quarles 2008). CASR has been shown to be an important drug target in the SHPT treatment due to its sensitivity to extracellular Ca^{2+} (Trivedi et al. 2008). Three CASR SNPs in exon 7, G986T, A990G, and C1011G, have been described to have a role in the pathogenesis of SHPT (Yun et al. 2007). A study has shown that SNP A990G have a possible association with cinacalcet response, showing that individuals with the mutated allele 990G were more sensitive to cinacalcet (Rothe and Mayer 2006).

Calcitriol (1,25-(OH)2D), the active form of vitamin D, has pleiotropic physiological effects, and their action is mediated by VDR, an intracellular receptor-type protein phosphate. VDR expression is reduced in the parathyroid glands in CKD, and calcitriol metabolism and action are abnormal; thus, it has an important role in the development of HPTS and renal osteodystrophy. Renal production of calcitriol is decreased due to reduced renal mass and phosphate retention. Moreover, VDR SNPs may influence the biological action of calcitriol. The main VDR SNPs (located at chromosome 12, q13.11) are *Bsm*I, *Apa*I, *Taq*I (located at 3′-noncoding region), *Cdx*I (located at 5′-noncoding region), and *Fok*I (located at N-terminal region) (Erturk 2006).

VDR regulates intestinal calcium absorption and synthesis of PTH in the parathyroid glands. Furthermore, VDR binds calcitriol and this complex is involved in the transcriptional regulation of calcium homeostasis in the intestine, parathyroid, kidneys, and bone tissue. It belongs to the family of trans-acting transcriptional regulatory factors and shows a sequence of similarities to the steroid and thyroid hormone receptors (PharmGKB). VDR gene was cloned in 1988 by Baker et al. (1988), and its genomic structure firstly defined in 1997 by Miyamoto and Colleagues (1997), which showed that it consists of 9 exons with at least 6 isoforms of exon 1, located in chromosome 12q13.11, spans 63.5 kb, and encodes a 427 amino acid protein. Four common restriction fragment length polymorphisms (RFLP) in this gene have been related to a large number of diseases (Morrison et al. 2005). Although none of the VDR gene SNPs have a clear role in the Ca metabolism, previous studies suggested that *Bsm*I and *Fok*I VDR gene polymorphisms were associated with higher PTH and Ca levels in SHPT patients (Erturk 2006; Giannini et al. 2002).

- *Bsm*I has been associated with the severity of SHPT and high-turnover bone disease (Gago et al. 2005).
- *Fok*I is a mutation located at the first of the two start codons, ATG, changing the nucleotide sequence to ACG, leading to a translation of three codons downstream, and finally resulting in a protein that differs in length by three amino acids. It has been associated with higher PTH level and, consequently, with the SHPT pathophysiology (Gago et al. 2005).

The main function of the group-specific component of serum (GC-globulin) or vitamin D-binding protein (DBP or GC) is to transport vitamin D and its metabolites in the plasma. GC binds to vitamin D metabolites (e.g., 25-hydroxyvitamin D3, the main circulating metabolite, and 1.25-dihydroxyvitamin D3, the active form of vitamin D) through the sterol-binding domain; GC carries vitamin D to the liver, kidney, bone, and other tissues and stores and prolongs circulating vitamin D metabolites half-life (Fang et al. 2009).

GC is a member of the albumin, α-fetoprotein, and α-albumin/afamin gene family. The human GC is localized at the chromosome 4q11–13. The two most common genetic variations are D432E (rs7041–T1296C) and T436K (rs4588–C1307A) (Speeckaert et al. 2006; Fu et al. 2009). The main circulating vitamin D metabolite, 25-hydroxyvitamin D (25(OH) D), is the best short-term total vitamin D exposure biomarker and has the higher affinity to GC (about 90 % is binding to GC), and thus, lower 25(OH)D serum levels could be associated with these two GC polymorphisms (Speeckaert et al. 2006; Ahn et al. 2010; Sinotte et al. 2009; Wang et al. 2010; Riccardi and Martin 2008).

2.2 Cardiovascular Disease Risk in CKD: Homocysteine and Folate Metabolism

Hyperhomocysteinemia is a predictor of cardiovascular disease morbidity and mortality in patients with renal failure. Moderate hyperhomocysteinemia is present since the early stages of renal failure and increases in parallel to deterioration of renal function, and it is present in approximately 50–60 % of the patients. Major causes for fasting and post-methionine loading hyperhomocysteinemia are impairment of renal function, folic acid (FA), and vitamin B12 and vitamin B6 status (Domenici et al. 2007).

FA is a crucial nutrient that supports important physiological functions such as DNA synthesis, cell division, and substrate methylation. Low FA level caused by suboptimal intake, transport, and cellular utilization of FA is a risk factor for diseases such as spina bifida and cardiovascular diseases (PharmGKB).

Besides the exposed above, vitamin D deficiency could be related to CVD risk (Cormier et al. 2010).

2.2.1 Methylenetetrahydrofolate Reductase (MTHFR) and Reduced Folate Carrier (RFC1) Polymorphisms Related to Homocysteine and Folate Metabolism

Methylenetetrahydrofolate reductase (MTHFR) is a key enzyme in homocysteine (Hcy) and folate (FA) metabolism, whose gene is located at the chromosome 1, region 1p 36.3, and presents 11 exons ranging in size from 102 to 432 bp. The protein encoded by MTHFR gene catalyzes the conversion of 5,10-methylenetetrahydrofolate to 5-methyltetrahydrofolate, a co-substrate for the remethylation of Hcy to methionine. Genetic variations in this gene influence the susceptibility to vascular occlusive disease, neural tube defects, colon cancer, and acute leukemia, and mutations in this gene are associated with an enzyme deficiency. A common polymorphism in MTHFR gene, C677T (rs1801133), results in a substitution of C by T at nucleotide 677 changing alanine by valine at the position 222 of the amino acid sequence of the protein. This SNP turns the enzyme into a thermolabile and less active form. In patients with renal disease, this mutation is linked to elevated total plasma Hcy levels and with lower FA level in homozygous compared with heterozygous or normal individuals (OMIM) which are in turn linked to increased venous thrombosis and cardiovascular disease. Furthermore, MTHFR has been associated with cardiovascular risk in CKD patients (Jamison et al. 2009).

Another mutation in MTHFR gene, A1298C (rs1801131), consists of transition from A to C in nucleotide 1,298 and results in alteration of a glutamate (or glutamic acid) codon for alanine at position 429 of the amino acid sequence and also in enzymatic activity reduction, which is more pronounced in mutant homozygous individuals than in heterozygous. However, it has been less associated with the risk of vascular diseases than C677T mutation (Domenici et al. 2007; Jamison et al. 2009; Chatzikyriakidou et al. 2008).

The intracellular folate uptake is partially mediated by the transporter of reduced folate protein (RFC1), gene encoded by the human solute carrier family 19, member 1 (SLC19A1). RFC1 is a bidirectional transporter of 5-methyltetrahydrofolate and thiamine monophosphate (PharmGKB).

Reduced folate carrier (RFC1 or SLC19A1), located in chromosome 21q22.3, is responsible for FA transport into the cells (Chatzikyriakidou et al. 2008). A common non-synonymous SNP in exon 2 of the gene, Arg27His (80G>A), results in the substitution of a histidine for an arginine at residue 27 in the protein sequence. It has been associated with higher plasma FA levels in homozygous AA individuals in comparison to individuals who carried the G allele, and by the opposite association, Hcy levels are lower in AA patients (Chango et al. 2000).

3 Kidney Transplantation

3.1 Substitutive Renal Therapy

Most patients in stage 5 CKD need renal replacement therapy (RRT) to correct the accumulation of toxins, fluids, and electrolytes. The three basic choices of TRS are hemodialysis, peritoneal dialysis, and transplantation.

The TRS is the high-cost chronic therapy in specialized care. According to the World Health Organization (WHO), in 2008 the absolute

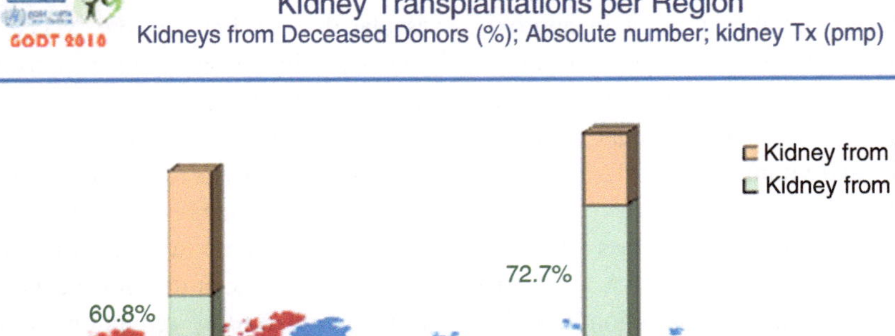

Fig. 34.2 Kidney transplantation distribution around world (Source: WHO)

number of kidney transplants in the world was 69.300 (WHO 2010). Figure 34.2 shows kidney transplant distribution according world regions by the WHO.

3.2 Calcineurin Inhibitors Immunosuppressive Therapy

3.2.1 Mechanism and Main Drugs (Cyclosporine and Tacrolimus)

Cyclosporine (Csa) and tacrolimus (Tac) are calcineurin inhibitors (CNI), which are cornerstone therapies in the prevention of rejection following solid-organ transplantation (Staatz et al. 2010). They act by inhibiting the phosphatase activity of calcineurin through binding of cyclosporine-cyclophilin and tacrolimus-FKBP12 complexes to calcineurin, which is an essential enzyme in the nuclear factor of activated T-cell (NFAT) pathway involved in T-cell activation. These two drugs have been successfully used in immunosuppressive protocol of more than 94 % of kidney transplanted recipients. However, their use is associated with toxicity beyond immunosuppression, such as metabolic alterations (hyperlipidemia and diabetes *mellitus*), hypertension, nephrotoxicity, and other problems (gingival hypertrophy and hypertrichosis). The main goal after transplantation is to maintain a reasonable balance between efficacy (avoid rejection), side effects of general immunosuppression (malignancies and infections), and specific toxicity, due to their narrow therapeutic range. Nevertheless, therapeutic drug monitoring (TDM) of CNIs is not the entire solution, since side effects are still observed with drug levels below the standard therapeutic range, while high peripheral blood levels of Csa or Tac are not fully protective against rejection (Naesens and Sarwal 2010).

One of the main limitations for the use of the immunosuppressive drugs in clinical practice is the association of major and unpredictable inter-individual variability in their pharmacokinetics, which leads to variations in drug exposure and a number of dose-related side effects. The CNIs have poor oral bioavailability; after oral administration, Csa and Tac absorption is approximately 30 %, but this may vary greatly. For example, the absolute bioavailability of Csa ranges from 10 to 60 %, and the oral bioavailability of Tac varies from 4 to 89 % among kidney and liver transplanted patients. There are clear ethnic differences in dose requirements for these drugs (Anglicheau et al. 2007).

Csa and Tac are both metabolic substrates for cytochrome P450 (CYP) 3A enzymes, in particular CYP3A4 and CYP3A5, and are transported out of cells by the P-glycoprotein (ABCB1) efflux pump. Different expression of CYP3A4, CYP3A5, and ABCB1 causes patient-to-patient variability in the absorption, metabolism, and tissue distribution of CNIs. This different expression is likely to be at least partially the result of mutations in the genes encoding for these enzymes and drug transporter, which may lead to different blood and at target sites drug concentrations, influencing drug efficacy, an individual's susceptibility to drug interactions or drug toxicity (Staatz et al. 2010).

3.2.2 Adverse Drug Reactions Associated to Calcineurin Immunosuppressor Treatment

One of the challenges in solid-organ transplantation is the tailoring of immunosuppressant therapy to specific requirements of individual transplanted recipients to optimize efficacy and minimize toxicity. Immunosuppressive therapy must be carefully titrated for the patient. Therapeutic drug monitoring of calcineurin inhibitors (CNIs) is routinely performed, with the drug dosage adjusted according to whole-blood drug concentrations and patient clinical response. However, TDM can only be performed after transplantation and is not useful for determining the optimal CNIs starting dose. Furthermore, achieving the recommended target range does not guarantee absence of drug toxicity or complete immunosuppressant efficacy because each transplanted individual responds differently to similar measured immunosuppressant concentrations (Staatz et al. 2010). The personalization and an increased understanding of drug therapy of the long-term graft biology changes are the best option available for improving transplant graft and patient survival and for reducing adverse drug reaction (ADR) (Burckart and Amur 2010).

A static long-term graft survival rate and a high incidence of severe ADR provide a situation that is ripe for intervention by complementary genomic technology. The population of patients is quite complex, and a major innovation in medical care is difficult to achieve but necessary (Burckart and Amur 2010). Candidate genes studies previously have provided a platform for creating a drug selection algorithm for transplant patients.

3.3 Pharmacogenetics of Calcineurin Immunosuppressors Cyclosporine and Tacrolimus

3.3.1 Drugs Target: ABCB1

Human multidrug resistance gene (ABCB1) is a ATP-binding cassette membrane transporter gene that contributes to the protection of the organism against the environment and certain drugs such as Tac, limiting its absorption by the intestine and promoting its excretion via urine and bile acids (Op den Buijsch et al. 2007). P-glycoprotein is encoded by the multidrug resistance gene (MDR1), also known as the ABCB1 gene, one of many ubiquitous adenosine triphosphate (ATP)-binding cassette (ABC) genes. ABCB1 is polymorphically expressed, with at least 50 SNPs identified to date. The most common and extensively studied ABCB1 SNP is the 3435C>T, in exon 26, a synonymous SNP (the altered genetic code does not lead to changes in the amino acid sequence of the encoded protein) that may decrease ABCB1 mRNA levels by decreasing mRNA stability or may affect the timing of folding and insertion of P-glycoprotein into the membrane, resulting in decreased substrate specificity. Its clinical function remains unclear; however, several studies have associated the mutated homozygous with lowered intestinal P-glycoprotein expression or activity. This would lead, in theory, to inefficient export of drugs from cells, leading to increased absorption of calcineurin inhibitors across intestine and higher systemic and intracellular drug concentrations (Staatz et al. 2010; Ieiri et al. 2004; Mourad et al. 2008). Furthermore, Naesens et al. (2009) reported that the effects of ABCB1 genotype are only evident when both donor and recipient are homozygous for the TT variant (Naesens et al. 2009). As it is known, up to 30 % of patients waiting for renal transplantation had a previous failed kidney allograft. This SNP has also been associated with acute graft rejection and survival rates (Krüger et al. 2008) and exhibits larger interethnicities allele frequency differences (CEU=54.3 %; Asian HCB=40 %; Asian JPT=47.7 %; YRI=11 %[1]) (Hodges et al. 2011).

3.3.2 Drug Metabolism: CYP3A5 and CYP3A4

CYP3A5 (Cytochrome P450, Family 3, Subfamily A, Polypeptide 5)

Along with CYP3A4, CYP3A5 is the main responsible to Tac metabolization (Staatz et al. 2010; Lamba et al. 2012). CYP3A5 is polymorphically expressed, with at least 11 SNPs identified to date. The most extensively studied involves an A to G transition at the position 6,986 within intron 3 of the gene (6986A>G). This SNP is unusual as its wild-type (WT) allele (*1) occurs at a lower frequency than the variant allele (*3), 11.7 and 88.3 % for Caucasian population, respectively (NIH). Homozygous or heterozygous carriers of the allele *1 produce high levels of full-length CYP3A5 mRNA and express similar high levels of functional CYP3A5 protein (CYP3A5 expressers), while homozygous carriers of *3 produce very low or undetectable levels of functional protein (nonexpressers). Individuals that carry at least one *1 allele should theoretically display higher clearance and lower oral bioavailability of drugs and should be more likely to experience a lack of efficacy at the standard dose. In contrary, individuals mutated homozygous (*3/*3) should experience lower clearance and higher bioavailability, and thus, they are more susceptible to suffer toxicity or an ADR and graft acute rejection. Furthermore, this SNP frequency is highly dependent on patient's ethnicity: CEU=94.2 %, Asian HCB=66.7 %, Asian JPT=75 %, and RI=15 % (*3 mutated allele frequencies reported by population) (HapMap; Staatz et al. 2010).

[1] HapMap population classification: CEU=European ancestry; Asian HCB=Han Chinese; Asian JPT=Japanese; YRI=Sub-Saharan African (Yoruban).

CYP3A4 (Cytochrome P450, Family 3, Subfamily A, Polypeptide 4)

CYP3A4 is the most abundant CYP enzyme, involved in metabolizing 45–60 % of all currently used drugs (Wang et al. 2011).

The CYP3A4 intron 6 C>T polymorphism is associated with altered Tac and Csa metabolism. CYP3A4 intron 6 C>T along with CYP3A5*3 (especially for Tac) pharmacogenetic testing performed just before transplantation may help in identifying patients at risk of CNI overexposure and contribute to limit CNI-related nephrotoxicity by refining the starting dose according to their genotype (Elens et al. 2011).

Gervasini et al. (2012) reported that CYP3A4*1B-CYP3A5*1 haplotype may have a more profound impact in tacrolimus pharmacokinetics than the CYP3A5*1 allele alone.

Table 34.2 summarizes the clinical relevance of these eight genes SNPs.

3.3.3 SNPs Relating to Drug Efficacy and Prescribed Doses of Drugs as Per Genotypes

Some authors have proposed a dosing equation for Tac using genetic variants and clinical factors (Passey et al. 2011, 2012). According to others, a biobank linked with electronic medical record seems to be useful to predict Tac dose requirements prior transplantation (Birdwell et al. 2012).

Thervet et al. (2010) have proposed a successful protocol to starting Tac dose determined by patient's genotype: CYP3A5 expressers (i.e., carriers of CYP3A5*1 allele) received 0.30 mg/kg/day, whereas CYP3A5 nonexpressers (CYP3A5*3/*3 genotype) received 0.15 mg/kg/day. The improved prediction of the required tacrolimus dose was associated with a lower number of dose modifications and a shorter delay between tacrolimus introduction and achievement of target blood concentration (Thervet et al. 2010).

The Royal Dutch Pharmacists Association – Pharmacogenetics Working Group has evaluated therapeutic dose recommendations for tacrolimus based on CYP3A5 genotype; however, they make no dosing recommendation (Swen et al. 2011).

Table 34.2 Genes clinical evidence

Gene	SNP (rs)	Allele variant	Medicines	Clinical evidence	References
Part I					
CYP3A5	776746	G6986A (*3)	Tacrolimus	Subjects with *CYP3A5*3/*3* genotype are considered to be CYP3A5 nonexpressers and need lower Tac initial dose. Nephrotoxicity risk	Li et al. (2012), De Meyer et al. (2012), Wehland et al. (2011)
	10264272	G14690A (*6)	Cyclosporine/tacrolimus	Allele T is associated with decreased dose of Tac but not with Csa in people with kidney transplantation as compared to allele C	Santoro et al. (2011)
	41303343	27131-27132insT	Tacrolimus	Tac initial dose	Santoro et al. ((2011)
CYP3A4	35599367	C>T (*22)	Tacrolimus	Tac overexposure (risk of supratherapeutic Tac concentrations) when combined with CYP3A5*3; Genotype AG is associated with decreased dose of tacrolimus in people with kidney transplantation as compared to genotype GG	Elens et al. (2011)
			Cyclosporine	Genotypes AA + AG are associated with decreased clearance of cyclosporine in people with kidney transplantation as compared to genotype GG	
ABCB1	1045642	C3435T	Tacrolimus	Pharmacokinetics, nephrotoxicity risk, increased incidence of acute rejection	Li et al. (2012), De Meyer et al. (2012), Shuker et al. (2012)
	2229109	G1199A	Cyclosporine/tacrolimus	Related to the donor, protective effect. Risk of developing calcineurin inhibitor-related nephrotoxicity	De Meyer et al. (2012)
	2032582	G2677T/A	Cyclosporine	Increased incidence of acute rejection. Possibly related to neurotoxicity	Shuker et al. (2012)
	1128503	C1236T	Cyclosporine	Increased incidence of acute rejection	Shuker et al. (2012)
Part II					
CaSR	1801725	G986T	Cinacalcet	Cinacalcet response and SHPT prognostic. G986T are associated with hypercalcemia	Rothe et al. (2005, 2008), OMIM
	1042636	A990G			
	1801726	C1011G			
VDR	1544410	*Bsm*I (A>G)	–	SHPT prognostic	Bu et al. (2010)
	2228570	*Fok*I (C>T)			
GC	7041	T1296G	–	Low levels of 25-(OH) D	Sinotte et al. (2009)
	4588	C1307A			
MTHFR	1801133	C677T	–	Risk for cardiovascular events, renal decline in African-Americans with hypertensive nephrosclerosis	Fung et al. (2012)
	1801131	A1298C			
RFC1	1051266	G80A	–	Hcy and FA levels	Galbiatti et al. (2011)

4 Translation of Pharmacogenetic Findings to the Clinical Practice Environment

Although this field has been extensively studied, there is still a need related to the implementation of pharmacogenomic/pharmacogenetic techniques into the clinical practice.

Even with medicine advances, optimal pharmacotherapy is still a clinical challenge that triggers continuous improvements. Patients are still suffering ADR, lack of efficacy, and the treatment monitoring is even still based on trial and error method. Interindividual variability in medication response is well known and depends, partially, on age, sex, liver and renal function, co-medication, etc. Furthermore, inherited variants in drug-metabolizing enzymes, transporters, receptors, and molecules of signal of transduction cascades may have a major impact on drug response (Tomalik-Scharte et al. 2008).

Pharmacogenetics (PGt) and pharmacogenomics (PGx) could be considered new disciplines in the clinical field. The approach of personalized medicine relies on decision-making based on a patient's genetic makeup, with the aim of optimize patient health outcomes; however, the translation into the clinical routine is still pending and needing more reliability.

Pharmacogenomics can be used as a tool for personalizing health care on the basis of individual genetic variations and may decrease the amount of time needed to identify the most beneficial drug and dosage for a patient, minimize exposure to ineffective treatments, reduce ADR, and improve the economic efficiency of the health-care system itself.

On the other hand, ADRs are a major health problem and its mechanism still remains poorly understood. Furthermore, most of them can be only detected after the drug has been administrated, and there is a lack of clinical reports about that. Pharmacovigilance's roles are to detect and assess risks of ADR prior to and during the marketing of medicines, to evaluate drugs in clinical use, to implement approaches to reduce risks, and in order to monitor the effectiveness. The application of PGt test moves toward ADR prevention (Alfirevic and Pirmohamed 2010).

PGt and PGx applicability must be based on genotype-phenotype valid biomarkers. Since 2004, FDA and EMA have been motivating pharmaceutical industries to perform PGt and PGx test in their clinical trials and submit the resulting information to the agencies (Goodsaid et al. 2010). Fundamentally, genotype data per se does not lead to the complete information applicable to the patient's care, and lack of knowledge regarding genotype-phenotype correlation is often considered as the major barrier delaying the PGt translation to clinical practice. GWAS could be indeed a good promise to reveal possible associations; however, large patients cohort without understanding the genotype implication in drug-response phenotype would not conclude any significant clinical association, even when the investment is highly costly (Gurwitz and Pirmohamed 2010).

Moreover, ethnic differences are a determinant issue in the response variability to the same drug. The FDA recognizes that standard methods of defining racial and ethnic subgroups are necessary to ensure consistency in demographic subset analyses, to compare results across studies, and to assess potential subgroup differences in safety and effectiveness. The inclusion of different ethnicity studies in the medication label is encouraged by FDA. An example of ethnicity-related safety and efficacy information included in the label is Tac, an older marketed drug. The label notes that a retrospective comparison in kidney transplant patients suggested that black (African-American) patients required higher Tac doses to reach trough concentrations similar to those observed in white patients; however, whether this phenotype depends only on ABCB1 and CYP3A5 polymorphisms remains unclear (Yasuda et al. 2008; Huang and Temple 2008).

Another point of interest that should be discussed is the economic advantage that pharmacogenetics could bring compared to current standard of health care. Firstly, more studies demonstrating the links between genetic variation and response to medications in defined populations are required, along with development of valid tests to measure these specific variants and, thus, assure legitimate biomarkers. Secondly, studies should be conducted to evaluate whether pharmacogenetic testing improves health outcomes for patients and if the decision to perform the pharmacogenetic test is cost-effective relative to usual care (Deverka and McLeod 2008). This type of analysis must be done with accuracy and robustness to assure its reliability (Payne and Shabaruddin 2010). Pharmacogenetic testing is likely to be cost-effective because it uses genomic information to improve drug effectiveness and reduce toxicity both in the drug development process and at the bedside (Deverka et al. 2010).

Ethical and social issues are priorities in pharmacogenetic context. Pharmacogenetic information and patient's anonymity must be under strict control and kept safe. Informed consent should be signed for a specific study purpose or biobank, and patient's information must be clear and easy to understand (Vijverberg et al. 2010).

A multidisciplinary team to elaborate the pharmacogenetic report is important because a junction of knowledge from different specialties is needed. Understanding the genetics basis and molecular mechanism of drug response is important prerequisite to PGt application, and continuous pharmacogenetic information update is extremely necessary (Giacomini et al. 2007; Gong et al. 2008; Thorn et al. 2010).

A recent meeting by the International Association of Therapeutic Drug Monitoring and Clinical Toxicology had reported that there is a diverse range of opinions from the provision of genotype only, leaving interpretation to the prescribing clinician, to the provision of an interpreted genotype with specific prescribing advice (Mac Phee 2010). In fact, a consensus of the information contained in the pharmacogenetic report (counseling) is necessary besides a guideline. External factors as diet, sun exposure, sex, and others could influence in the disease's and drug response's phenotypes; thus, clinical team must take this information into account in the interpretation of the pharmacogenetic test to make the best decision.

5 Future Perspectives

Patients with CKD-BMD and kidney transplanted will benefit with pharmacogenetic test application in the clinical routine, even if further studies about the relevant SNPs should be made. The most promising SNP up to now is CYP3A5 rs776746, as already has a clinical guideline being used in the routine, a step forward to starting dosing protocol (Ware 2012).

A range of companies are investing in the development of direct-to-consumer pharmacogenetic tests, though this has an uncertain value and needs to be deeply discussed (Chua and Kennedy 2012). Thus, the clinical pharmacist/pharmacologist has an important involvement in the multidisciplinary team interpreting pharmacogenomics/pharmacogenetics information, as it can optimize drug therapy, focus on disease and treatment-related outcomes, ensure compliance, and counsel patients about drug usage, and PGt seems to be a good ally to achieve this objectives (Stemer and Lemmens-Gruber 2010). PGt will implicate changes in clinical routine, as health professionals can access the patient's pharmacogenetic information before the prescription is being decided, assuring the patient's safety (Haga and Burke 2008).

Whether PGt and PGx will complement the actual standard of care is a question that will remain without answer for a while. Although divergent opinions about the clinical applicability of these technologies exist (Limdi and Veenstra 2010), PGt and PGx will progress as the new studies of emerging multicenter consortia will bring more reliable results.

References

Ahn J, Yu K, Stolzenberg-Solomon R, Simon KC, McCullough ML, Gallicchio L, Jacobs EJ, Ascherio A, Helzlsouer K, Jacobs KB, Li Q, Weinstein SJ, Purdue M, Virtamo J, Horst R, Wheeler W, Chanock S, Hunter DJ, Hayes RB, Kraft P, Albanes D (2010) Genome-wide association study of circulating vitamin D levels. Hum Mol Genet 9(13):2739–2745

Alfirevic A, Pirmohamed M (2010) Drug-induced hypersensitivity reactions and pharmacogenomics: past, present and future. Pharmacogenomics 11(4):497–499

Anglicheau D, Legendre C, Beaune P, Thervet E (2007) Cytochrome P450 3A polymorphisms and immunosuppressive drugs: an update. Pharmacogenomics 8(7):835–849

Baker AR, McDonnell DP, Hughes M, Crisp TM, Mangelsdorf DJ, Haussler MR, Pike JW, Shine J, O'Malley BW (1988) Cloning and expression of full-length cDNA encoding human vitamin D receptor. Proc Natl Acad Sci U S A 85(10):3294–3298

Bakris GL, Ritz E (2009) Hypertension and kidney disease, a marriage that should be prevented. Kidney Int 75:449–452

Birdwell KA, Grady B, Choi L, Bian A, Denny JC, Jiang M, Vranic G, Basford M, Cowan JD, Richardson DM, Robinson MP, Ikizler TA, Ritchie MD, Stein CM, Haas DW (2012) The use of a DNA biobank linked to electronic medical records to characterize pharmacogenomic predictors of tacrolimus dose requirement in kidney transplant recipients. Pharmacogenet Genomics 22(1):32–42

Bu FX, Armas L, Lappe J, Zhou Y, Gao G, Wang HW, Recker R, Zhao LJ (2010) Comprehensive association analysis of nine candidate genes with serum 25-hydroxy vitamin D levels among healthy Caucasian subjects. Hum Genet 128:549–556

Burckart G, Amur S (2010) Update on the clinical pharmacogenomics of organ transplantation. Pharmacogenomics 11(2):227–236

Cabrera SS (2004) Definición y clasificación de los estadios de la enfermedad renal crónica. Prevalencia. Claves para el diagnóstico precoz. Factores de riesgo de enfermedad renal crónica. Nefrología 24(6):27–34

Chango A, Emery-Fillon N, de Courcy GP, Lambert D, Pfister M, Rosenblatt DS, Nicolas JP (2000) A polymorphism (80G->A) in the reduced folate carrier gene and its associations with folate status and homocysteinemia. Mol Genet Metab 70(4):310–315

Chatzikyriakidou A, Vakalis KV, Kolaitis N, Kolios G, Naka KK, Michalis LK, Georgiou I (2008) Distinct association of SLC19A1 polymorphism –43TNC with red cell folate levels and of MTHFR polymorphism 677CNT with plasma folate levels. Clin Biochem 41:174–176

Chua EW, Kennedy MA (2012) Current state and future prospects of direct-to-consumer pharmacogenetics. Front Pharmacol 3:152, Epub 2012 Aug 20

Cormier C, Courbebaisse M, Maury E, Thervet E, Souberbielle JC (2010) Effect of vitamin D deficiency on cardiovascular risk. J Mal Vasc 35(4):235–241

De Meyer M, Haufroid V, Elens L, Fusaro F, Patrono D, De Pauw L, Kanaan N, Goffin E, Mourad M (2012) Donor age and ABCB1 1199G>A genetic polymorphism are independent factors affecting long-term renal function after kidney transplantation. J Surg Res 178(2):988–995

Deverka PA, McLeod HL (2008) Harnessing economic drivers for successful clinical implementation of pharmacogenetic testing. Clin Pharmacol Ther 84(2):191–193

Deverka PA, Vernon J, McLeod HL (2010) Economic opportunities and challenges for pharmacogenomics. Annu Rev Pharmacol Toxicol 50:423–437

DiPiro JT et al (2005) Pharmacotherapy: a pathophysiologic approach, 6th edn. The McGraw-Hill Companies, New York. ISBN: 0-07-141613-7

Domenici FA, Vannucchi MTI, Simões-Ambrosio LMC, Vannucchi H (2007) Hyperhomocysteinemia and polymorphisms of the methylenetetrahydrofolate gene in hemodialysis and peritoneal dialysis patients. Mol Nutr Food Res 51:1430–1436

Drüeke TB (2004) Modulation and action of the calcium-sensing receptor. Nephrol Dial Transplant 19(Suppl 5):v20–v26

Elens L, van Schaik RH, Panin N, de Meyer M, Wallemacq P, Lison D, Mourad M, Haufroid V (2011) Effect of a new functional CYP3A4 polymorphism on calcineurin inhibitors' dose requirements and trough blood levels in stable renal transplant patients. Pharmacogenomics 12(10):1383–1396

Erturk S (2006) Gene polymorphism association studies in dialysis: bone and mineral metabolism. Semin Dial 19(3):232–237

Fang Y, van Meurs JBJ, Arp P (2009) Vitamin D binding protein genotype and osteoporosis. Calcif Tissue Int 85(2):85–93

Frase WD (2009) Hyperparathyroidism. Lancet 374: 145–158

Fu L, Yun F, Oczak M, Wong BYL, Vieth R, Cole DEC (2009) Common genetic variants of the vitamin D binding protein (DBP) predict differences in response of serum 25-hydroxyvitamin D [25(OH)D] to vitamin supplementation. Clin Biochem 42:1174–1177

Fung MM, Salem RM, Lipkowitz MS, Bhatnagar V, Pandey B, Schork NJ, O'Connor DT, AASK Study Investigators (2012) Methylenetetrahydrofolate reductase (MTHFR) polymorphism A1298C (Glu429Ala) predicts decline in renal function over time in the African-American Study of Kidney Disease and Hypertension (AASK) Trial and Veterans Affairs Hypertension Cohort (VAHC). Nephrol Dial Transplant 27(1):197–205

Gago EV, Cadarso-Suárez C, Perez-Fernandez R, Burgos RR, Mugica JD, Iglesias CS (2005) Association between vitamin D receptor FokI polymorphism and serum parathyroid hormone level in patients with chronic renal failure. J Endocrinol Invest 28:117–121

Galbiatti AL, Ruiz MT, Rezende PD (2011) A80G polymorphism of reduced folate carrier 1 (RFC1) gene and head and neck squamous cell carcinoma etiology in Brazilian population. Mol Biol Rep 38(2):1071–1078

Gervasini G, Garcia M, Macias RM, Cubero JJ, Caravaca F, Benitez J (2012) Impact of geneticpolymorphisms on tacrolimus pharmacokinetics and the clinical outcome of renal transplantation. Transpl Int 25(4):471–480

Giacomini KM, Brett CM, Altman RB, Benowitz NL, Dolan ME, Flockhart DA, Johnson JA, Hayes DF, Klein T, Krauss RM, Kroetz DL, McLeod HL, Nguyen AT, Ratain MJ, Relling MV, Reus V, Roden DM,

Schaefer CA, Shuldiner AR, Skaar T, Tantisira K, Tyndale RF, Wang L, Weinshilboum RM, Weiss ST, Zineh I (2007) The pharmacogenetics research network: from SNP discovery to clinical drug response. Clin Pharmacol Ther 81(3):328–345

Giannini S, D'Angelo A, Nobile M, Carraro G, Rigotti P, Silva-Netto F, Pavan S, Marchini F, Zaninotto M, Dalle Carbonare L, Sartori L, Crepaldi G (2002) The effects of vitamin D receptor polymorphism on secondary hyperparathyroidism and bone density after renal transplantation. J Bone Miner Res 17(10):1768–1773

Gong L, Owen RP, Gor W, Altman RB, Klein TE (2008) PharmGKB: an integrated resource of pharmacogenomic data and knowledge. Curr Protoc Bioinformatics 23:14.7.1–14.7.17

Goodman WG, Quarles LD (2008) Development and progression of secondary hyperparathyroidism in chronic kidney disease: lessons from molecular genetics. Kidney Int 74:276–288

Goodsaid FM, Amur S, Aubrecht J, Burczynski ME, Carl K, Catalano J, Charlab R, Close S, Cornu-Artis C, Essioux L, Fornace AJ Jr, Hinman L, Hong H, Hunt I, Jacobson-Kram D, Jawaid A, Laurie D, Lesko L, Li HH, Lindpaintner K, Mayne J, Morrow P, Papaluca-Amati M, Robison TW, Roth J, Schuppe-Koistinen I, Shi L, Spleiss O, Tong W, Truter SL, Vonderscher J, Westelinck A, Zhang L, Zineh I (2010) Voluntary exploratory data submissions to the US FDA and the EMA: experience and impact. Nat Rev Drug Discov 9(6):435–445

Grzela T, Chudzinski W, Lasiecka Z, Niderla J, Wilczynski G, Gornicka B, Wasiutynski A, Durlik M, Boszczyk A, Brawura-Biskupski-Samaha R, Dziunycz P, Milewski L, Lazarczyk M, Lazarczyk M, Nawrot I (2006) The calcium-sensing receptor and vitamin D receptor expression in tertiary hyperparathyroidism. Int J Mol Med 17(5):779–783

Gurwitz D, Pirmohamed M (2010) Pharmacogenomics: the importance of accurate phenotypes. Pharmacogenomics 11(4):469–470

Haga SB, Burke W (2008) Pharmacogenetic testing: not as simple as it seems. Genet Med 10(6):391–395

Hodges LM, Markova SM, Chinn LW, Gow JM, Kroetz DL, Klein TE, Altman RB (2011) Very important pharmacogene summary: ABCB1 (MDR1, P-glycoprotein). Pharmacogenet Genomics 21(3):152–161

Huang SM, Temple R (2008) Is this the drug or dose for you? Impact and consideration of ethnic factors in global drug development, regulatory review, and clinical practice. Clin Pharmacol Ther 84(3):287–294

Ieiri I, Takane H, Otsubo K (2004) The MDR1 (ABCB1) gene polymorphism and its clinical implications. Clin Pharmacokinet 43(9):553–576

International HapMap Project. Available in: http://hapmap.ncbi.nlm.nih.gov/. Accessed in Sept 2012

Jamison RL, Shih MC, Humphries DE, Guarino PD, Kaufman JS, Goldfarb DS, Warren SR, Gaziano JM, Lavori P, Investigators VAS (2009) Effect of the MTHFR C677T and A1298C polymorphisms on survival in patients with advanced CKD and ESRD: a prospective study. Am J Kidney Dis 53(5):779–789

Kasper DL, Braunwald E, Hauser S, Longo D, Jameson JL, Fauci AS (2005) Harrison's principles of internal medicine, 16th edn. Copyright© 2005 by The McGraw-Hill Companies, Inc., New York. ISBN-13: 978-0071402354

Kidney Disease: Improving Global Outcomes (KDIGO) CKD–MBD Work Group (2009) KDIGO clinical practice guideline for the diagnosis, evaluation, prevention, and treatment of chronic kidney disease–mineral and bone disorder (CKD–MBD). Kidney Int 76(Suppl 113):S1–S130

Krüger B, Schröppel B, Murphy BT (2008) Genetic polymorphism and the fate of the transplanted organ. Transplant Rev 22:131–140

Laaksonen MML, Outila TA, Kärkkäinen MUM, Kemi VE, Rita HJ, Perola M, Valsta LM, Lamberg-Allardt CJ (2009) Association of vitamin D receptor, calcium-sensing receptor and parathyroid hormone gene polymorphisms with calcium homeostasis and peripheral bone density in adults finns. J Nutrigenet Nutrigenomics 2:55–64

Lamba J, Hebert JM, Schuetz EG, Klein TE, Altman RB (2012) PharmGKB summary: very important pharmacogene information for CYP3A5. Pharmacogenet Genomics 22(7):555–558

Li Y, Hu X, Cai B, Chen J, Bai Y, Tang J, Liao Y, Wang L (2012) Meta-analysis of the effect of MDR1 C3435 polymorphism on tacrolimus pharmacokinetics in renal transplant recipients. Transpl Immunol 27(1):12–18

Limdi NA, Veenstra DL (2010) Expectations, validity, and reality in pharmacogenetics. J Clin Epidemiol 63(9):960–969

MacPhee IAM (2010) Use of pharmacogenetics to optimize immunosuppressive therapy. Ther Drug Monit 32(3):261–264

McCann LM, Beto J (2010) Roles of calcium-sensing receptor and vitamin D receptor in the pathophysiology of secondary hyperparathyroidism. J Ren Nutr 20(3):141–150

Miyamoto K, Kesterson RA, Yamamoto H, Taketani Y, Nishiwaki E, Tatsumi S, Inoue Y, Morita K, Takeda E, Pike JW (1997) Structural organization of the human vitamin D receptor chromosomal gene and its promoter. Mol Endocrinol 11(8):1165–1179

Morrison NA, George PM, Vaughan T, Tilyard MW, Frampton CM, Gilchrist NL (2005) Vitamin D receptor genotypes influence the success of calcitriol therapy for recurrent vertebral fracture in osteoporosis. Pharmacogenet Genomics 15(2):127–135

Mourad M, Wallemacq P, De Meyer M, Malaise J, De Pauw L, Eddour DC, Goffin E, Lerut J, Haufroid V (2008) Biotransformation enzymes and drug transporters pharmacogenetics in relation to immunosuppressive drugs: impact on pharmacokinetics and clinical outcome. Transplantation 85(7 Suppl):S19–S24

MTHFR in OMIM. Online c. Available in: http://www.ncbi.nlm.nih.gov/omim/607093. Accessed in Aug 2012

Naesens M, Sarwal MM (2010) Monitoring calcineurin inhibitor therapy: localizing the moving target. Transplantation 89(11):1308–1309

Naesens M, Lerut E, de Jonge H, Van Damme B, Vanrenterghem Y, Kuypers DR (2009) Donor age and renal P-glycoprotein expression associate with chronic histological damage in renal allografts. J Am Soc Nephrol 20(11):2468–2480

National Institute of Health (NIH). Available in: http://www.ncbi.nlm.nih.gov/projects/SNP/. Accessed in Sept 2012

National Kidney Foundation Kidney Disease Outcomes Quality Initiative (2002) K/DOQI clinical practice guidelines for chronic kidney disease: evaluation, classification, and stratification. Available in: http://www.kidney.org/professionals/kdoqi/guidelines_ckd/p4_class.htm. Accessed in July 2012

National Kidney Foundation Kidney Disease Outcomes Quality Initiative (2002) K/DOQI clinical practice guidelines for bone metabolism and disease in chronic kidney disease. Available in: http://www.kidney.org/professionals/kdoqi/guidelines_bone/index.htm. Accessed in Sept 2012

Op den Buijsch RAM, Christiaans MHL, Stolk LML, de Vries JE, Cheung CY, Undre NA, van Hooff JP, van Dieijen-Visser MP, Bekers O (2007) Tacrolimus pharmacokinetics and pharmacogenetics: influence of adenosine triphosphate-binding cassette B1 (ABCB1) and cytochrome (CYP) 3A Polymorphisms. Fundam Clin Pharmacol 21:427–435

Passey C, Birnbaum AK, Brundage RC, Oetting WS, Israni AK, Jacobson PA (2011) Dosing equation for tacrolimus using genetic variants and clinical factors. Br J Clin Pharmacol 72(6):948–957

Passey C, Birnbaum AK, Brundage RC, Schladt DP, Oetting WS, Leduc RE, Israni AK, Guan W, Matas AJ, Jacobson PA (2012) Validation of tacrolimus equation to predict troughs using genetic and clinical factors. Pharmacogenomics 13(10):1141–1147

Payne K, Shabaruddin FH (2010) Cost–effectiveness analysis in pharmacogenomics. Pharmacogenomics 11(5):643–646

PharmGKB, Pharmacogenomics Knowledge Base. Available in: http://www.pharmgkb.org/index.jsp. Accessed Sept 2012

Riccardi D, Martin D (2008) The role of the calcium-receptor in the pathophysiology of secondary hyperparathyroidism. NDT Plus 1(Suppl 1):i7–i11

Rodríguez M, Nemeth E, Martín D (2005) The Calcium-sensing receptor: a key factor in the pathogenesis of secondary hyperparathyroidism. Am J Physiol Renal Physiol 288:F253–F264

Rothe H, Mayer G (2006) Clinical importance of calcium-sensing receptor gene polymorphism Arg990Gly in the age of calcimimetic therapy. Curr Pharmacogenomics Person Med 4:153–156

Rothe H, Shapiro W, Sun WY, Matalon A (2008) CaSR polymorphism Arg990Gly and response to calcimimetic agents in end-stage kidney disease patients with secondary hyperparathyroidism and in cell culture. Per Med 5(2):109–116

Rothe HM, Shapiro WB, Sun WY, Chou SY (2005) Calcium-sensing receptor gene polymorphism Arg990Gly and its possible effect on response to cinacalcet HCl. Pharmacogenet Genomics 15:29–34

Santoro A, Felipe CR, Tedesco-Silva H, Medina-Pestana JO, Struchiner CJ, Ojopi EB, Suarez-Kurtz G (2011) Pharmacogenetics of calcineurin inhibitors in Brazilian renal transplant patients. Pharmacogenomics 12(9):1293–1303

Sinotte M, Diorio C, Bérubé S, Pollak M, Brisson J (2009) Genetic polymorphisms of the vitamin D binding protein and plasma concentrations of 25-hydroxyvitamin D in premenopausal women. Am J Clin Nutr 89:634–640

Shuker N, Bouamar R, Weimar W, van Schaik RH, van Gelder T, Hesselink DA (2012) ATP-binding cassette transporters as pharmacogenetic biomarkers for kidney transplantation. Clin Chim Acta 413(17–18):1326–1337

Speeckaert M, Huang G, Delanghe JR, Taes YEC (2006) Biological and clinical aspects of the vitamin D binding protein (Gc-globulin) and its polymorphism. Clin Chim Acta 372:33–42

Staatz C, Goodman LK, Tett SE (2010) Effect of CYP3A and ABCB1 single nucleotide polymorphisms on the pharmacokinetics and pharmacodynamics of calcineurin inhibitors: Part I. Clin Pharmacokinet 49(3):141–175

Stemer G, Lemmens-Gruber R (2010) Clinical pharmacy services and solid organ transplantation: a literature review. Pharm World Sci 32:7–18

Stenvinkel P (Division of Renal Medicine, Department of Clinical Science, Intervention and Technology, Karolinska Institutet, Stockholm, Sweden) (2010) Chronic kidney disease: a public health priority and harbinger of premature cardiovascular disease (Review). J Intern Med 268:456–467

Swen JJ, Nijenhuis M, de Boer A, Grandia L, Maitland-van der Zee AH, Mulder H, Rongen GA, van Schaik RH, Schalekamp T, Touw DJ, van der Weide J, Wilffert B, Deneer VH, Guchelaar HJ (2011) Pharmacogenetics: from bench to byte–an update of guidelines. Clin Pharmacol Ther 89(5):662–673

Thervet E, Loriot MA, Barbier S, Buchler M, Ficheux M, Choukroun G, Toupance O, Touchard G, Alberti C, Le Pogamp P, Moulin B, Le Meur Y, Heng AE, Subra JF, Beaune P, Legendre C (2010) Optimization of initial tacrolimus dose using pharmacogenetic testing. Clin Pharmacol Ther 87(6):721–726

Thorn CF, Klein TE, Altman RB (2010) Pharmacogenomics and bioinformatics: PharmGKB. Pharmacogenomics 11(4):501–505

Tomalik-Scharte D, Lazar A, Fuhr U, Kirchheiner J (2008) The clinical role of genetics polymorphisms in drug-metabolizing enzymes. Pharmacogenomics J 8:4–15

Trivedi R, Mithal A, Chattopadhyay N (2008) Recent updates on the calcium-sensing receptor as a drug target. Curr Med Chem 15(2):178–186

U.S. Pharmacist. Postgraduate Health Care Education (2007) The role of the pharmacist in the identification and management of secondary hyperparathyroidism. Available in: http://www.uspharmacist.com/continuing_education/ceviewtest/lessonid/105514. Accessed in Aug 2012

Vijverberg SJH, Pieters T, Cornel MC (2010) Ethical and social issues in pharmacogenomics testing. Curr Pharm Des 16(2):245–252

Waller S (2011) Parathyroid hormone and growth in chronic kidney disease. Pediatr Nephrol 26(2):195–204

Wang TJ, Zhang F, Richards JB, Kestenbaum B, van Meurs JB, Berry D, Kiel DP, Streeten EA, Ohlsson C, Koller DL, Peltonen L, Cooper JD, O'Reilly PF, Houston DK, Glazer NL, Vandenput L, Peacock M, Shi J, Rivadeneira F, McCarthy MI, Anneli P, de Boer IH, Mangino M, Kato B, Smyth DJ, Booth SL, Jacques PF, Burke GL, Goodarzi M, Cheung CL, Wolf M, Rice K, Goltzman D, Hidiroglou N, Ladouceur M, Wareham NJ, Hocking LJ, Hart D, Arden NK, Cooper C, Malik S, Fraser WD, Hartikainen AL, Zhai G, Macdonald HM, Forouhi NG, Loos RJ, Reid DM, Hakim A, Dennison E, Liu Y, Power C, Stevens HE, Jaana L, Vasan RS, Soranzo N, Bojunga J, Psaty BM, Lorentzon M, Foroud T, Harris TB, Hofman A, Jansson JO, Cauley JA, Uitterlinden AG, Gibson Q, Järvelin MR, Karasik D, Siscovick DS, Econs MJ, Kritchevsky SB, Florez JC, Todd JA, Dupuis J, Hyppönen E, Spector TD (2010) Common genetic determinants of vitamin D insufficiency: a genome-wide association study. Lancet 376:180–188

Wang D, Guo Y, Wrighton SA, Cooke GE, Sadee W (2011) Intronic polymorphism in CYP3A4 affects hepatic expression and response to statin drugs. Pharmacogenomics J 11(4):274–286

Ware N (2012) The role of genetics in drug dosing. Pediatr Nephrol 27:1489–1498

Wehland M, Bauer S, Brakemeier S, Burgwinkel P, Glander P, Kreutz R, Lorkowski C, Slowinski T, Neumayer HH, Budde K (2011) Differential impact of the CYP3A5*1 and CYP3A5*3 alleles on predose concentrations of two tacrolimus formulations. Pharmacogenet Genomics 21(4):179–184

World Health Organization (2010) Global observatory on donation and transplantation. Available in: http://www.transplant-observatory.org/pages/home.aspx. Accessed in Sept 2012

World Health Organization (WHO). International Classification of Diseases (ICD-10) (2007) Available in: http://apps.who.int/classifications/apps/icd/icd10online/. Accessed in July 2012

Yasuda SU, Zhang L, Huang SM (2008) The role of ethnicity in variability in response to drugs: focus on clinical pharmacology studies. Clin Pharmacol Ther 84(3):417–423

Yun FHJ, Wong BYL, Chase M, Shuen AY, Canaff L, Thongthai K, Siminovitch K, Hendy GN, Cole DE (2007) Genetic variation at the calcium-sensing receptor (CASR) locus: implications for clinical molecular diagnostics. Clin Biochem 40(8):551–561

Zaza G, Granata S, Sallustio F, Grandaliano G, Schena FP (2009) Pharmacogenomics: a new paradigm to personalize treatments in nephrology patients. Clin Exp Immunol 159:268–276

ERRATUM

Section V

Personalized Medicine: Metabolic Disorders

D. Barh et al. (eds.), *Omics for Personalized Medicine*,
DOI 10.1007/978-81-322-1184-6, © Springer India 2013

DOI 10.1007/978-81-322-1184-6_35

Please note that the title of Section V on page 559 has been incorrectly published. The correct section title should read as below:

Personalized Medicine: Metabolic Disorders and Infectious Diseases

The online version of the book can be found at
http://dx.doi.org/10.1007/978-81-322-1184-6

Editors Bios

Debmalya Barh is the founder and president of the Institute of Integrative Omics and Applied Biotechnology (IIOAB), India, and a consultant biotechnologist having international repute. He is an active researcher in multidisciplinary "omics" fields that include biomarkers, target and targeted drug discovery, pharmacogenomics, nutrigenomics, neuroscience, cancer, and cardiovascular, infectious, and metabolic diseases. He works with nearly 400 researchers from 30–35 countries and has published more than 100 papers. He is a well-known editor for many research reference books in the foremost domains of "omics" published by top-ranked international publishers. He also serves as an editorial and review board member for a number of highly professional international research journals.

Dipali Dhawan is one of the promising researchers in the field of pharmacogenomics and induced pluripotent and cancer stem cells. A postgraduate in Molecular and Genetic Medicine from University of Sheffield, UK, and a PhD in Biotechnology from Nirma University, India, Dr. Dhawan is presently associated with the Institute of Life Sciences, Ahmedabad University, Ahmedabad, India. She has published many research works in her field and has also been awarded CSIR and DST fellowships.

D. Barh et al. (eds.), *Omics for Personalized Medicine*,
DOI 10.1007/978-81-322-1184-6, © Springer India 2013

Nirmal Kumar Ganguly, *M.D., D.Sc. (hc), FAMS (India), FINSA (India), FNASc (India), FIASc (India), Fellow of Imperial College, Faculty of Medicine (London), FIWAS (Italy), FRC Path (London), Fellow FICAS (Canada)*, is a Distinguished Biotechnology Research Professor, Department of Biotechnology (DBT), Government of India. He is President of the Jawaharlal Institute of Postgraduate Medical Education and Research (JIPMER) as well as that of the Asian Institute of Public Health, Bhubaneswar, Odisha. He is the former Director General, Indian Council of Medical Research (ICMR), New Delhi; former Director, PGIMER (Chandigarh); and former Director, National Institute of Biologicals (Noida).

Prof. Ganguly has published 757 research papers and has supervised 130 PhD theses as Supervisor/Co-supervisor. His major areas of research have been tropical diseases, cardiovascular diseases, and diarrheal diseases. His interest encompasses the disciplines of immunology, biotechnology, and public health. He is member of the Advisory Group, Cholera Vaccine Investment Case Preparation, International Vaccine Institute, Seoul, South Korea. He is also Member of Global Access Advisory Committee, University of Western Ontario, Canada; Worldwide Antimalarial Resistance Network (WWARN); Centre for Tropical Medicine, University of Oxford, UK; and Asian AIDS Vaccine Network Task Force, WHO, Geneva, Switzerland. He is member of Editorial Board of Molecular and Cellular Biochemistry, Institute of Cardiovascular Sciences, University of Manitoba, Canada. He is Advisory Board Member of Grand Challenges (Canada). He is president of the Asian Conference on Diarrhoeal Diseases and Nutrition (ASCODD), Yogyakarta, Indonesia. Prof. Ganguly is an Honorary Global Health Research Fellow and Adjunct Professor at Boston University, USA. He is also an adjunct professor of Environmental Health, School of Public Health, University of Minnesota, USA. He is also a member of the Scientific Board, Grand Challenges, Bill & Melinda Gates Foundation.

He has received 116 awards, including 6 international and 110 national awards. He has been honored with the prestigious **Padma Bhushan Award** by Her Excellency the President of India on 26th January 2008 in the field of "medicine."

Index

D. Barh et al. (eds.), *Omics for Personalized Medicine*, DOI 10.1007/978-81-322-1184-6, © Springer India 2013

D

E

F

G

H

I

J

K

L

Q

R

S

T